Handbook
of Food
Analysis

FOOD SCIENCE AND TECHNOLOGY

A Series of Monographs, Textbooks, and Reference Books

1. *Flavor Research: Principles and Techniques,* R. Teranishi, I. Hornstein, P. Issenberg, and E. L. Wick
2. *Principles of Enzymology for the Food Sciences,* John R. Whitaker
3. *Low-Temperature Preservation of Foods and Living Matter,* Owen R. Fennema, William D. Powrie, and Elmer H. Marth
4. *Principles of Food Science*
   Part I: Food Chemistry, edited by Owen R. Fennema
   Part II: Physical Methods of Food Preservation, Marcus Karel, Owen R. Fennema, and Daryl B. Lund
5. *Food Emulsions,* edited by Stig E. Friberg
6. *Nutritional and Safety Aspects of Food Processing,* edited by Steven R. Tannenbaum
7. *Flavor Research: Recent Advances,* edited by R. Teranishi, Robert A. Flath, and Hiroshi Sugisawa
8. *Computer-Aided Techniques in Food Technology,* edited by Israel Saguy
9. *Handbook of Tropical Foods,* edited by Harvey T. Chan
10. *Antimicrobials in Foods,* edited by Alfred Larry Branen and P. Michael Davidson
11. *Food Constituents and Food Residues: Their Chromatographic Determination,* edited by James F. Lawrence
12. *Aspartame: Physiology and Biochemistry,* edited by Lewis D. Stegink and L. J. Filer, Jr.
13. *Handbook of Vitamins: Nutritional, Biochemical, and Clinical Aspects,* edited by Lawrence J. Machlin
14. *Starch Conversion Technology,* edited by G. M. A. van Beynum and J. A. Roels
15. *Food Chemistry: Second Edition, Revised and Expanded,* edited by Owen R. Fennema
16. *Sensory Evaluation of Food: Statistical Methods and Procedures,* Michael O'Mahony

Handbook
of Food
Analysis

(in two volumes)

V O L U M E 1

PHYSICAL

CHARACTERIZATION

AND

NUTRIENT

ANALYSIS

edited by

Leo M. L. Nollet

Hogeschool Gent
Ghent, Belgium

Marcel Dekker, Inc.

New York • Basel • Hong Kong

Library of Congress Cataloging-in-Publication Data

Handbook of food analysis / edited by Leo M. L. Nollet.
    p. cm. — (Food science and technology)
    Includes index.
    Contents: v. 1. Physical characterization and nutrient analysis — v. 2
Residues and other food component analysis.
    ISBN 0-8247-9682-9 (vol. 1 : hardcover : alk. paper) —
    ISBN 0-8247-9683-7 (vol. 2 : hardcover : alk. paper)
    1. Food—Analysis—Handbooks, manuals, etc. 2. Food contamination —
Handbooks, manuals, etc. I. Nollet, Leo M. L. II. Series: Food science
and technology (Marcel Dekker, Inc.) ; v. 77.
    TX541.H36 1996
    664'.07—dc20
                                         96-17851
                                              CIP

The publisher offers discounts on this book when ordered in bulk quantities. For more information, write to Special Sales/Professional Marketing at the address below.

This book is printed on acid-free paper.

Marcel Dekker, Inc.
270 Madison Avenue, New York, New York 10016

Current printing (last digit):

10 9 8 7 6 5 4 3 2 1

PRINTED IN THE UNITED STATES OF AMERICA

Preface

The science of food analysis has developed rapidly in recent years. The number of articles and papers is increasing daily. New analysis techniques are being developed and existing techniques optimized. Analysis systems have been fully automated. A compilation of these new analytical methods is needed by food chemists.

Analytical and qualitative testing of product composition is one part of the analysis of food compounds. Another part is the guarantee of product quality as productivity increases. In the food industry there is need for analysis of components in both raw and processed products. All sorts of analysis techniques are necessary in the development of food products and in controlling food safety.

Knowledge of the chemical and biochemical composition of foods is required. What structural changes may occur during the processing and storage of foods? What are the effects of the use of agrochemicals? Detection of additives and toxins is necessary. Food product labeling must include significant ingredient and nutrient information. Directives (EEC) or Acts (U.S.) are developed for that reason. To get ingredient and nutrient information, food has to be analyzed or tested. Knowledge of chemical composition is important to the health, well-being, and safety of the consumers.

Most chapters in this book have been built using the same structure. The physical and chemical properties are enumerated first: What are the properties in food? Why must one detect the different components? What are the regulations in the food industry? Next, sample preparation and/or derivatization techniques are described. What extraction and clean-up methods are used?

Detection methods are then fully detailed. Brief accounts of the historical methods are presented. More attention is given to classical methods and even more to recently developed and automated methods. What are the future trends in the analysis of the different food components? What are the advantages and/or disadvantages of the different methods? What are the detection limits, the accuracy, the duration of analyses, the costs, or the reliability? Each chapter ends with the enumeration of reviews.

Data are summarized in charts and tables, structured through the volume in the same way. An abundance of such material, along with graphs and examples, makes this manual a solid reference book.

The handbook contains 48 chapters covering all topics of food analysis. It is divided into five parts. In the first part of Volume 1, sample preparation and chemometrics are discussed. Chapter 1 deals with sample preparation. Techniques and types of sampling, homogenization and preparation are discussed. Statistics and chemometrics are essential to the modern food analyst, and Chapter 2 provides applications of multivariate techniques as well as of conventional data analysis routines.

The next four chapters deal with the physical characterization of food. Methods for measuring characteristics such as dry matter, ash content, viscosity, elasticity, and so on, are fully detailed in these chapters.

The third part of Volume 1, on nutrient analysis, discusses different food components. Chapters 7 to 25 give methods of sample preparation, analysis, and detection of those components. The reader can find information on amino acids, peptides, proteins, enzymes, lipids, phospholipids, carbohydrates, alcohols, fat-soluble vitamins, water-soluble vitamins, organic acids, organic bases, phenolic compounds, bittering substances, pigments, aroma compounds, and dietary fiber.

In Volume 2 of the handbook, methods of detection of residues in food are summarized. Residues originate from different sources. The possible contamination of foodstuffs by fungi is well known. Antimicrobial agents are valuable for livestock productivity, but the residues present a great concern for public health.

Other residues of different origins are also of concern. In 15 chapters methods for detecting various residues are discussed: mycotoxins, phycotoxins, antibacterials, residues of growth promoters, residues of various pesticides, fungicides, and herbicides, food packaging residues, dibenzo-*p*-dioxins, dibenzofurans and biphenyls, *n*-nitroso compounds, polycyclic aromatic hydrocarbons, metal contamination.

In the fifth part miscellaneous components are discussed and a final chapter on instrumentation and technique is given. Many substances are added to foodstuffs to improve or preserve the quality and/or to prevent microbial growth. Topics such as nonenzymic browning, colorants, preservatives, synthetic food antioxidants, and intense sweeteners are discussed.

Anions and cations occur naturally in foods and can also be added as preservatives. Analytical methods to detect anions and cations are summarized in Chapter 46. Recently developed methods of identification of irradiated foodstuffs are given in the following chapter. Finally, Chapter 48 describes most of the cited techniques, separation, and detection methods.

This is a handbook of methods of analysis, and it offers the reader detailed descriptions of step-by-step procedures as well as a great number of references summarized in tables. This volume may be used as a primary textbook for undergraduate students in the techniques of food analysis. Furthermore, it is intended for use by graduate students and all scientists involved in the analysis of food components.

The contributing authors from throughout the world are strong leaders in all fields of food analysis in both industry and academic institutions. I congratulate them and thank them for their excellent efforts.

This work is dedicated to my wife.

Leo M. L. Nollet

Contents of Volume 1: Physical Characterization and Nutrient Analysis

NUTRIENT ANALYSIS

A cumulative index appears in Volume 2.

Contents of Volume 2: Residues and Other Food Component Analysis

Contributors to Volume 1

Claudia Abrigo Department of Analytical Chemistry, University of Torino, Torino, Italy

Mauro Amelio Quality Control Laboratory, Fratelli Carli SpA, Imperia, Italy

Leon Baert Department of Applied and Physical Chemistry, University of Ghent, Ghent, Belgium

George F. M. Ball Independent Consultant, Windsor, Berkshire, England

Robertino Barcarolo Instrumental Analyses Laboratory, Dairy and Food Biotechnology Institute of Thiene, Thiene, Italy

Jeffrey H. Baxter Department of Strategic Research—Medical Nutrition, Ross Products Division, Abbott Laboratories, Columbus, Ohio

Jean A. Bézard Nutrition Research Unit, University of Burgundy, Dijon, France

Paola Biglino Department of Analytical Chemistry, University of Torino, Torino, Italy

Carlos Calvo Institute of Agrochemistry and Food Technology, High Council of Scientific Research (CSIC), Valencia, Spain

Ennio Campi Department of Analytical Chemistry, University of Torino, Torino, Italy

Pierino Casson Instrumental Analyses Laboratory, Dairy and Food Biotechnology Institute of Thiene, Thiene, Italy

Giuliana Drava   Institute of Food and Pharmaceutical Analysis and Technology, University of Genoa, Genoa, Italy

Reinhard Eder   Department of Chemistry, Federal Research Institute and College for Viticulture and Pomology, Klosterneuburg, Austria

Michele Forina   Institute of Food and Pharmaceutical Analysis and Technology, University of Genoa, Genoa, Italy

Maria Carla Gennaro   Department of Analytical Chemistry, University of Torino, Torino, Italy

Donatella Giacosa   Department of Analytical Chemistry, University of Torino, Torino, Italy

D. Blanco Gomis   Department of Physical and Analytical Chemistry, University of Oviedo, Oviedo, Spain

Dolores González de Llano   Dairy Products Institute of Asturias, High Council of Scientific Research (CSIC), Villaviciosa (Asturias), Spain

Tomás Herraiz   Department of Food Chemistry, Institute of Industrial Fermentations, High Council of Scientific Research (CSIC), Madrid, Spain

A. Huyghebaert   Department of Food Technology and Nutrition, University of Ghent, Ghent, Belgium

Hyoung S. Lee   Citrus Research and Education Center, Florida Department of Citrus, Lake Alfred, Florida

Michael J. Lichon   Department of Plant Science, University of Tasmania, Hobart, Tasmania, Australia

David Madigan   Research Centre, Guinness Brewing Worldwide, Dublin, Ireland

J. J. Mangas Alonso   Departmento di Sidra, Centro de Investigación Aplicada y Technologia Agroalimentaria (CIATRA), Villaviciosa (Asturias), Spain

Ian McMurrough   Research Centre, Guinness Brewing Worldwide, Dublin, Ireland

John A. Monro   Food Science and Technology Division, New Zealand Institute for Crop and Food Research Limited, Palmerston North, New Zealand

R. Muñoz   Department of Plant Biology, University of Murcia, Murcia, Spain

Young W. Park   Cooperative Agricultural Research Center, Prairie View A&M University, The Texas A&M University System, Prairie View, Texas

Giovanni Parolari   Experiment Station for Food Processing and Preservation, Parma, Italy

Miguel Peris-Tortajada   Department of Chemistry, Polytechnic University of Valencia, Valencia, Spain

M. Carmen Polo Institute of Industrial Fermentations, High Council of Scientific Research (CSIC), Madrid, Spain

A. Ros Barceló Department of Plant Biology, University of Murcia, Murcia, Spain

L. Faye Russell Department of Agriculture and Agri-Food Canada, Centre for Food and Animal Research, Ottawa, Ontario, Canada

Boukaré G. Semporé Nutrition Research Unit, University of Burgundy, Dijon, France

D. W. Stanley Department of Food Science, University of Guelph, Guelph, Ontario, Canada

Ashley P. Stone Department of Food Science, University of Guelph, Guelph, Ontario, Canada

Marvin A. Tung Department of Food Science, University of Guelph, Guelph, Ontario, Canada

Cristina Tutta Instrumental Analyses Laboratory, Dairy and Food Biotechnology Institute of Thiene, Thiene, Italy

John Van Camp Department of Food Technology and Nutrition, University of Ghent, Ghent, Belgium

Jan Vanderdeelen Department of Applied Analytical and Physical Chemistry, University of Ghent, Ghent, Belgium

Paul Van der Meeren Department of Applied Analytical and Physical Chemistry, University of Ghent, Ghent, Belgium

Bill W. Widmer Citrus Research and Education Center, Florida Department of Citrus, Lake Alfred, Florida

1

Sample Preparation

Michael J. Lichon
University of Tasmania, Hobart, Tasmania, Australia

I. INTRODUCTION

Almost without exception, food is a complex inhomogeneous mixture of a staggering range of chemical substances. The isolation and measurement of individual chemical compounds in food usually represents a difficult task. Even with powerful modern techniques of separation and identification, such as high-performance liquid chromatography (HPLC), rarely is it possible to directly load a syringe with a food matrix and inject to obtain a sensible result. Perhaps surprisingly, it is not unusual to find analytical methods published with precision data reflecting repeated direct assays of standard solutions. This tells the reader little about the practicality of the methods to real world samples. Procedures for preparation of the sample should be developed, evaluated, and published as an integral part of any analytical method.

There are three steps involved in sample preparation for chemical analysis of foods: (a) *sampling*, i.e., obtaining a sample for the laboratory; (b) *homogenization* of the laboratory sample to enable the taking of test portions; and (c) *sample preparation*, i.e., physical and chemical manipulation of the test portion prior to analytical measurement. It should be appreciated that elements of these steps may occasionally occur in the reverse order or as combined operations. The fourth and final step of the analysis is the actual determinative assay procedure. Paradoxically, although the purpose of each of the three steps is to increase the accuracy and precision of the analysis so the test portion reflects the composition of the bulk, each step also introduces inherent errors. The error contributions of these steps for a typical food analysis scheme are shown in Fig. 1. Analyte concentration is limited at one end by detection limit and at the other by overloading of preparation stages or the measurement instrument by either analyte or matrix. The significance of the contributions of these steps to the total error for the analysis are mathematically described by the relative standard deviation (RSD) relation, Eq. (1).

$$RSD_{total} = [RSD^2_{sampling} + RSD^2_{homogenization} + RSD^2_{sample\ preparation} + RSD^2_{analytical}]^{1/2} \quad (1)$$

The equation shows the errors to be additive. Either a greater number of preparative steps or larger errors in a smaller number of steps may create a worse overall error. Furthermore, the errors

1

of each step are not only cumulative but strictly irreversible. Any error generated, by sampling (for example), cannot be compensated for by any subsequent treatment.

From the equation it is clear that if any one of the contributing factors is significantly greater than the others, it is futile to attempt to reduce any of the other contributors, as the total error will be disproportionately dictated by the most dominant factor being squared. A good example is aflatoxin analysis, where as long as the sampling error contributes 90% of the error there is little justifiable incentive to improve the analytical precision (1) beyond any professional interest.

It may often be clear to the experienced analyst what the approximate proportions of the contributing errors are of the total for a familiar analysis. If this is not so, then these may be defined by rigorous assays of replicates and recoveries testing the effects of each successive step of the analysis. This chapter has been divided into three sections to consider problems encountered in each of these respective steps.

II. SAMPLING

There has been great concern among analysts over the validity of analytical methods. Attempts to rigorously define precision and accuracy of methods include such measures as international collaborative trials. For all the benefits of implementing such expensive measures there is one important oversight, i.e., the issue of sampling is not examined. Experience demonstrates that sampling can often be the greatest source of error in chemical analysis, particularly for food matrices (Fig. 1). Frequently lack of attention to sampling is the hidden factor behind "unexpected" analysis results.

> The classic example of incorrect sampling procedure and its ridiculous consequences is given by the fable of the blind men and the elephant. The consequences are sometimes no less ridiculous for incorrect chemical sampling [W. J. Blaedel and V. W. Meloche, quoted in (2)].

Prescribed undergraduate texts often form the basis of analysts' future attitudes. The spectrum of emphasis on sampling ranges from serious but brief mentions (3–5), through a good treatment

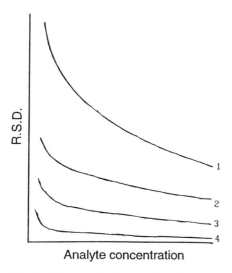

Fig. 1 Typical relative standard deviation of error components for an inhomogeneous food matrix. 1 = sampling, 2 = homogenization, 3 = sample preparation, 4 = assay procedure.

but at chapters at the end of the book (2), to an integral treatment from the beginning (6). This choice resides with course designers and coordinators.

Unfortunately, in the real world the benefits of correct sampling come at a cost. However, if one considers the assay of a sample that does not represent the bulk, or "population," of interest as nearly useless, then to obtain meaningful analysis one must be prepared to pay for it. If the samples, individually and collectively, cannot provide the required information, then they are seldom worth the time and expense of analysis. Correct sampling by competent personnel may require an investment of anything up to, or even more than, the cost of the laboratory assays (7). Planning for informative sampling must be an integral part of any study. Subsequent occasions of analysis of the same situation may cost somewhat less on account of experience. The degree of justifiable investment often depends on the purpose of the assay: For assay leading to the determination of a market price of a shipment, it may well be acceptable to spend 1% of the market value of the commodity on reliable sampling. While this may at first seem a high cost, it may be low when regarded as an insurance policy against larger losses. If a pesticide content of an export shipment of beef is suspected of approaching the acceptable threshold, the sampling could justify a far higher expense than, for example, when routinely examining for other contaminants known to be far below regulatory limits.

The best method of sampling in a given situation will depend on such issues as, What information is sought? What resources are available? What is the accessibility of the target population? Is the population heterogeneous? If so, is the variation general, localized, or stratified? Is there temporal variation as well as spatial variation? What is the required turnaround time? What is the perishability of the food and the analyte? Should the population be sampled critically or representatively, randomly or systematically? What are the criteria for acceptability and sampling correctness? Are samples to be pooled or replicated? Should analysis be performed separately on different portions of the sample? Is the surface to be included in the bulk? What monitoring should occur to prevent contamination and abuse?

The difficulties of obtaining good representative sampling depend on the nature of the "population." Discrete populations are relatively easy to sample. In the case of a shipment of pallets of canned food, the units are discrete and may be mathematically modeled and sampled; the only inconvenience then is the necessity of dismantling some of the pallets to extract the desired target cans. This correct procedure is often bypassed, and cans are taken from the top corner of the most accessible pallets to give a highly biased sampling.

Sampling of continuous populations is by far more common. Most difficult sampling problems usually involve a 3D bulk population: direct sampling is rarely possible (7). If at all possible in practice, the most correct approach is to convert the bulk into a 1D flowing stream. The stream is then divided systematically, randomly, or by a combined formula of random delineation of systematic segments, into target segments in time, such that any part of the bulk has an equal probability of scrutiny (7). The entire cross-section segments (strictly delineated by parallel planes, regardless of whether a 1D sample is flowing or not, Fig. 2) of the stream are sampled at the target times, then taken either separately or pooled for analysis. In a factory situation, the best sampling points are often conveyor belts, from which segments may be easily extracted. Another method of treatment of a 3D bulk is to convert it to a flattened 2D configuration. The 2D bulk may then be sampled by cylindrical corer, at random or systematic points on the surface, ensuring that the entire thickness of the bulk is recovered. In practice, this method often fails to strictly recover a strictly delineated cylindrical core or to adequately expose the very bottom of the bulk to scrutiny by virtue of the sampling device being unable to retrieve the lowest part of the cylinder. The 2D bulk may be fashioned into a long narrow pile and cross-sampled in space as a 1D bulk (Fig. 3), with randomly or systematically located segments delineated by strict parallel planes (Fig. 2). The classical method of coning and quartering can rarely be done correctly: operator bias often fails

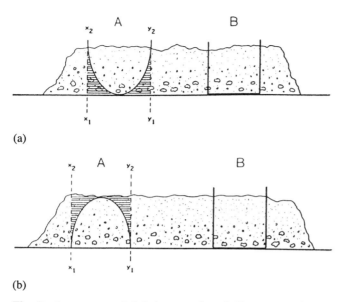

(a)

(b)

Fig. 2 Importance of defining samples of 1D segments by parallel planes. (a) A: incorrect increment delimitation introduced by a scoop with round section; B: correct increment delimitation performed by a scoop with square section. (b) A: incorrect increment delimitation introduced by a scoop with no side walls to prevent material from falling; B: correct increment delimitation performed by a scoop with correct side walls. (From Ref. 7.)

the process (7), and fundamental problems of unmixing (both spontaneously and when being moved or agitated) of flowable materials come into play (see Section III.B).

It may on some occasions be desirable to analyze unrepresentative target, or extreme portions of a population, as well as a representative sample. It may be of interest to find extreme values for vitamins, to find the worst case for environmental contamination, or to link distribution with a causative factor. It may be desirable to select specimens with a view to breeding selection.

To consider some additional variables in real sampling problems, take the example of analysis of peas: Sample variables include size distributions, position of individual peas in their pods, height of pods up the individual vine, individual plant genetics, cultivar, time of planting, efficiency of

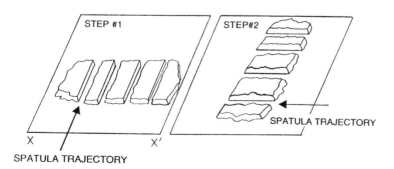

Fig. 3 Sampling 3D bulk by flattening into a 2D slab. Cross-sampling of slab to obtain 1D lots. The 1D lots are then cross-sampled again. Correct increment delimitation is thus possible. (From Ref. 7.)

pollination, watering history, soil type and underlying geology, previous crop history, soil fertilization, crop maturity, disease and pestilence attack, length of time, handling and storage conditions since harvest. Some further questions follow: Will either the sample deteriorate or the data expire before the sampling, analysis, and reporting is complete? Are the peas to be cooked? If so, how and how long? Is the analysis to represent data for a single plot, farm, locality, or national database? Does the sampling need to be certified for legal or economic reasons?

The compilation of food composition data presents considerable sampling problems. Typically nutrient contents frequently reflect limited localized sampling (8). The task of nationwide representative sampling is difficult to organize and resource-intensive.

Environmental monitoring frequently involves the sampling of foods, such as trout or mollusks, which accumulate heavy metals from the environment. Other foods, such as milk, through which children may be particularly exposed to some pollutants, are often the subject of monitoring programs (9).

Production sampling may seek to determine a market price based on nutritional content (such as oil in seeds) or to select on the basis of contaminant content (aflatoxins, pesticides, etc.).

These issues are discussed to varying degrees in the sampling literature; they are mostly self-explanatory or specific to situations. In the latter case discussion is usually found in sections of literature specific to the situation or remains the unrecorded know-how of specialists. These specialists must be encouraged to include such details in their publications.

Literature concerning sampling is well dispersed and generally not easy to locate. A useful survey (10) lists over 60 references of possible use to food analysts, some general references (11–15) being recommended. Several others included various mathematical treatments (16–18) of both general and specific problems. A mathematical consideration (19) of chemical analysis was not included, presumably because it concentrates on assessment of data quality rather than the practical aspects of sample planning. The American Chemical Society has published guidelines (20) regarding sampling. International standards are complete for sampling of fruits and vegetables (21), meat (22), and oilseeds (23).

More recent discussions of sampling (24–27) have attempted to integrate sampling approaches with laboratory practices, sample characteristics, and analytical problems.

Pitard (7) explains the exhaustive sampling correctness theory of Gy, which remains the most authoritative to date. Despite the context and examples of this work centering on the problems of the mining industry, it takes little imagination to relate the applications to the problems of food analysis. Minkkinen (28) devised a computer program based on Gy's theory to perform various calculations, e.g., to estimate minimum sample size to meet precision criteria given certain particle size.

The importance of sampling was recently highlighted by problems encountered in analysis of aflatoxins in peanuts. Several workers have written papers specifically addressing this type of sampling situation (1,29–32). In these cases of highly inhomogeneous distribution of analyte in the matrix, the proportion of total analytical error attributable to sampling is commonly greater than 90%. Similar distribution inhomogeneity may present problems in sampling individual chocolate chip cookies, cans of sausages and beans, and croutons in a soup packet. The best strategy may often be to take a very large initial sampling quantity prior to preliminary homogenization. Another approach is to take many replicate samples, say 20–200, for analysis to show individual unit variability. The differing approaches yield different information that may meet different requirements. Pooling of several individual units may give better definition of the true mean of assay results (Fig. 4). However, individual unit variation may better reflect single-serving variability for the consumer. The example of β-carotene content in canned stew in Fig. 4 is a reflection of statistical sampling of units containing relatively few, large particles of carrots in a small discrete sampling bulk, from a large population. By pooling 12 cans together for assay, the manufacturer

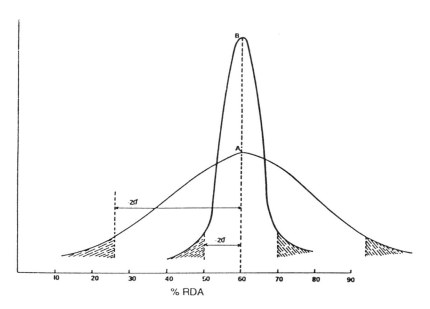

Fig. 4   Distribution curves for vitamin A content of canned food with carrot chunks. A: single-can samples; B: 12-can pooled samples. (From Ref. 13.)

is able to meet specifications for vitamin A content using less of the ingredient, simply by the reduction of the RSD.

All too often the analyst has little influence over the taking of the sample. A widening of appreciation of the importance of sampling may serve to rectify this problem. Ideally the analyst should appraise the problem and take the samples personally. Failing that, the analyst should endeavor to thoroughly brief the sampler as to the most appropriate methods for each situation. A specific instance of such difficulty concerned an untrained sampler being assigned to learn about the traditional indigenous foods of Australian aboriginals. He was briefed to collect specimens for nutritional analysis but was apparently little influenced by the analysts. Constraints apart from the lack of training were the hot climate, lack of refrigeration, desirability of taking pocket size samples, and remoteness from the laboratory, some 3000 km and several days freight away. All of these factors contributed to the degradation of sample integrity. Many of the samples analyzed consisted, for example, of three individual thawed fruits totaling 15 g. However, while the limitations of the nutritional assays on such unrepresentative samples are obvious, the data are nevertheless at least indicative and as such still useful in this context (33–36), where previously nothing was documented in the way of nutrient content of these foods.

III.   HOMOGENIZATION

The complex structure and composition of food substrates necessitates homogenization prior to most chromatographic analysis. Variable texture, structure, and viscosity as well as the presence of immiscible phases and hygroscopic or hydrophobic matter all contribute to the difficulty of this operation meeting with success. The observation that collaborative test results for food materials often show greater coefficients of variation than other matrices (37) is therefore not surprising.

The classical wet chemistry methods of analysis held sway for decades. They rarely involved the manipulation of test portions <1 g and measurement precision was variable. Hence the old conventional methods of homogenization usually provided adequate test portion sampling of common foods. More recently, with the development of analytical instrumentation, detection limits and test portions decreased dramatically, while sensitivity to contaminants and artifacts increased. Simultaneously, as food matrices became more technologically diverse, so did the diversity of analytes of interest. The need arose to review existing methods for homogenization and develop new techniques to adequately meet the new demands.

Problems encountered with test portion sampling, particularly for semimicrocombustion analysis (38), led to the writer's investigations concerning homogenization methods used for food samples (39). This paper surveys nine conventional methods and three cryogenic methods of homogenizing numerous food samples, condensed into seven categories of matrix type. Several methods were subjected to more rigorous examination. Other workers have considered more restricted ranges of (conventional) methods applied to limited food types (13,40–43). For assays using test portions of 1 g or more, several of the conventional methods prove satisfactory with compatible matrices (39). Many method–food category combinations proved to be incompatible, some unexpectedly so. Conventional techniques usually operate by cutting and shearing for samples of wet consistency and by shearing and impact for dry samples. Weaknesses of these techniques include the reliance of the inertia of the supporting fluid, or surrounding particles, as the "anvil" for the grinding action. Examples of failure of these methods include the persistence of corn kernel skins and long meat fibers. For techniques such as hammer milling (39) the serrated body of the grinder serves this purpose. The CO_2 milling technique (39) uses dry ice as inert mass particles, which also serve to embrittle the food particle. The liquid N_2 milling technique (39) similarly uses cryogenic embrittlement as well as a metal–metal impacting action. For assays using smaller test portions or requiring stringent homogeneity of very heterogeneous foods the cryogenic treatments proved well worth the extra effort after a conventional pretreatment and freeze-drying. This dual treatment reduced particle sizes to below 60 μm (97% below 10 μm) for one of the most difficult matrices (muesli bars) with sufficient mixing to take reproducible test portions of 1 mg. The average RSD of semimicrocombustion protein assays for a range of foods was 1.33%, performed on test portions of 2–5 mg (39). The number of particles included in each 2-mg test portion following N_2 milling was approximately 1×10^5.

The required size of the test portion and the sample's characteristics will dictate the degree and type of homogenization required. If several different assays are to be performed on a sample then whichever has the most stringent requirements will often dictate the homogenization requirements. Experience shows that it is often prudent to sequentially use two or more homogenization techniques. It may be desirable to split the sample after an initial wet basis homogenization treatment; analyzing the first part for labile vitamins directly with further rigorous homogenizing after freeze-drying before subjecting the second part to other analysis. A generalized example scheme for homogenization of samples for nutritional analysis is shown in Fig. 5. This scheme includes approximations for quantities and particle sizes at each stage and what types of assays are amenable to the products of each stage. However, it must be reiterated that each sample will have different characteristics that will require different homogenization treatments. Table 1 illustrates a summary of example assays and the consequences of using three example homogenization treatments. The main conclusion that should be drawn is that each sample should be homogenized by methods that have proven effectiveness with the particular matrix, either from experience, from the literature, or by experiment, to a degree that meets the test portion requirements and other facets of the assay procedures. This must be confirmed by the precision of replicate assays, blanks, and recoveries.

A. Particle Size Reduction

There are two functions of homogenization: reduction of particle size and mixing. Reduction of particle size involves cutting, shattering, and shearing. The various homogenization devices

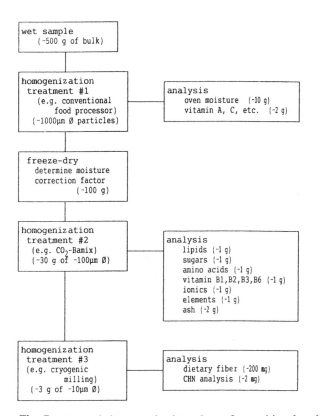

Fig. 5 A sample homogenization scheme for nutritional analysis of food. ø = diameter. (From Ref. 78.)

achieve these in different ways and to different degrees. This necessitates judicious choice of homogenization methods that have demonstrated applicability for use on particular matrices. The efficacy of a method may be observed by microscopic examination or sieving of the product. The importance of particle size reduction is intuitive; the smaller the particles, the greater the number of particles included in a given test portion, and hence the greater the probability of sampling parts of the original bulk. Quantifying this statistical notion is more difficult. A simplified treatment with graphed relationships (14) is recommended reading for nonmathematical analysts. The computer program (28) mentioned previously may have application as well. The standard deviation plot in Fig. 6 clearly shows the improvement in precision that results from particle size reduction. For reasonable precision in assaying a complex food, the sample should be ground sufficiently such that the smallest test portion should contain at least 500 particles of the least common ingredient.

Apart from the aspect of test portion sampling, particle size may influence other facets of the analytical procedure, where surface area and/or particle radius are critical to reagent exposure and penetration. Hence the degree of particle size reduction may affect the accuracy as well as the precision of the analysis (Table 1) in ways that may not be readily revealed by recovery data. A typical case is the accessibility of the food matrix to enzymatic digestion. The efficiency of enzymatic digestion is proportional to the surface area (or degree of homogenization) of the food substrate, i.e., inversely proportional to particle size (radius). Large particle size may inhibit enzyme access to the whole of the food; or specific portions may be encapsulated and rendered unavailable for subsequent extraction; such problems have been encountered in thiamine (44),

Table 1 Consequences of Using Various Homogenization Treatments Prior to a Selection of Food Assays

Assay and test portion	Type of homogenization treatment, approx. particle size		
	Conventional, 1–2 mm	CO_2 mill, 100–200 μm	N_2 mill, 10–20 μm
Fat, 1–5 g	Poor extraction, OK sampling	Better	Best extraction but possible for emulsion problems
Fiber—AOAC, 1 g	Poor sampling, OK digestion	Best	Losses through 90-μm pore filter
Fiber—Englyst, 200 mg	Sampling too poor	OK	Best
Moisture, 2–10 g	Best	NA—process changes moisture	
Vitamins, 1–2 g	Best for labile vitamins	Better sampling (if vitamins stable after grinding treatment)	
Bomb calorimeter, 0.5 g	Poor	OK	Preferred
GFAAS direct, 1–10 mg	Useless	Poor sampling	Preferred
CHN, 2–5 mg	Useless	Poor sampling	Preferred

GFAAS, graphite furnace atomic absorption spectroscopy; CHN, semimicrocombustion CHN elemental analysis.

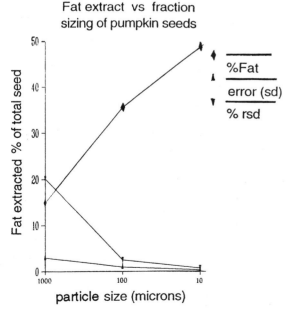

Fat extract vs fraction sizing of pumpkin seeds

%Fat
error (sd)
% rsd

Fig. 6 Fat extraction from pumpkin seeds homogenized to three particle sizings by conventional treatment, CO_2 milling, and N_2 milling. Total extractible fat and error (both absolute and relative standard deviation) improves dramatically with reduction of particle size. This figure is derived from data (used with permission) from the writer's research at Defence Science and Technology Organization, Materials Research Laboratory, Tasmania.

lipid (45), moisture, and dietary fiber (46) analysis. Efficient thiamine extractions were only possible after digestion of finely homogenized foods. In the second example, digestion of a well-homogenized substrate was required for complete release of lipids. Efficient solvent extraction was found to be possible only with small particle radius (Fig. 6). However, if some foods were too finely ground (with very large reactive surface area), problems with emulsion formation hampered the efficiency of the subsequent liquid–liquid extraction. Moisture determinations are similarly affected by large particle size (7). Large-radius particles often fail to quantitatively yield water during the procedure, as the innermost molecules are trapped by the depth of surrounding material. Moisture–time curves for such samples are shallow, poorly defined, and do not reach recognizable plateaus. In contrast, Englyst dietary fiber determinations (46) on poorly homogenized samples give spuriously high results due to incorporation of undigested starch in the fiber fractions. Analysis of total dietary fiber by the AOAC method (47) has different problems: if the sample is too finely ground then there is the risk of low results through losing fiber through the 90-μm porosity filter, even with the use of filter aids; if the sample is too coarse then there will be high results from insufficient enzymatic digestion of other components, and a loss of precision from poorer test portion sampling of larger particles (Table 1).

B. Mixing

While particle size reduction helps to improve sampling at the test portion by increasing the number of particles sampled, the sample must be well mixed to achieve the gain in accuracy and precision.

Mixing is not easy—ask anyone who has attempted to blend home-made muesli thoroughly enough to prevent naughty children from "segregating" their breakfast serving to maximize their favorite ingredient content at the expense of siblings!

Most sampling and homogenization processes for foods lead to powdered samples. Simple observation of powders leads to the definition of two broad categories: free flowing and cohesive (48). The distinct properties of each lead to important consequences for analysis of test portions.

Free-flowing powders, such as rice or coarse sugar, exhibit a smooth flow, have an attractive nondusty appearance, and adhere little to container walls. In broad terms, particle size for free flow needs to be above 50 μm, given average surface properties and relative dimensions of the particles. This cutoff size increases with particle elongation or stickiness. On the face of it, flowable powders are attractive for homogenization of food samples, as they are easy to mix and handle. However, the flowable properties lead to mixtures unmixing as easily as mixing. As individual particles are free to move, they will move and settle differentially according to individual characteristics such as mass, density, shape, angularity, angle of repose, size, and surface properties. Some examples of simple unmixing causes are shown in Fig. 7. Therefore, any induced motion will lead to classification and segregation. The muesli example clearly falls into this category. Segregation mechanisms include projection, percolation, and elutriation. Coarse particles, by virtue of momentum, are projected farther than finer particles. Smaller, denser particles have a greater ability and mobility to percolate through a loose mass of particles. Percolation may take place when moving down an incline; rolling and tumbling; pouring into a pile; or when subject to vibration. Elutriation is the loss of fines as dust. This commonly occurs when filling a container, i.e., the displacement air blows dust out of the mixture; or when stirring or mixing in an open vessel. This loss of fines may be critical to analysis, as the composition of the dusts often differs markedly from that of the coarser particles. For example, dust in peanuts may be disproportionately rich in mycotoxins. The counters to these drawbacks of flowable powders are to reduce particle size by homogenization into a cohesive powder, or to ensure that any mixing is active, not passive. Mixing should involve active displacement of particles rather than allow them to choose their own paths. A rolling, tumbling mixer gives individual particles freedom and is undesirable, whereas a paddle mixer pushes and relocates groups of particles and is preferred.

Cohesive powders, such as flour, have an erratic stick-slip flow, are often dusty, and stick to container walls. Smaller particle size (<50 μm) ipso facto enhances the potential for mixing to meet test portion precision requirements. Furthermore, the cohesive structure inhibits segregation. Cohesive powders lack mobility as light individual particles are held in a structure and cannot easily move independently. Hence mixing tends to behave irreversibly, with the structure preventing classification. Thorough mixing of a cohesive powder is more difficult due to poor mobility, is slower, and involves a more intensive process. Many of the mixing properties of cohesive powders are shared with thick fluids, typical of many wet basis food samples.

Mixing procedures need attention to two areas: adequate shearing and inclusion of the entire sample. The former is necessary to overcome cohesive forces, to break up and disperse aggregates into the mixture; it is necessary to repeatedly break individual particles free from their structuring neighbors. For many powders gentle rolling and breaking of a tumbling mixer may cause sufficient restructuring. For more cohesive aggregates, it may be necessary to put in more energy in the form of high-speed impactors to create a high shear zone in the mixer. However, for food care it is required not to overheat the mixture either locally or as a whole. The latter area requiring attention is to ensure the inclusion of the whole sample in the mixing process. The cohesion of finer powders lends them to be held up in dead spaces in the mixer and be isolated from the mixing action. Efficient scraping and sweeping of the entire mixing volume is essential.

There are several means that may prevent adequate mixing of foods: classification, agglomeration, and phase separation. Causes include differentiated particle sizes and shapes, surface adhesivity, electrostatic charging, disruption of stable structures maintaining surface tension, destruction of encapsulating structures, and various hydrophobic-hydrophilic interactions. Typical practical food examples include oil separation in finely ground nuts and classification of whole-grain flour. Some samples of brans and flours circulated for dietary fiber collaboration were found impossible to mix (46); the only means of foiling the classification of these flowable powders was to further grind the samples into cohesive powders. One common means of inducing classification error is the use of a vibrating spatula. This device should only be used for pure compounds. Mixing may be thwarted by the presence of a few large particles, especially if those contain the analyte of interest. Such a mixture needs to be reground to eliminate the large particles.

The surest way of avoiding such problems is selecting appropriate homogenization methods through experience and learning, and very often by intelligent trial-and-error experimentation. The CO_2-Bamix homogenization method (39), for example, usually produces a powder with particle size around 100 μm, yet is usually cohesive in mixing properties. The shearing action of this technique tends to produce irregular elongate particles, and food particles are often adherent, giving cohesive properties. The cohesion of the food–CO_2 mixture requires the systematic shaking and occasional scraping of the assembly during grinding to ensure inclusion of "dead" pockets. Mechanical mortar and pestle (39) grinding similarly requires regular scraping, in addition to that provided by the mechanical scraper. This method is very successful on some foods; however, the method is unsuitable for fatty foods because phase separation becomes a problem.

C. Examination of Particle Size and Mixing

Mixing may be more difficult to examine than particle size reduction. Experience suggests visual inspection of color and texture is very useful, but not necessarily rigorous, especially in the case of a sample consisting of components of similar appearance. The combined effect of reduction and mixing may be examined by performing assays on replicate test portions. For this examination it may be prudent in some circumstances to run simple, cheap assays rather than use the actual target

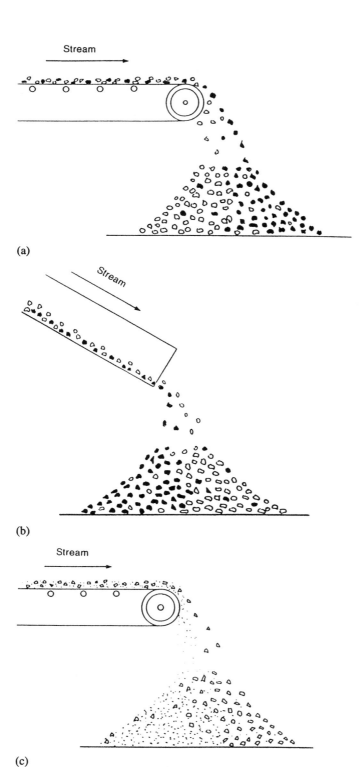

Fig. 7 Some causes of particle classification. (a) differences in momentum due to density variation, (b) differences in distribution by settling within flowing stream due to density differences, (c) differences in momentum due to size variation, (d) percolation of fines into pile while coarse fragments roll down the outside, (e) segregation generated by differences in angle of repose, (f) segregation by differences in friction rates. (From Ref. 7.)

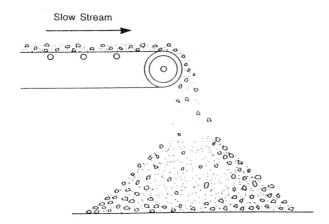

(d)

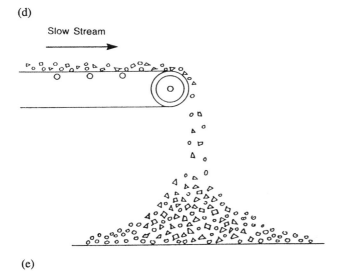

(e)

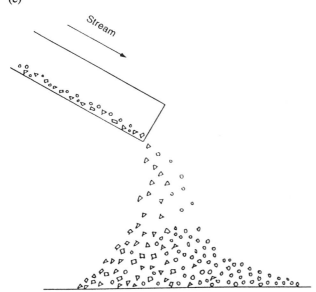

(f)

assay. The use of an assay more sensitive to variation than the target assay may be warranted. For a complex mixture, the confirmation of adequate randomization of one ingredient does not necessarily ensure that other components are mixed. It may often be safer to carry out separate statistical analyses for at least the key different types of analytes associated with different ingredients.

Another method of potential is statistical image analysis of the homogenized test portions under the microscope. The determination of particle size distribution is relatively straightforward. It may be necessary to color-label some components prior to treatment to aid differentiation of the constituent particles and/or observe under different wavelengths. A simple example is the addition of iodine to differentiate starch particles in a light-colored mixture.

Fortunately, most recommended homogenization techniques are found to simultaneously reduce particle size and mix samples with at least reasonable efficiency for both.

IV. SAMPLE PREPARATION

Sample preparation includes any operation performed to the test portion prior to the analytical measurement: storage, preservation, weighing, dilution, cleanup, extraction, digestion, purification, separation, derivatization, etc. Descriptions of these are usually included in publications, although frequently lack background information, "tricks of the trade," finer detail, or rigorous theoretical or practical error analysis.

A. Apportionment and Preparation of Foods and Storage of Laboratory Samples

Careful thought and often some background research is required to decide what parts of the sample are to be analyzed. For example, is fresh produce to be washed prior to pesticide analysis? If so, how? What constitutes the "edible portion?" Examples of such separations include removal of outer leaves, peels, and pips from fresh fruits and vegetables; removal of bones and trimming of excess fat from meat; exclusion of brines from canned vegetables but inclusion of liquid from canned fruit. Many such choices are subject to debate, such as the inclusion or otherwise of seeds in a sample of blackberry jam. It may be desirable to analyze both portions. In any case it is standard practice to weigh the separate portions, analogous to the determination of moisture when drying a sample.

Several workers have emphasized that for nutritional evaluation it is desirable to prepare the food in the same way as it is commonly consumed (13,24) prior to analysis. What are the customary methods and times of cooking of meals? What constitutes complete preparation of powdered soups and hot beverages? An example of inappropriate preparation is the analysis of such beverages for vitamins B_1 and C, where the assay of the beverage powder was assumed to be the measure of dietary intake (49,50). In reality, these vitamins degrade significantly when the powders are stirred into boiling water, especially if the water has been sterilized by iodine agents, as is recommended practice in this context. Vitamins in cooked foods may be reduced by as much as 80% when compared with the raw ingredient vegetables and meats (51,52). Food held for serving in Bains-Marie suffers further losses. When bread is baked, 30% of thiamine is lost; if alkaline baking powder is used, the loss can increase to 80%.

Storage and identification of samples requires meticulous attention and control. Analyzing the wrong samples is worse than any error generated by sampling, homogenization, preparation, or assay technique! Storage conditions will be determined by the perishability of the sample/analyte, delay before assay, and duration of sample archival. Foods are generally frozen in sealed plastic after preliminary homogenization, or freeze-dried before and after further homogenization treatments and stored in desiccators.

B. Mass and Volume Measurements

A practically universal step in the manipulation of the test portion is the weighing step. Fortunately, the precision of modern balances is commonly six significant figures or better. Judicious choice of balance can maintain this precision for a large range of test portion size, typically from 10 g down to 1 mg for food analysis. However, the analyst must ensure that the operation is performed accurately; moisture variation of food samples requires particular attention. Electrostatic charging may cause problems with dry samples and may be eliminated through the use of ionizing tools, such as "antistatic guns" sold in audiophile stores for discharging vinyl records. These same errors must be considered when drying samples and determining moisture correction factors. Once dry, samples should be stored in a desiccator. Weighing of dry powders should occur with desiccant present in the weighing chamber of the balance; boats of silica gel are adequate for general use, phosphorus pentoxide being recommended for operations with more hygroscopic materials. Care is needed to avoid moisture adsorption problems with weigh containers and tools. It is prudent to store these in a desiccator after oven drying.

A frequent operation contributing to the sample manipulation error is the volumetric dilution. For headspace sampling, the size, pressure, equilibration time, and temperature of the space are all critical. Volumetric errors are inherently orders of magnitude greater than those for mass measurements. Volumetric errors include the delivery volume of gas chromatography syringes and the like.

C. Chemical Manipulations

Digestion, extraction, and derivatization should all be quantitative. Efficiency may be enhanced, e.g., by application of microwaves for HPLC (53), Kjeldahl (54), and metal (55–58) analysis, but only if sample integrity is maintained. Minor elements may be determined after bomb calorimeter analysis in preference to oven ashing and digestion (59). Confirmatory tests should be used in doubtful cases. A typical simple test is the testing for residual starch with drops of iodine to confirm completion of amylase digestion (46).

The efficiency of extraction should be shown to be quantitative. Reextraction of residues should be blank. Recovery of added spikes should be 100%, though this does not conclusively mean that the (bound) sample analyte is necessarily fully extracted. Better assurance may be achieved by the use of two independent methods or the use of standard reference materials. As intimated earlier, degree of homogenization (particle size) may influence the efficiency of extraction to a greater (Fig. 6) or lesser (60) degree.

Recent developments in chromatographic analysis include the use of supercritical fluid extraction (SFE) (61,62). The use of CO_2 reduces potential for contamination and eliminates the problems associated with toxic solvents. Another novel cleanup method is the use of dialysis to remove all but small molecules prior to chromatography (63). Large molecules may also be digested by enzyme treatment, e.g., prior to amino acid (64) or nitrate analysis (65).

The methods of ensuring good recoveries while using solid phase absorption columns prior to liquid chromatography are straightforward. The use of preconcentration methods in gas chromatography (66) can be fraught with complications of volatility differences, reactivity, and adsorption. These issues are generally adequately discussed in methodology papers alongside chromatography details.

D. Degradation and Contamination

Degradation of the sample and analyte integrity may take place at any stage, from the taking, transport, and storage of the sample; through drying, homogenization; and sample preparation; to the injection of the manipulated test portion. Addition of contaminants, exposure of samples to heat, warmth (microbiological activity), moisture, oxygen, visible and ultraviolet light, reagent fumes can all compromise accuracy. These problems are considerable in vitamin analysis. Consider some examples:

Riboflavin is sensitive to ultraviolet light. Vitamins A, B_6, D, E, and folic acid are light-sensitive. Laboratory manipulations are usually performed using low-actinic glassware and preferably in the dark (44,67,68).

Plant tocopherols are sensitive to oxidation. Losses during homogenization may be minimized by addition of pyrogallol (69).

Ascorbic acid is particularly sensitive to degradation by oxidation, especially when exposed to atmospheric oxygen, heat, or high pH. Analysis schemes aim to reduce manipulation and turnaround time to an absolute minimum, making use of various stabilizing agents. A number of methods incorporate metaphosphoric acid at the homogenization stage. The 10-fold variation in ascorbic acid found in *Terminalia ferdinandiana*, a native plum found in northern Australia rich in this vitamin (33,34), is at least partly due to degradation during lengthy transport on different occasions. A less obvious hazard is contamination by traces of copper, which catalyzes the oxidation reaction.

Metal contamination, particularly by trace elements, is a common laboratory problem. Metals may be of interest from a nutrient point of view or as environmental contaminants. Many laboratories redistill acids on site, as well as using elaborate water purifiers. Recent work quantified the metal increment of using laboratory blenders (70,71). This increment could be reduced by chemical treatment of the blender surfaces (71). Another approach has been to design (72) or seek out homogenization devices with inert working surfaces. For laboratories dedicated to trace metal work, a clean room is a good investment to minimize laboratory environmental contamination (73). High purity, smooth surfaces, filtered ventilation, and positive pressure airlocks are typical features. Precautions may include the use of overclothing and hair confinement as well as strict hygienic practices analogous to those used in infectious bacteriology. Care is taken to avoid transfers via all contact surfaces, hands and implements, air and dust. Such practices are not confined to metal analysis but to any trace and environmental analysis whereby contamination of the samples or personnel is a potential hazard. Introduction of metal contamination may be serious beyond the simple raising of metal content. The presence of metal may promote degradation reactions compromising sample integrity, as mentioned above, but may also interfere with extraction, cleanup, and enzymatic digestion steps of sample preparation procedures.

Contaminants may be introduced by reagents. The development stages of a method for enzymatic digestive release of lipids (45) revealed that several commercial enzyme preparations contained unacceptably significant amounts of ether-extractable contaminants, adding to determination of fat.

Contamination and the potential complications in sample preparation procedures are highlighted by the gas chromatography of alditol acetates in the Englyst dietary fiber determination. Plasticizers may contaminate food samples at literally any point from the farm to the chromatograph. Pure samples of ubiquitous plasticizer contaminants were found to chromatograph at similar times to some of the analytes (74) but could be resolved from analyte peaks by capillary columns. More recent investigations found that exposing plasticizers to the derivatization procedure used to form alditol acetates yielded multiple and broad peaks that potentially interfered with the analytes (Fig. 8) (46). This is an additional artifact caused by the sample preparation technique fundamental to the analysis. This led to the elimination of certain plastics (such as polyvinylchloride) from the analytical procedure. Sample preparation artifacts have similarly been observed in fatty acid methyl ester analysis (75), whereby compounds synthesized by the derivatization interfere with the gas chromatography peaks of interest. A recent monograph (76) deals specifically with such analytical

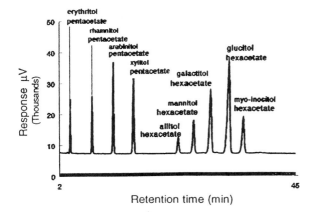

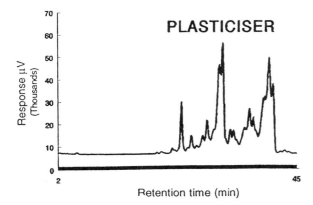

Fig. 8 Artifact peaks from "derivatized" contaminant plasticizers (lower chromatogram) can interfere with measurement of alditol acetate analyte peaks (upper chromatogram) for dietary fiber fractionation analysis. Chromatograms derived from data (used with permission) from the writer's research at Defence Science and Technology Organization, Materials Research Laboratory, Tasmania.

artifacts, with considerable attention to problems with gas chromatography–mass spectrometry, a technique widely regarded as definitive with respect to analyte specificity.

A novel approach to derivatization of fatty acid mixtures is to react the sample with reagent in the gas chromatography injection port (77), thus eliminating bench chemistry and attendant problems of sample contamination, errors, and limited control of reaction conditions.

All elements of the gamut of test portion manipulations performed are potentially significant error contributors. The consistent use of observation, replication of test portions, recoveries and reference materials should highlight problem areas. These may be reduced, or at least quantified, using the skill of the analyst.

V. CONCLUSION

Analytical errors usually cost money in terms of wrong decisions based on erroneous results, lost business, false security, litigation, and so forth. The analyst wishing for accuracy and precision must focus on all elements in each of the four steps of analysis: sampling, homogenization, sample preparation, and analytical technique. Resources and attention should be allocated to every one of these steps. While the last step is often routine in laboratories, the former steps are often less so despite their frequent domination of experimental error. Critical examination should reveal weaknesses where sample integrity may be compromised. The greatest effort should be expended to reduce contributions in the error-dominating steps. Authors should be encouraged to include all experimental and practical details of the first three steps in their publications. Progress in this direction may be enhanced by critical feedback from peers, referees, and editors.

REFERENCES

1. W. Horwitz and J. W. Howard, *NBS Special Publication 519*: 231 (1979).
2. W. E. Harris and B. Kratochvil, *An Introduction to Chemical Analysis*, W. B. Saunders, Philadelphia, 1981.
3. J. S. Fritz and G. H. Schenk, *Quantitative Analytical Chemistry*, Allyn & Bacon, Boston, 1978.
4. W. F. Pickering, *Modern Analytical Chemistry*, Marcel Dekker, New York, 1971.
5. A. I. Vogel, *A Textbook of Quantitative Inorganic Analysis*, Longman, London, 1978.
6. B. W. Woodget and D. Cooper, *Samples and Standards*, John Wiley and Sons, Chichester, 1987.
7. F. F. Pitard, *Pierre Gy's Sampling Theory and Sampling Practice*, CRC Press, Boca Raton, 1989.
8. L. F. Russell, L. W. Douglass, and J. T. Vanderslice, *J. Food Compos. Anal. 5*: 224 (1992).
9. L. H. Keith, *Environmental Sampling and Analysis: A Practical Guide*, Lewis, Chelsea, 1991.
10. B. G. Kratochvil and J. K. Taylor, *NBS Tech. Note 1153* (1982).
11. W. F. Kwolek and E. B. Lillehoj, *JAOAC 59*: 787 (1976).
12. R. C. Tomlinson, in *Comprehensive Analytical Chemistry*, Vol. 1A (W. L. Wilson and D. W. Wilson, eds.), Elsevier, Amsterdam, 1959, p. 36.
13. H. G. Lento, *NBS Special Publication 519*: 243 (1979).
14. W. E. Harris and B. Kratochvil, *Anal. Chem. 46*: 313 (1974).
15. B. Kratochvil and J. K. Taylor, *Anal. Chem. 53*: 924A (1981).
16. E. L. Bauer, *A Statistical Manual for Chemists*, Academic Press, New York, 1971.
17. W. G. Cochran, *Sampling Techniques*, John Wiley and Sons, New York, 1963.
18. W. J. Youden, *Statistical Methods for Chemists*, John Wiley and Sons, New York, 1959.
19. K. Eckschlager and V. Štěpánek, *Information Theory as Applied to Chemical Analysis*, John Wiley and Sons, New York, 1979.
20. Am. Chem. Soc., *Anal. Chem. 52*: 2242 (1980).
21. International Standards Organization, *Fresh Fruits and Vegetables—Sampling*. ISO 874 (1980).
22. International Standards Organization, *Meat and Meat Products*, Parts 1 and 2, ISO 3100/1&2 (1975).
23. International Standards Organization, *Oilseeds—Sampling*, ISO 874 and ISO 664 (1980).
24. F. M. Garfield, *JAOAC 72*: 405 (1989).
25. J. A. Springer and F. D. McClure, *JAOAC 71*: 246 (1988).
26. W. Horwitz, *JAOAC 71*: 241 (1988).
27. J. H. Cunningham, *Food Aust. 42*: S16 (1990).
28. P. Minkkinen, *Anal. Chim. Acta 196*: 237 (1987).
29. A. D. Campbell, T. B. Whitaker, A. E. Pohland, J. W. Dickens, and D. L. Park, *Pure Appl. Chem. 58*: 305 (1986).
30. D. L. Park and A. E. Pohland, *JAOAC 72*: 399 (1989).
31. N. Apro, S. Resnik and C. Ferro Fontan, *An. Asoc. Quim. Argent. 75*: 501 (1987).
32. S. Hisai, *Kogai to Taisaku 15*: 417 (1979).
33. K. W. James, P. J. Tattersall, and M. J. Lichon, *Proc. 5th ANZAAS-AIST Conf. Science & Technology, "Technology Today and Tomorrow"*, *July 1986*, Sydney, 165 (1986).
34. K. W. James, A. T. Hancock, M. J. Lichon, and L. Robertson, *Chem. International Food Forums Proc. August 1989*, Brisbane, 245 (1989).
35. P. M. A. Maggiore, *Proc. Nutr. Soc. Aust. 15*: 220 (1990).

36. J. Brand Miller, K. W. James and P. M. A. Maggiore, *Tables of Composition of Australian Aboriginal Foods*, Aboriginal Studies Press, Canberra, 1993.
37. Y. Malkki, *JAOAC, 69*: 403 (1986).
38. M. J. Lichon and K. W. James, *Proc. Govt. Food Analysts 2nd Meeting, October 1985*, Sydney, 195 (1985).
39. M. J. Lichon and K. W. James, *JAOAC 73*: 820 (1990).
40. R. B. H. Wills, N. Balmer, and H. Greenfield, *Food Tech. Aust. 32*: 198 (1980).
41. R. A. Beebe, E. Lay, and S. Eisenberg, *JAOAC 72*: 777 (1989).
42. J. D. Pettinati, S. A. Ackerman, R. K. Jenkins, M. L. Happich, and J. G. Phillips, *JAOAC 66*: 759 (1983).
43. J. W. Dorner and R. J. Cole, *JAOAC 76*: 983 (1993).
44. K. W. James and A. T. Hancock, *Chem. International Food Forums Proc. August 1989*, Brisbane, 327 (1989).
45. M. J. Lichon, P. J. Tattersall, and K. W. James, *Proc. 9th Aust. Symp. Anal. Chem.*, Vol. 1, April 1987, Sydney. *1*: 282 (1987).
46. M. J. Lichon and K. W. James, *JAOAC Int.*, in press.
47. Assoc. Off. Anal. Chem. *Official Methods and Analysis*, Arlington, VA, Sec. 985.29, 1990.
48. N. Harnby, M. F. Edwards, and A. W. Neinow, *Mixing in the Process Industries*, Butterworth-Heinemann, Oxford, 1992.
49. K. W. James, M. J. Lichon, P. J. Tattersall, G. F. Thomson, and A. T. Hancock, Laboratory evaluation of Australian ration packs, Technical Note, MRL-TN-540, Defence Science & Tech. Organ., Aust., 1988.
50. K. W. James, G. F. Thomson, A. T. Hancock, G. J. Walker, R. A. Coad, and M. J. Lichon, Laboratory evaluation of Australian ration packs, MRL Report, MRL-TR-92-30, Defence Science & Tech. Organ., Aust., 1993.
51. C. A. Kwiatkowska, P. M. Finglas, and R. M. Faulks, *J. Hum. Nutr. Dietet. 2*: 159 (1989).
52. R. A. McCance and E. M. Widdowson, *The Composition of Foods*, HMSO, London, 1978.
53. K. Ganzler, A. Salgó, and K. Valcó, *J. Chromatogr. 371*: 299 (1986).
54. C. L. Suard, M. H. Feinberg, J. Ireland-Ripert, and R. M. Mourel, *Analusis 21*: 287 (1993).
55. H. M. Kingston and L. B. Jassie, *Anal. Chem. 58*: 2534 (1986).
56. M. D. Mingorance, M. L. Perez-Vazquez, and M. Lachica, *J. Anal. At. Spectrom. 8*: 853 (1993).
57. H. M. Kuss, *Chem. Labor Biotech. 42*: 11 (1991).
58. A. Krushevska, R. M. Barnes, and C. Amarasiriwaradena, *Analyst 118*: 1175 (1993).
59. G. Schwedt, *Dtsch. Lebensm.-Rundsch. 87*: 223 (1991).
60. M. Takeda, H. Sekita, K. Otsuki, H. Tanabe, Y. Kawai, Y. Mishima, K. Kogo, and Y. Katayama, *Shokuhin Eiseigaku Zasshi 14*: 569 (1973).
61. L. J. D. Myer, J. H. Damian, P. B. Liescheski, and J. Tehrani, *ACS Symp. Ser. 488*: 221 (1992).
62. F. David, *Chem. Mag.* (Ghent) *17*(4): 12 (1991).
63. F. Verillon, F. Qian, and P. Rasquin, *Anal. Sci. S7*: 1511 (1991).
64. M. Hauck, *Dtsch. Lebensm.-Rundsch. 86*: 12 (1990).
65. R. Gromes and T. Siegl, *GIT Fachz. Lab. 35*: 623 (1991).
66. W. G. Jennings and M. Filsoof, *J. Agric. Food Chem. 25*: 440 (1977).
67. D. J. Aulik, *JAOAC 57*:1190 (1974).
68. M. H. Bui, *JAOAC 70*: 802 (1987).
69. T. Ujiie, A. Tanaka, A. Kondo, R. Hiroe, and T. Tetsushige, *Bitamin 66*: 101 (1992).
70. C. Schlage and B. Wortberg, *Z. Lebensm.-Unters.-Forsch. 145*: 97 (1971).
71. I. B. Razagui and P. J. Barlow, *Food Chem. 44*: 309 (1992).
72. V. W. Bunker, H. T. Delves, and R. F. Fautley, *Ann. Clin. Biochem. 19*: 444 (1982).
73. G. E. Batley, *Trace Element Speciation: Analytical Methods and Problems*, CRC Press, Boca Raton, 1989.
74. R. J. Henry, P. J. Harris, A. B. Blakeney, and B. A. Stone, *J. Chromatogr. 262*: 249 (1983).
75. C. F. Moffat, A. S. McGill, and R. S. Anderson, *J. High Resolut. Chromatogr. 14*: 322 (1991).
76. B. S. Middleditch, *Analytical Artifacts*, Elsevier, New York, 1986.
77. E. L. Nimz and S. L. Morgan, *J. Chromatogr. Sci. 31*(4): 145 (1993).
78. M. J. Lichon, *J. Chromatogr. 624*: 3 (1992).

2

Chemometrics

Michele Forina and Giuliana Drava
*Institute of Food and Pharmaceutical Analysis and Technology, University of
Genoa, Genoa, Italy*

In the past 30 years the information made available by chemical analysis has increased greatly due to both the increased number of analyzed samples and the number of analytes measured in each sample. At the same time, the availability of desktop computers with continuously increasing speed and low cost has permitted the handling of information using the tools of multivariate statistics and applied mathematics.

Two disciplines, experimental design and chemometrics, began to be part of the background of chemists, the first one apparently coming from agronomy and the second one from the older biometry.

The special needs of chemistry, in particular food chemistry, and the continuous exchange of ideas and techniques among the two disciplines had a strong influence on their evolution. As a result, we now prefer to redefine chemometrics to include experimental design or, better, according to R. Phan-Tan-Luu, to use the name "Methodology of Experimental Research" in Chemistry.

Chemometrics as a chemical discipline has the following objectives:

To plan the chemical experiment in order to obtain the maximum chemical information with the minimum cost

To optimize the analytical process, especially when a chemical quantity is computed from a measured physical quantity

To extract useful chemical information from the computed chemical quantities

In this way, three moments of chemometrics are defined: before, during, and after the experiment.

Generally, chemical analyses are performed to give information relevant to solve a given chemical problem; in the case of the absence of problems, when chemical data are obtained just to fill a lot of sheets of a useless report destined to be buried on a dusty shelf, chemometrics is useless, maybe pernicious, because it reduces the volume of the data to their essential significance, sometimes a few rows or a few words.

Before the analytical experiments, we collect samples. Sampling is done (a) to study the effect of some factors, (b) for quality control, and (c) for process control.

In the first case, samples have to be selected to span the space of the studied factors and the noise. Some factors (e.g., temperature, time, pressure) can frequently be fixed to selected values; other factors (e.g., solvent, additive) cannot be expressed by a defined specific measurable quantity and cannot be fixed to a selected value. Here the selection of samples representative of the studied factors is the main problem.

In the second case, the quality of a product is defined by many characteristics, and samples are drawn from the production line to verify that the product satisfies the quality requirements. The frequency of the sampling and the consequent cost of the control are the most important problems.

In process control, samples are drawn (or physical quantities measured) at certain points. Where and when should samples be collected?

The classical experimental design, the design on the principal properties, and sampling theory give answers to these problems.

During the analytical experiment we first select the suitable analytical technique. Then we select the operative conditions. We do calibration. Finally, we obtain the chemical quantity with a certain confidence interval. Frequently the analytical chemist can perform these operations without chemometrics. Sometimes, however (e.g., choice of a mixture of solvents or of a complex gradient in chromatography, computation of a calibration model in multivariate calibration) chemometrics is necessary.

After the analytical experiment, the amount of collected chemical information must be applied to the problem, sometimes with information obtained from different sources (e.g., sensory data, climatological or geographic information). Is the information relevant, sufficient, redundant? How can the chemical information be used to give the required answer to the problem? Even an excellent table of analytical data cannot by itself give the required answer. From data we must extract rules, i.e., mathematical models to compute from the original data the required quantities. The rules must be verified; the importance of the measured variables must be evaluated. Anomalous samples and nonrelevant variables can be detected; by deleting these objects and variables, the model can be improved.

I. TOOLS AND VALIDATION PROCEDURES OF CHEMOMETRICS

A lot of techniques have been developed by statisticians and mathematicians and then applied to chemical problems. Some of these techniques are of great importance in chemistry; some have been modified according to the chemical problem. These techniques are the tools of chemometrics.

A special characteristic of chemometrics is that several techniques are used to predict some characteristics of new samples, so that the evaluation of the prediction ability of a model is its most important characteristic. This prediction ability is evaluated by means of validation procedures.

Here a list of some important families of chemometric tools is presented, with a short description of the chemical problem for which the tools are used.

A. The Tools

1. *Experimental Design*

The general application is for the selection of an optimal subset of the combinations of values of the factors having influence on the experiment, to obtain information about their effect on the result (one or more characteristics of the experiment) with the minimum cost, and, as a consequence, to fix the factors in order for the experiments to give optimal results.

Experimental design is used in sampling (to select the minimum number of samples representative of the studied analytical system), in blending of foods (to select the optimum mixture of some original products), in the selection of experimental conditions of an analytical method (e.g., the combination of temperature, % organic solvent, pH in chromatography), and in calibration (selection of standards). Experimental design includes:

1. *Optimal design*: General strategies (1).
2. *Factorial fractional design*: Strategy used in the case of large number of factors (1).
3. *Screening*: Case of very large number of factors, to select relevant factors.
4. *Mixtures*: Special strategies to be applied in the optimization of mixtures (e.g., to select the optimum percentage of the components of a solvent mixture for chromatography) (2).
5. *Design on principal properties*: Used when the factors are molecules or substituents. Molecules can be described by a lot of properties (more or less correlated), as molecular weight, melting point, dipole moment, etc. The problem can be to select, e.g., for a chemical reaction, some solvents representative of all of the available solvents (3).

2. Pretreatment

There are three possible objectives: (a) to eliminate the scale effect among different variables (it can have heavy effects on the performance of the techniques reported below); (b) to obtain variables having normal distribution; (c) to reduce the dimensionality of data; (d) to reduce noise.

1. Scaling
2. Transforms
3. Fourier transforms

3. Visualization

Visualization is a very important family of techniques used in multivariate data analysis to display in a plane (bidimensional) the maximum amount of information contained in the multidimensional space of the variables.

1. Principal components
2. Biplots
3. Nonlinear mapping
4. Projection pursuit

4. Factor Analysis

Factor analysis encompasses techniques used to reduce noise in measured data, to find significant interpretable factors responsible for the differences between objects, to separate overlapping signals in hyphenated techniques, and to identify pure components (with their percentage) in a set of mixtures.

1. Abstract factor analysis
2. Orthogonal rotations
3. Oblique rotations
4. Target factor analysis
5. Iterative target factor analysis
6. Evolving factor analysis

5.   *Clustering*

Clustering refers to the methods applied to search for groups of similar samples or variables, of special usefulness in explorative data analysis; also used as classification techniques (see below).

1.   Hierarchical methods
2.   K means
3.   Minimal spanning tree

6.   *Classification and Class Modeling*

These are the techniques used in identity problems (genuine–adulterate, cultivar categories, technological categories, origin, etc.)

1.   Linear discriminant analysis (LDA)
2.   Quadratic discriminant analysis (QDA) and UNEQ
3.   Potential functions methods
4.   K nearest neighbors (KNN)
5.   Soft independent modeling of class analogr (SIMCA)
6.   Artificial neural networks (ANN)

7.   *Regression*

Regression refers to the techniques used to study the relationships between two variables (one predictor variable and one response variable) or between blocks of variables. The very large field of applications and future development includes multivariate calibration (e.g., near infrared spectroscopy), quantitative study of property–molecular structure relationships, image analysis, relationship between composition or treatment and panel scores, transfer of spectra from server to master instrument, relationships among panelists, etc.

1.   Ordinary least squares (OLS)
2.   Principal component regression (PCR)
3.   Partial least squares (PLS)
4.   Alternating conditional expectations (ACE)
5.   Robust methods
6.   Other (ANN, ridge regression, procrustes analysis, consensus PLS)

8.   *Feature Selection*

Feature selection represents the techniques used to reduce the number of variables, either because some variables give only noise or, in the case of redundant information, to obtain a more economical model. Frequently feature selection is part of a classification or regression technique.

9.   *Time Series*

Time series refers to techniques used to study regularities in a time-dependent phenomenon, e.g., product quality control, chromatography of series compounds.

B.   The Validation Procedures

Hard models are based on theory, and the objective of data analysis is to estimate the parameters of the model with a minimum of error.

Chemometrics deals with soft models; both classification and regression models are developed with a very empirical objective, i.e., the prediction, with a minimum of error, of one or more quantities (a class index, the value of one or more response variables). Sometimes from the soft models theoretical models can be developed, but this is a secondary objective.

So a model must be evaluated on the basis of its predictive ability; the measure of the predictive ability is a fundamental step in data analysis.

Validation procedures have been developed to measure the predictive ability (4). These procedures can be very simple or somewhat complex, according to the requirements of the technique to be validated and of the computing time required.

1. Single Evaluation Set

This procedure is the simplest and quickest validation method. We use it to introduce some elements of basic terminology.

A sample, the nth sample in a collection of N samples, is described by V measured or computed variables; the ordered values of the variables for this sample are the nth row of a data matrix (N rows and V columns). The sample (or molecule, or panelist) is the

nth object

of the matrix $_N\mathbf{X}_V$.

To develop a model—a rule for either classification or regression—we need some more information: in the case of classification, a column vector $_N\mathbf{c}$ (N rows) with the class index of each object; in the case of regression, a column vector of known values of a response variable $_N\mathbf{y}$ or a matrix $_N\mathbf{Y}_M$ of known values of M response variables.

In the validation procedure we divide the N objects into two sets. One is the *training* set (with I objects), used to compute the parameters of the model; the other is the *evaluation* set (with $J = N - I$ objects), used to compute the predictive ability.

In the case of a classification technique, the objects in the training set are used to obtain the parameters of a model that computes the class of whichever object. When the model has been obtained, we can use the model to compute the class of the objects in the same training set. The percentage of correct decisions gives us the *classification ability*. Then we use the model to predict the class of the objects in the evaluation set: the percentage of correct decisions gives us the *prediction ability*.

In the case of regression, the regression rule is used to compute the value of the response variable for the objects in the training set, the vector $\hat{\mathbf{y}}$. The *fitting ability* can be measured by the fraction of the explained variance, or R^2:

$$R^2 = 1 - \frac{\sum_I (y_i - \hat{y}_i)^2}{\sum_I (y_i - \bar{y})^2} \tag{1}$$

where y_i is the measured value of the response variable for the ith object; $\bar{y}$ is the mean of the response variable computed on the I objects of the training set; $\hat{y}_i$ is the value computed by the regression model; $(y_i - \bar{y})$ is the before-regression error; $(y_i - \hat{y}_i)$ is the after-regression error for the objects in the training set, i.e., the error of fitting.

On the objects of the evaluation set we compute a *prediction ability*, as the *validated* fraction of the explained variance R_v^2:

$$R_v^2 = 1 - \frac{\sum_J (y_j - \hat{y}_j)^2}{\sum_J (y_j - \bar{y})^2} \tag{2}$$

In this case, the after-regression error $(y_j - \hat{y}_j)$ is the error of prediction.

With the use of more complex models it is even possible to increase up to 100% the classification ability or R^2; on the contrary, the optimum prediction ability is obtained with a model of suitable complexity; a more complex model can noticeably reduce the predictive ability.

The evaluation set is usually obtained by random choice of objects. Usually 25–33% of the objects are assigned to the evaluation set.

The disadvantages of the procedure are as follows:

1. The casualness of the choice can produce significant overestimate or underestimate of the true predictive ability.
2. A high percentage of objects in the training set originates an evaluation set that is not representative; a low percentage of objects in the training set decreases the validity of the computed model, sometimes preventing us from using a selected technique, when a minimum ratio between the number of objects and the number of variables is required.

2. Cross-validation

Cross-validation is the common validation procedure (5).

We divide the N objects into G *cancellation groups*. The objects are assigned to a cancellation group by their index n (position in the data matrix): the first object is assigned to cancellation group 1, the second to group 2, the gth to group G. Then the $(g + 1)$th object is assigned again to group 1, and so on.

The model is computed G times. Each time the objects in the corresponding cancellation group form the evaluation set; the objects in the other groups are the training set, used to compute the model parameters. At the end of the procedure, each object has been $(G - 1)$ times in the training set and once in the evaluation set.

The predictive ability (in the case of a regression technique) is obtained as R^2_{cv} (cross-validated R^2):

$$R^2_{CV} = 1 - \frac{\sum_N (y_n - \hat{y}_n)^2}{\sum_N (y_n - \bar{y})^2} \tag{3}$$

The numerator refers to the value of the response variable computed by the model when the nth object was in the evaluation set; $\bar{y}$ is the mean of the response variable computed with all of the objects. Figure 1 shows the before-regression errors in a simple case with only one predictor variable X; the generalized centroid is the point with coordinates $\bar{x}, \bar{y}$, mean values computed with all of the objects.

The number of cancellation groups usually ranges from 3 to N. In the last case we have as many cancellation groups as objects, so that the model is computed N times, with $(N - 1)$ objects in the training set and only one object in the evaluation set; cross-validation with N cancellation groups is also known as the leave-one-out procedure. It has the advantage of being unique for a given dataset, whereas when $G < N$ the result depends on the order of the objects.

However, the leave-one-out procedure produces a very small change in the training set. The complete dataset is rarely a "perfect" representation of the whole population of possible objects: generally it has a "representation error." A small perturbation has the consequence that in each cancellation cycle we have a training set with about the same representation error. The measure of the predictive ability can be overly optimistic. On the contrary, with a small number of cancellation groups, the training sets are very different, with different representation errors, and the measure of the predictive ability is not optimistic, perhaps pessimistic.

So we suggest performing cross-validation many times, with a different number of cancellation groups, from the leave-one-out procedure up to three cancellation groups. Moreover, for a given number $G < N$ of cancellation groups, it is advisable to repeat the validation with a different order

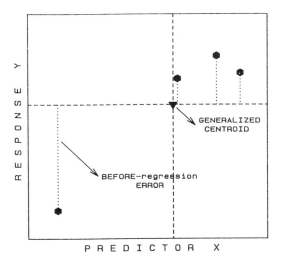

Fig. 1 Generalized centroid and before-regression errors for a small dataset with only one predictor.

of the objects, in order to have a different composition of the cancellation groups. The validation procedure becomes time consuming, but it ensures a significant measure of predictive ability.

3. Repeated Evaluation Set

In this procedure, several evaluation sets are created by means of random generation, with different number of objects and with different objects. Each object is many times in the evaluation set, and each time in combination with different companions. A total of $T \gg N$ predictions are performed and the predictive ability is evaluated by means of

$$R^2_{RE} = 1 - \frac{\Sigma_T (y_t - \hat{y}_t)^2}{\Sigma_T (y_t - \bar{y})^2} \qquad (4)$$

4. Full Validation

All validation procedures have as a common characteristic the prediction on one or more evaluation sets; it is very important that no information from the evaluation set be used to build the model; this is the fundamental characteristic of "full validation" (6).

A very common case is that of a simple pretreatment, as column centering (where to each variable its mean is subtracted), done with all the objects, and followed by the use of a selected technique and cross-validation.

Some techniques use the centroid of data (computed in column centering) as an important parameter, but the centroid contains information from the future evaluation sets, and an overestimate of the prediction ability can be obtained.

Figure 2 shows the wrong procedure. In the prediction of the response variable for the deleted object, the regression line (computed only with data from the objects in the training set) through the centroid (the new origin after column centering) computed with all the objects gives a prediction error that is small in comparison with the corresponding before-regression error in Fig. 1. The position of the generalized centroid contains information from the deleted object: it "shows" to the three retained objects in the training set the position of the deleted object.

Figure 3 shows the correct procedure. In the prediction of the response variable for the deleted

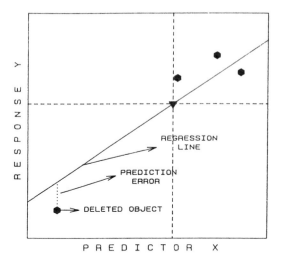

Fig. 2 Example of uncorrect validation. The prediction error is small because the deleted object (excluded from the computation of the regression line) has not been used for computation of the centroid.

object, a "cancellation group centroid" is computed with the objects in the training set. The regression line through this centroid gives a very high prediction error for the deleted object: really, the three objects in the training set have not information sufficient to predict the response variable for the deleted object.

Frequently chemometric techniques are used as black boxes, executable programs whereby it is not possible to observe the details of the procedure. The above wrong procedure is present in some commercial packages; despite the fact that in many cases the consequences are without importance (low error in the estimate of the prediction ability), in some cases the conclusion can

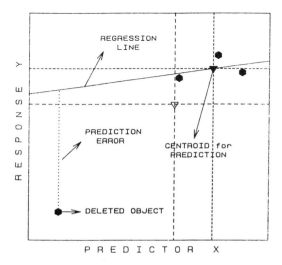

Fig. 3 Example of correct validation. The prediction error is very high because the deleted object has not been used for computation of the centroid.

be that the block of predictor variables X has a sufficient predictive ability whereas it has no predictive ability at all; it contains only useless information.

5. *Optimization*

An important case in which the fundamental rule of full validation is frequently violated is the case of techniques characterized by a model whose parameters (model parameters, as regression coefficients) depend on some hidden parameters that do not appear explicitly in the model. By changing these hidden parameters several models, sometimes an infinity of models, can be obtained by the same technique. An example regards the *weights* assigned to the variables. This weight can be a two-level parameter (selected-unselected variable) or a continuous parameter.

In this case usually the values of hidden parameters are selected so that the maximum predictive ability is obtained; the values of hidden parameters are changed and for each combination the predictive ability is evaluated by means of a validation method, e.g., cross-validation. The techniques of experimental design are used to detect the combination of hidden parameter values producing a model with the maximum predictive ability.

Obviously, in this way the objects in the evaluation set give an information used to build the model.

The correct procedure, in this case where the model parameters are optimized by action on the hidden parameters, involves three sets:

Training set
Optimization set
Evaluation set

The optimization set is used to select the optimum values of the hidden parameters; the true predictive ability of the optimized model is computed on the evaluation set, which objects were not used to compute, directly or through the hidden parameters, the model.

The procedure with these three sets can be that of cross-validation: in each of the G_1 cancellation cycles some objects are assigned to the evaluation set. Then the remaining objects are assigned to G_2 cancellation groups for the cross-validated selection of the optimum values of the hidden parameters. So the procedure is repeated $G_1 \times G_2$ times [$N \times (N - 1)$ times in the case of leave-one-out procedure], and in the G_2 inner cycles many models are computed searching for the optimum model. The computing time can be very high, but it is necessary to avoid model being forced toward an illusory high predictive ability.

II. TECHNIQUES

A lot of analytical techniques produce multivariate information: chromatography, spectroscopy, hyphenated techniques, etc.

Chemometric techniques apply to multivariate data, whereby objects are described by many variables, from 3 (rarely), to 1000 (as in near-infrared spectroscopy), to many thousands (as in image analysis or in structure–activity relationships).

However, a chemometric technique can be better explained by means of examples whereby the objects are described by only two variables, so that it is possible to visualize the information on the plane of a figure, as shown in Fig. 4.

The range of the variables defines the space of the information, a square in the case of Fig. 4. Sometimes the objects are scattered more or less uniformly through the whole information space. In this case (Fig. 4A) the only information we can obtain is about the *similarities* among objects: two close objects are similar because they are described by close values of all (both) variables.

In Fig. 4B the objects fill only a part of the information space (inner space); we can single out

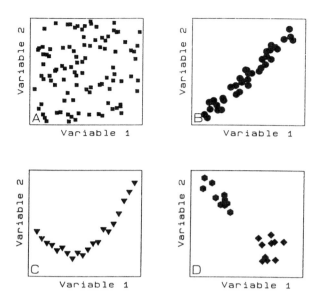

Fig. 4 Examples of datasets. (A) Objects uniformly scattered in the plane of two variables. (B) Data having a linear structure. (C) Data having a nonlinear structure. (D) Data having a two-block structure.

a linear structure in the data. A structure gives us the possibility of prediction, that of the value (with a more or less great error) of one of the variables from the value of the other variable. At the same time a structure means that the two variables have information in common: from a certain perspective this information is redundant (so that it can be canceled without loss of useful information); from another perspective the common information is synergic.

Figure 4C shows a case of nonlinear structure.

Finally, Figure 4D shows a two-block structure, and the two blocks of objects can be separated because they have a different position on the linear structure of the whole dataset. Moreover, each block has its own structure. The first block has a linear structure; the second block fills only a small part of the space of the information, i.e., the space of the block.

A. Principal Components

Principal components analysis (PCA) is the basic tool for data analysis. It is used to visualize the information contained in a complex data matrix, to reduce the dimensionality of a too large data matrix, and it is also the basis of important techniques of class modeling (SIMCA) (7) and of regression (Principal components regression, or PCR, and partial least squares, or PLS (8)).

1. Scaling

Information is variability; no information can be obtained from a constant "variable." (Some might not agree with this statement, but they are keeping in mind at least a second level of the same quantity, i.e., in their problem the quantity is a variable, not a constant.)

Variability is measured by the variance: for the variable x measured on N objects the variance is given by

$$s^2 = \frac{\sum_n (x_n - \bar{x})^2}{N - 1} \tag{5}$$

and a generic residual $(x_n - \bar{x})$ is the distance between the point x_n and the centroid $\bar{x}$. The numerical value of a distance depends on the unity used to measure the variable, so that the variance depends on the square of the unit scale. That dependence makes it difficult to compare the information from variables of a different nature.

For this reason, frequently the original variables are modified in order to have the same or approximately the same variance. The more frequently used procedure is autoscaling, i.e., the use of the Student transform:

$$t_n = \frac{x_n - \bar{x}}{s} \tag{6}$$

and in this case the new variable t has unity variance (and mean zero, i.e., we made both a centering and a standardization). In the case of more variables, after autoscaling the variables have the same variance, i.e., they yield the same amount of information.

This is a very good strategy when we do not have a priori information about the variables because autoscaling gives the same starting importance to all of the variables, whatever their original magnitude. However, frequently chemists have some a priori knowledge about the variables. That is, when we compare spectra of different chemical species and no species absorb in a given region, the differences between spectra in this region are without importance, due exclusively to noise. So when we apply autoscaling to a series of spectra (each variable in this case is the absorbance at a selected wavelength), we give the same importance both to the noise and to the wavelengths of the meaningful region of the spectra. For this reason, after autoscaling, we often must weigh each variable by multiplying it by a factor inversely proportional to its relative noise.

Moreover, autoscaling involves a centering of each variable: in the subsequent data analysis a piece of information will be lost, i.e., that related to the centroid $\bar{x}$ (the vector of the means of the variables, barycenter of the information in the hyperspace of the variables), Frequently this information is without importance (we are interested in comparing samples using differences between samples rather than the absolute values of the variables); in some cases, however, this information is important.

There are many scaling procedures other than autoscaling: column centering, standardization, range scaling, global centering, etc. Column centering and standardization are the two steps that jointly produce autoscaling; range scaling consists of subtraction of the minimum from each value of a variable followed by division by the range of the variable. The transformed variable has a minimum of 0 and a maximum of 1. Global centering consists of subtraction of the mean from all of the data in the data matrix; the mean is computed on all objects and all variables. Global centering is frequently used in multivariate calibration.

Independently of the scaling procedure, we will continue to indicate with $_N X_V$ the data matrix as usual but also with the specification of the performed transform.

2. Rotations and Visualization

In the upper part of Fig. 5 a dataset of 9 objects and 2 autoscaled variables, X and Y, is reported. The two variables are clearly positively correlated. We can rotate, e.g., clockwise, the axes, as shown in the lower part of Fig. 5. Here we can see the original axes with the objects in the space of the new coordinates. These new coordinates are uncorrelated variables, so that they give independent information. Moreover, the new abscissa is the direction with the maximum possible variance, and the new ordinate is the direction with the minimum possible variance.

With this rigid (orthogonal) rotation we have obtained some interesting results:

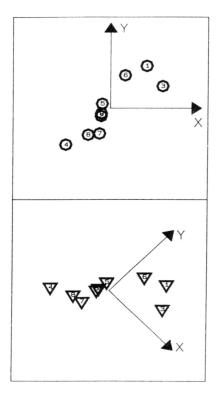

Fig. 5 Upper: Objects described by two correlated variables. Below: After rotation of the axes, the new two variables are uncorrelated.

1. Uncorrelated variables starting from correlated variables.
2. New axes ranked in order of variance; the variances on the new axes are the *eigenvalues*.
3. The possibility of representing the directions of the original variables and to reason in the space of the new axes in terms of the original variables.

The new axes are the principal components (PCs) of the autoscaled data (generally the term principal components is related to autoscaled data; PCs are the eigenvectors of the matrix of the correlation coefficients).

The results with two variables are much more interesting in the case of many variables:

1. Because of the uncorrelation, a plot with two PCs does not show duplicate information.
2A. When we observe the first two PCs we see the maximum amount of information (variance) available on a plane in the multidimensional space of the original information.
2B. The high-order PCs have low variances, i.e., low information. Moreover, this information is generally useless, i.e., noise; the useful information is generally concentrated on the first (one to five) PCs.
3. We can represent in the space of the PCs the directions of the original variables, so that we can interpret a great amount of information in terms of the original variables. A plot whereby both objects in the space of two PCs and the directions of the original variables are represented is called biplot.

Figure 6 shows the biplot on the two first principal components of a set of 324 extravirgin olive oils described by eight variables (fatty acids). The oils come from 9 Italian regions (9), namely,

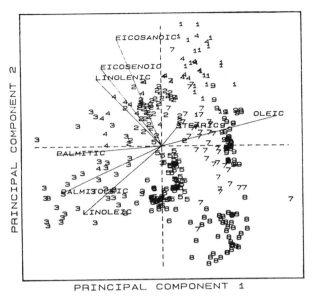

Fig. 6 Biplot on the two first principal components of a set of 324 Italian extravirgin olive oils described by eight fatty acids: the samples are divided into nine categories corresponding to the region of origin.

North Apulia, Calabria, South Apulia, Sicily, Inland Sardinia, Coastal Sardinia, East Liguria, West Liguria, and Umbria, so that the information is stored in a data matrix $_{324}X_8$ and in a vector $_{324}c$ containing the class number of the samples. Data were autoscaled so that the total variance was 8. PC rotation produces 8 PCs, the first with eigenvalue 3.393 (42.4% of the information), the second with eigenvalue 1.863 (23.3% of the information), so that in the plane of Fig. 6 65.7% of the information is reported.

The figure shows:

How these chemical data can be used to classify an oil sample in one of the nine regional classes;

The main characteristics of the oils of a class, e.g., oils of class 1 have a relatively high content in oleic, linolenic, eicosanoic, and eicosenoic acids;

How samples in some classes have very similar characteristics (e.g., class 9) and samples in other classes have a wide range of composition (e.g., class 3);

The main correlations among the variables (e.g., eicosanoic, eicosenoic, and linolenic acids are positively correlated);

The relative importance of the variables in the plane of the two components (e.g., stearic acid is represented with a short line), meaning that the plane of the two first PCs has a strong inclination on the axis of stearic acid (this acid is very important for the third component).

The rotation from the space of the original variables to the space of PCs is done by means of an orthogonal rotation matrix:

$$_N S_V = {_N}X_V \, {_V}L_V \tag{7}$$

Here **S** is the matrix of *scores*, coordinates of the N objects in the space of PCs, and **L** is the rotation matrix, where each element is the cosine of the angle between a variable and a principal component (the *loading* of the variable on the principal component). The smaller the angle, the larger is the cosine and also the coordinate of the variable on the component in the biplot, as in the case of oleic acid in Fig. 6, having a large cosine on the first component and a small cosine on the second component.

3. The Space of Significant Information and Target Factor Analysis

Numerous techniques have been developed to evaluate the number of significant components; among these techniques the so-called double-cross validation (10), based on the predictive ability of each component, is the one most used in chemometrics.

The significant components define the space of structured information, the *inner space*; the information in the outer space, the space of nonsignificant components, has no predictive ability; it represents noise.

We can study the information in a few dimensions of the inner space, where relationships that are easily understandable in the case of two variables are still valid.

The first example concerns a matrix $_{50}X_{101}$; the variables are the absorbances at 101 wavelengths and the objects are mixtures of some pure components.

In Fig. 7 the spectra of four pure constituents A, B, C, and D (at a given concentration) are reported, corresponding to the row vectors $_A x_{101}^T$, $_B x_{101}^T$, $_C x_{101}^T$, and $_D x_{101}^T$ (x^T is for the row vector transposed of a column vector). Note that there are not specific wavelengths.

Each of the 50 objects is a mixture of A, B, C, and D:

$$_n x_{101}^T = k_A \, _A x_{101}^T + k_B \, _B x_{101}^T + k_C \, _C x_{101}^T + k_D \, _D x_{101}^T$$

The problem requires determination of the following:

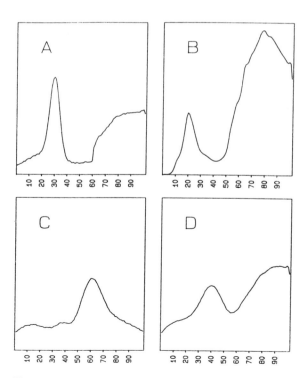

Fig. 7 Spectra of four pure constituents of a mixture.

1. How many pure constituents have been mixed to obtain the 50 objects, i.e., how many coefficients k_A, k_B, etc., are different from zero
2. What these coefficients are, i.e., what pure constituents have been mixed
3. The values of the coefficients for each object

The number of possible pure constituents, here 4, can be very high in practical problems; the pure constituents can also be chemical mixtures (e.g., when they are pollution sources and the problem is to detect how many and which possible pollution sources are responsible for the pollution on a territory where some samples have been collected and analyzed) or solutions.

Figure 8 shows that two pure constituents, X_1 and X_2, can be mixed to result in composition x_{1i} and x_{2i}.

It is:

$$x_{1i} = a\,x_{1A} + b\,x_{1B}$$
$$x_{2i} = a\,x_{2A} + b\,x_{2B}$$

with a and b nonnegative, and $a + b = c$. When $c = 1$, i.e., in the case of a mixture without dilution, the representative point of the mixture is on the line joining the points that represent A and B in the plane X_1, X_2 (triangles in Fig. 8). When $c < 1$ (mixture with dilution) the mixture is between this line and the origin, in the sector defined by the origin and the points A and B (hexagons in Fig. 8). When $c > 1$ (mixture with concentration) the mixture is farther away from the origin than the line $\overline{AB}$ (diamonds in Fig. 8). Figure 9 shows that the coefficients a and b can be obtained from the coordinates on the oblique axes represented by the lines $\overline{OA}$ and $\overline{OB}$.

This is easily understandable in the case of mixtures described by only two variables, but similar reasoning can be applied in the case of the significant eigenvectors.

In our example, data were not autoscaled because centering produces a loss of the information about the origin (necessary to apply the above considerations) and scaling gives too great an importance to noisy wavelengths. Here the eigenvectors are those of the matrix of the second-order moments (variances and covariances about the origin).

By means of double-cross validation it was found that there are two significant components.

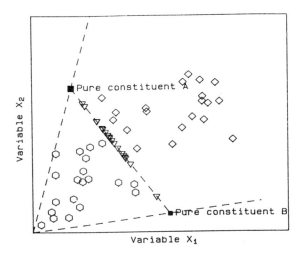

Fig. 8 Two pure constituents A and B, described by two variables; mixtures without dilution (triangles); mixtures with dilution (hexagons); mixtures with concentration (diamonds).

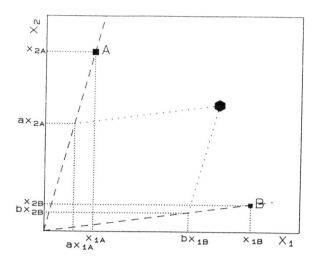

Fig. 9 Geometric representation of the computation of *a* and *b*.

The eigenvectors are abstract mathematical factors, but the number of real factors (here the pure constituents) must be the same, i.e., two. We answered the first question: Within the noise of our experimental data only two pure constituents contribute to the composition of the 50 objects.

We can observe all the significant information in the plane of the first two eigenvectors (Fig. 10). Here we can also represent the four pure constituents (obviously they have not been used to compute the loadings of the eigenvectors). Only two constituents (B and D) identify a sector that includes all of the objects, so that B and D must be the two real factors. Their mixtures have a spectrum **x**, linear combination of the spectra of B and D:

$$_n\mathbf{x}^T_{101} = k_{\mathrm{B}} \,_{\mathrm{B}}\mathbf{x}^T_{101} + k_{\mathrm{D}} \,_{\mathrm{D}}\mathbf{x}^T_{101}$$

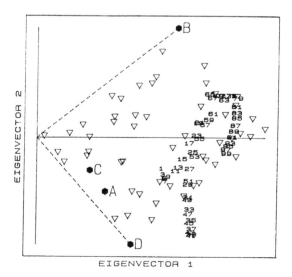

Fig. 10 Representation of 50 objects (triangles), mixtures of pure constituents B and D (filled hexagons), described by 101 variables (absorbances at 101 wavelengths) in the plane of eigenvectors 1 and 2.

This is the equation of a plane (the plane of the first two eigenvectors) in the space of the 101 original variables; the plane contains all of the possible combinations of B and D, and so also the two special objects with $k_B = 1$, $k_D = 0$ and $k_B = 0$ and $k_D = 1$, i.e., the pure constituents B and D.

However, due to noise, experimental spectra are not exactly on the plane. In Fig. 11 we can see the noise, i.e., the deviation from the plane of the structured information. The noise is represented here only by the scores on the third eigenvector, which is the most important component of the outer space, the space of noise. Eigenvectors 3 and 1 are slightly correlated because the eigenvector rotation was made around the origin (the covariance about the origin is zero). The figure shows the extent of noise in the 50 objects and how the constituents B and D can be considered within the noise on the same plane of the 50 objects. To the contrary, the constituents A and C are very far from the plane of the structured information; their distance cannot be explained by the noise; they cannot be the real factors.

Coming back to Fig. 10 we can see that the objects do not have a structure. To the contrary, the variables, represented by their loadings on the two eigenvectors, have a structure which can be interpreted using the information in Fig. 7. For instance, the constituent D in Fig. 10 is in the direction identified mainly by the variables 33–47, corresponding to the main relative difference between the two spectra.

In some cases the objects also have a structure. Figure 12 describes a hypothetical elution profile of a mixture B + D. A sample is the spectrum recorded during the chromatogram. Figure 13 shows that in this case the objects projected on the two significant eigenvectors have a characteristic structure reflecting the chromatogram. The structure of the variables is the same as in Fig. 10 (the same spectra); in Fig. 13 it is reflected on the axis of the ordinates (the sense of the eigenvectors is without importance). Results such as those in Fig. 13 generally occur in the case of hyphenated techniques with overlapping chemical species. Eigenvector analysis allows detection of the overlapped constituents; then, by means of oblique rotation (as that shown in Fig. 9), the contribution of each constituent to an object can be obtained, i.e., the elution profiles as those in Fig. 12.

The procedure described is known as target factor analysis (11), an example of the rotations in

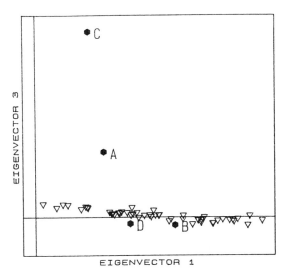

Fig. 11 Representation of the 50 mixtures and of the 4 pure constituents in the plane of eigenvectors 1 and 3. The third component contains mainly noise.

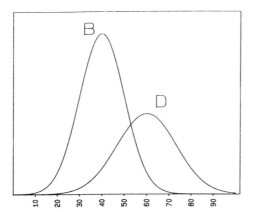

Fig. 12 Hypothetical elution profile of a mixture B + D.

the inner space of the significant information that allows one to obtain real factors (as the chemical constituents) from the abstract mathematical factors (the eigenvectors).

B.   Classification and Class Modeling with SIMCA

Classification techniques are used when the objective is to assign one or more objects to one of several well-defined categories. For instance, a classification technique can be used to assign samples of oil to one category among olive oil, soy oil, peanut oil, sunflower oil, etc.

Frequently the practical problem is to decide if a certain sample satisfies some requirements, as some characteristics define only one category, as in the case of quality control. Here there are no other defined categories. Likewise, it is not very important to describe all of the possible adulterations when the problem is to decide if a product is genuine or not. In this case the

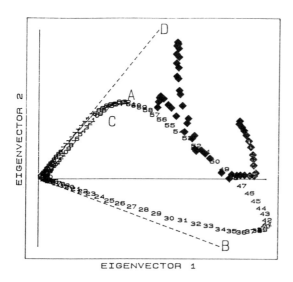

Fig. 13 Representation of the spectra during the chromatogram in the plane of the first two eigenvectors; variables are shown as filled diamonds.

classification problem is between the target category and the universe of all other possible categories. Class modeling techniques build a model of the studied category: samples that fit the model, i.e., that satisfy the required characteristics, are classified in the target category; the other objects are classified as rejected by the target category.

Principal components are the basis of some class modeling techniques. One of these, SIMCA (Soft Independent Modeling of Class Analogies) (7), was the first class modeling technique used in chemometrics.

Each category in the problem (frequently only one category) is described by a PC model. The objects of the training set of the studied category are used to develop the model. First, to compute means and standard deviations to perform an autoscaling called "separated scaling," it means that objects in other categories (when present) are also scaled using the dispersion parameters of the studied category. Then the PCs are computed and the number A of significant PCs is determined by double-cross validation (10). This number A defines the dimensionality of the inner space of the category, the space of the model. This space includes all of the significant correlations among the variables; in the outer space there is only uncorrelated noise. Sometimes the significant PCs, abstract factors defining the inner space, can be interpreted in terms of real factors (e.g., ripening, preparation temperature, etc.).

The scores of objects of the training set on each significant PC are within a range, called normal range.

In the case where the number of objects in the training set is small, the normal range is an underestimate of the true range; the normal range is extended to define the range of the class model. So the class model is defined by A principal components and by a range for each of them.

Because of the noise, both the objects in the training set and those to be checked (in validation or in the real use of the model) have a distance from the model, both in the outer space of the noise and in the inner space, but outside the boundaries of the model defined by the range of the components.

The residual standard deviation (RSD) from the model computed on the objects in the training set is used to define a maximum allowed distance from the model (at a selected probability level, as in the usual confidence intervals). This distance defines around the class model the class space; objects in the class space are accepted by the model of the category. Objects outside the class space are rejected by the model.

Figure 14 shows two class models. Both are built with only one principal component, and the class spaces are obtained by the distances of the objects of the category from the model. In the case of two (or more) categories the spaces of two categories can overlap, and in this case a part of the data space is common to the two categories. The objects in this part of the space fit both category models. Other objects can fall in the space of only one category. Finally, other objects are very far from both models, so that neither category accepts them.

Figure 14 refers to a bidimensional case, very different from the practical cases where the number of variables can be very high, frequently much more than the number of the objects of a category. (In that case classification or class modeling techniques other than those based on principal components cannot be used.) So to obtain a graphical representation of the modeling in the case of two categories, Coomans diagrams (12), as shown in Fig. 15, are used. Here the distances from two models are reported. The critical level, used to define the boundaries of the class space, divides the Coomans plot into four parts. Objects with a distance smaller than the critical distance from the model of class A are accepted by this model (almost all of the filled squares in Fig. 15). Objects with a distance smaller than the critical distance from the model of class B (almost all of the triangles in Fig. 15) are accepted by this model. Some objects have too high a distance from model A and model B; they are rejected from both models, and are outliers. Three objects in Fig. 15 are outliers, but one is so close to the class boundary that it can be accepted.

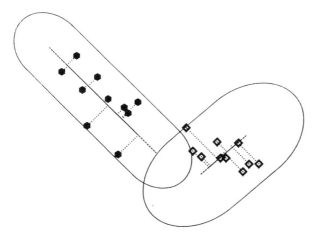

Fig. 14 Class models obtained from SIMCA (one principal component). Solid lines are the two models (normal range), broken lines are the extensions of the normal range, dotted lines are the distances of the objects from the model. For both classes the class space around the model is shown.

Finally, some objects are accepted by both models: they are in the lower left-hand corner of the figure (eight objects in Fig. 15). When SIMCA is used as a classification technique, the objects are assigned to the class with the minimum distance, i.e., using as class boundary the hypersurface (in the space of the variables) where the condition of equal distance from the class models is satisfied (in the Coomans plot this surface is represented by the diagonal line).

SIMCA gives a lot of interesting parameters: the interclass distance, which is a measure of the separation between the models of two classes; for each class, the modeling power of the variables

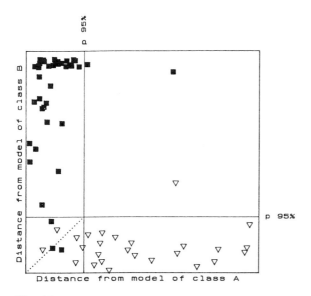

Fig. 15 Coomans plot of the objects of class A (filled squares) and of class B (triangles), showing the distance of each object from the model of A and of B. The critical distance from the model of each class is shown.

(their contribution to the class model, obtained by the loadings on the significant PCs of the model); the discriminating power of the variables (a measure of the multivariate importance of a variable in the separation of the classes). Moreover, each class model can be refined by deleting outliers. Also, the variables can be weighted, after class-separate autoscaling, to take into account information from measurement error or of nutritional importance or based on some other knowledge of the user.

This quality and the possibility of changing the model range make SIMCA a very flexible technique, which is a great advantage when it is used by people with great experience both in the problem and in chemometrics.

There are many other classification and class modeling techniques, frequently so rigid that the user can gain advantage from his experience with the problem in no other way than by deleting variables. Linear discriminant analysis (LDA) (13) is the oldest multivariate classification technique, based on the use of multivariate probability distribution, under the hypothesis of normal distribution with the same variance–covariance matrix in all classes in the problem. Quadratic discriminant analysis (QDA) uses the hypothesis of LDA, but the variance–covariance matrix is different, computed for each class. QDA has been transformed in a class modeling technique with the name UNEQ (14). Here the class model is the class centroid, and the class box boundary has the form of a confidence hyperellipsoid, the limit of the confidence interval of the normal multivariate distribution. Figure 16 shows the projections on the two first PCs of the confidence hyperellipsoids, boundaries of class models of the same nine classes of the biplot in Fig. 6. PCs are used in this case to give an approximate idea of the degree of overlapping of the class models.

When the distribution is not normal, a probabilistic nonparametric technique, the potential function probability technique (15), can be used as both a classification and a class modeling technique, but it requires a very great number of objects.

Modified Coomans plots can be used to compare the performances of models of the same class obtained with different techniques or with the same technique in different conditions (different number of variables, different number of PCs in SIMCA models). Figure 17 shows the use of the Coomans plot in the comparison of SIMCA and UNEQ models (class 1 of the olive oils of the previous example). In this case the UNEQ model accepts all of the samples, whereas the SIMCA

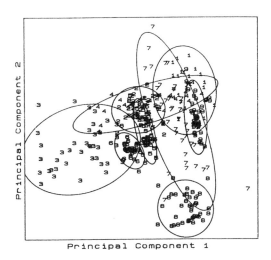

Fig. 16 UNEQ applied to the dataset of 324 Italian extravirgin olive oils belonging to nine classes (regions): projection of the boundaries of the nine class models on the plane of the two first PCs.

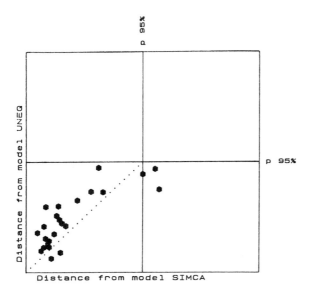

Fig. 17 Modified Coomans plot. Comparison between the performances of two modeling techniques (SIMCA and UNEQ) for the samples of class 1 of the dataset of Italian extravirgin olive oils.

model rejects four samples (one at a very low significance level, not shown in the figure). The percentage of the objects used to develop the model and accepted by the model measures the model sensitivity. In order to compare different models the specificity must also be considered, as measured by the fraction of objects of a different class correctly rejected by the model.

C. Latent Variables and Regression

Regression is a fundamental activity in analytical chemistry: univariate linear regression is generally used to compute the parameters of the relationship between a chemical quantity (known in a series of standards) and the physical measured quantity. Then, by the inverse relationship, the chemical quantity is obtained by the physical quantity measured on samples of unknown composition. The parameters (slope and intercept) of the linear model are estimated so that the sum of the squared residuals (differences between the measured quantity and its estimate by the regression line) is minimum: this is the basis of the least squares technique.

Recently, analytical chemists have used all the implications in the statistical model to correctly compute the confidence interval for the unknown.

Even more recent is the application of robust regression techniques as the simple single-median regression, which can detect outliers and linearity intervals.

1. Univariate Regression in Consensus Problems

The field of application of regression techniques is much wider than usually recognized. Here an example is reported about the use of simple least squares regression in consensus problems.

Data were organized in a table of 12 objects (members of a panel) described by 26 variables (scores given by the experts to 26 wines). These wines were of high quality, so that also for experts the differences in quality are quite difficult to detect. Only one attribute, the total quality, was judged by the 12 experts.

Table 1 shows the correlation coefficients among the experts, ranging from a value .31 (not

Table 1 Correlation Coefficient Matrix of 12 Experts in a Wine Panel (Overall Quality, 26 Wine Samples)

	A	B	C	D	E	F	G	H	I	L	M	N
A	1.00	0.50	0.65	0.76	0.47	0.70	0.66	0.66	0.58	0.58	0.49	0.71
B	0.50	1.00	0.77	0.77	0.80	0.85	0.37	0.71	0.51	0.84	0.56	0.74
C	0.65	0.77	1.00	0.79	0.62	0.82	0.59	0.76	0.57	0.75	0.42	0.74
D	0.76	0.77	0.79	1.00	0.70	0.80	0.46	0.87	0.63	0.79	0.62	0.94
E	0.47	0.80	0.62	0.70	1.00	0.74	0.31	0.68	0.61	0.77	0.57	0.69
F	0.70	0.85	0.82	0.80	0.74	1.00	0.57	0.75	0.52	0.89	0.68	0.82
G	0.66	0.37	0.59	0.46	0.31	0.57	1.00	0.46	0.59	0.39	0.35	0.47
H	0.66	0.71	0.76	0.87	0.68	0.75	0.46	1.00	0.75	0.75	0.68	0.87
I	0.58	0.51	0.57	0.63	0.61	0.52	0.59	0.75	1.00	0.55	0.44	0.65
L	0.58	0.84	0.75	0.79	0.77	0.89	0.39	0.75	0.55	1.00	0.61	0.81
M	0.49	0.56	0.42	0.62	0.57	0.69	0.35	0.68	0.44	0.61	1.00	0.74
N	0.71	0.74	0.74	0.94	0.69	0.82	0.47	0.87	0.65	0.81	0.74	1.00

significantly different from 0) to .94. Experts D and N have the maximum correlation. Figure 18 shows the regression of the scores of expert N on those of expert D, with the confidence interval of the regression line. In spite of the reduced range of scores given by expert N, there is a very good correlation between the two experts. To the contrary (Fig. 19), expert G has about the same range as expert D, but with a very low correlation coefficient (.46).

Because the two experts D and N show the maximum agreement, one of them can be selected as a reference. The regression of a generic expert e on a reference expert, e.g., D, gives the regression line as

$$\hat{x}_e = a_e + b_e x_D \tag{8}$$

and the residuals are

$$r_e = x_e - \hat{x}_e = x_e - a_e - b_e x_D \tag{9}$$

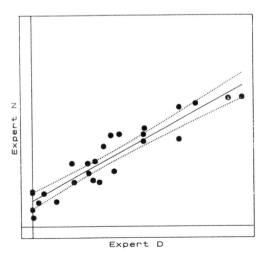

Fig. 18 Scores assigned to 26 wines by two experts. The regression line and its confidence intervals show a very good agreement between expert D and expert N.

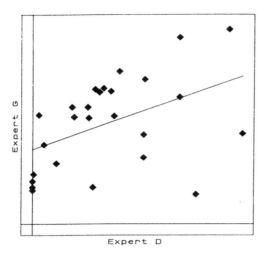

Fig. 19   Scores assigned to 26 wines by two experts. Very low correlation between expert D and expert G.

Scores of expert e are modified in the following way:

$$x_{me} = x_D + r_e = x_D + x_e - a_e - b_e x_D$$

so that the systematic part of the difference with expert D is corrected by the regression line, and the random part, measured by the residual, is retained.

Figure 20 shows the use of PCs in the visualization of the effect of the correction described. PCs were computed on an enlarged data matrix with the 12 original rows of scores (triangles in Fig. 20), with 11 rows of the experts corrected against expert D (filled diamonds in Fig. 20), and the 11 rows of the experts corrected against expert N (filled hexagons in Fig. 20). The original dispersion, given by the triangles, can be compared with the final consensus configuration obtained

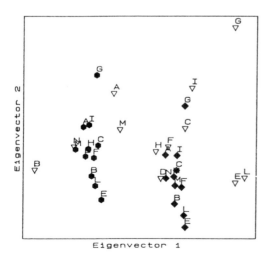

Fig. 20   Plot on PCs 1 and 2 of 12 panelists: experts before consensus (triangles), experts after consensus with expert D as reference panelist (filled diamonds) and after consensus with expert N as reference panelist (filled hexagons).

with reference to expert D or N. The more anomalous expert G is still far from the cluster of the other experts, but the degree of agreement obtained among the other 11 experts is very high.

2. *Multivariate Regression*

Multivariate regression gives the relationship between more variables (predictors) and one or more other variables (response variables), whose direct measure is difficult or requires too much time or money.

Ordinary least squares (OLS) regression, or multiple regression (16), is the multivariate technique analogous to the method used in the univariate case. It is widely applied in experimental design, when the factors under study are independent; however, in practice, it is frequently used also in cases of many correlated predictors, where one of the fundamental assumption of OLS (uncorrelation of predictors) fails.

The relationship between the response variable Y and the predictor variables X_v is presented in the form

$$\hat{y} = x^T b \tag{10}$$

where x^T is the row vector of the V predictors augmented with a term 1, for a total of $M = V + 1$ elements; b is the column vector of the regression coefficients; the mth coefficient (corresponding to 1 in the x^T vector) is the intercept; $\hat{y}$ is the estimate of the response variable as obtained by the regression equation.

The regression coefficients are estimated with a training set of I objects, a matrix $_IX_M$ of I rows (objects, samples) and $M = V + 1$ predictor variables, and a vector y of I rows. Vector b is obtained as

$$b = (X^TX)^{-1}X^Ty \tag{11}$$

In the case of univariate regression ($M = 2$), this equation corresponds exactly to the equation used to obtain the intercept and the slope of the straight line; in particular, for the slope

$$b = \frac{\Sigma_I \, (x_i - \bar{x}) \, (y_i - \bar{y})}{\Sigma_I \, (x_i - \bar{x})^2} \tag{12}$$

Very useful diagnostics (17) have been developed for OLS, including the confidence interval of each regression coefficient b_v, the leave-one-out statistics corresponding to the validation with the leave-one-out technique, the leverage of each object. This is a measure of the importance of the object in the equation of the regression model: an object very far from the centroid gives a large contribution to the elements of the dispersion matrix X^TX and draws on itself the regression plane; an object with a very high leverage due to a heavy error can dramatically affect the regression equation.

The OLS model searches for the regression plane with the minimum sum of the squared residuals:

$$\Sigma_I \, (y_i - \hat{y}_i)^2$$

and uses too much the noise in the data. Sometimes a small improvement in the fitting (a smaller sum of squared residuals) is obtained at the expense of a great error in the estimate of regression coefficients.

Figure 21 shows this effect in the case of univariate regression. Both the upper and lower plots show the same model (the dotted line, the theoretical relationship we can obtain only with infinite objects) and two different estimates, in both cases with three experimental points, in both cases with the same error on the first, second, and third points. The large deviation of the model estimated

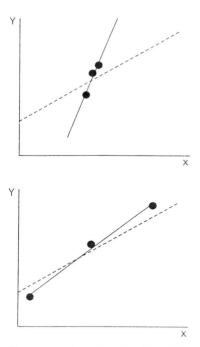

Fig. 21   Effect of the location of the experimental points on the regression line. Dotted line is the "true" regression line, solid line is the computed regression line.

from the true model in the upper part of the figure is due to the closeness of the three points, which means a very low value of the denominator of the equation for the estimation of the slope b, $\sum_I (x_i - \bar{x})^2$.

The value of this denominator is, in the case of multivariate OLS, the determinant of the dispersion matrix $\mathbf{X}^T\mathbf{X}$. A large value of the determinant ensures a low uncertainty of the regression coefficients (and generally a low error of the regression equation); the value of the determinant depends on the range of the predictors (as in Fig. 21), usually fixed by experimental requirements, and (very important) decreases with the increasing correlation between the predictors. The determinant has the significance of the dispersion in the V-dimensional space of the predictors: when the correlation between the predictors is very high, the objects are in a $(V - 1)$–dimensional space (a line in a two-dimensional space) and the dispersion in the V-dimensional space is null.

An obvious procedure is to compute the PCs of the predictors and then apply OLS on the scores in the space of the significant components: this procedure is known as principal component regression (PCR). Principal components are uncorrelated variables, so that their determinant is simply the product of the eigenvalues, the terms on the diagonal of the dispersion matrix.

PCR works very well in practice. However, it is a two-step technique, and presently a similar technique is of general use in regression problems, with the same performances as PCR, or better, and with an easier interpretation. This technique is PLS (8), i.e., partial least squares regression, or projection on latent structures. A latent structure, a latent variable, is a direction in the space of the predictors with some analogy to principal components. Really, as PCs are the eigenvectors of the matrix $\mathbf{X}^T\mathbf{X}$ (autoscaled predictors), latent variables are the eigenvectors of the matrix $\mathbf{X}^T\mathbf{y}^T\mathbf{y}\mathbf{X}$: the value of the response variable weighs each row of the predictor matrix.

The mathematical formalism, however, cannot help to make comprehensible the real significance of these latent variables. Here a small numerical example is used to introduce both the PLS algorithm and the meaning of latent variables.

Let us study the regression between one response variable y and one predictor x. Be x measured on 10 samples, twice for each sample. Data are drawn from a population with relation $y = 2x$ and normally distributed noise added to x.

Because of the noise, different values of x are obtained in the two replicates (Table 2).

In this case, we know that the mean of the two replicates is a better estimate of the true value of the predictor because the variance of the mean is half as large as the variance of the single repetition. So the correct procedure is to compute the mean $\bar{x}$, then to use the univariate least squares regression, y vs. $\bar{x}$.

The following result is obtained:

$$\hat{y} = -2.266 + 2.059\bar{x}$$

The statistical characteristics are as follows:

Slope: $b_{\bar{x}} = 2.059$ $s_b = 0.046$ Conf. int. $= 0.106$
Intercept: $a_{\bar{x}} = -2.266$ $s_a = 5.6$ Conf. int. $= 12.9$

The standard deviation of the error is 8.12.

Since the theoretical equation has intercept zero, the relatively large confidence interval for the intercept is not surprising, with the consequent acceptance of the hypothesis of intercept zero.

Alternatively, we can consider the two repetitions x_1 and x_2 as two distinct predictors and apply multivariate regression; we have not recognized the presence of a "latent variable," i.e., the mean.

OLS gives the following results:

$$\hat{y} = -2.034 + 0.412x_1 + 1.633x_2$$

with the following statistical characteristics:

$b_1 = 0.412$ $s_{b_1} = 0.344$ Conf. int. $= 0.814$
$b_2 = 1.633$ $s_{b_2} = 0.336$ Conf. int. $= 0.795$
$b_3 = -2.034$ $s_{b_3} = 4.954$ Conf. int. $= 11.7$

The fitting is as follows:

Table 2 Numerical Example for OLS and PLS

Obj. index	Value of the variables			
	x_1	x_2	$\bar{x}$	y
1	18.69	21.52	20.107	40
2	46.50	38.54	42.521	80
3	58.44	63.80	61.118	120
4	78.59	77.50	78.046	160
5	102.82	102.59	102.704	200
6	113.12	113.36	113.243	240
7	126.95	134.80	130.876	280
8	147.69	163.30	155.497	320
9	170.88	177.96	174.421	360
10	204.87	197.30	201.084	400

Standard deviation of the error $= 7.18$
Mean absolute error $= 4.90$
$R^2 = 0.9973$

The prediction (leave-one-out) is as follows:

$R^2_{CV} = 0.9948$
CV RSD $= 8.75$
CV Mean prediction error $= 7.22$

Note that:

1. The regression coefficients b_1 and b_2 are very different; it seems that x_2 is four times more important than x_1, which is completely absurd because the two variables are repetitions of the same measurement.
2. The confidence interval of both slopes is very high; the first slope b_1 seems not significantly different from zero, so that we can consider this variable x_1 as a useless quantity.
3. The sum of the two slopes, 2.045, is a good estimate of the true regression coefficient 2.

In this case, where one or more regression coefficients are not significantly different from zero, usually a stepwise technique is applied, which selects one variable at a time to build an OLS regression model. On the basis of a statistical F test, new variables are accepted or not. Applied to the present example, stepwise OLS accepts only one variable, which means that the information given by the second repetition of the measurement is lost.

PLS computes first the slopes of the partial regressions: from the regression of y on x_1 it obtains the partial slope $b_{P1} = 2.0716$; from the regression of y on x_2 it obtains $b_{P2} = 2.0315$.

These slopes are divided by the square root of the sum of their squares, the Euclidean norm. These normalized slopes are $w_1 = 0.71398$ and $w_2 = 0.70016$. They are called PLS weights.

PLS latent variable is defined as

$$t = w_1 x_1 + w_2 x_2 \tag{13}$$

Because of the normalization, the sum of the squares of the weights is 1: so the weights have the significance of director cosines of the latent variable.

The latent variable is almost exactly proportional to the mean of the two variables. PLS uncovered that not the original variables but rather this function t must be used in regression. In fact, PLS regression is then performed as the regression of y on the latent variable t. The regression equation is:

$$\hat{y} = -2.265 + 1.456t$$

with the following statistical characteristics:

Slope $b_t = 1.456$ $s_b = 0.033$ Conf. int. $= 0.076$
Intercept $a_t = -2.265$ $s_a = 5.6$ Conf. int. $= 13.0$

Standard deviation of the error $= 8.16$

Note that the RSD and the relative confidence interval of b_t are the same as those of $b_{\bar{x}}$, and about the same are the estimates of the standard deviation of the error, of the intercept and its standard deviation.

The next step of PLS is the substitution of the latent variable with its function of the original variables. The final form (closed form of PLS regression) is obtained as:

$$\hat{y} = -2.265 + 1.0395x_1 + 1.0194x_2$$

and the regression coefficients indicate that the two predictors have the same importance; their sum is the same as $b_{\bar{x}}$.

In this case the following statistical characteristics for PLS regression are not so favorable; the cross-validated prediction is slightly worse than that one obtained with OLS. In real multivariate problems with many predictor variables generally PLS predicts much better than OLS, and this is its principal advantage together with the use of the latent variables.

The fitting is as follows:

Standard deviation of the error = 7.70
Mean absolute error = 5.98
$R^2 = 0.9960$

The prediction (leave-one-out) is as follows:

$R^2_{CV} = 0.9948$
CV RSD = 9.50
CV Mean prediction error = 7.57

3. Multivariate Calibration

PLS and PCR are widely applied in a lot of regression problems. PLS has more flexibility than PCR: with the name PLS-2, it can also be applied to the regression of a block of response variables on a block of predictor variables, and in this case also latent variables for the block of responses are defined; moreover PLS can be applied as Consensus-PLS, to many blocks problems. This happens when many experts in a panel evaluate a number of samples for many attributes: in this case each expert corresponds to a data matrix with as many rows as samples and as many columns as attributes (and this number can be different, as in free-choice evaluation of foods), so that the number of blocks is the same as the number of experts.

In analytical chemistry one of the most interesting fields of application of multivariate regression is multivariate calibration (18), where the value of one (or more) chemical quantities is obtained as a function of more physical nonspecific quantities. The specificity is obtained as a result of the regression procedure.

Multivariate calibration is widely applied in near-infrared spectrometry (NIRS). However, it was also applied in X-ray axial tomography, and in principle it can be applied to many techniques whereby the single signal is nonspecific but the complex of signals under different conditions can give useful analytical information, e.g., fluorescence, nuclear magnetic resonance (NMR), ultraviolet (UV), etc.

The advantage of multivariate calibration is in the application to fast methods of analysis of complex matrices, where the time and the cost of the treatment needed for the usual analytical procedure are too high for application in quality control and when the analytical information must be obtained in a very short time, as in the case of perishable samples.

In the following example (19), the response variable is moisture in soy samples, and the predictor variables are 19 reflectances from a filter NIRS instrument. Sixty samples of soy were analyzed, which was well representative of the Italian production.

Tables 3 and 4, and Figs. 22 and 23, show the results of multivariate regression with OLS and PLS, respectively. PLS has not a diagnostics as complete as OLS, so that the uncertainty on regression coefficients is not reported. Results of PLS are shown as a function of the number of latent variables. After one latent variable has been computed, its information is canceled by the matrix of the predictors, and a new latent variable is computed with the residuals of the predictors. The number of latent variables used defines the complexity of the PLS regression model. The optimum complexity is that giving the minimum prediction error. As the number of latent variables

Table 3 Results of OLS Applied to 60 Samples of Soy

Wavelength	Slope	SD	95% CI
1	−218.07416	339.7122	686.4904
2	405.79183	393.0893	794.3549
3	−134.62146	328.4933	663.8192
4	79.20624	496.5756	1003.4799
5	424.29168	595.1891	1202.7581
6	−737.52700	617.9301	1248.7130
7	1402.23132	580.2750	1172.6196
8	−64.71404	442.2657	893.7305
9	−1030.86676	830.0522	1677.3695
10	−267.85859	187.9339	379.7769
11	−37.18117	496.3503	1003.0247
12	−692.72045	905.3195	1829.4696
13	−267.24236	314.8850	636.3195
14	335.05786	959.4461	1938.8485
15	299.57042	167.8608	339.2130
16	965.36018	772.1568	1560.3743
17	−1028.68378	884.5394	1787.4771
18	71.19596	219.0959	442.7489
19	531.34858	337.1648	681.3425
Intercept	−7.96317	10.9649	22.1578

SD	1.18
Mean absolute error	0.76
R^2	0.905

Prediction (leave-one-out)	
CV RSD	1.57
CV Mean prediction error	1.20
R^2_{CV}	0.752

increases, the PLS model tends to the result of OLS regression. The larger the number of latent variables beyond the optimum complexity, the more noise that is picked up by the regression model; the fitting ability increases because a complex model is able to fit noise better than a simple model, but the prediction ability decreases. In the example regarding the soy, the optimum complexity is obtained with four latent variables, but the prediction error (measured with the leave-one-out procedure) is almost stable from three to eight latent variables, and it is about 20% less than the error obtained by OLS. With cross-validation and five cancellation groups, the results in Table 5 have been obtained. The optimum complexity for the PLS model is 4, unchanged; the estimate of the prediction ability is worse for OLS, almost the same obtained with the leave-one-out procedure for PLS.

Fitting ability of PLS model increases with the number of latent variables and approaches that of OLS.

Figure 24 shows the regression coefficients of the variables (reflectances at 19 wavelengths) computed by OLS and PLS. The absolute values of OLS regression coefficients (triangles in Fig. 24) are much larger than those of PLS (circles). The value of the response variable is obtained as the sum of a negative and a positive contribution (negative and positive regression coefficients); the same value (approximate) of the response variable is obtained by OLS as the sum of two large

Table 4 Results of PLS Applied to Soy Samples

A. Fitting

Component	RSD	Expl. variance (%)	R^2 (%)
1	1.6044	73.627	74.074
2	1.3540	81.212	81.849
3	1.2344	84.379	85.173
4	1.1745	85.853	86.812
5	1.1586	86.230	87.397
6	1.1212	87.101	88.413
7	1.1144	87.251	88.764
8	1.1027	87.512	89.206
9	1.0794	88.031	89.857
10	1.0717	88.195	90.196
11	1.0737	88.146	90.356
12	1.0793	88.017	90.454
13	1.0884	87.808	90.495
14	1.0995	87.553	90.507
15	1.1114	87.276	90.511

B. Prediction (leave-one-out)

Component	CV RSD	CV expl. variance (%)	Mean Error
1	1.6393	72.475	1.3165
2	1.4360	78.880	1.1369
3	1.3388	81.641	1.0208
4	1.2705	83.468	0.9665
5	1.3145	82.303	0.9926
6	1.4352	78.902	1.0394
7	1.3384	81.652	0.9789
8	1.3931	80.122	1.0175
9	1.4679	77.930	1.1244
10	1.5292	76.049	1.1765
11	1.5477	75.466	1.1913
12	1.5767	74.539	1.2144
13	1.5809	74.402	1.2088
14	1.5811	74.394	1.2067
15	1.5742	74.617	1.1998

contributions, by PLS as the sum of two small contributions. So the noise at a selected wavelength, because of the generally larger coefficients, produces an error on the response variable larger with OLS than with PLS.

4. Experimental Design in Multivariate Calibration

Frequently there are no standards available for analytical problems where multivariate calibration is applied. To compute the model parameters it is necessary to select some natural samples, then

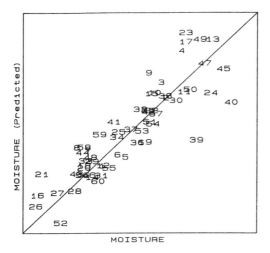

Fig. 22 Relationship between moisture measured on 60 samples of soy vs. moisture predicted by OLS. The predictors are reflectances at 19 wavelengths.

analyze them by means of a reference technique, which is generally expensive and time consuming.

To reduce the cost of calibration, it is possible to select an optimal subset of samples among a lot of possible samples for calibration, analyze only the samples in this subset, and then use these samples to compute the regression model.

The optimal subset is selected according to the rules of experimental design. The procedure here described can be applied to a lot of practical cases where the result of the experiment depends on uncontrollable and correlated factors (as in the case of the selection of some representative solvents among all of the possible solvents in chromatography or in organic synthesis).

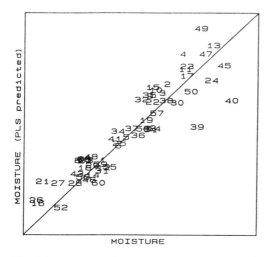

Fig. 23 Relationship between moisture measured on 60 samples of soy vs. moisture predicted by PLS. The predictors are reflectances at 19 wavelengths.

Table 5 Prediction with OLS and PLS,
Five Cancellation Groups

	OLS	PLS[a]
CV RSD	1.86	1.3
CV Mean prediction error	1.41	1.03
R^2_{CV}	0.650	0.802

[a]Four latent variables.

Here the samples are described by a spectrum, with many (sometimes more than 1000) highly correlated wavelengths; the spectra are those of the many possible samples for the selection of the optimal subset.

We can compute the principal components of the spectra. The PC plots can show how well the whole set of the samples covers the experimental space; when samples are scattered regularly on the PC plot they explore well the experimental space. To the contrary, when they cluster in some regions of the plot, they are not representative of the (more or less well-known) factors determining the variability.

One of the techniques of experimental design is to select the optimal subset so that the determinant of the dispersion matrix $\mathbf{X}^T\mathbf{X}$ will be maximum (D-optimal design) (20). In Fig. 25 the PCs of the 60 soy spectra are shown. Filled hexagons indicate the spectra selected for the subset. In this case we decided to select nine samples and to take into account the possibility of nonlinear relationships between spectrum and composition. The original data matrix was augmented with squares and cross-products of the variables in order to select the subset on the hypothesis of a quadratic relationship. Then we can select some other samples for the validation of the PLS model.

In Fig. 25 it is possible to verify that the 60 samples are well distributed on the space of the two PCs (only one sample is described by a large score on the second component). The selected samples are located at the edge of the cluster of points and in the center. When a sample selected by the optimal design must be discarded for some other reason, it is possible to substitute with a sample

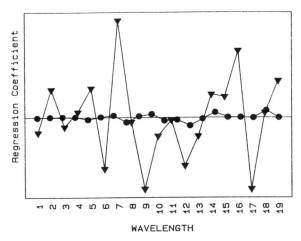

Fig. 24 Regression coefficients computed by OLS (triangles) and by PLS (circles) for the reflectances at 19 wavelengths.

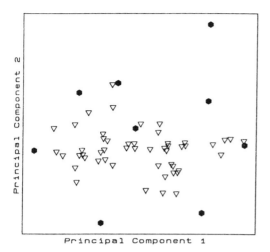

Fig. 25 Plot on PCs 1 and 2 of 60 NIR spectra (soy samples). Filled hexagons indicate the spectra selected for an optimal subset for calibration.

very close to it in the plot without a noticeable worsening of the design performances. Obviously, the reduced number of samples used for calibration generally produces a model with a little worse predictive ability—the price of economy.

5. Multivariate Regression in Calibration Transfer

Another problem, typical of multivariate calibration, is that the response of different instruments is not exactly the same, so that we can have a problem of consensus among instruments. The most common case is that calibration is made on a master instrument of high quality and it cannot be applied on a series of slave instruments.

Multivariate regression offers the possibility of transferring both the signal (the spectra) and the regression equation from one instrument to another. We can compute from the spectra recorded on a slave instrument the spectra that the master instrument would obtain on the same sample. Then we can apply the regression model computed on the master instrument. Otherwise, we can transfer the regression equation from the master instrument to the slave instrument, and the slave instrument can use its own spectrum with this predicted regression equation.

The overall schema is shown in Fig. 26. Here a possible intermediate step is shown with the projection of the original spectra on their principal components (generally the eigenvectors of global centered data are used because autoscaling tends to give too much importance to wavelengths without useful information). The use of scores is suggested to reduce the dimensionality of data and to cancel some noise from the original data.

A multivariate regression technique, applied to samples in the training set, can be used to compute all of the possible relationships between blocks of variables (between the two blocks X of physical variables, between the two blocks S of scores obtained from the original blocks X, between each block X and the vector of the response variable y, between S and y). With the use of the relationship (given by the matrix of loadings) between S and the corresponding X, and by combining the different regression models, it is possible to obtain other results: the equation linking a block X to the response variable via the scores S, the equation connecting a block X to the response variable y via the other block X, etc.

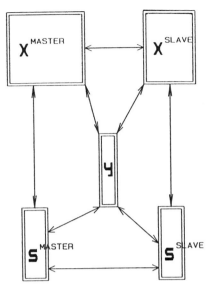

Fig. 26 Schema of calibration transfer between a master instrument and a slave instrument. Regression is applied to compute the relationship between spectral variables (X) or scores (S) and chemical variable (y) of master and of slave, and between spectral variables (or scores) of master and of slave.

An extreme example of the transfer of spectrum is shown in Fig. 27: the (slave) instrument has 19 filters, the (master) instrument is an NIR spectrometer at 700 wavelengths—two very different instruments, two very different spectra. PLS was applied to the scores on the first 10 eigenvectors (almost all of the variance of the original data is retained in these eigenvectors), and the predicted spectrum (leave-one-out procedure) is practically equal to the spectrum recorded on the master instrument on the spectrum scale. Figure 27 shows also the errors (both in fitting and in prediction) on an expanded scale.

When the two instruments are of comparable quality, the transfer procedure gives a spectrum with less noise than that recorded on the master instrument. Obviously the transfer procedure cannot reproduce noise, and the transferred spectrum can be considered as the average between the spectrum of the slave instrument and the spectrum of the master instrument as interpreted by the regression equation between the two X or S blocks.

This example of the transfer of calibration between instruments shows the flexibility and the power of regression techniques. Surely chemists have a lot of problems to which similar strategies can be advantageously applied.

III. COMPUTER PROGRAMS

Many chemometric tools are available on general statistics software. However, there are some packages written by chemometricians that are oriented to chemical problems with special attention to the most recent chemometrics literature. A very short review of the best known packages is given here.

Arthur (21) was the first package born in chemometrics in the mid-1970s. It is one of the chemometric software products commercialized by InfoMetrix. The upgrade of Arthur is Pirouette (22), which combines exploratory data analysis (clustering and PCA) with classification and class modeling techniques (KNN and SIMCA) as well as regression techniques (PCR, PLS). The

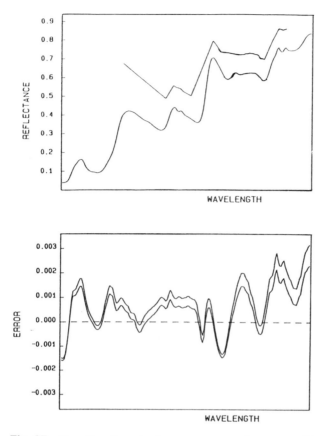

Fig. 27 Transfer of spectra between a master instrument (700 wavelengths) and a slave instrument (19 filters). Upper: comparison between spectrum recorded by the slave instrument and spectrum of the master instrument. The predicted spectrum is perfectly overlapping on the plot scale. Below: errors in fitting and in prediction (expanded scale).

package is now available in a Windows version: it is specially designed to simplify data analysis and creation of chemometric models by means of many data management features and a friendly graphical environment. InStep (23), designed for Windows, was created for routine prediction applications, working alongside an analytical instrument system. Ein*Sight (24) is an exploratory data analysis system that includes only clustering techniques and PCA; it can be considered as an introduction to the Pirouette system.

SIMCA (25) is a package for multivariate data analysis with special attention to PC and PLS analysis for different kinds of problems. It is equipped with predictive tools for optimizing products and processes. MODDE (26) is a package for experimental design, handling both qualitative and quantitative factors. OLS and PLS can be used for model fitting. Graphical user interface, Windows facilities, and powerful diagnostics are common features of both packages.

Unscrambler (27) contains PC analysis, PC and PLS regression, SIMCA class modeling, artificial neural networks, and experimental design techniques. Very powerful weighting and preprocessing of variables, validation methods, outlier detection, and excellent presentation of results are other important characteristics of this package, which is oriented to multivariate calibration.

SCAN (28) includes the best known data exploration, preprocessing, classification, class modeling, and regression techniques. It also has many clustering techniques, both hierarchical and nonhierarchical as K means and Jarvis–Patrick clustering. Ridge regression and alternating conditional expectations (ACE) are special regression techniques; ACE is a very powerful nonlinear regression technique.

Sirius (29) includes experimental design techniques for screening and optimization; extensive preprocessing capability; techniques for exploration, classification, and regression of multivariate data; and Windows on-line help.

NEMROD (30) is a special package for experimental design. It allows a choice among many experimental strategies for solving different problems: screening of influent factors, study of the effect of the factors, optimization, searching of the best mixture in formulation problems, etc. Multilinear regression, statistical tests, graphical tools, and response surfaces are available for analysis of the results.

Principal Components (31) is an educational set with four videos and a double software package for the introduction to the use of PC analysis and related methods in chemometrics. The first software is a tutorial version of SPECTRAMAP (32), a performant commercial software for methods derived from PC analysis. The second is a list of MATLAB programs.

Finally, PARVUS (33) was the first package especially written for personal computers. It includes most preprocessing, clustering, class modeling (SIMCA, UNEQ, and potential functions methods) and regression techniques (from univariate regression to PLS and ACE), factor analysis (Varimax rotations, evolving factor analysis), with special attention to validation procedures. Color plots done on a Hewlett-Packard plotter can be obtained for PCs, latent variables, and dendrograms obtained from clustering techniques.

REVIEWS

D. L. Massart, B. G. M. Vandeginste, S. N. Deming, Y. Michotte, and L. Kaufman, *Chemometrics: A Textbook*, Elsevier, Amsterdam, 1988.

J. E. Jackson, *A User's Guide to Principal Components*, John Wiley and Sons, New York, 1991.

R. G. Brereton, *Chemometrics: Application of Mathematics and Statistics to Laboratory Systems*, Ellis Horwood, Chichester, 1990.

M. Meloun, J. Militky, and M. Forina, *Chemometrics for Analytical Chemistry: PC-Aided Statistical Data Analysis*, Ellis Horwood, Chichester, 1992.

REFERENCES

1. G. E. P. Box, W. G. Hunter, and J. S. Hunter, *Statistics for Experimenters*, John Wiley and Sons, New York, 1978.
2. J. A. Cornell, *Experiments with Mixtures*, John Wiley and Sons, New York, 1990.
3. R. Carlson, *Design and Optimization in Organic Synthesis*, Elsevier, Amsterdam, 1992, p. 337.
4. I. E. Frank and R. Todeschini, *The Data Analysis Handbook*, Elsevier, Amsterdam, 1994, p. 212.
5. M. Stone, *J. Roy. Stat. Soc., Ser. B 36*: 111 (1974).
6. S. Lanteri, *Chem. Int. Lab. Sys. 15*: 159 (1992).
7. S. Wold, *J. Pattern Recog. 8*: 127 (1976).
8. S. Wold, C. Albano, W. J. Dunn III, U. Edlund, K. Esbensen, P. Geladi, S. Hellberg, E. Johansson, W. Lindberg, and M. Siostrom, *Chemometrics: Mathematics and Statistics in Chemistry* (B. R. Kowalski, ed.), Reidel, Dordrecht, 1984.
9. M. Forina and E. Tiscornia, *Ann. Chim. 72*: 143 (1982).
10. S. Wold, *Technometrics 20*: 397 (1978).
11. E. R. Malinowski, *Factor Analysis in Chemistry*, 2nd Ed. John Wiley and Sons, New York, 1991.
12. D. Coomans, PhD thesis, Vrije Universiteit Brussel, 1982.
13. R. A. Fisher, *Ann. Eugenetics 7*: 179 (1936).

14. M. P. Derde, D. Coomans, and D. L. Massart, *Anal. Chim. Acta 141*: 187 (1982).
15. M. Forina, C. Armanino, R. Leardi, and G. Drava, *J. Chemometrics 5*: 435 (1991).
16. N. R. Draper and H. Smith, *Applied Regression Analysis*, 2nd Ed. John Wiley and Sons, New York, 1981.
17. D. A. Belsley, E. Kuh, and R. E. Welsch, *Regression Diagnostics: Identifying Influential Data and Sources of Collinearity*, John Wiley and Sons, New York, 1980.
18. H. A. Martens, *Multivariate Calibration*, Dr. techn. thesis, Technical University of Norway, Trondheim, 1985.
19. M. Forina, G. Drava, C. Armanino, R. Boggia, S. Lanteri, R. Leardi, P. Corti, P. Conti, R. Giangiacomo, C. Galliena, R. Bigoni, I. Quartari, C. Serra, D. Ferri, O. Leoni, and L. Lazzeri, *Chem. Int. Lab. Sys. 27*: 189 (1995).
20. T. J. Mitchell. *Technometrics 16*: 203 (1974).
21. *Arthur 4.1*, InfoMetrix Inc., Seattle, 1986.
22. *Pirouette*, InfoMetrix Inc., Seattle, 1993.
23. *InStep*, InfoMetrix Inc., Seattle, 1993.
24. *Ein*Sight 3.0*, InfoMetrix Inc., Seattle, 1991.
25. *SIMCA for Windows*, Umetri AB, Umeå, 1993.
26. *MODDE for Windows*, Umetri AB, Umeå, 1993.
27. *Unscrambler II* (Extended Memory Version 5.0), CAMO A/S, Trondheim, 1994.
28. R. Todeschini, U. Cosentino, I. E. Frank, and G. Moro, *SCAN: Software for Chemometric Analysis* (Rel. 3.1), JerIl Inc., Stanford, 1992.
29. *SIRIUS for Windows*, Pattern Recognition Systems A/S, Norway, 1991.
30. D. Mathieu and R. Phan-Tan-Luu, *NEMROD: New Efficient Methodology for Research Using Optimal Design*, LPRAI, Aix-Marseille, 1992.
31. D. L. Massart and P. J. Lewi, *Principal Components*, Elsevier, Amsterdam, 1994.
32. *SPECTRAMAP 1.0*, Janssen Pharmaceutica, Beerse, Belgium, 1988.
33. M. Forina, S. Lanteri, C. Armanino, R. Leardi, and G. Drava, *PARVUS 1.3: An Extendable Package of Programs for Data Explorative Analysis, Classification and Regression Analysis*, Istituto di Analisi e Tecnologie Farmaceutiche ed Alimentari, Genova, 1994.

3

Determination of Moisture and Ash Contents of Food

Young W. Park
Cooperative Agricultural Research Center, Prairie View A&M University,
The Texas A&M University System, Prairie View, Texas

I. ANALYSIS OF MOISTURE CONTENTS

A. Introduction

Water is undoubtedly the most ubiquitous substance in nature. Three-fourths of the earth's surface is covered with water. It is contained in all existing materials in earth and serves as the fundamental basis for life-supporting systems of every living creature. Even the atmosphere contains a small percent of water as vapor (1). In addition to its ubiquity, water is unusually reactive due to its high polarity (2).

The terms "water content" and "moisture content" have been used interchangeably in the literature to designate the amount of water present in foodstuffs and other substances. The abundance and chemical reactivity cause moisture and the determination of moisture to be of great concern to many industries such as food, paper, and plastics, in which acceptable levels of moisture vary among materials and in some cases minute quantities of moisture can adversely affect product quality (1).

Moisture determination is an important and widely used analytical measurement in the processing and testing of food products (3). The amount of moisture is a measure of yield and quantity of food solids, and is frequently an index of economic value, stability, and quality of food products. Therefore, it is of paramount importance for the food industry as well as researchers to have simple, rapid, and accurate methods for moisture determination in food products. The availability of such analytical procedures underlies compliance with statutory requirements of regulatory agencies, quality control in manufacturing of food products, and good business management (4,5).

An ideal method for moisture assay has been suggested (6). It should (a) be rapid, (b) be applicable to the broadest range of materials, (c) be performable by any, preferably nontechnical, person with brief training, (d) use readily available apparatus of low initial investment and low cost per test, (e) have reasonable accuracy and good precision, and (f) present no operational hazards. Analytical methods of moisture determination are usually selected for either rapidity or accuracy, although both goals are simultaneously sought.

The accurate determination of moisture is frequently one of the most difficult tasks that the food chemist encounters, even if it is in some respects a very simple analytical procedure. This is largely attributable to the difficulty in achieving complete separation of all water from a food sample without causing simultaneous decomposition of the product. The production of water by decomposition and loss in weight would affect the accuracy of the determination. The loss of volatile constituents from the food is another difficulty involved in moisture determination. The complexity of water determination will depend on the condition in which it is present and the nature of other substances present (5,7).

B. Properties of Water (in Food System)

Water can exist physically in three forms: gas, liquid, and solid. It exists in the gaseous state as the monomolecular water vapor, in the liquid state largely as dihydrol in which two molecules of water are bound by hydrogen bond forces, and in several solid forms as ices varying in degree of association (8).

Water inside a foodstuff exists in two forms: free and bound. However, water can be classified in at least three forms (9–11): First, a certain amount exists as *free water* in the intergranular spaces and within the pores of the material. Such water serves as a dispersing medium for hydrophilic macromolecules such as proteins, gums, and phenolics to form molecular or colloidal solutions, and a solvent for the crystalline compounds. Second, part of the water is *adsorbed* as a very thin, mono- or polymolecular layer on the internal or external surfaces of the solid components (i.e., starches, pectins, cellulose, and proteins) by molecular forces or by capillary condensation. This water is closely associated with the absorbing macromolecules by forces of absorption, which is attributed to van der Waals forces or to hydrogen bond formation. Third, some of the water is in chemical combination as water of hydration, so-called *bound water*. Carbohydrates such as dextrose, maltose, and lactose form stable monohydrates, and salts such as potassium tartrate also form hydrates. Water of hydration can be clearly observed from gels of proteins or polysaccharides in which the bound water is firmly held by hydrogen bonds.

The state of water in colloidal systems and the nature of bound water is still not clear. It has been suggested that water found in biological material may exist as (a) occluded water, (b) capillary water, (c) osmotic water, (d) colloidal water bound by physical forces, and (e) chemically bound water (12).

Most researchers have defined bound water as the water that remains unchanged when food is subjected to a particular heat treatment. For example, a certain proportion of the total water present in the biocolloids are not readily separated by freezing (even at $-230°C$) or drying. All the free water is usually frozen at $-125°C$ and the remaining bound water is not frozen at considerably lower temperatures (11,12). Such bound water concentration varies from one food to another. It is the bound water and not the free water that is related to the ultimate accuracy of an analytical method for moisture determination. On drying, part of the water in a sample is retained for longer times at higher temperatures than the remainder. The range of bound water in foods is less than 9.5% to over 30% of the total water present, corresponding to 0.1–2.2 g/g total solids (11,13). Except for surface water, all water may be considered as bound to a variable extent. The most tightly bound water is the BET monolayer water (the model of Brunauer, Emmett and Teller) (5). For most foods, the BET monolayer values range from a few percentage points to about 12% (wet basis) of the food or food component (14).

It has been shown that small changes in water content exert a large influence on storage stability in storage of low-moisture foods. Irreversible changes in the texture of foods also occur during freezing and freezing storage. Therefore, removal of free water rather than bound water from dried foods has been proposed to improve storage stability (15). Likewise, decrease in free water contents of foods to be preserved by freezing, concentration, partial dehydration, or addition of sugar is believed to improve storage stability of frozen foods and food products (11).

Water displays abnormally high values of certain physical constants such as specific heat, specific gravity, heat of infusion, heat of evaporation, surface tension, and viscosity. These special constants may be derived from its remarkable and variable solvent power, high dielectric constant, dissolving and ionization ability, and its own molecular aggregation tendency (16). The characteristics of water in chemical reactivity, volatility, solvent power, electrical properties (high dielectric constant, conductivity, and magnetic resonance absorption), thermal conductivity, and light scattering and absorption have been applied to the determination of moisture content in foods and other materials.

C. Mechanism of Drying (Moisture Loss)

Water is the largest single constituent of all living things. Water is present in most natural foods to the extent of 70% of their weight or greater. Moisture content greatly affects food quality, value, and freshness. Quantification of moisture content is directly affected by drying rate of a food sample. It is desirable to understand the mechanism of drying or moisture removal in moisture determination of a sample. The mechanism of moisture loss during the drying process has been demonstrated in two distinct stages (17,18), as shown in Fig. 1. In the first stage, the drying rate remains constant and equals that of evaporation from a free liquid surface. This stage is called a *constant rate* drying period. This phase of constant rate continues as long as water reaches the surface of the material as fast as evaporation takes place. There is a sudden drop in the drying rate at the end of the constant rate period, in which the point is termed "critical moisture content". This sudden fall of drying rate is caused by the physicochemically bound water. The second stage of drying begins at the point of the critical moisture content and extends to the final moisture content. This stage is called the *falling rate* drying period, which is divided into two subperiods. The first subperiod of falling rate corresponds to that part of the drying cycle at which less than the entire surface is wetted. For the second subperiod, the falling rate curve will asymptotically approach the absolute water content of the material. This second subperiod begins when the entire surface is completely dry. At this point, the internal diffusion becomes the controlling factor and changes in external conditions no longer affect the rate of drying. In practice, if 90% of water in a product is removed in 4 hr in the first stage, it may require another 4 hr to remove most of the remaining 10% at the falling rate period (18). Since the removal rate asymptotic, zero moisture is never reached under practical operating conditions.

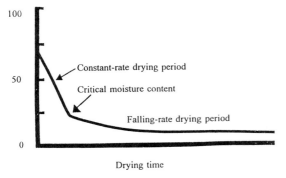

Fig. 1 Diagrammatic representation of the phases in a food drying process. (From Ref. 17.)

As shown in Fig. 2, it is more likely that decomposition takes place with flour or other sugar-containing products at about 180°C as depicted by the discontinuity in the straight line. Extrapolation of the main straight line to 250°C gives the "true" moisture content (19). The rate of loss of moisture can be increased by drying under reduced pressure. Vacuum drying is particularly helpful with foods that decompose at comparatively low temperatures. For instance, food samples that contain much sugar and fruit products are recommended for drying in a vacuum oven at around 70°C.

D. Factors Affecting Rate of Moisture Removal

1. Temperature

Temperature differential is the most integral part of the drying or evaporation process. The greater the temperature difference between the heating medium and the food, the greater will be the rate of heat transfer into the food. This condition provides the driving force for moisture removal. If the heating medium is air, temperature plays a second important role. As water is evaporated from the sample as vapor, it must be driven away or it will create a saturated atmosphere at the surface of the sample. The higher the temperature of the air, the more moisture it will hold before becoming saturated.

The loss of weight in the food sample will be increased due to volatization as the drying temperature increases. From Fig. 3 it is apparent that a maximum value is eventually reached at each different drying temperature. The greater the temperature, the shorter the time to be reached at the maximum drying. The differences in the maximum weight loss as well as drying time can be easily noticed by the differentials of three different temperatures applied. Although the proportion of bound water decreases as the drying temperature increases, it is extremely difficult to remove all of the moisture in the sample.

2. Air Velocity

Heated air takes up more moisture from the sample than cool air does. In addition, air in motion, i.e., high-velocity air, will take moisture away from the drying surface of food, which prevents the moisture from being in a saturated atmosphere. This is explained by the fact that clothes dry more

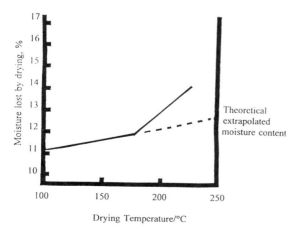

Fig. 2   Graph showing increase in moisture lost on drying with increase in drying temperature. (From Ref. 19.)

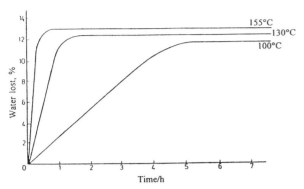

Fig. 3   Rate of loss of moisture from wheat flour at various drying temperatures. (From Ref. 19.)

rapidly on a windy day. Obviously, a greater volume of air can take up more moisture than a lesser volume in a limited oven space.

3. Atmospheric Pressure and Vacuum

The atmospheric pressure surrounding the food sample also has a significant influence on drying rate. At a pressure of 1 atm (760 mm Hg), water boils at 100°C. As the pressure is lowered, the boiling temperature decreases. At a constant temperature, a drop in atmospheric pressure would increase the rate of boiling or drying. This principle is applied to the vacuum drying process. A food sample in a heated vacuum chamber will lose more moisture at a lower temperature than it would be in a chamber at atmospheric pressure. It is important for a heat-sensitive food to reduce drying temperature in time to prevent a possible decomposition.

4. Humidity

The drying rate of a food sample would undoubtedly be affected by the humidity of the drying medium by which the sample is placed. As the humid air approaches saturation, it would hold and absorb less additional moisture than dry air. The drier the air of the medium, the faster the rate of drying for a given sample.

   The humidity at which the produce neither loses nor gains moisture is the equilibrium relative humidity (ERH). Diagramic plots of such data yield water sorption isotherms as shown in Fig. 4. Each food has its own ERH. Below the atmospheric humidity level, food can be dried further, whereas above this humidity it may pick up moisture from the atmosphere. The ERH at different temperatures can be measured by exposing the dried food sample to different humidity atmospheres in bell jars and weighing the sample after several hours of exposure (18). As illustrated with potato in Fig. 4, the product comes into equilibrium at 4% moisture at 100°C and 40% RH. Similarly, if the product dries down to 2% moisture, the equilibrium is attained at 15% and 100°C. Similar water sorption isotherms have been established for a wide variety of food products.

5. Drying Time

Drying time is a function of temperature during evaporation process. Drying period is inversely correlated with the temperature applied to the sample for the evaporation of moisture. As water evaporates from a surface, it cools the food surface. This cooling is largely the result of absorption by the water of the latent heat of phase change from liquid to gas (18). The heat of vaporization

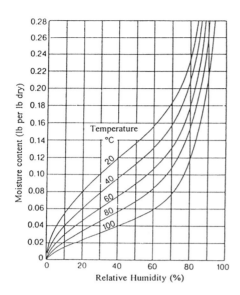

Fig. 4 Water sorption isotherms of potato. (From Ref. 18.)

goes from water to vapor, where the vaporization time is delayed if the amount of latent heat is reduced. With few exceptions, drying processes that employ high temperatures for short times cause less damage to the food sample than those employing lower temperature for longer times.

E. Sampling for Moisture Determination

It is absolutely essential that a representative sample of the food in question be carefully prepared prior to any analysis. No method of moisture determination is of value unless the samples are representative of the food or other material to be analyzed. Foodstuffs and food ingredients are relatively heterogeneous materials, whereby it is difficult to obtain a single absolutely representative sample for laboratory analysis. It is possible that sampling errors can sometimes be greater than the experimental error of analysis.

Although it is not possible to give general methods of sampling that are applicable to all foods, random sampling is the most recommended fundamental concept. Random sampling is applicable for relatively homogeneous food samples, whereas stratified random sampling is employed for heterogeneous food samples. For the stratified random sampling, the sample population is subdivided into small groups that may be treated as homogeneous. Too large or numerous subdivision of the sample should be carefully considered due to increase in cost of sampling.

In order to obtain precise analytical results of moisture determination, the laboratory sample should be as homogeneous as possible. The homogenization method utilized would depend on the type of food sample to be analyzed. Reduction of the size of food particles and thorough mixing of the food products can be efficiently performed using a number of electrical mechanical devices. Blenders, mincers, graters, homogenizers, powder mills, and grinders are essential pieces of equipment for homogenization of dry, moist, and wet samples in a food analysis laboratory.

Dry foods should usually be passed through an adjustable hand or mechanical grinder and then mixed thoroughly with a spoon or spatula. In some cases, it is advisable to pass the powder through a sieve of suitable mesh size. The size of dry or powder bulk samples can be reduced by the process known as quartering. This process involves tipping the bulk food into a uniform pile on a large sheet of glazed paper, glass, or the surface of clean laminated bench or table top. The pile is divided into four equal parts by separating quarter segments. Two quarters are rejected and the other two are thoroughly mixed, and the process is repeated until a suitable size of laboratory samples (i.e., 200–400 g) is obtained, as shown in Fig. 5.

Moist foods such as meat and fish products as well as vegetables are best homogenized by chopping rather than mincing, using a modern domestic food processor or blender. The process is repeated at least once more before transferring the mixture to a stoppered jar and storing it under refrigeration.

Wet foods, i.e., those containing fruits and vegetables such as pickles, sauces, and canned products, are best treated in a high-speed blender. Care must be taken to ensure that there is no fat separation by blender due to emulsions of samples such as salad cream or cream-based soups. Fluid foods are best emulsified by top-drive or bottom-drive blenders.

Oil and fats are easily prepared by gentle warming and mixing. The heated sample should be filtered. If the sample is filtered at too high a temperature, an antioxidant in the sample may be lost.

Fatty mixed-phase products such as cheese, butter, margarine, and chocolate are difficult to homogenize. Cheese and chocolate are best grated followed by hand mixing of the grated material. Butter and margarine should be warmed to 35°C to melt the fat in a screw-capped glass jar and shaken.

Sampling of dry foods should be performed rapidly because of possible grain or loss of moisture from the atmosphere during preparation. The prepared samples need to be quickly placed in dry rigid plastic or glass containers with tight closures and clear labels, and then stored at suitable cold temperatures prior to chemical analysis.

Solid foods have an additional complexity of sampling not encountered with liquids, which is the minimum weight of the sample. Sampling of a uniform size solid such as cereal grains, presents fewer minimum weight problems than that of a nonuniform size and heterogeneous solid foods such as frozen meat. The official method of analysis (20–22) described minimum weight problems and conditions to be considered in sampling for moisture measurement as summarized in Table 1.

F. Methods of Moisture Determination

Methodologies of moisture determination have been reviewed by many authors (2,5,7,10,11,23, 24). In his comprehensive monograph, Pande (7,23) published an extensive review in four volumes, *Handbook of Moisture Determination and Control.*

Analytical methods of moisture determination can be classified in two ways as shown in Table 2. One way is by the four major analytical methods; drying, distillation, chemical, and physical. The other is by direct and indirect procedures based on underlying scientific theory. With the direct

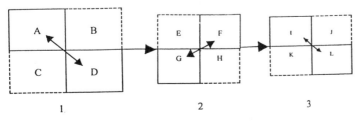

Fig. 5 Preparation of samples by quartering. (From Ref. 19.)

Table 1 Factors Affecting Sampling Conditions for Moisture
Determination

1. Type of food sample
       Solid vs. liquid
       Homogeneous vs. heterogeneous
2. Particle size and shape of sample
3. Sample preparation
       Homogenization, blending, mixing, grating, milling, rending mortar,
          sieve size, heat of mechanical device to sample
4. Representative sampling
       Random sampling
       Stratified random sampling
       Quartering
5. Sample contamination
       Chemical
       Bacterial (microorganism)
       Atmospheric (moisture, dust)
6. Aging of sample
       Oxidation
       Decomposition
       Environmental relative humidity
       State of hysteresis
          Adsorption
          Desorption
          Equilibrium moisture content
       Ratio of "free" to "bound" water
8. Sample storage
       Storage period
       Storage temperature
       Time to be analyzed

methods, moisture is normally removed from the solid food samples by drying, distillation, etc., and its quantity is measured by weighing, titration, and so forth. With the indirect methods, moisture is not removed from the sample, but instead properties of the wet solid that depend on the amount of water or number of hydrogen atoms are measured. These indirect methods must be calibrated against standard moisture values that have been precisely determined by using one or more of the direct methods. Therefore, the accuracy of indirect methods is dependent on the values of direct measurements against which they are calibrated.

Direct methods usually yield accurate and even absolute values for moisture determination, but they are mostly manual and time consuming. On the other hand, indirect methods are rapid, nondestructive, and offer the possibility of automation for continuous determination. For the convenience of presentation, methods of moisture determination are classified into two divisions: direct and indirect. Advantages and disadvantages of each individual method under the two classifications are described in Tables 3 and 4.

1. Direct Methods

a. Air-Oven Drying Owing to its convenience, air-oven drying is the most common and widely used method for routine moisture determination in standard laboratories around the world. Either convection-type ovens or forced-draft ovens can be used. Modern drying ovens, usually

Table 2 Classification of Analytical Methods for Moisture Determination

Classification by four major principles	Classification by direct/indirect procedures
1. Drying methods	1. Direct methods
Oven drying	Gravimetric methods
Vacuum drying	Oven drying
Freeze-drying	Air oven
Chemical desiccation	Vacuum oven
2. Distillation method	Freeze-drying
Azeotropic distillation	Thermogravimetric analysis
3. Chemical methods	Chemical desiccation
Karl Fischer titration	Distillation method
Generation of acetylene	Chemical titration method
Heat on mixing with H_2SO_4	(Karl Fischer)
4. Physical methods	Extraction methods
Infrared absorption	Gas chromatography
Near-infrared reflectance	Refractometry
Gas chromatography	2. Indirect methods
Nuclear magnetic resonance	Spectroscopic methods
Refractometer	Infrared absorption
Neutron scattering	Near-infrared reflectance
Electrical	Nuclear magnetic resonance
Microwave absorption	Gas chromatography
Dielectric capacitance	Electrical–electronic methods
Conductivity	Microwave absorption
	DC and AC conductivity
	Dielectric capacitance
	Sonic and ultrasonic method
	Neutron and γ-ray scattering

heated by electrically or infrared heaters, are equipped with built-in balances for routine and fast analysis of any stable solid materials. Because the principle of oven drying is based on weight loss, the sample needs to be thermally stable and should not contain a significant amount of volatile compounds (5).

The general operational procedures for the conventional method of moisture determination using drying oven and analytical balance involve sample preparation, weighing, drying, cooling, and reweighing. The general steps and principles of the procedures follow. The information has been basically adapted from the officially accepted procedures (20,21,25).

APPARATUS
1. Dishes. Nickel, stainless steel, aluminum, or porcelain. Metal dishes should not be used when the sample may have a corrosive action.
2. Analytical balance. 0.1-mg sensitivity.
3. Desiccator. Containing efficient desiccant such as phosphorus pentoxide, calcium sulfate, calcium chloride, etc.
4. Atmospheric oven.
5. Blender. Oster, Waring, or equivalent for high-moisture samples.
6. Grinder and mill for low moisture samples.
7. Spatula or plastic spoon.
8. Steam bath, predrying for high-moisture samples such as dairy products.

Table 3 Comparison of Advantages and Disadvantages Among Direct Methods of Moisture Determination

Method	Advantages	Disadvantages
A. Oven drying	1. Standard conventual method 2. Convenient 3. Relative speed and precision 4. Accommodates more samples 5. Attains the desired temperature more rapidly	1. Variations of temp due to particle size, weight of sample, position in the oven, etc. 2. Difficult to remove bound water 3. Loss of volatile substances during drying 4. Decomposition of sample (i.e., sugar)
B. Vacuum oven drying	1. Heating at low temperature 2. Prevents samples (i.e., sugar containing) from decomposition 3. Good for volatile organic samples 4. Uniform and constant heating and evaporation	1. Drying efficiency low for high-moisture foods 2. Often cannot analyze as many sample as oven dry
C. Freeze-drying	1. Excellent for sensitive, high-value liquid foods 2. Preserved best for textural and appearance attributes 3. No foaming 4. No case hardening 5. No oxidation 6. No bacterial changes during drying	1. High unit cost 2. Longer drying time than oven dry 3. Sample must be frozen before drying 4. Applicability limited to low-moisture samples
D. Azeotropic (distillation) method	1. Determines water directly and not the loss in weight 2. Apparatus simple to handle 3. Frequently more accurate than oven drying method 4. Takes relatively short time (30 min to 1 hr) to determine 5. Prevents oxidation of sample 6. Not affected by environmental humidity	1. Low precision of measuring device 2. Immiscible solvents, such as toluene, may be a fire hazard 3. Can have higher results due to distillation of water-soluble components (substances such as glycerol and alcohol 4. Unclean internal surface of the apparatus may cause erroneous results with the adhered water 5. Reading accuracy of water volume is limited 6. Solvents may be toxic (i.e., benzene)
E. Karl Fischer (chemical method)	1. One of the standard methods for moisture assay 2. Accuracy and precision higher than other method 3. Useful for determining water in fats and oils by preventing samples from air oxidation 4. Once apparatus is set up, determination takes a few minutes	1. Chemicals of the highest purity to be used for preparing the reagent 2. Titration endpoint may be difficult to determine 3. Reagent is unstable and standardization before usage 4. Titration apparatus must be protected from atmospheric moisture due to extreme sensitivity of reagent to moisture 5. Use of pyridine in reagent is highly reactive

F. Chemical desiccation	1. Can serve as reference standards against other rapid procedures 2. Can be done at room temperature 3. Good for measuring moisture in tea and spices containing volatile substances	1. Requires a lengthy time to achieve constant dry weight 2. Moisture equilibrium depends on chemical reactivity of desiccant
G. Thermogravimeric analysis	1. Semiautomatic and fully automatic method over standard oven drying 2. (Sample is not removed from oven) Weighing error is minimal 3. Sample size is small	1. Excellent for research but not practical
H. Refractometry	1. Determination takes only ⌣ .(. min (rapid) 2. Does not require complex and expensive instrumentation 3. Simple and reasonable accuracy 4. Excellent method for fruit, milk, and carbohydrate-rich products	1. Temperature-sensitive 2. Requires tissue slurry preparation (i.e., meat) 3. Requires uniformity of type of samples
I. Gas chromatography	1. Analysis is rapid (5–10 min per sample) 2. Reliable method for meat, cereal, and fruit products	1. Unit cost per sample may be higher than oven dry

9. Crucible tongs.
10. Thermometer, 0–110°C.

PROCEDURE

1. The empty dishes should be washed thoroughly, rinsed, and dried in an oven for several hours at 100°C. Store in a clean desiccator at room temperature before use.
2. Mix the prepared sample thoroughly and weigh approximately 2–3 g sample (1 g if low moisture) into a preweighed dish on an analytical balance, to the nearest mg.
3. Place the dish with cover on the metal shelf in the atmospheric oven, avoiding contact of the dish with the walls. Consult Table 5 for steam bath requirement, oven temperature, and drying time in oven for dairy and other food products.
4. After specified time in oven, use tongs to press cover tightly into dish, remove dish from oven, and place dish in desiccator for at least 30 min to cool to room temperature.
5. Weigh the dish on analytical balance and calculate moisture loss.

CALCULATIONS

$$\% \text{ moisture} = \frac{\text{loss in weight} \times 100}{\text{weight in sample}}$$

$$\% \text{ solids} = 100 - \% \text{ moisture}$$

A number of factors influence the accuracy of any particular drying procedure for the determination of moisture. Erroneous results in moisture determination by oven drying may be obtained due to variations in weighing samples, oven conditions, drying conditions, and postdrying treatments. Sample weighing is influenced by adsorption of atmospheric vapor, length of weighing, spillage, and balance accuracy. Oven conditions are affected by temperature, air velocity, pressure, and relative humidity, whereas drying conditions are changed by size and shape of sample container, type of heating element and its location, drying period, scorching, loss of volatile

Table 4 Comparison of Advantages and Disadvantages Among Indirect Methods of Moisture

Method	Advantages	Disadvantages
A. Infrared analyses		
a. Infrared absorption	1. Can perform multicomponent analysis 2. Most versatile and selective 3. Nondestructive analysis	1. Accuracy on calibration against reference standard 2. Temperature-dependent 3. Dependent on homogenizing efficiency of sample 4. Absorption band of water is not specific
b. Near-infrared reflectance spectroscopy	1. Rapid 2. Precise 3. Nondestructive 4. No extraction required 5. Minimum sample preparation	1. Reflectance data is affected by sample particle size, shape, packing density, and homogeneity 2. Hydroxyl group interferes with amine group 3. Temperature-dependent 4. Refractive index is changed by level of fluorescence 5. Equipment is expensive
B. Microwave absorption meter	1. Nondestructive 2. No extraction required 3. More accurate than low-frequency resistance or capacitance meters	1. Possible leakage of microwave energy during measurement 2. Has relatively low sensitivity and limited range for moisture determinations 3. Depends on the fluctuation of the material density in the volume measured 4. Results affected by such factors as particle size, temperature, soluble salt contents, bonded water, polarization and frequency of sample, etc.
C. Dielectric capacitance	1. Has high sensitivity due to large dielectric constant of water 2. Convenient to industrial operations with the continuous measurement system 3. Electrode system can be modified so that the method has universal applicability	1. Affected by texture of sample, packing, mineral contents, temperature, and moisture distribution 2. Conductivity considerably affected by acid salts in sample 3. Calibration difficulty for the meter beyond sample pH 2.7–6.7 4. Difficult to measure bound water at high frequencies
D. AC and DC conductivity	1. Measurement is instantaneous 2. Nondestructive 3. Precise	1. Measures only free water, and the conversion charts is needed for the fixed amount of bound water to specific samples 2. Accuracy and precision temperature, electrolyte content, and contact between electrode and samples are affected 3. Difficult to maintain calibration of the equipment

E. Sonic and Ultra-sonic absorption	1. Bound water can be determined in aqueous solution of electrolytes and nonelectrolytes 2. Nondestructive	1. Dependent on the type of medium for sound passes 2. Total water not measured
F. Mass spectroscopy	1. Can simultaneously analyze a large number of components separately out of a complex mixture of chemical compounds 2. No electrical leakage problem due to low potentials applied to the beam tube	1. High variation between theoretical moisture values and hydrated substances 2. Major instrumental problem encountered with memory effect from the preceding sample
G. NMR spectroscopy	1. Analysis is quick (a matter of seconds) 2. Accurate 3. Nondestructive 4. Can analyze raw foods as well as finished products 5. Can differentiate between free and bound water 6. Particle size and packing of granular samples have no effect on signal absorption (advantage over other electrical methods)	1. Cost of equipment is high 2. Separate calibration curves required for different substances 3. Applicability limited to fat and oil foods 4. Constant and correct sample weight required
H. Neutron and γ-ray scattering method	1. Used for soil moisture assay 2. The absolute error is claimed to be less than $\pm 0.5\%$ 3. Density and moisture content can be measured simultaneously in large heterogeneous objects	1. Applicable to only substances that are relatively proton-free 2. Expensive

compounds, and decomposition. Postdrying factors such as final temperature at weighing, desiccator efficiency, loss of dried sample, balance buoyancy effect, and so forth may also contribute to erroneous data. Advantages and disadvantages of air-oven drying methods as well as those of other direct method should be obtained from Table 3.

Different operating oven temperatures and drying times have been recommended for different types of samples by the Association of Official Analytical Chemists (AOAC) (20,21). The conditions for steam bath requirement, oven temperature, and length of time in oven for various products are exemplified in Table 5. Some high-water–containing foods such as milk products require partial drying on steam bath before the samples can be placed in an air oven. Maintaining drying temperature below 70°C is important to prevent samples from decomposition or loss of volatile substances by using steam bath or infrared radiation.

b.   Vacuum Oven Drying   Vacuum oven drying is generally the standard and most accurate method of moisture determination for most foods. Some of the drawbacks of the air-oven drying method can be overcome by this method, where volatile organic materials are heated at lower temperatures (60–70°C). Even though it is impossible to obtain an absolute moisture level by

Table 5 Atmospheric Oven Temperatures and Time Settings for Oven Drying of Milk and Other Foods

Product	Dry on steam bath[a]	Oven temp. ($^0C \pm 2$)	Time in Oven (hr)
Buttermilk, liquid	X	100	3
Cheese, natural type only		105	16–18
Chocolate and cocoa		100	3
Cottage cheese		100	3
Cream, liquid and frozen	X	100	3
Egg albumin, liquid	X	130	0.75
Egg albumin, dried	X	130	0.75
Ice cream and frozen desserts	X	100	3.5
Milk–whole, low-fat, and skim	X	100	3
Condensed skim		100	3
Evaporated milk		100	3
Nuts almonds, peanuts, walnuts, etc.		130	3

[a]X indicates that samples must be partially dried on steam bath before placing in oven.
Source: Ref. 25.

drying methods, moisture measurement by drying in a vacuum is a close and reproducible estimate of true moisture content (7,10).

Several types of vacuum ovens can be developed. Recent models of laboratory vacuum ovens can be attached to a vacuum line and electrically heated. The front door can be made air-tight by grinding into position with emery powder, or a gasket may be used. A vacuum of 100–600 mm Hg is maintained inside the sample chamber. The rate of drying can be increased by lowering the vapor pressure in the air via vacuum. If no air were let into the vacuum oven during drying, the pressure of water vapor in the oven would eliminate the usefulness of the vacuum oven, especially in samples with high moisture content. The AOAC (21,22) procedure recommends that moisture content of samples be determined with heating in the vacuum oven at 70°C for 6 hr at a pressure not exceeding 100 mm Hg. This technique is especially desirable for the analysis of sugar-containing products, particularly fructose-containing foods as these substances have a tendency to decompose during the drying process. The advantages and disadvantages of vacuum drying are also described in Table 3.

The general apparatus and procedure for vacuum drying are as follows:

APPARATUS
1. Vacuum oven. Thermostatically controlled and connected with a vacuum pump capable of maintaining the pressure in the oven below 25 mm Hg. The oven should be provided with an air inlet connected to a sulfuric acid gas-drying bottle and trap for releasing the vacuum.
2. Dishes. Metal dishes with close-fitting lids and flat bottoms to provide maximum area of contact with the heating plate.
3. Other apparatus and equipments are the same as those for air-oven drying.

PROCEDURE
1. Wash, dry, cool in desiccator and weigh the metal dish to the 0.1-mg scale.
2. Weigh 3.0- to 5.0-g sample into a preweighed dish on an analytical balance. Distribute sample evenly over the bottom of the dish, where possible. Some samples, require predrying as described for the air-oven drying to prevent decomposition.

3. Place the sample dishes in the vacuum oven, partially uncover the dish, evacuate the oven, and dry the sample at the specified temperature and vacuum pressure shown in Table 9. During drying, admit into oven a slow current of air (about 2 bubbles/sec) dried by passage through sulfuric acid washing bottles.
4. Shut off the vacuum pump after 5 hr and slowly readmit dry air into the oven. Press cover lightly into dish using tongs, transfer the dish to desiccator to cool, and reweigh.
5. Dry for another hour to ensure that constant weight has been achieved.

CALCULATIONS

$$\% \text{ moisture} = \frac{\text{loss in weight} \times 100}{\text{weight in sample}}$$

$$\% \text{ solids} = 100 - \% \text{ moisture}$$

c. Freeze-drying   No drying method is better than freeze-drying for preserving dried samples for freshness and textural appearance. Freeze-drying can be utilized to dehydrate sensitive, high-value liquid foods such as coffee and juices, but it is especially suited to drying solid foods of high value such as strawberry, whole shrimp, chicken dice, mushroom slices, and sometimes food pieces as large as steaks and chops (18). Freeze-drying has been developed to a highly advanced state in recent years. The unit cost of drying may be two to five times greater per weight of water removed than other common drying methods. Therefore, much of the development work has focused on the optimization of the process and equipment to reduce drying cost. On the other hand, many advantages of the freeze-drying method have been demonstrated (26), as some of them are shown in Table 3.

The primary principle of the freeze-drying method is based on the sublimation characteristics of water vapor. When a material can exist as a solid, a liquid, and a gas, but goes directly from a solid to a gas without transforming into the liquid phase, the material is said to sublime. Freeze-drying is achieved in the situation where water is evaporated from ice without melting the ice under certain conditions of low vapor pressure. Ice (frozen water) will sublime if it is placed at 0°C or below in a vacuum chamber of a freeze dryer at a pressure of 4.7 mm Hg or less. Under such conditions the water remains frozen and its molecules can leave the ice block at a faster rate than water molecules from the surrounding atmosphere reenter the frozen block (18). Heat is applied to the frozen food samples to speed sublimation within the vacuum chamber of the dryer. When the pressure of the vacuum is maintained sufficiently high (i.e., 0.1–2 mm Hg) and the heat is controlled just short of melting the ice, the water vapor will sublime at a nearly maximum rate.

As drying progresses, the surface of the ice front continues to recede toward the center of the frozen sample, then the final ice sublimes when the moisture content of the food becomes less than 5%. A typical freeze-drying curve is displayed using an asparagus sample shown in Fig. 6. It illustrates the curve characteristics of the temperatures of the heating plates and those of the food surface during the drying process. At the beginning of the drying stage, no moisture has yet been removed, and the temperature of the frozen sample remains somewhat below –30°C at its center and surface. As the sublimation continues, the temperature of the product surface rises, and approximately after 2½ hr, the moisture content of the asparagus drops to 50%. The chamber is evacuated to a pressure of 150 μm and the heating platen is set at 120°C. Sufficient heat is required to provide the driving force to induce fast sublimation without melting the ice in the receding ice core. The drying may be completed in 8 hr or longer.

d. Azeotropic Distillation Method   This method utilizes a well-known property of water called "azeotropy." It is based on the principle that water is simultaneously distilled with an immiscible liquid at a constant ratio. There are two main types of distillation procedures. In the first, water is distilled from an immiscible liquid of high boiling point. The sample suspended in

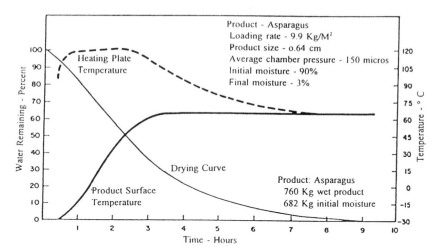

Fig. 6 Drying curve and significant temperature changes during freeze dehydration of asparagus. (From Ref. 18.)

a mineral oil having a flash point much above the boiling point of water is heated to a predetermined temperature in a suitable apparatus (5,7,10,19). The water that distills off condenses and is collected in a suitable measuring cylinder. In the second type, the mixture of water and an immiscible solvent such as toluene and xylene distills off and is collected in a suitable measuring appartus where water separates and its volume is measured. Although the respective boiling points of water and toluene as single components are 100 and 110°C, the boiling point of the binary mixture of water/toluene is 85°C with the ratio of relative quantities at the boiling point being 20/80, respectively (5,7).

Distillation with a boiling liquid provides an effective means of heat transfer; the water is distilled rapidly and the measurement is made in an inert atmosphere, which minimizes the danger of oxidation. Figure 7 illustrates the Dean and Stark distillation apparatus for azeotropic distillation. The apparatus consists of a heating source under the flask, a 250- or 500-ml round-bottom boiling flask, a Bidwell–Sterling receiver, and condenser, which can simultaneously determine water, fat, and residue in a food sample.

The laboratory data obtained from azeotropic distillation have consistently been shown to approach the value of the theoretical water content to within 0.1%. This value represents the normal magnitude of error for the measurement of the distillation method. Several difficulties may be encountered in the determination of moisture by the azeotropic method. However, it has many advantages over the other direct methods, which are listed in Table 3.

Azeotropic distillation procedures are as follows:

APPARATUS
1. Dean and Stark distillation apparatus (see Fig. 7)
2. Heating mantle

REAGENTS
1. Xylene or toluene

PROCEDURE
1. Weigh 10- to 20-g sample (usually fatty foods containing significant amounts of volatiles) and place in a 250-ml flask.

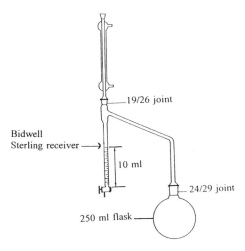

- 19/26 joint
- Bidwell Sterling receiver →
- 10 ml
- 24/29 joint
- 250 ml flask

Fig. 7 Dean and Stark distillation apparatus.

2. Add 100 ml of xylene or toluene to the flask.
3. Assemble Dean and Stark apparatus by connecting the flask to the condenser.
4. Run the cooling water through the condenser and heat the flask by heating mantle to boil the contents.
5. Adjust the heating apparatus so that the flask contents are just kept boiling, and continue heating for at least 1½ hr.
6. Turn off the heating mantle, and allow the apparatus to cool, especially the side arm.
7. Record the volume of water in the side arm.

CALCULATIONS

$$\% \text{ moisture } = \frac{\text{loss in weight} \times 100}{\text{weight in sample}}$$

e. Karl Fischer Chemical Method The Karl Fischer titration has become a standard method for the moisture measurement in liquids and solids (27). It is the method of choice for the absolute determination of water because of its selectivity, high precision, and speed (28,29). The method is particularly applicable to foods that give erratic results with methods of heat and vacuum application. This chemical technique has been found to be the preferred method for determination of water in many low-moisture foods such as wheat flour, cocoa powder, molasses, malt extract, tea, spices, margarine, oils, and fats. This method is especially superior in determining moisture of sugar-rich foods, or foods rich in both reducing sugars and proteins (10,21). The method, however, is seldom used for water determination in structurally heterogeneous, high-moisture foods such as fresh fruits and vegetables. The accuracy and precision have been found to be higher than by air-oven drying and other methods.

This titration method is based on the reaction demonstrated by Bunsen in 1853 showing the reduction of iodine by sulfur dioxide in the presence of water:

$$2H_2O + SO_2 + I_2 \rightarrow H_2SO_4 + 2HI$$

Karl Fischer (30) modified the procedure and established the new quantitating conditions. The titration reagent comprises a mixture of iodine, pyridine, and sulfur dioxide in methanol solution. The reactions involved take place in two steps (31) as follows:

and

As shown by the above reactions, for each mole of water, 1 mole of iodine, 1 mole of sulfur dioxide, 3 moles of pyridine, and 1 mole of methanol is required. The endpoint is indicated by the appearance of the brown color of free iodine. Iodine is consumed as long as there is any water present in the solution. The merits and demerits of Karl Fischer method is to be consulted with Table 3. The general apparatus, reagents, and procedure of Karl Fischer titration method are as follows:

APPARATUS
1. Burette. Automatic filling type, all glass, fully protected against moisture ingress.
2. Titration vessel. Having agitation device, such as magnetic stirrer or dry inert gas (N_2 or CO_2) maintained a small positive pressure to exclude air.
3. Eletrometric apparatus and galvanometer, suitable for "dead-stop" endpoint technique.

REAGENTS
1. Methanol. Anhydrous.
2. Sodium acetate trihydrate.
3. Karl Fischer reagent. To minimize losses of active reagent from side reactions, many laboratory suppliers provide the Karl Fischer reagent as two solutions: (a) a solution of iodine in methanol and (b) sulfur dioxide in pyridine. The solutions are mixed shortly before use.

PROCEDURE
1. Weigh an amount of sample containing approximately 100 mg water into a predried round-bottom 50-ml flask.
2. Add 40-ml methanol into the flask; quickly place it on the heating range and connect the reflux condenser.
3. Boil the contents of the flask gently under reflux for 15 min.
4. Stop heating with the condenser atttrached and let it drain for 15 min.
5. Remove and stopper the flask.
6. Pipette a 10-ml aliquot of the extract into the titration vessel, titrate with the Karl Fischer reagent to the dead-stop endpoint and record volume of titrant used.
7. Run a blank flask without sample as the same procedures described above.

CALCULATION

$$\% \text{ water of sample} = \frac{0.4 \times F \times (\text{ml reageant used for sample} - \text{ml reageant used for blank})}{\text{weight of sample(g)}}$$

where F = standardization factor of reagent (mg water/ml)

f. Chemical Desiccation This method of moisture determination in dried foods is carried out by desiccation in an evacuated desiccator over dehydrating or moisture-absorbing chemicals. The amount of water removed from the sample is dependent on the strength of the drying agent employed. Desiccation of the sample is usually performed at room temperature. With few exceptions, desiccation techniques are a lengthy procedure, often requiring weeks and even months for the sample to achieve constant weight. The moisture equilibrium depends strongly on the force of chemical reactivity of the sample with moisture. Nevertheless, results obtained by this method often serve as reference standards against which data obtained by more rapid procedures are calibrated.

Relative efficiencies of drying agents have been studied by several workers in considerable detail. Results reported by the International Critical Tables (32), Bower (33), and Trusell and Diehl (34) are summarized in Table 6. The most commonly used agents are chosen on the basis of residual water per liter of air. The most practical desiccating agents recommended by the AOAC (20,22) are phosphorus pentoside, barium monoxide, and magnesium perchlorate. An example of the practical usage of chemical desiccation is the moisture determination in tea and spices which contain volatile substances other than moisture.

    g. Thermogravimetric Method Thermogravimetric analysis (TGA) is a semiautomatic or fully automatic improved method of the standard oven drying. The microthermogravimetric analysis is a more advanced TGA technique, which is particularly useful for differential microcalorimetry or differential thermal analysis.

This TGA method is equipped with the special feature of a thermobalance, which automatically measures and records the weight changes of a food sample as a function of time and temperature while the sample is being heated. Moisture is continuously evaporated from the sample and the weight loss is recorded until the sample has reached a constant weight. The recorder curve known as a thermoweighing curve or thermogram is produced in reference to the particular temperature programming. The weighing error of the TGA becomes minimal because the sample is not taken out from the furnace or placed in a desiccator before weighing. The water content measured by TGA is comparable to the Karl Fischer method.

This method is automatic and requires small amounts of sample. The TGA is highly applicable

Table 6 Relative Efficiencies of Drying Agents

	Residual water, μg/L gas		
Substance	Truell and Diehl (34) (25°C)	Bower (33) (30.5°C)	ICT[a] (32) (25°C)
Cooling to -190°C	—	—	1.6×19^{-20}
$Mg(CIO_4)_2$, anh.	0.2	2	<0.5
$Mg(CIO_4)_2 \cdot 1.48H_2O$	1.5	—	—
BaO	2.8	0.65	—
Al_2O_3	2.9	5	3
P_2O_5	3.5	—	<0.02
H_2SO_4, anh.	—	—	3
Molecular sieve 5A	3.9	—	—
$Mg(CIO_4)_2 \cdot 3H_2O$	—	30	<2
$CaCl_2$, anh.	67	360	360
$CaSO_4$, anh	67	5	—
Silica gel	70	30	—
$NaOH \cdot 0.03H_2O$	513	—	—
CaO	650	3	200
MgO	750	—	—
H_2SO_4, conc.	—	—	300
$CaCl_2$, gran.	—	1500	140–1250
$CuSO_4$, anh.	—	2800	1400

[a]International Critical Tables.

in connection with moisture equilibrium isotherm determinations and is used to study the bound water in solids. A standard TGA instrument designed for calibrating and certifying moisture meters was used to determine moisture in industrial and agricultural products, thus providing a national standard for quantifying moisture in solids (35).

h.   Refractometry   This optical method for determining moisture content has been used for moisture analysis of dry fruit (36) and meat (37). The method depending on the refractive index of tissue slurry requires uniformity of type and preparation of samples. Density or specific gravity and refractive index measurements can be utilized for rapid estimations of several percent water in a variety of liquids and inert solids. Such techniques are suitable for binary or other liquid solutions in which only variation in the water content will affect measurements.

The apparatus for these measurements varies from simple laboratory units such as hydrometers and Abbé or dipping refraction meters to automated process stream analyzers (24). In refractometry, stable temperature control is essential because both density or specific gravity and refractive index are temperature-sensitive. Differential refractometry does not require temperature control because the instrument records the difference between sample and reference.

The critical angle refractometer was used for moisture analysis of petroleum and food industries (38), and the instrumentation is demonstrated in Fig. 8. Since measurement becomes strictly a surface phenomenon, the technique can be used on opaque liquids and some suspensions as well as transparent liquids. With this method, the foodstuff such as meat is homogenized with an anhydrous solvent (i.e., isopropanol) and equilibrated to a predetermined temperature, then the refractive index is measured. The percent moisture is calculated in reference to the values of standard solutions. The procedure takes only 5–10 min to complete and does not require expensive complex instrumentation with reasonable accuracy.

i.   Gas Chromatography   The versatility of gas chromatography (GC) in analytical chemistry has been applied to moisture determination. The moisture content is quantitated by determining the peak areas of water and solvent, and comparing them with the standard curve for a known amount of water.

The GC method of moisture determination involves moisture extraction with a solvent followed by quantitative separation of the water–solvent mixture with the instrumentation (39,40). A weighed

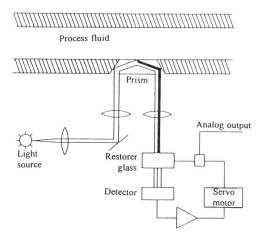

Fig. 8   Components of critical angle refractometer. (From Ref. 38.)

amount of food (i.e., meat) is homogenized in a known quantity of anhydrous methanol, ethanol, or isoporpanol, using an ultrasonic homogenizer. The extract is then analyzed by a GC equipped with a column such as Poro-pack-Q and a thermal conductivity detector. The analysis takes only about 5–10 min. The GC method provides a decent result with a higher standard deviation than the oven method but smaller then the toluene distillation (39), as shown in Table 7.

j. Application of Direct Methods in Moisture Analysis Air-oven drying methods are usually used for the least temperature-sensitive foods, whereas vacuum oven drying is applied to temperature-sensitive foods such as sugar-containing products. Because of decomposing and volatile components in the sample, the vacuum oven method uses the lower-than-atmospheric pressures to lower the boiling point of water. Azeotropic distillation methods are performed under atmospheric pressure and completed in short time (30 min). Toluene distillation is only applied to solid foods, where the temperature of the water–toluene azeotrope is 85°C.

Karl Fischer titration and vacuum desiccation methods are performed at low temperatures for special situations of moisture determination. The Karl Fischer method remains a laboratory technique, which requires a considerable degree of dexterity and care from the analyst, even if it is a rapid method. On the other hand, the vacuum desiccation technique takes too long to reach moisture equilibrium to be considered for the method of product quality control.

In comparison with four direct methods for determining moisture in malt, a temperature gradient was found in the oven that caused a 0.1% weight difference between low- and high-temperature zones (41). The values by the water oven method were consistently about 0.5% lower than those by vacuum oven, Karl Fischer, and toluene distillation (Table 8). The official methods for moisture determination in various foods recognized by the AOAC and the International Standards Organization (ISO) have been summarized in Table 9. The time and temperature revealed an inverse relationship, especially in the oven drying method, which was explained in Figs. 1, 2, and 4.

2. Indirect Methods

a. Infrared Absorption Spectroscopy The infrared spectrum of a chemical compound is probably the most characteristic physical property of that compound, which is known as its

Table 7 Moisture Content of Meat Samples Obtained by Conventional and GC Methods

	% Moisture[a]	
Samples	GC	Conventional
Fat	6.9	6.6[b]
Pork jowl	27.2	27.1[b]
Pork trim	36.6	37.2[b]
Navels	38.2	38.0[b]
Salami	40.5	39.3[b]
Emulsion, frankfurter[d]	51.1	51.0[b]
Cow meat	55.4	54.9[c]
Bull meat	69.4	70.6[c]
Cheek meat	71.4	69.9[c]
Turkey	75.9	76.3[c]

[a]Mean of five determinations.
[b]Toluene distillation.
[c]Oven drying at 105°C for 24 hr.
[d]Emulsion obtained immediately prior to extrusion.
Source: Ref. 39.

Table 8   Moisture Content of Malts Determined by Four Methods

	Sample no. (wt %)						
Method	1	2	3	4	5	6	7
Water oven, nominal 98°C, 3–4 hr	2.91	2.94	2.87	3.03	2.87	2.80	2.90
Vacuum oven, 40°C, 0.5 mm	3.46	3.64	3.68	3.39	3.56	3.56	3.59
Karl Fischer reagent	3.75	3.69	3.70	3.60	3.61	3.90	3.70
Toluene distillation	3.5	3.6	3.4	3.7	3.9	3.8	3.9

Source: Ref. 41.

fingerprint (23). The technique of infrared absorption spectroscopy is one of the most versatile methods for the measurement of moisture content of a large number of substances in solids, liquids, or gases by employing suitable wavelengths at which maximum absorption is expected to occur.

Analysis of fluid milk by infrared (IR) is based on absorption of IR energy at specific wavelengths by carbonyl groups in ester linkages of fat molecules (5.723 μm), by peptide linkages between amino acids of protein molecules (6.465 μm), and by OH groups in lactose molecules (9.610 μm) (25). Total solids can be measured by addition of an experimentally determined factor to the percentage of fat, protein, and lactose; water content is then measured by subtracting total solids from 100.

The IR beam is passed through an optical filter that transmits energy at the wavelength of maximum absorption for the component being measured. The filtered beam then passes through the sample cell and then a detector. Signals from the detector are then used to compare the absorption at the two wavelengths and hence to the concentration of the component (25).

The spectral region for water from 0.7 to 2.4 μm has been investigated for measurement of the moisture content in cereal grains. Absorption bands for water have been shown to occur at 0.76, 0.97, 1.118, 1.45, and 1.94 μm (5). The moisture content in a sample can be determined by comparison of the depth of the band of interest with that of the same band for standard concentrations of water.

b.   Near-Infrared Reflectance Spectroscopy   The high resolving power of reflectance spectra in the near-infrared (NIR) range (0.8–2.5 μm) recently was utilized to develop NIR spectroscopy. Mid-IR range (2.5–24 μm) has high resolution in the absorption spectrum and can absorb IR radiation effectively from many compounds, but resolution is poor in the reflectance spectrum (42).

During the last two decades or so, NIR has assumed immense economic importance as a rapid, integrated form of multicomponent testing of a wide range of organic products grown or manufactured. The NIR technique is widely used to predict the moisture, oil, and protein content of grains and oil seeds. The speed of analysis is the primary advantage of the method, and this technique does not require wet chemical analysis. However, the accuracy of the NIR instrumentation depends entirely on wet chemical analysis data of calibration standard samples. If the calibration sample set does not adequately represent all future unknown samples, then NIR will not be successful.

The reflectance (R) spectra from a Cary model 14 monochromator operated in a single- or double-beam mode can be recorded into a multipurpose computer (43). Ground samples are packed into a sample holder maintaining direct contact with a concentric IR transmitting quartz window. The reflected radiation signals of the diffused spectra from the glass window are collected with four lead sulfide detectors equally spaced around the incident beam. The signal from the detectors is amplified with a logarithmic response amplifier digitized and fed to a digital computer. The wavelength range from 1100 to 2500 nm is scanned at every 2 nm (or 0.5 nm) width of the

Table 9 Comparison of Direct Methods on Moisture Determination of Foods Recommended by AOAC [20,21]

Moisture assaying method	Drying temp. (°C)	Pressure (mm Hg)	Sample weight (g)	Time required (hr)	Food products
Air oven drying	100–102	760	2–6	Until constant weight	Butter, cacao powder, sugar-containing products
	105	760	2–10	16–18	Meat products
	115	760	2	8	Various foods
	125	760	2	4	Various foods
	130	760	2–5	1–2	Cheese, wheat flour
	135	760	2	2	Various foods
Vacuum oven drying	60–70	50–100	2–5	2–6	Dried fruits, honey, syrup, modified starch
	98–100	25	2–5	5	Coffee, tea, wheat flour, gelatine, pasta products
	100	100	1–3	4–5	Dried and malted milk, cheese, meat products
	125	100	5	5	Fats and oils
Toluene distillation	85	760	10	0.5–1	Cheese and meat products, instant coffee products
Karl Fischer method	—	—	2–2.5	A few min (if apparatus is set up)	Molasses, dried vegetables, tea, spices, cocoa products
Vacuum desiccation	25–30	10	2–5	Until constant weight	Tea, spices

reflectance curve (44). The infrared reflectance (R) curves are recorded as the second derivative of the original log ($1/R$) absorption curve.

Chemical analysis data of calibration samples are recorded in the computer, which analyzes them with a stepwise multiple linear regression method to develop prediction equations by regression analysis of IR spectral data against chemical data (43–45).

Commercial NIR instruments are manufactured on the basis of three geometries as to the method of collecting the reflectance as (a) integrating sphere, (b) large solid angle detector, and (c) small detector. Each of these types has advantages and disadvantages. The large solid angle detector offers good collection efficiency, simplicity of construction, and minimum interference from specular reflectance as shown in Fig. 9 (46).

The calibration results for water in beef samples shown in Fig. 10 demonstrate a correlation coefficient of .998 and a standard error of 0.61% water. This was obtained by using a second derivative at 1975 nm divided by a second derivative at 1608 nm (46). The reproducibility of the reflectance technique can be evaluated from these data. The computed standard deviations of repeated analysis were 0.42% for fat and 0.40% for water. These errors include both the instrumental and sampling error.

  c. Microwave Absorption Method   The electric dipole relaxation of water in centimeter wave band and has been widely employed as a nondestructive test for water content (47). The principle of microwave instrumentation is based on the property that the dipole water molecule absorbs several thousand times more microwave energy than a similar volume of any perfectly dry substance (5).

At normal temperatures in the frequency range 1–30 GHz, the dielectric constant of water is between 40 to 80 and the loss tangent ranges from 0.15 to 1.2. Most dry materials have a dielectric constant ranging from 1 to 5 and a loss tangent between .001 and .05, unless some conducting material is present (23).

The microwave moisture meter consists of (a) a constant source of microwave radiation (2450 to 10680 MHz), (b) waveguide terminating in a horn and associated components, (c) a microwave detector, (d) a microwave attenuator and amplifier, and (e) an indicating meter (23). The meter also comprises centimeter range wavelength transmitting and receiving antennas that are placed at either side of the sample. The accuracy of the microwave measurement is affected by leakage

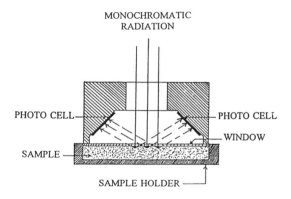

Fig. 9   Large, solid angle detector geometry for reflectance. (From Ref. 46.)

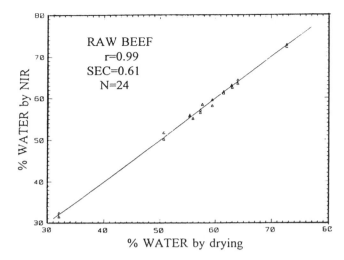

Fig. 10 Calibration plot for near-infrared measurement of water content of raw beef. (From Ref. 46.)

of microwave energy, sample temperature and particle size, polarization of different material, soluble salt content, bonded water and frequency, etc. At higher frequencies, the presence of electrolytes has a negligible effect on moisture readings. Bound water displays lower microwave absorption than free water due to the restricted dipole mobility. The microwave instruments are shown to be more accurate than the low-frequency capacitance or resistance meters.

d. Dielectric Capacitance The dielectric constant of water at 20°C is 81, whereas that of most organic bone-dry materials ranges from 2.2 to 4.0. In case of a 1% increase in water content of any substance, it will elevate the dielectric constant of the system by approximately 0.8 unit. This principle is applied to quantify the moisture content of foods such as cereals. The variation of dielectric constant of water-containing materials is nearly linear over a limited range from low to about 30% moisture content at room temperature.

The heart of the dielectric instrument is a capacitance cell in which two metal plates of a given size are at a fixed distance. Measurements can be made over the range −20°C to 180°C. The two plates have equal but opposite charges that are reversed at fixed frequencies to give an alternating current field. If a sample is placed between two electrodes, polarization takes place that increases the charge of the plates and their capacitance. The capacity change introduced into the electrode system due to moisture level in the sample is converted to a current and voltage, which in turn is converted to moisture content for instantaneous readout.

It is determined by comparing the capacitance C of a capacitance cell filled with the tested dielectric substance to the capacitance C_0 of an empty (vacuum) condenser, such as

$$C = EC_0$$

where E is a dimensionless constant that is affected by temperature and frequency (10). Capacitance meter measurement is influenced by moisture distribution, presence of electrolyte, temperature, and sample density. The advantages and disadvantages of this and other indirect methods are listed in Table 4.

e. DC and AC Conductivity Method In conductivity methods, conductivity and resistance are measured when an electrical circuit is introduced into a food sample. There is a distinct relationship between the moisture content of materials and their direct current or indirect current conductivity or resistance. Electrical resistance decreases with increasing moisture content in the sample. The logarithm of the resistance has practically a linear function of the moisture content on the range of about 5–30% moisture.

Several instruments, widely used by the industry for rapid routine moisture determination, are based on measuring the electrical resistance. In the Universal meter, the electrical resistance is indicated on an ohmmeter of the dynanometer type. A correction scale for temperature is provided. The instrument requires neither battery nor power supply. The Marconi instrument is battery-operated (10,23).

For determination of moisture by conductivity meter, the sample is held between two electrodes and the current flowing through the sample is measured by the change of electrical resistance. The determination is nondestructive and instantaneous. The accuracy and precision are affected by uniform moisture distribution, temperature (20–25°C), uniform electrolyte content, and good contact between the electrodes and samples. Conductivity meters can measure only free water and a fixed amount of bound water, where the conversion charts are needed due to variations of bound water content among samples. The DC and AC conductivity values are similar at higher moisture levels in the samples, whereas DC conductivities are significantly lower than the AC ones at lower moisture levels.

f. Sonic and Ultrasonic Absorption   The degree of absorption of sound energy is dependent on the medium through which it is transmitted (23). Sonic and ultrasonic absorption also depends on the quantity of water present in the sample. Based on this principle, ultrasonic velocity measurements have been used for determining bound water in aqueous solutions of electrolytes and nonelectrolytes (48).

For moisture determination, the sample is positioned between an energy generator and a microphone. The energy output of the sample is amplified to become an voltmeter reading, which in turn is converted to moisture content. Owing to their high frequency (short wavelength), ultrasonic waves have special properties of reflection, refraction and absorption.

The amount of bound water measured by ultrasonic method is shown to be 0.43, 0.23, and 0.40 g for glucose, maltose, and dextrin, respectively. The reports indicate that one molecule of water is bound to each free OH group for glucose, whereas there is a decrease in bound water for maltose possibly due to a result of intermolecular hydrogen bonding (5,7).

g. Mass Spectroscopy   Mass spectrometry has recently been utilized as a successful tool of chemical analysis including moisture determination even if the technique was originally developed for the measurement of isotope abundance ratios and other relevant quantities.

In a mass spectrometer, the sample is first introduced to a vacuum chamber and reduced to low vapor pressure by suitable heating. The vapor formed is ionized by passing it through a beam of electrons, where ions are accelerated by an electric field and passed into the magnetic field. Ions of various mass-to-charge (m/e) ratios can be focused on a collector plate and grounded by means of varying either the velocity of the particles (by changing accelerating voltage) or changing the magnetic field strength. The resulting current is amplified and recorded by a suitable electrometer circuit (23). The plot of mass vs. intensity is known as mass spectrum, which is reduced to a mass pattern by dividing each of the peak intensities by the intensity of the largest or base peak.

The main importance of the mass spectrometry lies in being able to analyze simultaneously a large number of components separately out of a complex mixture of chemical compounds. This technique is relatively rapid, with an hour or two being sufficient for the analysis of a complicated mixture. However, mass spectroscopy is often considered unsuitable for the quantitative measurement of water, whereas it gives fairly reliable results over a wide range of concentrations.

h.   Nuclear Magnetic Resonance Spectroscopy   Nuclear magnetic resonance (NMR) spectroscopy is a nondestructive method of moisture determination discovered in 1946, in which the nuclear magnetism of the hydrogen atom is utilized. It is based on the nuclear properties of the hydrogen atoms in water rather than on properties of the water molecule itself (23).

The instrument consists of four main units: a permanent magnet, a magnetic field sweep system, a radiofrequency generator, and a radiofrequency receiver (5). In addition to mass and charge, most nuclei possess the extra characteristic of spin. This spin property of the atom is the basis of the phenomenon of NMR. When the radiofrequency radiation is passed through the magnetized sample, it absorbs the energy. There is a linear relationship between the indicator readout in millivolts and the moisture content.

The application of NMR to moisture measurement is an excellent effort to meet the long-sought industrial requirement for a quick, accurate, and nondestructive method of measurement and control of moisture of raw materials and finished products. Although the NMR instrument measures primarily free water, it can differentiate free and bound water by using a high-resolution NMR apparatus. The NMR techniques have been applied to determination of moisture in virtually unlimited types of foods, including corn, wheat flour, rice, sugar, candy, starch and its derivatives, cheeses, and many others.

Most of NMR instruments are suitable for a variety of materials for a moisture range of 5–100% with an accuracy of 0.2%. This technique has a great advantage over the electrical methods because its signal absorption is not affected by particle size and packing for granular samples. However, it requires a constant and correct weight of sample, and analysis can be done in a matter of seconds. The advantages and disadvantages of the NMR method are given in Table 4.

i.   Neutron and γ-ray Scattering Methods   The neutron was discovered by Chadwick (49) in 1931. The neutron scattering method of measuring water content of soil or other materials is founded on the physical principles concerning the interaction between neutrons and the medium (23).

Energetic neutrons or γ-ray are scattered by nuclei and lose their energy. This moderation process forms the basis of a determination of moisture. Due to its large cross-section for collision, hydrogen is the most effective moderator for neutrons (2). For moisture determination, the sample is bombarded with fast neutrons and the density of slow neutrons produced is measured. The slowing down of neutrons is directly affected by the number of hydrogen molecules, which is proportional to the amount of water present in a sample. This method is applicable to substances that are relatively proton-free.

The neutron-scattering instrumentation consists of four major components: (a) probes containing a source of fast neutrons, (b) a detector for slow neutrons, (c) an instrument to determine the count rate from the detecting device, and (d) a suitable standard to verify the performance of the equipment.

Neutron and γ-ray scattering methods are applied to moisture determination of soils, coal, ceramics, coke, pulse and cellulose, etc. A neutron device that was developed to measure moisture content of blast furnace coke in the range 0–10% water (50).

II.   ANALYSIS OF ASH CONTENTS

A.   Introduction

Ash is the inorganic residue from the incineration of organic matter. The ash content is determined from the loss of weight that occurs during complete oxidation of the sample at a high temperature (usually 500–600°C) through volatilization of organic materials. The ash obtained is not necessarily of exactly the same composition as the mineral components present in the original food because there may be losses via volatilization or some interaction between constituents.

For complete ashing, the heating is continued until the resultant ash is uniform in color, white or gray, occasionally green or reddish, and free from particles of unburned carbon and fused lumps. Ashing may be performed by incineration over an open flame, in a muffle furnace, in a closed system in the presence of oxygen, or by wet combustion in the presence of sulfuric acid, nitric acid, and perchloric acid alone or in mixtures.

The determination of ash content is of value in the analysis of food for various reasons. The ash content can be regarded as a general measure of quality in certain foods such as tea, flour, and edible gelatin, and often is a useful criterion in identifying the authenticity of a food (51). The ash analysis has been chiefly used for the determination of adulteration of certain foods. A high-ash figure suggests the presence of an inorganic adulterant, and this condition is advisable to determine the acid-insoluble ash. The presence of large amounts of ash in finished products such as sugar, starch, gelatin, fruit acids, or pectin is objectionable. The ashing of vegetable and plant materials, particularly cut sections, is recognized as a useful tool in determining the nature and distribution of mineral constituents of plants (52). The ash content is an index of the quality of feedstuffs used for poultry and cattle. The ash content serves as a reliable index of the metabolism of yeast.

Total ash content is a useful parameter of the nutritional value of many foods and feeds. It is helpful with many foods to quantify not only the total ash but also the ash soluble and insoluble in water, the alkalinity of the soluble ash and of the total ash, and the proportion of ash insoluble in acids. High levels of acid-insoluble ash indicate the presence of sand or dirt in the sample.

B. Methods of Ashing

There are two major procedures of ashing, which are eventually utilized to determine mineral contents of the sample. These methods are dry ashing and wet ashing techniques: in dry ashing the organic matter of the sample is oxidized by complete incineration at a high temperature in the presence of oxygen, whereas in wet ashing the sample is oxidized with a mixture of concentrated strong acid. In addition to the two direct methods, there are indirect ashing techniques, such as conductometric methods, that can determine the total electrolyte content of foods. The nature of the ashing procedure is determined by the purpose for which the ash is prepared, the particular constituents to be determined, and the method of analysis to be used.

1. Dry Ashing

Dry ashing is the most standard method to determine the ash content of a sample. The total ash content of a foodstuff is the inorganic residue remaining after the organic matter has been burnt away. In dry ashing, the sample is ignited at 550–600°C to oxidize all organic materials without flame. The inorganic material that does not volatilize at that temperature is called ash. The required equipments and general procedure of dry ashing method for total ash are described in the following:

EQUIPMENT AND APPARATUS
1. Muffle furnace (electric), thermostatically controlled
2. Porcelain crucible, Coors #1 or platinum dish
3. Analytical balance, 0.1-mg sensitivity
4. Desiccator, charged with efficient desiccant
5. Tongs, muffle
6. Marking ink, permanent type for crucibles

Additional equipments for wet samples:

1. Bunsen burner
2. Hot plate thermostatically controlled
3. Tripod, iron
4. Steam bath
5. Atmospheric oven
6. Wash bottle
7. Double-deionized water

PROCEDURE

1. Place new or clean marked crucibles in a muffle furnace at 600°C for 1 hr. Turn off the furnace, transfer cooled crucibles from furnace to a desiccator, and cool to room temperature.
2. Weigh them as quickly as possible to prevent moisture absorption. Use metal tongs to move the crucibles after they are ashed or dried (crucible preparation).
3. Weigh accurately 2 g of dry foods or 10 g of wet samples into the prepared, preweighed crucible. If alkalinity of ash is to be determined, calculations are simplified if exactly these weights are used.
4. For wet sample, evaporate the wet sample to dryness on a hot plate or steam bath or in a atmospheric oven at 100°C for 1 hr. Omit this step for dry samples.
5. Carbonize the samples under the hood. Place the sample crucibles on iron tripod over a Bunsen burner or hot plate and slowly char for about 30 min. Heat cautiously to prevent spattering and add a few drops of olive oil to reduce spattering.
6. Place the charred crucibles in muffle furnace set at 550°C. Burn off the samples until they become completely free from carbon to be a light gray or white ash.
7. If black carbon spots persist in the sample crucible due to incomplete oxidation of tough samples (i.e., animal tissues), turn off the furnace, add a few drops of concentrated nitric acid directly onto the unburnt spots, and switch on the furnace again. Raise the temperature to 600°C and leave for 1–2 hr depending on the incompleteness of the ashing.
8. Transfer the ashed sample crucible to a desiccator and cool to room temperature. When cooled, weigh the crucible as quickly as possible to prevent moisture absorption.
9. Save the ash sample if mineral determinations are to be made.

CALCULATIONS

$$\% \text{ ash} = \frac{\text{weight of residue} \times 100}{\text{weight of sample}}$$

2. Wet Ashing

Wet ashing is used primarily for the digestion of samples for determination of mineral elements. The wet oxidation method is perferable to the dry ashing procedures for the decomposition of organic matter prior to the determination of mineral constitutents because they obviate difficulties resulting from losses of more volatile constituents during dry ashing and slow solution of the residue after ashing.

The use of a single acid is desirable but usually not practical for the complete decomposition of organic material. Nitric acid alone is a good oxidant, but it usually boils away before the sample is completely oxidized. The use of perchloric acid with nitric acid or nitric-sulfuric acid mixtures has been suggested (53) for the rapid decomposition of many organic compounds that are difficult to oxidize.

In wet ashing, the organic matter of the sample is oxidized with concentrated nitric and perchloric acids. The acids are partially removed by volatilization and the soluble mineral constituents remain dissolved in nitric acid. Any silica present is dehydrated and made insoluble. The equipment, reagents, and general procedures of wet ashing methods are as follows:

EQUIPMENT
1. Fume hood, constructed for safe exhaustion of perchloric acid and nitric acid fumes
2. Hot plate, thermostatically controlled
3. Volumetric flasks: 100 ml, 250 ml, Pyrex or equivalent quality made of borosilicate glass
4. Wash glass
5. Goggles (protective glasses)
6. Glass rod, for sample stirring and transfer
7. Boiling beads
8. Air oven
9. Steam bath

REAGENTS
1. Nitric acid, 69–71% concentrated.
2. Digestion acid, add 1 vol perchloric acid (60–62% $HClO_4$) to 4 vol nitric acid (69–71% HNO_3).
3. Hydrochloric acid, approximately 2 M. Add 1 vol hydrochloric acid (36% HCl) to 5 vol deionized water.

PROCEDURE
1. Weigh accurately 1 or 2 g dry and ground sample into a 250-ml beaker, and add 20 ml digestion acid and three glass boiling beads under a fume hood. Cover the beaker with a watch glass. Wear the goggles.
2. For liquid foods such as wine, juices, and beverages an aliquot of the sample should first be dried in a hot-air oven or steam bath; then dry residue should be ashed with digestion acid.
3. After addition of the digestion acid, allow the reaction to proceed in the hood at room temperature for 3–4 hr with occasional swirling using hands and glass rod, or leave overnight.
4. Place the sample beaker on a hot plate maintained at the liquid in the beaker just simmers. When the initial reaction has subsided, increase the temperature of the hot plate to 180–200°C.
5. Continue the digestion until the fluid in the beaker becomes clear and no visible particles remain. If the solution darkens when the volume is reduced, remove the flask from the hot plate, add 1 or 2 ml nitric acid, and continue the digestion with occasional swirling.
6. Raise the temperature of the hot plate to 240°C and evaporate the digestion acid until dense white fumes are formed within the beaker.
7. Remove the beaker from the hot plate. Transfer the solution quantitatively to a 100-ml volumetric flask with deionized distilled water.
8. Stopper the flask, thoroughly mix the solution, and leave overnight. The water- and acid-insoluble silica will settle to the bottom and aliquots of the solution are drawn from the top, filtered, and stored in small polyethylene bottles for later analyses.
9. The sample solution prepared can be used for the determination of the various macro and trace minerals by atomic absorption spectrophotometry or other methods.
    Caution: Perchloric acid is a violent oxidant and explosive. Therefore, exercise special care by using protective glasses and fume hood.

3. Conductometric Method

Conductometric methods are indirect methods for measuring the total electrolyte content of foods. Conductometric procedures provide a simple, rapid, and accurate means of determining the ash content of sugars (54). Foods containing high sugar (i.e., sucrose, syrup, or molasses) are generally low in minerals and require oxidation of a large amount of samples that have strong foaming carbohydrates.

The principle of conductometric methods is based on the fact that the mineral matter that constitutes the ash of the sugar dissociates in a solution, whereas the sucrose, a nonelectrolyte, does not dissociate. Therefore, the conductance of the solution is an index of the concentration of the ions present, which is the mineral or ash content of the sample.

4. Comparison of Dry Ashing with Wet Ashing

There are quite a few advantages and disadvantages of dry ashing in the determination of ash and mineral compared with those of wet ashing. Dry ashing is the most commonly used procedure to determine the total mineral content of foods. Dry ashing is used to measure water-soluble, water-insoluble, and acid-insoluble ash. The advantages and disadvantages of the two direct ashing methods are summarized in Table 10.

5. Soluble and Insoluble Ash

A finally prepared ash sample is composed of soluble and insoluble portions in water and/or acids. To determine water-insoluble ash, the ash in the crucible (dish) is solubilized with about 25 ml

Table 10 Comparison of Advantages and Disadvantages of Dry Ashing with Wet Ashing Method

Ashing method	Advantage	Disadvantage
Dry ashing	1. Simple 2. No attention required during ashing 3. Generally no reagent and no blank substraction required 4. Can handle routine large number of samples 5. Standard method for determining total ash content 6. Can assay water-soluble, water-insoluble and acid-insoluble ash content	1. High temperature required 2. Equipment is expensive 3. Volatilization loss of minerals 4. Interaction occurs between mineral components or receptacle materials 5. Absorption of trace elements by porcelain crucibles or other silica vessels 6. Poor utility for analysis of certain minerals such as Hg, As, P, Se 7. Excessive heating makes certain mineral compounds insoluble (i.e., those of tin) 8. Difficulty in handling ash residue for subsequently analysis due to high hygroscopicity, lightness, and fluffiness of the ash
Wet ashing	1. Relatively low temperature required 2. Apparatus is simple 3. Oxidation is rapid 4. Maintained in liquid conditions, which is convenient for mineral analysis 5. Equipment is inexpensive 6. Less volatilization loss of minerals	1. Requires large amounts of corrosive reagents 2. Explosive acids (i.e., perchloric requires attention 3. Requires the correction from the reagents (for calculation). 4. Evolves hazardous fumes continuously 5. Handling routine large number of samples is difficult 6. Procedure is tedious and time consuming

deionized water after weighting the total ash content. A watch glass is covered over the dish to avoid loss by spattering and is then heated to near boiling. The ash solution is filtered through an ashless filter paper and washed with an equal volume of hot water. The filter paper and residue are again placed in the crucible, ignited, and weighed. From that weight, the water-insoluble ash is calculated and the water-soluble ash is determined by the difference.

The acid-insoluble ash can also be measured in the same way as the above procedure of water-insoluble ash by replacing deionized water with 10% hydrochloric acid (sp. gr. 1.050) as solvent.

6. Alkalinity of Ash

The alkalinity of the ash is ascribed to the presence of the salts of such acids as citric, tartaric, or malic, which are converted to the corresponding carbonates on incineration (11). Therefore, the alkalinity of the water-soluble ash from grape products is potassium carbonate derived from naturally occurring cream of tartar, whereas the alkalinity of the water-insoluble ash results from calcium and magnesium carbonates derived from calcium salts of fruit acids. It is known that the ash of fruit and vegetable products is alkaline in reaction, whereas that of meat products and certain cereals is acid (11).

The alkalinity of the ash can be used in determining the acid–base balance of the food and in detecting adulteration of the food with minerals. The concentrations of the acid-forming elements (P, S, Cl) and the base-forming elements (K, Na, Ca, and Mg) directly influence the acid–base balance in a food material. The quantity of these elements in a given weight of food material is converted into milliliters of 1 N acid or base, respectively, equivalent to the amount of acid or base element present (11,55). The general procedure and calculation for determination of alkalinity are as follows:

PROCEDURE
1. Moisten an ashed sample with a small amount of deionized water and quantitatively transfer the sample into a 400-ml beaker.
2. Add 50 ml of 0.1 N HCl to the beaker. Cover the sample beaker with a watch glass and boil gently for 5 min. Cool the beaker and rinse the watch glass with freshly boiled deionized water and drain rinsings back into the beaker.
3. Add 30 ml of 10% $CaCl_2$ solution, stir gently, cover with a watch glass, and let stand for 10 min.
4. Add 10 drops of 1% phenolphthalein indicator (1.0 g in 100 ml 95% ethyl alcohol) and titrate with 0.1 N NaOH to a faint pink color persisting for 30 sec.

CALCULATIONS

$$\text{ml 0.1 N HCl/100 g} = \frac{1000 \times 50 \times (\text{N of HCl}) - (\text{N of NaOH} \times \text{ml NaOH})}{\text{weight of sample}}$$

REVIEWS

AOAC, *Official Methods of Analysis*, 15th Ed., Association of Official Analytical Chemists, Washington, DC 1990.

A. Pande, *Handbook of Moisture Determination and Control*, Marcel Dekker, New York, Vol. I (1974), Vols. II, III, IV (1975).

J. Mitchell, Jr., and D. M. Smith, *Aquametry*, John Wiley and Sons, New York, Part I (1977), Part II (1980), Part III (1984).

Y. Pomeranz and C. E. Meloan, *Food Analysis: Theory and Practice*, AVI, Wesport, 1978.

M. A. Joslyn, *Methods in Food Analysis*, 2nd Ed., Academic Press, New York, 1970.

REFERENCES

1. R. L. Hassel, *Am. Lab.* (*1976*): 33.
2. J. W. Pyper, *Anal. Chim. Acta*, *170*: 159 (1985).
3. B. Makower, *Advances in Chemistry*, Vol. 3, Am. Chem. Soc., Washington, DC, 1950.
4. J. D. Pettinati, C. E. Swift, and E. H. Cohen, *JAOAC 56*: 544 (1973).
5. E. Karmas, *Food Technol. 34*: 52 (1980).
6. C. W. Everson, T. Keyahian, and D. M. Doty, Am. Meat Inst. Found. Bull. No. 26, Washington, DC, 1955.
7. A. Pande, *Handbook of Moisture Determination and Control*, Vol. 1, Marcel Dekker, New York, 1974, p. 1.
8. N. E. Dorsey, *Properties of Ordinary Water Substance in All Its Phases: Water Vapor, Water and All the Ices*, 2nd ed., Reinhold, New York, 1950.
9. A. G. Ward, *Recent Advances in Food Science*, Vol. 3 (J. M. Leitch and D. N. Rhodes, eds.), Butterworths, London, 1953.
10. Y. Pomeranz and C. E. Meloan, *Food Analysis: Theory and Practice*, AVI, Westport, 1978, p. 521.
11. M. A. Joslyn, *Methods in Food Analysis*, 2nd Ed., Academic Press, New York, 1970, p. 67.
12. J. Kuprainoff, *Fundamental Aspects of the Dehydration of Foodstuffs*, Macmillan, New York, 1958, p. 14.
13. O. R. Fennema, *Principles of Food Science*, Vol. 1 (O. R. Fennema, ed.), Marcel Dekker, New York, 1976, p. 13.
14. H. A. Iglesias and J. Chirife, *Lebensm.-Wiss.-Technol. 9*: 107 (1976).
15. H. Salwin, *Food Technol. 13*: 594 (1959).
16. R. A. Horne, *Surv. Progr. Chem. 4*: 1–43 (1968).
17. S. E. Charm, *The Fundamentals of Food Engineering*, 2nd Ed., AVI, Westport, 1971.
18. N. N. Potter, *Food Science*, 4th Ed., Reinhold, New York, 1986, p. 56.
19. D. Pearson, *Laboratory Techniques in Food Analysis*, John Wiley and Sons, New York, 1973, p. 27.
20. AOAC, *Official Methods of Analysis*, 12th Ed. (W. Horwitz, ed.), Washington, DC, 1975.
21. AOAC, *Official Methods of Analysis*, 14th Ed., Washington, DC, 1984.
22. AOAC, *Official Methods of Analysis*, 15th Ed., Washington, DC, 1990.
23. A. Pande, *Handbook of Moisture Determination and Control*, Vols. 2–4, Marcel Dekker, New York, 1975, p. 313.
24. J. Mitchell, Jr., and D. M. Smith, *Aquametry*, John Wiley and Sons, New York, Part I (1977), Part II (1980), Part III (1984).
25. G. H. Richardson, *Standard Methods for the Examination of Dairy Products*, 15th Ed. (G. H. Richardson, ed.), Am. Public. Health Assoc., Washington, DC, 1985, p. 373.
26. E. W. Flosdorf, *Freeze-Drying*, Reinhold, New York, 1949.
27. A. S. Bobrow, *Am. Lab.* (*1983*): 92.
28. F. E. Jones, *Anal. Chem. 53*: 1955 (1981).
29. Annual Book of ASTM Standards, *ASTM E 203-75, Part 30*, Am. Soc. Test. and Materials, Philadelphia, 1981, p. 803.
30. K. Fischer, *Angew, Chem. 48*: 394 (1935).
31. J. Mitchell, *Anal. Chem. 23*: 1069 (1951).
32. *International Critical Tables*, Vol. 3, McGraw-Hill, New York, 1928, p. 385.
33. J. H. Bower, *J. Res. Natl. Bur. Stds. 12*: 241 (1934).
34. F. Trusell and H. Diehl, *Anal. Chem. 35*: 674 (1963).
35. A. P. Chaladze and V. E. Melkumyan, *Izmeritel'naya Tekhnika* (*1980*); *Meas. Tech.* (USSR) *23*: 346 (1980).
36. H. R. Bolin aNd F. S. Nury, *J. Agric. Food Chem. 13*: 590 (1965).
37. P. R. Addis, G. A. Reineccius, and A. S. Chudger, 32nd Annual Institute of Food Technologists Meeting, Minneapolis, 1972.
38. F. W. Karasek, *Research/Development 20*: 68 (1969).
39. G. A. Reineccius and P. B. Addis, *J. Food Sci. 38*: 355 (1973).
40. A. Khayat, *Can. Inst. Food Sci. Technol. J. 7*: 25 (1974).
41. A. Bennett and J. R. Hudson, *J. Inst. Brewing*, 60, 35 (1954).

42. Y. W. Park, *PhD dissertation*, Utah State Univ., Logan, 1981, p. 11.
43. J. S. Shenk and M. R. Hoover, Infrared Reflectance Spectro-computer Design and Application, 7th Technicon Int. Congr., New York, 1976.
44. Y. W. Park, M. J. Anderson, J. L. Walters, and A. W. Mahoney, *J. Dairy Sci. 66*: 235 (1983).
45. K. H. Norris, R. F. Barnes, J. E. Moore, and J. S. Shenk. *J. Anim. Sci. 43*: 889 (1976).
46. K. H. Norris, in *Modern Methods of Food Analysis* (K. K. Stewart and J. R. Whitaker, eds.), AVI, Westport, 1984, p. 167.
47. J. B. Hasted, *Progress in Dielectrics*, Heywood, London, 1961, p. 102.
48. K. F. Herzfeld and T. A. Litovitz, *Absorption and Desorption of Ultrasonic Waves*, Academic Press, New York, 1959, p. 353.
49. Chadwick, *Proc. Roy. Soc. Ser. A136*, 1931, pp. 692–703.
50. A. K. Stroikovskii, A. A. Pershkin, and A. N. Sheikin, *Izmeritel'naya Tekhnika* (*1980*): 54; *Meas Tech.* (*USSR*) *23*: 351 (1980).
51. R. S. Kirk, and R. Sawyer, *Pearson's Composition and Analysis of Foods*, 9th Ed., Longman, London, 1991, p. 2.
52. F. M. Uber, *Bot. Rev. 6*: 204 (1940).
53. G. F. Smith, *Anal. Chem. Acta. 5*: 397 (1953).
54. T. R. Gillet, *Anal. Chem. 21*: 1081 (1949).
55. J. Davidsion and J. A. LeClere, *J. Biol. Chem. 108*: 337 (1935).

4

Mechanical Properties of Food

D. W. Stanley, Ashley P. Stone, and Marvin A. Tung
University of Guelph, Guelph, Ontario, Canada

I. INTRODUCTION

For food analysts whose intent is to quantify the quality attribute of texture, chemical testing has proved to be of little use in predicting the response of commercial products whereas measuring mechanical properties has been much more beneficial. This is because it is not only the chemical composition of a food that dictates texture but how these chemical components are organized into a structure that is important in determining mechanical properties and, hence, texture. Indeed, one definition of texture is the way in which the structural components of a food are arranged into a structure and the external manifestations of this structure (1). Thus it becomes clear that measurements of the mechanical properties of food are made in order to acquire some sense of its underlying structure and to obtain an objective assessment of texture. These data also aid the food scientist in developing suitable process parameters, packaging materials, and storage conditions for food products.

Until recently, workers in this area have been constrained by the use of large deformation tests to quantify texture and glass lenses to examine structure. While these results often correlate significantly with sensory analysis, they frequently fail to respond to relatively minor changes in composition or processing variables. However, in this era of food fabrication it is critical to know how these effects alter the way food components interact and how structure formation occurs in foods. The development of small deformation testing instruments and electron lenses suitable for and affordable by a food analysis laboratory has allowed the study of these properties and led to a greater understanding of the mechanical properties of food. This chapter will provide an overview of the classical area of food texture and rheology and explore some of the new tools available to study this subject. Emphasis will be placed on textural evaluation of food materials, by which analysts can measure properties of importance to food quality and acceptance, and rheological evaluation, wherein structural information can be obtained. Examples will be given of how these methods can be applied to food systems.

II. TEXTURE AND RHEOLOGY OF FOODS

In order to understand how to design and interpret mechanical tests it is necessary to have some appreciation of the fundamental principles from which they are derived. The following discussion is of necessity abbreviated and simplified; it should be augmented by referring to the literature cited at the end of the chapter. It will be noted that the behavior of certain polymers has been used to provide a model for foods, although their suitability for this is far from universal.

A. Theoretical Considerations

Rheology is concerned with deformation and flow in materials and the influence of time on these properties. Classical rheology begins with a consideration of two ideal materials; the elastic solid and the viscous liquid. An elastic solid is defined as a material with a definite shape which, when deformed within certain limits by an external force, will return to its original dimensions upon removal of that force. A viscous liquid has no definite shape and will flow irreversibly upon application of an external force, including gravity. As would be expected, the preponderance of food materials have rheological properties somewhere between these two model representations and are classified as viscoelastic or, as termed in the food industry, soft solids. Since most foods are viscoelastic, this subject will be covered in more detail in a subsequent section.

For the simplest case of an ideal viscous liquid (Newtonian), a shear stress (σ, force/unit area, N m^{-2} or Pa) acting throughout a liquid contained between two parallel plates will result in deformation or strain (γ) at a particular rate ($d\gamma/dt$, where this derivative of strain with time is often shown as $\dot{\gamma}$, sec^{-1}, "gamma dot") related to the magnitude of the stress and resistance to flow offered by the liquid. For a Newtonian liquid this relationship is a direct proportionality between shear stress and shear rate:

$$\sigma = \eta\dot{\gamma} \tag{1}$$

where the proportionality constant (η) is the viscosity, which represents the resistance to flow in the liquid, and may be expressed as Pascal second, Pa·sec, or poise, P, where 1 P equals 0.1 Pa·sec. Figure 1 shows this in diagrammatic form. In most food materials the relationship between shear stress and shear rate is nonlinear (non-Newtonian).

For the simplest case of an ideal elastic solid (Hookean), a shearing stress (σ) applied to the body produces an instantaneous shearing strain (γ, dimensionless):

$$\sigma = G\gamma \tag{2}$$

The proportionality constant (G, N m^{-2} or Pa) is an elastic modulus also called the shear modulus (Fig. 2). It is likewise possible to determine a compression or tension modulus (E or Young's modulus) and a bulk modulus (K) for hydrostatic or triaxial deformation. These two equations point out the analogy between Hooke's law for an elastic solid and Newton's law for viscous liquids, but note that for solids the stress is linearly related to the strain whereas for liquids the stress is linearly related to the rate of change of strain or strain rate.

B. Viscosity of Fluid Foods

Measurements of viscosity are the most frequently undertaken attempts at characterizing the rheological properties of foods by mechanical means. Food processors use viscosity data to test incoming raw materials and standardize final products whereas the development of new foods often relies on attaining the proper consistency in order to match sensory requirements.

A

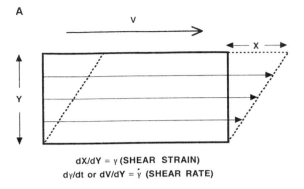

dX/dY = γ (SHEAR STRAIN)
dγ/dt or dV/dY = $\dot{\gamma}$ (SHEAR RATE)

B

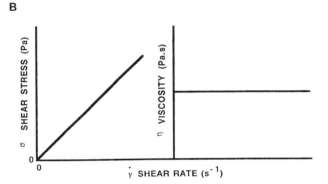

Fig. 1 Simple shear applied to an ideal viscous or Newtonian fluid. (A) A liquid is contained between two parallel plates. The bottom plate is fixed while the upper plate is moving with a velocity (V, cm sec^{-1}) caused by a shear stress (σ, N m^{-2} or Pa) acting over time (t, sec) that produces a displacement or shear strain (γ, dX/dY, dimensionless) per unit time equal to the shear rate ($\dot{\gamma}$, sec^{-1}, $d\gamma/dt$ or dV/dY). The proportionality constant between σ and $\dot{\gamma}$ is the viscosity (η, Pa-sec or P). (B) Flow curves or rheograms for a Newtonian fluid. (Adapted from Refs. 1 and 85.)

1. Measuring viscosity: Instrumentation

Essentially, only two motions are available by which stress can be applied to any sample, linear and rotational, although the rotational stress may be steady (continuous), periodic (as in sinusoidal oscillation), or in a number of transient ways such as step or ramp programs.

    a. Measuring System Geometry. In order to perform rotational measurements, samples are tested in one of several systems including cone and plate, parallel plate, and concentric or coaxial cylinder geometries (Fig. 3). All of these cells can be obtained in different sizes and with different gaps between the surfaces and different angles for cones. As a rule, the larger the sample area the lower the stress per unit torque and as the gap increases the strain per unit displacement decreases as does the shear rate per unit velocity. Thus, cell diameter determines stress and gap determines strain, within the limits imposed by the sample. The gap size between parallel plates and between concentric cylinders is often dictated by the nature of the material, especially if it is particulate. Truncated cones may be used if the sample contains large particles but these particles must be small (<10%) in relation to the minimum separation between the fixture surfaces. Cone and plate geometry is particularly useful for non-Newtonian systems since it provides a constant rate of

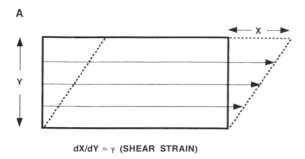

dX/dY = γ (SHEAR STRAIN)

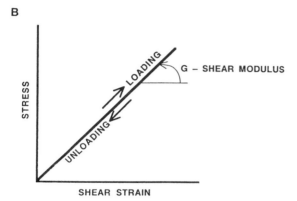

Fig. 2 Simple shear applied to an ideal elastic or Hookean solid. (A) A solid is sheared by the application of a shear stress (σ, N m^{-2} or Pa) that produces an instantaneous displacement or shear strain (γ, dX/dY, dimensionless). The proportionality constant between σ and γ is the shear modulus (G, N m^{-2} or Pa). (B). Stress/strain plot of an ideal elastic solid. (Adapted from Refs. 1 and 85.)

strain across the sample and is also applicable if only small amounts of sample are available. However, positioning of the two fixture components requires great care to ensure that the apex of the cone coincides with the surface of the plate. Truncated cones are also commonly employed to prevent friction due to contact between the components; a gap of several micrometers is normally used. Cone and plate fixtures are used for constant temperature studies and are not recommended for temperature ramping experiments because the sample gap dimensions are sensitive to expansion or contraction of the fixture assemblies by changing temperature. Parallel plate fixtures are

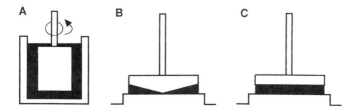

Fig. 3 Measuring system geometries. (A) Concentric or coaxial cylinders. (B) Cone and plate. (C) Parallel plates. (Adapted from Ref. 2.)

beneficial in studies involving changing temperature and very narrow gap settings can be avoided. Additional information on measurement geometries may be obtained from Shoemaker et al. (2).

   b.   Viscometers.   In measuring the viscosity of fluid foods it is important to select a range of shear rates that are appropriate for the material being tested. As mentioned previously, these can vary over a wide range depending on both the sample and the application. A further critical parameter in measuring viscosity is temperature control; this subject is discussed in a following section. Viscosity measurements of food can be obtained in a variety of ways using any of a multitude of commercial instruments. The stress or shear rate used in viscosity testing can result from gravity, such as in the capillary-type and falling ball viscometers, and therefore be stress-controlled, or from motor-driven apparatus. With modern instrumentation, controlled speed or controlled shear rate devices are popular because it is electronically possible to generate constant or controlled speed and then to measure the torque required to maintain this defined shear rate. Such an instrument can characterize Newtonian and non-Newtonian liquids over the range of shear rates available, this variable being limited by the operating speed range of the instrument. Further practical information on the theory and measurement of viscosity in food may be obtained from Refs. 3–6.

2.   Non-Newtonian Behavior in Fluid Foods

It was mentioned previously that most food substances exhibit nonideal behavior and are said to be non-Newtonian. Non-Newtonian behavior applies to all fluids except ideal viscous fluids, i.e., any fluid with a flow curve, at a given temperature and pressure, which is not linear and does not pass through the origin on arithmetic coordinates of shear stress vs. shear rate. Nonlinearity indicates that viscosity changes with shear rate. Three broad categories of fluids may be considered as non-Newtonian: (a) time-independent fluids for which the shear stress at any position within the fluid is dependent only on the shear rate at that location, e.g., dilute purees and colloidal suspensions; (b) time-dependent fluids for which the shear stress is a function of both the magnitude and duration of shear, and possibly of the time lapse between consecutive applications of the shearing treatments, such as egg albumen and gelatinized starch suspensions that have a shear sensitive secondary structure; (c) viscoelastic fluids that show partial elastic recovery upon removal of a deforming shear stress, thus demonstrating properties of both elastic and viscous materials as in gels and high-internal-volume emulsions.

   Although these categories provide a convenient system for discussing rheological phenomena, some real foods show properties under various conditions that include a combination of classes. A gel, for example, may be viscoelastic in showing partial elastic recovery when the deforming shear stress is removed. With greater magnitude and duration of shear, the elastic response disappears and time-dependent behavior is observed with shear stress decreasing as secondary structure is destroyed. Eventually, at a certain time, or at high temperatures, all secondary structure could be gone and the behavior becomes time-independent. Additional information on non-Newtonian flow can be obtained from most rheological volumes; Tanner (7) discusses this subject from an engineering perspective.

   a.   Time-independent fluids.   Along the flow curve of a non-Newtonian fluid the ratio of shear stress to shear rate is not constant; thus the term *apparent viscosity* is applied to this ratio. Since the non-Newtonian shear stress/shear rate ratio changes along the flow curve, a value for apparent viscosity has no meaning unless accompanied by the corresponding shear stress or shear rate. Another important consideration when setting out to measure rheological behavior of a fluid food is determining viscosity at shear rates that will be similar to those which exist in the practical application under consideration (e.g., shelf stability, sensory evaluation, mixing, etc.). Apparent viscosity may vary more than 100-fold over shear rates from 1 to 1000 sec^{-1} for fluid foods such as ketchup, salad dressings, or applesauce.

PSEUDOPLASTIC FLUIDS   The most common type of time-independent non-Newtonian behavior is pseudoplastic flow in which the fluid exhibits shear thinning (reduced viscosity with increasing rate of shear) over a wide range of shear rates. These fluids may be dispersions containing asymmetrical particles or high molecular weight polymers in which the particles are randomly oriented and entangled when the fluid is at rest. This condition of random structure follows the tendency of such systems to maximize entropy through Brownian motion. The material behavior when sheared may be imagined to change over different shearing conditions as follows.

At very low shear rates the entangled particles separate and intertwine in a dynamic fashion as the disrupting influence of shearing is balanced by Brownian motion to preserve an essentially random structure. In this low shear rate range, resistance to shearing is proportional to the shear rate and the fluid exhibits a Newtonian-like constant viscosity, η_0, called the zero shear viscosity. This region of constant viscosity at low shear rates may not be observed for all pseudoplastic fluids under common testing conditions because relatively few instruments have adequate sensitivity for these measurements at extremely low shear rates (e.g., 0.001–0.00001 sec^{-1}).

As the shear rate is increased the asymmetrical particles tend to align themselves with the shear planes such that frictional resistance is reduced. In this manner the random structure at ultralow shear rates gives way to a shear oriented structure at higher shear rates. The progressively decreasing resistance to flow at higher shear rates is observed as a decreasing apparent viscosity. This shear thinning is usually the only phenomenon observed over a practical range of shear rates. However, for any given pseudoplastic system it is reasonable to expect that at some high shear rate the particles will be fully aligned along the laminar shear planes and that no further streamlining is possible. This will be the limit of shear rate thinning for the fluid. An important consequence of this reduced resistance to flow at high shear rates is that pseudoplastic fluids may be pumped, mixed, or otherwise processed at high shear rates with a lower power requirement than at low shear rates.

At still higher shear rates the shear stress may again be proportional to shear rate and a second Newtonian-like region may be evident. This constant viscosity at very high shear rates is called the infinite shear viscosity, η_∞. In practice the high-shear Newtonian-like plateau may not be demonstrated easily because of the high shear rates required to achieve this fully aligned condition with food dispersions.

As the shear rate is decreased, the shear-oriented structure will progressively disappear as the influence of Brownian motion causes a return to the random structure at lower rates of shear. Thus, the pseudoplastic shear rate thinning phenomenon is instantaneously, completely reversible, and dependent only on the rate of shear in a given fluid at a given temperature. Recall that in this model the orientation of particles is random. The shear thinning phenomenon in fluids with more complex structures will be discussed later.

An example of pseudoplastic flow is shown in Fig. 4A as shear stress vs. shear rate on arithmetic coordinates. These types of plots are called rheograms and show relationships among rheological quantities. Near the origin the σ vs. $\dot{\gamma}$ relationship is essentially linear and the slope of this segment corresponds to the low-shear Newtonian-like apparent viscosity, η_0. As the shear rate is increased the σ vs. $\dot{\gamma}$ ratio steadily decreases but at high shear rates the rheogram again becomes linear and exhibits Newtonian-like flow characterized by η_∞, the infinite shear rate apparent viscosity. When plotted on logarithmic coordinates as apparent viscosity vs. shear rate (Fig. 4B), these three regions are evident. For many pseudoplastic fluids the intermediate shear thinning behavior follows a power function which appears as a straight line with negative slope on the logarithmic plot (see

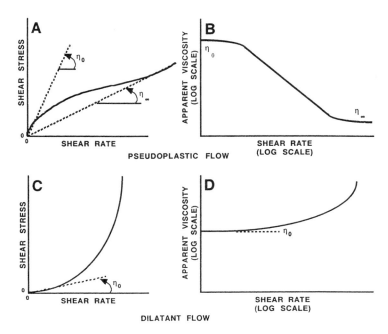

Fig. 4 Rheograms for non-Newtonian fluids. Pseudoplastic flow: (A) Arithmetic coordinates. (B) Logarithmic coordinates. Dilatant flow: (C) Arithmetic coordinates. (D) Logarithmic coordinates. η_0, zero shear apparent viscosity. η_∞, infinite shear rate apparent viscosity.

subsequent discussion). In terms of food operations, pouring occurs in the shear rate region of ~10 sec^{-1}, chewing at ~50–100 sec^{-1}, and spreading at ~1000 sec^{-1}.

DILATANT FLUIDS   A comparatively small group of fluids exhibit an increased resistance to flow with increasing shear rate and this time-independent non-Newtonian behavior is called *dilatant flow*. The increasing apparent viscosity at higher rates of shear may or may not be accompanied by a volume increase, which would be termed *volumetric dilatancy*. Rheological dilatancy or shear rate thickening has been observed in some concentrated suspensions; however, it is not restricted to dispersions of this type.

An explanation of shear rate thickening in concentrated suspensions centers on the alterations required within the system to achieve flow. At rest, the solid particles of the suspension are consolidated in a close-packed structure with the liquid continuous phase occupying the void spaces. When small shear stresses are applied to the system, particles in relatively widely separated planes can experience slip in the direction of the applied stress as the particles glide past one another in the thin film of adhering liquid. For small stresses the deformation rate or shear rate appears to be governed by the continuous phase liquid and is proportional to the shear stress; thus a Newtonian-like behavior will be noted over a small range of low shear rates.

As the shear rate is increased, movement may occur in planes that are more closely spaced and an increase in void volume is required to permit this large-scale interparticle movement. Since the volume of continuous phase liquid present is insufficient to provide surface wetting of all particles as the sample volume increases, the surfaces of adjacent particles come into direct contact to produce a high internal friction. This internal friction increases rapidly with increasing shear rate, thereby accounting for the increase in apparent viscosity with increasing shear rate. A reduction in shear rate will allow a return to the close-packed structure with improved surface wetting or lubrication leading to a lower apparent viscosity. Accordingly, the dilatant flow increase is instantaneously reversed when shear rates are lowered.

Dilatant flow may occur in a wide variety of fluid systems, even in dilute dispersions. A mechanism based on the formation of intermolecular junctions during shearing is sometimes used to explain dilatant behavior in a dilute polymer solution. Higher shear rates increase the collision frequency of reactive sites on differing molecules and the resulting network provides an elastic structure of increasing apparent viscosity. An example of dilatancy in food materials may be found in relatively concentrated aqueous starch suspensions. Figure 4C and D shows rheograms of dilatant flow.

MODELS FOR TIME-INDEPENDENT NON-NEWTONIAN FLOW   Several models have been proposed to describe relationships for time-independent non-Newtonian fluids. Some of these flow models are based on theoretical concepts whereas others are strictly empirical.

The power law model is an empirical relationship and is the simplest and most popular model applied to non-Newtonian flow:

$$\sigma = m\dot{\gamma}^n \tag{3}$$

where m and n are parameters called the consistency coefficient and the flow behavior index, respectively. Comparison of this relationship with the Newtonian model indicates that Newtonian fluids would be a special case in which $n = 1$ with η in place of m. The flow behavior index (n) provides a convenient identification of shear thinning ($n < 1$) and shear thickening ($n > 1$) types of flow. The power law model can also be shown as the apparent viscosity as a function of shear rate:

$$\eta = m\dot{\gamma}^{n-1} \tag{4}$$

The consistency coefficient (m) is numerically equal to the shear stress or apparent viscosity at a shear rate of 1 $\sec^{-1}$ and has units of Pa $\cdot$ $\sec^n$.

In order to calculate the m and n parameters, the equations can be cast in a logarithmic form that will be linear:

$$\log \sigma = \log m + \log n \log \dot{\gamma} \tag{5}$$
$$\log \eta = \log m + (n - 1)\log \dot{\gamma} \tag{6}$$

Thus data may be plotted directly on logarithmic coordinates and a straight line calculated. Figure 5 shows examples of power law rheograms for a food dispersion over a range of shear rates. The material is shear rate thinning or pseudoplastic and follows the power law model closely since the rheogram is essentially linear on a logarithmic plot.

The Casson model was originally developed to describe the flow of printing inks but has been found suitable for some food systems, e.g., molten chocolate. The equation is:

$$\sigma^{0.5} = k_0 + k_1\gamma^{0.5} \tag{7}$$

This equation describes a linear rheogram when the square root of shear stress is plotted against the square root of shear rate. For Casson fluids, extrapolation of the rheogram to zero shear rate provides an estimate of the yield value, σ_y, which can be defined as the maximum shear stress just before flow begins. The concept of yield value or yield stress will be discussed subsequently. An example of a food material displaying Casson flow is shown in Fig. 6. The estimated yield stress is found by extrapolating the flow curve to zero shear rate and taking the square of the intercept value. The common problem of initiating flow from a bottle with a small opening containing such materials is a consequence of this yield stress. The presence of a yield value indicates a form of plastic flow which is a consequence of a solid-like structure when undisturbed. At shear stresses

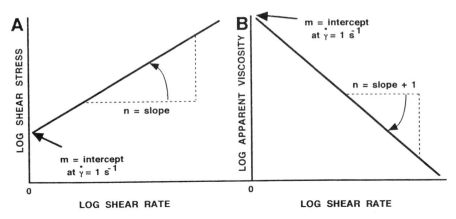

Fig. 5 Examples of power law model rheograms for a food product on logarithmic coordinates. (A) Shear stress vs. shear rate. (B) Apparent viscosity vs. shear rate.

above the yield stress the material will behave as a fluid and exhibit non-Newtonian flow. An obvious example of this type of product is tomato ketchup.

The power law plastic or Herschel-Bulkley model (8) contains three parameters and is a modification of the power law to include a yield value for non-Bingham plastic flow in the form:

$$\sigma = \sigma_y + m'\dot{\gamma}^{n'} \tag{8}$$

The parameters σ_y, m', and n' are analogous to their descriptions in earlier sections. The flow behavior index, n', can be used to identify shear thinning ($n' < 1$) and shear thickening ($n' > 1$) and when $n' = 1$ the equation describes Bingham plastic flow with m' in place of η_p.

In order to evaluate the flow parameters for a given set of viscometric data it is necessary to first have an estimate of the yield value. This may be available directly from the viscometric test or may be estimated by applying the Casson model to the data at low shear rates. With the estimate for the yield stress, the logarithmic form of the model:

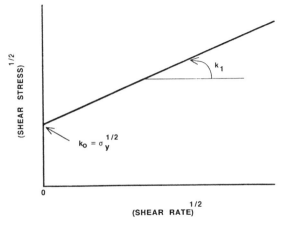

Fig. 6 Example of Casson flow model rheogram for a food product.

$$\log (\sigma - \sigma_y) = \log m' + n' \log \dot{\gamma} \tag{9}$$

can be used to provide a straight line relationship when values of $\sigma - \sigma_y$ vs. $\dot{\gamma}$ are plotted on logarithmic coordinates. The consistency coefficient, m', would be the ordinate value at a shear rate of 1 sec^{-1} and the slope of the rheogram is the flow behavior index, n'.

A rheogram for power law plastic flow is shown in Fig. 7. It is necessary to first obtain an estimated value for yield stress in order to plot such a flow curve. In some cases, plastic flow is more accurately described by one model or the other (Casson or power law plastic) and for some materials both models apply equally well.

CHOOSING A RHEOLOGICAL MODEL   Many other rheological models are available to describe time-independent non-Newtonian flow of fluid foods, but from the variety of potentially useful forms it is necessary to choose the most suitable for a particular situation. Non-Newtonian fluid behavior may be expressed by equations that differ in complexity. A suitable model should accurately represent flow behavior over a range of shear stresses or shear rates pertinent to the performance requirements of the fluid, and viscometric evaluations would necessarily include these conditions. The power law model generally fits the flow curve over the central shear rate range, whereas other more elaborate models not discussed here, such as the Cross and Sisko models, cover a wider scope of shear rate. Rheological relationships must be defined in terms of practical utility.

For materials known to have yield values, potential models would include the Bingham plastic, power law plastic, and Casson relationships. Fluid foods that exhibit shear rate thinning and thickening over intermediate shear rates are often described accurately by the power law model. Where information on rheological behavior at ultralow and/or ultrahigh shear rates is relevant, the use of more complicated multiparameter models may be required to accurately characterize these fluids. One example of very high and low shear rate applications would be in the preparation of an emulsion and its stability against separation after formation. Shear rates in colloid mills or homogenizers may exceed 10^5 sec^{-1}, whereas droplet migration during storage of the emulsion would involve near-zero shear rates.

In general, simpler models are preferred over more complex ones on the bases of convenience and validity. A suitable model must provide an accurate fit to the experimental data; this may be determined statistically. The model should agree closely with the data over the entire flow curve, which may be verified by visual examination of the data plotted along with the least squares fitted function. Restrictions on the choice of potentially suitable models may be imposed by the types

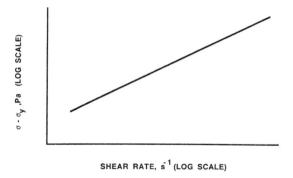

Fig. 7   Example of power law plastic flow model rheogram for a food product.

of available instrumentation and computing facilities. The most commonly used industrial equipment is not suited for measurement at very low or very high shear rates, and yield values, if present, may not be detected because this characteristic depends on both the instrument used and the methodology followed. In spite of recent advances in rheometer design and data-processing devices, these facilities are not always accessible.

   b. Time-Dependent Fluids.   In the previous section fluids were considered for which the shear stress was dependent only on the shear rate. However, some fluids exhibit time-dependent flow in which there are decreasing or increasing effects on shear stress and apparent viscosity with time at a constant shear rate and temperature. In some cases this change is reversible and the fluid will recover its original condition with time at rest, or the change brought about by shearing may be irreversible. Time dependency is thought to result from structural alterations in the material such as the adhesive or weak secondary interaction of dispersed particles to form three- dimensional structures that will reversibly break with time in a continuous flow (thixotropy); if the loss of structure is not recoverable the behavior is termed rheodestructive.

   An understanding of the time-dependent nature of flow in some fluid foods is of considerable importance in their processing or handling. The duration of pumping or mixing operations, for instance, must be carefully controlled to assure desired apparent viscosity in time-dependent foods such as custard, margarine, or liquid egg albumen. Conversely, variation in shearing time may be used as a means of altering flow characteristics.

THIXOTROPIC FLUIDS   A thixotropic system will experience a reversible decrease in shear stress and apparent viscosity at a constant shear rate and temperature. Fluids of this type are thought to consist of asymmetrical dispersed particles or molecules that interact with adhesive or weak secondary bonding forces to form a network or aggregated structure at rest. When continuous flow is imposed on such a system at a constant shear rate, some of the interparticle bonds break with time, thereby offering less resistance to flow, and the shear stress relaxes to some constant value over a period of time. Since the bonds are thought to vary in strength, many weak bonds will be broken near the onset of shearing, resulting in a relatively rapid loss of structure at first, with gradual slowing as the weaker bonds are depleted until the remaining structure is due to bonds of greater strength than the shearing forces applied in the system.

   The extent of structural breakdown to achieve this equilibrium condition will depend on the rate of steady shear imposed, with higher shear rates resulting in a more rapid breakdown of structure. Since the shear rate is constant during this shear stress relaxation period, this may be interpreted as an apparent viscosity decay. When the shearing treatment is stopped, the dispersed particles can once again form the network or aggregated structures as random Brownian motion gradually restores the particles to positions where interparticle interactions can occur. Over sufficient time the original density of linkages is achieved, and thus the process is reversible.

   The major difference between the structural shear rate sensitivity of thixotropic and pseudoplastic systems is the time required to reach a stable shear-induced condition and to restore the original structure when shear stresses are removed. This change and reversal is thought to be instantaneous in pseudoplastic fluids, whereas a relatively long time is required in thixotropic fluids. For some thixotropic materials the shear stress relaxation may span several minutes before an equilibrium condition is established, and complete recovery may require several hours.

RHEODESTRUCTIVE FLUIDS   In some fluids, the time-dependent loss of structure is not recoverable and fluids that experience permanent loss of structure may be called rheodestructive. The loss of structure is envisaged in the manner described earlier for thixotropy; however, there is no tendency to reform the structural elements when shearing has stopped. Rheodestruction is compared with thixotropy in Fig. 8 as time-dependent viscosity decay curves during steady shear, followed by differing behavior when the fluids are at rest.

Not all time-dependent shear thinning systems fall into one or the other purely thixotropic or totally rheodestructive categories. In some materials, partial recovery of structure is noted but the original conditions are not restored even after long periods of rest. Such systems may be described as partially rheodestructive or as a combination of thixotropic and rheodestructive.

RHEOPECTIC FLUIDS   If a fluid system exhibits a reversible increase in shear stress or apparent viscosity when sheared at a constant rate and temperature, it is said to be rheopectic. This relatively rare form of rheological behavior is the opposite of thixotropy and is referred to as antithixotropy or rheopexy. Rheopexy is explained in much the same way as dilatancy except that the developing resistance to flow occurs over a measurable time whereas the dilatant response is instantaneous. Similarly, rheopectic materials require a considerable period of time before there is a complete return to the original structure of low apparent viscosity.

MEASUREMENT OF TIME-DEPENDENT FLOW   Time-dependent flow in a given fluid may be characterized by subjecting the sample to steady shear at a constant rate while the fluid temperature is controlled. Several mathematical models have been proposed to describe the nonlinear relation of apparent viscosity to time. Different rates of shear will provide different shear stress relaxation or apparent viscosity decay relationships. When time-dependent flow is pertinent to the performance of a fluid system, as in pumping or mixing, this behavior should be examined over a range of shear rates and times encountered in that application.

Another technique for characterizing time-dependent flow consists of a viscometric test in which the shear rate is steadily increased from zero to a given maximum value and then back to zero again. For a particular experiment the acceleration and deceleration rates may be equal and the test will span a certain range of shear rates; however, these may be varied to provide a spectrum of testing conditions. In these tests, apparent viscosity will be a function of shear rate and time.

For a time-dependent fluid, the increasing and decreasing rheograms will not coincide, i.e., they will show hysteresis. The general shape of the curves and the extent of hysteresis may be useful to characterize time-dependent flow behavior. Examples of idealized flow curves for time-dependent non-Newtonian systems are shown in Fig. 9. The arrows indicate the increasing and decreasing shear rate curves and the concavity of the lines will identify time-dependent thinning (concave to shear rate axis) or thickening (convex to shear rate axis). As well, thixotropic

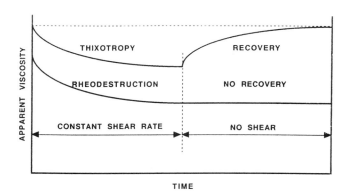

Fig. 8   Idealized rheograms for thixotropic and completely rheodestructive dispersions.

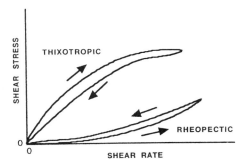

Fig. 9 Idealized rheograms of time-dependent flow behavior.

and rheopectic behavior can be characterized empirically by measuring the area within the loops seen in Fig. 9 as a measure of structure breakdown or buildup.

3. Temperature Effects of Flow Behavior

The influence of temperature on rheological behavior of fluid foods is of prime importance because many food processing unit operations involve heating or cooling, e.g., cooking of starch-based sauces, production of margarine, sterilization of aseptically canned puddings, and chilling of milk products. Processors must also be able to relate initial or processing flow behavior to desirable textural attributes of the fluid product at serving temperature. Moreover, elevated temperatures are often used in the food industry to alter flow characteristics so as to facilitate such operations as pumping, mixing, or container filling. It follows from this that strict temperature control is necessary when attempting viscosity measurements.

The viscosity of many liquids decreases with increasing temperature according to the Arrhenius-type relationship:

$$\eta = A e^{\Delta E/RT} \tag{10}$$

where T is the absolute temperature, R is the universal gas constant, A is a parameter, e is the natural logarithm base, and ΔE is the activation energy for viscous flow. The parameter ΔE expresses the energy barrier that must be overcome before the elementary flow process can occur and is related to the coherence of the molecules in the liquid. A high value for ΔE indicates that the liquid is highly associated and that a large amount of energy is required to bring about the dissociation resulting in flow. The activation energy of viscous flow, ΔE, for a Newtonian liquid may be evaluated from viscometric data over a range of temperatures. Data are then plotted on semilogarithmic coordinates with viscosity on the logarithmic ordinate and the reciprocal of the absolute temperature on the arithmetic abscissa (Fig. 10). The slope of the line is $\Delta E/2.303R$. Thus, the activation energy can be calculated from the rheogram slope and expressed in joules per mole of liquid.

It may occur that a slight curvature, concave to the ordinate, will be seen in a graph like Fig. 10. This is not uncommon for Newtonian liquids with substantial secondary bonding. When a liquid deviates from the linear relationship over a wide temperature range it may be necessary to fit the relation over smaller intervals and express the viscosity–temperature relationship as several pairs of ΔE and A values with associated temperature ranges. With this information it is possible to predict the viscosity of a Newtonian fluid at any temperature within that range. However, since ΔE reflects molecular association within a fluid, this may be a useful tool in the study of interactions in fluid systems and their changes with the application of heat.

Since the apparent viscosity of a non-Newtonian fluid is a function of shear rate or shear stress,

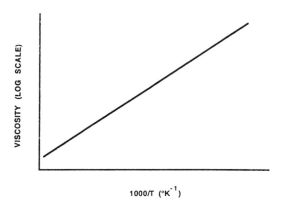

Fig. 10 Example of an Arrhenius-type plot of viscosity data to show temperature dependence of a Newtonian liquid.

the relationship with temperature must contain either σ or $\dot{\gamma}$ as an additional parameter. This may be done by expressing the apparent viscosity–temperature relationship while holding shear rate or shear stress at a constant value. For example, apparent viscosity could be determined at a selected shear rate over a range of temperatures and treated in much the same way as for the Newtonian fluid data. This may be repeated for a number of different shear rates to provide information over a range of conditions pertinent to a given application.

It would also be expected that temperature would influence time-dependent behavior of fluids. Higher temperatures would provide additional kinetic energy in the sample to aid in the breakdown of structure under the influence of shear. A direct approach to the examination of the effect of temperature on non-Newtonian flow would be to evaluate the change in flow model parameters with temperature, using equations suitable for the flow behavior of the fluid over the temperature range of interest. Another procedure for evaluating the effect of temperature will be presented when viscoelastic properties are discussed.

4. Concentration Effects on Flow Behavior

Rheological properties of fluid food dispersions are often sensitive to changes in concentration. Consequently, the flow behavior of many food solutions, suspensions, and emulsions may be modified by simply changing the amounts of one or more ingredients. Early theoretical work relating viscosity to volume concentration for ideal dilute suspensions of rigid spheres was done by Einstein who proposed that

$$\eta = \eta_s(1 + a\phi) \tag{11}$$

where η_s is the viscosity of the suspending liquid, ϕ is the fractional volume concentration of the particles, and the parameter a has a value of 2.5. Other assumptions in the derivation are that no particle interactions occur, particles are uncharged, there is no slip between the particles and suspending liquid, and viscosity arises only from the dissipation of energy by modification of fluid motion near the particle surfaces. This equation is valid only for dilute suspensions where $\phi < 0.05$. For concentrated suspensions of spheres, short fibers, and other crystalline forms such as cubes of narrow size distribution in Newtonian liquids, a simple empirical equation can be used:

$$\eta = \eta_s[1 - (\phi/A)]^{-2} \tag{12}$$

The magnitude of the constant A ranges from 0.68 for spheres to 0.18 for short fibers with a length/diameter ratio of 27.

For suspensions of higher volume concentration, there would be increasing interaction among particles as the flow patterns of the surrounding liquid around the particles come closer together. This hydrodynamic interaction without flocculation of particles increases the observed viscosity and results in a more complex function of dispersed phase concentration. In real systems of increasing volume concentration, particle collisions would occur and clusters of particles would be expected to form and disperse. Observed viscosities for suspensions of nonspherical particles are greater than predicted for spheres when Brownian motion is sufficient to maintain a random orientation of the particles. Conversely, the viscosity is less than predicted for these systems when Brownian motion is small as would be the case at low temperatures and for high solvent viscosities. A further limitation of models for concentration dependence is that they apply to Newtonian-like flow behavior at low volume concentration and/or low shear rate conditions.

For some fluid systems, empirical relationships between viscosity and concentration have been determined. For example, in sucrose solutions:

$$\log \eta = a + bc \tag{13}$$

where c is concentration in g/100 g water. This is shown in Fig. 11.

With non-Newtonian fluids such as polymer solutions, suspensions, and emulsions, η vs. $\dot{\gamma}$ rheograms show increasing values of apparent viscosity along the curve when concentrations of the dispersed phase are increased. When the flow curves at different concentrations follow the same flow model it may be possible to evaluate the concentration effect on each parameter of the model. If the relationships among flow model parameters and concentration are known over a range of concentrations, the parameter values at any intermediate concentration can be evaluated and the corresponding flow curve drawn. An example is given in Fig. 12 with the power law consistency coefficient shown as a semilogarithmic function of solids content for aqueous dispersions of tomato solids. Using the corresponding variation of the flow behavior index with concentration, flow curves can be constructed for any given solids content. Moreover, by knowing the effects of concentration on flow behavior for a particular dispersed system it may be possible to formulate a product to provide desired flow properties.

Rheological properties of fluids may be influenced by several factors in addition to those already mentioned. These are either externally imposed, such as pressure, or factors involved with fluid composition, such as the nature of the continuous and dispersed phases in a dispersion, ionic strength, pH, the presence of interactions among components, and so forth.

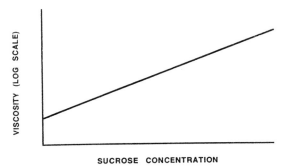

Fig. 11 Example of rheogram for sucrose in water showing viscosity as a function of concentration.

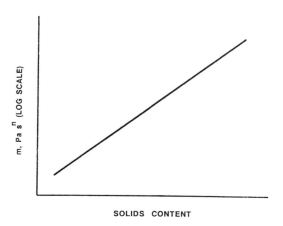

Fig. 12 Example of rheogram for tomato solids dispersed in water showing power law consistency coefficient as a function of solids content.

C. Rheological Evaluation of Solid Foods

The mechanical behavior of solid foods may be determined in several ways. If the test employed is used to measure a strictly defined rheological parameter it is classified as fundamental but, and much more likely, if the test is designed to obtain an attribute that has been found through practice to reflect textural quality, then it is classified as empirical. Traditionally, these latter tests employ large deformations and take the sample beyond the point of failure. Since sensory tests use a similar approach these two methods are frequently found to correlate highly, allowing the partial replacement of sensory panels with more objective instrumental measurements. Even though empirical tests are widely used in the food industry, an understanding of the principles on which they are at least loosely based is necessary.

1. Principles of Testing Solid Foods

A force acting externally on a solid leads to one or more of the following: tension (application of uniaxial parallel force to cause extension), compression (application of uniaxial parallel force to cause flattening), or shear (application of uniaxial tangential force to cause separation or cutting); bending involves tension, shear and compression; torque involves shear; and so forth (Fig. 13). When a material is subjected to one or more of these forces, dimensions, such as length, will change by some amount (ΔL), i.e., a deformation will occur. Alternatively, stress, the intensity factor of force, expressed as force per unit area (F/A) will produce strain or linear deformation ($\Delta L/L$). Thus, either force/deformation or stress/strain results can be obtained. For certain ideally elastic solid bodies the relationship between stress and strain is linear and passes through the origin of the resulting graph. As described previously, the ratio of stress to strain is the slope and is a constant termed the modulus of elasticity. Examples are given in Table 1. Applying a constant stress will result in a constant strain over time, and when the stress is removed the strain will return to zero.

These theoretical considerations are useful in contemplating rheological measurements of food. Unfortunately, there are several underlying assumptions about these models that are rarely, if ever, fulfilled with foods: (a) It is assumed that specimen dimensions are known throughout the application of force, since stress is a function of the area through which it acts. However, materials subjected to force change dimensions both along the axis through which it is applied and also

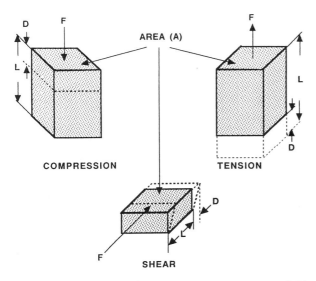

Fig. 13 Actions of forces on solids. (Adapted from Ref. 3.)

along the axis perpendicular to which it is applied. (b) The models assume that the stress and resulting strain are uniformly distributed across and throughout the sample i.e., the sample is completely homogeneous—a most unlikely case with food. (c) It is assumed by classical elastic theory that deformations are small. Since most foods are tested to failure, this is not the case. Thus there are large deviations from the ideal state. On the other hand, useful information can be obtained for solid foods through the application of deforming forces. For example, the texture profile analysis procedure of Bourne (5) used compression–decompression cycles to predetermined deformations for quantifying such secondary parameters as hardness, fracturability, cohe-

Table 1 Modulus of Elasticity and Poisson's Ratio for Various Materials

Material	Elastic modulus (Pa)	Poisson's ratio
Aluminum	6.2×10^{10}	
Apples	$0.6–4.2 \times 10^{7}$	0.29–0.21
Brass	9.3×10^{10}	
Concrete	$0.22–1.7 \times 10^{10}$	
Glass	$6.0–7.0 \times 10^{10}$	0.24
Ice	9.7×10^{9}	
Meat		0.197–0.262
Nylon	2.6×10^{9}	
Plexiglas	3.0×10^{9}	
Polystyrene	3.1×10^{9}	
Potatoes	$0.3–1.4 \times 10^{7}$	0.49–0.45
Rubber	$8.0 \times 10^{5} – 1.9 \times 10^{7}$	0.45
Silver	7.6×10^{10}	
Sweet potatoes		0.25–0.45
Water	2.3×10^{9}	

Data from Refs. 10, 11, and 28.

siveness, etc. Kaletunic et al. (9) described methodology to determine the degree of elasticity in solid foods using repeated compression–decompression cycles.

When an ideal elastic solid is subjected to hydrostatic or triaxial compression, the volume will decrease but the shape will remain the same, whereas for a shearing stress the shape will change but the volume remains the same. In a tensile test, solids change shape and some change in volume. Thus another parameter, the Poisson ratio, was introduced to relate this change in volume to the change in shape. This parameter (μ) is the ratio of lateral strain to axial strain for an elastic solid in a tensile or compression test or μ = change in width per unit width/change in length per unit length. Typical values are given in Table 1. For example, a material like cork or sponge is a mixture of solid structure and air. When a cylinder of such a material is compressed, air may be squeezed out and the total volume may be so reduced without sideways expansion. The Poisson ratio of such a material, which can be compressed with no change in diameter, is zero. For a material where no volume change takes place when it is stretched or compressed the Poisson ratio is 0.5. Anisotropic materials, those whose mechanical properties differ depending on sample orientation, will have more than one ratio. Mitchell (10) described methodology for measuring the Poisson ratio of foods.

If an ideal elastic solid is tested in tension or compression, the loading and unloading curve is a single straight line on a stress/strain plot (Fig. 2). Structurally, ideal elastic behavior is explained by intermolecular bonding mechanisms that provide coherence within the sample. An applied force will cause a mutual displacement of the molecules which absorbs energy in the form of internal tensions that balance the applied forces. When the applied force is removed, the internal forces are dissipated by restoring the molecules to their original dimensions.

2. Testing Procedures

Numerous empirical tests are used in evaluating the mechanical properties of foods. Some examples of tests that rely on linear motion include shearing, puncture, extrusion, bending, compression, tension, and cutting. These tests have in common that the results are strongly dependent on such operating factors as sample mass and geometry, cell geometry, test speed, etc. Due to the irregular shape of food samples, the measurement of mechanical properties of solid foods often requires sample cell geometries that are specifically designed for a particular material. Design of measuring systems for solids was covered extensively by Mohsenin (11).

It was noted previously that a force acting on a solid will lead to tension, compression, shear, or a combination of these "pure" stresses. However, even though the theory of these ideal bodies exist, have well-defined engineering characteristics, and are useful in contemplating food behavior, they are rarely applied to texture measurement. Thus, fundamental properties are infrequently measured; rather, forces and deformations are observed during an arbitrary test period. Although some instruments are still being used that give only one reading, usually maximum force, most of these are either being replaced or being connected to recorders to give a graph of the complete deformation process, as a function of either distance or time. Such a force–deformation plot (Fig. 14) differs markedly from a true stress–strain curve and obviously contains a great deal of information. Much care must be taken in the interpretation of these curves because it is quite easy to misjudge the significance of certain aspects of the data or to ignore useful parts of the curve. A wealth of textural information may be hidden in such characteristics as yield point, initial angle of force uptake, work of rupture, compression or elongation required to produce yield, number of peaks observed, etc. A particularly useful procedure is to examine the structure of the sample during deformation; this can be done macro- or microscopically and aids the experimenter in knowing the nature of the material studied. Works on this subject by Stanley (12) and Aguilera and Stanley (13) give examples and methodology useful in measuring food structure. For greater

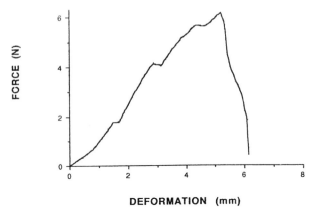

Fig. 14 Force–deformation curve resulting from puncture testing a canned green bean sample. Plot represents 326 data points. Unpublished data.

detail on instrumentation, techniques, and interpretation of data from the testing of solid foods, readers are directed to Refs. 3–5 and 14–17.

C. Selecting a Test Method

From a practical viewpoint, perhaps the most important decision to be made by a food analyst is the selection of an instrumental method. Several key factors play a role in this determination (5,18). The first consideration is the purpose of the test. The aim may be to develop a quality control procedure. If this is the case then a test should be sought that accurately measures the quality characteristic chosen, that satisfactorily duplicates sensory evaluation, that can be used routinely during all steps of the process, and that can be performed within the necessary time limits for process control. The choice of instruments will be influenced by ruggedness and versatility within the constraint of cost. If, on the other end of the scale, the purpose is to undertake research on food texture, then cost becomes secondary to precision, accuracy, and sophistication in instrument selection.

D. Viscoelastic Behavior of Foods

Many other factors also influence method selection, too numerous and diverse to discuss fully. A valid approach is to judiciously apply all of the structural and compositional information available and, taking into account the nature of the food, the objective of the test, and the instrumentation available, select a group of tests that can be implemented using what is accessible. It is then necessary to evaluate these preliminary tests over a broad spectrum of textures. In quality control situations there are usually sensory evaluations involved and, even though it is unlikely that a single instrumental test can replace integrated sensory responses, comparisons can be made during the selection process. Paralleling sensory analysis, instrumental data can often be reduced by multivariate statistical analysis to a few key independent measurements, thus allowing subsequent testing to concentrate only on those of importance. Upon the adoption of one or more instrumental tests it is recommended that sensory panels not be disbanded, if for nothing else than to intermittently ensure the continued validity of instrumental data. It should be added that sensory analysis can be of great use to those studying the mechanical properties of food since this technique allows the experimenter to determine which of these properties are important in judging acceptability and quality. A full discussion of this topic is not possible in the present work but it is suggested strongly that this area be examined thoroughly; a suitable starting point is Bourne (5).

Since what is most usually required in the instrumental analysis of food texture is a test or tests that can accurately and quickly estimate sensory evaluations, it follows that as long as the attribute selected correlates highly with texture it does not necessarily have to be a mechanical property that is measured. Thus, it has been found that quite a few physical properties, measured by a nonmechanical device, can be used to predict food texture. Chemical properties, such as compositional analysis, are also known to sometimes be associated with texture but the time, expertise, and cost required to obtain the data make this option unpopular. Some examples of nonmechanical physical properties related to texture are given in Table 2. Nonmechanical devices can be useful if a relationship with texture can be established. Readers interested in a more complete discussion of physical characterization of foods are directed to the volume edited by Gruenwedel and Whitaker (19).

D. Viscoelasticity

The classical theories given previously describe ideal material behavior. Hookean elasticity and Newtonian viscosity rarely extend to real materials such as food. Only under conditions of infinitesimal strain or rates of strain do solids or liquids lend themselves to interpretation by these explanations. When finite strain and rates of strain are employed, typical of those encountered in processing (mixing, pumping, etc.) and mastication, then departure from ideal behavior is observed.

Alternatively, deviations may arise through stress–strain relations exhibiting time dependency. When this occurs at an infinitesimal strain, i.e., when the ratio of stress to strain is constant, the behavior is termed linear viscoelastic. The strain limit of this linear viscoelastic region (LVR) is dependent on the material structure; the significance that this limit holds for foods will be discussed later. To describe rheological behavior within the LVR, constitutive equations of the general form

$$\sigma = f(\gamma, \dot{\gamma}, t, \ldots T, P, \ldots) \tag{14}$$

are used where shear stress (σ) is a function of strain (γ), strain rate ($\dot{\gamma}$), and time (t); higher terms such as temperature (T) and pressure (P) might also be incorporated. In essence, the constitutive equation is a combination of those used to describe viscous [Eq. (1)] and elastic [Eq. (2)] bodies

Table 2   Examples of Nonmechanical Physical Properties Related to Texture

Property	Parameter(s) measured	Examples(s)
Geometric	Particle size and shape	Ground coffee, salt, milk powder
	Particle shape and orientation	Textured vegetable protein, tomato juice
	Volume	Loaf volume
Optical	Light diffraction	Sarcomere length of muscle tissue
	Reflected visible light	Ripeness of fruit
	X ray, transmission	Lettuce quality, hollow heart in potatoes
	X ray, diffraction	Crystal form of fat crystals
Thermal	Enthalpy change, transition temperature	Thermomechanical properties of meat
Electrical	Dielectric constant	Fruit texture and maturity
	Conductivity	Bean hardness
Auditory	Sound amplitude	Crispness. fruit ripeness
	Resonance frequency	Fruit and vegetable quality

Data from Ref. 13.

with either strain and time or strain rate and time being held constant. The use of constitutive equations to describe nonlinear behavior (i.e., when finite strains and strain rates are applied) is complicated through aberrations in stress–strain relationships, which is to say that mechanical properties are not only a function of time but also the stress magnitude. Since many food materials are inherently nonlinear, attempts to describe behavior by derivation of suitable models have increased in recent years. The theoretical concept of nonlinearity will not be discussed here; instead the reader is advised to consult Refs. 20–24 for further explanations.

An important aspect of linear constitutive equations central to the theory of linear viscoelasticity is that the effect of mechanical history (i.e., sequential changes in strain) are linearly additive; this is known as the Boltzmann superposition principle. Assuming that a material's behavior is linear, the Boltzmann superposition principle allows a prediction of strain behavior over a range of different stresses from a single set of stress–strain data.

The principles of viscoelasticity are covered by authors to varying degrees; for a theoretical approach, Ferry's (25) is a suitable text; for a more applied treatment relevant to food materials, refer to Refs. 10 and 26–28).

Additive constitutive equations lead to a mechanical model for viscoelastic behavior in which a spring is used to represent a Hookean ideal elastic solid having a proportionality constant (G) and a dashpot, a cylinder filled with liquid of viscosity (η) is used to represent a Newtonian ideal viscous liquid. These two elements can be combined in various ways to represent analogs of rheological behavior and can be expressed as mathematical equations. Thus, a spring and dashpot in parallel is termed the Voigt or Kelvin model (Fig. 15). This model has been likened to the suspension system of a car in which the spring is analogous to a coil spring or torsion bar and the dashpot is analogous to a shock absorber. Another model useful in considering rheological experiments is the Maxwell model, consisting of a spring and dashpot in series (Fig. 15). Many other more complex mechanical models have been advanced to describe viscoelasticity, as well as models based on molecular theories in which viscoelastic polymers are considered as a series of springs in a viscous solution. These models are sometimes useful in explaining and predicting the behavior of materials under various circumstances. More complex models and their response to constant and oscillatory forms of deformation will be discussed subsequently.

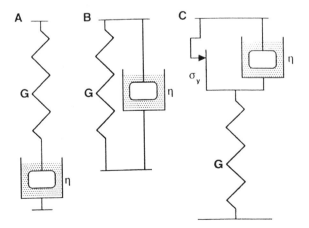

Fig. 15 Mechanical models for (A) Maxwell model; (B) Kelvin or Voigt model; (C) Bingham body. A spring is used to represent a Hookean ideal elastic solid with a proportionality constant (G); a dashpot, a cylinder filled with liquid of viscosity (η), is used to represent a Newtonian ideal viscous liquid; a friction element is used to represent yield stress (σ_y). (Adapted from Refs. 1 and 86.)

Viscoelastic materials often demonstrate a third component; along with viscous and elastic behavior, it is also necessary to consider the concept of plasticity. Plastic materials do not undergo a permanent deformation until they are stressed beyond their yield point, when they flow as a liquid. Thus, elastic deformation is exhibited at stresses below the yield value and plastic flow at higher stresses. This behavior is explained by the presence of a relatively weak long-range structure that may be overcome by stresses above the yield stress (σ_y). The mechanical model element used to represent a plastic body is the St. Venant body or friction element. A simple model of viscoelastic material with a plastic component is the Bingham body, shown in Fig. 15. When the stress applied to this model is below the yield stress it reacts as an elastic body but at stress values beyond the yield stress two components function, one is constant (friction element) while the other is proportional to the shear rate (viscous flow element).

Yield stress is of practical importance in many industrial situations involving foods such as pumping and mixing, and to the quality of many products such as whipped cream and chocolate. However, theoretical rheologists are becoming inclined to consider this value at least partially a function of instrument sensitivity in that any material will flow under only the force of gravity if sufficient time is allowed (8; also see subsequent discussion of Deborah number). Although yield stresses are most often obtained by extrapolation of rheograms, Steffe (28) points out that instrumentally obtained yield stresses are defined by the technique employed and the assumptions implicit in the methodology. It is possible to differentiate static yield stress, obtained from measurement of samples at rest, from dynamic yield stress, often a lower value, obtained from stressed samples. Techniques useful for obtaining yield stress data are described by Tung et al. (29) and Steffe (30).

1. Effect of Time

Rheology also is concerned with time effects. As emphasized previously, the results of stressing a viscoelastic material depend on the amount of stress applied and the time for which it is applied; small stresses and short times may cause slow and reversible flow whereas larger stresses over longer periods can lead to nonelastic or permanent deformation. Note the difference between time effects in non-Newtonian fluid flow and viscoelastic materials. The former are inelastic and the time-dependent behavior results from structural changes but in the latter the stress–strain relationship is not instantaneous; it is possible for real materials to be both time-dependent and viscoelastic (28). In order for rheological tests to be valid the elastic limit of the material cannot be exceeded, i.e., it must be in the LVR. This requires that strain must be linearly related to the applied stress, the stress–strain ratio must be a function of the time for which the stress is applied and not the magnitude of the stress, and the strain after a stress is applied depends on the shear history of the sample (Boltzman superposition principle). While single-point static tests involving large deformations have been used widely in food texture studies, mainly due to time constraints and instrument cost, it is now being appreciated that the behavior of viscoelastic materials cannot be adequately described by these data alone. Selecting the operational parameters to be used for large deformation measurements must be done with care since the rate and degree of force application can have marked effects on the results. On the other hand, these tests have frequently been shown to correlate with sensory analysis. Small deformation testing has theoretical advantages for certain materials and yields data amenable to the generation of predictive equations. This subject will be discussed more thoroughly in a subsequent section.

Another aspect of time effects on rheology relates to the time scale of the experiment. Sample characteristics are defined by this parameter in that if an experiment is relatively slow, a viscoelastic sample will appear to be more viscous, but if the experiment is relatively fast, elastic behavior is observed. This raises the question as to whether a particular material is a "solid" or a "liquid." This issue is best resolved in terms of the Deborah number (D), a term coming from the observation that if a long enough observation period is used, everything will flow. The Deborah number is the ratio of relaxation time of a material to the time of observation. If D is small (short relaxation time, long observation time) the material flows and behaves like a liquid. If D is large (long relaxation time, short observation time) the material appears to be a solid.

There continues to be discussion on the degree of deformation that should be used in the testing of food materials. Large deformations are characterized by nonlinear effects and can lead to fracture; however, data from such studies can often be related to commercially important properties and masticatory studies. The degree of deformation is of particular importance in food testing since with increased deformation comes increased structural breakdown. Compressing an ideal elastic solid produces a deformation proportional to the force applied. When stress is removed the material returns to its original state because no permanent damage has occurred and the experiment can be performed repeatedly with the same results. The slope of the ideal stress–strain curve is thus a constant (the elastic modulus). With foods, however, recovery is often incomplete, indicative of permanent (plastic) deformation. The experiment cannot be repeated, so that the slope in not a constant because the proportionality limit has been exceeded. Therefore, the use of the elastic modulus is most often inappropriate with food materials, where the curve is frequently not linear above a deformation of as little as 0.1–1%. Above this point the data represent simultaneous events such as bond breaking and bond reformation, with structural breakdown being the ultimate outcome.

Another confounding result of working with foods is their heterogeneity. If foods are considered structurally to be areas or regions held together by a combination of long- and short-range bonding forces, then the common occurrence of incongruities cannot be overlooked. As well, foods can exist in differing geometries and many exhibit the property of anisotropy, which means that their mechanical properties differ depending on the orientation.

2. Viscoelasticity and Mechanical Analogs

Linear viscoelastic materials may be characterized by three forms of measurement: transient (i.e., creep and stress relaxation), dynamic (oscillatory) shear, and normal stress measurements. To help conceptualize behavior using these techniques, mechanical analogs composed of springs and dashpots similar to those developed for Hookean and Newtonian bodies are used; the simplest combinations are the Maxwell (in series) and Kelvin–Voigt (in parallel) elements (Fig. 15).

a. Transient Experiments. Two widely used static methods of measurement designed to study the behavior of viscoelastic materials as a function of time are relaxation and creep. In the former, the sample is subjected to constant strain and the decay of stress is measured over time. Creep may be defined as the slow deformation of a material, usually measured under constant stress. In creep analysis the sample is subjected to the instantaneous application of a constant load or stress and strain is measured over time.

STRESS RELAXATION   An instantaneous strain is applied (γ_0) and held constant, and the resultant decay in stress required to maintain this strain is measured as a function of time (Fig. 16). In a simple shear experiment, the viscoelastic response at time t is represented by the stress relaxation modulus $G(t)$:

$$G(t) = f(t) = \sigma(t)/\gamma_{constant} \tag{15}$$

$G(t)$ having units of Pa (Pascal) or N m^2. If the mode of deformation is either compression or tension then $E(t)$ is used. When a perfectly elastic solid is exposed to sudden deformation the stress remains constant, with $G(t)$ equal to the shear modulus (i.e., $G(t) = G = \sigma/\gamma$, Fig. 16). For a

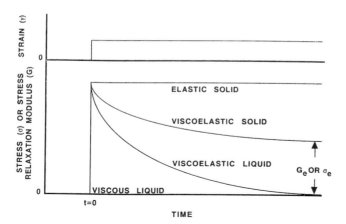

Fig. 16 Stress relaxation curves for an elastic solid, viscoelastic solid, viscoelastic liquid, and viscous liquid. (Adapted from Ref. 28.)

viscoelastic solid, the stress will decay to a finite value, the equilibrium stress value ($\sigma_e > 0$); subsequently, the shear modulus G will also reach equilibrium at $G_e > 0$. By contrast, for a viscoelastic liquid sufficient viscous flow allows the stress to decay to zero within the time frame of the test (Fig. 16). Viscous liquids are unable to maintain any stress in the absence of motion; subsequently, the stress relaxes instantaneously.

Alternatively, materials may be characterized by using an experimental value termed the relaxation time; defined as the time necessary for the stress to decay to σ/e (where e is the base of the natural logarithm) or 36.7% of its original value. Since this empirical measurement permits only limited interpretation of molecular and structural events, it is normally used for comparative assessment only. Each relaxation spectrum is characteristic for a particular material, the phenomenon of relaxation arising from molecular and structural reorientation as a consequence of an applied deformation. To help interpret stress relaxation data recorded for viscoelastic liquids, the Maxwell model is often used. In dealing with a single Maxwell element its response to constant strain is described by:

$$G(t) = G \exp(-t/\sigma_{rl}) \tag{16}$$

where σ_{rl} is the relaxation time. This simplistic approach is generally insufficient for describing viscoelastic behavior in real materials. A better interpretation is given when a number (i) of Maxwell elements arranged in parallel are used (Fig. 17). The stress relaxation is now given by:

$$G(t) = \Sigma_i\, G_i \exp(-t/\sigma_{irl}) \tag{17}$$

where G_i and σ_{irl} are the shear modulus and the relaxation time of the ith Maxwell element. In the case of a viscoelastic solid, a similar expression is used except that for one of the elements the relaxation time must be infinite such that the stress decays to σ_e (i.e., G_e) instead of zero.

From the number of elements (i) required to fit the data and the values calculated for each component (the moduli for springs and viscosity for dashpots) certain inferences can be made about structure. If, for example, the data are adequately described by three Maxwell elements in parallel, the assigned values for which are vastly different, then it is likely that each element represents a separate structural event. Alternatively, if the values are similar, the individual elements may represent separate stages of a multiphase process. The Kelvin–Voigt element (Fig.

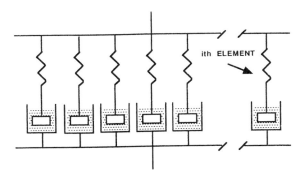

Fig. 17   Generalized Maxwell (Weichert) model with elements in parallel.

15) is considered unsuitable for stress relaxation studies because it cannot be made to deform instantaneously.

CREEP   A constant stress is applied instantaneously and the resulting increase in strain (creep) is monitored as a function of time (Fig. 18). Viscoelastic behavior (simple shear) is described in terms of a creep compliance function $J(t)$:

$$J(t) = f(t) = \gamma(t)/\sigma_{constant} \tag{18}$$

which has units of Pa^{-1} or $(N m^2)^{-1}$. The creep compliance of a perfectly elastic solid is theoretically the reciprocal of the shear modulus $(1/G)$, although experimentally this does not hold as a consequence of differing experimental time patterns. Again, differences in spectra are apparent for viscoelastic solids and liquids; in the former J will attain an equilibrium value J_e, whereas for the latter, J becomes a linear function of time. On removal of the stress at t_1, J decays to zero for a viscoelastic solid, whereas for a viscoelastic liquid a finite value is reached (Fig. 18). This permanent deformation arises through molecular rearrangement. From the degree of recovery we can measure the function $J_r(t)$ (i.e., creep compliance recovery):

$$J_r(t) = J(t) - J(t - t_1) \tag{19}$$

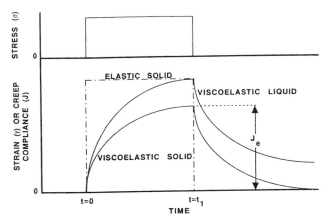

Fig. 18   Creep compliance curves for an elastic solid, viscoelastic solid, and viscous liquid. (Adapted from Ref. 10.)

As for the stress relaxation measurement, material creep can be represented by mechanical models:

$$J(t) = t/G + t/\eta \tag{20}$$

is used to describe a single Maxwell element and

$$J(t) = 1/G\,(1 - \exp{-t/\sigma_{rt}}) = 1/G\,[1 - \exp(-t/\sigma_{rt})] \tag{21}$$

for a single Kelvin–Voigt element, where σ_{rt} is the retardation time (i.e., η/G). Similarly, the latter may be expanded for a number (i) of Kelvin–Voigt elements in series (Fig. 19):

$$J(t) + \Sigma_i\, 1/G_i\,[1 - \exp(-t/\sigma_{irt})] \tag{22}$$

giving a better representation of the response of a real body to constant stress.

Although use of a discrete number (i) of elements gives a satisfactory interpretation of viscoelastic behavior, further improvements toward a more realistic response may be achieved by considering an infinite number of elements. Equation (17) can now be written as:

$$G(t) = G_e + \int_{-\infty}^{+\infty} H\,(\sigma_{rt})\,\exp(-t/\sigma_{rt})\,d\ln\sigma_{rt} \tag{23}$$

where $H(\sigma_{rt})$ is the relaxation spectrum and G_e the equilibrium shear modulus for the spring element that does not decay to zero (i.e., $G_e > 0$). For a more detailed account of the derivation of these and other mechanical models, the text by Ferry (25) is recommended.

The mechanical analog approach to quantitatively elucidating material structure is strongly enhanced when coupled with data from complementary techniques such as microscopy. Shama and Sherman (31) were early proponents of this; their characterization of ice cream by creep analysis is often used as an example of the potential of rheological methods. They derived a six-element model, assigning each element to a particular structure, e.g., fat crystals and stabilizer

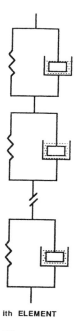

ith ELEMENT

Fig. 19   Generalized Kelvin–Voigt (Becker) model with elements in series.

gel. By adjusting temperature and overrun, Shama and Sherman were able to calculate values for each of the constants associated with the six elements. Although this procedure allows quantification of the time scales of molecular interactions, the use of an infinite number of elements as described in Eq. (22) may be considered a more realistic approach because few molecular motions are truly independent.

For both types of transient experiments the stresses and deformations used should be kept small so as to ensure that behavior is from within the LVR, a point of consideration that is easily overlooked. The advent of reliable controlled stress or strain rheometers has made creep and stress relaxation testing more appropriate for food materials. Also, computerized methods for analyzing creep behavior has reduced the calculation procedure from hours to minutes (32).

It is noted that for these two methods the loading is linear, i.e., a single mode of deformation is used. This one-dimensional approach is often not adequate to deal with the complex nature of viscoelasticity in nonideal materials. Creep and stress relaxation tests usually cover a frequency range of only ~1 Hz (cycle/sec) and below. Also, it is now recognized that in order to obtain a thorough understanding of viscoelasticity, data must be obtained over a wide range of deformation times during which biological changes can occur in the sample.

b. Dynamic Experiments. The development of instrumentation capable of applying a dynamic sinusoidal stress or strain makes it possible to obtain a more thorough understanding of viscoelastic behavior during a reasonably short experimental time and so obtain data that complement creep tests that only provide results over a longer period. In considering the application of sinusoidal stress or strain to a viscoelastic material it will be remembered that for an ideal Hookean solid, stress is directly proportional to strain and vice versa. This usually holds for viscoelastic food materials if the deformation is small, i.e., if it is in the LVR. In dynamic testing, commonly referred to as periodic or oscillatory, the test material is exposed to a sinusoidally varying stress or strain so that deformation varies harmonically with time (Fig. 20). Whether the controlled input is stress or strain depends on the type of instrument used. The resultant strain or stress amplitude and phase angle (δ, the phase shift between input and output; degrees or radians) are measured and from these separate elastic and viscous components are derived. The frequency (ω) of oscillation (measured in Hz or rad; $\sec^{-1}$) is analogous to the shear rate in a steady-shear experiment and can be varied to access information over a wide range of time scales. Compared with transient (or nonperiodic) experiments where data are collected over comparatively long periods of time, ranging from minutes to hours, dynamic testing is able to access information over considerably shorter time scales (at 10 Hz the time period for one oscillation is 0.1 sec). Qualitatively, data from a periodic experiment run at a frequency ω are equivalent to a transient (nonperiodic) test at time $t = 1/\omega$.

Dynamic measurements yield a number of viscoelastic parameters and the derivation of these will now be considered. For a strain-controlled rheometer one part of the fixture is stationary and the other surface is subjected to oscillatory motion where the variation in strain with time (i.e., simple shear) can be described by:

$$\gamma = \gamma_0 \sin \omega t \tag{24}$$

where γ_0 is the maximum strain amplitude (Fig. 20). The corresponding stress (σ generated within the material is determined by:

$$\sigma = \gamma_0 (G' \sin \omega t + G'' \cos \omega t) \tag{25}$$

The resultant stress comprises two frequency-dependent functions, one corresponding to the amount of stress in phase (i.e., $\delta = 0°$) with the strain [$G'(\omega)$] and the other out of phase (i.e., $\delta = 90°$) with the strain [$G''(\omega)$]. These two parameters are respectively known as the storage (G') and

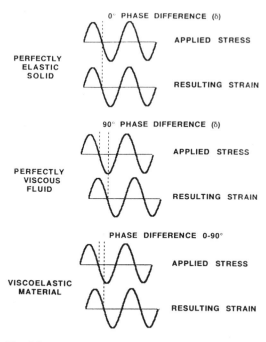

Fig. 20 Schematic response of an elastic, viscous, and viscoelastic material of dynamic (oscillatory) shear. (Adapted from Ref. 10.)

loss (G'') moduli. These terms represent the amount of energy elastically stored and recovered per cycle of deformation (elastic component, G') and the amount of energy lost per cycle by viscous dissipation (G''). For a perfectly elastic body the stress will be in phase with the strain (i.e., $\delta = 0°$) since all energy is stored and recovered; consequently G'' will be zero. By comparison, for a perfectly viscous (Newtonian) body where the stress is out of phase (i.e., $\delta = 90°$) the opposite applies; all energy is dissipated viscously and as a result $G' = 0$. Viscoelastic bodies lie between these two extremes with the phase angle between 0° and 90° (Fig. 20), with the amount of energy stored being a function of the materials structure.

Equation (14) can now be rewritten incorporating the phase angle (δ) by which the stress precedes the strain:

$$\sigma = \sigma_0 \sin(\omega t - \delta) \tag{26}$$

or

$$\sigma = \sigma_0 \cos \delta \sin \omega t - \sigma_0 \sin \delta \cos \omega t \tag{27}$$

where σ_0 is the maximum stress amplitude (Fig. 20). From Eqs. (14) and (16), G' and G'' can be derived:

$$G' = (\sigma_0/\gamma_0) \cos \delta \tag{28}$$
$$G'' = (\sigma_0/\gamma_0) \sin \delta \tag{29}$$

where $G''/G' = \tan \delta$; $\tan \delta$ is commonly referred to as the loss tangent. This is a measure of energy lost compared with energy stored per cyclic deformation; for a predominantly elastic material $\tan \delta < 1$ whereas for a liquid $\tan \delta > 1$. Storage and loss moduli are measured in units of Pascals or N m^2. Additional parameters which can be calculated include the complex modulus (G^*):

$$G^* = \frac{\sigma_0}{\gamma_0} = \sqrt{(G')^2 + (G'')^2} \qquad (30)$$

which is the ratio of the total stress to the strain (both in-phase and out-of-phase stress), and the complex viscosity (η^*):

$$\eta^* = \frac{G^*}{\omega} = \sqrt{(\eta')^2 + (\eta'')^2} \qquad (31)$$

which is the ratio of the total stress to the frequency of oscillation. The in-phase (η') and out-of-phase (η'') dynamic viscosity components of η^* are given by:

$$\eta' = G''/\omega \qquad (32)$$
$$\eta'' = G'/\omega \qquad (33)$$

Both dynamic viscosities and the complex viscosity have units of Pa · s or N m^2·sec.

As with transient experiments, mechanical analogs can be fitted to dynamic data. The viscoelastic response in terms of G' and G'' for the Maxwell and Kelvin–Voigt elements to sinusoidally varied strain is given by:

$$G'(\omega) = G\omega^2\sigma_{rl}^2/(1 + \omega^2\sigma_{rl}^2) \qquad (34)$$
$$G''(\omega) = G\omega\sigma_{rl}/(1 + \omega^2\sigma_{rl}^2) \qquad (35)$$

and for a single Maxwell element and for a Kelvin–Voigt element by:

$$G'(\omega) = G \qquad (36)$$
$$G''(\omega) = \omega\eta \qquad (37)$$

An advantage of using such mechanical analogs to describe experimental data is that it permits the establishment of a framework within which measurements, in this case viscoelastic, can be compared for real systems. Nevertheless, there are flaws to this approach; models tend to fall short when it comes to linking mechanical elements with specific structural and molecular components. Generally, model parameters obtained through curve fitting, such as the decay constants assigned to an exponential stress relaxation process, reveal little about actual structural events. An alternative approach is to compare the behavior of real with those of model systems (e.g., crosslinked synthetic polymers) in which molecular interactions and structure have previously been established through complementary techniques. For example, Ferry (25) evaluated the viscoelastic behavior of synthetic polymers with various degrees of branching and crosslinking. Based on this work, Morris and Ross-Murphy (33) extended this technique to biopolymers such as gels as well as dilute and concentrated solutions. Viscoelastic responses of these systems were illustrated using mechanical spectra, i.e., frequency dependence plots of G', G'', and η^*. Mechanical spectra, along with their structural interpretation, will be dealt with subsequently.

Mechanical spectra, typically recorded as a log-log plot, cover three to four decades of frequency and most commercial instruments operate from around 0.01 to 10 Hz. It is possible to extend this range by applying the principle of time–temperature superposition, a qualitative approach based on the principle of viscoelastic corresponding states. For constitutive equations [e.g., Eq. (14)] inclusion of higher terms such as temperature and pressure is possible. Normally, for a frequency-dependent test, these terms along with strain would be held constant. However, since temperature and frequency of oscillation are related by the above principle, one can be effectively controlled through the other. In practical terms, lowering the sample temperature is equivalent to increasing test frequency. The result is a plot in terms of reduced variables that when used correctly allows prediction of the viscoelastic response over many decades of frequency. The application of the time–temperature superposition principle, however, is somewhat limited to materials that do not undergo compositional or conformational changes as a result of temperature shifts. Unfortunately, this excludes the bulk of food materials, so that this approach tends to be reserved for synthetic polymers for which it was originally developed. In spite of this problem, attempts to apply temperature superposition to foods have met with some degree of success. The Dea et al. (34) study of ice cream is one example.

Frequency dependence measurements should be performed at a strain (in the case of a strain-controlled rheometer) or stress (in the case of a stress-controlled rheometer) within the LVR of the material. In order to establish the extent of this region the strain or stress dependence of the material must first be established. If the subsequent mechanical spectra are to cover several or more decades of frequency, then strain or stress dependence measurements should be conducted at the extremes of this range to verify linear behavior. Within the LVR, G' and G'' remain constant (Fig. 21); departures from linearity at excessive stress or strain, as in the case of G' decreasing suddenly, are viewed as a deviation from the previous level response. The strain limit (or critical strain) of the LVR is characteristic of a material; for colloidal dispersions it extends to 0.01 (1%), while for gels and biopolymer solutions it is not uncommon to observe linear behavior up to 0.5–1.0 (50–100%) strain (26, 27). An example of the significance that the critical strain holds in itself for food material functionality is the spreadability of margarine. At low γ (i.e., during storage), G' is high and important for shape retention, whereas at higher γ, beyond the LVR, G' is low and necessary for spreading (26). Some typical strain limits for a variety of foodstuffs are given in Table 3.

It is worthwhile to reemphasize the necessity of varying the frequency when attempting to characterize the viscoelastic behavior of an unknown material. If this is not done, the situation is similar to measuring viscosity at only one shear rate or shear stress. The need to characterize materials using multipoint, i.e., calibrated over a range of shear rates or frequency of oscillation, rather than single-point methods is an important concept in rheology. Figure 22 shows the reason for this graphically. If only one rate of deformation is used it is quite likely that incorrect conclusions will be made because foods most often exhibit non-Newtonian (i.e., shear rate–, shear stress–, or frequency-dependent) behavior.

To summarize, dynamic or oscillatory testing consists in essence of applying a small perturbation in the form of a sinusoidal stress or strain pattern and analyzing the resulting sinusoidal strain or stress output by examining the degree to which it is out of phase with the input wave. The objective is to test various structures within the sample nondestructively. By varying the frequency of the oscillatory input it is possible to probe the numerous elements contributing to structure. Modern

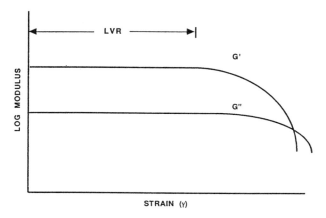

Fig. 21   Strain dependence experiment at constant frequency to determine LVR. Unpublished data.

Table 3 LVR Strain Limits for Various Food Materials

Material description	Strain limit of LVR
Ovalbumin gel (10% w/w)[1]	0.02
Butter (5°C)[2]	0.001
Cocoa liquor (40°C)[3]	0.002
Dextran solution (15% w/w)[3]	0.10
Dextran solution (20% w/w)[3]	0.06
Ice cream (−10°C)[3]	0.0002
Salad dressing[4]	0.05
Whipped cream (10°C)[3]	0.0005

Sources: [1](64); [2](81); [3]unpublished data; [4](82).

dynamic testing procedures are of growing importance to the food analyst because (a) they provide information not possible using large deformation tests since these methods are particularly sensitive to such events as glass transitions, crosslinking, phase separation, molecular aggregation, etc.; (b) they are nondestructive in nature, give results in basic units, and theory exists to which these results can be compared. This is in contrast to the vast majority of textural evaluations carried out on food, which are of an empirical nature (10); (c) they are well suited to viscoelastic food materials because instrumental output can be separated into a viscous and an elastic component and the sample can be tested over a wide frequency range; (d) they have predictive capabilities and are therefore of interest in research and development. Some studies have shown significant correlations between dynamic measurements and sensory analysis [see Richardson et al. (35) for an example]. Also, the physical quantities measured can lead to structural interpretation [see Rha (36) for examples] and processes such as gel formation can be monitored in real time.

    c. Normal Stress Measurements. The lack of suitable commercial rheometers in the past meant that normal stress measurements were somewhat overlooked. However, the advent of more

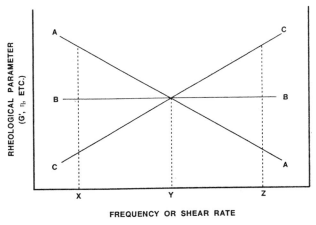

Fig. 22 Influence of changing deformation rate on non-Newtonian or nonlinear viscoelastic behavior. If three different materials (A, C—nonideal; B—ideal) are tested at three different rates of deformation (frequency, shear rate, etc.) in order to determine a rheological parameter (storage modulus, viscosity, etc.), then the perceived result for the nonideal materials will depend on the deformation rate selected (X, Y, Z). This emphasizes the necessity of using multipoint rather than single-point tests. (Adapted from Ref. 37.)

sophisticated instrumentation has generated interest in normal stress behavior, although for food materials examples of such measurements are still uncommon. Normal stresses occur during finite deformation under conditions of simple shear acting at right angles to the direction of flow. In the case of a cone-and-plate geometry where the cone is rotated (steady shear), the lower plate will experience a downward force in addition to the induced torque. In viscoelastic materials it is the elastic component that is responsible for the development of normal stresses. Characterization of this effect is by the normal stress coefficient (Ψ) and normal stress (N_1) measurement. For a cone-and-plate geometry they are described by:

$$\Psi = 2F/(\pi R^2 \dot{\gamma}^2) \tag{38}$$
$$N_1 = \dot{\gamma}\Psi(\dot{\gamma}) \tag{39}$$

where F is the normal force and R the cone radius. Two frequently used examples of normal stress behavior are the Weissenberg effect, where a viscoelastic liquid is observed to climb a stirring rod, and die swell, where upon exiting a die an extruded material is seen to expand suddenly. The Weissenberg effect has been reported for numerous food materials including cake batter, aged condensed milk, wheat flour dough, malt extract, and eucalyptus honey (37). Kokini and Plutchok (24) applied the measurement of normal stresses to foods.

3. Effect of Temperature on Viscoelastic Behavior

At this point it is important to address the action of temperature on the mechanical behavior of viscoelastic materials. Because the science of rheology was developed primarily for the testing of polymers, many of the terms found in the literature refer to these materials. In particular, polymers are classified as either glassy, brittle solids (sometimes defined as capable of supporting their own weight against flow due to the force of gravity), or in a transition state characterized as elastic or rubbery, or as viscous liquids that will flow in real time. A single polymer can exhibit all of these mechanical behaviors depending on the temperature and time scale of measurement. It is often useful to consider polymers as models for real foods, even though there are many obvious fallacies in equating these two disparate materials. Perhaps the most widely recognized thermal event occurring in polymers is the glass transition, which takes place at a temperature characteristic of a given material (T_g). This is the point below which flexible polymer molecules become inflexible and conformational changes can occur only slowly as a result of a marked change in the thermal expansion coefficient and free volume. The latter term can be used as the basis for explaining the glass transition. The specific volume of a polymer (total volume/g) consists of a large portion of occupied volume and a small portion of free volume not occupied by the polymer. This free volume represents holes of molecular dimensions or smaller voids associated with packing irregularities. While the occupied volume increases linearly with temperature, free volume undergoes a rapid expansion at T_g. The T_g term is also associated with abrupt changes in other physicochemical properties such as refractive index, permeability, dielectric constant, thermal conductivity, and specific heat.

In considering the influence of temperature on the rheological behavior of polymers, data for storage compliance (J') can be obtained as a function of temperature and frequency. Examination of such an experiment shows that the shape of the compliance–frequency curve changes with temperature. At low temperatures the compliance is low, suggesting a glassy structure, and changes little with frequency, whereas at high temperatures the compliance is high, suggesting a soft rubbery solid, and again changes little with frequency. In the intermediate temperature zone, viscoelastic behavior is observed and there is usually a pronounced frequency-dependent behavior, characteristic of conformational change. From these data a so-called master curve may be constructed by empirically shifting data obtained at different temperatures along a logarithmic frequency axis to a selcted temperature. In order to convert data obtained at one frequency and temperature to another frequency and temperature, a correction or shift factor (a_T, the ratio of relaxation times at two different temperatures) is applied. Thus, from the original data in which rheological responses are seen as a function of both temperature and frequency, it has been possible to separate these two effects into frequency and temperature alone. It is important to remember that this transformation and the interpretation of data derived from it are predicated on the assumption that no structural or compositional changes occur over the temperature range in question. With food materials this premise is often incorrect.

An important observation from experiments of this type is that the influence of temperature is similar for many amorphous polymers and glass-forming liquids. This led to the development of the WLF (Williams, Landel, Ferry) equation (25):

$$\log a_T = \frac{C_1(T - T_S)}{C_2 + (T - T_S)} \tag{40}$$

where C_1 and C_2 are constants and T_S is the reference temperature. It has been shown subsequently that the original WLF equation [Eq. (40)] can be cast in terms of the glass transition temperature:

$$\log a_T = \frac{C_3(T - T_g)}{C_4 + (T - T_g)} \tag{41}$$

where C_3 and C_4 are new constants and T_g is the glass transition temperature. It will be appreciated that changes in temperature would be expected to have more of an effect on the viscosity component than the elastic component. By ignoring the elastic modulus, the shift factor can be equated to the ratio of the viscosities at the experimental temperature to the glass transition temperature, i.e., $a_T = \eta_T / \eta_{T_g}$. The WLF equation is important because it has been found that Arrhenius kinetics do not apply to polymer behavior such as viscosity in the viscoelastic state. Rather, a larger temperature dependence is observed and this is described by the WLF equation. Figure 23 illustrates the discontinuity in Arrhenius kinetics for a typical polymer. While Arrhenius kinetics are appropriate below T_m and above T_g, the range between these two temperatures is characterized by WLF kinetics with a larger temperature dependence. This effect is discussed in greater detail by Ferry (25).

4. Instrumentation for Measuring Viscoelasticity

A new generation of rheometers has made measuring the mechanical characteristics of food much easier and more precise. This section will attempt to describe some of the basic features of the devices that are currently available.

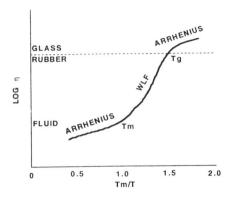

Fig. 23 Viscosity (h) as a function of reduced temperature (T_m, transition temperature from rubberlike to glasslike consistency; T_g, glass transition temperature) for typical polymer. (Adapted from Ref. 87.)

a. Measuring System Geometry.   Raw data generated by rheometers generally take the form of force and deformation measurements. In order to derive the previously described fundamental rheological parameters (e.g., storage and loss moduli) the force/deformation ratio must be converted to stress/strain using form factors. Since the calculation of these factors has been discussed elsewhere (25), the following will only include modifications to existing form factors so that they are applicable to viscoelastic techniques.

As with steady shear experiments, the three basic measuring system geometries (i.e., concentric or coaxial cylinders, cone-and-plate, and parallel plate fixtures) are employed in viscoelastic studies on liquid and semisolid food materials (Fig. 3). The same restrictions for steady shear imposed by the nature of the sample (e.g., particulate material bridging the geometry gap) apply to viscoelastic measurements. However, when sample thickness becomes greater as is the case in large-diameter fixtures and when greater separation distances are used, secondary flows can take place; flow patterns have been demonstrated to be extremely complex even for the limiting cases of low frequency and high viscosity. A thorough examination of such secondary flow patterns is covered by Walters (38). Further modification to equations of stress and strain may be required when used in the oscillatory mode, since geometry, instrument inertia, and material inertial effects are potential problems that can preclude obtaining accurate measurements. Good rheometric practice requires that tests be repeated in two or more fixture configurations, with corrections applied to the results as necessary and testing of their concurrence before pooling the findings.

b. Rheometers.   Currently, there are a number of commercial rheometers designed for measurement of viscoelastic parameters. These machines fall into one of two categories in their mode of operation; they are either strain- or stress-controlled. In the case of the former the extent of deformation applied to a material is regulated and the resulting induced stress is measured, whereas for the latter the reverse applies. Examples of strain-controlled machines include the Weissenberg rheogoniometer, the Rheometrics mechanical spectrometer, and the Bohlin VOR (viscous–oscillatory rheometer) while the stress-controlled version is represented by the Carri-Med CS and the Bohlin CS (controlled stress) and Krusse (formerly Rheo-Tech International) rheometers.

Both types of machine are essentially similar in design using a rotating fixture or rotor to transmit a shearing deformation/force to the material, with the resulting force/deformation generated within the sample being measured through a stationary fixture. Where the instruments do differ, however, is in the mechanism by which rotational motion is controlled. In the strain-controlled variant the rotor is either connected directly by a drive shaft to a motor in which case a torsion bar/spring is attached to the other fixture (Fig. 24) or, alternatively, when motion and torque sensing are combined, a torsion dynamometer is incorporated within the drive shaft. By comparison, the stress-controlled instrument employs a low-friction air bearing to support the drive system so that all the torque, which is applied through an induction motor, is also transmitted to the sample (Fig. 24). Deformation may be measured by a proximity transducer. Both types of machine drive systems are capable of running in oscillatory mode. However, transient tests are specific to the mode of operation; a strain-controlled rheometer is used for stress relaxation and the stress-controlled version for creep. Additional information on rotational rheometers can be obtained from the reviews by Shoemaker et al. (2) and Brownsey (39).

III. APPLICATION OF VISCOELASTIC MEASUREMENTS TO FOODS

Because so many foods and processing operations associated with foods are related to viscoelasticity, and as a result of previously mentioned advances in both instrumentation and theory to deal with viscoelasticity in foods, many researchers are now considering viscoelastic measurements.

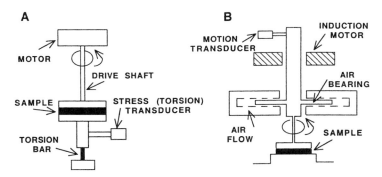

Fig. 24 Schematic cross-sectional representation of (A) strain and (B) stress-controlled rheometers with parallel plate geometry. (Adapted from Ref. 88.)

The following section will provide examples of various fields to which these procedures are applicable. Other areas of practical interest not mentioned include subjects such as mixing, pumping, coating, sagging, spreading, filling and dispensing.

A. Gelation Studies

The field of protein and carbohydrate gelation has received much attention in recent years, primarily due to advances in rheometer design and increased commercial availability. Theoretical treatises describing polymer gelation were published in the 1940s and 1950s by Flory (40,41) and Stockmeyer (42,43). Subsequent studies relating the mechanisms and kinetics of network formation along with equilibrium gel structure to fundamental rheological measurements were conducted in the following years by numerous researchers (25,44–48). Protein and polysaccharide macromolecules were often the choice of gelling species in developing these models. Theoretical aspects of gelation are beyond the scope of this chapter and the reader is advised to consult the comprehensive reviews of Clark and Ross-Murphy (49), Ziegler and Foegeding (50), and Clark (51).

Advances in viscoelastic measurement methodology have prompted researchers to investigate experimentally the process of gelation in addition to final gel characterization. Of the three rheological measurements (transient, dynamic, and normal), dynamic measurements are favored in monitoring gelation because the strain (or stress) can be kept sufficiently small, minimizing mechanical input and the effect it has on network development. Two types of oscillatory shear experiments are typically used for this: isothermal (constant) temperature and dynamic (scanning) temperature.

1. Isothermal Experiments

In this case samples are held at a constant temperature while changes in shear moduli (i.e., G', G'') are recorded as a function of time. The frequency, typically set at 1 Hz for practical reasons, and strain (or stress), set within the LVR, are also held constant. The bulk of earlier studies were predominantly of this type, also known as gel cure experiments. A typical plot of storage and loss moduli (G' and G'', Pa) against time for a constant temperature experiment is shown in Fig. 25. Initially $G'' > G'$, as would be expected for a viscoelastic fluid. After some time the bonding forces in the matrix of macromolecules prevent movement and G' increases rapidly (increased elastic response), such that $G' > G''$; this point is often referred to as the gel point, the lag period prior to this event being termed the gel time (t_g). Various authors have considered theoretical determination of the gel point; see, for example, Refs. 52–55. Gel time (t_g), along with the magnitude of G' in

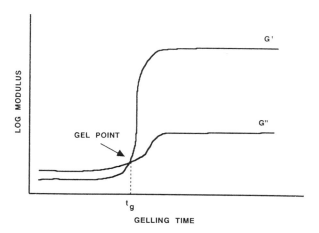

Fig. 25 Isothermal or constant temperature experiment. Changes in storage (G') and loss (G'') moduli with time at constant strain and frequency. (Adapted from Ref. 49).

the plateau region, is concentration-dependent, becoming shorter and larger, respectively, with increased concentration of the gelling species (56).

The G' plateau or equilibrium modulus value and its concentration dependence has been considered by numerous authors who have attempted to adapt it to classical theories of gelation and rubber elasticity (44,45,57). From these theories it is possible to determine the critical concentration required before continuous network formation can proceed.

Isothermal experiments have been used to determine kinetic parameters. Oakenfull and co-workers (58,59) used a kinetic approach to determine the molecularity of a gelling species from gel time to concentration data. Improvements to this model were later made by Clark (60) and Ross-Murphy (56). Other kinetic studies are reported by te Nijenhuis (61) and Djabourov and her co-workers (62,63).

2. Dynamic Temperature Experiments

In these experiments, temperature is increased at a predetermined fixed rate while rheological parameters are monitored. One of the main advantages of this technique over isothermal conditions is that transition temperatures can be singled out. However, such transitions cannot be considered absolute because insufficient time (dictated by the heating rate) is allowed for equilibrium to be established at any one temperature. When comparing transition values from other sources or with different techniques (e.g., DSC) attention should be paid to the heating rate, if different. As with isothermal experiments, the frequency and strain (or stress) are held constant. When setting the strain for both isothermal and dynamic experiments, magnitudes must be such that conditions of linearity are met at all stages of network development from the initial viscoelastic fluid through denaturation (in the case of proteins), aggregation, and finally gelation. Since the point of gelation and the arising network development is usually the focus of these experiments, strain (or stress) levels are set so as to minimally influence these events. These strains are, however, not optimal for detecting viscoelastic changes in the heated sol, but this is usually of secondary importance. Figure 26 is a generalized mechanical thermogram (G' and tan δ) for heat-set myofibrillar proteins. Thermal transitions, in particular the onset of network development, can usually be more accurately detected from these data by using a derivative plot of the rheological parameter with respect to temperature (i.e., dG'/dT).

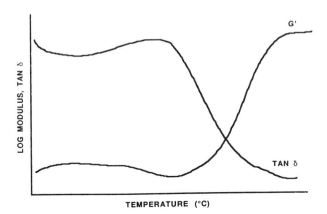

Fig. 26 Dynamic or scanning temperature experiment. Changes in storage modulus (G') and tan δ with increasing temperature at constant strain and frequency for a myofibrillar protein preparation. Unpublished observations.

As with isothermal experiments, equilibrium G' values for cooled gels have been used in evaluating network structure. One example is that of Van Kleef et al. (64) who, using theories developed by Flory (41) and Ferry (25), calculated the number of crosslinks per molecule for ovalbumin gels. Similar studies on meat and milk gels were conducted by Blanshard and Derbyshire (65) and Culioli and Sherman (66), respectively, applying the theory of rubber elasticity developed by Treloar (67).

3. Final Gel Characterization

Once gel formation has reached equilibrium and crosslink formation has ceased, either at the temperature of formation or in a cooled system, gels can be characterized using both transient and dynamic experiments. For the latter, the strain and frequency dependence of the gel is measured. For most biopolymer gel systems the LVR rarely extends beyond $\gamma = 1$ before nonlinearity is observed. The magnitude of G' within the LVR and the strain limit of this region are influenced by gel structure, e.g., number and nature (covalent or physical) of molecular interactions.

Plots of rheological parameters as a function of frequency are known as mechanical spectra and some typical values for G', G'', and tan δ for various gels are given in Table 4. For a permanently crosslinked gel, both moduli show little if any frequency dependence, G' being several orders of magnitude greater than G'' (Fig. 27). The degree of this dependence is again characteristic of gel structure, particularly the nature of molecular crosslinking. For example, in a gel in which covalent bonds (e.g., disulfide bridges) predominate, G' and G'' vary only slightly with the frequency of oscillation as crosslinks remain intact even over long time scales. By comparison, a physical gel (one in which physical interactions such as H–bonds crosslink molecules) will display increased dependence, particularly at low frequencies, as gradual network rearrangement through new bond formation occurs in response to deformation (68).

Egelandsdal et al. (69) and, later, Stading and Hermansson (70) used the following simple relationship to determine the type of bonding:

$$\log G' = p \log \omega + q \tag{42}$$

where q is a constant and p is the slope of the log G' – log ω plot. For a covalent gel $p = 0$, whereas a physical gel has $p > 0$. Examples of values for p in various gelling species are given in Table 5.

Table 4 Rheological Parameters (G', G'', and tan δ) Taken from
Mechanical Spectra for Various Food Gels (Frequency 0.1 Hz)

Gelling species	G'(Pa)	G''(Pa)	tanδ
Agar (1% w/w)[1]	50,000	3,500	0.07
Starch gels (6.5% w/w)[2]:			
Maize	141	7.9	0.056
Barley	53.9	10.4	0.192
Potato	16.7	5.5	0.331
Ovalbumin (10% w/w)[3]	2,240	110	0.05
Myofibrillar protein (7% w/w)[4]	2,500	320	0.128

Sources: [1](47); [2](83); [3](64); [4](84).

Transient experiments such as stress relaxation and creep have been used to test the perma-
nence of physical interactions, with time scales for the application of constant stress or strain of
around 10^2–10^6 sec typically being used. In addition to dynamic shear techniques, Van Kleef et
al. (64) examined the stress relaxation behavior of ovalbumin gels. Calculation of the degree of
molecular crosslinking using the equilibrium stress relaxation modulus, G_t, taken after 60 h gave
better values than those obtained from G' (storage modulus) from dynamic measurements. This
was attributed to a contribution from noncovalent bonds that persist over short time scales but
are negated when a sample is deformed, as in the case of a transient experiment, for a considerable
length of time.

B. Biopolymer Solutions

Biopolymers such as polysaccharides are often used as thickeners due to the profound effect they
exert on solution behavior even at low concentrations. This desirable attribute has prompted many
researchers to focus their efforts on the field of hydrocolloid rheology. As with polymers in general,
much of the groundwork for establishing theoretical treatises governing the hydrodynamic
behavior of polymers in solution was conducted by scientists in the plastics industry. Over the last
20 years or so, these theories have been extended frequently to food polymers, particularly

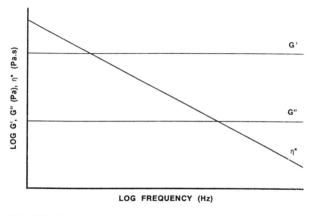

Fig. 27 Typical mechanical spectrum for a biopolymer gel. G', G'', and η^* plotted against frequency.
(Adapted from Ref. 68.)

Table 5 Frequency Dependence (p Value) of Various Gels

Gelling species	Conditions	p value
β-Lactoglobulin (12% w/w)[1]	pH 3.0	~.038
	pH 5.5	~.057
	pH 7.5	~.040
Myosin (1% w/w, 0.6 M)[2]	50°C	~.075
	60°C	~.095
	70°C	~.038
Myofibrillar protein (0.01 M, pH 7.4)[3]	3.8% w/w	.105
	4.3% w/w	.101
	4.8% w/w	.059
	7.0% w/w	.082

Sources: [1](70); [2](69); [3](84).

polysaccharides; the writings of Morris (68), Mitchell (71), and Robinson et al. (72) are but a few examples worth consulting.

The majority of these theories have been developed for streamline flow situations such as those encountered in steady shear. An example of this is the Mark–Houwink relationship, $[\eta] = KM\alpha$, where $[\eta]$ is the intrinsic viscosity (measured by very low steady shear methods), M the molecular weight, and K and α are parameters related to chain stiffness (73,74). Similarly, for a solution of a random coil polymer, the concentration at which coil overlap occurs (i.e., the critical concentration, c^*) is determined using steady shear measurements and is estimated from the change in slope of a log-log plot of specific viscosity (η_{sp}) against concentration. At the critical concentration a transition from dilute to concentrated behavior ensues. Certain polysaccharides, such as dextran, also exhibit a semidilute region (75).

This transition at c^* is also evident from frequency dependence plots of G', G'', and η^*. Figure 28 represents generalized mechanical spectra for dilute and concentrated hydrocolloid solutions. In these solutions, storage of energy is mainly a result of elastic deformation of the polymer chains. Dilute behavior appears essentially Newtonian; η^* remains constant over the frequency range accessed, whereas G' and G'' increase linearly with frequency. At low frequencies $G'' \gg G'$, energy being predominantly lost through molecular translational motion, whereas at higher frequencies G' approaches G'' as a greater proportion of energy is stored via strained conformational changes to the polymer chain. It should emphasized that there are limitations to using oscillatory techniques with dilute solutions, the most commonly encountered problem being detection of the extremely low stresses induced in the material.

Turning to the concentrated solution, it is seen that η^* is initially constant but decreases as higher values of ω are approached. This is analogous to the low-shear Newtonian plateau and the subsequent shear-thinning behavior observed in steady shear experiments (Fig. 4). In fact, the two curves (i.e., η^* or η' vs. ω, and η vs. $\dot{\gamma}$) are closely superimposable, an observation first reported by Cox and Merz from whom this frequently used empirical approach now takes its name (76). However, as with most rules, there are exceptions, e.g., gum kara (24); also Bistany and Kokini (77) indicated that a number of different fluid food materials did not obey this relationship.

Examination of the storage and loss moduli (Fig. 28) indicates that at low frequencies the behavior is characteristic of that of a dilute solution, i.e., $G'' \gg G'$; when the time scale of deformation is sufficiently long, chain entanglements are disrupted through translational motion. By contrast, at higher values of ω these entanglements remain in place as there is insufficient time for molecular rearrangement within the period of one oscillation. The outcome is a response resembling that of a crosslinked system (i.e., $G' > G''$) such as a gel, where both moduli appear almost

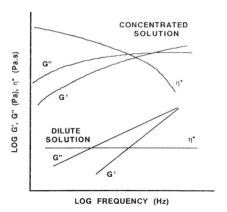

Fig. 28 Typical mechanical spectra for dilute ($c < c^*$) and concentrated ($c > c^*$) biopolymer solutions. G', G'', and η^* plotted against frequency. (Adapted from Ref. 68.)

independent of frequency. The point of crossover between G'' and G' has itself been used as a parameter for comparing systems, shifting to lower frequencies as the level of coil overlap increases.

Previously it was demonstrated how the accessible frequency range of a rheometer could be extended by using the time–temperature superposition principle. Similarly, by adjusting solution concentration a master curve may be constructed spanning many decades of frequency. This approach, referred to as frequency–concentration superposition, entails rescaling the moduli by dividing by concentration and laterally shifting the data along the frequency axis relative to the lowest concentration. The result is a log-log plot of reduced frequency (ω_r) against G'/c and G''/c (Fig. 29). Use of this method of reduced variables allows a prediction of moduli response at frequencies beyond the scope of the rheometer but that might be encountered, for example, during processing.

C. Dispersions and Suspensions

There are numerous examples of food systems in which one phase is dispersed or suspended in another, the most common being the suspension of particulate matter in an aqueous phase, e.g.,

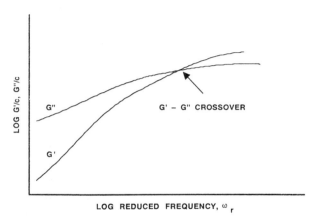

Fig. 29 Frequency–concentration superposition. Log–log plot of reduced frequency (ω_r) against G'/c and G''/c. (Adapted from Ref. 26.)

starch pastes, fruit and vegetable purees and juices, soups, and salad dressings. The viscosity of the suspending medium (continuous phase) in aqueous suspensions may also be modified by use of high molecular weight macromolecules such as hydrocolloids to enhance product stability and texture. The term *dispersion* also extends to encompass foams (air in liquid), doughs (starch granules in gluten), emulsions (oil droplets in water, e.g., butter, mayonnaise, and margarine), and solid particles in fat systems (e.g., chocolate).

Instability in a dispersed system may manifest itself as phase separation and particulate flocculation, and is contingent on volume concentration of the dispersed phase (volume fraction) particulate interactions, particle size, size distribution, and particle deformability, as well as the viscosity, pH, and electrolytic content of the continuous phase. Dispersion stability can be controlled, in the case of particle–particle interactions (e.g., van der Waals attractive forces), by masking these interactions through steric and electrostatic barriers. In order to study the effect of these and other forms of stabilization, the rheological behavior of the rest structure of the system (i.e., structure assumed in the absence of an applied force other than gravity) must be examined. Rheological analysis of these rest structures requires that the imposed strain be minimal (i.e., within the LVR); for electrostatically stabilized systems critical strains range from 0.001 to 0.005 whereas for a sterically stabilized structure they are of the order 0.01 to 0.05. The increased strain level for the latter is due to steric barriers exerting their influence over greater distances. The critical strain in conjunction with moduli values from within the LVR is also influenced by the volume fraction; higher concentrations lead, as would be expected, to increased moduli values but also to increased strain dependence. This nonlinear behavior has been attributed to the gradual breakdown of rest structures, being less apparent at low concentrations where rest structures are of diminished consequence.

The experimental prerequisite of minimal strain precludes the use of steady shear experiments where such critical strain levels are exceeded within the first seconds of the test even at low (i.e., 0.01 sec^{-1}) shear rates. Only dynamic experiments are suited to the examination of these rest structures. Some examples of food dispersions and their strain limits are given in Table 3.

Like gels and biopolymer solutions, the viscoelastic behavior of dispersions can be characterized by their frequency dependence. The mechanical spectra for dispersions bear close resemblance to those for biopolymer solutions (Fig. 28); the frequency–concentration superposition principle also applies (Fig. 29). The interpretation of these spectra are, however, slightly different for suspensions. At low-volume fractions (i.e., low concentrations of the disperse phase) the dispersed particles form an equilibrium configuration in which Brownian forces predominate and the resulting behavior is predominantly viscous (i.e., Newtonian, $G'' \gg G'$). As the volume fraction is increased so that particle–particle interactions contribute to particle arrangement, a degree of elasticity is introduced to the response ($G'' \rightarrow G'$) and behavior comparable to that of a concentrated solution is observed. Further increases in phase volume lead to solid-like ($G'' \ll G'$) behavior, with both moduli showing reduced frequency dependence. At these high concentrations, the dynamic response is almost entirely due to steric (or electrostatic) layer compression, Recommended references for the subject of suspension rheology include those by Sherman (78) and Krieger (79).

D. Challenges Associated with Rheological Analysis of Food Materials

Mechanical properties of foods were first described using empirical methods. It is now recognized that if the industry is to progress, a fundamental understanding of material behavior is necessary. The first and probably most important consideration when assessing a food material is the choice of test. As described previously, selection should be based on the information required, the nature of the material, financial constraints, and practical considerations. Poor selection at this point can easily lead to misinterpretation of the mechanical properties.

Additional challenges associated with any analysis of foods include the complex nature of the material and inadequate sampling methods. These lead to sources of error and poor reproducibility. Physical phenomena such as phase separation, segregation, and dehydration can also lead to erroneous data. Various authors have suggested that rheological testing is unlikely to yield reproducible data unless the sample is deaerated, either by vacuum or centrifugation. However, use of these techniques may result in unwanted artifacts. Compositional and/or conformational changes (i.e., protein denaturation, water loss in gels) should also be avoided.

Some specific concerns are peculiar to dynamic rheological experiments. These include nonlinear effects and, although already mentioned, it must be emphasized that linearity checks are necessary during testing. This is of particular importance because nonlinear effects can "resonate" under certain circumstances; however, these effects may themselves be of considerable interest and yield important information about material structure. Another source of difficulty is the natural frequency (ω_0) associated with the measuring system. This is the frequency at which free oscillation occurs. At ω_0, G' and G'' have no meaning because the amplitude ratio is 1 and the phase angle is 0. Consequently, use of linear equations is rendered inconclusive. Typically, data at the ω_0 are excluded from analysis. Measurements recorded near ω_0 display discontinuity in curves for η' vs. ω and G' vs. ω.

IV. CONCLUSIONS

Not long ago, a book chapter on the topic of material properties of food would have dealt mainly with viscosity measurement of liquids and large-scale deformation testing of solids as related to food texture. Now food scientists are becoming more engaged in determining the viscoelastic characteristics of food using small deformation testing. The reasons for this shift are an increasing awareness that food texture is a direct manifestation of structure and that, as well as traditional microscopic methods, small deformation testing yields results that aid in understanding structure formation and breakdown. New instrumentation is available for obtaining these data that allow testing over a wide range of conditions.

Results from studies of this sort have stimulated the field of food materials science. Researchers in this area of food science undertake studies aimed at measuring the size and distribution of elements in complex mixtures, understanding how these elements interact and behave under mechanical stress, and developing processes to maximize the formation and retention of useful structures (80).

Finally, the authors trust they have demonstrated that a knowledge of the mechanical properties of food is necessary to undertake the controlled transformation of native materials to forms suitable for consumption. Evidence of the economic importance of mechanical properties is found in legal regulations setting standards for government grades. These standards often take the form of instrumental specifications; an example is tomato products, for which definite measured consistencies are required.

Readers interested in supplementing their understanding of mechanical properties beyond that offered in this necessarily brief chapter are directed to several recent general texts (28,89,90) and two dealing specifically with the techniques of measurement (91,92).

ACKNOWLEDGMENTS

Original research described herein was supported by the Natural Sciences and Engineering Research Council of Canada and the Ontario Ministry of Agriculture and Food.

REVIEWS

J. F. Steffe, *Rheological Methods in Food Engineering*, Freeman Press, East Lansing, 1992.

M. A. Rao and S. S. H. Rizvi, *Engineering Properties of Foods*, Marcel Dekker, New York, 1986.

M. A. Rao and J. F. Steffe, *Viscoelastic Properties of Foods*, Elsevier, Barking, 1992.

A. A. Collyer and D. W. Clegg, *Rheological Measurement*, Elsevier, Barking, 1988.

R. W. Whorlow, *Rheological Techniques*, 2nd Ed., Ellis Horwood, New York, 1992.

REFERENCES

1. J. M. deMan, in *Rheology and Texture in Food Quality* (J. M. deMan, P. W. Voisey, V. F. Rasper, and D. W. Stanley, eds.), AVI, Westport, 1976, pp. 8–27.
2. C. F. Shoemaker, J. I. Lewis, and M. S. Tamura, *Food Technol. 41*(3): 80 (1987).
3. P. W. Voisey, in *Rheology and Texture in Food Quality* (J. M. deMan, P. W. Voisey, V. F. Rasper, and D. W. Stanley, eds.), AVI, Westport, 1976, pp. 79–141.
4. P. W. Voisey and J. M. deMan, in *Rheology and Texture in Food Quality* (J. M. deMan, P. W. Voisey, V. F. Rasper, and D. W. Stanley, eds.), AVI, Westport, 1976, pp. 142–243.
5. M. C. Bourne, *Food Texture and Viscosity*, Academic Press, New York, 1982.
6. J. D. Dziezak (ed.), *Food Technol. 45*(7): 82 (1991).
7. R. I. Tanner, *Engineering Rheology*, Clarendon Press, Oxford, 1985.
8. H. A. Barnes, J. F. Hutton, and K. Walters, *Introduction to Rheology*, Elsevier, New York, 1989.
9. G. Kaletunic, M. D. Normand, E. A Johnson, and M. Peleg, *J. Food Sci. 56*: 950 (1991).
10. J. R. Mitchell, in *Food Analysis, Principles and Techniques*, Vol. 1, *Physical Characterization* (D. W. Gruenwedel and J. R. Whitaker, eds.), Marcel Dekker, New York, 1984, pp. 151–220.
11. N. N. Mohsenin, *Physical Properties of Plant and Animal Materials*, 2nd Ed., Gordon and Breach, New York, 1986.
12. D. W. Stanley, in *Food Texture* (H. R. Moskowitz, ed.), Marcel Dekker, New York, 1987, pp. 35–64.
13. J. M. Aguilera and D. W. Stanley, *Microstructural Principles of Food Processing and Engineering*, Elsevier, London, 1990.
14. M. C. Bourne, in *Rheology and Texture in Food Quality* (J. M. deMan, P. W. Voisey, V. F. Rasper, and D. W. Stanley, eds.), AVI, Westport, 1976, pp. 244–274.
15. J. H. Prentice, *Measurements in the Rheology of Foodstuffs*, Elsevier, London, 1984.
16. V. N. Mohan Rao and G. E. Skinner, in *Engineering Properties of Foods* (M. A. Rao and S. S. H. Rizvi, eds.), Marcel Dekker, New York, 1986, pp. 215–254.
17. M. Peleg, in *Food Texture* (H. R. Moskowitz, ed.), Marcel Dekker, New York, 1987, pp. 3–33.
18. J. M. deMan and D. W. Stanley, in *Food Analysis: Principles and Techniques*, Vol. 1, *Physical Characterization* (D. W. Gruenwedel and J. R. Whitaker, eds.), Marcel Dekker, New York, 1984, pp. 221–245.
19. D. W. Gruenwedel and J. R. Whitaker (eds.), *Food Analysis: Principles and Techniques*, Vol. 1, *Physical Characterization*, Marcel Dekker, New York, 1984.
20. J. S. Dodge and J. M. Krieger, *Trans. Soc. Rheol. 15*: 589 (1971).
21. F. J. Lockett, *Non-linear Viscoelastic Solids*, Academic Press, London, 1972.
22. Y. Chen and J. Rosenburg, *J. Text. Stud. 8*: 477 (1977).
23. M. van de Temple, in *Rheometry: Industrial Applications* (K. Walters, ed.), John Wiley and Sons, New York, 1980, pp. 179–207.
24. J. L. Kokini and G. J. Plutchok, *Food Technol. 41*(3): 89 (1987).
25. J. D. Ferry, *Viscoelastic Properties of Polymers*, 3rd Ed., John Wiley and Sons, New York, 1980.
26. S. B. Ross-Murphy, in *Biophysical Methods in Food Research* (H. W.-S. Chan, ed.), Blackwell Scientific, Oxford, 1984, pp. 138–199.
27. S. B. Ross-Murphy, in *Food Structure: Its Creation and Evaluation* (J. M. V. Blanshard and J. R. Mitchell, eds.), Butterworths, London, 1984, pp. 387–400.
28. J. F. Steffe, *Rheological Methods in Food Engineering*, Freeman Press, East Lansing, MI, 1992.
29. M. A. Tung, R. A. Speers, I. J. Britt, S. R. Owen, and L. L. Wilson, in *Engineering and Food*, Vol. 1 (W. E. L. Speiss and H. Schubert, eds.), Elsevier, New York, 1990, pp. 79–89.

30. J. F. Steffe, in *Advances in Food Engineering* (R. P. Singh and M. A. Wirakartakusumah, eds.), CRC Press, Boca Raton, 1992, pp. 363–376.
31. F. Shama and P. Sherman, *J. Food Sci. 31*: 699 (1966).
32. M. Balaban, A. R. Carrillo, and J. L. Kokini, *J. Text. Stud. 19*: 171 (1988).
33. E. R. Morris and S. B. Ross-Murphy, *Carbo. Metab. B310*: 1 (1981).
34. I. C. M. Dea, R. K. Richardson, and S. B. Ross-Murphy, in *Gums and Stabilizers for the Food Industry*, *Vol. 2* (G. O. Phillips, D. J. Wedlock, and P. A. Williams, eds.), Pergamon Press, Oxford, 1984, pp. 357–366.
35. R. K. Richardson, E. R. Morris, S. B. Ross-Murphy, L. J. Taylor, and I. C. M. Dea, *Food Hydrocoll. 3*: 175 (1989).
36. C. K. Rha, *Food Technol. 33*(10): 71 (1979).
37. M. J. Lewis, *Physical Properties of Foods and Food Processing Systems*, Ellis Horwood, Chichester, 1987.
38. K. Walters, *Rheometry*, Chapman and Hall, London, 1975.
39. G. J. Brownsey, in *Rheological Measurement* (A. A. Collyer and D. W. Clegg, eds.), Elsevier, London, 1988, pp. 405–431.
40. P. J. Flory, *J. Am. Chem. Soc. 63*: 3083 (1941).
41. P. J. Flory, *Principles of Polymer Chemistry*, 7th Ed., Cornell Univ. Press, Ithaca, NY, 1953.
42. W. H. Stockmeyer, *J. Chem. Phys. 11*: 45 (1943).
43. W. H. Stockmeyer, *J. Chem. Phys. 12*: 125 (1944).
44. J. Hermans, Jr., *J. Polym. Sci. 3*: 1859 (1965).
45. C. A. L. Peniche-Covas, S. B. Dev, M. Gordon, M. Judd, and K. Kajiwara, *Faraday Dis. Chem. Soc. 57*: 165 (1974).
46. A. Clark, R. K. Richardson, G. Robinson, S. B. Ross-Murphy, and J. M. Stubbs, *Macromolecule 16*: 1367 (1983).
47. D. A. Oakenfull, *J. Food Sci. 49*: 1103 (1984).
48. A. Clark, R. K. Richardson, G. Robinson, S. B. Ross-Murphy, and A. C. Weaver, *Prog. Food. Nutr. Sci. 6*: 149 (1982).
49. A. H. Clark and S. B. Ross-Murphy, *Adv. Polym. Sci. 83*: 57 (1987).
50. G. R. Ziegler and E. A. Foegeding, in *Advances in Food and Nutrition Research*, Vol. 34 (J. E. Kinsella, ed.), Academic Press, New York, 1990, pp. 203–298.
51. A. H. Clark, in *Physical Chemistry of Foods* (H. G. Schwartzberg and R. W. Hartel, eds.), Marcel Dekker, New York, 1992, pp. 263–306.
52. C. Y. M. Tung and P. J. Dynes, *J. Appl. Polym. Sci. 27*: 569 (1982).
53. F. Chambon and H. H. Winter, *J. Rheol. 31*: 683 (1987).
54. H. H. Winter, *Polym. Eng. Sci. 27*: 1698 (1987).
55. G. Cuvelier, C. Peigney-Nourry, and B. Launay, in *Gums and Stabilizers for the Food Industry*, *Vol. 5* (G. O. Phillips, D. J. Wedlock, and P. A. Williams, eds.), IRL Press, Oxford, 1989, pp. 549–552.
56. S. B. Ross-Murphy, in *Food Polymers, Gels and Colloids* (E. Dickinson, ed.), Royal Soc. Chem., London, 1991, pp. 357–368.
57. A. H. Clark and S. B. Ross-Murphy, *Brit. Polym. J.*, *17*: 164 (1985).
58. D. A. Oakenfull and A. Scott, in *Gums and Stabilizers for the Food Industry*, *Vol. 3* (G. O. Phillips, D. J. Wedlock, and P. A. Williams, eds.), Elsevier, London, 1986, pp. 465–475.
59. D. A. Oakenfull and V. J. Morris, *Chem. Ind.* (March): 201 (1987).
60. A. H. Clark, in *Food Polymers, Gels and Colloids* (E. Dickinson, ed.), Royal Soc. Chem, London, 1991, pp. 322–368.
61. K. te Nijenhuis, *Colloid Polym. Sci. 259*: 522 (1981).
62. M. Djabourov, J. Marquet, H. Theveneau, J. Leblond, and P. Papon, *Br. Polym. J. 17*: 169 (1985).
63. M. Djabourov, J. Leblond, and P. Papon, *J. Physique 49*: 333 (1988).
64. F. Van Kleef, J. Boskamp, and M. Van den Tempel, *Biopolymer 17*: 225 (1978).
65. J. M. V. Blanshard and W. Derbyshire, in *Meat* (D. J. Cole and R. A. Lawrie, eds.), Butterworths, London, 1975, pp. 269–283.
66. J. Culioli and P. Sherman, *J. Text. Stud. 9*: 257 (1978).
67. L. R. G. Treloar, *The Physics of Rubber Elasticity*, 3rd Ed., Clarendon Press, Oxford, 1975.

68. E. R. Morris, in *Gums and Stabilizers for the Food Industry* (G. O. Phillips, D. J. Wedlock, and P. A. Williams, eds.), Pergamon Press, Oxford, 1984, pp. 57–78.
69. B. Egelandsdal, K. Fretheim, and K. Samejima, *J. Sci. Fd. Agric. 37*: 944 (1986).
70. M. Stading and A-M. Hermansson, *Food Hydrocoll. 4*: 121 (1990).
71. J. R. Mitchell, in *Polysaccharides in Food* (J. M. V. Blanshard and J. R. Mitchell, eds.), Butterworths, London, 1979, pp. 51–72.
72. G. Robinson, S. B. Ross-Murphy, and E. R. Morris, *Carbo. Res. 107*: 17 (1982).
73. H. Mark, *Der Fester Körper*, Hirzel, Leipzig, 1938.
74. R. Houwink, *J. Prakt. Chem. 157*: 15 (1940).
75. R. D. McCurdy, H. D. Goff, D. W. Stanley, and A. P. Stone, *Food Hydrocoll.* (submitted) *8*: 609 (1994).
76. W. P. Cox and E. H. Merz, *J. Polym. Sci. 28*: 619 (1958).
77. K. I. Bistany and J. L. Kokini, *J. Text. Stud. 14*: 113 (1983).
78. P. Sherman, *Industrial Rheology*, Academic Press, New York, 1970.
79. I. M. Krieger, in *Physical Properties of Foods* (M. Peleg and E. Bagley, eds.), AVI, Westport, 1983, pp. 385–398.
80. D. W. Stanley, *Food Res. Int. 27*: 135 (1994).
81. H. Rohm and K.-H. Weidinger, *J. Text. Stud. 24*: 157 (1993).
82. J. Munoz and P. Sherman, *J. Text. Stud. 21*: 411 (1990).
83. L. Lindahl and A.-C. Eliasson, *J. Sci. Food Agric. 37*: 1125 (1986).
84. A. P. Stone and D. W. Stanley, *Food Res. Intl. 27*: 155 (1994).
85. A. Dinsdale and F. Moore, *Viscosity and Its Measurement*, Chapman and Hall, London, 1962.
86. I. M. Ward, *Mechanical Properties of Solid Polymers*, 2nd Ed., John Wiley and Sons, New York, 1983.
87. L. Slade and H. Levine, *CRC Crit. Rev. Food Sci. Nutr. 30*: 115 (1991).
88. C. F. Shoemaker, in *Viscoelastic Properties of Foods* (M. A. Rao and J. F. Steffe, eds.), Elsevier, New York, 1992, pp. 233–246.
89. M. A. Rao and S. S. H. Rizvi, *Engineering Properties of Foods*, Marcel Dekker, New York, 1986.
90. M. A. Rao and J. F. Steffe, *Viscoelastic Properties of Foods*, Elsevier, New York, 1992.
91. A. A. Collyer and D. W. Clegg, *Rheological Measurement*, Elsevier, New York, 1988.
92. R. W. Whorlow, *Rheological Techniques*, 2nd Ed., Ellis Horwood, New York, 1992.

5

Optical Properties

Carlos Calvo

Institute of Agrochemistry and Food Technology, High Council of Scientific Research (CSIC), Valencia, Spain

I. INTRODUCTION

The various parameters that define the quality of food products can be divided into three groups: visual parameters, texture parameters, and flavor and aroma parameters (1). By visual parameters we understand those that we perceive with the sense of sight. The various properties perceived by the sense of sight can be further divided into three subgroups: optical properties, properties depending on physical form, and properties depending on the method of presentation (2).

By optical properties we understand those that depend on geometric or chromatic modification of the light striking the product in question. These properties are color, gloss, translucency, and their uniformity on the surface of the product.

If the reader should ask why we measure color and what importance it has, the answer is that with the sense of sight we obtain 83% of the information we receive about our environment, whereas a mere 17% is obtained from the other senses. Also, visual evaluation of appearance precedes the evaluation of other parameters (flavor, aroma, and texture), so that it may have an excluding effect. If we reject a food product on the basis of a visual examination, we do not go on to evaluate its other properties, however excellent they may be. This importance is recognized by food legislation in all countries, where color always features in the description of any foodstuff and also features as a quality factor to distinguish among different quality levels for a particular product.

The first and longest part of this chapter is devoted to color, followed by a description of the measurement of gloss and translucency.

II. SAMPLE PREPARATION

A. Solid

The surface of the sample to be analyzed must be as flat as possible, whether one is measuring with a colorimeter or making a visual comparison with a color dictionary.

In the case of canned foods or foods that are naturally soft, all that is needed is to press the sample gently onto the plate of optical glass that covers the window of the colorimeter. If one is comparing with color dictionary cards, the problem of sample curvature is less than with colors that are printed on the pages of the dictionary.

With viscous products the procedure is to fill the cuvette to the brim, smooth the surface with a spatula, and cover it with an optical glass taking care not to leave any bubbles.

If the product is in powder form, it is pressed into the cuvette and covered with an optical glass in the same way as for doughy products, so that the surface is smooth and homogeneous.

If the product is granular, such as wheat, rice, millet, etc., the cuvette is filled and the surface smoothed, and if possible a colorimeter with a large measurement area is used.

B. Liquid

If the liquid is clear and transparent, there are no problems. All that is needed is to place it in the quartz cuvette in the spectrophotometer or colorimeter and read by transmission. If the liquid contains small particles in suspension, it is filtered, provided that this operation does not alter the color. Finally, if the liquid is cloudy, the best approach is to use the Kubelka–Munk formulas (Sec. V.B).

C. Calibration Standards

Both tristimulus colorimeters and spectrophotometers are supplied with black and white plates for calibration. These plates are assayed by the manufacturer of the apparatus.

In the case of spectrophotometers equipped with a reflecting device, they are supplied with porcelain or plastic blanks and calibration lists giving the reflection percentage (every 5 or 10 nm) in relation to the $BaSO_4$ reference blank. The blanks supplied need to be recalibrated periodically against $BaSO_4$ in the user's laboratory. The procedure is to make a piece or tablet of $BaSO_4$ from the product in powder form, compressing it in a suitable device so as to form the tablet. This needs to be handled with great care, as it is quite fragile and tends to crack and split, changing its reflecting power. So it is best to use it soon after it has been made.

D. Lighting and Illuminants

The color of food—as with any other product—can vary according to the way it is lit and the illuminant employed. Consequently, special attention must be given to this aspect.

In order to standardize lighting conditions, in 1931 the Commission Internationale de l'Eclairage (CIE) selected the illuminants A, B and C; and later (in 1966), D65. These illuminants are defined by their color temperature (Table 1) and by their spectral distribution curve. Thus, illuminant A has an unequal spectral distribution, with more energy in the area of the reds than in that of the blues, and so gives a reddish hue. Illuminants B and C come closer to an imitation of sunlight while D65, which has a fairly uniform spectral distribution and is closest to natural daylight, is the one most frequently used nowadays.

In addition to the international illuminants quoted there are others for particular purposes or for commercial use, such as TL84, F2, D50, D60, and D75. TL84 is used by large commercial chains; F2 is a fluorescent light; and D50, D60, and D75 are variations of D65 with modifications of the color temperature.

If we wish the light reflected by the sample to give us information about color, it must be diffuse. To achieve this, the sample is lit with a beam of white light at an angle of 45° and this is reflected at right angles to the surface of the sample to be viewed. This is known as 45/0 lighting. If the lighting and viewing angles are reversed we have 0/45 lighting. According to Helmholtz's

Table 1 Standard Illuminants

Illuminant	Color temp. (k)
A	2854
B	4870
C	6770
D65	6500

reciprocity principle, the two lighting systems can be interchanged without affecting the result of the measurements.

The lighting may be in the form of a single or double beam, or a circumferential beam, which evens out small irregularities in the surface of the sample. Another way of obtaining diffuse light is by means of an integrating sphere. This consists of a hollow sphere, painted white on the inside with $BaSO_4$. In this way the light that enters is reflected in all directions and reaches the sample as diffuse light.

III. COLOR MEASUREMENT

A. Color Specification

1. 1931 CIE System

In order to standardize color measurement, in 1931 the CIE met in Paris to define a physical color space based on the theory of trichromatic perception. The construction of this space was based on matching colors by mixing appropriate quantities of three basic stimuli, using a defined standard observer and a particular system of coordinates (3). In the *CIE* system color is specified in terms of three primary values—X (red), Y (green), and Z (blue), the so-called tristimulus values. Tables have been produced to show the values of $\bar{x}$, $\bar{y}$, $\bar{z}$ corresponding to the quantities of the primaries X, Y, Z needed by the standard observer to match each wavelength in the visible spectrum with the same energy. These tables were published by Hardy, Judd and Wyszescky, MacAdam, and Wright (3). The tristimulus values are calculated from the equations:

$$X = \Sigma \bar{x}_\lambda E_\lambda R_\lambda \Delta_\lambda \qquad Y = \Sigma \bar{y}_\lambda E_\lambda R_\lambda \Delta_\lambda \qquad Z = \Sigma \bar{z}_\lambda E_\lambda R_\lambda \Delta_\lambda$$

where E_λ is the relative energy and R_λ the percentage of reflectance or transmittance, as appropriate. The tristimulus values are obtained by drawing a graph of the equations with the wavelength as the abscissas and measuring the area enclosed. The calculation can be made by the weighted ordinates method or the selected ordinates method (4).

Once the tristimulus values X, Y, Z have been obtained, the proportion of each can be calculated:

$$x = \frac{X}{X + Y + Z} \qquad y = \frac{Y}{X + Y + Z} \qquad z = \frac{Z}{X + Y + Z}$$

These new variables are called chromaticity coordinates or trichromatic coefficients. Since $x + y + z = 1$, it is only necessary to use x, y to define chromaticity. Therefore colors are generally specified in terms of Y, x, y.

The representation of the colors in the spectrum in terms of x, y produces a horseshoe-shaped graph known as the *chromaticity diagram* (Fig. 1). The values of x, y give the chromaticity of the sample; together with the value of Y, the percentage of luminosity, they form a three-dimensional

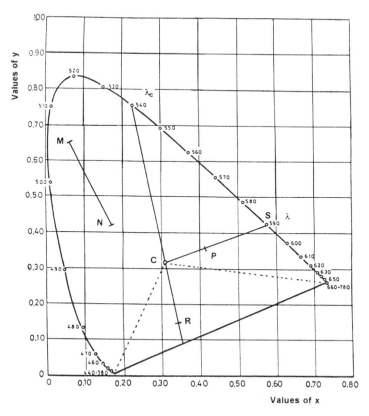

Fig. 1   CIE chromaticity diagram.

diagram. In practice, however, we work with only two dimensions (x, y). The chromaticity diagram gives extensive information on the color thus specified, as we shall see.

If one has only a spectrophotometer and not a colorimeter, the chromaticity coordinates can be calculated by the selected ordinate method. Table 2 shows the calculation of these parameters for a specific example.

From the reflectance values of the sample, read within the visible spectrum, we can draw the spectral reflectance curve, which shows percentage refeflectance values as ordinates and wavelength in nanometers as abscissas. From this graph we can read off and note down the values of R that correspond to the wavelengths selected in columns X, Y, and Z in Table 2. In our case, a wavelength of 435.5 corresponds to an R value of 0.9, 489.4 corresponds to 1.0, and so on. The R values are then totaled for each of the three columns and the result is multiplied by a constant factor for each of the three totals. This gives us the values of X, Y, and Z.

In the diagram the values of x, y for a given color provide us with its chromaticity, i.e., its predominant wavelength and its purity. The *predominant wavelength* is a psychophysical magnitude which corresponds to the sensory perception of hue. To calculate it graphically a straight line is drawn connecting the illuminant, point C, to the color, represented by point P; the line is extended until it cuts the border of the diagram, giving the predominant wavelength (Fig. 1). If the point defining the color is located in the area of colors known as "purple," i.e., in the 400-nm–C–700-nm triangle, the procedure is as above, but the line is extended toward point C until it cuts the curve; this point corresponds to a complementary wavelength represented as λ_c.

Table 2 Calculation of CIE Chromaticity Coordinates for a Tomato Ketchup by 10-Selected-Ordinates Method

Ordinate no.	X	R (%)	Y	R (%)	Z	R (%)
1	435.5	0.9	489.4	1.0	422.2	0.9
2	461.2	0.9	515.2	1.1	432.0	0.9
3	544.3	1.5	529.8	1.2	438.6	1.0
4	564.1	2.2	541.4	1.5	444.4	1.0
5	577.4	3.0	551.8	1.5	450.1	1.0
6	588.7	4.0	561.9	2.0	455.9	1.0
7	599.6	5.1	572.5	2.5	462.0	1.0
8	610.9	5.8	584.8	3.7	468.7	1.0
9	624.2	7.0	600.8	5.1	477.7	1.0
10	645.9	_7.4_	627.2	_7.3_	495.2	_1.1_
Total		37.8		26.9		9.9
Factors	0.098		0.100		0.118	

$$X = (37.8)(0.098) = 3.704$$
$$Y = (26.9)(0.100) = 2.690$$
$$Z = (9.90)(0.118) = 1.168$$
$$\text{Total } (X + Y + Z) = 7.562$$
$$x = (3.704)/(7.562) = 0.489$$
$$y = (2.690)/(7.562) = 0.355$$

The psychophysical magnitude known as *colorimetric purity* corresponds to the sensory perception of color saturation. In the diagram this is represented by the relative distance of the point representing the color from the border. Its graphical measurement is given by: purity = (CP/CS)/100 (Fig. 1); and it can be calculated from the coordinates of the points.

The x, y chromaticity diagram is flat, but it becomes three-dimensional with the addition of the perpendicular Y axis, which passes through the center C, giving us the color *luminosity*. In the CIE system the values of Y can vary from 0 to 100. When representing a color in a diagram by its x, y coordinates, the Y value is frequently given as a reference point.

Neutral or achromatic colors, as their name indicates, have no chromaticity and are located in the middle of the diagram, at the point representing the illuminant, through which the Y axis passes. Black, white, and shades of gray are represented on this axis.

If we have two colors given in the diagram by points M and N, all of the possible *mixtures* of them are represented by the segment MN (Fig. 1). The midpoint corresponds to equal parts of both colors and the other points to relative proportions of each color.

The colors located on opposite sides of any straight line passing through the center of the diagram are called *complementary colors* because when added together they give a gray color more or less affected by the color of one of them according to its distance from point C and the brightness level (Y) that applies.

The spectral reflectance curve of the colors known as *purple* has maxima in the area of the blues and reds and a minimum in the area of the greens. These colors do not exist in the visible spectrum as monochromatic light but are obtained, as has been indicated, by an additive mixture of radiations from the extremes of the visible spectrum. If they are represented in the chromaticity diagram, they are located in the 400-nm–C–700-nm triangle.

2. Hunter System

In parallel with the development of tristimulus colorimeters, the Hunter L, a, b system (5) was developed from Hering's theory of opposite colors.

With these new coordinates a Cartesian space is defined in which L corresponds to luminance or brightness and a and b to chromaticity. Specifically, a defines the red–green component: red for positive values and green for negative values; the b parameter defines the yellow–blue component: yellow for positive values and blue for negative values. As the distance increases between the center of the graph and the points defining a color, the color saturation also increases (Fig. 2).

The relationship between the values of L, a, b and the tristimuli X, Y, Z is given by the following equations:

$$L = 10Y^{1/2}$$

$$a = \frac{17.5(1.02X - Y)}{Y^{1/2}} \qquad b = \frac{7.0(Y - 0.847Z)}{Y^{1/2}}$$

The relative cheapness and rapid response of tristimulus colorimeters by comparison with conventional spectrophotometers contributed significantly to the adoption of the L, a, b system. Its popularity was also due to the intuitive nature of the graphical representation and the fact that the system was not simply a convenient idea produced by the authors but was based on the theory of opposite colors.

This is the system most commonly employed in Hunter, Gardner, and similar colorimeters, and it is much used for food analysis. However, one of its drawback is its lack of uniformity in the blue range.

3. CIELAB System

In 1971 the CIE put forward a new chromatic space called CIELAB, created by nonlinear transformations of the 1931 CIE system (or CIEXYZ). In the new system a space is defined within rectangular coordinates (L^*, a^*, b^*) together with another space in cylindrical coordinates (L^*, H^*, C^*) (6).

The relationship between the CIELAB and CIEXYZ coordinates is as follows:

$$L^* = 116(Y/Y_0)^{1/3} - 16$$
$$a^* = 500(X/X_0)^{1/3} - (Y/Y_0)^{1/3}, \qquad \text{where } Y/Y_0 > 0.01$$
$$b^* = 200(Y/Y_0)^{1/3} - (Z/Z_0)^{1/3}$$

where X, Y, Z are the tristimulus values of the sample and X_0, Y_0, Z_0 are the values of the achromatic point corresponding to the illuminant employed.

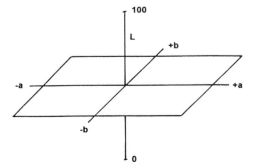

Fig. 2 Hunter diagram.

In this system the difference in color between two samples is given by the experssion:

$$\Delta E^* = [(\Delta L^*)^2 + (\Delta a^*)^2 + (\Delta b^*)^2]^{1/2}$$

For representation in cylindrical coordinates the variables used are L^*, C^*, H^*, where L^* is luminance, C^* chroma (saturation), and H^* hue, as defined by the following equations:

$$C^* = (a^{*2} + b^{*2})^{1/2} \qquad H^* = \arctan \frac{b^*}{a^*}$$

A representation using the CIELAB coordinate system is given in Fig.3.

B. Sensory Evaluation

1. Human Sight

The easiest and most elementary way of evaluating the color of a food product is by visual observation. The human eye has two very valuable properties: (a) its ability to discriminate, i.e., its ability to distinguish between very similar colors, and (b) its ability to produce a single integrated response for the color of a surface with slight stains or discolorations or a group of items where some are of a different color or are in shadow. These advantages are offset by two

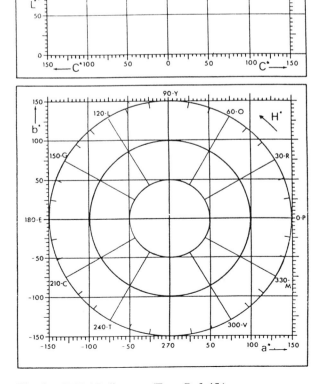

Fig. 3 CIELAB diagram. (From Ref. 45.)

disadvantages: (a) the lack of a color memory to recall a previously seen color and (b) the inadequacy of existing language for describing colors.

If we say that some peas are bright green or some asparagus is a light cream color, this gives us very little information to form an idea of the wide range of colors that may be covered by these words. There is also the fact that not all human beings perceive color in the same way. Natural variations in the human population cause the perception of certain colors to vary. This is compounded by biases mainly attributable to cultural or regional preferences (7). Finally, there may be defects in color vision: protanopia (red deficiency), deuteranopia (green deficiency), and tritanopia (blue deficiency). For all of these reasons sensory measurement of color should be performed under standardized lighting conditions (Sec. II.D) with a suitably selected panel of judges (Sec. III.B.2).

2. Selection of Judges: Farnsworth–Munsell Test

A frequently used method for selecting judges is to test their discriminative ability with a ranking test. The Farnsworth–Munsell test (8) is available for this purpose, consisting of arranging a group of standard color records (obtainable from Munsell Color Co., PO Box 950, Newburgh, NY 12550), selected from the range of greens, blues, reds, etc., according to the samples to be analyzed.

The test should be carried out under standardized conditions in a light chamber with illuminant C or else done at midday with light coming from the north. The color records should be lit at an angle of 90° to the work plane and viewed at an angle of 60°. Each set of colors to be arranged is stored in a long box containing 24 color records. The two extreme colors are attached to the two ends of the box, and the others can be moved; on the back of each is a number indicating its correct position. The potential tester recives the colors out of order and has to arrange them in the right order in a maximum time of 2 min.

To rank the candidates a table is constructed like that in the example (Table 3). The first row shows the correct order of the colors, the second the order given by the candidate, and the third the score, which is the difference between the correct number for each color and the numbers of the two colors that have been placed on either side. The candidate's score is the total of all of the numbers in the score row. A score of 0–16 is considered to show high discrimination, 17–100 is medium, and over 100 is low. In our example the total is 120, which shows poor discrimination. Usually one starts with more candidates than are needed and, as most have medium discrimination, they are ranked by score and the first n needed are selected.

3. Comparison with Standards

Because color is perceived with the sense of sight, the best way to evaluate it is by actually using the sense of sight. In Sec. III.B.1 we saw that human sight has certain characteristics. Let us consider one good feature and one bad one: our great discriminative ability and our poor color memory. So, with the aid of a reference the observer can compare a food product and classify it as better than, equal to, or worse than the reference, a task that otherwise would be practically impossible. In other words, with the reference we make up for our deficient color memory and

Table 3 Farnsworth–Munsell Test: Example of Recording Data

Theoretical	1	2	3	4	5	6	7	8	9	10	11	12	13	14	15	16	17	18	19	20	21	22	23	24
Order		2	3	5	4	6	7	8	11	9	13	10	12	15	20	17	14	19	19	22	21	18	23	
Score		2	3	3	3	3	2	4	5	6	7	5	5	8	8	6	8	8	9	7	4	8	6	

poor color vocabulary, taking advantage of our powerful discriminative ability. There are two kinds of references: dictionaries and plates.

   a.   Color Dictionaries   A color dictionary or atlas is a collection of colors arranged according to a specific criterion or system, based in many cases on the three-parameter nature of color, specifically on the sensory attributes of hue, luminance, or brightness and saturation or intensity. Part of the difficulty lies in having to make a two-dimensional reproduction, on the pages of an atlas, of a three-dimensional space. Hence the existence of various systems.

   The following survey of color atlases and collections takes in the main systems but does not claim to be exhaustive. The systems most commonly used in the area of food analysis are described in greater detail.

MUNSELL SYSTEM   This is a system for classifying real color samples. The basis for its nomenclature is a description of a continuous three-dimensional color space where equal spaces—insofar as the geometry of the system allows—correspond to adjacent entries for each of the three attributes: hue, value, and chroma. These three variables are arranged so that value is the vertical axis with white at the top, black at the bottom, and grays in between. The hues are arranged in circles perpendicular to the axis. These circles are divided into 100 parts containing the five basic hues (red, yellow, green, blue, and purple) and the five secondary hues (red–yellow, yellow–green, green–blue, blue–purple, and purple–red), thus giving the 10 basic hues (Fig. 4). For each hue, chroma is represented by the distance from the central axis on the corresponding radius.

   The solid space of the Munsell system can be compared to an orange where the vertical axis corresponds to value, with white at the top and black at the bottom. Each slice cut at right angles to the axis has the most saturated colors at the periphery. Each segment of the orange corresponds to a hue, i.e., to one of the pages of the atlas. Each page has an indication of the hue in the top right corner. Each row on the page consists of colors of equal value, increasing from the bottom of the page ($V = 1$) to the top of the page ($V = 9$). The chroma values increase by two Munsell units at a time, progressing from the central axis (value) toward the outside (Fig. 5).

   In current editions of the atlas (9), four pages are assigned to each of the 10 basic colors, so that there is a difference of 2.5 hue units from one page to the next. The brighter colors are at the top of each page and the darker colors at the bottom. The achromatic colors (from white to black, ranging through shades of gray) are in the area nearest to the central axis, and the colors that are most saturated with chroma in the area furthest from the axis.

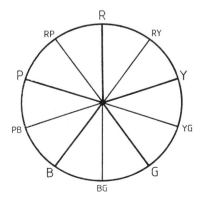

Fig. 4   Organization of hue Munsell circle.

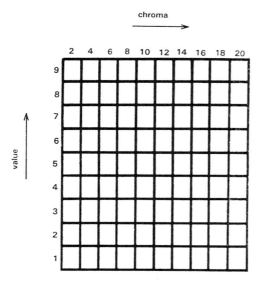

Fig. 5 Design of a page of Munsell color dictionary.

One advantage of the Munsell system is that if a new real color appears that has a more saturated hue than the example in the atlas, it can be incorporated without any alteration to the structure or nomenclature of the system. This is not the case with mixture systems, such as the Ostwald system and others.

Another practical advantage, having to do with handling, is that the colors are printed on small cards with the notation printed at the bottom, fitted into slots provided in the atlas. Thus it is not necessary to move the sample to the atlas, with the danger this might create in the case of moist foods; instead, the card can be placed next to the sample and then returned to its position in the atlas without any risk of confusion because each card has its Munsell notation printed at the bottom.

This notation consists of three parts, referring to the three color attributes. For example, if a color is given as 5R 2.6/13, this means that the hue is 5R, the value is 2.6, and the chroma is 13. In other words: H, V/C.

The Munsell system is one of the most frequently used systems in the field of food analysis and there are many references to it in the literature. It has also been adopted in the legislation of various countries (USA, Canada, France, etc.) as the reference for defining the color of canned tomato derivatives.

The Munsell system defines the relationship between the Y variable in the CIE system and V (value) in the Munsell system by the equation:

$$Y = 1.2219V - 0.23111V^2 + 0.23951V^3 - 0.021009V^4 + 0.0008404V^5$$

For the attributes of hue and chroma there is no simple mathematical formula relating the CIE and Munsell values. However, there are tables and computer programs with which information can be converted from one system to the other.

DIN SYSTEM   The DIN system (Deutsche Industrienormen) defines a color space in terms of three attributes: hue, chroma, and darkness (10). These three attributes are the three coordinates of the DIN space and are represented by T for hue (Farbton), S for saturation (Sättigungsstufe), and D for degree of darkness (Dunkelstufe).

With these three coordinates any color can be specified, e.g., 8.0:4.2:3.8 (*T:S:D*), just as *H*, *V/C* is used in the Munsell system.

The brightness scale or degree of darkness goes from white to black, ranging from 0 to 10. For hue the basis taken is the Ostwald circle, giving 24 equally spaced hues with the maximum possible saturation. These hues, unlike those in the Munsell system or Ostwald's full colors, are not empirical colors (obtained by subtractive mixing: paint, ink, etc.), but are taken from the color spectrum.

The way in which the DIN dictionary is presented has certain similarities to the Munsell atlas. There are 25 double pages into which 25 sets of color cards are inserted. Each page corresponds to a particular hue, which is indicated; there are 8 rows and 7 columns into which the color cards are inserted; those with the lightest colors are at the top and the darkest are at the bottom. The saturation increases from left to right (Fig. 6). The color cards are inserted into specially provided slots and have the DIN identification on the back. Each of the 25 double-page spreads has the cards on the right and on the left their equivalent of the DIN nomenclature in the CIE system in terms of *X, Y, Z* and also *x, y, z*. Of the 25 pages, 24 are devoted to the various colors and the last page is simply for the achromatic colors, with a gray scale ranging from white to black. As with the Munsell dictionary, in the DIN system not all of the spaces on each page are occupied. There are more cards in the middle of the page and fewer at the top and bottom, near white and black.

There are no equations for converting DIN units to the CIE system. In practice this is done by consulting specially designed tables, by reading a graph, or by means of a computer program.

MAERZ AND PAUL DICTIONARY   This dictionary is based on color mixtures (10). The first edition of Maerz and Paul's *A Dictionary of Color* dates from 1930 and was made by printing small dots of three separate colors. The dots were too small to be differentiated by the human eye, so that they gave the sensation of a color that was a mixture of the three elements used: one achromatic and two chromatic. With this printing method it is possible to produce a color atlas with a large number of reference colors, suitably graduated, at a modest price.

The atlas contains a total of 7056 color standards. The order in which they are presented follows the order of the visible spectrum, which they divide into seven main groups: from red to orange, from orange to yellow, from yellow to green, from green to turquoise, from turquoise to blue, from

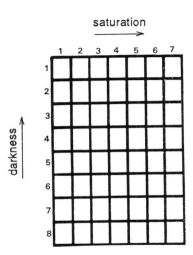

Fig. 6   Design of a page of DIN color dictionary.

blue to red, and from red to purple. By mixing these pairs of basic colors it is possible to produce a staggered series of intermediate colors with very small differences between them. Each series takes up eight consecutive pages in the atlas. The first has colors of the greatest purity, printed on white paper, and the following seven pages use progressively darker gray paper that ends up by being almost black. Each page of the atlas is divided into 144 compartments arranged in 12 columns and 12 rows, identified by numbers and letters (Fig. 7).

To represent the mixtures of two colors, e.g., yellow (Y) and orange (O), increasing quantities of ink Y are used from left to right, from column B to column L. With ink O printing starts in row 2 and increasing quantities are used from top to bottom of the page, finishing in row 12. Thus square 1A remains white because no ink is printed in it. Row 1 contains increasing quantities of ink Y with no ink O, and square 1L shows the typical color Y (yellow). Similarly, column A has increasing quantities of ink O and square 12A shows the typical color O (orange). The diagonal 2B–12L corresponds to equal mixtures of the two colors, and the colors in between are arranged in order on the page.

The atlas is designed so that when it is opened the colors printed as described are on the right-hand page and the left-hand page has a similar grid where, instead of printed colors, there is the name of those colors that have a specific name.

The identification of each color is given by two numbers and a letter. For example: 9.D.2 means the color on page 9, column D, row 2; on the left-hand page its name is given as "cream."

To complete the atlas, there is an index of names or denominations of colors in alphabetical order together with the corresponding identifying code so that they can be located and inspected on the appropriate page.

This atlas has been widely adopted in the area of food analysis to the extent that Canadian legislation defines the color of canned tomatoes in terms of it, as was mentioned in Sec. IV.A.3. However, because of its empirical basis there is no way of converting these colors to the CIE system except by measuring with a colorimeter.

The Villalobos Atlas, produced in Argentina and containing over 7000 reference colors, is

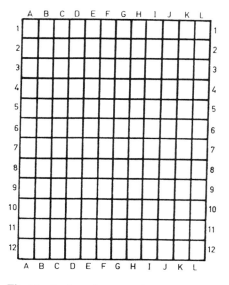

Fig. 7   Design of a page of Maerz and Paul color dictionary.

based on the same principle. In it the standards are 1-cm squares with a circular hole in the middle, 4 mm in diameter, to facilitate comparison.

OTHER COLOR DICTIONARIES (11)   The Ridgway *Color Standards and Color Nomenclature Atlas* and the *Color Harmony Manual*, published in the United States, are based on the Ostwald system. In addition, there are the OSA-UCS and ISCC-NBS systems, also from the USA; the atlases of the Royal Horticultural Society (RHS) (12) and Imperial Chemical Industries (ICI) in the United Kingdom; the SCOTDIC and Chroma Cosmos atlases in Japan; the Natural Color System (NCS) in Sweden (13); and the Coloroid system in Hungary (14).

b.   Color Plates   There are cases where color comparison does not need a whole dictionary but only a part of it. Moreover, in other cases it is sufficient to have a color reference that marks the boundary between two quality levels or defines the borderline between acceptance and rejection. As these colors do not generally coincide with one of the colors in a dictionary, the method adopted is to manufacture small plates of durable material—usually plastic—which are extremely convenient to use both for the manufacture and inspection of food products. Section IV.A.3 comments on particular cases of the use and application of these color plates.

4.   Visual Colorimeters

One way of overcoming the problem of the discrete color references provided by atlases is to use visual colorimeters. With these instruments the detector is still the human eye. A standardized illuminant is connected to the apparatus, but the reference colors can be varied continuously or almost continuously by mixing standard colors. There are two kinds of mixing: (a) additive color mixing, using red, green, and blue as primaries; and (b) subtractive color mixing achieved by eliminating part of the white light with the aid of suitable filters until a match for the problem is found. In this case the primaries are magenta, yellow, and cyan.

There is a clear difference between additive and subtractive colorimetry. If we mix red, green, and blue light we obtain white light, but if we place red, yellow, and blue filters in the way of white light we get black.

Accordingly, visual colorimeters can be classified as additive or subtractive (4).

a.   Additive Colorimeters   Additive colorimeters can use a mixture of colored lights or a mixture of sectors from a Maxwell disk. Wright and Guild colorimeters belong to the first category. For foods, however, disk colorimeters are used more commonly. They are equipped with two or more disks constructed so that each sector of the surface of the circle is covered by one of the colors of the mixture. The disks are placed on a platform connected to a motor which turns at great speed, so that to the eye the appearance is of a uniform color, which can thus be compared with the sample in question. The resulting color is expressed as a percentage of each of the components.

The original idea for disk colorimetry predates the adoption of the CIE system in 1931, and so for a long time the specification for the disks was given in Munsell notation. Nowadays, however, the CIE equivalent in terms of X, Y, Z is also given.

The simplicity of the apparatus has contributed to its general popularity. With commercial models the motor, platform, and disks are contained inside a chamber fitted with standardized illumination. This kind of colorimeter is much used for tomatoes and tomato derivatives, as will be seen in Sec. IV.A.4.

b.   Subtractive Colorimeters   As has been said, with subtractive colorimeters portions of white light are eliminated by means of filters until the sample is matched.

The most representative instrument in this group is the Lovibond tintometer. It was designed over 150 years ago (in 1833) by Joseph William Lovibond, a brewer in Salisbury, England, to measure the color of his beer.

Although it has been modified a certain number of times, the basic concept has not changed. The sample and the standard are observed through a viewer constructed with prisms and mirrors so that to the operator the visual field appears as a circle divided into two equal parts. To measure a color, the two parts have to be matched. A series of numbered filters—red, green, and blue—can be placed in front of a white block, so that it is possible to form a series of graduated hues ranging from very bright to very dark (from 0 to 10), until the color of the sample is matched. For each color there is a wide range of intensities. Each filter is marked with its value in Lovibond units in relation to the unit to which it is equivalent. If two or more filters of the same color are used, the values are added. The units for the filters are arbitrary, but they are related to the three basic standards. When the color filters that come closest to matching the sample appear matt by comparison, the sample is defined as glossy and neutral tint filters are placed in front of the sample. The value of the neutral tint filter(s) added gives a measurement of the gloss of the sample.

This apparatus is very widely used in the food industry, where the number of filters can be reduced considerably when it is used for specific purposes with a single product: oil, margarine, malt, beer, sugar solutions, etc. A factor that has contributed to its popularity is the fact that the tintometer company has managed to keep the calibration of its filters constant over the years.

The instrument was supplied together with a set of nomograms to convert Lovibond units into X, Y, Z units. This operation, which used to be fairly awkward, has been greatly simplified with the aid of computer programs.

The main advantages of the Lovibond colorimeter are (a) the fact that it uses the human eye as the detector, which is important as color is a human sensation; (b) the consistency of its filters over the years; (c) the simplicity of the apparatus.

The disadvantages to be considered are (a) the natural fatigue of the observer after a certain number of measurements, as occurs with any instrument using a human detector; (b) for a given sample, the results can vary from one day to another and from one observer to another.

C. Instrumental Measurement

1. Spectrophotometers

Spectrophotometers are instruments designed to measure the transmission of light through liquid samples. To measure the light reflected by opaque solid samples they are connected to a suitable accessory. In this way they can measure the quantity of light reflected from the surface, over the whole visible spectrum, as a fraction of the light reflected by the white standard illuminated in the same way. This fraction is called the *reflectance*.

The standard blank used is supplied with the apparatus. It tends to be white marble or a hard white plastic. In either case, periodically it should be calibrated with barium sulfate to test its stability and to take account of possible variations that may have occurred with the passage of time.

Measurements performed with these instruments provide a detailed reflectance spectrum from which complete information about the color of the sample measured can be obtained. The old disadvantage of the slowness of the measuring process has been overcome in modern instruments, which supply color information in a few seconds, displaying the spectral reflectance curve and data for the spectral sweep on the screen. These data can, of course, be directed to the printer.

There is still the problem of price and a further difficulty that has not yet been mentioned—that of placing the sample in the apparatus. With a spectrophotometer it is easy to measure the color of a piece of paper, plastic, cloth, cardboard, etc., but it is difficult and often impossible to insert samples such as an apple, or asparagus, or a pimiento.

2. *Tristimulus Colorimeters*

One of the instruments most commonly employed for measuring the color of food products is the tristimulus colorimeter. With this apparatus the integration procedures that incorporate the standard observer curves (weighted ordinates and selected ordinates) are replaced by the use of various suitable filters that simulate the response.

A diagram of the tristimulus colorimeter is shown in Fig. 8. A beam of white light from an illuminant strikes the sample at an angle of 45° and is reflected at right angles to the surface of the sample (diffuse light). The beam of light then passes through the three X, Y, Z filters and is measured by a photocell. Another way of obtaining diffuse light is by means of an integrating sphere. The sphere is hollow, painted white on the inside with $BaSO_4$. Consequently, the light that enters is reflected in all directions, striking the sample as diffuse light.

The instrument is calibrated, as we said, with a white plate and a black plate. Recalibration with a plate of a similar color to the product to be measured improves the accuracy of the measurement considerably (15).

The kind of light used has less importance here because when it is modified by a filter and a detector the result is similar to the response of the human eye when it sees an object illuminated by a standard light source.

Commercial Wolfram lamp x suitable filter x spectral sensitivity of detector = standard illuminant x response of standard observer

The key lies in achieving a lamp–filter–photocell combination that exactly reproduces the response of the human eye. The lighting system in these colorimeters is basically produced in one of two ways: (a) with a Wolfram incandescent lamp or (b) with a xenon flash lamp.

Wolfram lamps need to be heat-stabilized for a certain time and must then be kept on until the measurements are finished. Consequently, their useful life is only a small percentage of the time that they remain on. The lighting provided by these lamps may be a light beam that is concentrated by a lens before striking the sample or circular lighting, where the light from the lamp is reflected onto the sample by a ring of mirrors.

The light beam system is suitable for products with a smooth surface for taking readings, such as tomato purée, peanut butter, canned pimientos, etc., whereas the mirror ring system is useful when the reading might vary with the orientation of the sample or when there is some danger of shadows, e.g., peas, rice, beans, etc.

Examples of colorimeters that use a Wolfram lamp include Hunter, Gardner, Elrepho, and

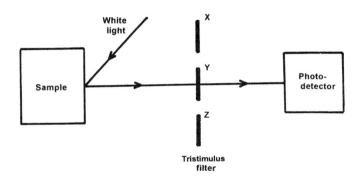

Fig. 8 Essential elements of tristimulus colorimeter.

MOM (a list that should not be considered complete or commercial, or to show a preference for any particular manufacturer).

The use of xenon flash lamps offers the advantage that there is no "dead time" when they are used. The lamp is only on for the moment of measurement. To measure color, the sample is connected to the apparatus, lit with the flash, and the reflected light is read. A necessary condition if all the measurements are to be comparable is that all of the flashes should be identical. Examples of these colorimeters are the instruments produced by Dr. Lange and Minolta.

The first colorimeter models only gave readings in terms of L, a, b, but with present day computer-controlled models the color is expressed in various scales, and with some models even in Munsell notation.

Tristimulus colorimeters have been very successful since they first appeared on the market. Their advantages by comparison with classical spectrophotometers are that (a) they are cheaper, (b) they respond faster, (c) they are easier to handle, and (d) they measure so quickly that the problem of sample heating is eliminated.

The main disadvantage of these instruments is that they do not give exact measurements. A color cannot be defined in terms of a measurement by a tristimulus colorimeter. However, their measurements of color differences are exact.

The increasingly widespread use of computer-controlled spectrophotometers has not finished off tristimulus colorimeters, which still have their particular field of applications, and this has even been expanded by the use of portable instruments.

3. Spectrocolorimeters

As their name indicates, these instruments measure color from its reflection spectrum. They could have been included in the group of spectrophotometers, but because of their particular characteristics we have chosen to keep them in a separate section.

They are designed to measure the color of opaque bodies on the basis of a simplified spectral sweep, every 20 or at most 10 nm within the visible spectrum. As these instruments are computer-controlled, they take only a few seconds to calculate any colorimetric parameter from the spectral reflectance curve. Normally readings are given in four scales: L, a, b; X, Y, Z; x, y, Y; L^*, a^*, b^*. They draw the spectral reflectance curve and give numerical data for the spectral sweep and they can reference the color at two viewing angles with various illuminants: A, C, D65, etc. They can also calculate a huge number of colorimetric indices and of course store a large quantity of data in memory and fix tolerances in relation to a color or sample established as a standard. As for convenience of use, some have a number of diaphragms for measuring, other are fitted with an optic fiber so that the unit can be moved to the sample, etc.

These instruments started off as simplified spectrophotometers, taking readings every 20 nm to deduce the spectral reflectance curve, but nowadays there are fully computer-controlled models which give a reading every nanometer and are backed up by a vast array of programs.

By way of example there are Labscan and ColorQuest, manufactured by Hunter; Spectrograd from Gardner; Color Eye from Macbeth; the instruments produced by Minolta; and Dr. Lange's Xeno-Color.

IV. FOOD ANALYSIS APPLICATIONS

A. Color Requirements in Quality Regulations

1. Verbal Description

The vast majority of regulations simply describe the color of food products verbally, which is a very imperfect way of doing so because language has a limited vocabulary, which gives very little idea of the varied range of colors that may be covered by these words.

2. Reference to Dictionaries

One way to overcome the shortcomings just mentioned is by comparing the color of the food product in question with standards in a color dictionary (16).

The comparison of sample and standard must be made in standardized illumination, which can be achieved by taking the measurements in a standardized lighting chamber or cabin. This consists of a large box without a front. The surfaces are painted neutral gray on the inside and there is a standardized illumination that includes at least C or D65 illuminants. The illumination/observation conditions should always be 45/0 or 0/45.

As an example we could mention the USDA regulations for sauerkraut (17). These regulations describe the color of sauerkraut by reference to the "olive buff" and "dark olive buff" standards in Ridgway's "Color Standards and Nomenclature" (Table 4).

Another example of the use of a dictionary is the measurement of the color of canned peeled tomatoes according to Canadian regulations (18). In this case the color is described by reference to plates 3, I-2 and 4, F-12 in Maerz and Paul's *Dictionary of Color* (10) (Table 5).

We could also quote the case of canned tuna, where the quality depends on how light or dark the color is. United States legislation (19) distinguishes quality levels in terms of Munsell values (color dict.): *white*, over 6.3; *light*, between 6.3 and 5.3; and *dark*, below 5.3.

To measure the color, the tuna sample is first pressed through a quarter-inch sieve. The sieved product is mixed well and placed in the measuring cuvette, and the surface is pressed down and flattened. After 10 min it is compared with the Munsell standards, the sample and the standard being of equal area and not greater than 2 in. across. They are placed together in a chamber painted matt gray to avoid reflections. They are lit at an angle of 45° with 100-W tungsten lamps and the sample is viewed perpendicularly.

3. Reference to Standards

A more convenient and practical method than the use of color dictionaries is that of developing specific standards for particular foods. These standards can be printed on a plate to define the different levels or they can be pieces of plastic in the form of strips or tubes, depending on the circumstances.

The comparison of sample and standard should be made under standardized conditions such as those described above in Sec. IV.A.2.

An example of these standards being quoted in U.S. legislation is the case of frozen French-fried potatoes, for which there is a strip showing the colors of dark and light fried potatoes. There are also plates to grade canned pimientos (20) and others for peanut butter (21). Finally, we could quote the standards used for measuring the color of orange juice. United States regulations (22) classify the color of orange juice by means of six plastic test tube–shaped standards, 1 in. in diameter. The tubes range in color from intense orange to light yellow, coded as OJ 1 to OJ 6.

The juice being studied is placed in a tube that is the same size as the standards. The comparison

Table 4 Sauerkraut: USDA Standards (1985)

Grade	A	C	Sstd
Points	15–13	12–10	9–0
Normalized color standard (Ridgway)	Equal or lighter than "olive buff"	Not darker than "dark olive buff"	Darker or pink tinted color

Table 5 Canned Tomatoes: Canadian Standards (1985)

Maerz and Paul's "Dictionary of Color	"Tomato red" (plate 3, I-2)	(Orange reddish) (plate 4, F-12)
"Practically uniform good red color"	Not less than 95% of surface areas	Not more than 5% of surface areas
"Fairly good red color"	Not less than 75% of surface areas	Not more than 25% of surface areas
"Reasonably good red color"	—	The comminuted drained solids from any one can are not less red

is made by placing both the standards and the sample on a rack, so that the lighting is at 45° and observation is perpendicular to the tubes.

4. Reference to Colorimeters and Spectrophotometers

The colorimeter most generally used for measuring the color of food products is the disk colorimeter, particularly with reference to tomato products. The color of tomato concentrate and ketchup is defined by reference to Munsell colors (18,23,24). The standard colors are the result of mixing four colors—5R 2.6/13, 2.5YR 5/12, N1, and N4—according to specified proportions (Table 6). The regulations in France make the same requirements (25). For tomato juice and peeled tomatoes the U.S. regulations use the same standards (26).

Legislation in the United States (27) also defines the color of black olives in terms of measurements made with disk colorimeter but using the standards 5R 4/14, 2.5Y 8/12, and N1 (Table 7).

Another instrument quoted in legislation is the spectrophotometer. For canned tuna, U.S. legislation (19) indicates that its reflectance can be measured at a wavelength of 555 nm, given that a luminous reflectance of 33.7% or over corresponds to the *white* classification, under 22.6% to *dark*, and *light* falls between the two. All of these data relate to magnesium oxide.

B. Special Proposals

In addition to the official color standards quoted, there are a great many other nonofficial ones. Some are proposed by various organizations and others are simply developed by researchers in order to provide objective references for the color control of different foods.

Among the proposals made by various nonofficial organizations we can mention first the Munsell disks, developed for various foods but not featured as a compulsory requirement in the legislation in any country. The disks would apply to tomato products, egg yolks, beans, green peas,

Table 6 Tomato Ketchup USDA Standards (1992)

Grade	A	C	Sstd
Points	25–21	20–17	17–0
Red (%)	65	53	—
Orange (%)	21	28	—
Black + Gray (%)	14	19	—

Table 7 Canned Ripe Olives: USDA Standards (1977)

Grade	A	B	C
Points	30–27	26–24	23–21
Red (%)	3.5	6	6
Yellow–red (%)	3.5	6	6
Black (%)	93.0	88	88
% units equal or darker	90	80	60

beef, carrots, milk, beets, asparagus, semolina, macaroni, etc., and are obtainable from the Munsell Color Co., PO Box 950, Newburgh, NY 12550.

Among the reference colors developed at the Instituto de Agroquímica y Tecnología de Alimentos CSIC (Valencia, Spain) are those for pimientos (28), peeled tomatoes (29), asparagus (30), peas (31), tomato concentrate (32), and green beans (33).

In Japan standards have been developed, on cards, for various kinds of fresh fruit (34).

At the Instituto de la Grasa CSIC (Seville, Spain) standards have been developed to compare the color of olive oil (35) by mixing together the three following solutions in different proportions:

KH_2PO_4 (1/15 M)
$Na_2HPO_4 \cdot 2H_2O$ (1/15 M)
Bromothymol blue   (0.04%)

In addition to the standards used for comparison, whether reference plates or liquids, there is a series of numeric indices obtained from instrumental measurements that can be used to classify the color of food products.

For paprika there is the index developed by the American Spice Trade Association, known as the ASTA value (36), which is given by the expression:

$$\frac{(A_{460})(16.4)(\mathrm{IF})}{(\mathrm{g\ of\ sample})}$$

where A_{465} is the absorbance at 465 nm and IF an instrument correction factor for a filter supplied by the National Bureau of Standards (NBS SRM 2030):

$$\mathrm{IF} = \frac{A_{465}\ \text{given by the NBS}}{A_{465}\ \text{measured each time}}$$

For raw tomato the food industry in California frequently uses the Agtron grade, which is given by the expression:

$$G = 276\,\frac{(X_G - 0.7)}{(X_R - 0.7)}$$

where G = Agtron grade or TCI (tomato color index)
    X_G = reflectance at 546 nm (green)
    X_R = reflectance at 640 nm (red)

The value of G can be calculated from the formula or obtained directly by means of an Agtron colorimeter (37). These colorimeters were designed to measure the color of tomatoes and tomato derivatives. Basically they measure the spectral reflectance at two wavelengths, giving the nondimensional relationship known as the Agtron grade.

There is also a color index for orange juice known as the color number (CN), which used to be defined in terms of the colors known as citrus red (CR) and citrus yellow (CY), determined with a Hunter citrus colorimeter. As this apparatus is no longer manufactured, the CN value is deduced from the values of X, Y, Z measured with a Minolta 2002 from the equation (38):

$$CN = 56.5X/Y - 18.4Z/Y + 48.2/Y - 8.57$$

A color index has also been proposed for green olives (39):

$$i = \frac{4R_{635} + R_{590} - 2R_{560}}{3}$$

by which they can be classified into different grades.

More recently, values of $a*$ have been deduced that can be used to distinguish between different kinds of quince jelly (40).

Black olives can be color-classified in terms of their reflection percentage at 700 nm (41) and can thus be sorted into five different color grades.

Recently a flour color index (FCI) has been proposed, defined as $L* - b*$, which corresponds very closely to sensory evaluation (42).

V. OTHER VISUAL PARAMETERS

A. Gloss

The apparent color of an object can vary when the viewing angle varies. The cause of this variation is gloss, which can be defined as the fraction of light reflected in the specular direction or, expressed more simply, the degree to which it resembles the surface of a mirror.

The appropriate instrument for measuring gloss is a goniophotometer. It is designed to measure the quantity of light reflected when the viewing angle changes (37). The graphs representing the change in reflectance with the change in viewing angle are called goniophotometric curves and measure luminous directional reflectance in relation to the angle of illumination and the viewing angle. A goniophotometric curve is produced by fixing the light source at a particular angle and measuring the light reflected at different viewing angles, although it is also possible to use reciprocal conditions by fixing the viewing angle and varying the angle of illumination. The goniophotometric curve of a surface is the basic determination of gloss, just as the spectrophotometric curve determines color.

Examples of goniophotometric curves are given in Fig. 9. These curves show the quantity of light reflected in various directions with the sample illuminated at 45°. The reflectance factor is plotted on the vertical axis on a logarithmic scale and the angle of reflection is plotted on the horizontal axis. The sample with the highest gloss is the one that produces the highest peak. The light reflected at an angle of −45° represents the light reflected in the specular direction and gives an indication of the gloss of the sample.

In practice it is generally not necessary to obtain the whole goniophotometric curve and it is sufficient to take a few measurements for various selected angles of illumination and vision. For this purpose there are simplified goniophotometers called glossmeters, which measure at three angles of reflection (20, 60, 85°) that correspond to equal values for the angles of incidence. The angle of 20° is used for very glossy surfaces, 60° for those with normal gloss, and 85° for those with low gloss. The gloss data in this method are obtained by comparing the reflectance of the sample with a standard glass. A plate of highly reflecting flat glass is used as a calibration standard, and this is given a value of 100 for each geometry. The plate comes with the apparatus, but also the NBS supplies standards for calibrating these instruments.

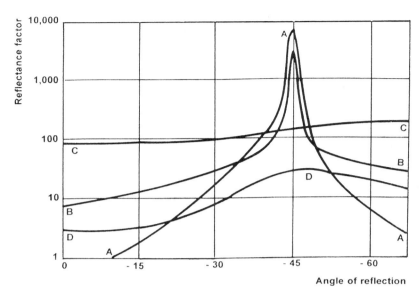

Fig. 9 Goniophotometric curves for four samples. Spectral reflection is indicated at 45° viewing angle.

However, the measurement of gloss and its application in the field of food technology is not widespread despite its undoubted importance in relation to the appearance of many products (5).

B. Translucency: Kubelka–Munk Equations

In the food industry there are many translucent products, which are neither totally opaque nor completely transparent, whereby the light is absorbed, reflected, transmitted, and dispersed or scattered. For this kind of product the objective measurement of color presents serious difficulties and if is carried out in the conventional way it is very likely that a good sensory/instrumental relationship will not be obtained. The problem may be due to the fact that one is not measuring the same thing with the optical instrument as with the eye.

A colorimeter cannot evaluate visual appearance. It simply records the proportion of light striking the photocell in relation to the standard. The correct detection of differences between samples, whether visual or instrumental, is considerably affected by the way in which the sample is presented, and this can make measurement easier or more difficult. Factors to be taken into account are (a) the light source, (b) the size of the sample, (c) the container in which the sample is placed, and (d) the color of the bottom of the cuvette containing the sample. If a good sensory/instrumental correlation is still not found, this may be because one or both of the evaluations are incorrect. One way to check the validity of the method is by progressively modifying the parameter being studied (hue, saturation, or luminance) in a controlled way.

The instrumental problems posed by the definition and measurement of the color of cloudy or translucent food products can be overcome by applying Kubelka–Munk thin-layer analysis (43). These authors propose a measurement for the reflection of a thin sample against a white background and a black background by relating it to a coefficient of absorption (K) and a coefficient of scattering (S):

$$\frac{K}{S} = \frac{(1 - R_\infty)^2}{2R_\infty}$$

where R_∞ is the reflectance of a layer, described as infinitely thick, which is so thick that if the thickness is increased there is no change in reflectance. The value of this parameter is obtained from measurements made with the black and white backgrounds using the equations:

$$R_\infty = a - b \qquad a = 1/2 \frac{R + (R_0 - R + R_g)}{R_0 R_g} \qquad b = (a^2 - 1)^{1/2}$$

where

R = reflectance of the sample against a white background
R_0 = reflectance of the sample against a black background
R_g = reflectance of the white background

In many cases it may be practical to replace each of the R values with one of the corresponding tristimulus values X, Y, Z, or L. In certain cases better correlations with the sensory evaluation have been obtained by using K/S values calculated in this way (44).

ACKNOWLEDGMENTS

This work was supported by Comision Interministerial de Ciencia y Tecnologia (Spain). Projet ALI94-0731.

REVIEWS

F. J. Francis, in *Engineering Properties of Foods* (M. A. Rao and S. H. Rizvi, eds.), Marcel Dekker, New York, 1995, p. 495.
T. J. Mabon, *Cereal Food World 38*, 21 (1993).

REFERENCES

1. A. Kramer and B. A. Twigg, *Fundamenals of Quality Control for the Food Industry*, 2nd Ed., AVI, Westport, 1970, p. 16.
2. J. B. Hutchings, *Sensory Properties of Foods* (G. G. Birch, J. G. Brenan, and K. J. Parker, eds.), Applied Science, London, 1977, p. 45.
3. D. L. MacAdam, *Color Measurement: Theme and Variations*, Springer-Verlag, Berlin, 1985, p. 9.
4. G. MacKinney and A. C. Little, *Color of Foods*, AVI, Westport, 1962, pp. 97, 196.
5. R. S. Hunter and R. W. Harold, *The Measurement of Appearance*, 2nd Ed., John Wiley and Sons, New York, 1988, pp. 141, 273.
6. K. MacLaren, *The Color Science of Dyes and Pigments*, 2nd ed., Adam Hilger, Ltd., Bristol, 1986, p. 134.
7. T. J. Mabon, *Cereal Foods World 38*: 21 (1993).
8. D. Farnsworth, *The Farnsworth–Munsell 100-Hue Test*, Munsell Color Company, Baltimore, 1957.
9. Munsell Color Company, *Munsell Book of Color*, Munsell Color Company, Baltimore, 1966.
10. A. Maerz and M. Paul, *A Dictionary of Color*, 2nd Ed., McGraw-Hill, New York, 1950.
11. C. Calvo, *Rev. Agroquím. Tecnol. Aliment. 29*: 15 (1989).
12. D. H. Voss, *HortScience 27*: 1256 (1992).
13. A. Hard and L. Siuik, *Color Res. Appl. 6*: 128 (1981).
14. A. Nemecsics, *Color Res. Appl. 12*: 135 (1987).
15. M. Kent and S. Porretta, *Industria Conserv. 65*: 9 (1990).
16. C. Calvo, *Rev. Agroquím. Tecnol. Aliment. 32*: 589 (1992).
17. USDA. *United States Standards for Grades of Sauerkraut*, May 13, Washington, DC, 1963.
18. Ministry of Supply and Services Canada, Canada Agricultural Products, *Standards Act. Processed Products Regulations*, Ottawa, 1985, p. 105.

19. FDA. *Food and Drug Administration Standards of Identity. Canned Tuna*, Dec. 12, Washington, DC, 1990.
20. USDA. *United States Standards for Grades of Canned Pimientos*, October 23, Washington, DC, 1967.
21. USDA. *United States Standards for Grades of Peanut Butter*, February 5, Washington, DC, 1972.
22. USDA. *United States Standards for Grades of Orange Juice*, Dec. 10, Washington, DC, 1982.
23. USDA. *United States Standards for Grades of Canned Tomato Puree (Tomato Pulp)*, May 1, Washington, DC, 1978.
24. USDA. *United States Standards for Grades of Tomato Catsup*, August 31, Washington, DC, 1953.
25. Ministre de L'Agriculture, Normes de fabrication des Conserves de Produits Agricoles. Conserves de Légumes. Décision No. 8. Conserves de Purée de tomates. Centre Technique des Produits Agricoles, Paris, 1983.
26. USDA. *United States Standards for Grades of Canned Tomatoes*, April 13, Washington, DC, 1990.
27. USDA. *United States Standards for Grades of Canned Ripe Olives*, July 24, Washington, DC, 1977.
28. E. Primo, L. Duran, A. M. Bermell, and N. McLean, *Rev. Agroquim. Tecnol. Aliment. 10*: 258 (1970).
29. N. McLean, A. M. Bermell, and L. Duran, *Rev. Agroquim. Tecnol. Aliment. 10*: 84 (1970).
30. N. McLean, and L. Duran, *Rev. Agroquim. Tecnol. Aliment. 9*: 396 (1969).
31. M. J. Alcedo, L. Duran, and A. M. Bermell, *Rev. Agroquim. Tecnol. Aliment. 13*: 124 (1973).
32. M. J. Alcedo, L. Duran, and A. M. Bermell, *Rev. Agroquim. Tecnol. Aliment. 13*: 583 (1973).
33. E. Primo, L. Duran, and A. M. Bermell, *Rev Agroquim. Tecnol. Aliment. 10*: 550 (1970).
34. T. Yamazaki and K. Suzuki, *Bull. Fruit Tree Res. Stn. (Japan) A7*: 19 (1980).
35. R. Gutierrez Gonzalez-Quijano and F. Gutierrez Rosales, *Grasas y Aceites 37*: 282 (1986).
36. J. E. Woodbury, *JAOAC 60*: 1 (1977).
37. F. J. Francis and F. M. Clydesdale, *Food Colorimetry: Theory and Applications*. The Avi Pub. Co., Inc. Westport, Conn. (1975) p. 186, 108.
38. B. S. Buslig, *Proc. Fla. State Hort. Soc. 106*: 262 (1993).
39. A. H. Sanchez Gomez, L. Rejano Navarro, and A. Montaño Asquerino, *Grasas y Aceites 36*: 258 (1985).
40. C. Calvo and A. Gambaro, "Puesta a punto de un metodo objetivo para clasificar dulces y cremas de membrillo en funcion de su color," Proceedings II Congreso Nacional de Color. Paterna, Valencia, Spain, 1991, p. 7.
41. M. J. Fernandez Diez and A. Garrido Fernandez, *Grasas y Aceites 22*: 193 (1971).
42. J. R. Oliver, A. B. Blakeney, and H. M. Allen, *Cereal Chem. 69*: 546 (1992).
43. D. B. Judd and G. Wyszeki, *Color in Business, Science and Industry*, John Wiley and Sons, New York, 1967, p. 387.
44. C. Calvo, *Rev. Agroquím. Tecnol. Aliment. 33*: 597 (1993).
45. J. M. Artigas, J. C. Gil, and A. Felipe, *Rev. Agroquím. Tecnol. Aliment. 25*: 316 (1985).

6

Sensory Evaluation Techniques

Giovanni Parolari
Experiment Station for Food Processing and Preservation, Parma, Italy

I. INTRODUCTION

Since their first applications in grading goods for the armed forces and hospitals, sensory analysis and product testing have become popular in marketing research as well as in quality assurance and in research and development. The advances from primitive, empirically conducted organoleptic evaluations to modern sensory science have been outlined by Pangborn (1). Today the widespread interest in sensory disciplines is witnessed by the increasing number of articles published in journals dealing with food science and in the three sensory-related *Journal of Sensory Studies*, *Food Quality and Preference*, and *Journal of Texture Studies*. Several straightforward and comprehensive books and papers are available to researchers wishing to start practicing sensory techniques; they range from the mechanism of perception (2,3) to methodology and standardized testing procedures (4–15) up to experimental design and statistic or chemometric methods for the interpretation of experimental data (16–22). Getting acquainted with the principles of sensory evaluation is recognized to be a requirement for minimizing the biases and errors that are likely to occur in everyday sensory work; in fact, as stated by Meilgaard et al. (7), though sensory science has become a science over the years, it is still an inexact science.

II. PRODUCT TESTING AND SENSORY ANALYSIS

In a broad sense, the ultimate goal of sensory evaluation of a food product is to predict its acceptance. As long as product testing pertained to marketing management, a number of simple appraisal techniques were conducted to answer questions dealing with products of the competition or to forecast the success of a new item, as compared with a traditional one. For these purposes, a few so-called affective tests have long been used with consumer panels made up of untrained people, usually drawn from the local market. In such tests the terminology in use is naive, resembling the household vocabulary for grading goods. Typically, respondents in consumer tests are asked to record overall liking or intent of purchase and they need not be trained because consumers, who are the ultimate target of testing, are not trained to taste prior to purchase. For these reasons, a consumer team must be somewhat numerous to compensate for the bias typical of untrained judges.

From the late 1960s, in a rather new and original approach to product testing, expert panels have been used to describe food properties through selected attributes and rate them accordingly. A basic principle underlying this type of appraisal is that human senses are to be regarded as instruments for the measurement of specific odor, flavor, or texture stimuli peculiar to the product. Consequently, subjects to enter an expert panel should be selected, trained, and periodically checked just as an instrument is calibrated from time to time. The expert panel is expected to be highly reproducible and consistent in the long run; thus only a few members will suffice but will be chosen from a wide pool of candidates, usually recruited from clerical personnel within the company or the department where the analyses will be carried out. The main differences between expert and consumer panels are outlined in Table 1.

Results from trained panels may contain huge amounts of information and require specific statistics to interpret; multivariate analysis proves helpful in this sense, for sensory problems are generally multivariate problems. Moreover, in many sensory applications experiments have to be carefully designed to obtain meaningful data and to achieve some major objectives such as product optimization, category classification, or to find correlations between sensory and physical or chemical properties of foods. As sensory analysts have made use of experimental design and statistical methods a new and rigorous class of sound testing procedures have become available to practitioners dealing with characterization and analysis of food products. Such procedures look quite different from previous grading schemes, which still thrive in marketing simulation tests and rely on numerous unexperienced respondents. The term *sensory analysis* is currently used to denote the standardized and codified methodology, as opposed to trade-oriented consumer or product testing. Scientifically conducted sensory evaluation is increasingly acknowledged to be a powerful tool of food analysis, capable of providing as original and relevant information as well-established instrumental techniques.

According to the subject of this book, the present chapter will deal with the application of sensory analysis in research, development, and in the quality control of foods; readers interested in consumer testing should refer to specific books and papers (23,24) as well as to the trade-oriented *Journal of Marketing Research* and *Journal of Consumer Research.*

Whichever the application in flavor and texture analysis, the development and use of a sensory panel is essentially a multistep problem (Table 2). Each step has been deeply studied in the last decades and originally different techniques were eventually developed by pioneering scientists and consolidated into more integrated available procedures. Some major procedures and guidelines are described in the following paragraphs with the aid of examples based on original or literature data.

Table 1   Profile of Expert vs. Consumer Panel

Factor	Expert	Consumer
Recruitment	Usually from the company or department	Local community
Members	At least 6	At least 40
Training time	At least 1 month	Little or null
Usage of item investigated	Not needed	Must use and accept
Terminology	Selected attributes	Scanty
Motivation	High, if properly sustained (participation, discussion and share of results, etc.)	High, if rewarded

Table 2 Flow Sheet for Panel Development and Use

Recruitment of panelists
Screening of panelists
Selection of attributes
Panel training
Performance evaluation
Method validation
Method application

Specific selection and training tests have been designed to fit the product to be assessed, the kind of analysis (texture, flavor, taste, odor), and the type of test (descriptive, difference, ranking, etc.). Nevertheless the procedures that can be followed for the preliminary panel recruitment and screening are of general validity. The guidelines contained in the ASTM Special Technical Publication 758 *Guidelines for the Selection and Training of Sensory Panel Members* (25) and in the ISO *Guide for the Selection and Training of Assessors* (26) are generally accepted. As will be shown in the following paragraphs, the recommendations for the selection of personnel overcome the traditional individual check based on recognition of the four basic tastes—acid, salty, bitter, and sweet—and tend to include health and behavioral aspects that were neglected in the past.

III. RECRUITMENT (27–29)

To allow complete selection of candidates, the panel leader ought to prescreen twice as many or more people than those entering the final panel. Candidates are often recruited from the internal personnel; in this case, one should take into account problems of time allocation between participation in the panel and other work within the company; such conflicts are very likely to occur in the long term, impairing the assessor's performance; therefore, panelists involved in an extended sensory task should be recruited from external participants, as is commonly done with consumer tests (23).

The following prerequisites should be met by the candidate prior to the screening and training stages.

A. Health

The criterion is that no disease should be related to the sensory properties being evaluated. In particular, dentures and food allergies may affect flavor and texture assessment, as do chronic cold and rhinitis with odor and fragrances. Hypertension, diabetes, and metabolic disorders are possible drawbacks and the effect of ingredients and additives should be inspected in this respect.

B. Individual Attitude

Subjects must be interested in participation and must be available throughout training and for at least 80% of the test sessions. If they are to be involved in consumer tests, they should declare their acceptance of and acquaintance with the food item under examination. In most cases, the information at the last two points is gathered by telephone or by simple questionnaires.

IV. SCREENING (30,31)

Candidates having passed the recruitment tests shall qualify for subsequent acuity tests based on detection and discrimination of standard substances at given concentrations. Whichever their use in flavor analysis, the panelists will have first to recognize the four basic tastes. Those successful will eventually undergo discrimination tests based on the same references or on actual products for given attributes. While passing the preliminary four-taste check is a requisite that cannot be set aside, some degree of tolerance is allowed in the latter stages, as described in the following paragraphs. Typically, screening is a two-step procedure, aimed at the candidate's ability to detect threshold amounts of reference substances and to rank different concentrations of the same compounds. Table 3 reports standards that may be used for this purpose.

A. Detection of Basic Tastes

Reference solutions will be used to check individual attitudes to detect standard substances at the threshold level.

1. Test Procedure

Present participants with four three-digit coded identical glasses or cups containing the same volume of threshold solutions. Serve in random order and provide a form as shown in Table 4. Ask subjects to taste the samples and fill in the form following instructions therein. The test may be repeated several times.

2. Test Results

In principle, subjects should be thoroughly correct to qualify; however, since the four-taste check is being considered less informative by modern sensory analysts, the project leader may plan to reject people only after a second, easier test based on more concentrated standards (typically, 5 × threshold levels in Table 3). Poor candidates unsuccessful with the latter check will be dropped from the screening process. The others pass to the following.

B. Ranking Test Procedure

Present coded solutions for each of the four tastes and ask candidates to rank them according to increasing perceived intensity. Use the form as given in Table 5. Reject tasters unable to order correctly. Some panel leaders allow mistaking one sample for the adjacent one.

V. TRAINING (32–35)

This is the kernel of the sensory project, representing the true investment in human resources. During this step, prescreened panelists are familiarized with the test procedure and get acquainted with product properties and descriptors. Depending on the complexity of subsequent applications,

Table 3 Standard References for Recognition and Ranking Tests

Taste	Substance	Concentration[a]				
		Threshold	Ranking			
Sweet	Sucrose	8.0	9.0	18.0	26.0	72.0
Acid	Citric acid	0.20	0.25	0.50	1.0	2.0
Salty	Sodium chloride	0.25	1.0	2.0	4.0	8.0
Bitter	Caffeine	0.25	0.3	0.6	1.2	2.4

[a]In grams per liter of tasteless water.

Table 4 Score Sheet for Detection Test

Taste and assign the coded samples to the four basic tastes

Sweet: Code #_____   Acid: Code #_____   Salty: Code #_____   Bitter: Code #_____

Date_____   Signature_____

training may take days or months; in any case, assessors passing training and related tests will join and augment the company's potentiality to control manufacturing and develop new items. Basic instruction of panelists will deal with the following steps.

A. Training Individuals for Descriptive Analysis

It is a principle of modern sensory analysis that food properties can be described in terms of sensory attributes and that attributes can be used for both qualitative and quantitative analysis. When properly trained, panelists prove able to detect and describe the perceived sensory attributes in a consistent and reliable manner. For this purpose, they must become familiarized with scaling, terminology, and testing procedures.

1. Scaling

Individual reactions to flavor, odor, or texture stimuli can be recorded with the aid of numbers and words to report the intensity of perceived attributes. The measurement of sensory responses has long been based on *classification* of samples into nominally differing groups and on *ranking* by ordinal scales. As long as descriptive techniques have been accepted in product development and maintenance, the sensory analyst has turned to apparently more complex but also more informative scaling based on numerical scales (*rating*) and line scales (*scoring*) relying on equally spaced value intervals (Fig. 1).

   a. Rating   Also called numerical or category scaling, rating is a measurement technique in which the sample attributes are assigned a value on a numerical scale. The scale width has to span the full range of attribute intensities and allow sufficient sensitivity by adequate segmentation. This is normally achieved by choosing a category scale from 0 to 9, but scale ranges from 0 to 5 and 0 to 15 are also encountered, depending on the number of steps the assessor will be able to afford. As with numeral scales, the two extremes are anchored with minumum and maximum reference stimuli.

   b. Scoring   The attribute intensity is recorded by placing a mark on a 15-cm or 6-in. linear scale anchored either at the extremes of 1.5 cm from both ends. The use of line scales is

Table 5 Score Sheet for Ranking Test

Order samples for increasing perception of the four tastes.			
		Taste	
	Sweet	Acid	Salty	Bitter
Intensity	Code #	Code #	Code #	Code #
Low	_____	_____	_____	_____
	_____	_____	_____	_____
	_____	_____	_____	_____
High	_____	_____	_____	_____

a- rating

Number	Anchor term
0	No salt at all
1	Slight
2	Moderate
3	Distinct
4	Strong
5	Extremely salty

b- scoring

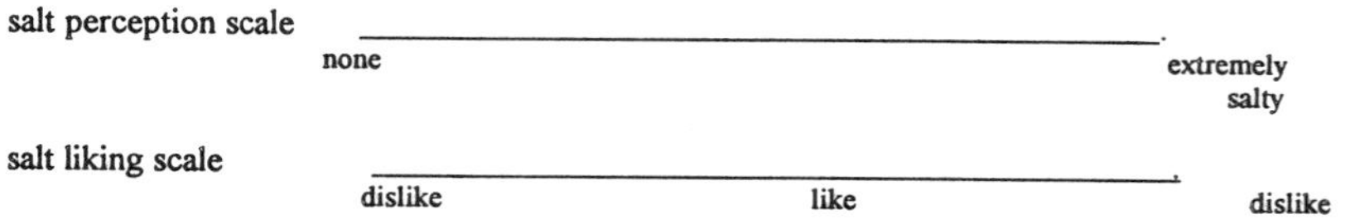

salt perception scale

none   extremely
                                                                            salty

salt liking scale

      dislike   like   dislike

Fig. 1   Sensory evaluation of saltiness by (a) numerical and (b) line scales.

straightforward, with the left end corresponding to absence and the right end to very strong perception of the stimulus. However, depending on the type of attribute, the two extremes may not correspond to diametrically opposed perceptions, as shown in Fig. 1b, where salt liking is maximum at about intermediate salt perception. In these cases, to improve understanding, one might consider splitting the scale into two one-way scales. The use of line scales means further burden to the analyst, who must convert marks in the line to numbers for subsequent use and interpretation.

Thanks to their ability to provide quantitative sensory data as do other instrumental methods, scaling techniques have gained in popularity among sensory scientists, who enjoy the opportunity to treat the results with a wide class of procedures available with computer statistical packages. A major pitfall of numerical and, to a lesser extent, line scales is that panelists tend to use only part of the scale, namely, the middle one. Typically, assessors are reluctant to assign a sample extreme scores such as 0 or 9 in order to keep them for hypothetically forthcoming lowest and highest intensities. Since such extreme scores are unlikely to occur with most samples, the scale might end up being misused, resulting in an overcrowding of samples around intermediate scores.

The aforementioned drawback, as well as the need for consistent scale usage over time, is usually tackled by training assessors with meaningful reference material, prepared by the leader or, better, picked up from the market. The comprehensive list of standard references contained in Spectrum intensity scales represents a milestone for the analyst coping with the training of panelists for flavor, odor, and texture evaluation. As shown in the excerpted list in Table 6, the intensity values suggested for the four basic tastes can be easily explained with the aid of well-defined products available over-the-counter.

To improve the trainee's expertise with the product of interest, the team leader may also use some relevant references, obtained by, for example, selecting or even spiking the product itself with substances recalling the attributes to be evaluated. Table 7 lists a few references that may be used as upper-end anchors for the rating of raw sausage. Tasters are familiarized with terms and instructed on the usage of standard references. This is a typical roundtable procedure, where subjects are encouraged to interact and discuss. Discussion proves useful to harmonize the term usage but becomes essential when the panel has to develop its own vocabulary of attributes.

Table 6 Spectrum Intensity Scale Values (0–15) for the Four Basic Tastes

Term	Sweet	Salt	Sour	Bitter
American cheese (Kraft)		7	5	
Applesauce, natural (Mott)	5		4	
Applesauce, regular (Mott)	8.5		2.5	
Big Red gum (Wrigley)	11.5			
Bordeaux cookies (Pepperidge Farm)	12.5			
Celery seed				9
Chocolate bar (Hershey)	10		5	4
Coca Cola Classic	9			
Endive, raw				7
Fruit punch (Hawaiian)	10		3	
Grape juice (Welch's)	6		7	2
Grape Kool-Aid	10		1	
Grapefruit juice, bottled (Kraft)	3.5		13	2
Kosher dill pickle (Vlasic)		12	10	
Lemon juice (ReaLemon)			15	
Lemonade (Country Time)	7		5.5	
Mayonnaise (Hellman's)		8	3	
Orange (fresh squeezed juice)	6		7.5	
Soda (Orange Crush)	10.5		2	
Frozen orange concentrate (Minute Maid) reconstituted	5.5		5	
Potato chips (Frito Lay)		9.5		
Potato chips (Pringles)		8.5		
Ritz cracker (Nabisco)	4	8		
Soda cracker (Premium)		5		
Spaghetti sauce (Ragu)	8	12		
Sweet pickle (Vlasic)	8.5		8	
Tea bags/1-hr soak				8
Triscuit (Nabisco)		9.5		
V-8 vegetable juice (Campbell)		8		
Wheatina cereal		6		2.5

Source: Ref. 7.

B. Selection of Terms for Descriptive Analysis (36–39)

It is well established that for a descriptive test to be reliable in the long run the descriptors must be meaningful. In other words, even a timely and accurate training will prove useless if relevant descriptors are missing. Quite often, subjects are made acquainted with a list of attributes previously chosen by the project leader on the basis of personal experience, or literature, or simply prompted by the marketing management. In these cases, the panel will be taught the meaning of attributes and how to use them with the aid of given references. In a few cases, as happens with new formulations or with sensory profiling, the assessors may be asked to take part in the development of terminology. This is usually accomplished through thorough discussion, protracted up to full agreement on which descriptors will enter the score sheet and how to use them. It is not uncommon for a score sheet development stage to require up to months. Two newly developed techniques—*free-choice profiling* (FCP) and the *repertory grid* method (RGM)—look promising in terms of saving time and generating relevant attributes.

Table 7 Standard References Corresponding to Maximum
Intensities of Raw Sausage Attributes (Original Data)

Attribute	Standard	Value
Aged	10-month-old salami, salt = 5–5.5%, pH = 5–5.5	7
Pungent	Salami, salt = 3–3.5; proteolysis = 20–22%; acetic acid > 0.1%; moisture/protein > 2	7
Rancid	dry-cured ham, external fat	7
Buttery	250 ppm diacetyl in water	7
Sweetish	salami, fat/protein > 2	7
Salty	salami, salt = 7–7.5%	13
Acid	salami, pH < 4	13

[a]On a 0 to 7 scale; attributes salty and acid rated on bipolar 1 to 13 scale.

1. Free-Choice Profiling (40–43)

This technique enables each assessor to select attributes by allowing him or her develop a score card and rate samples accordingly. Members are presented with samples belonging to the category being examined and encompassing the widest array of flavor, odor, or texture features and are instructed to use a common scale; discussion is not needed and is even undesired because subjects might be led astray by reciprocal suggestion; for the same reason training is unnecessary. This is why selection of attributes via FCP will eventually result in dramatic time saving, compared with the number of sessions required for conventional procedure based on group discussion. A major drawback with the FCP approach is that untrained people may feel distressed to retrieve and use terms developed during previous sessions; in view of this, it is useful to present the panel with replicate samples so as to disclose the individual's attitude to recognize his or her own vocabulary and use it consistently.

2. Repertory Grid Method (44–46)

To reduce fatigue associated with term development, the project leader may present each subject with two or three samples of the same class of product and ask him or her to describe the product in terms of differences and similarities between the samples. The panel leader must himself take part in this stage to help assessors dissect their own perceptual space using the interview technique. The repertory of attributes collected from each individual through presentation of several sample couples or terns (hence the name *dyad* or *triad repertory grid*) will then be analyzed to focus on analogies among descriptors obtained.

It is still a matter of study as to which procedure provides better profiling; FCP or RGM. Both are regarded as valuable instruments for term generation and time saving. However, it is a fact that with RGM the interviewer must spend considerable time in the development phase.

3. Analysis of FCP and RGM Data

At the end of both FCP and RGM sessions, the panel leader often faces the problem that panelists may have used differing words in response to the same stimuli, and vice versa. Advances in sensory research have shown that the problem may be solved by treating the data with the computing technique known as generalized procrustes analysis (GPA) (47). This method, which belongs to the class of multivariate analysis techniques, has a major advantage over conventional methods in that it allows the handling of uneven data matrices as are those usually originating from FCP and RGM sessions (Fig. 2a). By application of procrustes analysis individual attributes

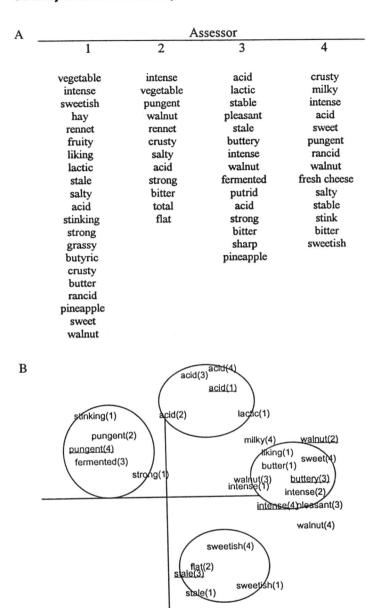

Fig. 2 Two steps in attribute development of matured cheese by combined free choice profile–generalized procrustes analysis (GPA). (A) Each member develops a score sheet for taste evaluation. (The figure reports only the first four panelists, generating 12–21 terms.) (B) GPA facilitiates the grouping of individual terms into classes of synonyms (groups are circled, subjects are denoted by number in parentheses). After discussion, the attributes underlined enter the common scoresheet. Analysis of original data by the Procrustes PC v.2.2 program (Oliemans Punter & Partners BV, Utrecht, The Netherlands, 1991).

are eventually grouped in classes of terms representing each member's words or expressions used to describe similar perception (Fig. 2b) (48–51). Next in score sheet development the project leader will pick up from each class of words the terms he or she believes necessary to describe the product and will encourage the panel to recognize and use them as synonyms for each one's terms.

This approach is rather appealing being less coercive than traditional sessions whereby people are forced to use terms of which they might be unaware; weighed down with fewer contraints, the panel may be expected to feel more comfortable and prove consistent over time. However, one should not be overconfident of substantially untrained assessors and regular checking has to be programmed to prevent bias and inaccuracy to impair sensory work based on FCP- and RGM-developed attributes (52).

C. Further Screening with Attributes Selected

Once individuals are in possession of the attributes of concern, they shall undergo additional screening, based on recognition and application of terms previously selected. In most cases, candidates are checked for their ability to discriminate actual samples for given descriptors.

1. ANOVA of Rating Data

Analysis of variance or ANOVA (53), applied to individual rating data, is a choice technique for screening purposes because it provides an easily understandable criterion for checking each subject's performance. Typically, the procedure relies on 6–12 samples, chosen so as to provide wide value ranges for attributes selected. Testing is spread over the number of sessions required to allow three or four replicates of each sample. The order of presentation is random and three-digit coded samples are examined by candidates in a testing room equipped with sensory facilities (see Sec. VI.A). The following example applies to panelists and attributes selected through previous free-choice profiling of cheese.

Example

TESTING   Six cheese samples are analyzed for buttery, acid, and intense tastes on a 0–9 scale anchored at extremes with given references. Samples (each replicated four times) are served on white plastic dishes coded with three-digit random numbers. All six samples are presented at each session, for a total of four sessions, held over 4 days. The order of presentation is random. Panel comprises 16 trained candidates (it is advisable to screen twice as many people as those needed for panel working); they are asked to taste about 10 g of each sample and chew it for at least 20 sec before swallowing. Five-minute intervals are allowed between samples; water and fresh carrots are provided to clean the mouth.

ANALYSIS OF DATA   Data from each candidate are treated by one-way ANOVA to search for differences among the eight cheeses (treatments). In Table 8 the results in the form of F ratios can be compared with the critical F value of 8.4 (at the .05 significance level), showing that subject 2 is poor for all of the attributes and subject 3 for buttery only. Based on these data, candidates can be ordered by performance rank (Table 9), indicating (see figures in the sum row) that assessor 2 is by far the last and suggesting rejection for this panelist.

The panel leader can himself take part in the test, provided he is unaware of the samples being examined. It is not uncommon, as reported by Cross et al. (34) in an ANOVA-based screening procedure for meat texture, that the team leader is overcome by other candidates; due to the test consequences, one had rather not rely on one evaluation to state who will or will not enter the panel.

Table 8 Screening Candidates for Discrimination of
Eight Cheese Samples by Three Taste Attributes: ANOVA
of Rating Data for the First Six Candidates

Taste	F ratio Panelist					
	1	2	3	4	5	6
Buttery	9.3	3.4	3.1	12.4	7.9	13.2
Acid	11.0	2.0	9.2	10.0	8.9	15.2
Intense	8.0	1.9	8.3	13.2	10.1	7.9

As shown in this example, the higher the F ratio the greater the panelist's ability with a given attribute; however, one might wonder for how many attributes the candidate has to prove successful in order to join the panel. Powers et al. (54) suggest following the binomial procedure as a criterion of acceptance/rejection of candidates tested for discrimination of samples through sensory attributes. Thus if candidates are tested for rating samples with N descriptors, they will be analyzed by ANOVA for each term and the number of significant or successful attributes will be compared with the critical value in binomial tables; let the critical value be M of N trials at a given significance level; those candidates unable to discriminate samples for equal/more than M attributes will be dropped.

2. Sequential Analysis (55–57)

To cut the overall number of trials required to check each individual's ability to discriminate, the candidates may be screened by triangle tests arranged in a sequential procedure. The triangle test is the most popular discriminating test; its use is familiar to every practitioner (see Sec. VIII.C.1a).

Sequential analysis relies on four parameters that the panel leader must select prior to testing:

P_0: maximum proportion (0–1) of correct responses for a candidate to be rejected
P_1: minimum proportion of correct responses for a candidate to be accepted
α: probability of choosing a candidate that should be rejected
β: probability of dropping a candidate that should be accepted

Because with the triangle test the proportion of correct responses by chance is 0.33, one has to assign P_0 a value exceeding 0.33 and P_1 a value exceeding P_0; α and β are usually assigned values of .05 and .10, respectively.

Table 9 Ranking of Candidates by F Ratios

Attribute	Candidate					
Taste	1	2	3	4	5	6
Buttery	3	5	6	2	4	1
Acid	2	6	4	3	5	1
Intense	4	6	3	1	2	5
Sum	9	17	13	6	11	7
Order	3	6	5	1	4	2

Two equations can be derived from these parameters (58) that will represent the upper and lower boundaries in the sequential graph (Fig. 3) where the cumulative number of correct decisions for each candidate is plotted vs. the total number of tests. Subjects falling in the accept or reject zone will not continue testing, whereas those in the intermediate will go on until they overstep either border. This will result in overall time saving. Samples for the triangle test will be prepared to provide couples differing by 2–3 points (on a 9-point scale) for the attributes being evaluated.

The two tests illustrated in this section (discrimination by rating and triangle test) represent powerful tools for panel selection. Since the employment of unsatisfactory members is very likely to yield unreliable results, it is recommended to start with a large number of participants in order to ensure an adequate panel size after rigorous selection. A minimum size of six subjects is commonly established, but larger panels are strongly suggested because small panels are excessively dependent on each member's behavior.

VI. PERFORMANCE EVALUATION

As with any instrument of measurement, the performance of the panel involved in sensory practice needs to be evaluated to make sure that the requirements for reproducibility and accuracy are constantly met. In fact, despite prolonged training procedures, human senses are prone to bias or changes that might be clueless even to a skilled analyst; for this reason the sensory response is averaged over several judges, just as an analytical measurement is replicated to control instrumental variability. As with instruments, the more precise the panel, the higher its ability to discriminate between samples.

The panel's performance is often checked by running tests based on sample discrimination and analyzing results by ANOVA. Samples are commonly arranged following either a *complete block* or a *balanced incomplete block* design. In the first case, panelists are presented with the whole set of samples in a randomized order and the test is replicated over sessions so that a "session" effect besides a "sample" or treatment effect can be calculated. When the sample size is large and the onset of fatigue is a concern, it is advisable that the experimental design be of the incomplete type. With such design, individuals are presented at each session with a subgroup of the entire set of samples, selected so that every sample is rated the same number of times and all possible sample pairs are served an equal number of times. So shaped, the balanced incomplete block enables finding differences, if any, between blocks (the session or day effect) and to regard such effects as a result

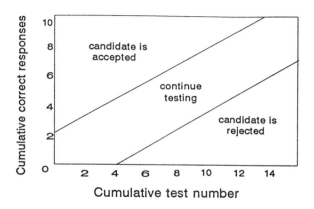

Fig. 3 Sequential chart for acceptance/rejection of candidates.

of individual inconsistency. Complete and incomplete block design studies are described in books dealing with experimental design (59) and procedures are available to the analyst wishing to balance the effect of presentation order and first-order carryover from one sample to the next (60).

Example A few months after training, the panel leader determines to check the panel's ability to recognize the same three attributes given in Table 8 and use them to differentiate cheese samples. Since a major objective is to evaluate the panel's reproducibility in time, an experiment based on 10 samples replicated 4 times over 8 sessions held on 4 different days (2 sessions per day, covering the entire set of 10 samples) is conducted following an incomplete balanced block design (59). The samples are chosen to represent adequate value ranges in the selected attributes.

Results from ANOVA are summarized in a typical two-way ANOVA table (Table 10) where the two effects (treatments or samples and blocks or days) are reported as individual F ratios for each attribute. Well-performing, consistent judges shall be high in treatment (able to discriminate samples) and low in day effects (stable over time). In this case, nonsignificant day effects demonstrate that the panel is reliable and significant F ratios for all tastes indicate satisfactory performance and sensitivity.

One of the main advantages of expert over nonexpert panels is that only a few participants are necessary because training always yields a homogeneous as well as consistent group. A minor source of inhomogeneity is due to individual use of differing portions of the scale, resulting in shifted lines in the plot of scores vs. samples (Fig. 4A). This problem witnesses differing attitudes toward scaling and can be easily accounted for by the analyst, who regards this problem as an assessor "main effect" (in an ANOVA sense). In contrast, the panel homogeneity is likely to be severely affected if scaling attitude changes with samples, as is shown graphically in Fig. 4B, where lines intersect, indicating that two panelists are rating samples in a way dramatically different from that of the rest of the team. This latter problem, often referred to as treatment–panel interaction, will result, by averaging scores over the team, in flattened scores and hence in a drop in sensitivity. In order to check out any source of interaction, a new two-way ANOVA test may be carried out as described below.

Example Eight panelists are presented with a set of four samples in four sessions. The whole set, in a randomized order, is served at each session and judges are asked to rate samples for the three attributes given in the previous example. ANOVA enables splitting total variance into three effects, namely, the "samples" and "panelists" main effects and the effect due to sample–panelist interaction. High treatment or sample F ratios will indicate the panel's discriminating ability, high panelist F ratios will suggest constant differences between members, whereas significant interac-

Table 10 ANOVA Testing of Eight Panelists Assessing 10 Samples (Treatments) Over Eight Sessions Held in 4 Days[a]

| | | F ratio | | | | | | | |
| | | Panelist | | | | | | | |
Attribute	Factor	1	2	3	4	5	6	7	8
Buttery	Treatment	4.3	4.9	6.1	5.0	6.6	4.8	4.7	8.0
	Day	0.3	0.9	0.9	1.1	0.5	0.9	0.9	1.0
Acid	Treatment	5.6	4.9	8.0	4.9	5.0	4.9	4.9	7.9
	Day	0.5	0.8	1.1	1.5	0.5	1.0	0.3	0.4
Intense	Treatment	6.1	4.7	7.7	6.0	8.9	5.9	4.9	10.0
	Day	0.9	0.7	1.8	1.0	0.4	1.6	0.9	0.6

[a]Treatment F-ratios significant at $P < .01$ for all panelists. Day F ratios nonsignificant at $P = .05$.

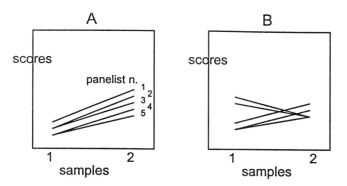

Fig. 4 Panelist–sample plots. (A) Additive effect. (B) Interaction effect.

tions will suggest that some major variability is occurring within judges. Table 11 shows a significant treatment effect and a panelist effect for the attribute buttery, whereas all interaction *F* ratios are below the significance threshold.

Although ANOVA-related tests are made easy by most statistical packages providing experimental design and analysis of variance routines (however, one should make sure that his or her package fits the right type of experimental design), simpler criteria are also followed to check the panel's performance in the sensory practice. A rule of thumb is that no panelist should vary by more than ±15% from the mean response. Thus, if a product is assigned a mean distance of 100 mm on a 150-mm scale, each judge shall fall within an 85–115 range. This criterion is straightforward and requires only a pocket calculator; however, one should be aware that at the scale extremes scores tend to be less spread out and that some traits, such as visual features, are easier to assess than, say, olfactory properties. Therefore it would be advisable that tolerance ranges be adapted from case to case.

Finally, the aforementioned ANOVA-based tests belong to the class of univariate techniques. Thus even with a properly designed experiment the relationships between attributes are lost, wasting otherwise available information. Multivariate techniques such as MANOVA, cluster, and factor analysis (61,62) may help inspect the panel–attribute linkage.

VII. METHOD VALIDATION (63–68)

In sensory practice it is implicitly assumed that the method developed through panel training and attribute selection is "on the target," i.e., judges, descriptors, and procedure are appropriate to the

Table 11 Two-Way ANOVA Table for the Search of Treatment × Panelist Interactions: F Ratios by Taste Attribute

	F-ratio		
Attribute	Treatment	Panelist	Panelist × Treatment
Buttery	9.5[a]	5.6[b]	0.5
Acid	8.0[a]	2.1	0.6
Intense	8.8[a]	1.0	0.9

[a]Significant at $P < .01$.
[b]Significant at $P < .05$.

problem of concern. It would be frustrating to discover, after heavy selection and training, that the sensory project is reliable and accurate but valueless because it is irrelevant to the product or to the marketing needs; in other words, that the panel isn't hitting the mark.

For a sensory team to be credible it needs to be validated; this is usually accomplished by matching the panel with other panels for the same samples, or relating its response to another profile, obtained by, for example, chemical or physical measurements. If independent assessments lead to a similar conclusion, then one feels more confident when using an expert panel, even if it is made up of few in-house people. The test method by far most used in validity studies is the analysis of correlation between blocks of data, such as those achieved by experts using sensory attributes and consumers rating the same samples for a few overall traits.

Example To validate an in-house expert panel dealing with taste traits of mint drops, the marketing management designs a study based on 10 samples encompassing a variety of flavor properties and assessed by a 7-member expert team as well as by a 100-consumer group. The attributes are as follows.

Experts: sweet, taste strength, taste length, intense, aftertaste. They are rated on a 0–9 scale.
Consumers: sweetness liking, freshness, overall liking. They are rated on a 0–100 scale.

Results from the two blocks undergo correlation analysis, to measure the extent to which variables are related to each other. The correlation coefficient, r, is the well-known measure of the strength of relationship; it ranges from -1 to 1, the two extremes indicating perfect inverse and perfect direct linkage, respectively. Correlation coefficients for the drop experiment are summarized in Table 12, indicating significant relationships between the consumers' ratings and 4 of 5 experts' traits.

Sensory strength, length, and intensity of taste correlate with a major property of mint drops, i.e., freshness; sweetness liking is expectedly linked with sweet intensity and overall liking seems to be well described in terms of four sensory attributes. The result is satisfactory and one may be confident that he or she won't be astray when using in-house profiling of mint drops for quality control or marketing purposes.

Correlation coefficients are straightforward but may be misleading if used to infer conclusions other than how two variables covary. Figure 5A shows what happens if one regards the sweetness relationship in Table 12 as a clue for existing linear dependence of sweetness liking on sweetness intensity: liking increases, then decreases, following a parabola-type curve. Further increase of sweetness beyond the breakpoint would impair sweetness preference, unlike what one might

Table 12 Linear Correlation Between Consumer Liking and Sensory Attributes Rated by Experts

	Consumers		
Experts	SL[a]	FR[a]	OL[a]
Sweet	0.61	−0.14	0.69[b]
Strength	−0.36	0.89[b]	0.80[b]
Length	−0.15	0.59[b]	0.85[b]
Intense	−0.24	0.91[b]	0.59[b]
Aftertaste	−0.15	−0.10	−0.05

[a]SL, sweetness liking; FR, freshness; OL, overall liking.
[b]Denotes significance at $P < .05$.

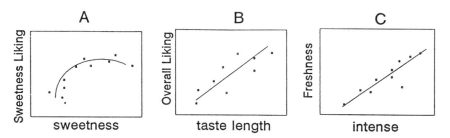

Fig. 5 Plots of consumer response vs. sensory panel evaluation of mint drops.

conclude based on the r value. In contrast, overall liking appears to be more clearly related to length (Fig. 5B) and freshness to taste intensity (Fig. 5C). In these two cases the high correlation coefficients contained in Table 12 are very likely to disclose linear interdependences.

The search for equations describing quantitatively a dependent variable as a function of other factors is a major concern in food science. To solve this problem it is worth turning from univariate linear relationships to nonlinear and often to multiple linear and nonlinear equations. For example, adding a quadratic term (sweetness)^2 to the linear term (sweetness) would linearize the relationship in Fig. 5A, resulting in a greater amount of explained variance (or squared r). Overall liking would likewise be explained better in terms of length, intensity and aftertaste, all used at the same time, to the first power, as shown by the following equation, obtained by multiple regression analysis:

Overall liking = −34.9 + 8.3 [length] + 6.1 [intense] −1.2 [aftertaste], with multiple regression coefficient $R^2 = 0.96$.

Predictive equations are attractive tools in quality control and development but may hide some pitfalls; those interested in linear or nonlinear equations as well as simple and multiple relationships should become acquainted with the theoretical background of regression techniques (69), prior to analyzing data with one of the widely available computer programs.

VIII. METHOD APPLICATION

For a correct use of the panel in sensory work some basic requirements must be met; they concern how samples are to be presented and assessed and where the panel is to be held. The following guidelines apply to the sensory laboratory and testing procedures.

A. The Sensory Laboratory (70–73)

The testing area must be designed to keep the test under control and minimize any source of bias. A minimum requirement is three independent rooms: a preparation area, a discussion area, and a testing ambient equipped with individual booths to prevent reciprocal suggestion. As a rule, the three rooms should be arranged as described in Fig. 6, supplied with furniture and tools for sample storage and serving. Each booth should accommodate one subject and allow sample serving through a hatch located in the front panel (Fig. 7).

It is essential to ensure adequate air circulation and use absolutely odorless and tasteless cups, dishes, and any other serving material. The testing area should be easily accessible and kept at about 22°C or 72°F, at a relative humidity of 50%.

To provide the testing environment with odorless air, a slightly positive are pressure inside the booth is useful, as are activated charcoal filters to treat incoming air and air from the preparation and cooking area. One should also make sure that no cleansing traces are present inside the booth.

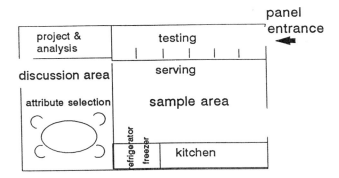

Fig. 6 Schematic layout of the various facilities within the sensory test laboratory.

Odorless cleansing agents are to be preferred. Since a major drawback with sensory work is fatigue, it is fundamental that tasters feel at their ease in the working environment: avoid noise to prevent annoyance and consequent performance drop and provide restful lighting by equipping booths with direct, uniform, shadowless light. Furthermore, because differing light intensities may be required depending on the product under evaluation, controllable intensity lighting is recommended; either incandescent and fluorescent sources will cover the widest range of applications, though special lighting may be necessary for special purposes. Typical fluorescent lights range from the so-called cool white, to warm white, to north daylight. Colored sheet filters may be used to generate colored lighting where shape or color is thought to bias the judge having to focus on traits such as flavor or texture.

B. Sample Presentation

The set of kitchen utensils must be chosen to minimize any source of bias due to odor-releasing material, difficulty in equipment usage, etc. Monouse plastic kitchenware is cheap and easy to dispose of, but some practitioners warn against possible foreign off-flavors and recommend odor-resistant chinaware, glass, and steel. Cups and dishes should be marked with odorless ink or, better, with printed labels, coded with two- or three-digit random numbers.

If the kind and number of samples enables presentation of the entire set without fatigue, the

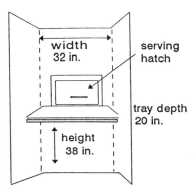

Fig. 7 Description of a typical booth area in a sensory evaluation laboratory (not drawn to scale).

panel leader may follow a complete block design (59), where independently replicated samples are served in a separately randomized order for each assessor. With large sample sizes, incomplete block designs have to be adopted by balancing the order of presentation (59,74–75), which means that within each session every sample will have the same position an equal number of times.

As a rule, serving should reflect the same conditions in use for the product under examination. Serving temperature is of prime importance in this view and one should make sure that besides temperature, size, shape, and weight are the same as well. Depending on the type of product, the leader may instruct judges on how to sniff or taste, how much to taste, number of bites, time of chewing, whether to swallow or discard, and number of samples per session. Suitable designs (76) have been suggested to control the carryover effect or bias between samples; this effect is commonly subdued by providing water or, in the case of fat matrices, unsalty crackers or diced carrots or apples. Fatigue is prevented by limiting the number of samples to four to six units per session and allowing few-minute intervals between samples because it is commonly believed that performance decreases sharply with time. In contrast, Moskowitz (77) claims that boredom is the main cause for poor performance and that a major drawback is too little time between samples. Quoting the same author: "In a conventional 20 min overall test, the first 5 min, or 25% may not generate as good data as the remaining 15 min. In contrast, a 4-hour test, allowing for 10 min orientation in the beginning and 10 min at the end, generates a total of 220 or more than 90% production time."

It seems advisable that the number of samples fit the difficulty of the test, the complexity of the score sheet, and the flavor properties of samples; with mild or delicate foods one may present the judges with a greater number of samples, provided adequate resting is allowed in between.

C. Type of Test

Two groups of procedures, difference and descriptive tests, cover most applications of practical use in sensory analysis. The former class comprises well-established procedures that are the basis of everyday work with both trained and untrained subjects, whereas descriptive tests pertain essentially to expert panels. Both classes set up the basic tools every sensory professional should be fitted out with for any kind of application.

1. Difference Tests

They are used to test two or more samples for difference as expressed by overall judgments or by selected attributes. In this last case the test is commonly referred to as attribute difference analysis. Two major objectives with difference tests are as follows: (a) Is there a difference between samples? (b) Which sample is preferred for a given descriptor? Test statistics will disclose significant differences (if any), but the analyst must be careful not to consider the absence of significant difference as evidence for sample equality.

a. Triangle Test (78–81)   This is by far the most popular among difference tests, historically adopted by all branches of the manufacturing industry for a variety of purposes, such as evaluating the effectiveness of an ingredient, establishing the effect of processing changes vs. traditional manufacturing, comparing the sensory quality of standard produce with a competitor's one, and so on. The triangle test is sensitive and can be successfully adopted in decision making even with less numerous panels; but it is also fatiguing and should be avoided when strong tastes or carryover effects are concerned. The typical procedure is described below.

Example   The R&D laboratory of a tomato-canning factory is asked to study the effectiveness of citric acid to control the acidity of tomato paste and increase protection against some microbial strains. The laboratory is also required to investigate the consumer acceptability of the product after acid treatment. A sensory test is planned in this way: a 2-kg batch of untreated tomato paste and a 2-kg batch of treated product are each subdivided into 18 samples and the two sets are presented to 12 trained panelists. Each assessor is presented with three samples, set out in the following balanced order:

Assessor	Left	Center	Right
1	U	T	T
2	T	U	T
3	T	T	U
4	U	U	T
5	U	T	U
6	T	U	U
7	U	T	T
8	T	U	T
9	T	T	U
10	U	U	T
11	U	T	U
12	T	U	U

T, treated; U, untreated

Each subject is informed that his or her triad contains two identical and one different or odd sample and is asked to taste samples from left to right and pick up the odd one; the score sheet is given in Fig. 8. After the test, 8 of 12 subjects turned out to be correct; by comparing the 8/12 couple (number of correct responses/number of respondents) with the critical values for the triangular test (see figures in Table 13 for up to 20 answers at the 5% and 1% significance levels), one could conclude that it was possible for assessors to distinguish between the treated and untreated tomato paste. To reduce fatigue associated with triangle test two easier procedures may help detect differences.

    b. Paired Comparison Test (78,79,82)   Among difference tests the paired comparison test is considered less fatiguing and proves suitable when the panel has to deal with complex sensory traits or strong stimuli resulting in possible carryover effects. A paired test is effective in much the same situations as depicted for a triangle test; however, since each member is presented with two samples, the probability of assessors giving the correct answer by chance is half, i.e., greater

Instructions.
You are given three samples of tomato paste. Two are identical, one is different. Taste the samples and pick up the different one. Report its code number in the 'odd sample' column. Guess in case of uncertainty.

Sample code   Odd sample

___   ___   ___   ___

Name: Date:

Fig. 8   Score sheet for a triangle test of tomato paste.

Table 13   Critical Values for the Triangle Test

No. of sets of 3 samples (respondents)	Significance level[a] 5% (.05)	1% (.01)
5	4	5
6	5	6
7	5	6
8	6	7
9	6	7
10	7	8
11	7	8
12	8	9
13	8	9
14	9	10
15	9	10
16	9	11
17	10	11
18	10	12
19	11	12
20	11	13

[a]Figures are the minimum number of correct responses needed for significance at the selected level for the corresponding number of respondents.

than with the triangle test; therefore the minimum number of correct responses required for significance will be higher and larger panels will be needed.

Typically, an equal number of sample pairs A/B and B/A is presented to a 20-member (or larger) panel. In the case of a two-sided test the panelists will be asked if they find any difference between samples; if the problem is one-sided or directional (e.g., which sample has stronger taste or firmer consistency), the question will be of the type: which sample differs from the other for the given attribute?

Table 14 contains critical values for up to 35 pairs in both the two-sided and one-sided mode; in comparison with the triangle test it stands out for the high proportion of correct responses required for significance.

c.   Duo-Trio Test (83,84)   Judges are confronted with three samples—two identical, one odd, just as with the triangle test. In this case one of the twin samples is identified as reference and panelists are asked to select the sample that differs from the reference. In most applications the reference is drawn from actual production and its properties are well known to a panel of trainees who have been instructed as to how to recognize the reference; in the less frequent cases where both samples are unknown or panelists are inexperienced a balanced procedure is preferred, consisting of presenting either sample as the reference, at random. Being less selective than other methods (correct response by chance has an associated probability of as much as .5), the duo–trio test requires at least a 15- to 20-member panel but a much larger size of 30 or more is recommended if some ticklish decision has to rely on difference test data. The procedure is straightforward, as outlined by instructions in Fig. 9, but with strong flavors or difficult traits the duo–trio test becomes as fatiguing as the triangle test. For data analysis and interpretation one may refer to critical values in use for one-sided paired comparison test (Table 14).

Table 14   Critical Values for Two-Sided and
One-Sided Paired Comparison Test

No. of pairs (respondents)	Two-sided 5% (.05)	Two-sided 1% (.01)	One-sided 5% (.05)	One-sided 1% (.01)
15	12	13	12	13
16	13	14	12	14
17	13	15	13	14
18	14	15	13	15
19	15	16	14	15
20	15	17	15	16
21	16	17	15	17
22	17	18	16	17
23	17	19	16	18
24	18	19	17	19
25	18	20	18	19
26	19	20	18	20
27	20	21	19	20
28	20	22	19	21
29	21	22	20	22
30	21	23	20	22
31	22	24	21	23
32	23	24	22	24
33	23	25	22	24
34	24	25	23	25
35	24	26	23	25

The header for the significance columns reads: Significance Level[a], divided into Two-sided and One-sided, each with 5% (.05) and 1% (.01).

[a]Figures are the minimum number of correct responses at
5% and 1% significance level for the corresponding
number of respondents.

   d.   A–Not A Test (85,86)   To compare test samples with a well-accepted standard reference
when sensory fatigue would discourage the recourse to traingle or duo–trio test, the A–not A
test may be the choice procedure. The field of application is the same depicted for the other
difference test, i.e., inspection of differences resulting from changes in recipe, manufacturing,
storage, and in general for tracking competitive product changes. As with the other tests, best

Instructions.
Samples from left to right are a reference and two test samples. One of the two
is equal to the reference, one is different. Pick up the sample matching the
reference and report its number in the code column.

Samples: ____   ____   Code:____

Name: Date:

Fig. 9   Score sheet for duo–trio test.

results are achieved if judges are skilled in recognition of the reference product. Depending on fatigue level each member may receive one sample or more; samples should be arranged in a balanced order, so that the number of A samples equals that of not A. Results are usually analyzed with the aid of the χ^2 statistic (tables are reported in most statistics handbooks) with data summarized as follows:

Panel response	Actual values		Total	Mean = row total/2
	A	not A		
A	$N1$	$N2$	$T1$	$M1 = T1/2$
not A	$N3$	$N4$	$T2$	$M2 = T2/2$
Total	$T3$	$T4$	T	

hence

$$\chi^2 = \frac{(N1 - M1)^2}{M1} + \frac{(N2 - M1)^2}{M1} + \frac{(N3 - M2)^2}{M2} + \frac{(N4 - M2)^2}{M2}$$

Although all four tests described so far find recommended use in specified areas and for the purposes summarized in Table 15, they provide a universal tool for selection and training of subjects for the testing of differences, which is by far the most popular branch among sensory applications. As a rule, the sensory professional should choose the method that combines maximum selectivity and minimum fatigue. Since selectivity is inversely related to the probability of guessing the correct answer by chance, the four methods given in the present section do not qualify for better than is provided by triangle test, which has an associated 33% probability.

By further reducing that probability one enjoys a considerable increase in selectivity, as is the case with the "two-of-five" test (87), having a probability of 10%, allowing significant differences to be found even by panels of as few as five or six members. Being strongly fatiguing, this test is confined to very easy evaluations, such as visual or tactile, and is commonly carried out by small in-house skilled teams.

Table 15 Main Features of Sensory Difference Tests

Test	Fatigue	Minimum panel size	Application
Paired comparison	Low	15–20	Performable with any sensory attribute in both one-sided (directional or preference) and two-sided modes
A–not A	Low	25	Used to match a sample with a well-known reference product. Increased efficiency with skilled judges
Duo–trio	Low-medium	15	Useful to inspect differences from a reference product even with little or no previous experience
Triangle	Medium	6–8	Any kind of task. Avoid or limit sample sets if strong flavors or difficult attributes are considered. Little training required
Two-of-five	High	5	Used to focus differences in attributes producing little sensory fatigue

2. Descriptive Analysis (88–92)

Although widely accepted for their ability to answer basic questions arising in quality control, difference tests fail to extract the plentiful information that is otherwise available with descriptive analysis. In principle, difference tests might be regarded as an application of concepts relying on deep knowledge of sensory properties of foods. Finding an overall difference between competitive samples may seem exhaustive to the marketing manager but may be less helpful to the R&D leader if he or she is unaware of the reasons for such difference. In a broad sense, descriptive analysis has two valuable applications: (a) It reveals which properties are mainly affected by ingredients, processing, packaging, and any other technological factor. (b) In marketing research general assessments such as intent of purchase and overall liking may be related to more detailed product attributes, enabling one to focus on those properties that are actually responsible for product acceptance.

These applications would themselves warrant the widest application in most branches of food analysis, but the need for very skilled personnel, heavy training, and maintenance still restrict extended recourse to attribute testing.

Finally, descriptive analysis yields a great deal of data that require acquaintance with suitable interpretation techniques. This means that not only does the project leader need to be skilled in testing procedure but he or she must also be familiarized with experimental design as well as with basic and advanced statistical methods. In those cases where statistical advice is unavailable from in-house facilities, the sensory laboratory shall turn to an external consultant, whose reward will further increase the project costs for personnel. In spite of the amount of time and money required, descriptive sensory methods have eventually joined other analytical techniques, mainly within large companies, where specified procedures have been developed and fitted on the products of interest. Being proprietary, most such techniques are usually unavailable from the literature; nevertheless, efforts in the last two decades from independent professional groups toward standardized descriptive procedures have led to a few well-recognized methods that, though proprietary or patented, make up the common knowledge of sensory descriptive techniques by this time. Two of these methods deserve further details.

a. Quantitative Descriptive Analysis (QDA) (93) In the mid-1970s Stone et al. (94) argued that sensory profiles, as they were first developed (95–97), needed a more quantitative approach and statistical analysis of data. They introduced a new method having three basic steps in screening and training of panelists and development of a common list of sensory attributes for subsequent sample testing. Screening is based on product recognition and has to be followed by further check of panel's reliability and by panel maintenance (see Sec. IV and V). Supervised by the leader, whose role is substantially limited to assistance and sample preparation, the panel sets apart preliminary discussion sessions to the ballot development and settles the testing procedure. Among procedures, the use of balanced block designs and of line scales with anchors at the extremes and, above all, the spider representation of data are peculiar aspects of QDA. The plot in Fig. 10, drawn from the wealth of literature on this subject, shows that attribute intensities are easily depicted through the spider web graph, with each attribute corresponding to one spoke and the distance to the midpoint being proportional to the attribute intensity, averaged over the panel. The resulting plot is straightforward, disclosing main sample differences at the first glance, thus allowing the product manager, even with little or no experience in sensory analysis, to feel more comfortable with tasks such as tracking of product changes and shooting of troubles. From its appearance, QDA has gained success among practitioners, providing considerable help to the spreading of descriptive analysis.

b. Spectrum Descriptive Analysis (7) This method, which originates from the broad experience gained by Gail Civille and other sensory professionals with a wide array of food, textile, and

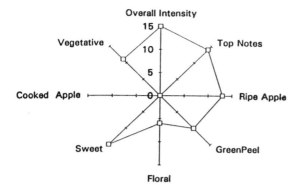

(a) Golden Delicious

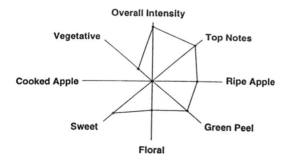

(b) Winter Banana

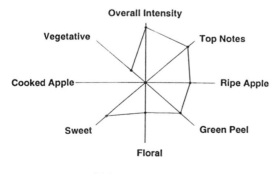

(c) Red Delicious

Fig. 10 Aroma intensity of juice obtained from three apple varieties (a–c). (From Ref. 88, pp. 332–3).

personal care products, has been conceived to provide the user with those tools that are actually needed to carry out most descriptive procedures. In a very pragmatic and comprehensive fashion, the panelists are first familiarized with principles underlying the specific features under investigation. For example, if the problem deals with texture analysis, the assessors will be taught the mechanical, geometric, or rheological properties of the product in question, according to the texture characteristics to be evaluated. Selection and training are timely and panelists must qualify for reliability in recognition and use of selected attributes. For training made extensive use of reference materials is made, based quite often on actual products. The standards provided by the method cover most product categories of potential use and represent a meaningful step forth in sensory analysis. A few selected references given in Table 6 show that human senses can be calibrated with the aid of commercially available materials just as a measurement instrument is calibrated with a standard. Attributes are rated by the panel in the order in which they are perceived and intensities are recorded with a 15-cm linear scale, or a numeric scale, or other scales, depending on the project task.

Descriptive methods are applied in current practice for most purposes, and papers have become available on the visual, flavor, and textural properties of meat and poultry (98–106), alcoholic beverages (107–116), dairy products (117–121), fruits and vegetables (122–129), as well as of other foods (130–133). Their relationships with analytical or instrumental measurements have also been investigated. It is noteworthy, however, that current texture analysis, though sharing the same principles underlying the two methods above, results from procedures originally devised to provide a rheological definition of food products. Such procedures have eventually joined into an accepted and codified method, often referred to as the texture profile method (134).

3. Multivariate Statistics for Descriptive Analysis

Many interesting issues may be stated when dealing with descriptive analysis. They range from sample discrimination to panel behavior, from classification by food categories to sensory-instrumental correlation, and so on. Whatever the objective, the analyst should be aware that because the problem with sensory attributes is generally of the multivariate type it has to be approached accordingly; it means that unless those cases where univariate methods such as ANOVA (53) are appropriate, the interpretation techniques are to be multivariate, in order to extract all of the possible information contained in the data.

There are some basic questions usually addressed to multivariate analysis of sensory data:

1. Are panelists consistent in the use of attributes? Should one or more subjects be excluded from successive assessments and computation?
2. Are the samples similar or do they differ, and for which variables?
3. If regression, classification, or optimization models are being developed, which variables are relevant to the problem and which could otherwise be neglected with little or no adverse effect on the model performance?

A number of multivariate procedures help answer these questions. A few interesting applications arise from recent advances in chemometrics, due to independent groups of scientists endeavoring interpretation of multivariate data in many branches of experimental sciences. Since it is beyond the objectives of this chapter to delve into details of multivariate statistics, the reader is addressed to the available works describing the theoretical (135) and practical (136–138) aspects of this discipline; this section will focus on a few essential guidelines.

Practitioners approaching multivariate statistics should be conscious that, though powerful, these techniques can't provide a solution to any problem and that best results are achieved if the interpretation method and the experimental design are chosen contextually. Analysts skilled in

Table 16　Main Features of Some Multivariate Techniques of Frequent Use in Sensory Analysis

Technique	Main goals/main output	Requirements and drawbacks
Principal components analysis (PCA) and factor analysis (FA)	Outlook of original data; reduction of the original set of variables to a smaller set of transformed variables (called factors or components); usable for classification purposes. The sample plot is used for the search of outliers and of sample pattern; the variable loading plot (or co-efficient plot) for correlation between variables and for redundant variables; the amount of total variation explained by each factor and by each variable is a criterion to retain factors and variables.	Minimum requirements, depending on the algorithm adopted. Usable with as few cases as the number of variables.
Cluster analysis (CA)	Group cases or variables according to a criterion of similarity; classify samples by group membership. The dendrogram shows which samples are grouped together and the distance among groups. It helps define the number of actual clusters.	Heavy computation time and memory with large databases. Interpretation of results may be lengthy.
Discriminant analysis	Separate samples belonging to a given class from samples belonging to other classes as well as possible; classify unknown samples according to a maximum likelihood criterion. The table of probability of classification into different groups provides a quantitative criterion for sample classification.	Classes of products are to be actual and relevant; they have to be defined through meaningful samples described in terms of relevant attributes; ensure a number of cases much greater then the number of variables.
Multidimensional scaling (MDS)	Represent stimuli as points in a reduced rank space akin to low-dimensional space in PCA; find similarities and individual differences in taste and smell perception; reveal basic stimuli in the true dimensional space. Representation of samples and variables on two-dimensional maps provides an easy way to interpret criterion of similarity.	High number of comparisons per individual.

Technique	Description	Notes
Multiple regression analysis (MRA)	Estimate a relationship between one dependent variable and two or more dependent variables. Predict acceptance from sensory attributes or sensory traits from instrumental measurements, etc. The regression equation depicts the role of independent variables in the predictive model. The multiple R^2, the F statistic in the stepwise procedure, and the plots of residuals are typical tools for checking the model.	The number of cases must be higher than the number of regressors. MRA is affected by strong correlations between the independent variables; it is prone to overfitting when the number of cases is low.
Latent variable regression (PCR and PLS) techniques	Same goals as MRA; used to overcome the computational problems arising from intercorrelation between variables. Two main offspring techniques: 1. Principal component regression (PCR), based on PCA of regressors followed by MRA of the dependent variable on the PCA factor scores. 2. Partial least squares regression (PLS), where the two previous steps are accomplished simultaneously. May be used to handle one (PLS1) or several dependent variables as well (PLS2).	Requirements are less stringent then with MRA, but a low number of cases may easily result in overfitting.
Procrustes analysis	Output substantially similar to that of PCA. In the sample and in the variable plots reports the ordinate is the dependent variable (PCR and PLS1) or a combination of several dependent variables (PLS2). Match each panelist's response to the average panel configuration. Compare data matrices of different origin (e.g., sensory and instrumental measurements on the same samples). Develop a common vocabulary of terms for panelists profiling the same set of samples in the free choice fashion. The individual sample plots are matched to the consensus plot and the individual correlation plots are matched to the average correlation plot to check each subject's behavior. The consensus and residual variation table provides additional information with this respect.	Similar to PCA and practically as easy to interpret, but less available on computer programs than principal components and related techniques.

Table 16 (*Continued*)

Technique	Main goals/main output	Requirements and drawbacks
Response surface method (RSM)	Explore data relationships to establish optimum conditions from a limited and feasible number of experimental tests. Typically used in product development to predict the best combination of ingredients, processing, and storage conditions. The predictive equation relating the response variable (typically, an overall rating, or a sensory attribute) to the independent variables is represented in graphic form through a response surface plot or a contour plot enabling easy interpretation of results.	In general, same requirements as MRA. Proper experimental design needed. RSM programs not always available within multipurpose statistical packages.

multivariate analysis warn against abuse of computational routines that are easily available through statistical or chemometrical packages for computers (139–144); to prevent misleading conclusions users not in possession of a sound statistical background are advised to follow a few recommendations (145):

Inspect data by all possible bivariate two-dimensional plots to facilitate a preliminary understanding of results.

Check out outlying and influential observations; these cases might have an abnormal effect on subsequent calculations leading to poor interpretation of data or to models unrepresentative of the bulk of data. Examine the database through simple representation (such as frequency techniques) or with the aid of multivariate procedures (principal components, cluster analysis) to identify possible outliers. Use the leverage methods included in residual analysis procedures available with regression techniques to disclose influential cases.

Run a preliminary principal components analysis and plot the data on the first two factors; check the plot for unexpected data patterns, then choose the procedure that seems appropriate.

In model building, chiefly with regression techniques, an enhancement of predictivity is often obtained by increasing the number of independent variables; however, this is likely to result in overfitting modeling, i.e., models fitting the original data rather than having general validity. Guard against this major fault by resorting to the cross-validatory methodology (146). As a rule, test the method by means of external cases, i.e., observations not employed for calculation. If cross-validation is unavailable, split the data into a training set and a more tiny test set to be used to validate the model relying on the training data.

Finally, when checking the original observations, be careful to remove only those cases that deserve rejection for ascertained measurement errors or that turn out to be clearly foreign to the dataset. Adjusting the data following one's expectation would result in a self-complying interpretation.

A few multivariate statistics of frequent use in sensory science are schematically compared in Table 16.

REVIEWS

J. R. Piggott, *Sensory Analysis of Foods*, 2nd Ed., Elsevier, London, 1988.

H. Stone and J. L. Sidel, *Sensory Evaluation Practices*, Academic Press, Orlando, 1993.

M. Meilgaard, G. V. Civille, and B. T. Carr, *Sensory Evaluation Techniques*, 2nd Ed., CRC Press, Boca Raton, 1991.

H. R. Moskowitz, *Product Testing and Sensory Evaluation of Foods, Marketing and R&D Approaches*, Food and Nutrition Press, Westport, 1983.

REFERENCES

1. R. M. Pangborn, Sensory evaluation of food: a look backward and forward, *Food Technol.*, *18*, 1309 (1964).
2. J. R. Piggott, *Sensory Analysis of Foods*, 2nd Ed., Elsevier, London, 1988.
3. F. A. Geldard, *The Human Senses*, 2nd Ed., John Wiley and Sons, New York, 1972.
4. H. Stone and J. L. Sidel, *Sensory Evaluation Practices*, Academic Press, Orlando, 1993.
5. R. M. Pangborn, Sensory techniques of food analysis, in *Food Analysis: Principles and Techniques*, *Vol. 1, Physical Characterization* (D. W. Gruenwedel and J. R. Whitaker, eds.), Marcel Dekker, New York, 1984.
6. D. Laming, *Sensory Analysis*, Academic Press, Orlando 1986.

7. M. Meilgaard, G. V. Civille, and B. T. Carr, *Sensory Evaluation Techniques*, 2nd Ed., CRC Press, Boca Raton, 1991.
8. D. Carter, *Methods and Data Handling Techniques for Sensory Analysis*, technical memorandum, Campden Food Preservation Research Association, No. 430, 1986.
9. American Society for Testing and Materials, *Basic Principles of Sensory Evaluation*, STP 433, ASTM, Philadelphia, 1968.
10. British Standard, BS 5929, *Methods for Sensory Analysis of Food, Part 1, General Guide to Methodology*, United Kingdom, British Standards Institution, 1986.
11. H. R. Moskowitz, *Applied Sensory Analysis of Foods*, Vols. I and II, CRC Press, Boca Raton, 1988.
12. N. T. Gridgeman, Tasting panels: sensory assessment in quality control, in *Quality Control in the Food Industry*, 2nd ed. (S. M. Hersschdorfer, ed.), McGraw-Hill, New York, 1984.
13. A. A. Williams and R. K. Atkin (eds.), *Sensory Quality in Foods and Beverages: Its Definition, Measurement and Control*, Ellis Horwood, Chichester, 1983.
14. Anonymous, Sensory evaluation guide for testing food and beverage products, *Food Technol. 35(11)*, 50–59 (1981).
15. International Organization for Standardization, *Sensory Analysis: Methodology General Guidance*, ISO 6658, 1985.
16. M. O'Mahony, *Sensory Evaluation of Food: Statistical Methods and Procedures*, Marcel Dekker, New York, 1986.
17. M. C. Gacula and J. Singh, *Statistical Methods in Food and Consumer Research*, Academic Press, Orlando, 1984.
18. A. V. A. Resurreccion, Application of multivariate methods in food quality evaluation, *Food Technol. 42(11)*, 128, 130, 132–134, 136 (1988).
19. J. R. Piggott (ed.), *Statistical Procedures in Food Research*, Elsevier, London, 1986.
20. J. L. Sidel and H. I. Stone, Experimental design and analysis of sensory tests, *Food Technol. 30*, 32, 34, 36–38 (1976).
21. T. Aishima and S. Nakai, Chemometrics of flavor research, *Food Rev. Int. 7*, 33 (1991).
22. M. C. Gacula, Jr., *Design and Analysis of Sensory Optimization*, Food and Nutrition Press, Westport, 1992.
23. H. R. Moskowitz, *Product testing and Sensory Evaluation of Foods: Marketing and R&D Approaches*, Food and Nutrition Press, Westport, 1983.
24. H. L. Meiselman, Consumer Studies of Food Habits, in *Sensory Analysis of Foods* (J. R. Piggott, ed.), Elsevier, London, 1984.
25. American Society for Testing and Materials, *Guidelines for the Selection and Training of Sensory Panel Members*, STP 758, Philadelphia, 1981.
26. International Organization for Standardization, *Sensory Analysis, Choosing and Training Assessors*, ISO 8586, Paris, 1985.
27. R. J. Winger and C. G. Pope, Selection and training of panelists for sensory evaluation of meat flavours, *J. Food Technol. 16*, 661 (1981).
28. S. L. Martin, Selection and training of judges, *Food Technol. 22*, 22, 24 (1973).
29. H. Stone, The selection of judges for sensory testing, *Food Technol., 31(11)*, 50 (1977).
30. J. J. Powers, Uses of multivariate methods in screening and training sensory panelists, *Food Technol. 42(11)*, 123, 126, 136 (1988).
31. M. McDaniel, L. A. Henderson, B. T. Watson Jr, and D. Heatherbell, Sensory panel training and screening for descriptive analysis of the aroma of Pinot noir wine fermented by several strains of malolactic bacteria, *J. Sensory Stud. 2*, 149 (1987).
32. K. P. Rutledge and J. M. Hudson, Sensory evaluation: method for establishing and training a descriptive flavour analysis panel, *Food Technol. 44(12)*, 78 (1990).
33. E. Chambers IV, J. A. Bowers, and A. D. Dayton, Statistical designs and panel training/experience for sensory analysis, *J. Food Sci. 46*, 1902 (1981).
34. H. R. Cross, R. Moen, and M. S. Stanfield, Training and testing of judges for sensory analysis of meat quality, *Food Technol. 32*, 48 (1978).
35. G. V. Civille and H. T. Lawless, The importance of language in describing perceptions, *J. Sensory Stud. 1*, 203 (1986).
36. G. V. Civille, Development of vocabulary for flavour descriptive analysis, in *Flavour Science and*

Technology (M. Martens, G. A. Dalen, and H. Russwurm Jr., eds.), John Wiley and Sons, New York, 1987, p. 357.

37. J. R. Piggott, Selection of terms for descriptive analysis, in *Sensory Science Theory and Applications in Foods* (H. Lawless and B. Klein, eds.), Marcel Dekker, New York, 1991, p. 339.

38. A. C. Noble, R. A. Arnold, J. Buechsenstein, E. J. Leach, J. O. Schmidt, and P. M. Stern, Modification of a standardized system of wine aroma terminology, *Am. J. Enol. Vitic. 38*, 143 (1987).

39. M. O'Mahony, L. Rothman, T. Ellison, D. Shaw, and L. Buteau, Taste descriptive analysis: concept formation, alignment and appropriateness, *J. Sensory Stud. 5*, 71 (1990).

40. J. Benedict, E. M. Steenkamp, and H. C. M. Van Trijp, Free-choice profiling in cognitive food acceptance research in *Food Acceptability* (D. M. H. Thomson, ed.), Elsevier, London, 1988, p. 363.

41. R. J. Marshall and S. P. J. Kirby, Sensory measurement of food texture by free-choice profiling, *J. Sensory Stud. 3*, 63 (1988).

42. A. A. Williams and S. P. Langron, The use of free-choice profiling for the evaluation of commercial ports, *J. Sci. Food Agric. 35*, 558 (1984).

43. A. A. Williams and G. M. Arnold, A comparison of the aromas of six coffees characterised by conventional profiling, free-choice profiling and similarity scaling methods, *J. Sci. Food Agric. 3*, 204 (1985).

44. D. M. H. Thomson, Recent advances in sensory and affective methods, *Food Science and Technology Today, 3* (1989).

45. F. M. Scriven, N. Gains, S. R. Green, and D. M. H. Thomson, A contextual evaluation of alcoholic beverages using the repertory grid method, in *Int. J. Food Sci. Technol. 24*, 173 (1989).

46. J. A. McEwan and D. M. H. Thomson, An investigation of the factors influencing consumer acceptance of chocolate using the repertory grid method, in *Food Acceptability* (D. M. H. Thomson, ed.), Elsevier, London, 1988, p. 347.

47. J. C. Gower, Generalized procrustes analysis, *Psychometrica 40*, 33 (1975).

48. G. M. Arnold and A. A. Williams, The use of generalized procrustes techniques in sensory analysis, in *Statistical Procedures in Food Research* (J. R. Piggot, ed.), Elsevier, London, 1986, p. 233.

49. D. C. Oreskovich, B. P. Klein, and J. W. Sutherland, Procrustes analysis and its applications to free-choice and other sensory profiling, in *Sensory Science Theory and Applications in Foods* (H. Lawless and B. Klein, eds.), Marcel Dekker, New York, 1991, p. 353.

50. J. R. Piggott and M. P. Watson, A comparison of free-choice profiling and the repertory grid method in the flavor profiling of cider, *J. Sensory Stud. 7*, 133 (1992).

51. I. N. Wakeling, M. M. Raats, and H. J. H. MacFie, A new significance test for consensus in generalized procrustes analysis, *J. Sensory Stud. 7*, 91 (1992).

52. G. Parolari, Taste quality of Italian raw ham in a free-choice profile study, *Food Qual. Pref. 5* 129 (1994).

53. G. E. P. Box, W. G. Hunter, and J. S. Hunter, *Statistics for Experimenters*, John Wiley and Sons, New York, 1978.

54. J. J. Powers, Uses of multivariate methods in screening and training of sensory panelists, *Food Technol. 38(6)*, 74 (1984).

55. J. R. Gacula, L. A. Parker, J. J. Kubala, and J. Reaume, Data analysis: a variable sequential test for selection of sensory panel, *J. Food Sci. 39*, 61 (1974).

56. A. Vie and M. O'Mahony, Triangular difference testing: refinements to sequential sensitivity analysis for predictions for individual triads, *J. Sensory Stud. 4*, 87 (1989).

57. M. O'Mahony and M. Odbert, A comparison of sensory difference testing procedures: sequential sensitivity analysis and aspects of taste adaptation, *J. Food Sci. 50*, 1055 (1985).

58. American Society for Testing and Materials, *Manual on Sensory Testing Methods*, STP 434, Philadelphia, 1968.

59. W. G. Cochran and G. M. Cox, *Experimental Designs*, John Wiley and Sons, New York, 1957.

60. M. Meilgaard, G. V. Civille, and B. T. Carr, *Sensory Evaluation Techniques*, 2nd Ed., CRC Press, Boca Raton, 1991, p. 303.

61. T. M. M. Malundo and A. V. A. Resurreccion: a comparison of performance of panels selected using analysis of variance and cluster analysis, *J. Sensory Stud. 7*, 63 (1992).

62. J. J. Powers, K. Shinholser, and D. R. Godwin, Evaluating assessors' performance and panel homogenehity using univariate and multivariate statistical analysis, in *Progress in Flavor Research 1984* (J. Adda, ed.), Elsevier, New York, 1985, p. 193.

63. R. Griffin and L. Stauffer, Product optimization in central-location testing and subsequent validation and calibration in home-use testing, *J. Sensory Stud. 4*, 231 (1991).
64. A. Howard, Taste panel technique. I. Reproductibility, reliability and validity. *Food Res. Quart. 32*, 80 (1973).
65. H. G. Schutz, Sources of invalidity in the sensory evaluation of foods. *Food Technol. 25*, 249 (1971).
66. A. A. Williams, C. A. Rogers, and A. J. Collins, Relating chemical/physical and sensory data in food acceptance studies, *Food Qual. Pref. 1*, 25 (1988).
67. N. Daget, Understanding and interpreting consumer answers in the laboratory, in *Sensory Quality in Foods and Beverags: Its Definition, Measurement and Control* (A. A. Williams and R. K Atkin, eds.), Ellis Horwood, Chichester, 1983.
68. H. G. Schutz, Predicting preference from sensory and analytical data, in *Flavour Science and Technology* (M. Martens, G. A. Dalen, and H. Russwurm Jr., eds.), John Wiley and Sons, New York, 1987, p. 399.
69. N. Draper and H. Smith, *Applied Regression Analysis*, 2nd Ed., John Wiley and Sons, New York, 1981.
70. American Society for Testing and Materials. *Physical Requirements Guidelines for Sensory Evaluation Laboratories*, STP 913, Philadelphia, 1986.
71. International Organization for Standardization, *Sensory Analysis: General Guidance for the Design of Test Rooms*, ISO 8589, 1988.
72. B. A. Billmeyer and G. Wyman, Computerized sensory evaluation system, *Food Technol. 45(7)*, 100 (1991).
73. G. F. Russell, Some basic considerations in computerizing the sensory laboratory, *Food Technol. 38*, 67 (1984).
74. M. Meilgaard, G. V. Civille, and B. T. Carr, *Sensory Evaluation Techniques*, 2nd Ed., CRC Press, Boca Raton, 1991, p. 257.
75. D. G. Mayer and J. C. Mulder, Factors influencing the efficiency of incomplete block designs in sensory evaluation experiments, *J. Sensory Stud. 4*, 121 (1989).
76. H. J. MacFie and N. Bratchell, Designs to balance the effect of order of presentation and first-order carry-over effects in Hall tests, *J. Sensory Stud. 4*, 129 (1989).
77. H. R. Moskowitz, *New Directions for Product Testing and Sensory Analysis of Foods*, Food and Nutrition Press, Westport, 1985, p. 114.
78. C. James, Paired-comparison and triangle sensory methods compared for use in product improvement, *J. Food Qual. 9*, 175 (1986).
79. E. B. Roessler, R. M. Pangborn, J. L. Sidel, and H. Stone, Expanded statistical tables for estimating significance in paired-preference, paired-difference, duo-trio and triangle tests, *J. Food Sci. 43*, 940 (1978).
80. United Kingdom, British Standards Institution, *Method for Sensory Analysis of Food. III. Triangular Test*. British Standard, BS 5929 Part 3 (1984).
81. International Organization for Standardization, *Sensory Analysis. Mechodology. Triangular Test*. ISO 4120, 1983.
82. United Kingdom, British Standards Institution, *Methods for Sensory Analysis of Food. II. Paired Comparison Test*. British Standard, BS 5959, 1982.
83. K. Mullen, D. M. Ennis, E. De Doncker, and J. A. Kapenga, Models for the duo–trio and triangular methods, *Biometrics 44*, 1169 (1988).
84. International Organization for Standardization, *Sensory Analysis. Methodology. Duo–Trio Test*. ISO/DIS 10399, 1990.
85. International Organization for Standardization, *Sensory Analysis. Methodology. A–Not A Test*. ISO 8588, 1987.
86. United Kingdom, British Standards Institution, *Methods for Sensory Analysis of Food. Part 5. A–Not A Test*, British Standard, BS 5929, 1988.
87. M. Meilgaard, G. V. Civille, and B. T. Carr, *Sensory Evaluation Techniques*, 2nd Ed., CRC Press, Boca Raton, 1991, p. 67.
88. M. E. Einstein, Descriptive techniques and their hybridization, in *Sensory Science Theory and Applications in Foods* (H. T. Lawless and B. P. Klein, eds.), Marcel Dekker, New York, 1991, p. 317.
89. M. Gillette, Applications of descriptive analysis. *J. Food Prot. 47*, 403 (1984).
90. J. J. Powers, Current practices and applications of description methods, in *Sensory Analysis of Foods* (J. R. Piggott, ed.), Elsevier, London, 1988, p. 187.

91. J. R. Piggott and P. R. Canaway, Finding the word for it: methods and uses of descriptive sensory analysis, in *Flavour '81* (P. Schreier, ed.), Walter de Gruyter, Berlin, 1981, p. 33.

92. M. O'Mahony, L. Rothman, T. Ellison, D, Shaw, and L. Buteau, Taste descriptive analysis: concept formation, alignment and appropriateness, *J. Sensory Stud. 5*, 71 (1990).

93. H. Stone and J. Sidel, *Sensory Evaluation Practices*, 2nd Ed., Academic Press, Orlando, 1993, p. 204.

94. H. Stone, J. Sidel, S. Oliver, A. Woolsey, and R. C. Singleton, Sensory evaluation by quantitative descriptive analysis, *Food Technol. 28*, 24 (1974).

95. J. F. Caul, The profile method of flavor analysis, *Adv. Food Res. 7*, 1 (1957).

96. M. A. Brandt, E. A. Skinner, and J. A. Coleman, Texture profile method, *J. Food Sci. 28*, 404 (1963).

97. A. S. Szczesniak, Classification of textural characteristics, *J. Food Sci. 28*, 404 (1963).

98. B. W. Berry and G. V. Civille, Development of a texture profile panel for evaluating restructured beef steaks varying in meat particle size, *J. Sensory Stud. 1*, 15 (1986).

99. P. J. Shand, Z. J. Hawrysh, R. T. Hardin, and L. E. Jeremiah, Descriptive sensory assessment of beef steaks by category scaling, line scaling and magnitude estimation, *J. Food Sci. 50*, 495 (1985).

100. M. Careri, A. Mangia, G. Barbieri, L. Bolzoni, R. Virgili, and G. Parolari, Sensory property relationship to chemical data in Italian-type dry-cured ham, *J. Food Sci. 58*, 968 (1993).

101. F. Shahidi, L. J. Rubin, and L. A. D'Souza, Meat flavor volatiles: a review of the composition, techniques of analysis, and sensory evaluation, *CRC Crit. Rev. Food Sci. Nutr. 24*, 141 (1986).

102. F. Sinesio, E. Risvik, and M. Rodbotten, Evaluation of panelist performance in descriptive profiling of rancid sausages: a multivariate study, *J. Sensory Stud. 5*, 33 (1990).

103. B. G. Lyon and C. E. Lyon, Texture profile of broiler pectoralis major as influenced by postmortem deboning time and heat method, *Poultry Sci. 69*, 329 (1990).

104. D. Rajalakshmi, S. Dhanaraj, Nagin-Chand, and V. S. Govindarajan, Descriptive quality analysis of mutton, *J. Sensory Stud. 2*, 93 (1987).

105. S. L. Beilken, L. M. Eadie, P. L. Jones, and P. V. Harris, Sensory and other methods for assessing salami quality, *CSIRO Food Res. Quart. 50*, 54 (1990).

106. B. C. Paterson and F. C. Parrish, Jr., A sensory panel and chemical analysis of certain beef chuck muscles, *J. Food Sci. 51*, 876 (1986).

107. J. X. Guinard and M. Cliff, Descriptive analysis of Pinot noir wines from Carneros, Napa, and Sonoma, *Am. J. Enol. Vitic. 38*, 211 (1987).

108. H. Heymann and A. C. Noble, Descriptive analysis of commercial Cabernet Sauvignon wines from California, *Am. J. Enol. Vitic. 38*, 41 (1987).

109. A. A. Williams, S. P. Langron, and A. C. Noble, Influence of appearance on the assessment of aroma in Bordeaux wines by trained assessors, *J. Inst. Brew. 90*, 250 (1984).

110. A. C. Noble, A. A. Williams, and S. P. Langron, Descriptive analysis and quality ratings of 1976 wines from four Bordeaux communes, *J. Sci. Food Agric. 35*, 88 (1984).

111. W. O. Kwan and B. R. Kowalski, Data analysis of sensory scores. Evaluations of panelists and wine score cards, *J. Food Sci. 45*, 213 (1980).

112. M. A. Amerine and E. B. Roessler, *Wines: Their Sensory Evaluation*, W. H. Freeman, San Francisco, 1976.

113. J. R. Piggot and S. P. Jardine, Descriptive sensory analysis of whisky flavour, *J. Inst. Brew. 85*, 82 (1979).

114. N. Gains and D. M. H. Thomson, Sensory profiling of canned lager beers using consumers in their own homes, *Food Qual Pref. 2*, 39 (1990).

115. F. R. Sharpe, Assessment and control of beer flavour, *J. Inst. Brew. 94*, 301 (1988).

116. C. Morten, C. Meilgaard, and J. E. Muller, Progress in descriptive analysis of beer and brewing products, Technical Quarterly, Master Brewers' Association of the Americas, Vol. 24, 1987, p. 79.

117. H. Tuorila, Sensory profiles of milks with varying fat contents, *Lebensm. Wissenshaft Technol. 19*, 344 (1986).

118. L. M. Poste and C. F. Patterson, Multidimensional scaling: sensory analysis of yoghurt, *Can. Inst. Food Sci. Technol. J. 21*, 271 (1988).

119. E. Casiraghi, M. Lucisano, and C. Pompei, Correlation among instrumental texture, sensory texture and chemical composition of five Italian cheeses, *Italian J. Food Sci. 1*, 53 (1989).

120. J. A. McEwan, J. D. Moore, and J. S. Colwill, The sensory charcteristics of cheddar cheese and their relationship with acceptability, *J. Soc. Dairy Technol. 42*, 112 (1989).

121. A. Vangtal and E. G. Hammond, Correlation of the flavor characteristics of Swiss-type cheeses with chemical parameters, *J. Diary Sci. 69*, 2982 (1986).
122. W. Pfannhauser, Sensorische und instrumentelle analytische Untersuchungen des Kiwi-Aromas, *Zeitschrift für Lebensmittel Untersuchung Forschung 187*, 224 (1988).
123. E. L. Wilson, M. G. Hogg, and D. J. W. Burns, Sensory evaluation of clarified kiwifruit juices, *J. Food Qual. 7*, 43 (1984).
124. M. Martens, H. Martens, and S. Wold, Preference of cauliflower related to sensory descriptive variables by partial least squares (PLS) regression, *J. Sci. Food Agric. 34*, 715 (1983).
125. R. Stanley, *Sensory quality assessment of fresh strawberries*, Technical memorandum, Campden Food Preservation Research Association, No. 479, 1988.
126. A. E. Watada, J. A. Abbott, R. E. Hardenburg, and W. Lusby, Relationships of apple sensory attributes to headspace volatiles, soluble solids and titratable acids, *J. Am. Soc. Horticultural Sci. 106*, 130 (1981).
127. A. G. H. Lea and G. D. Ford, Instrumental and sensory characterization of commercial apple juice, *Flüssiges Obst. 58*, 29 (1991).
128. S. Dhanaraj, S. M. Ananthakrishna, and V. S. Govindarajan, Apple quality: development of descriptive quality profile for objective sensory evaluation, *J. Food Qual. 4*, 83 (1980).
129. E. Guichard, P. Schlich, and S. Issanchou, Composition of apricot aroma: correlations between sensory and instrumental data, *J. Food Sci. 55*, 735 (1990).
130. F. C. F. Galvez and A. V. A. Resurreccion, Comparison of three descriptive analysis scaling methods for the sensory evaluation of noodles, *J. Sensory Stud. 5*, 251 (1990).
131. T. H. Sanders, J. R. Vercellotti, K. L. Crippen, and G. V. Civille, Effect of maturity on roast color and descriptive flavor of peanuts, *J. Food Sci. 54*, 475 (1989).
132. P. A. Prell and F. M. Sawyer, Flavor profiles of 17 species of North Atlantic fish, *J. Food Sci. 53*, 1036 (1988).
133. U. Hellemann, H. Tuorila, H. Salovaara, and L. Tarkkonen, Sensory profiling and multidimensional scaling of selected Finnish rye breads, *Int. J. Food Sci. Technol. 22*, 693 (1987).
134. A. Muñoz, Development and application of texture reference scales, *J. Sensory Stud. 1*, 55 (1986).
135. K. V. Mardia, J. T. Kent, and J. M. Bibby, *Multivariate Analysis*, Academic Press, Orlando, 1982.
136. D. M. Ennis, Multivariate sensory analysis, *Food Technol. 42*(11), 118 (1988).
137. S. L. Bieber and D. V. Smith, Multivariate analysis of sensory data: a comparison of methods, *Chem. Senses 11*, 19 (1986).
138. H. J. H. MacFie, Data analysis in flavour research: achievements, needs and perspectives, in *Flavour Science and Technology* (M. Martens, G. A. Dalen, and H. Russwurm Jr., eds.), John Wiley and Sons, New York, 1987, p. 423.
139. J. A. McEwan, Review of statistical software for sensory evaluation, *Food Qual. Pref. 1*, 179 (1989).
140. M. Cliff and J. S. Wild, Development of SAS registered computer programs for sensory graphics, *J. Sensory Stud. 4*, 249 (1990).
141. E. Risvik and R. Rogers, Sensory analysis: a view on the use of computers, *Food Qual. Pref. 1*, 81 (1989).
142. A. A. Williams and P. Brian, The scope of the microcomputer in sensory analysis, *Chem. Ind. 4*, 118 (1986).
143. M. R. Savoca, Computer applications in descriptive testing, *Food Technol. 38*(9), 74 (1984).
144. P. L. Brady, Computers in sensory research, *Food Technol. 38*(9), 81 (1984).
145. H. Martens, S. Wold, and M. Martens, A layman's guide to multivariate data analysis, in *Food Research and Data Analysis* (H. Martens and H. Russwurm Jr., eds.), Applied Science, Barking, 1983, p. 492.
146. S. Wold, Cross validatory estimation of the number of components in factor and principal components models, *Technometrics 20*, 397 (1978).

7

Amino Acids

Jeffrey H. Baxter
Ross Products Division, Abbott Laboratories, Columbus, Ohio

I. INTRODUCTION

A. General Background

Amino acids are components of essentially all foods, though there is substantial variability in content. They are found primarily as a component of proteins. Digestion breaks down protein to free amino acids and small peptides, which are absorbed by the body. Dietary amino acids are used to produce the proteins needed for cellular and tissue function and structure, as well as neurotransmitters, hormones, and so forth.

The amino acids are α-amino, carboxylic acid derivatives (the one notable exception, proline, is an α-imino acid) having the general structure shown below:

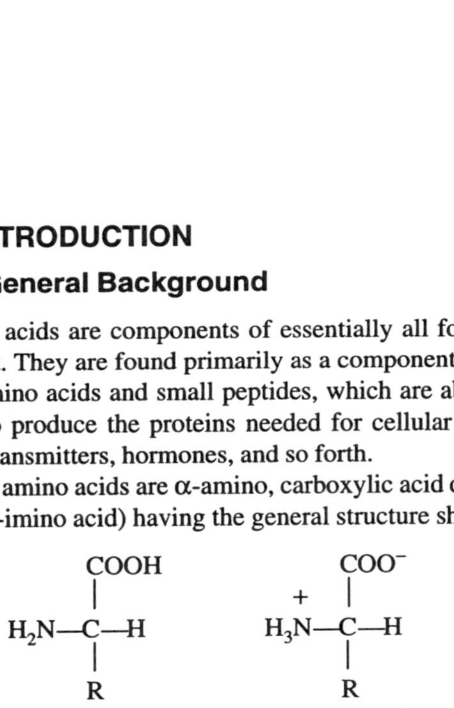

$$
\begin{array}{cc}
\text{COOH} & \text{COO}^- \\
| & + \quad | \\
\text{H}_2\text{N}-\text{C}-\text{H} & \text{H}_3\text{N}-\text{C}-\text{H} \\
| & | \\
\text{R} & \text{R} \\
\text{Unionized form} & \text{Zwitterion form}
\end{array}
$$

In nature, the stereochemistry around the anomeric (α) carbon is almost always L-, as shown above. D-Amino acids are found in nature, but are very specialized; for example, they are often associated with toxins. The R group can vary, having dramatic effects on the overall chemical and physical properties of the molecule. Fortunately for the analyst, there are only 20 commonly occurring variants, which are listed in Table 1. It should be noted that these molecules are almost always found in the zwitterionic form—even in the solid. This fact has significant effects on their physical properties. While there are very large differences in their structure (and thus their chemical and physical properties) there are similarities as well, and it is these that are used in general analytical approaches (e.g., methods designed to determine all or most of the common amino acids). Since

Table 1 Chemical Structures of 20 Naturally Common Amino Acids

Name	Abbrev.	Structure	Name	Abbrev.	Structure
Glycine	Gly (G)	COO^- $\mid$ $^+H_3N-C-H$ $\mid$ H	Aspartate	Asp (D)	COO^- $\mid$ $^+H_3N-C-H$ $\mid$ CH_2 $\mid$ COO^-
Alanine	Ala (A)	COO^- $\mid$ $^+H_3N-C-H$ $\mid$ CH_3	Glutamate	Glu (E)	COO^- $\mid$ $^+H_3C-C-H$ $\mid$ CH_2 $\mid$ CH_2 $\mid$ COO^-
Valine	Val (V)	COO^- $\mid$ $^+H_3N-C-H$ $\mid$ $C-H$ $/\backslash$ $H_3C\ CH_3$	Asparagine	Asn (N)	COO^- $\mid$ $^+H_3N-C-H$ $\mid$ CH_2 $\mid$ $CONH_2$
Leucine	Leu (L)	COO^- $\mid$ $^+H_3N-C-H$ $\mid$ CH_2 $\mid$ $C-H$ $/\backslash$ $H_3C\ CH_3$	Glutamine	Gln (Q)	COO^- $\mid$ $^+H_3N-C-H$ $\mid$ CH_2 $\yen$ CH_2 $\mid$ $CONH_2$
Isoleucine	Ile (I)	COO^- $\mid$ $^+H_3N-C-H$ $\mid$ $CH-CH_3$ $\mid$ CH_2 $\mid$ CH_3	Lysine	Lys (K)	COO^- $\mid$ $^+H_3N-C-H$ $\mid$ CH_2 $\mid$ CH_2 $\mid$ CH_2 $\mid$ CH_2 $\mid$ $^+NH_3$
Phenylalanine	Phe(F)	COO^- $\mid$ $^+H_3N-C-H$ $\mid$ CH_2 $\mid$ ⬡			

Tryosine   Tyr (Y)

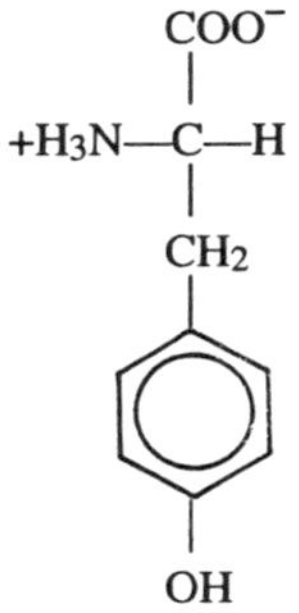

$$
\begin{array}{c}
\text{COO}^- \\
| \\
{}^+\text{H}_3\text{N}-\text{C}-\text{H} \\
| \\
\text{CH}_2 \\
| \\
\bigcirc \\
| \\
\text{OH}
\end{array}
$$

Tryptophan   Trp (W)

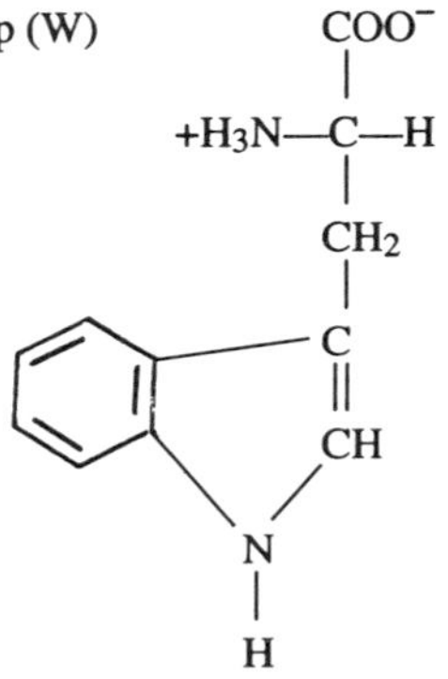

$$
\begin{array}{c}
\text{COO}^- \\
| \\
{}^+\text{H}_3\text{N}-\text{C}-\text{H} \\
| \\
\text{CH}_2 \\
| \\
\text{C} \\
\| \\
\text{CH} \\
| \\
\text{N} \\
| \\
\text{H}
\end{array}
$$

Cysteine   Cys (C)

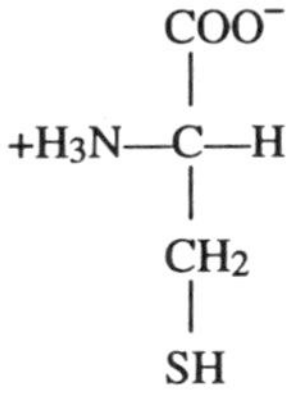

$$
\begin{array}{c}
\text{COO}^- \\
| \\
{}^+\text{H}_3\text{N}-\text{C}-\text{H} \\
| \\
\text{CH}_2 \\
| \\
\text{SH}
\end{array}
$$

Methionine   Met (M)

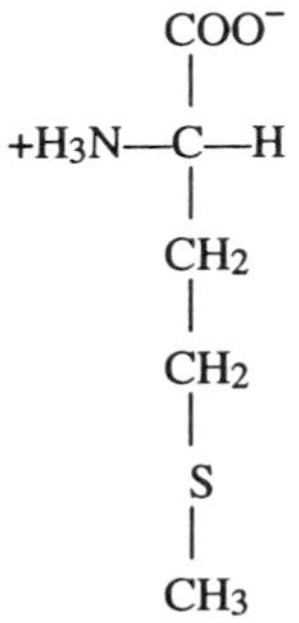

$$
\begin{array}{c}
\text{COO}^- \\
| \\
{}^+\text{H}_3\text{N}-\text{C}-\text{H} \\
| \\
\text{CH}_2 \\
| \\
\text{CH}_2 \\
| \\
\text{S} \\
| \\
\text{CH}_3
\end{array}
$$

Arginine   Arg (R)

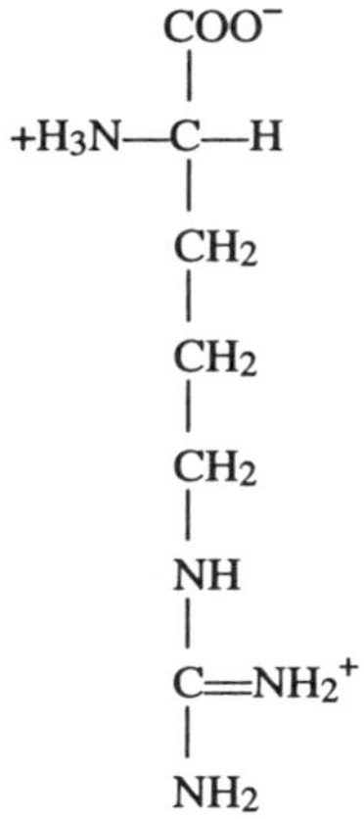

$$
\begin{array}{c}
\text{COO}^- \\
| \\
{}^+\text{H}_3\text{N}-\text{C}-\text{H} \\
| \\
\text{CH}_2 \\
| \\
\text{CH}_2 \\
| \\
\text{CH}_2 \\
| \\
\text{NH} \\
| \\
\text{C}=\text{NH}_2^+ \\
| \\
\text{NH}_2
\end{array}
$$

Histidine   His (H)

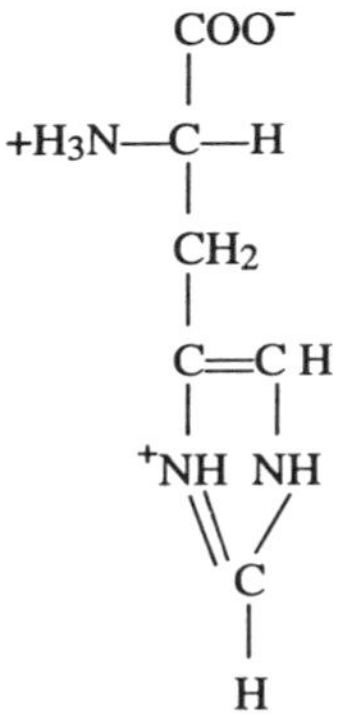

$$
\begin{array}{c}
\text{COO}^- \\
| \\
{}^+\text{H}_3\text{N}-\text{C}-\text{H} \\
| \\
\text{CH}_2 \\
| \\
\text{C}=\text{CH} \\
| \quad | \\
{}^+\text{NH} \quad \text{NH} \\
\backslash\!\!/ \\
\text{C} \\
| \\
\text{H}
\end{array}
$$

Serine   Ser (S)

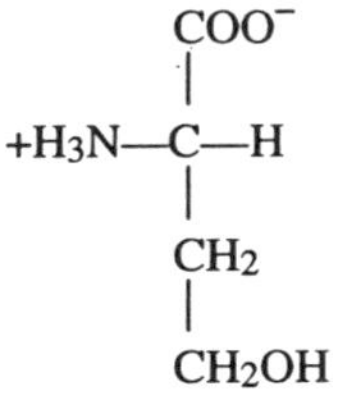

$$
\begin{array}{c}
\text{COO}^- \\
| \\
{}^+\text{H}_3\text{N}-\text{C}-\text{H} \\
| \\
\text{CH}_2 \\
| \\
\text{CH}_2\text{OH}
\end{array}
$$

Threonine   Thr (T)

$$
\begin{array}{c}
\text{COO}^- \\
| \\
{}^+\text{H}_3\text{N}-\text{C}-\text{H} \\
| \\
\text{H}-\text{C}-\text{OH} \\
| \\
\text{CH}_3
\end{array}
$$

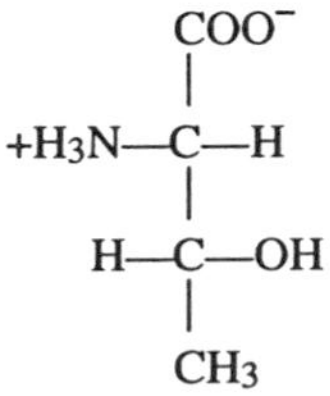

Prolene   Pro (P)

$$
\begin{array}{c}
\text{COO}^- \\
| \\
\overset{+}{\text{H}_2\text{N}}-\text{C}-\text{H} \\
\end{array}
$$

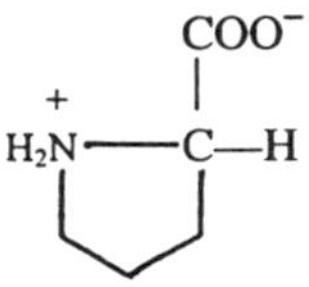

there are no particularly attractive absorbance or fluorescence properties associated with all of the amino acids, some form of derivative formation is usually needed for analysis. In their free form, all of the amino acids have at least one carboxylic acid functionality and (again with the exception of proline) at least one primary amine. Due to the universal occurrence of the carboxylic acid group, one would suppose that its use is indicated in any derivative formation chemistry. However, reagents for amino group labeling were much more common during the early development of amino acid analysis, and they still predominate. While methodologies exist for forming derivatives of carboxylic acids, they have almost no history of application to amino acid analysis. It is a fairly large task to establish the validity of any method of analysis, and often the historical is the path of least resistance. There are several good reviews and reference books written on the subject of amino acid analysis (e.g., 1–7).

Amino acids are nonvolatile, crystalline white solids in their pure form. They have no defined melting points, but decomposition occurs at temperatures ranging from 185°C to 342°C. Almost all are optically active (glycine is the exception). They are all soluble to some extent in water (though some are only sparingly so and can be problematic in certain extractions). As expected, water solubility is minimized at the isoelectric point of the molecule. Amino acids have limited to no solubility in even the polar organic solvents (only proline and hydroxyproline have appreciable solubility in ethanol). All are capable of forming salts, and it is these that are usually most cheaply obtained and most stable. A summary of some of these and other properties of the 20 common amino acids is found in Table 2. Others have compiled data concerning the less common derivatives of amino acids found in very specialized proteins or in their free form (e.g., 8).

The fact that most amino acids in foods are present as components of proteins presents a unique challenge to the analyst. To measure all of the individual amino acids in the protein component, it is necessary to hydrolyze the amide bonds linking these amino acids without destroying the amino acids themselves. Since several of the amino acids are relatively labile to the chemical treatments needed to break the amide bond, a variety of approaches have been devised. However, it should be noted that, to date, it is not possible to simultaneously analyze for all of the 20 common amino acids in a food matrix. While less often encountered, measurement of free amino acid content in foods is sometimes necessary. This is often much easier than total amino acid profile determinations since one can avoid complications from hydrolysis. However, interference associated with small peptides and so forth may be problematic. Any extraction process (usually needed for free amino acid determinations) will need to take account of the differences in solubility and dissolution kinetics of the various amino acids, as well as any matrix structural hindrance to extraction. It should be noted that quantitative extraction processes are *not trivial*, and the analyst should carefully validate any approaches.

B. Why Analyze for Amino Acid Content?

Food must provide the nutrients needed for growth and maintenance of the organism. Normally, most living things can synthesize at least some of the amino acids (*nonessential*) from other food components and/or other amino acids, but those that cannot be produced internally (*essential*) must be provided in the diet. There are also certain disease states, or special needs circumstances (e.g., after surgery, etc.), whereby other amino acids become essential (*conditionally essential*). Included in this category are those amino acids that are needed only during early development. Table 3 shows the amino acids currently recognized in each of these categories for humans, though the conditionally essential category is not claimed to be complete. As biological and medical scientists investigate the amino acids more thoroughly and in more stress, injury, or other nonnormal situations, it is likely that the beneficial effects of particular amino acids will be more obvious.

Table 2 Properties of the 20 Common Amino Acids[a]

Amino acid	Formula	Mol wt	Decomp. T. (°C)[b]	Solubility[c]	pKa values	pI[d]	Specific rotation $[\alpha]_D$[e]
L-Alanine	$C_3H_7O_2H$	89.09	297	166.5 (25°C)	2.34, 9.69	6.01	+2.8 (w, c = 6, 25°C)
L-Arginine	$C_6H_{14}O_2N_4$	174.20	244	148.7 (20°C)	2.01, 9.04, 12.48	10.76	+12.5 (w, c = 3.5, 20°C)
L-Asparagine	$C_4H_8O_3N_2$	132.12	234–235	35.3 (28°C)	2.02, 8.80	5.41	−5.42 (w, c = 1.3, 20°C)
L-Aspartic acid	$C_4H_7O_4N$	133.10	270–271	4.5 (20°C)	2.10, 3.86, 9.82	2.98	+4.36 (w, c = 1, 20°C)
L-Cysteine	$C_3H_7O_2NS$	121.16	240	160 (20°C)	1.71, 8.27, 10.78	5.02	+9.8 (w, c = 1.3, 30°C)
L-Cystine	$C_6H_{12}O_4N_2S_2$	240.30	260–261	0.112 (25°C)	1.04, 2.05, 8.0, 10.25	5.02	−223 (1 N HCl, c = 1, 20°C)
L-Glutamic acid	$C_5H_9O_4N$	147.13	247–249	8.64 (25°C)	2.10, 4.07, 9.47	3.08	+31.4 (6 N HCl, c = 1, 22°C)
L-Glutamine	$C_5H_{10}O_3N_2$	146.15	185–186	26.0 (18°C)	2.17, 9.13	5.65	+6.5 (w, c = 2, 25°C)
Glycine	$C_2H_5O_2N$	75.07	233–290	250 (25°C)	2.35, 9.78	6.06	Not Active
L-Histidine	$C_6H_9O_2N_3$	155.16	287	41.9 (25°C)	1.77, 6.10, 9.18	7.64	−39.7 (w, c = 1.13, 20°C)
L-Isoleucine	$C_6H_{13}O_2N$	131.17	284	41.2 (25°C)	2.36, 9.68	6.02	+11.29 (w, c = 3, 20°C)
L-Leucine	$C_6H_{13}O_2N$	131.17	293–295	24.3 (25°C)	2.36, 9.60	5.98	−10.8 (w, c = 2.2, 25°C)
L-Lysine	$C_6H_{14}O_2N_2$	146.19	224.5	>1000 (25°C)	2.18, 8.95, 10.53	9.47	+14.6 (w, c = 6.5, 20°C)
L-Methionine	$C_5H_{11}O_2NS$	149.21	280–282	53.7 (20°C)	2.28, 9.21	5.74	−8.2 (w, c = 1, 25°C)
L-Phenylalanine	$C_9H_{11}O_2N$	165.19	283	29.6 (25°C)	1.83, 9.13	5.48	−35.1 (w, c = 2, 20°C)
L-Proline	$C_5H_9O_2N$	115.13	220–2	1620 (25°C)	2.00, 10.60	6.30	−80.9 (w, c = 1, 20°C)
L-Serine	$C_3H_7O_3N$	105.09	228	359.7 (20°C)	2.21, 9.15	5.68	−6.8 (w, c = 10, 20°C)
L-Threonine	$C_4H_9O_3N$	119.12	255–257	90.3 (20°C)	2.71, 9.62	6.16	−28.3 (w, c = 1.1, 26°C)
L-Tryptophan	$C_{11}H_{12}O_2N_2$	204.22	290–292	11.4 (25°C)	2.38, 9.39	5.88	−31.5 (w, c = 0.5, 20°C)
L-Tyrosine	$C_9H_{11}O_3N$	181.19	342–344	0.453 (25°C)	2.20, 9.11, 10.07	5.63	−10.6 (1 N HCl, c = 4, 22°C)
L-Valine	$C_5H_{11}O_2N$	117.15	315	88.5 (25°C)	2.32, 9.62	5.96	+22.9 (20%HCl, c = 0.8,23°C)

[a]Data compiled from: 1. *CRC Handbook of Chemistry and Physics*, 54th Edition, CRC Press, Cleveland, OH. 2. *Merck Index*, 11th Edition, Merck & Co., Rahway, N.J. 3. *Chemistry and Biochemistry of the Amino Acids*, G. C. Barrett, Ed., Chapman and Hall, New York. 4. *Ulmann's Encyclopedia of Industrial Chemistry*, Vol. A2, 1985, p. 62.
[b]Decomp. T. decomposition temperature.
[c]Solubility, grams per 100 grams water, at specified temperature.
[d]Isoelectric point.
[e]Concentration data (c =) are expressed in % (w/v). Solvents and temperature are indicated. w = water.

Table 3 Amino Acids Essential, Nonessential, and Conditionally Essential for Humans

Essential	Nonessential	Conditionally essential[a]
Histidine	Alanine	Arginine[b]
Isoleucine	Arginine	Glutamine[c]
Leucine	Asparagine	Cysteine[d]
Lysine	Aspartate	Tyrosine[e]
Methionine	Cysteine	Ile, Leu, Val[f]
Phenylalanine	Glutamate	Taurine[g]
Threonine	Glutamine	
Tryptophan	Glycine	
Valine	Proline	
	Serine	
	Tyrosine	

[a]No attempt is made at completeness in this list.
[b]May be essential in infant nutrition or in total parenteral nutrition.
[c]Improves rate of healing in gut trauma.
[d]Essential in premature infants, usually sulfur amino acids in infants.
[e]Essential in patients with PKU.
[f]In hepatic disease.
[g]In infant animals, particularly cats, for proper neurological development. Maybe developmentally important in humans, but this is not well established.

This list is therefore incomplete and is intended only to inform of the possibility of nutritional necessity in particular situations.

Amino acid profile is not of overriding concern to the food industry in general. The normal adult diet is usually quite varied and provides sufficient amounts of the needed amino acids, and limitations in amino acid uptake (within reason) are not normally necessary. In most cases, a total protein determination is all that is needed or required. There are, however, many cases where amino acid analysis (partial or complete profile data) is either mandatory or highly advisable.

1. Dietary Restriction/Sole Source Nutritional

When dietary restriction is encountered, or in conditions whereby the entire diet is supplied from one source, it is critical that the food material provide all of the biological needs. Amino acid profile is often determined for infant formulas and adult nutritional products designed for sole-source use.

2. Intentional Addition of Amino Acid(s)

Another case where analysis is indicated is when one or more amino acids are intentionally added to the food. This is especially true when a specific nutritional effect is desired, e.g., enhancing the nutritional quality of the protein source or targeting some unusual metabolic need. Examples in this area include fortification of soy-based infant formulas with methionine and medical nutritional products such as those fortified with glutamine (9,10) or arginine (11–13) for metabolically stressed patients. It is incumbent on the manufacturer to assure correct addition as a general quality assurance practice. It is even more important when a specific nutritional effect is targeted, as in some nutraceutical-type products.

3. Absence (or Established Levels) Nutritionally Important

Another reason to undertake amino acid analyses is to demonstrate absence (to below predetermined limits) of one or more amino acids. This may seem odd, but there are special disease states (inborn errors in metabolism) where strict limiting of specific amino acids is absolutely critical to health. Such diseases include phenylketoneuria (PKU) where phenylalanine must be limited, and maple syrup urine disease (MSUD) where the branched chain amino acids (isoleucine, leucine, and valine) are problematic. Analytical data in the case of foods designed for these special needs becomes a very serious matter, with both patient health and legal ramifications. Foods designed for these persons must be analyzed to verify an upper limit content of those analytes. Many manufacturers are providing label information with either a general warning (e.g., this product contains phenylalanine) or a statement of specific levels of certain amino acids in the product as a service to those customers benefited by limiting amino acid intake.

Another example in this latter category is the recent public health concern over monosodium glutamate (MSG) which has prompted the U.S. Food and Drug Administration (FDA) to propose labeling (and thus testing) regulations concerning the levels of free glutamic acid in foods. While it is not yet certain that these regulations will be adopted, this is an instance where specific free amino acid analysis challenges could present themselves.

4. Avoidance of Formulation Problems

Problems with the final product can be averted by commodity testing, including certain kinds of amino acid testing. One case involves the estimation of the relative Maillard browning potential of various protein ingredient candidates. For intact protein systems, such browning is usually related to the ε-amino group of lysine (14,15). But with the more recent popularity of protein hydrolyzates, there is a need for "available" amine determinations (free amino acids + N termini of peptides + ε-amino group of lysine) to estimate the potential for this sometimes serious food processing problem. Maillard reaction products have been associated with a variety of problems ranging from inhibition of digestion (e.g., 16) to mutagenesis (e.g., 17,18). Therefore, minimizing or preventing their formation is a concern for the food industry.

5. Ingredient Quality Assurance Testing

Occasionally it is necessary to analyze a relatively pure amino acid commodity, i.e., one obtained in a food grade form to be used in the manufacture of a food product. Some amino acids and their derivatives are used as flavor enhancers (e.g., MSG, a form of free glutamic acid). It is occasionally desirable to fortify with one or more of the amino acids to boost the protein quality in an otherwise useful protein source (i.e., cheap, easy to formulate, bland flavor, less allergenic, better tolerance, etc.). Examples of this are soy protein (fortification with methionine) or rice protein (fortification with lysine and threonine).

6. Legal Issues

There are often legal requirements for assurances concerning amino acid content. For example, the Infant Formula Act of 1980 (USA) mandates regular assurance of adequate protein quality. This requirement is usually met by total amino acid testing and periodic protein efficiency ratio (PER) testing. The FAO/WHO also recommends suitable amino acid patterns for infant, youth, and adult diets (19), and these recommendations are often heavily considered by regulatory agencies. Certainly, all cases of intentional addition of free amino acids would usually require analytical verification for good quality assurance. Since legal requirements are somewhat different from country to country and often change, no comprehensive effort is made to cover them in this work. However, in general, if the nutrient is intentionally added (as in amino acid fortification), or if the finished food is claimed as a sole-source nutritional product (as in infant formulas or liquid diets designed for tube feeding), or if a claim is made concerning the amino acid content either on the package or in literature used as sales aids, one can expect some legal requirements for amino acid testing. Usually, applicable laws also mandate that substantial efforts be made to assure method validity (an expectation of good analytical work). Thus, proof that the method detects the analyte in a quantitative manner in the matrix to be tested and that the result does not vary substantially from day to day or analyst to analyst is needed. For amino acid profile analysis, methods which deliver 95–105% or better overspike recovery and less than 5% relative standard deviation for all analytes tested are usually deemed appropriate. Method performance should be better than this if only one amino acid is determined; fewer compromises are necessary and more flexibility is available to the analyst. As is the case for any analysis, the use of high-quality, well-characterized reference standards is mandatory. System suitability checks (e.g., minimum peak height to width ratios, several required retention time windows, etc.) and control charting to track method performance on a day-to-day basis are also very good practices and are often required.

Finally (as a comment only), one of the current trends is away from "simple" analysis of complete amino acid profile, and more toward analyses that estimate *bioavailable* amino acid profile, especially for foods designed as sole-source nutritionals. The suggestion of a protein digestability–adjusted amino acid score as a more appropriate test (20) and the inclusion of this technique in the latest edition of the *AOAC Official Methods of Analysis* (21) is a clear indication of things to come, and the analyst would do well to keep this in mind.

II. ANALYSIS

Amino acid analysis can be divided into two major categories: free amino acid analysis and amino acid analysis of peptides and proteins. These require quite different sample workup and yet utilize the same or similar approaches in derivative formation and separation. Another consideration in the food industry is that both commodity materials (amino acids, relatively pure proteins, etc., that are used in food material production) and in-process or finished food materials themselves need to be analyzed. These different sample types both present their own quite different challenges.

A. Free Amino Acids as Commodities

The simplest of these assays is the analysis of an amino acid commodity acquired to fortify a food product. These are usually fairly pure (90% or higher by weight). The certificate of analysis from the manufacturer is usually accepted as to purity, although in heavily regulated industries purity is often verified anyway. It is standard practice to establish a series of two or three simple, quick, and inexpensive tests to check for the errors one can anticipate, e.g., shipping errors or mislabeling. Usually, optical rotation under well-defined conditions, color checks, ultraviolet or infrared spectral comparisons, and/or a test such as thin layer chromatography (TLC) with specific staining (e.g., ninhydrin or fluorescamine spray; e.g., 22) or TLC after derivative formation (e.g., 23) are sufficient for this purpose. If a batch of material is suspect, more sophisticated analysis and follow-up with the supplier will usually resolve the issue.

B. Sample Preparation

Samples which contain significant quantities of materials other than free amino acids (e.g., food samples or protein/peptide commodities) must be prepared for application to typical amino acid analysis systems. The amino acids need to be in their free form and in solution. This is often not a trivial task.

1. Free Amino Acids in Products

Free amino acid analysis is often needed for finished product or relatively complex in-process samples. In such cases, sample preparation for determination of free amino acid content usually involves two steps. First the material must be rendered easily extractable. For liquids and most powders, this is usually already done, e.g., the sample can be efficiently extracted without prior preparation. For solids (food bars, meat products, etc.), the material must be homogenized (wet solid, such as a meat sample) or ground/powdered (dry solid) to allow more efficient extraction. In other words, seeds, meat products, and other samples with significant physical structure (or even large particle size) usually require breakdown of this structure to allow efficient extraction. There are very few published procedures for food materials where free amino acid analysis is done; it has not been deemed a widely necessary determination to date. Although some analyses have been reported (e.g., seeds and wine), the sample type is limited. Some of these specific examples will be covered in a later section of this chapter. The numerous publications on small molecule analysis in biological tissues and fluids, particularly the methods used for homogenization, could prove useful to the analyst needing to produce such information (e.g., 24).

Once the sample is ready for extraction, stirring in a suitable solvent is usually sufficient to solublize the free amino acids. However, leucine and cystine are often problematic, requiring relatively acidic conditions and relatively longer stirring times to completely solubilize. Also, it is known that some amino acids bind rather tightly to proteins. The most notable example of this is tryptophan in blood serum (25) and human milk (26) samples. This is a clinical laboratory problem; serum free amino acid profiles are run for diagnostic purposes. It should be assumed that other samples might present this difficulty as well, and appropriate care during method development is required to avoid errors.

One solvent of particular utility is dilute hydrochloric acid. In our laboratories, we extract a complete powdered nutritional product containing free amino acids as the only protein source by stirring the sample in 1.5 N HCl for at least 15 min at room temperature. Subsequently, the pH is adjusted, and a suitably diluted sample is subjected to a standard amino acid analysis. This procedure gives suitable recovery of all of the amino acids in the product. Excessive heat should be avoided if applying this procedure; several of the amino acids are labile to acidic conditions at elevated temperatures. Development of extraction procedures should be done carefully, and extraction efficiency should be verified through both kinetics and overspike recovery experiments.

2. Total Amino Acid Profile

By far the most common amino acid analysis for food materials is total amino acid profile. This requires that the proteins and peptides in the sample be hydrolyzed. In most reviews of amino acid analysis, it is acknowledged that the most formidable problems facing the analyst in total amino acid analysis involve hydrolysis. The primary problems involve tryptophan and cyst(e)ine; tryptophan is degraded in hot strong acid (particularly in the presence of carbohydrates), and the cyst(e)ine sulfur is often oxidized. In some cases, degradation of serine and threonine (presumably by dehydration) is observed. Also, in any method involving hydrolysis, the amides glutamine and asparagine are converted to glutamic and aspartic acid, respectively.

A fairly large variety of methods for protein hydrolysis have been developed, each with limitations and advantages. A listing of the more commonly used hydrolysis methods, their applications, and their advantages and disadvantages is given in Table 4A. It is important to note that there is no known method for hydrolysis which gives quantitative recoveries of all nutritionally

Table 4A Some Common Protein Hydrolysis Procedures: Standard Hydrolysis Methods[a]

Agent	Conditions	Ref.	Notes
6 N HCl (with or without protectant, e.g., 2% thioglycolic acid)	16–72 hrs, 110°C 4 hrs, 145–155°C microwave (minutes)	27, 28 29, 30 31, 32	Multiple hydrolysis times and extrapolation needed for accurate quantitation. Single-time analysis results often acceptable. Microwave digestion is fast. Inexpensive. Automated systems available. Most frequently reported in literature.
6 N HCl (with tryptamine)	22 hrs, 110°C	33, 34, 35	Attempt to recover TRP while determining AA profile.
4 N methanesulfonic acid (with or without protectant, e.g., 3-(2-aminoethyl)indole	22–24 hr, 115°C	36, 37, 38	Sensitive to carbohydrate. Requires peptide/protein cleanup prior to application to foods—losses of small peptides and free amino acids can be problematic in sample processing to remove carbohydrate. Useful in protein commodity analysis.
Propionic + hydrochloric acid	50/50 v/v	39	Used for hydrolysis of resin-bound peptide in synthesis; not applied to foods.
3 N p-Toluenesulfonic acid	22 hr, 110°C	28, 40	Alternate acid to HCl. Attempt to obtain Trp and Cys with one hydrolysis preparation. Extensive application to foods not done.
3 N mercaptoethanesulfonic acid	22 hr, 110°C	40	See 3 N p-toluenesulfonic acid, above.

[a]Almost without exception requiring inert atmosphere or vacuum.

important amino acids in a food matrix. Of course, glutamine and glutamic acid are determined as GLX, and asparagine and aspartic acid are determined as ASX in any system which hydrolyzes peptide bonds (there is no significant difference between the side chain amides found in asparagine and glutamine and the peptide bond itself). Even allowing for this, however, all of the conditions in the reasonable analytical methods developed thus far are compromises; the chemical stability of the various amino acids are different enough that a few of them (especially tryptophan, cyst(e)ine, serine, and threonine) degrade to some extent under conditions needed to guarantee complete hydrolysis of the more stable peptide bonds (e.g., those involving isoleucine and leucine). For this reason, when performing amino acid analysis on isolated proteins, the early experiments were done using several hydrolysis times and extrapolating, either back to zero time (for labile amino acids) or to infinite time (for hard-to-hydrolyze bonds). While this may still be suitable (and even necessary) for research, routine analysis of foods utilizing this approach is not feasible (time and cost); thus the compromise. Even so, the complete amino acid profile analysis in complete food matrices involves application of several techniques. This is so because acid hydrolysis of a food matrix (which would normally contain carbohydrate), while suitable for most of the amino acids, usually destroys substantial amounts of tryptophan and cyst(e)ine). It is routinely necessary to use alternative methods for the determination of these amino acids. There are several hydrolysis methods which have been developed for analysis of either of these amino acids; some of the more widely applied are presented in Table 4B. While some more direct methods for both of these amino acids, (e.g., fluorescence analysis of tryptophan (50) and DTNB (5,5′–dithiobis(2–nitrobenzoic acid)) reaction for cyst(e)ine (51), have been utilized for protein analysis without hydrolysis, these methodologies are not generally applicable to food matrices due to interferences.

Despite all of the effort applied to the hydrolysis problem in amino acid analysis over the last 30–40 years, the most widely used hydrolysis conditions are still 6 N HCl at 105–120°C for 18–24 hr. This provides a sample suitable for determination of the primary amino acid profile and is

Table 4B   Some Specialized Hydrolysis Methods[a]

Agent	Conditions	Reference(s)	Notes
For tryptophan:			
4.2 N NaOH (+50 mM ascorbate)	110°C, 22 hr	41, 42	Need inert hydrolysis vessel; 85–95% recoveries
4 N Ba(OH)$_2$	110°C, 22 hr	43	Inert vessel and atmosphere
	125°C, 16 hr	44	Inert vessel and atmosphere
4 N LiOH	110°C, 22 hr	45	Inert vessel and atmosphere
	145°C, 4–8 hr	45	
	Microwave heating, 50–60 min	46	With internal standard 5-methyltryptophan, quantitative results
Pronase digestion	50°C, 6 hr	47	Background correction needed, good recoveries
For cyst(e)ine and/or methionine:			
Performic acid oxidation/6 N HCl	See references	48	Cyst(e)ine →cysteic acid (measured)
		49	Methionine →methionine sulfone (measured)
			Problem: oxidation is a normal route of degradation
			Corrosive fumes from oxidation. Multistep process

[a]Usually target one or only a few amino acids.

supplemented by base hydrolysis of samples (NaOH, LiOH, BaOH, 105–120°C for 18–24 hr) for tryptophan and performic acid oxidation/acid hydrolysis to determine cyst(e)ine (as cysteic acid). There are many variations on these classical approaches, e.g., elevated temperatures, microwave heating, and addition of antioxidants which afford savings in sample processing time and/or recovery in particular matrices. The typical analyst will usually resort to the classical combination of hydrolysis conditions for treatment of samples to determine the complete amino acid profile. One unusual approach is hydrolysis using proteases; while this is not a new idea, the increased availability of a large variety of highly purified proteases (and a better understanding of their action) may make this more serviceable today. Proteolytic digestions may allow determination of protein glutamine and asparagine, which today is usually not feasible (without protein sequencing). However, the enzymatic digestions thus far applied are not complete (e.g., 52), a commonly encountered problem with this approach. Other methods have been developed for the amide amino acids, e.g., chemical conversion followed by acid hydrolysis (e.g., 53), but have not yet been applied to food matrices.

C. Sample Pretreatments

Several sample pretreatments may be applied prior to derivative formation and separation in any analysis of foods. Operations such as filtration, concentration, etc., may be required based on the characteristics of the sample. One cannot inject particulates onto a modern high-performance liquid chromatography (HPLC) column, and there are limits to the determination of minute amounts of amino acids, even with derivative formation. Therefore, treatments to remove particulates and to concentrate the sample are common. Filtration can be accomplished using any of a number of commercially available filter manifolds. Filter papers should be of high quality since contamination of the sample may be an issue. Whatman paper or glass fiber disks are generally acceptable, and no doubt many others could work as well for gravity filtration of a sample. Such filtrations can be slow, and centrifuge-driven filtration using centrifuge cones (e.g., those from Amicon) can be advantageous. Syringe-mounted HPLC filters are also commonly utilized, but only for relatively clean samples since they plug easily if particulates are extensive.

Free amino acid determinations often cite deproteinization treatments as a significant part of sample cleanup. Commonly used treatments include acidification, e.g., with sulfosalicylic acid (54) or perchloric acid (55), or use of organic solvents, e.g., acetonitrile (56), and, after a suitable incubation, centrifugation or filtration to remove precipitated protein.

Another concentration/cleanup step involves the use of minicolumns (e.g., Sep-Pak). Such a procedure, utilizing packings such as Rexyn 101(H) (e.g., 57), Amberlite IR120 (e.g, 58), etc., is commonly applied, particularly with gas chromatography (GC) analyses. This involves binding the amino acids on a strong cation exchanger under highly acidic conditions, such as those used to hydrolyze proteins. Elution is accomplished using a strong, volatile base (e.g., ammonia) and the amino acids are recovered by evaporation of the solvent. There are some examples of application of this technology in HPLC-based analyses as well. The Pico-Tag system originally suggested use of C-18 Sep-Pak columns to remove high molecular weight contaminants and lipids (59,60), but this procedure has been questioned (61).

In some types of samples (e.g., cheese and fatty seeds) it is necessary to remove the high fat content to prevent unduly rapid loss of column performance in HPLC separation systems. This is usually done using an organic solvent extraction [hexane, 3:1 acetone/chloroform, etc. (e.g., 56,62)]. However, unless the sample is very high in fat content this is usually not necessary.

Many of these types of sample manipulations are useful and often necessary to apply a particular analytical system to a given sample. However, as with all manipulations involving the possibility of analyte loss and/or degradation, great care must be taken during method development to assess analyte recovery and stability. It is generally true that the more steps in the sample preparation, the greater the risk of error, and the more demanding the process is on the analyst. Control charting of several parameters and the use of internal and external calibration is highly recommended in all analysis. This becomes critical in the case of analyses involving multiple cleanup steps and complex derivative formation and separation schemes. It is a good general rule that if you cannot demonstrate a requirement for a cleanup step, don't use it—it adds not only potential error but cost (in both supplies and operator time).

D. Derivative Formation Chemistries

While the aromatic amino acids have useful ultraviolet (UV) absorbance characteristics (usable absorbance from 250 to 280 nm, and tryptophan is even fluorescent (excitation 295 nm, emission 345 nm), most of the amino acids do not have useful absorption), fluorescent, or electrochemical properties. And none of them are volatile enough for GC analysis. It is therefore almost always necessary to produce a derivative of some kind. Derivative formation chemistries differ widely, as might be expected, but can be divided into two general classes: (a) where the primary goal is to ease detection (by chromophore, fluorophore, or electroactive product formation) and (b) where the primary goal is to produce a volatile adduct.

1. Derivative Formations as Detection Aid

Derivative formation chemistries which have the primary aim of rendering the amino acid detectable are used for separations with liquid chromatography—almost exclusively HPLC though some of the same chemistries have been applied to capillary zone electrophoresis (CZE) and to more exotic, but still liquid-based, separations. The three standard detectors typically used in amino acid analysis are absorbance (UV or visible), fluorescence (including laser-induced), and electrochemical. By far the most frequently encountered are the absorbance and fluorescence detectors, with ultraviolet absorbance probably the most common today. There are a very large number of derivative formation chemistries which at one time or another have been used in amino acid analysis. Some of these are listed in Table 5 (with selected references and comments for each). Details of each chemistry, including detection limits usually associated with the derivatives, mode of detection, and so forth, are listed in Table 6.

There are two analytical objectives which are affected by the choice of derivatizing chemistry: (a) analyte separation and (b) easing detection. Either of the basic operations—separation or derivative formation—can be (and has been) done first. Reaction before separation is called *precolumn* derivative formation whereas after separation is *postcolumn* derivative formation.

a. Precolumn. These chemistries offer the advantage of concomitantly rendering the amino acid more hydrophobic—and thus more amenable to reverse phase HPLC (RP-HPLC) separations. Only a few underivatized amino acids are retained (and those only weakly) using standard RP-HPLC columns and solvents. RP-HPLC provides high resolution, is relatively common, and utilizes widely available, relatively inexpensive columns and versatile equipment. The general approach of precolumn derivative formation suffers from the fact that interfering substances may also be derivitized and present separation problems. However, since there are no serious restrictions on reaction time (unlike postcolumn reactions; see below), a very large variety of chemistries are available (again, see Tables 5 and 6). This is a distinct advantage, because derivative formation can be tailored to the problems peculiar to the matrix and to the available equipment.

There are currently three major chemistries utilizing this approach which are commercially available as kits: PITC, OPA/FMOC, and AQC. While relatively expensive (compared to reagents

Table 5 Some Derivitization Chemistries Used in Amino Acid Analysis

Reaction	Representative refs.	Comments
PITC (phenylisothiocyanate)	32, 35, 54, 56, 59, 63–68	Initially used in protein sequencing. Application to amino acid analysis later. Now the basis for the Pico-Tag prepackaged amino acid analysis system. Stable derivative, UV-active. Some toxicity of agent.
FMOC (9-fluorenylmethyloxycarbonyl chloride)	69, 70	Side reaction products a separation problem. Stable derivative, UV active. Relatively expensive agent.
Naphthylene 2,3-dicarboxaldehyde (NDA)	71, 72, 73	No significant literature. Very little used. Relatively exotic separation system, and electrochemical detection, or fluorescence used. Requires cyanide as coreactant presenting safety and disposal issues.
o-Phthaldialdehyde (OPA)	60, 74–77	Rapid reaction, fluorescent product, TLC or test tube assays have been used. Pre- or postcolumn use possible. Requires mercaptan coagent. Agent not fluorescent by itself. Derivative not stable. Does not quantify Cys or Pro. Some separation and stability changes with mercaptan selected. Electrochemical detection possible. Inexpensive, and available as premixed solution.
Fluorescamine	78–81	Rapid reaction, fluorescent product, excess reagent breaks down in water to nonfluorescent byproduct. Needs organic solvent (dry!) and very rapid and through mixing for quantitative use. Expensive agent, not much applied in amino acid analysis except as TLC spray or testtube assay for total primary amine estimation. Could be applied postcolumn.
Ninhydrin (2,2 Dihydroxy-1,3 indanedione)	82	Very widely used. Most history, but being replaced by precolumn agents. Strictly postcolumn, will detect all amino acids, though Pro needs different wavelength for sensitive detection. Produces visible absorbing chromophore. Can also be used for TLC spray or test tube detection for amine. Reagent stability is an issue. Many prepackaged commercial applications, esp. with classical amino acid analyzers.
DABITC	83, 84	
Dansyl chloride	23, 62, 85	Stable derivative, fluorescent. Side reaction products also fluorescent, presenting a separation problem. Strictly precolumn. Has been used on TLC plates.

Dabsyl chloride	86–89	Relatively simple derivatization process, but several drying steps. Postcolumn derivatization only. Multiple reagent byproducts problematic. Complex gradients used for resolution. UV-active.
NBD-F (4-fluoro-7-nitrobenzo-2-oxa-1,3-diazole)	90	Pre- or postcolumn, relatively stable, fluorescent derivative. Not much used in food analysis.
1-Naphthylisocyanate	55	Precolumn, fluorescent derivative. Relatively stable, but not much applied to food analysis.
OPA/FMOC combined method	91–94	Fluorescent derivatives. Attempt to overcome the limitations of OPA. Will detect secondary amines (FMOC). Basis for the AminoQuant amino acid analysis package.
FMOC/ADAM combined method	95	Relatively stable, fluorescent derivative. Eliminated the largest problem of FMOC—interfering byproduct—by producing highly hydrophobic adduct with ADAM, which elutes after the amino acid derivatives. Not much used in food analysis.
AQC (6-aminoquinolyl-N-hydroxysuccinimidylcarbamate)	96, 97, 98	Relatively new, stable, fluorescent derivative (also useful UV absorbance). Basis for AccQ Tag amino acid analysis system. Not much literature yet.
FDNB (1-fluoro-2,4-dinitrobenzene)	99	Available from, e.g., Sigma. Multiple reaction byproducts can be problematic. Slow reaction, stable derivatives, if light-shielded. UV-active.
DNPAA (1-fluoro-2,4-dinitrophenyl-5-L-alanine amide)	100	Residual reagent problematic. Available from, e.g., Pierce. Relatively complex derivatization procedure. No significant literature. Relatively stable derivative. UV-active.

Table 6 Characteristics of Some Derivatization Techniques for Amino Acids

Agent	DL[a]	Detection[b]	Stable[c]	Commercial kit	Kinetics[d]	2° Amines	Cyst(e)ine
PITC	50–100 pmole	A (254–280)	Yes	Yes (PicoTag)	Moderate	Yes	Yes
FMOC	"high fmole"	F (265/313)	Yes	No	Fast	Yes	Yes
NDA	0.1–0.5 fmole	E (0.8–1 V) F (420,440/490)	Int.	No	Moderate	No	No
OPA	5–10 fmole	F (330/455) E (0.4–0.7 V)	No	See OPA/FMOC	Very Fast	No	Yes[e]
Fluorescamine	50–100 pmole	F (390/475)	Int.	No	Very Fast	No	Yes
Ninhydrin	50–100 pmole	A (570+440)	Int.	Yes (see Table 7)	Fast	Yes (440 nm)	Yes
DABTC	1–20 pmole	A (436)	Yes	No	Slow	Yes	Yes
Dansyl Chloride	10–50 pmole	F (340/510)	Yes	No	Moderate	Yes	Yes
Dabsyl Chloride	"pmole"	A (425, 436)	Yes	No	Moderate	Yes	Yes
NBD-F	10–3000 fmole	F (470/530)	Yes	No	Fast	Yes	Yes
1-Napthylisocyanate	"low pmole"	F (238,305/385) A (222)	Yes	No	Moderate	Yes	Yes
OPA/FMOC	"high fmole"	F (see above) (OPA & FMOC)	No (OPA)	Yes (AminoQuant)	Slow[f]	Yes	Yes[g]
FMOC/ADAM	~50 fmole	F (265/313)	Yes	No	Slow[f]	Yes	Yes[g]
AQC	"fmole"	F (250/395) A (254)	Yes	Yes(AccQ Tag)	Fast	Yes	Yes
FDNB	10 pmole	A (365)	Yes	No	Slow	Yes	Yes
DNPAA	"pmole"	A (340, 414)	Yes	No	Slow	Yes	Yes

[a]DL = detection limit.
[b]A = absorbance (wavelength(s), nm); F = fluorescence (excitation/emission wavelengths, nm); E = electrochemical.
[c]Int. = intermediate stability, slow degradation is observed.
[d]Very fast < 1 min, fast = 1–10 min, moderate = 10–60 min, slow > 60 min.
[e]Can be detected, but only in the presence of a reducing agent, such as chloramine T.
[f]Several sample derivatizing steps, reagent addition slow.
[g]Only when sulfhydryls are reduced and alkylated prior to derivatization. Assumes hydrolysis can release without destroying.

"made from scratch"), they offer the analyst a reasonable alternative to optimizing the chemistry. Now it is only necessary to test the prepackaged system and then concentrate on separations in the matrix. Also, keep in mind the time for reagent preparation in the made-from-scratch reagents; someone has to pay for this as well, and reagent preparation can be time consuming. The author does not particularly endorse any of these prepackaged systems, but only indicates that the reagent/supplies cost is often more than offset by savings in method development time and periodic reagent preparation. It should not be surprising that these chemistries account for a very large percentage of reported amino acid analyses utilizing precolumn derivative formation, especially for industrial applications.

b. Postcolumn. These chemistries offer the advantage of separation prior to rendering the molecule UV/Vis- or fluorescence-active. Sample preparation is almost always significantly simpler since derivative formation manipulations are eliminated. Separations are also somewhat less demanding since it is unlikely that fluorescent or highly UV/Vis-active molecules (the only ones detected) will elute with the underivatized amino acids. Further, the derivatizing agent recognizes only compounds that have a selected chemical functionality (most often an amine). Thus, separations only have to be complete from matrix compounds having this same functionality; others will be transparent to the detector. The postcolumn reaction approach suffers in several ways. First, it adds significant postseparation volume and equipment cost since a reactor is needed. This is typically a second oven, similar to a column oven. This must have a large enough volume to allow for thorough mixing of eluant and reactant, and a long enough residence time to assure that the reaction is reasonably complete. All of this adds volume between the separating column and the detector, resulting in loss of resolution (band broadening) and increasing the time necessary for the instrumental analysis. Second, the separations which may be applied [cation exchange (mainly) and ion pair chromatography (limited applications thus far)] have somewhat less flexibility in run parameters, allowing less optimization. After all, the analyst is restricted to the systems which can separate the amino acids in their underivitized form. The original automated amino acid analysis was, however, of this type; it used a postcolumn reaction with ninhydrin. The ninhydrin reaction and the reaction involving o-phthaldialdehyde in the presence of mercaptan are by far the major postcolumn chemistries used. A third limitation, perhaps obvious to the reader from the discussion above, is that the reaction needs both to be very fast and to use easily miscible solvents. However, in spite of all of the advances in this field, a modernized version of the cation exchange and postcolumn derivative formation approach is still very often used. It is relatively rugged, is not terribly matrix-sensitive, and is commercially available as prepackaged systems which are usually easy to install and validate. They are, however, expensive to obtain and run. These systems, while still in extensive use, seem to be phasing out in favor of the precolumn approach (see above).

2. Volatile Derivatives

Separations utilizing gas–liquid chromatography (GLC) require that the compounds be rendered volatile. The majority of derivative formation schemes applied to GLC-based amino acid analyses seem to follow the same general approach. Acidified alcohol is used to produce the ester, followed by N-acylation using any of a number of acid anhydrides. Some of the more commonly used acidified alcohols and acid anhydrides as well as representative references are listed in Table 7. The recent trends in this area are focusing more on simplifying the derivative formation process or highlighting the major strength of GLC analysis—its very high resolution. One paper (111) cites a single-step derivative formation process utilizing BSTFA [Bis(trimethylsilyl)trifluoroacet-amide], but this procedure has not been widely applied. A very recent paper (110) illustrating the very high resolving power of the GLC approach describes the use of isobutylchloroformate

Table 7 Derivatization Chemistries Useful for GLC Analysis of Amino Acids

Target functionality	Representative Refs.
Esterification Agents	
3 N HCl/isopropanol	57, 101–103
3 N HCl/isobutanol	104–107
3 N HCl/n-butanol	58, 107
3 N HCl/n-propanol	108
3 N HCl/methanol	109
3 N HCl/isoamyl alcohol	109
Isobutylchloroformate	110
N-Acetylation Reagents[a]	
N-methyl-N-(tert-butyldimethylsily)trifluoroacetamide	110
Pentafluoropropionic anhydride	101, 108
Trifluoroacetic anhydride	58, 108
Heptafluorobutyric anhydride	57, 102, 103, 105–107, 109

[a]Fluorinated derivatives allow use of electrochemical detectors as well as flame ionization detectors. This is thus a common element in the acetylation reagents.

followed by *t*-butyldimethylsilylation to allow GLC separation of 53 different free amino acids from seed samples.

E. Separation Systems

After a sample has been treated in such a way as to extract (free amino acids) or liberate (hydrolysis for total profile) the amino acids, and appropriate derivatives have been formed, one must separate and quantitate the desired analytes. In the case of a "complete" amino acid profile (total of 18 analytes) this is a nontrivial task. There are, however, many different methods available to solve the problem, all with their strengths and weaknesses. This section will focus on those separation/detection systems that are reasonably well developed and reasonably modern. (Immobilized enzyme reactors, microbial growth tests, etc., which are either not widely used/developed or are generally considered as outdated, will be dealt with exclusively in the last section of the chapter.) Even with these limits, a very large amount of material is covered, and no real attempt at completeness is claimed.

The choice of separation system seems mainly an empirical matter dependent on limitations in the sample matrix or equipment available or even personal preferences. There are always a number of approaches that are suitable for every analytical problem, and each has advantages and drawbacks, depending on the sample matrix, the sample load, budgetary constraints, etc. A summary of the various general approaches regularly used (and some not so frequently encountered) is provided in Table 8. Some consideration of the advantages and limitations usually associated with each approach is given, and a very limited number of representative references are cited. As indicated before, the HPLC-based methods very much dominate reported separations in the food industry, and the advantage in this general area goes to precolumn derivative formation and RP-HPLC. Figures illustrating separations possible with several of the more frequently used systems are included in the food applications section below. Whatever the approach used, careful consideration to method control utilizing system suitability checks, proper equipment maintenance, and appropriate, well-characterized reference standards is essential to assure the quality of the data generated.

Table 8 Major Separations Systems Applied to Amino Acid Analysis

Separation system (general)	Advantages	Drawbacks	Ref.[a]
Cation exchange/Postcolumn derivatization	Classical—very large literature Commercial Instruments: *Beckman *Biotronik (Eppendorf) *Dionex *LKB *Pickering *Etc.	Expensive hardware (high maintenance) Limited derivatization chemistries *Rapid reaction necessary *Miscible solvent needed *OPA or ninhydrin used Limited run parameter flexibility	5, 38, 82
RP-HPLC/precolumn derivatization	Commercial prepackaged systems: *Waters PicoTag and AccQTag *Hewlett–Packard AminoQuant *Etc?. Large literature—research and food appl. Less costly, more versatile hardware Sample handling reduced (vs precolumn derivatization) RP-HPLC resolution w/limited sample preparation (derivatization postcolumn)	Higher sample handling time Interfering substances derivatized	66, 93, 96
Ion pair chromatography/postcolumn derivatization		Derivatizing chemistries limited (see Cation exchange, above) Postcolumn reactor: *Band broadening and increased cost Complex (?) mobile phases Small literature—esp. in food analysis	112, 113
Gas–liquid chromatography	Very high resolution Inexpensive operation—no solvent disposal Simple detection—chromophore not needed Rugged, moderate cost instrument	Volatile derivative required Multiple derivatization common Small literature—esp. in food analysis	114–117
Capillary zone electrophoresis	Very high resolution—comp. to GLC Fast Ultrahigh sensitivity	Not fully developed, esp. in food analysis Very small sample injections	118–120

[a]See Table 5 for references dealing with specific derivatization chemistries.

Method control is absolutely essential in any analytical technique. Basically, system suitability for the assay must be demonstrated on a regular enough basis to assure proper operation during analysis. This is done by establishing and running system suitability tests. These are usually done on standards injections. Measurement of such parameters as retention time for several peaks (these must fall within certain "windows"), peak shape (ratio of peak height to width at half height, or asymmetry measurement, if the software or integrator handles this), peak area for a set amount of standard injected, resolution of the "worst separated" pair of analyte peaks, and so forth are often done. Some of these are illustrated in Fig. 1. These are short cuts to the more classical resolution and plate count calculations cited in good analytical methods texts (e.g., 121,122). It is almost always the case that several of these, measured at least once in the analysis run (or with a specified frequency), are used to monitor system suitability. Notice that this is a use test; things like measuring flow rate, detector function, etc., do not assure the full utility of the system *for that particular analysis.* Use of the system to analyze a known solution and measurement of parameters in the results is always necessary. This can be done after the run, as long as the data are collected during the run, or periodically, as required (established during method development). Charting the results of these system suitability tests can detect a drift from usability—caused by deterioration, for example, of column performance or detector lamp intensity—before bad data are reported.

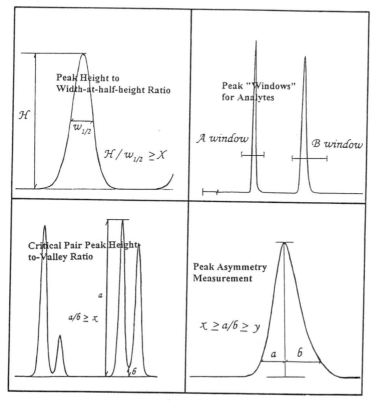

Fig. 1   Examples of system suitability shortcuts. All of these measurements are empirical. Suitable value limits must be established for each assay during method development and validation.

F. Quantitation and Reporting

Calibration of the system response is always required. This is a different problem from system suitability. Well-characterized analytical reference standard materials are required and may be obtained from any of several sources. Of course, reference standards may be established in-house, but keep in mind that this is not a trivial task. Not only must the identity of the material be established, but its purity, contaminants, and so forth, must be established and periodically reevaluated as well. All of this must be painstakingly recorded, and these records must be kept, often for long periods. It is usually easier and more cost-effective to obtain reference standard materials from a reputable commercial supplier—referred to as a *compendial* source (e.g., National Institute of Standards and Technology, e.g., NIST (old NBS), U.S. Pharmacopeial Convention (USP), British Pharmacopoeia (BP), European Pharmacopoeia (EP), World Health Organization (WHO). Analysis of a standard to calibrate instrument response should be done with all analytical runs—and sometimes may be required after a certain number of sample injections because of possible system drift (derivatizing agent aging, small changes in sparged mobile phase, etc.). The shelf life of prepared standards and the required frequency of use during analysis need to be established during method development.

Internal standard materials are often added to samples during sample preparation. There are several reasons for doing this. Once the stability of the internal standard material to the conditions of sample preparation/derivative formation has been established, its addition compensates for less than perfect quantitative technique by the analyst and internally corrects for small variations in the injection volume. One needs only to know the exact sample mass (or volume) and the exact mass of internal standard to calculate the result. Both of these can be measured very accurately in all analysis situations, and dilutions then become necessary only to assure that the detector response falls within the established linear range of the method.

Another reason for internal standard addition is to correct for analyte instability through the sample preparation. This has been used in the case of tryptophan analysis. Addition of 5-methyltryptophan as an internal standard in tryptophan analysis (usually done using base hydrolysis of the sample) results in near-quantitative recovery of tryptophan *relative to the internal standard*. To allow use of this approach, it is necessary to rigorously establish that the stability of the analyte and the internal standard are identical or nearly so. Including a system suitability check verifying this would be advisable. As is obvious from this (admittedly brief) discussion of internal standards, there is no such thing as a "perfect" internal standard. The intended purpose of the internal standard directs its desirable qualities, which can be conflicting. Correction for less than perfect quantitative technique, injector imperfections, etc., absolutely requires very high stability, whereas compensation for analyte degradation during sample preparation requires instability closely matching the analyte of interest! A listing of some of the more commonly used internal standards (and some references) is provided in Table 9.

Finally, quantitation can be done by any of several techniques. The two major approaches are based on chromatographic peak height or peak area. There are advantages and disadvantages associated with each method, and the reader is referred to any of a number of very good method development books (e.g., 121,122) or any good textbook on modern instrumental analysis. Measurement of peak area or height is done almost exclusively using chromatography software, but choosing the right parameters can be critical to the success of the analysis, and understanding the basic principles is essential. Most amino acid analyses (particularly those done with ion exchange) use the peak area method of quantitation, while peak height is frequently used for GLC quantitation. While much of this work is now done by computers, it is always necessary for the analyst to check the data; there are many things that can go wrong, and simply looking at the chromatograms and/or comparing analytical results to those obtained previously for similar samples can often reveal problems.

Table 9 Some Compounds Used as Internal Standards for Various Amino Acid Analyses

A. *HPLC Analysis*

Chemical name	Ref.
Norleucine	55, 59, 66, 68, 71, 87, 96, 99, 123
Norvaline	92, 95, 124
α-Amino-*n*-butyric acid	56, 64, 66, 68, 73, 96, 98
Sarcosine	92
Piperidine 4-carboxylic acid	92
β-Amino-*n*-butyric acid	125
Homoserine	125
Homocystine	126
Aminohexanoic acid	63
Diaminobutyric acid	62
Ethanolamine	60, 127
γ-Amino-*n*-butyric acid	112

B. *GC Analysis*

Chemical name	Ref.
Aminopimelic acid	107
Norvaline	110, 117
3,4-Dimethylbenzoic acid	110
n-Butylstearic acid	58, 111, 128
Norleucine	102, 106, 109, 129–133
Cycloleucine	103
Tranexamic acid	134
γ-Amino-*n*-butyric acid	101
Ornithine	135
Pipecolic acid	104, 136

G. Food Applications

Analysis of amino acids in food matrices is done almost exclusively with liquid chromatography. While very high resolution is possible with GC separations, the derivative formation reactions applied are often complex and time consuming. Other technologies, such as capillary zone electrophoresis (CZE) are still in the development stage, and their main utility lies in analysis of very small sample sizes, not usually a concern in food analysis. In the past, microbiological assays were used. However, these are very labor-intensive compared to the separatory methods and have largely been eliminated. They do, however, have the advantage of using only the L-amino acids. The methods commonly used for routine amino acid profile do not distinguish between D- and L-amino acids. This is not usually a problem in foods, because protein does not contain D-amino acids and processing does not typically produce racemic mixtures. This may not be the case with free amino acids, however. Methods (both HPLC- and GLC-based) are now available for determination of mixtures of D- and L-amino acids (e.g., 101,108,137,138).

Liquid chromatographic methods applied to foods are almost exclusively divided into two types (though there are other systems, these are rare): ion exchange procedures modeled after the original work of Moore and Stein (27) and precolumn derivative formation followed by RP-HPLC. The separations obtainable from these two types of systems are roughly similar, though the edge in speed goes to the RP-HPLC. A classical system applied to foods, and the one most often used until the last 5–10 years, is acid hydrolysis followed by cation exchange separation (typically using a step gradient) and postcolumn derivative formation with ninhydrin. Using this approach, the chromatograms presented in Fig. 2 were generated.

One of the more commonly used precolumn derivative formation/RP-HPLC systems uses phenylisothiocyanate. This may be more common because it was one of the first systems of this

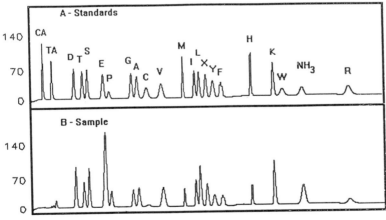

Fig. 2 Amino acid profile analysis using cation exchange with postcolumn ninhydrin reaction. A sample of a liquid nutritional product containing approximately 50 mg protein was hydrolyzed under nitrogen with 6 N HCl for 22 hr at 110°C and an aliquot evaporated via rotary evaporation. The residue was dissolved in a pH 2.2 buffer and an aliquot injected onto a Beckman 6300 amino acid analyzer. The standard three-buffer hydrolyzate amino acid profile elution program as supplied by the manufacturer was used. Postcolumn reaction with ninhydrin and monitoring the sum of 440-nm and 570-nm absorbance allowed quantitation when peak areas (chromatogram B) are compared to reference standards (chromatogram A). Standard one-letter abbreviations for the amino acids are used; CA = cysteic acid, X = norleucine (internal standard), and TA = taurine. Total run time illustrated was 75 min. Data supplied by Bart Jacoby, Ross Products Division of Abbott Laboratories, Columbus, OH.

type available as a prepackaged commercial system (PicoTag). An example of chromatograms generated using this approach is shown in Fig. 3. One very new, commercially available, prepackaged system is based on the AQC chemistry. This has also been applied to food matrices, and an example of the separations possible is shown in Fig. 4. The RP-HPLC–based systems can usually be run very quickly (run times of 10–15 min) with resolution adequate for routine quantitation. For complete resolution, however, runs of 30–60 min are often required, which compares with the ion exchange systems. A listing of amino acid analyses of food matrices (illustrative but by no means complete) is provided in Table 10.

III. HISTORICAL PERSPECTIVE/FUTURE TRENDS

Amino acids, both as separate entities and as components of proteins, have been known for a very long time. Glycine and leucine were the first amino acids discovered (in 1820) whereas the last

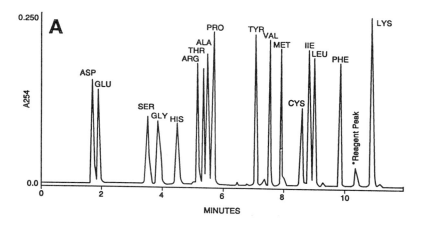

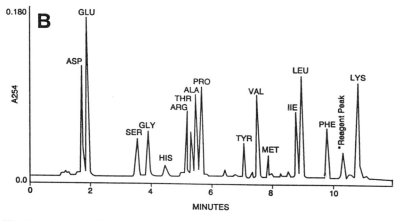

Fig. 3 Amino acid profile analysis using precolumn derivative formation with PITC and RP-HPLC separation. A powder sample of a soy-protein–based nutritional product was hydrolyzed under vacuum using the Pico-Tag workstation with 6N HCl. Derivative formation was done with PITC, and separation was by RP-HPLC, utilizing a complex gradient, and the separations solvents provided with the PicoTag amino acid analysis package. Column temperature was 37°C and separation flow rate was 1 ml/min. Detection was at 254 nm. Chromatogram A resulted from injection of standards and chromatogram B from a hydrolyzed sample of a soy protein–based nutritional product. (From Ref. 158.)

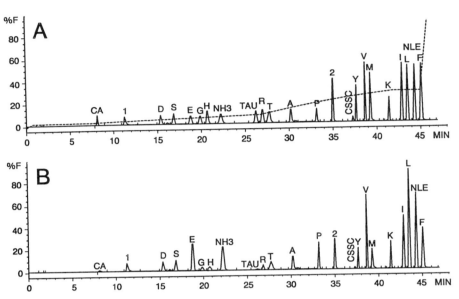

Fig. 4 Amino acid profile analysis using precolumn derivative formation with AQC and RP-HPLC separation. A protein sample was hydrolyzed as described in Fig. 2. This sample and a standards mixture were derivatized and separated using the AccQTag amino acid analysis kit, as recommended in the instructions accompanying the kit. An aliquot was injected onto the column and the gradient indicated in the figure eluted the analytes. Initial gradient condition is 2%B and final is 100%B. A = 0.175 M sodium acetate, 0.02 M triethylamine, 0.01% (w/v) sodium azide, pH 5.0; B = 60:40 acetonitrile/water. Chromatography was done at 32°C and 1 ml/min flow, and fluorescence detection was used, as recommended. Standard one-letter amino acid abbreviations are used. 1 = AMQ byproduct, 2 = hydroxynorvaline (internal standard), CSSC = cystine (cysteine dimer), and NLE = norleucine. Chromatogram A is from a reference standards injection, Chromatogram B is from a liquid nutritional product. Data provided by Stephen D. Smith, Ross Products Division of Abbott Laboratories, Columbus, OH.

of the 20 basic amino acids to be isolated was threonine (in 1935). Analysis of particular amino acids has been done using a wide variety of approaches over the years, and a brief discussion of these in a historical time frame can give some appreciation of the field. The original analyses of these compounds were done using isolation techniques. Amino acids liberated (often via acid hydrolysis treatments) were laboriously extracted and differentially purified from other compounds present—a herculean task often taking months. Flow charts were developed for these kinds of chemical analyses (e.g., 147). Briefly (during the 1940s) some analyses of amino acid content were done using microorganisms. Mutant strains (e.g., *Lactobacillus* mutant strains produced by irradiation) were selected by identifying and isolating those that grew only when supplied with a particular amino acid. There were problems with reversions to wild type (thus providing false-positive signal), and the usual uncertainties associated with microbiological growth produced rather large standard deviations both within lab and (especially) between labs (148–153). Another problem with this approach was the need of a bacterial mutant for the particular amino acid(s) of interest; not all were easily found. However, some of these methods are still used (e.g., 154, where results from a microbiological and an HPLC method for tryptophan are compared).

Other methods for analysis of amino acid content involved the application of wet chemical analysis. These approaches suffered from the need for specific tests for each of the different amino acids (not all of which were or are available), as well as the significant effect of interfering

Table 10 Food and Biological Samples Analysis Examples

Sample(s)	Chemistry	Reference(s)
Total Amino Acid Profiles		
Foods		
Infant formulas	PITC (PicoTag)	1, 59, 68, 139, 140
Cheese	OPA/FMOC (AminoQuant)	92
	PITC–RP-HPLC	56
Beer polypeptides	Cation exchange/ninhydrin	141
African oil beans	PITC (PicoTag)	143
Casein, beef, potatoes, peas, lentils, wheat	PITC	64
Mustard seed	GLC of *N*-heptafluorobutyrylisobutyl esters	104
Conophor nut	GLC of heptafluorobutyryl esters	129
Soy flour	PITC–RP-HPLC	56
Animal feeds		
Various	OPA and RP-HPLC	60, 76
	GLC of *N*-trifluoroacetyl *n*-butyl esters	111
	PITC (PicoTag)	63
Wheat/poultry feed	PITC–RP-HPLC	56, 142
Fish meals	Several GC and ion exchange chromatography methods	144
Feed grains	AQC (AccQTag)	96, 98
Whey protein isolate, soya molasses	PITC (PicoTag)	59
Physiological Samples		
Human urine	GLC of *N*-trifluoroacetyl *n*-butyl esters	111
Liver and muscle tissue	OPA and RP-HPLC	60
Other Sample Types		
Fossil bone	PITC	146
Fossil brachiopods and black shales	GLC of *N*(*O*)-heptafluorobutyrylisobutyl esters	106
Free Amino Acid Profiles		
Foods		
Grapes	1-Fluoro-2,4dinitrophenyl-5-L-alanine amide	145
Beer and beer wort	PITC–RP-HPLC	56
	OPA–RP-HPLC	75
Soy sauce	PITC–RP-HPLC	56
Cheese	GLC of *N*-heptafluorobutyryltrifluoroacetyl ester	102
Wine	GLC of isopropyl *N*-heptafluorobutyryl esters	103
	PITC or OPA/FMOC and RP-HPLC	65
Orange juice	GLC of *N*(*O*)-pentafluoropropionyl ester or *N*(*O*)-trifluoroacetyl 1-propyl ester	108
Honey	GLC of *N*-heptafluorobutyryl isobutyl esters	130
Potato	GLC of *N-heptafluorobutyryl isopropyl esters*	57
Nuts, seeds	GLC of *N*(*O,S*)-isobutyloxycarbonyl-*t*-butyldimethylsilyl esters	110
	Cation exchange/ninhydrin	126

Physiological Samples

Plasma, blood, serum	OPA and RP-HPLC	60, 124, 125, 127
	FMOC chloride and RP-HPLC	124
	PITC and RP-HPLC	54, 124
	Dansyl chloride and RP-HPLC	124
	GLC of N-trifluoroacetyl n-butyl esters	111
	1-napthylisothiocyanate-RP-HPLC	55
	PITC and RP-HPLC or cation exchange/ninhydrin	123
	Ion pair chromatography-OPA postcolumn	112
	Dabsyl chloride–RP-HPLC	89
	GLC of N-acetyl n-propyl esters	133
Urine	PITC and RP-HPLC or cation exchange/ ninhydrin	123
	GLC of N-acetyl n-propyl esters	133
	Dabsyl chloride–RP-HPLC	89
	PITC–RP-HPLC	54
Amniotic fluid	PITC and RP-HPLC or cation exchange/ninhydrin	123
Brain tissue	1-naphthylisothiocyanate–RP-HPLC	55
	OPA and RP-HPLC	127
Cerebrospinal fluid	1-naphthylisothiocyanate–RP-HPLC	55
	OPA and RP-HPLC	127
Liver tissue	Dabsyl chloride–RP-HPLC	89
	PITC–RP-HPLC	54
Muscle tissue	Dabsyl chloride–RP-HPLC	89

Other Sample Types

Plant tissue extracts	GLC of N-acetyl n-propyl esters	133
Sea water	GLC of N-heptafluorobutyrylisobutyl esters	132
Lake and river water	GLC of N-heptafluorobutyrylisobutyl esters	107
Ascitic fluid	PITC–RP-HPLC	54
IV Solutions	6-Aminoquinolyl-N-hydroxysuccinimidylcarbamate (AccQTag)	96

substances—a serious problem in food applications. Some isolation of the protein and/or amino acids from the matrix were needed prior to application of these methods. This could introduce analytical errors.

A major breakthrough in amino acid analysis came with the advent of cation exchange separations and postseparation colorimetric reactions. This work was pioneered by Moore and Stein (27,82) whose automation of this analysis was quite elegant. However, this approach still took many hours for one injection (sample, control sample, standards, etc.). In other words, a routine food analysis for use in supporting label claims for amino acid content today would have taken several days just to do the chromatography—if nothing went wrong! Of course that didn't include the hydrolysis time involved in sample preparation. The advent of HPLC in the 1970s and the development of a large arsenal of derivative formation chemistries through the late 1970s to date has resulted in rather routine analysis of the amino acid content of food proteins with 24-hr turnaround times possible. Modern instruments based on HPLC cation exchange separations (very similar to the original Moore and Stein system) are still in use and allow separations taking 30–60 min per injection.

It is somewhat surprising that the application of GLC approaches to this analytical problem have not been more popular. But, initially, the chemistries involved in rendering the amino acid volatile—an absolute requirement for GLC—were cumbersome (see Table 7), and the ease of use of HPLC seemed to dominate in the field over those few who tried to apply GLC. Now that the "standard" methods are so thoroughly entrenched, even the development of a simple derivative formation scheme would probably receive minimal application.

One of the more recent advances that may significantly impact our ability to perform amino acid determinations is the use of microwave heating. Using this simple modification, hydrolyses that used to take 22–24 hr can now be done in as many minutes (31,32), though the degradation of serine and threonine can be problematic, requiring careful balance of hydrolysis vessel pressure and heating time, and limiting the time savings realizable. Still, the time saved can be substantial, justifying the cost of the microwave digestor.

Another significant area of interest is the application of enzymes. This is actually a rather old approach. Amino acid decarboxylases were used, by measuring released CO_2 with a Warburg manometer back in 1945 (155). This approach was limited to the availability of the enzymes. One more recent example is the use of proteases to very gently hydrolyze a protein sample to release sensitive amino acid(s) (e.g., 47, for tryptophan analysis). However, these proteases may not work for all protein samples encountered, and careful validation for each protein source is essential. Another problem is that proteolytic digestions rarely (if ever) completely release all of the amino acids in a protein. There are some small peptides remaining under any but the longest of digestion times. Proteases also tend to digest themselves as well, requiring the use of an enzyme blank and subtraction for proper quantitation. Immobilized enzymes may help but they are considerably more expensive to use. Another example of interest is the use of specific enzymes as amino acid detectors (e.g., 156, where L-amino acid oxidase is used as a postcolumn detector). Some more recent work reports on the development of amino acid–specific electrodes, using enzymes specific for degradation of a particular amino acid, and measuring one of the products, e.g., ammonia (157). In cases of fortification of a food product with a specific amino acid, this approach would be particularly useful since essentially no sample preparation and minimal analysis time is involved. However, we still have to find the right enzyme, immobilize it without affecting its activity, and solve some potential problems with diffusion time, enzyme shelf life, etc.

The advances of the last 30–40 years really have revolutionized the approach to amino acid analysis. Not long ago, such analyses were a rather cumbersome research tool. Now we execute hundreds if not thousands of them in routine work in a food company in any given year. There are a number of interesting approaches in development now—only a few of which were illustrated above—that are very likely to make what we now consider quick and easy just as cumbersome and slow as we consider the original amino acid analyzer today. It is certain that work currently being done will render this writing obsolete in a relatively short number of years, but it will be exciting to watch these developments take place.

ACKNOWLEDGMENT

The author thanks Dr. Phillip Davis and Ms. Maureen Snowden for their careful review of the manuscript.

REVIEWS

G. Sarwar and H. G. Botting, *J. Chromatogr. 615*: 1 (1993).
Joint FAO-WHO Expert Consultation, Protein Quality Evaluation, Food and Nutrition Paper 151, Food and Agriculture Organization of the United Nations, Rome, 1991.
G. C. Barrett (ed.), *Chemistry and Biochemistry of the Amino Acids*, Chapman and Hall, New York, 1985.
Z. Deyl, J. Hyanek, and M. Horakova, *J. Chromatogr. 379*: 177 (1986).
A. P. Williams, *J. Chromatogr. 373*: 175 (1986).
I. M. Moodie, G. S. Shephard, and D. Labadarios, *J. High Res. Chromatogr. 12*: 509 (1989).
J. M. Rattenbury (ed.), *Amino Acid Analysis*, John Wiley and Sons, New York, 1981.

REFERENCES

1. G. Sarwar and H. G. Botting, *J. Chromatogr. 615*: 1 (1993).
2. Joint FAO-WHO Expert Consultation, Protein Quality Evaluation, Food and Nutrition Paper 151, Food and Agriculture Organization of the United Nations, Rome (1991).
3. G. C. Barrett (ed.), *Chemistry and Biochemistry of the Amino Acids*, Chapman and Hall, New York, 1985.
4. Z. Deyl, J. Hyanek, and M. Horakova, *J. Chromatogr. 379*: 177 (1986).
5. A. P. Williams, *J. Chromatogr. 373*: 175 (1986).
6. I. M. Moodie, G. S. Shephard, and D. Labadarios, *J. High Res. Chromatogr. 12*: 509 (1989).
7. J. M. Rattenbury (ed.), *Amino Acid Analysis*, John Wiley and Sons, New York, 1981.
8. S. Hunt in *Chemistry and Biochemistry of the Amino Acids* (G. C. Barrett, ed.), Chapman and Hall, New York, 1985, p. 55.
9. J. M. Lacey and D. W. Wilmore, *Nutr. Rev. 48*: 297 (1990).
10. T. R. Ziegler, K. Benfell, R. J. Smith, L. S. Young, E. Brown, E. Ferrari-Baliviera, D. K. Lowe, and D. W. Wilmore, *JPEN 14*(4): 137S (1990).
11. S. J. Kirk and A. Barbul, *JPEN 14*: 226S (1990).
12. A. Barbul, S. A. Lazarou, D. T. Efron, H. L. Wasserkrug, and G. Efron, *Surgury 108*: 331 (1990).
13. H. Saito, O. Trocki, S.-L. Wang, S. J. Gonce, S. N. Joffe, and J. W. Alexander, *Arch. Surg. 122*: 784 (1987).
14. R. F. Hurrell and P.-A. Finot, *Experientia Suppl. 44*: 135 (1983).
15. J. M. O'Brien and P. A. Morrissey, *Bull. Int. Dairy Fed. 238*: 53 (1989).
16. R. E. Oste, A. Daahlquist, H. Sjostrom, O. Noren, and R. Miller, *J. Agric. Food Chem. 34*: 355 (1986).
17. W. D. Powrie, C. H. Wu, and V. P. Molund, *Environ. Health Persp. 67*: 47 (1986).
18. C. A. Krone, S. M. J. Yeh, and W. T. Iwaoka, *Environ. Health Persp. 67*: 75 (1986).
19. Joint FAO/WHO/UNU Expert Consultation, Energy and Protein Requirements, WHO Tech. Rept. Ser. No. 724, World Health Organization, Geneva, 1985.
20. G. Sarwar and F. E. McDonough, *J. Assoc. Off. Anal. Chem. 73*(3): 347 (1990).
21. Method 960.48, *AOAC Official Methods of Analysis*, Sections G–I, 15th Ed. by the Association of Official Analytical Chemists, Arlington, VA, 1990.
22. A. Neiderwieser, *Meth. Enzymol. 25*: 60 (1972).
23. R. Bhushan and G. P. Reddy, *Biomed. Chromatogr. 3*(6): 233 (1989).

24. C. H. Suelter, *A Practical Guide to Enzymology*, John Wiley and Sons, New York, 1985, p. 64.
25. J. D. Fernstrom, *J. Neural Transm. Suppl. 15*: 217 (1979).
26. V. Zanardo, G. Bacolla, M. Biasiolo, and G. Allegri, *Biol. Neonate 56*: 57 (1989).
27. S. Moore and W. H. Stein, in *Methods of Enzymology* Vol. 6 (S. P. Colowick and N. O. Kaplan, eds.), Academic Press, New York, 1963, p. 819.
28. T.-Y. Liu and Y. H. Chang, *J. Biol. Chem. 246*: 2842 (1971).
29. P. E. Hare, *Meth. Enzymol. 47*: 3 (1977).
30. R. W. Zumwalt, J. S. Absheer, F. E. Kaiser, and C. W. Gehrke, *J. Assoc. Off. Anal. Chem. 70*: 147 (1987).
31. L. B. Gilman and C. Woodward, *Current Research in Protein Chemistry*, Academic Press, San Diego, 1990, p. 23.
32. E. Tatár, M. Khalifa, G. Záray, and I. Molnár-Perl, *J. Chromatogr. A 672*: 109 (1994).
33. L. C. Gruen and P. W. Nichols, *Anal. Biochem. 47*: 348 (1972).
34. V. Fábián, M. P. Szakács, and I. M. Perl, *J. Chromatogr. 520*: 193 (1990).
35. I. Molnar-Perl, M. Pintér-Szakács, and M. Khalifa, *J. Chromatogr. 632*: 57 (1993).
36. A. S. Inglas, D. T. W. McMahon, C. M. Roxburgh, and H. Takayanagi, *Anal. Biochem. 72*: 86 (1976).
37. R. J. Simpson, M. R. Neuberger, and T.-Y. Liu, *J. Biol. Chem. 251*: 1936 (1976).
38. K. S. Lee and D. G. Drescher, *Int. J. Biochem. 9*(7): 457 (1978).
39. F. Westall and H. Hesser, *Anal. Biochem. 61*: 610 (1974).
40. B. Penke, R. Ferenczi, and K. Kovacs, *Anal. Biochem. 60*: 45 (1974).
41. T. E. Hugli and S. Moore, *J. Biol. Chem. 247*: 2828 (1972).
42. J. L. MacDonnald and M. C. Allred, *J. Assoc. Off. Anal. Chem. 71*: 603 (1988).
43. S. Delhaye and J. Landry, *Anal. Biochem. 159*: 175 (1986).
44. J. Landry, S. Delhaye, and G. Viroben, *J. Agric. Food Chem. 36*: 51 (1988).
45. B. Lucas and A. Sotelo, *Anal. Biochem. 109*: 192 (1980).
46. J. Kivi and S. Smith, Ross Laboratories, Columbus, OH, personal communication.
47. S. E. Garcia and J. H. Baxter, *J. Assoc. Off. Anal. Chem. Int. 75*: 1 (1992).
48. J. L. MacDonald, M. W. Krueger, and J. H. Keller, *J. Assoc. Off. Anal. Chem. 68*(5): 826 (1985).
49. E. Schram, S. Moore, and E. J. Bigwood, *Biochemistry 57*: 33 (1957).
50. P. Pajot, *Eur. J. Biochem. 63*: 263 (1976).
51.. P. W. Riddles, R. L. Blakeley, and B. Zerner, *Meth. Enzymol. 91*: 49 (1983).
52. F. F. Shih and A. D. Kalmar, *J. Liq. Chromatogr. 7*(6): 1169 (1984).
53. D. Fouques and J. Landry, *Analyst 116*: 529 (1991).
54. V. Fierabracci, P. Masiello, M. Novelli, and E. Bergamini, *J. Chromatogr. 570*: 285 (1991).
55. A. Neidle, M. Banay-Schwartz, S. Sacks, and D. S. Dunlop, *Anal. Biochem. 180*: 291 (1989).
56. B. A. Bidlingmeyer, S. A. Cohen, T. L. Tarvin, and B. Frost, *J. Assoc. Off. Anal. Chem. 70*(2): 241 (1987).
57. A. Golan-Goldhirsh, A. M. Hogg, and F. W. Wolfe, *J. Agric. Food Chem. 30*: 320 (1982).
58. A. I. Ihekoronye, *J. Sci. Food Agric. 36*: 1004 (1985).
59. J. A. White, R. J. Hart, and J. C. Fry, *J. Automat. Chem/J. Clin. Lab. Automat. 8*(4): 170 (1986).
60. C. R. Krishnamurti, A. M. Heindze, and G. Galzy, *J. Chromatogr. 315*: 321 (1984).
61. R. J. Hart and J. A. White, *J. Chromatogr. 368*: 164 (1986).
62. A. P. Thio and D. H. Tompkins, *J. Assoc. Off. Anal. Chem. 72*(4): 609 (1989).
63. R. G. Elkin and A. M. Wasynczuk, *Cereal Chem. 64*: 226 (1987).
64. S. R. Hagen, B. Frost, and J. Augustin, *J. Assoc. Off. Anal. Chem. 72*: 912 (1989).
65. M. Calull, J. Fábregas, R. Marcé, and F. Borrull, *Chromatographia 31*(5/6): 272 (1991).
66. S. A. Cohen and D. J. Strydom, *Anal. Biochem. 174*: 1 (1988).
67. B. A. Bidlingmeyer, S. A. Cohen, and T. L. Tarvin, *J. Chromatogr. 336*: 93 (1984).
68. G. Sarwar, H. G. Botting, and R. W. Peace, *J. Assoc. Off. Anal. Chem. 71*(6): 1172 (1988).
69. S. Einarsson, B. Josefsson, and S. Lagerkvist, *J. Chromatogr. 282*: 609 (1983).
70. S. Einarsson, *J. Chromatogr. 348*: 213 (1985).
71. M. D. Oates and J. W. Jorgenson, *Anal. Chem. 61*: 432 (1989).
72. M. C. Roach and M. D. Harmony, *Anal. Chem. 59*: 411 (1987).
73. F. Lai and T. Sheenan, *Biotechniques 14*(4): 642 (1993).
74. P. Bohlen, *Meth. Enzymol. 91*: 17 (1983).

75. L. A. Allison, G. S. Mayer, and R. E. Shoup, *Anal. Chem. 56*: 1089 (1984).
76. R. G. Elkin, *J. Assoc. Off. Anal. Chem. 67*(5): 1024 (1984).
77. M. R. Euerby, *J. Chromatogr. 454*: 398 (1988).
78. M. Weigle, S. L. DeBarnardo, J. P. Tengi, and W. Leimgruber, *J. Am. Chem. Soc. 94*: 5927 (1972).
79. S. Udenfriend, S. Stein, P. Böhlen, W. Dairman, W. Leimgruber, and M. Weigele, *Science 178*: 871 (1972).
80. S. Stein, P. Böhlen, J. Stone, W. Dairman, and S. Udenfriend, *Arch. Biochem. Biophys. 155*: 203 (1973).
81. K. Sogowa and K. Takahashi, *J. Biochem. (Tokyo) 83*: 1783 (1978).
82. D. H. Spackman, W. H. Stein, and S. Moore, *Anal. Chem. 30*: 1190 (1958).
83. J. Y. Chang, A. Lehmann, and B. Whittmann-Liebold, *Anal. Biochem. 102*: 380 (1980).
84. A. Lehmann and B. Wittmann-Liebold, *FEBS Lett. 176*(2): 360 (1984).
85. Y. Tapuhi, D. E. Schmidt, W. Lindner, and B. L. Karger, *Anal. Biochem. 115*: 123 (1981).
86. J. K. Lin and C. H. Wang, *Clin. Chem. 26*: 579 (1980).
87. V. Stocchi, G. Piccoli, M. Magnani, F. Palma, B. Biagiarelli, and L. Cucchiarini, *Anal. Biochem. 178*: 107 (1989).
88. J. Lammens and M. Verzele, *Chromatographia 11*: 376 (1987).
89. J.-K. Lin and S.-Y. L. Shiau, *J. Chinese Biochem. Soc. 12*(2): 47 (1983).
90. K. Imai, Y. Watanabe, and T. Toyo'oka, *Chromatographia 16*: 214 (1982).
91. D. T. Blankenship, M. A. Krivanek, B. L. Ackerman, and A. D. Cardin, *Anal. Biochem. 178*: 227 (1989).
92. U. Bütikofer, D. Fuchs, J. O. Bosset, and W. Gmür, *Chromatographia 31*(9/10): 441 (1991).
93. H. Godel, P. Seitz, and M. Verhoef, *LC-GC Intl. 5*(2): 44 (1992).
94. A. G. Schepky, K. W. Bensch, P. Schulz-Knappe, and W.-G. Forssmann, *Biomed. Chromatogr. 8*: 90 (1994).
95. I. Betnér and P. Földi, *LC-GC 6*(9): 832 (1988).
96. S. A. Cohen and K. M. De Antonis, *J. Chromatogr. A. 661*: 25 (1994).
97. S. A. Cohen and D. P. Michaud, *Anal. Biochem. 211*: 279 (1993).
98. H. J. Liu, *J. Chromatogr. A 670*: 59 (1994).
99. R. C. Morton and G. E. Gerber, *Anal. Biochem. 170*: 220 (1988).
100. S. Kochhar and P. Christen, *Anal. Biochem. 178*: 17 (1989).
101. S. Bunjapamai, R. R. Mahoney, and I. S. Fagerson, *J. Food Sci. 47*: 1229 (1982).
102. L. C. Laleye, R. E. Simard, C. Gosselin, B. H. Lee, and R. N. Giroux, *J. Food Sci. 52*(2): 303 (1987).
103. A. M. P. Vasconcelos and H. J. Chaves das Neves, *J. High Res. Chromatogr. 13*(7): 494 (1990).
104. S. L. MacKenzie and D. Tenaschuk, *J. Chromatogr. 97*: 19 (1974).
105. S. L. MacKenzie and L. R. Hogge, *J. Chromatogr. 132*: 485 (1977).
106. D.-M. Liu, *J. High Res. Chromatogr. 12*: 239 (1989).
107. G. Bengtsson and G. Odham, *Anal. Biochem. 92*: 426 (1979).
108. H. Brückner and M. Lüpke, *Chromatographia 31*(3/4): 123 (1991).
109. P. Felker and R. S. Bandurski, *Anal. Biochem. 67*: 245 (1975).
110. C.-H. Oh, J.-H. Kim, K.-R. Kim, D. M. Brownson, and T. J. Mabry, *J. Chromatogr. A 669*: 125 (1994).
111. C. W. Gehrke, R. W. Zumwalt, and K. Kuo, *J. Agric. Food Chem. 19*(4): 605 (1971).
112. T. Hayashi, H. Tsuchiya, and H. Naruse, *J. Chromatogr. 274*: 318 (1983).
113. T. A. Walker and D. J. Pietrzyk, *J. Liq. Chromatogr. 8*(11): 2047 (1985).
114. S. L. MacKenzie, *Meth. Biochem. Anal. 27*: 1 (1981).
115. S. L. MacKenzie, *J. Assoc. Off. Anal. Chem. 70*(1): 151 (1987).
116. D. Labadarios, I. M. Moodie, and G. S. Shephard, *J. Chromatogr. 310*: 223 (1984).
117. P. Hušek, *J. Chromatogr. 552*: 289 (1991).
118. Z. Deyl and R. Struzinsky, *J. Chromatogr. 569*: 63 (1991).
119. M. V. Novotny, K. A. Cobb, and J. Liu, *Electrophoresis 11*: 735 (1990).
120. M. Albin, R. Weinderger, E. Sapp, and S. Moring, *Anal. Chem. 63*: 417 (1991).
121. L. R. Snyder and J. J. Kirkland, *Introduction to Modern Liquid Chromatography*, John Wiley and Sons, New York, 1979.
122. L. R. Snyder, J. L. Glajch, and J. J. Kirkland, *Practical HPLC Method Development*, John Wiley and Sons, New York, 1988.
123. J. F. Davey and R. S. Ersser, *J. Chromatogr. 528*: 9 (1990).

124. P. Fürst, L. Pollack, T. A. Graser, H. Godel, and P. Stehle, *J. Chromatogr. 449*: 557 (1990).
125. G. A. Qureshi and A. R. Qureshi, *J. Chromatogr. 491*: 281 (1989).
126. H. F. Marshall, G. P. Shaffer, and E. J. Conkerton, *Anal. Biochem. 180*: 264 (1989).
127. B. N. Jones and J. P. Gilligan, *J. Chromatogr. 266*: 471 (1983).
128. C. W. Gehrke, R. W. Zumwalt, and L. L. Wall, *J. Chromatogr. 37*: 398 (1968).
129. A. O. Ogunsua, *Food Chem. 28*: 287 (1988).
130. A. Pirini, L. S. Conte, O. Francioso, and G. Lerker, *J. High Res. Chromatogr. 15*(3): 165 (1992).
131. R. J. Pearce, *J. Chromatogr. 136*: 113 (1977).
132. R. J. Siezen and T. H. Mague, *J. Chromatogr. 130*: 151 (1977).
133. R. F. Adams, *J. Chromatogr. 95*: 189 (1974).
134. F. E. Kaiser, C. W. Gehrke, R. W. Zumwalt, and K. C. Kuo, *J. Chromatogr. 94*: 113 (1974).
135. A. Golan (Goldhirsh) and F. H. Wolfe, *Can. Inst. Food Sci. Technol. J. 12*(3): 123 (1979).
136. J. P. Zanetta and G. Vincendon, *J. Chromatogr. 76*: 91 (1973).
137. N. Nimura, A. Toyama, and T. Kinoshita, *J. Chromatogr. 316*: 547 (1984).
138. R. H. Buck and K. Krummen, *J. Chromatogr. 315*: 279 (1984).
139. A. P. Barba de la Rosa, J. Gueguen, O. Paredes-López, and G. Viroben, *J. Agric. Food Chem. 40*: 931 (1992).
140. G. Sarwar, H. G. Botting, and R. W. Peace, *J. Assoc. Off. Anal. Chem. 72*(4): 622 (1989).
141. C. J. Dale, T. W. Young, and S. Brewer, *J. Inst. Brew. 95*: 89 (1989).
142. R. W. Beaver, D. M. Wilson, H. M. Jones, and K. D. Haydon, *J. Assoc. Off. Anal. Chem. 70*(3): 138 (1987).
143. E. O. Udosen and E. T. Ifon, *Food Chem. 36*: 155 (1990).
144. E. L. Miller, J. M. Juritz, S. M. Barlow, and J. P. H. Wessels, *J. Sci. Food Agric. 47*: 293 (1989).
145. V. G. Dourtaglou, N. G. Yannovits, V. G. Tychopoulos, and M. M. Vamvakias, *J. Agric. Food Chem. 42*(1): 338 (1994).
146. L. R. Gurley, J. G. Valdez, W. D. Spall, B. F. Smith, and D. D. Gillette, *J. Protein Chem. 10*(1): 75 (1991).
147. A. C. Chibnall, M. W. Rees, and E. F. Williams, *Biochem. J. 37*: 372 (1943).
148. J. G. Wooley and W. H. Sebrell, *J. Biol. Chem. 157*: 141 (1945).
149. S. Shankman, M. S. Dunn, and L. D. Rubin, *J. Biol. Chem. 150*: 477 (1943).
150. S. Shankman, M. S. Dunn, and L. D. Rubin, *J. Biol. Chem. 151*: 511 (1943).
151. K. A. Kuiken, W. H. Norman, C. M. Lyman, F. Hale, and L. Blotter, *J. Biol. Chem. 151*: 615 (1943).
152. J. R. McMahan and E. E. Snell, *J. Biol. Chem. 152*: 83 (1944).
153. R. D. Greene and A. Black, *Proc. Soc. Exp. Biol. Med. 54*: 322 (1943).
154. H. S. Kim and G. Angyal, *J. Assoc. Off. Anal. Chem. Int. 76*(2): 414 (1993).
155. E. F. Hale, *Biochem. J. 39*: 46 (1945).
156. N. Kiba and M. Kaneko, *J. Chromatogr. 303*: 396 (1984).
157. A. A. Suleiman, R. L. Villarta, and G. G. Guilbault, *Bull. Electrochem. 8*(4): 189 (1992).
158. N. Astephen and T. Wheat, *Am. Lab. February T159 (1993)*.

8

Peptides

Dolores González de Llano
Dairy Products Institute of Asturias, High Council of Scientific Research (CSIC), Villaviciosa (Asturias), Spain

Tomás Herraiz and M. Carmen Polo
Institute of Industrial Fermentations, High Council of Scientific Research (CSIC), Madrid, Spain

I. INTRODUCTION

Peptides are present naturally in foods, arising mainly from the partial degradation of protein polypeptide chains. Some peptides, like the dipeptides carnosine, anserine, and balenine in vertebrate muscle or glutathione in fruits, are nonproteinic in origin. In other cases peptides are present in foods because they are used as additives (sweeteners, flavor enhancers, bulking agents in light drinks, etc.).

Enzymatic hydrolysis of food proteins yields peptides that are of great interest to the food industry and are utilized for various purposes, e.g., improving the functional properties of foods and flavoring agents, in dietetic food, for parenteral feeding (casein hydrolyzates) or milk protein substitutes in cases of intolerance.

The multiple functions of peptides in foods (antioxidants, antimicrobial agents, interfacial agents) and their role in the development of characteristic flavors (sweetness, bitterness), as well as the information they can provide about the genuineness of foods, makes peptide analysis a necessity.

Both producers and government laboratories have considerable interest in the study of peptides, both for research purposes and for the control of raw materials or manufactured foods. For this reason, substantial attention is now being focused on the development of analytical techniques designed to separate, characterize, and quantify peptides.

II. PROPERTIES OF PEPTIDES

A. Physicochemical Properties

Peptides are a very heterogeneous group of compounds that share the common characteristic of being made up of chains of amino acids joined by an amide bond, also known as a peptide bond. The large number of different amino acids that exist in nature and the large variations in the number of amino acids in a given peptide mean that peptides can be highly diverse.

Peptides are termed di-, tri-, tetra-, etc., peptides, according to the number of amino acids; the term *oligopeptides* is used for low molecular weight (MW) of fewer than 10 residues, while the term *polypeptides* is used for peptides with higher MW. The dividing line where peptides stop and proteins start has not been sharply defined, and proteins are normally considered to have at least 100 residues (MW >10 kD). The authors have also made this assumption in the present chapter.

The acid–base behavior of peptides (1) is determined by the free α-amino group on the N-terminal residue, by the free α-carboxyl group on the C-terminal residue, and by the ionizable R groups located at intermediate positions. The pK values of the terminal α-carboxyl groups are somewhat higher and those of the α-amino groups somewhat lower than those of the corresponding free amino acids (Table 1).

The terminal amino and carboxyl groups react in the same way as the corresponding amino acids (acylation, esterification, etc.). Some of the reactions of the amino groups (with ninhydrin, orthophthaldehyde, fluorescamine, etc.) are used for peptide quantification and detection purposes, as will be discussed later. Peptides also react in ways that free amino acids do not, like the classic biuret reaction, which consists of the formation of a colored complex with a transition metal (Cu, Ni, etc.) in an alkaline medium and which can be measured quantitatively by spectrophotometry. This reaction is used in peptide quantification.

In the presence of carbonylic compounds like aldehydes, certain peptides take part in the Maillard reaction, leading to the formation of melanoid pigments that contribute to the development of undesirable flavor and coloration in foodstuffs. Conversely, there are also lysine-containing peptides that retard browning reactions with glucose, and these may be appropriate for fortifying the lysine content of sugar-containing foods to be cooked by heating.

Peptides also play a very important role in determining the rheological properties of foods. Hydrolysis of proteins to peptides during the processing (fermentation, ripening, cooking, etc.) of foods may modify food texture.

B. Sensory Properties

Peptides are tasteless or bitter, with the exception of certain dipeptides of glutamic acid and aspartic acid, which are sweet. Because refined sugar, the principal nutritive sweetener, can cause dental

Table 1 Dissociation Constants and Isoelectric Points of Various Peptides (25°C)

Peptide	pK_1	pK_2	pK_3	pK_4	pK_5	pI
Gly-Gly	3.12	8.17				5.65
Gly-Gly-Gly	3.26	7.91				5.59
Ala-Ala	3.30	8.14				5.72
Gly-Asp	2.81	4.45	8.60			3.63
Asp-Gly	2.10	4.53	9.07			3.31
Asp-Asp	2.70	3.40	4.70	8.26		3.04
Lys-Ala	3.22	7.62	10.70			9.16
Ala-Lys-Ala	3.15	7.65	10.30			8.98
Lys-Lys	3.01	7.53	10.05	11.01		10.53
Lys-Lys-Lys	3.08	7.34	9.80	10.54	11.32	10.93
Lys-Glu	2.93	4.47	7.75	10.50		6.10
His-His	2.25	5.60	6.80	7.80		7.30

Source: Ref. 2

decay and is suspected of contributing to obesity and adult diseases such as diabetes mellitus and hypertension, many peptide sweeteners have been developed.

The peptide aspartame, L-aspartyl-L-phenylalanine methyl ester, has been the foremost alternative to sugar but presents many problems including low stability at extreme temperatures and at neutral pH, low solubility, sensitivity to enzyme digestion, and possible side-effects such as phenylketonuria and methanol accumulation. To resolve these problems other dipeptide analogs have been devised and several studies carried out to relate the structure and taste of analogs of this dipeptide (3). Tsang et al. (4) reported that the analogs at the lower end of the L-aspartyl-α-aminocycloalkanecarboxylic acid methyl ester series were sweet, the dipeptides containing α-aminocyclohexanecarboxylic acid methyl ester and α-aminocycloheptanecarboxylic acid methyl ester were bitter, and analogs containing α-aminocyclooctanecarboxylic acid methyl ester were tasteless. Rodríguez et al. (5) also found that a series of L-aspartyl-D-alanyl tripeptides went from sweet to bitter to tasteless as the size of the C-terminal amino acid increased. Thus, there is a close relationship between size and taste in these peptides.

Peptide sweeteners belonging to L-aspartyl-D-alanine amides seem to be good sugar substitutes, among them the peptide alitame, which is 2000-fold sweeter than sugar and according to the data accumulated until now is safe (6). Kim et al. (7) developed an HPLC method for determining a backbone compound of intense peptide sweeteners belonging to L-aspartyl-D-alanine amides.

Some peptides enjoy the property of masking the bitter taste of foods. Ohyama et al. (8) conducted sensory analyses using synthetic peptides and found that neutralized peptides consisting of aspartic acid and glutamic acid had a taste similar to that of a monosodium glutamate. They termed this "umami taste" or "relish."

Food may also contain bitter peptides, formed as a result of proteolytic reactions that take place in foods. These bitter peptides have been identified in soybean (9,10) and casein hydrolyzates (11,12) as well as in other foods, like cheese, that undergo fermentation or ripening processes with marked proteolytic activity. The bitter taste has been related to the hydrophobic amino acid content (13–17) and to chain length. Ney et al. (18) established a formula relating the bitterness of peptides to their amino acid composition and chain length. Too large a proportion of hydrophobic amino acids gives rise to bitterness, yet above a certain molecular weight bitterness is not perceptible even when there are hydrophobic amino acids. Proteins do not have a bitter taste even when they contain hydrophobic amino acids, yet hydrolysis of such proteins may indeed yield bitter peptides.

Otagiri et al. (19) used model peptides composed of arginine, proline, and phenylalanine to ascertain the relationship between bitter flavor and chemical structure. They reported that the presence of the hydrophobic amino acid at the C terminus and the basic amino acid at the N terminus brought about an increase in the bitterness of di- and tripeptides. They further noted a strong bitter taste when arginine was located next to proline and a synergistic effect in the peptides $(Arg)_l$–$(Pro)m$–$(Phe)n$ ($l = 1,2$; $m, n = 1,3$) as the number of amino acids increased. Birch and Kemp (20) related the apparent specific volume of amino acids to taste.

Diketopiperazines, which are cyclical dipeptides, are generally bitter and have been reported in many protein hydrolyzates and in foods such as cocoa and cheese. Recently, diketopiperazines have been observed in the aqueous fraction from Comté cheese together with other bitter compounds. However, the taste of whole fraction and cheese are hardly or not at all bitter (21).

Enzymatic activity is extremely important in the development of bitter flavors in foods. Trypsin does not degrade bitter peptides, although other peptidases may break down bitter peptides to nonbitter lower molecular weight peptides and amino acids.

Lower molecular weight peptides play an important role in the flavor intensity of meat and beef broth (22). However, warmed-over flavor in cooked-stored-recooked meat has been related to the higher hydrophobic peptide content in such meats as compared to cooked meats (23).

C. Functional Properties

Research on the functional properties of peptides began a few years ago in a search for new applications, but especially to achieve a better understanding of the structure–function relationships of proteins (24). With improved analytical methods, scientists are now able to detect more peptides in food and to gain greater knowledge on the influence of these peptides on the properties of proteins in foods. Moreover, more manufactured and formulated foods are being produced, thus placing an increased demand on ingredients with consistent and desirable functional attributes.

The functional properties of some proteins may be improved when they are enzymatically hydrolyzed and their hydrolyzates can be used as additives to enhance food quality. These hydrolyzates have less secondary structure, more solubility near the isoelectric point, less viscosity, and better foaming, gelling, and emulsifying properties than native proteins. Peptide size control is essential if optimum and reproducible changes in functional and sensorial properties are to be achieved (25). Extensive hydrolysis of food proteins often results in the formation of bitter-tasting peptides. The relationship between the degree of hydrolysis and the level of bitterness and solubility of casein hydrolyzates, produced using different proteases and different hydrolysis times, showed that limited hydrolysis to a preselected degree can give high solubilities and no (or little) development of bitter taste, depending on the enzymes used (26).

Certain protein hydrolyzates can be used for their antioxidant properties. Thus, a hydrolyzate of egg white slowed oxidation when added to a sample of cookies containing lard. The rise in stability was proportional to the amount of protein hydrolyzate added. Dipeptides were also found to have a great inhibiting effect on the autoxidation of linolenic acid than did mixtures of the corresponding free amino acids (27).

Adler-Nissen and Eriksen (28) established that the enzymatic hydrolyzates of soybean proteins conferred interesting functional properties on beverages. They pointed out that medium-size peptides had a marked bulking effect, while large peptides acted as foaming agents. Other researchers have also reported that polypeptides and proteins have an effect on foam stability (29). The isolation of these peptides, in order to determine their foaming activity, may be of great interest in relation to certain beverages like sparkling wines and beer, in which foam formation plays an important role.

Increases in the emulsifying capacity of various proteins have also been brought about by enzymatic hydrolysis. For the improvement of emulsifying properties by enzymatic hydrolysis, hydrophobicity and large molecular weight (>20 residues) of resulting peptides are reported to be important factors. Larger peptides can be produced by limiting the degree of protein hydrolysis or by separation of hydrolyzates using an ultrafiltration (UF) reactor system. Also, enzyme specificity is important to peptide functionality as it strongly influences the nature of peptides produced. For example, tryptic peptides from whey proteins have better emulsifying and interfacial properties than chymotryptic peptides (30). Peptides generally have better emulsifying properties at acid pH, indicating that environmental conditions exercise an influence, and some synergistic effects of peptides on emulsifying properties have been noted, indicating that peptide behavior is complex (24,31).

Shimizu et al. (24) studied the emulsifying properties of peptides by isolating a peptide that contained 23 residues from the peptic hydrolyzate of α_{s1}-casein. This peptide fraction displayed an emulsifying capacity similar to that of α_{s1}-casein at a neutral pH and concentrations greater than 2%. Subsequently, using high-performance liquid chromatography (HPLC), Shimizu et al. (31) reported that this peptide fraction contained small quantities of other peptides that interfered with emulsifying activity. The emulsifying activity of these peptides would seem to be synergistic, since the purified peptide exhibited a lower emulsifying capacity.

Turgeon et al. (30) studied the interfacial properties of peptide fractions, obtained by tryptic hydrolysis in a UF reactor of β-lactoglobulin, to determine the physicochemical characteristics of those peptides and observed that the structure–function relationships of peptides are complex.

The functional properties of milk proteins and the peptides of corresponding hydrolyzates were studied in detail by Haque (32) and Haque and Mozaffar (33). Recently, Haque (34) published an interesting review on the influence of milk peptides in determining the functionality of milk proteins.

Various studies were carried out recently on peptides from protein hydrolyzates with poor functional properties as well as low solubility and digestibility with a view to their possible use in the food industry. The fraction of corn gluten meal is composed mainly of zein, a highly hydrophobic protein that in its native state possesses poor functionality. To enhance the utilization of corn proteins in human food products, investigation has been carried out into the use of enzyme hydrolysis to modify certain functional properties. Cassella and Whitaker (25) observed that the mixture of polypeptides produced by limited specific hydrolysis of zein with trypsin had substantially increased solubilities, emulsion stability, foamability, and foam stability. Mannheim and Cheryan (35) also reported that functional properties of zein in corn gluten meal can be improved by enzyme hydrolysis combined with membrane technology. Unconventional proteins of high quality are available in large amounts of fish and are transformed into fish meal for animal fodder because of their poor functional properties. Rebeca et al. (36) showed that fish protein hydrolyzates can be used in the food industry for their solubility and nutritional value.

The separation of peptides from protein hydrolyzates by size exclusion chromatography and sodium dodecyl sulfate–polyacrylamide gel electrophoresis (SDS-PAGE), in order to determine quantitative information on size distribution, may be of great interest in relation to certain functional properties.

D. Physiological Properties

Short food-derived peptides produced during processing or during digestion in vivo may have pharmacological or physiological activity. The potential for biological activity of diet-derived peptides is emphasized by known biochemical effects of peptides from protease-treated dietary proteins such as casein and gluten (37). These findings have introduced a new criterion for defining the nutritive value of food proteins: peptides that are inactive within the protein sequence can be released during digestion and may have physiological activity.

Milk protein is a rich source of biologically active peptides such as casomorphins and caseinophosphopeptides, which might have significant nutritional implications. The opioid peptides derived from bovine caseins are called α- and β-casomorphins or casoxins because their amino acid sequences indicated that they were fragments of α- and β-casein or K-casein, respectively. Those derived from whey proteins are called lactorphins. Since such peptides originate from exogenous food sources and exert naloxone-inhibitable opioid activities, they are termed exorphins (38). The major opioid peptides derived from milk proteins are fragments of β-casein (β-casomorphins). Morphiceptins, the amide forms of β-casomorphins 4 and 5, are the most active opioid peptides (38,39). Among hydrolyzates of whey protein at least α-lactorphin exerted weak but continuous opioid properties (40). Caseinophosphopeptides have a potent ability to form soluble complexes with Ca and trace elements, and improve the utilization of minerals. The opioid β-casomorphin-11 and the α_{s1}-caseinophosphopeptides-9, isolated as in vivo digestion products of bovine caseins, have been characterized by Meisel and Frister (41,42).

Other peptides isolated from K-casein, fragment 106–116, and lactotransferrin, fragment 39–42, were active on platelet functions. Their behavior was compared to that of fibrinogen peptides possessing similar effects (43).

In addition to the possible liberation of bioactive peptides during digestion, peptides with nutritional and physiological implications may be released during the manufacture of milk or food products. Peptides from hydrolyzates of milk whey proteins have lower antigenicity than the native proteins and could be used for hypoallergenic infant formulas (44). Hydrolyzates of milk proteins have also been used for clinical applications in parenteral nutrition. Preparation of peptides with high content in branched chain amino acids and low content in aromatic amino acids may be used in producing physiologically functional food because these peptides have been related to improvement of a liver disease (45). Conversely, other peptides released during digestion could produce a food intolerance reaction. For example, gluten-derived peptides could be particularly toxic to coeliac patients. Wieser et al. (46) isolated a pure peptide from tryptic digest of whole gliadin by ultrafiltration and chromatography which shows activity for all celiac patients in very low concentration.

III. SAMPLE PREPARATION AND CLEANUP OF FOOD PEPTIDES

Analytical determination of peptides in foods, like that of any other chemical component present in a complex matrix, requires sample preparation involving their isolation from the rest of the components, their separation with high-resolution techniques, and optimal, sensitive detection. HPLC or capillary electrophoresis analysis of peptides requires a prior step of isolation, purification, and concentration. This involves, on the one hand, concentrating the peptide sample for suitable detection and eliminating components from the matrix which may interfere on the analysis, on the other. Table 2 summarizes the most common procedures for preparative fractionation of a peptide sample.

A. Peptide Extraction and Cleanup

Generally, a first step of extraction followed by deproteinization is performed. For extraction, which is essential when analyzing solid samples, it is possible to use homogenization in water, solutions of acetic acid or sodium chloride, or also mixtures of water and methanol or ethanol, or even a mixture of chloroform and methanol. Aqueous mixtures in an acid medium are used for hydrophilic peptides, while highly hydrophobic peptides are extracted with mixtures of organic and aqueous solvents. Water has been used for the extraction of peptides from cheese (49,55) and legumes (51). Saline solutions have been used in the extraction of peptides from meat (50,52–54) and from flour (56). The hydrophobic peptides from cheese have been extracted with mixtures of chloroform and methanol (2:1) or methylene chloride and methanol (1:1) (47,48,57). Tieleman and Warthesen (77) compared various extraction procedures with water and mixtures of methanol and methylene chloride to characterize cheese proteolysis. Extraction with methylene chloride/methanol 2:1 v/v is sometimes used in meat (23) or with hexane in cheese (55) in order to eliminate lipid components.

Deproteinization is usually performed by means of protein precipitation in solutions of ethanol, methanol, acetone, or trichloroacetic acid (TCA), sulfosalicylic acid, phosphotungstic acid (PTA), or picric acid, or also by means of addition of salts (ammonium sulfate) or adjusting the pH to the isoelectric point of the protein. After precipitation, filtration or centrifugation is performed to separate proteins from soluble peptides and/or dialysis in order to eliminate salts. At times this step results in the fractionation of peptides on the basis of their solubility since the soluble peptide fraction differs according to the precipitant agent and its proportion. Yvon et al. (59) found differences in the solubility of peptides from hydrolyzates of caseins in function of the percentage of TCA used. González de Llano et al. (58) detected only small peptides (lower than 700 Da) and amino acids when precipitating the water-soluble fraction from cheese with 5% PTA. Bican and Spahni (60) compared up to nine different analytical procedures, based on the solvents mentioned

Table 2 Procedures for Preparative Fractionation of Peptide Mixtures in Foods

Procedure	Material used	Property exploited	Application	Ref.
Extraction	H_2O, diluted acids, NaCl, H_2O/methanol, $CHCl_3$/methanol	Peptide solubility	Peptide solution and removal of insoluble interferences	47–57
Deproteinization and subsequent centrifugation, filtration, or dialysis	Organic solvents (ethanol, acetone), acids (trichloroacetic, sulfosalicilic, phosphotungstic, picric), salts	Protein in solubility	Removal of proteins	58–60
Ultrafiltration	Membranes of several materials and cutoffs	Fractionation by molecular size	Removal of large molecules and prefractionation of peptides by size	46,61–66
Preparative low-pressure gel filtration chromatography	Sephadex, BioGel of different porosity	Fractionation by molecular size	Removal of very small and large molecules and prefractionation of peptides by size	23,58,61,65–69
Preparative low-pressure ion exchange chromatography	Charged polymers (celluloses, Sephadex, polystyrene derivatives)	Fractionation by charge and some influence of polarity	Removal of uncharged and also other charged molecules	46,56,63,68,70
Affinity chromatography	Matrix-binding peptide receptors (lectins, antibody-Sepharose)	Specific noncovalent interaction	Specific isolation of bioactive peptides	71
Ligand exchange chromatography	Sephadex-Cu^{2+}, Chelex-Cu^{2+}	Exchange of ligands	Peptide separation from amino acids	69,70,72
Solid phase extraction	Disposable columns with polar, nonpolarized, and charged matrices	Nonpolar, polar, or ionic interaction	Prefractionation and purification of peptides	23,66,73–76

earlier, for the determination of peptides formed during cheese ripening. These pointed out the need to optimize sample extraction and deproteinization processes.

B. Fractionation of Peptides by Ultrafiltration

Ultrafiltration is a simple and highly useful and selective method for fractionating peptides and eliminating proteins in foods. A wide variety of membranes of several different materials, filtration surfaces, and cutoff sizes are commercially available. Kaiser et al. (62) fractionated the soluble peptides at pH 4.6 from cheddar cheese into four ranges of molecular mass (I < 3000 Da, II > 30,000 Da, III > 10,000 Da, IV > 3000 Da) using ultrafiltration into different molecular mass cutoffs. Turgeon et al. (63) used it to fractionate tryptic peptides from β-lactoglobulin which have surface-active properties and Bouhallab et al. (78) used it to isolate bioactive peptides from casein. The filtration cross-flow procedure has been used in the isolation of bioactive peptides (64).

C. Isolation and Fractionation of Peptides Using Low-Pressure Liquid Chromatography

The isolation and concentration of peptides via separation from other components of foods such as sugars, organic acids, amino acids, phenols, salts, and others can be carried out using different classic preparative chromatographic techniques such as molecular exclusion chromatography, ion exchange chromatography, ligand exchange chromatography, affinity chromatography, etc.

Low-pressure molecular exclusion chromatography on column fractionates peptides according to their molecular size. This is traditionally very useful as an initial step in the fractionation of complex mixtures. Two main types of material have been used for preparative molecular exclusion, crosslinked dextran (Sephadex) or polyacrylamide (BioGel P). The choice of pore size is determined by the range of molecular weights of the peptides from the mixture. This type of chromatography enables the separation of low molecular weight peptides from proteins or from high molecular weight peptides by selecting the suitable pore size of the polymeric material. At the same time it facilitates the separation of peptides from other smaller molecules such as organic acids, carbohydrates, salts, phenols, and amino acids. Among the eluants used to elute peptides from the column we can mention: (0.05–0.1)M NH_4HCO_3 solution, water, diluted ammonia, 5% formic acid, 30% acetic acid, 0.1% trifluoroacetic acid, and pyridine–acetic acid buffers (79).

This procedure has been widely used in the study of peptides in foods. Lemieux and Amiot (69) used classic molecular exclusion chromatography on Sephadex G-25 as the first step in the fractionation of hydrolyzates of casein, and Spanier and Edwards (23) used it to fractionate peptides extracted from meat. Sephadex G-10 with water as eluant has been used to separate peptides from amino acids in the 5% PTA fraction soluble from cheese (58,67), as is shown in Fig. 1. Sephadex LH-20 has also been used to fractionate peptides in wine (65).

Preparative isolation of peptides from other charged or noncharged components can be performed by ion exchange chromatography on low-pressure columns using anion or cation exchange resins. Hydrophilic materials such as cellulose or crosslinked dextrans or materials based on crosslinked organic polystyrene polymers (Dowex-type resins) are now available. Hydrophilic materials are very useful for separating large peptides. With cation exchange cellulose such as CM-cellulose or anion exchange cellulose such as DEAE-cellulose, linear eluting gradients with ammonium acetate (pH 5–6) and ammonium carbonate (pH 8) buffers, respectively, are used. Crosslinked polystyrene materials (Dowex cation and anion resins) have been used for small peptides, where separation is based on the peptide's charge and polarity. In this case, elution occurs with buffers containing pyridine in order to eliminate hydrophobic interactions with the resin. BioRex membranes (AG1, AG50), crosslinked anion and cation exchange resins enmeshed in polytetrafluoroethylene may be used as an alternative to classic ion exchange resins. They achieve

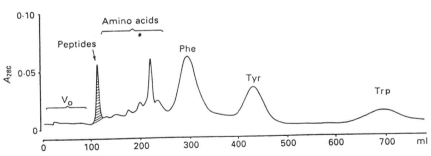

Fig. 1 Elution profile of the 5% phosphotungstic acid–soluble nitrogen of a blue cheese on a Sephadex G-10 column using Milli-Q water as eluant. (From Ref. 67.)

the same effectiveness but analysis times are shorter (80). Cation and anion exchangers can be used to complement each other. For example, Collar et al. (56) used prefractionation on Dowex strong anion exchange columns in order to retain anionic compounds which interfere with the polypeptide fraction of bread dough during fermentation. Next they used cation exchange chromatography with Dowex 50Wx2 resin to separate the retained peptides from noncharged components and from carbohydrates and anions. Turgeon et al. (63) purified fractions by anion exchange chromatography of surface-active peptides prior to RP-HPLC.

Affinity chromatography is based on the isolation of peptides containing very specific functional groups by using receptors covalently bonded to an agarose matrix (Sepharose) and by exploiting biospecific reactions. To our knowledge it has not been used to isolate peptides from foods but has been used in other fields to isolate glycopeptides joined to immobilized lectins, or peptides isolated with the antigen–antibody reaction (81). Recently, Saito et al. (71) used lectin affinity chromatography to characterize the glycopeptide from bovine colostrum.

When working with small molecular size peptides, it is advisable to separate them from amino acids. It is possible to use ligand exchange chromatography on stationary phases modified with copper (Cu^{2+}) (82). Chelex resins are used to eliminate the metals in the peptide eluate. Rothenbühler et al. (83) described a method based on copper-Sephadex columns in which separation is based on the different stabilities of the copper complexes with Sephadex, peptides, and amino acids and apply it to the separation of peptides from amino acids in cheese and yeast extract. Peptides elute first, while the amino acids are retained and elute in later fractions. Acedo and Polo (72) use the Sephadex copper system followed by Chelex 100 to eliminate the Cu^{2+}, for the separation of peptides and amino acids in wine, Prieto (70) for peptides in bread dough, and Lemieux and Amiot (69) for hydrolyzates of casein.

D. Isolation and Fractionation of Peptides by Solid Phase Extraction

One of the techniques which has been most developed in recent years in procedures for sample preparation is sold phase extraction. The popularity and versatility of this technique have increased significantly with the commercial development of disposable cartridges and columns with several bonded phases. Three methods of chromatographic separation can be used to separate peptides: normal phase polar interaction, reverse phase hydrophobic interaction, and ion exchange (85,86). Bennett (87) fractionated a mixture of peptides using cation exchange cartridges (Accell Plus CM) and anion exchange cartridges (Accell Plus QAM) in series. This obtains three fractions—neutral peptides, acid peptides, and basic peptides—which significantly simplifies later HPLC analysis. The octadecyl cartridges and columns, which separate by reverse phase hydrophobic interaction, are highly useful in the isolation and purification of peptides and are often used in a stage prior to analysis by high-resolution techniques (88). By this means a peptide-enriched and protein- and salt-free fraction is obtained. However, experimental conditions of retention and elution of each peptide must be optimized (75,89). C-18 cartridges have been used recently for the isolation and fractionation of peptides in milk (73,74), cheese (66,76), meat (23), and tryptic hydrolyzates of lactic proteins (75). Fig. 2 shows an HPLC chromatogram of peptides in cheese prefractionated by extraction on octadecyl silica and elution with different proportions of acetonitrile (66). The peptides are retained in the octadecyl cartridges while interfering compounds are eluted.

It is often necessary to use several combined procedures of sample preparation in order to isolate and concentrate peptides from foods. Following are some examples: Wieser et al. (46) used ultrafiltration, gel filtration, cation exchange chromatography, and anion exchange chromatography to separate tryptic peptides from gliadin. Spanier and Edwards (23) fractionated peptides from meat by means of molecular exclusion chromatography on Sephadex G-25 and obtained two groups of polypeptides with molecular weights of 1500 and 3000. The fractions were lyophilized and passed through a C-18 cartridge. The peptides were eluted with water/acetonitrile 20:80 prior to their chromatographic injection. Singh et al. (66) used ultrafiltration, chromatography on Sephadex G-25, and solid phase extraction with C-18 cartridges prior to HPLC analysis in the fractionation of peptides in cheese. Dale and Young (61) isolated peptides from beer using ultrafiltration and Sephadex columns prior to HPLC analysis. Zevaco and Ribadeau-Dumas (68) used a Sephadex G-50 column to purify glycomacropeptides previously separated by DEAE-cellulose.

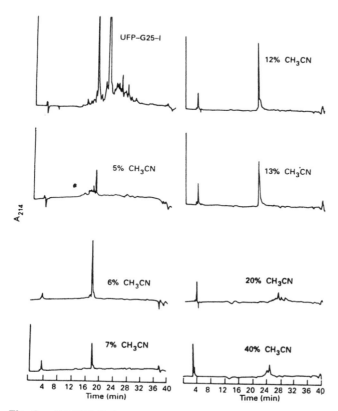

Fig. 2   RP-HPLC chromatograms of peptides fractionated on Sep-Pak C-18 cartridges. (From Ref. 66.)

Forecasts for the future are for a generalization of the use of automatization for sample preparation, using above all solid phase extraction. At present robots are used in the laboratory for the automatization of solid phase extraction and chromatographic analysis (90,91).

IV. ANALYTICAL SEPARATION OF PEPTIDES

The procedures of preparative fractionation described so far for sample isolation, preparation, and cleanup provide poor resolution of complex mixtures of peptides even when long fractionation times are used. It is necessary to use high-resolution techniques: reverse phase high-performance liquid chromatography (RP-HPLC), ion exchange high-performance liquid chromatography (IE-HPLC), size exclusion high-performance liquid chromatography (SE-HPLC), hydrophobic interaction chromatography, high-performance displacement chromatography, perfusion chromatography, slab gel electrophoresis, and capillary electrophoresis (CE) among others. In this chapter we shall concentrate on those most widely used in peptide analysis: RP-HPLC, SE-HPLC, IE-HPLC, slab gel electrophoresis, and CE.

A. Separation of Peptides by HPLC

HPLC is characterized by the use of uniform microparticles (5–10 µm) as chromatographic support on small-diameter (≤5 mm for analytical separation and ≥10 mm for semipreparative and preparative separations) and short-length (<300 mm) columns, and by the use of advanced instrumentation. The use of narrow-bore (≤2 mm i.d.) columns increases resolution and sensitivity and reduces loading capacity and eluant consumption. HPLC displays high speed, high resolution, accuracy, and good detectability as compared with low-pressure techniques (92). HPLC has taken the place of the classic methods as a technique for peptide separation and purification. It enables small amounts of peptide material to be handled, thus reducing sample losses, and uses nondestructive detection systems. Recovery of biological activity after HPLC is quite good with most low molecular weight peptides. Table 3 lists the characteristics and factors in HPLC separation and references of its application in foods.

1. RP-HPLC Separation of Peptides

RP-HPLC is the most commonly used method for separating complex mixtures of peptides (108). The stationary phases of the RP columns most often used in peptide separation are *n*-alkyl hydrophobic chains of octadecyl, octyl, or butyl covalently bonded to silanol groups of the silica support (solid matrix). RP columns based on organic polymeric materials (polystyrene divinylbenzene) (macroporous polymer resins) are also available. The latter have a certain advantage over columns based on silica in that they are stable at high values of pH, which enables separation of peptides over wide ranges of pH (1–13) (110). These columns are highly suitable for analysis of peptides more soluble in slightly alkaline solutions than in acid ones and are becoming increasingly popular (111,112).

The separation mechanism in RP-HPLC is based on the hydrophobic interaction of peptides with the nonpolar stationary phase. Elution occurs on a gradient of a mixture of a polar eluent, normally water, and an organic modifier, generally acetonitrile, methanol, ethanol, propanol, or isopropanol. Generally peptides demonstrate increasing resolution with increasing flow rate and decreasing gradient rate (113). Since hydrophobic interaction is the major principle of peptide retention on the stationary phase, their retention time is inversely proportional to the concentration of the organic modifier up to approximately 40% concentration (114).

a. Factors Influencing Separation of Peptides by RP-HPLC. RP-HPLC is a highly flexible technique in that it enables manipulation of parameters such as the type of stationary and mobile

Table 3 HPLC Types and Factors Influencing Separation of Peptides

HPLC type	Separation mechanism	Characteristics	Influencing factors	Ref. of applications
RP-HPLC	Hydrophobic interaction of peptides with the stationary phase	High resolution High speed Standard separation procedures Prediction of retention time Good detection and reproducibility Quantitative recovery using volatile buffers Moderate capacity in analytical columns	Polarity of peptide *Mobile phase*: pH, counterion, organic modifier, *Stationary phase*: alkyl or aryl chain, packing porosity, pore size.	23,55,58,61,62,65, 67,74,77,93–97
IE-HPLC	Ionic interaction of peptides with anion exchangers or cation exchangers as stationary phase	Moderate resolution High speed High capacity Prefractionation of peptides by charge Wide range of pH (acid and alkaline) Further desalting may be necessary	Net charge and polarity of peptides Type of matrix and ionic exchanger pH, ionic strength of the sample and eluents Organic modifier	69,98–103
SE-HPLC	Size exclusion of peptides on hydrophilic polymers as stationary phase	Moderate or low resolution High speed Very high capacity Desalting Good prefractionation by size Molecular weight determination Presence of nonspecific interactions	Molecular size of peptides Packing porosity and pore size Mobile phase to avoid nonspecific interactions	69,98,104–109
Chiral HPLC	Peptide interaction with chiral stationary phase or chiral mobile phase	Separation of enantiomeric peptides	Type of stationary phase Type of chiral modifiers	

phase, mobile phase pH, mobile phase counterion, and temperature. The type of silica, the pore, the particle size, the type and extension of end capping, and above all the type of alkyl chains affect surface hydrophobicity (114). Large peptides (>4000 Da) are analyzed with RP columns containing particles with a greater pore diameter (300 Å) than conventional columns so that the molecules pass freely through the pores and have greater access to the alkyl chains. For smaller molecular size peptides conventional pore diameters of 60 and 100 Å are used. The nature of the stationary phase allows the possibility of changing selectivity and resolution. Butyl (C4) stationary phase is recommended for large and hydrophobic peptides, while octadecyl and octyl (C18 and C8) are recommended for small and hydrophilic, or not very hydrophobic, peptides. Columns with cyanopropyl stationary phase are more polar than those mentioned previously and can be used for highly hydrophobic peptides. The most widely used organic modifier for peptide separation and elution is acetonitrile due to its high transparency in UV detection at around 200–220 nm, its low viscosity and high volatility, which facilitates further elimination. Isopropanol or n-propanol alone or mixed with acetonitrile may also be used as organic modifiers to elute large or hydrophobic peptides. Acetonitrile displays greater resolution and selectivity and is the best solvent for most practical purposes (115).

Peptide retention and separation in HPLC improves when the negative charges of the silanol groups in the stationary phase and the polar nature of the peptides are reduced (116). In order to block the silanol groups, salts or strong acids are added to the eluant. As well as blocking the silanol groups, the addition of strong acids diminishes the pH and the polar nature of peptides since the carboxyl groups of the amino acid C terminal and of the aspartic acid and glutamic acid are present in a nondissociated form. For this reason, most peptide separations are performed at pH <3, although excellent separations can also be performed at neutral pH. Mant and Hodges (117) studied the effect of pH on the separation of a mixture of basic peptides. The ionic interactions with the silica depend on the pH and affect the resolution of basic peptides.

In peptide elution, together with water and an organic modifier, it is common to use an ionic modifier (counterion) which is an ion-pairing reagent. At acid pH values the amino groups of the basic amino acids histidine, lysine, and arginine and of the amino acid N terminal are charged. Thus, RP-HPLC separation is improved if the mobile phase is an anionic buffer so that it forms pairs of ions with the basic groups of these amino acids. Since its introduction by Bennett et al. (88,118), trifluoroacetic acid (TFA) has been commonly used as an ion pair–forming reagent for peptide separations at acid pH and at a concentration generally around 0.1%. TFA has numerous advantages: it is transparent to ultraviolet light, it is an excellent polypeptide solvent, it is a strong ion-pairing reagent, and it does not block the amino terminal groups and is miscible with most organic modifiers. Since TFA is volatile, it can be easily eliminated by lyophilization. Other ion-pairing reagents, such as heptafluorobutyric acid (HFBA) (119,120), phosphoric acid (121), hydrochloric acid, formic acid (122), acetic acid (123), ammonium acetate (114), and tertiary alkylamines, particularly triethylammonium phosphate (124), can also be used to modify selectivity (117,125,126). HFBA is suitable for increasing the retention of polar peptides (108,118,120, 126,127). HFBA is also volatile and transparent to UV light at low concentrations. Phosphoric acid is useful in hydrophobic peptide separations (128–130). The effect of the ion pair–forming reagent on the resolution of a mixture of peptides can be observed in Fig. 3, where the use of TFA, phosphoric acid, and HFBA is compared. Peptide retention increases as the hydrophobicity of the ion pair–forming reagent increases and this effect is more marked in peptides with basic residues.

The use of tertiary alkyl amines (cationic ion–pairing reagents), particularly triethylammonium phosphate (108,131,132), has the disadvantage that generally elution is carried out at pH >4 (pH greater than the pKa of the acid groups of the lateral chains of the amino acids), and with nonvolatile mobile phases. A later step of desalting becomes necessary although it enables detection at 200–220 nm. Thus, although cationic ion–pairing reagents can allow good control of

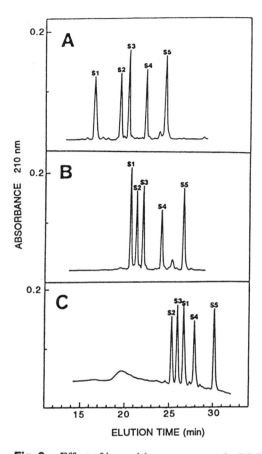

Fig. 3 Effect of ion-pairing reagents on the RPC separation of a mixture of synthetic peptide standards. Mobile phase: linear gradient 1% B/min. eluant A) water, eluant B) acetonitrile, both containing 0.1% H_3PO_4 (panel A), 0.1% TFA (panel B), or 0.1% HFBA (panel C). (From Ref. 108.)

retention behavior, it is best to avoid their use if possible (108). However, they may be suitable for the separation of acid peptides. Recently, Calderan et al. (133) were able to separate acid peptides using tetrabutylammonium as counterion. The order of elution is parallel to the hydrophobicity of the ion pair reagent/peptide complex, which increases with the number of acid functions present in the amino acid sequence. Ammonium acetate is a volatile basic counterion which has the advantage of being easily eliminated and has been used for the RP-HPLC purification of acid peptides which are insoluble in strong acid such as TFA (134).

   b. Prediction of Peptide Retention in RP-HPLC. Due to its undoubted interest, in the past few years there has been an attempt to predict the retention of peptides separated by RP-HPLC in terms of the composition of their amino acids. For this purpose it has been common to use as mobile phase a linear gradient from 0.1% TFA in water to 0.1% TFA in acetonitrile. The prediction models should help to avoid long periods of empirical optimization. Each amino acid's contribution to retention is calculated by means of linear regression using the observed retention times of standard peptides of known amino acid composition. The predicted retention time (t) of a peptide on a particular column will be equal to the sum of the contributions of its amino acids (retention coefficients (ΣRc), plus retention time of a nonretained solute (t_0), plus a correcting time (t_s) given by the difference between observed retention times (obs) and those calculated for a standard peptide (std) injected on that column (113,115).

$$t = \sum Rc + t_0 + t_s$$
$$t_s = (tr^{obs}std - (\sum Rc^{std} + t_0))$$

Since 1980 numerous researchers have established the retention coefficients of the different amino acids (113,115,127,135–137). A computer program called ProDigest-LC has recently been designed for the prediction of the retention time of peptides based on the retention coefficients of the amino acids described above (138,139). In peptides with more than 20 residues the secondary structure and shape can have a significant influence on retention. For these, retention time is equal to the sum of the contributions of their amino acids and of factors due to chain length (140). Chabanet and Ybon (141) studied the retention of 104 peptides in order to establish the contribution of chain length to peptide retention. The contribution of each amino acid to peptide retention is not constant but rather a decreasing function of peptide length.

Peptides with similar hydrophobicities can elute as a single peak in RP-HPLC. In order to avoid this disadvantage other modes of separation, such as ion exchange or size exclusion chromatography, can be employed.

2. IE-HPLC Separation of Peptides

Ion exchange chromatography has played an important role in peptide separation and purification (99,142). The development of HPLC has contributed to the improvement in peptide separation by ion exchange, although it has been used to a lesser extent than RP-HPLC (99,100,143). This technique uses columns 5–30 cm long and with an internal diameter of 3–5 mm, containing cation or anion exchangers as stationary phase. The stationary phase contains ionized groups with the opposite charge to the peptides to be separated. These groups are generally fixed to a silica or polymeric material support (polystyrene-divinylbenzene, PS-DVB) (92,99). PS-DVB columns must have a lower percentage of crosslinking and are mainly suitable for small peptides. Silica admits lower pH values (2–8) and polystyrene-derived resins allow for alkaline pH (up to pH 12).

The peptide separation mechanism in IE-HPLC is based on differences in net charge (Table 3). Peptide retention is generally directly proportional to the charge it has (98,138). Moreover, retention and elution involve the substitution of certain ionic species in the stationary phase by others. Normally, a particular peptide may be chromatographed on both anion exchangers and cation exchangers depending on the pH of the sample and mobile phase and whether the surface of the exchanger is suitably charged at that pH (100,103). At present there are numerous anion and cation exchangers available, both strong and weak, which contain $-SO_3H$ (sulfo), $-COOH$ (carboxy), $-OPO_3H_2$ (phospho) $-N^+H$ (primary, secondary, tertiary ammonium) groups (Table 4). Peptides can be successfully chromatographed over a wide range of pH values, ionic strengths, buffer types, organic solvents, temperatures, and matrix types (92).

The mobile phase controls peptide retention in terms of its ionic strength, buffer pH, and the possible presence of an organic modifier (Table 3). If the buffer's ionic strength is too weak, the peptides may remain in the stationary phase indefinitely. Elution is frequently performed on a gradient of sodium or potassium chloride in phosphate, in tris, or in citrate (108,125). Control of the pH is important due to its influence on the charge of the peptides and weak exchangers, both cationic and anionic. In contrast, the effect is less in strong exchangers or for highly ionized solutes (144).

Both the cation and anion exchanger methods appear to be extremely useful for resolving many separation problems. Acid peptides are separated better in anion exchangers (99), whereas neutral or basic peptides are separated better in cation exchangers (92). Strong cation exchange columns

Table 4 Ion Exchangers Employed in Peptide
Separation [92]

Exchanger	Characteristics
Anion	
Polyamino	Weak
Diethylaminoethyl	Moderate
Quaternary ammonium	Strong
Cation	
Carboxymethyl	Weak
Phospho	Moderate
Sulfo	Strong

have been used for peptide separation with volatile pyridine–acetic acid buffers (103), sodium phosphate buffers (145), and lithium citrate buffers (146). Equally, weak cation exchangers have been used for separation of highly basic peptides using 50 mM KH_2PO_4 buffer and a gradient of KCl (100,102), and also strong cation exchangers using phosphate buffer and a gradient of salt (98). Weak anion exchangers prepared on porous silica ($-CH_2-CH_2-CH_2-^+NH_2-CH_2-CH_2-^+NH_3$) were used for HPLC separation of dipeptides and diastereoisomeric peptides (147) using mixtures of triethylammonium acetate (TEAA) and acetonitrile as eluants for separation and detection over the range 210–225 nm (99). Since TEAA is volatile, it enables peptides to be recovered easily. This method has been applied to the separation of mixtures of bioactive peptides and tryptic digestion of proteins (99,101). Closely related acidic and neutral peptides can be successfully resolved by HPLC on a weak anion exchange bonded phase with good results in terms of sensitivity, peak symmetry, reproducibility, high capacity, and recovery (69).

Although IE-HPLC has been used to a lesser extent than RP-HPLC for peptide separation, it has the advantage that separation can be performed at an alkaline pH and with 10–100 times more sample than that for RP-HPLC. One of the disadvantages is that most of the buffers used for elution absorb at low wavelengths. Moreover, RP-HPLC and IE-HPLC are used in a complementary manner for the separation of complex mixtures of peptides (99,108).

3. SE-HPLC Separation of Peptides

SE-HPLC separates peptides on the basis of their molecular size (Table 3). The material used as stationary phase provides pores with a controlled size which filter the components of the sample. Smaller solutes have access to a greater proportion of the total pore volume and are retained longer than larger molecules. The theory and applications of SE-HPLC were recently reviewed (148). In recent years a wide range of organic and inorganic packing materials with different pore sizes have been manufactured. These HPLC columns can be used with a great variety of eluants, both aqueous and organic. Stationary phases based on inorganic packing materials such as silica or alumina have the drawback of adsorbing charged molecules and so must be deactivated, e.g., with organosilanes. Among the organic materials used, the following can be mentioned: methacrylate-glycerol, sulfonated styrene-divinylbenzene, and polyesters with hydroxyl groups (149). By means of developing small, rigid particles of uniform size, packed uniformly in <10-mm-i.d. and 200–500-mm-long HPLC columns, a high number of plates per unit of length have been obtained (150).

The mobile phase used in SE-HPLC should be merely a means of transport since separation occurs only in terms of molecular size and should have no interaction with the column, the sample, or the packing. Mant et al. (153) have studied the most suitable chromatographic conditions for peptide separation by SE-HPLC. Aqueous solvents and buffers containing 100–400 mM salt are commonly employed as mobile phase in SE-HPLC in order to eliminate nonspecific interactions (108,138). Small concentrations of an organic modifier help to eliminate nonspecific interactions (153).

SE-HPLC has been more widely used in protein analysis, but it has also been used for peptide separation. Vijayalakshmi et al. (151) separated peptides with molecular weights between 6000 and 250 Da using a hydroxylated polyester column (TSK-SW-2000) eluting with phosphate 50 mM containing 0.1% TFA and 35% methanol. Lemieux and Amiot (69,104,105) used a similar column to fractionate peptides from hydrolyzates of phosphorylated and dephosphorylated casein which were then separated by RP-HPLC. This column has also been used to separate soluble peptides from milk and to quantify the glycomacropeptide (106,107). Lemieux et al. (57) separated the astringent (84,000–18,000 Da) and bitter (5000–110 Da) fractions of cheddar cheese made with *Lactobacillus* to accelerate ripening.

Richter et al. (152) described a method for suitable determination of the molecular weight of small peptides (1000–10,000) by SE-HPLC. Peptide separation can be achieved with a non-denaturing volatile mobile phase such as 0.1% TFA or in highly denaturalizing conditions which are more suitable for predicting behavior of peptides in SE-HPLC of peptides in the determination of molecular weight (50 mM KH_2PO_4, pH 6.5, 0.5 M KCl, 8 M urea) (108,153). SDS and guanidine hydrochloride are used also in denaturing phases. Saito et al. (71) used 6 M guanidine hydrochloride to determine the molecular weight of glycopeptides from bovine colostrum.

The practical value of molecular exclusion columns is somewhat limited for peptide resolution when compared to RP-HPLC (108,151). The columns are very expensive and deteriorate quickly. The behavior of peptide separation is often unpredictable due to the fact that complex mechanisms of size exclusion, hydrophobic adsorption, and electrostatic interactions are involved (151,153, 154). These disadvantages make it less suitable than other types of HPLC (108,117). In many cases, its use is almost limited to the separation of large, highly hydrophobic peptides, which can present problems in RP-HPLC or for preliminary peptide separation before using RP-HPLC, which enables long peptide fragments or nondigested proteins to be eliminated.

4. Combination of HPLC Modes

The complete separation of all of the components of a mixture of peptides may require the combination of different separation modes (multiple HPLC): IE-RP, SE-IE-RP, SE-RP, SE-RP-IE-RP (101,108,155,156). SE-HPLC and IE-HPLC can be performed principally in order to obtain a good, rapid resolution of peptide fractions for further RP-HPLC analysis. RP is generally used as the last purification step for the elimination of salts and also for its greater selectivity and resolution capability (Table 3). Prior to use of SE enables the elimination of long fractions or nondegraded proteins which may be accompanying the peptides of interest and which would complicate the other types of chromatography (125,157). In these cases it is useful to employ volatile eluants such as 0.1% TFA, HFBA, acetonitrile, and ammonium acetate, which are easily eliminated and do not leave any salts. This is important when IE-HPLC is performed afterward. Some studies have combined the three techniques of SE-, IE-, and RP-HPLC for the complete resolution and identification of hydrolyzates of β-casein (69,98) and to identify hydrolyzates of hemoglobin (158). Takahasyhi et al. (159) describe an automated tandem HPLC system with step gradient elution from an anion HPLC exchanger SpheroGel-TSK IEX-540 DEAE column and direct application of the eluate fractions onto a RP-HPLC column which they applied to the analysis of the peptide map of genetic variants of albumin from human serum (160).

5. Separation of Peptide Stereoisomers by HPLC

One of the most important aspects in peptide separation by HPLC is stereoisomer (enantiomer and diastereoisomer) resolution. These peptides differ in the configuration (L, D) of their amino acids.

Diastereoisomers can be separated by HPLC due to their different physical and chemical properties on conventional phases (reverse phase or ion exchange). In diastereoisomers small structural differences result in different hydrophobicities. However, enantiomeric peptides have identical physical properties and cannot be separated in an achiral medium, and so their separation in HPLC is complicated. Their separation is of great interest in different fields, including of course food science, since organoleptic receptors distinguish between stereochemically different compounds. Optically active peptides exhibit particular biological and taste activities (3,4).

Enantiomeric peptides are separated on chiral phases or are transformed into diastereoisomers by reacting with a chiral reagent and are then separated on archiral phases. Peptide separation on chiral phases is not yet a technique that can be considered routine (161,162). Pirkle et al. (163) and Hyun et al. (164) separated enantiomeric peptides (methyl, esters of N-3,5-dinitrobenzoyl derivatives) on chiral stationary phases (CSPs) derived from (R)-α-arylalkylamines, (S)-N-(2-naphthyl)valine, or (S)-1-(6,7-dimethyl-1-naphthyl)isobutylamine. These researchers separated four peaks for each dipeptide derivative corresponding to the two enantiomeric pairs $(R,R)/(S,S)$ and $R,S)/(S,R)$. In the past few years cyclodextrins have been used as chiral stationary phase for the separation of enantiomeric peptides or amino acids (165–168), or have been added to the mobile phase (168,169). In these cases optical resolution seems to be greater if the hydrophobic amino acids have a chiral center and can form inclusion complexes in the hydrophobic interior of the cyclodextrin (170). Zukowski et al. (171) studied the mechanism of retention and optical resolution of derivatized peptides on cyclodextrin-bonded phases and came to the conclusion that as well as the formation of inclusion complexes, hydrogen bridges are also necessary for optical resolution. Enantiomeric and diastereoisomeric dipeptides were recently chromatographed on chiral stationary phases based on immobilized α-chymotrypsin (172).

HPLC separation of diastereoisomeric peptides is more common. Separations of diastereoisomeric di- and tripeptides have been performed on RP columns. Cahill et al. (173) separated derivatized diastereoisomeric peptides using esters of N-hydroxysuccinamide of t-butylcarbonyl-L-amino acid on C-18 and C-8 columns. Linder et al. (174) separated dipeptide derivatives on RP-C8 columns, adding a metal chelate. Mixtures of DL and LD dipeptides can be separated by RP-HPLC into two peaks, one containing the LL and DD isomers, and the other the LD and DL isomers. The separation of these dipeptides by weak anion exchange HPLC also gave rise to two peaks, but the order of elution was the opposite to that in RP-HPLC (99).

The separation of enantiomeric peptides has been mainly applied in the study of peptides with biological activities (enkephalins, endorphins, hormones) but is not yet common in food science. However, we can expect significant developments in the future for studying bioactive peptides and peptides with sensory properties.

B. Electrophoretic Analysis of Peptides

1. Slab Gel Electrophoresis

Electrophoresis (EF) has been universally adopted for research on food proteins and large insoluble peptides since the introduction of polyacrylamide as a support (175,176). The technique has progressed from separations based only on charge/mass (PAGE) to those based on molecular mass (SDS-PAGE) or isoelectric point (isoelectric focusing, IEF). These electrophoretic techniques have good resolution, but at the same time are relatively inexpensive and simple to set up and use, reproducible, reliable, and relatively rapid. The combination of two of the methods into a two-dimensional separation (2D-EF) gives unparalleled resolution but is more time consuming.

The recent introduction of precast ultrathin minigels and an automated system for performing electrophoresis, staining, and destaining (Phast System, Pharmacia) has reduced analysis time and improved detection limits. Also the introduction of electroblotting and recent improvements in the membranes available and in transfer technique have overcome many of the limitations of EF as a preparative method in analyzing peptides because after separation the peptides can be transferred to a matrix compatible with further analysis.

Electrophoresis has been widely employed for separations of hydrophobic and large peptides since these peptides can be solubilized with detergents and EF is compatible with detergents unlike RP-HPLC. In the majority of studies published, large insoluble peptides are analyzed together with proteins using protein electrophoretic methods (177,178). Procedures of electrophoresis under nondenaturing conditions (PAGE and urea-PAGE in alkaline or acid buffers) are used for peptide and protein analysis (179). In PAGE, peptides are negatively or positively charged, depending on buffer conditions, and migrate with mobilities related to their charge/mass ratio. Small peptides can be separated according to size and charge in high-percentage (40–50%) polyacrylamide gels without SDS. Two buffer systems were described: an acid gel with an operative pH of 3–3.1 for basic peptides and an alkaline gel system with an operative pH of 9.1–9.2 for acid peptides (180,181).

However, electrophoresis in the presence of SDS, a method for estimation of molecular weights and determination of purity, is one of the most frequently used techniques in food analysis to separate proteins and peptides. Among the electrophoretic procedures, the most common one is probably the Laemmli method (182). This is a powerful and useful method for separation of large peptides and molecular weight analysis of most of them, but it provides unsatisfactory results, especially for analysis of small peptides. Peptides with a molecular mass below about 15,000 are not well resolved on uniform concentration polyacrylamide gels, even on 15% polyacrylamide, when SDS is the sole dissociating agent.

Thus, in order to analyze low molecular weight (LMW) peptides, alternative procedures for electrophoresis have been developed. One way of making conditions suitable for peptide analysis is to decrease the effective pore size of the gel by adding urea and/or using highly crosslinked gel. Swank and Munkres (183) showed that separation of low molecular peptides can be considerably improved by using 12.5% polyacrylamide gels prepared with a high ratio of bisacrylamide crosslinker and the inclusion of 8 M urea in a continuous buffer system. Later, to improve resolution, Hashimoto et al. (184) introduced the use of concentration gradient gel electrophoresis in the presence of SDS and urea, and Anderson et al. (185) used a discontinuous SDS-PAGE system that provides superior resolution of polypeptides in the weight range of 2500–90,000. Although these conditions (higher concentration of both crosslinker and urea) give good separations for LMW peptides, the presence of urea in gels and sample solution sets limitations, since it can carbamylate peptides and salt out in the gel.

The other ways of optimizing SDS-PAGE for peptide analysis are modification of buffer systems in gel and/or running buffer without using urea. Several procedures for the separation of LMW peptides in the absence of urea have been developed. Increasing tris concentration in the separating gel and running buffer of the Laemli system improve resolution of LMW peptides (186). Schägger and Von Jagow (187) introduced the use of tricine as trailing ion, instead of glycine, which allows the separation of polypeptides in the range of 1–100 kDa at lower acrylamide concentrations than in the glycine-SDS-PAGE system. The use of tricine has other advantages: the gels can be run with and without urea, there is no need for casting gradient gels, overloading effects are reduced, and the omission of glycine and urea prevents disturbances in the course of subsequent amino acid sequencing. In the modified Laemmli method reported recently by Okajima et al. (188), high-molarity buffers were used in the separating gel and in the running buffer to provide superior resolution for peptides as small as MW 5000.

IEF has had limited utilization in peptide analysis. This technique only applies to amphoteric compounds and more precisely to good ampholytes with a steep titration curve around their pI, a necessary condition for any compound to focus as a narrow band. This is a problem with short peptides that need to contain at least one acid, or basic amino acid residue, in addition to the -NH$_2$ and -COOH terminal. Peptides which have only these terminal charges are isoelectric over the entire range from approximately pH 4 to pH 8 and do not focus. Another limitation with short peptides is encountered at the level of detection methods: carrier ampholytes are reactive to most peptide stains (189).

Table 5 shows some of the main electrophoretic methods developed to analyze peptides. Although good electrophoretic methods exist for the separation of LMW peptides, some limitations exist. Due to limited peak capacity, separation of complex peptide mixtures cannot be obtained and selectivity cannot be changed in a simple manner (192). Special nonstandard procedures must be used for separation, and quantification is poor due to difficulties associated with staining (193). LMW peptides do not bind SDS in a characteristic, stoichiometric fashion as do proteins, and so slab gel electrophoresis is not commonly used for the MW determination of peptides.

Thus resolving power, reproducibility, and quantification of LMW peptides by slab gel EF are poor, electrophoretic analysis of LMW peptides is purely analytical and is not as routinely performed as other techniques such as RP-HPLC and, more recently, CE.

2. Capillary Electrophoresis

During the past few years, CE has captured the attention of bioanalytical chemists and biochemists as a tool for the separation and quantification of peptides. Peptides have traditionally been analyzed by HPLC, TLC, or EF. CE is rapidly becoming an accepted complement to these methods because it provides an alternative selectivity to conventional techniques, combined with inherently high separation efficiency, simplicity, high speed, good quantitative results, and very small sample requirements. The general principles of CE, theory, instrumentation, and detection as well as a broader account of numerous applications of CE have been described by several authors (194–196).

Colburn (193) demonstrated how CE has been used to assess the purity of peptides, extracted structural information about peptides, discussed the basis for CE peptide separations, and described the problems encountered in CE of peptides and their solutions. Recently, the separation and analysis of peptides and proteins by CE was reviewed (170,197). Schöneich et al. (170) focused on the various advances and principles that have been incorporated into methodologies for CE applied to the separation and analysis of peptides and proteins. Various approaches to peptide and protein separation using CE are discussed by Schwartz et al. (197), who reported that whereas the separation of small peptides is relatively straightforward and well understood, it appears that no single strategy is applicable for large peptides and proteins, due largely to the diversity and complexity associated with these molecules. In the past decade, numerous studies have been published on peptide analysis by CE, but the majority are performed with model peptides or hydrolyzates of proteins with physiological activities (hormones, myoglobin, β-endorphin, etc.) or with hydrolyzates of lactic proteins (β-lg, β-CN), since the structures of these compounds are well known. However, in the field of foods there are few specific applications, although it is to be expected that it will become a routine technique in the near future. Several commercial instruments have been brought out lately, which points to the great strides made by this method as a complement to HPLC.

Capillary electrophoresis separations can be categorized according to the following separation modes: capillary zone electrophoresis (CZE), capillary isoelectric focusing (CIEF), capillary gel electrophoresis (CGE), micellar electrokinetic capillary chromatography (MECC), and capillary isotachophoresis (CITP).

CZE, also known as free solution capillary electrophoresis (FSCE), is the simplest form of CE.

Table 5 Main Procedures Used in Peptide Analysis by Gel Electrophoresis

Electrophoretic techniques	MW	Stacking gel	Resolving or separating gel	Running buffer	Detection staining	Ref.
SDS-PAGE	1200–10,000	—	12.5%T; 10%C; 8 M urea SDS-tris-phosphate	SDS-tris-phosphate pH 6.8	Coomassie blue R-250	183
SDS-PAGE	2500–90,000	3.125%T; 20%C	8%T; 5%C; 8 M urea SDS-tris-SO$_4$ pH 7.8	Anode: SDS-tris-SO$_4$ pH 7.8 Cathode: SDS-tris-HCl pH 7.8	Coomassie blue R-250	185
Non-urea-DS-PAGE	2000–200,000	—	Gradients of acrylamide (9–26%); bis (0.18–1.05%) glycerol (0–0.8%) tris-phosphate pH 6.8	SDS-tris phosphate pH 6.8	Silver	190
Tricine-SDS-PAGE	1000–100,000 1000–70,000 5000–20,000 <5000	4%T; 3%C Stacking + spacer 4%T; 3%C + 10%T; 3%C 4%T; 3%C + 10%T; 3%C 4%T; 3%C + 10%T; 3%C	10%T; 3%C 16.5%T; 3%C 16.5%T; 6%C 16.5%T; 6%C, 6 M urea SDS-tris-HCl pH 8.45	Anode: Tris-HCl pH 8.9 Cathode: SDS-tris-tricine-HCl pH 8.25	Serva blue G	187
Non-urea SDS-PAGE	5000–100,000	4.5%T; 2.7%C	20%T; 2.7%C + NaCl or LiCl SDS-tris-HCl pH 8.85	SDS-tris-glycine-HCl pH 8.3	Coomassie blue R-250	188
Acid PAGE	Small peptides 100–15,000	3.3% Acrylamide; 0.16% bis	50% acrylamide; 0.03% bis, 6 M urea Acetic acid-ammonium hydroxide pH 3.5 Operative pH resolving gel 3.0–3.1	Acetic acid-glycine pH 2,9	Autoradiography and fluorography	191
Alkaline PAGE	Small peptides 100–15,000	3.3% Acrylamide; 0.16% bis	40% Acrylamide; 0.037% bis, 6 M urea, tris-HCl pH 8.9, Operative pH resolving gel 9.1–9.2	Tris-glycine pH 8.3	Autoradiography and fluorography	180

Ideally, the separation mechanism of CZE is due to the differing electrophoretic mobilities (based on solute size and charge at a given pH) of the various components of a mixture. Since the separation is solely due to differing electrophoretic phenomena, CZE is a very promising alternative to traditional slab gel electrophoresis for the analysis of peptides, and as compared to LC results in differing selectivity, which is often complementary to the information derived from LC. Applications of FSCE to the analytical scale separation of peptides were already described by Grossman et al. (195). These authors demonstrate the suitability of FSCE for the analysis of purity of peptide samples, the homogeneity of peptide samples prior to sequencing, the identity of peptides by using electrophoretic mobility values, and the reduction of an intrachain disulfide bridge. They also evaluated FSCE as a complement and/or alternative to HPLC for peptide mapping and reported that FSCE can provide valuable complementary information to RP-HPLC, since the mechanisms of separation of both techniques are different, but increased selectivity is necessary for trace level assays. CE in free solution has been successfully applied to peptide separation.

Separation in capillary isoelectric focusing is based on the differences in the pI values of molecules. The mixture of compounds to be resolved is added to a solution of ampholytes, which encompass the pH range of interest. During the focusing step, an electric field is applied and a pH gradient is established ranging from a low pH at the anode terminus to a high pH at the cathode end. Each peptide will emigrate to a region within the capillary where the pH is equivalent to the pI of the peptide, at which point the electrical neutral molecule ceases to emigrate further. Establishment of linear pH range and the slope of the gradient are important factors to be considered for resolution optimization. Utilization of CIEF in peptide mapping or analysis has been limited, although CIEF introduces yet another dimension in selectivity in addition to LC and CZE since the separation order corresponds to the pI of the digested fragments (198).

The mechanism in capillary gel electrophoresis is based on size exclusion (sieving effects). The size distribution of the channel within the crosslinked network may be controlled by the percentage of crosslinking reagent. Separations of peptides based solely on molecular weight were achieved by complexation of the denatured peptide with SDS, which occurs at a constant ratio of 1.4 g of SDS/g of peptide. The utilization of gels in a capillary format offers the opportunity for very high separation efficiencies and resolving power for peptides and proteins. Cohen and Karger (199) reported on the separation of peptides by SDS-PAGE in a capillary format and observed that the log molecular weight vs. mobility gives a linear relationship.

The addition of surfactants to the running electrolyte in capillary electrophoresis provides micellar electrokinetic capillary chromatography (MECC) with an additional mechanism for selectivity. Separations by MECC are based on differential interactions with a micelle that allows the separation of neutral molecules or analytes of similar charge. While initially advocated for the separation of neutral analytes that are not resolved by CZE, this technique is finding use for the separation of charged small molecules and, more recently, peptides. The ability of MECC for separating structurally similar model peptides and tryptic digest is demonstrated by Liu et al. (200). The micelle formatting compounds—cyclodextrins and the anionic (SDS) and cationic surfactants (dodecyltrimethylammonium and hexadecyltrimethylammonium bromide)—used under different analytical circumstances exhibited certain beneficial effects for peptides with similar net charges but different hydrophobicities. These authors also reported that to achieve higher sensitivity of peptides, fluorescence measurements become feasible after derivatization with a suitable fluorogenic reagent, ortoftaldialdehyde (OPA), fluorescamine, and 3-(4-carboxybenzoyl)-2-quinoline carboxaldehyde (CBQA), and when utilizing cyclodextrins as buffer additives they have noticed significant increases in the measured fluorescense intensity for CBQA-tagged peptides.

Capillary isotachophoresis (CITP) was the earliest capillary electrophoretic technique, for which a commercial instrument, the LKB tachophor, was introduced in the mid-1970s. Separations occur in the gap between the leading and terminating electrolytes based on the individual mobilities of peptides. A broad and detailed review (with 148 references) of CITP in peptide analysis was described by Kasicka and Prusik (201).

V. DETECTION OF PEPTIDES

Table 6 shows the most important methods used in the detection of peptides. Some of them are based on the direct detection of peptides, while others are based on pre- or postcolumn reaction with a derivatizing agent and subsequently detection of the reaction product by absorbance or fluorescence in order to increase sensitivity or specificity. The derivatization reaction should be rapid and simple and quantitative yield should be obtained. The derivatizing reagent should generate a strong signal and the reaction byproduct should not complicate the analysis. Table 7 compares some of the advantages and disadvantages of pre- and postcolumn derivatization (206,207,232,233). It is always preferable to use more than one detection system in order to obtain the maximum information from the peptide sample.

A. Absorbance Detection of Peptides in HPLC

The most widely applied method for HPLC peptide detection is absorbance detection in far-UV. The wavelength selected is a compromise between the highest absolute sensitivity using the absorbance of the peptide bond (<210 nm) and the need to reduce background absorbance of the solvent and its impurities (generally more significant at lower wavelengths). Wavelengths between 210 and 220 nm are the most frequently used for detection of peptides at concentration between 100 and 1000 ng (121). Detection limit at 214 nm is generally between nanomoles and picomoles (202) and decreases with the use of narrow-bore columns. The aromatic groups of some amino acids such as tyrosine, phenylalanine, histidine, and tryptophan contribute to absorbance and thus the response is not proportional to the concentration of peptide bonds. At low wavelengths many components of foods and solvents absorb, therefore the samples must be carefully purified and solvents transparent to ultraviolet light must be used.

Peptides with aromatic residues (Phe, Tyr, Trp) can be detected at 254 nm. Detection is performed at 280 nm in some cases in peptides containing Tyr or Trp. The use of multiwavelength detection has the advantage of enabling the identification of peptides containing tyrosine and tryptophan. Second-order derivatives of the spectrum of Phe, Tyr, and Trp present characteristic absorption minima at 257, 280, and 290 nm. In addition, each aromatic amino acid has other minima of lower intensity: 250 and 264 nm for Phe, 272 nm for Tyr, and 268 and 278 nm for Trp (203). In peptides composed of only Tyr and Phe, the spectral contribution of each aromatic amino acid to the derivative of the peptide spectrum can be distinguished quickly by the Phe (257 nm) and Tyr (278 nm) absorption minima. However, identification of Tyr in the presence of Trp is unclear because of spectral interference by the Trp minimum. The ratio between absorbance at the two wavelengths 255:270 nm or 255:265 nm is also a useful aid in identifying aromatic amino acids or combinations thereof (204). At present, the use of computers and software enables a large number of spectra to be stored and then used and interpreted at a later time.

Absorbance detection of peptides can be carried out by forming derivatives which absorb at higher wavelengths, thus avoiding the disadvantages of working at wavelengths below 220 nm. The most commonly used derivatizing reagents are ninhydrin, dansyl chloride, phenyl isothiocyanate, and phthaldialdehyde.

Ninhydrin reacts with primary amino groups of peptides to generate a blue color (absorbance at 570 nm). The method enables detection of almost all peptides except those which contain proline at the N terminal or have the N terminal blocked. ε-NH_2 groups of lysine also react and contribute

Table 6 More Usual Methods Employed for Detection of Peptides Separated by Chromatography

Method of Detection	Application	Ref.
Absorbance		
UV (<220 nm)	Most common method for detection in HPLC analysis of peptides	121, 202
UV (254 and 280 nm)	Detection of peptides with aromatic amino acids (Phe, Tyr, Trp)	203, 204
Precolumn reaction with dansyl chloride (UV 254 nm)	HPLC analysis of amino acids and peptides	67, 150, 205, 206
	Detection of the N-terminal amino acid of peptides	
	Possibility of additional detection by fluorescence	
Precolumn reaction with phenylisothiocyanate (PITC) (UV 254 nm)	Detection of N-terminal amino acid of peptides.	70, 207–210
	Peptide sequence determination	
	Peptide detection in HPLC	
	Analysis of amino acids by HPLC	
Reaction with ninhydrin (570 nm)	Postcolumn detection of peptides and amino acids	211, 212
	Detection in TLC	
Fluorescence		
Natural fluorescence of Trp and Tyr	Detection of peptides containing Trp and Tyr	213
Pre or postcolumn reaction with fluorescamine (390 nm excitation and 475 nm emission)	HPLC detection of peptides	214–217
	TLC detection of peptides	
Pre or postcolumn reaction with OPA/thiol (340 nm excitation and 455 nm emission)	HPLC detection of peptides	67, 74, 206, 218–221
	Standard method for amino acid analysis	
	TLC detection of peptides	
	Additional detection by absorbance, amperometry, and chemiluminescence is possible	
Precolumn reaction with naphthalene-2,3-dicarboxaldeyde/cyanide (420 nm excitation and 490 nm emission)	HPLC detection of peptides	202, 222, 223
	Amino acid analysis	
Reaction with benzoin (325 nm excitation and 435 nm emission)	Detection of arginine-containing peptides	224
Mass Spectrometry	Structural information for identification of peptides.	79, 96, 225–231
	Detection in HPLC and CE	

Table 7 Advantages and Disadvantages of Precolumn and Postcolumn Derivatization of Peptides

Precolumn (dansyl chloride, phenylisothiocyanate, OPA, fluorescamine, NDA/CN)		Postcolumn (ninhidrin, OPA, fluorescamine)	
Advantages	Disadvantages	Advantages	Disadvantages
Easy automatization	Possible formation of artifacts	Separation of peptides in their original form	Less time for reaction
No time limit for derivatization	Separation of derivatives and reactives	Standard procedures of separation	Additional pumps required
	Instability of derivatives during separation	Lower artifact formation	Higher cost of instruments and reactives
	Various peaks from a single peptide	No necessity of complete reaction if enough reproducibility	Less efficiency due to broader chromatographic peaks
		Higher reproducibility	Excessive reactive present
		Easy automatization on-line	
		Double possibility of detection	

to the development of color. The reaction of proline with ninhydrin does not give a blue color (absorbance at 570) but absorbs at 420 nm. This reaction is successfully used for postcolumn detection of peptides in foodstuffs (211,212) and for detection on TLC plates.

Dansyl chloride reacts with amino groups of peptides in basic medium (pH 9.5) forming sulfonamide derivatives which are highly stable to hydrolysis and are photosensitive. The precolumn formed derivatives can be separated by HPLC and are detected by UV at 254 nm and also by fluorescence (excitation 320–360 nm, emission 430–470 nm). Reaction conditions must be optimized in order to avoid the formation of byproducts which may interfere with detection (234–236). Dansylsulfonic acid is formed as a reaction byproduct and the excess reagent reacts with the dansyl derivatives to form dansyl amide. Dansyl chloride has been used for the determination of the N-terminal residue in peptides after acid hydrolysis (67,150,205). Precolumn derivatization with dansyl chloride is useful for peptide mapping, since the resolution of the dansylated peptides is satisfactory and picomoles can be detected (206). One of the disadvantages of the use of this reagent is that it also reacts with hydroxyl, thiol, and imidazole groups (150).

Phenylisothiocyanate (PITC) reacts with primary and secondary amino groups to form phenylthiocarbamyl (PTC) derivatives. The reaction is complete and reproducible, and peptide derivatives are stable for weeks and adsorb over a broad region of the spectrum with a maximum of 254 nm. This method is used for the precolumn derivatization of amino acids (207,209) and peptides (208), and it becomes possible to detect picomoles. The PITC peptides can be fractionated and purified and are suitable for microsequence analysis. PITC is used to determine the peptide sequence by Edman degradation. When phenylthiocarbamyl derivatives are treated with an acid in an organic solvent, cyclization takes place and phenylthiohydantoin amino acid is formed; this can be separated from the rest of the chain, which remains intact, and further identified chromatographically. The process can then be repeated. PITC is also used for peptide sequencing in combination with colored Edman's reagent 4-N,N'-dimethylaminoazobenzene-4'-isothiocyanate (DABITC) (237). DABITC derivatives of formed peptides can be separated by RP-HPLC on a C-18 column and detected at 436 nm. It has the drawback that the reaction is slow and incomplete and several different derivatives may be obtained from the same peptide.

Using dansyl-Edman degradation, Leadbeater and Bruce-Ward (205) identified 60 tryptic peptides of β-casein. Colilla et al. (208) showed the effectiveness of this method by microsequencing of phenylthiocarbamyl peptides isolated from tryptic digests of α-lactoalbumin and β-lactoglobulin. PITC derivatization has been applied to the analysis of acid hydrolyzates of purified proteins and peptides in foods (70,209,210).

B. Fluorescence Detection of Peptides in HPLC

Peptides can be detected by means of the natural fluorescence of the aminoacids Tyr and Trp. The highest emission is obtained at 320–360 nm when peptides are excited at 220–280 nm. The strong native fluorescence of these residues can make this method several times more sensitive than detection at 254 nm (213).

In order to increase detection sensitivity and selectivity, fluorescent derivatives can be obtained. Fluorescamine, o-phthaldehyde (OPA); and naphthalene 2,3-dicarboxyaldehyde/cyanide or anthracene 2,3-dicarboxyaldehyde are the reagents most often used for derivatization. The sensitivity of fluorescence can be three times greater than that achieved by UV light (121).

Fluorescamine (214) reacts instantaneously with primary amino groups (α- or ε-amino of peptides) giving rise to a fluorescent derivative when excited at 390 nm and with emission at 475 nm. One of the disadvantages of this reagent is that the secondary amines do not give fluorescent products with fluorescamine. Moreover, it is insoluble in water and is commonly prepared with acetone. However, fluorescamine and its degradation products are not fluorescent, which is an important advantage. The fluorescence intensity is not affected at pH values between 4 and 10 but is affected by the composition of the solvent (215) and depends on the structure of the peptide. However, it enables at least picomoles to be detected. Peptide detection with fluorescamine can be automatized for postcolumn detection (216,238), with the disadvantage of the high cost of the reagent, or for precolumn detection (217), applied less often due to the fact that the derivatives are unstable.

OPA reacts with primary amines at an alkaline pH (9–11) in the presence of a thiol such as 2-mercaptoethanol, 3-mercapto-1-propanol, ethanothiol, N-acetyl-L-cysteine, or N,N-dimethyl-2-mercaptoethylammonium chloride to give fluorescent derivatives (218–220). The isoindole, which is formed between OPA and the primary amine, is unstable but can be stabilized in an acid medium and can be detected by fluorescence at 455 nm when excited at 340 nm. Neither the reagent nor the reaction products are fluorescent, which is undoubtedly an advantage as in the case of fluorescamine. Sensitivity is of the same order as for fluorescamine, but OPA derivatives are not hydrolyzed in an aqueous solution. The reaction takes place in approximately 1 min, which makes it suitable for precolumn peptide derivatization (206) and for postcolumn on-line derivatization (67,74). Gluta-thione, a common peptide in foods, which contains an SH group in its molecule, does not require the use of any thiolic reagent in order to give rise to a fluorescent derivative with OPA (74,239).

OPA reacts both with the NH_2 of the N terminal of peptides and with the ε-amine of lysine. One disadvantage is the lack of reaction when the N terminal is proline since it does not react with secondary amines. For reaction with these it is possible to use the derivatizing agent 9-fluorenyl-methylchloroformate (233). Some authors (74,221,240) have shown that fluorescent response with OPA depends on the peptide and is generally less than the response obtained by absorbance at 214 nm except in peptides with lysine, which provide much greater fluorescence intensity (up to 50 times more) than peptides without lysine. In peptides with lysine a detection limit of picomoles can be achieved. The authors (67,74) have shown the usefulness of detection by absorbance and postcolumn on-line OPA derivatization of peptides separated by HPLC in blue cheeses and in peptides from skim milk. Selectivity in the analysis of peptides in foods is increased using fluorescence detection coupled to absorbance, as shown in Fig. 4 for skim milk peptides.

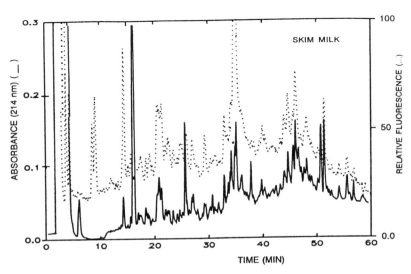

Fig. 4   RP-HPLC analysis of peptides from skim milk detected by absorbance and OPA-fluorescence on-line. (From Ref. 74.)

It must be pointed out that *o*-phthaldehyde reaction products can also be detected by absorbance (241), amperometry (242,243), and chemiluminescence (244).

The sensitivity of the precolumn fluorescent derivatization procedure of peptides can be improved by using naphthalene 2,3-dicarboxyaldehyde (NDA) or anthracene 2,3-dicarboxyaldehyde and cyanide (CN) instead of OPA (222,223,245). Patel et al. (202) described a detection limit with NDA/CN of 50–100 fmoles for angiotensins using excitation at 420 nm and emission at 490 nm. On the other hand, Ohno et al. (224) described the on-line postcolumn fluorescence detection of arginine containing peptides separated by HPLC, following reaction of the HPLC eluate with benzoin. Picomole quantities of arginine-containing peptides were detected by emission at 435 nm with excitation at 325 nm. Less sensitively, UV absorption could also be used.

The sensitivity of fluorescence can be increased by applying laser-induced detection instead of conventional fluorescence, giving rise to detection limits two orders of magnitude below conventional fluorescence.

C. Mass Spectrometry (MS) Detection of Peptides

From all the detectors available, MS is perhaps the most universal. In addition to detecting the presence of a peptide, MS provides information on its structure and thus enables its identification. The use of mass spectrometry in protein and peptide analysis was recently reviewed by Burlingame et al. (247) and McCloskey (248), while Schöneich et al. (170) focus on the use of on-line detection with a separation technique.

MS was initially used as an off-line technique in peptide determination. However, in the past few years, the difficulties involved in the liquid chromatography/mass spectrometry (LC-MS) interface has been gradually overcome through the development of new techniques of sample introduction, e.g., direct liquid introduction (DLI), moving belt, and thermospray (TSP); and ionization, e.g., continuous flow (or dynamic) fast atom bombardment (FAB), atmospheric pressure ionization systems (nebulized, electrospray [ESI], and nebulized electrospray) (249). Generally, the use of interfaces limits chromatographic conditions such as flow rate, buffers, and flow splitting. The moving belt interface is not very sensitive for peptide analysis and produces some thermal decomposition (250). Use of DLI reduces sensitivity because the flow is split before analysis (detection limit of approximately 10 µg). Also, variable tuning occurs, which complicates the analysis because the analyst does not have a suitable compound to tune on (249).

Table 8 lists the most widely used LC-MS interfaces for peptide analysis. The thermospray interface was initially quite popular and sensitivity for small peptides is in the range of 100–500 pmol for full-scan spectra. As in DLI, the tuning conditions are crucial and some analytes may not be observed whereas others are (250), which may be due to thermal decomposition. TSP produces spectra with the molecular ion which sometimes include characteristic fragments of the structure in sufficient intensity for structural elucidation (251).

The application of the above interfaces to peptide and protein LC-MS analysis has been limited in comparison with FAB and ESI (227,260). CF-FAB is a soft ionization technique which produces MH^+ or $(M-H^-)$ ions of a peptide or small protein. This technique is a method for the exact determination of the molecular weights of peptides (error of ±0.2 amu) (255,226). It can provide certain information on the peptide sequence (254) and has been optimized in order to increase fragmentation (229,255). CF-FAB has traditionally competed with TSP-MS in being the most popular technique for MS peptide analysis. The detection limit is less than 10 ng for peptides of <2000 amu (atomic mass unit) using microbore columns (261). CF-FAB provides high-intensity MH^+ ions, a good MS-MS spectrum, and tolerates a wide range of buffers (228).

ESI produces multiple charged ions of proteins and peptides at room temperature and atmospheric pressure. These multiple charged ions enable the exact determination of the molecular

Table 8 LC-MS techniques for the Analysis of Peptides

LC-MS interface	HPLC flow rate	Spectral information	Sensitivity	MW range (amu)	Ref.
Thermospray (TSP)	1–2 ml/min	MH^+, $(M-H)^-$ ions and fragments for possible structural elucidation Thermal degradation may occur	100–500 pmol	<1000	249–251
Dynamic fast atom bombardment (FAB)	1–10 µl/min	MH^+, $(M-H)^-$ ions. Some fragments are produced that are useful for structural elucidation No thermal degradation	5–50 pmol	<10,000	225, 226, 228, 253–255
Electrospray (ESI)	1 µl/min–<1 ml/min	$(M + nH)^{n+}$, $(M-nH)^{n-}$ for peptides >1000 amu MH^+, $(M-H)^-$, and some fragmentation for small peptides No thermal degradation	5–50 pmol	<200,000	170, 228, 247, 252, 256, 258
LC-MS/MS	<1 ml/min	MH^+ subject to collision-induced decomposition (CID) that produces fragmentation Peptide sequence and structural information	—	—	229, 256, 257, 259

weight of molecules with higher molecular weight than the upper mass limit of the spectrometer (256). Sensitivity depends on the compound introduced and may be very similar to CF-FAB (5–50 pmoles) when peptide digests are analyzed (228). Good full-scan spectra of peptides with molecular weights >1000 on samples smaller than 100 ng can be obtained. The ESI methodology has brought about a revolution in protein MS. Current research has gone from molecular weight information of proteins to attempts to study secondary/tertiary structure or enzyme–substrate complexes (247). The advantage of this technique is that it covers a wide range of molecular weights than FAB and shows high sensitivity for compounds of high molecular weight such as peptides and proteins (252,258). The upper mass limit of the technique is usually cited as about 100–150 kDa. From the standpoint of flexibility, convenience, sensitivity, cleanliness of the mass spectrometer, and ease of maintenance, the method of LC-ESI is an effective and practical interface that allows a wide range of conditions in flow rate and solutions (262).

As well as enabling exact and sensitive determination of the molecular weight of peptides and proteins, ESI-LC-MS also facilitates peptide mapping. If sequence information is sought, peptide molecular ions (usually MH^+) can be subjected to collision-induced decomposition (CID) in a tandem mass spectrometer to provide sequence-specific information (252,256,257,259). A large amount of structural information may be deduced very rapidly, such as location of glycosilated or disulfide-bridged peptides. This combination further enhances usefulness and flexibility and provides information that was difficult or impossible to obtain by other methods (256).

As has been the case with HPLC, CE has also been studied for its linkage to MS (CF-FAB and ESI). According to Suter et al. (228), the two separation techniques provide good results for the analysis of the peptides of a tryptic hydrolyzate of the same protein. Sensitivity is of the order of tens of picomoles, but CZE is preferred if the available sample is a limiting factor. In CZE the peaks are very narrow and the small amount of sample and the narrow-packed capillaries fix MS performance requirements (228,263). The microbore HPLC system is the best method when large amounts of sample are available and trace compounds are to be detected. The coupling of CZE-MS with the ESI interface has been explored with tryptic peptides of bovine serum apotransferrin (264).

The use of MS as a detector limits HPLC separation conditions with regard to eluants (mobile phase) which can be used and flow rates or splitting. Generally, volatile buffers and small flow rates (less than 100 µl/min) are required for CF-FAB and ESI. The increase in microbore columns in LC is accompanied by an increase in the use of LC-MS systems.

MS has generally been applied to the study of endogenous biological peptides (enkephalin, neurotensin, angiotensin, bradykinin, insulin, glucagon (265). It is often used in peptide mapping for the purpose of determining the primary sequence (256,266) and is particularly useful in the identification of posttranslationally modified residues. Although to a lesser extent, it is also beginning to be used in the study of peptides from foodstuffs or enzymatic hydrolyzates of proteins from foodstuffs which produce bioactive peptides (96,230). Huang and Henion (257) used LC-MC and LC/MS/MS (electrospray) with tryptic digest of β-lactoglobulin A and B to obtain information on the sequence. Addeo et al. (96) characterized the oligopeptide of parmigiano-reggiano cheese using FAB on the basis of the apparent molecular mass (MH^+). Van der Greef et al. (231) identified and quantified diketopiperazines (cyclic peptides) in extracts of cocoa powder. Identification of bioactive peptides from casein has been carried out by Bouhallab et al. (78).

D. Detection Methods Used in Electrophoresis

Techniques for detection and quantification of bands of slab gels are an important aspect of electrophoretic analysis. Direct photometric scanning of unstained gels is unsuitable for most applications. The usual method is to increase sensitivity by reacting the peptide bands with an easily visualized reagent (179). The most common approaches involve the use of stains, autoradiography, and fluorography.

Unlabeled peptides are stained with the dyes, Coomassie blue and silver, used usually in electrophoretic analysis of proteins. Amido black 10B has been largely superseded by Coomassie blue R-250 and G-250, which are more sensitive. Coomassie blue R-250 is widely used to stain peptide bands. The dye is usually dissolved in methanolic solution containing glacial acetic acid and the excess of dye removed from the gel by destaining (183,185,188,263,267,268). In other described methods (269), this dye is dissolved in TCA and destaining is not necessary. Thus these methods are more rapid and furthermore TCA is a much better fixative than acetic acid and so may be preferred to detect LMW peptides. Coomassie blue G-250 has similar sensitivity to R-250; usually 0.2-0.5 µg of any protein in a sharp band can be detected. This dye is usually prepared in solutions of methanol and TCA. Blaskeley and Boezi (270) described a staining method in which destaining is not necessary.

A more sensitive staining procedure is silver staining, which is over 100-fold more sensitive than Coomassie blue. However, this dye has a number of drawbacks: high background, expensive, laborious, some bands stain poorly, other compounds are detected (e.g., DNA, polysaccharides). Many variations of silver staining procedures have been described to avoid these disadvantages (271,272). Since the sensitivity of detection using simple organic stains like Coomassie blue is sufficient for many applications, it is advisable to use silver staining only when superior sensitivity is required, e.g., after peptide mapping. Silver staining has been used in peptide analysis by several authors (273).

There are major limitations in the fixation of small peptides in staining procedures and possible improvements are still necessary. Quantification of stained bands is achieved with the use of a scanning densitometer. Laser densitometers give higher resolution than conventional ones.

Unlabeled peptides separated by EF can also be transferred by capillary blotting or electroblotting to an immobilizing membrane for subsequent detection.

If the molecular sample contains radioactive atoms or can be tagged with radioisotopes, then radiolabeled peptides can be detected by autoradiography or fluorography. The first consists of placing the gel next to X-ray film whereupon radioactive emissions cause the production of silver atoms within the halide crystals and these are visualized after developing the film. The second approach involves slicing the gel after electrophoresis and determining the radioactive protein content of each slice by scintillation counting (179). These procedures are more sensitive than staining methods for unlabeled peptides and have been widely used in peptide mapping (180,275).

Detection in CE analysis is achieved as in HPLC, as described above (V.A–V.C). UV detectors are the most often used for CE of peptide analysis because they are reasonably robust, inexpensive, and versatile. However, other detectors such as fluorescence have been employed because of the need for increased sensitivity, since only small amounts of sample can be loaded. Tagging of peptides with fluorescence moieties becomes one of the logical strategies for achieving such improvements (200). Indirect fluorescence detection has been used for the analysis of trace quantities of mixtures of peptides separated by CZE (276,277) because it enables high sensitivity typical of fluorescence detection for nonfluorescent analytes without the need for sample derivatization and without altering the structure of analytes. Another detector of great interest in CE peptide analysis in mass spectrometry, although the identification of unknown peaks becomes very difficult in the small samples loaded and many studies have attempted the interfacing of CE with MS (278,289).

E. Other Methods of Detection of Peptides

Quantitative analysis of peptides in column fractions is possible by amino acid analysis after hydrolysis of samples with 6 M HCl for 24 hr at 110°C. Hermann et al. (280) combined HPLC-radioimmunoassay (RIA) for the characterization and quantitative measurement of neuropeptides. Glycopeptides can also be detected by testing the content in carbohydrates by reaction with sulfuric acid/phenol (281).

VI. APPLICATIONS OF PEPTIDE ANALYSIS IN FOODS

A. Applications of HPLC to the Analysis of Peptides in Foods

Up to the present day, HPLC has been the most widely used technique for peptide separation in foods. Table 9 shows examples of its recent use together with the analytical separation and detection conditions. In most cases RP-HPLC is used with octadecyl columns and elution gradient with mixtures of water and acetonitrile in the presence of TFA. Detection is generally by absorbance at low wavelengths (lower than 230 nm), in a few cases accompanied by fluorescence obtained by OPA derivatization or by mass spectrometry. It has been extensively applied in the study of peptides involved in cheese ripening (55,60,62,66,67,76) in the study of proteolysis in milk (74,95,287), in bioactive peptides (78,284), in bitter peptides from caseins (282), and in peptides with tensioactive properties from β-lactoglobulins (283). It has also been used in the determination of peptides from meat (54), in alcoholic beverages (61,65), and in fermented cocoa seeds (286).

Peptide mapping using HPLC is a topic of great interest since it reveals small differences between individual proteins and is capable of detecting posttranslational modifications of amino acids of genetic variants. HPLC has proven valuable when preparing, separating, and analyzing peptides for amino acid sequencing studies (288). In this regard, the proteins from dairy products are those which have been most thoroughly studied (149,205,282,289,290). Lemieux and Amiot (57,69,104,105) used various types of HPLC to characterize peptides from hydrolyzates of phosphorylated and dephosphorylated casein. Erhardt et al. (291) isolated and completed primary sequence of a new ovine wild-type β-lactoglobulin C by using RP-HPLC of tryptic peptides. Figure 5 shows RP-HPLC chromatograms of cheese peptides during ripening.

HPLC peptide analysis has also been used to characterize foods by Carnegie et al. (52,53). They separated dipeptides on the muscles of various species of meat in order to identify the origin of the meat product. Medina and Phillips (50) analyzed the peptide pattern obtained by the enzymatic hydrolysis of proteins isolated from several sources such as beef, pork, chicken, and soybean by TLC and HPLC to identify proteins. RP-HPLC has been applied to differentiate mixtures of milk in cheese, after tryptic hydrolysis of milk proteins (292,293). Pham and Nakai (94) correlated RP-HPLC of peptides with cheese age. Olieman and van den Bedem (109) and Vreeman et al. (107) developed methods of SE-HPLC for detecting glycomacropeptide resulting from the action of rennet on the K-caseins during cheese manufacturing in order to measure the amount of whey in powdered skim milk. Later, Olieman and Van Riel (97) developed a complementary procedure based on RP-HPLC for determination of glycomacropeptide in proteolyzed milk. Quantitation of degradation products of raw milk proteins by HPLC may furnish information on the shelf life of subsequently prepared UHT milk. Mottar et al. (106) applied HPLC to determine the specific proteolytic components which provide information concerning the presence and activity of psychrotropic bacteria.

Table 9 Recent Examples of Chromatographic Separation of Peptides in Foods by HPLC

Column	Eluants for gradient	Detection	Food	Ref.
RP-ODS (24 × 0.46 cm)	0.005 M triethylammonium formide (pH 4.6) (A); 0.05 M triethylammonium formide in 80% acetonitrile (B)	220 nm	Cheese ripening	62
RP-ODS (25 × 0.4 cm)	0.1% TFA in water, 0.1% TFA in acetonitrile (B)	214 nm and OPA fluorescence	Wine	65
RP Zorbax ODS	0.1% TFA in water (A); 0.05% % TFA in acetonitrile (B)	Diode array (200–340 nm)	Meat	54
RP Hypersil ODS (25 × 0.46 cm)	0.01 M ammonium acetate (pH 6.5) (A); 0.01 M ammonium acetate (pH 6.5) in 50% acetonitrile (B)	230 nm	Casein bitter peptides	282
Ultrasphere ODS (25 × 0.4 cm)	0.1% TFA in water (A); 0.1% TFA in 75/25 acetonitrile/water (B).	230 nm and OPA fluorescence	Blue cheese	67
Ultrasphere ODS (25 × 0.46 cm)	0.1% TFA in water (A). 0.1% TFA in acetonitrile (B)	214, 254, and 280 nm	Cheese	66
RP NovaPak C-18 (15 × 0.39 cm)	0.1% TFA in water (A); 0.1% TFA in acetonitrile (B)	214 nm and postcolumn OPA fluorescence on-line	Milk	74
RP Vydac C-18	0.115% TFA in water (A); 0.1% TFA in (60:40) acetonitrile/water (B)	214 and 280 nm	Tensioactive peptides from β-lactoglobulin	283
RP Vydac C-18 (25 × 0.46 cm)	0.1% TFA in water (A). 0.1% TFA in water/acetonitrile (10/90) (B)	220, 254, and 280 nm	Cheese proteolysis	77
RP Vydac C-18	0.05% TFA in water (A); 0.05% TFA in 60% acetonitrile (B)	220 and 260 nm	Beer	61
RP Vydac C-18. (25 × 0.46 cm)	0.1% TFA in water (A). 0.1% TFA in 70% acetonitrile (B)	215 nm	Proteolysis in cheese	76
RP μBondapak C-18 (25 × 0.4 cm)	0.1% %TFA in water (A). 0.1% TFA in acetonitrile (B)	220 nm and MS detection off-line	Cheese	96
RP μBondapak C-18	0.115% TFA in water (A).; 0.1% TFA in 60% acetonitrile (B)	214 nm	Peptides from casein	69, 104, 105
RP DeltaPak C-18 (15 × 0.39 cm)	0.1% TFA in water (A); 0.1% TFA in 70% acetonitrile (B)	214 and 280 nm	Bioactive peptides from casein	284
RP Nucleosil C-18	0.1% TFA in water (A); 0.1% TFA in acetonitrile 90%. (B)	220 and 280 nm	Milk	73

Table 9 (*Continued*)

Column	Eluants for gradient	Detection	Food	Ref.
RPC HR 5/5 (5 × 0.5 cm)	0.1% TFA in water (A); 0.1% TFA in methanol (B)	214 nm	Cheese	55
RPSC Ultrapore (7.5 × 0.46 cm)	0.1% TFA in water (A); 0.1% TFA in acetonitrile (B)	210 nm	UHT milk proteolysis	285
RP Ultrasphere ODS (25 × 0.46 cm)	0.11% TFA in water (A); 0.1% TFA in acetonitrile (B)	210 nm	Fermented cocoa seeds	286
RPSC Ultrapore	0.1% TFA in 15% acetonitrile (A); 0.1% TFA in 30% acetonitrile (B)	210 nm	Rennet whey in milk	97
SE-TSK 2000 SW	0.1 M buffer phosphate (pH 6.0); 0.15 M Na_2SO_4	205 nm	Milk proteolysis	106
SE-TSK 2000 SW	0.1 M KH_2PO_4; 0.01 M K_2HPO_4; 0.15 M Na_2SO_4	205 nm	Rennet whey in milk	109

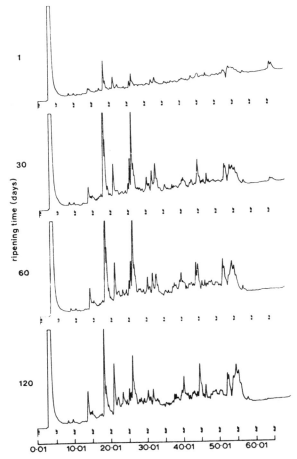

Fig. 5   HPLC elution profiles of appenzeller cheese extracts. (From Ref. 76.)

B.   Application of Slab Gel Electrophoresis to the Analysis of Peptides in Foods

Slab gel electrophoresis has been widely adopted for research on proteins and large insoluble peptides of foods, whereas it has been less applied for LMW peptide analysis. Table 10 shows some applications of analysis of peptides by EF in foods.

SDS-PAGE electrophoretic profiles of seed proteins from 22 different peanut cultivars revealed the presence of a 36-kDa major polypeptide in most of the cultivars and breeding lines with poor blanchability (294). The presence *vs.* absence of this polypeptide could be used as a reliable indicator of blanchability and it may also serve as an important criterion in the assessment of the quality of peanuts for processing purposes.

Shate et al. (268) analyzed 31 soybean varieties using two electrophoretic procedures: the Laemmli (182) method and the method modified by Fling and Gregerson (186). Those observed that the modified method afforded better resolution and visualization of polypeptides in the MW range of 29,000–32,000 and the 10,000-Da region, which may permit identification of varietal differences. Two-dimensional gel electrophoresis, IEF in the first dimension followed by SDS-PAGE in the second dimension, also permits high resolution of polypeptides (301) but is more time

Table 10 Applications of Gel Electrophoresis to Peptide Analysis in Foods

Sample type	Fraction studied	Electrophoretic modes	Applications	Ref.
Peanuts	Extracted with urea, mercaptoethanol (β-ME)	SDS-PAGE, IEF	Quality of peanuts: blanchability	294
Soybean	Extracted with Tris-HCl, β-ME, with or without urea	SDS-PAGE, Nonurea-SDS-PAGE	Identification of unique seed varieties	268
Meat	Hydrolyzed with CNBr	SDS-PAGE, 2D-EF	Accelerate conditioning in bovine meat	295
Fishery products	Digested enzymatically	SDS-PAGE, 2D-EF	Fish species identification	273
Cheese	pH 4.6 insoluble fraction (caseins, casein peptides)	SDS-PAGE, urea PAGE	Proteolysis during ripening of cheese	296
Cheese	pH 4.6 insoluble fraction	SDS-PAGE	Proteolysis during storage of UHT milk	287
Milk, cheese	pH 4.6 soluble and insoluble fractions	SDS-PAGE, urea PAGE	Effects of thermal treatments in milk proteins	297
Caseins	pH 4.6 insoluble fraction	Urea-PAGE and electroblotting	Proteolytic specificity of plasmin	298
Cheese	Ultrafiltrate permeate of the water soluble nitrogen fraction	Urea PAGE	Fractionation of cheese nitrogen and identification of principal peptides	299
Soy protein	Digested enzymatically	SDS-PAGE	Protein digestibility	300

consuming. Two-dimensional SDS-PAGE was also used by Stanton and Light (295) to investigate the potential of prerigor lactic acid injection to accelerate conditioning (tenderizing) in bovine meat.

In the dairy sector, EF has been widely used by many researchers to monitor the proteolysis or formation of peptides during ripening and storage of cheese and milk; to determine the activity of residual coagulant, plasmin, and proteinases of starter on caseins of cheese; to study effects of thermal treatments on milk proteins (287,296–299). Figure 6 shows the SDS-PAGE pattern of enzyme-treated caseins and casein fraction of stored UHT whole and skim milk. HPLC and EF applications in determinations of the proteins and peptides in dairy products were reviewed by Strange et al. (289).

Kim and Barbeau (300) evaluated a SDS-PAGE method for estimating protein digestibility and reported that this technique had two advantages: it allows estimation of the molecular weight of polypeptides/peptides remaining in protein hydrolyzates and visualization of the process of protein digestion from densitometric analysis.

C. Peptide Mapping by Slab Gel Electrophoresis

Traditionally, the confirmation of protein identity involves the fragmentation of proteins into peptides, which can then be analyzed. Fragmentation is most often accomplished by digesting the protein enzymatically or chemically. A high degree of specificity in peptide bond cleavage is perhaps less easy to achieve with chemical methods of hydrolysis than with enzymatic methods. The most popular proteinases used for enzymatic cleavage of polypeptide chains tend to be trypsin, *Staphylococcus aureus*, V8 proteinase, chymotripsin, papain, and pepsin. The digestion of protein results in a complex mixture of peptides, which are resolved to form a fingerprint or map characteristic for that protein. Slab gel electrophoresis and RP-HPLC have been the most widely used methods for analysis of that mixture. Presently, CE can also provide a complementary method for generating the map. There are a great many possible ways of doing peptide mapping, according to the enzyme and method of analysis chosen.

Peptide mapping using electrophoretic techniques is described in detail by Andrews (274). As sample proteins are not readily available in a purified form, instead of performing some preliminary purification scheme, it is often convenient to apply the samples directly to an electrophoresis gel and, once the components have been separated into their respective zones, to perform the

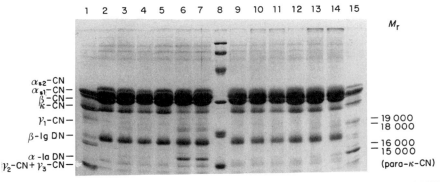

Fig. 6 SDS-PAGE pattern of enzyme-treated caseins and casein fractions of stored UHT milks. Casein with added plasmin (1); direct UHT whole milk: 0d (2), 42d (3), 78d (4); direct UHT skim milk: 0d (5), 42d (6), 78d (7); molecular mass standard (8); indirect UHT whole milk: 0d (9), 42d (10), 78d (11); indirect UHT skim milk: 0d (12), 42d (13), 78d (14); casein with added proteinase from *Pseudomonas fluorescens* B52 (15). CN, casein; DN, denatured; d, days. (From Ref. 287.)

hydrolysis to peptides and map the peptides in a second electrophoresis stage. Various methods (PAGE, SDS-PAGE, and IEF) can be used for the initial separation of the protein component in the sample, but the SDS-PAGE method is the one most often employed in the final stage to separate the peptides into a one-dimensional map. If the sample is relatively pure, it will be preferable to generate the peptide mixture in vitro before electrophoretic analysis because it is then easier to define and control the hydrolysis process. In electrophoretic peptide mapping, it is important to avoid methods which generate large numbers of very small peptides, since these may be difficult to resolve or even lost from the gel during processing. When chemical methods are used, it may be necessary to remove reagents and/or reduce the ionic strength and change the pH before analysis. In such cases it may be best to use other nonelectrophoretic methods to separate peptides.

Although there are many good and much earlier electrophoretic peptide mapping procedures, the most quoted method is that described by Cleveland and Fisch (302). It involves partial proteolytic cleavage of proteins separated by PAGE, without elution of protein followed by a second SDS-PAGE separation during which the peptide products separate on a size basis.

Although the concentration of gradient gels depends on the molecular mass of the original protein and size of the peptide fragments produced, gels with linear $T = 5$–20% or 10–25% gradients are often suitable for peptide mapping.

Peptide mapping represents a useful supplement to SDS-PAGE and IEF in fish identification because, in addition to molecular weight and charge of the proteins, the amino acid sequence is used as a third characteristic for the differentiation of proteins. Fish species of seafood made of washed fish flesh (e.g., imitation crab meat from surimi) were identified by peptide mapping of the myosin heavy chain (MHC) (273). This author concluded that peptide maps of MHC of raw muscle of various fish are species-specific and in accordance with the maps for surimi of the respective species (Fig. 7). Admixtures of other proteins to surimi having the same molecular weight as MHC, e.g., shrimp muscle, was detectable by the change in the pattern of the peptide map. Peptide mapping was also applied to elucidate differences in the primary structure of MHCs from white and red muscle of fish (303) and from muscle of fish acclimatized to different temperatures (304).

D. Applications of Capillary Electrophoresis

CE is particularly suitable for peptide mapping because of its very high efficiencies and because complete resolution can often be obtained for analytes differing by only one amino acid substituent, enabling mutations and posttranslational modifications to be detected. Table 11 shows applications of peptide mapping in foods by CE.

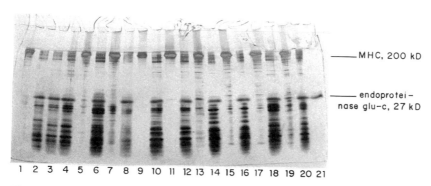

Fig. 7 Peptide maps of the myosin heavy chain (MHC) from various fish and fish products. (From Ref. 273.)

Table 11 Applications of Capillary Electrophoresis to Peptide Mapping in Foods

Compounds	Capillary electrophoretic modes	Detection	Applications	Ref.
Tryptic peptides, β-casein	CZE	UV (215 nm)	Peptide mapping	305
Tryptic peptides, β-lactoglobulin A	CZE	UV (220 nm)	Purity assays: to screen samples before sequencing analysis	306
Tryptic peptides, serum albumin bovine	CZE	UV (200 nm)	Study the utilization of coated and uncoated capillaries in CZE analysis	307
Tryptic peptides, β-casein	CZE	Indirect fluorescent detection	Analysis of trace quantities of peptides	276
Tryptic peptides, β-casein and serum albumin bovine	CZE	Indirect fluorescent detection	Analysis of trace quantities of peptides	277

Peptide maps of β-casein tryptic digest using CE and microcolumn LC were reported by Cobb and Novotny (305). Peptide maps from phosphorylated and dephosphorylated forms of β-CN have been readily distinguishable using both separation methods, demonstrating an ability to detect a single modification in a protein. The CZE peptide mapping was found to be one to two orders of magnitude more sensitive, the time analysis was significantly shorter by CZE (20 vs. 70 min), and relative standard deviations were less than 1% and better than by HPLC. Figure 8 shows the peptide maps of phosphorylated and dephosphorylated forms of β-casein, obtained with CZE.

Zu et al. (307) used coated capillaries to minimize adsorption and observed that coated capillary at alkaline pH exhibits good performance for the separation of tryptic fragments of bovine serum albumin by CZE.

Indirect fluorescence detection has been used to increase sensitivity in peptide mapping. Hogan and Yeung (276) used indirect fluorescence detection–CZE (IFD-CZE) to observe the separation of subfemtomole quantities of β-casein digest mixtures in 3 min. The quantitative and qualitative data were reproducible and the mass limit was 180 and 3000 times lower than that obtained with UV-CE and HPLC, respectively. Another procedure of IFD-CZE has been used to analyze trace quantities of cationic peptides derived from the tryptic digest of β-casein (β-CN) and bovine serum albumin (BSA) (277).

CE has been used to screen the purity of the fractions of tryptic peptides of β-lg A isolated by RP-HPLC. Fraction A and C appear as a single peak on the chromatogram (HPLC) but as three and two peaks, respectively, on the electrophoretogram (CZE). Subsequent sequence analysis of these fractions gave results consistent with the presence of three and two peaks (306).

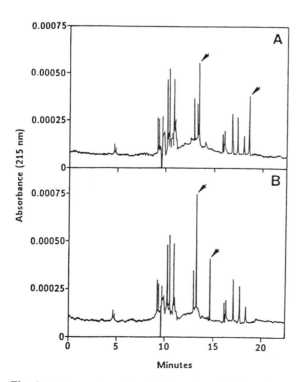

Fig. 8   Comparison of tryptic digest from (A) phosphorylated and (B) dephosphorylated forms of β-casein, separated by CZE. Arrows point to the two peaks which exhibit different migration times between the two forms. (From Ref. 305.)

Recently, Tsuda et al. (308) investigated optimum separation conditions and the identification of LMW peptides in cheese using computer simulation systems of capillary isotachophoresis. They reported that this technique can contribute to the analysis of LMW peptides and amino acids, which can be expected to improve the ripening conditions and better characterize the flavor of cheese.

VII. FUTURE DEVELOPMENTS

In the future, HPLC-MS and CE-MS will become generalized as routine methods of peptides and proteins in foods as well as in hydrolyzates for sequence determination. The additional structural information provided by obtaining a mass spectrum associated with a peak eluting from a high-resolution technique (HPLC and CE) makes separation-MS approaches highly attractive despite the expense and complexity of the instrumentation.

Minaturization of separation techniques (microbore and narrow-bore columns) is now pushing detection limits on MS instruments, and future breakthroughs in utility will come from adapting the most sensitive ionization–analyzer combinations with very high-resolution separation techniques. MS will become an ever more important analytical tool for studying food systems. Although numerous methods of detection exist, it would be useful to develop new, more specific methods of on-line peptide detection in CE and HPLC.

Robotization in sample preparation and its on-line coupling (mainly in solid phase extraction) with the advanced high-resolution and detection techniques is another point for future development in food peptide analysis. On the other hand, it is to be expected that the separation of enantiomeric peptides by HPLC and CE will be widely applied more in the identification of food peptides with specific properties.

ACKNOWLEDGMENTS

The authors thank the financial support given by the Spanish Comision Interministerial de Ciencia y Tecnologia (Projects ALI 93-0873-C02-02 and ALI 94-0737).

REVIEWS

A. L. Burlingame, T. A. Baille, and D. H. Russel, Mass spectrometry, *Anal. Chem. 64*: 467R (1992).

W. G. Kuhr and C. A. Moning, Capillary electrophoresis, *Anal. Chem. 64*: 389R (1992).

M. C. Polo, D. González de Llano, and M. Ramos, Derivatization and liquid chromatographic separation of peptides, in *Food Analysis by HPLC* (L. M. L. Nollet, ed.), Marcel Dekker, New York, 1992, pp. 117–140.

E. D. Strange, E. L. Malin, D. L. Van Hekken, and J. J. Basch, Chromatographic and electrophoretic methods used for analysis of milk proteins, *J. Chromatogr. 624*: 81 (1992).

C. Schoneisch, S. K. Kwok, G. S. Wilson, S. R. Rabel, J. F. Stobaugh, T. D. Williams, and D. G. Van der Velde, Separation and analysis of peptides and proteins, *Anal. Chem. 65*: 67R (1993).

REFERENCES

1. A. L. Lehninger, in *Bioquímica Las Bases Moleculares de la Estructura y Función Celular*, Omega, S. A. Barcelona, 1988, pp. 97–127.
2. H. D. Belitz and W. Grosch, *Food Chemistry*, Springer-Verlag, New York, 1986, p. 26.
3. M. Kawai, M. Chorev, J. Marin-Rose, and M. Goodman, *J. Med. Chem.*, 23: 420 (1980).
4. J. W. Tsang, B. Schmied, R. Nyfeler, and M. Goodman, *J. Med .Chem.*, 27: 1663 (1984).
5. M. Rodriguez, J. M. Bland, J. W. Tsang, and M. Goodman, *J. Med. Chem.*, 28: 1527 (1985).
6. T. M. Freeman, *Cereal Foods World, 34*: 1013 (1989).

7. N. Kim, D. Y. Kwon, H. D. Hong, C. Mok, Y. J. Kim, and Y. J. Nam, *Food Chem., 47*: 407 (1993).
8. S. Ohyama, N. Ishibashi, M. Tamura, H. Nishizaki, and H. Okai, *Agric. Biol. Chem., 52*: 871 (1988).
9. S. Arai, M. Yamashita, H. Kato, and M. Fujimaki, *Agric. Biol. Chem., 34*: 729 (1970).
10. M. Fujimaki, M. Yamashita, Y. Okazawa, and S. Arai, *J. Food Sci., 35*: 215 (1970).
11. S. Visser, K. J. Slangen, and G. Hup, *Neth. Milk Dairy J., 29*: 319 (1975).
12. E. Minagawa, S. Kaminogawa, F. Tsukasaki, and K. Yamauchi, *J. Food Sci., 54*: 1225 (1989).
13. A. B. Cliffe and A. J. Law, *Food Chem., 36*: 73 (1990).
14. N. Ishibashi, Y. Arita, H. Kanehisa, K. Kouge, H. Okai, and S. Fukui, *Agric. Biol. Chem., 51*: 2389 (1987).
15. N. Ishibashi, K. Sadamori, O. Yamamoto, H. Kanehisa, K. Kouge, E. Kikuchi, H. Okai, and S. Fukui, *Agric. Biol. Chem., 5*: 3309 (1987).
16.. N. Ishibashi, T. Kubot, M. Chino, H. Fukui, I. Shinoda, E. Kikuchi, H. Okai, and S. Fukui, *Agric. Biol. Chem., 52*: 95 (1988).
17. N. Ishibashi, I. Ono, K. Kato, T. Shigenaga, I. Shinoda, H. Okai, and S. Fukui, *Agric. Biol. Chem., 52*: 91 (1988).
18. K. H. Ney and G. Retzlaff, in *The Shelf Life of Foods and Beverages* (G. Charalambous, ed.), Elsevier, Amsterdam, 1986, pp. 543–550.
19. K. Otagiri, Y. Nosho, I. Shinoda, H. Fukui, and H. Okai, *Agric. Biol. Chem., 49*: 1019 (1985).
20. G. G. Birch and S. E. Kemp, *Chem. Senses, 14*: 249 (1989).
21. F. Roudot-Algaron, D. Le Bars, J. Einhorn, J. Adda, and J. C. Grippon, *J. Food Sci., 58*: 1005 (1993).
22. M. I. Cambero, I. Seuss, and K. O. Honikel, *J. Food Sci., 57*: 1285 (1992).
23. A. M. Spanier and J. V. Edwards, *J. Liq. Chromatogr., 10*: 2745 (1987).
24. M. Shimizu, S. W. Lee, S. Kaminogama, and K. Yamauchi, *J. Food Sci., 49*: 1117 (1984).
25. M. L. A. Casella and J. R. Whitaker, *J. Food Biochem., 14*: 453 (1990).
26. G. E. Vegarud and T. Langsrud, *J. Dairy Res., 56*: 375 (1989).
27. M. Fujimaki, *Rev. Agroquim. Tecnol. Aliment., 21*: 299 (1981).
28. J. Adler-Nissen and S. Eriksen, in *The Shelf Life of Foods and Beverages* (G. Charalambous, ed.), Elsevier, Amsterdam, 1986, pp. 551–567.
29. F. R. Sharpe, D. Jacques, A. G. Rowsell, and A. L. Whitear, Rapid methods of measuring the foam-active nitrogenous components of worts and beers, *Proceedings of the 18th EBC Congress*, Copenhagen, 1981, pp. 607–614.
30. S. L. Turgeon, S. F. Gauthier, and P. Paquin, *J. Agric. Food Chem., 39*: 673 (1991).
31. M. Shimizu, S. W. Lee, S. Kaminogama, and K. Yamauchi, *J. Food Sci., 51*: 1248 (1986).
32. Z. U. Haque, *J. Dairy Sci., 71*: 96 (1988).
33. Z. U. Haque and Z. Mozaffar, *Food Hydrocolloids, 5*: 559 (1992).
34. Z. U. Haque, *J. Diary Sci., 76*: 311 (1993).
35. A. Mannheim and M. Cheryan, *J. Am. Oil Chem. Soc., 69*: 1163 (1992).
36. B. D. Rebeca, M. T. Peña-Vera, and M. Díaz-Castañeda, *J. Food Sci., 56*: 309 (1991).
37. E. N. C. Mill, M. J. C. Alcocer, and M. R. A. Morgan, *Trends Food Sci. Technol., 3*: 66 (1992).
38. H. Meisel and E. Schlimme, *Trends. Food Sci. Technol., 1*: 41 (1990).
39. J. L. Maubois and J. Lenoil, *Lait, 69*: 245 (1989).
40. P. Antila, I. Paakkari, A. Järvinen, M. J. Mattila, M. Laukkanen, A. Philanto-Leppälä, P. Mäntsälä, and J. Hellman, *Int. Dairy J., 1*: 215 (1991).
41. H. Meisel and H. Frister, *Biol. Chem. Hoppe-Seyler, 369*: 1275 (1988).
42. H. Meisel and H. Frister, *J. Dairy Res., 56*: 343 (1989).
43. A. M. Fiat, S. Levy-Toledano, J. P. Caen, and P. Jollès, *J. Diary Res., 56*: 351 (1989).
44. T. Nakamura, H. Sado, Y. Syukunobe, and T. Hirota, *Milchwiss., 48*: 667 (1993).
45. S. Adachi, Y. Kimura, K. Murakami, R. Matsuno, and H. Yokogoshi, *Agric. Biol. Chem., 55*: 925 (1991).
46. H. Wieser, H. D. Belitz, A. Ashkenazi, and M. Idar, *Z. Lebensm. Unters. Forsch., 176*: 85 (1983).
47. V. R. Harwalkar and J. A. Elliot, *J. Dairy Sci., 54*: 8 (1971).
48. W. A. McGugan, D. B. Edmons, and E. Larmond, *J. Dairy Sci., 62*: 398 (1979).
49. C. N. Kuchroo and P. F. Fox, *Milchwiss, 37*: 331 (1982).
50. A. B. Medina and J. G. Phillips, *J. Agric. Food Chem., 30*: 1250 (1982).
51. S. H. Wu, *J. Chromatogr., 245*: 268 (1982).

52. P. R. Carnegie, M. Z. Ilic, M. O. Etheridge, and M. G. Collins, *J. Chromatogr., 261*: 153 (1983).
53. P. R. Carnegie, M. G. Collins, and M. Z. Ilic, *Meat Sci., 10*: 145 (1984).
54. A. M. Spanier, J. V. Edwards, and A. P. Duphy, *Food Technol., 42*: 112 (1988).
55. A. J. Cliffe, D. Revell, and B. A. Law, *Food Chem., 34*: 147 (1989).
56. C. Collar, J. A. Prieto, A. F. Mascaros, and C. Benedito de Barber, *Rev. Agroq. Tecnol. Aliment., 29*: 266 (1989).
57. L. Lemieux, R. Puchades, and R. E. Smard, *J. Food Sci., 54*: 1234 (1989).
58. D. Gonzalez de Llano, M. Ramos, and M. C. Polo, *Chromatographia, 23*: 764 (1987).
59. M. Yvon, C. Chabanet, and J. P. Pelissier, *Int. J. Peptide Protein Res., 34*: 166 (1989).
60. P. Bican and A. Spahni, *Lebensm.-Wiss. Technol., 24*: 315 (1991).
61. C. J. Dale and T. W. Young, *J. Inst. Brew., 95*: 35 (1989).
62. K. P. Kaiser, H. D. Belitz, and R. J. Fritsch, *Z. Lebensm. Unters. Forsch., 195*: 8 (1992).
63. S. L. Turgeon, S. F. Gauthier, D. Mollé, and J. Léonil, *J. Agric. Food Chem., 40*: 669 (1992).
64. H. Schluter, N. Krivoy, A. Hurster, A. Ingendoh, M. Karas, and W. Zidek, *J. Chromatogr., 625*: 3 (1992).
65. M. I. Acedo, E. Pueyo, and M. C. Polo, *Am. J. Enol. Vitic., 45*: 167 (1994).
66. T. K. Singh, P. F. Fox, P. Hojrup, and A. Healy, *Int. Dairy J., 4*: 111 (1994).
67. D. Gonzalez de Llano, M. C. Polo, and M. Ramos, *J. Dairy Res., 58*: 363 (1991).
68. C. Zevaco and B. Ribadeau-Dumas, *Milchwiss., 39*: 206 (1984).
69. L. Lemieux and J. Amiot, *J. Liq. Chromatogr., 13*: 4023 (1990).
70. J. A. Prieto, Composición cualitativa y cuantitativa de la fracción nitrogenada de harina y masa panaria y sus cambios durante el amasado, la fermentación y la cocción, PhD dissertation, Universidad de Valencia, 1989.
71. T. Saito, T. Yoshida, and T. Itoh, *Int. Dairy J., 3*: 129 (1993).
72. M. I. Acedo and M. C. Polo, Cromatografia de liquidos de alta eficacia de péptidos de bajo peso molecular. Aplicación al estudio del vino, in *Abstracts from Reunion Cientifica del Grupo de Cromatografia y Tecnicas Afines,* San Sebastian, 1991, p. 51.
73. A. Voirin, J. F. Letavernier, and B. Sebille, *J. Chromatogr., 553*: 155 (1991).
74. T. Herraiz, V. Casal, and M. C. Polo, *Z. Lebens. Unters. Forsch. 199*: 265 (1994).
75. T. Herraiz and V. Casal *J. Chromatogr. A780*: 209 (1995).
76. P. Bican and A. Spahni, *Int. Dairy J., 3*: 73 (1993).
77. A. E. Tieleman and J. J. Warthesen, *J. Dairy Sci., 74*: 3686 (1991).
78. S. Bouhallab, D. Mollé, and J. Leonil, *Biotechnol. Lett., 15*: 697 (1993).
79. G. Allen, Sequencing of proteins and peptides, in *Laboratory Techniques in Biochemistry and Molecular Biology* (R. H. Burdon and P. H. Van Knippenberg, eds.), Elsevier, Amsterdam, 1989, pp. 129–132.
80. R. M. Pérez-Castellanos, *Quimica 2000, 41*: 66 (1989).
81. S. Ostrove, *Meth. Enzymol., 182*: 357 (1990).
82. S. Fazakerley and D. R. Best, *Anal. Biochem., 12*: 290 (1965).
83. E. Rothenbuhler, R. Waibel, and J. Solms, *Anal. Biochem., 97*: 367 (1979).
84. M. Salmona, M. Mussini, F. Pocchiari, and G. Beggolin, *J. Chromatogr. 246*: 334 (1982).
85. B. Tippins, *Nature, 334*: 273 (1988).
86. F. J. Al-Shammary, *J. High Resol. Chromatogr., 13*: 309 (1990).
87. H. P. Bennett, *J. Chromatogr., 359*: 383 (1986).
88. H. P. J. Bennett, A. M. Hudson, C. McMartin, and G. E. Purdon, *Biochem J., 168*: 9 (1977).
89. T. Higa and D. M. Desiderio, *Int. J. Peptide Protein Res., 33*: 250 (1989).
90. J. Pivnichny, A. A. Lawrence, and J. D. Strong, *J. Chromatogr. Sci., 25*: 181 (1987).
91. G. J. Schmidt and M. W. Dong, *Am. Lab. (Fairfield) 19*: 62 (1987).
92. M. P. Henry, in *High Performance Liquid Chromatography in Biotechnology* (W. S. Hancock, ed.), John Wiley and Sons, New York, 1990, pp. 205–261.
93. E. Li-Chang, L. Kwan, and S. Nakai, and G. F. Amantea, *J. Food Sci., 57*: 350 (1992).
94. A. M. Pham and S. Nakai, *J. Dairy Sci., 67*: 1390 (1984).
95. R. Lopez-Fandiño, M. I. Acedo, and M. Ramos, *J. Dairy Res., 60*: 117 (1993).
96. F. Addeo, L. Chianese, A. Salzano, R. Sacchi, U. Cappucio, P. Ferranti, and A. Palorni, *J. Dairy Res., 59*: 401 (1992).

97. C. Olieman and J. M. A. Van Riel, *Neth. Milk Dairy J., 43*: 171 (1989).
98. C. T. Mant and R. S. Hodges, *J. Chromatogr., 327*: 147 (1985).
99. M. Dizdaroglu, *J. Chromatogr., 334*: 49 (1985).
100. M. Dizdaroglu, in *High Performance Liquid Chromatography in Biotechnology* (W. S. Hancock, ed.), John Wiley and Sons, New York, 1990, pp. 263–278.
101. M. Dizdaroglu and H. C. Krutzsch, *J. Chromatogr., 264*: 223 (1983).
102. P. J. Cachia, J. van Eyk, W. D. McCubbin, C. M. Kay, and R. S. Hodges, *J. Chromatogr., 343*: 315 (1985).
103. A. D. Radhakrishnan, S. Stein, A. Licht, K. A. Gruber, and S. Udenfriend, *J. Chromatogr., 132*: 552 (1977).
104. L. Lemieux and J. Amiot, *J. Chromatogr., 473*: 189 (1989).
105. L. Lemieux and J. Amiot, *J. Chromatogr., 519*: 299 (1990).
106. J. Mottar, R. Van Renterghem, and J. De Vilder, *Milchwiss., 40*: 717 (1985).
107. H. J. Vreeman, S. Visser, C. J. Slangen, and J. A. M. Van Riel, *Biochem. J., 240*: 87 (1986).
108. C. T. Mant and R. S. Hodges, *J. Liq. Chromatogr., 12*: 139 (1989).
109. C. Olieman and J. W. Van den Bedem, *Neth. Milk Dairy J., 37,*: 27 (1983).
110. D. P. Lee, *J. Chromatogr., 443*: 143 (1988).
111. Y. Kato, S. Nakatani, T. Kitamura, Y. Yamasaki, and T. Kashimoto, *J. Chromatogr., 502*: 416 (1990).
112. V. A. Davankov, A. A. Kurganov, and K. K. Unger, *J. Chromatogr., 500*: 519 (1990).
113. D. Guo, C. T. Mant, A. K. Taneja, and R. S. Hodges, *J. Chromatogr., 359*: 519 (1986).
114. F. Lottspeich and A. Henschen, in *High Performance Liquid Chromatography in Biochemistry* (A. Henschen, K. P. Hype, F. Lottspeich, and W. Voelter, eds.), VCH, Weinheim, 1985, pp. 177–178.
115. D. Guo, C. T. Mant, A. K. Taneja, J. M. R. Parker, and R. S. Hodges, *J. Chromatogr., 359*: 499 (1986).
116. F. E. Regnier, *Meth. Enzymol., 91*: 137 (1983).
117. C. T. Mant and R. S. Hodges, *Chromatographia, 24*: 805 (1987).
118. H. P. J. Bennett, C. A. Browne, and S. Solomon, *J. Liq. Chromatogr., 3*: 1353 (1980).
119. D. R. K. Harding, C. A. Bishop, M. F. Tarttelin, and W. S. Hancock, *Int. J. Peptide Protein Res., 18*: 214 (1981).
120. H. P. J. Bennett, *J. Chromatogr., 266*: 501 (1983).
121. W. S. Hancock and D. R. K. Harding, in *Handbook of HPLC for the Separation of Amino Acids, Peptides and Proteins*, Vol. 1 (W. J. Hancock, ed.), CRC Press, Boca Raton, 1984, pp. 189–192.
122. D. J. Poll and D. R. K. Harding, *J. Chromatogr., 539*: 37 (1991).
123. B. S. Welinder and H. H. Sorensen, *J. Chromatogr., 537*: 181 (1991).
124. J. E. Rivier, *J. Liq. Chromatogr., 1*: 343 (1978).
125. C. T. Mant and R. S. Hodges, in *High Performance Liquid Chromatography of Biological Macromolecules: Methods and Applications* (K. Gooding and F. Reigner, eds.), Marcel Dekker, New York, 1988.
126. M. Guo, C. T. Mant, and R. S. Hodges, *J. Chromatogr., 386*: 205 (1987).
127. C. A. Browne, H. P. J. Bennet, and S. Solomon, *Anal. Biochem., 124*: 201 (1982).
128. M. J. O'Hare and E. C. Nice, *J. Chromatogr., 171*: 209 (1979).
129. B. Grego, F. Lambrou, and M. T. W. Hearn, *J. Chromatogr., 266*: 89 (1983).
130. H. Gaertner and A. Puigsever, *J. Chromatogr., 350*: 279 (1985).
131. M. T. W. Hearn and B. Grego, *J. Chromatogr. 266*: 75 (1983).
132. G. E. Tarr and J. W. Crabb, *Anal. Biochem., 131*: 99 (1983).
133. A. Calderan, P. Ruzza, O. Marin, M. Secchieri, G. Borin, and F. Marchiori, *J. Chromatogr., 548*: 329 (1991).
134. L. Lozzi, M. Rustici, A. Santucci, L. Bracci, S. Petrini, P. Soldani, and P. Neri, *J. Liq. Chromatogr., 11*: 165 (1988).
135. J. L. Meek, *Proc. Natl. Acad. Sci. USA, 77*: 1632 (1980).
136. J. L. Meek and L. Rossetti, *J. Chromatogr., 211*: 15 (1981).
137. M. C. J. Wilce, M. I. Aguilar, and M. T. W. Hearn, *J. Chromatogr., 548*: 105 (1991).
138. R. S. Hodges, J. M. R. Parker, C. T. Mant, and R. R. Sharma, *J. Chromatogr., 458*: 147 (1988).
139. C. T. Mant, T. W. Burke, N. E. Zhou, J. M. R. Parker, and R. S. Hodges, *J. Chromatogr., 485*: 365 (1990).
140. C. T. Mant, T. W. L. Burke, J. A. Black, and R. S. Hodges, *J. Chromatogr. 458*: 193 (1988).
141. C. Chabanet and M. Yvon, *J. Chromatogr., 599*: 211 (1992).

142. R. L. Hill and R. Delaney, *Meth. Enzymol., 11*: 339 (1967).
143. J. A. Smith and R. A. McWilliams, *Am. Lab. (Fairfield), 12*: 25 (1980).
144. D. H. Robertson, in *Analysis of Foods and Beverages* (G. Charalambous, ed.), Academic Press, New York, 1984, p. 118.
145. T. P. Bohan and J. L. Meek, *Neurochem. Res., 3*: 367 (1978).
146. H. Nakamura, C. L. Zimmerman, and J. J. Pisano, *Anal. Biochem., 93*: 423 (1979).
147. M. Dizdaroglu and M. G. Simic, *J. Chromatogr., 195*: 119 (1980).
148. H. G. Barth and B. E. Boyes, *Anal. Chem., 64*: 428R (1992).
149. M. C. Polo, D. Gonzalez de Llano, and M. Ramos, in *Food Analysis by HPLC* (L. M. L. Nollet, ed.), 1992, p. 117.
150. G. Allen, in *Laboratory Techniques in Biochemistry and Molecular Biology: Sequencing of Proteins and Peptides* (R. H. Burdon and P. H. Van Knippengerg, eds.), Elsevier, Amsterdam, 1989, pp. 64, 134.
151. M. A. Vijayalakshmi, L. Lemieux, and J. Amiot, *J. Liq. Chromatogr., 9*: 3559 (1986).
152. W. O. Richter, B. Jacob, and P. Schwandt, *Anal. Biochem., 133*: 288 (1983).
153. C. T. Mant, J. M. R. Parker, and R. S. Hodges, *J. Chromatogr., 397*: 99 (1987).
154. H. Yoshida and S. Najo, *Anal. Biochem., 159*: 273 (1986).
155. S. A. Margolis and M. Dizdaroglu, *J. Chromatogr., 322*: 117 (1985).
156. S. A. Margolis and P. L. Konash, *Anal. Biochem., 134*: 163 (1983).
157. C. T. Mant and R. S. Hodges, *J. Chromatogr., 326*: 349 (1985).
158. J. M. Piot, D. Guillochon, Q. Zhao, G. Ricart, B. Fournet, and D. Thomas, *J. Chromatogr. 481*: 221 (1989).
159. N. Takahashi, N. Ishioka, Y. Takahashi, and F. W. Putnam, *J. Chromatogr., 326*: 407 (1985).
160. N. Takahashi, Y. Takahashi, N. Ishioka, B. Blumberg, and F. W. Putnam, *J. Chromatogr. 359*: 181 (1986).
161. S. Hara and A. Dobashi, *J. Liq. Chromatogr., 2*: 883 (1979).
162. A. Dobashi, Y. Dobashi, and S. Hara, *J. Liq. Chromatogr., 9*: 243 (1986).
163. W. H. Pirkle, D. M. Alessi, M. H. Hyun, and T. C. Pochapsky, *J. Chromatogr. 398*: 203 (1987).
164. M. H. Hyun, I. K. Baik, and W. H. Pirkle, *J. Liq. Chromatogr., 11*: 1249 (1988).
165. I. Florance and Z. Konteatis, *J. Chromatogr., 543*: 299 (1991).
166. S. Li and W. C. Purdy, *J. Chromatogr., 543*: 105 (1991).
167. C. A. Chang, H. Ji, and G. Lin., *J. Chromatogr., 522*: 143–152 (1990).
168. S. Friebe, B. Hartrodt, K. Neubert, and G. J. Krauss, *J. Chromatogr. A 661*: 7 (1994).
169. K. Schimida and K. Hirakata, *J. Liq. Chromatogr., 15*: 1763 (1992).
170. C. Schoneisch, S. K. Kwok, G. S. Wilson, S. R. Rabel, J. F. Stobaugh, T. D. Williams, and D. G. Van der Velde, *Anal. Chem., 65*: 67R (1993).
171. J. Zukowski, J. Pawlowska, M. Nagatkina, and D. W. Armstrong, *J. Chromatogr., 629*: 169–179 (1993).
172. P. Jadaud and I. W. Wainer, *J. Chromatogr., 476*: 165 (1989).
173. W. R. Cahill Jr., E. P. Kroeff, and D. J. Pietrzyk, *J. Liq. Chromatogr., 3*: 1319 (1980).
174. W. Linder, J. N. Le Page, G. Davies, D. E. Seitz, and B. L. Karger, *J. Chromatogr., 185*: 323 (1979).
175. B. J. Davis, *Ann. NY Acad. Sci., 121*: 404 (1964).
176. L. Ornstein, *Ann. NY Acad. Sci., 121*: 321 (1964).
177. B. D. Hames and D. Rickwood, *Gel Electrophoresis of Proteins: A Practical Approach*, 2nd ed., IRL Press, Oxford, 1990.
178. A. T. Andrews, *Electrophoresis: Theory, Techniques and Biochemical Clinical Applications*, 2nd ed., Oxford University Press, New York, 1986.
179. B. D. Hames, in *Gel Electrophoresis of Proteins: A Practical Approach*, 2nd ed. (B. D. Hames and D. Rickwood, eds.), IRL Press, Oxford, 1990, pp. 103–108.
180. M. H. P. West, R. S. Wu, and W. M. Bonner, *Electrophoresis, 5*: 133 (1984).
181. C. A. Dadd, R. G. Cook, and C. D. Allis, *Biotechniques, 14*: 266 (1993).
182. U. K. Laemmli, *Nature, 227*: 680 (1970).
183. R. T. Swank and K. D. Munkres, *Anal. Biochem., 39*: 462 (1971).
184. F. Hashimoto, T. Horigome, M. Kanbayashi, K. Yoshida, and H. Sugano, *Anal. Biochem., 129*: 192 (1983).

185. B. L. Anderson, R. W. Berry, and A. Telser, *Anal. Biochem., 132*: 365 (1983).
186. S. P. Fling and D. S. Gregerson, *Anal. Biochem., 155*: 83 (1986).
187. H. Schägger and G. Von Jagow, *Anal. Biochem., 166*: 368 (1987).
188. T. Okajima, T. Tanabe, and T. Yasuda, *Anal. Biochem., 211*: 293 (1993).
189. P. G. Righetti, E. Gianazza, C. Gelfi, and M. Chiari, in *Gel Electrophoresis of Proteins: A Practical Approach*, 2nd ed. (B. D. Hames and D. Rickwood, eds.), IRL Press, Oxford, 1990, p. 152.
190. D. B. De Wald, L. D. Adams, and J. D. Pearson, *Anal. Biochem., 154*: 502 (1986).
191. M. H. P. West, R. S. Wu, and W. M. Bonner, *Electrophoresis, 5*: 133 (1984).
192. F. Lottspeich, *Chromatographia, 28*: 89 (1989).
193. J. C. Colburn, in *Capillary Electrophoresis: Theory and Practice* (P. D. Grossman and J. C. Colburn, eds.), Academic Press, San Diego, 1992, pp. 237–271.
194. B. L. Karger, A. S. Cohen, and A. Guttman, *J. Chromatogr., 492*: 585 (1989).
195. P. D. Grossman and J. C. Colburn, *Capillary Electrophoresis: Theory and Practice*, Academic Press, San Diego, 1992, pp. 237–271.
196. W. G. Kuhr and C. A. Moning, *Anal. Chem., 64*: 389R (1992).
197. H. E. Schwartz, R. H. Palmieri, J. A. Nolan, and R. Brown, *Introduction to Capillary Electrophoresis of Proteins and Peptides*, Beckman Instruments, Inc., Fullerton, CA, 1993.
198. I. S. Krull and J. R. Mazzeo, *Nature, 357*: 92 (1992).
199. A. S. Cohen and B. L. Karger, *J. Chromatogr., 397*: 409 (1987).
200. J. Liu, K. A. Cobb, and M. Novotny, *J. Chromatogr., 519*: 189, (1990).
201. V. Kasicka and Z. Prusik, *J. Chromatogr., 569*: 123 (1991).
202. H. B. Patel, J. F. Stobaugh, and C. M. Riley, *J. Chromatogr., 536*: 357 (1991).
203. B. Grego, E. C. Nice, and R. J. Simpson, *J. Chromatogr., 352*: 359 (1986).
204. F. Nyberg, C. Pernow, U. Moberg, and R. B Eriksson, *J. Chromatogr., 359*: 541 (1986).
205. L. Leadbeater and F. Bruce-Ward, *J. Chromatogr., 397*: 435 (1987).
206. E. Mendez, R. Matas, and F. Soriano, *J. Chromatogr., 323*: 373 (1985).
207. S. A. Cohen and D. J. Strydom, *J. Chromatogr., 336*: 93 (1988).
208. F. J. Colilla, S. P. Yadav, K. Brew, and E. Mendez, *J. Chromatogr., 548*: 303 (1991).
209. B. A. Bidlingmeyer, S. A. Cohen, T. L. Tarvin, and B. Frost, *J. Assoc. Off. Anal. Chem., 70*: 241 (1987).
210. R. G. Elkin and A. M. Wasynczuk, *Cereal Chem., 64*: 226 (1987).
211. N. M. Griffiths and M. J. Billington, *J. Assoc. Publ. Analyst, 21*: 89 (1983).
212. I. B. Agater, K. J. Briant, J. W. Llewellyn, R. Sawyer, F. J. Bailey, and C. H. S. Hitchcock, *J. Sci. Food Agric., 37*: 317 (1986).
213. T. D. Schlabach and C. T. Wehr, in *HPLC of Proteins and Peptides* (T. W. Hearn, F. E. Regnier, and C. T. Wehr, eds.), Academic Press, London, 1983, p. 221.
214. S. Udenfriend, S. Stein, O. P. Bohlen, W. Dairman, W. Leimgruber, and M. Weigele, *Science, 178*: 871 (1972).
215. R. V. Lewis, in *Handbook of HPLC for the Separation of Amino Acids, Peptides and Proteins*, Vol. 1 (W. S. Hancock, ed.), CRC Press, Boca Raton, 1984, p. 193.
216. S. Stein and S. Udenfriend, *Anal. Biochem., 136*: 7 (1984).
217. V. K. Boppana, C. Miller-Stein, J. F. Politowski, and G. R. Rhodes, *J. Chromatogr., 548*: 319 (1991).
218. M. Roth, *Anal. Chem., 43*: 880–882 (1971).
219. S. S. Simons and D. F. Johnson, *J. Anal. Chem. Soc., 98*: 7098 (1976).
220. H. Frister, H. Meisel, and E. Schlimme, *Fresenius. A. Anal. Chem., 330*: 631 (1988).
221. T. M. Joys and H. Kim, *Anal. Biochem., 94*: 371 (1979).
222. P. De Montigny, J. F. Stobaugh, R. S. Givens, R. G. Carlson, K. Srinivasachar, L. A. Sternson, and T. Higuchi, *Anal. Chem., 59*: 1096 (1987).
223. H. Koning, H. Wolf, K. Venema, and J. Korf, *J. Chromatogr. 533*: 171 (1990).
224. M. Ohno, M. Kai, and Y. Ohkura, *J. Chromatogr., 392*: 309 (1987).
225. M. Barber, R. S. Bordoli, R. D. Sedgwick, A. N. Tyler, and E. T. Whalley, *Biomed. Mass Spectrom., 8*: 337 (1981).
226. M. Barber, R. S. Bordoli, G. J. Elliot, R. D. Sedgwick, A. N. Tyler, and B. N. Green, *J. Chem. Soc. Chem. Commun., 16*: 936 (1982).
227. R. M. Caprioli, *Anal. Chem., 62*: 477A (1990).

228. M. J. F. Suter, B. B. DaGue, W. T. Moore, S. N. Lin, and R. M. Caprioli, *J. Chromatogr., 553*: 101 (1991).
229. D. B. Kassel, B. D. Musselman, and J. A. Smith, *Anal. Chem., 63*: 1091 (1991).
230. P. Petrilli, P. Pucci, J. P. Pelissier, and F. Addeo, *Int. J. Peptide Protein Res., 29*: 504 (1987).
231. J. Van der Greef, A. C. Tas, L. M. Nijssen, J. Jetten, and M. Hoehn, *J. Chromatogr., 394*: 77 (1987).
232. U. A. Th. Brinkman, R. W. Frei, and H. Lingeman, *J. Chromatogr., 492*: 251 (1989).
233. R. Schuster, *J. Chromatogr., 431*: 271 (1988).
234. I. Tapuhi, D. E. Schmidt, W. Lindner, and B. L. Karger, *Anal. Biochem., 115*: 123 (1981).
235. B. Grego and M. T. W. Hearn, *J. Chromatogr., 255*: 67 (1983).
236. P. Martin, C. Polo, M. D. Cabezudo, and M. V. Dabrio, *J. Liq. Chromatogr., 7*: 539 (1984).
237. J. Y. Chang, *Biochem. J., 199*: 557 (1981).
238. R. Newcomb, *LC-GC, 10*: 34 (1992).
239. P. Leroy, A. Nicolas, M. Wellmann, F. Michelet, T. Oster, and G. Siest, *Chromatographia, 36*: 130 (1993).
240. T.D. Schlabach and C. T. Wehr, *Anal. Biochem., 127*: 222 (1983).
241. M. K. Radjai and R. T. Hatch, *J. Chromatogr., 196*: 319 (1980).
242. L. A. Allison, G. S. Mayer, and R. E. Shoup, *Anal. Chem., 56*: 1089 (1984).
243. K. Bratin and P. T. Kissinger, *J. Liq. Chromatogr., 5*: 881 (1982).
244. S. Kobayashi, J. Sekino, K. Honda, and K. Imai, *Anal. Biochem., 112*: 99 (1983).
245. M. C. Roach and M. D. Harmony, *Anal. Chem., 59*: 411 (1987).
246. J. Liu, K. A. Cobb, and M. Novotny, *J. Chromatogr., 519*: 189 (1990).
247. A. L. Burlingame, T. A. Baille, and D. H. Russell, *Anal. Chem., 64*: 467R (1992).
248. J. A. McCloskey, *Methods in Enzymology*, Vol. 193, Academic Press, San Diego, 1990.
249. K. B. Tomer and C. E. Parker, *J. Chromatogr., 492*: 189 (1989).
250. D. E. Games and E. D. Ramsey, *J. Chromatogr., 323*: 67 (1985).
251. T. R. Covey and J. D. Henion, *Anal. Chem., 55*: 2275 (1983).
252. M. H. Hail, S. Lewis, I. Jardine, J. Liu, and M. Novotny, *J. Microcol. Sep., 2*: 285 (1990).
253. R. M. Caprioli, T. Fan, and J. S. Cottrell, *Anal. Chem., 59*: 2949 (1986).
254. M. E. Hemling, G. D. Roberts, W. Johnson, S. A. Carr, and T. R. Covey, *Biomed. Environ. Mass Spectrom., 19*: 677 (1990).
255. D. B. Kassel, K. P. Williams, B. D. Musselman, and J. A. Smith, *Anal. Chem., 63*: 1978 (1991).
256. K. Mock, *Peptide Res., 6*: 100 (1983).
257. E. C. Huang and J. D. Henion, *J. Am. Chem. Soc. Mass Spect., 1*: 158 (1990).
258. E. D. Lee, J. D. Henion, and T. R. Covey, *J. Microcol. Sep., 1*: 14 (1989).
259. A. Cappiello, P. Palma, I. A. Papayannopoulos, and K. Biemann, *Chromatographia, 30*: 477 (1990).
260. R. D. Smith, J. A. Loo, R. R. Ogorzalekloo, M. Bushman, and H. R. Udseth, *Mass Spectrom. Rev., 10*: 359 (1991).
261. D. W. Hutchinson, A. R. Woolfitt, and A. E. Ashcroft, *Org. Mass Spectrom., 22*: 304 (1987).
262. P. J. Arpino, in *Mass Spectrometry in the Biological Science: A Tutorial* (M. L. Gross, ed.), Kluwer Academic, Dordrecht, 1992, p. 253.
263. M. A. Moseley, L. J. Deterding, K. B. Tomer, and J. W. Jorgenson, *J. Chromatogr., 516*: 167 (1990).
264. P. Thibault, C. Paris, and S. Pleasance, *Rapid Commun. Mass Spectrom., 5*: 484 (1991).
265. D. M. Desiderio, in *Handbook of HPLC for the Separation of Amino Acids, Peptides and Proteins*, Vol. 1 (W. S. Hancock, ed.), CRC Press, Boca Raton, 1984, p. 197.
266. D. J. Bell, M. D. Brightwell, W. A. Neville, and A. West, *Rapid Commun. Mass Spectrom., 4*: 88 (1990).
267. K. Weber and M. J. Osborn, *J. Biol. Chem., 244*: 4406 (1969).
268. S. K. Shate, G. G. Lilley, A. C. Mason, and C. M. Weaver, *Cereal Chem., 64*: 380 (1987).
269. W. Diezel, G. Kopperschläger, and E. Hoffmann, *Anal. Biochem., 48*: 617 (1972).
270. R. W. Blaskeley and J. A. Boezi, *Anal. Biochem., 82*: 580 (1977).
271. C. R. Merril, D. Golddman, S. A. Sedman, and M. H. Ebert, *Science, 211*: 1437 (1981).
272. J. Heukeshoven and R. Dernick, *Electrophoresis, 9*: 28 (1988).
273. H. Rehbein, *Electrophoresis, 13*: 805 (1992).
274. A. T. Andrews, in *Gel Electrophoresis of Proteins: A Practical Approach*, 2nd ed. (B. D. Hames and D. Rickwood, eds.), IRL Press, Oxford, 1990, pp. 301–319.

275. C. A. Dadd, R. G. Cook, and C. D. Allis, *Biotechniques, 14*: 66 (1993).
276. B. Hogan and E. S. Yeung, *J. Chromatogr. Sci., 28*: 15 (1990).
277. L. Gross and E. S. Yeung, *Anal. Chem., 62*: 427 (1990).
278. R. D. Smith, J. A. Loo, C. J. Barinaga, C. J. Edmonds, and H. R. Udseth, *J. Chromatogr., 480*: 211 (1989).
279. M. A. Moseley, L. J. Deterding, K. B. Tomer, and J. W. Jogerson, *Anal. Chem., 63*: 109 (1991).
280. K. Hermam, R. E. Lang, Th. Unger, C. Bayern, and D. Ganten, *J. Chromatogr., 312*: 273 (1984).
281. M. Dubois, K. A. Gilles, J. K. Haulton, P. A. Rebers, and F. Smith, *Anal. Chem., 28*: 350 (1956).
282. E. Bumberger and H. D. Berlitz, *Z. Lebensm. Unters. Forsch., 197*: 14 (1993).
283. S. L. Turgeon, S. F. Gauthier, D. Mollé, and J. Léonil, *J. Agric. Food Chem., 40*: 669 (1992).
284. A. P. Lappälä, P. Antila, P. Mäntsälä, and J. Hellman, *Int. Dairy J., 4*: 291 (1994).
285. R. Lopez-Fandiño, A. Olano, C. San-Jose, and M. Ramos, *J. Dairy Res., 60*: 111 (1993).
286. J. Voigt, B. Biehl, H. Heinrichs, S. Kamaruddin, G. Gaim Marsoner, and A. Hugi, *Food Chem., 49*: 173 (1994).
287. R. Lopez-Fandiño, A. Olano, N. Corzo, and M. Ramos, *J. Dairy Res., 60*: 339 (1993).
288. B. L. Jones and G. L. Lookhart, *Cereal Chem., 62*: 89 (1985).
289. E. D. Strange, E. L. Malin, D. L. Van Hekken, and J. Bash, *J. Chromatogr., 624*: 81 (1992).
290. M. Dalgalarrondo, J. M. Chobert, E. Dufour, C. Bertrard-harb, J. P. Dumont, and T. Haertle, *Milchwiss., 45*: 212 (1990).
291. G. Erhardt, J. Godavac-Zimmermann, and A. Conti, *Biol. Chem. Hoppe Seyler, 370*: 757 (1989).
292. M. H. Tobler, B. Windemann, and B. Baumgartner, *Mitt 'Gebiete Lebensm. Hyg., 74*: 132 (1983).
293. K. P. Kaiser and I. Krause, *Z. Lebensm. Unters. Forsch., 180*: 181 (1985).
294. E. H. Shokraii, A. Esen, and R. W. Mozingo, *J. Agric. Food Chem., 33*: 1114 (1985).
295. C. Stanton and N. Light, *Meat Sci., 27*: 141 (1990).
296. D. González de Llano, M. Ramos, A. Rodriguez, A. Montilla, and M. Juarez, *Int. Dairy J., 2*: 121 (1992).
297. M. M. Calvo, L. Amigo, A. Olano, P. J. Martin, and M. Ramos, *Food Chem., 32*: 99 (1989).
298. P. L. H. McSweeney, P. F. Fox, J. A. Lucey, K. N. Jordan, and T. M. Cogan, *Int. Dairy J., 3*: 613 (1993).
299. T. K. Sing, P. F. Fox, P. Hojrup, and A. Healy, *Int. Dairy J., 4*: 111 (1994).
300. Y. A. Kim and W. E. Barbeau, *J. Food Sci., 56*: 1082 (1991).
301. B. Hu and A. Esen, *J. Agric. Food Chem., 30*: 21 (1982).
302. D. W. Cleveland and S. G. Fisch, *J. Biol. Chem., 252*: 1102 (1977).
303. I. Martinez, R. Ofstad, and R. L. Olsen, *FEBS Lett., 265*: 23 (1990).
304. G. C. Hwang, S. W. Watabe, and K. Hashimoto, *J. Comp. Physiol. B., 160*: 233 (1990).
305. K. A. Cobb and M. Novotny, *Anal. Chem., 61*: 2226 (1989).
306. P. D. Grossman, J. C. Colburn, H. H. Lauer, R. G. Nielsen, R. M. Riggin, G. S. Sittampalam, and E. C. Rickard, *Anal. Chem., 61*: 1186 (1989).
307. M. Zu, R. Rodriguez, D. Hansen, and T. Wehr, *J. Chromatogr., 516*: 123 (1990).
308. T. Tsuda, M. Yamada, and Y. Nakazama, *Milchwiss., 48*: 74 (1993).

9

Proteins

John Van Camp and A. Huyghebaert
University of Ghent, Ghent, Belgium

I. INTRODUCTION

A large variety of food proteins, either from animal (milk, meat, eggs, blood, fish) or plant (seeds, cereals) origin, are nowadays available in the food industry. To illustrate the discussion on the determination of proteins in foods and food products, an important representative from each product class (milk proteins for animal species and cereal proteins for plants) was chosen. In Tables 1 and 2 an overview is given of the molecular structure and basic properties of the major protein fractions present in both substrates. Specific attention is drawn to those properties which might be of importance for their determination in foods and food products. Milk proteins are subdivided into random coiled caseins, which can be precipitated by acidification of raw skim milk to pH 4.6 at 20°C, and into more globular whey proteins, which remain in the serum after precipitation of the caseins. Cereal proteins are primarily composed of prolamins (e.g., gliadins for wheat, zeins for maize, hordeins for barley, and avenins for oats) and glutelins (e.g., glutenins for wheat), which can be isolated from cereal grains and flours by the classical Osborne fractionation procedure (1,2). More detailed information on both types of food proteins can be found in the reviews of Whitney (3) and Schofield et al. (4).

Apart from their nutritional value, food proteins serve in foods and in food products as important functional ingredients. Functional properties of proteins are defined as those physical and chemical properties which affect the behavior of proteins in food systems during processing, storage, preparation, and consumption (5). Important classes include surface-active properties (emulsification, foaming), rheological properties (viscosity, gelation), and structural properties (e.g., extrudability, dough formation). More detailed information on the structure–function relationship of food proteins was given by Kinsella (5) and Cheftel et al. (7). Specific information on the subject with regard to milk proteins has been published by Kinsella (8), and for cereal proteins by Kinsella (5) and Schofield et al. (4).

As illustrated by the examples given in Table 3, the determination of proteins in foods and food products has an important nutritional, functional, and technological significance. A regulation and

Table 1 Molecular Structure and Properties of the Major Milk Proteins

Food protein	Protein component	% of total protein	Molecular weight	Isoelectric point	Molecular structure
Caseins	•α_{s1}-Caseins	32–40	22,068–23,724	4.44–4.76	•Primarily random coiled •Rich in Pro and hydrophobic amino acids •Strong tendency for association by hydrophobic interactions
	•α_{s2}-Caseins	8–10	25,230	4.20–4.60	•Primarily random coiled •Most hydrophilic of major caseins
	•β-Caseins	24–29	23,944–24,092	4.83–5.07	•Either highly asymmetrical or random coiled •Most hydrophobic of major caseins •Strong tendency for association by hydrophobic interactions
	•κ-Caseins	5–10	19,007–19,039	5.30–5.80	•Random coiled, containing distinct hydrophobic (para-κ-casein) and hydrophilic (macropeptide) regions •Tendency for association by disulfide bonds and hydrophobic interactions
Whey proteins	•β-Lactoglobulin (β-Lg)	8–17	18,205–18,363	5.35–5.49	•Globular protein, containing one-cysteine and two-cysteine residues •Formation of dimers in pH range 5.2–7.5; octamerization may occur at pH 3.5–5.2
	•α-Lactalbumin (α-La)	2–7	14,147–14,175	4.2–4.5	•Compact globular protein, containing four cystine residues •High stability between pH 5.4–9.0
	•Bovine serum albumin (BSA)	2	66,267	4.7–4.9	•Rod-shaped protein, containing one cysteine and 17 cystine residues •Partial unfolding at low (<4) and high (>8) pH values
	•Immunoglobulins (Ig)	2–4	150,000–1.10^6	5.5–8.3	•Monomers or polymers of four-chain molecule containing two light (MW 20,000) and two heavy (MW 50,000–70,000) chains linked by disulfide bonds

Source: Data from Refs. 2 and 152.

Table 2 Molecular Structure and Properties of the Major Wheat Proteins

Protein component(s)	% of total protein	General characteristics
Albumins and globulins	12–20	•Extractable by water or dilute salt solutions (e.g., 0.1–0.5 M NaCl) •Contains enzymes (amylases, proteases, lipoxygenases) •Rich in Lys, Asp, Ala, Cys, Leu, and Arg; low in Met content
Gliadins	35	•Soluble in aqueous solutions of alcohols (e.g., 70% ethanol or 55% isopropanol) •Primarily globular proteins (MW 30,000–40,000) stabilized by intramolecular disulfide bonds •Rich in Glu, Pro, and hydrophobic amino acids •Low content of basic amino acids; Glu and Asp primarily in amide form; low net charge
Soluble glutenins	15–35	•Extractable in dilute acid or alkali (e.g., 0.05 N acetic acid) •High MW (100,000 →3,000,000) complexes containing low MW (40,000 →100,000) subunits linked by disulfide bonds; extra stabilization by hydrogen and hydrophobic interactions •Rich in Glu, Pro, and hydrophobic amino acids •Low content of basic amino acids; Glu and Asp primarily in amide form; low net charge
Insoluble glutenins ("residue protein")	10–40	•Not extractable by dilute acid or alkali •Primarily composed of high MW aggregated glutenin molecules containing several intermolecular disulfide bonds

Source: Data from Refs. 4–6.

standardization of the protein content in food matrices can be obtained by the use of official and statutory methods, which have been approved by international organizations. For milk and cereal proteins, the official protein determination methods currently available include Kjeldahl, colorimetric (Biuret, dye binding), and the more recently developed infrared techniques (Table 4).

II. SAMPLE PREPARATION

For most protein determination methods, proteins have to be solubilized in order to be determined. An exception to this rule is given by the Kjeldahl method and by the recently developed near-infrared (NIR) reflection spectrophotometry, which allow the analysis of unsoluble and eventually almost intact food samples. Prior to protein quantification, cereal grains are usually grinded to a meal having a similar particle size distribution to flour. Since a loss of moisture can occur during this process, a simultaneous moisture determination should be performed for accurate protein determination to occur, IR spectrophotometry techniques are advantageous to Kjeldahl as they allow the determination of moisture and protein in a single detection run (9). In addition, for various dye binding procedures developed to quantify the protein content in cereal grains, protein compounds can interact with the dye without being fully solubilized (9,39).

A suitable sample pretreatment is required when proteins in unsoluble samples are to be quantified using the Biuret method. Samples containing unsoluble proteins (e.g., collagen or creatine) can be oxidized in the presence of an excess hydrogen peroxide, followed by drying and solubilization in an alkalinized sodium dodecyl sulfate (SDS) solution at elevated temperatures prior to treatment with the Biuret reagent (40,41). Alternatively, in some applications (e.g., for wheat protein determinations) the alkaline copper solution is used not only as a reagent but as an extractant for protein (40,42). The interference of food components like starch and lipids, which

Table 3 Motivations for Determination of Milk and Cereal Proteins in Foods and Food Products: Examples

Motivation for protein determination	Applications in food analysis	Ref.
Formulation of food and feed products	•Production of balanced food diets with standardized protein content	9, 10
Control of overall performance of food products	•Evaluation of raw milk for UHT processing by determination of protein breakdown	11, 12
	•Heat classification of skim milk powders	11, 13, 14
	•Determination of casein content in milk for prediction of cheese yield	15
	•Evaluation of beer brewing process by determination of the protein content of barley	9, 16
	•Deduction of quality of baked goods via protein content of cereal grains and flours	9, 17, 18, 19
Improvement of yield and quality of foods	•Use of genetic variants of milk proteins to evaluate milk yield and composition	20, 21
	•Cultivar identification and discrimination of cereal proteins	22, 23, 24, 25
Tracing of adulterations in food products	•Detection of rennet whey and buttermilk solids in milk powders	11, 26, 27
	•Detection of cow's milk in sheep or goat's milk used for cheese production	28, 29, 30
	•Detection of milk proteins in meat products	31, 32
	•Identification of milk-clotting enzymes in cheese making	33
	•Detection of soft wheat in flour sold for hard wheat meal	34
	•Detection of maize or rice as alternative source of starch in malted barley	34
Optimization of protein fractionation and determination methods	•Electrophoretic characterization of prolamins fractionated by HPLC	17, 18, 35, 36
	•Electrophoretic evaluation of the fractionation of whey proteins by FPLC	37, 38

are extractable together with the proteins in the alkaline Biuret solution, can be significantly reduced by pretreatment of the samples with organic solvents, which retard solubilization of the compounds in the Biuret reaction mixture (43). Turbidity of whole milk is reduced by complete solubilization of casein micelles and fat globules in the presence of n-butylamine (44–46).

The influence of various compounds interfering during the Lowry procedure (cf. Sec. III) can be successfully decreased by the usage of a suitable sample pretreatment. A frequently employed technique based on the specific isolation of the protein fraction prior to analysis is the treatment of the sample with trichloroacetic acid (TCA), followed by resolubilization of the precipitate and quantification of the protein in solution with the Lowry reagent (47). The quantitative recovery of protein from the TCA precipitate at low protein concentrations can be successfully achieved by application of a suitable detergent (e.g., sodium deoxycholate) in addition to TCA during precipitation (48).

For the chromatographic and electrophoretic characterization of cereal proteins present in one of the extracts obtained during the Osborne fractionation process (cf. Sec. I), it can be sufficient to clarify the solution by centrifugation and/or filtration prior to analysis (49). The unsoluble glutelin fraction might be solubilized by the addition of reagents for reduction (e.g., 10 mM β-mercaptoethanol, β-ME, or 0.1–1% dithiothreitol, DTT; 25,50,51), denaturation (2–8 M urea; 16,51), and alkylation (4-vinylpyridine; 24,50,52). Alternatively, an almost complete protein

Table 4 Standards for Determination of Milk and Cereal Proteins in Foods and Food Products (Editions 1994)

Organization	Standard code	Last update	Standard title
International Organization for Standardization (ISO)	1871	1975	Agricultural food products; general directives for the determination of nitrogen by the Kjeldahl method
	5542	1984	Milk; determination of protein content; amido black dye binding method (routine method)
	5549	1978	Caseins and caseinates; determination of protein content (reference method)
International Dairy Federation (IDF)	20A	1986	Milk; total nitrogen content (Kjeldahl method)
	25	1964	Determination of the protein content of processed cheese products
	29	1964	Determination of the casein content of milk
	92	1986	Caseins and caseinates; determination of protein content (reference method)
	98A	1985	Milk; protein content (amido black dye binding method)
	141A	1990	Whole milk; milk fat, protein, and lactose content; guide for the operation of mid-infrared instruments
Eurofood monitor (EEC)	503	1985	Acid caseins, rennet caseins, caseinates; determination of protein content by the Kjeldahl method
American Association of Cereal Chemists (AACC)	38-10	1982	Gluten; hand washing method
	38-11	1961	Gluten; machine washing method
	39-10	1989	Infrared reflectance method for protein and oil determination
	46-08	1982	Crude protein; automated Kjel-Foss method
	46-09	1982	Crude protein; automated colorimetric method
	46-10	1985	Crude protein; improved Kjeldahl method
	46-11A	1985	Crude protein; improved Kjeldahl method, copper catalyst modification
	46-12	1986	Crude protein; Kjeldahl method, boric acid modification
	46-13	1986	Crude protein; micro Kjeldahl method
	46-14A	1985	Crude protein; Udy dye method (acid orange 12)
	46-15	1986	Crude protein; 5 min Biuret method for wheat and other grains
	46-18	1982	Table: crude protein, calculated from percentage of total nitrogen, in feeds and feedstuffs
	46-19	1982	Table: crude protein, calculated from percentage of total nitrogen, in wheat and flour
	46-20	1982	Undenatured whey protein nitrogen
	46-21	1982	Whey protein, Harland and Ashworth method
	46-22	1982	Whey protein, Rowland method

extraction from wheat flour without chemical reduction of disulfide bounds could be obtained by applying mechanical shear with an ultrasonic probe in a 2% SDS buffer at pH 6.9 (11,53). Milk proteins are not included in a solid matrix and normally do not demand a rigorous sample preparation procedure prior to analysis. Acid-precipitated caseins can be resolubilized in distilled water at neutral pH (54). Whey proteins can be analyzed without modification (13,38), although if necessary complete denaturation and masking of charges can be achieved by boiling the solution for 3–5 min in the presence of 2–5% SDS and 0.1 M β-ME or 20 mM DTT (55).

When proteins are immunochemically detected in foods and food products, care should be taken not to induce excessive modification within the antigenic structure of the protein during the extraction procedure. A specific discussion regarding the influence of the extraction procedure during the immunochemical detection of milk proteins in meat products was given by Sinell and Mentz (56). More general information can be found in the review of Daussant et al. (34).

III. DETERMINATION OF PROTEINS IN FOODS

A. Historical Evolution

In the Dumas procedure, originally developed in 1831, the nitrogen-containing components of the sample are converted to oxides at elevated temperatures and subsequently reduced with copper to gaseous nitrogen, which is quantified and used as a measure for the protein content of the sample (40). Since only small sample sizes (5–50 mg) in the dry state can be used, significant errors may occur when heterogeneous materials are analyzed (40,57). The Kjeldahl method was introduced in 1883 and has gained much more importance in the field of food analysis in comparison to the Dumas procedure (58).

Colorimetric protein determination methods became available in the beginning of the century and have since been further developed for the analysis of proteins in foods and food products. The Biuret procedure was first applied to wheat and flour in 1949 by Pinckney (9). A dye binding procedure using orange G was proposed by Udy in 1956 (9), and the Folin–Ciocalteau method originally developed in 1927 was further modified and optimized by Lowry et al. in 1951 (59). The use of NIR reflectance spectrophotometry for the quantification of protein in cereals was first reported by Norris, who originally invented a moisture meter based on the NIR principle in 1964 (9).

Column chromatography became a powerful tool in food analysis since the high-pressure liquid chromatography (HPLC) technique was introduced in the early 1970s (60). As an alternative to chromatography, the free or moving boundary electrophoresis reported in 1930 by Tiselius developed into the electrophoresis in porous gel media, which are nowadays frequently applied in food protein analysis (61). The principle of immunochemical analysis, which involves the interaction between a soluble antigen and a soluble antibody resulting in an insoluble precipitate, was reported by Kraus in 1897. Shortly afterward immunoprecipitation in gel media became known, which forms the basis for several immunoassays currently used in food protein analysis (62).

B. Classical and Recently Developed Techniques

1. Quantification of Total Protein

The quantification of total protein in food products can be performed directly, or by determination of the total organic nitrogen followed by conversion of total nitrogen to crude protein content using a suitable conversion factor. An overview of various techniques available in both classes of protein determination methods is presented in Table 5. In spite of the large number of techniques presented,

Table 5 Methods for Total Protein Determination in Foods and Food Products

Method classification	Determination method	Ref.
Total nitrogen determination, followed by conversion to crude protein content	•Kjeldahl	9, 40, 57, 63, 64, 65
	•Dumas/Pregl-Dumas	9, 40, 65
	•Neutron activation	9, 57
	•Proton activation	9, 57, 66
Direct protein determination	•Formol titration	65
	•Direct distillation, Kofranyi	9, 63, 65
	•UV spectrophotometry	
	UV absorption	40, 57, 65, 67
	UV fluorescence	57, 65
	•Visible region spectrophotometry (colorimetry)	
	Biuret	44, 45, 46
	Folin–Ciocalteau/Lowry	47, 59
	Bicinchroninic acid	67
	Dye binding	9, 40, 57, 65
	Silver staining	67
	•Infrared spectrophotometry	
	IR absorption	65
	NIR reflectance	9
	NIR transmittance	9
	•Pulsed nuclear magnetic resonance (NMR) spectrophotometry	68, 69
	•X-ray photoelectron spectrophotometry	57
	•Turbidimetry/nephelometry	40, 65
	•Refractometry	40
	•Polarography	40
	•Radioactivity (liquid scintillation counter)	9, 70

few seem of interest for protein quantification in different types of food products (e.g. Kjeldahl, colorimetry, and IR spectroscopy). Specific attention will be given to these methods in the following discussion.

Conversion of nitrogen to protein can only occur accurately if the nitrogen content of the protein fraction is known and if the food product includes no other nitrogen-containing matter besides protein. A list of useful conversion factors for purified proteins and protein-containing food materials is given by Gáspár (40) and by Kirk et al. (63). If no specific information is available, an approximative value of the protein content can be obtained by multiplying the nitrogen content by 6.25, a factor corresponding to an average protein nitrogen content of 16%. The accuracy of selecting a predescribed conversion factor is discussed in more detail by Haidelbaugh et al. (10), who compared for 68 different food products the protein content derived from the Kjeldahl nitrogen by different conversion factors to those calculated from the amino acid content as determined by chemical analysis.

The presence of nonprotein nitrogen (NPN) compounds in foods, like amino acids, peptides, ammonia, urea, and trimethylamine oxide, overestimates their true protein nitrogen content as derived from the current nitrogen determination methods. Separation of NPN from true protein nitrogen can be achieved by the addition of a protein precipitating agent (e.g., trichloroacetic acid or perchloric acid). Since the composition and the amount of NPN varies according to the process conditions applied during precipitation, it is necessary to specify type and concentration of precipitating agents used (64). Alternative techniques such as dialysis and gel filtration are probably more accurate to remove the NPN fraction, but they remain unacceptable for routine analysis. A review on NPN determination methods in cow milk, and on aspects regarding the composition of the NPN fraction, is given by Wolfschoon-Pombo et al. (71,72).

In direct protein determination methods, the response varies with the amino acid composition of the investigated protein(s). As a consequence, if accurate absolute measurements are required, these methods have to be calibrated against a purified sample of the protein under investigation or against a protein with similar response characteristics to the one determined. In addition, possible interferences from nonprotein components like carbohydrates, salts, lipids, and detergents have to be taken into account. Due to the instability of reagents used or to the time dependency of the response measured, a precise timing for reagent addition, mixing with sample, and response determination must be followed if accurate and repeatable results are to be obtained.

Table 6 summarizes the advantages and disadvantages of the methods currently used for the analysis of total protein in food products. Kjeldahl, Biuret, and dye binding procedures are classical methods which remain important in modern food protein analysis. Lowry is more frequently applied in biochemical research than in food analysis but can be of importance for specific applications where a high sensitivity is required. UV absorption is particularly suited when used in combination with a chromatographic separation technique (cf. Sec. III.B.2), while IR spectro-photometry has become increasingly important during the last few years as a fast and nondestruc-tive method for the quantification of total protein in food products.

a.   The Kjeldahl Method.   The Kjeldahl method maintains its position as the most frequently used technique for the determination of organic nitrogen in food products. The technique is integrated in various official and statutory methods, and is currently applied as a reference for the evaluation of alternative protein determination techniques. In the overall process, two consecutive steps can be discerned: (a) decomposition of organic matter by heating in the presence of concentrated sulfuric acid and (b) determination of ammonia released from the digest.

During the decomposition step, dehydration and carbonization of organic matter occurs, combined with oxidation of the liberated carbon to carbon dioxide. Organic nitrogen is transformed to ammonia, which is retained in solution as ammonium sulfate. Acceleration and completion of the decomposition step can be achieved by the addition of salts, which raise the overall decompo-sition temperature (e.g., potassium sulfate), by the addition of oxidants (e.g. hydrogen peroxide, perchloric acid, persulfates, and chromic acid), which additionally minimize foaming, and by application of a catalyst. For the latter mercuric oxide has been reported as being most effective, although mercury–ammonia complexes are formed during digestion which limit the release of ammonia into solution. Additional information on the usage of salts or oxidants in the digestion process is given by Gáspár (40), while suggestions for the selection of a suitable catalyst have been reported by Lakin (57). In order to guarantee a quantitative conversion of nitrogen to ammonia, care must be taken to include a sufficiently long after-boiling time, i.e., the digestion time during which the amount of nitrogen transformed approaches asymptotically the absolute nitrogen value of the sample. Its purpose is to ensure that more resistant intermediate components of the decomposition process are fully transformed into ammonia, water, and carbon dioxide. By using one of the accelerating agents mentioned above, process times of 45 min or lower have been reported for the total digestion step.

Ammonia is liberated from the digest by distillation in the presence of alkali. If mercury was used as a catalyst in the digestion step, addition of potassium sulfide or sodium thiosulfate can destabilize the mercury–ammonia complexes formed during digestion by precipitation of the mercury. A total recovery of ammonia from the digest can be obtained within 5–20 min by direct distillation with moderate heating; if a steam distillation is applied, process times do not exceed

Table 6 Advantages and Disadvantages of Methods Used for Determination of Total Protein in Foods and Food Products

Method	Advantages	Disadvantages
Kjeldahl	• Appropriate for various (soluble and insoluble) types of food products • High reliability • Included in methods approved by international organizations	• Interference by nonprotein nitrogen compounds • Excessive foaming during digestion • Use of toxic and/or expensive catalysts • Choice of conversion factor; low sensitivity • Time consuming; length of after boiling time
Biuret	• No interference by free amino acids • Little influence of amino acid composition on color development • Simplicity of operation, ease of handling large number of samples	• Interference from ammonia, detergents, and buffer salts (e.g., tris) • Low sensitivity (concentration: 1–6 mg/ml; amount of protein needed: 0.05–5 mg)
Lowry	• High sensitivity (amount of protein needed: ±5–100 µg) • Simplicity of operation; ease of handling large number of samples	• Moderate dependence of color development on amino acid composition • High number of interfering compounds • Instability of Folin–Ciocalteau reagent at alkaline pH • Nonlinear standard curve at high protein concentrations • Large dilutions prior to measurement necessary
Dye binding	• High sensitivity • Rapid, easily automated • Simplicity of operation; suitable for large number of samples • Included in methods approved by international organizations	• Adsorption of protein–dye complex to glass surfaces • Dependence of protein–dye interaction on amino acid composition • Interference from laboratory chemicals (e.g., Triton X-100, SDS, colorants, buffer salts) • Insolubility of analyzed components in acid medium • Variations of binding capacity for different batches of commercial grade dyes
Near-UV absorption	• Rapid, nondestructive • No addition of reagents required	• Low sensitivity (0.05–2 mg of protein needed) • Interference by UV absorbing compounds (especially nucleic acids and nucleotides) • Strong influence of amino acid composition on signal response

Table 6 (*Continued*)

Method	Advantages	Disadvantages
Far-UV absorption	•Rapid, nondestructive •No addition of reagents required •High sensitivity (0.01–0.05 mg protein needed) •Low dependence of signal response on amino acid composition •Low interference from nucleic acids and nucleotides	•Necessity of using very clean sample cells and relatively new deuterium lamps •Spectral interference by UV absorbing compounds (e.g., buffer salts)
Infrared absorption	•Rapid, multicomponent analysis •No sample preparation •Nondestructive	•Strong interference by water •Influence from lipids and sample particle size •Complex calibration procedure
Near-infrared reflection	•Rapid, multicomponent analysis •Cheap and robust instrument components •High signal-to-noise ratio •Applicable to solid materials •Protein quantification in the presence of water •Limited sample preparation	•Interference by starch and lipids •Displacement of reflectance spectrum by moisture content and particle size of sample •Complex calibration procedure

10 min. The quantification of distilled ammonia is generally achieved by titration with a colorimetric endpoint detection. Frequently used is the absorption of ammonia in an excess of boric acid, followed by titration against standard acid in the presence of a suitable indicator (e.g., Tashiro indicator: 0.2% methyl red and 0.1% methylene blue). The minimum amount of nitrogen to be quantified is significantly improved when ammonia is directly determined in the digest by color-inducing compounds such as ninhydrin, indophenol, or the Nessler reagent. Unfortunately, the color development by ammonia is influenced by various components used during the digestion step (e.g., catalysts). By using gas-sensing probes, which measure potentiometrically the partial pressure of gases in solution, the direct analysis of ammonia in the Kjeldahl digest has been performed within 5–10 min. Complete conversion of ammonia to the gaseous form is achieved by an alkalinization of the digest to pH 11–13. The method has been adapted for measurement of the ammonium content of Kjeldahl digest solutions in a continuous flow system at a rate of 60 samples per hour (23).

The determination of protein in cereal grains has been optimized on a microscale using a colorimetric method based on the Berthelot reaction (salicylate-nitroprusside reagent) for determination of nitrogen in the Kjeldahl digests (29). Nitrogen concentrations in the range 0.2–0.8 µg N/ml for standard solutions and 0.3–1.3 µg N/ml for the sample digest could be analyzed within 35–40 min. On a macro scale, modern Kjeldahl apparatus allow the analysis of nitrogen in the range 0.2–160 mg N/sample. Various degrees of automation are available including an automated digestion and distillation followed by manual titration; a fully automated digestion, distillation, and titration; and the use of block digesters and autosamplers for the unattended analysis of maximum 60 samples per batch. Depending on the analysis procedure used, the scale of operation applied, and the degree of automation installed, the analysis time of the procedure could be further reduced, corresponding to frequencies of analysis up to 20 samples/hr (9,57,75,76).

b. The Biuret Reaction. The Biuret reaction comprises a one-step colorimetric assey of protein based on the binding of copper (II) to the peptide bonds in protein molecules at alkaline pH values. A stable violet complex is formed, which can be quantified by absorption at 310 nm or at 540–560 nm. Color development is believed to occur through coordination of one copper atom to four nitrogen atoms derived from peptide bonds, with an accompanying loss of a proton from each of the substituted amide groups (77). In addition, color formation may result from complex formation between copper and the free electron scales from nitrogen and oxygen atoms present in the peptide chains (44).

Although free amino acids were reported not to interfere with the reaction, a strong influence is noticed from tris buffers and ammonia. The sensitivity of the method is rather low (Table 6) but can be increased by separating the copper–protein complex from the lower molecular weight interfering compounds by gel filtration, followed by release of copper from the complex and determination of the element by colorimetry. Sensitivities ranging from 0.01 to 100 µg of protein/ml were obtained, depending on the type of color reaction used to determine the copper liberated from the complex. Alternatively, by measuring the absorbance of the Biuret complex in the near-UV region (260–330 nm), the negative interference from nucleic acids or from ammonia can be severely reduced, thereby shifting the working range of the method down to 0.01–3 mg of protein/ml. For protein quantification in cereals, an addition of propan-2-ol to the Biuret reagent was suggested to reduce possible interference from lipids and from starch and to facilitate the removal of unsolubilized particles by filtration prior to the colorimetric measurement (57). By combining the Biuret method with a liquid scintillation counter, the sensitivity of the method was increased for the determination of protein in whole milk (70).

After addition of the copper reagent, a sufficient equilibrium time is required to develop a stable and reproducible Biuret color. The Biuret method for protein determination in grains, as developed by Johnson et al. (78), has a total elapsed time per sample of 35–40 min, from which a minimum of 15 min is reserved for full development of the Biuret color. Various adaptions on the original concept of this method reduced the process time severely, e.g., by increasing the incubation temperature during color development, thereby obtaining process times of 10 min or lower (79–81). It must be specified though, that complete color development is probably not achieved when using short incubation periods. As a consequence, exact timing during incubation is essential to maintain a high overall reproducibility.

Developments of the Biuret method on a microscale have been reported (77,82). Bosset et al. (46) developed an automatic continuous flow analyzer for protein determination in milk according to the Biuret method, capable of analyzing up to 15 samples per hour. A similar rate of analysis was found for the semiautomated Biuret method applicable to cereal grains as described by Popineau et al. (83).

c. The Lowry Method. The Lowry method is based on a reduction of the Folin–Ciocalteau reagent, composed of a mixture of phosphomolybdic and phosphotungstic acid, by oxidation of tyrosine, tryptophan, and, to a lesser extent, cystine, cysteine, and histidine residues on the polypeptide side chains. The oxidation–reduction reaction is accompanied by the formation of a characteristic blue color with an absorption maximum at 745–750 nm. Copper chelates in the peptide structure which facilitates the electron transfer from the amino acid functional groups to the mixed acid chromatogen. The method is particularly suitable for the estimation of small amounts of protein in solution (Table 6). The color development is highly dependent on pH, which should be maintained between pH 10 and 10.5. Due to the instability of the Folin–Ciocalteau reagent at these alkaline pH values, a precise timing during the execution of each step in the overall procedure is required (67).

A high number of substances have been found to interfere with the method. An extensive list of these compounds, which include buffer salts, amino acids, ammonia, chelating agents, detergents, sugars, alcohols, and lipids, has been given by Dunn (67) and Peterson (47). The interference primarily results from an increase in the reagent blank, an effect which is usually accompanied by a reduction in color produced by the standard protein. In addition, substances like nonionic and cationic detergents, potassium, and lipids, form precipitates above a critical concentration which increase the apparent protein value due to turbidity or which react with the Folin–Ciocalteau reagent and disturb the final color development of the protein solution. Finally, compounds like the cryoprotectant polyvinylpyrrolidinone (PVP) interfere by destabilizing the molybdenum blue complex (47,84). Correction is possible by adding an equivalent amount of interfering substance to the reagent blank and to the standard protein solutions, or by removing interfering compounds present in the sample prior to analysis (cf. Sec. II). Additionally, several proposals were made to reduce the influence of specific interfering compounds by changing the reagent composition, the reagent concentration, and the operation temperature applied during incubation. A detailed overview is given by Peterson (47).

In the original method described by Lowry et al. (59), the total process time was set to approximately 40 min from which 30 min was reserved for the color development. By increasing the operating temperature during incubation and by changing the composition of the reagents required, Schacterle and Pollack (85) were capable of reducing the total assay time to 15 min.

A number of Lowry procedures have been optimized on a micro scale and/or were found suitable for automated analysis (86). Hegeman (87) developed an automated micro Lowry method capable of determining protein concentrations between 25 and 500 μg/ml in a sample volume as small as 0.4 ml. The continuous measurement of protein in milk using a modified Lowry method has been described by Anderson et al. (88).

d. Dye Binding Procedures. By carefully controlling the pH and the ionic strength of the medium, the acidic and basic functional groups of proteins can interact with oppositely charged groups of organic dyes. Frequently used colorants are sulfonated dyes like amido black and acid orange 12, which react at acid pH with the ε-amino group of lysine, the guanidine group of arginine, the imidazole group of histidine, and possibly also with the limited number of terminal α-amino groups. Furthermore, additional nonionic (mainly hydrophobic) interactions between the proteins and the nonionic moiety of the dye, and between the dye molecules in solution and those already bound to the protein, may occur.

The most successful dye binding method used for the determination of protein in foods and food products has been the indirect method based on the measurement of the excess dye remaining in solution after removal of the precipitated protein–dye complex. For the latter filtration has been preferred to centrifugation in view of its low price and speed of operation. However, care should be taken that none of the free colorant absorbs to the filter paper, thereby overestimating the protein content of the sample. In addition, the direct method introduced by Bradford (89), which benefits from its speed and simplicity of operation, may be of interest. In this assay, the binding of Coomassie brilliant blue G-250 to the protein causes a shift in the absorption maximum of the dye from 465 to 595 nm, which is quantified by monitoring the increase in absorption at 595 nm. The binding process is complete in approximately 2 min, while the usable working range of the method was set to 0.2–1.4 mg protein/ml for the standard assay and 5–100 μg protein/ml for the microassay procedure. Low concentrations of tris, acetic acid, 2-mercaptoethanol, sucrose, glycerol, EDTA, and detergents induce small interferences which can be eliminated by the use of a proper blank solution. Only detergents like SDS and Triton X-100 present in high concentrations were found to interfere excessively during determination.

The affinity of the colorants for the protein depends on the amino acid composition of the protein molecules. Whey proteins were found to adsorp 1.27 times more colorant than caseins. The NPN fraction is claimed not to interfere (65). In cereal products, both starch and bran were found to bind significant amounts of dye, thereby requiring separate calibration curves for different grades of flour (9). Also, it was found that for some colorants (e.g. amido black), different batches vary in their protein binding capacity due to a different degree of purification (9,65).

Care should be taken with food products containing colorants which interfere during the determination procedure (e.g., ice cream). Especially food colorants with an absorption spectrum equivalent to—or closely related to—those of the colorant added tend to underestimate the protein content of the sample and should be avoided when dye binding procedures are used (90). Erroneous results may also occur for foods with an increased amino group content due to proteolysis or with a decreased amino group content due to browning reactions.

Dye binding procedures with various degree of automation have been developed for protein determination in milk (57,91) and cereal products (9,57), allowing a full protein quantification in less than 5 min.

 e. Ultraviolet Spectrophotometry. Most proteins exhibit an absorption maximum at 280 nm which is attributable to the phenolic group of tyrosine and to the indolic group of tryptophan. Protein quantification based on absorption in the near-UV region has the advantages that no reagents are necessary and no sample destruction occurs during determination. If the solvent medium absorbs at 280 nm, an additional blank must be analyzed. The method is rather insensitive (Table 6) and suffers from the interference from compounds containing purine and pyrimidine rings (nucleic acids, nucleotides). For the latter, a suitable correction can be applied when the absorption is additionally measured at 260 nm. The measurement is also significantly influenced by the amino acid composition of the proteins. An overview of the molar absorptivities of various proteins, measured at different wavelengths, is given by Kirschenbaum (92).

The interference by nucleic acids and the dependence of the amino acid composition of the protein toward absorption can be reduced when measurements are performed in the far-UV region.

Although the peptide bonds in proteins exhibit a well-defined absorption maximum at 191–194 nm, the interference by oxygen limits the protein quantification at this wavelength. However, reliable and highly sensitive measurements can be made around 205–235 nm, especially when a correction is made for the tyrosine and tryptophan content by an additional absorption measurement at 280 nm (93).

UV absorption has gained significant importance for protein identification and quantification in the eluate from chromatographic columns (cf. Sec. III.B.2). For total protein determination in milk, an appropriate dilution step prior to the absorption measurement is necessary due to the extreme turbidity of the undiluted sample. In addition, demulsification of fat globules and complete solubilization of casein micelles is necessary to avoid scattering of light in the sample during the measurement. Reagents such as n-butylamine, EDTA, and acetic acid have been applied for this purpose (65).

    f. Infrared Spectrophotometry. Absorption of infrared (IR) light by protein molecules results from transitions between the vibrational energy levels. The IR light covers the wavelength region from 2.5 to 16 μm (i.e., 4000–625 cm^{-1}), which corresponds to transitions from the ground state to the first exitation level of molecular vibrations. The NIR light is situated between the IR and visible region with wavelengths between 0.75 and 2.5 μm (13,300–4000 cm^{-1}). Its higher energy content compared to IR light is suitable to induce transitions to higher vibrational levels and to cover the absorption bands related to combined interactive vibrations.

IR absorption can be used to determine the protein content in whole milk by measuring the intensity of the absorption band at the well-defined wavelength of 1548 cm^{-1}, which corresponds to the absorption maximum of the peptide bond. The measurements are fully automated, proceeding at a frequency of 100 up to 300 analyses per hour. Other components like lactose and fat, which display their own characteristic absorption bands at 1042 and 1745 cm^{-1}, can be quantified simultaneously.

Irradiation of food products with NIR light followed by measurement of the reflected radiation from the sample at different wavelengths produces a reflectance spectrum which contains quantitative information on different ingredients present in the sample. NIR spectroscopy has become a popular method for the determination of protein in cereal products, primarily due to its speed, simplicity of operation, safety, and low running costs. To avoid excessive interferences by starch, fat, and water, a wavelength of 2.18 nm—corresponding to a combined vibration of amide groups with wavenumbers 1650 and 1300 cm^{-1}—was chosen for quantification of the protein components. A measurement can be performed fully automatically in less than 1 min. In addition, a simultaneous determination of the oil and moisture content of the sample is possible.

Water absorbs significant amounts of energy between 1000 and 5000 cm^{-1}, thereby interfering excessively with the protein absorption bands in the IR spectrum. For protein determination in milk, the phenomenon could be neutralized by using a double-beam spectrophotometer with water in the reference cell and milk in the sample compartment (65). The fat globules in milk display a similar diameter as the wavelength of the incident light beam and have to be reduced in size by homogenization in order to reduce the degree of scattering during the measurement (65,94). In cereals, the change in NIR reflectance at 2.18 nm due to a change in protein content is somewhat low compared to the changes induced by variations in particle size and in starch, oil, and water content of different samples. These unwanted variations can be compensated for by using an adapted calibration procedure where specific absorption bands for protein, starch, oil, and water are taken into account (9).

2. Quantification of Specific Protein Components

Most currently used methods to detect and/or quantify specific protein components in foods and food products can be cataloged in the field of chromatography, electrophoresis, immunology, or a combination of them. Table 7 lists the various operation modes available in each method class, while Tables 8–10 specify the advantages and disadvantages of the methods which are currently used or which have been introduced recently for the determination and quantification of individual milk and cereal proteins in food matrices.

   a. Chromatography.   Within the area of chromatography, HPLC techniques show a growing importance for the analysis of specific protein components in foods and food products. The mixture of proteins to be separated is introduced into the mobile phase and separated on the basis of a difference in affinity between the protein components for the stationary phase in the column. Three

Table 7   Methods for Determination of Specific Protein Compounds in Foods and Food Products

Method classification	Determination method	Ref.
Chromatography	•Ion exchange (IE)	11, 95, 96, 97
	•Size exclusion (SE)	11, 97, 98, 99
	•Reverse phase (RP), hydrophobic interaction (HI)	11, 95, 97
	•Hydroxylapatite	65, 95, 97
	•Metal chelate	95
	•Covalent	95
	•Chromatofocusing	95
Electrophoresis (EF)	•EF under denaturing conditions (SDS-PAGE/acid urea PAGE)	55, 100
	•EF under native conditions (PAGE)	55, 100
	•Isoelectric focusing (IEF)	55, 100
	•Two-dimensional EF (2D-PAGE)	55
	•Capillary electrophoresis (CE)	101, 102
Immunology	•Passive agglutination	34
	•Precipitation reactions	
	Ring test technique	34
	Immunoturbidimetry/nephelometry	34
	Double diffusion (Ouchterlony, Kaminski)	34, 62
	Radial immunodiffusion (RID)	34, 62
	Immunoelectrodiffusion (IED)/ rocket immunoelectrophoresis	34, 103
	Electrosyneresis (counterimmunoelec- trophoresis, immunoosmophoresis)	34, 103
	Zone immunoassay (ZIA)	34
	Immunoelectrophoretic analysis (IEA)	34
	Crossed immunoelectrophoresis	34
	•Antigen/antibody labeling	
	Immunofluorescence	34, 62, 104
	Enzyme or protein A labeling of anti- bodies	34
	Enzyme-linked immunosorbent assay (ELISA)	34, 62, 105, 106
	Radioimmunoassay (RIA)	34, 62, 105, 107
	•Immunoadsorption	
	Affinity chromatography	91

Table 8 Advantages and Disadvantages of Chromatographic Methods Used for Determination of Specific Protein Components in Foods and Food Products

Method	Advantages	Disadvantages
Ion exchange chromatography (IEC)	•Mild separation process; limited protein denaturation; maintenance of biological activity •High-resolution power •Wide range of variables to influence retention and selectivity	•Laborious optimization procedure •Corrosive effects of salts •Interference by hydrophobic interaction with column packings
Size exclusion chromatography (SEC)	•Simple one-step analysis •Mild separation conditions; no protein denaturation	•Protein denaturation at extreme pH values •Peak broadening at large sample volumes •Decrease in resolution at high flow rates •Interference by hydrophobic or electrostatic interaction with column packings •Erroneous elution behavior of nonglobular and associated protein molecules
Reverse phase chromatography (RPC)	•High-resolution power •Suitable for protein analysis at low ionic strengths	•Risk of protein denaturation and loss of biological activity •Use of toxic solvents •Empirical separation process •Interference by hydrophobic contaminants •Difficulty of performing reproducible isocratic separations
Hydrophobic interaction chromatography (HIC)	•High number of potential elution conditions •High-resolution power •Less risk of protein denaturation compared to RP-HPLC •Suitable for protein analysis at high ionic strengths	•Corrosive effects from salts •Empirical separation process •Interference by hydrophobic contaminants

Table 9 Advantages and Disadvantages of Electrophoretic Methods Used for Determination of Specific Protein Components in Foods and Food Products

Method	Advantages	Disadvantages
Electrophoresis under denaturing conditions (SDS-PAGE)	• Applicable to unsoluble proteins • High resolution (e.g., 1 kDa for 100-kDa protein) and reproducibility	• Loss of enzymic and biological activity • No separation between proteins with identical MW or with MW <10,000 Da • Interference by nonprotein moieties (carbohydrates, lipids, phosphates) and by media with high ionic strength
Electrophoresis under native conditions (PAGE)	• No sample destruction; maintenance of enzymic and biological properties • Free selection of buffer composition and separation pH	• Limited to soluble proteins • Low resolution for dilute samples (e.g. <1 mg/ml) and for samples with high ionic strength
Isoelectric focusing (IEF)	• High-resolution power (0.02 pH units for CE and 0.001 pH units for IPG)	• Interference by charged detergents (e.g., SDS) and high salt concentrations • Long-term instability of pH gradients • High sensitivity toward protein micro-heterogeneity
Capillary electrophoresis (CE)	• On-line sample detection without need for staining procedures • Rapid, automated, and quantitative analysis • Wide application area (amino acids, peptides, nucleic acids, proteins, intact cells) • Small sample sizes • Improved heat dissipation during electrophoresis (high surface-to-volume ratio)	• Limited sensitivity of detection systems • Protein adsorption to capillary wall • Insufficient control of electroosmotic flow

Table 10 Advantages and Disadvantages of Immunological Methods Used for Determination of Specific Protein Components in Foods and Food Products

Method	Advantages	Disadvantages
Immunoturbidimetry/nephelometry	•Rapid, highly automated analysis •Suitable for routine analysis of large series of samples •High sensitivity and reproducibility	•High immune serum consumption/sample •Not suitable for polyspecific immune sera •Need for careful control of antigen/antibody ratio in reaction mixture
Single/double immunodiffusion	•Suitable for polyspecific immune sera •Means to evaluate monospecificity of immune sera	•Time consuming (2 days for double diffusion up to a week for single diffusion) •Semiquantitative
Immunoelectrodiffusion (IED)	•Suitable for polyspecific immune sera •Improved sensitivity and selectivity •Means to evaluate monospecificity of immune sera	•Time consuming (±1 day) •Complex experimental setup
Enzyme linked immunosorbent assay (ELISA)	•Low immune serum consumption •Suitable for routine analysis (high number of samples analyzed simultaneously) •Simplicity of operation, use of stable reagents •Simple separation of free and bound substances •High sensitivity and versatility	•Time consuming (±1 day) •Interference by enzyme inhibitors •Sensitive toward nonspecific reactions

basic separating principles can be discerned: ion exchange chromatography (IE-HPLC), size exclusion (gel permeation) chromatography (SE-HPLC), and reverse phase chromatography (RP-HPLC). A detailed overview of commercially available IE-, SE-, and RP-HPLC columns for separation of proteins, together with additional information on the type of support material, functional group, pore size, particle size, column dimensions, and manufacturers, is given by Johns (97). Fast protein liquid chromatography (FPLC) media only need one-third to one-tenth of the back pressure created in HPLC media during a comparable chromatography run (i.e., similar flow rate and column geometry). As a result, less risk of protein denaturation caused by shearing forces exists in FPLC columns compared to HPLC columns (11).

ION EXCHANGE HPLC   In ion exchange chromatography, proteins are separated on the basis of their net charge. The charge of protein molecules varies in function of the pH of the medium in which the protein is solubilized. At the isoelectric point (pI) of the protein, the amount of positive and negative charges are perfectly balanced and no net charge exists. Below the pI, proteins are positively charged and interact with negatively charged cation exchangers; above the pI, proteins are negatively charged and can bind to positively charged anion exchangers. Selective desorption of protein molecules is achieved by changing the pH or increasing the ionic strength of the elution buffer.

Modern IE supports are primarily based on silica or hydrophilic organic polymers, the latter being more stable at high pH values, which allows the use of more stringent regeneration procedures (e.g., strong bases) to extend the lifetime of the column. To allow a sufficient access of the high molecular weight proteins to the ion exchange sides, a pore size of 30–100 nm is recommended (97). While strong ion exchangers, e.g., those based on quaternary ammonium or sulfonic acid, are independent of pH between pH 2 and 12, the dissociation of weak ion exchange groups (e.g., diethylaminoethyl and carboxymethyl) is influenced by the pH of the eluting buffer (11). In selecting a suitable ion exchange column for protein analysis, properties like pH stability, pL, and molecular weight of the analyzed protein components have to be taken into account (95).

Unlike SE-HPLC and RP-HPLC, the application of IE-HPLC for the analysis of food proteins remains limited. Bietz (6) described the separation of albumin and globulin components, extracted with 0.1 M NaCl from wheat flour, on a carboxymethyl-300 weak cation exchange column at pH 7.0. Proteins were eluted from the column with a 0.02–0.31 M phosphate buffer gradient. By using a MonoQ anion exchange FPLC column, specific casein components (α_{s1}-, α_{s2}-, β-, and κ-casein) have been successfully separated and quantified in 20–30 min. The mobile phase contained 0.01 M imidazole-HCl buffer (pH 7.0), 3.3 M urea, and 0.01 M β-mercaptoethanol. A sample volume of 0.5 ml was injected onto the column, the adsorbed caseins were eluted with a linear NaCl gradient (0–0.5 M), and detection was achieved by measuring the UV absorbance at 280 nm. The method has been used to evaluate the fractionation of casein micelles with different size in skim milk on a Sephacryl S-1000 gel filtration column (54). The same column has also been used to study the effects of pH, type of counterions, and nature of the ionic strength gradient on the separation of whey proteins (37).

SIZE EXCLUSION (GEL PERMEATION) HPLC   In gel permeation chromatography, proteins are separated on the basis of their molecular size. Protein molecules small enough to enter the pores of the stationary packing material are distributed between the liquid mobile phase and the liquid situated in the pores of the matrix, and show longer retention times compared to larger proteins which cannot enter the pores of the matrix. For each column support material used, a specific working range is valid, given by a linear relationship between the retention time (or retention volume) and the logarithm of the molecular weight of the protein. Calibration is achieved by the use of purified proteins with known molecular weight (11,98).

The packing material in SE columns consists primarily of chemically modified silicas, modified in order to avoid excessive noncovalent interaction between the protein and the stationary phase. For steric reasons, bonding is often incomplete and residual silanol groups exhibit residual ion exchange properties. Addition of salts (e.g., 0.05–0.2 M phosphate buffer) is sufficient to neutralize the interference, although too high concentrations have to be avoided because they promote hydrophobic interactions (97).

Whey proteins are fairly well separated by SE-HPLC on a Zorbax GF-250 column (11,13), a TSK 3000 SW Spherogel column (14,108), and a Syncro-Pak GPC-100 column (109). Ten to twenty-five microliters of prefiltered whey can be directly analyzed within 15–20 min, using a slightly acidic to neutral mobile phase (pH 6–7.5), an elevated operating temperature (30–40°C), and a flow rate of 0.5–1.0 ml/min. Protein detection is usually achieved in the far- (220 μm) or near- (280 μm) UV region. By coupling two TSK 2000 SW columns in series, Olieman and van den Bedem (27) were able to accurately determine the characteristic glycomacropeptide (GMP) fraction of cheese whey. The method seems suitable to quanitfy the addition of rennet whey total solids in skim milk powder and sweet buttermilk powder with a detection limit of 0.8%.

SE-HPLC has also been proposed as a fast and nondestructive method for the separation of unreduced flour proteins into its glutenin, gliadin, and albumin + globulin components (19,53). Analyses were performed on a Waters Protein-Pak 300 column, using a 0.05 M sodium phosphate buffer (pH 6.9) with 0.1% SDS as mobile phase. The flow rate was set to 0.5 ml/min and protein determination was achieved by UV absorption at 210 nm. After protein extraction (cf. Sec. II), 20-μl samples containing 0.5–1 mg protein/ml were analyzed in less than 30 min. Replacement of the elution solvent by 50% (v/v) aqueous acetonitrile containing 0.1% TFA not only extends the lifetime of the column but significantly improves the overall resolution of the method (18).

REVERSE PHASE AND HYDROPHOBIC INTERACTION HPLC   In reverse phase chromatography (RPC), separation is based on differences in surface hydrophobicity between the protein molecules. Hydrophobic amino acid residues of proteins are bounded through hydrophobic interactions to a stationary matrix composed of a nonpolar surface (usually C2–C18 alkyl chains) and a silica-based core material. Protein retention is controlled by increasing the concentration of organic solvents like methanol, acetonitrile, and ethanol in solution, hereby altering the polarity of the aqueous mobile phase. In hydrophobic interaction chromatography (HIC), protein separation is based on the same principle but elution is promoted by a decrease in ionic strength of the mobile phase (i.e., use of a decreasing salt gradient).

The option of protein desorption by a decreasing salt gradient became possible through the use of less hydrophobic alkyl groups (e.g., C8) packed less densely on the stationary phase. Not only solvents can be avoided during analysis; the risk of protein denaturation also is severely reduced. Unreacted silanol groups might interfere with protein absorption and have to be neutralized by blocking agents (e.g., trimethylchlorosilane). In addition, wide-pore matrices (±30 nm) are required to maximize capacity and resolution.

Protein retention in RPC and HIC can be altered by adding aliphatic molecules with a charged group to the mobile phase which interact electrostatically with the protein (e.g., TFA). Alternatively, changes in pH might influence the degree of hydrophobic interaction: maximal protein hydrophobicity occurs around the pI; electrostatic repulsions might take place if protein and adsorbent carry similar charges. The interaction between a protein and the hydrophobic surface remains poorly understood, and predictions of retention on the basis of amino acid composition are not accurate.

The gliadin and glutenin fraction in cereal proteins has been investigated extensively by RP-HPLC. For gliadins, analysis are usually performed with a C18 column, at elevated temperatures (40–70°C), and using an increasing gradient of acetonitrile (20–60% acetonitrile in 0.1% TFA) to elute bound substances. The flow rate varies between 1 and 3 ml/min, and detection is usually achieved by absorption in the far-UV region (e.g., 210 nm) to avoid interferences by acetonitrile and TFA. A total analysis can be performed within 20–120 min, with sample sizes varying between 10 and 50 μl (11). In the case of glutenins, solubilization, denaturation, reduction, and alkylation of the high molecular weight complexes (e.g., by using dithioerythritol, urea, guanidine HCl) has been suggested to improve reproducibility and resolution, although an alkylation of the subunit mixture proved to be unacceptable for routine cultivar identification purposes (24). Dissociation of the glutenins by SDS was dissuaded in view of the inferior and nonreproducible results obtained (50). Concerning HIC, reduced and alkylated acetic acid–soluble glutenins were successfully fractionated using a phenyl-Sepharose CL-4B column (52).

A successful separation and quantification of whey proteins by RP-HPLC has been achieved in less than 7 min by the use of a Vydac 214 TPB10 column and a two-step linear gradient of acetonitrile/TFA/morpholine in water as mobile phase. The flow rate was set to 1 ml/min, and detection was performed in the UV region at 280 nm. Linear calibration curves were obtained between 8.5 and 170 μg/ml for BSA, 9.5 and 304 μg/ml for α-La, and 27.5 and 960 μg/ml for β-Lg. The method proved useful in separating the homologous whey proteins from cow, goat, and sheep milks, which in turn allowed the characterization of milk mixtures derived from these animal species (110). As an alternative to SE-HPLC, adulteration of skim milk powder and buttermilk powder can also be determined by RP-HPLC on a C-3 column with acetonitrile/TFA/water as mobile phase. The flow rate was set to 1 ml/min and detection was performed by UV absorption at 210 nm. The detection limit of the method was found to be 0.2% of rennet whey solids (26).

   b.   Electrophoresis.   The separation of proteins in an electric field on the basis of their charge density is a powerful and frequently used technique for the analysis of specific protein components in foods and food products. The separation is commonly achieved in a suitable support material (usually polyacrylamide gels) in order to counteract such effects as convection and diffusion during migration and to facilitate immobilization of separated proteins prior to their identification and quantification. Electrophoresis of proteins can be performed under native (PAGE) or denaturing (SDS-PAGE) conditions; applying homogeneous (fixed pore size) or gradient (increasing pore size) polyacrylamide gels; and using size (SDS-PAGE), net charge and size (PAGE), or charge (IEF) to induce protein separation (55,64,100).

   Addition of fixation agents (e.g., 20% w/v TCA or methanolic solutions of acetic acid) to the gel network after the separation process precipitates and immobilizes proteins and removes interfering nonprotein components prior to staining. The latter process is usually achieved by the use of Coomassie brilliant blue R-250 (CBB), which benefits from its speed of operation and absence of background staining but which is rather insensitive (0.5 μg protein/cm^2 gel). Silver staining techniques may increase this sensitivity with a factor of 20–200, although they are more expensive and time consuming than CBB staining. For both procedures, care should be taken with variations in degree of staining between different protein components (55).

   Modern apparatus for electrophoresis allows a full protein analysis within 1–2 hr, thanks to the use of small electrophoresis modules with a reduced separation distance, by cooling the migration bed in order to increase the applied field strength and by automatization of the staining procedure. An additional reduction in process time down to 10–30 min was achieved by the recently developed capillary electrophoresis techniques, which use an on-line detection system for the separated protein compounds instead of the time-consuming staining procedures.

ELECTROPHORESIS UNDER NATIVE (PAGE) AND DENATURING (SDS-PAGE) CONDITIONS   SDS-PAGE is a powerful determination technique for food proteins with low solubility. After reduction, different polypeptides bind the same amount of SDS on a weight basis (i.e., 1.4 g SDS/g polypeptide), producing polypeptide–SDS complexes with a constant net charge per unit mass. In an electric field, the migration of these complexes is proportional to the molecular weight of the protein, provided no excessive deviations occur in hydrodynamic forms between different proteins. Small polypeptides (MW < 10,000 Da) generate complexes with similar dimension and charge, making a separation by SDS-PAGE impossible. In addition, nonprotein moieties like carbohydrates (glycoproteins), lipids (lipoproteins), and phosphate (phosphoproteins) interfere with the binding of SDS to the protein molecule, thereby altering or decreasing its mobility in an electric field. As an alternative to SDS-PAGE, electrophoresis can also be performed in the presence of other denaturing compounds. For example, acidic urea PAGE can be used to separate low-solubility proteins on the basis of differences in charge density and mass (55,100).

Native PAGE techniques allow the characterization of soluble, undenatured proteins on the basis of their charge and molecular weight. Depending on the properties of the proteins analyzed (e.g., pI, stability toward pH), the electrophoresis can be performed under acidic or under basic conditions. An overview of the buffer systems available at various pH levels is given by Dunn (55) and Goldenberg (111). By using a discontinuous buffer system with a highly crosslinked stacking gel, sample proteins are concentrated into a narrow starting zone prior to separation, thereby increasing the resolution capacity of the method and allowing the use of dilute sample solutions (55). In some applications, unsoluble proteins can be analyzed under fairly native conditions in the presence of nonionic detergents (e.g., Triton X-100).

PAGE techniques proved to be extremely useful for the characterization of gliadin and glutenin compounds in cereal proteins. In the frequently used method described by Lookhart et al. (112), PAGE is performed in a homogeneous 6% acrylamide/0.3% bisacrylamide gel, using a 0.25% (w/v) aluminum lactate/lactic acid (pH 3.1) buffer. The staining solution contains 0.05% (w/v) CBB in 6% (v/v) ethanol and 11% (w/v) TCA; destaining is achieved in 12% (w/v) TCA. For SDS-PAGE analysis, predominantly discontinuous buffer systems with 3–4% acrylamide (AA) and 0.04–1.35% N,N'-methylenebisacrylamide (BIS) in the stacking zone and 12–17% AA and 0.08–1.35% BIS in the separating zone have been used. In addition, the suitability of linear gradients (16,36) and multistacking gels (113) has been examined.

Caseins and whey proteins are easily separated on PAGE and SDS-PAGE gels using a discontinuous buffer system (e.g., 4% and 15% AA crosslinked with 2.7% BIS in stacking and separating gel, respectively). However, care should be taken with an abnormal migration behavior of the individual casein components during SDS-PAGE (114). The phenomenon could be attributed to mutual interactions between the casein molecules, which disturbs their normal binding behavior toward SDS (0.9 g SDS/g for para-κ-casein, up to 3.4 g SDS/g for β-casein).

ISOELECTRIC FOCUSING (IEF)   Separations in IEF are achieved by means of a uniformly increasing pH gradient, created between both electrodes by incorporation of low molecular weight carrier ampholytes (CAs; polyamino- and polycarboxylic acids) in the PA gel matrix. Driven by the electric current, protein molecules migrate to—and are finally focused at—the point where the pH of the medium corresponds to the pI of the proteins. To increase the stability of the pH gradient, to improve the resolution and overall reproducibility of the separation process, and to reduce the sensitivity to high salt concentrations and protein microheterogeneity, buffering ampholytes were recently immobilized into the PA matrix (IPG). Due to the somewhat low conductivity of the system, the application of the technique is only recommended for very narrow pH gradients (<1 pH unit), unless low concentrations of synthetic CAs are added.

IEF has been used in both one- (36) and two- (115) dimensional separations of zein proteins. The analysis were performed in a 5% AA gel, containing 6 M urea and 2% ampholites buffering in the pH zone 6–8. In the study of Paulis et al. (36), it was clearly shown that several zein components containing a similar surface hydrophobicity (as revealed by RP-HPLC) varied largely in their pI. Reversely, similarities in amount and distribution of ionizable amino acids do not necessarily imply similarities in surface hydrophobicity.

IEF is an excellent means to separate and identify the genetic variants of caseins and whey proteins (21,116). In the study of Vegarud et al. (21), the α_{s1}, α_{s2}, and β variants of casein were best resolved in an ampholyte mixture with pH ranges 4.0–6.0, 4.5–5.0, and 5.0–7.0 in a ratio of 1:1:1. κ-Casein was preferentially analyzed in a 1:2 mixture of ampholytes with pH range 4.0–6.0 and 5.0–7.0, while variants of α-La and β-Lg could be identified in the broad pH range 3–9.

CAPILLARY ELECTROPHORESIS   Capillary electrophoresis is a recently developed separation technique for proteins which can be distinguished from classical electrophoresis techniques at two points: (a) electrophoresis is performed in narrow capillaries (usually fused silica with an inside diameter of 20–200 μm and a length ranging from 10 cm to more than 100 cm), and (b) supporting media (gels) are applicable but can be omitted (101). Both ends of the capillary are placed in separate electrolyte reservoirs, each containing an electrode in direct connection with a high-voltage power supply (20–30 kV). Similar on-line detection systems as those applied for HPLC analysis (e.g., UV absorption, fluorescence) can be used. Sample sizes do not exceed the nanoliter range, while detection volumes as low as 30 pl can be sufficient (102).

In CE, the low diffusion coefficients of large biomolecules (proteins) permit narrow elution peaks. However, chemical modification of the capillary wall in fused silica tubes (e.g., coating of the wall with polyethylene glycol) is necessary since slow adsorption/desorption kinetics during solute–wall interactions tend to stimulate peak broadening and loss of resolution. The overall sensitivity of the method is, in view of the small sample volumes required, basically determined by the detection system installed (101,102).

In a recent publication (128), the suitability of CE is discussed for the separation of the major whey proteins (β-Lg, α-La, and BSA) in whey samples derived from both raw and UHT milks. The fused silica capillary used (25 μm i.d., effective length 20 cm) was internally coated with crosslinked polyacrylamide to avoid interfering adsorption effects of proteins. The eluting buffer contained 10% (w/v) PEG 8000 and 0.1% (w/v) SDS; pressure-driven sample injection was carried out in the cathode, and detection was achieved by UV absorption at 214 nm. Compared to HPLC techniques, a separation between the protein components was obtained with a lower analysis time (9 vs. 35 min), a comparable reproducibility in migration time, but a significantly lower area reproducibility. The latter might have been caused by the pressure-driven injection technique, which is claimed to be less reproducible than the injection valves used in HPLC. Olieman (28) proved that CE has the potential to separate genetic variants of milk proteins and to detect the adulteration of cow milk with goat or sheep milk (and vice versa).

    c. Immunology.   Methods based on the interaction between an antigen and its corresponding antibody (immune serum) are a highly specific and sensitive means to identify and quantify minor protein components in complex matrices. In the case of foods and food products, the use of physicochemical and thermic procedures throughout the production process, and the application of rigorous isolation procedures for protein antigens prior to their identification and quantification, might change the original antigenic specificity of the protein components present in the food matrix.

A detailed overview of immunochemical methods used in food analysis has been given by Daussant and Bureau (34). Various types of determination methods are now available (cf. Table 7), of which especially immunoturbidimetry and immunonephelometry, single and double diffusion, immunoelectrodiffusion, and the enzyme-linked immunosorbent assay (ELISA) are regularly applied for the analysis of proteins in foods and food products.

IMMUNOTURBIDIMETRY/IMMUNONEPHELOMETRY   The immunocomplexes formed after reaction of the protein antigens with their equivalent antibodies can be quantified by measuring the scattered light (nephelometry) or the transmitted light that is not scattered (turbidimetry) by the immunocomplexes in suspension. Immunonephelometry has been found suitable for the rapid determination of immunoglobulin G in milk (34,65). After a 30-min incubation of antigen and immune serum, highly sensitive (order of ng) nephelometric measurements were performed at 633 nm within 15 sec, using 40 µl of immune serum for each determination. The method was presented as a sensitive and specific means to detect the presence of colostrum in milk. Gombocz et al. (118) optimized a method based on immunoturbidimetry in combination with a centrifugal analyzer for the rapid and highly automated determination of different types of proteins in food products. During rotation of the sample cell, extracted sample antigens and immune serum were mixed in a reproducible manner, followed by measurement of the turbidity at 290 nm. The system could analyze five different sample extracts within 10 min and allowed a protein determination in the concentration range of 5–65 mg/L for casein, 20–120 mg/L for whey proteins, and 1–4 g/L for gliadin.

SINGLE DIFFUSION/DOUBLE DIFFUSION   By allowing the immunochemical reaction to take place in a suitable support material (e.g., agarose gel), different antigen compounds in one sample extract can be separated and individually quantified during the same analysis. In the double-diffusion technique, the immune serum and antigen extract are placed in separate wells in the gel, after which both reactants are allowed to diffuse into the gel and form a precipitin band which can be stained (e.g., with CBB) and quantified. In the single-diffusion technique (also called radial immunodiffusion; RID), the immune serum is equally divided in the agarose gel after which the wells in the plate are filled with antigen extract and the antigen allowed to diffuse into the gel. The precipitation rings formed can be visualized by staining and the diameter of the rings used to quantify the amount of antigen in the sample extract.

By using the double-diffusion technique, Gombocz et al. (29) were able to identify on a semiquantitative basis the presence of cow's milk casein in sheep's cheese. The immunodiffusion step was performed in agarose gels saturated with antibovine-casein-serum and the immunoprecipitate formed was stained with amido black. The distance between the opposing precipitin bands was found to be linearly correlated with the concentration of standard casein in the range 10–100 mg/100 ml. The method should be capable of detecting a minimum addition of 5% cow's milk to the sheep's milk used for cheese production. The double-diffusion test has also been used by Sinell et al. (56), who demonstrated that liver proteins severely interfere with the immunological determination of foreign milk protein in liver-based products.

IMMUNOELECTRODIFFUSION (IED)   By combining immunochemistry with electrophoresis, a further increase in sensitivity and selectivity toward protein detection and quantification in foods can be achieved. In immunoelectrodiffusion (IED), the immune serum is equally divided within the agarose gel while the antigen solution is placed in the wells of the gel. By carefully selecting the electrophoresis conditions, the antibodies in the agarose gel can be held stationary while the antigens are forced to migrate into the gel network by both diffusion and electrophoresis. Both the height and the surface of the precipitin peaks can be used to determine the concentration of the antigens in the sample extract.

To quantify the amount of α-La and β-Lg in milk and whey, Dougherty (31) described a method based on the IED principle. The results were calculated from a graph obtained by plotting the peak heights of the standards (0.10–0.25 mg/ml for α-La and 0.15–0.40 mg/ml for β-Lg) as a function of the concentration. For the estimation of the whey protein concentrate (WPC) content in a commercialized meat product (frankfurters), α-La antiserum was used in view of its higher resistance to thermal denaturation compared to β-Lg. WPC concentrations of 3–7% were successfully determined in fortified frankfurters, although the use of a highly concentrated antiserum (10%) was necessary. Additionally, in view of the varying α-La content in WPC, a WPC identical to the one present in the sample had to be used as a standard. In the study made by Duranti et al.

(119), it was shown that the detection of α-La by IED has potential to evaluate the effect of heat treatment on whole milk.

ENZYME-LINKED IMMUNOSORBENT ASSAY   In the enzyme-linked immunosorbent assay (ELISA) test, protein antigens or antibodies are labeled with enzymes, after which one of the reactants is immobilized onto a support material. As soon as the immunochemical reaction has taken place, the unbounded substances are washed out and the bound material is quantified by measuring the activity of the enzyme spectrophotometrically. The immobilization is preferentially performed in the wells of polyvinylchloride or polystyrene microtiter plates, while the color-forming enzyme used is normally peroxidase, alkaline phosphatase, or glucose oxidase. A detailed overview of the various direct and indirect operating modes available for ELISA are given by Catty and Ragkundalia (106).

ELISA techniques benefit from their high sensitivity and simplicity of operation. The increasing availability of ready-made immunochemical reagents and microtiter plates with immobilized antigens or antibodies makes the technique particularly suitable for the routine analysis of large numbers of small-volume test samples.

Mao et al. (120,121) reported an ELISA procedure for the determination of α-La and β-Lg in bovine serum and skim milk. Protein determinations were performed with the peroxidase enzyme using tetramethylbenzidine (TMB) as a substrate. The cross-reactivity of the α-La and β-Lg antisera to other milk proteins was very low (<0.0003%), while the detection limit for the standard solutions was found to be 25 and 100 pg for α-La and β-Lg, respectively (i.e., 5 and 2 ng/ml for a 50-μl sample volume). When ready-made microtiter plates were used, the total analysis could be performed within a working day. The ELISA method has also been successfully applied for the determination of milk protein in meat products (32). After the immunochemical reaction, the peroxidase activity was monitored with 5-amino-2-hydroxybenzoic acid as a substrate. The method was capable of distinguishing 1% and 2% added milk protein in frankfurter sausages but proved to be highly dependent on the extractability of the protein to be determined.

IV. FUTURE DEVELOPMENTS

All protein determination methods described are not absolute and demand some form of calibration. The Kjeldahl method remains the only official method currently available for calibration purposes, although the determination technique itself has some disadvantages which limit its accuracy. Alternative techniques such as neutron and proton activation analysis are available, but it remains uncertain if a widespread use and acceptance of these techniques as official and statutory method will be granted. To improve and optimize the accuracy of the protein quantification in foods and food products, a further search for new and alternative absolute protein quantification methods is necessary.

An additional point of interest for future developments is the availability of rapid, nondestructive methods which can be applied for routine analysis of intact food materials. Infrared techniques have proven to be suitable for this purpose, although the method remains empirical and demands calibration against a reference method.

For the analysis of specific protein components in foods and food products, separation techniques which combine the chromatography process with electrophoresis and immunology are as yet not sufficiently explored. Capillary electrophoresis was found suitable for the rapid separation and quantification of individual food proteins, although some specific problems remain to be solved before a routine application of the technique becomes possible. Immunochemical ELISA techniques are undoubtedly promising for a widespread use in food protein analysis due to their high sensitivity, specificity, and simplicity of operation. However, the fact that protein

Table 11 Methods for the Determination of Milk Proteins in Foods and Food Products

Analysed component(s)	Sample type	Protein determination method(s)	Ref.
Total protein	• Whole milk	• Alkaline distillation, formol titration, refractometry vs. Kjeldahl	122
		• Biuret vs. Kjeldahl	44, 45, 46
		• Dye binding (amino black, orange G), Biuret and IR spectrophotometry vs. Kjeldahl	123
	• Whole milk, semiskimmed milk, dried skimmed milk, ice cream mix, chocolate drink, buttermilk	• Dye binding (acid orange 12) vs. Kjeldahl	124
	• Skimmed milk powders	• Dye binding (amido black) vs. Kjeldahl	125
	• Ice cream and frozen desserts	• Dye binding (amido black, orange G) vs. Kjeldahl	90
Total protein, caseins, whey proteins	• WPC powder	• Dye binding (CBB) vs. Kjeldahl	126
	• Whole milk	• Kjeldahl, Dumas, alkaline distillation, sialic acid analysis, formol titration, dye binding (amido black, acid orange 12), Lowry, Biuret, UV, and IR spectrophotometry, fluorescence, turbidimetry/nephelometry, electrophoresis, chromatography, immunology, action of peptidases	65
Total protein, total whey protein	• Whole milk, nonfat dry milk	• Modified Lowry vs. Kjeldahl	86
Total whey protein	• Whole milk	• Alkaline distillation, dye binding (amino black), Biuret vs. Kjeldahl	127
	• Whey	• Biuret vs. Kjeldahl	128
		• Dye binding (amino black) vs. Kjeldahl	129, 130
		• Immunoturbidimetry	118
Total casein	• Whole milk	• Refractometry and dye binding (amido black) vs. Kjeldahl	131
		• Dye binding (amido black) vs. Kjeldahl	129, 130
		• SE-HPLC (280 nm) vs. Kjeldahl	15
	• Sheep cheese	• Immunoturbidimetry	118
		• Double diffusion (Ouchterlony, amido black)	29

Protein	Sample	Method	Reference
Caseins, whey proteins	•Processed milks, whey protein concentrates	•SDS-PAGE (CBB)	114
	•Whey	•SDS-PAGE (CBB), acid and alkaline PAGE (CBB), IEF (CBB)	116
Ig, BSA, α-La, β-Lg	•Whole milk	•IEF (CBB)	21
	•Whey	•SE-FPLC (280 and 214 nm) and SDS-PAGE (CBB)	38
		•SE-HPLC (280 nm)	13, 108
	•Skimmed milk powder	•IE-FPLC (280 nm) and PAGE (CBB)	37
		•SE-HPLC (220 nm) and PAGE (amino black, CBB)	14
Ig, BSA, α-La, β-Lg, proteose peptone	•Freeze dried whey protein	•SE-HPLC (280 nm) and RP-HPLC (280 nm)	109
BSA, α-La, β-Lg	•Whey	•Capillary electrophoresis (214 nm) and RP-HPLC	117
		•RP-HPLC (280 nm) and IEF (CBB)	110
α-La, β-Lg	•Whole milk and whey, WPC powder	•IED (visible detection or CBB)	31
	•Whole milk	•Single diffusion	20
	•Sheep cheese	•Crossed immunoelectrophoresis	30
	•(milk)serum	•ELISA (peroxidase)	120, 121
α-La	•Whole milk	•IED (CBB)	119
	•Meat products (frankfurters)	•IED (visible detection or CBB)	31
Ig, casein	•Calf serum	•ELISA (peroxidase)	132
α-casein, β-casein, whole casein	•Meat products (liver homogenates, liver sausages)	•Double diffusion (Ouchterlony), IED (Supranol cyanine 6B), ring test	56
α-, β- and κ-caseins	•Whole milk	•IE-FPLC (280 nm) and RP-HPLC (280 nm), IEF (CBB)	54
		•PAGE (CBB)	20
α-casein	•Meat products (sausages)	•IED (amido black), double diffusion (Ouchterlony)	133
κ-casein glycomacropeptide	•Skimmed milk, animal foods	•IED	134
	•Skimmed milk powder and buttermilk powder	•RP-HPLC (210 nm) and SE-HPLC (210 nm)	26
hydrolyzed milk protein	•Meat products (sausages)	•SE-HPLC (205 nm)	27
		•ELISA (peroxidase)	32, 135

Table 12 Methods for the Determination of Cereal Proteins in Foods and Food Products

Analyzed component(s)	Sample type	Protein determination method(s)	Ref.
Total protein	• Wheat	• Biuret vs. Kjeldahl	81, 136
		• Dye binding (acid orange 12), Biuret vs. Kjeldahl	80
		• UV absorption vs. Kjeldahl	137
		• Kjeldahl, Dumas, neutron activation analysis, alkaline distillation, dye binding, Biuret, IR spectrometry	138
		• Lowry vs. Kjeldahl	139
		• NIR reflectance and dye binding (acid orange 12) vs. Kjeldahl	140
	• Wheat, barley	• Biuret vs. Kjeldahl	83, 141
	• Barley, malt	• Biuret, dye binding, alkaline distillation, IR spectrophotometry vs. Kjeldahl	142
	• Rye	• Dye binding (orange G) vs. Kjeldahl	143
	• Brown and milled rice	• Dye binding (acid orange 12), Biuret vs. Kjeldahl	39
	• Wheat, maize, barley, rice, oats, rye, triticale	• Kjeldahl, Dumas, Biuret, dye binding (acid orange 12, acid Green 25), alkaline distillation, neutron and proton activation analysis, IR spectrometry	9
Total protein, wheat protein fractions	• Small grains and corn	• Biuret vs. Kjeldahl	78
	• Whole wheat meal and flour	• Biuret vs. Kjeldahl	79
Albumins, globulins, gliadins, glutenins	• Wheat	• SE-HPLC (210–214 nm), SDS-PAGE (CBB)	18, 19, 53
Albumins, globulins, prolamins, glutelins	• Rice	• PAGE and SDS-PAGE (CBB), RP-HPLC (210 nm)	35
Gliadins, glutenins	• Wheat	• RP-HPLC (210 nm)	25
Zeins, glutelins	• Maize	• RP-HPLC (210 nm) and SE-HPLC (210 nm)	24, 50
		• RP-HPLC (210 nm) and SE-HPLC (210 nm)	6
			6

Prolamins, glutelins	•Wheat, barley, malt	•Immunoprecipitation in gels, enzyme and radioimmunoassays, immunofluorescence, immunonephelometry, immunoblotting	144
Albumins, globulins	•Wheat	•IE-HPLC (280 nm)	6
Gliadins	•Wheat	•RP-HPLC (210 nm)	22, 23, 49, 145
		•RP-HPLC (210 nm) and PAGE (CBB)	146
		•RP-HPLC (210 nm), SDS-PAGE and PAGE (CBB)	17
		•PAGE (CBB)	112
		•Immunoturbidimetry	118
α-Gliadin	•Bread dough	•ELISA (peroxidase)	147
α-Gliadin, total gliadin	•Rice, maize, oats, rye, barley, wheat, and wheat flours	•SDS-PAGE and 2D SDS-PAGE (CBB)	148
		•HIC (280 nm) and SDS-PAGE (CBB)	52
		•RP-HPLC (210 nm) and SDS-PAGE (CBB)	51
		•SDS-PAGE, 2-dim SDS-PAGE/PAGE followed by immunoblotting (Ponceau S)	149
Hordeins	•Barley	•SDS-PAGE (silver, CBB)	150
		•RP-HPLC (210 nm) and SDS-PAGE (silver)	16
Zeins	•Maize	•RP-HPLC (280 nm), SDS-PAGE and PAGE (CBB), IEF (silver)	36
		•2D IEF/PAGE (silver)	115
Avenins	•Oat	•RP-HPLC (210 nm) and PAGE (CBB)	151

antigens in food matrices or sample extracts rarely display their original native structure severely stresses the need for specific and adapted food protein immune serums to become available.

V. APPLICATIONS IN FOOD ANALYSIS

An overview of various methods used for the determination of milk and cereal proteins in foods and food products is given in Tables 11 and 12. Dye binding procedures are characterized by the type of colorant used. For the chromatography techniques the detection method is specified, while for the electrophoresis methods the type of staining procedure is given. Immunochemical methods are presented with their staining method (IED, double diffusion) or the type of enzyme used for protein quantification (ELISA).

REVIEWS

A. L. Lakin, *Developments in Food Analysis Techniques*, Vol. 1 (R. D. King, ed.), Elsevier, London, 1978, p. 43.

I. Kerese and R. A. Chalmers, *Methods of Protein Analysis*, Ellis Horwood Ltd, Chichester, 1980.

J. Daussant and D. Bureau, *Developments in Food Analysis Techniques*, Vol. 3 (R. D. King, ed.), Elsevier, Essex, England, 1984, p. 175.

H. Guillou, J. P. Pelissier, and R. Grappin, *Le Lait*, 66: 143 (1986).

B. G. Osborne, *Developments in Food Proteins*, Vol. 4 (B. J. F. Hudson, ed.), Elsevier, London, 1986, p. 247.

E. L. V. Harris and S. Angal, *Protein Purification Methods: A Practical Approach*, Oxford University Press, New York, 1989, p. 317.

J. Vervaeck and A. Huyghebaert, *Food Analysis by HPLC* (L. Nollet, ed.), Marcel Dekker, New York, 1992, p. 141.

REFERENCES

1. R. A. Orth and W. Bushuk, *Cereal Chem.*, 50: 191 (1973).
2. T. B. Osborne, *Carnegie Inst. Washington*, Washington DC (1907).
3. R. M. Whitney, *Fundamentals of Dairy Chemistry*, 2nd ed. (B. H. Webb, A. H. Johnson, and J. A. Alford, eds.), AVI, Westport, 1987, p. 81.
4. J. D. Schofield and M. R. Booth, *Developments in Food Proteins*, Vol. 2 (B. J. F. Hudson, ed.), Applied Science, Essex, England, 1983, p. 1.
5. J. E. Kinsella, *Food Proteins* (P. F. Fox and J. J. Condon, eds.), Applied Science, Essex, England, 1982, p. 51.
6. J. A. Bietz, *Cereal Chem.*, 62(3): 201 (1985).
7. J. C. Cheftel, J. L. Cuq, and D. Lorient, *Food Chemistry*, 2nd ed. (O. R. Fennema, ed.), Marcel Dekker, New York & Basel, p. 245 (1985).
8. J. E. Kinsella, *Crit. Rev. Food Sci. Nutr.*, 21(3): 197 (1984).
9. B. G. Osborne, *Developments in Food Protein*, Vol. 4 (B. J. F. Hudson, ed.), Elsevier, London, 1986, p. 247.
10. N. D. Heidelbaugh, C. S. Huber, J. F. Bednarczyk, M. C. Smith, P. C. Rambaut, and H. O. Wheeler, *J Agric. Food Chem.*, 23(4): 611 (1975).
11. J. Vervaeck and A. Huyghebaert, *Food Analysis by HPLC* (L. Nollet, ed.), Marcel Dekker, New York, 1992, p. 141.
12. J. Mottar, R. van Renterghem, and J. De Vilder, *Milchwissenschaft*, 40(12): 717 (1985).
13. J. W. van den Bedem and J. Leenheer, *Neth. Milk Dairy J.*, 42: 345 (1988).
14. W. Kneifel and F. Ulberth, *Milchwissenschaft*, 40(5): 265 (1985).
15. R. N. Carpender and R. J. Brown, *J. Dairy Sci.*, 68: 307 (1985).
16. B. A. Marchylo, J. E. Kruger, and D. Hatcher, *Cereal Chem.*, 63(3): 219 (1986).

17. G. L. Lookhart and L. D. Albers, *Cereal Chem.*, *65*(3): 222 (1988).
18. I. L. Batey, R. B. Gupta, and F. MacRitchie, *Cereal Chem.*, *68*(2): 207 (1991).
19. N. K. Singh, R. Donovan, and F. MacRitchie, *Cereal Chem.*, *67*(2): 161 (1990).
20. D. M. McLean, E. R. B. Graham, and R. W. Ponzoni, *J. Dairy Res.*, *51*: 531 (1984).
21. G. E. Vegarud, T. S. Molland, M. J. Broveld, T. G. Devold, P. Alestrøm, T. Steine, S. Rogne, and T. Langsrud, *Milchwissenschaft*, *44*(11): 689 (1989).
22. J. A. Bietz, T. Burnouf, L. A. Colb, and J. S. Wall, *Cereal Chem.*, *61*(2): 129 (1984).
23. J. A. Bietz and L. A. Colb, *Cereal Chem.*, *62*(5): 332 (1985).
24. B. A. Marchylo, D. W. Hatcher, and J. E. Kruger, *Cereal Chem.*, *65*(1): 28 (1988).
25. F. R. Huebner, J. A. Bietz, B. D. Webb, and B. O. Juliano, *Cereal Chem.*, *67*(2): 129 (1990).
26. C. Olieman and J. A. M. Van Riel, *Neth. Milk Dairy J.*, *43*: 171 (1989).
27. C. Olieman and J. W. van den Bedem, *Neth. Milk Dairy J.*, *37*: 27 (1983).
28. C. Olieman, *Voedingsmiddelentechnologie*, *26*(5): 10 (1993).
29. E. Gombocz, E. Helluig, and F. Petuely, *Z. Lebensm. Unters, Forsch.*, *172*: 178 (1981).
30. H. Elbertzhagen and E. Wenzel, *Z. Lebensm. Unters. Forsch.*, *175*: 15 (1982).
31. T. M. Dougherty, *J. Fd. Sci.*, *42*(6): 1611 (1977).
32. P. Teufel and V. Sacher, *Fleischwirtschaft*, *62*(11): 1474 (1982).
33. J. C. Collin, G. M. De Retta, and P. Martin, *J. Dairy Res.*, *49*: 221 (1982).
34. J. Daussant and D. Bureau, *Developments in Food Analysis Techniques, Vol. 3* (R. D. King, ed.), Elsevier, Essex, 1984, p. 175.
35. K. Khan, L. Huckle, and B. L. Jones, *Cereal Chem.*, *69*(3): 270 (1992).
36. J. W. Paulis and J. A. Bietz, *Cereal Chem.*, *65*(3): 215 (1988).
37. J. M. Girardet, D. Paquet, and G. Linden, *Milchwissenschaft*, *44*(11): 692 (1989).
38. A. J. R. Law, J. Leaver, J. M. Banks, and D. S. Horne, *Milchwissenschaft*, *48*(12): 663 (1993).
39. L. C. Parial and L. W. Rooney, *Cereal Chem.*, *47*(1): 38 (1970).
40. L. Gaspar, *Methods of Protein Analysis* (I. Kerese, ed.), Ellis Horwood Ltd., Chichester, 1980, p. 30.
41. I. Goshev and P. Nedkov, *Anal. Biochem.*, *95*: 340 (1979).
42. A. C. Jennings, *Cereal Chem. 38*: 467 (1961).
43. H. Mitsuda and T. Mitsunaga, *Agric. Biol. Chem.*, *38*(9): 1649 (1974).
44. J. Bosset, B. Blanc, and E. Plattner, *Analytica Chimica Acta*, *70*: 327 (1974).
45. J. Bosset, B. Blanc, and E. Plattner, *Analytica Chimica Acta*, *71*: 97 (1974).
46. J. Bosset, B. Blanc, and E. Plattner, *Analytica Chimica Acta*, *75*: 343 (1975).
47. G. L. Peterson, *Anal. Biochem.*, *100*: 201 (1979).
48. A. Bensadown and D. Weinstein, *Anal. Biochem.*, *70*: 241 (1976).
49. J. A. Bietz, T. Burnouf, L. A. Colb, and J. S. Wall, *Cereal Chem.*, *61*(2): 124 (1984).
50. F. R. Huebner and J. A. Bietz, *Cereal Chem.*, *64*(1): 15 (1987).
51. A. Kawka, P. K. W. Ng, and W. Bushuk, *Cereal Chem.*, *69*(1): 92 (1992).
52. E. M. Magnus and K. Khan, *Cereal Chem.*, *69*(6): 607 (1992).
53. N. K. Singh, G. R. Donovan, I. L. Batey, and F. MacRitchie, *Cereal Chem.*, *67*(2): 150 (1990).
54. P. Barrefors, B. Ekstrand, L. Fägestam, M. Larsson-Raznikiewics, J. Schaar, and P. Steffner, *Milchwissenschaft*, *40*(5): 257 (1985).
55. M. J. Dunn, *Protein Purification Methods: A Practical Approach* (E. L. Y. Harris and S. Angal, eds.), Oxford University Press, New York, 1989, p. 18.
56. H. J. Sinell and I. Mentz, *Fleischwirtschaft*, *62*(1): 99 (1982).
57. A. L. Lakin, *Developments in Food Analysis Techniques*, Vol. 1 (R. D. King, ed.), Elsevier Applied Science, London, 1978, p. 43.
58. A. Lillevik, *Methods in Food Analysis* (M. A. Joslyn, ed.), Academic Press, London, 1970, p. 601.
59. O. H. Lowry, N. J. Rosebrough, A. L. Farr, and R. J. Randall, *J. Biol. Chem.*, *31*: 964 (1951).
60. A. M. Siouffi, *Food Analysis by HPLC* (L. M. L. Nollet), Marcel Dekker, New York, 1992, p. 1.
61. W. G. M. Braam, *Scheidingsmethoden: elektroforese* (W. G. M. Braam, ed.), Wolters-Noordhoff, Groningen, 1987.
62. C. J. Smith, *Development and Application of Immunoassay for Food Analysis* (J. H. Rittenburg, ed.), Elsevier, Essex, England, 1990, p. 3.
63. R. S. Kirk and R. Sawyer, *Pearson's Composition and Analysis of Foods*, 9th ed. (R. S. Kirk and R. Sawyer, eds.), Longman, 1991.

64. J. M. Regenstein and C. E. Regenstein, *Food Protein Chemistry* (J. M. Regenstein and C. E. Regenstein, eds.), Academic Press, London.
65. H. Guillou, J. P. Pelissier, and R. Grappin, *Le Lait, 66*: 143 (1986).
66. D. A. Dohan and K. G. Standing, *Cereal Chem., 51*(1): 91 (1976).
67. M. J. Dunn, *Protein Purification Methods: A Practical Approach* (E. L. Y. Harris and S. Angal, eds.), Oxford University Press, New York, 1989, p. 10.
68. L. V. Coveny, *Scientific and Technical Surveys Leatherhead Food R. A.,* No. 123, 1980.
69. N. D. Shiralkar, H.-P. Harz, and H. Weisser, *Lebensm. Wiss. Technol., 16*: 18 (1983).
70. R. C. Noble, J. H. Shand, and I. G. West, *J. Dairy Sci., 64*: 14 (1981).
71. A. Wolfschoon-Pombo and H. Klostermeyer, *Milchwissenschaft, 36*(10): 598 (1981).
72. A. Wolfschoon-Pombo and H. Klostermeyer, *Milchwissenschaft, 37*(2): 80 (1982).
73. L. V. Coveny, *Scientific and Technical Surveys Leatherhead Food R. A.,* No. 118 (1980).
74. C. Nkonge and G. M. Ballance, *J. Agric. Food Chem., 30*: 416 (1982).
75. L. L. Wall, Sr., C. W. Gehrke, T. E. Neuner, R. D. Cathey, and P. R. Rexroad, *J. Assoc. Off. Anal. Chem., 58*: 811 (1975).
76. Tecator specification sheets Kjeltec System 1026 Distilling Unit, Kjeltec Auto 1030 Analyser, and Kjeltec Auto Sampler System 1035, Analyser/1038 Sampler (1993).
77. K. Kanaya and K. Hiromi, *Agric. Biol. Chem., 57*(7): 1885 (1987).
78. R. M. Johnson and C. E. Crany, *Cereal Chem., 48*(3): 276 (1971).
79. J. S. Noll, D. H. Simmons, and W. Bushuk, *Cereal Chem., 51*(5): 610 (1974).
80. W. T. Greenaway, *Cereal Chem., 49*(6): 609 (1972).
81. C. E. Craney, *Cereal Chem., 49*(4): 496 (1972).
82. R. F. Itzhaki and D. M. Gill, *Anal. Biochem., 9*: 401 (1964).
83. Y. Popineau and B. Godon, *Lebensm. Wiss. Technol., 9*: 38 (1976).
84. G. W. Pace, M. C. Archer, and S. R. Tannenbaum, *Anal. Biochem., 60*: 649 (1974).
85. G. R. Schacterle and R. L. Pollack, *Anal. Biochem., 51*: 654 (1973).
86. Y. W. Huang, R. T. Marshall, M. E. Anderson, and C. Charoen, *J. Food Sci., 41*: 1219 (1976).
87. G. D. Hegeman, *Anal. Biochem., 37*: 204 (1970).
88. M. E. Anderson and R. T. Marshall, *J. Food Sci., 40*: 728 (1975).
89. M. M. Bradford, *Anal. Biochem., 72*: 248 (1976).
90. M. Kroger, E. E. Katz, and J. C. Weaver, *J. Dairy Sci., 61*: 274 (1978).
91. D. L. Park and R. L. King, *J. Assoc. Off. Anal. Chem., 57*(1): 42 (1974).
92. D. M. Kirschenbaum, *Anal. Biochem., 90*: 309 (1978).
93. J. R. Whitaker and P. E. Granum, *Anal. Biochem., 109*: 156 (1980).
94. Grappin et al., *Le Lait, 56*: 498 (1979).
95. S. Roe, *Protein Purification Methods: A Practical Approach* (E. L. V. Harris and S. Angel, eds.), IRL Press, Oxford, 1989, p. 175.
96. M. P. Henry, *HPLC of Macromolecules: A Practical Approach* (R. W. A. Oliver, ed.), IRL Press, Oxford, 1988, p. 91.
97. D. Johns, *HPLC of Macromolecules: A Practical Approach* (R. W. A. Oliver, ed.), IRL Press, Oxford, 1988, p. 1.
98. G. W. Welling and S. Welling-Wester, *HPLC of Macromolecules: A Practical Approach* (R. W. A. Oliver, ed.), IRL Press, Oxford, 1988, p. 77.
99. A. Z. Preneta, *Protein Purification Methods: A Practical Approach* (E. L. V. Harris and S. Angel, eds.), IRL Press, Oxford, 1989, p. 293.
100. O. Takács and I. Kerese, *Methods of Protein Analysis* (I. Kerese, ed.), Ellis Horwood, Chichester, 1980, p. 87.
101. W. T. Kok and G. J. M. Bruin, *Eur. Chromatogr. News, 2*(5): 22 (1988).
102. M. J. Gordon, X. Huang, S. L. Pentoney, Jr., and R. N. Zore, *Science, 242*: 224 (1988).
103. J. H. Skerritt, *Development and Application of Immunoassay for Food Analysis* (J. H. Rittenburg, ed.), Elsevier, Essex, England, 1990, p. 81.
104. G. D. Johnson, *Antibodies, Vol. 2, A Practical Approach* (D. Catty, ed.), Oxford University Press, Oxford, 1989, p. 179.
105. J. H. Rittenburg, *Development and Application of Immunoassay for Food Analysis* (J. H. Rittenburg, ed.), Elsevier, Essex, England, 1990, p. 29.

106. D. Catty and C. Raykundalia, *Antibodies, Vol. 2, A Practical Approach* (D. Catty, ed.), Oxford University Press, Oxford, 1989, p. 97.

107. D. Catty and G. Murphy, *Antibodies, Vol. 2, A Practical Approach* (D. Catty, ed.), Oxford University Press, Oxford, 1989, p. 77.

108. C. V. Morr, *J. Food Sci.*, 55(4): 1177 (1990).

109. L. L. Diosady, I. Bergen, and V. R. Harwalkar, *Milchwissenschaft*, 35(11): 671 (1980).

110. M. de Frutos, A. Cifuentes, L. Amigo, M. Ramos, and J. C. Diez-Masa, *Z. Lebensm. Unters. Forsch.*, 195: 326 (1992).

111. D. P. Goldenberg, *Protein Structure: A Practical Approach* (T. E. Creighton, ed.), IRL Press, Oxford, 1989, p. 228.

112. G. L. Lookhart, B. L. Jones, S. B. Hall, and K. F. Finney, *Cereal Chem.*, 59(3): 178 (1982).

113. K. Khan and L. Huckle, *Cereal Chem.*, 69(6): 687 (1992).

114. J. J. Basch, F. W. Douglas, Jr., L. G. Procino, V. H. Holsinger, and H. M. Farrell, Jr., *J. Dairy Sci.*, 68: 23 (1985).

115. J. S. Wall, D. A. Fey, and J. W. Paulis, *Cereal Chem.*, 61(2): 141 (1984).

116. E. Erhardt, *Milchwissenschaft*, 44(3): 145 (1989).

117. A. Cifuentes, M. de Frutos, and J. C. Diez-Masa, *J. Dairy Sci.*, 76: 1870 (1993).

118. E. Gombocz, E. Helluig, and F. Petuely, *Z. Lebensm. Unters. Forsch.*, 172: 355 (1981).

119. M. Duranti, S. Pagani, S. Iametti, and A. Carpen, *Milchwissenschaft*, 44(3): 142 (1989).

120. F. C. Mao, and R. D. Bremel, *J. Dairy Sci.*, 74: 2946 (1991).

121. F. C. Mao, R. D. Bremel, and M. R. Dentine, *J. Dairy Sci.*, 74: 2952 (1991).

122. E. Renner and S. Ömeroglu, *Z. Lebensm. Unters. Forsch.*, 149(5): 267 (1972).

123. E. Renner and S. Ömeroglu, *Z. Lebensm. Unters. Forsch.*, 149: 329 (1972).

124. J. W. Sherbon, *J. Assoc. Off. Anal. Chem.*, 53(4): 862 (1970).

125. J. A. O'Connell and T. C. A. McGann, *Irish Agric. Cream. Rev.*, 25(11): 17 (1972).

126. J. P. Richard and P. Paquin, *Milchwissenschaft*, 45(2): 92 (1990).

127. E. Renner and S. Ömeroglu, *Z. Lebensm. Unters. Forsch.*, 150: 338 (1973).

128. H. El-Gazzer, *Agric. Res. Rev.*, 56(6): 171 (1978).

129. T. C. A. Mc Grann, A. Matthassen, and J. A. O'Connell, *Irish Agric. Cream. Rev.*, 26(7): 17 (1973).

130. E. Renner and S. Ando, *Z. Lebensm. Unters. Forsch.*, 155: 23 (1974).

131. E. Renner and S. Ömeroglu, *Z. Lebensm. Unters. Forsch.*, 150(5): 295 (1973).

132. H. Kiriyama, E. Harada, and B. Syuto, *J. Dairy Sci.*, 72: 398 (1989).

133. H. Brehmer and H. Gerdes, *Fleishwirtschaft*, 60(7): 1374 (1980).

134. A. Driou, G. Godbillon, and G. Linden, *Sciences des Aliments*, 5: 73 (1985).

135. C. Staak and U. Kämpe, *Fleishwirtschaft*, 62(11): 1477 (1982).

136. W. T. Greenaway and R. M. Johnson, *Bakers' Digest*, 48(2): 38 (1974).

137. E. Gabor, *Acta Alim.*, 18(1): 31 (1989).

138. Y. Pomeranz and R. B. Moore, *Bakers' Digest*, 49(1): 44 (1975).

139. B. Sebecic, *Nahrung*, 31(8): 817 (1987).

140. R. Biston and G. Clamot, *Cereal Chem.*, 59(5): 333 (1982).

141. H. Mitsuda and T. Mitsunaga, *Agric. Biol. Chem.*, 38(11): 2265 (1974).

142. Y. Pomeranz, R. B. Moore, and F. S. Lai, *J. Am. Soc. Brew. Chem.*, 35(2): 86 (1977).

143. H. D. Bock, J. Wuensche, and E. Borgman, *Nahrung*, 17(8):829 (1973).

144. J. H. Skerritt, *Adv. Cereal Sci. Techn.*, 9:263 (1988).

145. F. R. Huebner and J. A. Bietz, *Cereal Chem.*, 65(4): 362 (1988).

146. G. L. Lookhart, L. D. Albers, and J. A. Bietz, *Cereal Chem.*, 63(6): 497 (1986).

147. P. Meier, H. Windemann, and E. Baumgarthner, *Z. Lebensm. Unters. Forsch.*, 178(5): 361 (1984).

148. W. E. Werner, A. E. Adalsteins, and D. D. Kasarda, *Cereal Chem.*, 69(5): 535 (1992).

149. A. Curioni, A. D. B. Peruffo, G. Pressi, and N. E. Pagna, *Cereal Chem.*, 68(2): 200 (1991).

150. S. E. Heisel, D. M. Peterson, and B. L. Jones, *Cereal Chem.*, 63(6): 500 (1986).

151. G. L. Lookhart and Y. Pomeranz, *Cereal Chem.*, 62(3): 162 (1985).

152. B. Mertens, The properties of proteins in the formation and stabilization of o/w emulsions, doctoral thesis, University of Ghent.

10

Enzymes

R. Muñoz and A. Ros Barceló
University of Murcia, Murcia, Spain

I. INTRODUCTION

The history of enzymes is closely connected with our increased knowledge in the fields of foodstuff chemistry and technology. It started around 1526 with Paracelsus's studies of fermentation products. The assay of the enzyme characteristics of a certain food is carried out to determine its degree of freshness (as is the case with the oxidative enzymes found in vegetables), to detect particular treatments such as pasteurization (easily monitored in milk by measuring the levels of phosphatases and lactoperoxidase), or to see whether decay or microbial contamination has started.

Enzyme levels can be easily measured by observing the action they have on their substrates, since most of these enzyme-catalyzed reactions have an absolute specificity. The first satisfactory mathematical analysis of the course of an enzyme-catalyzed reaction was by Michaelis and Menten in 1913. These authors suggested that the rate of transformation of a substrate (v) is a function, over a certain range, of the concentration of substrate $[S]$ and the concentration of enzyme $[E]$:

$$v = \frac{k_{cat}\,[E\,][S\,]}{K_M + [S\,]} \tag{1}$$

where k_{cat} and K_M are constants. Assuming that this equation is valid for most enzyme-catalyzed reactions, the enzyme concentration (or activity) is proportional to the rate of the appearance of a product or disappearance of a substrate, and independent of substrate concentration, only when $[S] \gg K_M$, for which the above expression may be reduced to:

$$v = k_{cat}\,[E\,] \tag{2}$$

For this reason, it is desirable to know the K_M (Michaelis constant) for an enzyme using a particular substrate in a particular food. This can easily be calculated through the Lineweaver–Burk (double-reciprocal) plot of Eq. (1). This fact makes it difficult to describe a universal method of measurement for a certain enzyme since K_M values vary greatly for the same enzyme depending on the substrate used and the food analyzed. This means that if enzyme concentrations are to be accurately estimated, a previous knowledge of the K_M values is necessary.

311

From Eq. (2), it is not surprising that enzyme concentrations are given in katals, a unit of the SI system. The katal (kat) is defined as the amount of enzyme capable of transforming 1 mole of substrate in a product per second. Nevertheless, the concentration of enzymes in tissues normally varies between nkat (nanokatals) and pkat (picokatals) per g fresh weight.

Another factor to be taken into account in enzyme determinations is the pH, since most enzyme-catalyzed reactions show a strong dependence on the pH of the reaction medium. A well-known example is the case of grapevine peroxidase. Thus, whereas most plant peroxidases are assayed at pH 6.0–7.0, grapevine peroxidase is inactive at this pH, with its optimal activity occurring at pH 5.0. Strong variations in enzymatic activity according to pH are also observed in the case of lipoxygenase and polyphenol oxidase. It is therefore necessary to check the optimal pH for the enzyme being assayed in a particular food.

Together with pH, another factor which may mask enzymes in foods is the presence of endogenous inhibitors. For example, ascorbic acid is present in relatively large amounts in most fruits and vegetables, and is both a polyphenoloxidase and a peroxidase inhibitor. This makes enzyme purification necessary in some cases. The protocols for enzyme purification vary and in most cases depend on the food and on the grade of enzyme purity desired. In general, gel filtration, ion exchange chromatography, hydrophobic chromatography, isoelectric focusing, preparative electrophoresis, or high-performance liquid chromatography are useful procedures and are valid for most of the enzymes which are of interest in foodstuff chemistry.

Recently, two excellent books which describe the fundamentals of enzymology and explain the roles of enzymes in food, agricultural, and health sciences have appeared (1,2). For this reason, and due to the high number of enzymes used in foodstuff chemistry and the great number of assay methods which exist, this chapter cannot be, nor does it intend to be, exhaustive. Attention is only paid to the most widely used enzymes in foodstuff chemistry (oxidoreductases and hydrolases, which greatly affect the quality of processed food and their products) and to the most common methods for measuring them. In some cases, recently described methods that afford substantial improvements to existing methods are discussed.

II. OXIDOREDUCTASES

A. Amine and Amino Acid Oxidases

Amine and amino acid oxidases (EC 1.4.3.1–6) catalyze the following generic reaction:

Amine (amino acid) + H_2O + O_2 → aminoaldehyde (2-oxo acid) + NH_3 + H_2O_2

The importance of amine and amino acid oxidases for food chemists comes from the importance of amines and amino acids themselves, and of their oxidation products (aminoaldehydes, NH_3, and H_2O_2), in the quality of processed food and their products. Amines are pharmacologically and nutritionally important since they exhibit a wide range of physiological activities, which include the regulation of enzyme mechanisms, interaction with nucleic acids and membranes, bactericidal and pressor activities, neurotransmission, and toxicity. Thus, many cheeses are rich in tyramine and histamine, the products of bacterial action, which may cause several reactions in the consumer including hypertension, nausea, and vomiting. These symptoms are caused directly or indirectly by the vasoconstrictive and cardiostimulatory actions of food-derived precursor amines, as in the case of tyramine. Amines are also the direct or indirect causes of lathyrism and scombrotoxic fish poisoning.

In addition to the specific reactions described above, amine oxidases and amino acid oxidases can also affect the quality of foods in desirable or undesirable ways. Thus, amines add to the fuller flavor of some cheeses but are also responsible for the off-tastes of spoiling fish. Similarly, reaction products such as α-ceto acids may be involved in Maillard browning reactions, while other reaction products such as the semialdehydes may be involved in the modification of texture in certain formulated foods. H_2O_2 may also promote lipid peroxidation with changes in flavor and color.

1. Diamine Oxidase

Diamine oxidase [amine:oxygen oxidoreductase (EC 1.4.3.6)] catalyzes the following reaction:

$$\text{Diamine} + H_2O + O_2 \rightarrow \text{aminoaldehyde} + NH_3 + H_2O_2$$

This enzyme has been isolated from different sources and monoamines, diamines, polyamines, and certain amino acids are substrates of the enzyme. Diaminooxidases are enzymes which are widely distributed among living organisms, although their physiological role is not fully understood. They have often been considered as histamine scavengers in animal tissues, hence the trivial name "histaminase" attributed to them. In plants and bacteria, however, this hypothesis seems untenable. In plants, this enzyme is soluble and is easily extracted with phosphate aqueous buffers. In some cases (3), 0.5 M NaCl is added to the homogenization buffer to solubilize the tightly bound diaminooxidase fraction.

    a. Diaminooxidase Assays. The diaminooxidase assays are based on the measurement of either O_2 consumption or H_2O_2 production coupled to a peroxidase assay (Table 1). In all cases the addition of 1.0 mM aminoguanidine, a strong inhibitor of diaminooxidase, is used as a control for enzymatic activity. In the oxygen uptake assay, the reaction medium contains 50 μg catalase and a crude enzyme preparation in aerated 0.1 M sodium phosphate buffer, pH 7.0 (4). The reaction is initiated by the addition of putrescine (diaminobutane) to a final concentration of 1.0 mM. Oxygen uptake is monitored with an oxygen electrode, the activity being expressed in nmol O_2 consumed/min.

    In the peroxidase-coupled assay (5), the reaction mixture contained 0.1 M sodium phosphate buffer, pH 7.0, 0.5 mM guaiacol, 18.5 nkat peroxidase, and a sample of diaminooxidase. The reaction is started by the addition of putrescine (final concentration of 1.0 mM) and the increase in absorbance at 436 nm is recorded with time.

    In other spectrophotometric methods, the formation of Δ^1-pyrroline is used for the measurement of the enzymatic activity (6). In this case, the reaction is carried out in 0.1 M sodium phosphate buffer, pH 7.0, 10 mM putrescine, and the enzyme, in a total volume of 4.0 ml. After incubation at 37°C for 10 min, the reaction is stopped by adding 0.5 ml of 10% (w/v) trichloroacetic acid, followed by 0.05 ml o-aminobenzaldehyde (10 mg/ml) in ethanol. The absorbance at 430 nm is measured after removal of the protein by centrifugation. Enzymatic activities are calculated using an $\varepsilon_{430} = 1.86 \ 10^3 \ M^{-1} \ cm^{-1}$ for the Δ^1-pyrroline.

Table 1 Diaminooxidase Assays

Method	Reaction medium	Detection	Ref.
O_2 uptake	0.1 M Na phosphate pH 7.0 + 50 μg catalase + 1.0 mM putrescine	O_2 electrode	4
Peroxidase-coupled assay	0.1 M Na phosphate pH 7.0 + 0.5 mM guaiacol + 18.5 nkat peroxidase + 1.0 mM putrescine	Spectrophotometry 436 nm	5
$^1\Delta$-Pyrroline	0.1 M Na phosphate pH 7.0 + 10 mM putrescine + 10% trichloroacetic acid + 10 mg/ml o-aminobenzaldehyde	Spectrophotometry 430 nm	6

A widely used procedure for staining diaminooxidase activity based on a peroxidase-coupled assay (7) after the electrophoretic separation of isoenzymes consists of incubating the gels in 0.1 M sodium phosphate buffer, pH 7.0, containing 20 µg/ml peroxidase, 0.2 mg/ml benzidine, and 17 mM cadaverine.

B. Polyphenoloxidases

Polyphenoloxidases (EC 1.10.3.1–2) have received the continuous attention of food chemists because they are involved in the enzymatic browning of many edible plant products—especially fruits, vegetables, and mushrooms—as well as crustaceans such as shrimp, lobster, and crab. They are enzymes which catalyze the following reaction:

$$2 \text{ diphenol} + O_2 \rightarrow 2 \text{ quinone} + 2H_2O$$

The Enzyme Commission originally referred to two enzymes in this category: (a) catechol oxidase or o-diphenol:oxygen oxidoreductase (EC 1.10.3.1) and (b) laccase or p-diphenol:oxygen oxidoreductase (EC 1.10.3.2). Catechol oxidase is frequently referred to as phenolase, polyphenoloxidase, tyrosinase, catecholase, or cresolase. Catechol oxidase from animal tissues is relatively specific for tyrosine and dopa, while the fungal and higher plant enzymes act on a wide range of mono- and o-diphenols (8). On the other hand, laccase is a much less specific enzyme, and it is capable of oxidizing nonphenolic substances such as ascorbic acid, diaminofluorene, and hydroxyindoles.

1. Laccase

Laccase [p-diphenol:oxygen oxidoreductase (EC 1.10.3.2)] catalyzes the following reaction:

$$2 \, p\text{-diphenol} + O_2 \rightarrow 2 \, p\text{-quinone} + 2H_2O$$

The importance of laccase in food chemistry is, to date, restricted to its role as an indicator of *Botrytis* contamination in must and wines (9).

a. Sample Preparation and Laccase Assays. For the determination of laccase activity in must and wines (Table 2), phenolics should be removed by absorption on polyvinylpolypyrrolidone, and cleared must and wines must be used directly. Laccase activity may be monitored by oxygen consumption using 10 mM p-hydroquinone in a 0.1 M sodium acetate buffer, pH 5.0 (9) or by spectrophotometry using syringaldazine as substrate. This last assay is also carried out in sodium acetate buffer, measuring the increase in absorbance at 530 nm ($\varepsilon_{530} = 65{,}000 \text{ M}^{-1} \text{ cm}^{-1}$) of a reaction medium containing 10 mg/L of syringaldazine (9). Recently, assays of laccase activity in plant materials using hydroxyindoles (10) and diaminofluorene (11) as substrates have been described. Syringaldazine, hydroxyindoles, and aminofluorene may also be used to stain laccase isoenzymes after electrophoretic separation. Other reagents used for this purpose include

Table 2 Laccase Assays

Method	Reaction medium	Detection	Ref.
O₂ uptake	0.1 M Na acetate pH 5.0 + 10 mM p-hydroquinone	O₂ electrode	9
Syringaldazine	0.1 M Na acetate pH 5.0 + 10 mg/L syringaldazine	Spectrophotometry 530 nm	9

N,N,N',N'-tetramethyl-p-phenylenediamine (2.1 mg/ml) and α-naphthol (1.4 mg/ml) in 0.1 M sodium acetate buffer, pH 5.5 (12).

2. Catechol Oxidase

Catechol oxidase [o-diphenol:oxygen oxidoreductase (EC 1.10.3.1)] catalyzes two distinct reactions:

(i) the insertion of oxygen in an *ortho* position into an existing hydroxyl group, usually followed by oxidation of the diphenol to the corresponding quinone, and referred to as cresolase activity:

Monophenol + $O_2 \rightarrow o$-quinone + H_2O and (ii) the oxidation of an o-diphenol to o-quinone: 2 o-diphenol + $O_2 \rightarrow 2$ o-quinone + $2H_2O$ simply referred to as catecholase activity.

With the exception of that which occurs in fungi, in which catecholase activity has been reported as being secreted into the culture medium, plant catechol oxidases are generally located intracellularly, and are found in the soluble fractions and bound to membranes (8). This enzyme is responsible for most browning reactions in damaged fruits and vegetables, and so its control is of great importance. Although catechol oxidase is an enzyme widely distributed throughout the plant kingdom, in some cases it has been impossible to detect probably due to its latency and the occurrence of endogenous inhibitors.

a. Extraction and Purification of Catechol Oxidase. Soluble catechol oxidase may be extracted from most plant tissues by grinding in aqueous buffers. A homogenization buffer widely used is 0.1 M sodium phosphate buffer, pH 7.3, containing 10 mM ascorbic acid. In other cases, such as artichoke (13), Yali pear (14), and grape (15) acetone powders are used as an enzyme source for extraction. In these cases, soluble protein fractions are considered sources of enzyme.

The latency and activation of catechol oxidases has been widely studied (8). Its activation can be achieved by short exposure to acid (pH 3.0–3.5), alkali (pH 11.5), detergents, or by limited proteolysis with trypsin or carboxypeptidase (8). Latent catechol oxidase is also extracted from the debris resulting from the homogenization of tissues in aqueous buffers (16,17). For this, the debris are digested overnight at 4°C with 0.1 M sodium phosphate buffer, pH 7.3, containing 1.5% (w/v) Triton X-114 and 2% (w/v) polyvinylpolypyrrolidone, and then centrifuged at $25,000g_{max}$ for 55 min at 4°C. The supernatant is subjected to temperature phase partitioning via addition of Triton X-114 at 4°C up to 4% (w/v) (17). The mixture is kept at 4°C for 15 min and then warmed to 37°C. After 15 min, the solution is centrifuged at $5000g_{max}$ for 15 min at 25°C. The supernatant is used as an enzyme source, to which the protease inhibitors leupeptin and N-ethylmaleimide are added up to a final concentration of 20 µg/ml and 5 mM, respectively (17).

Procedures for purifying catechol oxidase include gel filtration (18), ion exchange chromatography on DEAE-cellulose (14,18,19), hydrophobic (affinity) chromatography on phenyl-Sepharose CL-4B (13,20), chromatography on hydroxyapatite (21) or on concanavalin A–agarose (20), high-performance liquid chromatography on a LiChrospher Si-100 diol column (15), and preparative isoelectric focusing (22).

b. Catechol Oxidase Assays. Catechol oxidase assays are based on the measurement of either O_2 consumption or phenol oxidation (Table 3). In the oxygen uptake assay, the reaction medium contains 10–50 mM catechol (or other phenolic substrates such as 4-methylcatechol, chlorogenic acid, or L-dopa at the appropriate concentration) and a crude enzyme preparation in 0.1 M sodium phosphate buffer, pH 4.5–7.0 (the optimal pH for catechol oxidase activity should be routinely checked). Oxygen uptake is monitored with an oxygen electrode.

Spectrophotometric assays for the measurement of catechol oxidase activity vary greatly, the most widely used being those which measure o-quinone formation in the 390- to 420-nm spectral range. For this purpose, the reaction media are similar to that described in the measurement of

Table 3 Catechol Oxidase and Cresolase Assays

Method	Reaction medium	Detection	Ref.
Catechol oxidase			
O_2 uptake	0.1 M Na phosphate pH 4.5–7.0 + 10–50 mM *o*-diphenol	O_2 electrode	See text
o-Quinone detection	i.d.	Spectrophotometry 390–420 nm	See text
Proline/catechol	0.1 M Na phosphate pH 4.5–7.0 + 30 mM catechol + 30 mM proline	Spectrophotometry 530 nm	23, 24
Cresolase			
p-Cresol	0.1 M Na phosphate pH 7.0 + 0.5–1.0 mM *p*-cresol	Spectrophotometry 400 nm	26

oxygen uptake. Recently, methods have been proposed for following catechol oxidase spectrophotometrically by using coupling reactions with the quinone formed. In such spectrophotometric assays, e.g., the proline-catechol assay (23,24), the reaction medium contains 30 mM catechol and 30 mM L-proline in 0.1 M phosphate buffer, pH 4.5–7.0, the absorbance at 530 nm being recorded as time elapses. Also based on a coupling reaction is a new spectrophotometric assay for catecholase activity which uses cysteine and several *o*-diphenols (25).

The cresolase activity of polyphenoloxidases is easily monitored using *p*-cresol (26). For this, the cresolase activity of enzyme preparations is assayed in a reaction medium containing 0.5–1.0 mM *p*-cresol in 0.1 M sodium phosphate buffer, pH 7.0, by measuring the increases in absorbance at 400 nm, and using an $\varepsilon_{400} = 1350 \, M^{-1} \, cm^{-1}$ for the 4-methyl-*o*-benzoquinone formed. Recently, it was proposed that cresolase activity of polyphenoloxidase can be measured spectrophotometrically by monitoring the adduct formed by reaction of the *o*-quinone endproduct with 3-methyl-2-benzothiazolinone hydrazone (27). All these coupled spectrophotometric assays have the advantage that they can also be used for staining both catechol oxidase and cresolase activity of polyphenoloxidases after their separation on polyacrylamide gels.

    c. Differentiation of Catechol Oxidase from Laccase and Peroxidase. At this point, it is extremely important to differentiate catechol oxidase from laccase and peroxidase, since both laccase and peroxidase are capable of oxidizing *o*-diphenols. This discrimination is rarely seen in the literature. Catecholase activity can easily be distinguished from peroxidase through the use of catalase as peroxidase inhibitor. In some cases even, peroxidase is capable of oxidizing *o*-diphenols in the absence of H_2O_2 and the use of catecholase inhibitors is necessary. For this, fusaric acid, salicylhydroxamic acid, or tropolone at 0.1–1.0 mM concentrations may be adequate. In these controls, the oxidation rate of *o*-diphenols sensitive to inhibitors should be considered as the true catechol oxidase activity.

    Catechol oxidase and laccase can easily be distinguished by (8) (a) the oxidation of *p*-diphenols, (b) the response to CO, and (c) the response to phenylhydrazine. Thus, laccase is not inhibited by CO or by phenylhidrazine, while catechol oxidase is incapable of oxidizing *p*-diphenols.

C. Catalases and Peroxidases

Catalases and peroxidases (EC 1.11.1.6–7) are enzymes widely distributed in the plant and animal kingdom, and are hemoproteins that use hydrogen peroxide as oxidizing substrate. Plant peroxidases and lactoperoxidase are of major interest to the food scientist.

1. Catalase

Catalase [hydrogen peroxide:hydrogen peroxide oxidoreductase (EC 1.11.1.6)] catalyzes the following reaction:

$$H_2O_2 + H_2O_2 \rightarrow O_2 + 2H_2O$$

The determination of catalase is of importance in the analysis of milk. Milk which is suitable for human consumption contains no significant amounts of catalase, whereas the presence of this enzyme indicates secretion disorders or diseases of the udder or colostrum, or bacterial contamination (28).

a. Preparation and Purification of Catalase from Milk. To obtain catalase from cow's milk, the cream is separated from fresh cow's milk. The cream is churned in a diamond-type metal churn to obtain buttermilk, which is freeze-dried (29). Partial purification of catalase using this buttermilk sample is then carried out by the method of Ito and Akuzawa (30), using *n*-butanol extraction, ammonium sulfate fractionation, and ethanol-chloroform fractionation.

b. Assays of Catalase in Milk. Catalase activity may be measured by following either the decomposition of H_2O_2 or the liberation of O_2 (Table 4). The preferred method for food analysis is the UV spectrophotometric method since the titrimetric methods are only suitable for comparative purposes.

In the spectrophotometric assay, catalase activity is monitored by the decrease in absorbance at 230 nm of a reaction medium containing 10–20 mM H_2O_2, 50 mM Na phosphate buffer, pH 7.0, and enzyme. Enzymatic units are calculated using an $\varepsilon_{230} = 67$ M^{-1} cm^{-1} (31).

In the titrimetric method, 60 mM Na perborate in 50 mM Na phosphate, pH 7.0, is preincubated for 10–20 min at 25°C and then added to the enzyme sample. After 30–60 sec of incubation, the remaining perborate is back-titrated with 0.05 N $KMnO_4$.

Also based on the decomposition of Na perborate to H_2O_2 and the subsequent evolution of O_2 due to catalase action, catalase activity may be monitored with an O_2 electrode (32). For this, 0.1 M Na perborate, standardized by titration with 0.1 N $KMnO_4$, in 50 mM Na phosphate buffer, pH 7.0, is incubated with catalase samples, and O_2 evolution is monitored with a Clark oxygen electrode. In all of the described cases, catalase acvtivity is taken as those sensitive to inhibition with 0.1 mM Na azide or with 20 mM 3-amino-1,2,4-triazole, two well-known catalase inhibitors.

2. Peroxidase

Peroxidase [donor:hydrogen peroxide oxidoreductase (EC 1.11.1.7)] catalyzes the following reaction:

$$Donor + H_2O_2 \rightarrow oxidized\ donor + 2H_2O$$

Table 4 Catalase Assays

Method	Reaction medium	Detection	Ref.
Peroxide assay	50 mM Na phosphate pH 7.0 + 10–20 mM H_2O_2	Spectrophotometry 230 nm	31
Titrimetric assay	50 mM Na phosphate pH 7.0 +60 mM Na perborate	Titration with $KMnO_4$	See text
O_2 evolution	50 mM Na phosphate pH 7.0 + 0.1 M Na-perborate	O_2 electrode	32

Although plant peroxidases oxidize a vast array of endogenous phenolics, these enzymes being involved in browning reactions, in the case of the lactoperoxidase system the donor is thiocyanate anion (SCN^-) (33), which plays a key role in the antibacterial system of the milk.

The determination of peroxidase is of importance in food analysis since it is involved in the ripening of fruits and thus affects the flavor, texture, and color of processed fruit products, although the relationship of peroxidase activity to off-flavors and off-colors in most raw and unblanched vegetables is still largely empirical (34). Likewise, the use of peroxidase activity as an indicator of blanching treatment is continuously being appraised in view of the higher temperatures required to inactivate peroxidases.

a. Extraction and Purification of Plant Peroxidases.   Peroxidases are easily extracted from plant tissues by grinding in aqueous buffers. Nevertheless, some peroxidase activity may be bound to cell structures, such as membranes and cell walls (35,36), and in this case it is necessary to add a solubilizing agent. Among these, the most widely used are KCl, NaCl, LiCl, and $CaCl_2$ at concentrations in the order of 0.1–2.0 M.

Due to the stability and solubility of plant peroxidases, their purification is easily achieved by gel filtration (37), ion exchange chromatography (37), hydrophobic chromatography (38,39), affinity-based reverse micellar extraction and separation (40), preparative isoelectric focusing (41), autofocusing (42), and high-performance liquid chromatography (43).

b. Peroxidase Assays.   Peroxidase activity is easily monitored spectrophotometrically by its action on their substrates when the oxidized donors are colored, and it is on this property that the assays of peroxidase activity are based (Table 5). In all cases, it is necessary to carry out controls for the enzymatic activity in the presence of catalase or in the absence of H_2O_2. Among the most widely used assays is the reaction with guaiacol. For this assay, the reaction medium contains 0.1 M Na phosphate buffer, pH 7.0, guaiacol 0.1–1.0 mM, and 0.1–1.0 mM H_2O_2. The reaction is initiated by the addition of enzyme and the increase in absorbance at 470 nm is monitored with time. Enzymatic activities are calculated using an $\varepsilon_{470} = 26.6 \times 10^3$ M^{-1} cm^{-1} (44) for the tetraguaiacol formed as endproduct. Another common method for the measurement of peroxidase activity is based on the measurement of the oxidation course of 3,3′-diaminobenzidine (DAB) (45).

4-Methoxy-α-naphthol can also be used for the determination of peroxidase activity and, together with DAB, has the advantage that it may be used for integrated spectrophotometric, zymographic, and cytochemical studies (46), since the colored product is partially stable and nondiffusible (46,47). For this, the reaction medium contains 0.1 mM 4-methoxy-α-naphthol, and 1.0 mM H_2O_2 in 0.1 M tris-acetate buffer, pH 5.0. The reaction is initiated by the addition of enzyme. Enzyme activity may be calculated using an $\varepsilon_{593} = 2.1 \times 10^4$ M^{-1} cm^{-1} for the dye product

Table 5   Peroxidase and Lactoperoxidsase Assays

Method	Reaction medium	Detection	Ref.
Peroxidase			
Guaiacol	0.1 M Na phosphate pH 7.0 + 0.1–1.0 mM guaiacol + 0.1–1.0 mM H_2O_2	Spectrophotometry 470 nm	44
4-Methoxy-α-naphthol	0.1 M tris-acetate pH 5.0 + 0.1 mM 4-methoxy-α-naphthol + 0.1 mM H_2O_2	Spectrophotometry 593 nm	46, 47
Lactoperoxidase			
ABTS	25 mM Na acetate pH 4.4 + 0.025–1.0 mM ABTS + 0.05–0.25 mM H_2O_2	Spectrophotometry 412 nm	48

(46). To stain peroxidase isoenzymes after electrophoretic separation, a reaction medium containing 1.0 mM 4-methoxy-α-naphthol, 1.0 mM H_2O_2 and 0.1 M tris-acetate buffer, pH 5.0, may be used (46).

In the case of lactoperoxidase, the most common method used is that of 2,2′-azinodi(3-ethylbenzthiazoline-6-sulfonic acid) (ABTS) (48). For this, the reaction medium contains 0.025–1.0 mM ABTS, 0.05–0.25 mM H_2O_2 in 25 mM acetate buffer, pH 4.4. The absorbance at 412 nm is followed with time once the reaction is initiated by the addition of enzyme.

D. Lipoxygenase

Lipoxygenase [linoleate:oxygen oxidoreductase (EC 1.13.11.12)] catalyzes the following reaction:

Unsaturated fat + O_2 → unsaturated fat peroxide

Thus lipoxygenase in the presence of dissolved oxygen catalyzes the formation of fatty acid hydroperoxides from the Z,Z-1,4-pentadiene portion of polyunsaturated fatty acids, the most common substrates being linoleic and linolenic acids (49). The hydroperoxides are broken down enzymatically and nonenzymatically into a variety of short-chain carbonyl compounds. These breakdown products, such as hexanal and E-hex-2-enal, have been implicated in aroma and taste changes in many processed foods (49). In the presence of linoleic and linolenic acid, lipoxygenase may also oxidize carotenoids and chlorophylls, resulting in the bleaching of colors in vegetables. Recently, lipoxygenase was also used as a blanching index for frozen soybeans (50).

1. Extraction and Assay of Lipoxygenase

Although lipoxygenase exists as a soluble enzyme in most plants and is easily extracted with aqueous buffers, in other cases, such as the kiwi fruit, a more complex protocol is needed (49). In this case, pericarp tissues are ground in liquid nitrogen to a fine paste, which is then extracted with 50 mM tris-HCl buffer, pH 7.0, containing 10 mM diethyldithiocarbamate, 1% Triton X-100, 5 g of polyvinylpolypyrrolidone, and 5 g Dowex 1. The mixture is stirred for 20 min, filtered through muslin cloth, and the filtrate centrifuged at $6000g_{max}$ for 60 min. The supernatant is then filtered through a 0.45-μm filter and passed through a C-18 reverse phase cartridge to remove the excess detergent. In other cases, hexane-defatted material (51) or acetone powder (52) is directly used as an enzyme source.

Lipoxygenase activity may be monitored polarographically or spectrophotometrically (Table 6). In the polarographic method (53), the reaction is carried out in a reaction medium containing 0.1–2.0 mM linoleic acid dispersed with 0.028% Tween 20 in 0.2 M tris-maleate buffer, pH 6.0–9.0. The reaction is initiated by the addition of enzyme and the oxygen uptake recorded continuously.

In the spectrophotometric method (54), increases in absorbance at 234 nm are followed with time, enzymatic units being calculated using an $\varepsilon_{234} = 25 \times 10^3$ M^{-1} cm^{-1} for the end product,

Table 6 Lipoxygenase Assays

Method	Reaction medium	Detection	Ref.
O_2 uptake	0.2 M tris-maleate pH 6.0–9.0 + 0.1–2.0 mM linoleic acid	O_2 electrode	53
Hydroperoxide	See text	Spectrophotometry 234 nm	54

hydroperoxylinoleic acid. For spectrophotometric measurements it is advisable to use linoleic acid solutions free of linoleic acid hydroperoxide. For this, the method of Tappel et al. (55) with some modifications (56) may be used. Stated briefly, 1 drop of Tween 80 is added to a mixture of 50 μl of linoleic acid and 50 μl of absolute ethanol in 10 ml of 50 mM tris-HCl buffer, pH 9.0 (deaerated with nitrogen), and the mixture was vortexed for 5 sec. The preparation is immediately poured into a cold beaker and maintained on ice.

III. HYDROLASES

A. Alkaline and Acid Phosphatases

Phosphatases (*ortho*-phosphoric monoester phosphohydrolase) catalyze the hydrolytic cleavage of phosphoric acid esters and are designated as either alkaline (EC 3.1.3.1) or acid (EC 3.1.3.2) phosphatases according to their optimal pH.

1. Alkaline Phosphatase

One of the most important features of alkaline phosphatase activity [*o*-phosphoric monoester phosphohydrolase (EC 3.1.3.1)] for food chemists arises from the usefulness of the phosphatase tests to check the temperature reached in milk during pasteurization processes (57,58). In addition, since milk is a proven vector for a number of pathogenic bacteria, the test is of very great significance to public health as a means of policing the thoroughness of heat treatments.

   a. Alkaline Phosphatase Assay.   The method is based (Table 7) on the colorimetric measurement of the *p*-nitrophenol produced in the enzymatic hydrolysis of *p*-nitrophenyl phosphate (59).
   To avoid fat, which interferes in the phosphatase assay, milk samples should be centrifuged at $300g_{max}$ for 10 min in order to separate the cream. The cream-free milk is taken off with a pipette and filtered through a small, moist, blue band filter.
   The alkaline phosphatase activity is followed in a reaction medium (1.1 ml) containing 50 mM glycine buffer, pH 10.5, 0.5 mM $MgCl_2$, 5.5 mM *p*-nitrophenyl phosphate, and 0.1 ml of filtered milk. After incubation for 1 hr at 37°C, add 5 ml of 0.02 N NaOH and measure the absorbance at 405 nm against a blank consisting of a medium to which NaOH is added prior to the sample. Enzyme activities are expressed as moles of phosphate ester hydrolyzed using a $\varepsilon_{405} = 18.5 \times 10^3$ M^{-1} cm^{-1} for *p*-nitrophenol.

2. Acid Phosphatase

Acid phosphatase [*o*-phosphoric monoester phosphohydrolase (EC 3.1.3.2)] is an enzyme widely distributed in citrus tissues, although it is much less abundant in the juice than in other parts of the

Table 7 Phosphate Assays

Method	Reaction medium	Detection	Ref.
Alkaline phosphatase			
p-Nitrophenyl phosphate	50 nM Na glycine pH 10.5 + 5.5 mM *p*-nitrophenyl phosphate + 0.5 mM $MgCl_2$	Spectrophotometry 405 nm	59
Acid phosphatase			
Phenyl phosphate	50 mM Na citrate pH 5.0 + 0.017 mM disodium phenyl phosphate	Colorimetry 605 nm	62

fruit. It decreases rapidly upon extraction and with storage of the juice (60). It has been proposed that acid phosphatase activity may be used as an indicator of the effectiveness of juice pasteurization (61). Similarly, residual acid phosphatase activity is an accepted marker for the assessment the core temperature reached during the pasteurization process of canned hams.

   a. Extraction and Assay of Acid Phosphatase.   Samples taken from the central area of the canned hams are ground through a 2-mm plate, and enzyme extraction is carried out by blending with 50 ml of 0.05 M citrate buffer, pH 5.4 (62).

The enzyme assay (Table 7) is based on the color which develops when 2,6-dibromoquinone chlorimide reacts with the phenol produced in the enzymatic breakdown of phenyl phosphate. For this, preincubate 10 ml of enzyme extract in a water bath at 37°C for about 15 min and then add 5 ml of 0.05 M disodium phenyl phosphate in 0.15 M citrate buffer, pH 5.0, prepared immediately before use. After 60 min, add 5 ml of 20% (w/v) trichloroacetic acid. Shake well and filter. To 5 ml of clear filtrate add 5.0 ml of 0.5 M sodium carbonate and 0.1 ml of 2,6-dibromoquinone chlorimide solution, prepared according to the *USDA Guidebook* (63). Stand in the dark for at least 15 min (not overnight) for color development and read the absorbance at 610 nm against a blank consisting of a similar medium in which the sample has been boiled for 30 min.

B. Peptidases and Proteinases

Peptidases (peptidyl hydrolases) are a broad family of enzymes which differ in the peptide bonds that preferably hydrolyze. Exopeptidases (carboxy- and aminopeptidases) cleave consecutive COOH or NH_2 terminal amino acids from the end of the polypeptide chain and endopeptidases or, more commonly, proteinases hydrolyze a specific bond without regard for its position in the polypeptide chain.

There are several uses for peptidase and proteinase activities in food processing and food quality control. It has been demonstrated, for example, that proteinase activity is responsible for undesirable changes in seafood products. Flesh degradation associated with proteolysis occurs in many commercially important species, including mackerel, cod, salmon, krill, prawn, and shrimp. The mushy texture which appears in the abdominal tail segments of prawns on cooking after ice storage seems to be caused by digestive enzymes diffusing into the tail segments from the disintegrating hepatopancreas. Baranowski et al. (64) and Nip et al. (65) suggested that collagenase (EC 3.4.24.3) is the main enzyme responsible for the mushiness phenomenon. Other crustacean digestive enzymes have been found to exhibit trypsinolytic, α-chymotrypsinolytic, and pepsinolytic activities in addition to collagenolytic activity (66,67). Histamine accumulation in mackerel and skip-jack tuna has been attributed to muscle and viscera proteases (68,69). In addition, proteolysis caused by alkaline proteases is thought to be involved in the deterioration of minced fish products during processing (70).

During the curing process, dry-cured hams develop a characteristic texture and flavor as a result of biochemical changes, including intense proteolysis. This proteolysis seems to be due to the action of muscle lysosomal cathepsins, which have the ability to degrade myofibrillar proteins and are active throughout the curing process (71–73).

The spoilage of milk and dairy products caused by psychrotropic bacteria is due in part to peptides and amino acids generated by milk proteolysis, which is known to cause a variety of tastes. Since hydrophobicity has been found essential for a bitter taste, casein, with relatively high hydrophobicity, yields a large amount of bitter peptides on proteolysis. Many aspects of the bitterness of peptides have been discussed (74) and no simple correlation has been found between enzyme specificity and the bitterness resulting from their action.

The psychrotropic spoilage of meat is caused by the bacterial flora, which are cold-tolerant organisms present on meat surfaces. Thus, it has been described that extracellular proteolytic enzymes produced by *Pseudomonas fragi* hydrolyze proteinaceous components of the meat and constitute one of the primary spoilage mechanisms (75,76).

It is known that the level of some essential amino acids and peptides in milk is too low to support the growth of lactic acid bacteria involved as starter in fermented milk products (77). Thus, the action of extracellular and cell wall–bound bacterial proteinases is responsible for the degradation of milk proteins, yielding transportable peptides which undergo further degradation by intracellular peptidases.

Finally, in the manufacture of cheese, chymosin from calf rennet, and chymosin substitutes of animal and plant origin are the milk-clotting enzymes because they catalyze the hydrolysis of casein.

1. Proteinase Extraction

The solubilization of proteinases from the digestive tract (stomach and intestine) of grass shrimp may be carried out by homogenizing the material with deionized water (1:3 w/v) followed by filtration with Celite and overnight dialysis against 10 mM phosphate buffer, pH 7.0. Further purification is achieved by ammonium sulfate precipitation followed by sequential chromatography on Sephadex G-75 and TSK DEAE-600 columns (78). Proteinases from the hepatopancreas of freshwater prawn are solubilized by blending with 0.05 M Tris-HCl buffer, pH 7.4, containing 5.0 mM $CaCl_2$, centrifugation at $2,000g_{max}$ for 30 min and filtration of the supernatant through glass wool (64).

The extraction of muscle cathepsins is carried out by homogenization of frozen muscle with 50 mM sodium citrate buffer, pH 5.0, containing 1 mM EDTA and 0.2% (w/v) Triton X-100 (1:10 w/v). The resulting homogenate is then stirred for 1 hr at 4°C and centrifuged at $30,000g_{max}$ for 20 min. The supernatant is filtered through glass wool (79).

2. Proteinase Assays

a. Trypsin and Chymotrypsin Assays. Trypsin and chymotrypsin activities are commonly measured by determining the trichloroacetic acid (TCA)–soluble peptides released in the reaction by using hemoglobin or casein as substrates (Table 8).

In the trypsin assay (80), 1.0 ml of the enzyme sample is added to 5 ml of denatured hemoglobin solution (1.67% in the reaction mixture). After incubation (10 min), the reaction is stopped by addition of 10.0 ml of 5% TCA. After shaking, allow to stand for 30 min, centrifuge at $4000g_{max}$ for 20 min, and measure the compounds soluble in TCA by the method of Folin and Ciocalteau with tyrosine as standard. The blank consists of a mixture to which TCA solution is added prior to sample solution. The denatured hemoglobin solution is prepared as follows: suspend 2 g of bovine hemoglobin with about 50 ml of doubly distilled water, add 36 g of urea and 8 ml of 1 N NaOH, and dilute to about 80 ml. Allow to stand for 30–60 min at room temperature to denature the hemoglobin, then add 10 ml of 0.05 M NaCl in 1 M boric acid, and, after shaking, add 4.4 ml of 5% $CaCl_2$ solution. Adjust to pH 7.5 with 1 N HCl and dilute to 100 ml. Centrifuge off any erythrocyte stroma at $4000g_{max}$ for 15 min.

A similar method can be used to measure chymotrypsin activity (81) in which, 1.0% (w/v) casein (0.5% final concentration in the reaction mixture), containing 0.055% $CaCl_2$ in 0.1 M borate buffer, pH 8.0, prepared by heating, is used as substrate solution. The products of enzymatic hydrolysis are estimated by measuring the absorbance at 280 nm of the supernatant after the TCA addition.

Trypsin-like (esterase) activity may be measured (Table 8) by using either benzoyl-L-arginine ethyl ester (BAEE) or *p*-toluenesulfonyl-L-arginine methyl ester (TAME) as substrate (78). In the BAEE assay, 0.2 ml of enzyme sample is added to 2.8 ml of 1 mM BAEE in 50 mM tris-HCl

Table 8   Peptidase and Proteinase Assays

Method	Reaction medium	Detection	Ref.
Trypsin			
Hemoglobin	1.67% denatured hemoglobin	Colorimetry (see text)	80
Chymotrypsin			
Casein	100 mM Na borate pH 8.0 + 0.5% casein + 2.5 mM $CaCl_2$	Spectrophotometry 280 nm	81
Trypsin-like activity			
BAEE assay	50 mM tris-HCl pH 8.0 + 1.0 mM BAEE + 20 mM $CaCl_2$	Spectrophotometry 253 nm	78
TAME assay	40 mM tris-HCl pH 8.1 + 1.25 mM TAME + 10 mM $CaCl_2$	Spectrophotometry 247 nm	78
Chymotrypsin-like activity			
ATEE assay	50 mM Na phosphate pH 7.0 + 1.0 mM ATEE	Spectrophotometry 237 nm	78
BTEE assay	40 mM tris-HCl pH 7.8 + 0.5 mM BTEE + 50 mM $CaCl_2$	Spectrophotometry 256 nm	78
Collagenase			
Hydroxyproline assay	50 mM tris-HCl pH 7.4 + 0.7% collagen + 5 mM $CaCl_2$	Spectrophotometry 560 nm	82, 83
Cathepsin B			
Fluorimetric assay	88 mM KH_2PO_4 + 12 mM Na_2HPO_4 (pH 6.0) buffer, + 0.01 mM Z-Arg-Arg-NHMec, + 1 mM EDTA + 2 mM cysteine	Fluorimetry (350–360 nm, excitation; 460 nm, emission)	84, 85
Cathepsin H			
Fluorimetric assay	50 mMKH_2PO_4 + 50 mM Na_2HPO_4 (pH 6.8) buffer, + 0.005 mM Arg-NHMec, + 1 mM EDTA, + 2mM cysteine	Fluorimetry (350–360 nm, excitation; 460 nm, emission)	84, 85
Cathepsin D			
Hemoglobin assay	170 mM Na citrate pH 3.7 + 0.63% denatured hemoglobin	Spectrophotometry 280 nm	87

buffer, pH 8.0, containing 20 mM $CaCl_2$ and the changes in absorbance at 253 nm with time are measured. For the TAME assay, add 0.3 ml of 10 mM TAME and 0.1 ml of sample to 2.0 ml of 46 mM tris-HCl buffer, pH 8.1, containing 11.5 mM $CaCl_2$ and read the changes in absorbance at 247 nm with time.

Chymotrypsin-like (esterase) activity can be estimated (Table 8) by using either N-acetyl-L-tyrosine ethyl ester (ATEE) or N-benzoyl-L-tyrosine ethyl ester (BTEE) as substrate (78). For the ATEE assay, 0.2 ml of enzyme solution is added to 2.8 ml of 1.0 mM ATEE in 50 mM phosphate buffer, pH 7.0, and the absorbance at 237 nm is continuously recorded. In the BTEE assay, add 1.5 ml of 80 mM tris-HCl buffer, pH 7.8, containing 0.1 M $CaCl_2$, 1.4 ml 1.07 mM BTEE in 50% methanol solution, and 0.1 ml of enzyme sample. Read the changes in absorbance at 256 nm with time.

b.   Collagenase Assay.   The method (82) is based on the measurement of the hydroxyproline content of soluble peptides released by the enzyme from insoluble collagen (Table 8). The reaction medium consists of 5–50 µl of enzyme extract and 1 ml of a mixture containing 7 mg insoluble collagen, 5 µmol CaCl$_2$, and 50 µmol tris-HCl, pH 7.4. A medium without enzyme extract is used as control. After incubation at 30°C, the reaction is stopped by the addition of 7 ml of cold water, and the collagen is removed by filtration and the supernatant passed through a 1.2-µm Millipore filter. An aliquot of 6 ml of the filtrate is freeze-dried, dissolved in 3 ml of 6 M HCl, and hydrolyzed for 12 hr at 108°C. Then the hydroxyproline content of the hydrolyzate is measured by the method of Neuman and Logan (83).

    c. Cathepsin Assays. The measurement of muscle cathepsin B (EC 3.4.22.1), L (EC 3.4.22.15), and H (EC 3.4.22.16) activities is based on the fluorimetric determination of 7-amino-4-methylcoumarin (NHMec) released from adequate fluorescent binding peptides (84) (Table 8). For this, cathepsins B, B + L, and H are assayed in a reaction medium containing 10 µM *N*-CBZ-L-arginyl-L-arginine, -7-amido-4-methylcoumarin (Z-Arg-Arg-NH Mec), 10 µM *N*-CBZ-L-phenylalanyl-L-arginine (Z-Phe-Arg-NHMec), a common substrate of cathepsins B and L, and 5 µM L-arginine-7-amido-4-methylcoumarin (Arg-NHMec), as substrates, respectively (85). The NHMec released in the reaction is continuously monitored by spectrofluorimetry. The NHMec produced is determined at 350–360 nm (excitation) and 460 nm (emission). An excitation scan should be performed before the assay. The instrument is calibrated by using pure NHMec (0.005–0.15 µM) (86).

    Cathepsin D (EC 3.4.23.5) may be determined by spectrophotometric measurement of TCA-soluble peptides using acid-denatured bovine hemoglobin as substrate. To measure muscle cathepsin D, the enzyme extract is added to 4 ml of 0.2 M sodium citrate buffer, pH 3.7, containing 0.75% acid-denatured bovine hemoglobin as substrate and incubated at 45°C for 1 hr. The control contains 40 µl of 3.0 mM isovalerylpepstatin, an inhibitor of cathepsin activity. At the end of incubation, 3 ml of 5% TCA is added. TCA-soluble peptides are measured at 280 nm against the control (87).

C. Pectic Enzymes

Pectic enzymes are widely distributed in higher plants and are also produced by microorganisms. The natural substrates of pectic enzymes are the pectic substances which occur as structural polysacharides in middle lamella and primary cell walls of plants. Pectic enzymes are responsible for important textural changes in fruits and vegetables during ripening, storage, and processing. Besides the involvement of pectic enzymes in these processes, industrially produced pectic enzymes from microorganisms constitute an important aid in food processing.

    Pectic substances are polymers with a α-D-1,4-polygalacturonan linear backbone, which contains 1,2-linked α-L-rhamnosyl residues in small amounts and branched regions mainly consisting of the neutral sugars D-galactose, L-arabinose, and D-xylose. In addition, carboxyl groups may be esterified by methanol or, to a greater or lesser degree, form calcium salts.

    Pectic enzymes are classified according to the bonds which split. Thus, polygalacturonases (poly-α-1,4-D-galacturonic acid glycanohydrolase, EC 3.2.1.15) catalyze the hydrolytic breakdown of glycosidic bonds next to free carboxyl groups; pectinases, also referred to as pectinesterase and pectin–methyl esterases (pectin pectyl hydrolase, EC 3.1.1.11), hydrolyze methyl ester linkages and yield pectins of a low degree of methylation or pectic acids; pectin lyase (EC 4.2.2.10) and pectate lyase (EC 4.2.2.2 and EC 4.2.2.9) catalyze the nonhydrolytic split of glycosidic linkages next to free carboxyl groups by β elimination. Pectin and pectate lyases differ in their substrates, being for the former pectins of a high degree of methylation and due to the latter pectins of a low degree.

    Pectinesterase in citrus fruit is responsible for the cloud loss in juices, one of the most important problems in citrus technology. The cloud is composed mainly of pectins, proteins, and lipids, and contains most of the characteristic flavor and color. Thus, if pectinesterase is not inhibited by heat inactivation or by freezing during processing, the enzymatically demethylated pectins will be coagulated by the calcium ions in the juice (88–90), this coagulation being accompanied by the formation of a sediment and a clear liquid unacceptable to the consumer. In relation to this problem, deesterified pectins in juice concentrates form calcium pectate and lead to gellification, and so no juice can be reconstituted from these concentrates (91).

Since pectic substances are in part responsible for the texture, one of the most important factors in determining fruit and vegetable quality is their pectic enzyme content. Pectinesterase, by catalyzing the demethylation of pectins, generates free carboxyl groups, which can then be crosslinked by calcium ions, thereby increasing the firmness of the tissue. On the other hand, the hydrolysis of pectates and pectins caused by polygalacturonase leads to a softening of the tissue. Finally, the involvement of pectin lyase in tissue softening seems to be restricted to microbial enzyme, which provokes a deterioration of fruit quality.

1. Extraction of Pectic Enzymes

Several procedures have been described for pectic enzyme extraction from fruits. Thus, pectinesterase and polygalacturonase can be extracted from several tropical fruits by homogenization with a cold solution of 12% poly(ethylene glycol) and 0.2% sodium bisulfite in water. The pellet after centrifugation at $7000g_{max}$ for 20 min is washed twice by resuspension in distilled water and the last residue is suspended in 0.5 M NaCl (92).

Fayyaz et al. (93) described a method for pectinesterase solubilization in which an extractant of high ionic strength is used: papaya pulp is homogenized with 2 M NaCl, pH 8.0, and incubated for 5 hr in a cold room. During incubation, the pH of the homogenate is maintained at pH 8.0 by adding either 2 M NaOH or 2 M HCl. At the end of incubation, the enzyme activity is recovered by filtration through Whatman No. 54 filter paper in a Buchner funnel with a slight vacuum.

Finally, pectinesterase from lemon fruits is solubilized by blending the chopped fruits with 1 M tris containing 1 M NaCl (approx. 1:1 w:v). The resultant slurry is allowed to stand for 90 min and filtered through muslin. After the addition of ammonium sulfate (to 25% saturation), the filtrate is centrifuged at $11,600g_{max}$ for 20 min. Further, ammonium sulfate (to 80% saturation) is added to the supernatant and the gelatinous pellet of centrifugation ($11,600g_{max}$ for 20 min) is dissolved in 0.1 M tris-HCl buffer, pH 7.0, containing 0.1 M NaCl, and left overnight to allow the pectates to precipitate. After centrifugation ($11,600g_{max}$ for 20 min) the pellet is washed with the last buffer. Finally, the pellet of a further ammonium sulfate precipitation (0–80% saturation) of the combined supernatant and washings is redissolved in the buffer (94).

2. Pectic Enzyme Assays

a. Polygalacturonase Assays. The most commonly used methods for polygalacturonase estimation are based on the colorimetric determination of the reducing groups released in the enzymatic reaction (Table 9). Additionally, a viscosimetric assay may be used.

A wide number of reaction media for polygalacturonase activity have been described (92,95, 96). The most common substrates are polygalacturonic acid and sodium polypectate at concentrations ranging from 0.25% to 1%, with the media generally buffered at acidic (4.0–5.0) pH values using acetate buffers (citrate buffers interfere with the Somogyi–Nelson assay). The presence in the reaction mixtures of NaCl (0.15 M) and the addition of sodium azide (in trace amounts) have been recommended in order to avoid microorganism growth during incubation. EDTA (2.0 mM) in the reaction mixtures inhibits contaminating pectate lyases. Medium controls contain boiled enzyme extract.

Table 9 Pectic Enzyme Assays

Method	Reaction medium	Detection	Ref.
Polygalacturonase			
Somogyi–Nelson assay	50 mM Na acetate pH 5.3 + 0.2% polygalacturonic acid, + 5 mM EDTA, + 100 mM NaCl, + sodium azide (trace)	Colorimetry (see text)	97
2-Cyanoacetamide assay	Acetate buffer pH 4.0–5.0 + 0.5–1% polygalacturonic acid or sodium polypectate, + 150 mM NaCl + 2 mM EDTA	Spectrophotometry 276 nm	92, 95, 96, 98
Viscosimetric assay	McIlvaine's buffer (pH 5.0), + 0.43% pectic acid	Viscosimetry	99
Pectinesterase			
Titrimetric assay	1.0–1.5% pectin (at the desired pH), + 100–150 mM NaCl	Titrimetry	93, 100, 101
Colorimetric assay	0.15 mM K-phosphate pH 7.5 + 0.33% pectin, + 0.0005% bromothymol blue	Colorimetry 620 nm	102
Pectate lyase			
4,5-Unsaturated product assay	50 mM tris-HCl pH 8.5, + 0.20% polygalacturonic acid, + 0.50 mM $CaCl_2$	Spectrophotometry 232 nm	97

In the Somogyi–Nelson assay (97), the reaction is stopped by addition of 0.5 ml of copper reagent to 0.5 ml of reaction mixture. After incubation for 10 min in a boiling water bath, the mixture is cooled to room temperature and 1.0 ml of arsenomolybdate reagent added. This is allowed to stand for 15–40 min at room temperature. Then the mixture is centrifuged and the absorbances of the supertanants are read at 500 nm. The absorbance values are used in reference to a D-galacturonic acid standard curve.

The copper reagent is prepared by dissolving 12 g of sodium potassium tartrate and 24 g of anhydrous sodium carbonate in 250 ml of water. The following reagents are then added in sequence while stirring: a solution of 4.0 g of cupric sulfate pentahydrate in 100 ml of water, 16 g of sodium hydrogen carbonate, and a solution of 180 g of anhydrous sodium sulfate in 500 ml of boiled water. The mixture is then brought to 1 L with water. The arsenomolybdate reagent is prepared by dissolving 25 g of ammonium molybdate in 450 ml of water and adding 21 ml of 96% sulfuric acid and a solution of 3.0 g of disodium hydrogen arsenate heptahydrate in 25 ml of water; the reagent is incubated at 37°C for 24 hr before use.

For the 2-cyanoacetamide assay (98), mix 1 ml of the enzyme reaction sample with 1 ml of 1% 2-cyanoacetamide and 2.0 ml of 0.1 M borate buffer (pH 9.0) in a test tube and heat the tube in a boiling water bath for 10 min. Cool the tube immediately by immersing in tap water and measure the absorbance of the reaction mixture at 276 nm within 2 hr. Read the carbohydrate concentration from the calibration curve obtained for the standard solutions of galacturonic acid by the same procedure.

The viscosimetric assay (99) is based on the determination of the decreases in viscosity of a pectic acid solution caused by polygalacturonase. In this method, 6 ml of 0.5% pectic acid solution in McIlvaine's buffer, pH 5.0, in an Ostwald viscosimeter, is incubated at 37°C for 3 min. After incubation, 1 ml of enzyme extract is added and the mixture is incubated at 37°C for 5 min. The rate of viscosity reduction (A) is calculated as follows:

$$A = \frac{(T_a - T)}{(T_a - T_0)} \times 100 \tag{3}$$

where T is the flow time (sec) of the reaction mixture, T_a is the flow time (sec) of pectic acid solution added to the heat-inactivated enzyme, and T_0 is the flow time (sec) of water added to the heat-inactivated enzyme. One unit of enzyme activity is defined as the activity reducing the viscosity by 50%.

b. Pectinesterase Assays. The methods for pectinesterase determination (Table 9) are based on titration of the carboxyl groups formed in the hydrolysis of pectins, the spectrophotometric determination of the changes in absorbance of a pH indicator added to the reaction mixtures, or the gas chromatographic determination of the methanol released during the enzymatic reaction.

In the titrimetric method (93,100,101), a pectin solution (in the range of 1–1.5%), often containing NaCl (0.1–0.15 M), at the desired pH (depending on the enzyme source) and enzyme extract is automatically titrated with a standard NaOH (0.01–0.02 M) solution. Enzymatic activity is expressed as moles of carboxylic groups released per time unit.

For the spectrophotometric method (102), mix 2 ml of 0.5% (w/v) citrus pectin (prepared in distilled water by heating to 40°C with continuous stirring) with 0.15 ml of 0.01% (w/v) bromothymol blue in 0.003 M potassium phosphate buffer, pH 7.5, and 0.88 ml of water. The reaction is monitored at 620 nm in a recording spectrophotometer. After determining the absorbance of the solution, the reaction is started by addition of enzyme extract and the decrease in absorbance is calculated from the progress curve of the reaction. A standard curve is plotted, by using different amounts of galacturonic acid instead of enzyme sample, in order to determine the equivalence between decreases in absorbance at 620 nm and micromoles of carboxylate groups. The enzymatic activity is expressed as microequivalents of released carboxylate groups per time unit. The assay must always be started at the same pH to ensure reproducible color changes; thus the reagents must be prepared in distilled water or weakly buffered solutions. The enzyme must be extracted by using unbuffered NaCl solutions.

In the GLC method (103), once the enzymatic reaction is stopped, the pH of the mixtures (3 ml) is adjusted to pH 7.0 and 330 µl of 25 mM isopropanol is added as an internal standard for gas chromatography. Samples (2 µl) are injected into a chromatograph fitted with a flame ionization detector and a 5% Carbowax 20 M, 80/120 Carbopack B column. The chromatogram is run at a constant temperature of 80°C. The chromatograms are analyzed with an integrator and the methanol concentration is calculated by comparing with the isopropanol internal standard. Enzyme activity is expressed as moles of methanol produced per time unit.

c. Pectate Lyase Assay. In this method (97), enzyme activity is measured by increases in absorbance at 232 nm of the reaction mixtures produced by the 4,5-unsaturated reaction products. For this, in a cuvette, mix 2.5 ml of 0.6 mM $CaCl_2$ and 0.24% (w/v) polygalacturonic acid in 60 mM tris-HCl buffer, pH 8.5, and 0.5 ml of adequately diluted enzyme sample and read changes in absorbance at 232 nm with time. To prevent the formation of an insoluble calcium–polygalacturonate complex during preparation of the substrate solution, first dissolve the polygalacturonic acid in 120 mM tris-HCl, pH 8.5, buffer, and then add the same volume of 1.2 mM $CaCl_2$ dissolved in water. The enzyme activity is expressed as micromoles of 4,5-unsaturated product per time unit for which an $\varepsilon_{232} = 4600$ M^{-1} cm^{-1} is used (104).

ACKNOWLEDGMENT

This work was supported in part by a grant from the CICYT (Spain) Project #ALI 573/93.

REVIEWS

J. R. Whitaker (ed.), *Principles of Enzymology for the Food Sciences*, Marcel Dekker, New York, 1994.
P. F. Box (ed.), *Food Enzymology*, Vol. 1, Elsevier, London, 1991.

REFERENCES

1. J. R. Whitaker (ed.), *Principles of Enzymology for the Food Sciences*, Marcel Dekker, New York, 1994.
2. P. F. Box (ed.), *Food Enzymology*, Vol. 1, Elsevier, London, 1991.
3. M. Scalet, R. Federico, and R. Angelini, *J. Plant Physiol.*, *137*: 571 (1991).
4. A. Rinaldi, G. Floris, and A. Finazzi-Agrò, *Eur. J. Biochem.*, *127*: 417 (1982).
5. P. Pec and I. Frébort, *Eur. J. Biochem.*, *209*: 661 (1992).
6. B. Holmsted, L. Larsson, and R. Tham, *Biochim. Biophys. Acta*, *48*: 182 (1961).
7. G. Floris, A. Giartosio, and A. Rinaldi, *Phytochemistry*, *22*: 1871 (1983).
8. A. M. Mayer and E. Harel, *Phytochemistry*, *18*: 193 (1979).
9. C. Grassin and D. Dubourdieu, *J. Sci. Food Agric.*, *48*: 369 (1989).
10. W. Cai, R. Martin, B. Lemaure, J. L. Leuba, and V. Pétiard, *Plant Physiol. Biochem.*, *31*: 441 (1993).
11. W. Bao, D. M. O'Malley, R. Whetten, and R. R. Sederoff, *Science*, *260*: 672 (1993).
12. R. Sterjiades, J. F. D. Dean, G. Gamble, D. S. Himmelsbach, and K. E. L. Eriksson, *Planta*, *190*: 75 (1993).
13. O. Leoni, S. Palmieri, V. Lattanzio, and C. F. Van Sumere, *Food Chem.*, *38*: 27 (1990).
14. H. W. Zhou and X. Feng, *J. Sci. Food Agric.*, *57*: 307 (1991).
15. O. Lamikanra, S. D. Kirby, and M. N. Musingo, *J. Food Sci.*, *57*: 686 (1992).
16. A. Sánchez-Ferrer, R. Bru, and F. García-Carmona, *Plant Physiol.*, *91*: 1481 (1989).
17. M. Jiménez-Atiénzar, M. A. Pedreño, and F. García-Carmona, *Biochem. Int.*, *25*: 861 (1991).
18. J. A. Ngalani, A. Signoret, and J. Crouzet, *Food Chem.*, *48*: 341 (1993).
19. F. C. Chilaka, E. O. Anosike, and P. C. Egbuna, *J. Sci. Food Agric.*, *61*: 125 (1993).
20. P. Wesche-Ebeling and M. W. Montgomery, *J. Food Sci.*, *55*: 1315 (1990).
21. M. Kidron, E. Harel, and A. M. Mayer, *Phytochemistry*, *16*: 1050 (1977).
22. M. Kidron, E. Harel, and A. M. Mayer, *Am. J. Enol. Vitic.*, *29*: 30 (1978).
23. T. Ohshima and F. Nagayama, *Bull. J. Soc. Sci. Fish.*, *46*: 1035 (1980).
24. X. Yan and K. D. A. Taylor, *Food Chem.*, *41*: 11 (1991).
25. F. Gauillard, F. Richard-Forget, and J. Nicolas, *Anal. Biochem.*, *215*: 59 (1993).
26. E. Valero, R. Varón, and F. García-Carmona, *J. Food Sci.*, *53*: 1482 (1988).
27. J. N. Rodríguez-López, J. Escribano, and F. García-Cánovas, *Anal Biochem.*, *216*: 205 (1994).
28. J. Schormüller, *Methods of Enzymatic Analysis* (H. U. Bergmeyer, ed.), Academic Press, New York, 1974, p. 71.
29. O. Ito, S. Kamata, N. Kaki-Ichi, Y, Suzuki, M. Hayashi, and K. Uchida, *J. Food Sci.*, *55*: 1172 (1990).
30. O. Ito and R. Akuzawa, *J. Dairy Sci.*, *66*: 967 (1983).
31. A. C. Maehly and B. Chance, *Meth. Biochem. Anal.*, *1*: 357 (1954).
32. M. R. S. Clavero, Y. C. Hung, L. R. Beuchat, and T. Nakayama, *J. Food Prot.*, *56*: 55 (1993).
33. L. M. Wolfson and S. S. Sumner, *J. Food Prot.*, *56*: 887 (1993).
34. F. S. Burnette, *J. Food Sci.*, *42*: 1 (1977).
35. A. Ros Barceló, R. Muñoz, and F. Sabater, *Physiol. Plant.*, *71*: 448 (1987).
36. A. Ros Barceló, R. Muñoz, and F. Sabater, *Plant Physiol. Biochem.*, *26*: 575 (1988).
37. D. J. Sessa and R. L. Anderson, *J. Agric. Food Chem.*, *29*: 960 (1981).
38. J. J. Jen, A. Seo, and W. H. Flurkey, *J. Food Sci.*, *45*: 60 (1980).
39. C. Chavez and W. H. Flurkey, *J. Chromatogr.*, *298*: 169 (1984).
40. V. M. Paradkar and J. S. Dordick, *Biotechnol. Progr.*, *9*: 199 (1993).
41. M. Morales, M. A. Pedreño, R. Muñoz, A. Ros Barceló, and A. A. Calderón, *Food Chem.*, *48*: 391 (1993).
42. T. Dobránsky, O. Sova, and M. Teleha, *J. Chromatogr.*, *474*: 430 (1989).
43. R. Janzen, K. K. Unger, H. Giesche, J. N. Kinkel, and M. T. W. Hearn, *J. Chromatogr.*, *397*: 81 (1987).
44. B. Chance and A. C. Maehly, *Meth. Enzymol.*, *2*: 764 (1955).
45. V. Herzog and H. D. Fahimi, *Anal. Biochem.*, *55*: 554 (1973).

46. M. A. Ferrer, A. A. Calderón, R. Muñoz, and A. Ros Barceló, *Phytochem. Anal.*, *1*: 63 (1990).
47. G. G. Guilbault and D. N. Kramer, *Anal. Chem.*, *36*: 2494 (1964).
48. K. K. Mäkinen and J. Tenovuo, *Anal. Biochem.*, *126*: 100 (1982).
49. O. S. Boyes, C. Perera, and H. Young, *J. Food Sci.*, *57*: 1390 (1992).
50. S. C. Sheu and A. O. Chen, *J. Food Sci.*, *56*: 448 (1991).
51. P. R. Bhirud, F. W. Sosulski, and K. Sosulski, *J. Food Sci.*, *58*: 1090 (1993).
52. C. Ganthavorn and J. R. Powers, *J. Food Sci.*, *54*: 371 (1989).
53. H. M. Henderson, N. A. M. Eskin, and J. Borsa, *Food Chem.*, *36*: 97 (1990).
54. K. Surrey, *Plant Physiol.*, *39*: 65 (1964).
55. A. L. Tappel, P. D. Boyer, and W. O. Lundberg, *Arch. Biochem. Biophys.*, *42*: 293 (1953).
56. A. P. Kulkarni, A. Mitra, J. Chaudhuri, J. Z. Byczkowski, and I. Richards, *Biochem. Biophys. Res. Commun.*, *166*: 417 (1990).
57. H. D. Kay and W. R. Graham, *J. Dairy Res.*, *6*: 191 (1935).
58. R. Aschaffenburg and J. E. C. Mullen, *J. Dairy Res.*, *16*: 58 (1949).
59. E. Bernt, *Methods of Enzymatic Analysis* (H. U. Bergmeier, ed.), Academic Press, New York, 1974, p. 868.
60. J. H. Bruemmer and B. Roe, *Proc. Fla. State Hort. Soc.*, *88*: 300 (1975).
61. W. B. Sinclair, *The Biochemistry and Physiology of the Lemon and other Citrus Fruits* (W. B. Sinclair, ed.), University of California, Oakland, 1984, p. 444.
62. L. Körmendy, G. Zsarnóczay, and V. Mihalyi, *Food Chem.*, *44*: 367 (1992).
63. J. Lind, *USDA Chemistry Laboratory Guidebook*, 1987, p. 3.49.
64. E. S. Baranowski, W. K. Nip, and J. H. Moy, *J. Food Sci.*, *49*: 1494 (1984).
65. W. K. Nip, C. Y. Lan, and J. H. Moy, *J. Food Sci.*, *50*: 1187 (1985).
66. A. S. Eisen, K. O. Henderson, J. J. Jeffrey, and R. A. Bradshaw, *Biochemistry*, *12*: 1814 (1973).
67. A. Z. Eisen and J. J. Jeffrey, *Biochim. Biophys. Acta*, *191*: 517 1969).
68. T. Kawabata and S. Suzuki, *Bull. Jap. Soc. Sci. Fish.*, *25*: 481 (1959).
69. S. B. Pan, J. M. Kuo, L. J. Luo, and H. M. Yang, *J. Food Biochem.*, *10*: 305 1986).
70. H. Makinotan, *Gyonikumeriseihin, Minced Fish Product*, Koseisha, Tokyo, 1984, p. 34.
71. F. Toldrá, *New Technologies for Meat and Meat Products* (F. J. M. Smulders, F. Toldrá, J. Flores, and M. Prieto, eds.), Audet, Nijmegen, The Netherlands, 1992, p. 209.
72. F. Toldrá and D. J. Etherington, *Meat Sci.*, *23*: 1 (1988).
73. F. Toldrá, M-J. Motilva, E. Rico, and J. Flores, Enzyme Activities in the Processing of Dry-Cured Ham, Proceedings 37th Int. Congress of Meat Science and Technology, Vol. I, Kulmbach, Germany, 1991, pp. 954–957.
74. J. Adler-Nissen, *Enzymic Hydrolysis of Food Proteins*. Elsevier, London, 1986.
75. R. Y. Yada and B. J. Skura, *J. Food Sci.*, *46*: 1766 (1981).
76. R. H. Dainty, B. G. Shaw, and E. S. J. Scheps, *J. Appl. Bacteriol.*, *39*: 73 (1975).
77. T. D. Thomas and O. E. Mills, *Neth. Milk Dairy J.*, *35*: 255 (1981).
78. S.-T. Jiang, M. W. Moody, and H.-S. Chen, et al., *J. Food Sci.*, *56*: 322 (1991).
79. D. J. Etherington, M. A. J. Taylor, and E. Dransfield, *Meat Sci.*, *20*, 1 (1987).
80. W. Rick, *Methods of Enzymatic Analysis* (H. U. Bergmeier, ed.), Academic Press, New York, 1974, p. 1013.
81. W. Rick, *Methods of Enzymatic Analysis* (H. U. Bergmeier, ed.), Academic Press, New York, 1974, p. 1006.
82. P. Lindner, S. Angel, Z. G. Weinberg, and R. Granit, *Food Chem.*, *32*: 19 (1989).
83. R. Neuman and M. A. Logan, *J. Biol. Chem.*, *184*: 299 (1950).
84. A. J. Barrett, *Biochem. J.*, *187*: 909 (1980).
85. M. Parreño, R. Cussó, M. Gil, and C. Sárraga, *Food Chem.*, *49*: 15 (1994).
86. D. J. Etherington and R. J. Wardale, *J. Cell Sci.*, *58*: 139 (1982).
87. E. Rico, F. Toldrá, and J. Flores, *Food Chem.*, *40*: 87 (1991).
88. W. Pilnik, *Berichte Wiss.-Techn. Komm. Intern. Fruchtsaft Union*, Juris Verlag, Zürich, Switzerland, 1958, p. 203.
89. M. A. Joslyn and W. Pilnik, *The Orange: Its Biochemistry and Physiology* (W. B. Sinclair, ed.), Univ. California, Riverside, 1961, p. 373.
90. J. J. P. Krop, PhD thesis, Agricultural University, Wageningen, The Netherlands, 1974.

91. A. H. Rouse and C. D. Atkins, *Food Technol.*, *7*: 221 (1953).
92. M. El-Zoghbi, *Food Chem.*, *49*: 33 (1994).
93. A. Fayyaz, B. A. Ashi, H. M. Ghazali, Y. B. Che Man, and S. Jinap, *Food Chem.*, *47*: 183 (1993).
94. H. M. Macdonald, R. Evans, and W. J. Spencer, *J. Sci. Food Agric.*, *62*: 163 (1993).
95. H. E. Hart, M. E. Parish, J. K. Burns, and L. Wicker, *J. Food Sci.*, *56*: 480 (1991).
96. H. Urbanek and J. Zalewska-Sobczak, *Biochim. Biophys. Acta*, *377*: 402 (1975).
97. A. Collmer, J. L. Ried, and M. S. Mount, *Methods in Enzymology*, Vol. 161 (W. A. Wood and S. T. Kellog, eds.), Academic Press, New York, 1988, p. 329.
98. S. Honda, Y. Nishimura, M. Takahashi, H. Chiba, and K. Kazehi, *Anal. Biochem.*, *119*: 194 (1982).
99. T. Sakai, *Methods in Enzymology*, Vol. 161 (W. A. Wood and S. T. Kellog, eds.), Academic Press, New York, 1988, p. 335.
100. A. H. Rouse and C. D. Atkins, *Univ. Fla. Agric. Exp. Stn. Bull. 570* (1955).
101. L. Wicker, *J. Food Sci.*, *57*: 534 (1992).
102. A. E. Hagerman and P. J. Austin, *J. Agric. Food Chem.*, *34*: 440 (1986).
103. R. F. McFeeters and S. A. Armstrong, *Anal. Biochem.*, *139*: 212 (1984).
104. C. W. Nagel and M. M. Anderson, *Arch. Biochem. Biophys.*, *112*: 322 (1965).

11

Analysis of Neutral Lipids: Fatty Acids

Boukaré G. Semporé and Jean A. Bézard
Nutrition Research Unit, University of Burgundy, Dijon, France

I. INTRODUCTION

In humans, lipids constitute an important part of the food intake since in Western countries they represent 30–40% of energy intake and since certain of the contituent lipid fatty acids, which cannot be synthesized de novo by tissues, should necessarily be present in food. In animals, lipid intake and requirement are less known and vary according to species. On the other hand, lipids are partly responsible for physical and chemical properties of food. Therefore, analysis of food lipids concerns such diverse groups as nutritionists and food manufacturers.

Lipids consist of a wide range of products which share the common property of being generally insoluble in water but soluble in organic solvents. However, this property is shared to different extents by the different classes of lipids, namely, phospholipids; mono-, di-, and triacylglycerols; esterified and free sterols; free fatty acids; and waxes. Properties of these lipids are in great part due to their component fatty acids.

It is out of the scope of this chapter to present in full detail the properties and analyses of the lipids concerned. More details can be found in comprehensive ancient and new books (see reviews).

II. FATTY ACIDS

Except for free sterols, the physical, chemical, and biological properties of the different classes of neutral lipids are principally those of their component fatty acids. On a molecular basis, fatty acids can represent, for example, more than 95% of triacylglycerols and they are thereby responsible for the main properties of these molecules.

A. Chemical Structure

Fatty acids constitute a wide range of molecules but all do not display the same importance as for the amount present in usual lipids.

A few different fatty acids represent by themselves about 95% of the total fatty acids present in food lipids for human or animal consumption. These major fatty acids comprise saturated, monounsaturated, and polyunsaturated fatty acids according to their chemical structure and properties. Besides, minor fatty acids are found in usual food lipids but in much smaller quantities, such as branched or odd carbon number fatty acids. Lastly, unusual fatty acids are present in less common food lipids, sometimes in high proportion, such as hydroxy or cyclic fatty acids.

Detailed information regarding the formulas and molecular structures of the different fatty acids are likely to be found in recent specialized chemical and biochemical books (12,8,10). Only brief information will be reported in this section. The nomenclature adopted was that proposed by the IUPAC-IUB (International Union of Pure and Applied Chemistry/International Union of Biochemistry) commissions (13,14).

All fatty acids comprise an alkyl (hydrocarbon) chain with a methyl group at one end and a carboxylic group at the other one. Most of them are unbranched and present an even number of carbon atoms, since they are synthesized by addition of two carbon units generally starting from acetate, in plants and animals. Fatty acids differ by their chain length and their degree of unsaturation.

In saturated fatty acids (Table 1), all of the carbon atoms of the hydrocarbon chain are saturated with hydrogen atoms. They differ by their chain length and can be divided into

Short-chain fatty acids (SCFA) from 2 to 8 carbon atoms
medium-chain fatty acids (MCFA) from 10 to 12 carbon atoms
long-chain fatty acids (LCFA) from 14 to 18 carbon atoms
Very-long-chain fatty acids (VLCFA) comprising 20 or more carbon atoms

This classification presents the advantage of taking into account their physiological properties which are related in great part to their physical and chemical properties, themselves related to their chain length and degree of unsaturation. For example, during digestion the less hydrophobic SCFA are totally absorbed via the portal blood, whereas the most hydrophobic LCFA and VLCFA are almost totally absorbed via the lymph. As for the MCFA, they are shared between the portal and the lymphatic route (15).

To illustrate the nomenclature of saturated fatty acids, the example of the most common fatty acid in food lipids can be taken. That is palmitic acid (trivial name) or hexadecanoic (official name) or 16:0 in short-hand representation, in which the number 16 indicates the total number of carbon atoms and 0 the absence of double-bonds.

Besides the even carbon number and straight chain saturated fatty acids, odd carbon number and branched fatty acids are encountered in animal food from bovine origin. For example, margaric acid (17:0) is present in bovine milk, butter, and cheese fats. In the most common branched fatty acids a methyl group is fixed on the second carbon atom of the aliphatic chain (when counted from the methyl end) to form iso fatty acids such as isopalmitic (iso 16:0). In anteiso fatty acids (anteiso stearic acid or anteiso 18:0) the methyl group is fixed on the third carbon.

In the monounsaturated fatty acids (monoenes) as those reported in Table 2, one double bond, generally of the cis (c) configuration, is present in the aliphatic chain. Its position is determined from the carboxyl end (Δ numbering) or from the methyl end (ω or n-x numbering). For example, oleic acid, one of the most common fatty acids in food lipids, or octadecenoic in official nomenclature, is represented by the numeric symbol 18:1 Δ9 or 9–18:1 or 18:1 ω9 (or n-9). Unless otherwise indicated, the double bond is of the cis configuration.

Elaidic acid (9t-18:1) is the trans (t) geometrical isomer of oleic acid and vaccenic acid (11t-18:1) is one of the positional isomers of elaidic acid. They are present in industrially (9t-18:1 in margarines) or biologically (11t-18:1 in butter) hydrogenated fats.

Other monoenes are of biological importance in vegetables and animals such as petroselinic

Table 1 Name, Selected Physical Properties and Sources of the Major Unbranched Saturated Fatty Acids Encountered in Foods

Chain length	Trivial name	Systemic name	Numerical symbol[a]	Molecular weight	Half-developed formula	Selected physical properties[b]				Main sources
						m.p.	b.p.	d	n	
1	Formic	Methanoic	1:0	46.03	HCOOH	8.6	100.8	1.220	1.3714	Vegetable fats
2	Acetic	Ethanoic	2:0	60.05	CH$_3$-COOH	16.5	118.1	1.0492	1.36976 (25)	Ruminant milk fats
3	Propionic	Propanoic	3:0	74.08	CH$_3$-CH$_2$-COOH	-22	140.9	0.992	1.38736 (19.9)	Ruminant milk fats
4	Butyric	Butanoic	4:0	88.11	CH$_3$-(CH$_2$)$_2$-COOH	-7.9	163	0.9587	1.39906	Cow milk fat
5	Valeric	Pentanoic	5:0	102.13	CH$_3$-(CH$_2$)$_3$-COOH	-34.5	186.4	0.9387	1.4086	Cow milk fat
6	Caproic	Hexanoic	6:0	116.16	CH$_3$-(CH$_2$)$_4$-COOH	-4	205	0.929	1.41635	Cow, sheep, goat milk fats
7	Enanthic	Heptanoic	7:0	130.19	CH$_3$-(CH$_2$)$_5$-COOH	-7.46	223	0.9216 (14)	1.42162 (19.8)	Ruminant milk fats
8	Caprylic	Octanoic	8:0	144.22	CH$_3$-(CH$_2$)$_6$-COOH	16	239	0.9088	1.4285	Milk fats and *Cuphea* Genus seed oils
9	Pelargonic	Nonanoic	9:0	158.24	CH$_3$-(CH$_2$)$_7$-COOH	12.3	254	0.9055	1.4330	Milk fats
10	Capric	Decanoic	10:0	172.27	CH$_3$-(CH$_2$)$_8$-COOH	31.3	269	0.8858	1.42855	Milk fats and *elus* seed oils
11	—	Undecanoic	11:0	186.30	CH$_3$-(CH$_2$)$_9$-COOH	28.5	284	0.8905	1.4294	Milk fats
12	Lauric	Dodecanoic	12:0	200.32	CH$_3$-(CH$_2$)$_{10}$-COOH	43.5	225 (100)	0.883	1.4183	Coconut oil, palm kernel oil
13	—	Tridecanoic	13:0	214.35	CH$_3$-(CH$_2$)$_{11}$-COOH	51	312.4	—	1.4249 (70)	Milk fats
14	Myristic	Tetradecanoic	14:0	228.38	CH$_3$-(CH$_2$)$_{12}$-COOH	54.4	250.5 (100)	0.8622 (94)	1.4308 (60)	Dairy products and plant seed fats (nutmeg butter)
15	—	Pentadecanoic	15:0	242.41	CH$_3$-(CH$_2$)$_{13}$-COOH	52.1	257 (100)	0.8423 (80)	1.4270 (70)	Milk and liver fats
16	Palmitic	Hexadecanoic	16:0	256.43	CH$_3$-(CH$_2$)$_{14}$-COOH	62.85	268.5 (100)	0.8487 (70)	1.4273 (79.8)	Practically all animal and vegetable fats
17	Margaric	Heptadecanoic	17:0	270.46	CH$_3$-(CH$_2$)$_{15}$-COOH	62	227 (100)	0.8579 (60)	1.4319 (70)	Milk and sheep fats
18	Stearic	Octadecanoic	18:0	284.49	CH$_3$-(CH$_2$)$_{16}$-COOH	69.6	298 (100)	0.9408	1.4299 (80.2)	Animal and vegetable fats
19	—	Nonadecanoic	19:0	298.52	CH$_3$-(CH$_2$)$_{17}$-COOH	69.4	287 (64)	—	—	Dairy products

Table 1 (*continued*)

Chain length	Trivial name	Systemic name	Numerical symbol[a]	Molecular weight	Half-developed formula	Selected physical properties[b]				Main sources
						m.p.	b.p.	d	n	
20	Arachidic	Eicosanoic	20:0	312.54	$CH_3-(CH_2)_{18}-COOH$	75.4	328	0.824 (100)	1.4250 (100)	Peanut oil
22	Behenic	Docosanoic	22:0	340.59	$CH_3-(CH_2)_{20}-COOH$	80	306 (60)	0.8221 (100)	1.4270 (100)	Peanut oil, waxes
24	Lignoceric	Tetracosanoic	24:0	368.65	$CH_3-(CH_2)_{22}-COOH$	84.2	297 (20)	0.8207 (100)	1.4287 (100)	Peanut oil, waxes. *Lophira alata*
26	Cerotic	Hexacosanoic	26:0	396.70	$CH_3-(CH_2)_{24}-COOH$	87.7	288 (8)	0.8198 (100)	1.4301 (100)	Waxes

[a]The first number indicates the total number of carbon atoms and 0 the absence of double bond.

[b]m.p., melting point in °C; b.p., boiling point in °C at 760 mm Hg or at the pressure indicated in parentheses; d, density measured at 20°C (unless otherwise indicated in parentheses) compared to water at 4°C; n, refractive index of sodium D line measured at 20°C (unless otherwise indicated in parentheses).

Source: Adapted from Refs. 8, 10, and 13.

Table 2 Name, Selected Physical Properties and Sources of the Major Unbranched Monoene Fatty Acids Encountered in Foods

Chain length	Trivial name	Systematic name[a]	Numerical Symbol[b]	Molecular weight	Half-developed formula	Selected physical properties[c]				Main sources
						m.p.	b.p.	d	n	
3	Acrylic	Propenoic	3:1 n-1	72.06	$CH_2 = CH\text{-}COOH$	13	141	1.062 (16)	1.4224	Cottonseed oil
4	Isocrotonic	*cis*-Butenoic	4:1 n-2	86.09	$CH_3\text{-}CH = CH\text{-}COOH$	15.5	169	1.0312 (15)	1.4457	Cottonseed oil
6	Isohydrosorbic	2-Hexenoic	6:1 n-4	114.15	$CH_3\text{-}(CH_2)_2\text{-}CH = CH\text{-}COOH$	32	217	0.9627 (40)	1.4601 (40)	Mint oil
10	Caproleic	9-Decenoic	10:1 n-1	170.25	$CH_2 = CH\text{-}(CH_2)_7\text{-}COOH$	—	145 (15)	0.9238 (15)	1.4488	Ruminant milk fats
12	Lauroleic	9-dodecenoic	12:1 n-3	198.31	$CH_3\text{-}CH_2\text{-}CH = CH\text{-}(CH_2)_7\text{-}COOH$	—	—	—	—	Cow milk fat
14	Myristoleic	9-Tetradecenoic	14:1 n-5	226.36	$CH_3\text{-}(CH_2)_3\text{-}CH = CH\text{-}(CH_2)_7\text{-}COOH$	—	—	0.9018	1.4549	Milk fat and liver fat
16	Palmitoleic	9-Hexadecenoic	16:1 n-7	254.42	$CH_3\text{-}(CH_2)_5\text{-}CH = CH\text{-}(CH_2)_7\text{-}COOH$	1	219	0.9003 (15)	—	Marine oils
18	Petroselinic	6-Octadecenoic	18:1 n-12	282.47	$CH_3\text{-}(CH_2)_{10}\text{-}CH = CH\text{-}(CH_2)_4\text{-}COOH$	32	209 (10)	0.8824 (35)	1.4535 (47)	Umbelliferae seed oils
18	Oleic	9-Octadecenoic	18:1 n-9	282.47	$CH_3\text{-}(CH_2)_7\text{-}CH = CH\text{-}(CH_2)_7\text{-}COOH$	13	286 (100)	0.895	1.45823	All animal and vegetable lipids
18	Elaidic	*trans*-9-Octadecenoic	t-18:1 n-9	282.47	$CH_3\text{-}(CH_2)_7\text{-}CH = CH\text{-}(CH_2)_7\text{-}COOH$	44.5	288 (100)	0.851 (79)	1.4405 (70)	Hydrogenated oils
18	Vaccenic	*trans*-11-Octadecenoic	t-18:1 n-7	282.47	$CH_3\text{-}(CH_2)_5\text{-}CH = CH\text{-}(CH_2)_9\text{-}COOH$	42.5	—	0.8560 (70)	1.4439 (50)	Ruminant depot and milk fat
18	*cis*-Vaccenic	11-Octadecenoic	18:1 n-7	282.47	$CH_3\text{-}(CH_2)_5\text{-}CH = CH\text{-}(CH_2)_9\text{-}COOH$	—	—	—	—	Bacterial lipids
20	Gadoleic	9-Eicosenoic	20:1 n-11	310.52	$CH_3\text{-}(CH_2)_9\text{-}CH = CH\text{-}(CH_2)_7\text{-}COOH$	24.5	—	—	—	Marine oils
20	Gondoic	11-Eicosenoic	20:1 n-9	310.52	$CH_3\text{-}(CH_2)_7\text{-}CH = CH\text{-}(CH_2)_9\text{-}COOH$	22	—	—	—	Rapeseed and mustard seed oils (cruciferae)
22	Cetoleic	11-Docosenoic	22:1 n-11	338.58	$CH_3\text{-}(CH_2)_9\text{-}CH = CH\text{-}(CH_2)_9\text{-}COOH$	32.5	—	—	—	Marine oils

Table 2 *(continued)*

Chain length	Trivial name	Systematic name[a]	Numerical Symbol[b]	Molecular weight	Half-developed formula	Selected physical properties[c]				Main sources
						m.p.	b.p.	d	n	
22	Erucic	13-Docosenoic	22:1 n-9	338.58	CH_3-$(CH_2)_7$-CH=CH-$(CH_2)_{11}$-COOH	33.5	281 (30)	0.860 (55)	1.4480 (64)	Rapeseed and mustard seed oils (cruciferae)
22	Brassidic	*trans*-13-Docosenoic	t-22:1 n-9	338.58	CH_3-$(CH_2)_7$-CH=CH-$(CH_2)_{11}$-COOH	61.5	282 (30)	0.8585 (57)	1.4347 (100)	Hydrogenated rapeseed oil
24	Nervonic	15-Tetracosenoic	24:1 n-9	366.63	CH_3-$(CH_2)_7$-CH=CH-$(CH_2)_{13}$-COOH	41	—	—	1.4535 (46)	Marine oils, bran lipids
26	Ximenic	17-Hexacosenoic	26:1 n-9	394.69	CH_3-$(CH_2)_7$-CH=CH-$(CH_2)_{15}$-COOH	45	—	—	—	*Ximenia americana* seed oil

[a]Unless specified the double bond is of the cis configuration. The number x indicates that the double bond starts on the xth carbon atom counted from the carboxyl end of the molecule.

[b]Unless specified the double bond is of the cis configuration. t = trans. In the symbol the first number indicates the total number of carbon atoms (n), the second number the presence of one double bond which starts on the yth carbon atom $(n-y)$ counted from the methyl end.

[c]As note b in Table 1.

Source: Adapted from Refs. 8, 10, and 13.

acid (18:1 n-12), palmitoleic acid (16:1 n-7), erucic acid (22:1 n-9) present in the former rapeseed oil, and nervonic acid (24:1 n-9) found in neural tissue lipids (sphyngomyelins).

In polyunsaturated fatty acids (polyenes) such as those reported in Table 3, several double bonds are present, more commonly ethylenic bonds of the cis configuration, each being separated from the next one by a methylene group.

In plants the most widespread methylene-interrupted polyene fatty acids are linoleic acid (18:2 n-6) and α-linolenic acid (18:3 n-3). Their presence in food is of great importance since they cannot be synthesized de novo by human and animal tissues and should thereby be provided with the diet. Their accurate analysis in food is a prerequisite to the preparation of diets which fulfill requirements in these two essential fatty acids (EFAs). In tissues these EFAs are converted to longer and more unsaturated fatty acids of the n-6 and n-3 families, such as arachidonic (20:4 n-6), eicosapentaenoic (EPA: 20:5 n-3), and docosahexaenoic (DHA: 22:6 n-3), which are present in meat and fish in relatively high amounts. Polyenes of the n-7 and n-9 families derived from palmitoleic acid (16:1 n-7) and oleic acid (18:1 n-9), respectively, are rarely encountered in animal food, except when animals were EFA-deficient (presence of 20:3 n-9). In plant and animal food, less common fatty acids may be present. Besides short-chain, odd-numbered, and branched chain fatty acids are present in ruminant milks and their derivatives, hydroxylated fatty acids can be found principally in plants lipids. Ricinoleic acid (12-hydroxyoleic) is the best known, being the major fatty acid in castor oil. Other oxygenated fatty acids, epoxy and keto fatty acids, are usual components of several seed oils. Conjugated, allenic, and acetylenic fatty acids do not occur to any significant extent in food lipids. Cyclic fatty acids are usual components of several seed oils. The cyclopropene fatty acids, malvalic and sterculic, are found in noticeable amounts in baobab and kapok seed oils and in trace amounts in cottonseed oil. They frequently are accompanied by cyclopropane fatty acids. The analysis of these cyclic fatty acids, which are consumed in different parts of Africa and India, is important since the innocuity of these acids in humans is far from being proved. Besides these naturally occurring cyclic acids, others are formed in trace amounts when oils containing trienoic acids (α-linolenic) are repetitively heated at relatively high temperature.

More detailed informations regarding these unusual dietary fatty acids can be obtained from specialized books (10).

B. Fatty Acid Properties

In this section physical and chemical properties which play a role in the analysis of fatty acids in food lipids will be emphasized.

1. Physical Properties

Several selected physical properties are reported in Tables 1–3.

  a.   Solid State Behavior.   Saturated fatty acids with an even number of carbon atoms crystallize in three different varieties termed A, B, and C, the C form being the stable crystal form with the highest melting temperature. Transition between the three forms is as follows:

A → B → C ↔ liquid.

All three crystal forms can be obtained by crystallization from solvents, whereas only the C form is obtained from the liquid state.

The saturated fatty acids with an odd number of carbon atoms crystallize in three forms termed A′, B′, and C′, related but not similar to the A, B, C forms. Differences in crystal forms were also observed with branched and polyunsaturated fatty acids (10).

  b.   Melting Point.   The melting point of the even-membered saturated fatty acids (Table 1) increases with chain length (16). There is a very pronounced alteration between even and odd

Table 3 Name, Selected Physical Properties, and Sources of the Major Unbranched Polyene Fatty Acids[a] Encountered in Foods

Chain length	Trivial name	Systematic name[b]	Numerical symbol[c]	Molecular weight	Half-developed formula	Physical properties[d]				Main sources
						m.p.	b.p.	d	n	
Family n-9										
20	"Mead" acid	5,8,11-Eicosatrienoic	20:3 n-9	302.46	$CH_3-(CH_2)_7-CH=CH-CH_2-CH=CH-CH_2-CH-(CH_2)_3-COOH$	—	—	—	—	Tissue lipids from essential fatty acid–deficient animals
Family n-6										
16	Hiragonic	4,7,10-Hexadecadienoic	16:3 n-6	250.38	$CH_3-(CH_2)_2-CH=CH-CH_2-CH=CH-CH_2-CH=CH-(CH_2)_2-COOH$	—	—	0.9288	1.4855	Marine oils
18	Linoleic	9,12-Octadecadienoic	18:2 n-6	280.45	$CH_3-(CH_2)_4-CH=CH-CH_2-CH=CH-(CH_2)_7-COOH$	-11	230	0.9025	1.4699	Vegetable oils
18	γ-Linolenic	6,9,12-Octadecatrienoic	18:3 n-6	278.44	$CH_3-(CH_2)_4-CH=CH-CH_2-CH=CH-CH_2-CH=CH-(CH_2)_4-COOH$	—	—	—	—	Evening primrose, borage, black currant seed oils
20	Arachidonic	5,8,11,14-Eicosatetraenoic	20:4 n-6	304.48	$CH_3-(CH_2)_4-CH=CH-CH_2-CH=CH-CH_2-CH=CH-CH_2-CH=CH-(CH_2)_3-COOH$	-49.5	—	—	1.8482	Animal fats (egg yolk, liver)
22	Adrenic	7,10,13,16-Dodecatetraenoic	22:4 n-6	332.52		—	—	—	—	Adrenal and liver lipids

Family n-3

18	α-Linolenic	9,12,15-Octa-decatrienoic	18:3 n-3	278.44	CH_3-CH_2-CH = CH-CH_2- CH = CH-CH_2-CH = CH-$(CH_2)_7$-COOH	−11	231 (17)	0.9046	1.4780	Vegetable oils (linseed, soya, walnut)
18	Stearidonic	6,9,12,15-Octa-decatetraenoic	18:4 n-3	276.42		—	—	0.9297	1.4911	Black currant seed oil
20	Timnodonic	5,8,11,14,17-Eicosapentaenoic	20:5 n-3	302.46	CH_3-CH_2-CH = CH-CH_2- CH = CH-CH_2-CH = CH-CH_2-CH = CH- CH_2-CH = CH-$(CH_2)_3$- COOH	—	—	—	—	Marine oils
22	Clupanodonic	4,7,10,13,16,19-Docosahexaenoic	22:6 n-3	328.52		−78	236 (5)	0.9290	1.4868	Marine oils

[a]Methylene-interrupted (nonconjugated) polyene fatty acids.
[b]The double bonds are of the cis configuration. The numbers indicate the location of the double bonds when counted from the carboxyl end of the molecule.
[c]The first number indicates the total number of carbon atoms (n), the second number of double bonds which start on the yth carbon atom (n–y) when counted from the methyl end.
[d]As note b in Table 1.
Source: Adapted from Refs. 8, 10, and 13.

members. The melting curve of the even members occurs at a higher temperature than the curve formed by the odd fatty acids, particularly for the short-chain fatty acids. These differences can be explained by the crystal structure of the C and C' forms due to differences in the packing of the methyl end groups. A dense packing is achieved in the C form of the even saturated fatty acids.

For unsaturated fatty acids (Table 2 and 3) there is a reduction in the melting point compared to the corresponding saturated fatty acids, and the higher the number of double bonds the higher the decrease for the same chain length. The melting point also depends on the position of the double bonds. In monoenes, it is lower when the double bond is near the carboxyl group, increasing successively toward the methyl end. The presence of a trans double bond increases the melting point when compared to the cis isomer.

   c. *Boiling Point.* The boiling point of the even saturated fatty acids (Table 1) increases with chain length (17). The boiling curve of the odd members is below that of the even fatty acids. The boiling point decreases at pressures lower than 760 mm Hg. For example, that of hexadecanoic (16:0) decreases from 353.8°C at 760 mm Hg to 153.6°C at 1 mm Hg (18). The presence of a cis double bond decreases the boiling point slightly, from 240°C for octadecanoic (stearic, 18:0) to 234°C for 9-octadecenoic (oleic, 18:1) at 15/16 mm Hg (17), for example. The effect is more accentuated when the unsaturation increases.

   The boiling point of the methyl esters is markedly lower than that of the free fatty acids, e.g., 148°C compared to 175°C at 1 mm Hg for hexadecanoic (palmitic, 16:0) (10). This property is widely used in the analysis of long-chain fatty acids by gas–liquid chromatography.

   d. *Solubility.* Solubility in water of fatty acids is generally low because of the hydrophobicity of the aliphatic chain in relation to the number of carbon atoms. The solubility increases with decreasing chain length. For example, the maximum concentration in bulk Krebs buffer was found to be 575 mM for 4:0 and only 23.5 µM for 16:0 and .015 µM for 22:0 (15,19). Even and odd fatty acids follow the same curve of solubility in water, that is there is no alternation. The solubility increases slightly with the presence of double bonds which bring about some polarity to the hydrocarbon chain. Solubility also increases with temperature. Conversely, the solubility of fatty acids in organic solvents is relatively high and increases with temperature. In these solvents, even and odd fatty acids alternate, the latter being more soluble. Methyl esters of fatty acids are more soluble in organic solvents than free fatty acids. For example, at 18.4°C the solubility of palmitic acid in acetone is 4.26 wt % whereas that of methylpalmitate at 18.1°C is 60.8 wt % (20). This property is useful in the analysis of fatty acid methyl esters by gas and liquid chromatography.

   e. *Other Properties.* Other physical characteristics of fatty acids will not be described in this section. Information regarding density, viscosity, refractive index, surface tension, and dielectric constant are reported in specialized books (1,8,10).

2. Chemical Properties

Several properties of fatty acids are related to the carboxyl group and others to the hydrocarbon chain.

   a. *Reaction of the Carboxyl Group.* One of the most important reactions in fatty acid analysis is esterification. Esters can be prepared by interaction of the acid with an alcohol or its equivalent and the reaction requires an acidic, basic, or enzymic catalyst (6). The procedures used will be more extensively developed in a next section because of the wide use of fatty acid esters, especially methyl esters, for chromatographic analyses.

   Soaps or metal salts of fatty acids, especially alkali-metal salts, are other important derivatives of fatty acids. Applications of soaps are related to their properties as surface-active agents. Soaps can be present in small amounts in food lipids, e.g., calcium salts.

Fatty acid chlorides and anhydrides are effective acylating agents used in the synthesis of acylglycerols and phospholipids. Because of their high sensitivity to moisture, they are not present in food.

Nitrogen-containing compounds, such as amides, are manufactured to a high extent, and their uses are based on their surface-active properties. Their presence in foods only occurs as contamination.

Other reactions of the carboxyl group can give rise to sulfonic acids, peroxy acids, and esters or to aldehydes.

b. Reaction of the Aliphatic Chain. The most important reactions occur on the double bonds. *Hydrogenation* is one of the most extensively studied because catalytic hydrogenation is an important industrial process used to harden liquid oils of vegetable or marine origin for margarine and shortening production. Partial hydrogenation of unsaturated fatty acids is accompanied by double-bond migration and stereomutation giving rise to a wide range of positional and configurational (trans) isomers. From a nutritional point of view analysis of these isomers in foods is important since it is still controversial as to whether *trans*-fatty acids exhibit harmful effects or not (27). Hydrogenation can also occur in the rumen of cows and sheep through the action of rumen bacteria. Fatty acids liberated from dietary triacylglycerols in the rumen undergo double-bond migration and partial or complete biohydrogenation. The main product is vaccenic acid (11t-18:1) recovered in milk and meat of ruminants.

Another naturally occurring reaction of unsaturated dietary fatty acids is nonenzymic and enzymic *oxidation*, occuring during storage and handling of food and producing several complex compounds. The primary oxidation products are hydroperoxides in which the double bonds may have moved and/or changed configuration from its original cis form. These hydroperoxides subsequently give rise to short-chain secondary products (aldehydes, ketones, alcohols, esters, hydrocarbons, cyclized products, etc.) which contribute to the odor and flavor of food, sometimes acceptable, sometimes undesirable (rancid fat). They also may be converted to higher molecular weight products by dimerization.

Other reactions are extensively used as laboratory or industrial procedures: *epoxidation* produces epoxidized oils used as stabilizer-plasticizer compounds; *hydroxylation* gives threo- and erythrodiols; oxidative cleavage with permanganate and/or periodate or *ozonolysis* is used in the laboratory to determine the location of the double bonds. *Halogenation* has held in the past a significant position in fatty acid chemistry for measuring total unsaturation, recorded as iodine value. *Oxymercuration* with mercuric acetate and methanol has been used in many ways, e.g., for measuring the cis/trans ratio of oils by their (^{1}H-nuclear magnetic resonance) spectrum. Details of these reactions can be found in recent specialized books (8,10).

III. ACYLGLYCEROLS

Acylglycerols are esters of glycerol and fatty acids. Triacylglycerols are the major constituents of natural fats and oils and represent more than 95% of lipid intake in humans. Mono- and diacylglycerols are important intermediates in the synthesis and metabolism of triacylglycerols and glycerophospholipids in plants, animals, and humans.

A. Chemical Structure

The carbon atoms of glycerol (propane-1,2,3 triol) are numbered stereospecifically in acylglycerol molecules. In a Fischer projection, the secondary hydroxyl group is at the left of the central prochiral carbon atom and the three carbon atoms are numbered 1, 2, 3 from top to bottom. Acylglycerols which are stereospecifically numbered in that way have the prefix "sn" immediately proceding the term "glycerol" in the name of the compound and the position of each acyl group is specified, e.g., 1-oleoyl-2-linoleoyl-3-palmitoyl-sn-glycerol. The sn-acylglycerol will be chiral when the acyl groups at the sn-1 and sn-3 positions are different since the central cabon atom of glycerol is asymmetrical, giving rise to optical isomers (enantiomers).

The prefix "rac" indicates that the compound is an equal mixture of both enantiomers (or antipodes), while the term "x" is used when the configuration is unknown or unspecified.

1.   Monoacylglycerols

These result from esterification of glycerol with only one fatty acid molecule. Two positional isomers exist according as one of the primary hydroxyl groups or the secondary hydroxyl group is esterified. These are 1(3)-monoacyl-sn-glycerol and 2-monoacyl-sn-glycerol. The 1(3)-mono-acyl-sn-glycerol presents two optical isomers according to the position of the acyl groups. These are 1-monoacyl-sn-glycerol and 3-monoacyl-sn-glycerol.

2.   Diacylglycerols

These are fatty acid diesters of glycerol. They occur in two positional isomeric forms; 1,3-diacyl-sn-glycerol and 1,2(2,3)-diacyl-sn-glycerol. The 1,2(2,3)-diacyl-sn-glycerol consists of two stereoisomers: 1,2-diacyl-sn-glycerol and 2,3-diacyl-sn-glycerol.

When the two constituent fatty acids are different the 1,3-diacyl-sn-glycerols occur in two stereoisomeric forms. The 1,2-diacyl-sn-glycerols also occur in two positional isomeric forms which are respectively the optical isomers of the two positional isomers of the 2,3-diacyl-sn-glycerols.

3.   Triacylglycerols

These are fatty acid triesters of glycerol. The constituent fatty acids may be all different, two different, or all alike. When the fatty acids are alike the triacylglycerol is named "monoacid" or "simple" triacylglycerol. When two are different, the term "diacid" is used and "triacid" or "mixed" when the three fatty acids are different. In the latter case, three positional isomers occur, each one having two stereoisomers, or six stereoisomers in total. Only three isomers occur in a "diacid" triacylglycerol.

B.   Physical and Chemical Properties

1.   Physical Properties

The 2 and 1(3) isomers of monoacylglycerols do not present the same properties. Regarding the solid state behavior, the 2-monoacyl-sn-glycerols have only one crystal form (β), in the solid state, while the 1(3) isomers show complex polymorphic changes when cooled from the melt, the β form being the last forming and more stable state. Table 4 shows that the β form of the two monoacyl-glycerol isomers also present different melting points, that of the 1(3) isomers being higher. Both isomers have higher melting points than the corresponding free fatty acids. In saturated monoacyl-glycerols the melting point increases with chain length and decreases with increasing unsaturation.

In rac-octadecenoylglycerol(18:1) the melting point of the trans isomers is higher than that of the cis isomer (Table 4). Monoacylglycerols are poorly soluble in water (less than 1% at 50°C for most of them), and their hydrophilic characteristic decreases with chain length, without any marked difference between the two isomers. They also poorly lower the surface tension of lipids.

Table 4 Melting Points of Saturated and Unsaturated Acylglycerols

Acylglycerols[a]	10:0	12:0	14:0	16:0	18:0		c-18:1	t-18:1	c,c-18:2
TAG	31.5	46.4	57.0	65.5	73.0				
rac-DAG	44.5	57.8	66.8	76.3	79.4	b			
1(3)-MAG	53.0	63.0	70.5	77.0	81.5	73.1	35.0	58.5	14.5
2-MAG	40.4	51.0	61.2	68.5	74.5				
FFA	31.3	43.5	54.4	62.8	69.6		13.0	44.5	−11

[a]TAG; triacylglycerols; DAG, diacylglycerols (rac = racemic); MAG, monoacylglycerols; FFA, free fatty acids.
[b]Racemic mixtures.
Source: Data from Ref. 8.

Table 4 shows that the melting point of racemic saturated *diacylglycerols* is lower than that of the corresponding 1(3)-monoacyl-sn-glycerol, but higher when compared to the 2 isomer. The melting point increases with the chain length of the constituent saturated fatty acids.

Crystallization of *triacylglycerols* involves different solid phases (α, β', β) with a successive improvement of the close packing of the molecules, the more stable being the β form. Regarding the phase transition in simple saturated triacylglycerols, the most extensively studied is described by the following scheme:

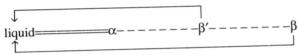

The same polymorphic forms also occur in triacylglycerols with different acyl chains and in natural fats, which are very complex mixtures of a large number of triacylglycerol molecules. Since melting properties of fats are important for industrial purposes, many reports on thermal data of triacylglycerols can be found in the literature (22). A few examples are given in Table 4 for saturated monoacid triacylglycerols. For a mixed triacylglycerol the melting point is roughly equal to the average value of the melting points of its component fatty acids, as a first approximation. Some data were published by Knoester et al. (23). The data in Table 4 show that the melting point of saturated monoacid triacylglycerols increases with the carbon number (total number of acyl carbon atoms). It is higher than the melting point of a saturated diacid or triacid triacylglycerol of the same carbon number. The presence of cis double bonds lowers the melting point. For example, for triacylglycerols of carbon number 54, the melting points in °C are 73, 42, 23, −3, −12, and −24 for 0, 1, 2, 3, 6, and 9 double bonds, respectively.

The presence of trans double bonds is less efficient in lowering the melting points. The melting point of mixed triacylglycerols depends on the position of fatty acids in the molecule. Transesterification, which distributes the component fatty acids in a random manner, generally modifies the melting point. Knowledge of other properties is useful in food analysis, e.g., solubility. Several reports have given extensive information on triacylglycerol solubilities in diethyl ether, petroleum ether, acetone, and ethanol (10). They show, for example, that solubilities of saturated monoacid triacylglycerols in 100% ethanol increase with decreasing carbon number and increase with temperature (24). The same was observed with mixed triacylglycerols (17). The solubilities in the four solvents tested increased in the order diethyl ether > petroleum ether >> acetone > ethanol.

As an example, the solubility at 25°C (g triacylglycerol/100 g solution) of 1-stearate, 2-myristate, 3-palmitate was 10.97, 7.59, 0.18, and 0.03 for the four solvents, respectively. For 1-stearate, 2-caprate, 3-laurate the corresponding solubilities were much higher, i.e., 192.13, 179.56, 13.49, and 0.39 (17). Other physical and optical properties are reviewed elsewhere (8,10).

2. Chemical Properties

In addition to the chemical properties of fatty acid constituents (see above), some properties of lipids are specific to the ester bond and will be briefly described in this section. Lipid *hydrolysis* liberates fatty acids and glycerol in the presence of water. The reaction is most conveniently effected with aqueous ethanolic alkali (saponification). Acidification of the hydrolyzate liberates fatty acids from the soaps. Fatty acids can be extracted with hexane while glycerol remains in the aqueous phase. The quantitative alkaline hydrolysis serves for determination of the saponification value.

Partial hydrolysis of food lipids occurs under the influence of lipases present in seeds or in animal tissues, so that most extracted lipids contain some free fatty acids and some partial acylglycerols, which is an undesirable change. Lipolysis can be minimized by inactivating lipases before extraction of lipids.

The reverse reaction of hydrolysis is *esterification*, whereby a carboxylic acid or an acyl derivative reacts with an alcohol or its equivalent. The reaction of an ester with an alcohol is named *alcoholysis*. The reaction of an ester with an acid is named *acidolysis* and with another ester *interesterification* or *transesterification*. For chromatographic purposes, fatty acids are frequently converted to their esters, most frequently to methyl esters by alcoholysis, without first isolating the free fatty acids. Procedures of fatty acid esterification will be precisely described in another section.

Interesterification is used industrially to improve the physical properties of fats. Natural fats are mixtures of triacylglycerols in which the fatty acids are generally distributed in a nonrandom manner. Under the influence of a catalyst the fatty acids are redistributed intra- and intermolecularly so that a complete random distribution is achieved (25). This procedure is generally used to harden fluid oils. Thus the melting point of soybean oil is raised from −7 to +6°C.

Enzymic interesterification using specific lipases is a recent development of considerable interest (25). The use of lipases in this reaction was recently reviewed by Eigtved (26).

IV. STEROLS AND STERYL ESTERS

Phytosterols, their esters, and cholesterol belong to the great family of steroids, which are compounds containing the nucleus perhydrocyclopentenophenanthrene (27). Steroids comprise a great variety of natural compounds such as sterols, biliary acids, steroid hormones, and several alcaloids. In this section, only sterols will be considered in detail because they can be present in relatively high amounts in foods from vegetable and animal origin.

Sterols are 3-monohydroxysteroids comprising in particular cholesterol widely present in animal foods and phytosterols principally present in vegetable foods. Besides the hydroxyl group in the 3 position, other functional groups can be present such as carboxyl, carbonyl (cetone), or additional hydroxyl groups (phenols) giving to the molecule several peculiar chemical properties.

A. Physical Properties

Steroids display some interesting *physical properties* useful for the structure characterization or for the dosage. Several of them present characteristical ultraviolet (220–280 nm) and infrared (2–25 μm) spectra. Their melting point is highly dependent on the crystallized form (~250°C). Their solubility is important to consider since it determines the quantitative extraction of sterols from foods and their chromatographic behavior in the separation of molecular species. Compounds non- or weakly ionized (cholesterol) are soluble in relatively unpolar solvents such as petroleum ether, ether, benzene, and chloroform. On the contrary, steroids with two or more hydroxyl groups are soluble in alcohols (methanol, ethanol). Steroids displaying ionized groups, such as carboxylic acid salts, glucuronides, or sulfo-conjugated acids are generally water-soluble.

B.   Chemical Properties

Chemical properties of steroids depend on the functional groups present in the molecule. The existence of a hydroxyl group in sterols give rise to acylated compounds. For example, cholesteryl esters are present in animals and acylated steryl glycosides are present in plants. The latter often contain glucose attached via a β-glycoside linkage to the 3 position of sterols, such as sitosterol, stigmasterol, campesterol, and cholesterol. The presence of a carboxylic group in a side chain gives rise to esters or to conjugated derivatives.

Phytosterols occur in plants in relatively high amounts, between 1 and 3 g/kg in such oils as groundnut oil (28) and soybean oil (29). β-Sitosterol is generally the major sterol in vegetable oils, followed by campesterol and stigmasterol. Brassicasterol is only present in rapeseed oil and cholesterol can be found in several oils and fats, such as cocoa butter, linseed oil, and palm kernel oil (30, 10) but in low amounts. On the contrary, cholesterol can be present in relatively high amounts in animal fats, for example 0.22–0.41 wt % in whole bovine milk lipids (31) as free cholesterol and cholesteryl esters. Quantitative analysis of cholesterol in animal foods is crucial because of involvement of this molecule in lipid metabolism disorders in humans.

V.   EXTRACTION OF LIPIDS FROM FOODSTUFFS

A.   Introduction

Most of the methods used in food lipid analysis cannot be directly applied to foods themselves because of their complexity. Lipids should first be extracted either for total lipid estimation or for further analysis to be performed. In all cases, lipids should be isolated quantitatively, free of nonlipid contaminants such as proteins and in their native state for accurate further analyses. For this purpose, care must be taken to ensure that lipolytic enzymes are deactivated. Precautions must also be taken to avoid or at least to minimize auto- or enzymic oxidation of polyunsaturated fatty acids.

The quantitative extraction of lipids often depends on whether the sample is of plant or animal origin as well as on the major component lipids. Neutral lipids (triacylglycerols) can easily be extracted by nonpolar solvents such as petroleum ether, hexane, or supercritical carbon dioxide. Many simple extraction procedures can thus be used for triacylglycerol-rich foods such as animal depot fats or oily seeds or fruits. On the other hand, if the sample contains polar near-water-soluble compounds (phospho- and glycolipids) polar solvents such as methanol must be used for quantitative determination, sometimes preceded by a chemical hydrolysis to break bonds formed between lipids and proteins.

A great variety of lipid analysis or extraction procedures have been developed. Several methods are not compatible with further analysis of the different lipid constituents. These are physical procedures in which lipids are not isolated, such as low-resolution ^1H NMR (32) or high-resolution ^{13}C NMR (33), and chemical procedures using solvent extraction with heating and acidic or alkaline hydrolysis. On the other hand, well-normalized procedures (American Oil Chemists' Society, or AOCS Aah-38) are used to determine the oil content of oily seeds (34). This is a refluxing extraction by petroleum ether in Soxhlet-type apparatus.

These methods will not be included in this section. Only methods which allow further analysis of the different lipids in addition to the total lipid extract measurement will be mentioned. These involve quantitative lipid extraction, generally by means of two miscible solvents, one unpolar (hydrocarbon) and one polar (alcohol). The latter, also breaks the hydrophobic or van der Waals forces, the hydrogen and ionic bonds, with the associated proteins or polysaccharides. A quantitative extraction of the unpolar and polar lipids, whatever the degree of polarity, is ensured by this binary solvent mixture, which, additionally, is selective since hydrophobic proteins and hydrophilic saccharides and peptides are not dissolved.

The extraction procedures involve four steps:

1. Extraction of lipids after grinding of the foodstuff and dispersion in the solvent.
2. Separation of the organic phase containing lipids from unextractable material by decantation or centrifugation and filtration.
3. Removal of nonlipid contaminants from the organic phase by washing.
4. Removal of solvent from extract by distillation or evaporation under nitrogen.

These different steps have been described in detail by Christie (35), together with very useful considerations regarding in particular storage of samples, preliminary treatments prior to extraction to minimize lipolysis and lipid peroxidation, factors affecting extractability of lipids by different solvent mixtures and artifacts encountered in extraction procedures. Readers are recommended to consult this excellent review. General considerations were also presented by Litchfield (7) and are reported in Table 5.

B. Recommended Extraction Procedures

A great number of extraction procedures have been described in the literature. They derived from the "Folch" procedure which to date remains the preeminent method (36). It was first proposed by the authors for the extraction of lipids from brain and animal tissues, but it is also convenient for vegetal tissues and any food sample. The operating conditions that the authors selected in the laboratory are as follows: The weighed sample (~1 g) is introduced into a 50-ml glass centrifuge tube and added with 13.3 ml, exactly measured, of chloroform-methanol (1:1, v/v). The sample is

Table 5   General Precautions for Preventing Contamination and Chemical Alteration of Lipid Samples During Analysis

---

To avoid contamination:
    1. Distill all solvents before use.
    2. Rinse all glassware with chloroform-methanol (2:1, v/v) immediately before use.
    3. Avoid sample contact with any rubber or plastic except Teflon.
    4. Store samples and solvents in glass containers with glass or Teflon-lined caps.
To prevent oxidation of polyunsaturated acids:
    1. Perform all laboratory manipulations under nitrogen.
    2. Bubble nitrogen or helium through solvents to purge dissolved oxygen.
    3. Evaporate solvents below 40°C under an inert atmosphere.
    4. Add 0.005% antioxidant (w/v) to solutions of stored samples.[a]
    5. Store samples or lipid extracts under nitrogen at −20°C.
To avoid chemical alteration:
    1. Extract lipids immediately after food sampling.
    2. Avoid prolonged sample contact with any alcohol, particularly under acidic or alkaline conditions.
    3. Remove the last traces of water by codistillation with pure ethanol.
    4. Do not leave lipids in the dry state. Cover with an inert nonalcoholic solvent such as hexane.

---

[a]Care should be taken to select an antioxidant that does not interfere with subsequent analyses. BHT (butylhydroxytoluene, or 2,6-di-*tert*-butyl-*p*-cresol) is often employed because it elutes in or near the solvent front during GLC of long-chain fatty acid methyl esters.
Source: Computed from Refs. 7 and 35.

homogenized for 5–10 sec by means of a grinder Ultraturrax which is rinced with exactly 6.7 ml of chloroform. The sample is homogenized in the presence of 20 ml of chloroform-methanol (2:1, v/v). The mixture is gently shaken to homogenize and left at room temperature for 2 hr for denaturation of proteins. Then 4.2 ml of 0.37 M potassium chloride is added, the mixture is homogenized and centrifuged at 3000 rpm for 5 min. The mixture partitions in two layers. The upper layer is composed of water-methanol-chloroform (47:48:3, in volumes) with denaturated proteins between the two layers. The lower layer containing the lipids is composed of chloroform-methanol-water (86:14:1, in volumes). The clear upper layer is drawn off by means of a Pasteur pipette under vacuum and replaced by 10 ml of "top phase," i.e., chloroform-methanol-water (47:48:3, in volumes). After homogenization, the tube is centrifuged at 3000 rpm for 5 min. The upper phase is eliminated by aspiration and, if necessary, a second washing is applied. The lower phase is filtered through a paper filter on a funnel to eliminate the denaturated proteins and collected in a 100-ml vessel. The centrifuge tube and the filter with the residue are rinsed with chloroform and with choloroform-methanol (1:1, v/v).

The solvent is removed under vacuum at 40°C using a rotating evaporator and the remaining water is eliminated by two successive codistillations with absolute ethanol. The lipids are redissolved in chloroform and quantitatively transferred into a 50-ml tared vessel. Chloroform is evaporated under vacuum and the vessel is kept under vacuum in a desiccator for several hours and weighed.

The lipids are stored at –20°C in a small volume of chloroform added with an antioxidant (BHT, butylhydroxytoluene). In most cases the complex polar lipids such as phospholipids and glycolipids are efficiently extracted.

A derived procedure, also commonly used, was developed by Bligh and Dyer (37) as an economical method of extracting lipids from large volumes of water-rich samples. In this procedure the endogenous water constitutes the ternary component of the extraction mixture and sufficient chloroform and methanol are added to give a single-phase system for homogenization. After filtration, the residue is rehomogenized with pure chloroform for neutral lipids to be completely extracted. The combined organic phases are washed with saline solution. The lower chloroform phase containing the lipids is quantitatively removed. This procedure is rapid and suited for routine analyses. It gives good recoveries of the major lipids. However, uncomplete recovery of minor acidic lipids has been sometimes reported.

A large variety of alternative binary, or even ternary solvent mixtures, have been proposed, principally because of solvent toxicity. In addition, chloroform and methanol are not completely stable and were found to give rise to acidic byproducts which could initiate esterification or transesterification of lipids. Such mixtures are as follows:

Dichloromethane-methanol (2:1, v/v) (38);
Dimethoxymethane-methanol (4:1, v/v) (39);
Chloroform-propan-2-ol (40);
Hexane-propan-2-ol (3:2, v/v) (47);
Hexane-methanol (1:1, v/v) (42);
Diethyl ether-ethanol (1:3, v/v) (43).

Different other mixtures of benzene or toluene with methanol or ethanol have been proposed for specific uses (for review, see Ref. 9).

Several studies dealing with comparison between extraction procedures have shown that efficiency of extraction increased with solvent polarity, while selectivity decreases.

Recently, supercritical fluids (SFs) have been evaluated as extractants for lipids (44). Carbon dioxide is by far the most commonly used SF because its low critical temperature (31.04°C) makes it suitable for extraction of thermally labile lipids. Additionally, the critical pressure (7.38 MPa) and critical density (0.47 g/cm^3) are high enough to give good solvating properties. To overcome restricted ability to extract polar compounds, a polar organic solvent, such as methanol, can be added in small amounts in supercritical CO_2 (45). This method seems to hold promise, although there appears to be little prospect for more general use.

As an alternative to conventional solvent extraction, solid phase extraction procedures have been described. In the method developed by Maxwell et al. (46) for extraction of lipids from animal tissues, a column is filled with anhydrous sodium sulfate and Celite 545 mixed with the ground food. Lipids are sequentially extracted, neutral lipids with dichloromethane and polar lipids with dichloromethane-methanol (9:1, v/v). Calcium hydrogenphosphate is then added to the lipid extracts to eliminate proteins. Wolff and Castera (47) proposed another procedure by solid phase extraction of lipids, in which a first extraction by hexane-propan-2-ol (3:2, v/v) is carried out. The lipid extract is dehydrated by passing through anhydrous Na_2SO_4 and purified through Celite 545. This simple and rapid procedure provides as good extractions as the Folch method and can be applied to a great variety of foods, in which the oil content can vary from 0.5 to 80 wt % and moisture from 0 to 90%. Furthermore, lipid fatty acids in hexane-isopropanol can be directly converted after acidification to isopropyl esters recommended in GLC analysis when short-chain fatty acids are present, as in dairy products. This method has been standardized and normalized Association Française de Normalisation; (AFNOR, NF V 83–30) as an extraction procedure also allowing characterization of lipid classes (48).

The use of a microwave oven has been recommended to facilitate extraction of grain and plant lipids (49).

When only the lipid content is to be determined in foods, the extraction procedures can also be preceded by an acidic or alkaline hydrolysis. The major methods used, and normalized, are reported in Table 6 together with the foods concerned.

C. Extraction of the Free Fatty Acid Fraction

In foods the free fatty acid fraction represents an important lipid fraction to be analyzed because it often reflects some degree of lipid hydrolysis. It can be isolated from a lipid extract by adsorption chromatography by means of one of the methods described in the next section in particular by TLC or by HPLC. However, when short-chain fatty acids, especially butyric, are present as free fatty acids in foods such as dairy products, the different operations of lipid extraction and lipid class separation cause an important loss of these acids because of their volatility and water solubility. In that case, it is preferable to immediately isolate the free fatty acid fraction from the lipid extract in such conditions as no loss of short-chain fatty acids occurs. Similar conditions should also be adopted for determining the fatty acid profile of this fraction by GLC. Several chromatographic methods have been proposed to isolate the free fatty acids from a lipid extract. They generally used a commercial prepacked column containing a stationary phase which may be silica gel (adsorption chromatography) or reverse phase materials such as octadecylsilyl (ODS) groups bonded to silica (partition chromatography) or aminopropyl moieties bonded to silica (ion exchange chromatography). These columns (or cartridges) are available from many manufacturers (e.g., Sep-Pak from Waters Associates). Ansari and Shoeman (50) proposed using a cartridge of silica gel to isolate the free fatty acids from brain lipid extracts. The acids were converted to ammonium salts by the addition of ammonium hydroxide to the extraction solvent. The simple lipids were eluted with chloroform saturated with the base and the unesterified fatty acid fraction was eluted with chloroform–acetic acid.

ODS cartridges were used by Pemkowiak (51) to adsorbe trace amounts of free fatty acids contained in seawater. The sorbed acids were eluted with 0.1 M ammonium hydroxide in water-methanol (1:1, v/v) followed by pure methanol. Such cartridges have also been used to

Table 6 Principal Methods Used to Determine the Lipid Content of Foods

Method	Foods concerned	Previous treatment	Extraction solvent	Ref.[a]
Direct extraction	Animal foods	Drying on anhydrous sodium sulfate	Ethyl ether	EEC Method A
	Oily seeds		Hexane	NF V 03 905
	Oil cakes		Hexane	NF V 03 924
	Oil cakes		Ethyl ether	NF V 03 925
	Meat		Hexane	NF V 04 403
	Miscellaneous		Isopropanol-hexane	NF V 83 30
	Miscellaneous		Dichloromethane-methanol	
Extraction after previous treatment	Aminal foods	HCl 3 N—100°C	Ethyl ether	EEC Method B
	Starch	HCl 4 N—100°C	Hexane	NF V 03 610
	Cocoa-chocolate	HCl 8 N—100°C	Hexane	EEC 16-10-75
	Casein	HCl 8 N—water bath	Ethanol–ethyl ether–hexane	NF V 04 382
	Cereals	HCl 8.4 N + formic acid—75°C	Ethanol–hexane	NF V 03 713
	Cream	NH₃ + ethanol	Ethyl ether–hexane	NF V 04 261
	Ice creams	NH₃ + ethanol	Ethyl ehter–petroleum ether	NF V 04 261
	Milks	NH₃ + ethanol	Ethyl ether–petroleum ether	NF V 04 214
	Milks (added with formol)	HCl 12 N—80–85°C	Ethyl ether	NF V 04 215
	Margarines	HCl N	Hexane	IUPAC
	Dairy products	HCl 4 N 100°C	Hexane	NF ISO 82 623
	Meats	HCl 4 N 100°C	Hexane	NF V 04 402
Physical methods	Oily seeds and oil cakes	Drying		Nuclear magnetic resonance, NF V 03 907

[a]EEC, European Economic Community; NF, Norme Française; IUPAC, International Union of Pure and Applied Chemistry.
Source: Data from A. Castera in Ref. 11.

isolate free fatty acids from serum by directly depositing the serum diluted with 10% acetic acid in water onto the column followed by water. The acids were recovered by elution with diethyl ether (52). However, lipid analysts have generally prefered using ion exchange chromatography to isolate the free fatty acid fraction from a lipid extract by means of cartridge with aminopropyl group chemically bonded to silica (53,54). With a sample of cheese (55) the extraction procedure was as follows: Cheese (1 g) was placed in a screw-capped tube and ground with 3 g anhydrous sodium sulfate, 0.3 ml 2.5 M sulfuric acid in water, and 1 ml internal standard solution (heptanoic 150 mg/L plus tridecanoic; 150 mg/L in hexane). Three milliliters diethy ether–hexane (1:1, v/v) was added and the mixture shaken for 3 min using a Vortex mixer. The supernatant was transferred to a screw-capped tube containing 1 g anhydrous sodium sulfate. This operation was repeated three times. The pooled lipid extracts were passed through an aminopropyl column Sep-Pak Plus, previously conditioned with 10 ml hexane. Simple lipids were eluted with 10 ml chloroform and the free fatty acids with 2.5 ml diethyl ether–formic acid (98:2, v/v). The diethyl ether solution was directly used for GC analysis of the fatty acids in the free form (see below).

VI. SEPARATION OF LIPID CLASSES

The complexity of natural lipid extracts, even from food lipids, is generally such that all of the lipid classes can rarely be separated in only one operation. It is therefore more convenient first to isolate pure lipid groups, such as "neutral" lipids and phospholipids, and second to separate the constituent lipid classes in each group.

However, methods have been developed in the past to separate all of the component neutral lipids but with the phospholipids in only one fraction. Conversely, by other methods phospholipid classes were separated with the neutral lipids in a single fraction.

More recently, the possibility of varying very uniformly and progressively the polarity of solvent has made it possible to separate in a lipid extract a wide range of lipid classes varying in polarity from cholesteryl esters to lysophospholipids in one run by HPLC with mass detection.

In this section we will consider different separation methods of simple lipid classes, using adsorption or partition chromatography, in different ways. The methods which have been to date the most utilized are column chromatography (CC) and thin-layer chromatography (TLC). However, high-performance liquid chromatography (HPLC), especially with mass detection, has the potential to replace the other methods.

A. Column Chromatography

The use of a column of relatively large volume allows high amounts of single lipids (several hundred milligrams) to be separated and isolated for further analysis. It has often been used to fractionate a pure triacylglycerol fraction prior to separation of molecular species by HPLC (56). The most commonly used adsorbants are as follows:

Florisil (magnesium silicate) deactivated by adding 7% wt water
Silicic acid containing 5–10% wt water
Silica gel (5% wt water)
Alumina (5% wt water).

In the first case (57) a glass column is filled with Florisil. The lipid extract in petroleum ether is deposited on the top of the column protected by a filter disk and the lipid classes are eluted first by 100 ml of hexane and then by hexane which is more and more enriched in diethyl ether and at least by a mixture of diethyl ether–ethanol. Details are given in Table 7 regarding the amount of adsorbant, the column size, the volume, and the composition of solvent mixtures used to elute the

Table 7 Some Examples of Simple Lipid Classes: Separation by Column Adsorption Chromatography

Adsorbent	Specifications[a]	Solvents (ml)			Eluted lipids	Ref.[b]
		Hexane	Diethyl ether	Methanol		
Florisil	30 g 60–100 mesh	100	—	—	Hydrocarbons	Carroll (1961) (57)
		95	5	—	Cholesteryl esters	
		85	15	—	Triacylglycerols + free fatty acids	
	Column: 20 × 170 mm	75	25	—	Free cholesterol	
		50	50	—	Diacylglycerols	
		—	90	10	Monoacylglycerols	
		Pentane	Diethyl ether	Methanol		
Silicic acid	55 g	291	9	—	Cholesteryl esters	Modified from MEAD (1961) (63)
		276	24	—	Triacylglycerols	
		160	40	—	Free fatty acids	
	Column: 35 × 500 mm	140	60	—	Diacylglycerols + free cholesterol	
		—	200	—	Monoacylglycerols	
		—	—	200	Phospholipids	
		Petroleum ether	Diethyl ether	Methanol		
Silicic acid	18 g 100 mesh	99	1	—	Cholesteryl esters	Day and FIDGE (1964) (69)
		96	4	—	Triacylglycerols + free fatty acids	
		92	8	—	Free cholestero	
	Column: 20 × 500 mm	75	25	—	Diacylglycerols	
		—	100	—	Monoacylglycerols	
		—	—	100	Phospholipids	

Table 7 (continued)

Adsorbant	Specifications[a]	Solvents (ml)				Eluted lipids	Ref.[b]
		Benzene	Diethyl ether	Methanol			
Silica gel	30 g	200	—			Triacylglycerols	IUPAC 2-321
	Column:	180	20			Diacylglycerols + free fatty acids	AOAC 965-35
	19 × 290 mm	—	200			Monoacylglycerols	AOAC 966-19
Silica gel	30 g	300	—			Triacylglycerols + 20% free fatty acids	
	Column:	225	25			Diacylglycerols + 60% free fatty acids	
	19 × 290 mm	—	200			Monoacylglycerols + 20% free fatty acids	

[a]Amount of adsorbant in grams (g), adsorbant particle size in mesh, internal diameter × length in millimeters.
[b]IUPAC, International Union of Pure and Applied Chemistry; AOAC, American Association of Analytical Chemistry.

different lipid classes (57). Several other methods have been published with slight modifications (58–62).

Procedures of simple lipid classes separation in the presence of silicic acid are very similar to those described for Florisil. The different lipid classes, cholesteryl esters, triacylglycerols, free fatty acids, free cholesterol, and diacylglycerols are eluted with petroleum ether or pentane containing increasing proportions of diethyl ether to increase the solvent polarity. Monoacylglycerols are eluted by pure diethyl ether and phospholipids by pure methanol. Several methods have been published (63–68, 208).

In the laboratory, the authors have generally used the method proposed by Fillerup and Mead (63), especially to fractionated partial acylglycerols and free fatty acids issued from lipolysis of triacylglycerols, because these lipid components were well separated and in sufficient amounts to proceed to further analyses. Details are given in Table 7. Under these conditions, diacylglycerols and free cholesterol are not easily separated. In a smaller column and with another sequence of solvents, Day and Fidge (69) achieved this separation, but triacylglycerols and free fatty acids eluted together. Separation conditions are reported in Table 7.

When partly coeluting with the triacylglycerol fractions, as with Florisil as adsorbant (57) or with silicic acid (69) free fatty acids can be separated on a second column containing Florisil or Sephadex (70) or aluminum oxide (77), or extracted by ion exchange (72,73) or by aqueous sodium carbonate (74).

Two methods of acylglycerol separation by column chromatography have been standardized (Table 7). Silica gel containing 5% wt of water is used as adsorbant and elution of mono-, di-, and triacylglycerols is by benzene or diethyl ether or a mixture of the two solvents.

In chromatography on columns packed with deactivated alumina (5% water), the solvents used are mixtures of petroleum ether–diethyl ether in proportion varying from 70:30 to 10:90 (v/v) (75).

Ion exchange column chromatography, on diethylaminoethyl (DEAE)-cellulose is used for the fractionation of particular groups of complex lipids in comparatively large amounts (200–400 mg on 30 mm i.d. × 500 mm column). The procedure can also be used on a smaller scale, with a smaller column, for isolation of polar plant lipids (3,76).

A complete separation by column chromatography takes in most cases 2–3 hr. Lipids which are present in the elution solvents are identified by analytical TLC. When only simple group separations are required, e.g., simple lipids or phospholipids for fatty acid GC analysis, simpler and more rapid small-scale procedures are more convenient (77,78). Prepacked cartridges of silicagel are now commercially available for these separations.

The procedure consists of applying about 30 mg of the lipid extract to this short column and eluting with 10 ml chloroform (or diethyl ether) to yield the simple lipids and 10 ml chloroform-methanol (1:1, v/v.) or even pure methanol to yield the more polar complex lipids.

B. Thin-Layer Chromatography

As for column chromatography (CC), the equipment required for TLC is simple, inexpensive, and flexible in that it can be used analytically (sample of less than 50 mg) and preparatively (79–82). By increasing the number of plates, the amounts of simple lipids which can be fractionated easily attain those obtained by column chromatography. Layers of silica gel G, containing calcium sulfate as binder, have been most frequently used for separation of lipid classes. Although precoated plates are now commercially available, TLC plates can be prepared in the laboratory as follows:

Depending on the thickness desired, 40–80 g of silica gel is homogeneously suspended, by vigorous stirring, in 80–160 ml of water and the mixture is spread on 200 × 200 mm glass plates at a thickness of 0.5–1.0 mm by means of a commercial device. The plates are dried in air, activated in an oven at 120°C for 1 hr, cooled, and stored in dry atmosphere. Lipids to be separated (5–50 mg) in chloroform solution are deposited by means of a Pasteur pipette or any other convenient device as a narrow, continuous band at about 20 mm from the lower edge of the plate. A mixture of standard lipids can be deposited at one extremity to help identification.

The plate is then developed in an appropriate solvent. The solvent elution systems we have the most commonly employed is hexane–diethyl ether–methanol–acetic acid (90:20:3:2, in volumes), or hexane–diethyl ether–acetic acid (90:30:1, in volumes) for the separation of acylglycerols and free fatty acids after lipolysis of triacylglycerols. The plate is then placed upright in a closed tank containing the solvent in such an amount that the level arrives a few mm below the lipid deposit. When the solvent has migrated up to 30 mm of the upper edge of the plate (after about 20 min for 200 × 200 mm plates), the plate is taken out and the solvent evaporated under a nitrogen flow. Bands are then conveniently detected by spraying with an 0.1% (w/v) solution of 2′,7′-dichlorofluorescein in 95% ethanol (or methanol) and viewing the dried plate under UV light (350 nm). The lipids appear as yellow bands. The elution order is from the solvent front:cholesteryl esters, triacylglycerols, free fatty acids, free cholesterol, diacylglycerols, monoacylglycerols, and total phospholipids as shown in Fig. 1. The order is that of increasing polarity of the lipid classes. The lipids adsorbed on the silica gel can be recovered. For this purpose, the bands are delineated under UV light and the adsorbant is scrapped off the plate, by means of razor blade, into a beacher or a small column. Simple lipids will be quantitatively eluted with chloroform or diethyl ether while elution of phospholipids will require chloroform-methanol (1:1, v/v) or the more polar solvent system chloroform-methanol-water (10:10:2, v/v). At this step, an internal standard can be added, such as an odd chain fatty acid not present in the sample (penta-, hepta-, or nonadecanoic) in the free fatty acid form or esterified in a triacylglycerol, to determine the amount of lipid, after GC analysis of fatty acids.

The procedure described above is convenient for simple lipids. To separate phospholipid classes, it is preferable to employ silica gel H as absorbant, with sodium carbonate as binder, and a more polar solvent elution system, such as chloroform–methanol–water–acetic acid (25:15:4:2, in volumes) (6,83).

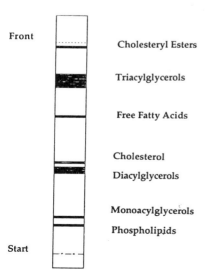

Fig. 1 Schematic separation of lipid classes of cheese by TLC on a precoated plate (Merck), developed with the eluant system of hexane–diethyl ether–acetic acid (70:30:1, in volumes).

C. High-Performance Liquid Chromatography

Separation of simple lipid classes by HPLC requires a sophisticated and comparatively expensive equipment, especially a suitable detector, contrary to the CC and TLC procedures described above. It is probably one of the reasons why HPLC is relatively less employed for this purpose. However, in many laboratories it is now supplanting TLC because it offers several advantages such as higher resolution, and easier and more accurate quantitation.

Many different types of column packing materials, solvent elution systems, and detectors have been used for lipid class separations by HPLC, as underlined by Christie (9), depending on the problem to be solved. The first problem which can be encountered by lipid analysts is to fractionate lipid classes from an extract for further analysis of their constituent fatty acids or of their different molecular species. In that case, a nondestructive detector, such as refractive index, UV, or IR detector, is recommended. If not, the detector should be sensitive enough for the major part of the column effluent to be diverted and recovered before detection.

The second problem encountered is to quantitatively determine the lipid class composition of the extract. In that case, a universal or mass-selective detector is the most convenient. If such a detector is not available, quantitation can be achieved by collecting the effluent corresponding to each lipid class, before (destructive detector) or after detection and adding an internal standard (odd chain fatty acid) to each fraction collected before GC analysis of the constituent fatty acids.

In both cases (determination of lipid class composition or fractination for subsequent analyses) the type of column and the type of elution solvent system must be selected in order to offer the best separations. In the case of simple lipid class separation, adsorption chromatography with columns of silica gel has been the most widely employed with a great variety of elution solvents and of detectors.

One of the first separations reported (84) was on a comparatively long and thin column (2.1 mm i.d. × 500 mm) packed with silica gel (MicroPak SH-10) with a gradient of ethanol into hexane-chloroform (9:1, v/v) as elution system. The detection was not by differential refractometry since the refractive index detector can only be used with isocratic elution, but by transport flame ionization. As shown in Fig. 2, a good separation of mono-, di-, and triacylglycerols, free fatty acids, and cholesterol was achieved in less than 15 min. By slightly modifying the elution conditions, cholesteryl esters could probably also be eluted ahead of triacylglycerols.

With the same type of column of silica gel (Ultrasil Si 5 μm), but with isooctane–tetrahydrofurane–formic acid (80:20:1, in volumes) as solvent and in isocratic mode, the elution order was somewhat different (85). In order of increasing retention time, it was cholesteryl esters, triacylglycerols, free fatty acids, free cholesterol, and diacylglycerols. Monoacylglycerols were not present in the sample. They probably would elute in last position. Since the detection was by a refractive index detector, the system could be used for collection of the lipid fraction at the outlet of the detector and for quantitative analysis, from peak areas, after calibration, or after internal standardization of the fractions collected (see below).

Reverse phase HPLC is also convenient for lipid class separation as shown by Palmer et al. (86) with a cyanopropyl-bonded phase, 0.05% isopropanol in hexane as elution solvent, and detection by UV light (210 nm) absorption. The elution order was cholesteryl esters, triacylglycerols, and cholesterol. Unfortunately, free fatty acids are more strongly absorbed on this type of column than on silicic acid columns. Moreover, detection by UV absorption used by the authors is of limited value for quantification purposes. For the same reason it has been rarely used with IR detection. With the recent development of mass detection, such as "light-scattering" detection, unsensitive to modifications of solvent composition, separation of not only simple lipids but of simple and complex lipids can easily be achieved in one step because the polarity of the solvent system can be progressively increased in the course of separations to elute the different lipid

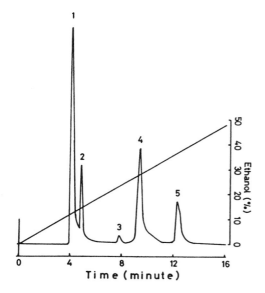

Fig. 2 Separation of a standard mixture of tristearin (1), distearin (2), cholesterol (3), palmitic acid, (4) and monopalmitin (5) by high-speed liquid chromatography on a Micropak SH-10 column eluted with a linear gradient solvent flow programming (see details in the text). (Reproduced from Ref. 84 by kind permission of Kiuchi et al. and the *Journal of Chromatographic Science.*)

components at the optimum polarity. Remarkable separations of very complex mixtures have been reported by Christie (87,88). In his first study (87), the author used a very small column (5 mm i.d. × 100 mm) packed with silica gel (Spherisorb, 3-μm particles) and a ternary gradient elution system. The three solvent mixtures used, in increasing polarity, were isooctane-tetrahydrofuran (99:1, v/v), isopropanol-chloroform (4:1, v/v), and isopropanol-water (1:1, v/v). At a flow rate of 2 ml/min, each of the main simple lipid (cholesteryl esters, triacylglycerols, free cholesterol) and phospholipid classes (phosphatidylethanolamine, phosphatidylinositol, phosphatidylcholine) from a rat liver lipid extract were clearly resolved in less than 20 min. Other components, such as mono- and diacylglycerols, free fatty acids, phosphatidylserine, and sphingomyelin, can also be separated under these conditions. Resolutions obtained by the author were of better quality and achieved in less time than those previously reported by Privett et al. (89). Furthermore these authors utilized a transport flame ionization detection device, less widely commercialized that the light-scattering detector.

In the future HPLC separation of lipid classes using this mass detector will certainly receive considerable attention from lipid analysts who want to determine qualitative and quantitative lipid class profiles of lipid extracts or individually collecting pure and clean lipid components for further analyses. The method will be particularly useful in food analysis.

VII. ANALYSIS OF FATTY ACIDS

A. Introduction

General information on the constituent fatty acids of fats can be obtained from different chemical analyses giving specific "values."

1. Saponification Value

Saponification value is the amount of alkali necessary to saponify a definite quantity of the lipid sample. It is expressed as the number of mg of potassium hydroxyde (KOH) required to saponify 1 g of the sample. The method of measure has been standardized and is thoroughly described in Official Methods (AOAC 920.160; AOCS Cd 3–25, Cd 3b-76, Cd 3c-91; AFNOR NFT 60–206) (90–92). The determination consists of saponifying the fat sample by refluxing with an excess of alcoholic KOH and titrating the unreacted KOH by hydrochloric acid with phenolphthalein as indicator. The saponification value gives information on the average chain length of the fat constituent fatty acids. It increases when the chain length decreases and conversely. For example, the saponification value of sunflower oil rich in C18 fatty acids is ~190, while that of coconut oil rich in lauric acid (C12) is ~260.

2. Iodine Value

Iodine value is expressed as the number of centigrams of iodine absorbed per gram of lipid sample. The method of measure has also been standardized and described in Official Methods (AOAC 920-158, 920-159; AOCS Cd 1–25, 1b-87; AFNOR NFT 60–203) (93–95). The determination consists of fixing iodine on the fatty acid double bonds using an excess of the Wijs solution (iodine chloride, ICl, in glacial acetic acid) and to titrate the unfixed iodine by sodium thiosulfate after adding potassium iodine, with starch as indicator. The iodine value gives information on the average degree of unsaturation of the constituent fatty acids. It increases with unsaturation. For example, the iodine value of sunflower oil rich in linoleic acid (18:2 n-6) is ~125 while that of coconut oil rich in saturated fatty acids is only ~8.

3. Hydroxyl Value

Hydroxyl value is defined as the number of milligrams of KOH equivalent to the hydroxyl content of the fat sample. Its determination is described in Official Methods (AOAC 965–32; AOCS Cd 4–40; AFNOR NFT 60–213) (96–98). The hydroxyl value gives information on the presence or absence of hydroxylated fatty acids in fats. For example, castor, which is rich in ricinoleic acid, has a hydroxyl value of ~150.

These chemical determinations are still used, e.g., in the control of food quality. However, the information brought about by these measures is general and does not indicate the specific fatty acids present in the fat and their proportion. That information can be obtained by GLC, which is probably still the most informative technique in fatty acid analysis.

B. Gas–Liquid Chromatography

1. Introduction

Gas liquid chromatograpy, or more simply gas chromatography (GC), was first developed by lipid analysts, who later contributed greatly to the development of the technique. Gas chromatography is based on the ideas of Martin and Synge (99). The concepts developed by these authors were applied by James and Martin (100) 10 years later, in the separation of short-chain fatty acids in the free form, with a column packed with Celite coated with paraffin, detection and quantitation being achieved by titration. The next report by the same authors (1956) dealt for the first time with the separation of methyl esters of fatty acids and automatic detection by the gas-density balance, first developed by Martin and James (101). The instrumentation then became more and more sophisticated, with the development of new detectors, new columns, and new ovens, the heating of which could be programmed according to the separations sought. Many reviews and books have regularly reported on the developments of the GC procedures and instrumentation (102–104). The more recent and comprehensive study was published by Christie (6).

It is beyond scope of this chapter to describe in detail the theory of GC and the instrumentation used which was developed since the basic work of James and Martin (100,101,105). We will principally focus on what was and is now the most widely used in GC analysis of fatty acids and the most useful for lipid analysts.

2. Derivatization of Fatty Acids

a. Introduction. Analysis of fatty acids in the free form by gas or liquid chromatography is practically reserved to analysis of the free fatty fraction of a lipid extract and only when volatile short-chain fatty acids are present in this extract, such as in cheese samples. This procedure avoids, or considerably limits, losses of short-chain fatty acids which are relatively volatile, especially as methyl esters, and partially water-soluble. When the short-chain fatty acids are esterified in triacylglycerols, their quantitative analysis can be carried out on the ester derivatives, in peculiar conditions exposed below.

For most fatty acid mixtures, use of ester derivatives, especially of methyl esters, is recommended for several reasons. The ester derivatives are more volatile than the correspondent free fatty acids and thereby more suited for their GC analysis in the gaseous form. They are much less polar and do not tend to adsorb on the support, even carefully "deactivated," or to dimerize, avoiding peak tailing, peak asymmetry, and peak shouldering. Thus, peak shape and resolution are considerably improved.

For the same reasons, it is recommended that other functional groups, such as hydroxyl groups, be masked to facilitate their analysis.

b. Fatty Acid Derivatives.

METHYL ESTERS   Methyl ester derivatives are by far the most widely used in gas and liquid chromatography. A great number of methods have been proposed to propare methyl ester derivatives of fatty acids, depending on the presence or absence of free fatty acids, in the lipid extract, or the presence of short-chain fatty acids, of highly polyunsaturated fatty acids or of particular fatty acids such as cyclopropene fatty acids (106–108).

Methylation can be acid- or base-catalyzed. In the first case the acidic catalyst used is hydrogen chloride (109,110) or sulfuric acid (111–113). or boron trifluoride (Lewis acid) (114,115). In the base-catalyzed procedures, use of strong bases such as sodium or potassium hydroxides (116) is recommended, or alkaline methylates such as sodium methylate (117,118), or other weak bases (119).

Several methods in neutral medium have been developed in recent years which avoid modifications of certain functional groups in the presence of an acidic catalyst when heating (120,121). The two methylation methods recommended in the AOAC Official Methods of Analysis (AOAC 969–33, IUPAC 2–301, AOCS Ce 2–66) (122–124) are as follows:

When fatty acids are esterified in acylglycerols, the procedure consists of a saponification step using a 0.5 N solution of sodium hydroxide in methanol followed by an esterification step with a solution of 12.5% (w/v) of boron trifluoride (BF_3) in methanol. The methyl esters are extracted with heptane added in the reaction medium.

When the fatty acids are in the free form, the saponification step is omitted and the fatty acids are directly esterified by the methanolic solution of BF3.

With a mixture of free fatty acids, diazomethane can also be utilized to prepare methyl esters in the presence of trace amounts of methanol as catalyst. The reagent is prepared by the action of a base (potassium hydroxide) on a nitrosamide (*N*-methyl-*N*-nitroso-*p*-toluenesulfonamide) and dissolved in diethyl ether, according to Schlenk and Gellerman (125). However, great care must be taken in the preparation and utilization of this reagent because of its toxicity and carcinogenicity.

The above methods are suited for plant and usual oils and fats only comprising the most common fatty acids. They cannot be used with lipids containing particular fatty acids in relatively high amounts, such as highly polyunsaturated fatty acids with acetylenic bonds or conjugated double bonds, or fatty acids containing the following groups: epoxy, hydroperoxy, cycloprophenyl, cyclopropyl, and possibly hydroxy. For the latter, specific procedures have been developed by Barford (126) for epoxy acids and by Bianchini et al. (127) for cyclopropenoic and cyclopropanic fatty acids. Metcalfe and Wang (119) proposed use of a weak base, the tetramethylammonium hydroxide, and Cecchi et al. (128) use of methyl *tert*-isobutyl ether as solvent.

The methylation methods recommended by the Association Française de Normalisation (AFNOR NF 60 233) depend on the lipid sample (129).

A general method is identical to that of the Association of Official Analytical Chemists (AOAC 969–33) (122). The other three methods are based on base-catalyzed reactions, using KOH-methanol or sodium methoxide–methanol as reagent. The first one, applicable to neutral lipids, is close to the Brockerhoff's method (116) except that in the latter the reaction takes place at room temperature. The second one is suited for acidic lipid samples and comprises a base-catalyzed methylation for esterified fatty acids followed by an acid-catalyzed esterification for free fatty acids. The last one, derived from the work of Luddy et al. (130) and Christopherson (131) is suited for mixtures containing short-chain fatty acids. In their laboratory the authors currently use two procedures for methylation of the most common esterified fatty acids, of chain length higher than 12 carbon atoms. For lower chain lengths they use butylesters. The first method, adapted from Metcalfe (132) and modified by Slover and Lanza (133), is suited for very small samples:

The lipid sample (~1 mg) in solution is introduced into a 4-ml tube with a Teflon-lined screwcap. After evaporation of the solvent, 0.5 ml of 0.5 N sodium hydroxide (NaOH) in methanol is added, and the tube is stoppered under nitrogen and heated at 80°C in an oven for 15 min. After cooling, 1 ml of 14% (w/v) BF_3 in methanol is added. The mixture is heated at 80°C for 15 min and cooled. For extracting the methyl esters, 1 ml of saturated NaCl solution in water is added together with 1 ml of isooctane or hexane. The mixture is vigorously stirred and after decantation the upper solvent phase is pipetted and passed through anhydrous sodium sulfate contained in a Pasteur pipette..

When only free fatty acids are present in the sample the procedure is the same except that the saponification step is omitted. According to the amount of sample treated, the amounts of reagents vary as shown in Table 8. The authors also currently use for neutral lipids a more rapid and simple method of methylation as follows: In a screw-capped 10-ml tube, 50–100 mg of lipids are added with 6 ml hexane and 0.25 ml 0.5 N sodium methoxide in methanol. The mixture is stirred

Table 8 Tube Capacity and Reagent Volumes Used for Methylation According to the Amount of Lipid Samples

Lipid sample (mg)	0.5	1	2–100
Tube volume (ml)	4	4	10
0.5 N NaOH/MeOH (ml)	0.3	0.5	1
14% w/v BF_3/MeOH (ml)	0.5	1	2
Saturated NaCl in water (ml)	0.5	1	2
Isooctane or hexane (ml)	1	1	1

NaOH, sodium hydroxide; MeOH, methanol; BF_3, boron trifluoride; NaCl, sodium chloride.

vigorously for 1.5 min and then centrifuged at 2000g for 3 min. The upper hexane phase containing the methyl esters is withdrawn and eventually diluted for GC analysis.

SHORT-CHAIN FATTY ACID DERIVATIVES   When the lipid sample contains short-chain fatty acids, the quantitative recovery of their methyl esters is questionable because of their volatility and partial solubility in water. The problem may be solved by preparing derivatives of higher molecular weight. As early as 1961 the authors have proposed to use butyl esters for GC analysis of complex mixtures of fatty acids from C_4 to C_{22} as those encountered in dairy products lipids (134). Several authors have later on confirmed this choice (135–137).

The procedure is very simple (138): approx. 1 mg of sample is heated in a 3-ml Teflon-lined screw-capped tube, at 100°C for 2.5 hr in the presence of 0.2 ml 2% (w/v) sulfuric acid in n-butanol. To the chilled reaction mixture are added 1 ml of 5% (w/w) potassium carbonate in water and 1 ml of hexane and the tube is vigorously shaken. After decantation (or centrifugation) the clear butanol-hexane upper phase is ready for GC analysis, without evaporation of solvent. The butanol, which is kept as solvent of the butyl esters for GC analysis, should be added in excess for esterification to be complete but not in too much excess so that the ester of butyric acid can be clearly separated from the not too broad butanol peak on the chromatogram.

In this method loss of short-chain fatty acid esters is avoided because of their comparatively low volatility and low water solubility because the alcohol is kept in totality as solvent of the esters and because there is no evaporation if hexane is added in suitable quantity to the mixture. When analyzed by GC, especially capillary GC, the butyl ester of butyric acid is very clearly separated from butanol. The presence of acetic acid can also be detected under these conditions.

The method can be adapted to very small amounts (10–15 µg) of lipid as demonstrated elsewhere (139). Now preparation of butyl esters may be carried out using a commercially available reagent constituted of a mixture of 14% (w/v) boron trifluoride in butanol. The use of sodium butoxide, prepared by reacting sodium on butanol, is also suited for preparing butyl esters from esterified fatty acids (134).

Esters of 2-chloroethanol (140) and of isopropanol (141) were also used to analyze mixtures of short- and long-chain fatty acids. Recently, Pina et al (142) proposed an elegant method to solve the problem of the presence of short-chain fatty acids in mixtures to be analyzed by GC. The method is based on the formation of tertiary alcohols from fatty acids esterified in triacylglycerols, under the influence of a Grignard reagent (alkyl magnesium bromide). If n is the number of carbon atoms of the alkyl group of the reagent, the teritary alcohol formed from a fatty acid has a number of carbon atoms higher of $2n$. The derivative of higher molecular weight is much less volatile.

Tertiary alcohols (dimethyl-, diethyl-, or dibutylcarbinols) are prepared at room temperature, in anhydrous diethyl ether, in a very short time (10 min). The excess of Grignard reagent is destroyed by 1 N hydrogen chloride in water, at 0°C. The upper diethyl ether phase contains the alcohols derived from the fatty acids.

GC analysis of these teritary alcohols showed that the dibutylalkylcarbinols prepared using butyl magnesium bromide afford the best results when compared to methyl, isopropyl, and other derivatives, especially when the fatty acid mixture contained butyric (4:0) caproic (6:0), or caprylic (8:0) acid. This method can only be applied to esterified fatty acids and not to free fatty acids which do not react with the Grignard reagent.

c.   Other Ester and Ether Derivatives.   For several reasons, other esters or ether derivatives of fatty acids may be required. Phenacyl ester derivatives have been used by Borch (143) for HPLC analysis because the aromatic group absorbs UV light, so the UV detectors are utilizable.

Fluorinated esters have been used for GC with electron capture detectors (144).

Trimethylsilyl esters of unesterified fatty acids have been prepared for GC analysis by the same method to derivatize the free hydroxyl group of mono- and diacylglycerols (see below) (145).

Specific ester derivatives are prepared for mass spectrometric (MS) analysis coupled with GC (or LC) analysis because they induce distinctive fragmentation useful for location of double bonds. Those are principally pyrrolidides and picolinyl esters, and to a lesser extent other amide derivatives, benzoxazole and triazolopyridine derivatives, or even trimethylsilyl or butyldimethylsilyl esters (6,146,147). The picolinyl esters are the most widely used because they offer specific mass spectrometric fragmentation pattern, of particular value in the double-bond location (146). They are prepared by first converting the fatty acids to acid chlorides with thionyl chloride followed by reaction with 3-pyridylcarbinol to give the ester 2- and 4-pyridylcarbinols. A precise description of the method was given by Harvey (146,148). Christie (6) prefers the method of Balazy and Nies (144) in which the free fatty acids first react with trifluoroacetic anhydride and then with 3-hydroxymethylpyridine and 4-dimethylaminopyridine. The procedure is milder and more convenient for sensitive fatty acids. The oxazoline derivatives are now more and more utilized for location of double bonds by GC-MS.

When fatty acids to be GC-analyzed contain polar hydroxyl groups, the latter should be masked by derivatization, after methylation of the carboxyl group, to improve resolution. The type of derivatives are the same as for mono- and diacylglycerols, namely, acetates, trifluoroacetates, or trimethylsilyl ethers or related compounds such as *t*-butylmethylsilyl ethers. Procedures will be described below for partial acylglycerols and are detailed in the book by Christie (6).

3. Instrumentation

a. Columns. The first columns used in gas chromatography have been *packed columns.* They were generally made of glass, which is a very inert material. Glass both allows one to see if the packing is correctly done and remains unbroken during use. However, because of its fragility, many analysts have preferred stainless steel columns over glass, despite the chemical reactivity of metal walls at high column temperatures. The columns were U-shaped in the first instruments and then coiled. This form required smaller ovens in which uniform temperature was easier to maintain. The solid support for the liquid phases was generally diatomaceous earths of uniform size (usually 80–100 or 100–120 mesh) deactivated by acid (HCl) washing and silylation before coating, to minimize adsorption of solutes on free sites of the support.

Coating the support with the liquid phase was a delicate operation since a very uniform film had to be obtained for high resolution. As experienced by the authors (150), the method recommended by Horning et al. (151) always gave good results. The liquid phase, say, diethylene glycol succinate (DEGS) 20–30 g, depending on the desired film thickness, was dissolved in excess (300 ml) of methylene chloride (CH_2Cl_2), with heating if necessary. The cooled solution was placed in a 1 L vacuum flask and added with 20–30 g of chromosorb W or Gas-Chrom Q (80–100 mesh). The suspension was then submitted to progressive vacuum for several minutes with gentle shaking to remove air from the support and then allowed to stand for several minutes. A second stronger vacuum was applied for several additional minutes and the suspension was filtered through fritted glass. The phase-impregnated support was heated overnight at 120°C to remove the remaining solvent. After washing the fritted glass with CH_2Cl_2, added to the remaining phase solution, the unfixed stationary phase was recovered by evaporating the solvent and weighed. The percentage of phase fixed on the support was determined by difference. It is the authors' experience that the application of vacuum on the suspension of the support in the phase solution, to remove air adsorbed on the poreous support, was very important and avoided tearing of the liquid film when the coated support was heated at ~180°C for fatty acid esters analysis. The same procedure of coating was used for triacylglycerol analyses (139).

Supports precoated with the desired liquid phases are now available at low cost. Their quality is not necessarily higher than laboratory-made coated supports.

A column is packed by introducing the dry coated support in small amounts through a funnel, while gently vibrating and applying vacuum to the exit end of the column. A glass-wool plug is placed at both ends of the column. Packing should neither be too loose nor too tight for good resolutions. The packed column must be conditioned before use for 1 or 2 days, flushed by the carrier gas, and at a temperature slightly higher than the routine analysis temperature. Packed columns are still used for fatty acid analysis in laboratories when precise information is not required. However, they tend to be replaced by *glass capillary columns* or wall-coated open tubular (WCOT) columns, especially since glass-drawing machines are commercially available because this type of column achieves much higher resolutions. However, the first capillary columns were made of stainless steel coated with the nonpolar phase Apiezon (152) or with the polar DEGS (153) and used for the separation of stereoisomers of unsaturated fatty acids. As early as 1969, with this type of column coated with the organosilicon polymer EGSS-X, stereoisomers of oleic and linoleic acids could be separated in a 50-m (0.254 mm i.d.) stainless steel column (154).

The capillary stainless steel columns are no longer used because the cost of empty columns is high, because the metal wall is not inert and can cause fatty acid rearrangement, and because coating was not easy to perform. On the contrary, glass capillary columns can be easily prepared, at very low cost, by drawing a heated glass tube with a drawing machine. Paradoxically, these columns are less fragile than glass-packed columns.

As experienced by the authors, the following procedure for preparing glass capillary columns has always given satisfaction: Tubes of two types of glass are used—an usual sodium/calcium glass for the polar phases (B44 from Cristalleries et Verreries, Choisy le Roy, France) and a Pyrex glass for apolar phases (Sovirel, France). Capillaries of up to 90 m (0.35 mm i.d.) are drawn using a commercial drawing machine (Spiral, Dijon, France). Before being coated, the inner surface of the column must be treated to favor adhesion of the phase. A great variety of procedures for treating the glass surface have been proposed (155), to render rougher and thus more wettable the smooth glass surface. For preparing polar columns, the gaseous etching process using hydrogen chloride (HCl) meets the preference of a great number of lipid analysts (156). The tube is filled with gaseous HCl and the two ends are sealed. The tube containing HCl is heated in a 350°C oven for 2 hr. The treatment probably induces some leaching of alkali metals from the surface together with inducing growth of sodium chloride crystals (the surface whitens), generating a rough surface with pits and crevices which greatly facilitate attachment of the phase. Following opening of the tube at its two ends, HCl is blown out with nitrogen and the column dried under vacuum. For preparing apolar columns (Pyrex glass), the tube is filled with 20% HCl in water, sealed, and heated at 170°C for 16 hr (157). Chlorhydric acid is then pushed with nitrogen and the column is washed with 1% aqueous HCl, then dried under vacuum at 250°C for 1 hr.

The treated surface must be deactivated to mask the glass active sites. For the sodium-calcium glass (polar columns) the tube is coated by a thin film of Carbowax 1000 by passing a 1% dichloromethane solution. The tube is dried under vacuum, filled with nitrogen, sealed, and heated at 280°C for 15 min, inducing further polymerization of Carbowax and some pyrolysis. The cooled tube is rinced with 15 ml dichloromethane and dried under vacuum. The Pyrex glass (apolar column) is deactivated by silanization. A solution of hexamethyldisilizane is passed through the tube, which is then sealed, heated at 380°C for 5hr, rinsed with diethyl ether, and dried under vacuum.

The last step is coating with the stationary phase by either dynamic or static technique. The dynamic technique is simpler and more rapid (158). A concentrated solution (~1 ml) of phase (10–20% in dichloromethane) is pushed by a mercury index under nitrogen pressure at low and constant rate. The solvent is then evaporated by flushing nitrogen. The static method (154) gives better results but is more difficult and takes a longer time. According to this procedure, the capillary column is filled with a very diluted (0.5%) solution of stationary phase in a volatile solvent. It is sealed at one end and the solvent is progressively evaporated by applying vacuum at the other end.

This operation takes 10–20 hr. To obtain a uniform coating the column should be maintained at constant temperature (20–30°C) and the phase solution carefully degassed before use. The authors' laboratory uses the dynamic procedure, which gives good analytical results and long life columns when carefully performed.

Now manufacturers propose *fused silica capillary columns,* which consist of a very inert amorphous silica coated with a chemically bonded phase. The phase polymeric molecules are additionally crosslinked to improve the stability at high temperature. These columns are more expensive than the classical glass columns but are very stable and have a long operating life, so that they are supplanting the other commercial columns.

b. Liquid Phases. The principal requirement of a stationary phase is to provide the separation desired. In addition, it should have an acceptable thermal stability to avoid high bleeding from the column and to ensure a reasonably long operating life of the column. Roughly two kinds of phases are used, i.e., nonpolar and polar. The most widely used nonpolar phases are silicone liquid phases, such as SE-30, OV-17, or JXR. These are very stable at high temperatures, with very low bleeding, easily supporting repeated temperature programmings. In the presence of these phases, separation of fatty acid esters occurs on the basis of their molecular weight on packed columns. However, with very efficient packed columns (low amount of stationary phase) or with capillary columns, saturated and unsaturated fatty acids of the same chain length can be separated, the latter eluting ahead of the saturated homologous (e.g., oleic acid and stearic acid). High molecular weight hydrocarbons, such as Apiezon L, are other nonpolar phases formerly in wide use. They can also separate saturated and unsaturated fatty acids of the same chain length, but unsaturated fatty acids differing only by the number of double bonds are very poorly or not separated (e.g., linoleic and oleic acids).

Polar polyester liquid phases are much more convenient for fatty analysis since they permit clear separations of fatty acids esters of the same chain length with a different number of double bonds. They are particularly useful in the analysis of the very polyunsaturated marine oils. With these phases geometrical and positional isomers of unsaturated fatty acids can also be separated.

The polar phases can be divided according to their polarity into four main groups:

The very high polar phases, such as alkylpolysiloxanes containing very polar substituents (nitrile groups, for example), commercialized under different appelations: Silar 10 C or OV-17.

The high polar phases, such as polyesters: polyethylene glycol succinate (EGS) polydiethylene glycol succinate (DEGS), or a copolymer of EGS with a methylsilicone (EGSS-X)

The medium polar phases, such as polyethylene glycol adipate (EGA), or polybutanediol succinate (BDS)

The low polar phases, such as Carbowax 20 M

The high polar DEGS or EGSS-X were frequently used in packed columns because of their relative high thermal stability and the good separations of polyunsaturated fatty acids they afford. The low polar Carbowax 20M is often used in glass capillary column because of its resistance to high temperatures and its low bleeding in temperature-programmed analyses. When fatty acids have to be analyzed by GC in the free form, such as the free fatty acid fraction of dairy products, an acidic component is added to the phase. The following phases are frequently used in that case: FFAP (Free Fatty Acid Phase) or Carbowax 20M–terephthalic acid (TPA).

c. Carrier Gases. As shown by the Van Deemter equation (160), column efficiency is inversely related to the carrier gas density. Therefore hydrogen and helium are more efficient than nitrogen or argon, especially the former. Hydrogen can be used in capillary column, using an injector without splitting and a flame ionization detector, because the flow rate of the carrier gas is very low (1–2 ml/min) and hydrogen is already used for this type of detector. However, analysts must be conscious of the risk of explosion when a leak of hydrogen occurs. Helium is much safer, also gives good results, but is more expensive.

With packed columns, nitrogen is suited as carrier gas. It provides good resolution and is cheaper than helium. Argon would be more convenient because it is more pure, but it is more expensive than nitrogen.

In all cases, the carrier gas should pass through an oxygen and moisture trap to prevent phase deterioration and afford baseline stability.

d. Ovens.   Reproducibility of fatty acids analysis in isothermal conditions and identification of the peaks on the chromatogram, by comparison with standards, requires an oven temperature that is well controlled and uniform in all regions. This is generally achieved in the present commercially available apparatus.

When the sample to be analyzed contains a wide range of fatty acids, with very different retention times, it is convenient to use temperature programming during the analysis. A classical example is butteroil fatty acids, comprising short- and medium-chain-length fatty acids together with long-chain-length homologous (see below). In that case the analysis is started at low temperature and then on at increasing temperature at an appropriate fixed rate up to a suitable final temperature which can be maintained or not. The optimal temperature program is such that the members of a homologous series (e.g., saturated fatty acids) elute at approximately equal time intervals.

Reproducibility in temperature programming and uniformity of temperature at any time, throughout the oven, are conditions of reproducible analyses for such mixtures. These are generally met in the present commercial instruments.

e. Detectors.   A great number of detectors have been developed for use in gas chromatography. In the past one of the most widely used because of its high sensitivity was the "argon detector" developed by Lovelock (161). The principle of this detector was that radioactive source (Sr-90) activated argon atoms, the energy of which was then utilized to ionize organic compounds emerging from the column and the ions were collected by an electrode to generate an ion current in a potential field. In spite of its high sensitivity it has been abandoned because of several drawbacks, in particular its comparatively high dead volume requiring a scavenge flow of argon, a comparatively expensive gas.

The flame ionization detector, which has by now been adopted in most laboratories, is not very different in its principle. In this detector the ions are generated by burning the organic compounds in a mixture of hydrogen and air. The ions are collected by a cylindrical electrode placed just above the flame and to which a voltage potential is applied to measure an ion current. As above, the ion current is linearly amplified and transmitted to a recorder. This detector displays a large number of qualities: high sensitivity, great stability even in temperature-programmed analyses, low dead volume avoiding peak broadening, short response time, and what is particularly important for quantitative analysis, i.e., linearity of response over a wide range of concentrations of the solutes in the carrier gas. However, response according to fatty acids has to be checked with quantitative standards.

f. Injectors.   With packed columns, the sample is simply injected in a small volume of solvent, by means of a microsyringe (10 μl) through a silicone septum, just onto the column packing. The injector temperature is generally higher than the column temperature to ensure rapid volatilization of the sample (e.g., 190–200°C for an isothermal analysis at 170–180°C). To avoid premature volatilization of the solvent before the needle is completely inserted into the injector, the sample can be completely drawn up into the syringe barrel before injection. It can be also useful to draw up a small amount of pure solvent ahead of the sample, to push the latter during injection. Whatever the technique, the syringe plunger must be pressed rapidly for injection to avoid peak broadening during separation.

With WCOT columns, the amount of sample to be injected onto the column is considerably lower than with packed columns to prevent overloading of the column. Several devices have been developed to introduce a small amount of sample in a small amount of solvent onto such columns. The problem which must be carefully solved is that the sample entering the column should have the same composition as the sample to be analyzed and contained in the injection syringe.

Four types of injectors have been developed to be used with glass capillary columns (162). The first is the splitter injector. In this device, after injection through a septum, the sample is vaporized and mixed with the carrier gas. A small fraction of the mixture is then carried onto the column, while the remainder is vented out by means of a stream splitter. A second gas stream controlled by a pressure regulator is connected to the splitter exhaust line to maintain a constant inlet pressure for the capillary column. The split ratio can be varied, commonly from 1:10 to 1:100. Because of the carrier gas vented out, the gas flow through the whole injector is very high. However, this injector is perfectly convenient for the most common fatty acid mixtures in which the range of components is very narrow. But with mixtures such as those encountered in lipids from dairy products in which short-chain fatty acids are present together with long-chain fatty acids, this type of injector discriminates against the less volatile components and the analysis is generally not quantitative. The "on-column" injection procedure should be preferred in those cases.

Another type of detector, the *splitless injector*, is used with dilute solutions or in cases where a limited amount is available. The sample solvent used must have a high boiling point with respect to the column temperature and a low boiling point with respect to the most volatile sample components (generally 30°C above the column temperature). In that case, the vaporized solvent recondenses in the column inlet to form a temporary thick-film stationary phase, trapping the sample components into a narrow band at the top of the column. After a purge time the gas flow is redirected to the bottom of the inlet where the flow is divided. One part continues forward to serve as carrier gas while the other one sweeps the residual volatiles and excess solvent from the injection chamber and vents them to the atmosphere. The oven temperature is then rapidly raised to begin the analysis process, and the solvent condensed at the top of the column again vaporizes and reconcentrates the sample components.

A third type injector, the *solid injector*, first developed by Ros (163), is particularly useful for high-boiling components such as fatty acids and steroids. A schematic diagram is given in Fig. 3. The injection is made by means of a glass needle (4), brought into the "loading position" by a small magnet (3). The sample is slowly deposited at the top of the needle at ambient temperature from a syringe through a septum (12), via the injection side arm. The solvent is allowed to evaporate and is vented out through the restriction at the top (1). The magnet is then moved down manually to the "injection position." The needle tip comes to a few millimeters above the capillary column inlet (11), in the heated zone of the injection port. The solid compounds condensed at the tip of the needle evaporate and are swept into the column by the carrier gas. The needle is then retracted using the magnet. Cold trapping can be used to concentrate the sample as for the splitless injector.

It is the authors' experience that this type of injector gives excellent results for the most common plant and animal fatty acid mixtures and for sterols. However, because of the previous evaporation of the solvent, even at room temperature, very short-chain fatty acid methyl esters are also partly blown off. The injector discriminates to the detriment of the most volatile components.

In the analysis of fatty acid mixtures containing short-chain fatty acids, in particular butyric acid, the *on-column injector* is recommended. It is designed to allow a liquid sample to be directly injected into the capillary column. The utilization of the cold septumless on-column technique has

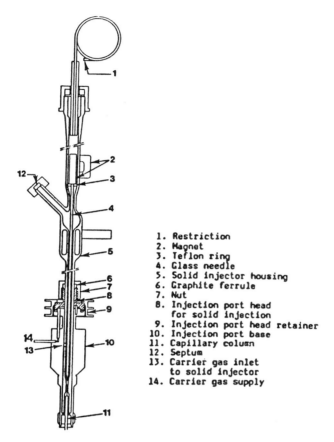

1. Restriction
2. Magnet
3. Teflon ring
4. Glass needle
5. Solid injector housing
6. Graphite ferrule
7. Nut
8. Injection port head
   for solid injection
9. Injection port head retainer
10. Injection port base
11. Capillary column
12. Septum
13. Carrier gas inlet
    to solid injector
14. Carrier gas supply

Fig. 3 Schematic diagram of a typical solid injector for capillary columns. (From Ref. 163.)

been shown to yield important advantages compared to conventional injection techniques. Discrimination caused by selective vaporization or by the stream splitter cannot occur. Thermal and catalytic decomposition are minimal because the sample is injected into a cold column preventing premature evaporation of the sample. The cold on-column injection is generally used in GC analysis of fatty acids from dairy products because it provides a high degree of accuracy and reproducibility, particularly when butyl esters are used as fatty acid derivatives (138).

4. Analysis of Fatty Acids

a. Qualitative Aspects of Fatty Acid Analysis. The first problem encountered in GC analysis of fatty acid derivatives (or free fatty acids) is the identification of peaks on the chromatogram. This can be achieved more or less accurately but rapidly with conventional methods and more accurately by specific methods such as mass spectrometry.

CONVENTIONAL METHODS   Conventional methods are based on comparison between the retention times (or volumes) of known fatty acids in a synthetic or natural mixture and of the components to be identified in the analyzed mixture. Equal retention times are in favor of identity of the two fatty acids. Standard mixtures of known methyl esters of saturated, mono-, and polyunsaturated fatty acids are commercially available (e.g., from Nu-Chek Prep, Elysian, MA). Natural mixtures whose composition have been well documented in the literature are also likely to be used. They generally contain a wider range of fatty acids than synthetic standard mixtures. In the laboratory the authors commonly use methyl ester mixtures prepared from peanut oil for the most common fatty acids (16:0, 18:0, 18:1, 18:2) and because peanut oil contains several long-chain saturated fatty acids (20:0, 22:0, 24:0). They also use the more complex mixture prepared from rat liver lipids containing several polyunsaturated long-chain fatty acids (20:4 n-6, 22:6 n-3) and in the analysis of marine oils, they use fatty acid methyl esters prepared from cod liver oil or salmon oil containing very different highly polyunsaturated n-3 fatty acids. A mixture of the two latter mixtures together with a small amount of linseed oil fatty acids containing 18:3 n-3 in high amount affords a wide range of n-6 and n-3 polyunsaturated fatty acids, likely to be encountered in the most lipid samples.

In the analysis of lipid fatty acids samples from dairy products, analyzed as butyl esters, a reference mixture can be obtained from a bovine butteroil. When the free fatty acid fraction of dairy products is analyzed without previous esterification, a standard mixture of free fatty acids can be prepared from individual fatty acids of commercial sources.

Retention times utilized must be normally corrected from the dead volume of the column. This is particularly true for short-chain fatty acids whose retention time is low in comparison to the column dead volume. If the air peak does not exist on the chromatogram to indicate the end of the dead volume, which is generally the case, the retention times must be measured from the very start of the (volatile) solvent.

Retention times are frequently expressed relative to one fatty acid, commonly occurring in the mixtures, namely, palmitic acid or stearic acid because the relative retention times are much less sensitive than absolute values to changes in the operating conditions or in the column characteristics. The values can be used to identify unknown acids if analysis is carried under the same isothermal operating conditions.

Another means of identifying fatty acids of a homologous series is to use a well-known property of their elution in *isothermal* conditions, i.e., the linear relationship between the logarithms of the corrected retention times and the number of carbon atoms of the aliphatic chain. Such a relationship is illustrated in Fig. 4 for the series of straight-chain saturated fatty acids from C10 to C22 analyzed as butyl esters on a packed polar column (134). This relationship exists for any type of ester derivatives, in particular for methyl esters. With an unknown mixture, the straight line can be drawn from only two points, from commonly occurring fatty acids, palmitic and stearic acids, or for a more precise tracing from myristic acid (C14), often present in samples from animal origin and stearic acid or even arachidic (C20) acid often present in trace amounts. The retention times of the other members of the homologous series can be deduced from this graph and, conversely, their identification can be tentatively deduced from their measured retention times.

Evidently, this type of relationship exists when analysis is carried out with capillary columns when operating under isothermal conditions. It also exists for other homologous series of saturated fatty acids, such as series of the branched iso- or anteisoacids. The straight line is parallel to that obtained for the straight-chain fatty acids and located under this line because of slightly shorter retention times.

A similar relationship can be graphically illustrated for homologous series of cis monounsaturated fatty acids and used for identification of these acids in a mixture. However, in this case, retention times depend on the position of the double bond (164, 165). However, the straight line can be drawn from the retention times of palmitoleic acid (16:1 n-7) and oleic acid (18:1 n-9), generally present in the most common fatty acid mixtures and utilized to identify the other homologous monounsaturated fatty acids with the double bond located at the $\Delta 9$ position (14:1, 20:1, 22:1, 24:1). In packed columns, the retention times of close positional isomers, e.g., $\Delta 9$ and $\Delta 11$ 18:1, are very closed (165) and some misidentification is possible. The same is true for positional isomers of 18:2 n-6 (linoleic acid).

In Fig. 4, the straight line for $\Delta 9$ monounsaturated fatty acids was drawn starting from the

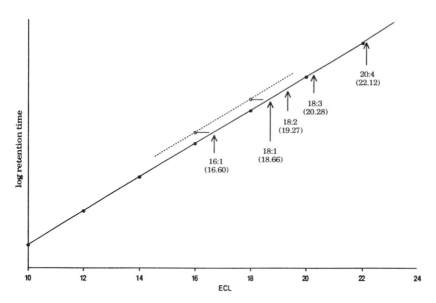

Fig. 4 Plot of the retention times of the butyl esters of a homologous series of straight chain saturated fatty acids from C10 to C22, vs. the number of carbon atoms in the aliphatic chain (equivalent chain lengths, ECLs) and of two monounsaturated fatty acids (16:1 n-7 and 18:1 n-9), on a packed column of EGS. The ECL (in parentheses) of several other unsaturated fatty acids (18:2 n-6, 18:3 n-3, and 20:4 n-6) are indicated. The corrected elution time of C20 was 20 min and of C22 40 min under the reported analytical conditions. (Based on data from Ref. 134.)

retention times of butyl esters of palmitoleic and oleic acids taking in abscissis the number of carbon atoms of the chain (carbon number), i.e., 16 and 18, respectively. It is parallel to the line for straight-chain saturated fatty acids.

From the graph obtained with the series of straight-chain saturated fatty acid esters (plain line in Fig. 4), an equivalent chain length (ECL) can be obtained from the retention time of any unsaturated fatty acids. It is the chain length of a hypothetical saturated straight-chain fatty acid which would be eluted at the same time as the considered fatty acid. It is easily graphically deduced from the graph in Fig. 4. For the reported unsaturated fatty acids the respective ECLs were 16:1 n-7 (16.60), 18:1 n-9 (18.66), 18:2 n-6 (19.27), 18:3 n-3 (20.28), and 20:4 n-6 (22.12).

The linear relationship between fatty acid carbon numbers and log (retention times) under *isothermal* operating conditions is very useful for tentative identification of unknown components. However, it does not offer any certainty and overall it cannot be utilized in temperature-programming analyses of mixtures comprising a wide range of fatty acids differing both in chain length and unsaturation. Use of standard mixtures remains the simpler and more rapid method in those cases. However, a safe and definitive identification can only be obtained from other, more sophisticated methods such as those indicated below.

SPECIFIC METHODS

Mass spectrometry

Introduction

Mass spectrometry coupled to GC (GC-MS) is probably the most powerful tool for the identification of fatty acids separated by gas chromatography. Mass spectrometer can also be used as a mass detector, since in general the molecular ion is present in the spectrum. It is beyond the scope of this chapter to develop in detail all of the principles and applications of the technique to fatty acid structural analysis. The reader should consult one of the authoritative reviews on the subject for more complete information (6,146,147,166–169).

The principle of mass spectrometry, in its simplest form, is that derivatized fatty acids in the gaseous effluent of the column are bombarded with accelerated electrons (electron impact = EI) and form positive ions (molecular ions M^+ from the parent molecules). These positive ions can fragment to give smaller ionized ions which are directed to an intense magnetic (or electrostatic) field, in which they are separated according to the mass-to-charge ratio (m/z). They are collected in sequence as the ratio increases and the amplified ion currents are registered. The total ion current generated from the column is recorded continuously and the trace obtained resembles a chromatogram obtained from a classical detector. Fragment ion currents are also continuously recorded and to each peak can correspond a spectrum.

Since fragmentation does not occur arbitrarily but at the weakest sites of the molecule, characteristic fragments are produced from which the structure of the molecule can be theoretically deduced. Of course, this description is oversimplified and the reality is more complex, particularly since the introduction of different ionization systems, such as the chemical ionization procedure. But the principle of the technique is the same in any case.

Application to structural analysis of fatty acids

Saturated fatty acids

Mass spectrometry (MS) is particularly suitable for locating the double bonds in polyunsaturated fatty acids. However, electron impact MS can also be utilized to identify *saturated* fatty acids because their spectra are characterized by a prominent molecular ion M^+ (or MH^+ with chemical ionization) giving the molecular weight M of the acid. Methyl esters, picolinyl esters, pyrrolidines, and oxazolines were the derivatives used for this purpose. The presence of a methyl group (14 mass units) on one of the aliphatic carbon atoms can be detected by a 14-mass-unit gap in the spectrum at the site of branching (148). However, when the methyl group is located at the extremity of the aliphatic chain (iso-isomers), which is the most commonly occurring situation, the identification is harder (6). The methyl ester derivatives have been the most widely used for the identification of branched chain fatty acids in natural mixtures. But pyrrolidides and picolinyl esters are more and more used for this purpose.

Unsaturated fatty acids

With simple derivatives, such as methyl esters or trimethylsilyl ethers, the position of double bonds is poorly exhibited in unsaturated fatty acids mass spectra by the conventional EI spectrometry. The reason is the pronounced tendency of the double bonds to migrate along the aliphatic chain. The solutions adopted mostly rely on "fixing" the double bonds. The first solution to the problem was to chemically modify the double bonds by ozonization or oxidation followed by GC-MS of the resulting compounds after derivatization (167). These procedures were successfully applied to monounsaturated fatty acids but were not suitable for highly unsaturated molecules. For these polyunsaturated fatty acids, the terminal carboxyl group was derivatized to pyrrolidines, to picolinyl esters, or, more recently, to oxazolines. The latter, in particular the 2-alkenyl-44-dimethyloxazolines, seem the most valuable (170).

Analysis by GC-MS of all *monoenoic* acids as oxazoline derivatives shows distinctly different spectra, but all containing an intense molecular ion. Besides a series of smaller ions, from which the location of the double bond can be safely deduced from the empirical rule edicted by Zangh et al (170): "If a mass separation of 12 instead of the regular 14 units is observed between two neighboring even-mass homologous fragments containing $n-1$ and n carbon atoms of the original acid moiety, a double bond exists between carbon n and $n+1$ in the chain." Otherwise speaking, "the point of unsaturation is indicated by the fragment, where an interruption of the 14-mass spaced homologous ion series occurs."

For example, in the spectrum of 18:1 Δ9 oxazoline (Fig. 5) there is a mass separation of 12 (CH$_2$–2H) (196–208) between two mass separations of 14 (CH$_2$) (182–196 and 208–222). Thus breaking occurred between carbon 9 (n) and carbon 8 (n–1) and the double-bond is located between carbon 9 (n) and carbon 10 (n+1), that is in Δ9 position. A mass separation of 40 (14+12+14) (182–222) can also be used to locate the double-bond. Moreover the presence of the two more intense fragments, 182 and 236, is also indicative of the location of the double-bond, according to Zangh et al. (170), who pointed out that "if the intense peaks containing n–2 and n+2 carbon atoms of the fatty acid residue are present, the double bond is at the carbon n n+1".

Thus in Fig. 5 (18:1 Δ9), the difference between fragments 236 (n+2) and 208 (n) is 2 units (14 × 2 = 28). The double bond is between carbons n (9) and n+1 (10), i.e., in the Δ9 position. The same rules can be applied to methylene-interrupted di-, tri-, and other polyenoic fatty acids, the most commonly occurring fatty acids in natural mixtures, when analyzed by GC-MS as their oxazoline derivatives. Typical 12-unit intervals take place in the proximity of the carbon atoms where the double bonds are located, as can be seen in Fig 6.

The figure reports the spectrum obtained by GC-MS of a trans isomer of arachidonic acid (O. Berdeaux et al., submitted). Two isomers were separated by GC on a BPX 70 fused silica column (50 m × 0.25 mm i.d.) with helium as carrier gas and identified as 20:4 ccct (c for cis; t for trans; double bonds are counted from the carboxyl end) and as cctc. Figure 6 reports the spectrum obtained with 20:4 ccct. M is the molecular ion, which permits one to identify the fatty acid oxazoline derivative. Four mass separations of 12 can be seen in Fig. 6, at C4–C5, C7–C8, C10–C11, and C13–C14, indicating the the four double bonds were located at the Δ5, Δ8, Δ11, and Δ14 positions. The positions of the double bonds can also be deduced from the four mass separations of 40 indicated with arrows in Fig. 6.

The configuration cis or trans of the double bond located at Δ14 position cannot be identified by the spectrum. It was identified by capillary GC of a pure 20:4 n-6 trans isomers isolated by successive fractionations (concentration by urea adduct formation, reverse phase HPLC, and silver ion TLC of the methyl ester derivatives).

These two examples demonstrate that in unsaturated fatty acids the positions of the double bonds are univocally identified by the GC/MS method.

Other fatty acids

Structural study of *cyclic* fatty acids can also be performed by GC-MS. These acids are naturally present in special oils such as baobab seed oil (177) or are produced in oxidized oils. The naturally

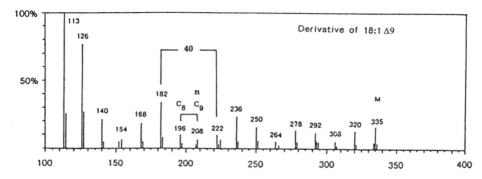

Fig. 5 Mass spectrum of 2-substituted, 4,4-dimethyloxazoline derived from oleic acid (18:1 Δ9) recorded on a MAT 711 mass spectrometer coupled with an SS 166 data system. (Reproduced by kind permission of the authors, Zhang et al. (1988) (170) and of *Biomedical and Environmental Mass Spectrometry*.)

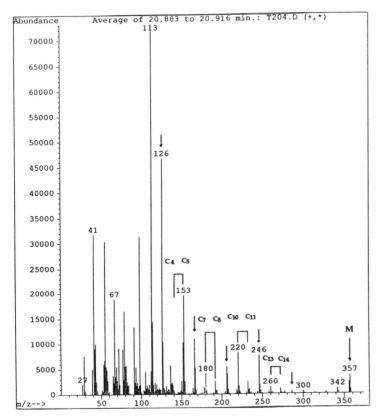

Fig. 6 Mass spectrum of 2-substituted 4,4-dimethyloxazoline derived from a trans isomer of arachidonic acid (20:4, Δ4c, 7c, 12c, 13t) recorded on a mass spectrometer MSD 5970 B Hewlett Packard coupled to a gas chromatograph 5890 A Hewlett-Packard. (The spectrum was kindly obtained by O. Berdeaux, INRA, Dijon.)

occurring cyclic fatty acids are principally cyclopropanoid and cyclopropenoid acids. Mass spectrometry of their methyl esters, pyrrolidine, or picolinyl ester derivatives have been used for location of the ring in the molecule since characteristic fragments are formed from either side of this ring (167). Cyclic fatty acids produced in oxidized oils have also been subjected to structural analysis by GC-MS (172).

Location of specific functional groups in the fatty acid aliphatic chain is likely to be determined by GC-MS since specific fragments are formed in the vicinity of the group. It is the case with keto fatty acids analyzed as methyl esters as well as with epoxy fatty acids (173).

Spectroscopic and chemical methods

Although simple GC-MS systems are becoming less costly, more reliable, and easier to use, many laboratories will not have the opportunity or the means to set up such systems in the next few years. Fortunately, other methods have long been available to study the structure of fatty acids and can be used for this purpose. Generally, the first step consists of isolating pure individual fatty acid derivatives by chromatographic techniques, more often after previous isolation of fatty acid families.

Isolation of fatty acid fractions

One method is fractionated *crystallization* of fatty acid methyl esters dissolved in acetone at low temperature. A fraction enriched in polyunsaturated fatty acid methyl esters is readily obtained by this method by separating, through a Buchner funnel, the cooled solution from the solid crystals of saturated and monounsaturated fatty esters.

Fractionated *distillation* of the methyl esters can also be utilized for this purpose, though this is less simple and requires more expensive equipment.

The formation of *urea adducts* is another means to isolate the most unsaturated fraction from a fatty acid ester mixture, since saturated esters form solid urea inclusion complexes with urea in solution in methanol. The experimental procedure is described by Christie (6).

Silver ion chromatography, by TLC or HPLC, is a valuable technique to isolate fatty acid methyl esters according to unsaturation. Most of the data have been obtained to date by TLC. Usually plates of silica gel impregnated with 5–10% by weight of silver nitrate are used and developed with hexane–diethyl ether as mobile phase. Since the double bonds of fatty acids form complexes with the silver ions the migration of unsaturated fatty acids is retarded, increasingly with an increased degree of unsaturation. In convenient experimental conditions, it is possible to separate into distinct fractions the fatty acid esters comprising zero to six double bonds. When visualized under UV light after spraying with a 0.1% 2′,7′-dichlorofluorescein in methanol, the bands can be delineated, the silica gel scraped off the plate, and the methyl esters extracted by chloroform-methanol for further analysis.

Positional isomers of unsaturated fatty acid methyl esters are also readily separated (174).

Now argentation TLC is beginning to be supplanted by argentation HPLC, since stable columns can be prepared by retaining the silver ions by sulfonic acid residues bonded to the silica gel solid phase, as described by Christie (175).

The use of a stream splitter at the outlet of the column allows the separated fractions to be collected for further analysis by GC or for further fractionation of the individual esters by reverse phase HPLC.

Gas chromatography, in a semipreparative scale, was formerly used to collect fatty acid ester fractions. Specific devices have been developed (150) and used for determining the location of the double bond in fatty acids derived from erucic acid metabolism by a chemical procedure (176). This fractionation method is now little used and advantageously replaced by reverse phase HPLC.

Spectroscopy of fatty acids

Applications of spectroscopic procedures other than MS in fatty acid analysis have been reviewed in detail elsewhere (177) and summarized by Christie (6).

Fatty acids may be subjected to IR absorption (at wavelengths of 2.5–15 μm) preferably as methyl ester derivatives and in tetrachloride solution. Specific absorption bands are indicative of double bonds (of cis and trans configuration) and of functional groups such as free hydroxyl groups. IR spectroscopy is most often used to detect the presence of and tentatively quantify the *trans*-fatty acids in lipid mixtures.

Specific bonds are also found for trans double bonds by *Raman* spectroscopy (178) and the method can be used to detect these isomers of cis-unsaturated fatty acids.

UV spectroscopy, at 200–400 nm, is generally used to detect the presence of conjugated double bonds (177).

Nuclear magnetic resonance (NMR) spectroscopy presents a wider field of applications. The principle of the method is that the nucleus of certain isotopes is continuously spinning, generating a magnetic field, and can resonate between different energy levels when an external oscillating magnetic field is applied to the nucleus. The resonance occurs at specific frequencies, absorbing some of the energy applied. This change in energy level is detected, amplified, and registered. The trace obtained is the NMR spectrum. The hydrogen isotope ^1H displays this property, as does carbon isotope ^{13}C but to a much lesser extent.

These last years, ^1H and ^{13}C NMR have been applied to the detection and the location of double bonds in the methyl ester fatty acid chain, as reported by Chapman and Goni (177), since nearly each carbon atom displays a distinct signal which is very specific in the neighboring of double bonds. A wide range of unsaturated fatty acids have been submitted to proton (^1H) spectroscopy, including cis and trans isomers of the n-6 and n-3 families, methylene-interrupted and conjugated polyenes, and acetylenic compounds (6,179). But a lot of studies have also been carried out with saturated fatty acids and those containing hydroxyl group or oxo moieties, as pointed out by Christie (6).

Chemical procedures

Chain length determination

When analyzing the structure of an unsaturated fatty acid, the first step is to determine its chain length. This is readily achieved by catalytic hydrogenation and identification of the saturated fatty acid formed. This is usually carried in a test tube in the presence of hydrogen and a platinum oxide as catalyst. However, this can be more easily and more rapidly effected by means of a special device fitted at the outlet of the GC column (180). It consists of a short, stainless steel column containing 0.5% platinum deposited on gas-Chrom P (100–120 mesh), maintained at 200°C, and tightly fixed at the outlet of the detector (katharometer). The fatty acid esters are separated on a GC column with hydrogen as carrier gas, individually hydrogenated inside the device and condensed on glass wool contained in a glass cartridge. The saturated fatty acid esters are eluted by passing pentane through the glass wool and identified by GC.

Figure 7 shows the chromatograms obtained in the analysis of walnut oil fatty acid butyl esters before and after hydrogenation by means of the device. The three unsaturated fatty acids in C18 (oleic, linoleic, and linolenic acids) were completely hydrogenated, with no formation of hydrogenolysis products, which would elute very early. The four C18 fatty acid esters together represented 93.6% before hydrogenated and the only C18 (stearic) peak after hydrogenation amounted to

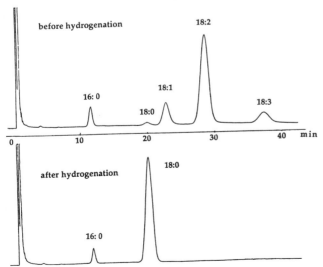

Fig. 7 Separation of walnut oil butyl esters before and after hydrogenation by the device developed by Lecerf and Bézard (1966) (180) on a U-shaped glass column (2 m × 4 mm i.d.) packed with 20% DEGS on Chromosorb W (80–100 mesh) with argon as carrier gas. (Reproduced by the kind permission of the *Revue Française des Corps Gras.*)

93.5%, proving a very good recovery. This device was successfully utilized to determine the chain length of the unsaturated fatty acid esters present in six fractions of fish oil fatty acid butyl esters separated by GC and individually hydrogenated and identified by means of their retention times.

Location of double bonds

In the chemical procedures, the position of the double bonds is usually determined after cleavage by oxidation and identification of the products formed. The two oxidation procedures generally used are those with permanganate-periodate or ozone as oxidative reagents. The former method yields mono- and dibasic acids as the products while ozonolysis followed by cleavage of the ozonides generates acids and aldehydes. Both procedures have been reviewed by Ackman et al. (181).

The permanganate-periodate procedure is probably the most largely used because it requires inexpensive equipment and chemicals and because overoxidation, which is negligible in the ozonolysis procedure, can be minimized here.

The oxidation operating conditions have been successfully used by the authors (182) to determine the proportion of the different isomers of hexadecenoic and octadecenoic acids in rat lung lipids. Briefly, butyl esters of these acids (~2 mg) were first isolated by argentation-TLC in the monoene fraction and then collected after GC separation at the outlet of the column (150). They were oxidized in tertiary butanol by sodium metaperiodate and potassium permanganate, with potassium carbonate as buffer for 1 h at room temperature. The solution was then acidified by sulfuric acid and the oxidative reagents destroyed with sodium bisulfite. The acids formed were extracted with hexane and converted to sodium soaps by $n/100$ sodium methoxide to avoid loss of the short-chain acids. The solvent was evaporated and the soaps converted to butyl esters by acidic butanol (134) in a sealed tube at 100°C for 3 hr. The chilled reaction mixture was added with 5% potassium carbonate and with pentane and shaken. After decantation the upper layer of butanol-pentane was directly used for GC analysis of the butyl esters, without evaporation of pentane.

Control analyses of different mixtures of palmitoleic, oleic, and vaccenic acids have demonstrated the quantitative recovery of the products formed by oxidation. The results showed that lung phospholipids contained a high proportion of the Δ7 isomer (~27%) of palmitoleic acid (Δ9) and of the Δ11 isomer (~18%) of oleic acid (Δ9). These isomers are readily separated in high-resolution capillary gas chromatography. The permanganate-periodate oxidation procedure is also suitable for monoenes and dienes. For more polyunsaturated fatty acids more sophisticated procedures are used (6).

Location of other functional groups

In general, spectroscopic methods are used for this purpose, rather than chemical methods. However, the latter can be an aid to spectrometric identification. For example, epoxyl groups are directly cleaved with periodic acid and the position of the oxygenated ring is deduced from the identity of the products formed as determined by GC analysis of their butyl esters, as above.

The ring of cyclopropene fatty acids can be disrupted by permanganate-periodate oxidation and the β-diketo compound can be safely identified by MS. The ring of cyclopropane fatty acid reacts with boron trifluoride methanol reagent to produce methoxy derivatives which are likely to be characterized by MS. The chemistry of these cyclic fatty acids has been reviewed by Christie (183). In branched fatty acids, the site of branching by means of a keto compound formed by acidic permanganate oxidation can be identified by GC (184).

b.   Quantitative Aspects of Fatty Acid Analysis.   In the quantitative analysis of fatty acids by GC, especially in the analysis of lipids containing lower fatty acids (dairy products), two aspects should be given special attention. The first is the quantitative recovery of the free fatty acid fraction from the product when this one will be analyzed in the free form or the quantitative conversion of fatty acids to esters and the quantitative recovery of the esters for GC analysis. This aspect has been examined in the section devoted to the preparation of fatty acid derivatives and was studied by Badings and De Jong (185).

The second aspect is the accurate analysis of the free fatty acids or the fatty acid esters by GC. This aspect was also examined by Badings and De Jong (185) in the capillary GC analysis of methyl esters. It is highly dependent on the equipment utilized. At first, during the injection process, it is extremely important that no loss of the most volatile compounds of the mixture occur. For the most common fatty acid mixtures, split or splitless injectors are convenient. In the laboratory, use of the "solid injector" (163) is satisfactory for fatty acid and sterol mixtures.

However, when the fatty acid mixture comprises short-chain fatty acids, in particular butyric acid, or a fortiori acetic acid (present in milk triacylglycerols), loss of the shortest fatty acids occurs with split injectors and this loss is higher with the more volatile methyl esters than with higher molecular weight derivatives. For such mixtures the use of the on-column injector is highly recommended.

A second possible cause of nonquantitative analysis is some loss in the chromatographic columns by thermal degradation, by adsorption on the solid support, by chemical modifications or by any other cause. The use of glass columns (and injector), deactivated solid support, and pure carrier gas (passing through oxygen and moisture traps) highly minimizes specific fatty acid losses. It was sometimes reported that there were some losses of polyunsaturated fatty acids in columns coated with polyester phases. The phenomenon was in part attributed to transesterification of the methyl esters with the polyester liquid phase due to catalyst residues. Now with the use of fused silica capillary columns coated with phase chemically bonded to the inner surface, such losses should not be significant. At last, if we suppose that the electronic integrator, which is the only method suitable for quantification in analyses with capillary columns, delivers signals at any time proportional to the detector signals, the last factor of quantitative analysis is the detector.

Since in a fatty acid mixture the components are generally in different proportions, the signal delivered by the detector for a given compound should be rigorously proportional to the amount to be detected, for a wide range of concentrations in the carrier gas. A second condition for quantitative results to be obtained is that for the same amount (mass) detected the signal should be independent of the compound detected. This is especially true with the short- and long-chain saturated fatty acids and for the polyunsaturated and saturated fatty acids, with the flame ionization detector (FID) the most widely used by lipid analysts. To check the capacity of the analytical system to give accurate and reproducible analyses, several assays should be made. To check the linearity of the detector response to the quantity of compound to be detected, in a wide range of quantities, the following experiment can be carried out. Since it is not easy to inject precisely known amounts of a given methyl ester, it is preferable to use a mixture of three methyl esters in low, medium, and high proportion, respectively, and to inject increasing amounts of this mixture. The relative proportion of the three esters calculated from peak areas determined by electronic integration should not change with the amount injected in a range as to detect the components in trace amounts. To check the reliability of the GC equipment to give quantitative analyses of complex methyl ester fatty acid mixtures containing a wide range of saturated and unsaturated fatty esters, standard quantitative mixtures should be repetitively analyzed. The mean compositions determined from peak areas should precisely correspond to the known compositions. Such standard mixtures are commercially available.

If only small differences systematically appear, response factors have to be determined and applied to correct the experimental data. These factors are easily established from the ratio of the known proportions of the test mixture components to the proportions calculated from peak areas. To express the results in fatty acid percent and not in fatty acid ester percent, the response factors should take into account the conversion of the methyl esters to the free fatty acids starting from their molecular weight.

In most analyses of fatty acids, the results are expressed as a percentage of each component in the mixture. However, in some instances, the absolute amount of each component or of the total fatty acid sample, needs to be measured. This is best accomplished by adding a precise quantity of fatty acid methyl ester not present in the mixture—generally an odd chain fatty acid (15:0, 17:0, or 19:0)—as internal standard. If a more complete and precise standardization is desired, the internal standard (as triacylglycerol, for example) should be added to the sample from which the lipids are extracted, to also take into account the extraction yield, but more commonly to the lipid extract to check the derivatization yield together with GC apparatus response.

When the mixture of fatty acids to be analyzed comprises short-chain fatty acids, it is the authors' experience that higher derivatives than methyl esters should be preferred, even if the recommendations of methyl ester preparation by Badings and De Jong (185) are carefully followed because the detector response for butyl esters, for example, is higher than that for methyl esters. This is shown in Table 9. The table reports results obtained in the analysis of methyl and butyl esters of a standard triacylglycerol mixture containing a wide range of short-, medium-, and long-chain saturated fatty acids and unsaturated fatty acids (138). It can be seen that from 10:0, the response factor is very close to unity for both types of esters, generally a little more for butyl esters than for methyl esters. But for short-chain fatty acids (40:0, 6:0, and 8:0) the response factor is much closer to unity for butyl esters than for methyl esters, particularly for butyric acid, which is in comparatively high proportion in milk lipids. This fatty acid is thus more easily detected and its proportion more accurately measured. Table 9 also reports the fatty acid composition, as mol %, of a sample of butterfat, determined by GC analysis of their butyl esters. The low standard errors of the means of six analyses prove the good reproducibility in the GC analysis of such complex mixtures.

In routine analyses it is important to establish regular checks with defined standard mixtures

Table 9 Molar Response Factors[a] for Acid Methyl and Butyl Esters Determined from a Standard Triacylglycerol (TG) Mixture and Composition (mol %) of Major Fatty Acids of Total Butterfat Triacylglycerols Analyzed as Butyl Esters

Fatty acid	Response factor		Total butterfat TG
	Methyl esters	Butyl esters	
4:0	1.669 ± 0.018	1.116 ± 0.010	9.61 ± 0.05
6:0	1.341 ± 0.017	1.155 ± 0.018	4.85 ± 0.04
8:0	1.147 ± 0.013	1.061 ± 0.012	2.17 ± 0.01
10:0	1.052 ± 0.011	1.031 ± 0.005	3.87 ± 0.01
12:0	0.998 ± 0.004	0.994 ± 0.003	3.71 ± 0.01
14:0	0.973 ± 0.006	0.984 ± 0.003	11.37 ± 0.02
14:1	0.985 ± 0.004	1.007 ± 0.004	1.08 ± 0.01
16:0	0.952 ± 0.008	0.969 ± 0.004	24.78 ± 0.03
16:1	0.966 ± 0.002	0.996 ± 0.002	1.51 ± 0.02
18:0	0.988 ± 0.011	1.028 ± 0.007	11.36 ± 0.02
18:1	0.929 ± 0.008	0.960 ± 0.006	23.02 ± 0.03
18:2	0.919 ± 0.007	0.949 ± 0.002	1.70 ± 0.04
18:3	0.942 ± 0.015	0.970 ± 0.010	0.83 ± 0.01
20:0	1.015 ± 0.014	1.050 ± 0.012	0.14 ± 0.01

[a]Values are means ± standard errors of means (SEM) of six GC analyses.

to ensure that the different procedures and GC equipment function correctly. This will guarantee good quality in GC fatty acid analysis.

  c. *Separation of Different Mixtures of Fatty Acids.*   To illustrate the separation of fatty acid esters that can be achieved by GC, a few examples of applications have been selected. The analyses were performed on capillary glass columns, generally laboratory-made, coated with polar phases, which are used almost universally for fatty acid analysis.

  The first example, illustrated in Fig. 8, is the separation of the methyl esters of fatty acids of a commercial sunflower oil on a usual laboratory-made glass capillary column (30 m × 0.30 mm i.d.) coated with Carbowax 20M a moderate polar polyglycol phase, in isothermal operating conditions (198°C), with helium as carrier gas at a rate of 3 ml/min. For the five major fatty acid methyl esters, the analysis is achieved in less than 10 min with clear separation of the components, in particular peaks 4 and 5. The five fatty acid esters were easily identified by their retention times since this profile is very usual in vegetable oils. Standard mixtures of close fatty acid composition can be used to confirm the identification. The components in trace amounts are also detected but not easily quantified in this analysis. They were identified by comparison with standard mixtures or by reference to known data such as equivalent chain lengths, for example.

  The determination of fatty acids in edible oils and fats by capillary GLC had been standardized in Official Methods (AOCS Ce le-91) (186). Another standard method (AOCS Ce-1d-91) is to be applied to the methyl esters of n-3 and n-6 unsaturated fatty acids obtained from vegetable oils

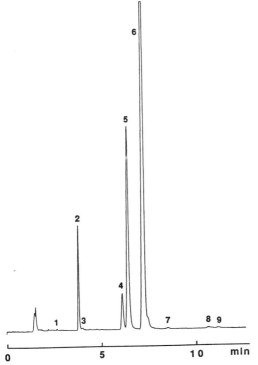

Fig. 8   Separation of a commercial sunflower oil fatty acid methyl esters on a Carbowax 20M glass column (30 m × 0.30 mm i.d.) at 198°C. Fatty acids: (1) 14:0, (2) 16:0, (3) 16:1 n-7, (4) 18:0, (5) 18:1 n-9, (6) 18:2 n-6, (7) 18:3 n-3, (8) 20:0, (9) 20:1 n-9. (The chromatogram was kindly obtained by J. Gresti.)

(187). It reports the statistical results of interlaboratory studies at the international level which are of great interest to appreciate the repeatability and the reproductibility of the method.

The second example, illustrated in Fig. 9, concerns the separation of the methyl esters of the fatty acids of cod liver oil on the same type of column and equipment and the same operating conditions as above. The mixture analyzed was much more complex than precedingly, comprising a wide range of polyunsaturated fatty acids, the complete identification of which is not easy. In that case mass spectrometry is very useful, but useful information can also be obtained by using columns of different polarities and by comparison with published equivalent chain length data (165) and with published chromatograms (6,188). The comparison with the analysis of the fatty acid methyl esters of pig testis (6) and of menhaden oil (AOCS 1b-89) has been particularly useful (188). Because of the importance of fish oil n-3 polyunsaturated fatty acids in human nutrition, the identification and quantification of eicosapentaenoic acid (EPA: 20:5 n-3) and of docosahexaenoic acid (DHA: 22:6 n-3) in particular in human foods are of special importance. The same is true for arachidonic acid (AA:20:4 n-6) found in human foods from terrestrial animal origin.

The following example concerns the analysis of the methyl esters of the positional and geometrical isomers of mono- and polyunsaturated fatty acids. The importance of *trans*-fatty acids in human nutrition was recently underlined by Willett and Ascherio (189), and requires particular care in identifying and quantifying the fatty acids in human foods. Positional and geometrical fatty acid isomers are present in dairy products from bovine origin, formed by biohydrogenation in the rumen of cows, and also present in foods prepared with industrially hydrogenated oils.

The method for the fatty acid composition of hydrogenated vegetable fats and oils by capillary GC using an SP-2340 column has been normalized (AOCS Official Method Ce 1c89) (130).

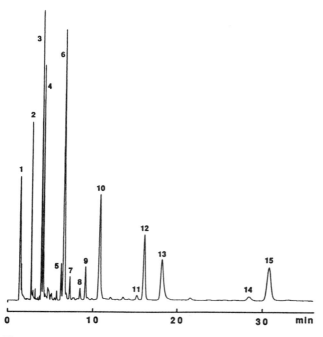

Fig. 9   Separation of a commercial cod liver oil fatty acid methyl esters on a Carbowax 20M glass column (30 m × 0.30 mm i.d.) at 198°C. Fatty acids: (1) solvent, (2) 14:0, (3) 16:0, (4) 16:1 n-7, (5) 18:0, (6) 18:1 n-9 + n-7, (7) 18:2 n-6, (8) 18:3 n-3, (9) 18:4 n-3, (10) 20:1 n-9, (11) 20:4 n-6, (12) 20:4 n-3, (13) 22:1 n-11, (14) 22:5 n-3, (15) 22:6 n-3. (The chromatogram was kindly obtained by J. Gresti.)

Figure 10 shows the separation of positional isomers of trans configuration of the monoenes of human milk. Milk lipid fatty acid isopropyl derivatives (141) were fractionated on silver ion TLC using hexane–diethyl ether (90:10, v/v) as eluant. The band containing the saturated and *trans*-monounsaturated fatty acid esters was isolated. The esters were extracted with methanol-hexane and analyzed by GC isothermally at 160°C on a CP Sil 88 fused silica capillary column (50 m × 0.25 mm i.d., 0.20 μm film thickness) with helium as carrier gas. The elution pattern was similar to that reported by Christie (165). The ECL increased as the double bond was nearer the methyl end of the molecule, reaching a maximum with 18:1 Δ16, which is clearly separated from its Δ15 isomer. The Δ11 isomer (*trans*-vaccenic acid) was predominant, as observed in bovine milk (197). The separation of the trans isomer of 18:1 from Δ6 to Δ16 was achieved within 4 min.

Figure 11 shows in part A the separation within 2 min of cis and trans isomers of linoleic acid found in human milk as methyl esters on a BPX 70 fused silica column (30 m × 0.32 mm i.d., 0.50 μm film thickness) with helium as carrier gas. The principal isomer was 18:2 cc, i.e., linoleic acid. It was accompanied by trans isomers, in very low amounts, with the isomer ct eluting ahead of tc. In part B is shown the separation within 4 min of the cis and trans isomers of α-linolenic acid on the same type of column. The all-cis isomer, i.e., α-linolenic acid (18:3 n-3), was by far the major fatty acid. However, trans isomers were present and eluted ahead of the cis isomer on this column. The isomer comprising two trans double bonds eluted ahead of those with only one. For the latter the elution order was cct. ctc, and tcc, i.e., the nearer was the trans double bond to the methyl end, the higher was the effect on the retention time.

Those two examples illustrate the high resolution that can be obtained with fused silica capillary columns. Despite that, to avoid too much overlapping between peaks, it is often useful to isolate

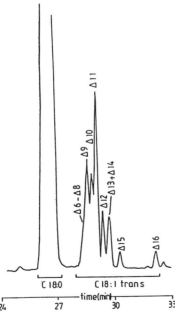

Fig. 10 Separation of postitional isomers of the *trans*-monoene methyl esters of human milk lipids on a CP Sil 88 fused silica column (50 m × 0.25 mm i.d.) at 160°C. (By the kind permission of J. M. Chardigny, INRA, Dijon.)

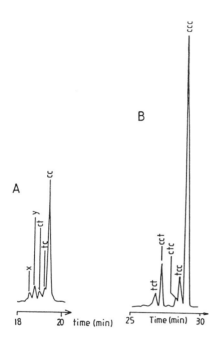

Fig. 11 Separation of *cis*-and *trans*-isomers of linoleic acid (18:2 n-6) and linolenic acid (18:3 n-3) of human milk lipids, on a BPX 70™ fused-silica column (30 m × 0.32 mm) at 160°C with helium as carrier gas. (By the kind permission of J. M. Chardigny, INRA, Dijon.)

simpler fatty acid mixtures from the original mixture, by TLC or HPTLC, before GC analysis, which was done with the human milk fatty acid isomers.

The next analysis, illustrated in Fig. 12, deals with another complex fatty acid mixture that is likely to be found in dairy products. It comprised a wide range of short-, medium-, and long-chain saturated fatty acids from C4 to C20.

As argued above, to avoid loss of short-chain fatty acids, butyric acid in particular, the analysis was performed on butyl ester derivatives and those were prepared in such a manner that butanol was kept as solvent of the esters, together with pentane or hexane, avoiding any extraction of the derivatives and any evaporation of the solvents. The GC equipment was different from that used for the more common fatty acid mixtures, with an on-column injector and a BPX 70 fused silica column (25 m × 0.32 mm i.d.), and the analysis was carried out in programmed temperature. The starting temperature (45°C) was maintained for 4 min. Then the temperature was raised to 165°C at a rate of 30°C/min (4 min), and again raised at 210°C at 4°C/min and maintained at this temperature until elution of the fatty acid butyl esters has been completed.

The chromatogram obtained is reported in Fig. 12. The separation was achieved in 25 min. A great number of peaks were detected, half of them present in trace amounts. The presence of branched and odd-numbered fatty acids are characteristics of ruminant lipids originating from bacterial metabolism in the rumen. On this type of column butyric acid butyl ester was very clearly separated from the butanol reagent and solvent. There is place for elution of butyl acetate between butanol and butyl butyrate. The octadecenoic fraction (16) is clearly separated from the butyl stearate (15). In this fraction, the trans isomer (vaccenic) and the cis isomer (oleic acid) coeluted.

This method, using butyl esters and particular esterification conditions, has proven to give quantitative results and is recommended for such analyses.

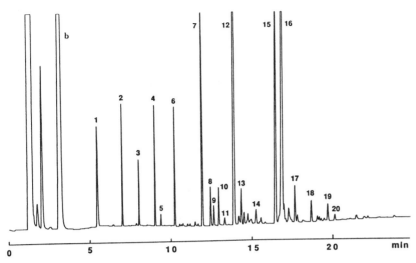

Fig. 12 Separation of butteroil fatty acid butyl esters on a BPX 70 fused silica column (25 m × 0.32 mm i.d.) at programmed temperature, with on-column injection. (b) Butanol (1) 4:0, (2) 6:0, (3) 8:0, (4) 10:0, (5) 10:1, (6) 12:0, (7) 14:0, (8) 14:1 n-5, (9) iso 15:0, (10) anteiso 15:0, (11) iso 16:0, (12) 16:0, (13) 16:1 n-7, (14) 17:0, (15) 18:0, (16) 18:1 n-9 + n-7, (17) 18:2 n-6, (18) 18:3 n-3, (19) conj. 18:2, (20) 20:0. (Reproduced by the kind permission of J. Gresti.)

The last illustration concerns GC analysis of a complex mixture of fatty acids, isolated from a cheese sample, and analyzed as free fatty acids on a fused silica capillary column coated with a polyethylene glycol phase added with terephthalic acid (Fig. 13).

The analysis was achieved in 70 min but the different major components were well separated. Peaks 4 and 8 correspond to the added standards 7:0 and 13:0, respectively; the presence of acetic acid was detected. The long-chain fatty acids were normally eluting although peak tailing was apparent, which is usual when analyzing fatty acids in the free form. Despite this tailing, monoenic

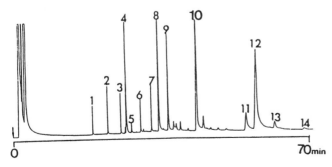

Fig. 13 Separation of cheese fatty acids in the unesterified form on polyethylene glycol TPA fused silica column (25 m × 0.32 mm i.d., 0.52-mm film thickness), with on-column injection and helium as carrier gas (2 ml/min). The analysis temperature was programmed from 60°C to 70°C at 1°C/min after a 5-min hold at 60°C and then to 220°C at 10°C/min and a 40-min hold at 220°C for complete elution. Fatty acids: (1) 2:0, (2) 4:0, (3) 6:0, (4) 7:0 as standard, (5) 8:0, (6) 10:0, (7) 12:0, (8) 13:0 as standard, (9) 14:0, (10) 16:0, followed by 16:1 n-7, (11) 18:0, (12) 18:1 n-9, (13) 18:2 n-6, (14) 18:3 n-3. (Reproduced from Ref. 55 by the kind permission of the authors and the *Journal of Food Science*.)

fatty acids were conveniently separated from the corresponding saturated fatty acid, e.g., oleic acid (12) from stearic acid (11).

In this type of analysis, great care must be taken to quantitatively extract the free fatty acid fraction and then to quantitatively analyze this fraction by GC. Checks with quantitative standard mixtures must be routinely used and correction factors applied if necessary.

C. High-Performance Liquid Chromatography

1. Introduction

Gas–liquid chromatography is the preferred technique for analysis of fatty acids because it offers very high resolution when using fused silica capillary columns, together with high sensitivity and very good reproducibility in quantitative analyses. However, high-performance liquid chromatography will supplant GLC with the development of highly resolutive capillary columns with a great variety of bonded phases of different polarities. But this is still to come.

For the moment HPLC is more often used for the isolation of specific fatty acid fractions, on a small scale, for further analysis. Because the separation is generally carried out at ambient temperatures, the isolation of fatty acids with labile functional group, such as cyclopropene rings, can be safely effected.

2. Fatty Acid Derivatives

Fatty acids can be separated in the unesterified form by HPLC in the reversed phase mode, provided that an acidic compound be added to the phase. However, the ester derivatives have been the most widely used to date. The methyl ester derivatives are the simplest and they have also proved to be of great value in the HPLC analysis of fatty acids. However, other alkyl ester derivatives have been proposed, in particular to increase the sensitivity of detection.

One category is represented by the derivatives which present a pronounced and specific absorption in the UV region of the spectrum (192). A large number of fatty acid esters containing UV-absorbing chromophores have been prepared. But phenacyl esters have probably been the most frequently used because of their relatively high molar extinction coefficient at the absorption maximum of 242 nm. The detection sensitivity is such that as little as 10 ng of phenacyl ester can be detected and quantified by this method. Additionally, since detection depends only on the alcohol moiety of the fatty acid derivative, the detector response is directly proportional to the molar amount of fatty acid and independent of the fatty acid. Quantitative analyses are thus easily achieved with this sort of derivative. One of the most comprehensive studies of phenacyl ester derivative separation by HPLC was performed by Borch (1975) and the advantages and drawbacks of these derivatives have been reviewed by Lam and Grushka (193).

Many other UV-absorbing fatty acid derivatives have been utilized, comprising substituted phenacyl derivatives, naphthacyl esters, and p-nitrobenzyl esters (Christie, 194).

The other category of fatty acid derivatives used to increase detection sensitivity is that of derivatives with fluorescent chromophores. The first and most widely used derivatives of this type are anthryl methyl esters. The range of detectability is the picogram range at 440 nm with fluorescent excitation at 360 nm, i.e., 10- to 100-fold better than in UV detection with the phenacyl esters. When used in HPLC-MS analysis, these derivatives give distinct molecular ions that permit detection of minor components.

Other workers have used coumarin esters, substituted phenanthrene derivatives, or quinoxalinonone derivatives of particularly low detectability (194). The preparation of these derivatives has been briefly described in the previous section. More details are likely to be found in the book by Christie (194).

3. Argentation-HPLC

The success of silver ion TLC prompted workers to use silical gel impregnated with silver nitrate in HPLC columns in the separation of fatty acid methyl esters.

The first columns prepared by mixing the adsorbant with a solution of silver nitrate suffered from silver ion leaching and for this reason did not have a long working life. This drawback has been overcome by using cation exchange resins as solid support, with sulfonic acid residues as functional groups. The silver ions are more strongly retained on the column and there is negligible bleed of these ions from the adsorbant.

The preparation of such a column containing sulfonic acid–bonded phase loaded with silver ion was described by Christie (134,135). The starting column is a commercial prepacked column (250 × 4.6 mm i.d.) of 5-μm silica particles (Nucleosil 5 SA) bearing bonded sulfonic acid groups. The column is loaded with silver ions simply by passing through the column a silver nitrate solution in water and then solvents of decreasing polarity to eliminate the excess of reagent. Such a column was used by Christie et al. (195) in the fractionation according to unsaturation of fatty acid methyl esters prepared from marine invertebrate lipids. Figure 14 illustrates the trace obtained with the invertebrate *Mytilus galloprovincialis*. The detection was by a mass light-scattering detector and a stream splitter was used to divert ~90% of the eluant from the column for collection of the fractions separated. The column was eluted with a gradient of methanol to methanol/acetonitrile (9:1, v/v) at a flow rate of 0.75 ml/min. The separation was achieved in 25 min.

As shown in Fig. 14, eight fractions were obtained and analyzed by capillary GC. The constituent fatty acids of each fraction were then identified by GC-MS as picolinyl esters. Data obtained showed that the first fraction (0) only contained saturated fatty acid methyl esters. The fraction (1) contained monoenes from C14 to C20 positional isomers. The fraction (2) of dienes was more complex, comprising 20:2 (minor first peak) and 22:2 (major second peak). Fraction (3) was composed of trienes from C16 to C22. Two tetraenes fraction (4) were eluted, of which the first contained predominantly n-6 isomers, while the second was mainly n-3 fatty acids. The pentaene (5) and hexaene (6) fractions contained the usual 20:5 n-3 and 22:6 n-3 found in marine lipids.

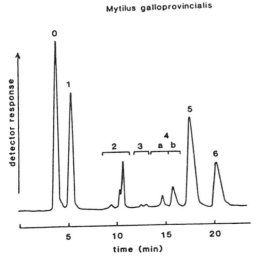

Fig. 14 Separation of the methyl esters derivatives of the lipid fatty acids of the marine invertebrate *Mytilus galloprovincialis* by silver ion HPLC. See the text for details on operating conditions. (Reproduced from Ref. 195 by kind permission of the authors and *Chemistry and Physics of Lipids*).

By combining fractionation of methyl esters by argentation-HPLC and analysis by GC and GC-MS of the fatty acids of each fraction, the authors succeeded in identifying more than 40 different fatty acids and quantifying 39 of them in the lipids of *M. galloprovincialis*, demonstrating the power of such combined techniques for investigation of fatty acid structural and compositional analysis.

Geometrical and positional isomers of unsaturated fatty acid have also been separated by silver ion HPLC as methyl esters on a column of silica gel impregnated with 5% silver nitrate into a slurry (196). The trans isomer of octadecenoic acid methyl ester eluted clearly ahead of the cis isomer, with hexane-tetrahydrofuran (100:0.6, v/v) as solvent, and the Δ7 isomer eluted ahead of the Δ9 isomer but less clearly. Nevertheless a good resolution was obtained for the four isomers likely to be present in hydrogenated oils. With columns prepared from cation exchange resins or from a silica support converted to aluminosilicate to hold silver ions by ionic bonds more strongly, such separations can be obtained (193).

Recently, this type of column was successfully used to separate methyl esters of cyclic fatty acids formed by oxidation of oil polyunsaturated fatty acids (Christie and Sebedio, personal communication).

Another form of silver ion chromatography is the incorporation of silver nitrate in the mobile phase and the use of octadecyl silyl (C18) stationary phase. In this case, since the complex formed between silver ions and cis double bonds is more stable than with trans double bonds, the cis isomers of unsaturated fatty acids elute ahead of the corresponding trans isomers (187). This type of argentation-HPLC is not much used because silver metal is likely to deposit in or corrode the HPLC equipment, especially the detector. It is so because stable columns loaded with silver nitrate can be easily prepared from prepacked Nucleosil columns.

4. Reverse Phase HPLC

Reverse phase HPLC is a very valuable technique for separation of the most commonly encountered fatty acids, more for collection of individual fatty acids or of fatty acid mixtures to be further analyzed than for analytical purposes.

By far the most widely used stationary phase consists of octadecylsilyl C18 groups covalently bonded to silica particles, although C8 phases are often found to be of some utility. Fatty acid methyl esters are frequently used as derivatives in HPLC analyses. With HPLC used in the reversed phase mode the unsaturated fatty acids are eluted ahead of the corresponding saturated fatty acids and roughly each double bond reduces the retention time by the equivalent of two carbon atoms.

The fatty acid derivatives can be characterized by the partition number (PN) of the fatty acid moiety. It can be calculated from the formula:

$$PN = CN - 2DB$$

in which CN is the total number of carbon atoms of the acyl chains and DB the total number of double bonds. PN can also be designated as ECN (equivalent chain length). This relation means that oleic (18:1) and palmitic (16:0) acids would coelute. However, with microparticulate phases and by selecting convenient operating conditions, these two fatty acids are readily separated. This is illustrated in Fig. 15. The separation of fatty acid methyl esters of a synthetic mixture was obtained with a 250 × 4.0 mm i.d. column containing 4-μm silica particles bonded with octadecylsilyl groups (Hibar LiChrocart Superspher RP 18) and isocratically eluted with acetonitrile-water as mobile phase (198). In part A the column temperature was maintained at 30°C and the separation was achieved in 40 min. Most of the fatty acid methyl esters present in the mixture were clearly separated except for the two positional isomers of 18:3 and for 20:2 n-6 and 16:0, forming a so-called critical pair of partition number 16. For the other critical pair, 18:1 n-9 and

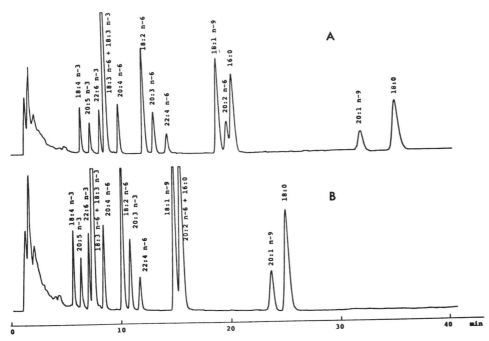

Fig. 15 Separation of a synthetic fatty acid methyl ester mixture by reverse phase HPLC on a octadecylsilyl (C18) column, isocratically eluted with acetonitrile-water, at 30°C (part A) and 40°C (part B). (Reproduced from Ref. 198 by kind permission of the authors and the *Journal of Chromatography*.)

16:0, the two members were clearly separated. The retention times could be highly reduced by increasing the analysis temperature, as shown in part B of Fig. 15. The same effect was obtained by decreasing the eluant polarity by decreasing the percentage of water in acetonitrile from 7% to 5%. However, in that case 20:2 n-6 tended to coelute with 18:1 n-9 and not with 16:0. This shows that separation of two particular fatty acids eluting in nearly the same time can be separated when selecting suitable operating conditions.

The method can be used in quantitative analyses provided calibration factors are applied if necessary. Table 10 shows such calibration factors calculated for the fatty acid methyl esters present in the mixture analyzed in Fig. 15 when a differential refractometer was used as a detector. In comparison to oleic acid (18:1 n-9) methyl ester, the unsaturated fatty acid esters were more highly detected (ratio lower than 1) than the saturated esters (ratio higher than 1).

An accurate fatty acid composition can be obtained by this method since the calibration factors are not too different from unity. However, the use of reverse phase HPLC as an analytical method for fatty acid mixtures is not widespread. Lipid analysts generally prefer for this purpose capillary GC, which offers higher resolution. But the possible development of capillary columns for use in HPLC as that proposed by Hirata and Sumiya (199) and their commercial availability can reorientate the choice of analysts in the future.

Up to now the privileged application field of reverse phase remains the isolation in a semipreparative scale (10–20 mg) of individual fatty acids or more often of fatty acids fractions of simple composition for further analysis. An example is given in Fig. 16. The purpose of the analytical work was to identify the trans isomers of arachidonic acid in liver lipids from rats fed a previously heated sunflower oil containing trans isomers of linoleic acid (18:2 n-6). In a first step, a fraction of the liver

Table 10 Calibration Factors for Quantitative Analysis of Fatty Acid Methyl Esters by HPLC with Detection by Differential Refractometry

	18:4 n-3	20:5 n-3	22:6 n-3	18:3 n-3, n-6	20:4 n-6	18:2 n-6	20:3 n-6	22:4 n-6	18:1 n-9	20:2 n-6	16:0	20:1 n-9	18:0
GC-RPLC[a] (n = 5)	0.90 ±0.03	0.84 ±0.04	0.86 ±0.04	0.92 ±0.03	0.93 ±0.04	0.95 ±0.02	0.92 ±0.02	0.88 ±0.03	1.00 ±0.03	1.01 ±0.02	1.08 ±0.03	1.01 ±0.02	1.08 ±0.03

[a]Ratios of the percentages of the fatty acid methyl esters present in the complex mixture previously analyzed (Fig. 15) as determined by GC after application of calibration factors, to the percentages of areas determined after analysis of the mixture by RPLC. Analytical conditions: temperature, 30°C; solvent, acetonitrile-water (93:7, v/v) at 1.0 ml/min. Values are means ± SE of five determinations.

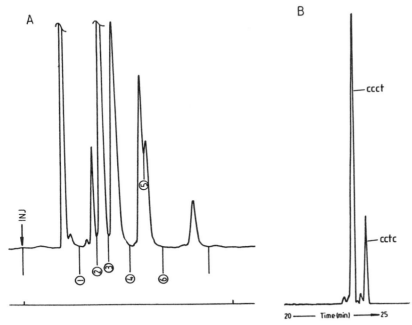

Fig. 16 Separation of a fraction of liver lipid fatty acid methyl esters enriched in unsaturated fatty acid derivatives by semipreparative reverse phase (C18) HPLC (A) and (B) separation of the trans isomers of arachidonic acid (20:4 n-6) isolated by silver ion TLC of the fraction 2 fatty acid methyl esters collected at the outlet of the HPLC column (A). In part B the two trans isomers are as indicated, ccct and cctc (c = cis- and t = trans double bond, counted from the carboxyl end). (Reproduced by the kind permission of O. Berdeaux, INRA, Dijon.)

lipid fatty acids as methyl ester derivatives, enriched in unsaturated fatty esters, was isolated by precipitating the saturated fatty esters as urea inclusion complexes from a cold solution of the esters in methanol in the presence of urea.

This fraction (20 mg) was then submitted to further fractionation by reverse phase HPLC on a semipreparative (250 × 10 mm i.d.) Nucleosil C18 column, eluted with acetonitrile-acetone (90:10, v/v) at a flow rate of 4 ml/min. Different fractions were separated . The authors were interested in analyzing fraction 2 containing more than 80% of arachidonic acid, which was collected at the outlet of the detector, the nondestructive differential refractometer. This fraction of 20:4 methyl esters was then further fractionated on silver ion TLC to isolate the trans isomers. GLC analysis of this fraction on a BPX 70 capillary glass column (50 m × 0.33 mm i.d.) with helium as carrier gas showed (Fig. 16B) that two major trans isomers were present, each having one trans double bond (t) but in different position and three cis double bonds (c).

Analysis by mass spectrometry of the first peak (ccct) gave the spectrum previously reported (Fig. 6), which showed that the double bonds were located at Δ5, Δ8, Δ11, and Δ14 positions, the same positions as in arachidonic acid, the all-cis isomer. This work illustrates perfectly the degree of analytical power the analyst can reach by using successive separations and fractionations to finally obtain separation by GC of individual isomers, the analysis of which is easy by MS. It also shows that in the analysis of complex mixtures of fatty acid isomers more than one separation method is necessary even with the powerful analytical GC-MS method.

Although the ester derivatives are the most used in HPLC analysis of fatty acids, several workers have separated fatty acids in their unesterified form, with octadecylsilyl columns, but it was necessary to add an acid in the mobile phase. Aveldano et al. (200) in their systematic study of the elution rules of fatty acids in reverse phase HPLC have used aqueous phosphoric acid added to acetonitrile. However, when the fatty acid fractions are to be collected to be further analyzed by TLC or/and GC, the absence of acid in the mobile phase and the presence of fatty acids as ester derivatives have to be preferred for commodity.

When the fatty acids to be analyzed have unstable functional groups such as a cyclopropene ring, HPLC gains interest over GLC because analysis is carried out at ambient temperatures and in the absence of oxygen, which is often present in trace amounts in the GC carrier gas. These unusual fatty acids can have adverse biological effects when consumed (201) and their detection and quantitation in human foods are of importance. Good separations of the methyl esters of malvalic and sterculic acids were obtained by Gaydou et al. (202) on a 250×4.6 mm i.d. column of LiChrosorb RP8 eluted with acetonitrile-water (170:30, v/v). Under these conditions, methyl sterculate, with a cyclopropene ring at the $\Delta9$ position and 19 carbon atoms, eluted just after methyl oleate. Sensitivity of detection of trace amounts of these acids in edible oils, such as cottonseed oil, can be increased by using phenacyl derivatives and UV detection.

Cyclopropene fatty acids from baobab seed oil were isolated on a semipreparative scale by reverse phase HPLC to use in biological experiments (203). It may be useful to recall that the methylation of these cyclic acids must be carried out in basic medium in the presence of sodium methoxide to avoid cleavage of the ring.

5.   Normal Phase HPLC

Normal phase partition HPLC using columns of silica gel has not been much employed for the analysis of usual fatty acids, either in unesterified or in derived form. Only a few separations have been published. On the contrary, this method has been widely used for less common fatty acids, such as hydroperoxides, as such or as hydroxides after reduction. These acids derived from usual unsaturated fatty acids by enzymatic oxidation or chemical autoxidation. For example, Chan and Levett (204) showed that hydroperoxide isomers could be separated on a column of silica gel, according to the position of the hydroperoxide groups and to the configuration cis or trans of the double bonds, both as hydroperoxides or as hydroxy derivatives after reduction.

Other oxygenated fatty acids were readily separated by adsorption HPLC, including hydroxy and epoxy fatty acids. Positional isomers of monohydroxy fatty acids were well resolved on a column of silica gel by Bandi and Reynolds (205). Di- and trihydroxy fatty acids and epoxy fatty acids were separated using different derivatives (4).

6.   Conclusion

Other types of HPLC have also been used for specific purposes, such as gel permeation to separate positional isomers of linoleic acid hydroperoxides (206) or to fractionate the methyl ester derivatives of the fatty acids liberated from polymerized triacylglycerols (207). However, reverse phase and silver ion HPLC remain by far the most widely used forms of HPLC, particularly for isolation of fatty acid fractions requiring other specific analyses.

REVIEWS

W. W. Christie, *Advances in Lipid Methodology*, Vol. 1, Oily Press, Dundee, 1992.
W. W. Christie, *Advances in Lipid Methodology*, Vol. 2, Oily Press, Dundee, 1993.
W. W. Christie, *Gas Chromatography and Lipids: A Practical Guide*, Oily Press, Ayr, 1992.

W. W. Christie, *High-Performance Liquid Chromatography and Lipids: A Practical Guide*, Pergamon Press, Oxford, 1987.

W. W. Christie, *Lipid Analysis*, 2nd ed, Pergamon Press, Oxford, 1982.

F. D. Gunstone, J. L. Harwood, and F. B. Padley, *The Lipid Handbook*, Chapman and Hall, London, 1994.

A. Karleskind, *Manuel des Corps Gras, Vols. 1 and 2, Technique et Documentation*, Lavoisier, Paris, 1992.

A. Kuksis, *Chromatography of Lipids in Biomedical Research and Clinical Diagnosis*, Elsevier, Amsterdam, 1987.

C. Litchfield, *Analysis of Triglycerides*, Academic Press, New York, 1972.

G. V. Marinetti, *Lipid Chromatographic Analysis*, 2nd ed. Vols. I, II, III, Marcel Dekker, New York, 1976.

J. L. Sebedio and E. G. Perkins, *New Trends in Lipid and Lipoprotein Analysis*, AOCS Press, Champaign, IL, 1995.

REFERENCES

1. C. Litchfield, *Analysis of Triglycerides*, Academic Press, New York, 1972.
2. G. V. Marinetti, *Lipid Chromatographic Analysis*, 2nd ed., Vols. I, II, III, Marcel Dekker, New York, 1976.
3. W. W. Christie, *Lipid Analysis*, 2nd ed., Pergamon Press, Oxford, 1982.
4. W. W. Christie, *High-Performance Liquid Chromatography and Lipids. A Practical Guide*, Pergamon Press, Oxford 1987.
5. A. Kuksis, *Chromatography of Lipids in Biomedical Research and Clinical Diagnosis*, Elsevier, Amsterdam, 1987.
6. W. W. Christie, *Gas Chromatography and Lipids. A Practical Guide*, Oily Press, Ayr, UK, 1992.
7. W. W. Christie, *Advances in Lipid Methodology*, Vol. 1, Oily Press, Dundee, UK 1992.
8. A. Karleskind, *Manuel des Corps Gras*, Vols. 1 and 2, *Technique et Documentation*, Lavoisier, Paris, 1992.
9. W. W. Christie, *Advances in Lipid Methodology*, Vol 2, Oily Press, Dundee, UK 1993.
10. F. D. Gunstone, J. L. Harwood, and F. B. Padley, *The Lipid Handbook*, 2nd ed., Chapman and Hall, London, 1994.
11. J. L. Sebedio and E. G. Perkins, *New Trends in Lipid and Lipoprotein analysis*, AOCS Press, Champaign, 1995.
12. J. F. Mead, R. B. Alfin-Slater, R. D. Howton, and G. Popjak, *Lipids: Chemistry, Biochemistry and Nutrition*, Plenum Press, New York, 1986.
13. IUPAC-IUB (International Union of Pure and Applied Chemistry/International Union of Biochemistry) Commission on Biochemical Nomenclature, *Eur. J. Biochem.*, 2: 127 (1967); *Biochem. J.*, 105: 897 (1967).
14. IUPAC-IUB Commission on Biochemical Nomenclature, *J. Lipid Res.*, 19: 114 (1978).
15. J. Bézard and M. Bugaut, *Absorption of glycerides containing short, medium and long chain fatty acids*, in *Fat Absorption*, Vol. 1 (A. Kuksis, ed.), CRC Press, Boca Raton, 1986, p. 119.
16. A. E. Bailey, *Melting and Solidification of Fats*, Interscience, New York, 1950.
17. W. S. Singleton, *Fatty Acids* (K. S. Markley, ed.), Interscience, New York, 1960.
18. T. E. Jordan, *Vapor Pressure of Organic Compound*, Interscience, New York, 1954.
19. H. Westergaard and J. M. Dietschy, *J. Clin. Invest.*, 58: 97 (1976).
20. A. E. Bailey, J. A. Harris, and E. I. Skau, *J. Am. Oil Chem. Soc.*, 47: 383 (1970).
21. W. C. Willett, M. J. Stampfer, J. E. Manson, G. A. Colditz, F. E. Speizer, B. A. Rosner, L. A. Sampson, and C. H. Hennekens, *Lancet*, 341: 581 (1993).
22. T. Malkin, *Prog. Chem. Fats Other Lipids*, 2: 1 (1954).
23. M. Knoester, P. De Bruijne, and M. Van Den Tempel, *Chem. Phys. Lipids*, 9: 309 (1972).
24. L. K. Arnold, B. B. R. Choudhury, and A. Guzman, *J. Am. Oil Chem. Soc.*, 4: 33 (1963).
25. A. R. Macrae, *J. Am. Oil Chem. Soc.*, 60: 291 (1983).
26. P. Eigtved, *Advances in Applied Lipid Research*, Vol. 1 (F. B. Padley, ed.), JAI Press, London, CT, 1992, p. 1.
27. N. Klyne, *La Chimie des Stéroïdes,* Gauthier-Villars, Paris, 1966.

28. J. B. Rossell and J. L. R. Pritchard, *Analysis of Oilseeds, Fats and Fatty Foods*, Elsevier, London, 1990.
29. D. R. Erickson, E. H. Pryde, O. L. Brekke, T.L. Mounts, and R. A. Falb, *Handbook of Soy Oil Processing and Utilization*, American Soybean Association and American Oil Chemists' Society, Champaign, IL, 1980.
30. T. Itoh, T. Tamura, and T. Matsumoto, *Oléagineux*, *29*: 253 (1974).
31. P. S. Dimick, R. D. McCarthy, and S. Patton, in *Physiology of Digestion and Metabolism of the Ruminant* (A. T. Phillipson, ed.), Oriel, Newcastle Upon Tyde, 1970.
32. M. C. M. Griebneau, *Trends Food Sci. and Technol.*, *3*: 186 (1992).
33. F. D. Gunstone, in *Advances in Lipid Methodology*, Vol. 2, (W. W. Christie, ed.), Oily Press, Dundee, UK, 1993.
34. American Oil Chemists' Society (AOCS), *Official Methods and Recommended Practices*, 4th ed. (D. Firestone, ed.), AOCS Press, Champaign, IL, AOCS Official Method Aa 4–38, 1994.
35. W. W. Christie, in *Advances in Lipid Methodology*, Vol. 2 (W. W. Christie, ed.), Oily Press, Dundee, UK, 1993, p. 195.
36. J. Folch, M. Lees, and G. H. S. Stanley, *J. Biol. Chem.*, *226*: 497 (1957).
37. E. G. Bligh and W. J. Dyer, *Can. J. Biochem. Physiol.*, *37*: 911 (1959).
38. I. S. Chen, C. S. J. Shen, and A. J. Sheppard, *J. Am. Oil Chem. Soc.*, *58*: 599 (1981).
39. J. L. Delsal, *Bull. Soc. Chim. Biol.*, *26*: 99 (1944).
40. R. M. Broekhuyse, *Clin. Chim. Acta*, *51*: 341 (1974).
41. A. Hara and N. S. Radin, *Anal. Biochem.*, *90*: 420 (1978).
42. S. Somersalo, P. Karunen and E. M. Aro, *Physiol. Plant.*, *68*: 467 (1986).
43. C. Entenman, *J. Am. Oil Chem. Soc.*, *38*: 534 (1961).
44. K. D. Bartle and T. A. Clifford, in *Advances in Applied Lipid Research*, Vol. 1 (F. B. Padley, ed.), JAI Press, London, 1992, p. 217.
45. P. Laakso, in *Advances in Lipid Methodology* Vol. 1 (W. W. Christie, ed.), Oily Press, Ayr, 1992, p. 81.
46. K. J. Maxwell, W. N. Marmer, M. P. Zubillaga, and G. A. Dalickas, *J. Assoc. Off. Anal. Chem.*, *63*: 600 (1980).
47. R. L. Wolff and A. Castera-Rossignol, *Rev. Fr. Corps Gras*, *34*: 123 (1987).
48. Association Française de Normalisation (AFNOR), *Recueil des Normes Françaises des Corps Gras, Graines Oléagineuses, Produits Dérivés*, 4th edition, edited by AFNOR, Paris, AFNOR Method NF V 83–30, 1988.
49. K. Ganzler, A. Salgo, and K. Valko, *J. Chromatogr.*, *371*: 299 (1986).
50. K. A. Ansari and D. Shoeman, *J. Chromatogr.*, *439*: 453 (1988).
51. J. Pempkowiak, *J. Chromatogr.*, *258*: 93 (1983).
52. T. Matsuzawa, K. Mishima, M. Nishii, and M. Ito, *Biochem. Int.*, *15*: 693 (1987).
53. M. A. Kaluzny, L. A. Duncan, M. V. Merritt, and D. E. Epps, *J. Lipid Res.*, *26*: 135 (1985).
54. C. De Jong and H. T. Badings, *J. High Resolut. Chromatogr.*, *13*: 94 (1990).
55. L. Lesage, A. Voilley, D. Lorient, and J. Bézard, *J. Food Sci.*, *58*: 1303 (1993).
56. G. Semporé and J. Bézard, *J. Chromatogr.*, *366*: 261 (1986).
57. K. K. Carroll, *J. Lipid Res.*, *2*: 135 (1961).
58. E. L. Jack, C. P. Freeman, L. M. Smith, and J. B. Mickle, *J. Dairy Sci.*, *46*: 284 (1963).
59. H. Chino and L. I. Gilbert, *Anal. Biochem.*, *10*: 395 (1965).
60. E. Kerkhoven and J. M. De Man, *J. Chromatogr.*, *24*: 56 (1966).
61. K. K. Carroll and B. Serdarevich, in *Lipid Chromatographic Analysis*, Vol. 1 (G. V. Marinetti, ed.), Marcel Dekker, New York, 1967, p. 205.
62. N. S. Radin, in *Methods in Enzymology*, Vol. 14 (J. M. Lowenstein, ed.), Academic Press, New York, 1969, p. 268.
63. D. L. Fillerup and J. F. Mead, *Proc. Soc. Exp. Biol. Med.*, *83*: 574 (1953).
64. J. Hirsch and E. H., Jr. Ahrens, *J. Biol. Chem.*, *233*: 311 (1958).
65. P. Quinlin and H. J. Weiser, *J. Am. Oil Chem. Soc.*, *35*: 325 (1958).
66. M. G. Horning, E. A. Williams, and E. C. Horning, *J. Lipid Res.*, *1*: 482 (1960).
67. M. R. Sahasrabudhe and D. G. Chapman, *J. Am. Oil Chem. Soc.*, *38*: 88 (1961).
68. M. R. Sahasrabudhe, J. J. Legari, and W. P. McKinley, *J. Assoc. Offic. Anal. Chem.*, *49*: 337 (1966).
69. A. J. Day and N. H. Fidge, *J. Lipid Res.*, *5*: 163 (1964).

70. R. F. Addison and R. G. Ackman, *Anal. Biochem.*, 28: 515 (1969).
71. R. G. Jensen, T. A. Marks, J. Sampugna, J. G. Quinn, and D. L. Carpenter, *Lipids*, 1: 451 (1966).
72. I. Hornstein, P. F. Crowe, and J. B. Ruck, *Anal. Chem.*, 39: 352 (1967).
73. T. F. Kelley, *J. Lipid Res.*, 9: 799 (1968).
74. J. C. Dittmer and M. A. Wells, in *Methods in Enzymology*, Vol. 14 (J. M. Lowenstein, ed.), Academic Press, New York, 1969, p. 513.
75. J. P. Wolff, in *Manuel d'Analyse des Corps Gras* (Azoulay, ed.), Paris, 1968.
76. G. Rouser, C. Kritchevsky, and A. Yamamoto, in *Lipid Chromatographic Analysis*, Vol. 1 (G. V. Marinetti, ed.), Edward Arnold, London, 1967, p. 99.
77. J. Bitman, D. L. Wood, M. Hamosh, P. Hamosh, and N. R. Mehta, *Am. J. Clin. Nutr.*, 38: 30 (1983).
78. P. Juaneda and G. Rocquelin, *Lipids*, 20: 40 (1985).
79. E. Stahl, *Pharmazie*, 11: 633 (1956).
80. J. M. Bobbitt, *Thin-Layer Chromatography*, Van Nostrand-Reinhold, Princeton, 1963.
81. V. P. Skipski and M. Barclay, in *Methods in Enzymology*, Vol. 14 (J. M. Lowenstein, ed.), Academic Press, New York, 1969, p. 530.
82. E. Stahl, *Thin-Layer Chromatography: A Laboratory Handbook*, 2nd ed., Allen and Unwin, London, 1969.
83. V. P. Skipski, R. F. Peterson, and M. Barclay, *Biochim. J.*, 90: 374 (1964).
84. K. Kiuchi, T. Otha and H. Ebine, *J. Chromatogr. Sci.*, 13: 466 (1975).
85. M. D. Greenspan and E. A. Schroeder, *Anal. Biochem.*, 127: 441 (1982).
86. D. N. Palmer, M. A. Anderson, and R. D. Jolly, *Anal. Biochem.*, 140: 315 (1984).
87. W. W. Christie, *J. Lipid Res.*, 26: 507 (1985).
88. W. W. Christie, *J. Chromatogr.*, 361: 396 (1986).
89. O. S. Privett, K. A. Dougherty, W. L. Erdhal, and A. Stolyhwo, *J. Am. Oil Chem. Soc.*, 50: 516 (1973).
90. Association of Official Analytical Chemists (AOAC), *Official Methods of Analysis*, 15th ed. (D. Firestone, ed.), Arlington, VA, AOAC Method 920–160, 1990, p. 957.
91. American Oil Chemists' Society (AOCS), *Official Methods and Recommended Practices*, 9th ed. (D. Firestone, ed.), AOCS Press, Champaign, USA, AOCS Official Methods Cd 3–25 and Cd 3b–76, AOCS Recommended Practice Cd 3c–91, 1994.
92. Association Française de Normalisation (AFNOR), *Recueil des Normes Françaises des Corps Gras, Graines Oléagineuses, Produits Dérivés*, 4th ed., edited by AFNOR, Paris, AFNOR Method NFT 60–206, 1988.
93. Association of Official Analytical Chemists (AOAC), *Official Methods of Analysis*, 15th ed. (D. Firestone, ed.), Arlington, VA, AOAC Methods 920–158, 1990, p. 955 and 920–159, p. 956.
94. American Oil Chemists' Society (AOCS), *Official Methods and Recommended Practices*, 4th ed. (D. Firestone, ed.), AOCS Press, Champaign, IL, AOCS Official Method Cd 1–25 and AOCS Recommended Practice Cd 1b-87, 1994.
95. Association Française de Normalisation (AFNOR), *Recueil des Normes Françaises des Corps Gras, Graines Oléagineuses, Produits Dérivés*, 4th ed., edited by AFNOR, Paris, AFNOR Method NFT 60–203, 1988.
96. Association of Official Analytical Chemists (AOAC), *Official Methods of Analysis*, 15th ed. (D. Firestone, ed.), Arlington VA, AOAC Method 965–32, 1990, p. 955.
97. American Oil Chemists' Society (AOCS), *Official Methods and Recommended Practices*, 4th ed. (D. Firestone, ed.), AOCS Press, Champaign, IL, AOCS Official Methods Cd 4–40 and Cd 13–60, 1994.
98. Association Française de Normalisation (AFNOR), *Recueil des Normes Françaises des Corps Gras, Graines Oléagineuses, Produits Dérivés*, 4th edition, edited by AFNOR, Paris, AFNOR Method NFT 60–213, 1988.
99. A. J. P. Martin and R. L. M. Synge, *Biochem. J.*, 35: 1358 (1941).
100. A. T. James and A. J. P. Martin, *Biochem. J.*, 50: 679 (1952).
101. A. J. P. Martin and A. T. James, *Biochem. J.*, 63: 138 (1956).
102. R. L. Grob, *Modern Practice of Gas Chromatography*, John Wiley and Sons, New York, 1977.
103. C. E. Poole and S. A. Schuette, *Contemporary Practice of Chromatography*, Elsevier, Amsterdam, 1984.
104. H. Jaeger, *Glass Capillary Chromatography in Clinical Medicine and Pharmacology*, Marcel Dekker, New York, 1985.

105. A. T. James and A. J. P. Martin, *Biochem. J.*, *63*: 144 (1956).
106. W. W. Christie, in *Topics in Lipid Chemistry*, Vol. 3 (F. D. Gunstone, ed.), Logos Press, London, 1972, pp. 171–197.
107. A. J. Sheppard and J. L. Iverson, *J. Chromatogr. Sci.*, *13*: 448 (1975).
108. J. D. Craske, C. D. Bannon, and L. M. Norman, *J. Am. Oil Chem. Soc.*, *65*: 262 (1988).
109. M. Kates, *J. Lipid Res.*, *5*: 132 (1964).
110. M. L. Blank, B. Verdino, and O. S. Privett, *J. Am. Oil Chem. Soc.*, *42*: 87 (1965).
111. K. V. Peisker, *J. Am. Oil Chem. Soc.*, *41*: 87 (1964).
112. B. Akesson, *Eur. J. Biochem.*, *9*: 463 (1969).
113. B. E. Phillips, C. R. Jr. Smith, and J. W. Hageman, *Lipids*, *4*: 473 (1969).
114. L. D. Metcalfe and A. A. Schmitz, *Anal. Chem.*, *33*: 363 (1961).
115. W. R. Morrison and L. M. Smith, *J. Lipid Res.*, *5*: 600 (1964).
116. H. Brockerhoff, *Arch. Biochem. Biophys.*, *110*: 586 (1965).
117. P. K. Raju and R. Reiser, *Lipids*, *1*: 10 (1966).
118. R. E. Anderson, N. R. Bottino, and R. Reiser, *Lipids*, *5*: 161 (1968).
119. L. D. Metcalfe and C. N. Wang, *J. Chromatogr. Sci.*, *19*: 530 (1981).
120. U. Schuchardt and O. C. Lopes, *J. Am. Oil Chem. Soc.*, *65*: 1940 (1988).
121. E. Schulte and K. Weber, *Fat Sci. Technol.*, *91*: 181 (1989).
122. Association of Official Analytical Chemists (AOAC), *Official Methods of Analysis*, 15th ed. (D. Firestone, ed.), Arlington, VA, AOAC Method 969–33, 1990, p. 963.
123. International Union of Pure and Applied Chemistry (IUPAC), *Standard Methods for the Analysis of Oils, Fats and Derivatives*, 7th Revised and Enlarged Edition, Blackwell Scientific, Boston, IUPAC Method 2–301, 1987.
124. American Oil Chemists' Society (AOCS), *Official Methods and Recommended Practices*, 4th ed. (D. Firestone, ed.), AOCS Press, Champaign, IL, AOCS Official Method Ce 2–66, 1994.
125. H. Schlenk and J. L. Gellerman, *Anal. Chem.*, *32*: 1412 (1960).
126. R. A. Barford, S. F. Herb, F. E. Luddy, P. Magidman, and R. W. Riemenschneider, *J. Am. Oil Chem. Soc.*, *40*: 136 (1963).
127. J. P. Bianchini, A. Ralaimanarivo, and E. M. Gaydou, *Anal. Chem.*, *53*: 2194 (1981).
128. G. Cecchi, S. Biasini, and J. Castano, *Rev. Fr. Corps Gras*, *32*: 163 (1985).
129. Association Française de Normalisation (AFNOR), *Recueil des Normes Françaises des Corps Gras, Graines Oléagineuses, Produits Dérivés*, 4th ed., edited by AFNOR, Paris, AFNOR Method NF 60–233, 1988.
130. F. E. Luddy, R. A. Barford, S. F. Herb, and P. Magidman, *J. Am. Oil Chem. Soc.*, *45*: 549 (1968).
131. S. W. Christopherson and R. L. Glass, *J. Dairy Sci.*, *52*: 1289 (1969).
132. L. D. Metcalfe, A. A. Schmitz, and J. R. Pelka, *Anal. Chem.*, *38*: 514 (1966).
133. H. T. Slover and E. Lanza, *J. Am. Oil Chem. Soc.*, *56*: 933 (1979).
134. G. Clément and J. Bézard, *C. R. Acad. Sci.*, Paris, *253*: 564, (1961).
135. B. M. Graig, A. P. Tulloch, and N. L. Marty, *J. Am. Oil Chem. Soc.*, *40*: 61 (1963).
136. J. Sampugna, R. E. Pitas, and R. G. Jensen, *J. Dairy Sci.*, *49*: 1462 (1966).
137. L. Marai, W. C. Breckenridge, and A. Kuksis, *Lipids*, *4*: 562 (1969).
138. C. Maniongui, J. Gresti, M. Bugaut, S. Gauthier, and J. Bézard, *J. Chromatogr.*, *543*: 81 (1991).
139. J. Bézard and M. Bugaut, *J. Chromatogr. Sci.*, *7*: 639 (1969).
140. K. Oette and E. H. Jr. Ahrens, *Anal. Chem.*, *33*: 1847 (1961).
141. R. L. Wolff and R. J. Fabien, *Le Lait*, *69*: 33 (1989).
142. M. Pina, C. Ozenne, G. Lamberet, D. Montet, and J. Graille, *Rev. Fr. Corps Gras*, *38*: 213 (1991).
143. R. F. Borch, *Anal. Chem.*, *47*: 2437 (1975).
144. A. Sonesson, L. Larsson, and J. Jimenez, *J. Chromatogr.*, *417*: 366 (1987).
145. A. Kuksis, J. J. Myher, L. Marai, and K. Geher, *Anal. Chem.*, *70*: 302 (1976).
146. D. J. Harvey, in *Advances in Lipid Methodology*, Vol. 1 (W. W. Christie, ed.), Oily Press, Dundee, UK 1992.
147. J. L. Le Quéré, in *Advances in Lipid Methodology*, Vol. 2, (W. W. Christie, ed.), Oily Press, Dundee, UK, 1993.
148. D. J. Harvey, *Biomed. Environ. Mass Spectrom.*, *11*: 340 (1982).
149. M. Balazy and A. S. Nies, *Biomed. Environ. Mass Spectrom.*, *18*: 328 (1989).

150. J. Bézard, P. Boucrot, and G. Clément, *J. Chromatogr.*, *14*: 368 (1964).
151. E. C. Horning, W. J. A. Vandenheuvel, and B. G. Creech, *Methods of Biochemical Analysis*, Vol. 2 (D. Glick, ed.), Interscience, New York, 1963.
152. S. R. Lipsky, R. A. Landowne, and J. E. Lovelock, *Anal. Chem.*, *31*: 852 (1959).
153. C. Litchfield, A. F. Isbell, and R. Reiser, *J. Am. Oil Chem. Soc.*, *40*: 302 (1962).
154. G. Lavoué and J. Bézard, *J. Chromatogr. Sci.* 7: 375 (1969).
155. R. R. Freeman, *High Resolution Gas Chromatography*, 2nd ed., Hewlett-Packard, 1981.
156. G. Alexander and G. A. M. F. Rutten, *Chromatographia*, *6*: 231 (1973).
157. K. Grob, G. Grob, and K. Jr. Grob, *Chromatographia*, *10*: 181 (1977).
158. G. Schomburg, H. Husman, and F. Weeke, *Chromatographia*, *10*: 580 (1977).
159. J. Bouche and M. Verzele, *J. Gas Chromatogr.*, *6*: 501 (1968).
160. D. H. Desty, *Gas Chromatography*, Buderworths, London, 1958.
161. J. E. Lovelock, *J. Chromatogr. 1*: 35 (1958).
162. K. Grob, in *Classical Split and Splitless Injection in Capillary Gas Chromatography*, 2nd ed., Hüthing, Heidelberg, 1988, p. 219.
163. A. Ros, *J. Gas Chromatogr.*, *3*: 252 (1965).
164. J. A. Barve, F. D. Gunstone, F. R. Jacobsberg, and P. Willow, *Chem. Phys. Lipids*, *8*: 117 (1972).
165. W. W. Christie, *J. Chromatogr.*, *447*: 305 (1988).
166. B. A. Anderson, *Progr. Chem. Fats Other Lipids*, *16*: 279 (1978).
167. D. E. Minnikin, *Chem. Phys. Lipids*, *21*: 313 (1978).
168. B. Schmitz and R. A. Klein, *Chem. Phys. Lipids*, *39*: 285 (1986).
169. N. J. Jensen and M. L. Gross, *Mass Spectrom. Rev.*, *6*: 49 (1987).
170. J. Y. Zhang, Q. T. Yu, B. N. Liu, and Z. H. Huang, *Biomed. Environ. Mass Spectrom.*, *15*: 33 (1988).
171. R. C. Badami and K. B. Patil, *Prog. Lipid Res.*, *19*: 119 (1981).
172. E. G. Perkins and W. T. Iwaoka, *J. Am. Oil Chem. Soc.*, *50*: 44 (1973).
173. G. W. Kenner and E. Stenhagen, *Acta Chem. Scand.*, *18*: 1551 (1964).
174. L. J. Morris, D. M. Wharry, and E. W. Hammond, *J. Chromatogr.*, *31*: 69 (1967).
175. W. W. Christie, *J. Chromatogr.*, *454*: 273 (1988).
176. P. Boucrot and J. Bézard, *Arch. Sci. Physiol.*, *27*: 1 (1973).
177. D. Chapman and F. M. Goni, in *The Lipid Handbook* (F. D. Gunstone, J. L. Harwood, and F. B. Padley, eds.), Chapman and Hall, London, 1988, p. 385.
178. J. E. D. Davies, P. Hodge, F. D. Gunstone, and M. S. F. Lie Ken Jie, *Chem. Phys. Lipids*, *15*: 48 (1975).
179. F. D. Gunstone and R. P. Inglis, in *Topics in Lipid Chemistry*, Vol. 2 (F. D. Gunstone, ed.), Logos Press, London, U.K., 1971, p. 287.
180. J. Lecerf and J. Bézard, *Rev. Fr. Corps Gras*, *13*: 455 (1966).
181. R. G. Ackman, J. L. Sebedio, and W. M. N. Ratnayake, *Meth. Enzymol.*, *72*: 253 (1981).
182. J. Lecerf and J. Bézard, *C. R. Acad. Sci.*, Paris, 272: 2104 (1971).
183. W. W. Christie, in *Topics in Lipid Chemistry*, Vol. 1 (F. D. Gunstone, ed.), Logos Press, London, 1970, p. 1.
184. N. Polgar, in *Topics in Lipid Chemistry*, Vol. 2 (F. D. Gunstone, ed.), Logos Press, London, 1971, p. 207.
185. H. T. Badings and C. De Jong, *J. Chromatogr.*, *279*: 493 (1983).
186. American Oil Chemists' Society (AOCS), *Official Methods and Recommended Practices*, 4th ed. (D. Firestone, ed.), AOCS Press, Champaign, IL, AOCS Official Method Ce 1e–91, 1994.
187. American Oil Chemists' Society (AOCS), *Official Methods and Recommended Practices*, 4th ed. (D. Firestone, ed.), AOCS Press, Champaign, IL, AOCS Official Method Ce 1d–91, 1994.
188. American Oil Chemists' Society (AOCS), *Official Methods and Recommended Practices*, 4th ed. (D. Firestone, ed.), AOCS Press, Champaign, IL, AOCS Official Method Ce 1b–89, 1994.
189. W. C. Willett and A. Ascherio, *Am. J. Public Health*, *84*: 722 (1994).
190. American Oil Chemists' Society (AOCS), *Official Methods and Recommended Practices*, 4th ed. (D. Firestone, ed.), AOCS Press, Champaign, IL, AOCS Official Method Ce 1c–89, 1994.
191. R. G. Jensen, in *Fatty Acids in Food and Their Health Implications* (C. K. Chow, ed.), Marcel Dekker, New York, 1992.
192. T. Jupille, *J. Chromatogr. Sci.*, *17*: 160 (1979).
193. S Lam and E. Grushka, *Sep. Pur. Meth.*, *14*: 67 (1985).

194. W. W. Christie, *J. High Res. Chromatogr., Chromatogr. Commun., 10*: 148 (1987).
195. W. W. Christie, E. Y. Brechany, and K. Stefanov, *Chem. Phys. Lipids, 46*: 127 (1988).
196. R. Battaglia and D. Frohlich, *Chromatographia, 13*: 428 (1980).
197. G. Schomburg and K. Zegarski, *J. Chromatogr., 114*: 174 (1975).
198. M. Narce, J. Gresti, and J. Bézard, *J. Chromatogr., 488*: 249 (1988).
199. Y. Hirata and E. Sumiya, *J. Chromatogr., 1267*: 125 (1983).
200. M. I. Aveldano, M. Van Rollins, and L. A. Horrocks, *J. Lipid Res., 24*: 83 (1983).
201. A. A. Andrianaivo-Rafehivola, J.-M. Cao, and E. M. Gaydou, *Rev. Fr. Corps Gras, 41*: 53 (1994).
202. E. M. Gaydou, J. P. Bianchini, and A. Ralaimanarivo, *Anal. Chem., 55*: 2313 (1983).
203. J. M. Cao, J.-P. Blond, and J. Bézard, *Biochim. Biophys. Acta, 1210*: 27 (1993).
204. H. W. S. Chan and G. Levett, *Lipids, 12*: 99 (1977).
205. Z. L. Bandi and E. S. Reynolds, *J. Chromatogr., 329*: 57 (1985).
206. H. Aoshima, *Anal. Biochem., 87*: 49 (1978).
207. O. N. Jensen and J. Moller, *Fette Seifen Anstrichm., 88*: 352 (1986).
208. L. G. Leeder and D. A. Clark, *Microchem. J., 12*: 396 (1967).

12

Analysis of Neutral Lipids: Triacylglycerols

Boukaré G. Semporé and Jean A. Bézard
Nutrition Research Unit, University of Burgundy, Dijon, France

I. EXPERIMENTAL METHODS

The development of analytical methods for the structural analysis of triacylglycerols (TAGs) from natural oils and fats starts with Chevreul (1915, 1923) (1,2), who identified natural fats as esters of glycerin and fatty acids, by crystallization and identification of the saponification products.

Later on, when glycerol was identified as a trialcohol [Berthelot, 1854 (3)] and when natural triacylglycerols were identified as mainly composed of triacid triacylglycerols [Pelouze and Boudet, 1838 (4); Bell, 1883 (5); Blyth and Robertson, 1889 (6); Heise, 1896 (7)], it appeared that oils and fats were very complex mixtures of triacylglycerols. To face this complexity, Hilditch and his collaborators (5) proposed to classify the triacylglycerol molecules according to their saturated (S) and unsaturated (U) fatty acid content. Between 1923 and 1950 they systematically applied the procedures of fractionated crystallization and of oxidative cleavage of fatty acid double bonds by potassium permanganate [Hilditch and Lea, 1927 (8)] to a large number of fats from vegetable and animal origin. Following this huge work, they reported the SSS, SSU, SUU, and UUU content of 1450 oils and fats starting from their fatty acid composition [Hildith and Williams, 1964 (9)]. It was the first comprehensive study on the triacylglycerol composition of natural fats.

The methods developed by Hilditch and his group were very time consuming (1–3 months for one sample). They required a high amount of fat (10–300 g) and results were at best semiquantitative. They were replaced from 1956 by chromatographic and enzymic methods, much less time consuming and less tedious, more accurate, and requiring only small amounts of fat (10–100 mg of fat were analyzed in a few hours).

The first method used was liquid–liquid partition chromatography, described independently by Dutton and Cannon (1956) (10), in the form of countercurrent distribution and by Priori (1956) (11), in the form of paper chromatography. In 1961 Huebner (12) demonstrated that gas–liquid chromatography (GLC) could be used to separate triacylglycerols in function of their molecular weight. Next year, De Vries (1962) (13) and Barrett et al. (1962) (14) developed silver ion thin-layer chromatography (TLC) for the separation of triacylglycerols in function of their degree of unsaturation.

The first work on the application of high-performance liquid chromatography (HPLC) in the reversed phase mode to the separation of simple medium-chain triacylglycerols was published by Pei et al. (1975) (15). From then on, considerable improvements have been brought about in the instrumentation and in the column technology. To-day this method is the one of choice in the analysis and fractionation of natural mixtures of triacylglycerols.

On the other hand, enzymic methods have been developed for the determination of the fatty acids located at the sn-1, sn-2, and sn-3 positions of the glycerol moiety and for the identification and quantification of all the triacylglycerol molecular species in natural mixtures.

From 1955, Mattson and Beck (1956) (16) and Savary and Desnuelle (1956) (17) developed an enzymic method, to identify the fatty acids esterified at the sn-2 position. This method was based on the specific property of pancreatic lipase to preferentially hydrolyze the ester linkages at the external sn-1 and sn-3 positions. Analysis of the 2-monoacyl-sn-glycerols formed permitted one to determine which fatty acids were located at the sn-2 position of the glycerol in triacylglycerols and in which proportion.

In 1965 Brockerhoff (18) proposed an enzymic method of stereospecific analysis of triacylglycerols, i.e., determination of fatty acids located at the stereospecifically distinct external sn-1 and sn-3 positions of the glycerol moiety. The method was based on the property of the snake venom phospholipase to specifically hydrolyze the fatty acid ester linkages at the sn-2 position of phospholipid-like molecules synthesized from diacylglycerols derived from partial chemical deacylation of triacylglycerols.

After 1985, two chromatographic methods have supplanted the Brockerhoff method for the identification of fatty acids in the sn-1 and sn-3 positions. They are based on the separation of diacylglycerol enantiomers by HPLC. The first method (Michelsen et al., 1985) (19) is based on the separation of diacylglycerol diastereoisomers on a silica column. The second method utilizes chiral stationary phases to separate urethane derivatives of diacylglycerols (Takagi and Itabashi, 1987) (20,21).

II. DETERMINATION OF TOTAL TRIACYLGLYCEROLS

After extraction of lipids from a food sample, the triacylglycerol fraction must be isolated by one of the methods described in a previous section. In general, for small size samples, adsorption TLC or HPLC is to be preferred. Column adsorption chromatography is suited for fractionation of large samples of triacylglycerols.

The second step is the quantitation of triacylglycerols. This can be made using several micromethods, each with advantages and disadvantages. The method of choice remains, in our opinion, the analysis by GC of the constituent fatty acids as methyl esters, if only long-chain fatty acids are present in the sample (otherwise as butyl esters), after addition of a known amount of an internal standard. Generally this standard will be an odd-numbered fatty acid (15:0, 17:0, or 19:0), absent, in most cases, in triacylglycerol samples. In the case of dairy products, triacylglycerols contain long-chain odd-numbered fatty acids together with short-chain fatty acids. The choice of shorter odd-numbered fatty acids, from 7:0, to 13:0, is recommended in that case.

For a better quantitation of the total triacylglycerols in a lipid extract, the internal standard can be added as triacylglycerol to the lipid extract (e.g., triheptadecanoin) before isolation of the triacylglycerol fraction. A possibly nonquantitative recovery of the triacylglycerol fraction from the lipid extract will not affect quantitation in that case. The weight of the methyl (or butyl) esters of the sample will be calculated from peak areas, or by weight percent of the fatty acid esters on the chromatogram, determined by an integrator-calculator, according to one of two formulas:

$$\text{Weight of sample esters} = \frac{\sum \text{corrected areas of sample esters}}{\text{Area of standard ester}} \times \text{weight of standard added}$$

$$\text{Weight of sample esters} = \frac{\sum \text{weight \% of sample esters}}{\text{weight \% of standard ester}} \times \text{weight of standard added}$$

The areas can be corrected after calibration of the detector response with quantitative standard mixtures of fatty acids.

From the fatty acid ester composition of the triacylglycerol sample, an average molecular weight (MW) of fatty acid esters is calculated by taking into account the MW of each component fatty acid ester and its percentage in the mixture. The MW of the free fatty acids is deduced from this value and utilized to calculate the average MW of triacylglycerols. The weight of triacylglycerols in the sample will be calculated using the formula:

$$\text{Weight of sample triacylglycerols} = \text{weight of sample esters} \times \frac{\text{average MW of triacylglycerols}}{3 \times \text{average MW of esters}}$$

To calculate the amount of sample triacylglycerols in mmol, the following formula is to be used:

$$\text{mmol of sample triacylglycerols} = \frac{\sum \text{mol \% esters}}{\text{mol \% standard}} \times \frac{\text{mmol standard}}{3}$$

If the fatty acid composition of the sample is not required, methods other than chromatographic may be more suited. In Table 1 (22) different methods are listed together with their sensitivity and their field of application.

III. DERIVATIZATION OF ACYLGLYCEROLS FOR STRUCTURAL ANALYSIS OF TRIACYLGLYCEROLS

A. Introduction

As a general rule, triacylglycerols are analyzed as unaltered molecules which can be collected for further investigations. However, triacylglycerol derivatives have sometimes been synthesized in order to improve chromatographic separations. Three functional groups in the triacylglycerol molecules can be used for preparation of derivatives: double bonds, ester linkage, functional groups such as hydroxy, epoxy, or keto. For example, double bonds have been used for hydrogenation, permanganate oxidation, ozonization, bromination, mercuration, etc.

Hydrogenation leads to saturated triacylglycerols; permanganate oxidation results in cleavage of double bonds to form triacylglycerols with dicarboxylic acids; ozonization, bromination, mercuration consist of the addition of ozone O_3, bromine, and mercuric acetate, respectively. Bromination of triacylglycerols changes their relative solubility and improves their separation characteristics. Ozonide derivatives of triacylglycerols can be separated by TLC on silicic acid according to their oxygen content. Mercuration increases the polarity of triacylglycerols proportionally to double-bond number and allows chromatographic separations according to this polarity.

Interesterification at the ester linkages, which gives rise to redistribution of acyl residues on the glycerol molecule, modifies the physical properties of triacylglycerols preserving their chemical, olfactory, and nutritional properties.

Triacylglycerols containing hydroxy, epoxy, or keto groups can be separated by adsorption chromatography without derivatization. However, specific derivatization can be necessary to stabilize the molecules or modify their physical properties. Litchfield (22) reviewed the reactions available for triacylglycerol derivative synthesis. He has described some of them and has evaluated their aptitude to quantitative analysis.

In the course of the structural analysis of triacylglycerols, nowadays the derivatization methods

Table 1   Micromethods for Quantitation of Triacylglycerols in a Food Sample

Methods	Approximate sensitivity (μg)	Observations	Ref.
Gas–Liquid Chromatography:			
Methyl esters Add internal standard; prepare fatty acids esters	0.1	Fastest method if fatty acid analyses are run by GLC	23, 24
Triacylglycerols Add internal standard	0.1	Fastest method if tracylglycerol carbon number distributions are determined by GLC	25, 26
Glycerol React with KOH or LiAlH$_4$. Recover glycerol, add internal standard such as butane-1,4-diol. Analyze free alcohols or their acetate or tri-methylsilyl ether derivatives	0.5	Slightly less accurate than GLC of methyl esters or triacylglycerols	27, 28
Spectrophotometry:			
Hydroxamic acid React with HONH$_2$HCl and ferric perchlorate to form ferric hydrox-amates. Measure absorption at 520 nm	20	Specific for ester linkages	29, 30
Chromotropic acid-acetylacetone-Phenylhydrazine Saponify. Acidulate. Convert glycerol to formaldehyde with NaIO$_4$ after: •Add NaAsO$_2$ or NaHSO$_3$, then chromotropic acid or acetylacetone. Heat and at last read at 570 nm for chromotropic acid or measure absorption or fluorescence at 405 nm for acetylacetone •Add phenylhydrazine, then potassium ferricyanide and HCl. Read color at 530 nm	5	Specific for 1,2-diols	31–34 35–37 38, 39
Glycerol kinase/glycerol dehydrogenase glycerol kinase/ pyruvate kinase/lactate dehydrogenase Saponify, acidulate, and remove fatty acids. Add buffer containing glycerol kinase, ATP, NAD, and MgCl$_2$ or buffer containing pyruvate kinase, lactate dehydrogenase, ATP, phos-phoenolpyruvate, NADH, and MgCl$_2$ •In the first case, incubate for 10 min at 30°C, then read absorption or fluorescence at 340 nm. Add glycerol dehydrogenase, wait 30 min, read again at 340 mn. Difference represents NADH from reduction of glycerol phosphate	5	Specific for glycerol	40–42 43–45

•In the second case, read absorption or fluorescence at 340 nm, add glycerol kinase, wait 10 min, read again at 340 nm. Difference represents NAD produced by sequence of enzymic reactions

Infrared	500	Specific for ester linkages; nondestructive	46

Measure ester bond absorption at 1745 cm^{-1} (5.73 μm)

(1) *3-Methylbenzothiazoline-2-One*; (2) *O-Aminophenol*; (3) *p-phenazobenzoyl chloride*; (4) *copper soaps*; (5) *dichromate/H_2SO_4 oxidation*; (6) *sulfophosphovanillin reaction*	2 or 5	Specific for 1,2-diols (1); glycerol (2); hydrophobic acids, aldehydes, and alcohols (3); hydrophobic acids produced by saponification of TG (4); all nonvolatile organic material (5); most lipids, but color intensity varies with fatty acid composition (6).	47–52

Titration:

Cerium perchlorate periodate	500	Specific for all alcohols (1) and 1,2-diols (2)	53, 54

Saponify, acidulate, and extract fatty acids.
•Titrate glycerol with cerium perchlorate (1).
•Add excess $NaIO_4$ to glycerol and back-titrate with $NaAsO_2$ (2).

KOH or NaOH	500	First method detects all acids. Second method specific for hydrophobic acids	55, 56

Saponify, then titrate excess KOH with HCl. Alternatively, saponify, acidulate, and extract fatty acids; titrate fatty acids with NaOH

Thin-Layer Chromatography:

Densitometry-fluorometry-GLC detector-spot area	1	Sample hydrogenation before analysis increases accuracy for densitometry and fluorometry. Measure of spot areas is less accurate than densitometry	57, 58, 59, 60, 61, 62, 63

•Spray plate with 50% H_2SO_4, 70% H_2SO_4 saturated with $K_2Cr_2O_7$, or other charring agent. Heat at 200°C and quantitate spots with densitometer
•Spray plate with rhodamine 6G and quantitate spots with scanning fluorometer
•Measure pyrolysis or combustion products with GLC detector
•Measure spot areas.

Radioactivity:

Transesterify with [^{3}H] or [^{14}C] methanol. Alternatively, saponify,acidulate, isolate acids, and react with ^{14}C-diazomethane, ^{63}Ni(NO_3)$_2$, or ^{60}Co(NO_3)$_2$. Isolate methyl esters and measure radioactivity	0.1	Inorganic salts not suitable for <C14 acids	64, 65

Table 1   (*continued*)

Methods	Approximate sensitivity (μg)	Observations	Ref.
Other Methods:			
Automatic column chromatography with: Refractive index detector (1) GLC detector (2)	20–100 (1) <1 (2).		66–68
Weight	500	Mostly for column chromatography work. Seldom used with TLC. Nondestructive	69–71

Source: Data from ref. 22.

are applied especially to monoacylglycerols and diacylglycerols in order to obtain more stable compounds and sometimes fluorescent or UV-absorbing derivatives. These derivatives simplify detection and quantification in HPLC analysis.

B.   Derivatization of Monoacylglycerols and Diacylglycerols

The most utilized derivatives of monoacylglycerols and diacylglycerols in the chromatographic analyses are the acetate and trimethylsilyl (TMS) ether derivatives. Many other derivatives have been prepared especially for reverse phase and chiral phase HPLC analyses. The preparation of the most important of them will be described.

1.   Acetate Derivatives of Monoacylglycerols and Diacylglycerols

The acetate derivatives preparation consists of blocking the free hydroxyl groups of mono- or diacylglycerols with an acetate group. The following procedure frequently used (72) is from Kuksis and Marai (73) and Kuksis et al. (74).

First 0.5 ml of a mixture of acetic anhydride and pyridine (10:1, v/v) is added to about 1 mg of lipids in a tube. The reaction mixture is heated at 60°C for 2 hr or left 12 hr at room temperature. To the medium is then added 1 ml of hexane and 1 ml of distilled water; the mixture is shaken and then centrifuged. The upper phase containing the mono- or diacylglycerol acetates is taken up and the lower phase is reextracted twice with hexane. Hexane is evaporated under nitrogen. The acetate derivatives are dried with absolute alcohol and dissolved in chloroform. If necessary, they can be purified by TLC on silicic acid using hexane–diethyl ether (80:20, v/v) or hexane–diethyl ether–acetic acid-methanol (90:20:2:3, in volumes) as developing solvent.

Acetate derivatives of monoacylglycerols were analyzed by GLC (75–77); diacylglycerol acetates were fractionated by argentation TLC (72–74,78–80), GLC (74,75,81,82), and HPLC (83–85).

2.   Trimethylsilyl Ether Derivatives

The trimethylsilyl (TMS) ether derivatives of mono- and diacylglycerols are more volatile but less stable than the acetate derivatives. They are slowly hydrolyzed on silica gel layers. The most frequently used reagent for preparing these derivatives is a mixture of hexamethyldisilazane-trimethylchlorosilane-pyridine (3:1:10, by volumes). It is used as follows (75,86):

To up to 10 mg of lipids is added 0.5 ml of pyridine, 0.15 ml of hexamethyldisilazane, and 0.05 ml of trimethylchlorosilane. The mixture is shaken for 30 sec; 5 min after it can be directly analyzed by GLC.

Alternatively, the mixture can be evaporated to dryness and the TMS ether derivatives redissolved in 5 ml of hexane. The solution obtained is washed with 1 ml of water, then dried through anhydrous sodium sulfate. Solvent is evaporated under nitrogen and the derivatives are redissolved in fresh hexane. They are stable at –20°C and can be stored for long periods.

Nowadays a great number of silylation reagents are commercially available. The most frequently used in lipid analysis is bis(trimethylsilyl)acetamide, or BSA. To 100 mg of lipids is added 100 hl of a mixture of heptane-BSA (10:1, v/v). After about 10 min, the mixture can be immediately analyzed by GLC (87). The derivatives can also be extracted as above.

Etherification with *t*-butyldimethylsilyl (t-BDMS) chloride (88) leads to derivatives 10^4 times as stable as TMS ether derivatives. The silylation reagent is a mixture of *t*-BDMS chloride (1 mmol), imidazole (2 mmol), and *N,N*-dimethylformamide (10 ml). Up to 10 mg of lipids is mixed with 0.5 ml of reagent. The mixture is heated at 60°C for 30 min. After cooling, 5 ml of hexane is added and the solution washed three times with 1 ml of water. It is then dried through anhydrous sodium sulfate, filtered, or decanted. The silylated derivatives are obtained after solvent evaporation.

Preparation and properties of TMS ether derivatives and related compounds are described elsewhere (89–91). These derivatives are used for GLC (87,92–94), TLC-AgNO3 (95), mass spectrometry (96–99), and HPLC (100,101) analyses.

It must be noted that monoacylglycerols have been transformed into trifluoroacetate (TFA) derivatives for GLC analyses (102). However these derivatives are rapidly hydrolyzed and render chromatographic columns unsuitable for other analyses. They also lead to losses of polyunsaturated compounds (103).

3. *Monoacylglycerol and Diacylglycerol Derivatives for HPLC Analyses*

Monoacylglycerols and diacylglycerols have been analyzed as underivatized compounds by HPLC (104–108). However, taking into account that they are better separated and sometimes detected and quantified as derivatives, numerous derivatives have been prepared essentially for the structural analyses of glycerophospholipids.

Thus monoacylglycerols have been analyzed as diacetate derivatives (109), *p*-nitrobenzoate derivatives (110), benzoate derivatives (111), urethane or 3,5-dinitrophenylisocyanate derivatives (108). Likewise, diacylglycerols have been separated using acetate derivatives (84,85), *tert*-butyldimethylsilyl ether derivatives (100,101,112), *p*-nitrobenzoate derivatives (110,113), *p*-anisoyl derivatives (114), dinitrobenzoate derivatives (115,116), benzoate derivatives (111,112,117–119), naphthylurethane derivatives (120–123), and dansylethanolamine phosphate derivatives (124,125).

In this chapter monoacylglycerols and diacylglycerols are considered as intermediaries in the structural analysis of triacylglycerols. The preparation of derivatives for stereospecific analysis of triacylglycerols will only be described.

a.   Preparation of Urethane Derivatives of Monoacylglycerols.   Monoacylglycerol 3,5-dinitrophenylurethane (DNPU) or dinitrophenylisocyanate derivatives are prepared according to the procedure described by Oi and Kitahara for derivatization of chiral alcohols (126) and adapted to monoacylglycerols by Takagi and Itabashi (20,127).

Twenty millimoles (~7 mg) of monoacylglycerols is dissolved in 450 µl of dry toluene in a 0.5-ml glass vial with Teflon-linked screwcap. To this solution is added 45 mmol (~10 mg) of 3,5-dinitrophenylisocyanate powder and 45 µl of dry pyridine. The mixture is heated at 70°C for 1 hr in an oven or left for 3 hr at room temperature, with occasional shaking. At the end of the reaction, the solution is cooled and the solvent removed under nitrogen. The resulting urethane derivatives are dissolved in 0.2 ml of chloroform and purified by TLC on silica gel 60 F254–precoated plates (20 × 20 cm, 0.25 mm thick) from Merck. The plates containing a fluorescence indicator are previously activated for 1 hr at 110–120°C in an oven. They are developed using a mixture of hexane–ethylene dichloride (or dichloromethane)–ethanol (40:10:3, v/v/v). The plates are dried under nitrogen. The monoacylglycerol derivatives are revealed under UV light (254 nm), delineated, and corresponding silica gel is scraped off the plate. The urethane derivatives are extracted from the adsorbant with diethyl ether.

Alternatively, the crude urethane derivatives of monoacylglycerols are purified by reverse phase HPLC instead of TLC (108). In this case, at the end of the derivatization reaction the mixture is left to decant. The clear upper phase is filtered through hyperfine glass wool. The solvent is evaporated under nitrogen and the monoacylglycerol derivatives are dissolved in chloroform for storage or in acetonitrile (or in the mixture corresponding to the HPLC mobile phase) for fractionation.

b. Preparation of 3,5-Dinitrophenylurethane (DNPU) or Isocyanate Derivatives of Diacylglycerols. This procedure is also derived from that employed by Oi and Kitahara (126) and adapted to diacylglycerols by Itabashi and Takagi (21,128).

Amounts of 1–5 mmol (0.6–3 mg) of diacylglycerols are dissolved in 4 ml of dry toluene in an 8-ml glass tube with a PTFE-lined screwcap. Ten to fifty millimoles (2–10 mg) of 3,5-dinitrophenylisocyanate (Sumitomo, Japan) and 40 µl of dry pyridine is added to the solution and the mixture left for 1 hr at ambient temperature with occasional shaking. At the end of the reaction, the solvent is removed under nitrogen and the residue is dissolved in 0.2 ml of chloroform. The diacylglycerol derivatives are isolated by TLC on silica gel 60 F254–precoated plates (Merck). The plates containing a fluorescence indicator were previously activated as above. The developing solvent is hexane–ethylene dichloride–ethanol (40:10:3, v/v/v). The remainder of the procedure, including the purification step by HPLC, is identical to that used for monoacylglycerol DNPU derivatives preparation.

c. Preparation of Naphthylethylurethane Derivatives of Diacylglycerols. The following procedure is that described by Laakso and Christie (129). One to two milligrams of diacylglycerols dissolved in 300 µl of dry toluene are reacted with 10 µl of (R)- or (S)-1-(1-naphthyl) ethylisocyanate in the presence of approximately 10 mmol of 4-pyrrolidinopyridine. The reaction mixture is heated at 50°C overnight. The products are extracted with hexane–diethyl ether (1:1, v/v) and washed with 2 N HCl and water. The organic layer is taken to dryness under nitrogen and the derivatives are purified on a short column of Florisil eluted with diethyl ether.

IV. STRUCTURAL ANALYSIS OF TRIACYLGLYCEROLS

A. Introduction

Natural oils and fats are very complex mixtures of triacylglycerol molecular species. Up to now, no physical or chemical method is able to separate all of the molecules present in these mixtures in only one step. Structural analysis must consequently proceed through several steps.

In a first step, triacylglycerols can be characterized by their component fatty acids, whatever their position at the 3 positions of the glycerol moiety. Such a "group" of triacylglycerols, more simply denominated "triacylglycerols," can be identified by using one chromatographic technique, or more often, a combination of two techniques. These chromatographic techniques will be described in the first section.

When such groups have been isolated and collected, study of the position of the constituent fatty acids at the 3 positions of the glycerol moiety can be undertaken. The study generally starts with the determination of fatty acids esterified at the *sn*-2 position and terminates with the determination of those esterified at the *sn*-1 and *sn*-3 positions. The techniques used for this purpose will be described in the second section.

Theoretically, the successive analyses result in the identification and quantitation of all of the molecular species of triacylglycerols present in the mixture.

Each group of triacylglycerols comprises six molecular species if the three constituent fatty acids are different. If total triacylglycerols contain N different fatty acids, the number of possible molecular species is N^3 while the number of "groups" in which the position of the constituent fatty acids is unspecified is only $(N^3 + 3N^2 + 2N)/6$. Their study is therefore easier.

B. Study of the Constituent Fatty Acids of Triacylglycerols

Several chromatographic methods can be used, alone or in combination, to identify and quantify the different groups of triacylglycerols present in a mixture.

1. Gas–Liquid Chromatography

Gas–liquid chromatography (GLC, or GC) was the first technique used to separate triacylglycerols as intact molecules, first by means of packed columns and now by means of the very efficient glass capillary columns. Several reviews have dealt with the method and its applications (22,75,102, 130–133).

a. Instrumentation.   The apparatus to be used in the GC analysis of triacylglycerols must be adapted to the separation of high-boiling-point compounds, i.e.:

An oven of which the temperature can be programmed and reach as high temperatures as 350–400°C
An on-column injector
A flame ionization detector or a "light-scattering" detector

b. Columns.   Two types of columns are used to separate triacylglycerols, packed columns, and wall-coated open tubular (WCOT) columns. *Packed columns* are made of glass or stainless steel, U-shaped in the past but now coiled—a form which is better adapted to small-volume ovens and to temperature programming. Both materials afford satisfactory results. However, Litchfield et al. (134) showed that glass columns were more efficient in the separation of such molecules. When used for analytical purposes, the internal diameter of the column generally is 2.5–3.5 mm while, according to sample, its length comprises 0.5–1 m not to give too high retention times.

The inert support, commercialized as Gas-Chrom Q, Chromosorb W, or Supelcoport consists of diatomaceous earth, deactivated by silanization of the silica active sites (22,135,136). For good resolutions, the particle size must be between very narrow limits, most commonly 60–80, 80–100, or 100–120 mesh. The peak resolution increases with decreasing particle size, but retention times are also increasing. A good equilibrium between column efficiency (depending on column length), particle size, other factors, and acceptable retention times must be achieved.

The support particles are impregnated by low levels of liquid phase, normally 1–3%, not to have high retention times while keeping good resolutions. The most frequently used stationary phases are silicone polymers of high thermal stability. They are, according to increasing polarity, OV-1 (137,138), OV-101 (138), JXR (134,140), SE-30 (141,142), Dexil-300 (143), OV-17 (144,145). The procedure for impregnating the support with the liquid phase is similar to that described in the fatty acid analysis section, recommended by Horning (146).

The column is packed as also described above for fatty acids and conditioned for several hours, generally 4 hr, at a temperature of 25°C higher than the utilization temperature to gain a good baseline stability. In the analysis of triacylglycerols, programming temperature is recommended, linearly from 180°C to 350°C for short columns at a rate of 2–5°C/min. Kuksis (147) prefers using nonlinear rate of temperature programming to improve peak resolution. For good resolution, the starting programming temperature must be 25–50°C lower than the ebullition point of the most volatile triacylglycerol.

With regard to the carrier gas, Litchfield et al. (134) showed that helium significantly improved peak resolution when compared to nitrogen, as predicted by the Van Deemter equations (148). However, nitrogen remains the most used carrier gas because of its lower cost. The carrier gas flow rate depends on the column size, the level of stationary phase fixed on the support, and the support particle size but is normally close to 100 ml/min. To obtain good separations, the sample, in solution in diethyl ether or hexane, or any other solvent, at 1–5% concentration, is deposited onto the head of the column, which is maintained at a temperature 40°C lower than the elution temperature of the most volatile triacylglycerol. The sample size is such that the amount of the major compound in the mixture is ~20 mg. According to Litchfield (22), the optimal operating conditions in the GC analysis of triacylglycerols from C30 to C60 (total acyl carbon number) are those reported in Table 2.

Several Offical Methods have been published for the GC analysis of triacylglycerols on packed columns (IUPAC-AOAC, 986-19; AOCS, Ce 5-86) (149,150). The specifications concerning the apparatus, the reagents, and the operating conditions are slightly but not basically different from those proposed by Litchfield (22) and reported in Table 2.

Figure 1 illustrates triacylglyerol separations obtained on a nonpolar packed column (151). In part A are short- and medium-chain triacylglycerols of coconut oil containing nearly 50% of lauric acid (12:0), 13% of caprylic acid (8:0), and 14% of myristic acid (14:0). Tritridecanoin (C39) used as internal standard to identify the peaks by their carbon number and to quantitate the total triacylglycerols in the sample was separated from the two adjacent peaks. Part B shows the medium-chain triacylglycerols of adipose tissue from rats fed coconut oil. The medium-chain fatty acids of the oil are associated with the usual long-chain fatty acids in the triacylglycerol molecules. These long-chain fatty acids are associated in adipose tissue triacylglycerols (part C) from rats fed their usual diet. The three major fatty acids (16:0, 18:1, and 18:2) were present in triacylglycerols of peak 52 (152).

In the most common natural oils and fats, the chromatogram obtained is very similar to that reported in part C of Fig. 1, since the major fatty acid chain length is 16 or 18. With the column (5 ft × 1/8 in. i.d.) the separations are achieved generally within 30 min, which is a reasonable time for routine analysis on packed columns.

Table 2 Optimal Operating Conditions in the GC Analysis of Triacylglycerols from C30 to C60[a]

Gas chromatograph	Instrument with on-column injector, equipped with flame ionization detector and temperature programming
Column	0.55 m × 2.5 mm i.d., glass or stainless steel
Packing	3.0% OV-1 on 100–120 mesh Gas-Chrom Q
Column conditioning	2–8 hr at 350°C
Carrier gas	Helium at 100 ml/min
Column temperature	170–350°C at 2–4°C/min
Flash heater temperature	300–350°C
Detector buse temperature	300–340°C
Sample size	10–20 mg triacylglycerol for each full-scale peak on a 1-mV recorder

[a]Total acyl carbon number.
Source: Data from Ref. 22.

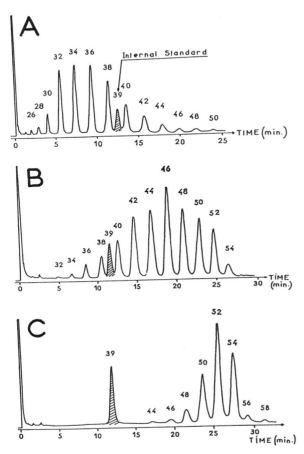

Fig. 1 Chromatograms recorded in the analysis by packed column gas chromatography of triacylglycerols from coconut oil (A), from rate adipose tissue after 6 weeks on coconut oil diet (B), from rat adipose tissue on chow diet (C). The triacylglycerol peaks are denoted by their total acyl carbon atoms (carbon numbers). Operating conditions: 5 ft × 1/8 in. o.d. stainless steel coiled column containing 1% JXR on 100/120 mesh Gas-ChromQ; temperature linearly programmed from 250°C to 360°C at a rate of 3°C/min; nitrogen, as carrier gas, at a constant flow rate of 50 ml/min; detection by flame ionization. The analytical column bleeding was compensated by a second similar column operating in the same conditions (dual column analysis). (Reproduced from Ref. 151.)

When high resolution is required in the GC analysis of triacylglycerols, *capillary columns* are preferred to packed columns. First made of glass (153), they are now in fused silica with the stationary phase chemically bonded to the inner wall and are thereby very stable to high temperatures. Their length usually varies from 5 to 25 m and the internal diameter from 0.20 to 0.32 mm depending on the efficiency desired. The stationary phase thickness is always very low, i.e., 0.10 to 0.12 mm.

The first stationary phases used were apolar, such as OV-101 (154) or Dexsil-300 (153). Later on, polar phases containing phenyl groups, such as the methylphenyl silicon polymer RSL-300 (155,156), were prepared and are now commercially available. The recommended instrumentation and operating conditions are as follows: The injection system is crucial. For accurate results, a cold on-column injector is necessary (102,157–160). As for packed columns, analyses must be carried out at programmed temperature, linearly or better, when possible, at a nonlinear (concave) rate to improve column efficiency. The initial and final temperatures of programming, as well as the rate of programming, have to be experimentally optimized according to the sample to be analyzed. Although helium can be used as carrier gas with capillary columns, hydrogen must be preferred, contrary to what was observed with packed columns. Hydrogen permits the elution of components at lower temperatures or with shorter retention times, thereby increasing the column working life (102).

Application of capillary columns in GC separations of triacylglycerols have been described (161,162). Figure 2 shows what kind of separation can be achieved on a medium-length (25-m) fused silica column but coated with a comparatively high polar phase in the analysis of butter oil triacylglycerols, which constitutes one of the most complex natural mixtures. A great number of triacylglycerols eluted together as a single peak on a packed column (Fig. 1) and exhibiting the same carbon number can be separated on a capillary column. For example, more than 10 triacylglycerols of carbon number 38 were totally or partially separated on this type of column. These peaks cannot be easily identified. However, several features have been previously established although on less efficient columns (163,164). For example, within a given carbon number, butyrates (4XX) were preceded by caproates (6XX), caprylates (8XX), and other triacylglycerols containing exclusively longer chain fatty acids. Separation also occurs according to unsaturation on this polar capillary column, saturated triacylglycerols being eluted ahead of unsaturated compounds. The use of such columns of very high efficiency must be reserved for the moment to establish the triacylglycerol profile of natural mixtures. To date a compositional determination cannot be considered since peak identification is very uneasy. On the other hand, this high efficiency can be very useful in the structural analysis of triacylglycerols, when utilized in the analyses of simpler fractions isolated by other means, such as reverse phase HPLC or argentation-HPLC. The combination of HPLC fractionation and GC analysis of butter oil triacylglycerols has proved to be very efficient in the determination of the molecular species of butter oil triacylglycerols (165).

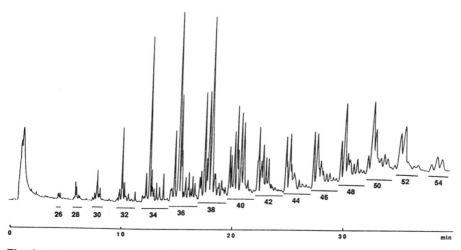

Fig. 2 Chromatogram registered in the analysis of butter oil triacylglycerols by capillary gas chromatography on a fused silica column (25 m × 0.25 mm i.d.) coated with TAP (0.1 μm thick). The temperature was programmed from 260°C to 360°C at 3°C/min. The carrier gas was hydrogen at 1 bar inlet pressure. The on-column injector was at a temperature of 380°C and the flame ionization detector at the same temperature. Triacylglycerols eluted inside groups denoted by their carbon numbers. (Analysis courtesy of J. Gresti.)

c. Identification and Quantitation. The triacylglycerols separated by GC are generally referred to in a simple nomenclature, by their total acyl carbon atoms, whatever the unsaturation, e.g., C54 for tristearin or triolein. This value is called the "carbon number." Natural triacylglycerol mixtures usually comprise homologous series of components, of which each term differs from the previous one from two carbon atoms. On chromatograms, triacylglycerols appear as regularly spaced peaks, each differing from the previous one by a two-carbon atom increment (Fig. 1).

To identify triacylglycerols on a chromatogram, a first method consists of analyzing a given amount of the sample and then on the same amount added with known standard triacylglycerols. The height of certain peaks is increased corresponding to the known added triacylglycerols. The other ones are identified by comparison with them (22). When triacylglycerols are analyzed in isothermal conditions, as fatty acids esters (Fig. 4, Chap. 11), a linear relationship exists between the logarithm of the retention time (or volume) and the total number of carbon atoms in a homologous series of triacylglycerols (e.g., triacylglycerols containing only straight-chain saturated fatty acids). From this relationship the identity of unknown triacylglycerols can be graphically deduced or their equivalent chain length (ECL) determined. However, since temperature programming is generally used, identification of the analyzed triacylglycerols from their retention time or volume is replaced by identification from the relative elution temperature (RET) (166), which is the ratio of the elution temperature of the unknown triacylglycerol to that of a known (standard) triacylglycerol (generally tripalmitin: C48). The RET is practically independent of the initial temperature and of the rate of temperature programming. It was found to be fairly reproducible for a triacylglycerol mixture from C6 to C54 analyzed in the presence of a silicone phase (167).

When the problem of peak identification on the chromatogram is solved, the next problem raised in GC analysis of triacylglycerols is that of *quantitation.* Several factors are likely to influence the quantitative detection of triacylglycerols. The first one is the detection response itself to increasing amounts of triacylglycerols emerging from the column. Using a flame ionization detector it has been shown with a mixture of triacylglycerols of C6 to C66 that the detector response was linearly related to the amount of triacylglycerols, up to 20 mg, except for triacetin (C6) and tripropionin (C9) (134,167,168). The second factor likely to influence the quantitative analysis of triacylglycerols is their retention on the column. It is supposed that the injection conditions are correctly selected (134). Under these conditions, the level of retention depends on the amount of triacylglycerol injected on the column, molecular weight, and the unsaturation of the varied triacylglycerols.

Different methods can be used to check the influence of these factors on the sample recovery at the outlet of the column and to determine correction factors to be applied to peak areas for quantitative results to be calculated.

The following method has been used to check the influence of the amount injected and of the unsaturation for four C54 triacylglycerols (152), as illustrated in Fig. 3.

To a given amount (0.5 mg) of tripalmitin (C48) was added increasing amounts (from 0.25 to 2.5 mg) of tristearin (000 in Fig. 3) or triolein (111) or trilinolein (222) or trilinolenin (333). A sample of each triacylglycerol mixture, such as each time it contained the same amount of C48 (4 mg), was injected onto the column. Thus, for a given C54 triacylglycerol, depending of the mixture analyzed the amount of C54 injected varied from 2 to 20 mg compared to 4 mg of C48.

Weight correction factors (fw) were calculated by comparing the C54 and C48 peak areas and the C54 and C48 amounts injected, using the formula

$$\frac{\text{Area C54} \times \text{ fw}}{\text{Area C48}} = \frac{\text{amount C54}}{\text{amount C48}}$$

or

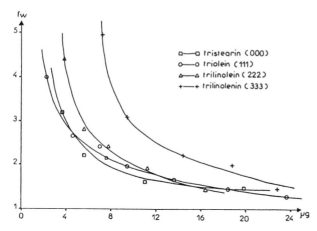

Fig. 3 Relative weight correction factors (fw) vs. amount injected (µg) of four simple triacylglycerols with same carbon number (54) and increasing number of double bonds (0–9), namely, tristearoyl-, trioleoyl-, trilinoleoyl-, and trilinolenoylglycerol. Tripalmitoylglycerol (48) present in the mixtures analyzed has been assigned an fw of 1. Operating conditions: stainless steel coiled column (80 cm × 1/8 in. o.d.) packed with 1% JXR on 100–120 Gas-Chrom Q; temperature- programmed from 240°C to 340°C at 3°C/min with a constant nitrogen flow of 45 ml/min; injector temperature 325°C; detection by flame ionization. (From Ref. 152.)

$$fw = \frac{\text{amount C54}}{\text{area C54}} \times \frac{\text{area C48}}{\text{amount C48}}$$

The second term of the equation (area C48/amount C48) was taken as 1, so that fw was expressed comparatively to C48 taken as unity. Results obtained, illustrated in Fig. 3, show that when low amounts of C54 triacylglycerol were injected, sample recovery at the outlet of the column was low, since the observed fw values were high. The losses on the column were higher for the most unsaturated triacylglycerol, i.e., trilinolenin (333). For the other three triacylglycerols, fw was relatively constant from 12 mg injected and slightly higher than 1, i.e., the fw of tripalmitin (C48).

The reason for the high fw when the C54 triacylglycerols were injected in low amounts could be that a somewhat constant amount was lost on the column, much more sensitive when taken up from a low injected amount.

This peculiarity was not observed with medium-chain triacylglycerols such as those found in coconut oil (169) and palm kernel oil (170) and with synthetic mixture of triacylglycerols from C30 to C42 (167).

To check the detector response according to the chain length of triacylglycerols, generally quantitative standard mixtures of simple saturated triacylglycerols were used from C24 to C54 (167) or from C24 to C66 (134). Such commercially available mixtures can also be used now. The sample size must be selected so that the amount of individual triacylglycerols injected is in the linear part of the curve of the detector response (Figure 3). The peak areas are now better measured by electronic integration, more accurately than by the triangulation method. Quantitative weight (fw) and molar (fm) calibration factors for individual triacylglycerol are calculated by the internal normalization method using the formulas

$$fw = \frac{\text{weight \%}}{\text{area \%}}$$

and

$$fw = \frac{mol \%}{area \%}$$

A value of 1.00 is assigned to fw and fm for a given triacylglycerol of intermediary molecular weight, generally trilaurin (C36) or tripalmitin (C48). From the data obtained a curve can be traced for fw and fm. Such curves, obtained by Litchfield et al. (134), are reported in Fig. 4. When compared to C36, the lower triacylglycerols showed a slightly higher fw. When the carbon number increased from C48 onward, fw increased rapidly. Similar results were obtained by Watts and Dils (167). Fw was shown to increase with the column chain length (134) especially for long-chain triacylglycerols because of longer retention volumes causing higher losses on the column. From the fw curve established for simple saturated triacylglycerols, fw values for mixed saturated triacylglycerols of intermediary carbon number are determined graphically. For unsaturated triacylglycerols the same internal normalization method can be applied to commercialized simple triacylglycerols such as those reported in Fig. 3, and the data so obtained for fw reported on the same graph, for each carbon number, like the fw of saturated simple triacylglycerols. For triacylglycerols of a given carbon number and for intermediary unsaturation, fw can be graphically deduced from the curves on the graph.

In the quantitative analysis of an unknown mixture on the same column, peak area of triacylglycerol of given carbon number and unsaturation must be multiplied by the corresponding fw (or fm) and that for each triacylglycerol before calculating the composition in weight % or mol %. This method of internal standardization when carefully applied gives very reproducible results and the accuracy is within 0.2–1.7% for C50 to C64 triacylglycerols (22). It is evident that the fw calculated are suited for analyses carried out on the same column. If another packed column is to be used, new fw have to be determined for this column.

With a capillary column coated with a nonpolar phase accurate determination of triacylglycerol composition depends on many factors (171). Determination of calibration factors is essential, more particularly with high molecular weight triacylglycerols such as triarachidin (C60), and for each column utilized. By optimizing the analysis operating conditions and the standardization, a very good reproducibility can be obtained.

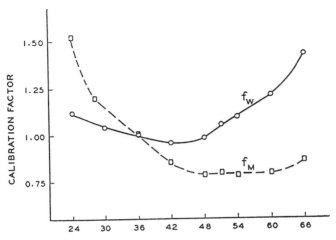

Fig. 4   Relative weight (fw) or molar (fm) calibration factors vs. carbon number for monoacid triacylglycerols. A value of 1.00 was assigned for trilaurin (carbon number 36). Operating conditions: 61 cm × 2.5–3.0 mm i.d. glass or stainless steel column packed with 3.0% JXR on 100/120 mesh Gas-Chrom Q. The column was programmed from 170°C to 325°C at 2-4°C/min with 100 ml/min of helium carrier gas flow. The flash heater was maintained at 320–325°C and the flame ionization detector at 300-340°C. (From Ref. 134.)

With capillary columns coated with a polar phase, accurate quantitative analyses are likely to be performed, following a careful standardization, even for high molecular weight triacylglycerols, with the exception of highly unsaturated oils such as marine oils (171,172). Geeraert (173) even reports that analysis of the most common oils, which do not comprise highly unsaturated triacylglycerols, such as soybean oil, butter oil, or cocoa butter, does not require determination of calibration factors. However, this interesting property should be checked before using a new column. If some loss of the higher triacylglycerols is observed to occur, calibration factors must be determined using the same procedure as for packed columns.

d. Application to Analysis of Triacylglycerol Mixtures

SEPARATION ON THE BASIS OF MOLECULAR WEIGHT Classically, gas–liquid chromatography has been utilized with nonpolar liquid phases such as JXR, OV-1, OV-101, SE-30, and DEXIL-300 for separation of intact triacylglycerols of various dietary products on the basis of molecular weight. The analyses have been performed both on packed conventional nonpolar columns (151,169,170,173–189) and on WCOT capillary nonpolar columns (157,164,172,190–197,199, 200). Thus natural oil triacylglycerols with short-, medium-, and long-chain fatty acids have been analyzed and the higher the number of fatty acids the greater the number of triacylglycerol peaks.

The separated peaks were identified by the total acyl carbon atoms. For example, the triacyl-glycerol C48 will be constituted of the triacylglycerol molecules containing 48 acyl carbon atoms, e.g., 16:0 16:0 16:0, 14:0 16:0 18:0, and 12:0 18:1 18:1.

On *packed columns* of 0.45–0.60 m long or more, mixtures of triacylglycerols differing by two carbon atoms are easily separated (168). For example (Fig. 5), the triacylglycerols of palm kernel oil (178, 179) containing short-, medium-, and long-chain fatty acids (12:0, 52%; 14:0, 16%; 18:1, 11%; 16:0, 7%; 8:0, 6%; and 10:0, 5%) were separated into 14 groups or peaks of triacylglycerols from C28 to C54 on a stainless steel column (120 × 0.3 cm i.d.) packed with JXR (2% on Gas-Chrom Q 100–120 mesh) with linear temperature programming from 220°C to 320°C at 3°C min, and nitrogen at 45 ml/min as carrier gas. The main peaks of triacylglycerol are C36, 27%; C38, 18%; C34, 11%; C40, 10%; C32, 9%; C42, 8%; C44, 5%.

On the other hand, the triacylglycerols from peanut oil of unsaturation 011 (Fig. 6) with

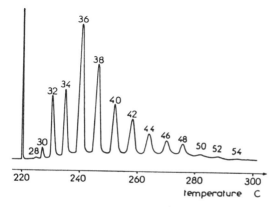

Fig. 5 GC elution pattern of palm kernel oil triacylglycerols. The triacylglycerols are denoted by their carbon number (total acyl carbon atoms). Operating conditions: 2.5 ft × 1/8 in. o.d. stainless steel column packed with 1% JXR on 100/120 mesh Gas-Chrom Q; linear temperature programming from = 220–320°C at 5°C/min; nitrogen flow, 90ml/min; injector temperature, 375°C; detection by flame ionization. (From Ref. 170.)

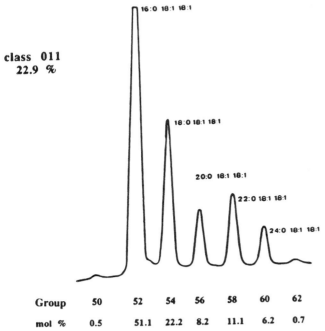

Fig. 6 Chromatogram recorded from gas chromatography of the peanut oil class 011 (22.9% of the total oil triacylglycerols) triacylglycerol fractionation by argentation-TLC under conditions reported in Fig. 7. Each peak corresponds to a group of triacylglycerols of same carbon number (total number of acyl carbon atoms) from 50 to 62. Their proportion (mol% of the total class 011 triacylglycerols) is indicated in the last line. Group 62 corresponds to 26:0 18:1 18:1 later identified. Operating conditions: 60 cm × 1/8 in. o.d. glass column packed with 100/120 mesh Gas-Chrom Q coated with 3% JXR; temperature programming from 220°C to 380°C at 4°C/min; injector temperature, 350°C; temperature of the flame ionization detector, 320°C. (From Ref. 180.)

very-long-chain fatty acids from 20 to 24 carbon atoms and as major fatty acids—18:1, 60%; 18:2, 18%; 16:0, 11%—were fractionated into C52 to C64 triacylglycerols (C54, 60%; C52, 23%; C56, 7%; and C58, 5%) (180). The analyses were carried out on a glass column (60 cm × 2 mm i.d., packed with JXR, 3% on Gas-Chrom Q 100–120 mesh; linear temperature programming from 220 to 380°C at 4°C/min; helium at 100 ml/min as carrier gas). Triacylglycerols differing by only one carbon atom have been separated on longer columns of about 1.83 m (168,176).

In principle, *capillary columns* coated with nonpolar liquid phases, 6 m long or less, provide the same separations as packed columns mentioned above, with separation according to carbon number (157,185,192–200).

With longer capillary columns, triacylglycerols having the same number of carbon atoms can be more or less partially resolved according to double-bond number or chain length (154,164,204–206) as also illustrated in Fig. 2.

SEPARATION OF TRIACYLGLYCEROLS ACCORDING TO NUMBER OF CARBON ATOMS AND DOUBLE BONDS These separations were obtained with *packed or capillary columns* in the presence of polar stationary phases (154,201–206). For a given carbon number, the polar liquid phases lead to separations according to number of double bonds (155,156,209). For example, triacylglycerols C52 and C54 of palm oil have been subdivided into seven and six fractions, respectively, with

RSL-300 fused silica column (25 m × 0.25 mm i.d., coated with a polymer of phenylmethylsilicon as stationary phase). The elution order observed for TG C54 on such liquid phase was as follows: 18:0 18:0 18:0 < 18:0 18:0 18:1 < 18:0 18:0 18:2 < 18:1 18:1 18:1 < 18:0 18:2 18:1 < 18:1 18:1 18:2 < 18:0 18:2 18:2 < 18:1 18:2 18:2 < 18:2 18:2 18:2 < 18:2 18:2 18:3. With these separations it can be seen that the elution of triacylglycerols is according to the number of double bonds of the constituent fatty acids and to distribution in the triacylglycerol molecules.

The same separations can be obtained with the commercially available liquid phases TAP-CB and with the columns WCOT fused silica and WCOT Ultimetal HT-TAP CB (25 m × 0.25 mm i.d.) from ChromPak. With the latter, using temperature programming from 340°C to 365°C, at 1°C/min and hydrogen as carrier gas, separations of triacylglycerols from soybean oil occurred in the following elution order: 16:0 18:1 16:0 < 16:0 18:2 16:0 < 16:0 18:1 18:0 < 16:0 18:1 18:1 < 16:0 18:2 18:0 < 16:0 18:2 18:1 < 16:0 18:2 18:2 < 16:0 18:2 18:3 < 18:0 18:1 18:0 < 18:0 18:1 18:1 < 18:1 18:1 18:1 < 18:0 18:2 18:1 < 18:1 18:1 18:2 < 18:0 18:2 18:2 < 18:1 18:2 18:2 < 18:2 18:2 18:2 < 18:2 18:2 18:3.

For a given carbon number, triacylglycerols have also been separated, particularly on polar capillary columns, according to fatty acid combinations between short- and medium- or long-chain fatty acids (161,207–209).

Other criteria of separation of triacylglycerols combining, hydroxy, and epoxy groups, for example, have been discussed by Litchfield (22).

GAS LIQUID CHROMATOGRAPHY AND TRIACYLGLYCEROL COMPOSITION OF NATURAL OILS AND FATS
GC separations of intact triacylglycerols according to carbon number and number of double bonds can lead to identification and quantification of individual triacylglycerols. An example is given with the analysis of soybean oil triacylglycerols on HT-TAP CB reported above. Besides the determination of an accurate triacylglycerol composition with only GC separations according to carbon number only, the most worked method for structural analysis of natural oils and fats needs combined techniques:

Analytical gas–liquid chromatography and preparative TLC-AgNO3;
Preparative and analytical gas liquid chromatography.

The combination of analytical gas liquid chromatography and preparative TLC-AgNO3 will be seen later.

In the procedure described by Bugaut and Bézard (211), which combines analytical and preparative GC, triacylglycerols were separated according to carbon number and quantified by analytical GC. Each peak was then collected by preparative GC and its fatty acid composition was determined by analytical GC.

To determine triacylglycerol composition of a peak, all possible combinations of three fatty acids, whose total acyl carbon atom was that of the peak, were established from the fatty acids detected in the peak. A system of equations was then established according to Bézard et al. (169) by stating that fatty acid composition calculated from the triacylglycerol composition was equal to the experimentally determined fatty acid composition. The solutions of the equations were the proportions of the triacylglycerols in the peak.

In conclusion, if GC separations of intact triacylglycerols according to carbon number and number of double bonds is very useful in basic research, separations uniquely according to carbon number are greatly utilized in industrial food controls (212,213). They allow one to distinguish between natural and fractionated fats and to identify transesterified fats. For example, the condensed fractions of butter have a TG C50 content superior to that of TG C38. It is the reverse in whole butter. The fluid fractions of palm oil almost completely relieved of tripalmitoylglycerol present in the whole oil have a C52/C48 ratio from 9.2 to 23.6, while this ratio is inferior to 5.9

for whole palm oil. In natural fats, fatty acids are not generally randomly distributed. It is the reverse in the case of interesterified fats.

e. Gas Chromatography of Partial Acylglycerols

MONOACYLGLYCEROLS Monoacylglycerols are preferably converted to ester or ether derivatives prior to gas–liquid chromatography. Thus they are analyzed most often as trimethylsilyl (TMS) ether derivatives (214), acetate derivatives (215), or trifluoroacetate (TFA) derivatives (216). Separations are carried out on packed (215,217) and capillary WCOT (218) columns containing nonpolar stationary phases (such as SP-100, OV-1, OV-3, JXR, SE-30) or polar stationary phases (such as polyester phases: EGS, EGSS-X, DEGS, EGA, PEGA; cyanosiloxane phases: Silar 5 CP, Silar 7 CP, Silar 10 C; dicyanoallylsilicone phase: OV-275).

Separation of monoacylglycerols according to molecular weight or carbon number

These separations are obtained on both polar and nonpolar columns (218–220). For example, Bézard et al. (221) and Morétain and Bézard (222) separated, according to carbon number, TMS ether derivatives of 2-monoacylglycerols prepared from palm kernel oil triacylglycerols by pancreatic lipase hydrolysis. A stainless steel column (1.20 m × 0.33 cm i.d.) of JXR (2% on Gas-Chrom Q 100–120 mesh) in programmed temperature from 120°C to 250°C at 3°C/min and with a constant flow rate of nitrogen of 45 ml/min was used in this instance.

2-Monoacylglycerols have also been resolved on the same stationary phase by Breckenridge and Kuksis (223,224) but as diacetate derivatives. However, the most interesting separations of monoacylglycerols for structural analyses of triacylglycerols are those carried out according to acyl carbon number, number of double bonds, and position in the glycerol moiety because they permit the relative proportion of sn-2-monoacylglycerols to be calculated comparatively to that of sn-1(3) isomers which can be subsequently resolved by chiral phase chromatography.

Separation of monoacylglycerols according to carbon number, unsaturation, and positional isomerism

On nonpolar capillary column (15–25 m) coated with SP-2100 or OV-3, with temperature programming from 80°C or 60°C to 360°C, TMS ether derivatives of monoacylglycerols can be resolved according to carbon number and position of fatty acids in the glycerol moiety (218). The sn-2 isomer elutes ahead of the sn-1(3) isomers.

The best separations are obtained on polar columns. On such columns, monoacylglycerols are separated according to carbon number and for a given carbon number the saturated and unsaturated compounds are separated. In addition, the sn-2- monoacylglycerol of each fatty acid is separated from the sn-1(3) isomers.

The cyanosiloxane liquid phases, which have high polarity and moderate thermal stability, have permitted the best separations obtained on polar columns. Thus Myher and Kuksis (75,215) using a glass column (180 cm × 3 mm i.d.) packed with Silar 5 CP and isothermal analyses have reported excellent separations of TMS ether and diacetate derivatives of monoacylglycerols based on carbon number, degree of unsaturation, and position of fatty acids. The elution order was that of increasing carbon atoms, and for each carbon number it was that of increasing double-bond number. While no resolution for the isomeric diacetates has been observed, the sn-2- monoacylglycerol and the sn-1(3)-monoacylglycerols as TMS ether derivatives eluted with satisfactory resolution, the sn-2-monoacylglycerol preceding the sn-1(3) isomers.

The general elution order of TMS ether derivatives of monoacylglycerols on this column was as follows: 2-MG-14:0 < 2-MG-16:0 <1(3)-MG-16:0 + 2-MG-16:1 < 1(3)-MG-16:1 < 1(3)-MG-18:0 + 2-MG-18:1 < 1(3)-MG-18:1 + 2-MG-18:2 < 2-MG- 18:4 < 2-MG-20:1 < 1(3)-MG-20:1 < 2-MG-20:3 < 2-MG-20:5 < 2-MG-22:0 < 1(3)-MG-20:5 < 2-MG-22:1 < 1(3)- MG-22:1 < 2-MG-22:6 < 1(3)-MG-22:6.

A comparative study of various phases: Silar 10 C (3 and 5%), Silar 7 CP (5%), Silar 5 CP (5%), OV-275 (5%) on 100–120 mesh Gas-Chrom Q, by Itabashi and Takagi (217) has shown in isothermal analyses (240–260°C, MG acetates, 170–220°C TMS ethers of MG) and temperature-programmed analyses (200–270°C, 1°C/min) that Silar 10 C provided the same type of separations as Silar 5 CP but with better resolution. While the monoacylglycerol acetate positional isomers were not resolved, the TMS ether derivatives of monoacylglycerols were separated in the following elution order: 2-MG-16:0 < 1(3)-MG-16:0 < 2-MG-18:1 + 1(3)-MG-18:0 < 1(3)-MG-18:1 < 2-MG-18:2 < 1(3)-MG-18:2 < 2-MG-18:3 < 1(3)-MG-18:3.

DIACYLGLYCEROLS The diacylglycerols have been usually submitted to gas–liquid chromatography as acetate or TMS ether derivatives. The gas–liquid chromatography has been carried out on polar or nonpolar liquid phases using packed or capillary columns with programmed or constant temperatures.

Diacylglycerol separations according to number of total acyl carbons

The best separations of diacylglycerol derivatives based only on the number of total acyl carbons have been obtained almost exclusively on thermostable nonpolar liquid phases, and usually under temperature programmed conditions, such as OV-1, JXR, SE-30 (93,102). According to Christie (102), optimum conditions for diacylglycerols analysis on *packed columns* are obtained with 1 m × 3 mm i.d. column packed with SE-30 (2% on silanized support), with a carrier gas flow of 100–200 ml/min, and linear temperature programming from 220°C to 300°C. However, this author has separated diacylglycerols differing by only one carbon atom with a 50 cm × 0.60 cm i.d. column packed with SE-30 (1%) used under the following conditions: programmed temperature of 250–300°C and constant nitrogen flow of 50 ml/min.

The diacylglycerol fractions separated by GC according to carbon number contain both the 1,3 and 1,2(2,3) isomers with different degrees of unsaturation (74,87,93).

Several workers have reported separations according to carbon number of acetate (225,226), TMS ether (179,227–229), and *tert*-butyldimethylsilyl (*t*-BDMS) ether (230,231) derivatives of diacylglycerols. The separations were obtained using stainless steel or glass columns (50–120 cm long, 2–3 mm i.d.) packed with JXR or OV-1 (2–3% on 100–120 mesh Gas-Chrom Q) using various temperature programs and nitrogen or helium as carrier gas, at 45 ml/min (179,225–231). These separations have been utilized

To study the stereospecific distribution of fatty acids in the triacylglycerols of milk fat (225,226)

To determine the *sn*-2-triacylglycerol compositions of saturated and unsaturated triacylglycerols of palm kernel oil (179,227)

For the structural analysis of peanut oil triacylglycerols (stereospecific distribution of fatty acids and triacyl-*sn*-glycerol composition (229)

To study the stereospecificity of lipoprotein lipase with respect to pancreatic lipase, using synthetic triacyl-*sn*-glycerols (228)

To develop a stereospecific analysis method (230,231)

Capillary columns coated with nonpolar liquid phases have also been used for analysis of diacylglycerol derivatives. Separations were based on carbon number and positional distribution of fatty acids in the diacylglycerol molecules. For example, the Official Method Cd-11b-91 (218) using a glass or fused silica capillary column (15–25 m, 0.25–0.35 mm i.d.) coated with SP-2100 or OV-3, a programmed temperature from 80°C to 360°C, 10°C/min, a helium flow rate of 5 ml/min, leads to separation of TMS ether derivatives of diacylglycerols according to carbon number and positional isomerism. For each carbon number, the 1,2(2,3) isomers are eluted ahead of the 1,3 isomer. Similar results have been obtained on a SE-30 column (232).

On the other hand, Myher et al. (233) in the stereospecific analysis of menhaden oil triacylglycerols used a fused silica column coated with the nonpolar phase SE-54 (8 m × 0.32 mm i.d.) and a programmed temperature of 40–350°C to separate TMS ether derivatives of diacylglycerols according to carbon number (234,235). Gaskell and Books (236) in studying phospholipids used isothermal analyses at 300°C to separate diacylglycerols according to carbon number on a column coated with OV-1 Silanox.

Another application of capillary GLC has been the separation of diacylglycerol diastereoisomers of short-chain fatty acids obtained on an SE-54 fused silica column (25 m) (237).

It should be noted that partial separations according to the degree of unsaturation of diacylglycerol fatty acids can sometimes be obtained on capillary nonpolar columns. They have to be converted to saturated compounds by hydrogenation prior to analysis.

Diacylglycerol separations according to carbon and double-bond numbers

Efficient separations of saturated and unsaturated diacylglycerols have been obtained on polar columns with both acetate and TMS ether derivatives. With *packed columns* cyanosiloxane liquid phases seem to be the best for this kind of separations. Thus, on a glass column (180 cm × 3 mm i.d.) packed with Silar 5 CP (3% on Gas-Chrom Q 100–120 mesh), Myher et al. (175,238) reported very good separations of TMS ethers of *sn*-1,2(2,3)- and 1,3-diacylglycerols which were analyzed separately. The separations are based on carbon number and number of double bonds for the same carbon number. For example, diacylglycerols C36 were separated in the following elution order: 18:0-18:1 < 18:1-18:1 < 18:1-18:2 < 18:2-18:2.

The general elution order of TMS ether derivatives of diacylglycerols on Silar 5 CP is as follows: 16:0-16:0 < 16:0- 18:0 < 16:0-18:1 < 16:0-18:2 < 16:0-18:3 < 16:0–20:4 < 18:0-18:0 < 18:0-18:1 < 18:1-18:1 < 18:0-18:2 < 18:1-18:2 < 18:0-18:3 < 18:2-18:2 < 18:1-18:3 < 18:2-18:3 < 18:0-20:4 < 18:3-18:3 < 18:0-20:0.

The efficiency of Silar 5 CP columns has been utilized in the structural analysis of peanut oil triacylglycerols (229) and in the study of lipoprotein lipase stereospecificity (228).

The same kinds of separations with higher separation factors for homologous series differing by one double bond but lower for those differing by two carbon atoms were obtained by Itabashi and Takagi (217) on a column packed with Silar 10 C.

Polar phases can provide diacylglycerol derivative separation based not only on carbon number and number of double bonds but on fatty acid combination according to chain length. Kuksis et al. (239) showed that, when analyzed as brominated acetate derivatives, the sn-2,3 enantiomers of diacylglycerols of same carbon number could be separated when they contained a butyrate (4:0), a caproate (6:0), or a caprylate (8:0) residue. The column used was a 180 cm × 3 mm i.d. column packed with 3% EGSS-X on 100-120 mesh Gas-Chrom Q and the analysis was isothermal at 250°C. The elution order observed was 10:0–6:0 < 12:0-4:0 < 12:0-6:0 < 14:0-4:0 < 15:0-4:0 < 12:0-8:0 < 14:0-6:0 < 16:0-4:0 < 17:0-4:0 < 16:0-6:0 < 18:0-4:0. It can be seen, for example, that the three diacylglycerols 12:0-8:0, 14:0-6:0, and 16:0-4:0 of the same carbon number 20 were resolved, under these conditions.

Positional distribution of fatty acids in the diacylglycerol molecules can also influence the separations when using polar liquid phases. Thus sn-1,2(2,3)- and sn-1,3- diacylglycerols as TMS ether derivatives were resolved on Silar 5 CP and Silar 10 C (217,238). The *sn*-1,2(2,3)-diacylglycerol preceded the sn-1,3 isomer for each diacylglycerol. The analyses of mixtures of isomers provided satisfactory separations on Silar 10 C (217) with a carrier gas flow rate of 20 ml/min. But some overlapping was observed on Silar 5 CP (75,238). The elution order on Silar 10 C is as follows: 1,3-16:0-16:0 < 1,2(2,3)- 16:0-18:1 < 1,3-16:0-18:1 < 1,2(2,3)-16:0-18:2 < 1,3-16:0-18:2 < 1,2(2,3)-18:1-18:1 < 1,3-18:1-18:1 < 1,2(2,3)-18:1-18:2 < 1,3-81:1-18:2 < 1,2(2,3)-18:2-18:2 < 1,3-18:2-18:2.

Polar liquid phases have also been utilized with *capillary columns* to separate TMS ether derivatives of diacylglycerols (240,233). Itabashi et al. (240) and Myher et al. (233) used such a column (15 m × 0.32 mm i.d. column coated with polar RT 2330, isothermally run at 260°C with hydrogen as carrier gas) to analyze the sn-1,2(2,3), sn-1,3, sn-1,2, and sn-2,3 isomers of diacylglycerols issued from menhaden oil, corn oil, cocoa butter, and lard triacylglycerols. Diacylglycerols were separated according to increasing carbon number and increasing number of double bonds for diacylglycerols of the same carbon number. This will be illustrated later when considering the applications of diacylglycerol separation in the stereospecific analysis of triacyl-sn-glycerols.

In the case of diacylglycerols produced by Grignard degradation of menhaden oil triacylglycerols (233), analyses on RT 2330 column have lead to elution of more than 70 molecular species of diacylglycerols, having 28–44 carbon atoms, 0–12 double bonds, and whose constituent fatty acids had 14–22 carbon atoms and 0–6 double bonds belonging to the n-9, n-6, n-4, n-3, and n-1 series for the unsaturated.

2. Adsorption Chromatography

In triacylglycerol analysis several alternative or complementary chromatographic techniques have been developed since 1962. The most widely used today are adsorption chromatography on sole silicic acid or, more often, on silicic acid impregnated with silver nitrate.

   a. Silicic Acid Chromatography. Adsorption chromatography is simple to set up but affords limited possibilities. It has been more often used to separate lipid classes, as illustrated in a previous section. However, in some cases it has been applied to the separation of triacylglycerols exhibiting great differences in polarity, such as triacylglycerols containing short-chain fatty acids or oxygenated fatty acids, which are strongly retained on silicic acid. Column and thin-layer chromatography are the most commonly used methods, generally using silicic acid as adsorbant but sometimes magnesium silicate (Florisil) or alumina oxide. Different types of separation were achieved.

SEPARATION ACCORDING TO MOLECULAR WEIGHT In a complex mixture, triacylglycerol molecules of high molecular weight are eluted ahead of low molecular weight molecules with petroleum ether–diethyl ether (80:20, v/v) as solvent (22). Clément et al. (241,242) used this property to separate three fractions of butter oil triacylglycerols by silicic acid column chromatography using as eluant pentane containing increasing amounts of diethyl ether. The triacylglycerols eluting ahead were poor in butyric acid (4:0) and rich in long-chain fatty acids (18:0, 18:1), while the reverse was true for the next two fractions, the fatty acid composition of which were very similar. Similarly, Blank and Privett (243) separated two fractions of butter oils triacylglycerols using petroleum ether–diethyl ether (97:3, v/v) as solvent. The first eluting fraction principally contained long-chain fatty acids, while the second one contained short-chain fatty acids.

Similar results were obtained by thin-layer chromatography (186,243–246). Triacylglycerols containing long-chain fatty acids and acetic or isovaleric acid have been separated according to their short-chain fatty acid content by TLC using petroleum ether–diethyl ether (92:8, v/v) (247) and hexane–diethyl ether–acetic acid (87:12:1, v/v/v) (175) as eluting solvent.

The presence of phytanic acid in a triacylglycerol molecule decreases its adsorption on silicic acid (248–250) permitting separation of triacylglycerols containing zero to three phytanyl residues. For this purpose, the solvent used was petroleum ether–diethyl ether–acetic acid (80:10:1, v/v/v) (251).

SEPARATION OF OXYGENATED TRIACYLGLYCEROLS Triacylglycerols containing polar hydroxyl, epoxyl, aldehyde, ozonide, or ester groups in their aliphatic chain are likely to be separated by adsorption chromatography on column or by TLC, according to the number of oxygenated fatty acids per molecule. Table 3 reports the solvent systems used according to the oxygenated triacylglycerols.

OTHER TRIACYLGLYCEROLS Triacylglycerols containing unsaturated fatty acids are more strongly adsorbed on silicic acid than those containing saturated fatty acids of the same chain length (274,275) and can be separated by silicic acid TLC.

Separation by silicic acid TLC of positional isomers of triacylglycerols containing two fatty acids of very different chain length has been reported. For example, β-PAP and β-PPA were resolved (P = 16:0, A = 2:0) (276–279) using the elution solvent petroleum ether–diethyl ether (80:20, v/v).

Separation of mercury derivatives prepared by addition of mercuric acetate on the double bonds has been carried out by TLC (280).

SEPARATION OF MONO- AND DIACYLGLYCEROLS Separations of the 1,2(2,3)-diacyl-*sn*-glycerols and 1,3-diacyl-*sn*-glycerol have been achieved by micropreparative TLC using petroleum ether–diethyl ether (50:50 or 60:40, v/v) as solvent (281–283) or chloroform-acetone (96:4, v/v). The silica plates can be impregnated with 8% of boric acid (282) to avoid acyl migration.

Separation of diacylglycerol molecular species has also been observed for triacylglycerols. Separation of *sn*-XX and *sn*-XA (X = 16:0 or 18:0, A = 2:0) has been achieved by TLC with the solvent system hexane–diethyl ether (70:30, v/v) (22). Epoxydiacylglycerols have been resolved in bands containing 0, 1, 2, epoxyacids by TLC using petroleum ether–diethyl ether–acetic acid (60:40:1, v/v/v) as solvent (284).

Monoacylglycerol isomers, 1(3)-monoacyl- and 2-monoacyl-*sn*-glycerols have been separated on TLC plates impregnated or not with boric acid. Different solvent systems have been used (285):

Binary systems constituting varied proportions of acetone in chloroform (chloroform–acetone 96:4, 92:8, 88:12, 50:50, v/v)
Tertiary systems of chloroform–acetone–methanol (71:25:4, 57:37:6, v/v/v) or of chloroform–acetone–acetic acid (95:5:4, v/v/v)
A quaternary system of chloroform–acetone–acetic acid–methanol (72.5:25:2:0.5, v/v/v/v)

b. Argentation Chromatography

INTRODUCTION Triacylglycerols, like fatty acid esters, can be separated according to unsaturation by adsorption chromatography in the presence of silver ions (22,102,286,287). The method is based on the existence of interaction between the silver ions and the π electrons of the double (or triple) bonds (288–291). On column in the non-HPLC mode the method is tedious, time consuming, and not very efficient. Location and identification of the fractions separated are uneasy. So the method had little development and is now rarely used. However, recently it was employed on a miniscale, using a Pasteur pipette column, to separate four fractions of triacylglycerols containing one, two, three, and four and more double bonds, with hexane–diethyl ether as solvent (292).

Nowadays commercial, ready-to-use columns are available (solid phase extraction, or SPE) that can be impregnated with silver nitrate (Bond-Elut SCX: propylbenzene sulfonic acid). A solution of silver nitrate (20 mg) dissolved in 0.25 ml of acetonitrile-water (10:1, v/v) is passed through such a column. The triacylglycerols (0.3 mg) dissolved in dichloromethane are eluted with 5 ml of dichloromethane (000: trisaturated triacylglycerols), then with 5 ml of dichloromethane–methyl acetate (85:15, v/v) for the monounsaturated triacylglycerols (001), then dichloromethane–methyl acetate (2:3, v/v) for 011 and 002 and 5 ml of acetone–acetonitrile (3:2, v/v) for the more unsaturated triacylglycerols.

The argentation procedure has also been applied to liquid–liquid partition chromatography on thin-layer plates or on paper by adding the polar mobile phase with silver nitrate (293,294). The

Table 3 Conditions of Oxygenated Triacylglycerol Separation by Column or Thin-Layer Chromatography

Oxygenated triacylglycerols	Chromatographic method	Eluting solvent systems	Observations	Ref.
Epoxytriacylglycerols	TLC	Petroleum ether-diethyl ether-acetic acid 80:20:1 (v/v)	Bands of triacylglycerols containing 0, 1, 2, 3 epoxyl groups per molecule	252–256
Hydroxytriacylglycerols	Column	Hexane-diethyl ether 80:20 (v/v)	Same separation	257
	TLC	Hexane-diethyl ether: 60:40 or 70:30 (v/v)	Bands of triacylglycerols containing 0, 1, 2, and even 3 hydroxyl groups	258, 259, 260, 261
		Benzene-diethyl ether 75:25 (v/v) Chloroform-methanol 99:1 (v/v)		
Ketotriacylglycerols	Column	Benzene-methanol 98:2 (v/v)	Same preparation	262–265
	Column	Benzene-methanol 98:2 (v/v)	Separation of triacylglycerols containing 0, 1, or 2 ketone groups	266
Ozonide triacylglycerol (double bonds converted to ozonides	TLC of ozonides	Petroleum ether-diethyl ether 80:20 (v/v)	Separation according to the number of epoxide groups per molecule	267
-CH-O-CH-O———O that may be reduced to aldehydes with triphenylphosphine or by hydrogenation)	TLC of aldehydes			268, 269
Oxidized triacylglycerols	TLC	Hexane-diethyl ether 60:40 (v/v)		270, 271
Triacylglycerols estolides	TLC		Separation according to the number of ester groups per molecule	272, 273

mobile phases used with methanol (293,294), methanol-water (295–297), and acetone-ethanol-water-acetonitrile (297a). They were saturated in silver nitrate. With such solvents, the separations were principally based on unsaturation or, in some cases, on unsaturation and chain length. This technique has not been much developed.

On the contrary, argentation TLC has been widely used in the separation of triacylglycerols (22) because of its simplicity, rapidity, good quality of the separations obtained, and facility of location and identification of the fractions separated. Allied with the low cost of the equipment required, it has become a classical technique in the separation of triacylglycerols from natural mixtures and is still widely used by the analysts.

However, in recent years argentation-HPLC has developed rapidly (298), particularly since stable columns can be prepared. It presents several advantages over argentation-TLC. Separations require short elution times. Unsaturated triacylglycerols are totally protected from oxidation and more complex separations can be performed. Unfortunately, its diffusion remains limited because of the high cost of the equipment and solvents. The method will be described in another section. This presentation will be limited to argentation-thin-layer chromatography.

ARGENTATION-TLC

The Plates

Glass plates, generally 200 × 200 mm, are covered with a thin layer of adsorbant, in most cases of silica gel G with calcium sulfate as binder. They can be prepared in the laboratory according to the procedure described for fatty acid esters. Silver nitrate is usually dissolved in water, but methanol, ethanol, acetone, acetonitrile, and even ammonium hydroxide are also suited.

The plates can also be prepared from commercial precoated plates by one of the following means:

The plates are dipped into a solution of silver nitrate in methanol, acetone (299), or acetonitrile (300–302).
The plates are sprayed with the silver nitrate solution.
The plates are developed using a 10–20% by weight of silver nitrate in acetonitrile (303,304).

In this latter case, a gradient of silver ions can be obtained which improves the separation of triacylglycerols.

The more suited ratio of silver nitrate to silica gel is 5% by weight (305). However, in the separation of highly polyunsaturated triacylglycerols (marine oils) or of the positional isomers of the constituent fatty acids, ratios of 8% (306) or even 23–30% (307,308) have been recommended.

The layer thickness selected is usually 0.25 mm for analytical purposes and 0.5–1 mm for semipreparative separations.

To improve efficiency, the plates should be activated for 2–4 hr at 100–120°C for laboratory-made plates and at 190–195°C for precoated plates (309,310). They are stored in darkness to avoid transformation of the silver ions to silver metal by daylight. To avoid the plates turning black, Muldrey (311) has proposed to use adsorbant impregnated with silver phosphate Zirconil insensitive to daylight and having the same efficiency as silver nitrate–impregnated plates.

The Eluant

Solvent systems used to elute triacylglycerols are generally binary and sometimes ternary mixtures in which petroleum ether, toluene, benzene, and chloroform are the most frequently found. Less usual solvents are diethyl ether, ethanol, methanol, acetone, and acetic acid. The mixture chloroform-methanol is the most often used. However, according to the unsaturation of triacylglycerols and of the operating conditions, the solvents and their relative proportion can be different (287,312,313). Litchfield (22) proposed varied operating conditions as reported in Table 4.

Table 4 Solvent Systems Recommended in the Separation of Triacylglycerols According to Unsaturation by Argentation–Thin Layer Chromatography

Number of double bonds per triacylglycerol molecule	Solvent systems (v/v) Chloroform: methanol or ethanol or acetic acid	Benzene: diethyl ether	Ref.
0–4	100:0 to 99:1	100:0 to 80:20	14, 314, 314a, 314b
1–6	99.2:0.8 to 98.5:1.5	90:10 to 80:20	308, 314c, 314d, 314e, 314f
5–9	97.5:2.5	0:100	273, 314c
7–12	94:6	—	306

In the fractionation of peanut oil triacylglycerols (180) the authors have used three different solvent systems in order to obtain clear fractions for accurate further analysis, two ternary and one binary:

A mixture of hexane–diethyl ether–methanol (79:20:1, v/v/v) for the separation of total triacylglycerols

A mixture of chloroform–methanol (99.5:0.5, v/v) for the separation of triacylglycerols 002 and 111 (the numbers 0, 1, 2 indicate the number of double bonds in the three constituent fatty acids of triacylglycerols)

A mixture of hexane–diethyl ether–methanol (77.5:20:2.5) for the separation of triacylglycerols 022, 122, and 222.

Figure 7 illustrates the separation so obtained. Nine bands were clearly separated and their triacylglycerols further analyzed, as will be seen later.

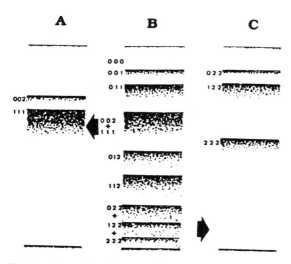

Fig. 7 Separation of peanut oil triacylglycerols by TLC on silica impregnated with silver nitrate. Triacylglycerols of the band (002 + 111) and of the band (022 + 122 + 222) were rechromatographed in conditions A and C, respectively, for better separation. Conditions for separation: silica layer thickness (µm): 250 (A and C), 500 (B); silver nitrate % (w/w): 20 (A and C), 10 (B); solvent: hexane-ethyl ether-methanol (v/v/v) (B: 158:40:2; C: 155:40:15) chloroform- methanol (A: 199:1, v/v). (From Ref. 180.)

Separation procedures

The triacylglycerol sample should contain no more than 40 mg when used at analytical scale and from 30 to 100 mg at semipreparative scale. It is applied as spots or as a narrow continuous band at the lower edge of the plate using a glass micropipette, a microsyringe, or an automatic device allowing a more uniform deposit.

The plates are developed in a glass tank, generally closed, in an atmosphere saturated with the mobile phase, most commonly at ambient temperature, sometimes at lower temperature, e.g., in the separation of triacylglycerol positional isomers (308).

To improve resolution, plates can be developed several times using either the same solvent (303,308,314) or different solvent systems, starting with the more polar (102). Development is generally stopped when the eluting solvent reaches the upper edge of the plate. However, the development is sometimes carried on (315) continuously, in an opened tank, with the solvent evaporating at the upper edge of the plate (316,317).

It is possible to detect the band migration and to identify the triacylglycerols by adding appropriate dyes to the sample before application (318,319): The decanal dinitrophenylhydrazone coelutes with tristearin, the 2c-7c-decadienal dinitrophenylhydrazone with triolein, and Soudan II containing two dyes migrates with triacylglycerols containing one or two cis double bonds.

Following development, the plates are dried under nitrogen and the bands are detected by means of a nondestructive or a destructive procedure:

> The most usual nondestructive procedure consists of spraying a solution of 0.05% (w/v) 2,7-dichlorofluorescein in methanol-water (50:50, v/v) or in ethanol. Under UV light (350 nm) the bands appear reddish. They can be delineated and the triacylglycerols can be extracted from the silica scraped off the plate. Alternatively, rhodamine 6G (320), dibromo-fluorescein (321), and sodium fluorescein (314) can be used instead of dichlorofluorescein.

> The destructive procedure is now rarely used. It consists of transforming the organic material into tar, by burning with a flame (322) or, better, by spraying the plates with an aqueous solution (50:50, v/v) of orthophosphoric acid or sulfuric acid (305) and heating in an oven at 200-400°C.

Quantification

Quantification of the triacylglycerol fractions separated by Ag^+-TLC is generally carried out as follows: The triacylglycerols are extracted from the silica scraped off the plate by first adding a solution of 1% by weight of NaCl in methanol-water (90:10, v/v) to break the complex Ag^+ double bonds, by fractions of 0.5 ml until disparition of the red color due to the complex Ag^+-dichlorofluorescein (323–326). The liberated triacylglycerols are then extracted with diethyl ether–methanol (90:10, v/v) containing 0.1 mg/L of BHT to prevent oxidation (323).

As experienced by the authors, the following procedure gives good results (152): The silica scraped off the plate is poured through a funnel in a tube containing chloroform-methanol (90:10, v/v). At this stage a primary standard of an odd chain triacylglycerol, generally tripentadecanoin, is added to the mixture. After a 2-hr contact, the mixture is filtered and the solvent evaporated to dryness. The residue in the tube is reextracted once with hexane and twice with hexane-water (1:1, v:v). The solutions are pooled in another tube, shaken, and the tube centrifuged. The upper hexane phase is washed three times with an equivalent volume of water. The hexane is evaporated and the extracted triacylglycerols are dried using absolute ethanol and dissolved in a known amount of chloroform. A second standard, generally triheptadecanoin, is added at this stage. It is recommended to perform the same extraction and washing phases on a blank which will serve as a control.

The amount of extracted triacylglycerols is then determined. For this purpose the method of choice is the GLC analysis of the component fatty acids (327–329) for fatty acid composition of the triacylglycerols and for the amount present in each band by means of the two internal standards. Comparison between the two standards allows one to measure the yield of extraction. The method is accurate since the absolute error to the mean of three or four analyses is ±0.9 for the percentages of the major components. The error is ±3 or 4 when only one analysis is carried out (327,330).

A direct quantification of the extracted triacylglycerols is likely to be performed by GLC (102,331) or HPLC. The amount of extracted triacylglycerols is determined for the standards (tripenta- and triheptadecanoin) and the composition in triacylglycerols separated according to the carbon number (GLC) or according to the partition number (reverse phase HPLC) is determined from peak areas. Other nondestructive methods of quantification have been proposed but are rarely used nowadays (314,332–341).

Quantification can also be performed directly on the plates, from the spot areas (342) or by photodensitometry of the tar in the spots or the bands (321,343,344). For this purpose good instruments are now commercially available (345). The triacylglycerols are visualized by spraying sulfuric acid (50–70% by weight in water) or other oxidative reagents (346–350). When heating at 220–250°C the lipids are oxidized and in the presence of silver ions an intense red color developed, independently of triacylglycerol unsaturation. The tar in spots is measured by densitometry at 400–450 nm. The densitometer is generally coupled to registers and integrators (347,351,352).

Argentation-TLC of triacylglycerols

Triacylglycerol separation according to cis double bonds

Elution order

After studying triacylglycerol mixtures containing saturated (0), monounsaturated (1), diunsaturated (2), and triunsaturated (3) fatty acids, Gunstone and Padley (353) established the elution order of triacylglycerols according to cis double-bond numbers on the chromatoplates coated with silica gel impregnated with silver nitrate. This elution order was as follows, from top to bottom: Top–000 > 001 > 002 > 111 > 012 > 112 > 022 > 003 > 122 > 013 > 222 > 113 > 023 > 123 > 223 > 033 > 133 > 233 > 333–Bottom. It has been confirmed by other workers (14,23,69,308) and can be predicted by attributing the following arbitrary values to the affinity of the silver ions for the double bonds of each fatty acid: saturated fatty acid = 0; monounsaturated fatty acid = 1; diunsaturated fatty acid = 2 + a; triunsaturated fatty acid = 4 + 4a. Thus the triacylglycerol 033 of six double bonds with an affinity silver ions double bonds of 8 + 8a is more retained on the plate than the triacylglycerol 223 with seven double bonds and an affinity silver ion double bond of 8 + 6a.

Ag$^+$-TLC of triacylglycerols

The resolution of the triacylglycerol bands on the chromatoplates decreases as the triacylglycerols become more unsaturated. Up to six double bonds, without linolenic acid (18:3), the mixtures of triacylglycerols from 000 to 222 can be fractionated and well resolved on a single plate (341,354), e.g., with chloroform-methanol (99:1, v/v) as developing solvent (102,298). Beyond six double bonds, the separations become more difficult (317,355). Two separations are often necessary using solvents of different polarity and the second separation can be performed with the same sample and, for example, a mixture of chloroform-methanol (95:5, v/v) to resolve the more unsaturated fractions (102,298). This procedure can be applied to natural oils or fats containing linolenic acid, whose triacylglycerols from 222 to 333 are in most cases partly resolved (318,353). The natural fat triacylglycerols containing fatty acids with four, five, or six double bonds are not well separated and only a few distinct bands can be observed (356).

Isomer separations

Ag$^+$-TLC allows the separation of four kinds of triacylglycerol isomers:

Triacylglycerols with the same number of double bonds but with different distributions of these bonds. For example, separations between 18:0 18:2 18:0 and 18:0 18:1 18:1 (020 and 011); 18:1 18:1 18:1 and 18:0 18:1 18:2 (111 and 012); 18:0 18:2 18:2 and 18:0 18:1 18:3 (022 and 013).

Triacylglycerols containing geometrical isomers (cis [c], trans [t]) of the same fatty acid. For example, the triacylglycerols containing stearic acid (18:0), oleic acid ($18:1_c$), and elaidic acid ($18:1_t$) are separated according to the following elution order: 18:0 18:0 18:0 > 18:0 18:0 $18:1_t$ > 18:0 $18:1_t$ $18:1_t$ > 18:0 18:0 $18:1_c$ > $18:1_t$ $18:1_t$ $18:1_t$ > 18:0 $18:1_c$ $18:1_c$ > $18:1_c$ $18:1_c$ $18:1_c$ (304,308,357).

Triacylglycerols containing isomers of fatty acid according to double-bond position in the hydrocarbon chain. For example, triacylglycerols containing oleic acid (18:1 cis Δ 9 = 0) and petroselinic acid (18:1 cis Δ 6 = Pe) are separated according to the following order: 000 > Pe00 > PePe0 > PePePe (308).

Positional isomers of triacylglycerols. Resolution of the following pairs of isomers (14,321,358, 359): β-001 and β-010; β-110 and β-101; β-002 and β-020; β-112 and β-121; β-220 and β-202; β-221 and β-212; β-102, β-012, and β-021 (0 = saturated fatty acids, 1 and 2 = mono- and diunsaturated fatty acids, respectively). Except for the isomers β-110 and β-101, in the other cases, symmetrical triacylglycerols (β-010, β-020, β-202, β-212) migrate ahead of the pair of isomers to be separated.

Separations of triacylglycerols with other functional groups having π electrons as, for example, triacylglycerols containing conjugated trienoic acids (360) or cyclopentenic acids (303,361) on Ag$^+$-TLC plates have been reported.

Utilization of argentation-TLC for the structural analysis of triacylglycerols

Ag$^+$-TLC has yielded much information on the triacylglycerol composition of many fats and oils (23,72,179–184,221,273,306,308,315,321,362–369).

It allows the determination of triacylglycerol class composition (000, 001, 011, 002, 111, etc.).

However, its greatest interest is to be complementary to the majority of other triacylglycerol analytical techniques for accurate determination of molecular species. For example, in the case of peanut oil triacylglycerol (180) the following procedure was used. Pure total triacylglycerols were fractionated into nine classes according to unsaturation, by argentation-TLC. An internal standard was added to each band. After extraction from the silica gel the bands or classes were analyzed by analytical gas chromatography for fatty acid and triacylglycerol groups according to carbon number. Such a chromatogram obtained for the class 011 triacylglycerols was previously illustrated in Fig. 6.

From the amount of the internal standard added and from the fatty acid composition in each band, the amounts and proportions of the bands in the total triacylglycerols were calculated.

From the fatty acid composition of each band the sum of the proportions of the fatty acids with the same degree of unsaturation allows the triacylglycerol class of each band to be identified. For example, the class 011 will have 33% of saturated fatty acids and 66% of monounsaturated fatty acids. The class 012 will have 33% of saturated fatty acids, 33% of monounsaturated fatty acids, and 33% of diunsaturated fatty acids, and so forth. All of these data are reported in Fig. 6.

From the fatty acid and the triacylglycerol groups (according to carbon number) compositions the accurate triacylglycerol composition of each band or class is obtained as follows: Three fatty acids from those found in each class or band are combined in every possible way with regard to the degree of unsaturation of the class. These triacylglycerols are then classified according to carbon number. The triacylglycerols for which the carbon number is not found in the triacylglycerol group composition of the class are dropped out. Two series of equations can then be established from the fatty acid and triacylglycerol group compositions and the list of triacylglycerols of the class. The formulation of these equations was reported elsewhere (169). Table 5 presents the method for the identification of the class of triacylglycerols and the method of the triacylglycerol composition calculation. The triacylglycerol composition of each class and the class proportions permit the calculation of the triacylglycerol composition of the total triacylglycerols.

As other practical applications, Ag^+-TLC is interesting for the determination of the trisaturated triacylglycerols (000), in margarines and in fats used for their preparation (31,370). Its capacity for resolving triacylglycerol positional isomers can lead to β- or sn-2-triacylglycerols identification and quantification without enzymic hydrolysis. Moreover, its capacity of resolving β-001, β-101, and β-002 isomers can permit the detection of adulterations of dietary fats. These isomers exist in negligible amount in vegetable oils but are normal constituents of interesterified fats (358,359).

Separations of cis–trans isomers of triacylglycerols by ag^+-TLC are also useful in the analysis of hydrogenated fats (371).

Argentation-TLC of diacylglycerols

Diacylglycerols and monoacylglycerols (321) produced by chemical deacylation or enzymatic hydrolysis of triacylglycerols can be fractionated by Ag^+-TLC. The best separations were obtained with sn-1,2(2,3)- or sn-1,3-diacylglycerols analyzed separately because the two kinds of isomers are differently adsorbed on Ag^+-TLC plates.

Free diacylglycerols can be used for these fractionations (310,372–375). However, because these compounds are unstable, they are often derivatized before Ag^+-TLC fractionation. Various derivatives have been thus prepared:

TMS ether derivatives (95)
t-BDMS ether derivatives (238,376)
Benzoyl ester derivatives (117)
Acetate derivatives (73,74,226,239,325,377,378)

Elution order of diacylglycerols

Diacylglycerol acetates have been the derivatives most frequently used in the fractionation by Ag^+-TLC. Studies by several workers (73,78–80,379–381) have permitted the establishment of the elution order of diacylglycerols according to unsaturation (0–6 double bonds in the constituent fatty acids). It is as follows from top to bottom of the plate: Top–00 > 01 > 11 > 02 > 22 > 03 > 13 > 04 > 14 > 24 > 05 > 06 > 15 > 55 + 56 + 66–Bottom.

Separation of diacylglycerols

In the work reported in Refs. 72 and 180, the separations obtained with diacylglycerol acetate prepared from total triacylglycerols and triacylglycerols of the class 011 from peanut oil are presented. A mixture of chloroform- methanol (99.4:0.6 and 199.5:0.5, v/v) was used as developing solvent. The total diacylglycerols were fractionated in six classes from 0 (00) to 4 (22) double bonds. According to Christie (286), diacylglycerol acetates from 0 to 6 double bonds can be fractionated on a single plate containing 10% (w/w) of $AgNO_3$, and chloroform-methanol, 99:1 (v/v) as developing solvent. Renkonen (80) separated a mixture of diacylglycerols having up to 12 double bonds, with a mixture of chloroform-methanol-water (65:25:4, in volumes) as developing solvent.

Separation of isomers

Many workers have reported diacylglycerol isomer separations as free diacylglycerols or acetate derivatives.

Table 5 Determination of the Triacylglycerol Composition of the Class 012 Triacylglycerols of Peanut Oil

Fatty acids	Mol%			Equations[d]
	a	b	c	
16:0	17.3	18.0	18.0	$a+b+d=18\times3$
18:0	5.7	6.0	5.9	$c+e+g=6\times3$
20:0	2.6	2.7	2.7	$f+h+j=2.7\times3$
22:0	4.1	4.3	4.4	$i+k+m+p=4.3\times3$
24:0	2.3	2.4	2.4	$1+n+o=2.4\times3$
26:0	Trace	Trace	0.3	
Σ	32.0			
16:1	0.4	0.4	0.4	$a+c+f+i+1=0.4\times3$
18:1	33.0	32.0	31.9	$b+e+h+k+n=32\times3$
20:1	0.7	0.7	0.7	$d+g+j+m+o=0.7\times3$
22:1	0.2	0.3	Trace	$p=0.2\times3$
Σ	34.3			
18:2	33.7	33.3	33.3	

Triacylglycerols[e]	Mol %[f]	Equations[g]	TG groups[h]	Mol %[i]	Solutions[j]
16:0 16:1 18:2	a	$a=1.2$	50	1.2	$a=1.2$
16:0 18:1 18:2	b	$b+c=49.7$	52	49.7	$b=50.7$
18:0 16:1 18:2	c	$d+e+f=19.7$	54	19.7	$c=$ trace
16:0 20:1 18:2	d	$g+h+i=8.5$	56	8.5	$d=2.1$
18:0 18:1 18:2	e	$j+k+l=12.8$	58	12.8	$e=17.5$
20:0 16:1 18:2	f	$m+n=7.2$	60	7.2	$f=$ trace
18:0 20:1 18:2	g	$o+p=0.9$	62	0.9	$g=$ trace
20:0 18:1 18:2	h				$h=8.2$
22:0 16:1 18:2	i				$i=$ trace
20:0 20:1 18:2	j				$j=$ trace
22:0 18:1 18:2	k				$k=12.2$
24:0 16:1 18:2	l				$l=$ trace
22:0 20:1 18:2	m				$m=$ trace
24:0 18:1 18:2	n				$n=7.2$
24:0 20:1 18:2	o				$o=$ trace
26:0 18:1 18:2	p				$p=0.9$

[a]Experimental fatty acid composition. The sum of the percentage of the fatty acids with the same unsaturation: Σ saturated = 32, Σ monounsaturated = 34.3, 18:2 = 33.7 indicates the nature of the class of triacylglycerols: class 012 whose triacylglycerols comprise one saturated (0), one monounsaturated (1), and one diunsaturated (2) fatty acid.

[b]Corrected by taking 33.3% of saturated, 33.3% of monounsaturated, and 33% of diunsaturated fatty acids.

[c]Calculated from the triacylglycerol composition in the last column.

[d]Equations derived from the fatty acid composition, established by writing that this composition calculated from the triacylglycerols a, b, c, etc., is equal to that experimentally determined.

[e]Triacylglycerols 012, formed by combining the fatty acids identified in the class, and classified according to carbon number.

[f]Percentages of the triacylglycerols in the class 012 which represent the unknowns to be determined.

[g]Equations derived from the triacylglycerol group composition. They mean that the sum of the percentages of the triacylglycerols with the same carbon number is equal to the percentage of this carbon number or group experimentally determined by GC.

[h]Groups of triacylglycerols separated according to carbon number by GC.

[i]Molar percentages of groups.

[j]Solutions obtained by combining the two series of equations, representing the percentages of the triacylglycerols of the class 012.

Source: Data from Ref. 180.

Separation of positional isomers of diacylglycerols: β-18:0-18:1 and β-18:1-18:0 (382)

Separation of stereoisomers of diacylglycerols (383,384)

Fractionation of diacylglycerols (issued from phosphatidylethanolamine) differing by double-bond configuration and fatty acid chain length, into seven fractions on chromatoplates containing 10% w/w AgNO3, using benzene-chloroform-methanol (98:2:0.1, in volumes) as developing solvent (385,386). The migration order in this case was as follows: $16:0\text{-}18:1_t$ > $16:0\text{-}16:1_t$ > $18:1_t\text{-}18:1_t$ > $18:1_t\text{-}16:1_t$ > $18:1_c\text{-}18:1_t$ > $16:1_c\text{-}18:1_t$.

The fractionation of diacylglycerols according to double-bond number has been used for the accurate determination of the proportions of the sn-2-triacylglycerols of several oils (72,180–184,362).

3. HPLC Analysis of Triacylglycerols

Partition liquid chromatography has been widely used in the past in the separation of triacylglycerols. The principle of the method is that molecules distribute according to their solubility between two nonmiscible solvents, one mobile and the other stationary. The method has been developed in the form of thin-layer, paper, column, and countercurrent chromatography (22). It has gained new interest since the introduction of HPLC, which may be considered now as the most efficient method for the analysis of lipids (65% of applications). In the reverse mode, an apolar stationary phase, most often an octadecyl silyl group, is chemically bonded to the support, generally silica, and the more polar mobile phase percolates through the column. Reverse phase HPLC has up to now been the method of choice in the analysis of triacylglycerols at the analytical and semipreparative scale.

a. Instrumentation.   The instrumentation used has been described in many reports (287,387–391). A complete equipment comprises the following:

A solvent distribution device, principally represented by pumps capable of delivering solvents at fixed and precise rates under pressure up to 6000 psi. The device should be able to generate solvent gradients according to programmed conditions.

An injection device comprising a valve in front of the column loop which allows the sample introduction onto the column without a pressure drop.

A stainless steel (rarely glass) column packed with an appropriate stationary phase fixed on a support (generally octadecylsilyl groups (C18 or ODS).

A device for detection and quantification of the compounds emerging from the column. In the analysis of triacylglycerols the most commonly used detectors are differential refractometers, UV and IR spectrophotometers, mass or light-scattering detectors, flame ionization detectors, mass spectrometers, and chemical detectors.

b. Columns.   Practically all of the triacylglycerol separations by reverse phase HPLC have been performed on columns packed with microparticular silica bonded with aliphatic groups, generally octadecylsilyl (392,393). Most of the commercial phases and columns have been used: μBondapak C_{18}, Supelcosil LC-18, LiChrospher 100-CH18, Nucleosil 5 C18, ChemCosorb I-5C18, μPorasil, Partisil ODS, Zorbax ODS, Ultrasphere ODS, Spherisorb ODS, HS 3 or 5 C18, M CH-10 RP-18, Hitachi Gel 3057 (394–397). Initially the silica particle size was 10 μm (398–402), but 5-μm particles rapidly appeared on the market and now 3-μm particles are increasingly utilized (395,396,403–406) since the smaller the support particle size, the greater the column efficiency.

The standard analytical columns, containing 5-μm silica particles, are 250–300 mm in length and 4–5 mm of internal diameter. Since efficiency increases with column length, several standard columns have sometimes been used in series, to reach in some cases 100 cm (395,407–413). For rapid separations 50- to 150-mm columns are used. Conversely columns of higher internal diameter are used for preparative separations (80 × 6.6 mm, 80 × 6.2 mm, 200 × 7.5 mm, 300 × 7.8 mm, 500 × 9.4 mm) since the separation capacity is proportional to the square of the diameter (397). Such columns are generally packed with particles of larger size (10–40 μm).

The high-efficiency columns packed with 3-μm particles are shorter, 100 × 4.6 mm i.d. The efficiency of such columns is comparable to that of standard columns, as for the number of theoretical plates, but the analysis time is shorter and less solvent is used. To improve resolution, several (up to three) short columns (3-μm particle size) have sometimes been mounted in series (395,403,405,406).

    c. Mobile Phases. The solvent system used as mobile phase is of great importance in the separation of triacylglycerols by HPLC. It is now well established that systems comprising acetonitrile as a weak solvent are the best suited (414). Addition of a strong solvent is often necessary to improve resolution by increasing the polarity and selectivity of the mobile phase. The range of solvents to be added to acetonitrile is wide.

Under isocratic conditions and with detection by refractive indices, the system acetone (AC)-acetonitrile (ACN) has been the most widely used (165,394,402,403,405,412,414–422). However, it is not suited when using detection by UV absorbance and above all trisaturated molecular species of high molecular weight are poorly soluble in this solvent system (412). It is the reason why other binary systems have been developed comprising, in addition to acetonitrile, methanol (MeOH) (414,415,423), ethanol (EtOH) (411,423–425), methylene chloride (MC) (404,412,426), Chloroform (CHL) (412,414), tetrahydrofuran (THF) (403,414), dichloromethane (DCM) (414), isopropyl alcohol (IPA) (414), diethyl ether (427). According to their efficiency in the separation of triacylglycerols, when studied together with other solvent systems using methanol as weak solvent [methanol-acetone (414,415) or methanol-isopropanol (414)], these alternative solvent systems can be classified as follows: DCM-ACN > CHL- ACN > IPA-ACN > THF-ACN > AC-ACN > IPA-MeOH > AC- MeOH. Several of these solvent systems have been used with detection by UV absorbance, namely, ACN-MeOH (423), ACN-EtOH (423,424), THF-ACN (414). Others are particularly efficient in the separation of very-long-chain saturated triacylglycerols. For example, tribehenin is eluted in 18 min with CHL-ACN (49:51, v/v) under isocratic conditions (412). Other solvents have been reported in the literature. Propionitrile is considered by Podlaha (408) to be more efficient than any other solvent systems based on acetonitrile. Other binary mixtures [MeOH-CHL (399,428–430) or systems comprising more than two solvents [ACN-THF-hexane (431)], IPA-AC-MeOH-ACN (394), ACN-AC-THF (418)] have also been proposed.

    Triacylglycerols analysis by HPLC has also been performed using solvent gradients, starting with solvent of low polarity and polarity increasing with linearly (or not) increasing concentration of another solvent of higher polarity. The procedure is suited for complex mixtures of triacylglycerols exhibiting a wide range of polarities, such as butter oil triacylglycerols or mixtures comprising saturated triacylglycerols of high molecular weight. In the simplest cases only two solvents are used and the gradient is linear, e.g., in addition to acetonitrile, 30–60% methylene chloride (404), 55–100% chloroform (412), 50–80%, 55%, 80%, and 85–99% acetone (412), 45–60% 2-propanol (432). Some authors have used nonlinear gradients, e.g., acetone and methyl *tert*-butyl ether in acetonitrile (433), 30–90% propionitrile in acetonitrile (200,434–437).

    Figure 8 illustrates the good resolution that can be obtained by using solvent gradients in the elution of butter oil triacylglycerols with detection by evaporative light-scattering detector (513). The two analytical columns (5 μm, 25 cm × 4.6 mm i.d.) connected in tandem were Spherisorb ODS-2 and Zorbax C18. Triacylglycerols were eluted using three solvents: acetonitrile as solvent

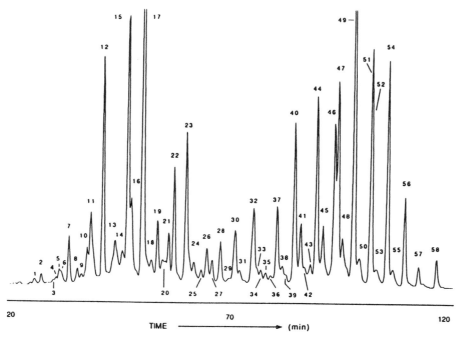

Fig. 8 Chromatogram produced by HPLC of milk fat triacylglycerols with evaporative light scattering detection. The columns (5 µm, 25 cm × 4-6 mm i.d.) were a Spherisorb ODS-2 and a Zorbax C18 connected in tandem with a Spherisorb ODS-2 precolumn (5 µm, 3.5 cm × 4.6 mm i.d.). Triacylglycerols were eluted using acetonitrile, solvent A; propionitrile, solvent B; and propionitrile-methylene chloride (72:28, v/v), solvent C. The elution involved two linear gradient steps at a flow rate of 0.7 ml/min. The first step was from 30% solvent A and 70% solvent B to 5% solvent A and 95% solvent B in 50 min. The second step was from 5% solvent A and 95% solvent B to 0% solvent B and 100% solvent C in 70 min. The evaporative light scattering detector (ELSD) was a model ELSD-MK IIA ELSD Varex (detector drift tube temperature 130°C; inlet pressure of nitrogen nebulizing gas 32 psi; detector exhaust temperature 81–82°C). A flow splitter was utilized to divide the column effluent between a fraction collector and the detector at a split ratio of 9:1. Peaks are numbered according to elution order. (From Ref. 513.)

A, propionitrile as solvent B, and propionitrile- methylene chloride (72:28, v/v) as solvent C. The elution involved two linear gradient steps at a flow rate of 0.7 ml/min. The first step went from 30% solvent A and 70% solvent B to 5% solvent A and 95% solvent B in 50 min. The second step went from 5% solvent A and 95% solvent B to 0% solvent A, 0% solvent B and 100% solvent C in 70 min. The analysis was carried out in 120 min but the milk fat triacylglycerols were resolved into 58 peaks which were collected at the outlet of the column and characterized by desorption chemical ionization mass spectrometry and tandem mass spectrometry.

When equipment permitting the use of solvent gradients is not available, an alternative solution is to proceed in two steps under different operating conditions (164). An example is given in Fig. 9 concerning analysis of butter oil triacylglycerols. In a first step the analytical conditions are those convenient for a good resolution of medium and high molecular weight triacylglycerols (fraction B). The low molecular weight triacylglycerols which eluted ahead of this fraction were collected (fraction A) and rechromatographed under more convenient conditions. A peak present in the two chromatograms (peak 27) permitted the quantitation of triacylglycerols in the two fractions.

The first analysis was run in 40 min and the second in 25 min. The butter oil triacylglycerols

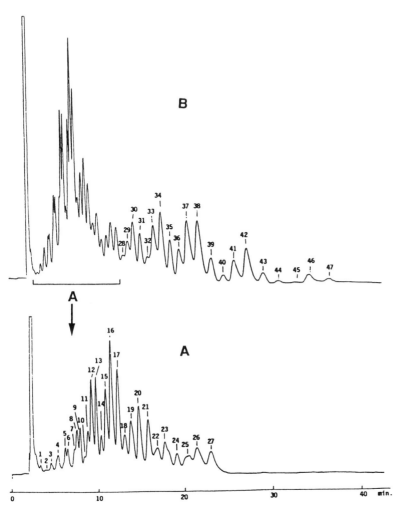

Fig. 9 Two-step reverse-phase HPLC of butterfat triacylglycerols. In (A) analysis at 30°C of the 27 fractions together collected in conditions (B), with acetone-acetonitrile (50:50, v/v) as solvent. In (B) analysis at 40°C using acetone-acetonitrile (55:45 v/v). The 250×4 mm i.d. LichroCart column was packed with LiChrospher 100 RP-18 (4 μm). The detection was by differential refractometry. The triacylglycerol fractions corresponding to the 47 peaks were collected at the outlet of the nondestructive detector for further analysis. (From Ref. 164.)

were resolved in 47 peaks which were collected at the outlet of the nondestructive detector and further analyzed by capillary GC for fatty acid composition and for triacylglycerol composition according to carbon number.

Besides binary gradient, more complex gradients have been preferred by several analysts: a ternary gradient of acetonitrile-methylene chloride-acetone from 85:15:5 to 20:60:20, v/v (438); a linear gradient from 100% acetonitrile to 100% acetonitrile-ethanol-hexane (40:40:20, w/w/w) (439); a mixture of two solvents A and B, in which A is composed of dioxane-triethylamine-iso-octane (1:0.1:98.9, v/v/v) and B is dioxane-triethylamine-(isooctane-methylene chloride, 50:50, v/v) (1:0.1:98.9, v/v/v) (440). Triacylglycerols eluted using these solvent systems were detected by UV absorbance (432,440), by flame ionization (404,433), by mass detection (439) or by light-scattering detection (412,438).

d. Operating Conditions. Before use, the mobile phase must be degassed under vacuum (287). The amount of triacylglycerols to be injected onto the column varies according to the detector sensitivity from 1 to 5 mg. Ideally, the solvent used to dissolve the sample must be identical to the mobile phase (441,442). When the dissolution is very poor, the sample can be dissolved in acetone or in acetone-tetrahydrofuran of polarity similar to that of the mobile phase or even in pure tetrahydrofuran. The volume to be injected is at most 5 ml. Most commonly, the analyses are carried out at ambient temperature. However, Jensen (431) and other authors (443) showed that lower temperatures increased retention times and improved resolution.

Figure 10 illustrates this property. The two chromatograms (a and b) concern two mixtures of three diacylglycerols, analyzed as derivatives, which can be considered as triacylglycerols. The analysis temperatures were 19°C in (a) and 10°C in (b), the other conditions being the same. In (b) peaks 1 and 2 were much better resolved than in (a), and pure fractions can be collected without any significant cross-contamination for further analysis. At 10°C the analysis time was not too much lengthened (17 min as compared to 12 min at 19°C).

When the sample triacylglycerols are saturated, several authors recommended proceeding at temperatures higher than ambient, 30–45°C to shorten retention times and to improve separation (395,405,423,444). Analyses at programmed temperature have sometimes been performed (423)

Figure 11 shows in comparison to Fig. 9 (chromatogram B) that resolution can be improved by temperature programming in analysis of butter oil triacylglycerols. Temperature was programmed from 10°C to 55°C at a rate of 1°C/min and maintained at 55°C until complete elution. Elution of the low molecular weight triacylglycerols was improved at lower temperature.

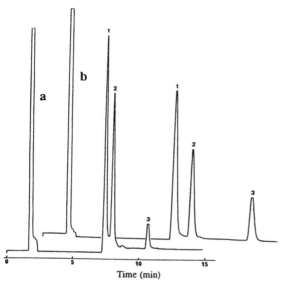

Time (min)

Fig. 10 Reverse phase HPLC separaton of 3,5-dinitrophenylurethane (DNPU) derivatives of 1,2(2,3)-diacyl-*sn*-glycerols originating from chemical deacylation of (a) peanut oil and (b) cottonseed oil palmitoyloleoyllinoleoylglycerol (16:0 18:1 18:2) at (a) 19°C and (b) 10°C. Other analytical conditions: stainless steel column (250 × 4 mm i.d.) packed with 4 μm LiChrospher 100 RP-18; eluant acetone-acetonitrile (40:60, v/v) at 1 ml/min; refractive index detection; isocratic analysis. (From Ref. 107.)

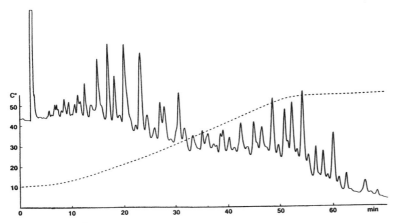

Fig. 11 RPLC profile of butterfat triacylglycerols using temperature programming from 10°C to 55°C at 1°C/min (dotted line). Other operating conditions: 250 × 4 mm i.d. LiChrospher 100 RP-18 column; eluant, acetone-acetonitrile (50:50 v/v) at 1 ml/min; refractive index detection. (From Ref. 164.)

e. Peak Identification. The most safe and accurate method to identify (and quantify) triacylglycerols separated by HPLC consists of collecting and analyzing each fraction at the outlet of the column. When using a nondestructive detector (differential refractometer, UV spectrometer), fractions are better collected at the outlet of the detector. With destructive detectors, the major part of the column effluent must be diverted for collection of peaks. After addition of an internal standard in each fraction collected—generally an odd chain simple triacylglycerol—the triacylglycerol component fatty acids are identified and quantified by GC. The triacylglycerols present in each fraction can generally be identified, more or less easily, from the fatty acid composition (165,399,402,405,421,423,428–430,432,433). This method is sometimes the only one that permits uncommon triacylglycerols to be identified with certainty. However, the method is time consuming. In routine analyses more rapid methods of peak identification are required. One of them is the use of partition number.

In reverse phase HPLC on a ODS column, triacylglycerols roughly elute according to increasing carbon number (CN) and decreasing unsaturation (DB), i.e., according to partition number (PN), which can be calculated by the formula:

$$PN = CN - 2DB$$

in which CN is the total number of acyl carbon atoms and DB the total number of double bonds in the triacylglycerol molecules (445–448). This elution property is verified with relatively poor resolutive columns. It permits prediction of the elution order. Two compounds displaying the same partition number, e.g., tripalmitin and triolein (PN = 48), are coeluted. They are called a critical pair (394). Nowadays, with the development of high efficiency columns, the formula defines a zone on the chromatogram inside which several triacylglycerols of the same PN are eluted, generally according to decreasing (287) unsaturation. With the very complex mixture of butter oil triacylglycerols, the chromatogram shows series of "quartets," with each quartet comprising four triacylglycerols of same PN (Fig. 9). With modern, efficient HPLC columns, more complex formulas are necessary to predict the elution order with accuracy.

El Hamdy and Perkins (394) proposed the concept of theoretical carbon number (TCN) as a means of identifying triacylglycerols. For each peak on the chromatogram, TCN can be experimentally determined from its capacity factor k':

$$k' = \frac{V_1 - V_0}{V_0}$$

in which V_1 is the retention volume of the peak and V_0 the retention volume of a triacylglycerol used as reference and from the linear relationship:

$$\text{TCN} = \text{NC}$$

established for a series of simple even chain length saturated triacylglycerols. It can also be calculated from the formula:

$$\text{TCN} = \text{ECN} - S^3{}_{1Ui}$$

in which ECN is the equivalent carbon number, i.e., the partition number (PN), and U_i the factor experimentally determined for each component fatty acid by using simple triacylglycerols. For example $U_i = 0.60–0.65$ for c18:1, 0.70–0.80 for cc18:2, 0.20 for t18:1, and of course 0 for saturated fatty acids. $S^3{}_1$ is the sum of the U_i of the three component fatty acids of a triacylglycerol molecule.

It is thus possible to identify triacylglycerols by means of their ECN or their NC and DB. Moreover, their component fatty acids can be determined from their k' and their TCN. For example, a triacylglycerol for which TCN = 44.8, ECN = 46.0, and $S_1{}^3 = 1.2$ is composed of two molecules of c18:1 and one molecule of 14:0. Additionally, when two triacylglycerols have identical CN and DB, the calculation of their TCN permits one to know if they can be separated (TCN different). The elution order is that of increasing TCN.

Podlaha and Toregard (408,449) developed another procedure of identification based on the ECN, with the meaning of ECL (equivalent chain length). The ECNs are deduced from the linear relationship:

$$\log t'_R = f(\text{CN})$$

in which t'_R is the reduced retention time and CN the carbon number of a series of simple saturated even chain length triacylglycerols. ECNs of triacylglycerols are graphically determined from their reduced retention times.

For homologous series of triacylglycerols displaying the same characteristics of unsaturation, a linear relationship exists between their ECN and their CN. The straight lines traced for series of triacylglycerols of different degrees of unsaturation are parallel. These straight lines traced for a reduced number of known triacylglycerols can be extrapolated for the identification of unknown peaks on the chromatogram. This is illustrated in the following results, obtained by Semporé and Bézard (405).

Peanut oil triacylglycerols were fractionated by reverse phase HPLC into 33 fractions, as shown in Fig. 12. Triacylglycerols of each fraction were analyzed by GC for fatty acid composition. In Fig. 13, 52 individual triacylglycerols are characterized by their ECN determined graphically and reported in the abcissa, and their carbon number (CN) calculated from that of the constituent fatty acids identified by GC analysis and reported in the ordinate. All the triacylglycerols with the same unsaturation characteristics (class) lie on a straight line and the different lines corresponding to nine classes are parallel. These lines can be traced from a limited number of model or natural triacylglycerols and then utilized to characterize the unsaturation and the chain length of the unidentified triacylglycerols present in the oil analyzed.

Determination of ECN is very reproducible. It can afford very informative indications on the component triacylglycerols of an unknown peak on the chromatogram such as total number of acyl carbon atoms, number and position of double bonds, and identity of the component fatty acids, since ECN of a triacylglycerol is the sum of the ECN of the component fatty acids. ECNs have

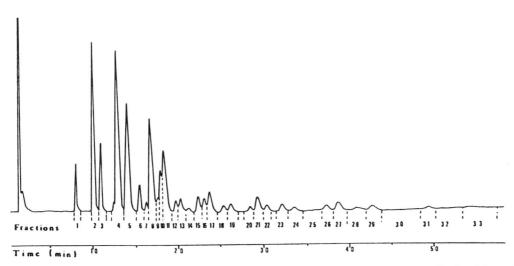

Fig. 12 Chromatograms recorded during the collection of the 33 fractions of peanut oil triacylglycerols separated by reverse phase HPLC on a 250 × 4 mm i.d. Hibar LiChrospher 100 CH-18 (3 µm) column eluted with acetone-acetonitrile (70:30 v/v) at 1.2 ml/min (isocratic conditions) and at 30°C. (From Ref. 405.)

proven to be very characteristic and useful elution parameters in modern reverse phase HPLC analysis of triacylglycerols.

A third method of identification has been described by Goiffon et al. (416,417). It is based on the use of selectivity of triacylglycerols that is the reduced retention times relative to that of trioleoylglycerol. Linear relationships exist between the number of double bonds of triacylglycerols and log α. These relationships have been observed by successive substitution of the identical component fatty acids, e.g., in the series 16:0 16:0 16:0, 16:0 16:0 18:1, 16:0 18:1 18:1, and 18:1 18:1 18:1. A diagram:

$$\log \alpha = f(\text{number of double bonds})$$

can thus be traced on a graph. Log α of a given triacylglycerol is equal to the sum of the log α of its three component fatty acids. Log α of each fatty acid is equal to one-third of the log α of the corresponding simple triacylglycerol: In the triacylglycerol ABC composed of the three fatty acids, A, B, and C, the relationship is:

$$\log \alpha \, (ABC) = \log \alpha \, (A) + \log \alpha \, (B) + \log \alpha \, (C)$$

with the other relationship:

$$\log \alpha \, (A \text{ or } B \text{ or } C) = 1/3 \log \alpha \, (AAA \text{ or } BBB \text{ or } CCC)$$

By calculation or by graphical determination, it is thus possible to know the retention time relatively to trioleoylglycerol of any triacylglycerol. Conversely, identification of a triacylglycerol the retention time of which is known can be graphically determined. Using this method, Perrin and Naudet (419) established the elution order of 120 triacylglycerols from the logarithm of their reduced retention times.

The method proposed by Goiffon et al. (416,417) is illustrated in Figure 14, which reports for 48 peanut oil triacylglycerols the relationship between the number of double bonds and log α. Points for 48 individual triacylglycerols lie on several series of straight parallel lines. One series

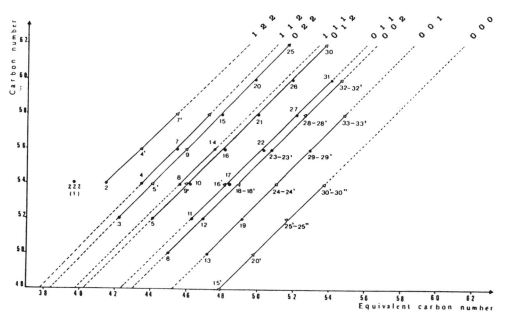

Fig. 13 Carbon number (CN) vs. equivalent carbon number (ECN) for 52 peanut oil triacylglycerols (TGs). ECNs were determined graphically from extrapolation of the linear relationship of CN vs. log α (definition in Fig. 14) for the five saturated simple TGs from trihexanoin (C18) to tritetradecanoin (C42). CNs were calculated from those of the three component fatty acids of the TGs after identification. The TGs are numbered as the collected fractions in which they were present (Fig. 12) and according to decreasing percentages in the case of several TGs in the same fraction. Twenty-eight TGs (plain circles) were immediately identified from the fatty acid composition of the fractions collected. Twenty-four TGs (open triangles) were identified later. Each straight line corresponds to one class of unsaturation (0, 1, 2 double bonds of the three component fatty acids). Numbers in the figure correspond to the following TGs in which L: 18:2, P: 16:0, S: 18:0, A: 20:0, B: 22:0, T: 24:0, H: 26:0, Po: 16:1, O: 18:1; Ao: 20:1, E: 22:1.

class 222 (1: LLL)
class 122 (2: OLL, 4′: AoLL, 7′: ELL)
class 112 (4: OOL, 7: OAoL)
class 022 (3: PLL, 5′: SLL, 9: ALL, 15: BLL, 20: TLL, 25: HLL)
class 111 (8: OOO, 14: OOAo)
class 012 (5: POL, 9′: PAoL, 10: SOL, 16: AOL, 21: BOL, 26: TOL, 30: HOL)
class 011 (11: POO, 16′: POAo, 17: BPoO, 22: AOO, 27: BOO, 31: BOAo)
class 002 (6: PPL, 12: PSL, 18: PAL, 18′: SSL, 23: PBL, 23′: SAL, 28: SBL, 28′: PTL, 32: STL, 32′: PHL)
class 001: 13: PPO, 19: PSO, 24: PAO, 24′: SSO, 29: PBO, 29′: SAO, 33: PTO, 33′: SBO)
class 000 (15′: PPP, 20′: PPS, 25′: PSS, 25″: PPA, 30′: PPB, 30″: PSA) (From Ref. 405.)

of parallel lines corresponds to triacylglycerols obtained by substituting each time one linoleic acid for palmitic acid, starting from a saturated triacylglycerol containing two or three molecules of palmitic acid. Another series of parallel lines is obtained by substituting oleic acid for palmitic. Other series of parallel lines can be obtained by substituting linoleic acid or oleic acid for the other saturated fatty acids. These straight and parallel lines can be traced out from a reduced number of known triacylglycerols, particularly from commercially available simple triacylglycerols or from triacylglycerols easily identifiable in a natural mixture as in peanut oil from this work.

The relationship log α = f(CN) of saturated triacylglycerols gives rise to two equations from

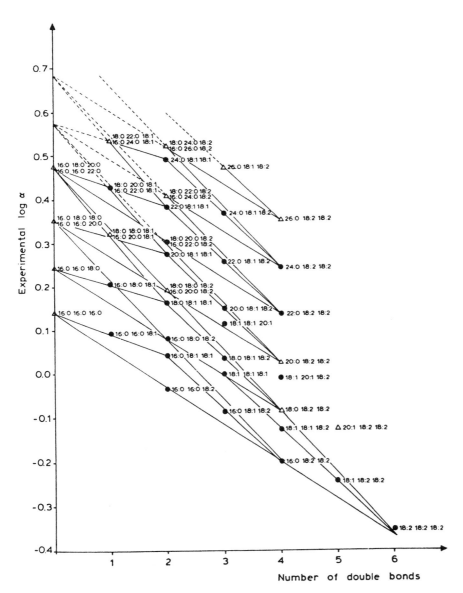

Fig. 14 Log α vs. number of double bonds. Log α is the logarithm of the reduced retention time of triacylglycerols relative to that of trioleoylglycerol (triolein). It was determined from the chromatogram reported in Fig. 12. The number of double bonds was calculated from that of the three component fatty acids of the triacylglycerols after identification. Twenty-seven triacylglycerols (plain circles) were immediately identified from the fatty acid composition of the fractions collected. The other 21 (open triangles) were identified later. (From Ref. 405.)

which calculation of log α from NC or of the ECN of Podlaha and Toregard (408,449) from log α is possible. In the work the authors have carried out on peanut oil triacylglycerols (405), the equations were:

$$\log \alpha = 0.057CN - 2.60$$

$$ECN = 17.54 \log \alpha + 45.60$$

f. Detectors and Quantification. Determination of the triacylglycerol composition of a sample analyzed by reverse phase HPLC is directly related to the detector used (287,445,450).

DETECTION BY THE REFRACTIVE INDICES

Differential refractometry has been used in some isocratic analyses. Quantitative determinations are based on peak area measurement or any other measurement relative to the registered peaks. In a first approach, peak areas were directly related to the mass of triacylglycerols and peak area percentages were taken as weight percentages of triacylglycerols (407,409,451,452). Practical data obtained were in good agreement with those later obtained by analysis of the constituent fatty acids of the collected fractions corresponding to the registered peaks (402,453).

A second, more complex approach consisted of determining calibration factors to improve accuracy of the analyses. These factors were experimentally determined by means of simple saturated and unsaturated triacylglycerols (405) or were calculated from the refractive indices of simple triacylglycerols (419).

Triacylglycerol composition calculated from peak areas by taking into account the calibration factors were very close to those determined after peak collection, addition of an internal standard, and GC analysis of the constituent fatty acids (405).

DETECTION BY UV ABSORBANCE

The UV detector response is highly influenced by the unsaturation of triacylglycerols at 205–210 nm. At 215 nm, using UV absorption of the ester linkages, peak areas represent a rather accurate measure of the amount of triacylglycerols especially under isocratic conditions (451,454). Shukla et al. (406) using a UV detector at 220 nm obtained acceptable quantitative data for natural oils. At this wavelength, UV absorption is due to the group C=O in the ester linkage.

DETECTION BY FLAME IONIZATION

Theoretically, it is the more suited detector in the analysis of triacylglycerols. When using the detector developed by Privett and Erdhal (404,455), there is proportionality between peak areas and amounts of triacylglycerols detected under isocratic conditions and with elution gradients. No calibration factors are needed. However, this detection displayed several drawbacks. It was difficult to handle (450) and overall was destructive. It has rarely been used.

LIGHT-SCATTERING DETECTOR

This kind of detector has been used by many analysts (410–412,441,442,456) because it is compatible with elution gradients (see Fig. 8). However, its response is not linearly related to the mass of triacylglycerols (456). But practically, in a mixture the relative proportion of the different triacylglycerols determined from peak areas is quite acceptable, even with elution gradients (456).

The laser light-scattering detector developed by Stolyhwo et al. (412,457–459) is well adapted to the quantitative analysis of triacylglycerols, although its response is not directly proportional to the mass injected but to the power x of this mass. However, it need not determination of calibration factors. The only precaution is to experimentally determine x, used to correct areas according to the formula:

Corrected area = (measured area)$^{1/x}$

Under these conditions, analysis of triacylglycerols without any other calibration factors gives results in good agreement with those obtained by GC analysis of the constituent fatty acids.

CHEMICAL DETECTION

A detection system involving chemical reactions of the compounds emerging from the column has been described by Kondoh and Takano (460). Triacylglycerols eluted are hydrolyzed by potassium hydroxide. The liberated glycerol is oxidized and converted to a derivative spectrophotometrically detected at 461 nm (460). The response is linearly related to the amount injected between 0.3 and 60 nmol.

    g. *Official Methods.*   Two methods of triacylglycerol analysis by HPLC are proposed by the American Oil Chemists' Society (AOCS):

    The method Ce 5c-93 (461), which is a method of separation, identification, and quantitative determination of individual triacylglycerols in vegetable and animal oils and fats containing only long-chain fatty acids (C16–C24)
    The method Ce 5b-89 (462), which is specific for liquid vegetable oils (olive oil, palm oil, peanut oil)

In both cases the recommended apparatus and reagents are as follows:

    A liquid chromatograph equipped with a device to control the column temperature.
    A differential refractometer, or a UV spectrophotometer, or a mass detector as detection system.
    A 250 × 4.5 mm i.d. column packed with 5-μm particles bonded with octadecylsilyl groups or two 150 × 4.5 mm i.d. columns of Spherisorb S3 ODS-2 (3-μm C18) mounted in series.
    A mobile phase depending on the detection system used, i.e., acetonitrile-acetone (differential refractometer) or acetonitrile-tetrahydrofuran (UV spectrophotometer at 220 nm) or a gradient of dichloromethane in acetonitrile (mass detector).
    Triacylglycerols to be injected are dissolved in acetone or in acetone-chloroform (1:1, v/v) and the sample injected is about 10 hl.

    h. *Determination of Triacylglycerol Composition.*   Nowadays with the reversed phase C18 columns, 5- or 3μm particles allowing the separation of individual compounds from the simplest natural mixtures of triacylglycerols, a detection system permitting the quantitation of peaks directly on the chromatograms (e.g., differential refractometer or light-scattering detector), a mobile phase leading to good resolution of triacylglycerols, and a suitable identification method, it is possible to obtain an accurate triacylglycerol composition of usual oils and fats in less than 1 hr. The rapidity and ease of operation (the extraction or purification step of the triacylglycerols is not always necessary) transform RP-HPLC, in that case, into a technique suitable for routine analyses. The triacylglycerol compositions of soybean, peanut, olive, palm, avocado, corn, palm kernel oils, cocoa butter, lard, and tallow were thus determined (404,418,419,438).

    The determination of an accurate triacylglycerol composition of the more complex mixtures always requires the combination of at least two different techniques because of the great number of triacylglycerols in each peak:

    Preparative or semipreparative RP-HPLC for collection of separated peaks, followed by GC analysis to obtain fatty acid composition or triacylglycerol group (according to carbon number) composition of each peak. The percentage of each peak can be calculated by means of peak areas or by standardization of the collected fractions (402,405). The accurate triacylglycerol composition of each peak is determined from its fatty acid composition and, if necessary, its triacylglycerol group (according to carbon number) composition. Table 6 reports the calculation of the triacylglycerol composition of an RP-HPLC peak of triacylglycerol (405). Wada et al. (399,428–430), Bézard and Ouedraogo (402), Semporé and Bézard (405), Maniongui et al. (164), Gresti et al. (165), Weber et al. (463, 464), and Barron et al. (465) performed such a work with cottonseed oil, peanut oil, soybean oil, fish oil, butter fat, and beef fat triacylglycerols, respectively.

Ag⁺-TLC and Ag⁺-HPLC to isolate triacylglycerols according to their number of double bonds. These fractions can then be analyzed by RP-HPLC and the separated peaks are either identified and quantified directly on the chromatograms, or collected and standardized before GC analysis of their fatty acids and eventually their triacylglycerol groups according to carbon number. This analytical scheme has been utilized especially with the first RP-18 columns (10-μm particles) in the study of the triacylglycerol composition of black cod lipids (429,430) and cottonseed oil (402).

ANALYSIS OF PARTIAL ACYLGLYCEROLS BY HPLC The pancreatic lipase hydrolysis of triacylglycerols produces 2- monoacyl-sn-glycerols and 1,2(2,3)-diacyl-sn-glycerols. The partial chemical deacylation by a Grignard reagent leads to a mixture of 1-, 2-, and 3-monoacyl-sn-glycerols and a mixture of 1,3- and 1,2(2,3)-diacyl-sn-glycerols. The analysis of these acylglycerols as part of structural analysis of triacylglycerols requires:

The separation of monoacylglycerols and diacylglycerols from the degradation products.
The analysis of these compounds, which can be preceded by a segregation of sn-2 and sn-1(3)-monoacylglycerols or sn-1,2(2,3)- and sn-1,3-diacylglycerols by TLC on silica gel impregnated with boric acid (285). Saturated and unsaturated compounds may be resolved by argentation TLC or HPLC.

The isolation of mono- and diacylglycerols from the degradation products and the separation of sn-2- and sn-1(3)-monoacylglycerols, sn-1,3- and sn-1,2(2,3)-diacylglycerols has been previously described. The resolution between saturated and unsaturated mono- and diacylglycerols will be seen in detail later.

Analysis of monocylglycerols

The first analysis of monoacylglycerols formed by pancreatic lipase hydrolysis or by Grignard deacylation requires separation of these compounds according to carbon number and number of double bonds of the fatty acid esterifying the glycerol molecule. It requires for the second analysis and in the case of monoacylglycerols produced by Grignard degradation the separation of the sn-2 isomer and the sn-1(3)-monoacylglycerols for each fatty acid.

A reverse phase HPLC method for the resolution of positional isomers of monoacylglycerols was described in 1985 by Kodali et al. (466). Separations were obtained with underivatized compounds and diacetate derivatives of monoacylglycerols:

Separation in this order of 2-monopalmitoyl-sn-glycerol and 1(3)-monopalmitoyl-sn-glycerol on Altex C18 column (250 mm × 4.6 mm i.d.) using as mobile phase a gradient of tetrahydrofuran-water-isopropanol- acetonitrile (from 7.6:23:5:64.4 to 3.2:0:70:26.8 in volumes).
Separation on a Nucleosil 7 OH column (250 mm × 5 mm i.d.) with isooctane-tetrahydrofuran (95:5, v/v) of a mixture of monopalmitoylglycerol and monooleoylglycerol as diacetates. The isomers were well resolved according to the following elution order:

Table 6   Determination of the Triacylglycerol-type Composition of Fraction 5 of Peanut Oil

Fatty acids			Triacylglycerol types				Mol %	Equations[e]	Solution[f]	
	Mol %									
Structure	Exptl[a]	Calc.[b]	CN^f	Class[c]	Log α[c]	Structure[d]			Fraction	TGs
16:0	28.20	29.80	54	022	−0.075	18:0, 18:2, 18:2	a	$b+d = 28.20 \times 3 = 84.60$	$a = 10.20$	1.38
18:0	3.42	3.53	52	012	−0.070	16:0, 18:1, 18:2	b	$a+c+e = 3.42 \times 3 = 10.26$	$b = 85.12$	11.55
			52	012	−0.063	18:0, 16:1, 18:2	c	$c+d+2e = 0.27 \times 3 = 0.81$	$c = 0$	0
			50	011	−0.058	16:0, 16:1, 18:1	d	$b+d+2f = 30.80 \times 3 = 92.40$	$d = 0.80$	0.11
16:1	0.27	0.27	50	011	−0.051	18:0, 16:1, 16:1	e	$2a+b+c+f = 37.31 \times 3 = 111.93$	$e = 0$	0
18:1	30.80	29.80								
						18:1, 18:1, 18:2	f		$f = 3.88$	0.53
18:2	37.31	36.59								13.57

[a] Determined experimentally by gas chromatography.

[b] Calculated from the TG-type composition reported in the last column.

[c] Carbon number (CN), class of unsaturation (class), and retention time relative to trioleoylglycerol (α) of the TG types present in the fraction.

[d] The last TG type represents a contaminant from the preceding fraction.

[e] Five equations corresponding to the five fatty acids present in the fraction and derived from the fatty acid composition of the TG types and from that of the fraction.

[f] Mol % of the component TG types in the fraction and in the total TGs of the oil.

Source: Data from Ref. 405.

1(3)-MG-18:1 < 2-MG-18:1 < 1(3)-MG-16:0 < 2-MG-16:0

The separation and quantitation of saturated (C6–C18) and unsaturated (C18:1 and C18:2) simple underivatized monoacylglycerols has been described by Maryuma and Yonese (104). The separations were obtained on a Unisil Q (C8, 250×4.6 mm i.d., 5-μm particles) using acetonitrile-water (7:3, v/v) as mobile phase.

The monoacylglycerol isomers separations by HPLC were later improved by Takano and Kondoh (105). These authors using a hypersil MOS (C8, 3-μm particles) column (150×4.6 mm i.d.) and acetonitrile-water (67:33 or 58:42, v/v) as mobile phase have succeeded in resolving a mixture of positional isomers of saturated (from C12:0 to C18:0) and unsaturated (C18:1 and C18:2) monoacylglycerols. These monoacylglycerols were separated according to the following elution order: 1(3)-MG-12:O < 2-MG-14:0 < 1(3)-MG-14:0 < 2-MG-18:2 < 1(3)-18:2 < 2-MG-16:0 < 1(3)-MG-16:0 < 2-MG-18:1 < 1(3)-MG-18:1 < 2-MG-18:0 < 1(3)-18:0. Unfortunately, the detector used to identify and quantify the compounds emerging from the column, the glyceride-selective postcolumn reactor detector (GS-PCRD), involved destruction of monoacylglycerols and did not allow their collection for further analysis.

Taking into account these previous results, Semporé and Bézard proposed in 1992 (108) a procedure of analysis and fractionation of monoacylglycerols as part of stereospecific analysis of triacylglycerols, permitting:

The RP-HPLC separation of complex mixtures of monoacylglycerols as underivatized molecules according to the nature of the constituent fatty acid (chain length and unsaturation) and to its positioning [sn-1(3)- and sn-2-monoacylglycerols]

The collection of the separated monoacylglycerols, especially of the sn-1(3) isomers for derivatization and enantiomer resolution

In the proposed method, the monoacylglycerols were separated on a LiChrospher 100 CH 18/II Super (4-μm particles), detected and quantified with a differential refractometer. The mobile phase was acetonitrile-water in various proportions (80:20, 85:15, 91:10, 95:5, v/v). The most suitable analytical conditions were as follows:

separations at ambient temperature (~19°C) and in the order rac-1-18:2 < sn-2-18:1 + sn-2-16:0 < rac-1-18:1 < rac-1-16:0 < rac-1-18:0.

separations at 12°C of the critical pair isomers MG-16:0 and MG-18:1 according to the order sn-2-18:1 < sn-2-16:0 < sn-1(3)-18:1 < sn-1(3)-16:0.

This procedure has been applied to the monoacylglycerols formed by Grignard deacylation of the triacylglycerols 16:0 18:1 18:1, 16:0 18:1 18:2, and 18:1 18:1 18:2 isolated from peanut oil, and 16:0 18:1 18:2 isolated from cottonseed oil.

Figure 15 illustrates the separations of the peanut oil 16:0 18:1 18:2 monoacylglycerols obtained under these conditions. The clear-cut resolution of the 1(3)-monoacyl-sn-glycerols (peaks 2, 5, 6) from the other peaks allowed collection of pure fractions without any cross-contamination for further analysis for enantiomer composition.

For the quantitative determination, the sn-2- monoacylglycerols formed by Grignard deacylation, isolated by borate-impregnated silica TLC (108), were analyzed for fatty acid composition by GC as methyl esters and by HPLC for monoacylglycerol composition determined from peak areas. The two series of determinations led to very similar results, since the variations did not exceed 4% showing that peak areas were representative of the amount of compounds detected by their refractive indices.

Analysis of diacylglycerols

Numerous methods have been developed for the analysis by reverse phase HPLC of diacylglycerols. In most cases, they were applied to 1,2-diacyl- sn-glycerols issued from phospholipase C

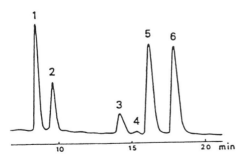

Fig. 15 Reverse phase HPLC separations of underivatized monoacylglycerols obtained by partial chemical deacylation of the peanut oil palmitoyloleoyllinoleoylglycerol (16:1 18:1 18:2). Mixture of 1 = sn-2-18:2; 2 = sn-1(3)-18:2; 3 = sn-2 18:1; 4 = sn-2-16:0; 5 = sn-1(3)-18:1; 6 = sn-1(3)-16:0. Operating conditions: 250 × 4 mm i.d. stainless steel column packed with 4 μm Lichrospher 100 CH-18 II; mobile phase, acetonitrile-water (85:15 v/v) at a flow rate of 1.2 mL/min; analysis temperature, 12°C; refractive index detection. (From Ref. 108.)

hydrolysis of phospholipids. Table 7 reports a certain number of these methods because the analytical conditions defined for these diacylglycerols can be applied to the analysis of the diacylglycerols formed by partial deacylation of triacylglycerols.

To date, reverse phase HPLC has rarely been applied to diacylglycerol mixtures as part of triacylglycerol structural analysis.

However, the study of Kondoh and Takano (106), who developed a method for simultaneous determination of mono-, di-, and triacylglycerols, must be cited. They separated both positional isomers and carbon number homologous of saturated simple standard diacylglycerols (differing by four carbon atoms, from C28:0 to C36:0) and saturated diacylglycerols (differing by two carbon atoms from C30:0 to C38:0) of an emulsifier in a commercial household product.

The analyses were carried out using a Hitachi gel 3057 (3-μm particles) and two mixtures of ethanol-acetonitrile-water (40:48:12 = solvent A and 65:35:0 = solvent B, in volumes) as mobile phase for gradient elution. The diacylglycerols were eluted according to increasing carbon number and for each of them the sn-1,3 isomer eluted ahead of the sn-1,2(2,3) isomer. However, as noticed in the monoacylglycerol analysis section, the detection method involved destruction of the partial acylglycerols and did not allow their collection for further analysis.

For the purpose of stereospecific analysis of triacylglycerol, Semporé and Bézard (107) studied the HPLC separation of complex mixtures of diacylglycerols according to the nature of the constituent fatty acids (chain length and unsaturation) and their positioning [sn-1,3- and sn-1,2(2,3)-diacylglycerols]. Analyses were carried out isocratically at ambient or subambient temperatures on a LiChrospher 100 CH-18/II Super (4-μm particles) column (250 × 4 mm i.d.) using acetone-acetonitrile (45:55 or 40:60, v/v) as mobile phase.

These analytical conditions have led to very good separations of the diacylglycerol positional isomers as urethane derivatives. Figure 16 illustrates the separations obtained with a mixture of sn-1,3- and sn-1,2(2,3)-diacylglycerols and of sn-1,2(2,3)-diacylglycerols alone, prepared by deacylation of natural triacylglycerols from peanut and cottonseed oils. The following conclusions can be drawn:

All of the sn-1,3-and sn-1,2(2,3)- diacylglycerols studied were clearly separated, even those of the critical pairs, i.e., the diacylglycerols presenting the same ECN or PN. This fact indicates that it is possible to analyze simultaneously the sn-1,3 and sn-1,2(2,3) isomers of diacylglyc-

Table 7 Conditions of Diacylglycerol (DG) Analysis

DG derivatives	Column	Mobile phase	Detector	Ref.
Acetates	Zorbax ODS (4.6 × 250 mm)	Acetonitrile-isopropanol-methyl-t-butyl ether (72:18:8:2, v/v), isocratic	UV 205 nm	3
Acetates	LiChrosorb RP-18	Acetonitrile-isopropanol-water (35:15:1, v/v), isocratic	Refractive indices	4
tert-Butyldimethylsilyl ethers	Supelco TM C18	Gradient of 30–90% of propionitrile in acetonitrile	Mass spectrometry	5, 6
p-Nitrobenzoates p-Methoxybenzoates	Brownlee RP-18	Isopropanol-acetonitrile (7:13, v/v)	UV, 254 nm	7, 8
p-Anisoyls	Spherisorb C6 (4.6 × 125 mm)	Gradient of 30–0% water in acetonitrile	UV 250 nm	9, 10
Dinitrobenzoates	Ultraspher ODS (4.4 × 250 mm)	Acetonitrile-isopropanol (4:1, v/v) or methanol-isopropanol (19:1, v/v)	UV	11, 12
Benzoates	Ultraspher ODS (4.4 × 250 mm)	Acetonitrile-isopropanol (70:30, v/v, isocratic) Gradient of 30–60% of isopropanol in acetonitrile	UV 230 nm	13–15 16
Naphthylurethanes	LiChrosorb RP-8 (10 µm) or RP-18 (5 µm)	RP-8: acetonitrile-water (83:17, 5 min then 87:13, v/v), RP-18, linear gradient of 10–4% water in acetonitrile, 1 ml/min; RP-18, linear gradient of 4–14% water in methanol	UV, 290 nm Fluorescence	17–20
5-Dimethylaminonaphthalene-1-sulfonyl (Dns)-phosphatidylethanolamines (dansylethanolamine)	RP C8 (250 × 4.5 mm)	Methanol-aqueous potassium phosphate pH 7 (83:17, v:v, 1 ml/min)	Fluorescence Excitation, 360 nm Emission >420 nm	21, 22

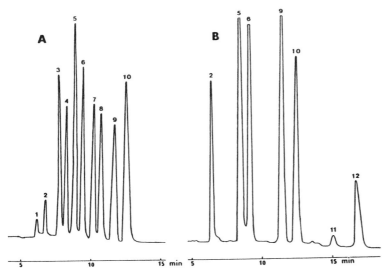

Fig. 16 Reverse phase HPLC separations of 3,5-dinitrophenylurethane (DNPU) derivatives of 1,3-diacyl-and 1,2(2,3)-diacyl-*sn*-glycerols formed by chemical deacylation of several peanut oil triacylglycerols. 1 = sn-1,3-18:2 18:2; 2 = sn-1,2(2,3)- 18:2 18:2; 3 = sn-1,3-18:1 18:1; 4 = sn-1,3-16:0 18:2; 5 = sn-1,2(2,3)-18:1 18:2; 6 = sn-1,2(2,3)-16:0 18:2; 7 = sn-1,3-18:1 18:1; 8 = sn-1,3-16:0 18:1; 9 = sn-1,2(2,3)- 18:1 18:1; 10 = sn-1,2(2,3)-16:0 18:1; 11 = sn-1,2(2,3)- 18:1 20:1; 12 = sn-1,2(2,3)-18:0 18:1. In (A) mixture of both sn-1,3-and sn-1,2(2,3) isomers. In (B) sn-1,2(2,3) isomers. Analytical conditions 250 × 4 mm i.d. stainless steel column packed with 4 µm LiChrospher 100 CH-18 II, isocratically eluted with acetone-acetonitrile (40:60, v/v) at a flow rate of 1.2 ml/min; analysis temperature (A) 20°C, (B) 19°C; refractive index detection. (From Ref. 107.)

erols without previous fractionation by TLC on borate- impregnated silica in the range of the diacylglycerols analyzed, which are those usually encountered in common oils and fats.

In both series of isomers, the diacylglycerols were eluted according to increasing ECN and when the ECN was the same the more unsaturated of the pair eluted earlier. For example, sn-1,2(2,3)-18:1 18:1 eluted earlier than sn-1,2(2,3)-16:0 18:1 (PN = 32).

For the diacylglycerols with the same two component fatty acids or the same ECN, the sn-1,3 isomers eluted earlier than the corresponding sn-1,2(2,3) isomers.

In each series of isomers, the *sn*-1,2(2,3)-diacylglycerols were better separated than the corresponding sn-1,3 isomers, exhibiting higher separation and resolution factors between two successive peaks. This property is important because in the stereospecific analysis of triacylglycerols the *sn*-1,2(2,3)-diacylglycerols have to be separated from the sn-1,3 isomers after deacylation and fractionated by reverse phase HPLC as derivatives before further separation of the two isomers on a chiral column. The good separation observed guarantees the purity of the fractionated diacylglycerols.

It has been observed that for a given series of isomers, a decrease in the analysis temperature considerably improved the separation of the different diacylglycerols (as previously illustrated in Fig. 10). The elution order of the diacylglycerols analyzed was as follows: sn-1,3-18:2-18:2 < sn-1,2(2,3)-18:2-18:2 < sn-1,3-18:1-18:2 < sn-1,3-16:0-18:2 < sn-1,2(2,3)-18:1-18:2 < sn-1,2(2,3)-16:0-18:2 < sn-1,3-18:1-18:1 < sn-1,3-16:0-18:1 < sn-1,2(2,3)-18:1-18:1 < sn-1,2(2,3)-16:0-18:1 < sn-1,2(2,3)-18:1-20:1 < sn-1,2(2,3)-18:0-18:1.

Different quantitative aspects of the diacylglycerol analyses have been studied and are reported in Table 8. The following should be noted:

Comparison between the fatty acid composition of the underivatized sn-1,2(2,3)-diacylglycerols and the composition reconstituted from fatty acid composition and the proportion of the RP-HPLC fractions of dinitrophenylurethane derivatives of these compounds shows a good agreement between the two series of data, the average deviation being only 2% with a maximum of 5.5%.

Comparison between the proportion of the different fractions of the sn-1,2(2,3)-diacylglycerols separated by RP-HPLC, calculated from peak areas (detection by differential refractometry) and from the fatty acid composition of the collected and standardized fractions, also shows a good agreement between the two series of data since the differences did not exceed 3.8%.

Comparison of the percentages of the sn-1,2(2,3)-diacylglycerol fractions calculated from the fatty acid composition of the sn-2-monoacylglycerols formed by pancreatic lipase hydrolysis and the percentages obtained from the fatty acid composition of the fractions collected after HPLC and standardized, or calculated from peak areas, shows that the series of data were very similar. The average difference was only 2.1% and the maximum was 3.7%.

These results show that (a) the method used generated representative sn-1,2(2,3)-diacylglycerols which can be accurately analyzed by RP-HPLC using differential refractometric detection and that (b) they can be fractionated by HPLC and collected for enantiomeric resolution by chiral phase

Table 8 Percentages of the sn-1,2(2,3)-Diacylglycerols Formed by Deacylation of Triacylglycerols and Fractionated as Urethane Derivatives by Reversed Phase HPLC

		Concentration (mol %)		
Triacylglycerols[a]	Diacylglycerols[b]	Peak areas[c]	Fatty acids[d]	sn-2-MG[e]
16:0 18:1 18:1	18:1 18:1	46.9	46.1	47.9
	16:0 18:1	53.1	53.9	52.0
16:0 18:1 18:2	18:1 18:2	49.2	48.1	49.1
	16:0 18:2	42.2	41.4	40.1
	16:0 18:1	8.6	10.5	10.8
18:1 18:1 18:2	18:1 18:2	77.9	77.6	78.1
	18:1 18:1	22.1	22.4	21.9
18:1 18:2 18:2	18:2 18:2	44.8	43.3	44.0
	18:1 18:2	55.2	56.7	56.0
16:0 18:1 18:2	18:1 18:2	50.5	48.4	48.5
	16:0 18:2	29.4	29.8	30.4
	16:0 18:1	20.1	21.8	21.1

[a]Triacylglycerols isolated from peanut oil (the first four) and cottonseed oil (the last) by combined argentation-TLC and reverse phase HPLC.
[b]sn-1,2(2,3)-Diacylglycerol fractions separated as derivatives and collected by reserve phase HPLC.
[c]Percentages calculated from peak areas registered in HPLC analysis of the mixture of diacylglycerol derivatives with refractometric detection (without any area correction according to fatty acids). Results are means of 3–5 different analyses.
[d]Percentages calculated from the fatty acid composition of the collected fractions after addition of a known amount of heptadecanoic acid.
[e]Percentages calculated from the percentages of sn-2-monoacylglycerols.
Source: Data from Ref. 107.

HPLC for stereospecific analysis of triacylglycerols, as will be illustrated. HPLC can also be used to separate enantiomers of monoacylglycerols (1- and 3-monoacyl-*sn*-glycerols) or enantiomers of diacylglycerols (1,2– or 2,3 diacyl–*sn*–glycerols), especially on chiral columns. The use of chiral phase HPLC will be developed and illustrated in the section dealing with the stereospecific analysis of triacyl-*sn*-glycerols.

4. Argentation-HPLC

a. Introduction. Argentation-TLC has several disadvantages: Unsaturated triacylglycerols are able to be oxidized; fractions collected are contaminated by silver ions, silica, and dye; and quantification is somewhat troublesome. The method is time consuming when carried out after previous extraction of the separated triacylglycerols from the silica and not very accurate for fractions amounting to less than 10% of the total (468).

This is why several authors have attempted to substitute argentation-HPLC, which is both more rapid and more quantitative. Three approaches have been used:

Addition of a silver salt in the mobile phase, in conventional reverse phase HPLC
Impregnation of the silica support with a silver salt
Linkage of silver ions by an ion exchange support

b. Addition of Silver Ions in the Mobile Phase. The first separations of triacylglycerols using this method were performed on reverse phase columns, e.g., Nucleosil 5 C18 with methanol-isopropanol (3:1, v/v) containing $5–10^{-2}$ M silver perchlorate (469). This study was followed by two others using the same approach (397,470) but with different columns (μ-Bondapak C_{18} and Zorbax ODS) and different elution solvents (acetonitrile-acetone, 2:1, v/v) containing 0.2 M silver. These methods did not present any advantage over conventional reverse phase HPLC (287) and the risk of corrosion and of silver deposit in the detection cells could not be avoided.

c. Columns Packed with Silver Nitrate–Impregnated Silica Gel. Better results were obtained with columns packed with silver ions adsorbed on the silica support. The supports used were Partisil 5 (347,468), Develosil 60-3 (316), Spherosil X OA 600 (471), and Nucleosil 100-3 (472). Impregnation of the support was achieved with a solution of silver nitrate in methanol (473), in acetonitrile (474), in water (475), in methanol-water (9:1, v/v) (476), or in commercial prepacked columns by injection of a solution of silver nitrate in a polar solvent, acetonitrile (304), or water (477,478).

The better ratio of silver nitrate/silica was shown to be around 10%. It represents a good compromise between acceptable resolution and relatively short retention times (468). Resolution is increased by decreasing the operating temperature, e.g., to 6–8°C in the separation of positional isomers of triacylglycerols (468). Use of small support particles (3 μm) improves resolution together while decreasing retention time (316,472). When using solvent gradients with detection by flame ionization it yielded the best separation of triacylglycerol positional isomers ever obtained by Ag$^+$-HPLC (472) such as separation of SOS and OSS, of SLS and SSL, of SOO and OSO, and of SLO and OSL. The mobile phases used were, under isocratic conditions, a mixture of toluene-hexane-ethyl acetate (30:20:1, v/v/v) (471) or benzene (316). In the latter case, very good resolutions of saturated and mono-, di-, and trisaturated triacylglycerols were achieved in analytical scale (250–300 × 4 mm i.d. columns, sample size up to 3 mg) as well as in semipreparative scale (200 × 7.5 mm i.d. columns, sample size up to 200 mg). Moreover, an accurate quantification was achieved using a differential refractometer as detector and by means of methyl santalbate (octadec-9-yne-11-*trans*-enoic acid methyl ester) as internal standard (468). Separation of triacylglycerol positional isomers on 3-μm silica particle size was achieved using a linear gradient of the two solvent systems, toluene-hexane (1:1, v/v) and toluene-ethyl acetate (9:1, v/v).

d. Cation Exchange Columns. Silver-loaded silica gel columns slowly lose silver ions in the mobile phase. This causes short operating life of the columns, deposit of silver, or even corrosion of the detection systems and contamination of the fractions collected at the outlet of the column. To circumvent those difficulties the most advantageous approach was to use a prepacked cation exchange column, containing benzenesulfonic acid chemically bonded to silica gel and retaining silver ions. The simplest method for preparing such a column (477,478) consists of injecting 1 ml 20% aqueous silver nitrate via the injector in 50-ml aliquots at 1-min intervals into an aqueous mobile phase in a standard Nucleosil 5 SA column (250 × 4.6 mm i.d.). The column is then washed with methanol for 1 hr and with 1,2-dichloroethane-dichloromethane (1:1, v/v) for another hour. Fifty- to eighty-milligram silver ions are bonded to the sulfonic groups. Such columns were reported to be stable for long periods of time since no leaching of silver ions occurred. This property is of special value when fractions are collected for further analysis. Resolutions achieved with these columns are very good. When resolutions deteriorate it is recommended that the column be flushed with methanol-acetonitrile or, if it does not improve the column efficiency, to repeat the silver ion loading procedure.

A column of this type is now commercialized by Chrompack International under the denomination of Chromspher Lipids. On this column, triacylglycerols are rigorously separated according to the number of double bonds when using a linear gradient of dichloromethane-dichloroethane (50:50, v/v) and acetone; retention increases with the degree of unsaturation. Under isocratic operating conditions using hexane-toluene-ethyl acetate as eluting solvent, resolution of positional isomers of unsaturated triacylglycerols is readily achieved on this column. The Nucleosil 5 SA column loaded with silver ions has been used with different solvent systems, more or less complex according to the sample and generally by means of solvent gradients (477–479).

Mixtures of triacylglycerols containing at most three double bonds, e.g., palm oil, have been resolved in six fractions containing from zero to three double bonds, without any cross-contamination and without any contamination by silver ions when collected, using a linear gradient of 1,2-dichloroethane-acetone, at 0.75 ml/min, in 30 min, and detection by a mass detector (477). Resolution of more complex mixtures, such as rat adipose tissue triacylglycerols (478), has required a gradient of two binary solvents: A—1,2-dichloroethane-dichloromethane (1:1, v/v) and B—acetone-acetonitrile (9:1, v/v) used with C—acetone. The solvent programming was as follows:

1. Linear gradient of A and B at 0.75 ml/min, for the separation of nine fractions of sheep subcutaneous adipose tissue, containing three major fatty acids (16:0, 26%, 18:0, 21%, 18:1n-9, 35%).
2. Gradient of 100% A to 50% A–50% B in 15 min then to 50% B–50% C in 25 min; the latter was maintained for 5 min. This program was used in rat adipose tissue or oils rich in linoleic acid. The solvent flow rate was constant and equal to 0.75 ml/min.
3. Gradient of 100% A to 50% A–50% B in 10 min, then to 70% B–30% C in 20 min, and finally 100% C for 30 min (C = acetone-acetonitrile, 4:1, v/v). This program was used with oils rich in linolenic acid and with fish oils, provided that elution by the last solvent was prolonged for an additional 10-min period (479). The triacylglycerol samples (0.25–0.8 mg) were dissolved in dichloroethane (5–10 ml) for injection onto the column.

The mobile phase recommended for the Chromspher Lipids are (A) dichloromethane-dichloroethane (50:50, v/v) and (B) acetone, with the following program for resolution of palm oil triacylglycerols strictly according to unsaturation: 100% A to 100% B at 1 ml/min in 40 min. Resolution of positional isomers of triacylglycerols is achieved under isocratic conditions with hexane-toluene-ethyl acetate (80:10:10, v/v/v).

e. Quantification.   The same methods as for conventional HPLC are suitable for quantification of triacylglycerol molecular species separated by Ag$^+$-HPLC. Quantitative data are derived from the detector signals, provided the detector response is calibrated. Calibration by *external standard* is carried out by analyzing equal sample volumes containing one or several compounds in known concentrations, within the expected concentrations of the samples to be analyzed. Calibration is considered to be accurate when the response is linearly related to the concentrations, starting from the zero point.

The method of *internal standardization* requires the use of a compound close to the type of molecule to be analyzed and eluting in the same area but distinctly. The internal standard is added in a precisely known amount before extraction or any other treatment.

The detector response for the analyzed compounds is compared to that of the internal standard and possibly used in the determination of calibration factors.

Methyl santalbate (methyl octadec-9-yne-11-*trans*-enoate) has been used as an internal standard for quantification of triacylglycerols SSS, SUS, and SSU with detection by differential refractometry (468). Trilaurin has been used in the quantification of palm oil triacylglycerols (16:0, 46%; 18:1, 40%) with detection by flame ionization (472).

The most accurate method of quantification, but one that is also time consuming, consists of *collecting the fractions* at the outlet of the nondestructive detector or at the outlet of the column by diverting the major part of the effluent, when using a destructive detector, and to add a precisely known amount of a standard methyl pentadecanoate (477) or nonadecanoate (478,479) before analyzing the constituent fatty acids by GC (477–481).

f. Separations of Triacylglycerols by Argentation–HPLC.   The elution of the triacylglycerols according to the degree of unsaturation of their three constituent fatty acids (zero, one, two, or three double bonds in each constituent fatty acid) is as follows: 000 > 001 > 011 > 002 > 111 > 012 > 112 > 022 = 003 > 013 = 122 > 113 > 023 = 222 > 123 > 033 > 223 > 133 > 233 > 333. This technique has been used to separate triacylglycerols of palm (477), sunflower, corn, safflower, linseed, and evening primrose oils, and subcutaneous fat of sheep (478), fishes (anchovies, pilchards, Atlantic and Baltic herrings), and sandeel (478,479). The fractions separated were collected manually by means of a stream splitter installed between the column and the mass detector, then standardized with methyl nonadecanoate. They were analyzed by GC as methyl esters for fatty acid composition.

Figure 17 reports an example of separations obtained with subcutaneous sheep triacylglycerols. It shows that resolution of triacylglycerols occurred according to unsaturation. It also shows the separation of triacylglycerols according to cis and trans configuration of one component fatty acid (18:1). Table 9 reports the fatty acid composition of the fractions collected after separation of sunflower oil triacylglycerols (478). It shows that separations were sufficiently precise to permit triacylglycerol composition of each fraction to be determined and triacylglycerol class to be defined from the proportions of fatty acids with the same degree of unsaturation.

However, in the case of oils or fats comprising a wide range of fatty acids with different chain lengths and degrees of unsaturation, such as fish oil (479), it is not possible to identify the individual components by this technique alone. After Ag$^+$-HPLC a complementary fractionation must be performed, e.g., by reverse phase HPLC.

5. *Mass Spectrometry of Triacylglycerols*

Mass spectrometry (MS) is an alternative or additional method of analysis which can be advantageously utilized to give qualitative information which cannot be obtained by other means of analysis, such as molecular weight, empirical formula, and structure. It can also give quantitative information on the triacylglycerol mixture composition. It consists in the study of molecular ions

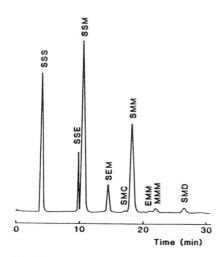

Fig. 17 Separation of triacylglycerols from sheep subcutaneous adipose tissue by HPLC with a silver ion column and mass detection. The triacylglycerol fractions are denoted by the unsaturation characteristic of the three component fatty acids. S = saturated; E = elaidic (*trans*-monoenoic); M = *cis*-monoenoic; C = conjugated dienoic; D = dienoic. Operating conditions: 250 × 4.6 mm i.d. Nucleosil 5 SA loaded with silver nitrate; elution with linear gradient of A (1,2-dichloroethane-dichloromethane, 1:1 v/v) to B (acetone) generated over 40 min, at a flow rate of 0.75 ml/min; detection by the ACS model 750/14 mass detector. (From Ref. 478.)

produced by ionization of molecules. This reaction is produced in the ion source of the instrument from the sample in the gaseous form, using electron impact procedure or chemical ionization or, more recently, field desorption and fast atom bombardment, which is a solid phase ionization technique utilizing production of charged particles from the surface of the condensed sample. Several comprehensive reviews deal with the equipment and techniques used in mass spectrometry of lipids (102,287,482–486).

Analysis of triacylglycerols by mass spectrometry was first carried out by directly introducing the sample into the heated ionization chamber (483,487–496). Mass spectrometers were then coupled to gas–liquid chromatographs. By means of an appropriate interface, the compounds emerging as vapors from the packed columns (497,498) or from the fused silica capillary columns (499,500) were directly introduced into the ionization chamber. The best mass spectra, showing intense molecular or quasi-molecular ions, were obtained using mild ionization procedures. Practically only the chemical ionization procedure can be used with gas chromatography on packed columns or on capillary columns. In the latter case, the technique has served in the analysis of triacylglycerols from algae (501) and from butter (502).

Now the mass spectrometer is also coupled to HPLC by means of different interfaces that are commercially available. Construction and properties of these interfaces have been reviewed elsewhere (102,287,503–506). Analysis and identification of triacylglycerol molecular species by HPLC-MS have principally been performed by the Kuksis group (112,200,507–509). The mass spectrometer used in the chemical ionization mode was equipped with an interface allowing introduction of the eluting solvent (acetonitrile or propionitrile) in the ionization chamber, serving as the reagent gas.

By means of HPLC-MS in the chemical ionization mode or by means of GC-MS using ammonia as the reagent gas to increase fragmentation of the parent ions (510–512), most of the major triacylglycerol molecular species from a great number of fats and oils of commercial interest were identified, as well as minor triacylglycerols, when conveniently separated (112,508).

Table 9 Fatty Acid and Triacylglycerol Composition (mol % of total) of Sunflower Seed Oil Total Triacylglycerol and of Fractions Obtained by Silver Ion HPLC

Fractions	1	2	3	4	5	6	7	8	9	Mol% of the total
TG classes	001	011	002	111	012	112	022	122	222	
%/class	0.6	2.3	2.8	3.9	10.2	9.3	17.7	26.8	26.6	
16:0	37.2	17.5	36.9	—	19.6	—	17.4	—	—	7.3
18:0	27.7	12.8	27.5	—	11.7	—	14.4	—	—	4.9
Σ saturated	64.9	30.3	64.4	—	31.3	—	31.8	—	—	21.9
18:1	35.1	69.6	—	100.0	34.2	61.9	—	31.9	—	21.9
18:2	—	—	35.5		34.5	38.1	68.2	68.1	100.0	65.9
TG										
16:0 16:0 18:1	14.6									0.09
16:0 18:0 18:1	85.4									0.51
16:0 18:1 18:1		57.8								1.33
18:0 18:1 18:1		42.2								0.97
16:0 16:0 18:2			14.6							0.41
16:0 18:0 18:2			85.4							2.39
18:1 18:1 18:1				100						3.90
16:0 18:1 18:2					62.6					6.39
18:0 18:1 18:2					37.4					3.81
18:1 18:1 18:2						100				9.30
16:0 18:2 18:2							54.7			9.68
18:0 18:2 18:2							45.3			8.02
18:1 18:2 18:2								100		26.80
18:2 18:2 18:2									100	26.60

Source: Data from Ref. 478.

The molecular ions (MH⁺) were indicative of the molecular weight of the triacylglycerol molecules, while the ions (MH-RCOOH)⁺ resulting from the loss of an acyl group gave information about the molecular structure. Via the identification of 1,2- and 2,3- diacylglycerols the structure of the parent molecule is likely to be unequivocally determined. For example, the mass spectrum of dipalmitoyllinoleoylglycerol (16:0-16:0-18:2) shows a MH⁺ ion at $m/z = 831$ and (MH-RCOOH)⁺ ions at $m/z = 551$ and 575 corresponding to dipalmitoylglycerol (16:0-16:0) and to palmitoyllinoleoylglycerol (16:0-18:2), respectively. The mass spectrum of trioleoylglycerol (18:1-18:1-18:1) shows an MH⁺ ion at $m/z = 885$ and a (MH-RCOOH)⁺ ion at $m/z = 603$. As an example, the characteristics of the diagnostic ions of the molecular species of corn oil are reported in Table 10 (508).

Figure 18 illustrates what kind of mass spectrum can be recorded when analyzing a triacylglycerol fraction separated by reverse phase HPLC. Fraction 12 of milk fat triacylglycerol was isolated after HPLC separation, as previously shown in Fig. 8, and submitted to positive ion desorption–chemical ionization–mass spectrometry (513). The mass spectrum reported in the figure shows the presence of a single triacylglycerol in this fraction, corresponding to the molecular ion (M + H)⁺ at $m/z = 611$, i.e., 34:0. Fragment ions correspond to elimination of butyric acid ($m/z = 523$), myristic acid ($m/z = 383$), or palmitic acid ($m/z = 355$) from the protonated molecule. The triacylglycerol can be easily identified as butyroyl-myristoyl-palmitoyl glycerol (4:0 14:0 16:0).

With more complex oils like peanut oil a previous fractionation by argentation-chromatography is necessary for identification of the compounds further separated by reverse phase HPLC. The differentiation between fatty acids at the sn-2 position from those at the sn-1(3) positions was achieved, but differentiation between the two enantiomers was not.

Table 10 Diagnostic Ions Present in the Mass Spectrum in the Chemical Ionization Mode of Triacylglycerols From Corn Oil

Triacylglycerol	MH⁺	(MH-R-COOH)⁺	
18:2-18:2-18:3	877	599	597
18:2-18:2-18:2	879	599	
18:1-18:2-18:3	879	601	597
16:0-18:2-18:3	853	575	597
18:1-18:2-18:2	881	601	599
16:0-18:2-18:2	855	575	599
18:1-18:1-18:2	883	603	601
18:0-18:2-18:2	883	603	599
16:0-18:1-18:2	857	577	601
16:0-16:0-18:2	831	551	575
18:1-18:1-18:1	885	603	
18:0-18:1-18:2	885	605	601
16:0-18:1-18:1	859	577	603
16:0-18:1-16:0	833	577	551
18:0-18:1-18:0	889	605	607
16:0-18:0-18:1	861	579	605

Source: Data from Ref. 508.

Quantification of triacylglycerols separated by HPLC-MS is not easily achieved (509). Yield in molecular ions and in ion fragments depends on the chain length, degree of unsaturation, and the position of the constituent fatty acids. However, when calibration is carefully worked out, quantification of triacylglycerols from natural oils and fats is quite possible.

6. Supercritical Fluid Chromatography

In supercritical fluid chromatography (SFC) the eluant is a gas compressed above its critical temperature (T_c) and critical pressure (P_c). Carbon dioxide is the most commonly used eluant for many reasons (514). SFC is situated between GC and HPLC. The method has gained great attention these last years, especially in the analysis of nonvolatile compounds including triacylglycerols.

a. Instrumentation. Details on instrumentation are likely to be found in comprehensive publications (514–517). Briefly, instrumentation comprises:

A system for delivery of the supercritical fluid, which allows fluid pressure and density programmation.

An injection device close to that used in HPLC.

Packed or WCOT columns (518–520). The packed columns are comparable to those used in HPLC (521,522), i.e., columns of 1–5 mm of internal diameter, packed with 3–10 μm silica and used in the normal or reverse mode. The solid phase used has also been polystyrene-divinylbenzene polymers (518,522). The WCOT columns are comparable to those used in GC but with a lower internal diameter (50–100 mm) and a stationary phase more tightly bound to column wall (523–525). The nonpolar columns are coated with methyl- or octylpolysiloxanes. The polar columns are coated with phenyl-, biphenyl-, or cyanopropylpolysiloxane.

A flow restrictor, attached to the outlet of the column which allows the maintaining of a constant

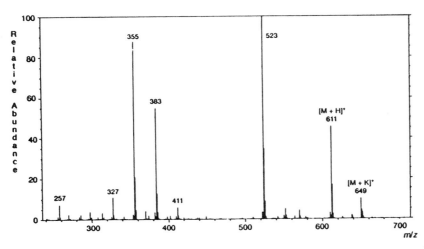

Fig. 18 Positive ion desorption chemical ionization (DCI) mass spectrum of fraction 12 (shown in Fig. 8) collected during reverse phase HPLC fractionation of milk fat triacylglycerols. Operating conditions: fraction collected through a flow splitter at the outlet of the HPLC column (split ratio = 9:1); isobutane positive ion DCI obtained using a JEOL JMS-HX11OHF double-focusing mass spectrometer equipped with a direct insertion probe; ion source temperature, 150°C. The presence of a single triacylglycerol (34:0) is indicated by the molecular ion (MH^+, m/z 611). Fragment ions correspond to elimination of butyric acid (m/z 523), myristic acid (m/z 383), or palmitic acid (m/z 355) from the protonated molecule. (From Ref. 513.)

density of the supercritical fluid all along the column and to control its linear velocity (526–529).

A detector comparable to that used in HPLC or GC (530).

  b. Analysis of Mono-, Di-, and Triacylglycerols.   Table 11 shows the different applications of SFC to the analysis of partial acylglycerols and of triacylglycerols (514). According to the experimental conditions, the triacylglycerols are separated according to chain length, degree of unsaturation, or both.

   Separation of triacylglycerols according to chain length was achieved using capillary columns coated with relatively nonpolar phases, DB-1, DB-5, or SE-54. For example, the triacylglycerols of palm oil or of milk chocolate were separated in the presence of the SE-54 stationary phase (5% phenyl-, 1% vinyl-, 94% methylpolysiloxane) with programmed CO_2 pressure (541) or programmed analysis temperature from 130 to 50°C (543). Other examples are analyses of triacylglycerols from soybean oil (542), butter oil (536,538,546), fish oil (536,539), or grain oils (534,536).

   SFC was also applied to the analysis of mixtures of mono-, di-, and triacylglycerols in the presence of dimethylsiloxane BP-10 stationary phase, with the elution order being mono-, then di-, and at last position triacylglycerols (543,545,554). When CO_2 added with methanol was used as mobile phase on a silica column, the elution order was reversed (553). Figure 19 shows part of a chromatogram (part A) recorded in SFC analysis of lipid classes from fungi (550). This part concerns the separation of triacylglycerols. In an attempt to characterize these triacylglycerols, standard triacylglycerols were analyzed under the same conditions. In part B of Fig. 19 is represented the retention factors of the fungi and the standard triacylglycerols.

   Comparative studies of milk chocolate triacylglycerols were carried out by GC using OV-1

Table 11 Applications of SFC to Analysis of Mono-, Di-, and Triacylglycerols

Acylglycerols	Column	Supercritical fluid	Detector	Ref.
Triacylglycerols	Packed	$CO_2 + CH_3OH$	LSD	531
		$CO_2 + ACN + IPA$	UV	532, 533
		CO_2	FID	522, 534
	Capillary	CO_2	FID	535–546
		CO_2	FT-IR	535
		CO_2	MS	538, 547, 548
Mono- and diacylglycerols	Capillary	CO_2	FID	549
Di- and triacylglycerols	Packed	CO_2-hexane or CH_3OH	UV	550
		CO_2	FID-UV	551
	Capillary	CO_2 or water saturated in CO_2	FID	552
Mono-, di-, and triacylglycerols	Packed	$CO_2 + CH_3OH$	LSD, UV, FT-IR	553
	Capillary	CO_2	FID	543–545, 553, 554

Supercritical fluids: CO_2 (carbon dioxide), CH_3OH (methanol), ACN (acetonitrile), IPA (isopropyl alcohol); *Detectors*: LSD (light-scattering detector), UV (ultraviolet detector), FID (flame ionization detector), FT-IR (fourier transform infrared detector), MS (mass spectrometer).
Source: Data from Ref. 514.

as stationary phase, an analysis temperature programmed from 290°C to 350°C at a rate of 20°C/min, and hydrogen as carrier gas and capillary SFC using the SE-54 phase, at 170°C and a linear programmation of the CO_2 pressure (544). Results showed that the proportions in the total of the major fractions were similar. But analysis by GC lasted 3 min as compared to 30 min by SFC.

Separation of triacylglycerols according to unsaturation (535,542,543,545) was carried out using more polar columns (25% cyanopropyl-, 25% phenyl-, 50% methylpolysiloxane) DB-225 (545), OV-225 (543), (50% cyanopropyl–50% methylpolysiloxane) 5B, cyanopropyl-50 (535), Carbowax 20M (542), and polyethylene glycol. In the presence of these stationary phases, triacylglycerols are separated according to their chain length and according to unsaturation within a given chain length with the more unsaturated being the more retained on the column.

SFC was adapted to argentation-chromatography of triacylglycerols (532,533). A capillary column of Nucleosil 551 loaded with silver nitrate and using CO_2-ACN-IPA (94:3:5.3:0.40, v/v/v or 92.8:6.5:0.7, v/v/v) as the supercritical fluid was utilized to analyze triacylglycerols from fish and grain oils (532), standard triacylglycerols and triacylglycerols from corn oil, linseed oil, sunflower oil, and palm oil (533). Results are obtained by Ag^+ reverse phase HPLC with, however, a different elution order since in Ag^+-SFC triacylglycerols are separated according to both unsaturation and chain length. SFC can also be coupled with SM. This coupling gave informations on the degree of unsaturation and the chain length of butter oil triacylglycerols in the presence of DB-5 (5% phenyl–95% methylpolysiloxane) as the stationary phase. Moreover, the fatty acids esterified at the 1(3) positions could be differentiated from those esterified at the sn-2 position (538).

With regard to identification and quantification of triacylglycerols separated by SFC, Proot et al. (541) studied the reproducibility of the retention times and the accuracy of the quantitative estimations. Following four successive analyses of a mixture of trilaurin-trimyristin-tripalmitin

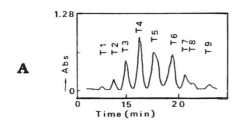

A

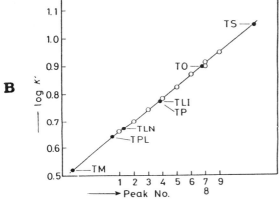

B

Fig. 19 In (A), separation of fungi triacylglycerols (T1 to T9) by supercritical fluid chromatography (SFC). Chromatographic conditions: Finepak Sil C18 (50 μm, 250 × 10 mm): mobile phase, supercritical dioxide with entrainer (hexane), at a flow rate of 3.0 ml/min; temperature, 40°C; pressure, 140 kg/cm^2; UV detector (195–350 nm). In (B), retention factor of the triacylglycerols peaks T1 to T9 (open circles) and of standard triacylglycerols (plain circles): TM = trimyristoyl-; TPL = tripalmitoeyl-; TLN = trilinolenoyl-; TP = tripalmitoyl-; TLI = trilinoleyl-; TO = trioleoyl-; TS = tristeraoylglycerol. (From Ref. 550.)

and tristearin, standard deviations lower than 0.16% and 0.8% for retention time and quantitative data, respectively, were observed.

C. Positional Distribution of Fatty Acids Between the Three Positions of the Triacylglycerols in Natural Oils and Fats

1. Determination of Fatty Acids Esterified in the sn-2 Position of Triacylglycerols

After the study of the constituent fatty acids (determination of their nature and proportion), the next step in the structural analysis of triacylglycerol mixtures of natural oils and fats is to determine the distribution of fatty acids between the primary and secondary positions of the glycerol molecule.

The study of the fatty acids esterifying the sn-2 position of triacylglycerols was rendered feasible owing to the investigations of Borgström (555), Mattson and Beck (16,556), and Savary and Desnuelle (17,557) on the specificity of pancreatic lipase. These authors and several other workers (558–564) showed that this enzyme preferentially hydrolyzes primary ester linkages in triacylglycerols to produce fatty acid residues from the external (sn-1 and sn-3) positions, sn-1,2(2,3)-diacylglycerols, and sn-2-monoacylglycerols. Thus the resultant sn-2-monoacylglycerols are indicative of the nature and proportion of fatty acids found in the 2 position of triacylglycerols and consequently of the nature and proportion of the β or sn-2 isomers of triacylglycerols.

Two Official Methods for hydrolysis of triacylglycerols using pancreatic lipase—separation of the formed sn-2-monoacylglycerols by TLC, methanolysis of these monoacylglycerols and analysis of their fatty acid methyl esters by gas chromatography—were adopted by the American Oil Chemists' Society (AOCS) and the International Union of Pure and Applied Chemistry (IUPAC), AOCS Official Method Ch 3-9 (565) and IUPAC Method 2-210 (566).

In these methods, which are not applicable to oils and fats containing substantial amounts of fatty acids with 12 or fewer carbon atoms, highly unsaturated fatty acids (more than four double bonds) containing 20 or more carbon atoms, or fatty acids containing oxygenated groups other than the acid group, the hydrolysis with pancratic lipase is performed as follows: To 100 mg of sample of triacylglycerols (dissolved in 0.2 ml of hexane with gentle warming if the sample is a solid fat) is added:

20 mg of pancreatic lipase (powder)
2 ml of a buffer solution, 1 M aqueous solution of Tris-hydroxymethylaminomethane (Tris), pH 8
0.5 ml of sodium cholate solution 1 g/L
0.2 ml of a calcium chloride solution 220 g/L

The digestion is allowed to proceed for 3 min:1 min in a thermostatted water bath maintained at 40°C with hand shaking, 2 min out of the water bath with agitation by means of an electric shaker. Lipolysis is terminated with 1 ml hydrochloric acid (HCl) and the products of lipolysis are extracted with 1 ml diethyl ether. The sn-2-monoacylglycerols are then isolated from the other lipids by preparative TLC, using n-hexane-diethyl ether-formic acid (70:30:1, v/v/v) as the developing solvent.

In the group of the authors (72) was used for the deacylation of triacylglycerols an adaptation of the semimicromethod of Luddy et al. (567) which has been widely used, with the reagents of Mattson and Volpenheim (568). To about 10 mg of triacylglycerols in a 5-ml glass tube is added 1 ml of 1 M buffer Tris pH 8 containing 5 mg $CaCl_2$, and 0.5 ml of 0.1% (w/v) taurocholate. The mixture is sonicated for 1 min in a cold bath and the tube is then placed and shaken in a bath at 40°C. When the mixture has reached this temperature, 100 μl of rat pancreatic juice (or 100 μl of purified porcine pancreatic lipase solution in buffer, 20 mg/ml) is poured. The reaction is stopped after 1 min (pH 1 with 1 N HCl) and the lipids are immediately extracted. The pancreatic juice is collected from a rat, with no bile at 0°C, according to Mattson and Volpenheim (569). Amounts of 0.5 ml are then distributed in sealed tubes, stored at −18°C. The tube is heated at 40°C for the experiment and only used one time. The lipids are extracted from the lipolysis medium three times with 1 ml of diethyl ether. They are recovered by evaporating ether under nitrogen and dried with absolute ethanol under reduced pressure. Then 0.2 ml chloroform is added and the lipolysis products are separated into monoacylglycerols (MGs), diacylglycerols (DGs), free fatty acids (FAs), and triacylglycerols (TGs) (according to elution order from bottom to top of the chromatoplate) by TLC on plates of 0.5-mm-thick Kieselgel G (Merck) containing 8% (w/w) boric acid to prevent acyl migration (282). The developing solvent is a mixture of hexane-diethyl ether-acetic acid-methanol (90:20:2:3, in volumes) as used by Brown et al. (570). The bands are visualized with 2′,7′-dichlorofluorescein (0.2% in absolute ethanol) under UV light (350 nm). After making their location, the silica gel zones are scraped off the plate in beakers containing a mixture of chloroform-methanol (90:10 in volumes). After 2 hr contact, the solution is filtered and the powder rinsed twice with chloroform-methanol. The solvent is then evaporated and the lipids are dissolved in pure chloroform.

Similar deacylation procedures have been described by other workers (270,325,568,571).

Likewise, enzymes other than pancreatic lipase can be used for the structural analysis of triacyl-glycerols, e.g., milk lipase, *Rhizopus arrhizus* lipase, *Geotrichum candidum* lipase specific for fatty acids with cis-9 unsaturation regardless of their position in the triacylglycerols. Some information and the utilization procedures concerning these enzymes were reviewed by Litchfield (22).

Pancreatic lipase hydrolysis has been widely used for determining the positional distribution of fatty acids in the internal position (572–579) and in the combined sn-1 and sn-3 positions of triacylglycerols (577,578,580–584).

Analysis of sn-2-monoacylglycerols by GC of the fatty acid methyl esters indicates the nature and the proportion of fatty acids esterified in the sn-2 position of triacylglycerols.

The fatty acid composition of the sn-1 and sn-3 positions can be found by analysis of the fatty acids released by the enzyme. Unfortunately, these fatty acids are generally not representative of the combined sn-1 and sn-3 positions of the original triacylglycerols. This is due to the following factors:

Fatty acid and triacylglycerol specificities of pancreatic lipase (22) cause the free fatty acid composition to vary with the extent of hydrolysis [the *sn*-1,2(2,3)-diacylglycerols may or may not be representative of the original triacylglycerols].

Some monoacylglycerols (1–2%) are completely hydrolyzed to glycerol and some fatty acids of the 2 position contaminate the free fatty acids.

When required, the positional distribution of the fatty acids at the combined sn-1 and sn-3 positions can be calculated as follows:

$$\% \text{ A at sn}{-}1,3 \text{ positions} = \frac{3[\% \text{ A TG}] - 2[\% \text{ A 2–MG}]}{2}$$

$$\% \text{ A at sn-1,3 positions} = 3[\% \text{ A TG}] - 2[\% \text{ A } sn\text{- } 1,2(2,3)\text{-DG}]$$

in which % A at sn-1,3 positions, % A 2-MG, % A *sn*-1,2(2,3)-DG, and % A TG represent the molar percentage of the fatty acid A at the combined sn-1,3 positions, in the 2- monoacylglycerols, in the *sn*-1,2(2,3)-diacylglycerols and in the original triacylglycerols, respectively.

The pancreatic lipase hydrolysis has also been used for determining the *sn*-2-triacylglycerol composition of natural oils or fats (72,179,181–184,221,273,339,362,364,585–590). According to the precision desired, more or less complex mixtures of triacylglycerols have been used as substrates for the lipolysis: total triacylglycerols, triacylglycerol classes having the same number of double bonds and isolated by TLC-AgNO$_3$ (181–184,362,578,585–587), triacylglycerol groups with the same number of carbon atoms and the same number of double bonds isolated by preparative GC of the triacylglycerol classes (179,221).

In most cases, the *sn*-2-monoacylglycerols alone were used in the calculation of the proportions of the sn-2 isomers (179,221,580). But for a precise *sn*-2-triacylglycerol composition, the authors (72) used the data provided by the *sn*-2-monoacylglycerols together with those provided by the *sn*-1,2(2,3)-diacylglycerols issued from pancreatic lipase hydrolysis of triacylglycerol classes isolated by TLC- AgNO$_3$. After controlling that representative diacylglycerols have been obtained by checking their fatty acid composition vs. that calculated (% A in sn-1,2(2,3)-diacylglycerols) from the fatty acid compositions of the original triacylglycerols (% A TG) and of the 2-monoacyl-glycerols (% A 2-MG), using the formula

$$\% \text{ A in } sn{-}1,2(2,3){-}\text{diacylglycerols} = \frac{3[\% \text{ A TG}] + [\% \text{ A 2–MG}]}{4}$$

these diacylglycerols were acetylated according to Kuksis et al. (73,74). They were then separated into classes according to number of double bonds by argentation-TLC. The *sn*-2-triacylglycerol compositions of each class of triacylglycerols and of the total triacylglycerols were determined by means of the triacylglycerol composition and of the sn-2-monoacylglycerol fatty acid composition and that of the diacylglycerol classes (Tables 12 and 13). This method has been applied to other oils and fats whose sn-2 isomers were completely identified (72,181–184,362) (Table 14).

2. Stereospecific Analysis of Triacylglycerols

a. *Enzymatic Methods of Stereospecific Analysis of Triacylglycerols.* For a complete analysis of triacylglycerol positional isomers, it is necessary to distinguish between the sn-1 and sn-3 positions, which are stereochemically distinct. The procedure leading to this distinction is called "stereospecific analysis" and was first introduced by Brockerhoff (18) who demonstrated in 1965 that fatty acids esterified in the sn-1 and sn-3 positions could be separately studied using the hydrolytic stereospecificity of the phospholipase A of snake venom toward phospholipid-like molecules synthesized from diacylglycerols derived from original triacylglycerols. Since this period, four enzymatic methods of stereospecific analysis have been developed: The sn-1,2(2,3)- and the sn-1,3-diacylglycerol methods of Brockerhoff (18,282,591–593), the method of Lands (594), and the method by Myher and Kuksis (230).

A detailed description is given of the sn-1,2(2,3)- diacylglycerol method of Brockerhoff, which has been the most utilized. The three other methods can be found in the original papers or in the book by Litchfield (22).

sn-1,2(2,3)-DIACYLGLYCEROL METHOD OF BROCKERHOFF The various steps of this method (18,282, 591–593) are as follows:

1. Incubation of the triacylglycerols with pancreatic lipase to obtain sn-2-monoacylglycerols
2. Simultaneous or subsequent deacylation of the triacylglycerols with a Grignard reagent (ethyl magnesium bromide) to obtain representative sn-1,2(2,3)-diacylglycerols (282,594)
3. Isolation of the sn-1,2(2,3)-diacylglycerols by preparative TLC
4. Reaction of the sn-1,2(2,3)-diacylglycerols with phenyldichlorophosphate to produce a mixture of sn-1,2-diacyl-3-phosphatidylphenol and sn-2,3-diacyl-1- phosphatidylphenol
5. Incubation with phospholipase A. The enzyme releases fatty acids from the sn-2 position of the sn-1,2-diacyl-3-phosphatidylphenol and leaves the sn-2,3-diacyl-1- phosphatidylphenol unhydrolyzed (595–598)
6. Separation and fatty acid analysis of the reaction products which allows calculation of the fatty acid composition of the sn-1, sn-2, and sn-3 positions in the original triacylglycerols

The fatty acid composition of the sn-1-position is given by the fatty acid composition of the lysophosphatides issued from phospholipase A hydrolysis. The fatty acids esterified in the sn-2 position are those of the sn-2-monoacylglycerols obtained from pancreatic lipase hydrolysis. They also correspond to the fatty acids released by phospholipase A hydrolysis. The fatty acid proportions in the sn-3 position are deduced from the fatty acid composition of the original triacylglycerols (TGs), of the monoacylglycerols formed by pancreatic lipase (2-MG), and of the lysophosphatides and the unhydrolyzed phosphatides after the phospholipase A hydrolysis, according to the formulas:

[A] sn-3 = (3 × [A]TG) – ([A] lysophosphatide) – ([A] 2-MG)

[A] sn-3 = (2 × [A] unhydrolyzed phosphatide) – ([A] 2-MG)

where [A] represents the percentage of the fatty acid A in the original triacylglycerols, the lysophosphatides, the 2- monoacylglycerols, and the unhydrolyzed phosphatides, respectively.

This method allows direct determination of the proportions of the fatty acids esterified in the

Table 12 Equations Derived from the Fatty Acid Composition of Monoglycerides and of Diglycerides 01[a] and 02[a] Issued from Lipolysis of Peanut Oil 012[a] Triglycerides

	Monoglycerides		Diglycerides 01[a]		Diglycerides 02[a]	
Fatty acids	Mol %	Equations[b]	Mol %[c]	Equations[d]	Mol %[c]	Equations[d]
16:0	1.7	$c + f + h = 1.7$	24.4	$a + c + d + f + h = 14.2$	29.2	$b + c + e + f + g + h = 42.2$
18:0	0.4	$k = 0.4$	8.7	$i + k = 5.1$	8.9	$j + k = 12.9$
20:0	0.2	$a = 0.2$	4.9	$l = 2.9$	3.3	$m = 4.8$
22:0	27.0	$d + i + l + n + p = 27.0$	8.2	$n = 4.8$	5.6	$o + r = 8.1$
24:0	70.7	$b + e + g + j + m + o + q + r = 70.7$	3.8	$p + 2.2$	3.0	$q = 4.3$
16:1			0.4	$a + c = 0.2$		
18:1			49.4	$d + f + i + k + l + n + p = 28.8$		
20:1			0.2	$h = 0.1$		

[a]Saturated (0), mono- (1) and diunsaturated (2) fatty acid; these diglycerides and triglycerides were fractionated by argentation thin-layer chromatography.
[b]These equations express that the summed percentage of the isomers which present a given fatty acid at position 2 is equal to the percentage this fatty acid presents in monoglycerides.
[c]The experimental values were slightly corrected; the percentage of 18:2 (50.0%) in diglycerides 02 was not used.
[d]The way in which these equations were established is presented in the text.
Source: Data from Ref. 72.

Table 13   Composition in 2-Isomers of the Triglycerides 012[a] from Peanut Oil

Mol%[b]	Triglyceride types[c]	2-Isomers[d]	Unknown	Solution 1[e]	Mol% Solution 2[e]	Mean
		16:0-16:1-18:2	a	0.2	0.2	0.2
1.2	16:0 16:1 18:2	16:0-18:2-16:1	b	1.0	1.0	1.0
		18:2-16:0-16:1	c	tr.	tr.	tr
		16:0-18:1-18:2	d	12.5	11.1	11.8
50.7	16:0 18:1 18:2	16:0-18:2-18:1	e	36.5	37.9	37.2
		18:2-16:0-18:1	f	1.7	1.7	1.7
2.1	16:0 20:1 18:2	16:0-18:2-20:1	g	2.1	2.1	2.1
		18:2-16:0-20:1	h	tr	tr.	tr.
		18:0-18:1-18:2	i	4.7	4.7	4.7
17.5	18:0 18:1 18:2	18:0-18:2-18:1	j	12.4	12.4	12.4
		18:2-18:0-18:1	k	0.4	0.4	0.4
8.2	20:0 18:1 18:2	20:0-18:1-18:2	l	2.9	3.4	3.1
		20:0-18:2-18:1	m	5.3	4.8	5.1
12.2	22:0 18:1 18:2	22:0-18:1-18:2	n	4.8	4.9	4.9
		22:0-18:2-18:1	o	7.4	7.3	7.3
7.2	24:0 18:1 18:2	24:0-18:1-18:2	p	2.2	2.9	2.6
		24:0-18:2-18:1	q	5.0	4.3	4.6
0.9	22:0 22:1 18:2	22:0-18:2-22:1	r	0.9	0.9	0.9

[a]Saturated (0), mono- (1) and diunsaturated (2) fatty acid; these triglycerides were fractionated by argentation thin-layer chromatography.
[b]As determined in a previous work (180).
[c]Nature of the component three fatty acids are known but not their positioning in the molecule.
[d]The fatty acid esterified at position 2 of glycerol is known; positions 1 and 3 are considered as equivalent.
[e]As determined from the equations listed in Table 12 solution 1 from diglycerides 01, solution 2 from diglycerides 02.
Source: Data from Ref. 72.

sn-1 and sn-2 positions. The sn-3 position can be obtained by calculation from two different experimental results.

The details of the procedures used along this method are as follows.

Chemical deacylation of triacylglycerols to representative diacylglycerols

The chemical deacylation procedure of the triacylglycerols used in the laboratory of the authors (599,600) is derived from an adaptation of the Brockerhoff method (282) for semimicro amounts of sample by Christie and Moore (594) whereby 0.5 ml of ethyl magnesium bromide (C_2H_5MgBr) solution (0.2 M in anhydrous diethyl ether, freshly prepared) is added to 20–40 mg of triacylglycerols dissolved in 1 ml of anhydrous diethyl ether. The mixture is shaken for 1 min at room temperature. The reaction is then stopped by adding 0.1 ml of glacial acetic acid and 1 ml of water. The reaction products are extracted with dilute aqueous $KHCO_3$ (5%, w/v) until neutrality. After removal of the solvent, the lipids are dried with absolute ethanol, dissolved in chloroform, and fractionated by preparative TLC on silicic acid impregnated with 8% (w/w) boric acid to prevent acyl migration (282). A mixture of hexane or petroleum ether-diethyl ether (50:50, v/v) is used as the mobile phase. From bottom to top of the chromatoplate the elution order is as follows: monoacylglycerols, *sn*-1,2(2,3)-diacylglycerols, *sn*-1,3-diacylglycerols, tertiary alcohols corresponding to free fatty acids, residual triacylglycerols.

The *sn*-1,2(2,3)-diacylglycerols (eventually the *sn*-1,3-diacylglycerols or the monoacylglycerols) are extracted as described in the pancreatic lipase hydrolysis method.

Table 14 sn-2-Triacylglycerol (sn-2-TG) Composition of Several Vegetable Oils (≥1 mol %)

sn-2-TG	Peanut oil	Olive oil	Sesame oil	Cottonseed oil	Shea butter
	Mol %				
18:1-18:1-18:1	24.7	34.8	8.7		5.8
16:0-18:1-18:1	11.2	19.5	5.3	2.5	2.4
18:1-18:2-18:1	9.5	6.5	5.9	2.7	
18:1-18:1-18:2	7.6	7.4	11.9	3.0	
16:0-18:2-18:1	5.2	4.7	5.2	8.1	
18:1-18:2-18:2	4.8	3.1	15.6	9.9	
18:0-18:1-18:1	4.7	4.3	3.6		23.9
22:0-18:1-18:1	2.5				
18:0-18:2-18:1	1.8		3.3	1.0	4.0
20:0-18:1-18:1	1.7				
16:0-18:1-18:2	1.7	2.9	3.8	7.6	
20:1-18:1-18:1	1.5				
16:0-18:2-18:2	1.4	1.5	6.3	24.5	
24:0-18:1-18:1	1.4				
16:0-18:1-16:0	1.2	3.2		2.9	
16:0-18:1-18:0	1.1	1.1	1.0		3.2
22:0-18:2-18:1	1.0				
16:1-18:1-18:1		1.2			
18:2-18:2-18:2			10.3	13.8	
18:2-18:1-18:2			5.3	3.2	
18:0-18:2-18:2			4.2	2.1	
18:0-18:1-18:2			2.2		1.1
16:0-18:2-18:0			1.4	1.2	1.7
18:0-18:1-18:0					42.1
18:0-18:2-18:0					4.6
16:0-18:0-18:1					2.1
18:0-18:1-20:0					1.5
16:0-16:0-18:2				1.1	
16:0-18:2-16:0				7.9	
18:1-16:0-18:2				1.4	
Fatty Acids					
14:0				0.7	
16:0	10.5	14.4	9.7	24.5	3.3
18:0	3.5	2.9	5.6	2.2	44.3
20:0	1.5	Trace	1.0		1.3
22:0	3.0				
24:0	1.6				
26:0	Trace				
16:1	0.3	1.2	0.2	0.9	
18:1	60.3	70.0	40.0	19.5	45.6
20:1	1.3	Trace			
18:2	18.0	11.6	43.3	52.2	5.5
18:3	Trace	Trace	0.2		

Palm kernel oil

Carbon number	sn-2-TG	Mol %
C32	8:0-12:0-12:0	7.1
C34	8:0-12:0-14:0	3.0
	8:0-14:0-12:0	1.5
	10:0-12:0-12:0	4.5
	12:0-10:0-12:0	1.1
C36	8:0-12:0-16:0	1.1
	8:0-14:0-14:0	1.1
	10:0-12:0-14:0	1.1
	12:0-10:0-14:0	1.6
	12:0-12:0-12:0	21.2
C38	12:0-10:0-16:0	1.2
	12:0-12:0-14:0	11.9
	12:0-14:0-12:0	3.6
C40	12:0-12:0-16:0	3.7
	12:0-14:0-14:0	2.4
	14:0-12:0-14:0	1.4
C42	12:0-12:0-18:0	1.0
	12:0-12:0-18:1	1.6
	12:0-18:1-12:0	1.0
C44	12:0-18:1-14:0	1.2
C46	12:0-18:1-16:0	2.0
Fatty acids		
6:0		0.3
8:0		5.5
10:0		4.7
12:0		51.6
14:0		16.3
16:0		7.1
18:0		1.7
20:0		Trace
16:1		Trace
18:1		11.3
18:2		1.5
18:3		Trace

Source: Data from Refs. 179–184, 221, 362.

Similar procedures of chemical deacylation of triacylglycerols have been published by other workers (310,601).

Chemical conversion of diacylglycerols into phospholipids

The chemical synthesis of phospholipids from diacylglycerols is carried out using a widely adopted adaptation of the original method by Brockerhoff (18). The reaction is quantitative for both sn-1,2(2,3)-diacylglycerols and sn-1,3-diacylglycerols (18,593).

The diacylglycerols (5–10 mg) are dissolved in 1 ml of anhydrous diethyl ether. To this sample is added, at 0°C, 0.56 ml of phenyldichlorophosphate solution (corresponding to a mixture of 0.01 ml of phenyl dichlorophosphate, 0.50 ml of chloroform, and 0.05 ml of pyridine). After 3 hr at room temperature, 0.1 ml of pyridine and five drops of distilled water to hydrolyze the second acyl chloride group which has not reacted with the free hydroxyl group of the diacylglycerol are again added at 0°C. The mixture is left for 1 hr at room temperature. Then 2 ml of chloroform, 2 ml of methanol, and 1.5 ml of distilled water is added. After shaking, the solution is decanted. The lower organic phase (chloroform-methanol) is recovered and neutralized with 0.05 ml of triethylamine. The solvent (chloroform-methanol) is evaporated under nitrogen and the lipids (phosphatidylphenols) are dissolved in 0.5 ml of chloroform. The phosphorylated diacylglycerols are purified by TLC on silicic acid using chloroform-methanol-water (65:25:4, v/v/v) as the developing solvent. The phenyl phosphate derivatives of diacylglycerols are extracted from the silica gel with a mixture of chloroform-methanol (50:50, v/v).

Similar methods of phosphorylation of diacylglycerols have been described by Akesson (310), Sampugna and Jensen (602), while Lands et al. (594) utilized an enzymatic synthesis employing a diacylglycerol kinase for the stereospecific conversion of sn-1,2-diacylglycerols to sn-1,2-diacyl-3-phosphatidate.

Hydrolysis of phospholipids by phospholipase A

This reaction utilizes the stereospecificity of phospholipase A from snake venom. This enzyme hydrolyzes fatty acid in the 2 position from an sn-1,2-diacyl-3-phosphatide, i.e., when the phosphate group is located in the 3 position (598,603), whereas the enantiomorphic sn-2,3-diacyl-1-phosphatide is not attacked (596–598,604). When presented to an sn-1,3- diacyl-2-phospholipid substrate, phospholipase A hydrolyzes the fatty acid in the sn-1 position (604,605).

Twelve hours before the experiment, two drops of a $CaCl_2$ solution (2.2%, w/v) is added to 1 mg of Ophiophagus hannah venom in order to activate the phospholipase. This mixture is then introduced in a flask containing the synthetic phospholipids (5–10 mg) dissolved in 0.7 ml of Tris buffer [1 M tris(hydroxymethyl)methylamine, pH 8]. The reaction mixture is incubated at 27°C for 6 hr with constant shaking (48). The reaction is stopped with 5 ml of absolute ethanol. The extracted lipids are then dried three times with absolute ethanol under reduced pressure at 40°C and dissolved in a mixture of chloroform-methanol (1:1, v/v). They are applied on a silica gel TLC plate which is developed in a mixture of chloroform-methanol-water (65:25:4, v/v/v). The elution order on the chromatoplate from top to bottom is as follows: free fatty acids, unhydrolyzed phosphatidylphenols, and lysophosphatidylphenols (hydrolyzed phospholipids).

After spraying with 2',7'- dichlorofluorescein solution (0.2%, w/v, in alcohol), the silica gel zones are detected under UV light (350 nm), scraped off the plate, and the lipids extracted with chloroform-methanol (1:1, v/v). The isolated products are converted to methyl esters for GLC analysis of fatty acids.

Other phospholipase A hydrolysis procedures have been reported by Brockerhoff et al. (592), Sampugna and Jensen (602), Akesson et al. (310).

APPLICATIONS: POSITIONAL DISTRIBUTION OF FATTY ACIDS AND SPECIFIC ISOMER DISTRIBUTION IN THE TRIACYLGLYCEROLS OF NATURAL OILS AND FATS The sn-1,2(2,3)-diacylglycerol procedure has been widely used by Brockerhoff and co-workers to determine the positional distribution of fatty acids between the sn-1, sn-2, and sn-3 positions of triacylglycerols in natural vegetable oils such as peanut, rapeseed, soybean, corn, olive oils, and cocoa butter (571) as well as in depot fats of mammals, birds, and various aquatic animals including freshwater and marine fishes, mammals, birds, amphibia, reptiles, and invertebrates (592,606,607). The same method has been used by several other workers. Most of the analyses have been carried out on the total triacylglycerols (571,592,606–608), sometimes on triacylglycerol classes isolated by TLC-AgNO$_3$ (609,610). Most of the results obtained have been reported on the basis of 100 mol % at each of the three positions of total triacylglycerols analyzed (22,611), without any information on the isomeric composition and therefore on the diacid and triacid structure.

The data obtained in the studies of the positional distribution of fatty acids in triacylglycerols by the pancreatic lipase hydrolysis and the phospholipase A stereospecific hydrolysis allowed the establishment of fatty acid distribution schemes in natural oils and fats, described in detail by Litchfield (22). Tables 15–17 report some results of stereospecific analysis of natural oils and fats. However, it has rarely been possible to determine the enantiomeric composition of natural oil triacylglycerols from the enzymatic stereospecific analysis methods. Only a few workers were successful in this way.

Sampugna and Jensen (602,613) isolated the monounsaturated triacylglycerol fraction of cocoa butter by preparative silver nitrate TLC. This fraction contains the triacylglycerols 16:0 16:0 18:1, 16:0 18:0 18:1, and 18:0 18:0 18:1. The purified triacylglycerol mixture has been hydrolyzed with pancreatic lipase and *Geotrichum candidum* lipase, the latter releasing oleic acid (18:1 Δ9) (613–616) whatever its position. The sn-1,3- and the sn-1,2(2,3)-diacylglycerols produced by the *G. candidum* lipase were analyzed as acetate derivatives by GC. They were also converted to phosphatidylphenols and subsequently incubated with phospholipase A. The fatty acid composition of the 3-monoacyl-2-*sn*-phosphatides, 1-monoacyl-3-*sn*-phosphatides, and 2,3-diacyl-1-*sn*-phosphatides produced by phospholipase A hydrolysis, the fatty acid composition of the sn-2-monoacylglycerols formed by pancreatic lipase, and the diacylglycerol composition of the sn-1,3- and sn-1,2(2,3)-diacylglycerol acetates have been used to calculate the proportions of the nine isomers (triacyl-*sn*-glycerols) present in the studied monounsaturated fraction.

Bézard et al. (599) also carried out the stereospecific analysis of the palmitoyl-oleoyl-linoleoyl glycerol (16:0 18:1 18:2) isolated from cottonseed oil triacylglycerols by combined argentation-TLC reverse phase HPLC, using the Brockerhoff sn-1,2(2,3)-diacylglycerol method (18,592). The distribution in the sn-2 position of the three component fatty acids 16:0, 18:1 and 18:2 is that observed in the fatty acid composition of the sn-2-monoacylglycerols issued from pancreatic lipase hydrolysis of the triacylglycerol. From the fatty acid composition of the sn-2-monoacylglycerols and from that of the sn-1-lysophosphatides formed after phospholipase A hydrolysis, which also represents the distribution of 16:0, 18:1 and 18:2 in the sn-1 position of the triacylglycerol, it has been possible to calculate the proportions of the six stereoisomers of the isolated 16:0 18:1 18:2 triacylglycerol. This result was obtained with a slight approximation by assuming that the relative proportion of 18:1 and 18:2 in the sn-1 position of the stereoisomers comprising 16:0 at internal position (2.3% of total isomers) was identical to that observed for all the six isomers. Results show that four stereoisomers were in high proportion, i.e., those comprising 18:2 and 18:1 in the sn-2 position.

b. Chromatographic Methods of Stereospecific Analysis of Triacylglycerols. Any compound which does not present a plan or a center of symmetry can exist under two different configurations, not superposable, symmetrical with regard to a plan, like an object and its image in a mirror. Such a compound is called *chiral*. Triacylglycerols are chiral compounds of which the glycerol moiety presents an asymmetrical center at the central carbon atom. The two nonsuperposable forms of a chiral molecule are called *enantiomers, enantiomorphs,* or *optical antipods.*

Table 15 Stereospecific Distribution of Fatty Acids in Triacylglycerols from Several Vegetable Oils

| Oils (ref.) | Position[a] | Fatty acid composition (mol %) | | | | | | | | | | | |
		14:0	16:0	16:1	18:0	18:1	18:2	18:3	20:0	20:1	22:0	22:1	24:0
Peanut	TAG	—	9.6	0.1	3.7	49.5	28.3	—	1.8	1.6	3.3	3.3	1.7
(North	sn-1		17.1	0.1	5.0	43.8	32.4	—	—	1.5	—	—	—
America)	sn-2		1.8	0.6	0.4	57.7	39.3	—	—	0.3	—	—	—
(231)	sn-3		11.4	0.2	6.3	50.2	11.3	—	4.9	3.0	8.0	—	4.8
Peanut	TAG	—	8.7	0.1	3.4	60.4	19.4	—	1.5	1.4	3.0	tr	1.7
(Africa)	sn-1	—	17.0	0.4	4.2	57.6	18.6	—	—	2.1	—	—	—
(231)	sn-2	—	2.1	0.6	0.5	67.2	29.4	—	—	0.2	—	—	—
	sn-3		8.8	0.2	4.7	56.5	10.1	—	4.2	2.7	7.8	—	5.1
Rapeseed	TAG	—	3.0	0.3	1.7	25.6	17.0	9.8	—	11.9	0.9	29.8	—
(571)	sn-1	—	4.1	0.3	2.2	23.1	11.1	6.4	—	16.4	1.4	34.9	—
	sn-2	—	0.6	0.2	—	37.3	36.1	20.3	—	2.0	—	3.6	—
	sn-3	—	4.3	0.3	3.0	16.6	4.0	2.6	—	17.3	1.2	51.0	—
Soybean	TAG	—	9.3	—	3.9	24.1	54.4	8.2	—	—	—	—	—
(571)	sn-1	—	13.8	—	5.9	22.9	48.4	9.1	—	—	—	—	—
	sn-2	—	0.9	—	0.3	21.5	69.7	7.1	—	—	—	—	—
	sn-3	—	13.1	—	5.6	28.0	45.2	8.4	—	—	—	—	—
Corn	TAG	—	11.3	0.5	1.6	27.5	58.1	0.9	—	—	—	—	—
(571, 593)	sn-1	—	17.9	0.3	3.2	27.5	49.8	1.2	—	—	—	—	—
	sn-2	—	2.3	0.1	0.2	26.5	70.3	0.7	—	—	—	—	—
	sn-3	—	13.5	0.1	2.8	30.6	51.6	1.0	—	—	—	—	—
Olive	TAG	—	10.4	0.8	2.2	76.2	9.6	0.9	—	—	—	—	—
(571)	sn-1	—	13.1	0.9	2.6	71.8	9.8	0.6	—	—	—	—	—
	sn-2	—	1.4	0.7	—	82.9	14.0	0.8	—	—	—	—	—
	sn-3	—	16.9	0.8	4.2	73.9	5.1	1.3	—	—	—	—	—
Cottonseed	TAG	0.7	24.5	0.9	2.1	19.6	52.2	—	—	—	—	—	—
(599)	sn-1	1.4	28.7	1.1	4.3	20.8	43.7	—	—	—	—	—	—
	sn-2	—	6.7	0.6	1.5	28.0	63.2	—	—	—	—	—	—
	sn-3	0.7	38.1	1.0	0.5	10.0	49.7	—	—	—	—	—	—
Linseed	TAG	—	5.9	0.2	3.4	16.2	16.7	57.5	—	—	—	—	—
(571)	sn-1	—	10.1	0.2	5.6	15.3	15.6	53.2	—	—	—	—	—
	sn-2	—	1.6	0.1	0.7	16.3	21.3	59.8	—	—	—	—	—
	sn-3	—	6.0	0.3	4.0	17.0	13.2	59.4	—	—	—	—	—
Cocoa butter	TAG	—	24.0	0.3	35.1	36.1	3.4	—	1.1	—	—	—	—
(571)	sn-1	—	34.0	0.6	50.4	12.3	1.3	—	1.0	—	—	—	—
	sn-2	—	1.7	0.2	2.1	87.4	8.6	—	—	—	—	—	—
	sn-3	—	36.5	0.3	52.8	8.6	0.4	—	2.3	—	—	—	—

[a]TAG = in total triacylglycerols. tr, traces.

For commodity, a numbering system called stereospecific, based on the nomenclature R/S (rectus/sinister) of the Cahn–Ingold–Prelog system, was adopted to better represent the stereochemistry of glycerolipids and to simplify their numbering (617,618). According to this stereospecific numbering, the secondary hydroxyl group of the glycerol molecule in the plane projection of Fisher is located at the left hand of the glycerol backbone and the carbon atoms are numbered downward. The prefix "sn" for stereospecifically numbered is included in the name of triacylglyc-

Table 16 Stereospecific Distribution of Fatty Acids in Triacylglycerols from Different Animal Fats

Animal fat (ref.)	Position[a]	4:0	6:0	8:0	10:0	12:0	14:0	14:1	15:0	15:1	16:0	16:1	17:0[b]	18:0	18:1	18:2	18:3	20:0	20:1	20:2[c]	22:1[d]	22:5	22:6
Cow milk (637)	TAG	8.8	5.0	2.5	5.3	5.2	13.8	1.7	1.0	0.2	28.0	2.3	0.8	8.3	14.1	1.1	0.9	—	—	—	—	—	—
	sn-1	—	—	0.2	1.4	3.4	12.7	1.4	1.5	0.3	41.2	2.7	1.3	15.2	16.2	1.2	0.7	—	—	—	—	—	—
	sn-2	—	0.1	2.9	7.8	8.1	23.9	2.4	1.8	0.4	36.7	3.3	0.7	3.0	7.2	0.6	0.3	—	—	—	—	—	—
	sn-3	26.4	14.8	4.4	6.7	4.0	4.9	1.2	0.3	—	6.1	1.0	0.4	6.7	19.0	1.6	1.7	—	—	—	—	—	—
Human milk (638)	TAG	—	—	—	1.5	7.8	8.0	0.2	0.4	—	24.0	2.2	0.3	7.4	31.2	13.8	1.3	0.1	0.5	0.8	0.2	0.1	0.2
	sn-1	—	—	—	0.4	2.3	3.5	0.3	0.2	—	12.4	1.6	0.4	15.2	46.4	14.6	—	0.2	0.5	0.7	0.4	—	—
	sn-2	—	—	—	0.6	7.8	12.5	0.2	0.8	—	51.2	2.4	0.3	1.5	11.5	8.5	—	0.2	0.3	0.8	0.3	0.2	0.3
	sn-3	—	—	—	2.3	13.9	10.7	0.7	0.4	—	11.7	3.2	0.2	5.2	31.8	16.7	—	—	0.3	0.9	0.2	—	0.1
Depot fat of:																							
Rabit (606)	sn-1	—	—	—	—	—	3	—	—	—	34	9	—	6	25	14	2	—	—	—	—	—	—
	sn-2	—	—	—	—	—	6	—	—	—	25	12	—	1	26	23	5	—	—	—	—	—	—
	sn-3	—	—	—	—	—	1	—	—	—	24	7	—	3	35	22	5	—	—	—	—	—	—
Beef (606)	sn-1	—	—	—	—	—	4	—	—	—	41	6	—	17	20	4	<1	—	—	—	—	—	—
	sn-2	—	—	—	—	—	9	—	—	—	17	6	—	9	41	5	<1	—	—	—	—	—	—
	sn-3	—	—	—	—	—	1	—	—	—	22	6	—	24	37	5	<1	—	—	—	—	—	—
Horse (606)	sn-1	—	—	—	—	—	3	—	—	—	39	7	—	6	27	5	11	—	—	—	—	—	—
	sn-2	—	—	—	—	—	7	—	—	—	9	10	—	1	29	17	25	—	—	—	—	—	—
	sn-3	—	—	—	—	—	3	—	—	—	30	6	—	7	37	5	11	—	—	—	—	—	—
Pig (606)	sn-1	—	—	—	—	—	2	—	—	—	16	3	—	21	44	2	<1	—	—	—	—	—	—
	sn-2	—	—	—	—	—	4	—	—	—	59	4	—	3	17	8	<1	—	—	—	—	—	—
	sn-3	—	—	—	—	—	<1	—	—	—	2	3	—	10	65	24	<1	—	—	—	—	—	—

[a] TAG = in total triacylglycerols.
[b] 17:0 + 17:1.
[c] 20:2 + 20:3 + 20:4.
[d] 22:1 + 22:2 + 22:4.

Table 17 Stereospecific Distribution of Fatty Acids in Triacylglycerols of Several Animal Depot Fats

Animal (ref.)	Position	14:0	16:0	16:1	18:0	18:1	18:2	18:3	20:1	20:5	22:1	22:5	22:6
Chicken	sn-1	2	25	12	6	33	14	2	—	—	—	—	—
(606)	sn-2	1	15	7	4	43	23	3	—	—	—	—	—
	sn-3	1	24	12	6	35	14	2	—	—	—	—	—
Duck	sn-1	1	26	7	8	43	14	2	—	—	—	—	—
(606)	sn-2	1	11	4	4	59	20	2	—	—	—	—	—
	sn-3	1	27	7	6	45	15	1	—	—	—	—	—
Turkey	sn-1	1	30	7	11	30	17	2	—	—	—	—	—
(606)	sn-2	2	13	5	4	44	39	2	—	—	—	—	—
	sn-3	1	27	8	8	33	19	2	—	—	—	—	—
Trout	sn-1	2	13	8	7	24	6	—	11	4	9	2	3
(25)	sn-2	3	6	14	1	35	11	—	7	4	2	2	9
	sn-3	4	13	8	8	25	5	—	12	6	9	2	1
Herring	sn-1	6	12	13	1	16	3	—	25	3	14	1	1
(25)	sn-2	10	17	10	1	10	3	—	6	18	5	3	13
	sn-3	4	7	5	1	8	1	—	20	4	50	1	3
Mackerel	sn-1	6	15	11	3	21	2	—	8	5	18	1	2
(25)	sn-2	10	21	6	1	9	1	—	5	12	5	3	20
	sn-3	2	5	4	2	21	2	—	19	10	24	1	5
Cod	sn-1	6	15	14	6	28	2	—	12	2	6	1	1
(25)	sn-2	8	16	12	1	9	2	—	7	12	5	3	20
	sn-3	4	7	14	1	23	2	—	17	13	7	1	6
Lobster	sn-1	3	13	10	3	22	2	—	11	8	7	<1	3
(25)	sn-2	4	12	7	1	17	2	—	10	13	2	<1	15
	sn-3	4	13	10	3	25	2	—	12	8	8	<1	5
Frog	sn-1	3	26	10	4	27	12	9	—	—	—	—	—
(25)	sn-2	2	17	8	1	30	25	13	—	—	—	—	—
	sn-3	2	18	8	3	35	16	9	—	—	—	—	—
Turtle	sn-1	18	24	7	12	12	1	—	5	1	5	1	1
(25)	sn-2	10	4	13	2	45	1	—	3	2	3	3	5
	sn-3	14	10	8	5	19	1	—	11	4	6	2	2

erols which are identified in this numbering system. Chirality of triacylglycerols have been reviewed by Smith (619).

The separation of the two enantiomers of a chiral molecule is difficult since, with the exception of the sign of the rotatory power, these isomers are entirely physically and chemically identical. Such a separation in most cases is based on the preparation of diastereoisomers by reacting the enantiomers with an optically active compound. These diastereoisomers, which are no longer images in a mirror, present more differentiated properties and are more easily resolved especially by HPLC.

In the stereospecific analysis of triacyl-*sn*-glycerols by the chromatographic methods (HPLC), the general procedure is as follows:

1. Total triacylglycerols of a natural oil or fat (129,620,621) or simpler mixtures, or individual triacylglycerols obtained from a double fractionation by argentation-chromatography (TLC or HPLC), followed by reverse phase HPLC (107) are submitted to partial chemical deacylation using ethyl magnesium bromide (594). This deacylation procedure is nonspecific, unlike the enzymic deacylation procedure (22).

2. The *sn*-2- and *sn*-1(3)-monoacylglycerols and the *sn*-1,3- and *sn*-1,2(2,3)-diacylglycerols are separately or together fractionated by TLC on boric acid–impregnated silica plates (285).

3. The monoacylglycerols are submitted to two types of fractionation according to the complexity of the mixture and to the further possibilities of separation. The first type is reverse phase HPLC. The monoacylglycerols are fractionated according to the component fatty acid and to the positional isomerism by means of an RP-18 column and acetonitrile-water (80:20, 85:15, 90:10, or 95:5, v/v) as eluting solvent (108). In that case, monoacylglycerols corresponding to the different fatty acids were separated and for each fatty acid the mixture of the two isomers sn-1 and sn-3 was separated from the sn-2 isomer. Figure 15 illustrates these good separations. The second type of fractionation is argentation-TLC in which saturated monoacylglycerols are separated from the unsaturated (622). Diacylglycerols can be submitted to a double fractionation according to the nature of constituent fatty acids (chain length and unsaturation) for the *sn*-1,2(2,3)-diacylglycerols or according to the nature of the constituent fatty acids and their positioning for the mixtures of *sn*-1,3- and *sn*-1,2(2,3)-diacylglycerols. These fractionations can be performed by reverse phase HPLC using an RP-18 column and acetone- acetonitrile (45:55 or 40:60, v/v) as mobile phase.

4. Resolution of the two enantiomers of the *sn*-1(3)-monoacylglycerol mixture or of the *sn*-1,2(2,3)-diacylglycerol mixture can be achieved according to two different methods. The first method, developed by Takagi and Itabashi (20,620,621,623), consists of a derivatization of the *sn*-1(3)-monoacylglycerols or of the *sn*-1,2(2,3)-diacylglycerols by reacting these compounds with an achiral reagent, the 3,5-dinitrophenylisocyanate, to improve the enantioselectivity and the sensitivity to UV detection (126,127). The dinitrophenylurethane (DNPU) derivatives of the optical isomers are resolved on columns packed with stationary phases composed of chiral selectors bonded to the silica support.

Three types of columns have been used:

The Sumipax OA-2100 column contains as chiral selector *N*-(*S*)-2-(4-chlorophenyl)- isovaleroyl-D-phenylglycine and as support γ-aminopropyl-silanized silica (5 μm)

The Sumipax OA-4100 column packed with *N*(*R*)-1- (*a*-naphthyl)ethylaminocarbonyl-(*S*)-valine chemically bound to γ-aminopropyl-silanized silica (5 μm)

The YMC-Pack-A-KO3 column packed with a polymeric phase composed of *R*-(*t*)-1-(1-naphthyl)ethylamine chemically bonded to spherical silica particles with wide pores (300 Å)

The monoacylglycerol enantiomers have been resolved on OA-2100 columns (250 × 4 mm i.d.) (20,127) and OA- 4100 (250 or 500 × 4 mm i.d.) (624) using hexane-dichloroethane-ethanol (40:12:3 and 40:10:1, v/v/v) as mobile phase (20,127,624). The elution order was sn-2, sn-1, and sn-3 with OA-2100 and sn-1, sn-2, sn-3 with OA-4100. In both cases, for each enantiomer, the retention time increased with unsaturation and decreased with chain length. Under the operating conditions of Takagi and Ando (624), the saturated monoacylglycerol enantiomers of 12–22 acyl carbon atoms were eluted in two groups of peaks, the first comprising the sn-1 isomers and the second the sn-3 isomers. Within each group, monoacylglycerols differing by two carbon atoms were separated. Those differing by one double bond (18:0, 18:1, 18:2, 18:3) were also resolved.

Separation of monoacylglycerol enantiomers has been used for the stereospecific analysis of triacyl-*sn*-glycerols (622,625–627). For this purpose, the *sn*-1(3)- and *sn*-2-monoacylglycerols previously isolated by boric acid–impregnated silica gel TLC were fractionated by Ag⁺-TLC into saturated and unsaturated compounds and converted to DNPU derivatives. The DNPU derivatives of *sn*-1(3)-monoacylglycerols were resolved on the basis of their chirality and chain length by

HPLC with OA-2100 as stationary phase. The proportion of each monoacylglycerol enantiomer, together with the percentage of fatty acids at sn-1(3) and sn-2 positions, allowed the stereospecific distribution of fatty acids in the three sn-1, sn-2, and sn-3 positions of triacylglycerols to be determined (622).

A method for stereospecific analysis of triacylglycerols from the monoacylglycerols using the OA-4100 chiral column was also described (625–627). In that case, no fractionation by Ag⁺-TLC was needed since this column satisfactorily resolved the complex mixtures of sn-1(3)-monoacylglycerols.

The authors have used the OA-4100 chiral column for the stereospecific analysis of triacylglycerols from a vegetable oil via the monoacylglycerol enantiomers (108,628). For this purpose, monoacylglycerols were isolated and quantified by reverse phase HPLC (108) as previously illustrated (Fig. 15). Fractionation was carried out by separating for each fatty acid the sn-2- and sn-1(3)-monoacylglycerols. The fractionated sn-1(3)- monoacylglycerols were resolved into sn-1 and sn-3 enantiomers as DNPU derivatives (628).

Figure 20 shows what kinds of separation can be performed on this column when analyzing mixtures of monopalmitoylglycerols. It can be seen in part A that the sequence of elution is sn-1, sn-2, sn-3. The sn-2 isomer is not completely separated from the sn-1 isomer eluted ahead, partly because of its high proportion in the mixture. Part B illustrates the very clear separation of the two sn-1 and sn-3 stereoisomers. Such a mixture can be obtained after fractionation of the monoacylglycerols obtained by chemical deacylation of the 16:0 18:1 18:0 triacylglycerol by reverse phase HPLC (fraction 6 in Fig. 15).

The proportion of sn-2- and sn-1(3)-monoacylglycerols as determined by reverse phase HPLC and the proportion of the two enantiomers of sn-1(3)-monoacylglycerols as determined by chiral phase HPLC were used to determine the stereospecific distribution of fatty acids in the three positions of the oil triacylglycerols. Data obtained were somewhat different from those determined

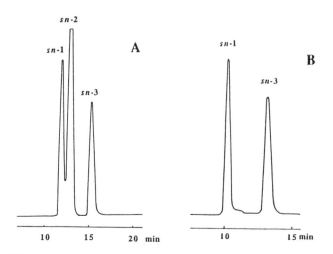

Fig. 20 HPLC isocratic analysis of monopalmitoylglycerol isomers as [di]3,5-dinitrophenylisocyanate (dinitrophenylurethane = DNPU) derivatives at ambient temperature and with UV detection (254 nm) on stainless steel (250 mm × 4 mm i.d.) chiral column. The column was packed with 5 μm of N (R)-1-(α-naph-thyl)ethylaminocarbonyl-(S)-valine chemically bonded to γ-aminopropyl- silanized silica (Sumipax OA-4100); mobile phase, hexane-ethylene dichloride-ethanol (40:12:3, v/v/v) at 1 ml/min. Part A: synthetic mixture of 1-, 2-, and 3-monopalmitoyl-sn-glycerols. Part B: racemic 1(3)-monopalmitoyl-sn-glycerols. (From Ref. 628.)

by using the *sn*-1,2(2,3)-diacylglycerols (629a) probably because of uncontrolled isomerization of monoacylglycerols.

To avoid or at least minimize the extent of partial acylglycerol isomerization, most analysts have first preferred using the diacyl-*sn*-glycerols instead of the monoacyl-*sn*-glycerols more sensitive to isomerization (22). The enantiomers of *sn*-1,2(2,3)-diacylglycerols were separated on the chiral Sumipax columns OA-2100 (21) and OA-4100 (128,623) and on the YMC-Pack-A-KO3 column (620,621). Several mobile phases were used under isocratic conditions to elute the optically active or inactive diacylglycerols, namely:

Hexane-ethanol (125:1, v/v) for three OA-2100 columns in series, i.e., 750 × 4 mm i.d. (15)

Several mixtures of hexane-dichloroethane-ethanol [80:20:1, v/v and 250:20:1, v/v (16) or 170:40:1, v/v, 170:20:1, 170:10:1, v/v (17)] with the OA-4100 columns (250 and 500 × 4 mm i.d.)

Hexane-dichloromethane (or dichloroethane)-ethanol (40:10:1, v/v) with the YMC-Pack A-KO3 column (250 × 4.6 mm i.d.)

The elution of diacylglycerol isomers was similar with the three types of columns and as follows: sn-1,3, sn-1,2, sn-2,3. With the OA-2100 column, the retention time of the simple diacylglycerol enantiomers C24 (dilauroylglycerol) and C36 (distearoylglycerol) ranged from 4 to 6 hr. In a homologous series of isomers, molecules differing from four acyl carbon atoms or two double bonds were separated (21).

According to the mobile phase (hexane-dichloroethane-ethanol: 80:20:1 and 150:20:1, v/v/v) the enantiomers *sn*-1,2-and *sn*-2,3-diacylglycerols generally eluted within 10 min and at the latest in 38 min on the OA-4100 column (128). With the first solvent, the optically inactive isomer, the *sn*-1,3-diacylglycerol coeluted with the sn-1,2 enantiomer whereas with the second solvent it eluted ahead of the two enantiomers. With the OA-4100 column, separations according to chirality and chain length or unsaturation were achieved (623). However, the retention times were particularly high, between 3 and 13 hr, which precluded any practical application.

The best resolutions of enantiomers were performed when using the YMC-Pack-A-KO3 column (620,621). It allowed separations of chiral diacylglycerol enantiomers with low retention times, i.e., 13 min for 1,2- and 2,3-dioleoyl-*sn*-glycerols (C36) and at most 40 min for the diacylglycerol optical isomers of linseed oil and fish oil triacylglycerols. On this type of column, the mixtures of *sn*-1,2(2,3)-diacylglycerols were resolved into two groups of peaks, each group corresponding to one enantiomer. Within each group the diacylglycerols were separated according to chain length and unsaturation, i.e., their partition number. The retention times of the optically inactive sn-1,3-diacylglycerols were comparable to those of the sn-1,2 isomers.

The high resolutive power of the OA-4100 column toward diacylglycerol enantiomers has been used by Semporé and Bézard in the detailed stereospecific analysis of triacylglycerols from a vegetable oil (107,629,630). The triacylglycerols were first fractionated into unsaturation classes by Ag⁺-TLC. Each class was then fractionated according to chain length by reverse phase HPLC into much simpler mixtures sometimes only comprising one individual triacylglycerol (629). These fractions were then submitted to partial chemical deacylation by means of the Grignard reagent ethyl magnesium bromide. The *sn*-1,2(2,3)-diacylglycerols isolated on silica plates impregnated with boric acid were converted to DNPU derivatives and fractionated by reverse phase HPLC according to their constituent fatty acids (107) as previously illustrated in Fig. 16. Each fraction so obtained was analyzed for enantiomer composition by chiral phase HPLC (630).

Figure 21 reports six chromatograms registered in the analysis on the OA-4100 chiral column of six mixtures of *sn*-1,2(2,3)-diacylglycerols of different chain length and unsaturation, obtained after chemical deacylation of natural oil triacylglycerols. The sn-1,2 isomer eluted ahead of the sn-2,3 isomer and the two stereoisomers were clearly resolved within 12 min in each case. The

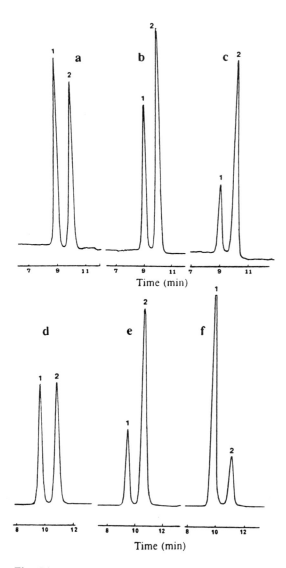

Fig. 21 HPLC separation of diacylglycerol enantiomers, as 3,5-dinitrophenylurethane (DNPU) derivatives on chiral OA-4100 column. Diacylglycerols derived from peanut oil and cottonseed oil triacylglycerols by chemical (b, c, e, f) and enzymatic (a, d) partial deacylation. They were previously fractionated by reverse phase HPLC as illustrated in Fig. 16. (a) 18:0-18:1; (b) 16:0-18:1; (c) 18:1-18:1; (d) 16:0-18:2; (e) 18:1-18:2; (f) 18:2 18:2. peaks: 1 = 1,2-diacyl-*sn*-glycerol and 2 = 2,3-diacyl-*sn*-glycerol in each chromatogram. Detection, UV absorption (254 nm); mobile phase, hexane-ethylene dichloride-ethanol (80:20:1, v/v/v) at a flow rate of 1 ml/min; analysis temperature, ambient (20°C). (From Ref. 630.)

composition in oil triacylglycerol molecular species was mathematically determined from the proportion of *sn*-1,2(2,3)-diacylglycerols and their enantiomer composition (629) as will be illustrated later on.

The YMC-Pack-A-KO3 column has been used for the analysis of the *sn*-1,3- and *sn*-1,2(2,3)-diacylglycerols from corn oil, linseed oil, and menhaden oil (620). It has been further used in the determination of molecular species of chiral diacylglycerols issued from partial chemical deacylation of cocoa butter, lard, and corn oil (621) as a prelude to stereospecific analysis of triacylglyc-erols (631). In this analysis, the *sn*-1,2(2,3)-diacylglycerols, as DNPU derivatives, were separated into two groups of peaks, each peak composed of pure sn-1,2 and sn-2,3 enantiomers. Each group of enantiomers was collected and the free diacylglycerols were regenerated by trichlorosilane (632). The latter were converted to TMS ethers using trimethylchlorosilanehexamethyldisila-zanepyridine (633). The TMS derivatives of diacylglycerols were then identified and quantified by capillary GC.

The second method to separate enantiomers of partial acylglycerols was initiated by Michelsen et al. (19) and principally applied to diacylglycerols. The latter were transformed into urethane (or carbamate) derivatives (diastereoisomers) by reacting with pure enantiomeric isocyanates. The most commonly used were (*S*)-(+)- or (*R*)-(−)-1-(1-naphthyl)ethylisocyanate. The diastereoisom-ers were resolved by HPLC on a silicic acid column. The diacylglycerols can be regenerated in the presence of trichlorosilane (632,634,635) for further analysis.

Laakso and Christie worked out the best resolutions of diacyl-*sn*-glycerol diastereoisomers as 1-(1-naphthyl)ethylurethane derivatives (129). They used a single column (250 × 4.6 mm i.d.) packed with 5-μm silica particles or, better, two columns in series of Hypersil, 3 μm (250 × 4.6 mm i.d.).

The most efficient mobile phases were:

Hexane-isopropanol (99.5:0.5, v/v) (84)

n-Propanol (2% water) in hexane or isooctane (636)

The latter gave the best separations with the most reproducible retention times. With the *S* derivatives the elution order was sn-1,3, sn-1,2, sn-2,3. The mixture of the urethane derivatives of the simple diacylglycerols containing 16:0, 18:0, 18:1, 18:2 was completely resolved so that all the sn-1,2 enantiomers eluted ahead of the sn-2,3. Inside each group the elution order was 18:1-18:1, 18:0-18:0, 18:2-18:2, 16:0-16:0. With the R derivatives the elution order was re-versed, with the sn-2,3 enantiomers eluting ahead of the sn-1,2.

APPLICATION OF HPLC TO THE STEREOSPECIFIC ANALYSIS OF TRIACYLGLYCEROLS Several groups have used the chromatographic separation of stereoisomers of partial acylglycerols, principally diacylglycerols, in the study of oil triacyl-*sn*-glycerols.

Method of Michelsen

Using the method initiated by Michelsen et al. (19) and thoroughly developed by Christie's group, several natural oils and fats were analyzed for distribution of fatty acids between the three positions of the glycerol moiety (129,636). For example, the triacylglycerols from corn oil, evening primrose oil, or egg yolk were partially deacylated by ethyl magnesium bromide. The diacylglycerols formed were fractionated on a preparative column of silica gel by HPLC, using hexane-tetrahy-drofurane-2-propanol (100:0.3:1.5, v/v/v) as eluting solvent. The isolated diacylglycerols were converted to 1-(1-naphthyl)ethyl urethane derivatives, and the diastereoisomers were resolved by HPLC on a silica gel column in three groups of peaks corresponding to the sn-1,3, sn-1,2, and sn-2,3 isomers. The diacylglycerols corresponding to each peak on the chromatogram were collected and their constituent fatty acids were analyzed by GC for identification.

The distribution of fatty acids in the three positions of the glycerol moiety of triacylglycerols was indirectly determined from the fatty acid composition of the three diacylglycerol isomers and from that of the starting triacylglycerols according to the following formulas:

(A) position *sn*-1 = 3 (A) TAG − 2 (A) *sn*- 2,3-DAG

(A) position *sn*-2 = 3 (A) TAG − 2 (A) *sn*- 1,3-DAG

(A) position *sn*-3 = 3 (A) TAG − 2 (A) *sn*- 1,2-DAG

in which (A) represents the percentage of a fatty acid "A" at the sn-1, sn-2, and sn-3 position, in the starting triacylglycerols (TAG) and in the diacylglycerol (DAG) isomers. Figure 22 shows the HPLC elution profile of the three groups of diacylglycerol isomers derived from palm oil triacylglycerols by chemical deacylation. Three groups of three peaks corresponding to the sn-1,3, sn-1,2, and sn-2,3 isomers were separated. All of the peaks in the three regions were collected together and analyzed by GC for fatty acid composition. Results for each of the three positions were calculated from the data for the composition of the intact triacyl-*sn*-glycerols and of 1,2- and 2,3-*sn*-diacylglycerols. They are reported in the upper part of Table 18. Positional distribution of fatty acids in palm oil triacylglycerols was also calculated from the data for 1,2- and 2,3-diacyl-*sn*-glycerols and those for position sn-2 obtained by pancreatic lipase hydrolysis (lower part of Table 18). The composition of position sn-2 obtained from the analysis of the 1,3-diacyl-*sn*-glycerols is reported in the last column.

Results obtained for position sn-2 from analysis of the *sn-1,2-* and *sn-2,3*-diacylglycerols were very similar to those obtained from pancreatic lipase hydrolysis (*e* in the table). But that obtained by utilizing the composition of the *sn-1,3*-diacylglycerols (*g* in the table) gave less reliable results, probably because palm oil triacylglycerols exhibit high asymmetrical distribution of fatty acids among the three positions.

Results reported in Table 18 show what has been established for many years for vegetable oils, that unsaturated fatty acids are preferentially found in sn-2 position. They also show that palmitic acid was more often in the sn-3 than the sn-1 position. The reverse was true for oleic and linoleic acid.

Method of Takagi and Itabashi: Method Based on Chiral Phase HPLC of Diacylglycerol Isomers

Using the method based on the use of chiral phase HPLC proposed by Takagi and Itabashi (21), the Kuksis group also analyzed positional fatty acid distribution among the three positions of triacylglycerols from natural oils and fats (620,621). They obtained very detailed results by

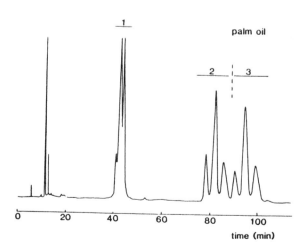

Fig. 22   HPLC resolution of (*S*)-(+)-1-(1-naphthyl)ethylurethane derivatives of diacyl-*sn*-glycerols derived from palm oil triacylglycerols by partial chemical deacylation. Two columns of silica gel (Hypersil, 3 μm, 25 cm × 4.6 mm i.d.) in series were utilized with 0.4–0.33% (v/v) 1-propanol (containing 2% water) in isooctane as mobile phase at a flow rate of 1 ml/min. Detection was by UV absorbance (280 nm). The numbers 1, 2, 3 above peaks refer to the 1,3-, 1,2- and 2,3-diacyl-*sn*-glycerol derivatives, respectively. (From Ref. 636.)

Table 18 Fatty Acid Compositions (mol % of total) of Palm Oil Triacylglycerols and of Positions sn-1, sn-2, and sn-3

Fatty		Position		
acid	TG	sn-1[a]	sn-2[b]	sn-3[c]
14:0	1.3	0.4	0.6	2.9
16:0	48.4	60.1	13.3	71.9
16:1	0.2	0.0	0.5	0.1
18:0	3.7	3.4	0.2	7.6
18:1	36.3	26.8	67.9	14.4
18:2	10.0	9.3	17.5	3.2

	Position			
	sn-1[d]	sn-2[e]	sn-3[f]	sn-2[g]
14:0	0.7	0.4	3.1	2.5
16:0	60.1	13.3	71.9	31.4
16:1	0.4	0.1	0.5	0.3
18:0	3.1	0.5	7.3	0.6
18:1	27.1	67.6	14.7	48.7
18:2	8.6	18.2	2.5	16.5

a = 3 × TG–2 × 2,3-DG; b = 3 × TG–(a + c); c = 3 × TG – 2 × 1,2-DG; d = 2 × 1,2-DG — e; e = via pancreatic lipase hydrolysis; f = 2 × 2,3-DG — e; g = 3 × TG – 2 × 1,3-DG
Source: Data from Ref. 636.

analyzing the *sn*-1,2- and *sn*-2,3-diacylglycerols fractionated by chiral phase HPLC by capillary GC of the TMS ethers after regeneration from the DNPU derivatives.

Figure 23 shows (A) that six major groups of diacylglycerols were clearly separated into two groups, which represented the sn-1,2 and sn-2,3 enantiomers. These two groups were isolated on a micromolar scale by preparative HPLC without cross-contamination as verified in (B) and (C). The chromatograms in (D) and (E) were obtained in the capillary GC analysis of the TMS ethers retrieved from the DNPU derivatives of these two groups of enantiomers. The diacylglycerol peaks, separated according to carbon number and number of double bonds, were identified by reference to standards and by comparison of the relative retention times to published data.

In addition to 16:0-18:1, 16:0-18:2, 18:1-18:1, 18:1-18:2, and 18:2-18:2 as major components, 18:0-18:1 and 18:0-18:2 were found as minor components.

Table 19 gives the quantitative composition of the diacylglycerols molecular species from corn oil triacylglycerols. All species found in this analysis were common to each enantiomer, but some differences were seen in proportions of the individual components, resulting from asymmetry in the distribution of fatty acids among positions sn-1 and sn-3. Table 19 also compares the composition of the original *sn*- 1,2(2,3)-diacylglycerols and those reconstituted from enantiomer composition. The good agreement is in favor of the accuracy of the procedures used for analyzing the diacylglycerol species. Fatty acid composition of positions sn-1, sn-2, and sn-3 can be calculated from these results and from the original triacylglycerol fatty acid composition using the formulas previously reported.

Using the method based on chiral phase HPLC of diacylglycerol isomers, the authors developed a method which permitted them to determine a detailed molecular species determination of peanut

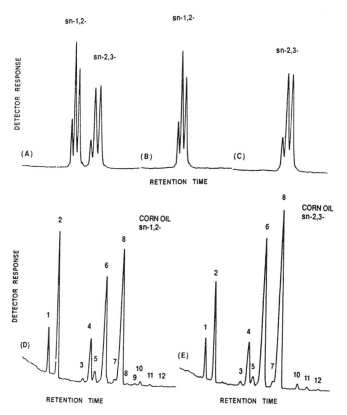

Fig. 23 Chiral phase HPLC of 3′-5′-dinitrophenylurethane (DNPU) derivatives of original corn oil *rac*-1,2-diacylglycerols (A) and of the *sn*-1,2 (B) and *sn*-2,3 (C) enantiomers after collection from the chiral column, along with polar capillary GC of TMS ethers of the *sn*-1,2- (D) and *sn*-2,3- (E) diacylglycerols regenerated from DNPU derivatives of the enantiomers. Peak identification: 1 = 16:0-18:1; 2 = 16:0-18:2; 3 = 18:0-18:1; 4 = 18:1 18:1; 5 = 18:0-18:2; 6 = 18:1-18:2; 8 = 18:2-18:3; 11 = 20:0-18:2. Operating conditions: for HPLC: chiral column (25 cm × 4.6 mm i.d.) containing *R*-(+)-1-(1-naphthyl)eth-ylamine polymeric phase bonded to 300-Å-wide pore spherical silica (YMC-Pack-A-KO3); isocratic solvent system, hexane- dichloromethane-ethanol (40:10:1) at a flow rate of 0.8 ml/min at an oven temperature of 28°C. For GLC: polar capillary column RT 2330 (15 m × 0.32 mm i.d.), isothermally operated at 260°C with hydrogen as the carrier gas (2 psi head pressure). (From Ref. 621.)

oil triacyl-*sn*-glycerols (629). The triacylglycerols were first fractionated into unsaturation classes by argentation-TLC. These classes were further fractionated according to chain length by reverse phase HPLC. Fractions isolated after these combined separations and comprising three known fatty acids were chemically deacylated. The *sn*-1,2(2,3)-diacylglycerols formed were analyzed and fractionated as DNPU derivatives by reverse phase HPLC. The fractions obtained were analyzed by chiral phase HPLC for enantiomer composition. From the *sn*-1,2(2,3)-diacylglycerol composition and the diacylglycerol *sn*-1,2 and *sn*-2,3 enantiomer composition, the individual molecular species of the major peanut oil triacylglycerols were identified and quantitated. An example is given for the triacylglycerol palmitoyl-oleoyl-linoleoylglycerol (16:0 18:1 18:2) from peanut oil.

Table 19 Composition of Enantiomeric Diacylglycerols Generated by Partial Grignard Degradation from Corn Oil Triacylglycerols (mol %)

Molecular species	Regenerated from DNPU			Original	Calculated[a]
	sn-1,2-	sn-2,3-	sn-1,2(2,3)-	sn-1,2(2,3)-	sn-1,2(2,3)-
16:0–18:1	6.80	4.72	5.77	5.64	5.76
16:0–18:2	21.86	11.80	17.38	16.85	16.83
18:0–18:1	0.63	0.58	0.59	0.65	0.61
18:1–18:1	9.04	7.99	8.36	8.14	8.51
18:0–18:2	2.41	1.73	1.99	2.02	2.07
18:1–18:2	25.19	32.49	28.58	28.02	28.84
18:2–18:2	34.07	40.69	37.33	38.68	37.38

[a](sn-1,2-Diacyl + sn-2,3-diacyl)/2.
Source: Data from Ref. 621.

In Table 20 are reported the sn-1,2(2,3)-diacylglycerol composition determined by reverse phase HPLC and the enantiomer composition determined by chiral phase HPLC.

In Table 21 is reported the mathematical method of determination of the percentages of the six molecular species found in the triacylglycerol. Three equations (1, 2, 3 in the lower part of the table) are derived from the diacylglycerol composition as determined by reverse phase HPLC. Three others (4, 5, 6) are derived from the enantiomer composition as determined by chiral phase HPLC. To solve this set of six equations comprising six unknowns, a slight approximation had to be made regarding the two molecular species comprising palmitic acid in position sn-2, which were in very low quantity.

Results show that the major molecular species was 1- palmitoyl-2-linoleoyl-3-oleoyl-sn-glycerol (16:0-18:2-18:1) amounting to more than 50% of the total. It was followed by 18:1- 18:2-16:0, also exhibiting 18:2 in the sn-2 position. Data obtained show that affinity of each fatty acid for the three positions was as follows:

16:0 : pos. 1 > pos. 3 >> pos. 2
18:1 : pos. 3 >> pos. 1 > pos. 2
18:2 : pos. 2 >> pos. 1 > pos. 3

Table 20 Enantiomer Composition of the sn-1,2(2,3)-Diacylglycerols Issued from Chemical Deacylation of Peanut Oil Triacylglycerols Isolated by Combined Argentation-TLC and Reverse Phase HPLC

Triacylglycerol	16:0 18:1 18:2					
sn-1,2(2,3)- Diacylglycerols[a]	18:1-18:2		16:0-18:2		16:0-18:1	
Mole %[b]	48.65		41.81		9.54	
Enantiomers	sn-1,2	sn-2,3	sn-1,2	sn-2,3	sn-1,2	sn-2,3
Mole %[c]	38.68	61.32	66.86	33.14	36.08	63.92

[a]The sn-1,2(2,3)-diacylglycerol fraction was separated from the sn-1,2 isomers in total diacylglycerols on borate-impregnated silica TLC.
[b]Mean of seven values determined in three different ways.
[c]Mean of three values determined in two different ways.
Source: Data from Ref. 629.

Table 21 Determination of the 1,2,3-Triacyl-sn-Glycerol Composition of Palmitoyloleoyllinoleoylglycerol Isolated from Peanut Oil

TG[a]	sn-2-TG[a]	sn-1,2,3-TG[a]	Mol %	sn-1,2-DG[b]	Mol %	sn-2,3-DG[b]	Mol %
16:0 18:1 18:2	16:0 18:1 18:2	16:0-18:1-18:2	a	16:0-18:1	a/3	18:1-18:2	a/3
		18:2-18:1-16:0	b	18:2-18:1	b/3	18:1-16:0	b/3
	16:0 18:2 18:1	16:0-18:2-18:1	c	16:0-18:2	c/3	18:2-18:1	c/3
		18:1-18:2-16:0	d	18:1-18:2	d/3	18:2-16:0	d/3
	18:1 16:0 18:2	18:1-16:0-18:2	e	18:1-16:0	e/3	16:0-18:2	e/3
		18:2-16:0-18:1	f	18:2-16:0	f/3	16:0-18:1	f/3

Equations[c]

$$(1)\ (a+b)+(c+d) = 48.65 \times 2 = 97.30 \qquad (4)\ \frac{b+d}{a+c} = \frac{38.68}{61.32}$$

$$(2)\ (c+d)+(e+f) = 41.81 \times 2 = 83.62 \qquad (5)\ \frac{c+f}{d+e} = \frac{66.86}{33.14}$$

$$(3)\ (a+b)+(e+f) = 9.54 \times 2 = 19.08 \qquad (6)\ \frac{a+e}{b+f} = \frac{36.08}{63.92}$$

Solutions

	A[d]	B[d]	Mean
a	5.57	6.00	5.78
b	10.81	10.38	10.59
c	54.09	53.66	53.88
d	26.83	27.26	27.05
e	0.88	0.88	0.88
f	1.82	1.82	1.82

[a] The triacylglycerol (TG) comprises three TGs whose fatty acid at the sn-2 position (sn-2-TG) is known, and six stereoisomers whose fatty acids at the 1, 2 and 3 positions (from left to right) are known (sn-1,2,3-TG).
[b] sn-1,2- and sn-2,3-Diacylglycerols generated by chemical partial deacylation.
[c] Equations (1), (2), and (3) are derived from the percentage of the individual sn-1,2(2,3)-diacylglycerols (Table 20, 3rd line); and Eq. (4), (5), and (6) from the percentage of the enantiomers. (Table 20, 5th line).
[d] Solution A is derived from Eq. (2), (5) and (1), (4) and solution B from Eq. (3), (6), and (1), (4).
Source: Data from Ref. 629.

Table 22   Composition in Molecular Species of Peanut
Oil Triacylglycerols (ô0.10mol %)

1,2,3-Triacyl- sn-glycerol	Exp. mol %	1,2,3-Triacyl- sn-glycerol	Exp. mol %
18:1 18:1 18:1	10.66	18:1 18:1 18:2	1.90
18:1 18:2 18:1	9.20	18:1 18:2 18:2	1.80
18:2 18:2 18:1	8.54	18:2 18:1 18:2	1.41
16:0 18:2 18:1	6.34	18:2 18:1 16:0	1.24
18:2 18:1 18:1	5.45	16:0 18:1 18:2	0.68
16:0 18:1 18:1	4.82	18:1 16:0 18:1	0.59
18:1 18:2 16:0	3.18	18:2 16:0 18:1	0.21
18:1 18:1 16:0	3.07	18:1 16:0 18:2	0.10
18:2 18:2 18:2	2.08		

Source: Data from Ref. 629.

In Table 22 are listed the major (≥0.10 mol % in the oil) molecular species of the peanut oil analyzed. This oil, of African origin, was rich in oleic acid and the major triacylglycerol was trioleoylglycerol, amounting to more than 10% of the total in the oil. The three next ones comprised linoleic acid in the sn-2 position and for the next ones this position was alternatively occupied by oleic or linoleic acid. This illustrates the high affinity of these two unsaturated fatty acids for the sn-2 position as for most vegetable oils. The molecules comprising palmitic acid in the sn-2 position were in very low percentages (<1%). Only nine molecular species were present at more than 2% in this oil and they represented more than 53% of the total. When taking into account the other eight, the total amounted to more than 61%. This method is time consuming but it can give a very detailed molecular species composition of any oil and fat.

Table 23   Comparison of Stereospecific Distributions of Acyl
Groups in Soybean Oil Triacyl-sn-Glycerols Obtained by HPLC
and Enzymatic Methods

		Composition (mol %)					
	Acyl	HPLC method[a]			Enzyme method[b]		
No.	group	sn-1	sn-2	sn-3	sn-1	sn-2	sn-3
1	14:0	0.1	0.1	0.1	0.3	0.3	−0.2
2	16:0	16.7	2.2	16.1	16.9	1.4	16.6
3	16:1n-7	0.1	0.1	0.1	0.1	0.1	0.0
4	17:0	0.1	0.0	0.1	0.1	0.0	0.0
5	18:0	5.4	0.3	4.6	5.2	0.3	4.7
6	18:1n-9	22.2	23.0	22.9	20.3	22.6	25.1
7	18:1n-7	2.1	0.4	1.7	2.0	0.4	1.8
8	18:2n-6	46.4	68.1	47.0	47.4	69.2	45.2
9	18:3n-6	6.4	5.7	7.0	7.5	5.7	5.9
10	20:0	0.5	ND[c]	0.4	0.2	ND	0.7

[a]Method presented here.
[b]Modified Brockerhoff method (286).
[c]*ND*, not detected.
Source: Data from Ref. 627.

Methods Based on Study of Monoacylglycerol Enantiomers

The methods of stereospecific analysis of triacylglycerols based on the study of the diacylglycerol enantiomers were initially preferred by the analysts because the monoacylglycerol isomers are very sensitive to acyl migration and results obtained from their study were not up-to-date or reliable. However, this problem of isomerization seems to have been solved since Takagi and Ando (627) reported stereospecific acyl distribution in different oil triacylglycerols calculated from data derived from monoacyl isomer analysis by chiral phase HPLC.

In Table 23 comparison can be made between the distribution of acyl groups in soybean oil triacyl-*sn*-glycerols obtained by chiral phase HPLC analysis of the monoacylglycerol isomers formed by the nonspecific chemical deacylation procedure of Grignard and the distribution observed when using the Brockerhoff method based on enzyme stereospecificity. There is good agreement between the two series of data, proving that the monoacylglycerol method can be used in the stereospecific analysis of triacyl-*sn*-glycerols.

REFERENCES

1. M. E. Chevreul, *Ann. Chim.* (Paris), *94*: 113 (1815).
2. M. E. Chevreul, *Recherches Chimiques sur les Corps Gras d'Origine Animale*, Levrault, Paris, 1823.
3. M. Berthelot, *Ann. Chim. Phys.*, *41*: 216 (1854).
4. J. Pelouze and F. Boudet, *Ann. Chim. Phys.*, *69*: 43 (1838).
5. J. Bell, *The Chemistry of Foods*, Vol. 2, Chapman and Hall, London, 1883, pp. 44–46.
6. A. W. Blyth and G. H. Robertson, *Proc. Chem. Soc., London*, *5*: 5 (1889).
7. R. Heise, *Arb. Kaiserl. Gesundheitsante*, *12*: 540 (1896).
8. T. P. Hilditch and C. H. Lea, *J. Chem. Soc.*, 3106 (1927).
9. T. P. Hilditch and P. N. Williams, *The Chemical Constitution of Natural of Natural Fats*, 4th ed., Chapman & Hall, London, 1964, pp. 358–527.
10. H. J. Dutton and J. A. Cannon, *J. Am. Oil Chem. Soc.*, *33*: 46 (1956).
11. D. Priori, Olii Miner, and Grassi Saponi, *Colori Vernici*, *33*: 23 (1956).
12. V. R. Huebner, *J. Am. Oil Chem. Soc.*, *38*: 628 (1961).
13. B. De Vries, *Chem. Ind., London*, 1049 (1962).
14. C. B. Barrett, M. S. J. Dallas, and F. B. Padley, *Chem. Ind., London*, 1050 (1962).
15. T. Pei, P. Henly and S. Ramachandran, *Lipids*, *10*: 152 (1975).
16. F. H. Mattson and L. W. Beck, *J. Biol. Chem.*, *219*: 735 (1956).
17. P. Savary and P. Desnuelle, *Biochim. Biophys. Acta*, *21*: 349 (1956).
18. H. Brockerhoff, *J. Lipid Res.*, *6*: 10 (1965).
19. P. Michelsen, E. Aronson, G. Odham, and B. Akesson, *J. Chromatogr.*, *350*: 417 (1985).
20. T. Takagi and Y. Itabashi, *Yukagaku*, *34*: 962 (1985).
21. Y. Itabashi and T. Takagi, *J. Chromatogr.*, *402*: 257 (1987).
22. C. Litchfield, *Analysis of Triglycerides*, Academic Press, New York, 1972.
23. M. L. Blank, B. Verdino, and O. S. Privett, *J. Am. Oil Chem. Soc.*, *42*: 87 (1965).
24. W. W. Christie, R. D. Noble, and J. H. Moore, *Analyst*, *95*: 940 (1970).
25. C. Litchfield, *Lipids*, *3*: 170 (1968).
26. A. Kuksis, W. C. Breckenridge, L. Marai, and O. Stachnyk, *J. Lipid Res.*, *10*: 25 (1969).
27. R. N. Roberts in *Lipid Chromatographic Analysis* (G. V. Marinetti, ed.), Vol. 1, Marcel Dekker, New York, 1967, pp. 447–463.
28. A. Rajiah, M. R. Subbaram, and K. T. Achaya, *J. Chromatogr.*, *38*: 35 (1968).
29. A. Antonis, D. S. Platt, and J. M. Thorp, *J. Lipid Res.*, *6*: 301 (1965).
30. R. G. Martinek, *J. Am. Med. Technol.*, *30*: 274 (1968).
31. C. Bandyopadhyay, *J. Chromatogr.*, *37*: 123 (1968).
32. M. Kraml and L. Cosins, *Clin. Biochem.*, *2*: 373 (1969).
33. H. P. Chin, S. S. A. El-Meguid, and D. H. Blankenhorn, *Clin. Chim. Acta*, *31*: 381 (1971).
34. P. Haux and S. Natelson, *Microchem. J. 16*: 68 (1971).
35. M. E. Royer and H. Ko, *Anal Biochem.*, *29*: 405 (1969).

36. W. Godicke and U. Gerike, *Clin. Chim. Acta, 30*: 727 (1970).
37. G. H. Mc Lellan, *Clin. Chim. Acta, 17*: 535 (1971).
38. F. G. Aletti, *Clin. Chim. Acta, 15*: 184 (1967).
39. M. R. Lloyd and R. B. Goldrick, *Med. J. Aust., 2*: 493 (1968).
40. H. Ko and M. E. Royer, *Anal. Biochem., 26*: 18 (1968).
41. S. S. Chernick, in *Methods in Enzymology*, Vol. 14 (J. M. Lowenstein, ed.), Academic Press, New York, 1969, pp. 627–630.
42. P. Belfrage, T. Wiebe, and A. Lundquist, *Scand. J. Clin. Lab. Invest., 26*: 53 (1970).
43. A. R. Timms, L. A. Kelly, J. A. Spirito, and R. G. Enstrom, *J. Lipid Res., 9*: 675 (1968).
44. J. L. Bell, S. M. Atkinson, and D. N. Baron, *J. Clin. Pathol., 23*: 509 (1970).
45. J. P. Mallon and C. Dalton, *Anal. Biochem., 40*: 174 (1971).
46. L. Notarnicola, *Riv. Ital. Sostanze Grasse, 44*: 72 (1967).
47. P. Malangeau and M. Pays, *Ann. Biol. Clin.* (Paris), *25*: 845 (1967).
48. D. Mendelsohn and A. Antonis, *J. Lipid Res., 2*: 45 (1961).
49. I. Katz and M. Keeney, *Anal. Chem., 36*: 231 (1964).
50. J. Sheath, *Aust. J. Exp. Biol. Med. Sci., 43*: 563 (1965).
51. J. Marsh and D. B. Weinstein, *J. Lipid Res., 7*: 574 (1966).
52. N. Zollner and K. Z. Kirsch, *Gesamte Exp. Med., 135*: 545 (1962).
53. J. R. Sand and C. O. Huber, *Talanta, 14*: 1309 (1967).
54. G. Jurriens, B. De Vries, and L. Schouten, *J. Lipid Res., 5*: 267 (1964).
55. M. J. Albrink, *J. Lipid Res., 1*: 53 (1969).
56. K. N. Jeejeebhoy, S. Ahmad, and G. Kozak, *Clin. Biochem., 3*: 157 (1970).
57. L. J. Nutter and O. S. Privett, *J. Chromatogr., 35*: 519 (1968).
58. G. Biernoth, *Fette Seifen Anstrichm., 70*: 402 (1968).
59. R. J. Nicolosi, S. C. Smith, and R. F. Santarre, *J. Chromatogr., 60*: 111 (1971).
60. F. B. Padley, *J. Chromatogr., 39*: 37 (1969).
61. E. Haahti, R. Vihko, I. Jaakonmaki, and R. S. Evans, *J. Chromatogr. Sci. 8*: 370 (1970).
62. G. Chlierf and P. Wood, *J. Lipid Res., 6*: 317 (1965).
63. R. J. Vander Wal, *J. Am. Oil. Chem. Soc., 42*: 754 (1965).
64. R. J. Ho and H. C. Meng, *Anal. Biochem., 31*: 426 (1969).
65. R. J. Ho, *Anal. Biochem., 36*: 105 (1970).
66. N. T. Werthessen, J. R. Beall, and A. T. James, *J. Chromatogr., 46*: 149 (1970).
67. G. Cavina, G. Moretti, A. Mollica, L. Moretta, and P. Siniscalchi, *J. Chromatogr., 44*: 493 (1969).
68. R. P. W. Scott and J. G. Lawrence, *J. Chromatogr. Sci., 8*: 65 (1970).
69. B. De Vries, *J. Am. Oil Chem. Soc., 41*: 403 (1964).
70. R. J. Komarek, R. G. Jensen, and B. W. Pickett, *J. Lipid Res., 5*: 268 (1964).
71. M. R. Subbaram and C. G. Youngs, *J. Am. Oil Chem. Soc., 41*: 445 (1964).
72. J. Bézard, G. Semporé, G. Descargues, and A. Sawadogo, *Fette Seifen Anstrichm., 83*: 17 (1981).
73. A. Kuksis and L. Marai, *Lipids, 2*: 217 (1967).
74. A. Kuksis, W. C. Breckenridge, L. Marai, and O. Stachnyk, *J. Am. Oil Chem. Soc., 45*: 537 (1968).
75. A. Kuksis, in *Lipid Chromatographic Analysis*, 2nd ed., Vol. 1 (G. V. Marinetti, ed.), Marcel Dekker, New York, 1976.
76. J. J. Myher and A. Kuksis, *Lipids, 9*: 382 (1974).
77. Y. Itabashi and T. Takagi, *Lipids, 15*: 205 (1980).
78. B. J. Holub and A. Kuksis, *Lipids, 4*: 466 (1969).
79. O. Renkonen, *Adv. Lipid Res., 5*: 359 (1967).
80. O. Renkonen, *Lipids, 3*: 191 (1968).
81. O. Renkonen, *Biochim. Biophys. Acta, 137*: 575 (1967).
82. A. Wood, W. J. Baumann, F. Snyder, and H. K. Mangold, *J. Lipid Res., 10*: 128 (1969).
83. T. Sasaki and H. Hasegewa-Sasaki, *Biochim. Biophys. Acta, 833*: 316 (1985).
84. Y. Nakagawa and L. A. Horrocks, *J. Lipid Res., 24*: 1268 (1983).
85. K. Itoh, A. Suzuki, Y. Kuroki, and T. Akino, *Lipids, 20*: 611 (1985).
86. R. D. Wood, P. K. Raju, and R. Reiser, *J. Am. Oil Chem. Soc., 42*: 161 (1965).
87. J. F. O'Brien and W. E. Klopfenstein, *Chem. Phys. Lipids, 6*: 1 (1971).
88. E. J. Corey and A. Venkateswaarlu, *J. Am. Oil Chem. Soc., 94*: 6190 (1972).

89. D. Droz, *J. Chromatogr.*, *113*: 303 (1975).
90. C. F. Poole, in *Handbook of Derivatives for Chromatography* (K. Blau and G. S. King, ed.), Heyden and Sons, London, 1978, pp. 152–200.
91. C. F. Poole and A. Zlatkis, *J. Chromatogr. Sci.*, *17*: 115 (1979).
92. G. Kresze, K. Bederke, and F. Z. Schauffelhut, *Anal. Chem.*, *209*: 329 (1965).
93. M. R. Sahasrabudhe and J. J. Lagari, *J. Am. Oil Chem. Soc.*, *44*: 379 (1967).
94. R. Watts and R. Dils, *J. Lipid Res.*, *10*: 33 (1969).
95. L. J. Morris, *Biochem. Biophys. Res. Commun.*, *20*: 340 (1965).
96. M. Barber, J. R. Chapman, and W. A. Wolstenholme, *Int. J. Mass Spectrom. Ion Phys.*, *1*: 98 (1968).
97. M. G. Horning, G. Casparrini, and E. C. Horning, *J. Chromatogr. Sci.*, *7*: 267 (1969); *Am. J. Clin. Nutr.*, *24*: 1086 (1971); *Anal. Lett.*, *1*: 481 (1968).
98. A. Morrison, M. D. Barratt, and R. Aneja, *Chem. Phys. Lipids*, *4*: 47 (1970).
99. E. G. Perkins and P. V. Johnston, *Lipids*, *4*: 301 (1969).
100. S. Pind, A. Kuksis, J. J. Myher, and L. Marai, *Can. J. Biochem. Cell. Biol.*, *62*: 301 (1984).
101. S. Pind, A. Kuksis, J. J. Myher, and L. Marai, *Can. J. Biochem. Cell. Biol.*, *63*: 137 (1985).
102. W. W. Christie, *Gas Chromatography and Lipids: A Practical Guide*, Oily Press, Ayr, 1992.
103. R. Wood and F. Snyder, *Lipids 1*: 62 (1966).
104. K. Maruyama and C. Yonese, *J. Am. Oil Chem. Soc.*, *63*: 902 (1986).
105. S. Takano and Y. Kondoh, *J. Am. Oil Chem. Soc.*, *64*: 1001 (1987).
106. Y. Kondoh and S. Takano, *J. Chromatogr.*, *393*: 427 (1987).
107. B. G. Semporé and J. A. Bézard, *J. Chromatogr.*, 547: 89 (1991).
108. B. G. Semporé and J. A. Bézard, *J. Chromatogr.*, 596: 185 (1992).
109. W. W. Christie and M. L. Hunter, *Biochem. J.*, *235*: 833 (1986).
110. M. Batley, N. H. Packer, and J. W. Redmond, *J. Chromatogr.*, *198*: 520 (1980).
111. J. T. Benert, N. K. Meredith, J. R. Akins, and W. H. Hannon, *J. Liq. Chromatogr.*, *8*: 1573 (1985).
112. A. Kuksis, L. Marai, and J. J. Myher, *J. Chromatogr.*, *273*: 43 (1983).
113. M. Batley, N. H. Packer, and J. W. Redmond, *Biochim. Biophys. Acta*, *710*: 400 (1982).
114. J. Kesselmeier and E. Heinz, *Anal. Biochem.*, *144*: 319 (1985).
115. M. Kito, H. Takamura, H. Narita, and R. Urade, *J. Biochem.* (*Tokyo*), *98*: 327 (1985).
116. T. Takamura, H. Narita, R. Urade, and M. Kito, *Lipids*, *21*: 356 (1986).
117. M. L. Blank, M. Robinson, V. Fitzgerald, and F. Snyder, *J. Chromatogr.*, *298*: 366 (1985).
118. M. L. Blank, E. A. Cress, M. Robinson, and F. Snyder, *Biochim. Biophys. Acta*, *833*: 366 (1985).
119. M. Robinson, M. L. Blank, and F. Snyder, *Arch. Biochem. Biophys.*, *250*: 271 (1986).
120. J. Kruger, H. Rabe, G. Reichmann, and B. Rustow, *J. Chromatogr.*, *307*: 387 (1984).
121. B. Rustow, D. Kunze, H. Rabe, and G. Reichmann, *Biochim. Biophys. Acta*, *835*: 465 (1985).
122. D. Kunze, B. Rustow, H. Rabe, and K.-P. Ullrich, *Clin. Chim. Acta*, *140*: 215 (1984).
123. M. Schlame, B. Rustow, D. Kunze, H. Rabe, and G. Reichmann, *Biochem. J.*, *240*: 247 (1986).
124. P. J. Ryan and T. W. Honeyman, *J. Chromatogr.*, *331*: 177 (1985).
125. P. J. Ryan, K. McGoldrick, D. Stichney, and T. W. Honeyman, *J. Chromatogr.*, *320*: 421 (1985).
126. N. Oi and H. Kitahara, *J. Chromatogr.*, *265*: 117 (1983).
127. Y. Itabashi and T. Takagi, *Lipids*, *21*: 413 (1986).
128. T. Takagi and Y. Itabashi, *Lipids*, *22*: 596 (1987).
129. P. Laakso and W. W. Christie, *Lipids*, *25*: 349 (1990).
130. A. Kuksis, in *Analysis of Lipid and Lipoproteins* (E. G. Perkins, ed.), American Oil Chemists' Society, Champaign, IL, 1975, pp. 36–62.
131. J. J. Myher, in *Handbook of Lipid Research: Fatty Acids and Glycerides*, Vol. 1 (A. Kuksis, ed.), Plenum Press, New York, 1978, pp. 123–196.
132. A. Kuksis, in *Chromatography: Fundamentals and Applications of Chromatographic and Electrophoretic Methods. B. Applications* (E. Heftmann, ed.), Elsevier, Amsterdam, 1983, pp. B.75–B.146.
133. P. Mares, in *Chromatography of Lipids in Biomedical Research and Clinical Diagnosis* (A. Kuksis, ed.), Elsevier, Amsterdam, 1987, pp. 128–162.
134. C. Litchfield, R. D. Harlow, and R. Reiser, *J. Am. Oil Chem. Soc.*, *42*: 849 (1965).
135. E. C. Horning, E. A. Moscatelli, and E. C. Sweeley, *Chem. Ind.*, *London*, 751 (1959).
136. W. R. Supina, R. S. Henly, and R. F. Kruppa, *J. Am. Oil Chem. Soc. 43*: 202A (1966).
137. J. A. Fioriti, M. J. Kanuk, and R. J. Sims, *J. Chromatogr. Sci.*, *7*: 448 (1969).

138. R. Wood, R. D. Harlow, and E. N. Lambremont, *Lipids*, *4*: 159 (1969).
139. K. L. Mikolajczak, C. R. Jr Smith, and L. W. Tjarks, *Lipids*, *5*: 812 (1970).
140. A. Kuksis and W. C. Breckenridge, *J. Lipid Res.*, *7*: 576 (1966).
141. C. G. Youngs and M. R. Subbaram, *J. Am. Oil Chem. Soc.*, *41*: 218 (1964).
142. A. Kuksis, *J. Am. Oil Chem. Soc.*, *42*: 269 (1965).
143. R. W. Finch, *Analabs Res. Notes 10*: 1 (1970).
144. A. Kuksis, L. Marai, and D. A. Gornall, *J. Lipid Res.*, *8*: 352 (1967).
145. R. Wood and R. D. Harlow, *Lipids*, *5*: 776 (1970).
146. E. C. Horning, W. J. A. Vandenheuvel, and B. G. Creech, *Methods of Biochemical Analyses*, Vol. 2 (D. Glick, ed.), Interscience, New York, 1963.
147. A. Kuksis, in *Lipid Chromatographic Analysis*, Vol. 1 (G. V. Marinetti, ed.), Marcel Dekker, New York, 1967, pp. 239–337.
148. D. H. Desty (ed.), *Gas Chromatography*, Butterworths, London, 1958.
149. Official Methods of Analysis, *Association of Official Analytical Chemists*, 15th ed., Arlington, VA, IUPAC-AOAC Method 986-19, 1990, p. 972.
150. Official Methods and Recommended Practices of the American Oil Chemists' Society, (D. Firestone, ed.) 4th ed., AOCS Press, Champaign, IL, AOCS Official Method Ce 5-86, 1994.
151. J. Bézard and M. Bugaut, *J. Chromatogr. Sci.*, *7*: 639 (1969).
152. J. Bézard and M. Bugaut, *J. Chromatogr. Sci.*, *10*: 451 (1972).
153. M. Novotny, R. Segura, and A. Zlatkis, *Anal Chem.*, *44*: 9 (1972).
154. E. Geeraert, P. Sandra, and D. De Schepper, *J. Chromatogr.*, *279*: 287 (1983).
155. E. Geeraert and P. Sandra, *J. High Res. Chromatogr.*, *Chromatogr. Commun.*, *7*: 431 (1984).
156. E. Geeraert and P. Sandra, *J. Am. Oil Chem. Soc.*, *64*: 100 (1987).
157. K. Grob, *J. Chromatogr.*, *178*: 387 (1979).
158. K. Grob, *J. Chromatogr.*, *251*: 235 (1982).
159. K. Grob, *J. Chromatogr.*, *279*: 225 (1983).
160. K. Grob and H. P. Newkom, *J. Chromatogr.*, *189*: 109 (1980).
161. E. Geeraert, in *Chromatography of Lipids in Biomedical Research and Clinical Diagnosis* (A. Kuksis, ed.), Elsevier, Amsterdam, 1987, pp. 48–75.
162. P. Mares, *Prog. Lipid Res.*, *27*: 107 (1988).
163. J. J. Myher, A. Kuksis, L. Marai, and P. Sandra, *J. Chromatogr.*, *452*: 93 (1988).
164. C. Maniongui, J. Gresti, M. Bugaut, S. Gauthier, and J. Bézard, *J. Chromatogr.*, *543*: 81 (1991).
165. J. Gresti, M. Bugaut, C. Maniongui, and J. Bézard, *J. Dairy Sci.*, *76*: 1850 (1993).
166. J. A. Schmit and R. B. Wynne, *J. Gas Chromatogr.*, *4*: 325 (1966).
167. R. Watts and R. Dils, *J. Lipid Res.*, *9*: 40 (1968).
168. R. D. Harlow, C. Litchfield, and R. Reiser, *Lipids*, *1*: 216 (1966).
169. J. Bézard, M. Bugaut, and G. Clément, *J. Am. Oil Chem. Soc.*, *48*: 134 (1971).
170. J. Bézard, *Lipids*, *6*: 630 (1971).
171. P. Mares and P. Husek, *J. Chromatogr.*, *350*: 87 (1985).
172. P. Mares, *Prog. Lipid Res.*, *27*: 107 (1988).
173. R. G. Ackman, C. A. Eaton, J. Kinneman, and C. Litchfield, *Lipids*, *10*: 44 (1975).
174. R. G. Ackman, C. A. Eaton, and C. Litchfield, *Lipids*, *6*: 69 (1971).
175. C. Litchfield, R. G. Ackman, J. C. Sipos, and C. A. Eaton, *Lipids*, *6: 674 (1971).*
176. C. Litchfield, R. D. Harlow, and R. Reiser, *Lipids*, *2*: 363 (1967).
177. C. Imai, H. Watanabe, N. Haga, and T. Ii, *J. Am. Oil Chem. Soc.*, *51*: 326 (1974).
178. C. Litchfield, E. Miller, R. D. Harlow, and R. Reiser, *Lipids*, *2*: 345 (1967).
179. J. P. Moretain and J. A. Bézard, *Rev. Fr. Corps Gras*, *24*: 303 (1977).
180. G. Semporé and J. Bézard, *Rev. Fr. Corps Gras*, *24*: 611 (1977).
181. G. Descargues and J. Bézard, *Riv. Ital. Sost. Grasse*, *58*: 613 (1981).
182. M. Ouédraogo and J. Bézard, *Rev. Fr. Corps Gras*, *28*: 473 (1981).
183. K. Sawadogo and J. Bézard, *Oléagineux*, *37*: 69 (1982).
184. M. A. Ouédraogo and J. Bézard, *Rev. Fr. Corps Gras*, *29*: 11 (1982).
185. M. L. Hunter, W. W. Christie, and J. H. Moore, *Lipids*, *8*: 65 (1973).
186. A. Kuksis, and W. C. Breckenridge, in *Dairy Lipids and Lipid Metabolism* (M. F. Brink and D. Kritchevsky, eds.), Avi, Westport, 1968, pp. 28–98.

187. N. C. Shanta and T. N. B. Kaimal, *Lipids*, *19*: 971 (1984).
188. J. Skorepa, P. Mares, J. Rublicova, and S. Vinogradov, *J. Chromatogr.*, *162*: 177 (1979).
189. R. Wood and R. D. Harlow, *Arch. Biochem. Biophys.*, *131*: 495 (1969).
190. E. Geeraert, in *Sample Introduction in Capillary Gas Chromatography*, Vol. 1 (A. Kuksis, ed.), Elsevier, Amsterdam, 1987, pp. 48–75.
191. R. P. D'Alonzo, W. J. Kozarek, and R. L. Wade, *J. Am. Oil Chem. Soc.*, *59*: 292 (1982).
192. K. Grob, H. P. Neukom, and R. Battaglia, *J. Am. Oil Chem. Soc.*, *57*: 282 (1980).
193. M. A. Amer, D. B. Kupranycz, and B. E. Baker, *J. Am. Oil Chem. Soc.*, *62*: 1551 (1985).
194. P. Kalo, K. Vaara, and M. Antilia, *J. Chromatogr.*, *368*: 145 (1986).
195. B. G. Muuse and H. J. Vander Kamp, *Neth. Milk Dairy J.*, *39*: 1 (1985).
196. P. P. Schmid, M. D. Muller, and W. Simov, *J. High Res. Chromatogr. Chromatogr. Commun.*, *2*: 675 (1979).
197. S. G. Wakeham and N. M. Frew, *Lipids*, *17*: 831 (1982).
198. J. J. Myher, A. Kuksis, L. Y. Yang, and L. Marai, *Biochem. Cell. Biol.*, *65*: 811 (1987).
199. J. J. Myher, A. Kuksis, L. Marai, and J. Cerbulis, *Lipids*, *21*: 309 (1986).
200. A. Kuksis, L. Marai, J. J. Myher, J. Cerbulis, and H. M. Farrell, *Lipids*, *21*: 183 (1986).
201. R. Aneja, A. Bhati, R. J. Hamilton, F. B. Padley, and D. A. Steven, *J. Chromatogr.*, *173*: 392 (1979).
202. A. Kuksis, in *Analysis of Lipids and Lipoproteins* (E. G. Perkins, ed.), American Oil Chemists' Soc., Champaign, IL, 1975, pp. 36–62.
203. T. Takagi and Y. Itabashi, *Lipids*, *12*: 1062 (1977).
204. T. Rezanka, P. Mares, P. Husek, and M. Podojil, *J. Chromatogr.*, *355*: 265 (1986).
205. H. Traitler and A. Prevot, *J. High Res. Chromatogr. Chromatogr. Commun.*, *4*: 109 (1981).
206. H. Traitler and A. Prevot, *Rev. Fr. Corps Gras*, *28*: 263 (1981).
207. J. V. Hinshaw and W. Seferovic, *J. High Res. Chromatogr. Chromatogr. Commun.*, *9*: 731 (1986).
208. M. Termonia, F. Munari, and P. Sandra, *J. High Res. Chromatogr. Chromatogr. Commun.*, *10*: 263 (1987).
209. E. Geeraert and P. Sandra, *J. High Res. Chromatogr. Chromatogr. Commun.*, *8*: 415 (1985).
210. J. J. Myher, A. Kuksis, L. Marai, and P. Sandra, *J. Chromatogr.*, *452*: 93 (1988).
211. M. Bugaut and J. Bézard, *J. Chromatogr. Sci.*, *8*: 380 (1970).
212. *AOCS Official Methods and Recommended Practices*, 4th ed., AOCS Press, Champaign, IL, AOCS Official Method Ce 5-86, 1994.
213. *IUPAC Standard Methods for the Analysis of Oils, Fats and Derivatives*, 7th Revised and Enlarged Edition, Blackwell Scientific, Boston, IUPAC-AOAC Method 986-19, 1990.
214. R. D. Wood, P. K. Raju, and R. Reiser, *J. Am. Oil Chem. Soc.*, *42*: 161 (1965).
215. J. J. Myher and A. Kuksis, *Lipids*, *9*: 382 (1974).
216. R. D. Wood and F. Snyder, *Lipids*, *1*: 62 (1966).
217. Y. Itabashi and T. Takagi, *Lipids*, *15*: 205 (1980).
218. *AOCS Official Methods and Recommended Practices*, 4th ed., AOCS Press, Champaign, IL, AOCS Official Method Cd 11b-9, 1994.
219. A. G. McInnes, N. H. Tattrie, and M. Kates, *J. Am. Oil Chem. Soc.*, *37*: 7 (1960).
220. V. R. Huebner, *J. Am. Oil Chem. Soc.*, *36*: 262 (1959).
221. J. Bézard, J. P. Moretain, and M. Bugaut, *Fette Seifen Anstrichm.*, *79*: 399 (1977).
222. J. P. Morétain and J. A. Bézard, *Rev. Fr. Corps Gras*, *24*: 303 (1977)
223. A. Kuksis, L. Marai, and D. A. Gornall, *J. Lipid Res.*, *8*: 352 (1967).
224. W. C. Breckenridge and A. Kuksis, *J. Lipid Res.*, *9*: 388 (1968).
226. L. Marai, W. C. Breckenridge, and A. Kuksis, *Lipids*, *4*: 562 (1969).
227. J. Bézard, J. P. Morétain, and M. Bugaut, *Fette Seifen Anstrichm.*, *79*: 399 (1977).
228. N. H. Morley, A. Kuksis, and D. Buchnea, *Lipids*, *9*: 481 (1974).
229. J. J. Myher, L. Marai, and A. Kuksis, *Lipids*, *12*: 775 (1977).
230. J. J. Myher and A Kuksis, *Can. J. Biochem.*, *57*: 117 (1979).
231. F. Manganaro, J. J. Myher, A. Kuksis, and D. Kritchevsky, *Lipids*, *16*: 508 (1981).
232. A. Prévot and J. J. Coustille, *Rev. Fr. Corps Gras*, *29*: 17 (1982).
233. J. J. Myher, A. Kuksis, and L. Y. Yang, *Biochem. Cell. Biol.*, *68*: 336 (1990).
234. J. J. Myher and A. Kuksis, *Can. J. Biochem.*, *60*: 638 (1982).
235. J. J. Myher and A. Kuksis, *Can. J. Biochem. Biophys. Methods*, *10*: 13 (1984).

236. S. J. Gaskell and C. J. W. Books, *J. Chromatogr.*, *142*: 469 (1977).
237. P. Michelsen and G. Odham, *J. Chromatogr.*, *331*: 295 (1985).
238. J. J. Myher and A. Kuksis, *J. Chromatogr. Sci.*, *13*: 138 (1975).
239. A. Kuksis, L. Marai, and J. J. Myher, *J. Am. Oil Chem. Soc.*, *50*: 193 (1973).
240. Y. Itabashi, A. Kuksis, and J. J. Myher, *J. Lipid Res.*, *31*: 2119 (1990).
241. G. Clément, J. Clément, and J. Bézard, *Arch. Sci. Physiol.*, *16*: 213 (1962).
242. G. Clément, J. Clément, J. Bézard, G. Di Costanzo, and R. Paris, *J. Dairy Sci.*, *46*: 1423 (1963).
243. M. L. Blank and O. S. Privett, *J. Dairy Sci.*, *47*: 481 (1964).
244. E. Fideli, *Riv. Ital. Sostanze Grasse*, *44*: 220 (1967).
245. R. L. Glass, R. Jenness, and L. W. Lohse, *Comp. Biochem. Physiol.*, *28*: 783 (1969).
246. A. H. Duthi and H. V. Atherton, *J. Chromatogr.* *51*: 319 (1970).
247. E. H. Jr. Gruger, D. C. Malins, and E. J. Gauglitz, Jr., *J. Am. Oil Chem. Soc.*, *37*: 214 (1960).
248. K. A. Karlsson, A. Norrby, and B. Samuelsson, *Biochim. Biophys. Acta*, *144*: 162 (1967).
249. K. A. Karlsson, K. Nilsson, and I. Pascher, *Lipids*, *3*: 389 (1968).
250. S. Laurell, *Biochim. Biophys. Acta*, *152*: 75 (1968).
251. G. Sezille, G. Biserte, J. Jaillard, and P. Scherpereel, *Eur. J. Clin. Biol. Res.*, *15*: 1122 (1970).
252. R. Kleiman, C. R. Jr. Smith, S. G. Yates, and Q. Jones, *J. Am. Oil Chem. Soc.*, *42*: 169 (1965).
253. J. A. Fioriti, N. Buide, and R. J. Sims, *J. Am. Oil Chem. Soc.*, *46*: 108 (1969).
254. J. A. Fioriti, N. Buide, and R. J. Sims, *Lipids*, *4*: 142 (1969).
255. B. E. Phillips, C. R. Smith, Jr., and J. W. Hagemann, *Lipids*, *4*: 473 (1969).
256. E. Vioque and M. P. Maza, *Grasas y Aceites*, *22*: 25 (1971).
257. W. H. Tallent, D. G. Cope, J. W. Hagemann, F. R. Earle, and I. A. Wolff, *Lipids*, *1*: 335 (1966).
258. R. W. Miller, F. R. Earle, I. A. Wolff, and S. Q. Jones, *J. Am. Oil Chem. Soc.*, *42*: 817 (1965).
259. F. D. Gunstone and M. I. Queshi, *J. Sci. Food Agr.*, *19*: 386 (1968).
260. E. Fideli, A. Tarenghi, and G. Jacini, *Riv. Ital. Sostanze Grasse*, *44*: 391 (1967).
261. K. Serck-Hanssen, *Acta Chem. Scand.*, *21*: 301 (1967).
262. K. L. Mikolajczak, C. R. Smith, Jr., and I. A. Wolff, *Lipids*, *3*: 215 (1968).
263. J. Pokorny, J. Hladik, and I. Zeman, *Pharmazie*, *23*: 332 (1968).
264. C. D. Evans, D. G. McConnell, R. L. Hoffmann, and H. Peters, *J. Am. Oil Chem. Soc.*, *44*: 281 (1967).
265. K. L. Mikolajczak and C. R. Smith, Jr., *Lipids*, *2*: 261 (1967).
266. C. Franzke, F. Kretzschmann, B. Rustow, and H. Rugentein, *Pharmazie*, *22*: 487 (1967).
267. O. S. Privett and M. L. Blank, *J. Am. Oil Chem. Soc.*, *40*: 70 (1963).
268. O. S. Privett and M. L. Blank, *J. Lipid Res.*, *2*: 37 (1961).
269. O. S. Privett, M. L. Blank, and J. A. Schmit, *J. Food Sci.*, *27*: 463 (1962).
270. G. Jurriens, *Anal. Character. Oils, Fats, Fat Prod.*, *2*: 237 (1968).
271. Y. Takahashi, *Yukagaku*, *17*: 492 (1968).
272. L. J. Morris and S. W. Hall, *Lipids*, *1*: 188 (1966).
273. F. D. Gunstone and F. D. Padley, *J. Am. Oil Chem. Soc.*, *42*: 957 (1965).
274. M. R. Sahasrabudhe and D. G. Chapman, *J. Am. Oil Chem. Soc.*, *38*: 88 (1961).
275. G. V. Marinetti, *J. Lipid Res.*, *7*: 786 (1966).
276. O. Renkonen, *Ann. Med. Exp. Biol. Fenn.*, *43*: 194 (1965).
277. R. Kleiman, R. W. Miller, F. R. Earle, and I. A. Wolff, *Lipids*, *1*: 286 (1966).
278. R. Kleiman, R. W. Miller, F. R. Earle, and I. A. Wolff, *Lipids*, *2*: 473 (1967).
279. B. Akesson, J. Elovson, and G. Arvidson, *Biochim. Biophys. Acta*, *210*: 15 (1970).
280. O. Hirayama, *Nippon Nogei Kagalu Kaishi*, *35*: 437 (1961).
281. P. Desnuelle and P. Savary, *J. Lipid Res.*, *4*: 369 (1963).
282. M. Yurkowski and H. Brockerhoff, *Biochim. Biophys. Acta*, *125*: 55 (1966).
283. W. W. Christie and J. H. Moore, *Biochim. Biophys. Acta*, *176*: 445 (1969).
284. J. A. Fioriti, N. Buide, and R. J. Sims, *46*: 108 (1965).
285. A. E. Thomas III, J. E. Scharoun, and H. Ralston, *J. Am. Oil Chem. Soc.*, *42*: 789 (1965).
286. W. W. Christie, *Lipid Analysis*, Pergamon Press, Oxford, 1982.
287. W. W. Christie, *High Performance Liquid Chromatography and Lipids: A Practical Guide*, Pergamon Press, Oxford, 1987.
288. O. K. Guha and J. Janak, *J. Chromatogr.*, *68*: 325 (1972).

289. S. D. Ittel and J. A. Iberis, in *Advances of Organometallic Chemistry*, Vol. 14 (F. G. A. Stone and R. West, eds.), Academic Press, New York, 1976, pp. 33–61.

290. C. L. De Ligny, in *Advances in Chromatography*, Vol. 14 (J. C. Giddings, E. Grushka, J. Cazes, and P. R. Brown, eds.), Marcel Dekker, New York, 1976, pp. 265–304.

291. P. H. Kasai, D. McLeod, and T. Watanabe, *J. Am. Oil Chem. Soc.*, *102*: 179 (1980).

292. K. Kemper, H. U. Melchert, K. Rubach, and H. Hoffmeister, *Fresenius Z. Anal Chem.*, *331*: 634 (1988).

293. A. G. Vereshchagin, *J. Chromatogr.*, *17*: 382 (1965).

294. A. G. Vereshchagin and S. V. Skortsova, *Dokl. Akad. Nauk, SSSR*, *157*: 699 (1964).

295. G. V. Novitskaya, A. V. Kaverina, and A. G. Vereshchagin, *Biokhimiya*, *30*: 1260 (1965).

296. G. V. Novitskaya and V. I. Maltseva, *Biokhimiya*, *31*: 953 (1966).

297. A. G. Vereshchagin and G. V. Novitskaya, *J. Am. Oil Chem. Soc.*, *42*: 970 (1965).

297a. W. O. Ord and P. C. Bamford, *Chem. Ind.*, *London*, 277 (1967).

298. B. Nikolova-Damyanova, in *Advances in Lipid Methodology*, Vol. 1 (W. W. Christie, ed.), Oily Press, Ayr, 1992, pp. 181–237.

299. J. K. Sliwowski and E. Caspe, *J. Steroid Biochem.*, *8*: 47 (1977).

300. L. J. Morris, in *Metabolism and Physiological Significance of Lipids* (R. M. C. Dawson and D. N. Rhodes, eds.), Wiley, New York, 1964, pp. 641–650.

301. R. Perron and M. Auffret, *Oléagineux*, *20*: 379 (1965).

302. C. Madelmont and R. Perron, *Bull. Soc. Chim. France*, 3442 (1967).

303. J. M. Cubero and H. K. Mangold, *Microchem. J.*, *9*: 227 (1965).

304. K. Aitzetmueller and L. A. G. Goncalves, *J. Chromatogr.*, *519*: 349 (1990).

305. L. J. Morris, *J. Lipid Res.*, *7*: 717 (1966).

306. N. R. Bottino, *J. Lipid Res.*, *12*: 24 (1971).

307. L. J. Morris, D. Wharry, and E. W. Hammond, *J. Chromatogr.*, *31*: 69 (1967).

308. H. Wessels and N. S. Rajagopal, *Fette Seifen Anstrichm.*, *71*: 543 (1969).

309. O. Renkonen, *Adv. Lipid Res.*, *5*: 329 (1967).

310. B. Akesson, *Eur. J. Biochem.*, *9*: 463 (1969).

311. J. E. Muldrey, *J. Am. Oil Chem. Soc.*, *43*: 138A (1969).

312. S. C. Rutan, P. W. Carr, W. J. Cheong, J. H. Park, and L. R. Snyder, *J. Chromatogr.*, *463*: 21 (1989).

313. L. R. Snyder, *J. Chromatogr. Sci.*, *16*: 223 (1978).

314. C. Litchfield, M. Farquhar, and R. Reiser, *J. Am. Oil Chem. Soc.*, *41*: 588 (1964).

314a. B. De Vries and G. Jurriens, *Fette Seifen Anstrichm.*, *65*: 725 (1963).

314b. M. Gold, *Lipids*, *4*: 288 (1969).

314c. M. M. Chakrabarty and D. Bhattacharyya, *J. Chromatogr.*, *31*: 556 (1967).

314d. F. D. Gunstone, F. B. Padley, and M. I. Qureshi, *Chem. Ind.* (*London*), 483 (1964).

314e. E. Gordis, *J. Clin. Invest.*, *44*: 1451 (1965).

314f. G. Jurriens and L. Schouten, *Rev. Fr. Corps Gras*, *12*: 505 (1965).

315. D. Chobanov, R. Tarandjiiska and R. Chobanova, *J. Am. Oil Chem. Soc.*, *53*: 48 (1976).

316. S. Takano and Y. Kondoh, *J. Am. Oil Chem. Soc.*, *64*: 380 (1987).

317. R. Tarandjiiska and H. Nguyen, *Riv. Ital. Sost. Grasse*, *66*: 99 (1989).

318. B. DeVries and G. Jurriens, *J. Chromatogr.*, *14*: 525 (1964).

319. G. Jurriens, *Chem. Weekbl*, *61*: 257 (1965).

320. O. S. Privett, M. L. Blank, D. W. Codding, and E. C. Nickell, *J. Am. Oil Chem. Soc.*, *42*: 381 (1965).

321. C. B. Barrett, M. S. J. Dallas, and F. B. Padley, *J. Am. Oil Chem. Soc.*, *40*: 580 (1963).

322. F. D. Gunstone, R. J. Hamilton, and M. I. Qureshi, *J. Chem. Soc.*, *London*, 319 (1965).

323. E. E. Hill, D. R. Husbands, and W. E. M. Lands, *J. Biol. Chem.*, *243*: 4440 (1968).

324. D. R. Husbands, *Biochem. J.*, *120*: 365 (1970).

325. O. S. Privett and L. J. Nutter, *Lipids*, *2*: 149 (1967).

326. J. N. Roehm and O. S. Privett, *Lipids*, *5*: 353 (1970).

327. M. L. Blank, B. Verdino, and O. S. Privett, *J. Am. Oil Chem. Soc.*, *42*: 87 (1965).

328. W. W. Christie, R. D. Noble, and J. H. Moore, *Analyst*, *95*: 940 (1970).

329. F. D. Gunstone, F. B. Padley, and M. I. Qureshi, *Chem. Ind. London*, 483 (1964).

330. E. Gordis, *J. Clin. Invest*, *44*: 1451 (1965).

331. A. Kuksis, W. C. Breckenridge, L. Marai, and O. Stachnyk, *J. Lipid Res.*, *10*: 25 (1969).

332. A. Rajiah, M. R. Subbaram, and K. T. Achaya, *J. Chromatogr.*, *38*: 35 (1968).

333. C. Bandyopadhyay, *J. Chromatogr.*, *37*: 123 (1968).
334. H. P. Kaufman and H. Wessels, *Fette Seifen Anstrichm.*, *68*: 249 (1966).
335. H. Grynberg, K. Ceglowska, and H. Szczepanska, *Rev. Fr. Corps Gras*, *13*: 595 (1966).
336. E. Vioque, M. P. Maza, and M. Calderon, *Grasas Aceites*, *15*: 173 (1964).
337. K. G. Berger and E. E. Akehurst, *J. Food Technol.*, *1*: 237 (1966).
338. G. Jurriens, B. De Vries, and L. Schouten, *J. Lipid Res.*, *5*: 267 (1964).
339. G. Jurriens, B. De Vries, and L. Schouten, *J. Lipid Res.*, *5*: 366 (1964).
340. F. Amat, E. Marquinez, R. M. Utrilla, and D. Martin, *Grasas Aceites*, *17*: 47 (1966).
341. H. P. Kaufman and H. Wessels, *Fette Seifen Anstrichm.*, *66*: 81 (1964).
342. R. J. Vander Wal, *J. Am. Oil Chem. Soc.*, *42*: 754 (1965).
343. E. Fideli and F. Camurati, *Riv. Ital. Sostanze Grasse*, *46*: 97 (1969).
344. H. P. Kaufman and K. D. Mukherjee, *Fette Seifen Anstrichm.*, *67*: 183 (1965).
345. J. C. Touchstone, *J. Chromatogr. Sci.*, *26*: 645 (1988).
346. M. Inomata, F. Takaku, Y. Nagai, and M. Saito, *Anal. Biochem.*, *125*: 197 (1982).
347. E. W. Hammond, *Chem. Ind. London*, 710 (1981).
348. D. Chobanov, R. Tarandjiiska, and R. Chobanova, *J. Am. Oil Chem. Soc.*, *53*: 48 (1976).
349. P. Ilinov, *Lipids*, *14*: 598 (1979).
350. L. V. Andreev, *J. High Res. Chromatogr. Chromatogr. Commun.*, *6*: 575 (1983).
351. C. B. Barrett, M. S. J. Dallas, and F. B. Padley, *J. Am. Oil Chem. Soc.*, *40*: 580 (1963).
352. B. Nikolova-Damyanova and B. Amidzhin, *J. Planar. Chromatogr. Mod. TLC*, *4*: 397 (1991).
353. F. D. Gunstone and F. B. Padley, *J. Am. Oil Chem. Soc.*, *42*: 957 (1965).
354. F. C. Denboer, *Z. Anal. Chem.*, *205*: 308 (1964).
355. R. Tarandjiiska and H. Nguyen, *Riv. Ital. Sost. Grasse*, *65*: 489 (1988).
356. N. R. Bottino, *J. Lipid Res.*, *12*: 24 (1971).
357. B. De Vries and G. Jurriens, *Fette Seifen Anstrichm.*, *65*: 725 (1963).
358. M. S. J. Dallas and F. B. Padley, *Lebensm. Wiss Technol.*, *10*: 328 (1977).
359. D. Gegiou and M. Georgouli, *J. Am. Oil Chem. Soc.*, *60*: 833 (1983).
360. F. B. Padley, PhD dissertation, University of St Andrews, 1965.
361. Z. L. Bandi and H. K. Mangold, *Segar. Sci.*, *4*: 83 (1969).
362. G. Semporé and J. Bézard, *J. Am. Oil Chem. Soc.*, *59*: 124 (1982).
363. J. Bézard and M. A. Ouédraogo, *J. Chromatogr.*, *196*: 279 (1980).
364. F. D. Gunstone and M. I. Qureshi, *J. Am. Oil Chem. Soc.*, *42*: 961 (1965).
365. H. Grynberg and K. Ceglowska, *Rev. Fr. Corps Gras*, *17*: 89 (1970).
366. M. Della Greca, P. Monaco, G. Pinto, A. Pollio, and L. Previtera, *Biochim. Biophys. Acta*, *1004*: 271 (1989).
367. N. Ozcimder and W. E. Hammers, *J. Chromatogr.*, *187*: 307 (1980).
368. T. W. Culp, R. D. Harlow, C. Litchfield, and R. Reiser, *J. Am. Oil Chem. Soc.*, *42*: 974 (1965).
369. P. Lund, *Milchwissenschaft*, *43*: 159 (1988).
370. U. Persmark and B. Toregard, *J. Chromatogr.*, *37*: 121 (1968).
371. J. W. E. Coenen, H. Boerma, B. E. Linsen, and B. De Vries, *Proc. 3rd Int. Congr. Catal.*, 1964, pp. 1397–1399 (1965).
372. B. Akesson, J. Elovson and G. Arvidson, *Biochim. Biophys. Acta*, *218*: 44 (1970).
373. J. B. Mudd, L. M. G. Van Golde, and L. L. M. Van Deenen, *Biochim. Biophys. Acta*, *176*: 547 (1969).
374. L. L. M. Van Golde and L. L. M. Van Deenen, *Biochim. Biophys. Acta*, *125*: 496 (1966).
375. L. L. M. Van Golde and L. L. M. Van Deenen, *Chem. Phys. Lipids*, *1*: 547 (1967).
376. J. J. Myher, A. Kuksis, and L. Marai, *Anal Chem.*, *50*: 557 (1978).
377. O. Renkonen, *J. Am. Oil Chem. Soc.*, *42*: 298 (1965).
378. A. Kuksis, L. Marai, W. C. Breckenridge, D. A. Gornall, and O. Stachnyk, *Can. J. Physiol. Pharmacol.*, *46*: 511 (1968).
379. O. Renkonen, *Ann. Med. Exp. Biol. Fenn.*, *44*: 356 (1966).
380. O. Renkonen, *Biochim. Biophys. Acta*, *125*: 288 (1966).
381. O. Renkonen, *Biochim. Biophys. Acta*, *152*: 114 (1968).
382. O. Renkonen, and L. Rikkinen, *Acta Chem. Scand.*, *21*: 2282 (1967).
383. L. M. G. Van Golde and L. L. M. Van Deenen, *Biochim. Biophys. Acta*, *125*: 496 (1966).
384. L. M. G. Van Golde and L. L. M. Van Deenen, *Chem. Phys. Lipids*, *1*: 157 (1967).

385. T. W. Keenan, Y. C. Awasthi, and F. L. Crane, *Biochim. Biophys. Res. Commun.*, *40*: 1102 (1970).
386. K. Yokota, R. Kanamoto, and M. Kito, *J. Bacteriol.*, *141*: 1047 (1980).
387. R. J. Hamilton and P. A. Sewel, *Introduction to High Performance Lipid Chromatography*, 2nd ed., Chapman & Hall, London, 1982.
388. R. Macrae, *HPLC in Food Analysis*, Academic Press, London, 1982.
389. N. A. Parris, *Instrumental Liquid Chromatography*, 2nd ed., Elsevier, Amsterdam, 1984.
390. L. R. Snyder and J. J. Kirkland, *Introduction to Modern Liquid Chromatography*, 2nd ed., John Wiley and Sons, New York, 1979.
391. V. R. Meyer, *J. Chromatogr.*, *334*: 197 (1985).
392. N. Cooke and K. Olsen, *Am. Lab.*, 45 (1979).
393. E. Grushka and E. J. Kikta, Jr., *Anal. Chem.*, *49*: 1005 A (1977).
394. A. H. El-Hamdy and E. G. Perkins, *J. Am. Oil Chem. Soc.*, *58*: 867 (1981).
395. H. F. Fieberg, *Fette Seifen Anstrichm.*, *87*: 53 (1985).
396. F. C. Phillips, W. L. Erdahl, J. D. Nadenicek, L. J. Nutter, J. A. Schmit, and O. S. Privett, *Lipids*, *19*: 142 (1984).
397. R. D. Plattner, *J. Am. Oil Chem. Soc.*, *58*: 638 (1981).
398. R. D. Plattner, G. F. Spencer, and R. Kleiman, *J. Am. Oil Chem. Soc.*, *54*: 511 (1977).
399. S. Wada, C. Koizumi, and J. Nonaka, *Yukagaku*, *26*: 95 (1977).
400. S. Wada, C. Koizumi, A. Takiguchi, and J. Nonaka, *Yukagaku*, *27*: 21 (1978).
401. A. H. El-hamdi and E. G. Perkins, *J. Am. Oil Chem. Soc.*, *58*: 49 (1981).
402. J. A. Bézard and M. A. Ouédraogo, *J. Chromatogr.*, *196*: 279 (1980).
403. M. W. Dong and J. L. DiCesare, *J. Am. Oil Chem. Soc.*, *60*: 788 (1983).
404. F. C. Phillips, W. L. Erdahl, J. A. Schmit, and O. S. Privett, *Lipids*, *19*: 880 (1984).
405. G. Semporé and J. Bézard, *J. Chromatogr.*, *366*: 261 (1986).
406. V. K. S. Shukla, W. S. Nielsen, and W. Batsberg, *Fette Seifen Anstrichm.*, *85*: 274 (1983).
407. B. Peterssen, O. Podlaha, and B. Toregard, *J. Am. Oil Chem. Soc.*, *58*: 1005 (1981).
408. O. Podlaha and B. Toregard, *J. High Res. Chromatogr. Chromatogr. Commun.*, *5*: 553 (1982).
409. O. Podlaha and B. Toregard, *Fette Seiffen Anstrichm.*, *86*: 243 (1984).
410. J. L. Robinson and R. J. Macrae, *J. Chromatogr.*, *303*: 386 (1984).
411. J. L. Robinson, M. Tsimidou, and R. Macrae, *J.Chromatogr.*, *324*: 35 (1985).
412. A. Stolyhwo, H. Colin, and G. Guiochon, *Anal Chem.*, *57*: 1342 (1985).
413. K. Takahashi, T. Hirano, M. Egi, and K. Zama, *J. Am. Oil Chem. Soc.*, *62*: 1489 (1985).
414. R. E. Pauls, *J. Am. Oil Chem. Soc.*, *60*: 1819 (1983).
415. B. Herslof, O. Podlaha, and B. Toregard, *J. Am. Oil Chem. Soc.*, *56*: 864 (1979).
416. J. P. Goiffon, C. Reminiac, and M. Ollé, *Rev. Fr. Corps Gras*, *28*: 167 (1981).
417. J. P. Goiffon, C. Reminiac, and D. Furon, *Rev. Fr. Corps Gras*, *28*: 199 (1981).
418. Y. Lozano, *Rev. Fr. Corps Gras*, *30*: 333 (1983).
419. J. L. Perrin and M. Naudet, *Rev. Fr. Corps Gras*, *30*: 279 (1983).
420. M. Tsimidou and R. Macrae, *J. Chromatogr.*, *285*: 178 (1984).
421. R. L. Kimmey and E. G. Perkins, *J. Am. Oil Chem. Soc.*, *61*: 1209 (1984).
422. K. Kinoshita, M. Kimura, K. Takahashi, and K. Zama, *J. Am. Oil Chem. Soc.*, *63*: 1558 (1986).
423. J. A. Singleton and H. E. Pattee, *J. Am. Oil Chem. Soc.*, *61*: 761 (1984).
424. J. A. Singleton and H. E. Pattee, *J. Am. Oil Chem. Soc.*, *64*: 534 (1987).
425. J. Kondoh and S. Takano, *Anal Chem.*, *58*: 2380 (1986).
426. N. A. Parris, *J. Chromatogr.*, *149*: 615 (1978).
427. E. Schulte, *Fette Seifen Anstrichm.*, *83*: 289 (1981).
428. S. Wada, C. Koizumi, A. Takiguchi, and J. Nonaka, *Yukagaku*, *27*: 579 (1978).
429. S. Wada, C. Koizumi, A. Takiguchi, and J. Nonaka, *Bull. Jpn. Soc. Sci. Fish.*, *45*: 611 (1979).
430. S. Wada, C. Koizumi, A. Takiguchi, and J. Nonaka, *Bull. Jpn. Soc. Sci. Fish.*, *45*: 615 (1979).
431. G. W. Jensen, *J. Chromatogr.*, *204*: 407 (1981).
432. J. D. Baty and N. W. Rawle, *J. Chromatogr.*, *395*: 395 (1987).
433. K. V. V. Nurmela and L. T. Satama, *J. Chromatogr.*, *435*: 139 (1988).
434. A. Kuksis, L. Marai, and J. J. Myher, *J. Chromatogr.*, *273*: 43 (1983).
435. A. Kuksis, J. J. Myher, and L. Marai, *J. Am. Oil Chem. Soc.*, *61*: 1582 (1984).
436. L. Marai, J. J. Myher, and A. Kuksis, *Can J. Biochem. Cell Biol.*, *61*: 840 (1983).

437. J. J. Myher, A. Kuksis, L. Marai, and F. Manganaro, *J. Chromatogr.*, *283*: 289 (1984).
438. J. L. Perrin and A. Prevot, *Rev. Fr. Corps Gras*, *33*: 437 (1986).
439. B. Herslof and G. Kindmark, *Lipids*, *20*: 783 (1985).
440. A. Karleskind, G. Valmalle, O. Midler, and M. Blanc, *Rev. Fr. Corps Gras*, *25*: 551 (1978).
441. M. Tsimidou and R. Macrae, *J. Chromatogr.*, *285*: 178 (1984).
442. M. Tsimidou and R. Macrae, *J. Chromatogr. Sci.*, *23*: 155 (1985).
443. E. Geeraert and D. Deschepper, *J. High Res. Chromatogr. Chromatogr. Commun.*, *6*: 123 (1983).
444. E. Deffense, *Rev. Fr. Corps Gras*, *31*: 123 (1984).
445. K. Aitzetmueller, *Prog. Lipid Res.*, *21*: 171 (1982).
446. R. J. Hamilton, in *Analysis of Oils and Fats* (R. J. Hamilton and B. J. Rossell, eds.), Elsevier, London, 1986, pp. 243–311.
447. E. G. Perkins, J. E. Bauer, N. Pelick, and A. El-Hamdy, in *Dietary Fats and Health* (E. G. Perkins and W. J. Wisek, eds.), American Oil Chemists' Soc., Champaign, IL, 1981, pp. 184–208.
448. R. D. Plattner, *Meth. Enzymol.*, *72*: 21 (1981).
449. O. Podlaha and B. Toregard, *J. Chromatogr.*, *482*: 215 (1989).
450. V. K. S. Shukla, *Prog. Lipid Res.*, *27*: 5 (1988).
451. L. Hernqvist, B. Herslof, and M. Herslof, *Fette Seifen Anstrichm.*, *86*: 393 (1984).
452. B. Herslof, M. Herslof, and O. Podlaha, *Fette Seifen Anstrichm.*, *82*: 460 (1980).
453. R. D. Plattner, K. Wade, and R. Kleiman, *J. Am. Oil Chem. Soc.*, *55*: 381 (1978).
454. B. Herslof, *J. High Res. Chromatogr. Chromatogr. Commun.*, *4*: 471 (1981).
455. O. S. Privett and W. L. Erdhal, *Anal Biochem.*, *84*: 449 (1978).
456. B. Herslof and G. Kindmark, *Lipids*, *20*: 783 (1985).
457. W. W. Christie, *J. Lipid Res.*, *26*: 507 (1985).
458. J. L. Perrin, A. Prevot, A. Stolyhwo, and G. Guiochon, *Rev. Fr. Corps Gras*, *31*: 495 (1984).
459. A. Stolyhwo, H. Colin, M. Martin, and G. Guiochon, *J. Chromatogr.*, *288*: 253 (1984).
460. Y. Kondoh and S. Takano, *Anal Chem.*, *58*: 2380 (1986).
461. American Oil Chemists' Society (AOCS), *Official Methods and Recommended Practices*, 4th ed. (D. Firestone, ed.), AOCS Press, Champaign, IL, AOCS Official Method Ce 5c-93, 1994.
462. American Oil Chemists' Society (AOCS), *Official Methods and Recommended Practices*, 4th ed. (D. Firestone, ed.), AOCS Press, Champaign, IL, AOCS Official Method Ce 5b-89, 1994.
463. K. Weber, E. Schulte, and H.-P. Thier, *Fat Sci. Technol.*, *9*: 341 (1988).
464. K. Weber, E. Schulte, and H.-P. Thier, *Fat Sci. Technol.*, *10*: 389 (1988).
465. L. J. R. Barron, T. G. Hiero, and G. Santa Maria, *J. Dairy Sci*, *57*: 517 (1990).
466. D. R. Kodali, T. G. Redgrave, D. M. Small, and D. Atkinson, *Biochemistry*, *24*: 519 (1985).
467. F. J. Heinze, M. Linscheid, and E. Heinz, *Anal. Biochem.*, *139*: 126 (1984).
468. E. C. Smith, A. Jones, and E. W. Hammond, *J. Chromatogr.*, *188*: 205 (1980).
469. B. Vonach and G. Schomburg, *J. Chromatogr.*, *149*: 417 (1978).
470. R. Berglund and K. Thente, *Int. Lab.* (Nov.), 34 (1983).
471. A. Monseigny, P. Y. Vigneron, M. Levacq, and F. Zwobada, *Rev. Fr. Corps Gras*, *26*: 107 (1979).
472. B. S. J. Jeffrey, *J. Am. Oil Chem. Soc.*, *68*: 289 (1991).
473. R. R. Hearth, J. H. Tomlinson, and R. E. Doolittle, *J. Chromatogr. Sci.*, *15*: 10 (1977).
474. R. R. Hearth, J. H. Tomlinson, R. E. Doolittle, and A. T. Proveaux, *J. Chromatogr. Sci.*, *13*: 380 (1975).
475. R. Battaglia and D. Froelich, *Chromatographia*, *13*: 428 (1980).
476. E. W. Hammond, *J. Chromatogr.*, *203*: 397 (1981).
477. W. W. Christie, *J. High Res. Chromatogr. Chromatogr. Commun.*, *10*: 148 (1987).
478. W. W. Christie, *J. Chromatogr.*, *454*: 273 (1988).
479. P. Laakso, W. W. Christie, and J. Pettersen, *Lipids*, *25*: 284 (1990).
480. W. W. Christie, *Fat Sci. Technol.*, *93*: 65 (1991).
481. B. Nikolova-Damyanova, W. W. Christie, and B. Herslof, *J. Am. Oil Chem. Soc.*, *67*: 503 (1990).
482. D. Chapman, *The Structure of Lipids*, Methuen, London, 1965.
483. R. Ryhage and E. Stenhagen, *J. Lipid Res.*, *1*: 361 (1960).
484. E. Stenhagen, *Anal. Character. Oils, Fats, Fat Prod.*, *2*: 26 (1968).
485. K. K. Sun and R. T. Holman, *J. Am. Oil Chem. Soc.*, *45*: 810 (1968).
486. J. L. Le Quere, in *Advances in Lipid Methodology*, Vol. 2 (W. W. Christie, ed.), Oily Press, Ayr, 1993, p. 215.

487. M. Barber, T. O. Merren, and W. Kelley, *Tetrahedron Lett.*, *18*: 1063 (1964).
488. M. Barber, J. R. Chapman, and W. A. Wolstenholme, *Int. J. Mass Spectrom. Ion Phys.*, *1*: 98 (1968).
489. W. M. Lauer, A. J. Aasen, G. Graff, and R. T. Holman, *Lipids*, *5*: 861 (1970).
490. R. A. Klein, *J. Lipid Res.*, *12*: 123 (1971).
491. C. B. Johnson and R. T. Holman, *Lipids*, *1*: 371 (1966).
492. A. L. Dolendo, J. C. Means, J. Tobias, and E. G. Perkins, *J. Dairy Sci.*, *52*: 21 (1969).
493. H. W. Sprecher, R. Maier, M. Barber, and R. T. Holman, *Biochemistry*, *4*: 1856 (1965).
494. R. A. Hites, *Anal. Chem.*, *42*: 1736 (1970).
495. H. M. Fales and G. W. A. Milne, *J. Am. Oil Chem. Soc.*, *48*: 333A (1971).
496. W. K. Rohwedder, *Lipids*, *6*: 906 (1971).
497. T. Murata and S. Takahashi, *Anal. Chem.*, *45*: 1816 (1973).
498. J. Blomberg, *Lipids*, *9*: 461 (1974).
499. P. A. Cranwell, N. Robinson, and G. Eglinton, *Lipids*, *20*: 645 (1985).
500. P. P. Schmid, M. D. Muller, and W. Simon, *J. High Res. Chromatogr. Chromatogr. Commun.*, *2*: 675 (1979).
501. T. Rezanka, P. Mares, P. Husek, and M. Podojil, *J. Chromatogr.*, *355*: 265 (1986).
502. O. Bertelsen and N. Dinh-Nguyen, *Fette Seifen Anstrichm.*, *87*: 336 (1985).
503. B. L. Karger and P. Vouros, *J. Chromatogr.*, *323*: 13 (1985).
504. N. M. M. Nibbering, *J. Chromatogr.*, *251*: 93 (1982).
505. O. S. Privett and W. L. Erdahl, *Meth. Enzymol.*, *72*: 56 (1981).
506. M. L. Vestal, *Science*, *226*: 275 (1984).
507. A. Kuksis, J. J. Myher, and L. Marai, *J. Am. Oil Chem. Soc.*, *61*: 1582 (1984).
508. L. Marai, J. J. Myher, and A. Kuksis, *Can. J. Biochem. Cell Biol.*, *61*: 840 (1983).
509. J. J. Myher, A. Kuksis, L. Marai, and F. Manganaro, *J. Chromatogr.*, *283*: 289 (1984).
510. T. Murata, *Anal. Chem.*, *49*: 2209 (1977).
511. T. Murata and S. Takahashi, *Anal. Chem.*, *49*: 728 (1977).
512. E. Schulte, M. Hohn, and U. Rapp, *Frezenius Z. Anal. Chem.*, *307*: 115 (1981).
513. G. A. Spanos, S. J. Schwartz, R. B. Van Breeman, and C.-H. Huang, *Lipids*, *30*: 85 (1995).
514. P. Laakso, in *Advances in Lipid Methodology*, Vol. 1 (W. W. Christie, ed.), Oily Press, Ayr, 1992, p. 80.
515. B. A. Charpentier and M. R. Sevenants, *Supercritical Fluid Extraction and Chromatography*, American Chemical Society, Washington, DC, 1988.
516. R. M. Smith, *Supercritical Fluid Chromatography*, Royal Society of Chemistry, London, 1988.
517. C. M. White, *Modern Supercritical Fluid Chromatography*, Huthig Verlag, Heidelberg, 1988.
518. P. J. Schoenmakers, in *Supercritical Fluid Chromatography* (R. M. Smith, ed.), Royal Society of Chemistry, London, 1988, pp. 102–136.
519. P. J. Schoenmakers and L. G. M. Vunk, *Adv. Chromatogr.*, *30*: 1 (1989).
520. H. E. Schwartz, P. J. Barthel, S. E. Moring, and H. H. Lauer, *LC-GC*, *5*: 490 (1987).
521. F. Pacholec, D. S. Boyer, R. K. Houck, and A. C. Roselli, in *Modern Supercritical Fluid Chromatography* (C. M. White, ed.), Hüthig Verlag, Heidelberg, 1988, pp. 17–43.
522. L. T. Talor and H.-C. K. Chang, *J. Chromatogr. Sci.*, *28*: 357 (1990).
523. M. L. Lee and K. E. Markides, *J. High Res. Chromatogr. Chromatogr. Commun.*, *9*: 652 (1986).
524. S. M. Fields, R. C. Kong, J. C. Fieldsted, M. L. Lee, and P. A. Peaden, *J. High Res. Chromatogr. Chromatogr. Commun.*, *7*: 312 (1984).
525. S. M. Fields, R. C. Kong, M. L. Lee, and P. A. Peaden, *J. High Res. Chromatogr. Chromatogr. Commun.*, *7*: 423 (1984).
526. R. D. Smith, J. L. Fulton, R. C. Petersen, A. J. Kopriva, and B. W. Wright, *Anal. Chem.*, *58*: 2057 (1986).
527. B. W. Wright and R. D. Smith, in *Modern Supercritical Fluid Chromatography* (C. M. White, ed.), Hüthig Verlag, Heidelberg, 1988, pp. 189–210.
528. T. L. Chester, D. P. Innis, and G. D. Owens, *Anal. Chem.*, *57*: 2243 (1985).
529. E. J. Guthrie and H. E. Schwartz, *J. Chromatogr. Sci.*, *24*: 236 (1986).
530. B. E. Richter, D. J. Bornhop, J. T. Swanson, J. G. Wangsgaard, and M. R. Andersen, *J. Chromatogr. Sci.*, *27*: 303 (1989).
531. P. Carraud, D. Thiebaut, M. Caude, R. Rosset, M. Lafosse, and M. Dreux, *J. Chromatogr. Sci.*, *25*: 395 (1987).

532. M. Demirbüher and L. G. Blomberg, *J. Chromatogr. Sci.*, *28*: 67 (1990).
533. M. Demirbüher and L. G. Blomberg, *J. Chromatogr.*, *550*: 765 (1991).
534. H. Kallio, A. Johansson, and P. Oksman, *J. Agric. Food Chem.*, *39*: 1752 (1991).
535. E. M. Calvey, R. E. McDonald, S. W. Page, M. M. Mossoba, and L. T. Taylor, *J. Agric. Food Chem.*, *39*: 542 (1991).
536. R. Huopalahti, P. Laakso, J. Saaristo, R. Linko, and H. Kallio, *J. High Res. Chromatogr. Chromatogr. Commun.*, *11*: 899 (1988).
537. H. Kallio and P. Laakso, *J. Chromatogr.*, *518*: 69 (1990).
538. H. Kallio, P. Laakso, R. Huopalahti, R. R. Linko, and P. Oksman, *Anal. Chem.*, *61*: 698 (1989).
539. H. Kallio, T. Vauhkonen, and R. R. Linko, *J. Agric. Food Chem.*, *39*: 1573 (1991).
540. R. P. Khosaa, in *Modern Supercritical Fluid Chromatography* (C. M. White, ed.), Hüthig Verlag, Heidelberg, 1988, pp. 155–187.
541. M. Proot, P. Sandra, and E. Geeraert, *J. High Res. Chromatogr. Chromatogr. Commun.*, *9*: 189 (1986).
542. B. E. Richter, M. R. Andersen, D. E. Knowles, E. R. Campbell, N. L. Porter, L. Nixon, and D. W. Later, in *Supercritical Fluid Extraction and Chromatography* (B. A. Charpentier and M. R. Sevenants, eds.), American Chemical Society, Washington, DC, 1988, pp. 179–190.
543. P. Sandra, F. David, F. Munari, G. Mapelli, and S. Trestianu, in *Supercritical Fluid Chromatography* (R. M. Smith, ed.), Royal Society of Chemistry, London, 1988, pp. 137–158.
544. P. Sandra, M. Proot, and E. Geeraert, in *Proceedings of the Seventh International Symposium on Capillary Chromatography* (P. Sandra, D. Ishii, and K. Jinno, eds.), Hüthig Verlag, Heidelberg, 1986, pp. 650–659.
545. C. M. White and R. H. Houck, *J. High Res. Chromatogr. Chromatogr. Commun.*, *8*: 293 (1985).
546. B. W. Wright and R. D. Smith, *J. High Res. Chromatogr. Chromatogr. Commun.*, *9*: 73 (1986).
547. J. Cousin and P. J. Arpino, *J. Chromatogr.*, *398*: 125 (1987).
548. B. W. Wright, H. T. Kalinoski, H. R. Udseth, and R. D. Smith, *J. High Res. Chromatogr. Chromatogr. Commun.*, *9*: 145 (1986).
549. T. W. Lee, E. Bobik, and W. Malone, *J. Assoc. Off. Anal. Chem.*, *74*: 533 (1991).
550. K. Sakaki, T. Sako, T. Yokochi, O. Suzuki, and T. Hakuta, *Yukagaku*, *37*: 54 (1988).
551. A. A. Nomura, J. Yamada, K. I. Tsunoda, K. Sakaki, and T. Yokochi, *Anal. Chem.*, *61*: 2076 (1989).
552. J. E. France, J. M. Snyder, and J. W. King, *J. Chromatogr.*, *540*: 271 (1991).
553. J. L. Perrin and A. Prévot, *Rev. Fr. Corps Gras*, *35*: 485 (1988).
554. T. L. Chester, *J. Chromatogr.*, *299*: 424 (1984).
555. B. Borgström, *Biochim. Biophys. Acta*, *13*: 491 (1954).
556. F. H. Mattson and L. W. Beck, *J. Biol. Chem.*, *214*: 115 (1955).
557. P. Savary and P. Desnuelle, *C.R. Acad. Sci.* (*Paris*), *240*: 2571 (1955).
558. N. H. Tattrie, R. A. Bailey, and M. Kates, *Arch. Int. Biophys.*, *78*: 319 (1958).
559. M. L. Karnovsky and D. Wolff, in *Biochemistry of Lipids*, Vol. 5 (G. Popjak, ed.), Pergamon Press, London, 1960, pp. 53–59.
560. B. Borgström, *J. Lipid Res.*, *5*: 552 (1964).
561. A. P. Levitskii, *Biokhimiya*, *30*: 45 (1965).
562. B. Entressangles, H. Sari, and P. Desnuelle, *Biochim. Biophys. Acta*, *125*: 597 (1966).
563. F. H. Mattson and R. A. Volpenheim, *J. Lipid Res.*, *9*: 79 (1968).
564. R. G. Jensen, R. E. Pitas, J. G. Quinn, and J. Sampugna, *Lipids*, *5*: 580 (1970).
565. D. Firestone (ed.), *AOCS Official Methods and Recommended Practices*, 4th ed., AOCS Press, Champaign, IL, AOCS Official Method Ch 3-91, 1994.
566. *IUPAC Standard Methods for Analysis of Oils, Fats and Derivatives*, 7th ed., Blackwell Scientific, IUPAC Method 2-210, 1987.
567. F. E. Luddy, R. A. Barford, S. F. Herb, P. Magidman, and R. W. Riemenschneider, *J. Am. Oil Chem. Soc.*, *41*: 693 (1964).
568. F. H. Mattson and R. A. Volpenheim, *J. Lipid Res.*, *2*: 58 (1961).
569. F. H. Mattson and R. A. Volpenheim, *J. Lipid Res.*, *7*: 536 (1966).
570. J. L. Brown and J. M. Johnston, *J. Lipid Res.*, *3*: 480 (1962).
571. H. Brockerhoff and M. Yurkowski, *J. Lipid Res.*, *7*: 62 (1966).
572. F. H. Mattson and E. S. Lutton, *J. Biol. Chem.*, *233*: 868 (1958).
573. F. H. Mattson and R. A. Volpenheim, *J. Lipid Res.*, *8*: 58 (1961).

574. F. H. Mattson and R. A. Volpenheim, *J. Biol. Chem.*, *236*: 1891 (1961).
575. F. H. Mattson and R. A. Volpenheim, *J. Lipid Res.*, *4*: 392 (1963).
576. F. D. Gunstone, R. J. Hamilton, F. B. Padley, and M. I. Qureshi, *J. Am. Oil Chem. Soc.*, *42*: 965 (1965).
577. N. R. Bottino, R. E. Anderson, and R. Reiser, *Lipids*, *5*: 165 (1970).
578. R. E. Anderson, N. R. Bottino, L. J. Cook, and R. Reiser, *Lipids*, *5*: 171 (1970).
579. L. M. Smith and S. Hardjo, *Lipids*, *9*: 713 (1974).
580. M. H. Coleman, *J. Am. Oil Chem. Soc.*, *38*: 685 (1961).
581. M. R. Subbaram and C. G. Youngs, *J. Am. Oil Chem. Soc.*, *41*: 595 (1964).
582. M. R. Subbaram and C. G. Youngs, *J. Am. Oil Chem. Soc.*, *44*: 425 (1967).
583. D. L. Carpenter, J. Lehmann, B. S. Mason, and H. L. Slover, *J. Am. Oil Chem. Soc.*, *53*: 713 (1977).
584. J. C. Hokes and R. E. Worthington, *J. Am. Oil Chem. Soc.*, *56*: 953 (1979).
585. K. A. Sawadogo and J. A. Bézard, *Oléagineux*, *37*: 247 (1982).
586. J. Bézard and K. A. Sawadogo, *Reprod. Nutr. Dev.*, *23*: 65 (1983).
587. G. Jurriens and A. C. J. Kroesen, *J. Am. Oil Chem. Soc.*, *42*: 9 (1965).
588. G. Jurriens and L. Schouten, *Rev. Fr. Corps Gras*, *12*: 505 (1965).
589. V. Koman and J. Kotuc, *J. Am. Oil Chem. Soc.*, *54*: 95 (1977).
590. V. Koman, J. Kotuc, and M. Csicsayova, *J. Am. Oil Chem. Soc.*, *54*: 95 (1978).
591. H. Brockerhoff, *Arch. Biochem. Biophys.*, *110*: 586 (1965).
592. H. Brockerhoff, R. J. Hoyle, and P. C. Huang, *Can. J. Biochem.*, *44*: 1519 (1966).
593. H. Brockerhoff, *J. Lipids Res.*, *8*: 167 (1967).
594. W. W. Christie and J. H. Moore, *Biochim. Biophys. Acta*, *176*: 445 (1969).
594a. W. E. M. Lands, R. A. Pieringer, S. P. M. Slakey and A. Zschocke, *Lipids*, *1*: 444 (1966).
595. G. Long and I. Penny, *Biochem. J.*, *65*: 382 (1957).
596. N. H. Tattrie, *J. Lipid Res.*, *1*: 60 (1959).
597. D. J. Hanahan, H. Brockerhoff, and E. J. Barron, *J. Biol. Chem.*, *235*: 1917 (1960).
598. L. L. M. Van Deenen and G. H. De Haas, *Biochim. Biophys. Acta*, *70*: 538 (1963).
599. J. Bézard, M. A. Ouédraogo, G. Semporé, and L. Schreiber, *Rev. Fr. Corps Gras*, *37*: 83 (1990).
600. J. Bézard, M. A. Ouédraogo, and G. Semporé, *Rev. Fr. Corps Gras*, *37*: 171 (1990).
601. R. Wood and F. Snyder, *Arch. Biochem. Biophys.*, *131*: 478 (1969).
602. J. Sampugna and R. G. Jensen, *Lipids*, *3*: 519 (1968).
603. G. Colacicco and M. M. Rapport, *J. Lipid Res.*, *7*: 258 (1966).
604. G. H. De Haas and L. L. M. Van Deenen, *Biochim. Biophys. Acta*, *106*: 315 (1965).
605. G. H. De Haas and L. L. M. Van Deenen, *Biochim. Biophys. Acta*, *84*: 467 (1964).
606. H. Brockerhoff, R. J. Hoyle, and N. Wolmark, *Biochim. Biophys. Acta*, *116*: 67 (1966).
607. H. Brockerhoff, R. H. Hoyle, P. C. Huang, and C. Litchfield, *Lipids*, *3*: 24 (1967).
608. W. C. Breckenridge and A. Kuksis, *Lipids*, *3*: 291 (1968).
609. W. W. Christie and J. H. Moore, *Biochim. Biophys. Acta*, *210*: 46 (1970).
610. J. J. Myher, L. Marai, A. Kuksis, and D. Kritchevsky, *Lipids*, *12*: 775 (1977).
611. H. Brockerhoff, *Lipids*, *6*: 242 (1971).
612. J. Sampugna and R. G. Jensen, *Lipids*, *4*: 444 (1969).
613. R. G. Jensen, J. Sampugna, J. G. Quinn, D. L. Carpenter, and T. A. Marks, *J. Am. Oil Chem. Soc.*, *42*: 1029 (1965).
614. R. G. Jensen, D. T. Gordon, N. H. Heimermann, and R. T. Holman, *Lipids*, *7*: 738 (1972).
615. R. G. Jensen, D. T. Gordon, and C. R. Scholfield, *Lipids 8*: 323 (1973).
616. R. G. Jensen, *Lipids*, *19*: 149 (1974).
617. IUPAC-IUB Commission on Biochemical Nomenclature, *Eur. J. Biochem.*, *2*: 127 (1967); *Biochem. J.*, *105*: 897 (1967).
618. IUPAC-IUB Commission on Biochemical Nomenclature, *J. Lipid Res.*, *19*: 114 (1978).
619. C. R. Smith, in *Topics in Lipid Chemistry*, Vol. 3 (F. D. Gunstone, ed.), Paul Elek, London, 1972, pp. 89–124.
620. Y. Itabashi, A. Kuksis, L. Marai, and T. Takagi, *J. Lipid Res.*, *31*: 1711 (1990).
621. Y. Itabashi, A. Kuksis, and J. J. Myher, *J. Lipid Res.*, *31*: 2119 (1990).
622. T. Takagi and Y. Ando, *J. Jpn. Oil Chem. Soc.*, *39*: 622 (1990).
623. T. Takagi and T. Suzuki, *J. Chromatogr.*, *519*: 237 (1990).
624. T. Takagi and Y. Ando, *Lipids*, *25*: 398 (1990).

625. Y. Ando and T. Takagi, *Inform*, 2: 353 (1991).
626. T. Takagi and Y. Ando, *Inform*, 2: 353 (1991).
627. T. Takagi and Y. Ando, *J. Jpn. Oil Chem. Soc.*, 40: 288 (1991).
628. B. G. Semporé and J. A. Bézard, *J. Liq. Chromatogr.*, 17: 1679 (1994).
629. G. Semporé and J. Bézard, *J. Am. Oil Chem. Soc.*, 68: 702 (1991).
629a. J. Bézard and G. Semporé, in *New Trends in Lipid and Lipoprotein analyses* (J. L. Sebedio and E. G. Perkins, eds.), AOCS Press, Champaign, 1995, pp. 106–132.
630. G. Semporé and J. Bézard, *J. Chromatogr.*, 557: 227 (1991).
631. L. Y. Yang and A. Kuksis, *J. Lipid Res.*, 32: 1173 (1991).
632. W. H. Pirkle and J. R. Hauske, *J. Org. Chem.*, 42: 2781 (1977).
633. J. J. Myher, A. Kuksis, L. Marai, and S. K. F. Yeung, *Anal Chem.*, 50: 557 (1978).
634. E. J. Corey and S. Hashimoto, *Tetrahedron Lett.*, 22: 299 (1981).
635. W. H. Pirkle and P. E. Adams, *J. Org. Chem.* 45: 4111 (1980).
636. W. W. Christie, B. Nikolova-Damyanova, P. Laakso, and B. Herlof, *J. Am. Oil Chem. Soc.*, 68: 695 (1991).
637. P. W. Parodi, *Lipids*, 17: 437 (1982).
638. J. C. Martin, P. Bougnoux, J. M. Antoine, M. Lanson, and C. Couet, *Lipids*, 28: 637 (1993).

13

Analysis of Neutral Lipids: Unsaponifiable Matter

Boukaré G. Semporé and Jean A. Bézard
Nutrition Research Unit, University of Burgundy, Dijon, France

I. INTRODUCTION

The unsaponifiable matter (UM) can be defined as those substances frequently found dissolved in oils and fats, which cannot be saponified by the usual alkaline treatment but are soluble in oil and fat solvents and thereby coextracted with the lipids.

UM generally represents a minor fraction of oils and fats, from 0.5% to 2%, except in some fats such as shea butter (7–10%), but it may contain a large number of compounds. Those comprise pigments, hydrocarbons, higher aliphatic alcohols, and sterols. After determination of its content, it is generally necessary to fractionate the UM to further characterize the different constituents.

II. DETERMINATION OF UNSAPONIFIABLE MATTER

The determination of the UM has been precisely described by several official methods. The AOCS method (Ca 6a-40) is applicable to normal animal and vegetable fats and oils, while a second method (Ca 6b-53) is applicable to fats and oils containing higher levels of UM and especially suited for marine oils. A normalized procedure has also been proposed by AFNOR (NF T60-205). It has been precisely described by Lozano et al. (1) as follows: the weighed lipid sample (~0.5 g) in a 20 × 100 mm screwcap tube is added with 1 ml of 0.1% betulin (Merck) ethanolic solution used as internal standard. Five milliliters of 1 N potassium hydroxide (KOH) in ethanol is added. The tube is capped and heated in a block heater for 1 hr at 95°C. After cooling, 10 ml of water is added and the solution is shaken and extracted three times with 10 ml of diethyl ether. The combined extracts are washed twice with water, once with 0.5 N KOH in ethanol and again with water until neutral. The organic layer is separated and dried over anhydrous sodium sulfate. (Na_2SO_4). After filtration, the solvent is evaporated to dryness under vacuum and the residue is weighed. This dry UM is then dissolved in 0.2 ml of chloroform and stored in capped vials until TLC or HPLC analysis.

The UM also contains the sterols originally present in the esterified form. The sterol esters, in particular cholesteryl esters, are slowly hydrolyzed and longer reflux times are necessary.

When a food lipid sample is saponified, the step in which the UM is extracted from the alkaline solution can be avoided. After acidification of the alkaline solution with hydrochloric acid in excess, the unsaponifiable materials and the free fatty acids can be isolated by adsorption thin-layer chromatography (TLC) or high-performance liquid chromatography (HPLC). Acidic and neutral materials can alternatively be separated by ion exchange chromatography in the presence of diethylaminoethyl (DEAE)-cellulose.

To gain further information on its constituents, the UM must be fractionated by adsorption chromatographic methods.

A.   Thin-Layer Chromatography

TLC is the traditional normalized method (AFNOR, NF T 60-232) using silica plates. Elution is made by a ternary solvent system hexane–diethyl ether–acetic acid (90:10:1, in volumes). The bands can be visualized by spraying a saturated solution of copper acetate and by heating 10 min at 150°C in an oven.

Figure 1 shows the separations obtained with unsaponifiable matter extracted from avocado oil (a) and sunflower oil (b). Avocado oil (c) and tocopherol (d) were used as controls (1).

Hydrocarbons were present at high level in the avocado and sunflower oil. They eluted clearly ahead of triacylglycerols (and cholesteryl esters). The two UMs also contained sterols, but Δ7 sterols were only present in the sunflower UM, contrary to Δ5 sterols present in both UMs.

Generally, other compound families can be detected, e.g., squalene, β-carotene, tocopherols, triterpene alcohols, methylsterols, and hydroxylated carotenoids. When the separated compounds have to be extracted from the silica, the bands are visualized at 366 nm after spraying with rhodamine B. They are then delineated and scraped off the plate and the compounds are extracted with dichloromethane.

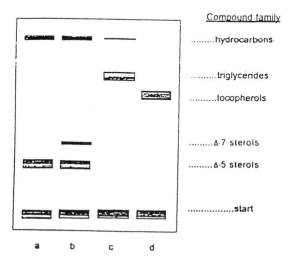

Fig. 1   Thin-layer chromatography of unsaponifiable matter (UM) extracted from avocado and sunflower oils (a = avocado UM; b = sunflower UM; c = avocado oil; d = tocopherol (1)). The silica TLC plates were eluted with a ternary solvent of hexane–ethyl ether-acetic acid (90:10:1, v/v/v). The plates were sprayed with a saturated solution of copper acetate and heated 10 min in an oven at 150°C. (Reproduced with kind permission of the authors and of the *Journal of American Oil Chemists' Society.*)

B.   High-Performance Liquid Chromatography

Fractionation of UM by TLC is somewhat tedious and time consuming. Analysts generally prefer using HPLC, which permits shorter separation and fraction collection times.

Mordret et al. (2) published an elegant and rapid method of fractionation of the sterol fraction from the sunflower oil UM by HPLC. They used a 250 × 4.6 mm i.d. column packed with silica (Si-60 5-μm particle size) and hexane-isopropanol (98.5:1.5, v/v) as eluting solvent. They separated in less than 15 min two clean sterol fractions, clearly separated, i.e., the Δ5 and the Δ6 sterols. These two fractions were collected and further analyzed by capillary gas chromatography with OV-1701 as stationary phase and the results obtained were compared to those obtained after fractionation by TLC. The two series of data were very similar. Since the HPLC method is much less time consuming, it must be preferred. Additionally, a simple analysis by HPLC permits one to determine the total sterols in the UM of a lipid extract and the ratio Δ5/Δ7 sterols. These data sometimes give enough information in food analysis.

Figure 2 illustrates the type of separation that can be obtained by HPLC (1). It concerns the UM extracted from sunflower oil in the following experimental conditions: the column was a 250 × 4.6 mm i.d. LiChrosorb Si-60 isocratically eluted with hexane-isopropanol (99:1, v/v) at 2 ml/min and 35°C. A fluorometric detector was used to analyze the tocopherol fraction. The outlet cell flow of this detector was then analyzed by a differential refractive index detector (RID) for quantitation of the sterol fractions. These fractions were hand-collected at the cell output for further GC analysis. Figure 2 shows the chromatogram recorded when the RID was used.

Several fractions were separated within 15 min. Fraction V corresponded to betulin added as internal standard before the saponification step in order to evaluate the recovery of UM during the

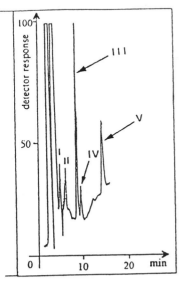

Fig. 2   High-performance liquid chromatography of unsaponifiable matter (UM) extracted from sunflower oil (1). I = fraction I not identified, II = fraction attributed to 4-monomethylsterols; III = Δ5-sterol fraction; IV = Δ7-sterol fraction; V = betulin added as internal standard. Peaks eluting ahead correspond to hydrocarbons, triacylglycerols, and tocopherols. Operating conditions: 250 × 4.6 mm i.d. LiChrosorb Si-60 column, isocratically eluted with hexane-isopropanol (99:1, v/v) at a flow rate of 2 ml/min at 35°C. The sterol fractions were detected by a differential refractive index detector. (Reproduced with kind permission of the authors and of the *Journal of American Oil Chemists' Society*.)

extraction step. The other fractions eluted ahead of betulin were those separated by TLC (Fig. 1), i.e., hydrocarbons, triacylglycerols, tocopherols, Δ5 and Δ7 sterols, in increasing retention time order. Peaks III and IV represented the Δ5 and Δ7 sterol fractions, respectively. Peak II was attributed to monomethylsterols. The different fractions of the UM can be collected and further analyzed.

III. ANALYSIS OF STEROLS

It is beyond the scope of this chapter to present in detail the analysis procedures for all of the UM fractions which can be separated by adsorption TLC or HPLC because, except for sterols in the free or esterified form, these constituents are not really considered as lipids, though extracted by the same organic solvents. Information is available in the report by Mordret (2). The authors will focus on the sterol fraction analysis in vegetable and animal foods, in particular on the determination of cholesterol present in animal foods in more or less amount, because this molecule, when in excess in blood, is repetitively associated with a high incidence of coronary heart disease in humans (3).

A. Analysis of Phytosterols

The total amount of sterols from a vegetable food sample is determined after isolation of the sterol fraction from the UM, separated by TLC or HPLC. Care should be taken to estimate the quantitative recovery of the UM by adding a nonreactive molecule as internal standard before saponification. Betulin and cholesterol are frequently used for this purpose. In their quantitative study, Lozano et al. (1) using betulin as internal standard showed that the recovery was only 76% and that losses took place during the extraction step. A correction factor of 1/0.76 had to be applied to the amount measured for a quantitative estimation.

After the sterol fraction is isolated from the UM, the different sterols have to be separated and quantitated. For this purpose, capillary GC is the most suitable method.

1. Capillary Gas Chromatography

Determination of the sterol composition of lipid extracts has been normalized by several organizations, e.g., AFNOR (NF ISO 6799). A typical procedure is described in the following example:

The unsaponifiable matter is extracted from the lipid extract and the sterols are separated by TLC as described above. After migration, the plate is sprayed with rhodamine 6G in methanol (0.05%, w/v) and the bands visualized under UV light and delineated. The silica containing the sterol fraction is scraped off the plate and the adsorbed sterols are extracted with 5 ml of chloroform, or hexane–ethyl acetate (80:20, v/v) or toluene-acetone (95.5 v/v), or pure ethyl acetate, or pure diethyl oxide in which rhodamine is insoluble. The mixture is heated in a water bath for 15 min under reflux. The chilled mixture is filtered on paper filter and the silica reextracted twice. The pooled solutions are evaporated under nitrogen and the residue is dissolved in the minimum amount (less than 1 ml) of ethyl acetate or diethyl oxide.

Before GC analysis, the sterols are generally derivatized as trimethylsilyl ethers or acetates.

The silyl ethers are prepared as follows: to a few milligrams of sterols, add 0.5 ml pyridine, 0.1 ml hexamethyldisilazane, and 0.04 ml trimethylchlorosilane. After 5 min of reaction the supernatant is ready for injection. The acetate derivatives are prepared as follows: to a few milligrams of sterols, add 0.1 ml acetic anhydride and 0.1 ml pyridine. After 1 hr at 70°C, add 2 ml ice water and 5 ml hexane. Shake vigorously and eliminate the water phase. Wash with 5 ml aqueous sodium bicarbonate (10 g/L) followed by 5 ml 0.5 N chlorhydric acid and at last with 5 ml water. The hexane phase dried on anhydrous sulfate is ready for GC analysis.

Analysis of the sterol derivatives by GC can be carried out on packed columns or preferentially on capillary columns. The packed column must be preferentially in glass to prevent decomposition of sterols at high temperature catalyzed by metals (except silver). The 180–200 cm × 2–4 mm i.d. glass column is silanized and packed with 2–5% SE30, JXR, OV-1, or OV-17 on 80–100 or 100–120 mesh diatomaceous earth. The analysis is run at 230–240°C with a nitrogen, helium, argon, or hydrogen flow rate of 30–50 ml/min with the injector and detector at 250–260°C.

The analysis will be preferentially carried out on glass capillary columns, under the following conditions: a 25 m × 0.32 mm i.d. glass column, coated with CP Sil 19 CB (0.44 μm thick); isothermal analysis at 265°C; carrier gas, hydrogen (0.6-bar head pressure); split/splitless or on-column injection system at 300°C; flame ionization detection at 320°C.

Such an analysis, on a glass capillary column, of sterols from sunflower oil as trimethylsilyl ethers is illustrated in Fig. 3. It shows that at least 12 sterols were present in the fraction. They were identified by means of standards or by mass spectrometry. Only trace amounts of cholesterol and of four other sterols were present, while β-sitosterol was the major sterol, followed by campesterol, campestenol, and Δ7-stimasterol. The sterol profile is generally highly characteristic of the oil analyzed and can serve to detect adulteration of edible oils.

2. Reverse Phase Liquid Chromatography

Capillary gas chromatography is by far the analytical method which offers the highest resolution in sterol analysis. However, the method fails to distinguish all of the sterols in a complex mixture and the use of a combination of chromatographic procedures is necessary. Reverse phase liquid chromatography offers the opportunity to separate sterol fractions for further analysis by GC alone or by GC-MS.

The following example (Fig. 4) illustrates this possibility. In this analysis, the UM of sunflower oil was fractionated on an alumina column (6). The dried UM, dissolved in carbon tetrachloride (5%, w/v), was fractionated using TLC by depositing 200 μl of the solution on a precoated silica gel plate and eluting with chloroform–diethyl ether (9:1, v/v). Cholesterol was used as standard for the identification of the sterol band and a rhodamine B spray for the detection at 366 nm. The zone corresponding to sterols was scraped off the plate and extracted with dichloromethane. The sterols were acetylated using acetic anhydride–pyridine (1:1, v/v) as reagents and the sterol acetate mixture was separated by micropreparative HPLC into five fractions analyzed by GC and GC-MS. The column used was a Hibar column (250 × 4.6 mm i.d.) packed with RP-8 LiChrosorb (5-μm particles) and eluted with acetonitrile-water (90:10, v/v) at room temperature.

Figure 4 shows the separation of five known sterol acetates. Separation of the sunflower sterols gave a somewhat different profile. But five fractions were also fractionated, the composition of which is given in Table 1. These five fractions were analyzed by GC using a WCOT glass capillary column (30 m × 0.36 mm i.d.) operated at 260°C. Analysis without previous fractionation by HPLC gave results reported in the second column of data in Table 1 showing that seven sterols were in low concentrations (<9%) and not easily identified. They were quantitated by GC-MS. On the contrary, these minor sterols were concentrated in certain fractions isolated by reverse phase HPLC, in particular in the first one, and could be more easily identified.

3. Conclusion

Accurate determination of sterols in edible oils or in food containing vegetable lipids is useful for detection of the presence of unexpected lipids. Indeed, several sterols are specific of certains oils and can be utilized as markers. For example, brassicasterol is specifically present in rape-

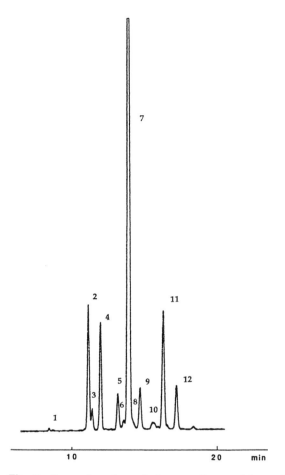

Fig. 3 Separation of sterols from sunflower oil by capillary gas chromatography. Operating conditions: glass column (25 m × 0.32 mm i.d.) coated with CP Sil 19 CB (0.44 μm thick); isothermal analysis at 265°C; hydrogen, as carrier gas (0.6 bar head pressure); ROS solid injector at 300°C; flame ionization detector at 265°C. Peak identification: 1 = cholesterol; 2 = campesterol; 3 = campestanol; 4 = stigmasterol; 5 = Δ7 campesterol; 6 = chlerosterol; 7 = β-sitosterol; 8 = stigmastanol; 9 = Δ5-avenasterol; 10 = Δ5-24 stigmastadienol; 11 = Δ7-stigmasterol; 12 = Δ7-avenasterol. (Chromatogram kindly produced by J. L. Coustille and F. Mordret, Iterg, Pessac, France.)

seed oil and its presence in a commercial sunflower oil proves an adulteration of this oil by rapeseed oil. In the same manner, Δ7-stigmasterol is specifically present in sunflower or safflower oil and its presence at high level in peanut oil proves an adulteration of one of these oils.

As another example, the presence of cholesterol is a good marker for the presence of animal fats in food and its determination is used for this purpose in food control laboratories.

The ratio of the concentrations of certain sterols is sometimes sufficient to detect an adulteration, e.g., the ratio β-sitosterol/(campesterol + stigmasterol) for olive oil.

Other constituents of the UM have also been used as markers in control analysis such as Δ-tocopherol for the presence of soybean oil, and tocotrienols for the presence of palm oil. The determination of sterols (and tocopherols) in soybean oil has been standardized (AOCS, Ce 3-74). Figure 5 illustrates the separation of both groups of tocopherols and sterols that can be obtained

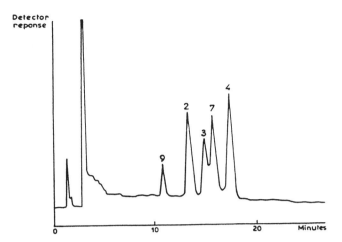

Fig. 4 Separation of five sterol acetates by reverse phase HPLC (5). Operating conditions: Hibar prepacked column (250 × 4.6 mm i.d.) slurry packed with RP-8 LiChrosorb (5 μm); eluant, acetonitrile-water (90:10, v/v); at a flow rate of 1 ml/min and at ambient temperature (20–22°C); detection by UV spectrophotometer. Peak identification: acetate of 9 = ergosterol, 2 = cholesterol, 3 = campesterol, 7 = stigmasterol, 4 = sitosterol. (Reproduced with kind permission of the authors and of the *Journal of American Oil Chemists' Society*.)

Table 1 Analysis (%) of the Sterol Fractions of Sunflower Seed Oil[a]

Sterol	α_2[b]	Direct GC[c]	Fractions[d]				
			1	2	3	4	5
Cholesterol	1.00	1.1	15.6	1.0			
Brassicasterol	1.14	0.4	11.2				
Campesterol	1.31	9.2	2.6	83.2	12.7		
Stigmasterol	1.43	9.1		7.3	78.3	8.4	
Δ7-Campestenol	1.55	0.9	10.3	3.1			
Sitosterol	1.63	60.5				28.8	91.8
Δ5-Avenasterol	1.81	1.8	18.9	2.1			
Δ7-Stigmastenol	1.94	12.4			7.4	59.6	6.7
28-Isoavenasterol	2.04	0.3	5.5				
Δ7-Avenasterol	2.15	2.8	29.6	2.1			
Unknowns		1.6	6.3[e]	1.2	1.6	3.2[f]	1.5

[a]Determined as sterol acetates from the unsaponifiable matter of the seed oil, after fractionation by TLC.
[b]Relative retention time of the sterol acetate on an OV-17 WCOT glass capillary column (cholesterol acetate 1.00) determined at 260°C.
[c]Without previous fractionation by HPLC.
[d]The five fractions were obtained by HPLC on a LiChrosorb RP-8 column with acetronitrile-water (90:10, v/v) as eluant.
[e]Among the other minor sterols, 24-methylenecholest-5-en-3β-ol (α_2 = 1.35) and 24E-ethylidenecholest-5-en-3β-ol (fucosterol, α_2 = 1.72) were tentatively identified.
[f]This fraction contained 1.5% of a sterol which was tentatively identified to 24ξ-ethyl-5α-cholest-7,E-22-dien-3β-ol (spinasterol, α_2 = 1.70).
Source: Data from Ref. 5.

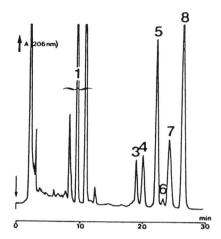

Fig. 5 Separation by reverse phase HPLC of a mixture of tocopherols, cholesterol, and rapeseed oil sterols (12). Operating conditions: 250×4.6 mm i.d. LiChrospher 100 CH-18 (4 μm); eluant, methanol- acetonitrile (40:60, v/v) at a flow rate of 1 ml/min and at thermostatted temperature of 35°C; detector by UV spectropho-tometer at 206 nm peak identification: 1 = tocopherols; 3 = Δ5 and Δ7 avenasterols; 4 = brassicasterol; 7 = campesterol; 8 = β-sitosterol. (Reproduced with kind permission of the authors and of the *Revue Française des Corps Gras.*)

by reverse phase HPLC on a LiChrospher 1000 CH-18 (250×4.6 mm i.d., 4-μm particles) at 35°C with the solvent system methanol-acetonitrile 40:60 v/v at 1 ml/min. The sample was a mixture of rapeseed oil tocopherols and cholesterol (12).

B. Analysis of Cholesterol

A high incidence of coronary heart disease is positively correlated to a high level of plasma cholesterol in humans (3,4). Any factor likely to raise blood cholesterol is considered to be unfavorable. Food cholesterol is generally considered to be one of these unfavorable factors. This underlines the importance of the determination of the amount of cholesterol in animal food. On the other hand, the cholesterol concentration in plasma is a diagnostic marker in diseases associated with dysfunctioning of lipid metabolism in humans. This explains why a great number of methods have been developed for the rapid and accurate determination of cholesterol alone by various procedures, in human plasma and in food.

Theoretically, free and esterified cholesterol can be measured during the separation of the lipid classes from a lipid extract by means of adsorption TLC or HPLC as previously seen.

However, enzymatic and colorimetric methods are preferred by analysts in most routine applications, in part because commercial kits are available for this purpose. But capillary GC analysis is also often used because of its higher precision and specificity and because the method permits detection of sterols other than cholesterol, in particular in foods containing animal and vegetable lipids. Moreover, molecular species of cholesterol esters can be separated by capillary GC, which can afford a great deal of information. At last, reverse phase HPLC is an alternative to GC in the measurement of sample lipid cholesterol.

1. Chemical Methods

The chemical method by Lieberman-Burchard (7) of cholesterol determination in a lipid sample has been widely used in the past and is still used because of its simplicity and low cost.

The measure is based on the development of a green coloration in the presence of acetic anhydride–concentrated sulfuric acid at room temperature, in the dark, after 30 min of reaction. The intensity of the coloration is measured by absorption spectrophotometry at 620 nm. The intensity is linearly related to cholesterol concentration between 100 and 600 µg, but it must be checked with control solutions of cholesterol of increasing concentrations for each series of measurements.

2. Enzymatic Methods

The colored reaction is characteristic of the sterol nucleus but not of cholesterol alone. It is the reason why the enzymatic method is preferred by more than 99% of analysts (8). It has been standardized (AFNOR NFT 60-243). This method is based on the oxidation of free cholesterol to Δ4-cholestenone with formation of hydrogen peroxide catalyzed by cholesterol oxidase. Since the enzyme is only active on cholesterol in the free form, the cholesteryl esters likely to be present in the lipid extract have first to be hydrolyzed in the presence of cholesterol esterase. The hydrogen peroxide (H_2O_2) formed by oxidation of cholesterol finally oxidizes a leukodye (such as triarylimidazole or the system aminophenazone-phenol) to a colored dye in the presence of peroxidase. The density of dye formed is proportional to the concentration of cholesterol. It is measured by absorption spectrophotometry.

In a lipid extract free and esterified cholesterol have been liberated from their complexes with proteins. It is not the case with a sample of plasma for which the cholesterol must be liberated using a nonionic detergent such as 3 mM sodium cholate in water.

Ready-to-use slides are now commercially available to determine cholesterol concentrations. They are multilayered elements coated on polyester supports. The spreading layer (baryum sulfate) contains a detergent (for plasma samples), cholesterol hydrolase, cholesterol oxidase, peroxidase, the leukodye, and buffered gelatin in that order from the top (deposit) to the bottom (support). As little as a 10-µl drop of lipid extract (or plasma) is deposited on the top of the slide. The density of the dye formed is measured by reflectance spectrophotometry at 540 nm (with triarylimidazole as leukodye).

If such slides are not available, dry mixtures of reagents are commercialized which must be added with a known volume of distilled water before using. Such a solution (Boehringer) contains:

In mmol/L: Tris buffer: 100, pH 7.7
Mg^{2+}: 50; 4-amino-phenazone: 1; sodium cholate: 10; phenol: 6; 3,4-dichlorophenol: 4
Fatty alcohol polyglycol ether: 0.6%
In international units/L: cholesterol esterase: 0.4; cholesterol oxidase: 0.25; peroxidase: 0.2.

Two milliliters of this reagent solution is added to 0.02 ml of the lipid extract (cholesterol concentration < 10 g/L) in a spectrophotometer cuvette (1 cm light path). The mixture is incubated at 37°C for 5 min or left at room temperature for 10 min and the absorbance is measured.

A test cholesterol solution is generally provided to assess the linearity of absorbance vs. the cholesterol concentration.

Instead of using the peroxidase to measure the hydrogen peroxide formed, other methods use the catalase as enzyme to form a yellow compound (lutidin) by coupling methyl aldehyde with acetylacetone. The hydrogen peroxide can also be quantitated by measure of the iodine formed from potassium iodide in the presence of molybdate.

3. Gas Chromatography

The above methods do not determine the individual sterols in mixtures such as those encountered in food samples often containing both phytosterols and cholesterol. For this purpose, chromatographic methods are preferred.

Gas chromatography on capillary columns is probably to date the most widely used method (9). The authors will describe in detail the method they currently use in the laboratory to measure total and free cholesterol in lipid extracts.

a. Material and Products. Teflon-lined screwcapped tubes (10 ml for saponification and 4 ml for silylation) are used. The silylation reagents are bis(Trimethylsilyl)trifluoroacetamide (BSTFA) and trimethylchlorosilane (TMCS). A mixture of BSTFA-TMCS (4:1, v/v) is prepared just before use. The saponification solution is an 11 N (616 g/L) solution of potassium hydroxide (KOH) in distilled water. The internal standard solutions are prepared from concentrated solutions of epicoprostanol (5β-cholestan-3α-ol) in absolute ethanol at a concentration of 100 µg/ml and 5α-cholestan in hexane (20 µg/ml) for estimation of extraction yield. These solutions have to be diluted before use according to sample cholesterol concentration in order to add in the reaction mixture 1 ml of epicoprostanol ethanol solution and 4 ml of cholestane hexane solution.

b. Saponification. For total cholesterol measurement (from 0.1 to 200 µg), esterified choles-terol must be liberated from the cholesteryl esters by saponification according to Abell et al. (10).

To a 10-ml tube add a known amount of the sample lipid extract such that the sterol content will be ~50–100 µg. Evaporate the solvent (chloroform reacts with KOH). Add 1 ml of the diluted ethanolic solution of epicoprostanol, 100 µl of aqueous KOH. Flush with nitrogen, cap the tube, and heat at 80°C for 1 hr. After cooling, add 4 ml of the diluted hexane solution of 5α-cholestane and shake. Add 2 ml distilled water and shake. Centrifuge and transfer the hexane upper phase containing the sterols into a 4-ml tube.

c. Derivatization. For the preparation of the trimethylsilyl (TMS) derivatives, the hexane is evaporated and the residue is added with 100 µl of the mixture BSTFA-TMCS (4:1, v/v) freshly prepared. The reaction mixture is heated at 80°C for 1 hr under nitrogen.

For measurement of only free cholesterol the saponification step is omitted. The sample lipid extract in the 4-ml vial (0.1–200 µg of free sterols) is directly submitted to derivatization as above.

If necessary, the reaction mixture is diluted with hexane and utilized for capillary GC analysis.

Figure 6 shows a typical chromatogram obtained with a human lipoprotein sample, only containing cholesterol as sterol and analyzed as TMS derivative. The analysis was carried out on a laboratory-made (10 m × 0.30 mm i.d.) glass capillary column coated with SE-30 as previously described for fatty analysis, with hydrogen as carrier gas (3 ml/min). The apparatus was a model 419 Packard chromatograph equipped with a solid injector maintained at 300°C and with a flame ionization detector. Peak areas were measured by an Enica 31 integrator calculator (Delsi).

The chromatogram shows that the cholesterol derivative was eluted in less than 5 min and was clearly separated from the two internal standards, cholestane and epicoprostanol. The amount of cholesterol present in the sample can be easily and accurately calculated from peak aeras, by comparison with the amount of epicoprostanol and by taking into account the extraction yield in terms of the amount of cholestan added. A previous calibration of the detector response with standard solutions of cholesterol, cholestan, and epicoprostanol is recommended for accurate quantitative analysis.

Similar results can be obtained with a very complex mixture of cholesterol and phytosterols by capillary GC of the acetate derivatives of sterols as shown by Kuksis et al. (11) and illustrated in Fig. 7. Such mixtures can be encountered in lipid extracts from human food containing animal and vegetable fats.

4. Gas Chromatography–Mass Spectrometry

Analysis of cholesterol after isotopic dilution by capillary gas chromatography coupled to mass spectrometry is the method of reference for the most accurate determination of cholesterol in a

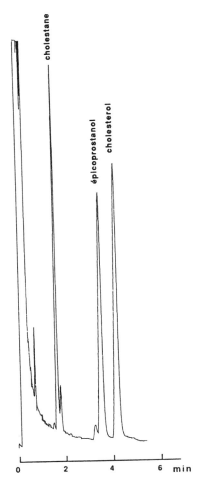

Fig. 6 Separation by glass capillary column gas chromatography of cholesterol and of two internal standards, cholestane and epicoprostanol. Operating conditions: laboratory-made glass column (10 m × 0.32 mm i.d.) coated with SE-30; carrier gas, hydrogen at a flow rate of 3 ml/min; isothermal analysis at 210°C; solid injection and flame ionization detection at 300°C. (Chromatogram kindly obtained from J. Gresti.)

sample. It is ideally suited as a "definitive" method, in particular in clinical chemistry, but also if necessary in food analysis.

The first methods proposed (13,14) involved gas chromatography on a packed column and the utilization of a [14]C- or deuterium-labeled cholesterol as internal standard and mass spectrometry in the electron impact ionization mode. A method displaying enhanced performance was developed by Gambert et al. (17) using a highly efficient glass capillary column and as cholesterol labeled by a non-radioactive isotope, namely, [3,4-^{13}C]cholesterol, instead of epicoprostanol as internal standard. The use of the chemical ionization mode in mass spectrometry led to a simpler fragmentation with a greater abundance of the high-mass ions.

Figure 8 illustrates the mass spectrum obtained with the trimethylsilyl ether of unlabeled cholesterol ($m/e = 458$) with methane as the ionizing agent. Three major fragments appear on the mass spectrum of unlabeled cholesterol. The OM-15 fragment corresponding to the loss of a

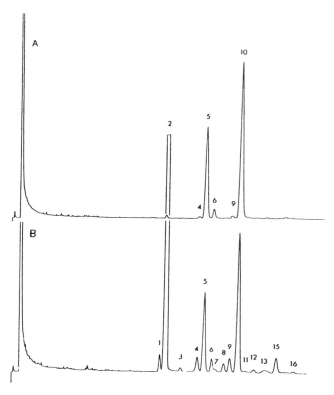

Fig. 7   GLC elution profile of sterol moieties of total steryl esters (11). (A) Common dietary plant sterols. (B) Plasma steryl esters from a patient with phytosterolemia. Peak identification: 1 = cholestanol; 2 = cholesterol; 3 = brassicasterol; 4 = campestanol; 5 = campesterol; 6 = stigmasterol; 7 = unknown; 8 = 24-metylenecholesterol; 9 = stigmasterol; 10 = β-sitosterol; 11–14 = unknown; 15 = avenasterol; 16 = unknown. Operating conditions: Supelcowax 10 on flexible quartz capillary column (15 m × 0.32 mm i.d.); temperature, 250°C isothermal; carrier gas, hydrogen, 5 psi inlet pressure; sample, sterylacetates. (Reproduced with kind permission of the authors and of *Lipids*.)

methyl group was chosen for cholesterol determination because of the absence of doublet, on the contrary present for the M + H – 90 fragment. The measure is based on the ratio of the abundance of the M – 15 fragment of cholesterol (m/e = 443.5) to that of the standard [3,4-^{13}C] cholesterol (m/e = 445.5).

The method was proven to be highly reproducible (0.5% variation between 28 analyses), very sensitive (0.125 pmol of cholesterol detected), linear within a wide range of concentrations, and as accurate as the GC method, which is less sensitive.

This method constitutes a reference method, but one which requires sophisticated and costly equipment. In most cases, simple GC remains the method of choice for sterol analysis in food because the sample to be analyzed is generally in sufficient amount that a very high detection sensitivity is rarely needed.

5.   Liquid Chromatography

As was shown in the first part of the chapter, free cholesterol can be directly determined by HPLC of a simple lipid extract on a Micropak SH-10 column in the adsorption mode. However, HPLC in the reverse phase mode is generally preferred. For example, Hurst et al. (15) deter-

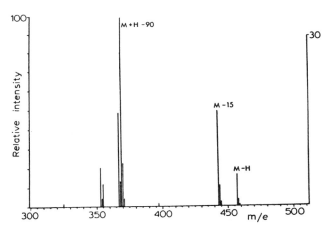

Fig. 8 Mass spectrum of trimethysilyl ether of unlabeled cholesterol (m/e = 458), eluted from SE-30 capillary column directly coupled to the ion source of a mass spectrometer by all-glass connection (17). The GC column temperature was programmed from 250 to 285°C at 1°C/min; nitrogen as carrier gas at 3 ml/min; injector temperature at 290°C. Mass spectrometer: Finnigan 3300 equipped with Interchim Data System. The chimical ionization mode was used with methane as the reagent gas at a pressure of 0.9 Torr. The M-15 fragment (m/e = 443.5) resulting from the loss of a methyl group from the molecular ion was selected for cholesterol determination. (Reproduced with kind permission of the authors and of the *Journal of Chromatography*.)

mined the total cholesterol concentration in milk fat from a semipurified unsaponifiable fraction by reverse phase HPLC on a octadecylsilyl (C18) Alltech column, using hexane-isopropanol (99.9:0.1, v/v) as eluting solvent.

Total cholesterol (free and esterified) can be determined after saponification, from the unsaponifiable fraction. Figure 7, previously reported, shows that cholesterol is relatively well separated from the other sterols. Although this separation was obtained with a peculiar sample, this situation can be encountered in foodstuffs.

Satisfactory results can be obtained with underivatized cholesterol. However, clearer separations or more sensitive detection can be achieved with specific derivatives. For example, a method has been described (16) for determining the cholesterol content in foodstuffs, in the form of benzoate ester derivative. The analysis was performed on a μ-Bondapak octadecylsilyl column with methanol as eluting solvent. The detection of this type of derivative by UV absorbance was very sensitive and as little as 10 ng of cholesterol benzoate could be detected.

Other derivatives, such as acetates or trimethylsilyls, are also likely to be used.

Reverse phase HPLC has also been used to separate molecular species of steryl esters. For example, Kuksis et al. (11) analyzed the steryl esters of plasma lipoproteins from a patient with phytosterolemia. This disease is characterized by the accumulation of plant sterols in the blood and tissues. Study of plasma sterols and steryl esters is useful in its diagnosis. The steryl esters isolated by TLC were analyzed by liquid chromatography coupled to mass spectrometry. The HPLC separation was carried out on a Supelcosil LC-18 column (250 × 2.5 mm i.d.) using a linear gradient of 30–90% propionitrile in acetonitrile as the eluting solvent at a flow rate of 1.5 ml/min and a temperature of 30°C. About 1% of the column effluent was admitted to a quadrupole mass spectrometer. The mass spectra were taken every 7 sec over the entire analysis. Figure 9 shows the elution profile of steryl esters of the low-density (LDL) and high-density (HDL) lipoproteins of plasma from this patient, recorded from the total ion current.

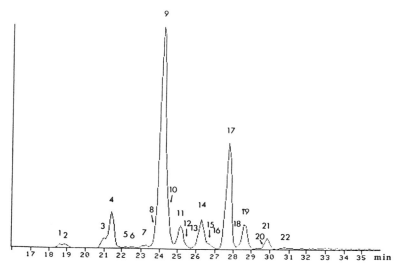

Fig. 9 Total ion current pattern recorded during liquid chromatography–mass spectrometry (LC-MS) elution of steryl esters of LDL (low-density lipoprotein) + HDL (high-density lipoprotein) fraction of plasma from a patient with phytosterolemia (11). Peak identification: 1 = cholesteryldocosahexaenoate (22:6); 2 = cholesteryleicosapentaenoate (22:5); 3 = cholesteryllinolenate (18:3); 4 = cholesterylarachidonate (20:4); 5 = unknown; 7 = avenasteryllinoleate (18:2); 8 = β-sitosterylarachidonate (20:4); 9 = cholesteryllinoleate; 10 = cholesterylpalmitoleate (16:1); 11 = campesteryllinoleate (18:2); 12–13 = unknown; 14 = β-sitosteryloleate (18:1); 15 = avenasteryloleate (18:1); 16 = unknown; 17 = cholesteryloleate (18:1); 18 = unknown; 19 = cholesterylpalmitate (16:0); 20 = unknown; 21 = β-sitosteryloleate (18:1); 22 = unknown. Operating conditions: reverse phase Supelcosil C18 (250 × 2.5 mm i.d.) LC column; eluant, a linear gradient of 30–90% propionitrile in acetonitrile at 1.5 ml/min and 30°C oven temperature. About 1% of the column effluent was admitted to Hewlett-Packard 5985B quadrupole mass spectrometer via a direct liquid inlet interface. (Reproduced with kind permission of the authors and of *Lipids.*)

It can be seen that, similar to fatty acid esters, the polyunsaturated steryl esters were eluted earlier and the less polar saturated esters later. Thus the elution order of cholesteryl esters were arachidonate (peak 4), linoleate (peak 9), oleate (peak 17), and palmitate (peak 19). Linoleates, oleates, and palmitates of the physterols, campesterol, and β-sitosterol were also present in the patient plasma. The authors also analyzed the LDL + HDL steryl esters by glass capillary column GC.

However, HPLC cannot easily compete with capillary GC for analytical purposes. It is more suited for isolating sterol fractions on a micropreparative scale for further analysis. But development of high resolutive columns which could be commercially available could reorientate the choice of the analysts.

6. Supercritical Fluid Chromatography

These last years, the applications of supercritical fluid chromatography (SFC) have been rapidly increasing because of its high separation efficiency and its applicability to the thermally labile compounds. On the other hand, supercritical fluid extraction (SFE) has also attracted interest because of its simplicity, efficiency, and selectivity.

In both SFC and SFE, carbon dioxide is frequently used as the mobile phase or as the extraction solvent because of its inert properties and because it is in a supercritical state at relatively low temperature and low pressure, making it a suitable choice from an instrumentation point of view.

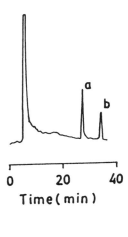

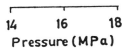

Fig. 10 Supercritical fluid chromatography (SFC) of cholesterol from egg yolk extracted by supercritical fluid extraction (SFE) (18). Peak identifications: a = cholesterol; b = cholesteryl chloroacetate as an internal standard. Operating conditions: cholesterol SFE at a constant pressure of carbon dioxide (17.7 MPa) and at a temperature of 45°C. SFC analysis: column, SE-52 fused silica capillary column (10 m × 0.1 mm i.d., coating thickness 0.45 μm); tapered restrictors rated nominally at 8 ml/min were connected at the column end for pressure reduction; injection valve, 1-μl loop; injection time, 1 sec; runs performed isothermally at 85°C and a pressure programmed from 14 to 20 MPa in 60 min. (Reproduced with kind permission of the authors and of the *Journal of Chromatography*.)

On the other hand, extraction using supercritical carbon dioxide avoids the use of dangerous or toxic organic solvents and the gas is easily removed by reducing the pressure.

SFE has been used in the extraction and SFC in the analysis of cholesterol in food or plasma samples. Ong et al. (18) applied SFE of cholesterol to egg yolk samples, using carbon dioxide at constant pressure (17.7 MPa) and at constant temperature (45°C) with an extraction time of 1 hr. They showed that the extraction efficiency was as high as with the conventional Soxhlet extraction procedure (~98%) and much less time consuming (1 hr instead of 7 hr).

Cholesterol was also quantitated by SFC in the egg yolk samples after addition of a selected internal standard, namely, cholesteryl chloroacetate added before extraction. The column was an SE-52 fused silica capillary column (10 m × 100 μm i.d., coating thickness 0.45 μm) and the mobile phase was supercritical carbon dioxide. The analysis was performed isothermally at 85°C using pressure programming for carbon dioxide from 14 to 20 MPa over 60 min. Detection was by flame ionization.

Figure 10 shows a typical chromatogram obtained by the authors from an extracted egg yolk sample. The amount of egg yolk used for extraction was 0.2 g and the amount of internal standard added was 0.03 g. The two compounds eluted between 20 and 40 min and were clearly separated. Peak area reproducibility was within 2% and retention reproducibility with 0.15% relative standard deviation. The detection limits for both cholesterol and cholesteryl chloroacetate were 25 ppm at a signal-to-noise ratio of 3, demonstrating the high sensitivity of the method.

Capillary SFC coupled to MS was also used in the determination of cholesterol (19).

REFERENCES

1. Y. E. Lozano, C. Mayer Dhuique, C. Bannon, and E. Gaydou, *J. Am. Oil Chem. Soc.*, 70, 561 (1993).
2. F. Mordret, in *Manuel des Corps Gras*, Vol. 2 (A. Karleskind, ed.), Lavoisier, Paris, 1992, p. 1171.
3. S. Renaud and M. De Lorgeril, *Repr. Nutr. Dev.*, 34, 599 (1994).
4. R. G. Jensen, in *Fatty Acids in Food and Their Health Implications* (C. K. Chow, ed.), Marcel Dekker, New York, 1992.
5. J.-P. Bianchini, E. M. Gaydou, J.-C. Sigoillot, and G. Terrom, *J. Chromatogr.*, 239, 231 (1985).
6. L. A. Swain, *Progr. Chem. Fats Other Lipids*, 5 (1958).
7. L. F. Fieser and M. Fieser, in *Steroids*, Reinhold, Chapman and Hall, Ltd, London, 1959, p. 31.
8. P. Trinder, *Ann. Clin. Biochem.*, 29, 1075 (1969).
9. H. Derks, A. van Heinigen, and H. Koedam, *Clin. Chem.*, 31, 691 (1985).
10. L. L. Abell, B. B. Levy, B. B. Brodie, and F. E. Kendall, *J. Biol. Chem.*, 195, 357 (1952).
11. A. Kuksis, J. J. Myher, L. Marai, J. A. Little, R. G. McArthur, and D. A. K. Roncari, *Lipids*, 21, 371 (1986).
12. J. L. Perrin and R. Raoux, *Rev. Fr. Corps Gras*, 35, 33 (1988).
13. I. Björkhem, R. Blomstrand, and L. Svensson, *Clin. Chim. Acta*, 54, 185 (1974).
14. R. Schaffer, L. T. White, L. T. Sniegoski, and Z. Breuer, *Anal. Chem.*, 279, 145 (1976).
15. W. J. Hurst, M. D. Aleo, and R. A. Martin, *J. Dairy Sci.*, 66, 2192 (1983).
16. D. R. Newkirk and A. J. Sheppard, *J. Assoc. Anal. Chem.*, 64, 54 (1981).
17. P. Gambert, C. Lallemant, A. Archambault, B. F. Maume, and P. Padieu, *J. Chromatogr.*, 162, 1 (1979).
18. C. P. Ong, H. K. Lee, and S. F. Li, *J. Chromatogr.*, 515, 509 (1990).
19. N. A. Porter, P. Dussault, R. A. Breyer, J. Kaplan, and J. Morelli, *J. Chem. Res. Toxicol.*, 3, 236 (1990).

14

Phospholipids

Paul Van der Meeren, Jan Vanderdeelen, and Leon Baert
University of Ghent, Ghent, Belgium

I. INTRODUCTION

Phospholipids are lipid-like natural surfactants. Figure 1 reveals that they contain a glycerol backbone onto which one or two fatty acids are esterified; saturated fatty acids mostly occupy the *sn* position 1, whereas unsaturated fatty acids are mainly present on *sn* position 2. On the third hydroxyl a phosphate residue is bound onto which different organic bases may be linked. The latter groups determine the type of phospholipid: phosphatidylcholine (PC), phosphatidylethanolamine (PE) and phosphatidylinositol (PI) are the most predominant natural phospholipids (1–3). Phosphatidic acid (PA), phosphatidylserine (PS), and phosphatidylglycerol (PG) are known as minor components. Lysophospholipids (LPC, LPE, etc.) contain only one acyl group per molecule. Sphingomyelin (SPH) is the major representative of the sphingophospholipids. It consists of a fatty acid linked by an amide bond to the amine group of sphingosine, a long-chain base, linked at position 1 to phosphorylcholine (Fig. 1). Sphingomyelin is a major component of animal complex lipids but does not occur in vegetable lipids.

Phospholipids are the main components of all natural membranes, so that they occur in all living organisms and the derived food products. The highest concentrations are encountered in egg yolk (12%) and brain (25%). In addition, phospholipids are widely used as food additives because of their emulsifying, baking improving and wetting properties in a whole range of products, among which margarines, chocolate, ice cream, animal food, and baking applications are the most important (3).

Besides their technological properties, phospholipids have a nutritional effect (3,4). Hence, it has been shown that they are a major source of choline. In addition, they can be a source of essential fatty acids. During the last years, this topic has been intensively investigated in fish nutrition (5). Finally, several authors have claimed that phospholipids may have pharmaceutical properties (3,6). In particular, their capability to reduce blood cholesterol levels and hence to cure arteriosclerosis has been studied quite intensively (3,7).

The major source of phospholipids is lecithin, which is mostly obtained as a byproduct during

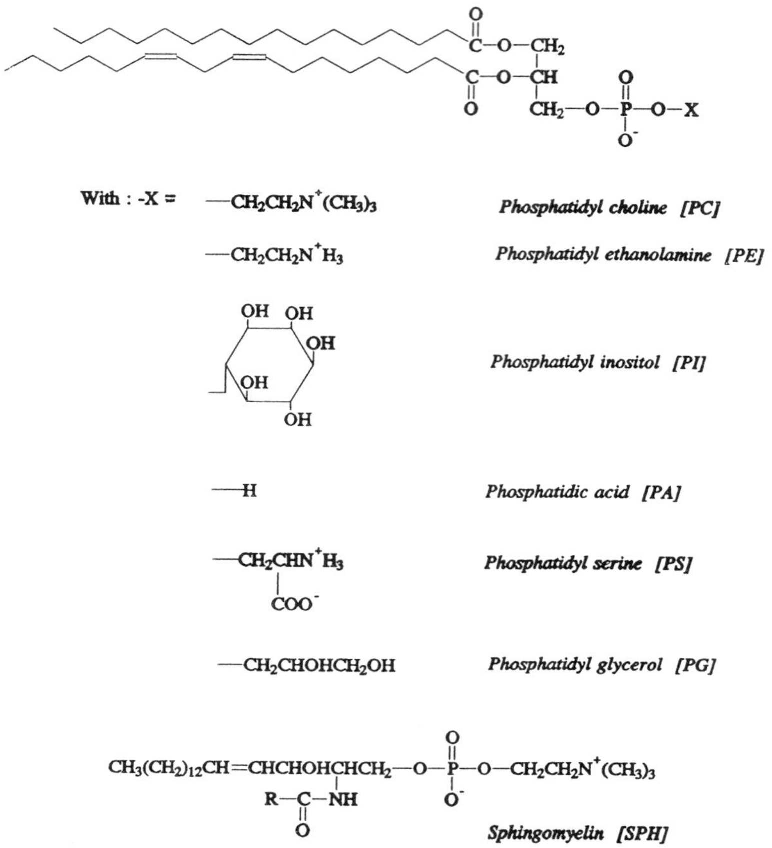

With : -X =   —CH₂CH₂N⁺(CH₃)₃   *Phosphatidyl choline [PC]*

—CH₂CH₂N⁺H₃   *Phosphatidyl ethanolamine [PE]*

Phosphatidyl inositol [PI]

—H   *Phosphatidic acid [PA]*

—CH₂CHN⁺H₃   *Phosphatidyl serine [PS]*

—CH₂CHOHCH₂OH   *Phosphatidyl glycerol [PG]*

Sphingomyelin [SPH]

Fig. 1   Chemical structure of the main phospholipids.

the refining of soybean oil (2). This crude soybean lecithin is viscous and contains about 60–70% of acetone insolubles (about 90% of them being phospholipids) as well as 27–37% of soybean oil (neutral lipids), 1.5–2.0% of water, and 0.5–2.0% of impurities. For the sake of completeness, it should be mentioned that egg lecithin also occurs, but this more expensive product is mainly used in pharmaceutical preparations such as parenteral nutrition. According to the American Food and Drug Administration (FDA) legislation, lecithins have been assigned the "Generally Regarded As Safe" (GRAS) label. According to the European Economic Community (EEC) legislation (8), lecithins, which were assigned E number 322, may be used as antioxidants, emulsifiers, and stabilizers. According to directive 78/664/EEC,

> Lecithins are mixtures of fractions of phosphatides obtained by physical procedures from animal or vegetable foodstuffs; they also include hydrolyzed products obtained through the use of harmless and appropriate enzymes. The final product must not show any signs of residual enzyme activity. The lecithins may be slightly bleached in aqueous medium by means of hydrogen peroxide. This oxidation must not chemically modify the lecithin phosphatides.

The same directive also includes some specific criteria of purity of lecithins related to the appearance, acid number, peroxide number, and composition of the lecithins. As far as the latter is concerned, the directive specifies that not less than 60% (w/w) of substances may be insoluble in acetone (56% for hydrolyzed lecithins), that not more than 0.3% (w/w) substances insoluble in toluene may occur and that not more than 2% (w/w) volatile matter may be present as determined by drying at 105°C for 1 hr. Methods to verify these criteria have been described in the directive 81/712/EEC. Besides, recommended methods have been proposed by some organizations (9,10) such as the American Oil Chemists' Society (AOCS) and the German Society for Fat Science (DGF). A summary of recommended methods related to phospholipids and lecithins is presented in Table 1.

Concerning all of the above-mentioned properties, it has been shown that they cannot solely be determined by the total amount of phospholipids since individual phospholipids may have largely different functional properties. Because of the antagonistic effect of various phospholipids in water, native lecithin doesn't have a general optimum efficacy. Hence, PC mainly stabilizes oil-in-water (O/W) emulsions, whereas PE, PS, PI, and PA are good stabilizers for water-in-oil (W/O) systems (11). In addition, the antispattering properties of lecithin are only observed when the PC/PE ratio is at least four (12). Besides, Rydhag and Wilton (13,14) observed large differences in emulsifying and emulsion stabilizing functionality of different lecithins; they ascribed the observed phenomena to differences in the content of charged phospholipids. Hence, separation and quantification methods have been sought.

II. SAMPLE PREPARATION

A. Extraction

As the structural formulas of phospholipids resemble those of triglycerides, they will generally be contained in the lipid fraction. The latter may be obtained either by solid–liquid extraction of dry solid materials or by liquid–liquid extraction of aqueous samples. Chloroform, di- and tetrachloromethane, petroleum ether, benzene, and isoamyl acetate are known as appropriate phospholipid solvents and hence may be used for solid–liquid extraction. Although the solubility behavior of

Table 1 Recommended Methods of the American Oil Chemists' Society (AOCS) and the German Society for Fat Science (DGF) Related to Phospholipids and Lecithins

Method	AOCS	DGF
Acetone-insoluble matter	Ja 4-46	F-I 5
Toluene-insoluble matter		F-I 4b
Hexane-insoluble matter	Ja 3-87	
Moisture	Ja 2a-46	
Moisture, Karl Fischer	Ja 2b-87	F-I 4
Phosphatides in oils, gravimetric		C-III 16
Phosphorus in oils, colorimetric	Ca 12-55	C-III 16a
Phosphorus in oils, atomic absorption	Ca 12b-87	
Phospholipids in vegetable oil	Ca 19-86	
Total phosphorus in lecithins	Ja 5-55	F-I 5a
Phospholipids in lecithin concentrates	Ja 7a-86, Ja 7b-91	F-I 6
Acid value	Ja 6-55	F-I 3
Peroxide value	Ja 8-87	F-I 3b
Iodine value	Cd 1-25	F-I 3a
Color	Ja 9-87	F-I 2
Viscosity	Ja 10-87, Ja 11-87	F-I 2a
Iron		F-I 4a

triglycerides and phospholipids roughly coincides, acetone is an important exception to this simple rule of thumb. As a further consequence, this solvent is frequently used for fractionation. Hence, soybean lecithin is deoiled by precipitating the phospholipids from a solution of crude lecithin by addition of acetone (1,2).

In most cases, however, it is preferable to add a somewhat polar solvent to break down lipid–protein complexes resulting from weak hydrophobic or van der Waals interactions as well as hydrogen bonds. Thus, it is known that lipid extraction using apolar solvents such as petroleum ether, hexane, or ethyl ether only yields the free lipids, including most of the neutral lipids but only part of the polar lipids (15); more polar solvents or mixtures are needed to recover the total lipids. Therefore, mixtures of chloroform and methanol (especially 2:1 v/v) have mostly been recommended. An exception to this general rule is used for the extraction of wheat lipids; i.e., water-saturated 1-butanol is generally considered to be the best solvent (16). This is due to the fact that inclusion complexes of (mainly lyso)phospholipids inside helical starch amylose molecules have to be destroyed.

For liquid–liquid extractions of aqueous samples, chloroform and methanol are also widely used. According to the Folch et al. (17) extraction procedure, methanol (10 ml/g of tissue) is added first and, after homogenization, 20 ml/g of chloroform is added. Subsequently, most of the nonlipid compounds are removed from the organic solution by adding one-quarter of the total volume of water or a dilute salt solution such as 0.88% KCl. The lower phase contains the lipids. In order to avoid lipid losses, it is important that the proportions in the final dispersion approximate 8/4/3 (v/v), so that the composition of further washing steps (if required) should be appropriately adjusted. The procedure of Folch et al. (17) was simplified by Bligh and Dyer (18), who proposed to extract wet tissues by a single-phase extractant consisting of chloroform and methanol as well as the water present in the sample in a 5:10:4 (v/v) ratio. Subsequently, water and chloroform were added to the extract to form two immiscible phases. The major advantage of this method is that it takes less time (typically about 10 min) and consumes less solvent.

The classical procedure of Folch et al. (17) is especially advisable for animal tissues. Using vegetable material, it is recommended to perform the extraction using 2-propanol in order to inactivate phospholipases. In the Nichols procedure (19), the sample is homogenized in 100% 2-propanol (100 ml/g). Subsequently, the 2-propanol-extracted residue is brought into a 1:1 (v/v) mixture of chloroform and 2-propanol (20 ml/g). These pretreatment procedures are followed by a Folch-type extraction of the evaporated extracts and of the residue by chloroform and methanol. However, in an extensive study involving 82 phospholipid determinations, using both the Folch et al. (17) and the Nichols (19) procedure, Meneses et al. (20) found that only six samples were significantly different, so that these authors were doubtful upon the presence of catabolic enzymes in the extract. They concluded that the problems encountered with extraction mostly rely on the sample handling before the extraction procedure rather than on enzymatic activity during the extraction itself.

Because of the potential health hazard in using chloroform, other solvent mixtures, such as hexane/isopropanol (21) and methylene chloride/methanol (22), have been proposed.

B. Fractionation

Mostly, a further fractionation of the lipid extract is necessary to get rid of most of the neutral and/or glycolipids. To this purpose, a modification of the Vorbeck and Marinetti (23) procedure has frequently been used. This column chromatographic method uses silicic acid as the stationary phase, which is preconditioned with diethyl ether and chloroform. Subsequently, the lipid sample, dissolved in chloroform, is poured on top of the column. Neutral lipids are eluted with chloroform (23,24) or with a mixture of hexane and diethyl ether (25); glycolipids are fractionated by acetone and the phosphatides are eluted with methanol (24). Hereby, additional fractions may be recovered by inserting elution steps using a combination of either chloroform and acetone or chloroform and methanol (Table 2). A similar setup with a completely different mode of action was proposed by Sen Gupta (26): washing a silica gel column containing a lipid sample with dry petroleum ether, phospholipid micelles eluted in the solvent front, whereas a 13:87 (v/v) mixture of petroleum ether–diethyl ether was needed to recover the (monomolecular) triglycerides.

Moreover, it should be emphasized that a whole range of methods have been described using a variety of other adsorbents, such as Florisil, DEAE-cellulose, Sephadex, and aluminum oxide. A review of these older methods is given by Christie (27) and by Zhukov and Vereshchagin (28). However, these open column chromatographic methods are very time consuming: following column preparation, several hours is usually required for separation by elution under gravity.

A similar, small-scale, and hence less time-consuming procedure was described by Christie (29). Applying 1 g of silica gel in a Pasteur pipette, about 30 mg of lipids may be separated into neutral lipids, glycolipids, and phospholipids using 10 ml of chloroform or diethyl ether, 10 ml of acetone, and 10 ml of methanol. To avoid phospholipid losses in the glycolipid fraction, some chloroform may be added to the acetone.

During the last years, numerous fractionation methods have been published using prepacked solid phase extraction columns (Table 3). Thus, Hurst and Martin (30) used silica gel Sep-Pak cartridges to clean up the lipid extract of chocolate in order to evaluate the PC content; after application of 10 ml of lipid extract dissolved in chloroform, the cartridge was washed with 10 ml of 7% petroleum ether in diethyl ether and the phospholipid fraction was recovered with 30 ml of methanol. A similar method has been described to remove most of the neutral lipids from milk fat samples (31). In this case, the lipid sample was dissolved in 2 ml of hexane; a 1/1 (v/v) mixture of hexane and diethyl ether was used to elute the neutral lipids whereas the complex lipids were recovered by elution with 20 ml methanol followed by 20 ml of a 3:5:2 (v/v) mixture of chloroform, methanol, and water. Juaneda and Rocquelin (32) proposed to elute a Sep-Pak silica cartridge with 20 ml of chloroform in order to remove the so-called nonphosphorous lipids (NPLs), whereas the phospholipids were subsequently recovered with 30 ml of methanol. Up to 100 mg of lipids was applied on a 25 x 10 mm i.d. column. In addition, these cartridges may be reused several times after washing with 20 ml of methanol and 20 ml of chloroform. Juaneda and Rocquelin (32) claim that this operation may be performed within 5 min, whereas the classical column chromatographic procedure lasts for several hours. Alternatively, Yao and Rastetter (33) proposed to use chloroform to recover neutral lipids, a 9:1 (v/v) mixture of acetone

Table 2 Composition (v/v/v) of Mixtures of Chloroform, Acetone, and Methanol Proposed by Vorbeck and Marinetti (23) for Fractionation of Total Lipid Extracts on Silicic Acid

Solvent composition (v/v/v)			
Chloroform	Acetone	Methanol	Fraction
100	0	0	Neutral lipids
50	50	0	ESG, MGDG, SG
0	100	0	SG, DGDG
95	0	5	PA + DPG
80	0	20	PE + PS
50	0	50	PI + PC
0	0	100	SPH + LPC

Table 3 Solid Phase Extraction Procedures to Recover Phospholipid Fraction from Total
Lipid Extracts

Stationary phase	Mobile phase	Fraction	Ref.
Sep-Pak silica	10 ml Chloroform solution		
	10 ml petroleum ether/diethyl ether (7:93)	NL	30
	30 ml methanol	PC	
Sep-Pak silica	40 ml hexane/diethyl ether (1:1)	NL	
	20 ml methanol		31
	20 ml CHCl$_3$/CH$_3$OH/H$_2$O (3:5:2)	PL	
Sep-Pak silica	20 ml chloroform	NPL	
	30 ml methanol	PL	32
Sep-pak silica	Chloroform	NL	
	Acetone/methanol (9:1)	Glycosphingolipids	33
	Chloroform/methanol (1:1)	PL	
Bond-Elut	Chloroform/2-propanol (2:1)	NL	34
aminopropyl	Acetic acid/diethyl ether (2:98)	FFA	
	Methanol	PL	
Sep-Pak silica	10 ml petroleum ether/diethyl ether (95:5)	TG	
	20 ml diethyl ether	Unsaponifiables	37
	10 ml methanol	PL	
Bond-Elut silica	30 ml hexane/diethyl ether (20:80)	NPL	
	30 ml methanol	PL	38

and methanol to collect glycosphingolipids, and a 1:1 (v/v) mixture of chloroform and methanol
for the phospholipids. Kaluzny et al. (34) separated up to 10 mg of total lipids into seven fractions
using three Bond Elut aminopropyl disposable columns (500 mg). Neutral lipids, free fatty acids,
and phospholipids were resolved on the first column by successive elution with a 2:1 (v/v)
mixture of chloroform and 2-propanol, 2% acetic acid in diethyl ether, and methanol. The second
and third columns were used for a further separation of the neutral lipid fraction into cholesteryl
esters, triglycerides, cholesterol, diglycerides, and monoglycerides. However, using the same
stationary phase, Christie et al. (35) obtained very poor recoveries of acidic phospholipids, such
as PI, which are known to be hardly soluble in methanol (36). Similarly, Nash and Frankel (37)
used Sep-Pak cartridges to recover the phospholipids from 1-g portions of crude oils; the former
represented only 0.5–1.0% of the total lipid fraction. The oil was applied as a 10% solution in
a 95:5 (v/v) mixture of petroleum ether and diethyl ether and neutral lipids (about 94–97%) were
eluted with 10 ml additional solvent; the so-called unsaponifiables (2–3%) were recovered using
20 ml diethyl ether and the phospholipid fraction was obtained with 10 ml of methanol. This
method was slightly changed by Melton (38): NPLs were removed with a 20:80 mixture of
hexane and diethyl ether.

Besides, more elaborate methods are available nowadays which make possible separation of
all individual lipid classes present. Thus, Prieto et al. (39) described the complete fractionation of
total lipids from wheat into eight neutral lipid, two glycolipid, and four phospholipid classes. In
addition to PC and LPC, N-acyl PE and N-acyl LPE were detected. However, two separate
stationary phases (silica and aminopropyl) as well as seven different mobile phases should be used
to achieve the separation of the phospholipids. Besides, about 14% of cross-contamination of PC
and LPC was observed and the recovery of the phospholipids was limited to about 85%. Therefore,
it should be concluded that solid phase extraction is a rapid, efficient, and versatile technique for
the preliminary fractionation of a limited number of groups of classes, but more powerful
alternatives should be used for the separation of the individual classes.

III. TOTAL PHOSPHOLIPID CONTENT

The classical methods for the determination of the total amount of phospholipids include the conversion of organic to inorganic phosphate, which may be performed either by dry ashing or by wet digestion in inorganic acids. Daun et al. (40) compared three chemical and two instrumental methods for determining phosphorus in crude and refined canola oils with the official American Oil Chemists' Society (AOCS) method Ca 12-55 (9). Hereby, a portion of polar lipids containing about 100 µg phosphorus is charred with 0.5 g of zinc oxide in a muffle furnace at 550–600°C for at least 2 hr. The ash is dissolved in 5 ml of water and 5 ml of HCl and heated to gentle boiling for 5 min. The solution is filtered and neutralized with KOH and then reacidified and made up to 100 ml. An aliquot is treated with molybdate and reducing agent to form the phosphomolybdenum blue complex, the absorbance of which is read at 650 nm. As this method is rather cumbersome and lengthy, alternative methods have been proposed. Rouser et al. (41) described a method including the heating of 20 mg of sample in a screw-capped tube in the presence of 0.4 ml of a 4:1 mixture of sulfuric and perchloric acid in a block heater at 180–200°C. After 30 min digestion, the sample is cooled and 5 ml of water, 1 ml of 2.5% ammonium molybdate, and 1 ml of 10% ascorbic acid is added. The tube is then heated for 20 min at 100°C and the blue color measured at 650 or 750 nm, depending on the phosphorus content. According to Daun et al. (40), this procedure gave consistently lower results than the AOCS procedure and, in addition, the reproducibility was poor between batches. Besides, the digests are often dark brown or yellow, and this might interfere with later colorimetric measurements. Alternatively, Prevot and Gente (42) proposed the use of atomic absorption spectroscopy (AAS). Their original method has been optimised to yield AOCS method Ca 12b-87 (9). Hereby, a sample of oil is diluted in a 1/1 (v/v) ratio with lanthanum/methylisobutylketone (MIBK) reagent consisting of 3.14 g of lanthanum 2,4-pentadionate and 20 g of 2-ethylhexenoic acid diluted to 100 ml with MIBK. Twenty microliters is introduced into the graphite furnace of an atomic absorption spectrophotometer equipped with an electrodeless discharge lamp using argon at 120 ml/min as the purge gas. The phosphorus absorption is measured at 213 nm. The results of AAS agreed within 10% with those obtained by the AOCS colorimetric method. The main advantage of this technique is that it is very fast and uses a minimum of sample. However, relatively expensive equipment is needed.

During recent years, some methods have been described to determine the total phospholipid content directly. Goh et al. (43) modified the Zinzadze reagent. Four milliliters of this reagent are added to the dried PL sample. After shaking for 30 min, the molybdenum blue complexes are extracted by shaking with 5 ml of hexane for 1 min. The absorbance is read at 711 nm against hexane as a blank. This direct method has the advantage that relatively large amounts of inorganic phosphate, present in crude oils, do not interfere as their molybdenum blue complexes are soluble in polar solvents, whereas these with phospholipids are soluble in apolar solvents such as hexane. The molar extinction coefficient of phospholipid–molybdenum blue complexes is about nine times smaller than this of inorganic phosphate complexes, but nevertheless, phospholipids at levels as low as 1 ppm can be determined (44). The results of the method agreed closely to these of a phosphorus assay of the ashings of purified phospholipids using the (modified) AOCS method.

An alternative direct quantitation method was introduced by Eryomin and Poznyakov (45), relying on the formation of complexes between phospholipids and the dyes Victoria Blue R and B. However, the response was largely dependent on the phospholipid class under consideration. More in particular, the method was very sensitive towards PA but could not be used for the choline-containing phospholipids, which are frequently the major phospholipids present.

IV. PHOSPHOLIPID COMPOSITION

A. Thin-Layer Chromatography

Until about 15 years ago, thin-layer chromatography (TLC) was the method of choice for the separation of individual phospholipid classes (27). As several samples could be run simultaneously on one single plate, the number of samples that could be handled daily largely exceeded that of the open column chromatographic alternative. The major advantage of TLC, however, relies on the existence of several staining sprays to perform specific analyses (27). In addition, universal phospholipid staining sprays have been developed that enable a very sensitive detection and quantification, even of minor components (46). Several universal detection reagents for lipids are based on a charring reaction. Typical examples are copper–acetate–phosphoric acid and sulfuric acid; the resulting color formation is dependent on the degree of unsaturation. Spillman et al. (47) observed that 8-anilino-1-naphthalene (ANS) and bromothymol blue stain all lipids indiscriminately; they are insensitive to varying degrees of unsaturation. A very promising feature of staining reagents is their selectivity, so that a complete resolution of all compounds present is not required. In phospholipid research, the molybdenum blue reagent, originally proposed by Dittmer and Lester (48) and modified by Ryu and MacCoss (49) is frequently used; as this stain only reacts with phosphate, it is specific for phospholipids. As a further consequence, the response is also independent of the fatty acid composition. Gustavsson (50) demonstrated that the molar absorptivities of most analyzed phospholipids, except from SPH, LPC, and PS, did not differ more than 10% from that of egg yolk PC. In addition, the Dragendorff stain (51) is specific for the choline-containing phospholipids PC, LPC, and SPH, whereas the ninhydrin stain (52) only reacts with amine-containing lipids. The usefulness of the availability of specific spray reagents has been exploited by Dieffenbacher and Bracco (53); they elaborated a method to detect a whole range of food emulsifiers. Thanks to the specificity of the sprays, a complete resolution of all emulsifiers, which could not be achieved anyway, was not needed. In this setup, the reagent of Dittmer and Lester was used as a specific reagent for organic phosphorous compounds, appearing as blue spots on a white background.

Previously, these stains were designed for visualization and to get a qualitative insight about the composition. A more reliable quantitative estimation was achieved after scraping the visualized phospholipid bands from the plate, followed by a P determination. Hence, TLC was considered to be an inexpensive technique with limited possibilities. Thanks to the advent of automated sample application systems and densitometric scanners, coupled with powerful image analysis and processing hard- and software, the accuracy and reproducibility have been highly increased, so that reliable quantitative information is extracted nowadays. Besides, the sensitivity is mostly better than that obtained by the evaporative light scattering detectors, widely used in high-performance liquid chromatography (HPLC).

The mobile phases of TLC are mostly based on chloroform, methanol, and water, to which acidic (formic or acetic acid) or alkaline (ammonia) modifiers may be added. The relevance of these modifiers becomes obvious from the two-dimensional phospholipid analysis of soybean lecithin (Fig. 2): The plate was developed in the first dimension using a 65:30:4 (v/v) mixture of chloroform, methanol, and 7 N ammonium hydroxide, whereas a 170:25:25:6 (v/v) mixture of chloroform, methanol, acetic acid, and water was preferred for the second dimension (54). Actually, this solvent system formed the basis of AOCS method Ja 7-86 (9). Figure 2 reveals that the zwitterionic phospholipids such as PC and LPC are hardly affected by the pH of the mobile phase, so that they reside near the bisectrix. On the other hand, the acidic phospholipids such as PI, PS, and (primarily) PA are largely affected. Thus, PA is rather immobile in the presence of ammonium hydroxide; its polarity is maximized by ionization. On the other hand, PA becomes protonated in the presence of acetic acid with a subsequent mobility increase. Similar effects are also observed for the free fatty acid (FFA) spot. This separation of 24 components out of soybean lecithin was achieved by the lipid

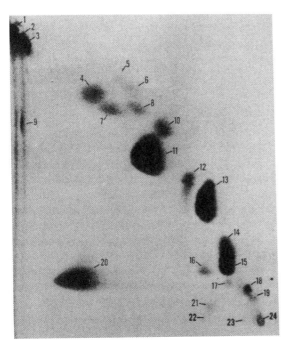

Fig. 2 Thin-layer chromatography of commercial lecithin on silica gel G. First dimension: chloroform/methanol/7 N ammonium hydroxide (65:30:4); second dimension: chloroform/methanol/acetic acid/water (170:25:25:6). 3 = N-acyl PE, 9 = FFA, 10 = PG, 11 = PE, 13 = PC, 14 = LPE, 15 = PI, 16 = PS, 18 = LPC, 20 = PA. (From Ref. 54.)

research group at the Hormel Institute at the Hormel Institute of the University of Minnesota, Austin, Minnesota, where very detailed separations of phospholipid mixtures were already accomplished more than two decades ago, both by TLC and liquid chromatography. Despite the huge technical progress, these results have not been surpassed in resolution up to now.

During the last decade, the design of TLC plates has been optimized in order to improve the resolution and to reduce the solvent consumption. These so-called HPTLC plates have smaller dimensions (5 cm instead of 10 or 12 cm) and consist of slightly thinner layers (0.20 instead of 0.25 mm) of smaller particles (7 instead of 12 μm) with a sharper particle size distribution. Hence, a better separation efficiency is achieved over a 50–60% shorter separation distance. As the analysis time is proportional to the length squared, HPTLC separations are at least four times faster. Besides, due to the improved separation efficiency and the reduced diffusional broadening, the detection limits are 10 to 15 times lower. Some typical HPTLC results using 5 x 5 cm Merck silica gel 60 HPTLC plates are shown in Fig. 3, giving proof of the influence of the composition of a mixture of chloroform, methanol, and water on the separation of vegetable phospholipids (55). Figure 3 reveals that even slight variations of the water content have a large effect on the separation. Besides, Lendrath et al. (55) observed that the resolution was improved by using acetate buffer (pH4) instead of water in the mobile phase. Hence, they proposed a 65:25:4.3 (v/v) mixture of chloroform, methanol, and 0.2 M acetate buffer. Besides, phosphoric acid impregnation of the silica gel layer was found to increase the stability and the intensity upon applying the Dittmer–Lester reagent, which is specific for phosphorus. Hereby, it was observed that the response was not significantly affected by the fatty acid composition of the phospholipid.

Similarly, Smiles et al. (56) were able to resolve seven phospholipid as well as seven glycolipid

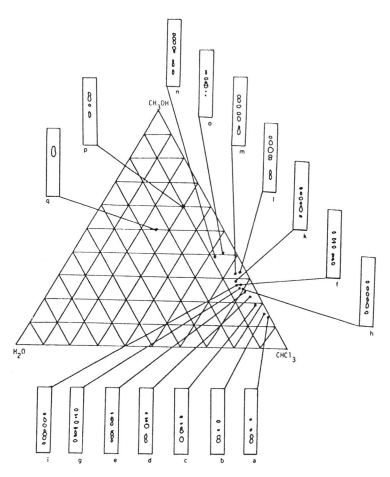

Fig. 3 Influence of the mobile phase composition on the separation of soybean lecithin phospholipids on 5 x 5 cm silica gel HPTLC plates. (From Ref. 55.)

bands within 30 min on LHP-K linear HPTLC plates (Whatman), using a 75:45:3:1 (v/v) mixture of chloroform, methanol, acetic acid, and water; the glycolipids were visualized by α-naphthol, PC, and LPC by Dragendorf and PE by ninhydrin, whereas a general spray reagent proposed by Vaskovsky et al. (57) was used for the detection of all phospholipids. A rapid multidevelopment HPTLC method capable of separating 18 different lipid components by means of three parallel chromatographic runs in different solvent systems was recently introduced by Pick and Kovacs (58). The whole procedure, including staining and densitometry, can be completed in 40 min. The first solvent mixture, consisting of a 23:29:11:10:12 (v/v) mixture of *n*-propanol, methyl acetate, methanol, ethanol, and 43 mM KCl was allowed to migrate over 8 cm and perfectly resolved the polar lipids, which part was cut from the remainings; phosphomolybdic acid was used for staining. After drying of the plate, the neutral lipid spot, which was now at the bottom of the plate (as the more polar lipids were cut), was resolved by a 300:54:25:25 (v/v) mixture of *n*-hexane, acetone, toluene, and chloroform. This part of the plate was stained by copper(II) sulfate.

In recent years, a significant improvement of chromatographic selectivity has been accomplished by automated multiple development (AMD) using a universal elution gradient (59,60).

Hereby the chromatogram is developed repeatedly in the same direction over migration distances increasing by 1–3 mm per run. As the strength of the solvents for each run decreases, a stepwise gradient is obtained. Thanks to this stepwise elution with solvents of decreasing elution power from 100% of methanol over 100% of dichloromethane to 100% of hexane, the fractions are refocused at each stage. As a further consequence, the peak width becomes independent of the migration distance and the width of the starting spot. The AMD process, which may consist of 15–30 runs, is carried out completely automated. In the method of de la Vigne and Jänchen (61), the spots are visualized by heating the plate for 20 min at 120°C after immersion for 5 sec in a solution containing 4 ml of concentrated sulfuric acid and 0.4 g of hydrated manganese chloride in 120 ml of water/methanol (1:1). The brown spots are densitometrically quantified by absorbance at 550 nm. Alternatively, the fluorescence at 366 nm excitation can be monitored; this approach enables a 10-fold decrease of the detection limit down to about 10 ng.

For the sake of completeness, it is mentioned that silica gel–sintered TLC plates may be used to enable the in situ mineralization and determination of phosphorus in phospholipids (62). These can withstand a treatment with strong acids at high temperatures. Hereby the digestion was accomplished by perchloric acid. The phosphoric acid liberated was determined by densitometry after visualization with brilliant green. This method is very sensitive; in the range from 10–100 ng of phosphorus, which represents about 0.25–2.5 μg of PL, the coefficient of variation was about 5–15%. Unfortunately, the individual calibration curves (on a phosphorus basis) were different for all phospholipids.

B. TLC-FID

A special mode of TLC is the application of thin layers of silica coated onto quartz rods (63,64). These so-called chromarods have been introduced to enable flame ionization detection, which is performed in a so-called Iatroscan (Iatron, Tokyo, Japan). The latter detection technique is universal and performs, at least theoretically, more or less as a mass sensitive detector. In practice, however, it has been reported that the response factors for the various lipids depend on variables such as sample volatility, amount of material analyzed, and fatty acid composition. Whitsett and Kennish (65) observed that the slopes of the calibration curves obtained using a series of molecular species of a given lipid class varied by as much as 50%; they concluded that accurate quantitation was not feasible using a single molecular species as the standard. Besides the response, the chromatographic behavior also depends on the fatty acid composition (66). Therefore, calibration should be elaborated by a mixture of molecular species closely resembling the actual sample rather than by a molecularly pure standard. As the response of some lipids is influenced by the composition of the mixture, it is important that the calibration solution consists of the analytes in approximately the same proportion as the sample (65).

The coefficient of variation typically amounts to about 5%, larger values being observed when the amount of lipid applied is below 1 μg or for large amounts whereby the separation capacity of the layer may be exceeded with incomplete combustion of some of the components (63). The sensitivity is also affected by instrumental parameters such as the chromarod movement rate through the flame, the distance of the ion collector above the rods, and the configuration of the ion collector. Thus, Przybylski and Eskin (67) modified the commercially available MKII flame ionization detector and obtained a 10-fold increase in detector response. Besides, the response for PS and PI was far better than reported previously. Whereas Du Plessis and Pretorius (68) found that the relative response for PC, PE, PA, and PI amounted to 7.7:7.6:3.2:1.0, Przybylski and Eskin (67) observed that the relative response for a whole range of phospholipids varied by a factor of 2 at most. Hence, from 0.1 up to 20 μg of phosphatides could be quantitated with a coefficient of variation of 4.4–7.2%. The procedure was adopted to determine the phospholipid composition of canola oils during the early stages of processing. Ohshima and Ackman (69) observed that older FID equipment, such as the Mark III and early Mark IV, gave nonlinear responses. On the other hand, the introduction of the current redesigned FID for later Mark IV models as well as of the Mark V Iatroscan has resulted in linear calibration curves over a reasonable working range (up to 25 μg). Besides, the sensitivity has been extended down to the nanogram scale.

Kaimal and Shantha (70) observed that the use of copper(II) sulfate–impregnated chromarods promoted ionization of organic compounds in the detector flame, thus increasing the FID response. However, Ranny et al. (71) found that impregnation of rods with divalent metals reduced the mobility of acidic phospholipids, so that charged phospholipids can hardly be resolved. According to the authors, this phenomenon is due to the formation of metal complexes with acidic phosphate groups. An alternative approach to increase the response consists of hydrogenating the lipid samples before analysis (69,72). As a result of the decreased number of molecular species, the peaks become sharper and both sensitivity and reproducibility are increased. Olsson (73) remarked, however, that hydrogenation decreased the resolution between PC and SPH because the differences in the hydrophobic part are largely removed, so that resolution almost entirely relies on the properties of the polar headgroup.

Several solvent systems have been described for the single-step separation of the main phospholipids (63). As expected from the similarity of this technique to TLC, mixtures of chloroform, methanol, and water are mostly used in this case too. The optimum chloroform-to-methanol ratio is roughly 2:1 (v/v); the solvent mixtures generally include about 3–4% of water. A decrease of the methanol content badly affects the separation of the fastest eluting compounds. Partial substitution of methanol by 2-propanol leads to an increase of the retention time and a decrease in the separation of SPH from PC, which is especially important for milk samples as their polar lipid fraction primarily consists of these compounds. The water content affects the separation of the more polar components to a high extent and depends on the methanol content. These systems are most useful for separating neutral ampholytic phospholipids. They are less useful for acidic lipids, such as PA, PI, and PS; broad tailing zones are formed whose R_F values depend on the concentration. This problem may be solved by the addition of ammonium hydroxide, which decreases the R_F value of acidic phospholipids, but this may cause some acidic phospholipids to split into two separate peaks; besides, the separation capacity of silica gel chromarods gradually decreases (63). On the other hand, the addition of acetic acid increases the mobility of acid PL.

Innis and Clandinin (74) proposed a double development scheme. Following the separation of the neutral lipids by a 85:15 mixture of petroleum ether and diethyl ether, the rods were partially scanned. Thus, the neutral lipids were removed, whereas the intact polar lipids remained at the origin. Subsequently, the polar lipids were separated by an 80:35:3 mixture of chloroform, methanol, and water, after which the chromarods were fully scanned. A similar approach was used by Du Plessis and Pretorius (68), Ratnayake and Ackman (75), and Whitsett and Kennish (65). Following the elution of neutral lipids to the upper part of the chromarod with acetone (68) or an 80:25:1.2 mixture of hexane, diethyl ether, and formic acid (65), they were removed by partial scanning. The phospholipids were then moved from the origin by a second development in a polar solvent system, such as an 80:35:3 mixture of chloroform, methanol, and water (65,68) and scanned for the quantification of the individual phospholipid classes.

C. High Performance Liquid Chromatography

Numerous methods have been described for the separation of phospholipids by HPLC (29,76,78). In general, it may be stated that the major challenge in HPLC is not to obtain a good resolution but to achieve a reliable and reproducible quantitative detection. Actually, both considerations affect each other. This is exemplified by the restrictions on the mobile phase composition imposed by the use of the UV detector, which was the detector of choice up to some years ago. Only solvents with a small UV absorbance could be selected. In addition, isocratic mobile phases were highly recommended in order to reduce baseline shift. Thus, mixtures of chloroform, methanol, and water, which were the most used solvents in TLC and open column chromatography, were excluded. Instead, mixtures of hexane, 2-propanol, and water, as well as mixtures of acetonitrile, methanol, and water or aqueous solutions of inorganic acids, have mostly been selected. An overview of these older solvent systems has been presented elsewhere (77). As the response of the UV detector is mainly affected by the degree of unsaturation and oxidation of the phospholipids (78), its sensitivity varies widely depending on the origin of the sample and its pretreatment. Therefore, other detection systems have been investigated whose response is more directly related to the amount of phospholipids present.

Thanks to the advent of these alternative detection systems, based on changes in the refractive index or on light scattering or flame ionization detection, the restrictions on the choice of the mobile phases were largely reduced, so that alternative solvent systems (Table 4) have been introduced. Thus, the construction of a home-made FID enabled the research group of Privett at the Hormel Institute to use a quaternary solvent gradient, whereby the separation of the phospholipid classes was realized by a gradient of chloroform and 8% NH_4OH in methanol (79). Using this solvent system, excellent separations were obtained in 1973. Erdahl et al. (54) demonstrated the resolution of 19 components of soybean lecithin. Unfortunately, this moving wire FID never became available commercially. In 1982, Tracor instruments developed a rotating disk FID for HPLC. Grieser and Greske (87) used this apparatus to determine the composition of corn and soy lecithin (Fig. 4). Thanks to the FID, chloroform could be incorporated in the mobile phases, in combination with methanol and 28.7% ammonium hydroxide (v/v). The latter improved peak shape and resolution but caused the phospholipid peaks to appear as doublets, even for synthetic species. According to the authors, the FID produces a linear response over at least a 25-fold concentration range, which is not affected by the variation in unsaturation of the fatty acid constituents in the phospholipids.

As can be seen from Table 4, UV-absorbing solvents can also be used in combination with a refractive index (RI) detector (80,84). However, as this detector can only be used with isocratic elution, its usefulness is limited to the quantitation of the most polar phospholipids such as PC, SPH, and LPC.

Mass spectrometry (MS) is a very promising technique combining detection and identification capabilities (88). In addition, the mass spectrum, recorded at any time, contains information on the molecular species present, so that their distribution may be judged during the elution of the phospholipid component. However, LC-MS cannot be considered as a routine analysis technique. Very experienced and skilled operators are needed. Moreover, since the equipment is very expensive, it is expected that this technique will mainly be used for academic research rather than for industrial food analysis applications.

During recent years, the evaporative light-scattering detector (ELSD) has become the detector of choice for quantitative lipid analysis by HPLC (92). The principle is based on the registration of the intensity of the light scattered by the solute particles remaining after evaporation of the mobile phase from the cloud of small droplets formed by a venturi (93). Van der Meeren et al. (94) showed that the response of the ELSD may be simulated if four phenomena are considered. By nebulization, the mobile phase containing the solute is divided into very small droplets. Next, the mobile phase is removed by evaporation. Subsequently, light scattering occurs as the solute particles move through a light beam. Finally, photomultiplier saturation determines the detection efficiency of the scattered light. On the basis of experimental data, it was shown that the profile of the response curves was largely determined by the light scattering phenomenon at small sample loads. As Rayleigh scattering prevails in this lower concentration range, the scattered intensity is

Table 4 Overview of Recent HPLC Phospholipid Separation Methods Using UV-Absorbing Solvents

Stationary phase	Mobile phase	Detection	Phospholipids	Ref.
Silicic acid minus 325 mesh	Chloroform 8% NH$_4$OH in methanol	FID	PG, PE, PC, PI, PA, LPC	54, 79
μ-Porasil	CHCl$_3$/CH$_3$OH/CH$_3$COOH/H$_2$O (14:14:1:1)	RI	PC	80
Spherisorb 3 μm	Isooctane/THF (99:1) 2-Propanol/CHCl$_3$ (4:1) 2-Propanol/H$_2$O (1:1)	ELSD	TG, C, MG, (D)PG, PE, PI, PS, PC, SPH	81, 82
Nucleosil 100 3 μm	CHCl$_3$/TFA (400:5) CHCl$_3$/CH$_3$OH/C$_7$H$_{16}$/H$_2$O/TFA (100:400:50:15:5)	ELSD	PG, PI, PS, PE, PC	83
μ-Porasil	CHCl$_3$/CH$_3$CN/CH$_3$OH/H$_2$O/85% H$_3$PO$_4$ (650:650:500:130:2)	RI	PC, SPH, LPC	84
Lichrosorb Si-60 5 μm	Hexane 2-Propanol/CHCl$_3$ (4:1) 2-Propanol/H$_2$O (1:1)	ELSD	PE, PI, PS, LPE, PC, SPH, LPC	85
Lichrospher Si-100 5 μm	CHCl$_3$ CH$_3$OH/H$_2$O/NH$_4$OH/CHCl$_3$ (92:5:2:1)	ELSD	MG, DG, FFA, PE, PI, PS, PC, SPH	86
Ultrasphere Si 5 μm	CHCl$_3$ CH$_3$OH/28.7% NH$_4$OH (86:14)	FID	TG, N-acyl PE, FFA, PE, PC, PI, LPC	87
Brownlee Si 5 μm	CH$_2$Cl$_2$/CH$_3$OH/H$_2$O (93:6.5:0.5) CH$_2$Cl$_2$/CH$_3$OH/H$_2$O/15M NH$_4$OH (65:31:4:0.2)	MS	PE, PC, PI, SPH, PS	88
Lichrosorb Si-60 10 μm	CHCl$_3$/THF (1:1) CH$_3$OH/NH$_4$OH/H$_2$O (92:7:1)	ELSD	PE, PI, PA, PC	89
Zorbax 7 μm	Isooctane/THF/2-propanol/CHCl$_3$/H$_2$O (415:5:446:104:30) (216:4:546:154:80)	ELSD	PE, PI + PA, PS, PC	38
Lichrosphere 100 Diol	Hexane/2-propanol/butanol/ THF/isooctane/H$_2$O (64.5:17.5:7:5:5:1) 2-propanol/butanol/THF/ iso-octane/H$_2$O (73:7:5:5:10)	ELSD	PE, PA, PC, PI, LPC	90, 91

proportional to the diameter of the particles to the sixth power. Hence, assuming the peak shape to remain constant and upon complete evaporation of the mobile phase, the response is proportional to the sample load squared. On the other hand, the characteristics of the photomultiplier tube become predominant at large sample loads. Due to oversaturation, the response levels off. As a consequence, a typical sigmoidal shape is mostly observed (81,83,86,94) having a fairly rectilinear part within the sample load range of 5–50 μg; as the response begins to fall off rapidly below 5 μg, the evaporative light-scattering detector is somewhat insensitive to minor components (92). Since the response is actually determined by the concentration at any given time, the calibration curves are largely affected by the peak shape (95). The major advantage of ELSD is that its response is not affected by the chain length and degree of unsaturation of the fatty acids present. Thus, Robinson et al. (96) found similar response curves for a range of triglycerides whose fatty

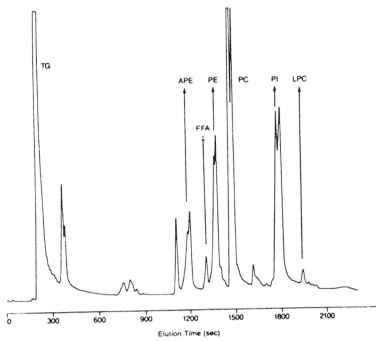

Fig. 4 HPLC separation of phospholipids in a typical soy lecithin and flame ionization detection. TG = triglycerides, APE = *N*-acyl PE. (From Ref. 87.)

acids contained from 12 to 18 carbon atoms with 0 to 3 double bonds. Representing their results on a double logarithmic scale, linear relationships were recorded. On the other hand, Robinson et al. (96) observed that the solvent composition largely affected the response. According to Charlesworth (97), this is due to a difference in particle size in the venturi and solvent evaporation, which is dependent on the molar volatility. According to Stolyhwo et al. (98), solvent vaporization is not complete at flow rates above 0.3 ml/min, so that noise emerges. To overcome this shortcoming, the authors proposed to use either small-bore columns or a stream splitter. Righezza and Guiochon (99) found that the nature of the solvent and solutes markedly affected the response of the ELSD. Hence they concluded that the ELSD was not a mass detector and calibration is required for quantitative analysis. However, it has been observed by many authors that the differences between calibration curves of different compounds are much smaller as compared to other detection techniques. Thus, Stockwell and King (100) compared the response factors for a range of 17 steroids using both light scattering and UV detection at 254 nm. Whereas the ratio of the lower to the highest value amounted to 0.82 for the former technique, it dropped to 0.01 for UV detection. The major conclusion is that it is simple to obtain semiquantitative information about the composition of the mixture under consideration. The larger peaks must be due to the major components. However, the accurate estimation of the actual concentration of the different components present requires a very tedious calibration. Van der Meeren et al. (95) demonstrated that even under these circumstances erroneous results may be obtained if the fatty acid composition of the standard does not match that of the sample. It was shown that the effect of the fatty acid composition was primarily due to its influence on the peak shape rather than its effect on the detector response.

The first successful applications of the ELSD for phospholipid separations were reported by

Christie (81,82,101). Originally, Christie et al. (101) used a gradient of hexane, 2-propanol, and water; at the end of the gradient these solvents were mixed in a 30:60:10 ratio. A typical result, showing the separation of the phospholipids from cows' milk, is shown in Fig. 5. In further experiments, dealing with animal tissues, a more complex gradient was introduced (81): three solvent mixtures, containing isooctane and tetrahydrofuran (99:1), 2-propanol and chloroform (4:1), and 2-propanol and water (1:1) were used. The presence of 1% of tetrahydrofuran (THF) had a beneficial effect on the rapid reequilibration of the column after an analysis. In a later study (82), electrolytes were added to the aqueous component to sharpen the resolution and to extend the column life. In the optimized solvent system, the water in the third solvent mixture contained 0.5 mM serine adjusted to pH 7.5 with ethylamine. Christie (82) mentioned that an old, abandoned column was completely restored by using this optimized solvent system. Arnoldson and Kaufmann (90) also promoted the use of ionic modifiers in the mobile phase; they proposed the addition of 180 mg of ammonium acetate per liter of both solvent mixtures used.

During recent years, the use of diol columns has been proposed (90,91,102–104). According to Christie (76), such columns offer a homogeneous surface of constant activity for adsorption chromatography. Components are eluted with somewhat less polar solvent than from silica gel.

For the sake of completeness, it should be noted that a chloroform-based solvent system was proposed more than one decade ago. Using a radioactivity detector, P-32-labeled phospholipids were baseline-separated by a gradient of three solvent mixtures, each consisting of chloroform, propanol, acetic acid, and water (105).

A promising alternative detection system was presented by Kaitaranta and Bessman (106,107) and by Islam et al. (108). By connecting the outlet of the HPLC column to an automatic phosphorus analyzer, a reliable quantitation is made possible. Besides, complete resolution of the neutral and

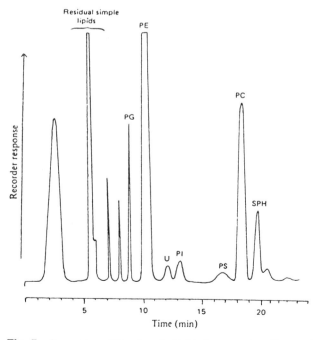

Fig. 5 Separation of the phospholipids from cow's milk on a 4.6 x 100 mm column of 3 μm Spherisorb by a gradient of hexane, 2-propanol, and water using evaporative light scattering detection. u = unidentified. (From Ref. 101.)

glycolipids from phospholipids is not required, so that a simple isocratic elution scheme seems to be effective. According to the authors, the range of 1–100 nmol (1–75 µg PL) may be regarded as a practical working range. The major disadvantage of this hyphenated technique is that the solvent flow rate is limited to about 0.4 ml/min by the digestibility of the phosphorus analyzer, so that large sample times are required. However, thanks to the introduction of smaller bore columns, this technique could show a revival during the coming years.

As a final remark on HPLC detection systems, it seems worthwhile to mention that UV detection may be very useful to quantify lipid peroxidation in foods, which is known to cause the appearance of unpleasant odor and taste, the emergence of toxic, potentially carcinogenic, substances, as well as nutritional damage. Terao et al. (109) revealed that normal phase HPLC is suitable for the estimation of the level of oxidation of each phospholipid class by using the ratio of the peak areas obtained by UV detection at 235 and 203 nm, respectively. As this ratio represents a relative value, it should primarily be used to evaluate the effect of different processing conditions on a given food sample.

D. P-31 Nuclear Magnetic Resonance

During the last years, there has been an increased interest in high-field P-31 nuclear magnetic resonance (NMR) for phospholipid quantification. First of all, this technique doesn't require a thorough sample pretreatment because nonphospholipid contaminants are not detected. Besides, the different phospholipid classes do not have to be separated; as the chemical shift of the phosphorus atom is dependent on its molecular environment, unique values are obtained for the different phospholipid classes. Another most interesting feature is that P-31 NMR spectra can be recorded in such a way that the peak areas are directly proportional to the phosphorus concentration. However, most of the phospholipid resonances are crowded in a very narrow chemical shift range of about 1.5 ppm wide. Hence, very narrow NMR resonances are needed. On the other hand, the P-31 spectra of phospholipids with relatively long acyl chains in aqueous dispersion consist of very broad lines due to the motional restriction associated with the lamellar bilayer structure. Hence, up to about 15 years ago, fairly broad P-31 NMR signals were always recorded. As a further consequence, resolution was limited and quantitation was limited to PC as compared to the remaining phospholipids. In order to reduce the peak width, two alternatives have been described. In 1979, London and Feigenson (110) demonstrated that the signals could be narrowed considerably through the use of detergents; the phospholipids are separated from each other by solubilization and reside in small mixed micelles of small bile salt aggregates. Alternatively, analysis of phospholipid headgroup composition becomes possible following lipid extraction and spectral acquisition in a solvent to obtain substantially narrower lines and hence an increased signal-to-noise ratio. The latter approach is mostly preferable to obtain a reliable quantitation of the different phospholipid classes present within a short time. This is mainly due to the fact that higher concentrations are generally obtained in the extract as compared to the aqueous sample. Besides, less molecular species resolution is observed in organic solvents; as a further consequence of the increased peak sharpness, higher peaks are obtained, which are easier to discern from the baseline noise. An overview of some ternary solvent systems used for the quantitative estimation of phospholipid classes is presented in Table 5.

Sotirhos et al. (111) managed P-31 NMR at 161.7 MHz (with proton decoupling) to analyze the phospholipid composition of egg lecithins. Hereby, 100 mg of phospholipids was dissolved in 3 ml of deuterated chloroform-methanol (2:1), containing triethyl phosphate as the internal standard; deuterated solvents are used to provide a deuterium reference signal for magnetic resonance field-frequency stabilization. In this study, it was observed that the temperature influenced the chemical shifts of the phospholipids, so that resolution between critical pairs could

Table 5 Composition of Ternary Solvent Mixtures, Obtained by Mixing Solvent A and Solvent B, Used for Quantitation of Phospholipid Classes by ^{31}P NMR

Solvent A		Solvent B		
Composition	Volume	Composition	Volume	Ref.
CDCl$_3$	2.0 ml	CH$_3$OH	1.0 ml	109
CHCl$_3$	2.0 ml	4 parts CH$_3$OH, 1 part 0.2 M Cs-EDTA in D$_2$O (pH 6)	1.0 ml	110
CDCl$_3$	3.5 ml	0.2 M Cs-EDTA in CH$_3$OH/D$_2$O (7:2)	1.0 ml	113
95% CHCl$_3$, 5% benzene-d$_6$, 0.075 mg TMP/ml	2.0 ml	4 parts CH$_3$OH, 1 part 0.2 M aq. Cs-EDTA (pH 6)	1.0 ml	17
CDCl$_3$, CH$_3$OH	400 μl 88 μl	0.2 M aq. Cs-EDTA (pH 6)	37 μl	111

be optimized by a change of temperature. Besides, addition of acids or bases was shown to largely affect the peak broadness and, to a lesser extent, the chemical shift. In addition, it was noticed that the chemical shift was slightly dependent on the phospholipid composition itself, which must be due to interactions between individual PL in organic solvents. On the other hand, differences in degree of unsaturation and chain length of the fatty acyl constituents present in phospholipid molecules do not significantly affect the chemical shift.

Meneses and Glonek (112) demonstrated that the resolution is affected by the solvent composition to a very high extent. Hereby, the presence of a univalent EDTA salt was essential. This is due to the fact that metallic cations interact with anionic phosphates forming coordination complexes, which alter the chemical shielding of the constituent phosphorus nucleus. Therefore, they recommended dissolving 0.01 to 100 mg of crude lipid extract in 2 ml of chloroform to which 1 ml of an aqueous methanol reagent was added. The latter contained one part 0.2 M (EDTA) in D$_2$O titrated to pH 6 with CsOH and four parts of methanol. Using this solvent system, a biphasic mixture is obtained; only the chloroform phase is sampled by the NMR. The small aqueous phase remaining on top prevents evaporation of the organic solvents, so that the solvent composition remains constant. Figure 6 contains a typical spectrum of a crude soybean phospholipid preparation. Besides the major phospholipids, their lyso derivatives, which are mostly not detected by TLC or HPLC, are also resolved. In a later study, Meneses et al. (20) experienced that the use of the above-mentioned Cs-EDTA–containing ternary solvent mixture could not always prevent peak broadening. Therefore, they recommended replacing KCl by K-EDTA (0.1 M) in the backwashing step of the classical Folch et al. (17) extraction procedure. The relevance of this small modification is shown in Fig. 7, representing the P-31 NMR spectra of extracted algal phospholipids. The upper trace was obtained following KCl backwashing, whereas the lower curve corresponds to the K-EDTA washed sample; the solvent composition is included in Table 5. Comparing lyso derivatives to their parent phospholipids, Meneses and Glonek (112) observed that the former are chemically shifted to lower magnetic fields by about 0.52∂. Hence, P-31 NMR is ideally suited to investigate hydrolytic degradation of phospholipids: diacyl phospholipids, 1-acyl lysophospholipids, 2-acyl lysophospholipids,; and glyceryl phosphorylcholine are completely resolved (113). Similarly, plasmalogens are shifted to lower magnetic fields by about 0.07∂ with respect to their parent compounds (112). This effect, although small, is sufficiently large to allow quantitation of the plasmalogens and their corresponding diacyl phospholipids. Besides, careful choice of the solvent system allows further resolution of classes defined by the various combinations of acyl, alkyl, and alkenyl hydrophobic chain linkages (114).

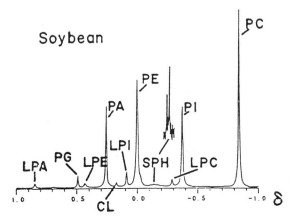

Fig. 6   P-31 NMR spectrum of crude soybean phospholipid preparation. The spectrum was accumulated over 24 hr to demonstrate signal stability. (From Ref. 112.)

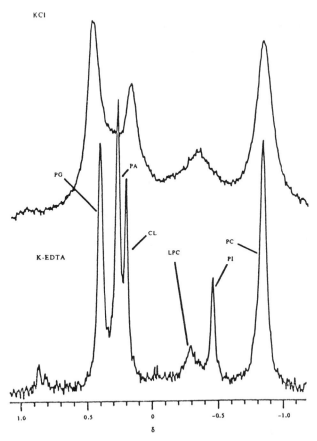

Fig. 7   P-31 NMR spectra of extracted algal phospholipids. The top spectrum was obtained after a simple Folch extraction procedure, using 0.1 M KCl in the backwashing step. The bottom spectrum was obtained using 0.1 M K-EDTA in the backwashing step. (From Ref. 20.)

Comparing P-31 NMR to HPLC, the former technique seemed to detect seven PL more than HPLC (112). In the HPLC profile, the unresolved signals were masked either by the void volume or by neighboring components, or not detected. By using the method of Meneses and Glonek (112), reliable and reproducible mole-fraction phospholipid compositions were obtained in about 15 min. The only drawback is that access to a powerful and sophisticated NMR spectrometer is required for high-resolution spectra or for dilute samples. Pearce et al. (115) slightly adjusted the solvent composition to resolve PS and SPH. The methanolic reagent consisted of 0.2 M Cs-EDTA in methanol and D_2O (7:2); 1 ml of this reagent was added to 3.5 ml of $CDCl_3$.

Pearce and Komoroski (116) observed much more complex NMR spectra when detergent solubilization was applied. Using synthetic PC standards, several empirical relationships could be drawn between acyl structure and P-31 NMR chemical shift. Thus, it was shown that the presence of double bonds in the acyl chains induced small but consistent upfield shifts relative to the corresponding disaturated acyl PC signals. However, species containing doubly unsaturated fatty acyls could not be resolved from singly unsaturated ones. Therefore the difference in chemical shift was about twice as large in sodium cholate as in deoxycholate. Besides, it was shown that resolution of molecular species (DPPC, POPC, and DOPC) was optimized at cholate to phospholipid ratios above 100 in 2% cholate. Besides, the chemical shift values of disaturated PC converge logarithmically to a value of about -0.50 as acyl chain length is increased. Considering a mixture of disaturated PCs, DLPC and DMPC were clearly resolved, whereas the more abundant DPPC, DSPC, and DAPC were hardly resolved. Besides, some other parameters were investigated from which it was deduced that consistent patterns of chemical shifts for diacyl PC molecular species in CHO were found over a wide range of conditions, but accurate reproduction of particular chemical shift values requires control of temperature, CHO/PL molar ratio, and CHO concentration. Considering natural phospholipids, the signal is partly resolved in three species groups defined by the number of acyl substituents (0, 1, or 2) that contain one or several double bonds. Chain length heterogeneity causes a further dispersion, typically an unresolved broadening. This is clearly exemplified by the P-31 NMR–expanded PC signal from amniotic fluid, egg yolk extract, soybean extract, and a mixture of synthetic PC molecular species. Figure 8 reveals that the soybean lecithin is much more unsaturated than egg yolk lecithin; whereas the former primarily contains diunsaturated molecular species (duPC), the latter almost exclusively contains molecular species with one saturated and one unsaturated fatty acyl group (suPC). Keeping account of the fact that differences in molecular species are more than 10 bonds away from the phosphorus nucleus, it seems unlikely that the observed effect is purely chemical, arising from a direct through-bond electronic effect. According to Pearce and Komoroski (116), the differences should be ascribed to the structure of the bile salt–phospholipid mixed micelles or the orientation and/or conformation of the phospholipids in the bile salt micelles.

For the sake of completeness, it should be mentioned that the resolution of groups of molecular species is certainly not a routine analysis. Data have to be accumulated over several hours at 121.65 MHz. However, sensitivity and resolution limitations can partially be overcome by multiple Lorentzian peak fitting. Hence, quantification may be accomplished without baseline resolution. As a further consequence, the sampling time may be reduced and less performant equipment can be used. Thus, Murcia and Villalain (117) described the quantification of the phospholipid composition of canned peas. The lipid extract was dissolved in 2 ml of 50 mg/ml sodium cholate containing 1 mM EDTA, 300 mg/g D_2O, and 50 µM inorganic phosphate as internal standard. Spectra were accumulated at 81 MHz. Although the accumulated NMR spectrum was badly resolved, displaying only two clear peaks besides several shoulders and humps, at least 11 Lorentzian-shaped bands could be discerned on multiple Lorentzian fitting, from which a detailed phospholipid quantification was deduced.

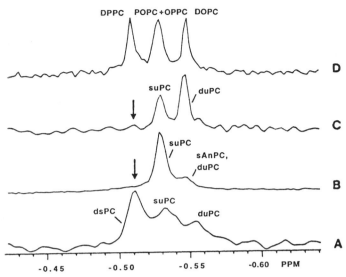

Fig. 8 P-31 NMR–expanded PC signal from (A) amniotic fluid, (B) egg yolk extract, (C) soybean extract, (D) model solution of standards. Spectra were acquired in 2% CHO, pH 8.0, 27°C, except (C) at 37°C. Arrows indicate the position of DPPC in spiked samples. (From Ref. 116.)

V.   APPLICATIONS IN FOOD ANALYSIS

By far the most important application of phospholipid analyses in food industry is the quality control of lecithin samples. As the price of lecithin is largely determined by its PC content (3), phospholipid quantitation is of utmost importance for the lecithin industry. Besides, the functional properties of the lecithins are highly dependent on their composition. Thus, alcohol fractionation of soybean lecithin yielded a PI- and a PC-enriched fraction; whereas the former was a good W/O emulsifier, the latter primarily promoted the formation of O/W emulsions (118). More recently, Kaufmann et al. (119) developed a chemometric strategy for the classification of lecithins according to their process performance. Based on the phospholipid composition and the fatty acid composition of the phospholipid classes, a relevant set of data was selected to predict the emulsifying properties of commercial lecithins. According to Chung et al. (120), the phospholipid content of wheat is related to bread-making (functional) properties. The experimentally determined highly significant correlations point to the potential usefulness of the phospholipid content and the neutral-to-phospholipid ratio for estimating loaf volume potential of hard red winter wheat flours. Hence, phospholipid analyses are essential for the lecithin customers as well.

   As a further consequence, there is an urgent need for a standard method for the determination of the phospholipid composition of lecithins, which would allow a better characterization of lecithin and phospholipid products, so that discussions on analytical data might be avoided in the future. Due to this need for accurate and reproducible data, the lecithin manufacturers founded the International Lecithin and Phospholipid Society (ILPS), whose analytical subgroup aims to develop a standardized methodology to quantify PC, PE, PI, PA, N-acyl PE, and LPC. Besides the choice of an appropriate analytical technique, the need for adequate calibration standards was recognized. According to De Kock (104), the lack of these standards on the market might be the greatest problem when comparing analytical results. Using P-31 NMR spectroscopy as an absolute method, ILPS-approved phospholipid standard mixtures are now available.

   From the above discussion, it follows that phospholipid analysis is especially important during the manufacturing of food products. Nevertheless, some applications exist on the use of phospholipid analyses of endproducts. First of all, Weihrauh and Son (4) mentioned the important dietetic properties of phospholipids. In response to frequent requests for data by researchers in nutrition and medicine, they studied the phospholipid composition of about 140 food products. Besides, phospholipid analyses may be used to detect food adulteration. Thus, Lechner (121) proposed to determine the total phospholipid content of buttermilk to detect adulteration with the (less expensive) cultured skimmed milk. As the former should contain over 100 mg/100 g whereas the latter contains only 15 mg/100 g, the amount of skimmed milk in commercial samples could be detected. Therefore the total amount of phospholipids was determined colorimetrically following ashing. However, Lechner (121) states that this test can easily be circumvented by the addition of soybean lecithin and proposes optimizing the test in order to discern milk phospholipids from those of other sources, which could be done by analysis of the phospholipid classes present. Actually, the latter approach was used by Badings and Koops (122) to detect adulteration of dried milk with dried sweet cream buttermilk. This test is based on the fact that the lipid fraction of dried milk contains about 1–2% of phospholipids, whereas only 0.15–0.25% of phospholipids are present in the lipid fraction of sweet cream buttermilk. As the adulteration may be masked by the addition of vegetable lecithins, the PC-to-SPH ratio was used as an additional criterion; the latter phospholipid only occurs in phospholipids of animal origin.

ABBREVIATIONS

AAS	atomic absorption spectroscopy
AMS	automated multiple development
ANS	8-anilino-1-naphthalene
CHO	cholic acid, sodium salt
DAPC	1,2-diarachidoylphosphatidylcholine
DLPC	1,2-dilauroylphosphatidylcholine
DMPC	1,2-dimyristoylphosphatidylcholine
DOPC	1,2-dioleoylphosphatidylcholine
DPG	diphosphatidylglycerol
DPPC	1,2-dipalmitoylphosphatidylcholine
DSPC	1,2-distearoylphosphatidylcholine
dsPC	1,2-disaturated acylphosphatidylcholine
duPC	1,2-di-unsaturated acylphosphatidylcholine
EDTA	ethylene diamine tetra–acetic acid
ELSD	evaporative light-scattering detector
FFA	free fatty acid
FID	flame ionization detector
GC	gas chromatography
HPLC	high-performance liquid chromatography
LC	liquid chromatography
LPC	lysophosphatidylcholine
LPE	lysophosphatidylethanolamine
MIBK	methylisobutylketone
MS	mass spectrometry
NL	neutral lipids
NMR	nuclear magnetic resonance
NPLs	nonphosphorous lipids
OPPC	1-oleoyl 2-palmitoylphosphatidylcholine

PA	phosphatidic acid
PC	phosphatidylcholine
PE	phosphatidylethanolamine
PG	phosphatidylglycerol
PI	phosphatidylinositol
PL	phospholipids
POPC	1-palmitoyl 2-oleoylphosphatidylcholine
PS	phosphatidylserine
RI	refractive index
sn	stereospecific numbering
SPH	sphingomyelin
suPC	1-saturated acyl 2-unsaturated acylphosphatidylcholine
TFA	trifluoroacetic acid
THF	tetrahydrofuran
TLC	thin-layer chromatography
TMP	trimethyl phosphate

REVIEWS

W. W. Christie, *Z. Lebensm. Unters. Forsch., 181*:171 (1985).

W. W. Christie, *HPLC and Lipids: A Practical Guide*, Pergamon Press, Oxford, 1987.

H. Pardun, *Die Pflanzenlecithine: Gewinnung, Eigenschaften, Verarbeitung und Anwendung pflanzlicher Phosphatidpräparate*, Verlag für chemische Industrie H. Ziolkowsky KG, Augsburg, 1988.

B. F. Szuhaj, *Lecithins: Sources, Manufacture and Uses*, American Oil Chemists' Society, Champaign, IL, 1989.

H. Pardun, *Fat Sci. Technol., 91*:45 (1989).

REFERENCES

1. H. Pardun, *Die Pflanzenlecithine: Gewinnung, Eigenschaften, Verarbeitung und Anwendung pflanzlicher Phosphatidpräparate*, Verlag für chemische Industrie H. Ziolkowsky KG, Augsburg, 1988.

2. B. F. Szuhaj, *Lecithins: Sources, Manufacture and Uses*, American Oil Chemists' Society, Champaign, Illinois, 1989.

3. H. Pardun, *Fat Sci. Technol., 91*: 45 (1989).

4. J. L. Weihrauh and Y. S. Son, *J. Am. Oil Chem. Soc. 60*: 1971 (1983).

5. A. Kanazawa, S. Teshima, and M. Sakomoto, *Z. Angew. Ichtyol., 4*: 165 (1985).

6. I. Hanin and G. Pepeu, *Phospholipids: Biochemical, Pharmaceutical and Analytical Considerations*, Plenum Press, New York, 1990.

7. W. Feldheim, *Phospholipids: Biochemical, Pharmaceutical and Analytical Considerations* (I. Hanin and G. Pepeu, eds.), Plenum Press, New York, 1990.

8. S. Amaducci, *Eurofood Monitor: European Community legislation on foodstuffs*, Agra Europe (London) Ltd, Tunbridge Wells, Kent, England, 1994.

9. D. Firestone, *Official Methods and Recommended Practices of the American Oil Chemists' Society*, 4th ed., American Oil Chemists' Society, Champaign, Illinois, 1990.

10. German Society for Fat Science, *German Standard Methods for the Analysis of Fats and Other Lipids*, Wissenschaftliche Verlagsgesellschaft mbH, Stuttgart, 1994.

11. H. Pardun, *Fette Seifen Anstrichm, 86*: 55 (1984).

12. R. Aneja, J. S. Chadcha, and R. W. Yoell, *Fette Seifen Anstrichm, 73*: 643 (1971).

13. L. Rydhag, *Fette Seifen Anstrichm, 81*: 168 (1979).

14. L. Rydhag and I. Wilton, *J. Am. Oil Chem. Soc., 58*: 830 (1981).

15. M. R. Sahasrabudhe and B. W. Smallbone, *J. Am. Oil Chem. Soc., 60*: 801 (1983).

16. W. R. Morrison, *Advances in ceral science and technology, Vol. 2* (Y. Pomeranz, ed.), American Association of Cereal Chemists, St. Paul, 1978, p. 221.
17. J. Folch, M. Lees, and G. H. Sloane-Stanley, *J. Biol. Chem., 226*: 497 (1957).
18. E. G. Bligh and W. J. Dyer, *Can. J. Biochem. Physiol., 37*: 911 (1959).
19. B. W. Nichols, *Biochim. Biophys. Acta, 70*: 417 (1963).
20. P. Meneses, J. N. Navarro, and T. Glonek, *Int. J. Biochem., 25*: 903 (1993).
21. A. Hara and N. S. Radin, *Anal. Biochem., 90*: 420 (1978).
22. I. S. Chen, C. S. J. Shen, and A. J. Sheppard, *J. Am. Oil Chem. Soc., 58*: 599 (1981).
23. M. L. Vorbeck and G. V. Marinetti, *J. Lipid Res., 6*: 3 (1965).
24. G. Rouser, G. Kritchevsky, G. Simon, and G. J. Nelson, *Lipids, 2*: 37 (1967).
25. A. Pollio, M. Della Greca, P. Monaco, G. Pinto, and L. Previtera, *Biochim. Biophys. Acta, 963*: 53 (1988).
26. A. K. Sen Gupta, *Fette Seifen Anstrichm., 78*: 111 (1976).
27. W. W. Christie, *Lipid Analysis—Isolation, Separation, Identification and Structural Analysis of Lipids*, 2nd ed., Pergamon Press, Oxford, 1982.
28. A. V. Zhukov and A. G. Vereshchagin, *Adv. Lipid Res., 18*: 247 (1981).
29. W. W. Christie, *HPLC and Lipids: A Practical Guide*, Pergamon Press, Oxford, 1987, p. 81.
30. W. J. Hurst and R. A. Martin, *J. Am. Oil Chem. Soc., 57*: 307 (1980).
31. J. Bitman, D. L. Wood, N. R. Mehta, P. Hamosh, and M. Hamosh, *Am. J. Clin. Nutr., 40*: 1103 (1984).
32. P. Juaneda and G. Rocquelin, *Lipids, 20*: 40 (1985).
33. J. K. Yao and G. M. Rastetter, *Anal. Biochem., 150*: 111 (1985).
34. M. A. Kaluzny, L. A. Duncan, M. V. Merritt, and D. E. Epps, *J. Lipid Res., 26*: 135 (1985).
35. W. W. Christie, R. C. Noble, and G. Davies, *J. Soc. Dairy Technol., 40*: 10 (1987).
36. P. Van der Meeren, J. Vanderdeelen, M. Huys, and L. Baert, *Phospholipids: Biochemical, Pharmaceutical and Analytical Considerations* (I. Hanin and G. Pepeu, eds.), Plenum Press, New York, 1990, p. 273.
37. A. M. Nash and E. N. Frankel, *J. Am. Oil Chem. Soc., 63*: 244 (1986).
38. S. L. Melton, *J. Am. Oil Chem. Soc., 69*: 784 (1992).
39. J. A. Prieto, A. Ebri, and C. Collar, *J. Am. Oil Chem. Soc., 69*: 387 (1992).
40. J. K. Daun, L. D. Davidson, J. A. Blake, and Wo Yuen, *J. Am. Oil Chem. Soc., 58*: 914 (1981).
41. G. A. Rouser, N. Siakotos, and S. Fleischer, *Lipids, 1*: 85 (1966).
42. A. Prevot and M. Gente, *Rev. Fr. Corps Gras, 24*: 491 (1977).
43. S. H. Goh, H. T. Khor, and P. T. Gee, *J. Am. Oil Chem. Soc., 59*: 296 (1982).
44. S. H. Goh, S. L. Tong, and P. T. Gee, *J. Am. Oil Chem. Soc., 61*: 1597 (1984).
45. V. A. Eryomin and S. P. Poznyakov, *Anal. Biochem., 180*: 186 (1989).
46. A. M. Siouffi, E. Mincsovics, and E. Tyihak, *J. Chromatogr., 492*: 471 (1989).
47. T. Spillman, D. B. Cotton, S. C. Lynn, and J. P. Bretaudiere, *Clin. Chem., 29*: 250 (1983).
48. J. C. Dittmer and R. L. Lester, *J. Lipid Res., 5*: 126 (1964).
49. E. K. Ryu and M. MacCross, *J. Lipid Res, 20*: 561 (1979).
50. L. Gustavsson, *J. Chromatogr., 375*: 255 (1986).
51. M. Kates, *Techniques of Lipidology: Extraction, Analysis and Identification of Lipids*, North-Holland, Amsterdam, 1972.
52. G. V. Marinetti, *New Biochemical Separations* (A. T. James and L. J. Morris, eds.), Van Nostrand, London, 1964, p. 349.
53. A. Dieffenbacher and U. Bracco, *J. Am. Oil Chem. Soc., 55*: 642 (1978).
54. W. L. Erdahl, A. Stolyhwo, and O. S. Privett, *J. Am. Oil Chem. Soc., 50*: 513 (1973).
55. G. Lendrath, A. Nasner, and Lj. Kraus, *J. Chromatogr., 502*: 385 (1990).
56. A. Smiles, Y. Kakuda, and B. E. MacDonald, *J. Am. Oil Chem. Soc., 66*: 348 (1989).
57. V. E. Vaskovsky, E. Y. Kostetsky, and I. M. Vasendin, *J. Chromatogr., 114*: 129 (1975).
58. J. Pick and L. Kovacs, *J. Planar Chromatogr., 4*: 91 (1991).
59. K. Burger, *Pflanzenschutz Nachrichten Bayer, 41*: 175 (1988).
60. H. Jork, *International Laboratory, 23*(6): 18 (1993).
61. U. de la Vigne and D. E. Jänchen, *International Laboratory, 21*(10): 22 (1991).
62. T. Terabayashi, T. Ogawa, Y. Kawanishi, and J. Ishii, *J. Chromatogr., 454*: 432 (1988).
63. M. Ranny, *Thin-layer Chromatography with Flame Ionisation Detection*, Reidel, Dordrecht, 1987.

64. R. G. Ackman, C. A. McLeod, and A. K. Banerjee, *J. Planar Chromatogr., 3*: 450 (1990).
65. J. F. Whitsett and J. M. Kennish, *J. Chromatogr., 435*: 343 (1988).
66. J. K. G. Kramer, R. C. Fouchard, and E. R. Farnworth, *Lipids, 20*: 617 (1985).
67. R. Przybylski and N. A. M. Eskin, *J. Am. Oil Chem. Soc., 68*: 241 (1991).
68. L. M. Du Plessis and H. E. Pretorius, *J. Am. Oil Chem. Soc., 60*: 1261 (1983).
69. T. Ohshima and R. G. Ackman, *J. Planar Chromatogr., 4*: 27 (1991).
70. T. N. B. Kaimal and N. C. Shantha, *J. Chromatogr., 228*: 177 (1984).
71. M. Ranny, J. Sedlacek, and C. Michalec, *J. Planar Chromatogr., 4*: 15 (1991).
72. N. C. Shanta and R. G. Ackman, *Lipids, 25*: 570 (1990).
73. N. U. Olsson, *J. Chromatogr., 624*: 11 (1992).
74. S. M. Innis and M. T. Clandinin, *J. Chromatogr., 205*: 490 (1981).
75. W. M. N. Ratnayake and R. G. Ackman, *Canad. Inst. Food Science Technol. J., 18*: 284 (1985).
76. W. W. Christie, *Z. Lebensm. Unters. Forsch., 181*: 171 (1985).
77. P. Van der Meeren, J. Vanderdeelen, and L. Baert, *Food Analysis by HPLC* (L. M. L. Nollet, ed.), Marcel Dekker, New York, 1992, p. 241.
78. F. B. Jungalwala, J. E. Evans, and R. H. McCluer, *Biochem. J., 155*: 55 (1976).
79. A. Stolyhwo and O. S. Privett, *J. Chromatogr. Sci., 11*: 20 (1973).
80. J. S. Rhee and M. G. Shin, *J. Am. Oil Chem. Soc., 59*: 98 (1982).
81. W. W. Christie, *J. Lipid Res., 26*: 507 (1985).
82. W. W. Christie, *J. Chromatogr., 361*: 396 (1986).
83. L. Breton, B. Serkiz, J. P. Volland, and J. Lepagnol, *J. Chromatogr., 497*: 243 (1989).
84. J. Scarim, H. Ghanbari, V. Taylor, and G. Menon, *J. Lipid Res., 30*: 607 (1989).
85. P. Juaneda, G. Rocquelin, and P. O. Astorg, *Proceedings of Eurolipid 1, Vol. 1*, Association Francaise pour l'étude des corps gras, 1989, p. 319.
86. A. Leseigneur, G. Gandemer, and D. Marion, *Proceedings of Eurolipid 1, Vol. 1*, Association Francaise pour l'étude des corps gras, 1989, p. 311.
87. M. D. Grieser and J. N. Geske, *J. Am. Oil Chem. Soc., 66*: 1484 (1989).
88. R. H. McCluer, M. D. Ullman, and F. B. Jungalwala, *Meth. Enzymol., 172*: 538 (1989).
89. T. L. Mounts, S. L. Abidi, and K. A. Rennick, *J. Am. Oil Chem. Soc., 69*: 438 (1992).
90. K. C. Arnoldsson and P. Kaufmann, *Chromatographia, 38*: 317 (1994).
91. K. C. Arnoldsson and U. Olsson, *Phospholipids: Characterization, Metabolism and Novel Biological Applications* (G. Cevc and F. Paltauf, eds.), in press.
92. W. W. Christie, *Advances in Lipid Methodology, Vol. 1* (W. W. Christie, ed.), Oily Press, Ayr, Scotland, 1992, p. 239.
93. T. H. Mourey and L. E. Oppenheimer, *Anal. Chem., 56*: 2427 (1984).
94. P. Van der Meeren, J. Vanderdeelen, and L. Baert, *Anal. Chem., 64*: 1056 (1992).
95. P. Van der Meeren, J. Vanderdeelen, G. Huyghebaert, and L. Baert, *Chromatographia, 34*: 557 (1992).
96. J. L. Robinson, M. Tsimidou, and R. Macrae, *J. Chromatogr., 324*: 35 (1985).
97. J. M. Charlesworth, *Anal. Chem., 50*: 1414 (1978).
98. A. Stolyhwo, H. Colin, M. Martin, and G. Guiochon, *J. Chromatogr., 288*: 253 (1984).
99. M. Righezza, G. Guiochon, *J. Liq. Chromatogr., 11*: 1967 (1988).
100. P. B. Stockwell and B. W. King, *Int. Chrom. Lab., 7*(10): 4 (1991).
101. W. W. Christie, M. L. Hunter, and R. C. Noble, *Biochem. Soc. Trans., 13*: 1221 (1985).
102. W. Kuhnz, B. Zimmermann, and H. Nau, *J. Chromatogr., 344*: 309 (1985).
103. B. Hersloef, U. Olsson, and P. Tingvall, *Phospholipids: Biochemical, Pharmaceutical and Analytical Considerations* (I. Hanin and G. Pepeu, eds.), Plenum Press, New York, 1990, p. 295.
104. J. De Kock, *Fat Sci. Technol., 95*: 352 (1993).
105. C. P. Blom, F. A. Deierkauf, and J. C. Riemersma, *J. Chromatogr., 171*: 331 (1979).
106. J. K. Kaitaranta, P. J. Geiger, and S. P. Bessman, *J. Chromatogr., 206*: 327 (1981).
107. J. K. Kaitaranta and S. P. Bessman, *Anal. Chem., 53*: 1232 (1981).
108. A. Islam, M. Smogorzewski, T. O. Pitts, and S. G. Massry, *Mineral Electrolyte Metab., 15*: 209 (1989).
109. J. Terao, M. Kawanishi, and S. Matsushita, *J. Agric. Food Chem., 35*: 613 (1987).
110. E. London and G. W. Feigenson, *J. Lipid Res., 20*: 408 (1979).
111. N. Sotirhos, B. Hersloef, and L. Kenne, *J. Lipid Res., 27*: 386 (1985).
112. P. Meneses and T. Glonek, *J. Lipid Res., 29*: 679 (1988).

113. J. M. Pearce, J. T. Krone, A. A. Pappas, and R. A. Komoroski, *Magn. Reson. Med., 30*: 476 (1993).

114. H. T. Edzes, T. Teerlink, M. S. van der Knaap, and J. Valk, *Magn. Reson. Med., 26*: 46 (1992).

115. J. M. Pearce, M. A. Shifman, A. A. Pappas, and R. A. Komoroski, *Magn. Reson. Med., 21*: 107 (1991).

116. J. M. Pearce and R. A. Komoroski, *Magn. Reson. Med., 29*: 724 (1993).

117. M. A. Murcia and J. Villalain, *J. Sci. Food Agric., 61*: 345 (1993).

118. *Kirk-Othmer Encyclopedia of Chemical Technology, 2nd ed., vol. 12*, John Wiley and Sons, New York, 1967, p. 353.

119. P. Kaufmann, U. Olsson, and B. G. Hersloef, *J. Am. Chem. Soc., 67*: 537 (1990).

120. O. K. Chung, Y. Pomeranz, and K. F. Finney, *Cereal Chem., 59*: 14 (1982).

121. E. Lechner, *Deutsche Milchwirtschaft, 33*: 1201 (1984).

122. H. T. Badings and J. Koops, *Neth. Milk Dairy J., 27*: 66 (1973).

15

Carbohydrates

Miguel Peris-Tortajada
Polytechnic University of Valencia, Valencia, Spain

I. INTRODUCTION

The analysis of food carbohydrates has been dealt with and reviewed by several authors, either in the form of books or chapters of books or as scientific papers in different journals. See the list of reviews.

A. Definition

Carbohydrates are one of the major classes of food components along with lipids and proteins. Although they are theoretically a well-defined group of compounds, some confusion often exists as to what constitutes a carbohydrate and what does not. Some definitions are as follows: (a) primary oxidation products of polyvalent alcohols, namely polyhydroxyaldehydes or polyhydroxyketones; (b) compounds containing carbon, hydrogen, and oxygen only and in which hydrogen and oxygen are in the same proportions as in water; and (c) compounds which contain carbon, hydrogen, and oxygen with the empirical formula $(CH_2O)_n$, n being 5 or 6, but including some compounds deficient in oxygen and others containing nitrogen and sulfur. Table 1 summarizes several different classifications of carbohydrates for the purpose of food analysis.

B. Characteristics and Properties

All foods, with few exceptions, contain carbohydrates, which may vary in form from a simple monosaccharide to a more complex polysaccharide. Table 2 summarizes the most important food carbohydrates along with their characteristics. It also briefly describes where they are usually found. Carbohydrates (sugars) are a major source of energy for the human body and their varied functional properties are utilized by the food industry to enhance, for instance, the palatability, acceptability and shelf life of foodstuffs. There is thus a continuing need to monitor levels of carbohydrates in foods in order to predict and control the properties of the food as well as the interactions of the components.

Table 1   Different Classifications of Food Carbohydrates

Group	Contributors
Total soluble carbohydrates	Glucose, fructose, galactose, lactose, maltose, sucrose
Total carbohydrates	Those mentioned above plus starch, gums, pectins, cellulose, and hemicellulose
Reducing sugars	Glucose, fructose, galactose, lactose, maltose
Available carbohydrates	All except fiber-type compounds
Aldehyde or ketone	
Pentose, hexose, heptose	

The analysis of carbohydrates frequently involves not only the determination of the total amount of sugar present in the sample but often the identity, configuration and conformation of the carbohydrate components as well. Therefore, since the interest of the food analyst in the analysis of carbohydrates is principally in their identification—foods usually contain a mixture of them and often their derivatives—and quantitative evaluation, some knowledge of the type of carbohydrate is required. Various excellent reviews have been written on the subject and the reader is directed to these for more detailed information (1,2). In any case, some simple qualitative tests may also be carried out.

II. SAMPLE PREPARATION

Regardless of the food product—and prior to the analysis—the sugars in the food must be dissolved in water and separation or fractionation of food components may be necessary to remove interfering substances. This involves the use of clarifying (clearing) agents, whose function is to eliminate colored compounds (which might interfere with the passage of light through the solution), all optically active nonsugar substances (tannins, glycosides, and amino acids), and other potentially interfering constituents. These clarifying agents should affect neither the type nor the composition of sugars in the final solution. When using reduction methods, it is also essential to remove colloidal materials (proteins), which would prevent the proper growth of Cu_2O precipitate, and obviously, nonsugar, copper-reducing materials.

Rapid filtration, minimum error, and efficiency of clarification are the main factors to be taken into account in the selection of clearing agents. In any case, the minimal reagents should be used, their amounts being governed by the intensity of color and the quantity of organic materials. Table 3 summarizes the clarifying agents most frequently used in carbohydrates analysis.

Finally, when starch is to be analyzed, the food is first extracted with 80% hot ethanol to eliminate low molecular weight sugars, followed by hot ethanolic KOH to remove interfering proteins and fats. As a last step, the food is reextracted with 80% hot ethanol and then heated to gelatinize the starch. The resulting glucose is then determined as described later.

III. METHODS OF ANALYSIS

Table 4 summarizes the most common methods for carbohydrate determination in foods.

A. Traditional Methods

Classically, carbohydrates in foods were determined "by difference" after ascertaining the content of other food components. However, this has become a redundant practice because the food

Table 2 Most Important Food Carbohydrates

Name	Structure	Redox characteristics	Remarks
Glucose	Monosaccharide	Reducing	Naturally very widespread in foods; also present in glucose syrups and invert sugars
Fructose	Monosaccharide	Reducing	Found in most foods in small amounts; higher concentrations in honey and fruit
Galactose	Monosaccharide	Reducing	Scarcely found in the free state in foods;
Sorbitol	Monosaccharide	Nonreducing	Reduction product of glucose; used mainly in foods for diabetics
Sucrose	Disaccharide (glucose + fructose)	Non-reducing	Widely used in foods
Lactose	Disaccharide (glucose + galactose)	Reducing	Found naturally only in milk and dairy products; used as a dusting powder for some baked goods
Maltose	Disaccharide (glucose + glucose)	Reducing	Found mainly in glucose syrups and malt extract
Raffinose	Trisaccharide (glucose + fructose + galactose)	Nonreducing	Found naturally
Starch	Polysaccharide (n times glucose)	Nonreducing	Used in foods as a filler and thickening agent; naturally present in many foods

Table 3 Clarifying Agents Most Commonly Used in Sugar Analysis

Name	Major applications	Remarks
Alumina cream	Fructose-rich products, e.g., honey	Not very efficient for highly colored solutions; recommended for colloidal suspensions
Lead acetate	Plant extracts, syrups, fruit and vegetable products, nonalcoholic beverages	Neutral form used when alumina cream fails to clarify sufficiently; basic form unsuitable for solutions of reducing sugars because of precipitation
Zinc ferrocyanide $[ZnAc_2 + K_3Fe(Cn)_6]$	In nearly all products	Good general clearing agent
Activated charcoal	In nearly all products	Tends to adsorb sugars; useful for qualitative analysis
Ion exchange resin	White potatoes, fresh leafy plant materials	Used to remove potential acidic or basic interfering substances
Phosphotungstic acid	Milk, cheese extracts	If large amounts of reducing substances are present, dilution may be required

Table 4 Major Methods of Carbohydrate Analysis

Method	Application field
Physical	
Refractometry	Total soluble carbohydrates
Polarimetry	All soluble carbohydrates
Hydrometry	Total soluble carbohydrates
Chemical	
Conventional	
Nelson–Somogyi	Reducing sugars
Picric acid	Reducing sugars, sucrose
Dinitrosalicylate	Reducing sugars
Copper reduction	Reducing sugars, sucrose
Phenol-sulfuric	All soluble carbohydrates
Anthrone	Free hexoses
Clegg–anthrone	Hexoses
Neocuproine	Reducing sugars
Ferricyanide	Reducing sugars
Triphenyltetrazolium	Reducing sugars
Iodometric	Glucose, lactose, galactose, and maltose
Automatic	
FIA	A great deal of carbohydrates
Physicochemical	
Chromatography	All soluble carbohydrates
Biochemical	
Enzymes	Most carbohydrates

industry demands more and more information on these versatile substances. Obviously, the accuracy of the by-difference method depends on the determination of the other food components and does not make any distinction between available and nonavailable carbohydrates. The complete analysis of foodstuffs may require the determination of several groups of compounds, e.g., simple sugars, reducing sugars, polysaccharides, and fiber, all of which may play an important role in the quality of the product. Developed to meet these demands, physical and chemical methods for carbohydrate analysis are well established and are a common practice in many laboratories.

1. Physical Methods of Analysis

Physical methods generally determine some overall feature of the sugars in the food, such as total carbohydrates. In addition to being used in routine analysis, they may be utilized for the on-line monitoring of certain manufacturing processes, since these techniques are usually nondestructive. Nevertheless, the results obtained may be affected by other compounds in the food contributing to the particular measurement. For example, proteins occasionally affect reducing sugar determinations or amino acids may contribute to optical rotation readings.

a. Refractometry. The ratio of the speed of the light in a vacuum to its speed in a substance is referred to as the *refractive index* of the substance. This speed is related to the temperature, composition, concentration, and purity of the substance. The index of refraction depends on the density of a liquid and thus on the concentration of solute in solution (3). Therefore the refractive index can be used to determine the concentration of carbohydrate solutions and the corresponding instruments (refractometers) may be calibrated with the refractive index or directly with sugar concentration scales (4). For example, they are often calibrated in °Brix, a parameter which is numerically equivalent to % sucrose w/w; in this case these instruments are utilized only with sucrose solutions. When only refractive index scales are available, tables are required to obtain carbohydrate concentration. Modern refractometers are electronically controlled and (theoretically) free from operator error, unlike older instruments which undoubtedly required certain dexterity for their use. Measurements are usually made at 20°C using the sodium D line as light source. Sucrose is most frequently used as the calibration medium for the instruments and the readings expressed as % sugar w/w "as sucrose" in the food. Obviously, such instruments are only accurate for pure sucrose solutions, although they are also widely used for foods containing other sugars such as glucose syrups or invert sugar, since the reading obtained may be corrected if required; usually the reading is used as a reference value to the sugar content (total soluble solids) of the solution. Corrections should also be applied if solutes other than carbohydrates occur, provided that their contribution to the reading is significant. Finally, as mentioned above, the refractive index is heavily influenced by changes of temperature and wavelength of the light source; therefore control is necessary. Nevertheless this drawback has been overcome by modern instruments because their reading is temperature-compensated.

b. Polarimetry. Compounds with a chiral center (e.g., an asymmetrical carbon atom) have the ability to rotate the plane of polarization of polarized light and the optical rotation is the angle through which the light is rotated when the light passes through a solution of the optically active compound. This is usually measured at 20°C using light from a sodium lamp ($\lambda = 589$ nm corresponding to the sodium D line). All sugars are optically active to a greater or lesser extent depending on structural considerations (5) and angular rotation is dependent on the concentration of the solution, length of light path, wavelength of light, and temperature. This rotation may be positive (+) or negative (-). Detailed descriptions of the instruments used (polarimeters) are given in the literature.

Specific rotation $[\alpha]_D$ is defined as follows:

$$[\alpha]\frac{T}{D} = \frac{100\alpha}{lc}$$

where T = temperature (°C)
 D = light from sodium D line (λ = 589 nm)
 α = observed angular rotation
 l = light pathlength in decimeters (dm)
 c = concentration of carbohydrate (g/100 ml or % w/v)

Specific rotation depends on the wavelength of the light and increases with decreasing wavelength. Different operators may obtain different results using the same polarimeter. This operator error can be eliminated through practice or by using more sophisticated instruments in which the optical rotation is shown on a digital display rather than depending on the observer first matching two areas for shade and then obtaining a reading from a scale observed down a telescope, with its related difficulties.

On the other hand, although specific rotation is theoretically a constant for a particular carbohydrate, it varies slightly for some sugars depending on the concentration used for the measurement. Glucose and sucrose have been taken as examples in Table 5 (6).

Temperature and the presence of inorganic species are two factors to be taken into account since they may also influence optical and specific rotations. The presence of inorganic species usually presents less of a problem. All foods contain metal ions and these will affect the optical rotation of the sugar solution to a greater or lesser extent depending on the species, although the effect is predictably greater at higher salt concentrations, which are not usually found in foods.

Temperature may cause more serious inaccuracies in measurements and must therefore be controlled. For example, with an invert sugar solution (glucose + fructose) using sodium light, specific rotation is 0 at 90°C and -20 at 20°C.

Attention must also be paid to the phenomenon called "mutarotation" (7), which takes place when freshly prepared solutions of some sugars change their optical rotation over a period of time until a constant value is reached. It is caused by the existence of α and β forms of the carbohydrate which constitute an equilibrium when one isomer is dissolved in water. Ammonia catalyzes this reaction and so does the heating to boiling of the solution, although mutarotation is essentially complete after 16–18 hr. Additionally, when open chain and ring structures attain equilibrium involving both pyranose and furanose forms, changes in optical rotation are also observed. Table 6 shows some examples of applications of polarimetry in carbohydrates analysis. Experimental details of the methods described above, details of polarimeters, and variations on the general principles of polarimetry such as saccharimetry may be found in the literature (8).

Table 5 Changes in Specific Rotation with Concentration

Carbohydrate	Specific rotation and concentration (g/ml)		
	0.01	0.1	0.2
Glucose[a]	+52.383	+52.711	+53.108
Sucrose	+66.464	+66.519	+66.542

Table 6 Several Cases of Practical Applications of Polarimetry in Sugar Analysis

Determination	Calculations	Remarks
Single carbohydrate in solution	$c_x(\%\ \text{w/v}) = \dfrac{\alpha_{\text{sol}} \cdot 100}{\alpha_x \cdot M} \times 100$	For example, fructose, glucose, sucrose (M = grams of solid sample in 100 ml of solution)
Sucrose in the presence of other carbohydrates	$c_s(\%\ \text{w/v}) = \dfrac{\alpha_{\text{b.i.}} - \alpha_{\text{a.i.}}}{Q}$	Clerget–Herzfeld method. Optical rotation of the solution is measured before and after inversion (enzymatic or acidic hydrolysis)
Sucrose and one other carbohydrate	$c_x(\%\ \text{w/v}) = \dfrac{100\alpha_{\text{sol}} - \alpha_s c_s}{\alpha_x}$	Concentration of sucrose previously determined by the Clerget–Herzfeld method. Then, $\alpha_{\text{sol}} = \alpha_s + \alpha_x$

c_x = concentration of the carbohydrate; α_{sol} = optical rotation of the solution; α_x = specific rotation of the carbohydrate; c_s = concentration of sucrose; $\alpha_{\text{b.i.}}$ = specific rotation before inversion; $\alpha_{\text{a.i.}}$ = specific rotation after inversion; Q = inversion divisor factor (depends on the clarifying agents used and the method of inversion); α_s = specific rotation of sucrose. Pathlength = 1 dm assumed in all instances.

    c. Hydrometry. Hydrometry is seldom used for the greatest accuracy. Rather, its usefulness depends on the case with which determinations of reasonable precision can be obtained. This method is then applied to the analysis of carbohydrate solutions (not to foods as such) provided that the precision of a refractometer is not required (9). Hydrometers are utilized to measure densities of liquids and are commercially available in a wide variety of scales and ranges, although the range of reading is generally small. For this reason, a set of hydrometers is always at the operator's disposal in every laboratory. They are often calibrated in wt % sucrose and the readings of these hydrometers are referred to as °Brix. Hydrometers graduated in degrees Baumé are also used, particularly when molasses or syrups containing both sucrose and glucose are concerned. Additionally, other applications may require other readings.

    Readings may be affected by several important factors such as temperature, surface tension, and presence of surface films. Temperature affects density and hence the readings; therefore a careful control of this variable is required for accurate results. Tables of corrections to be applied for temperatures other than 20°C are available. In general, tables showing the relation between density and % sucrose apply only to pure sugar solutions. However, much of the material that naturally occurs with sucrose has been found to have a similar density and these tables are used with relatively small error when working with mixtures of carbohydrate solutions. Table 7 shows a comparison of densities of 10% w/w solutions of several common sugars (10).

    Effects of surface tension result from the downward force on the stem exerted by the meniscus.

Table 7 Comparison of Densities (20/4°C) of Different Carbohydrate Solutions (10% w/w)

Carbohydrate	Density
Glucose	1.0377
Fructose	1.0385
Galactose	1.0379
Sucrose	1.0381
Maltose	1.0386
Lactose	1.0376
Raffinose	1.0375

The liquid adhering above the general liquid level has the same effect as adding to the mass of the hydrometer. Accordingly, two liquids having the same density but different surface tensions will give different readings. Hence it is necessary to specify the liquid for which a hydrometer is intended. Spontaneous formation of surface films may also cause problems. In such cases, frequent stirring or an overflow arrangement to renew the surface is required for precise readings.

2. Chemical Methods of Analysis

Unlike physical procedures, chemical methods are able to determine more specific features, e.g., reducing groups. Chemical analyses of foods in the past were frequently concerned with single-sample analysis but more recently automated methods have been developed to meet the requirements for multiple food analyses. However, chemical methods also have certain drawbacks insofar as they are destructive and time consuming. They may also suffer from additional problems such as involvement of other biological components with related structures.

Both quantitative and qualitative analyses may be required for carbohydrates in foods. Although the former are usually of greater importance, the qualitative determination of carbohydrates in foods can provide valuable information on possible interactions of occurring compounds; e.g., the presence of glucose may be indicative of browning problems. Nevertheless, qualitative methods of analysis are widely described in most organic chemistry books (11,12) and will therefore be omitted from this chapter.

Carbohydrates usually have characteristic structural features which can be utilized as a basis for identification or chemical analysis. Free ketone or aldehyde groups have reducing properties and are able to reduce alkaline solutions of metal (principally copper) salts to the free metal or to the oxide. Alternatively, a specific feature in the structure may exist, as with the amylose portion of starch and its helical structure which, with I_2, produces a blue color. Lack of specific groups can similarly provide some information; for example, lack of reducing power suggests the presence of nonreducing sugars—perhaps sucrose. A wide selection of chemical methods for carbohydrate analysis is currently available, although only those suitable for food will be described. These are fewer in number and are usually restricted to the determination of reducing sugars, sucrose, and starch. In general, two kinds of methods may be considered: those based on reducing properties (classical methods) and those involving colorimetric reactions.

a. Classical Methods. What we know as "classical" methods are related to the determination of reducing sugars. The Lane and Eynon titration method has customarily been used for this purpose (13). Equally to other methods, it involves reaction of reducing sugars with an alkaline Cu^{2+} complex solution. Unfortunately, it is not within limitations and a great deal of biological compounds interfere, although many of the drawbacks were eliminated through some modifications carried out by Khan (14) after a careful study of this method. The main reaction is based on the reduction of the cupric salt to cuprous oxide by the weak carbonyl group. Alternatively, other methods involved other metallic cations such as silver or mercury. By far the most widely used alkaline copper salt solution is Fehling's solution (or, more precisely, Soxhlet's modification), which actually consists of a mixture of two reagents—$CuSO_4$ solution (Fehling A) and NaOH and sodium potassium tartrate solutions (Fehling B)— in equal proportions. When the mixed Fehling's solution is boiled with a reducing sugar, the blue color (cupric tartrate) is discharged and a red precipitate forms (Cu_2O). According to the Lane and Eynon titration, the titration of the Fehling's solution with the sample containing the reducing sugar takes place in a two-stage process. First the bulk of the carbohydrate solution is added and then the remainder (usually less than 1 ml) is added dropwise. Due to the fact that the endpoint is not easily observed, an internal redox indicator is commonly used toward the end of the titration, with methylene blue being the usual choice. By reference to standard tables the concentration of various reducing sugars can be determined. Different tables are necessary for each reducing sugar since each one of these carbohydrates is degraded differently under the alkaline conditions of the reaction. Moreover, the official methodology clearly states that the different reducing sugars differ in their reducing power. Nevertheless some years ago the method was modified by the Laboratory of the Government Chemist (15), London, to avoid the use of tables.

The Lane and Eynon titrimetric procedure is the cornerstone of an empirical method developed to provide quantitative results and requires standardized experimental conditions according to which a volume of carbohydrate solution is titrated against 10 or 25 ml of Fehling's solution. Important variables such as concentration of reactants, heating time, and temperature affect the results and must therefore be carefully controlled. Additionally, sugar solutions must be near-neutral and neutralization is necessary after acid inversion; on the other hand, no other potentially interfering agents (i.e., reducing substances) should be present.

Although the Lane and Eynon titration only determines reducing sugars, a simple modification enables it to measure sucrose. After the initial determination of reducing sugars a sample of the sugar solution is hydrolyzed using either hydrochloric acid or invertase. This converts sucrose to a mixture of glucose and fructose (invert sugar), and by remeasuring reducing sugars the percentage of sucrose can be calculated. Alternatively, the cuprous oxide may be filtered off after the reaction is complete and further treated as in a conventional gravimetric procedure.

The basic method has also been slightly changed by making use of alkaline copper citrate rather than tartrate, with sodium carbonate providing the alkaline environment. It is then known as the Luff–Schoorl method. The modification is as follows: An excess of the reagent is added to the carbohydrate solution and, after the reduction reaction is complete, the excess copper citrate is reacted with excess KI and the liberated iodine is titrated with $Na_2S_2O_3$ using starch as indicator near the endpoint. By subtraction of a blank, the amount of copper solution reacted with the sugar solution can be obtained. This procedure has the advantage that fructose, glucose, and invert sugar are calculated from identical tables.

In the Munson and Walker method (16), a known volume of carbohydrate solution is heated under standard conditions with excess Fehling's solution. The Cu_2O precipitate that is filtered off may be washed, dried, and weighed giving percentage sugar by gravimetric calculation or, alternatively, titrated against either $Na_2S_2O_3$ or $KMnO_4$ after being dissolved in an appropriate solvent. In the first case the oxide is dissolved in HNO_3 where it is oxidized to cupric nitrate. Excess potassium iodide is added and the equivalent amount of iodide undergoes oxidation to iodine, which is then titrated with $Na_2S_2O_3$ as mentioned above (17). If the alternative of the permanganate titration is chosen, the precipitate is dissolved in acidified $Fe_2(SO_4)_3$ and then the Cu^+ ions reduce the ferric iron to ferrous, which is finally titrated with permanganate. Occasionally this method gives rise to unexpectedly low results because of the oxidation of Fe^{2+} under the acidic conditions, although this loss can be reduced if the ferric sulfate is acidified after addition of the Cu_2O.

In case of iodometric titrations (18), iodine oxidizes aldoses rather than ketoses at alkaline pH values, I_2 being transformed into IO^- and excess iodine titrated with thiosulfate. The reaction is nearly stoichiometric and may be used to determine glucose, lactose, and galactose in the presence of fructose.

Reducing sugars may also react with alkaline $Fe(CN)_6^{3-}$ (above pH 10.5) if heated forming a ferrocyanide derivative which reacts with Fe^{3+} salts to give rise to Prussian blue. Then the excess ferricyanide can oxidize potassium iodide to produce iodine which is titrated with sodium thiosulfate (19).

Jacobs reported other methods which may be utilized to determine reducing monosaccharides in the presence of reducing disaccharides, and fructose when aldoses and sucrose occur (20). Some authors have also proposed an alternative to measuring copper oxide based on the determination of unreacted copper using a copper ion–selective electrode; previously, a known excess of copper salt and reducing sugar are allowed to react. This procedure has been successfully applied to the determination of reducing sugars in several foodstuffs such as fruit juices and marmalade (21). Finally, the literature reports other modifications of the basic method (22) including an electrogravimetric procedure according to which the copper oxide is dissolved in HNO_3 and deposited on a platinum electrode, whose weight gain is related to the reducing sugar content of the sample solution (13).

   b. Colorimetric Methods. Copper(I) ions form with neocuproine (2,9-dimethyl-1,10- phenanthroline), an orange water-insoluble complex in the pH range 3–10, which is soluble in ethanol on brief shaking (23). The so-called neocuproine method is then based on the reduction of the Cu^{II}-neocuproine reagent [prepared by mixing an aqueous copper(II) sulfate solution with an ethanolic neocuproine solution] by reducing sugars, the resulting product (Cu^I complex) showing maximum absorption at 457 nm. The reaction of copper with neocuproine is specific and the determination is unaffected by the presence of many elements.

   According to the anthrone method, most carbohydrates—above all hexoses—react with anthrone (9,10-dihydro-9-oxoanthracene) under acidic conditions (concentrated H_2SO_4) to produce a blue–green color (24). The reaction takes place between the dehydration products of the sugar—furfural and hydroxymethylfurfural—forming condensation products with the anthrone. The reagent is most stable in strong (not quite concentrated) acid. The method can also determine different carbohydrates by merely changing the acid concentration and the temperature of heating. A modification to the standard technique, known as the Clegg–anthrone method, is based on the previous digestion of the food with perchloric acid. Thereafter, any disaccharides, trisaccharides, and higher oligomers will be hydrolyzed to their component reducing sugars and these react with the anthrone reagent (25).

   The Nelson–Somogyi method (26,27) carries out the determination of reducing sugars using arsenomolybdate-copper reagent. Copper(II) ions are initially reduced to the cuprous form by heating with the carbohydrate solution and the resulting Cu^+ further reduces the arsenomolybdate to molybdenum blue, which is then spectrophotometrically measured at 820 nm. Several species are known to interfere with the determination, e.g., ammonium ions. Previously, Shaffer and Somogyi used an iodimetric procedure to determine reducing sugars (28).

   Another typical colorimetric assay, the phenol-sulfuric method (29), involves two stages: dehydration of sugars to furfural and hydroxymethylfurfural with concentrated H_2SO_4 and condensation of these substances with phenol to produce a yellow color, its intensity being proportional to the carbohydrate concentration. This method has a high sensitivity and is therefore utilized to analyze eluates from paper chromatograms, e.g., where microgram quantities of sugar occur. The results obtained are given as total carbohydrates although different sugars give rise to different intensities of color depending on the way in which they are dehydrated by the acid.

   Reducing sugars also react with 3,5-dinitrosalicylate in alkaline solution giving red–brown colors which may be measured colorimetrically (30) or with picric acid in alkaline solution to produce red picramates which can also be monitored spectrophotometrically. The latter is known as the picric acid method (31).

   Finally, 2,3,5-triphenyltetrazolium chloride or bromide may react with reducing sugars at pH values above 12.5 to yield a pink/violet/blue color (depending on the sugar) as a result of the formation of triphenylformazon (32).

B.   Recent Methods

1.   Chromatography

Numerous books, reviews, and papers have been written on this subject and its application in food analysis. Therefore, it is not the author's intention to give an exhaustive review of the literature.

Rather, the reader can refer to the bibliographic sources mentioned at the beginning of this chapter if background information is required.

The earlier methods of paper chromatography (PC) and thin-layer chromatography (TLC) have now largely been replaced by instrumental methods such as high-performance liquid chromatography (HPLC) and gas–liquid chromatography (GLC), which in general provide more rapid analyses with greater precision and specificity. Instead, neither PC nor TLC is suited to automation and their resolution and speed of analysis are usually lower than those attained by HPLC and GLC techniques. Nevertheless, it must be borne in mind that for multiple samples, where only qualitative or semiquantitative data are required, PC or TLC may still be the most suitable option.

Table 8 shows the various factors involved in the comparison between HPLC and GLC methods, although the method of choice may ultimately depend on the nature of the sample (33). Both techniques need a careful sample treatment. In HPLC, this stage involves extraction (100% efficient or at least of known efficiency) and cleanup (interfering compounds must be removed but *not* the analyte), whereas GLC requires derivatization of the sugars (preparation of a volatile derivative in quantitative yield). In this last case, a survey of the literature indicates the predominance of the *o*-trimethylsilyl (TMS) derivatives for carbohydrate analysis, although both acetates and trifluoroacetates have also been used.

Tables 9 and 10 show a selection of some of the most important applications of HPLC and GLC to carbohydrate analysis. Further details may be found in the specialized literature.

2.   Biochemical Methods of Analysis

These procedures are based on the use of enzymes or on microbial assays using yeasts to ferment sugars such as sucrose or glucose. The latter methods are not so widely used although they do have some applications in the food industry, e.g., in glucose syrup analysis. Enzymatic methods are in theory very specific, but in practice their specificity depends on the purity of the enzymes used. For food analyses, where there are many different carbohydrates present, very pure enzymes are required to preclude interfering compounds. However, enzymes have been used indirectly in food analysis for many years, e.g., as hydrolyzing agents. Probably the best known of these cases is in the determination of sucrose, where invertase is utilized to hydrolyze this disaccharide into its constituents. Enzymic determinations require very small quantities of material and sampling is therefore important. Great care and cleanliness must be taken in the assays or the results may not be reliable.

In food analysis, enzymes are mainly used in two ways:
1.   In the analysis itself, to quantify the reaction products, e.g., by electrochemical or spectrophotometric methods
2.   To break down a substrate, with the reaction products being determined by conventional (chemical or physical) means

Table 8   Comparison Between HPLC and GLC for Carbohydrate Analysis

HPLC	GLC
More suitable for the determination of medium and high molecular weight polysaccharides	Far more sensitive
Shorter (50%) analysis times	Allows separation of α and β anomers
Higher recoveries of greater accuracy	Preferable for monosaccharides
Directly applicable to sugar samples	Derivatization required

Table 9 Examples of Applications of High-Performance Liquid Chromatography in Carbohydrate Analysis

Carbohydrate	Matrix	Extraction	Stationary phase	Mobile phase	Detector
1, 3, 4	Milk	80% CH_3CN	Aminex HPX 87-C	H_2O (80°C)	RID
1, 2, 3	Coffee	H_2O	Spherisorb-NH_2	H_2O/CH_3CN (16:84)	MD
1, 2, 5	Fruit	95% EtOH	Aminex HP-87	0.01% $CaCl_2$ (85°C)	RID
1, 2, 5, 6	Vegetables	H_2O	μ-Bondapak carbohydrate	H_2O/CH_3CN (25:75)	RID
1, 2, 5	Fruit juice	H_2O	Brownlee amine column	H_2O/CH_3CN (15:85)	UVS (λ= 190 nm)
1, 5, 7	Yogurt	70% EtOH	Spherisorb-NH_2	H_2O/CH_3CN (36:64)	RID
1, 2, 4, 5, 6	Beverages, cereals, cream	H_2O	μ-Bondapak carbohydrate	H_2O/CH_3CN (20:80)	RID
1	Starch hydrolyzates	H_2O	Dextropak	H_2O	RID
1, 2	Melon	80% EtOH	Radial-Pak silica (25:75) + TEPA	H_2O/CH_3CN	RID

1 = glucose; 2 = fructose; 3 = galactose; 4 = lactose; 5 = sucrose; 6 = maltose; 7 = raffinose; TEPA = triethylenephosphoramide; RID = refractive index detector; MD = mass detector; UVS = ultra violet spectrophotometer.

The first type is of greatest importance and is an area where a great deal of development work is being carried out to extend the list of sugars capable of being analyzed by this means. Unlike many chemical and physical methods, enzymic methods represent a specific, sensitive, rapid, and reproducible means of measuring food carbohydrates, although costs may be higher than for conventional techniques. Nevertheless, the use of immobilization techniques contribute to cheapen this type of analysis. Since enzymes are so specific, there is no need for elaborate cleanup procedures to remove interfering compounds. Trace metals may, however, cause problems. For

Table 10 Examples of Applications of Gas–Liquid Chromatography in Carbohydrate Analysis

Carbohydrate	Matrix	Stationary phase	Temp. (°C)	Detector
1, 6	Glucose syrup	1% OV-101 on 100/120 mesh HP Chromosorb G	120–350	FID
1, 3	Bread	3% SP-2330 on 100/120 mesh Supelcoport	215	FID
1, 2, 5, 6	Honey	2% OV-101 on 100/120 mesh HP Chromosorb G	120–300	FID
1, 3	Milk	10% E-30 on 100/120 mesh Diatomite CQ	100–230	FID
4, 5, 6	Bun mix	2% OV-101 on 100/120 mesh HP Chromosorb G	120–300	FID
1, 2	Syrups	1% OV-101 on 100/120 mesh Gas Chrom Q	125–330	FID
1, 3, 5, 7	Soybean	3% OV-1 on 80/100 mesh Chromosorb W	100–320	FID

1 = glucose; 2 = fructose; 3 = galactose; 4 = lactose; 5 = sucrose; 6 = maltose; 7 = raffinose; FID = flame ionization detector.

instance, heavy metal ions (Hg, Pg) and oxidizing agents (for –SH groups) can inactivate enzymes whereas other trace metals may act as enzyme activators. These interactions may be important when reaction rate methods are used to measure sugars.

Enzymic assays for quantitative carbohydrate analysis can be carried out according to the following alternatives (22):

1. Determine the rate of reaction whereby initial rate is proportional to substrate concentration. Since the rate is affected by temperature, pH, and reaction medium, strict control is necessary to obtain meaningful results. In this procedure, standards are usually run with each sample to eliminate errors to below 3%, although automation easily lowers this limit further.

2. Determine an endproduct of the enyzme–substrate interaction. The endproduct should be produced stoichiometrically from the carbohydrate to be measured and the reaction should preferably go to completion. If it does not, the concentrations of the reactants may be changed or endproducts may be removed to allow the reaction to proceed. Alternatively, equilibrium products may be measured and reference made to a calibration graph to obtain the concentration of the carbohydrate. Endproducts may either be measured directly or combined with a suitable indicator or dye to produce a colored complex which is then monitored spectrophotometrically. Examples of direct measurements include NADH production as a result of an oxidation reaction and optical rotation changes after invertase has been added to a sucrose solution. An example of the other case (indirect method) could be the formation of the colored complex in glucose determination by using β-D-glucose oxidase/peroxidase.

Table 11 shows major examples of the use of enzymes in carbohydrate analysis. Regarding less common sugars, analytical procedures are cited in the literature (34), as well as full experimental details for the above-mentioned enzymatic techniques (35,36). On the other hand, Table 12 summarizes the most important specific assays for the analysis of major sugars.

Combined instrumental-enzymic methods have also been developed for analysis of food sugars, especially glucose, galactose, sucrose, lactose, and starch (37). In particular, a quick method whereby hydrogen peroxide is measured by electrochemical oxidation at a platinum electrode is currently a common practice (38). Alternatively, oxygen uptake by sugar solutions can also be determined. The general reaction is:

$$\text{Carbohydrate} + O_2 \rightleftharpoons \text{oxidized carbohydrate} + H_2O_2$$

where an immobilized oxidase enzyme—contained within a suitable membrane support—is used. Each sugar requires a different enzyme/membrane and the content of resulting hydrogen peroxide is proportional to the sugar concentration. The oxygen is provided by the buffer solution in which the reaction takes place.

3. Flow Injection Analysis

Flow injection analysis (FIA) (39) requires relatively inexpensive materials for its excellent performance and has been the basis of a large number of methods for the analysis of carbohydrates (40,41), most notably reducing sugars, whose determination is a common analytical practice in many laboratories and industries. The analysis of these compounds is often required during the production process and quality control of several foods. There is a special problem due to heterogeneity of composition and concentration of carbohydrates in foods and the different reactivities of sugars. The evaluation of sugar content by means of a global index additionally causes difficulty of interpretation, since the use of a given standard (glucose, fructose, etc.) influences the results and complicates comparisons.

Maquieira et al. (42) propose and describe three automatic flow injection methods for the

Table 11 Examples of the Use of Enzymes in Carbohydrate Analysis

Carbohydrate	Reaction	Detection	Interferences
Glucose	Glucose + O_2 $\rightleftarrows$ gluconic acid + H_2O_2		Galactose
	H_2O_2 + indicator/dye acceptor $\rightleftarrows$ color + H_2O_2	UV-V	Mannose
			Ascorbic acid
	Glucose + ATP $\rightleftarrows$ glucose 6-phosphate	UV-V	Proteins
	Glucose 6-phosphate + NADP $\rightleftarrows$ 6-Phosphogluconate + NADPH		
	Glucose + O_2 $\rightleftarrows$ gluconic acid + H_2O	Electrochemical	
Fructose	Fructose + ATP $\rightleftarrows$ fructose 6-phosphate		
	Fructose 6-phosphate $\rightleftarrows$ glucose 6-phosphate	UV-V	
	Glucose 6-phosphate + NADP $\rightleftarrows$ 6-phosphogluconate + NADPH	Fluorimetry	
	Fructose $\rightleftarrows$ H_2O_2 + byproduct	Electrochemical	
Galactose	D-Galactose + NAD^+ $\rightleftarrows$ galactonic acid + NADH		
	D-Galactose $\rightleftarrows$ galactonic acid + H_2O		
Sorbitol	Sorbitol $\rightleftarrows$ fructose[a]		Xylitol
	Sorbitol + O_2 + NAD^+ $\rightleftarrows$ fructose + NADH	UV-V	
Sucrose	Sucrose $\rightleftarrows$ glucose + fructose		
	Glucose + fructose + ATP $\rightleftarrows$ hexose-6-phosphate		
	Fructose 6-phosphate $\rightleftarrows$ glucose 6-phosphate		Glucose
	Glucose 6-phosphate + NADP $\rightleftarrows$ 6-phosphogluconate + NADPH	UV-V	Fructose
	Sucrose $\rightleftarrows$ α-glucose + fructose	Electrochemical	Raffinose
	α-Glucose $\rightleftarrows$ β-glucose $\rightleftarrows$ H_2O		Mellibiose
			Glucose
Lactose	Lactose $\rightleftarrows$ glucose[b] + galactose[c]	UV-V	
	Lactose + O_2 $\rightleftarrows$ H_2O + modified substrate	Electrochemical	Fructose
			Erythrose
			Glycerol
			Mannose
Maltose	Maltose $\rightleftarrows$ glucose[b]		Sucrose
Raffinose	Raffinose $\rightleftarrows$ sucrose[d] + galactose[c]		Melibiose
			Stachyose
Starch	Starch $\rightleftarrows$ glucose[b]		
	Starch $\rightleftarrows$ β-D-glucose		
	β-D-glucose $\rightleftarrows$ H_2O		

[a] Determined as previously described for fructose.
[b] Determined as previously described for glucose.
[c] Determined as previously described for galactose.
[d] Determined as previously described for sucrose.

determination of reducing sugars in wine based on classical reactions (the picric acid method, the Cu^{II}-neocuproine method, and the Nelson–Somogyi method, all of them already mentioned above). Figure 1 shows the different configurations used in each case. The first one presents a dual-channel configuration, with pH being adjusted with NaOH at a confluence point. The same manifold is utilized for the Cu^{II}-neocuproine method, whereas the automation of the Nelson–Somogyi method makes use of a reversed FIA system, in which the sample flows continuously and then is mixed with the NaOH stream. In all cases, since the reduction of sugars requires drastic conditions for sufficiently fast development, the reactor in which the chemical reaction takes place is immersed in a thermostatic bath at a suitably high temperature. The usefulness of these methods

Table 12 Summary of Specific Enzymatic Assays for the Analysis of Carbohydrates

Carbohydrate	Immobilized enzyme	Detection principle
Glucose	GOD; membrane	O_2 or H_2O_2 probe
	GOD; membrane	H_2O_2 probe; longterm, *in vivo*
	GOD + mediator	Amperometric carbon probe
	Concanavalin A + fluorescein-labeled dextran	Optical fiber, fluorimetric (affinity sensor)
	Luciferase; reactor	Luminometer
	GOD; reactor	Enthalpimetric
	GOD; membrane	ISFET
	GOD; membrane	Micro-O_2 sensor
	GOD; reactor	Colorimetric or chemiluminescence
	ADH + GOD; nylon tubing	Colorimetric
	GOD + POD; reactor	F probe
	GOD; membrane	Externally buffered O_2 probe
	GOD; reactor	O_2 probe, bedside monitoring
	GOD; membrane	H_2O electrode, needle-type probe
	GOD + POD; dry reagent	Colorimetric
	GOD + catalase; reactor	Enthalpimetric; continuous monitoring
Lactose	Lactase + GOD; membrane	O_2 probe
Maltose	Glucan 1,4-glucosidase + GOD + POD; membrane	H_2O_2 probe
Sucrose	Invertase + GOD; membrane	H_2O_2 probe
	Invertase + mutarotase + GOD; membrane	O_2 probe
	Invertase; reactor	Enthalpimetric, continuous monitoring

GOD = glucose oxidase; ISFET = ion-sensitive field effect transistor; ADH = alcohol dehydrogenase; POD = peroxidase.

was tested by applying them to red and white wines, the results obtained being in good agreement with those provided by a batch standard method.

Peris-Tortajada et al. (43) enlarge this last work by successfully applying the Cu[II]-neocuproine method to the determination of reducing sugars in different food samples using a flow injection system. Two different sugars (glucose and fructose) were used as standards. Additionally, a dialysis procedure was assayed to avoid tedious decolorizations with charcoal (required in the work of Maquieira and co-workers), three types of membranes being tested for this purpose. The use of a dialysis unit (inserted between the injection valve and the reactor) also makes this method suitable for adaptation to on-line analysis.

All of these FIA methods proposed are straightforward with good sampling frequencies (over 40 samples per hour) and low sample and reagent consumption, and require affordable instrumentation.

Finally, different scientists are currently working on the automation of the above-mentioned enzymatic determinations of some carbohydrates (mainly glucose, fructose, lactose, maltose, and sucrose) by means of this technique (44–46). Several biosensors have also been designed—even commercialized—on this basis (47).

C. Future Trends

The future of carbohydrate analysis undoubtedly belongs to three analytical techniques—HPLC, enzymatic assays, and FIA—all of them having close relationships to one another.

(a)

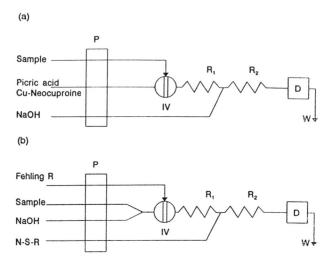

(b)

Fig. 1 (a) Normal FIA configuration for the picric acid and Cu(II)-neocuproine methods; (b) reversed FIA configuration for the Nelson–Somogyi method. N-S R = Nelson–Somogyi reagent; P = peristaltic pump; IV = injection valve; R_1 and R_2 = reactors; D = detector; W = waste. Courtesy of *The Analyst*, The Royal Society of Chemistry, U.K.

Considerable progress has been made in HPLC in such a way that it is now routinely and reliably used for quantitative analysis of food carbohydrates. Some separations remain to be improved, but the development of new bonded phases with various organic loadings (mainly from different aminoalkylsilanes) together with the establishment of the mass detector as an alternative to refractive index monitoring will offer much scope in improving the resolution of carbohydrates. Nevertheless, refractive index detectors are not expected to become "old-fashioned," provided that manufacturers succeed in reducing their noise level, so that greater sensitivity could be achieved.

On the other hand, a large number of highly purified enzymes is now commercially available for the specific detection and quantification of different food sugars, be they mono- or disaccharides or even polysaccharides. As foods become more complex and new sugars and their derivatives are used in their formulation, there is no doubt that enzymatic methods will be increasingly used to determine specific compounds. The widespread use of immobilized enzymes, including their use in enzyme electrodes, should allow the continuous assay of these essential components of foods, not only for the purpose of classical at-line analysis but with a view to on-line and in-line monitoring of food production processes. The latter is already a well-established analytical procedure and the growing availability of commercial biosensors will open a promising path. Additionally, the increasing stability of many enzymes that can be achieved by crosslinking and immobilization techniques will extend the range of such determinations.

Finally, FIA is expected to consolidate its leading position in the field of automated analysis, and the determination of food carbohydrates (especially reducing sugars) will not be an exception. Its coupling to enzymatic assays is likely to become a common analytical practice in most quality control laboratories, given its superb performances in terms of rapidity, precision, and accuracy. Some firms have been working in the development of FIA-based instrumentation especially designed for carbohydrate analysis in several matrices including wines and other sugar-rich products.

REVIEWS

Holme, D. J. and Peck, H. *Analytical Biochemistry*, Longman, London, 1993.

Coultate, T. P. *FOOD: The Chemistry of Its Components*, Royal Society of Chemistry, London, 1989, Chap. 2.

Mosbach, K (ed.), *Methods in Enzymology*, Vol. 137, Part D. Academic Press, New York 1988.

Birch, G. G. (ed.), *Analysis of Food Carbohydrate*, Elsevier, London, 1985.

Macrae, R. (ed.), *HPLC in Food Analysis*, Academic Press, London, 1982, Chap. 6.

King, R. D. (ed.), *Developments in Food Analysis Techniques*, Vol. 1, Applied Science, London, 1978, Chaps. 6, 9.

Southgate, D. A. T. *Determination of Food Carbohydrates*, Applied Science, London, 1976.

REFERENCES

1. T. W. G. Solomons, *Organic Chemistry*, John Wiley and Sons, New York, 1993.
2. El-Khadem, *Carbohydrate Chemistry: Monosaccharides and Their Oligomers*, London, 1988.
3. R.C. Weast, ed., *Handbook of Chemistry and Physics*, CRC Press, Boca Raton, 1984.
4. I.M. Kolthoff and P.J. Elving, *Treatise on Analytical Chemistry*, Part 1, Vol. 6, Interscience, New York, 1965.
5. F.J. Bates et al., *Polarimetry, Saccharimetry and the Sugars*, N.B.S. Circular C-440, Washington, DC, 1942.
6. F. Glover, J. Gaulden, *Nature*, *200*: 1165 (1963).
7. A. Vogel, *Textbook of Practical Organic Chemistry*, Longman, London, 1978.
8. H.A. Strobel, *Chemical Instrumentation*, Addison-Wesley, Reading, 1973.
9. F.D. Snell, *Encyclopedia of Industrial Chemical Analysis*, Vol. 18, John Wiley and Sons, New York, 1973.
10. I.M. Kolthoff and P.J. Elving, *Treatise on Analytical Chemistry*, Part 1, Vol. 7, Interscience, New York, 1965.
11. D.J. Pasto and C.R. Johnson, *Organic Structure Determination*, Prentice-Hall, Englewood Cliffs, 1980.
12. R.L. Schriner et al., *The Systematic Identification of Organic Compounds: A Laboratory Manual*, John Wiley and Sons, New York, 1980.
13. Association of Official Analytical Chemists (AOAC), *Official Methods of Analysis*, Washington, DC, 1984.
14. T.H. Khan, *Analyst*, *104*: 261 (1979).
15. H. Egan et al., *Pearsons Chemical Analysis of Foods*, Churchill Livingstone, London, 1981.
16. L.S. Munson and P.H. Walker, *J. Am. Chem. Soc.*, *28*: 663 (1906).
17. P.A. Shaffer and A.F. Hartmann, *J. Biol. Chem.*, *45*: 349 (1921).
18. C.L. Hinton and T. Macara, *Analyst*, *49*: 2 (1924).
19. R.L. Whistler and M.L. Wolfrom, *Methods in Carbohydrate Chemistry*, Vol. 1, Academic Press, New York, 1962.
20. M.B. Jacobs, *The Chemical Analysis of Foods and Food Products*, Van Nostrand Reinhold, New York, 1958.
21. D.S. Papastahopoulos, D.P. Nikolelis, and T.P. Hadjiioannou, *Analyst*, *102*: 852 (1977).
22. Y. Pomeranz and C.E. Meloan, *Food Analysis: Theory and Practice*, AVI, Westport, 1977.
23. R.A. Gahler, *Anal. Chem.*, *26*: 577 (1954).
24. J.H. Roe, *J. Biol. Chem.*, *212*: 335 (1955).
25. K.M. Clegg, *J. Sci. Fd. Agric.*, *1*: 40 (1956).
26. M. Somogyi, *J. Biol. Chem.*, *195*: 19 (1952).
27. J.P. Marais, J.L. de Wit, and G.V. Quiche, *Anal. Biochem.*, *15*: 373 (1966).
28. S. Ranganna, *Manual of Analysis of Fruit and Vegetable Products*, Tata–McGraw-Hill, New Delhi, 1977.
29. M. Dubois, K.A. Giles, J.K. Hamilton, P.A. Rebers, and F. Smith, *Anal. Chem.*, *28*: 350 (1956).
30. R.T. Bottle and G.A. Gilbert, *Analyst*, *83*: 403 (1958).
31. D.R. Osborne and P. Voogt, *The Analysis of Nutrients in Food*, Academic Press, London, 1978.

32. E.P. Cromwell and B.B. Burnett, *Anal. Chem., 39*: 121 (1967).
33. K.M. Brobst and H.D. Scobell, *Starch/Staerke, 34*: 117 (1982).
34. G.G. Guilbault, *Handbook of Enzymatic Methods of Analysis*, Marcel Dekker, New York, 1976.
35. H.U. Bergmeyer, *Methods of Enzymatic Analysis*, Vol. 1, Verlag Chemie, Weinheim, 1983.
36. BCL, *Catalogue*, Boehringer Corporation, London, 1984.
37. M. Mason, *JAOAC, 66*: 981 (1983).
38. P.J. Elving and J.D. Winefordner (eds.), *Immobilized Enzymes in Analytical and Clinical Chemistry*, John Wiley and Sons, New York, 1980.
39. M. Valcárcel and M.D. Luque de Castro, *Flow Injection Analysis: Principles and Applications*, Ellis Horwood, Chichester, 1987.
40. J. Ruzicka and E.H. Hansen, *Anal. Chim. Acta, 179*: 1 (1986).
41. J. Ruzicka and E.H. Hansen, *Flow Injection Analysis*, John Wiley and Sons, New York, 1988.
42. A. Maquieira, M.D. Luque de Castro, and M. Valcárcel, *Analyst, 112*: 1569 (1987).
43. M. Peris-Tortajada, R. Puchades, and A. Maquieira, *Food Chem., 43*: 65 (1992).
44. R.D. Schmid (ed.), *Flow Injection Analysis Based on Enzymes or Antibodies*, GBF Monographs, Vol. 14, Gesellschaft für Biotechnologische Forschung, Braunschweig, 1991.
45. K. Schügerl, L. Brandes, T. Dullau, K. Holzhauer-Rieger, S. Hotop, U. Huebner, X. Wu, and W. Zhou, *Anal. Chim. Acta, 249*: 87 (1991).
46. K. Schügerl, *Pure Appl. Chem., 64*: 205 (1992).
47. M. Alvarez-Icaza and U. Bilitewski, *Anal. Chem., 65*: 525A (1993).

16

Alcohols in Foods and Beverages

Mauro Amelio
Fratelli Carli SpA, Imperia, Italy

I. INTRODUCTION

The determination of alcohols in foods and beverages is of interest to many sectors, from market classification to quality control, from research of adulterations to understanding of the natural processes which control their formation or the effect of technological processes on products intended for human consumption.

The determination of the alcoholic content of wines and distilled spirits perhaps is the field which interested the analytical chemist first of all, given the huge importance of the production and commercialization of these products. Because of this, the producing countries have for a long time had regulations which control their production and commercialization, often with widely differing provisions. The current tendency leans to the greatest possible uniformity of regulations, in an attempt to tune them to ever wider markets (i.e., European Economic Community). The market classification of alcoholic drinks, given by the alcoholic content, is one of the most important parameters in establishing tax. The importance of the exact determination of alcoholic content and the risks of adulteration to which these products are exposed is then obvious.

However, interest in the alcoholic content of foods is also attributable to other reasons which are linked to their properties and which contribute to conferring the characteristics of that food. This is the case, for example, of alcoholic drinks whereby ethanol, besides improving preservation and giving sensory satisfaction, also has the role of emphasizing the aromatic components by increasing their volatility. A similar case is that of the higher alcohols, which in wines have an effect on the bouquet.

Glycerol, which is formed during the fermentation of musts, with its sweet taste contributes to the smoothness and the body of wines. However, in certain circumstances, it can indicate defects in harvested grapes (1). In alcoholic fermentation there is the formation of glycerol with yield of 2–3% from the fermented sugar. This content diminishes in acetic fermentation, but the best-quality vinegars contain larger quantities compared with ordinary ones. In fact, a very low amount of glycerol is a signal of possible adulteration.

Methanol, given its toxic properties which require a control of its concentration in drinks which may contain it, has a different importance. In wines it is naturally present in small quantities, as it originates during the fermentation of pectins in the grapes. In particular, in the preparation of distilled spirits, the methanol content must be kept under control, even though it is concentrated in the "head" of distillation, which is thrown away. Other alcohols are checked for different reasons. For example, 2-butanol in distilled spirits or in vinegars is indicative of bacterial spoilage in the mashes or wines used for distillation, whereas higher alcohols are indicators of wholesomeness and the botanical origin of the distillates.

There are also industrial reasons for determining alcoholic content, such as monitoring vinification processes and fermentation, production of low-alcohol beverages, etc. Since it is thought that the quality of white wines is in inverse relation to the content of higher alcohols (isobutanol, actamyl and isoamyl alcohols), studies to clarify their origin and conditions of formation have been carried out (2–4). In honey, on the other hand, the presence of not small quantities of ethanol is an indication of undesired fermentations which damage the product.

As was mentioned, another role is played by alcohols which contribute through their presence to the complex volatile fractions which give characteristics of taste and smell to a large number of foods. In milk, for example, these volatile fractions are often used as indicators of quality and their variation in composition is placed in relation to storage time and temperature, to thermal treatments, to exposure to light, etc. A study of these fractions also contributes to clarifying the metabolic mechanisms of their formation which, in the case of milk, depend on the animal which produces it. Besides, for certain foodstuffs, the aromatic component is decisive in characterizing the product and consents a specific market placing. In other cases (i.e., vegetable oils), the volatile component to which alcohols belong has been used to evaluate oxidation during storage.

The determination of alcohols is also used in the sector of adulteration controls, as in the case of olive oils. The determination of the content of fatty alcohols may be useful in recognizing mixtures between pressed olive oils and solvent-extracted oils (pomace oil) in samples under test. As certain technological refinement processes give rise to structural modifications of some triterpenic alcohols in relation to working conditions, their determination may be useful in recognizing fraudulent mixtures of virgin oils (i.e., oils obtained only by pressing) with elaborated oils.

In the sector of beers, the fraudulent addition of exogenous water (which means water other than that used during the photosynthesis of starch) to the fermentable juice or the type of cereal used may be discovered through suitable analytical determinations (nuclear magnetic resonance, NMR). Other fields of research may be interested in the determination of alcohols, such as, for example, certain sectors of agronomy. It is well known, for instance, that some herbivorous insects accept or reject plants according to their lipidic surface composition. The presence of long-chain fatty alcohols can affect the insects. Thus, a knowledge of the composition of these lipids may be useful in the fight against the parasites of certain fruit plants.

In a completely different sector, the physiology of nutrition, it has been demonstrated that octacosanol (C_{28}-OH) stimulates the conversion of lipids to energy, and a large number of patents has been issued for the use of these alcohols in the food and beverage industry (5). The search for the cheapest sources has inspired the study of plants which are rich in waxes containing the highest percentages of octacosanol, such as sugarcane (6).

A. Some Indications for the Reader

As can been seen from this brief introduction, there are many reasons for determining the content of alcohols in foods and beverages. These will appear more clearly in the applications section (V). The following short review only touches on the main aspects of so vast a subject. Therefore, brief coverage is given to the most frequently used analytical techniques, with special regard to the applications of analysis of alcohols. For further information, see the chapter on "Instruments and Techniques" in this book. To guide the reader, a few easily accessed texts are suggested. However, others in existence are just as valid.

It was not possible in the applications section (V) to describe analytical methods in depth, but only a few general details are given accompanied by bibliographical references to which the reader can refer for more detailed information. The same applies to very broad subjects such as wines—an exhaustive view can only be given in a more appropriate text. To the extent possible, recent works are treated, without however denying the validity of earlier published works.

Among the AOAC methods, some are classified as surplus; these are no longer in use. Nonetheless they are included because they are reported in the 15th edition (1990) to which previous reference was made.

Finally, regarding the chemical-physical characteristics of the alcohols studied, readers can consult the *Merck Index* (7).

II. SEPARATION OF THE ALCOHOLS FROM THE MATRIX AND CLEANUP

Given the complexity of the matrices in which alcohols are present in foods and beverages, some suitable procedures must be carried out in order to separate them in an appropriate form for later quality/quantity determinations. Each of them leads to a sample which is different depending on the type of determination to be made, the properties and concentration of alcohols to be determined, etc. Thus, a measure of density from which the ethanol content of a wine can be derived requires that a distillation be carried out with the intention of collecting a solution consisting of only water and ethanol. A determination of enzymatic type usually requires a very simple preparation of the sample (filtration, clarification, etc.) given the high specificity of the enzymes which therefore are not subject to much interference. On the other hand, the study of the alcoholic fraction of an aroma requires delicate techniques of stripping and recovery, and the use of instruments which are often very expensive. It is therefore not easy to compile an exhaustive review on the subject, and only an indication of the most common methods will be provided. For details, refer to individual works.

A. Distillation

In a great number of cases (i.e., wines, vinegars, extracts, etc.), *distillation* is the most suitable method for separating alcohols from the matrix. Distillation also collects other volatile fractions at those temperatures and which may recondense (e.g., methanol, higher alcohols, etc.). However, if ethanol is the principal alcohol, as in water-based products (e.g., wine), distillation is a suitable method for determining the alcoholic content through the measurement of the density of the distilled product. In the case of vinegars, the high concentration of acetic acid would interfere greatly in the measurement of density. The solution must therefore be neutralized prior to distillation by salification of the acetic acid. In several countries the law provides for such a step for these determinations. In the case of beverages containing large quantities of carbon dioxide (i.e., sparkling wines, beers, etc.), the solution must be decarbonized to avoid foaming during distillation. This operation must be carried out under controlled conditions to avoid evaporation of a part of the alcoholic content of the product. Usually the decarbonization is obtained by filtering on paper or by stirring with a glass rod. If alcohols with a higher boiling point are required, a *steam distillation* can be carried out. *Distillation with a Vigreaux column* prevents dragging, balances vapors better, and provides a very pure distillate due to its high number of plates. This is used for cordials, liqueurs, etc. In all distillations particular attention must be paid to the measurement of the volumes of solutions at the prescribed temperatures, given the high dilatation of hydroalcoholic solutions with rising temperatures. Besides, heat must be applied very carefully to avoid causing changes in the sample under examination.

Among the disadvantages of distillation are that it is time consuming and requires a relatively large volume of samples, which may not always be available (i.e., laboratory scale fermentations). In the study on aromatic components, the main disadvantages are artefact formation and the decomposition of individual components due to heating. *Vacuum distillation* is usually used in the study of aromatic fractions, both to collect the components with a high boiling point together with those with a low one, and to minimize changes which may be due to oxidation and the effect of higher temperatures.

As discussed at length in the chapter on aroma compounds, such distillations are carried out over a long period of time (several hours) in specifically designed equipment, since there is the need to collect suitable quantities of product for successive analyses (generally gas chromatography [GC] or gas chromatography–mass spectrometry [GC-MS]). In these investigations the volatile fraction must be collected in suitable traps. This is "cold trapping" (e.g., with liquid nitrogen) or *porous polymers* (e.g., Tenax) which, on the contrary, can work at room temperature. Their use is often associated, rather than with traditional distillation, with a *gas stripping by inert gas* (e.g., helium or nitrogen).

B.   Resins and Molecular Sieves

The use of resin extraction techniques has recently become known in the field of analysis of the aromatic component of wines. Hydrophobic resins (e.g., Amberlite XAD-2, C18, KS 112), which, however, do not give complete recovery for the presence of ethanol, are available. The use of hydrophilic resins (e.g., Extrelut) has been suggested which with suitable solvents releases these aromatic fractions to be subjected to further tests. In some cases the use of ionic exchange resins is used instead in order to carry out a cleanup of the solution to be tested (i.e., diethylene glycol [DEG] in wines). Molecular sieves have also been used as traps for volatile fractions (i.e., in dairy products), then analyzed in GC. In general, the principal disadvantages are the breakthrough of the lower boiling components and the accumulation of water in the trap, which interferes with further GC analysis.

C.   Solvent Extraction

A great number of works can be found in literature with descriptions of *extraction with solvents* which are also applied to research of the volatile components of various foodstuffs and drinks. The choice of solvents should be made taking into consideration types of analyte, types of matrix from which they are to be extracted, and the chosen type of analysis. It can usually be said that there is no ideal solvent or mixture of solvents which can quantitatively extract only the analytes of interest, in terms of the relationships of concentration. Their use may involve a simultaneous quantitative extraction of undesired substances which interfere in later determinations. It is usually advisable to evaluate each case in relation to the specific research demands.

D.   Microwave Extraction and SFE

Microwave extraction as a technique which may be used together with solvents has also been reported. Razungles et al. (8) applied it to the rapid analysis of volatile compounds in grapes and wines. The main advantage is speed (1–5 min) with respect to classic distillation (60–90 min) with good yields (over 50%) and with results comparable to those obtained with resins of the Amberlite XAD-2 type.

One kind of extraction which is becoming more and more important is *supercritical fluid extraction* (SFE) using gas, which in supercritical conditions shows liquid-like densities and solvating power along with very low viscosity. Supercritical carbon dioxide is by far the most used gas in this technique because of its chemical-physical characteristics, which make it very safe to use, and due to its chemical inertia. A brief introduction to SFE was written by Major (9).

Later on, in the review of applications, the use of SFE is found in the study of the volatile fraction of oils, where the authors find a considerable advantage compared with extractions with solvents in terms of increased sensitivity, less degradation of the analytes, and short analysis time. SFE, rather than a technique of extraction specific for alcohols, permits the separation of classes of compounds from various matrices to be subjected to successive separations and analyses. The possibility of varying the solvent strength of the supercritical fluid by acting on the pressure makes SFE extremely versatile in its applications.

E. Headspace–Gas Chromatography

Last to be mentioned is headspace gas chromatography (HS-GC), which has interesting applications in the study of volatile fractions (see relevant chapter for further details). The technique consists of subjecting to analysis the vapors which, after conditioning at a suitable temperature, are released from the sample. Naturally, the concentration of the analytes as vapors is proportional to their volatility, but it is well known that it also depends on their solubility in the matrix, on the nature of the other compounds present, and so forth. In this section, HS-GC is interesting as a means for separating the volatile fractions (and therefore the alcohols which are a part of them) from the matrix. Of the various HS techniques, perhaps the most common is dynamic HS, which provides for stripping of the analytes and, successively, concentration and analysis. In this case what has already been said about gas stripping by inert gas is valid.

The main advantage of this method is that a profile of the volatiles as they exist is obtained. On the other hand, if a concentration step does not exist, the number of compounds that can be detected is small.

III. DERIVATIZATION

As can be seen in the applications section (V) derivatization of alcohols during analysis of foodstuffs is not a normal procedure. It is often applied in the case of GC analyses to transform high-boiling alcohols in compounds with lower boiling point of reduced polarity, to make them more suitable for GC analysis. This is the case, for example, of diethylene glycol (DEG), which can be transformed into trimethylsilyl ether derivative (TMS), or fatty alcohols, also transformed into TMS derivatives.

In the frequent case of TMS derivatives, it should be observed that in polyhydroxylated alcohols (e.g., DEG), the main effect of their transformation into silyl ethers for GC analysis is that the boiling point (b.p.) is greatly lowered, and only secondarily the polarity of the molecule is reduced. In monofunctional alcohols, however, no appreciable variation in the b.p. is observed, but it is the reduced polarity which gives its advantages in GC analysis. Some derivatizing reagents introduce atoms of halogens, in particular fluorine, which make it possible for the molecule to be efficiently revealed by electron capture detector (ECD). The presence of fluorine is advantageous also in the use of FID, as it contributes to making the products of combustion, which tend to condense on the jet of the detector, more volatile. There are other situations where it may be necessary to transform alcohols into different derivatives, like acetyl derivatives, in relation to the specific demands of the analysis, like the stability of the derivative, type of analysis to which the sample is subjected, etc. For example, in the case of vegetable oils [see applications section (V)], derivatization with pivalic anhydride of alcohols and free sterols was suggested before subjecting the sample to on-line LC-GC-FID separation. The need to separate optical isomers while lacking chiral columns requires derivatization with suitable chiral reagents (i.e., chiral acids, acyl reagents, or isocyanates to form carbamate derivatives, etc.) to form diastereoisomer derivatives, which may be resolved chromatographically.

In the choice of derivatization method, it is also necessary to bear in mind that reactivity changes according to whether one is dealing with primary, secondary, or tertiary alcohols, and in relation to working conditions. There may be huge variations in the yield, and this may lead to erroneous quantifications if not kept in mind.

For a detailed description of the available methods of derivatization, for the theoretical considerations regarding them, and for an extensive bibliography, see the volumes mentioned in Refs. 10 and 11.

IV. DETECTION AND QUANTITATION

A. Physical Methods

There are, of course, a large number of methods of quantifying the content of alcohols in the test sample. The choice of the most suitable method is made by evaluating several aspects which depend on the information to be obtained from the analysis, on the analytes to be quantified, on the type of sample, on the method which may be used to separate the alcoholic component from the original matrix, from the degree of accuracy desired, and so forth.

Many methods proceed to quantification from alcoholic distillates (especially when the prevalent alcohol is ethanol) obtained from the product to be tested and which are generally water-based. It must be remembered that distillation collects not only ethanol but other substances of comparable volatility as well. Many methods exploit the physical properties of these distillates which change in relation to the alcoholic content. This is the case, for example, of wines, on whose alcoholic distillates *density* is measured at a certain temperature. From the density it is possible to work out the content in alcohols using tables. Usually measurement is carried out using a pycnometer. On the other hand, hydrometers (alcoholometers) give the content in alcohols directly on an incorporated scale. These methods require a particular approach and skill. Alcoholic degree may also be determined from a measurement of the refractive index of the hydroalcoholic solution, as, for example, in some AOAC methods (by Zeiss immersion refractometer). In the case of measurements of nondistilled liquid samples taken directly, the method is seen to be very fast and precise, but only for samples of similar composition. For example, in the brewing industry, several calibration curves are needed to keep account of the variations in the refractive index due to the composition.

With ebulliometric methods, work may be carried out both on alcoholic distillates and on wines, which, however, require a particular approach and which have particular limitations (e.g., sugars and salts lower the b.p., causing an error of excess on the alcoholic degree as in the case of sweet wines).

These are methods which are no longer in use but which may still be used for rapid screenings (precision is about 0.2° alcoholic for contents between 5° and 15°). The principle is that of measuring the boiling temperature of hydroalcoholic solutions, which is lowered with an increase in alcoholic content. Special tables give the alcoholic degree. It is always necessary to carry out a measurement with pure water in order to allow for variations in atmospheric pressure (zeroing of the apparatus).

Apparatus for taking these measurements are, for example, the Malligand ebullioscope, where the scale gives alcoholic degree immediately, or the Dujardin–Salleron ebulliometer, which measures the boiling temperature of wine. A graduated scale converts this to alcoholic degrees. Also available is Cachier's glass ebulliometer, with an electrically heated boiler.

Nowadays other physical methods which may be applied to wines only have historical value. Among these can be mentioned one which exploits the variations of *surface tension* which are inversely proportional to alcoholic content (Duclaux's dropping bottle), and one which uses the variations of *viscosity* (Otswald's viscometer).

All of these methods are dependent on temperature, and special care must be taken in this direction. There are correction tables which allow for the temperature at which measurements are taken.

B. Chemical Methods

The *Titrimetric* method consists of oxidizing the alcohol which has been separated from the tested product, with a known exceeding quantity of oxidizing agent (usually potassium dichromate or periodate), and of titrating the quantity which has not reacted. Oxidation with dichromate in analysis of wines was suggested by Martin (12) in 1924. Precision of this method is around +0.05°.

It must be observed, however, that the oxidation reaction is not specific, so other compounds also react, such as secondary alcohols, which form ketones. In fact, in the case of isopropanol determination, e.g., in lemon or cassia extracts, a successive distillation is carried out in order to collect dimethyl ketone and to make it react with an excess of iodine. Finally, a titration with thiosulfate is carried out.

Williams tubes take advantage of the different phase equilibrium reached by a suitable solvent (isoamylic alcohol/toluene/chloride acid = 70:28:2 v/v) and the alcoholic solution according to the alcoholic content. When settling is complete, the position of the meniscus between the two layers on the calibration mark shows the percentage (in volume) of alcohol.

C. Instrumental Methods

Of the *colorimetric* methods, the Roques method must be remembered for the determination of the higher alcohols. Concentration is evaluated by comparing the color which develops with that obtained from a series of standard solutions prepared with isobutanol which, on the other hand, gives a more intense color among higher alcohols. For this reason, the method could give lower results.

Also in use are *spectrophotometric* methods. In this case a substance must be obtained which has a specific absorbance, measurable at particular wavelengths and proportional to the alcohol to be quantified.

Generally the alcohol is oxidized and made to react to give colored derivatives. It is preferable for these derivatives to be stable, but this does not always happen, and the measurements must be taken quickly, at certain reaction times (i.e., certain methods for diols in wines). Usually, the measurements are taken at wavelengths in the visible spectrum (400–750 nm). These methods require calibration curves to be prepared with one single suitably chosen alcohol. However, in real samples more than one alcohol is present, and often it is desirable to quantify them together (i.e., higher alcohols). In this case, the results are less precise because each alcohol derivative has a different spectrophotometric response.

1. Gas Chromatography

Of the most widely diffused methods, gas chromatography is certainly the most common. The first method published for the determination of alcohols (higher alcohols) with GC (packed column) dates back to 1959 (13). During the 1960s and 1970s most of the analytical methods of alcohols in packed GC were developed, but the first satisfactory separations with capillary columns only took place at the end of the 1970s. [See Antonelli and Amati (14) for a brief review of this subject.] Available on the market is a wide choice of columns which are suitable for separation and quantification of alcohols, both at low and high temperatures.

The choice of the stationary phase depends on various factors, of which the first is the range of b.p. of the alcohols to be analyzed. The most low boiling are often also the most polar. As a consequence, the stationary phase will in this case have to be more polar than one which is designed to separate, for example, fatty alcohols containing over 20 atoms of carbon. Any derivatization will, however, change drastically its chemical-physical properties and will direct the choice of the stationary phase in a completely different way.

The most common stationary phases are of the polyglycol, polyalkylene glycol, or poly(cyano-propylphenyldimethylsiloxane) type. On the other hand, in the case of alcohols of high molecular weight, apolar stationary phases of the poly(diphenyldimethylsiloxane) type are used. In some cases porous polymer (mainly divinylbenzene or its copolymers) are also used for gas–solid chromatographic separations. Moreover, to research the enantiomeric composition of certain alcohols (i.e., 2-butanol), chiral stationary phases were used, of the type 2,3,6-trimethyl-β-cyclodextrin 5% vinyl. The market also offers the possibility of choosing between packed columns (packed and micropacked columns) generally made of glass or steel (diameters: from 1.8 mm i.d. or over; lengths: from a few decimeters to several meters), and capillary columns (diameters: 0.20–0.75 mm i.d.; lengths: from a few meters to over 100 m). The choice depends on the nature and quantity of the sample to be injected, on the concentration of the analytes to be quantified, etc. Examples of separations with various kinds of columns can be found in manuals, in literature, and in catalogs of chromatographic products and can help direct one's choice of method. It is to be remembered that a few couples of alcohols are often more difficult to separate, depending on the column used, e.g., the couple methanol–ethanol or methanol–acetaldehyde and amylic isomers. Many samples to be analyzed contain large amounts of water, which often constitute a cause of deterioration of the stationary phase and thus of the quality of the GC separation. Of the phase mentioned, poly(cyanopropylphenyldimethylsiloxane) is sensitive to moisture, whereas poly-alkylene glycol requires that water (and methanol) be vaporized before going into the column.

The accumulation of nonvolatile substances in the first part of the column may lower its quality and make the analysis less reliable. To lengthen the life of the column, it may be useful to rinse it from time to time, according to the manufacturer's instructions, or provide it with suitable precolumns. The choice of a bonded phase column is advisable for this aim besides giving greater thermal stability and less bleeding.

Quantification is usually carried out using calibration curves or by adding an appropriate internal standard. Quantifications using different methods are, however, possible, such as the addition of an external standard or the use of the method of standard additions. This is not the appropriate place to discuss criteria of choice or advantages and drawbacks of these quantification techniques, as books about general chromatography are easily available (i.e., 15).

The problem of quantification is often linked both to a good resolution of the peaks and to their shape. A common case is that of ethanol, which is often quantitatively prevalent and which, especially with packed columns, gives considerable tailings which make quantification difficult and interfere with successive peaks. The use of capillary columns of suitable polarity may greatly reduce these drawbacks. Simultaneously, particular care must be given to the technique of injection of the sample and, as a consequence, also to the choice of the quantification method. There are various techniques of injection of the sample into the column and the choice of one or the other naturally has advantages and drawbacks which must be considered in relation to each single case. Apart from analysis conducted with packed columns, where direct injection of the whole sample into the vaporizing chamber of the inlet takes place, in the case of capillary columns the most common but not the only technique is split injection. However, split injection causes considerable discriminations among the analytes which do in fact enter the column. [See Grob (16,17) for a closer examination of split-splitless and of on-column injection techniques, whereas Klee wrote a short review of the various injection techniques and relative inlets (18).]

The simultaneous separation of low-boiling and high-boiling alcohols is usually carried out using temperature-programmed runs. Usually they start from temperatures below 100°C (50°C or over) and reach temperatures of over 200°C. However, examples of isothermal separations abound, particularly with packed columns. Until now no explicit mention has been made of detectors which are available for gas chromatography. The characteristics and working principle of detectors for gas chromatography are described in GC manuals (i.e., 15) and in those of the manufacturers. Nonetheless, an overview may be obtained from a consultation of Buffington's booklet (19). It must also be remembered that the response of the detector is not usually the same for every analyte. If, for example, the internal standard method is chosen, the response factors (rf) of each single analyte to be quantified must be determined with respect to the internal standard. These rf values may vary with time, so it is preferable to determine them one or more times during a series of analyses.

It must be said that in most cases of analysis of alcohols, flame ionization detection is used. This detector gives one of the widest linear dynamic ranges and a satisfactory sensitivity for most applications. The mass selective detector (MSD) is particularly interesting. In this sector of analysis its use may prove fruitful in the identification of the peaks in very complex chromatograms, such as that of the analysis of the aromatic component (volatile) of several products (see Chapter 22).

2. High Performance Liquid Chromatography

Another chromatographic technique, which, however, is not as widely used in the analysis of alcohols in foods and beverages as gas chromatography, is high-performance liquid chromatography (HPLC). For a general introduction to this technique, see, for example, Refs. 20 and 21. In this case the choice of a suitable column is just as important as the choice of the most opportune mobile phase. Many analyses of alcohols with HPLC are applications to wines or musts and use various kinds of packing materials. Ion exchange columns are common. They give excellent results in the separation of mono- and disaccharides and sugar alcohols but are also suitable for the separation of low molecular weight alcohols. A study of the chromatographic behavior of 63 different compounds, 15 of which were alcohols, on columns of the strong cation exchange type may be seen in Ref. 22. These columns are made of polystyrene matrix resins which have sulfonic functional groups, usually in hydrogen or lead or calcium form. Separation takes place due to the effect of the hydroxyl coordination to the cation. In literature various applications may be found which are useful for directing the choice of the most suitable columns, but the manufacturers also provide useful indications in this.

From a general point of view, various different important elements must be considered for a good result of the analysis. Of these, readers will remember that the higher the crosslinkage, the more the resin is rigid and permeable to substances with a lower molecular weight, which are therefore held back longer. The resin ionic form contributes to selectivity due to interaction with the functional groups of analytes, whereas the choice of solvent, which is usually a diluted water solution of H_2SO_4, affects the resolution and symmetry of the peaks. Some authors, however, prefer to use phosphoric acid (H_3PO_4) as it is less corrosive on the components of HPLC (23). In some cases it may be advantageous to add to the mobile phase an organic modifier which greatly improves the resolution of the peaks. In deciding on the type and concentration of the modifier, it must be considered that the modifier may cause resin volume change and excessive back pressure (bed shrinkage and bed swelling). Very common is the use of acetonitrile in concentrations of between 5% and 30%. Also the temperature at which the chromatographic separation takes place is a determining factor for the resolution of the peaks since usually the efficiency of the column increases with the temperature and, correspondingly, a reduction of retention times is observed. The preparation of liquid samples for analysis is often limited to a simple filtration through a 0.45-μm filter. Other types of columns used in the HPLC separation of alcohols are the normal "direct phase" (silica) or "reverse phase" ones. The latter are prepared by bonding an alkylsilane reagent (methyl, octyl, octadecyl, or diphenyl) to silica.

There are several types of detectors for HPLC, but in the case of the analysis of alcohols, one of the most common is RI (refractive index). With this type of detector, a very accurate thermostatation of the cell is necessary, given the important effect of temperature on the RI. Electrochemical detectors, which function by applying suitable potentials at which the alcohols oxidize, are also available. The device keeps constant the potential across the cell by varying the current; then it is more properly termed "amperometric detectors." Le Fur et al. (24) examined their potentialities by applying them to the determination of flavor alcohols (mainly terpenols). The primary alcohols were quantified at concentration between 20 and 300 ppb, while the secondary and tertiary were quantified between 0.1 and 6 ppm. It also resulted that unsaturation in the alcohol drastically improves the limits of detection because of a better adsorption at the electrode surface.

With this type of detector it may be necessary to increase the conductivity of the solution by adding an electrolyte (i.e., NaOH, HClO4, etc.) at suitable concentrations. This postcolumn addition must be made by a reagent delivery module placed before the actual detector. This type of detector is very sensitive to bubbles, so the mobile phase must always be well degased. In particular, oxygen, which may interfere with the redox reactions, must be excluded, e.g., by helium sparging.

A detailed description of detectors for HPLC and the problems which may be encountered during their use may be found in Ref. 25.

The HPLC technique may be chosen not only for direct quantification of the alcohols but to separate them from complex matrices. This is the case of aliphatic alcohols and of triterpenic alcohols contained in fatty matter or in the unsaponifiable obtained from it [see applications section (V)] separated on a silica column (direct phase). In this case quantification takes place successively, using gas chromatography.

3. Supercritical Fluid Chromatography

Another chromatographic technique is supercritical fluid chromatography (SFC) which has attracted increasing interest during recent years. A mobile phase, consisting of a gas at temperature and pressure above its critical point, is used. Under these conditions the fluid has a density and solvating power similar to those of a liquid but the viscosity of a gas. For various reasons which will not be examined here, the gas most commonly used is carbon dioxide. However, it presents little ability to elute polar compounds and therefore the addition of small quantities of organic solvents (i.e., methanol, acetonitrile, etc.) may solve this problem. The solvent strength varies greatly with the density of the mobile phase and in a few SFC instruments it may be programmed by modifying radically the characteristics of the eluant. The detectors most commonly used are the FID and the light scattering detector (LSD). Since the elution of polar compounds with a mobile phase of carbon dioxide requires the addition of a polar solvent, the use of FID is not possible. In this case LSD may be a suitable choice. Nizery et al. (26) described the separation of standard solutions of fatty alcohols, on packed columns of the ODS (5–7 μm) or RP-18 (5 μm) type, and a mobile carbon dioxide–methanol phase (98.7:1.3, w/w). The analytes are revealed with an LSD whose working conditions they notably improve (nebulization and evaporation steps, before detection). Since the LSD response depends on the particle diameter in the cell of the detector, which is in relation to the concentration of the analyte, the use of a correction factor is suggested. However, the biggest drawback in using this detector is the nonlinear relationship between concentration and LSD response.

A combination of SFE and capillary SFC, applied to the analysis of saponified jojoba oil, is described by King (27). In this case an FID was used, and the fraction containing the fatty alcohols was well separated from the nonsaponified residue of oil. The advantage in using SFE is that one can carry out a selective extraction of the compounds of interest from complex matrices by using the solvent properties of the supercritical fluid, which are easily modifiable. In other words, it is possible to make "targeted" extractions. For a more detailed description, see the relevant chapter of this book as well as available texts (e.g., Ref. 28).

4. Thin-Layer Chromatography

Another chromatographic technique which in the field of analysis of alcohols has been applied in a very limited way is thin-layer chromatography (TLC). In the case of alcohols, of all of the TLC separation techniques, the most commonly used is "ascending TLC" in the common trough chamber, applied as single development.

The principles at the basis of TLC have been known for over a century, but only in the last 30 years has it been developed as a separative technique and codified in methods of analysis. The reader can consult specific texts (i.e., Refs. 29 and 30) for a description of the technique, which these days is becoming increasingly important in various fields of analysis, due in part to the huge progress which technology has made possible in providing quantitative, reproducible results. Devices have in fact been produced which make possible the nondestructive quantification of analytes separated with TLC. These are "thin-layer scanners," which can take measurements of absorption and fluorescence, take measurements at several wavelengths, carry out background correction, etc. All of these operations, widely automatized, are managed by means of microprocessors. With suitable calibration functions it is therefore possible to obtain quantitative results. Of the destructive analyzers, we remind readers of the chromarod-iatroscan, whereby a quartz rod is coated with silica gel (chromarod). Separation takes place along the chromarod which, at the end of the development, is scanned in the iatroscan by a flame which acts as an FID, for the whole length. The signal is recorded and elaborated by a microprocessor.

TLC has been used most of all as an efficient means of separating high molecular weight (MW) alcohols (aliphatic alcohols, triterpenic alcohols) from complex matrices and subjecting them to successive GC analysis. In some cases separation of high MW alcohols was preceded by their derivatization and then individual identification with HPLC was performed. In the case of low MW alcohols, however, it is necessary to derivatize the hydroxyl groups before TLC separation [formation of xanthates (31) or urethanes (32)], while the stationary phase and the mobile phase (eluant) are chosen in relation to the characteristics of polarity of the derivatives. On the other hand, in cases like determination of DEG in wines or of glycerol in vinegars, the sample can be applied as it is, without derivatization. Naturally, separation on silca gel layers requires a mobile phase of suitable polarity like, for example, mixtures of such solvents as acetone, chloroform, aqueous ammonia solutions, etc. So, in the separation of alcohols such as xanthates carried out on the stationary phase of microcrystalline cellulose, the eluant can be a mixture of n-butanol, water, and ammonia solution.

Once the chromatographic separation is over, the various analytes must be visualized for their qualitative and, perhaps, quantitative recognition. There is a large number of visualization reagents suited to various kinds of compounds. In the case of alcohols we can mention molybdatophosphoric acid, particularly suited for detection of reducing substances; potassium dichromate/H₂SO₄, which is a universal visualizer for its high oxidizing power (if necessary, the plate can be heated to 150°C); and dichlorofluorescein sodium salt, which is very efficient when the plate is exposed to ultraviolet light, where the bands take on a color that gives a distinct contrast with the fluorescence of the background. Quantification usually takes place in a removal of the analytes from the plate and by proceeding with other analytical techniques. But as it has already been said, in certain cases, the aforementioned devices may be used.

5. Nuclear Magnetic Resonance

One technique to be mentioned in passing, as it is adopted in the case of special studies, is nuclear magnetic resonance (NMR) [For a general description of the technique, see Chap. 46 as well as specific texts (i.e., 33).] Here the readers are reminded that the frequency at which the moment of spin "wobbles" permits the recognition of the elementary segments of which a molecule may consist.

This specificity allows the origin of ethanol in various alcoholic products to be identified, the quantity of added sucrose in wines to be measured, and even the effect of factors of climate on the biosynthetic mechanisms which take part in the formation of ethanol to be recognized, or the water resources and the manufacturing processing to be made known (fermentation, yeast, temperature cycle, etc.) (34,35).

6. Scintillation Spectrometry

The determination of synthetic ethanol (from petroleum or coal) in products which should not contain any (i.e., wines, vinegars) can be conducted by means of researching the level of radioactivity of the ^{14}C in ethanol. The method is rather complicated and makes use of scintillation spectrometry. For a complete description, please see the specific texts, as it is not usually used out of the context of this type of determination. From the number of disintegrations per minute per gram of C (dpm/g C) it is possible to find out whether there have been fraudulent additions. In fact, tables have been prepared which give the average values of radioactivity of ^{14}C in ethanol of safe origin, for each year, which constitute the comparable values for the tested samples.

7. Enzymatic Methods

Given their high specificity, enzymatic methods have been widely applied in the field of food analysis. These methods are based on a measurement of an absorbance in the UV spectrum or in the visible (colorimetric methods). In the first case, the variation of absorbance of the NADH coenzymes (reduced nicotinamideadenine dinucleotide) or the NADPH coenzymes (reduced nicotinamideadenine dinucleotide phosphate) is measured. If a spectrophotometer is available, the measurement is taken at the greatest absorbance (340 nm), whereas with a mercury lamp photometer the measurements are usually taken at 334 or 365 nm. During the last 10 years, thanks to immobilization techniques, a large number of biosensors have been perfected. These incorporate the specific enzymes for each analysis by means of adsorptive or covalent bonding. However, in some cases there is a crosslinking in which the enzyme itself, or after the addition of an inert protein (usually albumin), reacts forming a water-insoluble macromolecule. The biosensor is equipped with an electrode which in most cases is sensitive to O_2 or to H_2O_2, but there are others which are sensitive to NH_3, to pH, to CO_2, etc., depending on their uses. A review of this subject, with numerous bibliographical references, was published by Schwedt and Stein (36). In the case of alcohols, the most common biosensors measure pO_2, which diminishes during the oxidation of the alcohol (e.g., methanol and ethanol, respectively, to formaldehyde and acetaldehyde) because

O_2 is reduced to hydrogen peroxide (H_2O_2). Sensors with an optical detection method (fiberoptic biosensors) and O_2-sensitive fluorescing indicators (37) have also been described, and these find interesting applications in the field of fermentation processes (e.g., beer brewing). The advantage of biosensors in comparison with traditional enzymatic methods is mainly the possibility of constant monitoring of processes like industrial fermentations and that of taking immediate action. Naturally, the use of biosensors also requires that the temperature be controlled. Beyond certain values, for progressively longer times, the denaturing effect greatly lessens the sensor's response. The high specificity of the enzymatic methods usually allows a reduction in treatment of the sample, which may be cooled and/or deproteinized (with perchloric acid or by heating in hot water) to avoid interfering side reactions and which may, for spectrophotometric measurements, be decolored with products like polyvinylpolypyrrolidone or clarified with Carrez solutions. One of the most important things to remember when carrying out enzymatic analyses with kits which are available on the market is to prepare the reagents with bidistilled water on glass or "equivalent," better if freshly distilled. The presence of metal contaminants (which may come, for example, from metal distillers) of microorganisms, etc., may compromise the result of the analysis and the conservation of the reagents.

V. APPLICATIONS

A. Wines and Musts

1. Alcohol

a. AOAC Methods (38). Determination of alcohol (by volume) may be carried out by determining the specific gravity on the alcoholic distillate (method number 920.57), or by using a refractometer (method number 920.58), or by dichromate oxidation (method number 969.12). In this last case, a steam flow distillation is carried out (Figs. 1 and 2), and the distillate is collected in an Erlenmeyer which contains an acid solution of potassium dichromate. The excess of dichromate which has not reacted is titrated with $Fe(NH_4)_2(SO_4)_2$ using *o*-phenanthroline as indicator.

GC method (method number 983.13) requires the use of a glass-packed column and on-column injection. 2-Propanol is used as the internal standard. The determination of alcohol content (by weight) is always carried out on the alcoholic distillate by converting the value by volume into

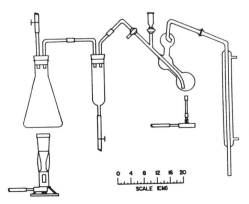

Fig. 1 Distillation apparatus for chemical determination of alcohol in wine. (From the *Official Methods of Analysis*, 15th ed., Chap. 28, method no. 969.12, Fig. 969.12A. Copyright 1990 by AOAC International.)

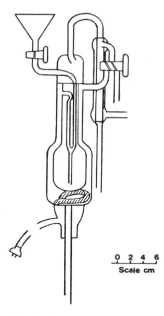

0 2 4 6
Scale cm

Fig. 2 Electric distillation apparatus for chemical determination of alcohol in wine. (From the *Official Methods of Analysis*, 15th ed., Chap. 28, method no. 969.12, Fig. 969.12B. Copyright 1990 by AOAC International.)

value by weight, by multiplying by the specific gravity and dividing by the specific gravity of the sample (method number 920.59).

b. EEC Methods (39).   The EEC regulations require the use of alcoholometers or densitometers according to the method described in (40).

c. Gas Chromatographic Methods.   Gerhards (41) describes a statical headspace GC method suitable for the routine determination of ethanol and methanol in wines. This is a method using dedicated apparatus which conditions samples and injects them automatically. The author suggests the use of *n*-butanol as an internal or external standard or the method of standard additions.

2. Methanol

a. EEC Methods (39).   Any GC method is suitable provided it is possible to reveal concentration less than 20 mg/L and the repeatability be less than 20 mg/L of 100% of methanol.

b. Other Methods.   The determination of methanol can be carried out following various methods, some of which are also common to liqueurs and alcoholic distillates (see V.D, V.E). In this section those which are studied specifically for wines are discussed. The Italian Ministry of Agriculture and Forestry issued a decree on the subject (42). Here two methods are described: one a colorimetric method providing for the oxidation of methanol with $KMnO_4$ and dose of formaldehyde formed colorimetrically after reaction with Schiff's reagent, the other a GC method with packed column, which uses *n*-butanol as the internal standard. The response factor must be determined. The result is expressed in ml/100 ml of total alcohol.

3. Ethanol

a. Gas Chromatographic Methods. Yusty et al. (43) suggested an isothermic GC separation on packed column (Porapak Q) of ethanol and glycerol in only 15 min. They also verified the effect of various possible interferents and quantified a CV% = 2.28 for ethanol and a CV% = 4.0 for glycerol. The method was then applied to 36 Spanish wines.

b. HPLC Methods. Pfeiffer and Radler (44) separated the main constituents of wine with a column cation exchange, a mobile phase of diluted H_2SO_4, and an RI detector. In particular, they obtained an excellent separation of ethanol and glycerol from the other components in a single run of 25 min. The authors also compared conventional methods, showing a good agreement in the results.

c. Enzymatic Methods. Belghith et al. (45) described an enzyme electrode for the on-line determination of ethanol and make a satisfactory comparison between the results obtained on a red wine with those of a classic enzymatic method and, on the distillate, between a chemical method and their biosensor. The biosensor measures pO_2 (oxygen partial pressure) and it has also been tested for the determination of methanol, but on standard solutions.

4. C1–C5 Alcohols and Glycerol

a. HPLC Methods. Pecina et al. (46) used fixed ion resin columns in hydrogen ion form to separate in a satisfactory way mixtures of low MW alcohols. At the same time, glycerol is also well resolved from other alcohols present. The authors worked on columns of 300 x 7.8 mm, at 50°C, using as eluant a 0.01 N H_2SO_4 solution, with a flow of 0.7 ml/min and RI detector.

For fast screenings, the use of shorter columns has been suggested (47), with an eluant which consists of a 0.002 N H_2SO_4 solution, greater flows (1.2 ml/min), thermostatation at 65°C, and RI detector.

5. Higher Alcohols

a. EEC Methods (39). This is a spectrophotometric method: the alcohols react with salicyl-aldehyde in the presence of hot sulfuric acid (reaction of Komarowsky). The solution absorbance is measured at 560 nm and compared to a previous calibration made with five 2-methylpropan-1-ol solutions (1, 2, 4, 6, 10 mg/L). The total aldehyde concentration has to be taken into account by a previous spectrophotometric measurement with a calibration with acetaldehyde solutions (from 1 to 10 mg/L). The method requires a repeatability of less than 2.5 mg/L of 100% ethanol.

6. Diethylene glycole

a. Gas Chromatographic Methods. Conte et al. (48) described a GC method for the determination of DEG and tried it on two wines (Sangiovese, red; Marino, white) spiked with increasingly large quantities of DEG. The method provides for the addition of 1,4-butanediol as the internal standard, an ion exchange with a suitable resin, and, on the filtrate concentrated with "rotavapor," extraction with diethyl ether and derivatization in pyridine with hexamethyldisyla-zane and trimethylchlorosylane before the programmed GC run. The authors tried the method using both capillary and packed columns and found a sensitivity level of 5 ppm.

b. HPLC Methods. With the same type of column used by Pecina et al. (46) (v. C1–C5 alcohols and gylcerol), determinations of DEG in wines were carried out (47). The eluant used was a 0.002 N H_2SO_4 solution at a flow of 0.6 ml/min, with the column at 65°C. In this case, too, the detector is the RI.

McMurrough and Byrne (49) reported a method which provides for the separation of DEG from wine by means of an Extrelut cartridge and a successive derivatization with anthracene-9-carbonyl chloride. HPLC separation takes place by means of a silica column (25 cm x 4.9 mm, 5μm), with a mobile phase of chloroform/hexane (20:80) at 35°C. They also use two detectors, for absorbance (252 nm) and fluorescence (ex. 360 nm, em. 460 nm) both for the best confirmation of the identity of the peak and for its greater sensitivity (less than 1 ppm).

7. 2,3-Butanediol

The Italian Ministry of Agriculture and Forestry issued a decree on the subject (42). With this method, the purified distillate obtained for the determination of glycerol is used. First, oxidation of 2,3-butanediol to acetic aldehyde with sodium periodate is carried out. Then sodium nitroprusside and piperidine are added. Finally, reading at 570 nm takes place. A calibration curve has to be prepared.

8. Glycerol

a. *AOAC Methods (38)*.   In wine and grape juice (method number 991.46), this is an HPLC method using a strong cation column at 65°C, with an RI detector and a mobile phase of degassed water by boiling or helium sparging. Before starting, the sample must be filtered (0.45 μm) and it is necessary to prepare different standard solutions (external standard) depending on whether one is dealing with wine or grape juices. In dry wines (method number 920.60), this is a surplus method. In sweet wines (method number 920.61), this is a surplus method.

b. *Gas Chromatographic Methods*.   See Ethanol, Yusty et al. (43).

c. *HPLC Methods*.   See Ethanol, Pfeiffer and Radler (44).

Caputi et al. (50) conducted a collaborative study for the determination of glycerol in wine and grape juice. This method uses a strong cation exchange type column (H^+ form), degassed water as the mobile phase, and RI detector. The glycerol was determined on samples simply after fitration (0.45 μm) and with an external standard. For wines, the authors obtained a repeatability of 1.25% and a reproducibility of 2.79%.

d. *Enzymatic Methods*.   Glycerol may be determined enzymatically with the method described in the Boehringer method collection (51). Only a preliminary dilution based on the estimated glycerol content is necessary. Red wines can be analyzed even without any previous decoloration. The reactions produce a quantity of NADH equivalent to the quantity of glycerol and absorbance is measured at 334 or 340 or 365 nm.

e. *Other Methods*.   The Italian Ministry of Agriculture and Forestry issued a decree on the subject (42). With this method, a distillation with a Vigreaux column is carried out, followed by a cleanup of the distillate on ion exchange columns. A part of it is again distilled and on the residue, oxidation of the glycerol to formaldehyde with sodium periodate, reaction with phloroglucinol, and reading at 460 nm are carried out. A calibration curve has to be prepared.

9. Alcohols in Volatile Fractions

Gomez et al. (52) researched the variation of volatile compounds due to the type of vinification on Monastrell-type grapes. They adopted the method of extraction described in (53): distillation at 35°C and 35 mm Hg of pressure. As internal standard, 2-octanol was added to the must before the isolation step. The distillation apparatus consisted of two traps, ice water and liquid nitrogen, to condense the compounds during a 2-hr distillation.

Furthermore, the aqueous condensed phase was solid–liquid–extracted (Chromosorb 101). Volatile compounds were eluted with diethyl ether and then concentrated by a stream of nitrogen.

The analyses were performed by high-resolution gas chromatography (HRGC), whereas identification was by GC-MS. The authors identified and quantified nine alcohols at different concentrations according to the type of fermentation (Tables 1–3).

To be cited is the work of Sefton et al. (54) on the volatile secondary metabolites of *Vitis vinifera* cv. Sauvignon blanc. For the analytical methods, the authors refer to preceding works and only point out extraction procedures with Freon F11, absorbant C-18 RP (reverse phase), and identification with GC-MS, ^{1}H NMR, and ^{13}C NMR. This work has a large bibliography.

Gerbi et al. (55) suggested the use of an Extrelut resin (Merck) to isolate the volatile fraction (Fig. 3) and compare this method with another method of extraction with solvent (pentane/dichloromethane, 7:3 for 72 hr continuously). They investigate a red wine, a sparkling "Asti" wine, and a white vinegar. The authors arrive at the conclusion that this method can be used as an alternative to the more laborious liquid–liquid method of extraction.

Di Cesare and Nani (56) also make use of a resin (KS 112) to adsorb the volatile fraction. In particular, they prepare a juice from "Isabella" grapes (from Catania, Italy) after enzymatic treatment for 90 min at 30–35°C and further extraction of volatiles in a glass climbing film evaporator. The extract was successively concentrated with KS 112 resin and then analyzed by GC-MS. The alcoholic fraction represents about 7% of the total area, the highest being 1-hexanol. They also found geraniol.

Romero et al. (57) researched the efficiency of extraction with solvents of the volatile fraction of 15 wines. Of this fraction, 9 alcohols are considered: 2-methylpropanol, 2-phenylethanol, 2-methylbutanol, 3-methylbutanol, 3-pentanol, 2-pentanol, 1-pentanol, and 1-butanol. Heptanol and octanol were used as internal standard. In particular, they investigate the dichloromethane-pentane (2:3, v/v) system and find that efficiency increases with concentration and molecular weight of the volatiles. They also find a CV = 5% in 60% of cases and not more than 10% in the other cases. Identification and quantification of the components were carried out by means of capillary or packed GC.

Stashenko et al. (58) studied the formation of volatile during fermentation of must using a purging/extraction apparatus devised by Umano and Shibamoto (59). It uses purified air followed by nitrogen to purge the headspace and dichloromethane for extraction. To identify the volatile, the authors used a gas chromatograph equipped with an FID or an FPD (flame photometric detector) interfaced to a mass spectrometer. They mainly found ethanol, isopentanol, pentanol, and hexanol.

Uzochukwu et al. (60) researched the volatile fraction of palm wine and pasteurized palm sap with the intention of identifying the components responsible for their typical aroma. They trapped the volatiles on Tenax GC by bubbling helium through 2 ml of the sample at 25°C. Following this, the authors carried out GC and GC-MS analysis. Of the alcohols identified, they found ethanol, propanol, 2-methylpropanol, 3-methylbutanol, esanol, octanol, and, after full fermentation, dodecanol and hexadecanol. Moreover, where possible, they correlate every component with an odor comment (Fig. 4).

    a. Other Determinations. Mazzoleni and Colagrande (61) investigated more than 200 sparkling wines and 16 volatile components were determined on a smaller number of samples. They also determined *ethanol* according to Ref. 42. The % vol was comprised between 10.82 and 13.72 with a CV% = 3.3%. *Higher alcohols* (88 wines) were analyzed according to Ref. 62. This is a programmed temperature GC method, which uses a packed column with Porapak Q and a precalibration with standards at different concentration. The *volatiles* were analyzed by HRGC after continuous extraction in pentane/methylene chloride (2.1 v/v) according to Ref. 63. The CV

Table 1 Changes in Alcohols During Red Wine Vintification (mg/L) (In the present work, the alcohol section only has been reproduced from the original table.)

Component	1 (1989)	1 (1990)	3 (1989)	3 (1990)	5 (1989)	5 (1990)	10 (1989)	10 (1990)	15 (1989)	15 (1990)	45 (1989)	45 (1990)	90 (1989)	90 (1990)
Alcohols														
1-propanol	—[a]	—	—	—	0.02	—	0.02	0.14	0.08	0.11	0.07	0.07	0.06	0.06
Isobutanol	0.06	—	0.55	0.04	0.51	2.93	3.48	3.44	5.26	5.48	5.34	2.20	4.57	2.23
1-butanol	0.05	0.08	0.07	0.01	0.11	0.08	0.10	0.26	0.17	0.20	0.11	0.09	0.07	0.12
Isoamyl alcohol	0.28	0.19	2.89	0.33	137.23	72.18	125.80	120.60	186.18	187.76	184.25	177.39	182.22	179.20
(E)-2-hexenol	0.61	0.59	0.09	0.71	—	—	—	—	—	—	—	—	—	—
Hexanol	0.39	0.19	1.31	0.60	1.27	1.08	1.14	1.34	1.22	1.43	1.10	1.34	1.08	1.04
Amyl alcohol	—	0.01	0.02	0.02	0.07	0.09	0.06	0.15	0.08	0.21	0.08	0.08	0.09	0.07
1-decanol	—	0.02	—	0.02	1.82	1.64	1.89	1.28	1.62	0.38	1.12	0.19	0.46	0.22
2-phenylethanol	0.10	0.03	0.9	0.13	23.31	12.76	22.37	12.44	21.61	20.31	26.35	15.08	35.20	14.33
Sum of alcohols	1.49	1.11	5.83	1.86	164.34	90.76	154.86	139.65	216.22	215.88	218.42	196.44	223.75	197.27

Source: Institute of Food Technologists, Chicago.

Table 2 Changes in Alcohols in Rose Vinification (mg/L) (In the present work, the alcohol section only has been reproduced from the original table.)

Component	1 (1989)	1 (1990)	3 (1989)	3 (1990)	5 (1989)	5 (1990)	10 (1989)	10 (1990)	15 (1989)	15 (1990)	45 (1989)	45 (1990)	90 (1989)	90 (1990)
Alcohols														
1-propanol	—a	0.01	—	0.07	—	0.13	0.12	0.03	0.18	0.21	0.14	0.05	0.19	0.19
Isobutanol	0.17	—	0.25	1.15	1.16	1.19	5.55	2.34	3.38	2.50	4.24	3.67	8.03	2.78
1-butanol	0.07	—	0.11	0.10	0.12	0.12	0.25	0.13	0.08	0.06	0.06	0.05	0.08	0.07
Isoamyl alcohol	1.27	0.29	1.57	31.11	12.60	83.62	121.22	144.00	135.81	124.00	122.93	75.20	146.23	113.00
(E)-2-hexenol	0.73	1.49	0.64	—	—	—	—	—	—	—	—	—	—	—
Hexanol	0.58	0.46	0.61	1.91	2.00	1.60	2.86	1.72	1.52	1.71	1.53	1.99	1.61	2.27
Amyl alcohol	0.09	—	—	0.06	0.05	0.16	0.05	0.02	0.05	0.06	0.06	0.06	0.06	0.06
1-decanol	0.02	—	0.09	2.63	0.63	5.12	0.69	5.93	7.88	0.87	2.75	1.04	2.15	1.22
2-phenylethanol	0.30	0.05	0.68	1.33	0.62	4.12	8.02	16.26	11.92	22.16	13.21	9.39	10.71	14.35
Sum of alcohols	3.23	2.30	3.95	38.36	17.18	96.06	138.76	170.43	160.82	151.57	144.92	91.45	169.06	133.94

Source: Institute of Food Technologists, Chicago.

Table 3 Changes in Alcohols in Carbonic Maceration (mg/L) (In the present work, the alcohol section only has been reproduced from the original table.)

	Days									
	1		3		5		15		90	
Component	1989	1990	1989	1990	1989	1990	1989	1990	1989	1990
Alcohols										
1-propanol	0.05	0.05	0.06	0.20	0.14	0.34	0.14	0.06	0.20	0.16
Isobutanol	4.20	3.53	4.84	4.42	6.54	7.17	2.89	2.47	4.65	1.64
1-butanol	0.12	0.03	0.13	0.42	0.18	0.67	0.96	0.23	0.15	0.17
Isoamyl alcohol	40.06	51.87	100.00	206.00	173.89	184.00	132.23	128.00	161.80	110.00
(E)-2-hexenol	–[a]	–	–	–	–	–	–	–	–	–
Hexanol	0.70	1.48	0.66	1.02	0.79	1.14	0.63	1.21	0.85	1.15
Amyl alcohol	0.06	0.02	0.06	0.17	0.22	0.11	0.05	0.09	0.06	0.11
1-decanol	0.30	2.30	3.44	3.41	2.05	3.80	3.25	1.27	0.75	0.26
2-phenylethanol	4.00	4.53	5.71	10.00	15.65	10.30	22.64	14.28	14.40	13.44
Sum of alcohols	49.49	63.91	114.90	225.64	199.46	207.57	162.79	147.61	182.86	126.63

Source: Institute of Food Technologists, Chicago.

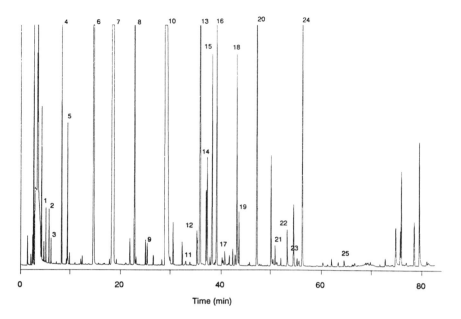

Fig. 3 Gas chromatogram of the aromatic fraction of a white vinegar neutralized with ammonia and extracted with Extrelut resin. 2, 2-Butanol; 3, 1-propanol; 4, 2-methyl-1-propanol; 6, 3-methyl- + 2-methyl-1-butanol; 9, 3-ethoxy-1-propanol; 13, (-)-2,3-butanediol; 15, (*R, S*)-2,3-butanediol; 23, benzyl alcohol; 24, 2-phenylethanol. [Only the alcoholic fraction has been referenced in the present work. From *Ital. J. Food Sci., 4*: 259 (1992).]

are between 29.7 (1-butanol) and 44.8 (2-phenylethanol). *Glycerol* was analyzed according to Ref. 64. Concentration ranges from 3.82 g/L up to 9.56 g/L, with a CV% = 23.2%.

B. Vinegars

1. Alcohol

a. AOAC Methods (38).   A sample of 100 ml faintly alkaline (NaOH) is distilled and the alcholic distillate is transferred to the pycnometer so that its density may be measured (method number 930.35Q). The result is expressed as % by volume from 913.02.

b. Other Methods.   In Ref. 65, there is the description of a method which is different from the AOAC 930.35Q method due to the fact that the alcoholic distillate is carried to volume, slightly acidified (H_2SO_4, 1:4); and redistilled. The second distillate is once more carried to volume and its density measured with the pycnometer.

2. Methanol

The Italian Ministry of Agriculture and Forestry issued a decree on the subject (42). See Methanol, Other methods, in Wines and Musts (V.A).

3. Ethanol

a. Enzymatic Methods.   The Boehringer enzymatic method (51) does not require the neutralization of the sample of tested vinegar if it is already greatly diluted. Otherwise the sample must be filtered and neutralized before it can be subjected to analysis.

Plessi et al. (66) used the enzymatic method to determine the ethanol contained in the traditional

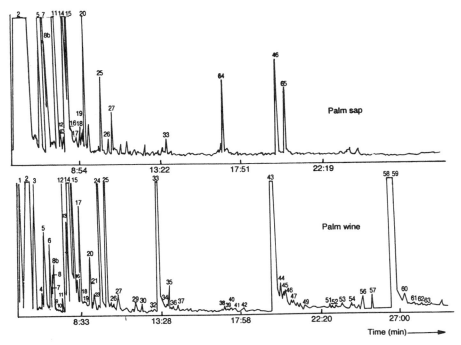

Fig. 4 Gas chromatograms of pasteurized palm sap (above) and palm wine (below). 1, ethanol; 4, unknown + propanol; 5, 2-methylpropanol; 14, 3-methylbutanol; 36, hexanol; 38, octanol [Only the alcoholic fraction has been referenced in the present work. From Volatile constituents of palm wine and palm sap, by S.V.A. Uzochukwu, E. Balogh, O.G. Tucknot, M.J. Lewis, and P.O. Ngoddy, *J. Sci. Food Agric.*, *64*: 405 (1994).]

Modena balsamic vinegar and in apple vinegar by means of the oxidation to acetaldehyde which is further oxidized to acetic acid. The concentration of NADH produced is measured at 340 nm.

4. Nonbiogenic Ethanol (65)

This method is based on the measurement of natural radioactivity of ^{14}C. The measurement is made with a scintillation spectrophotometer in liquids, and the result is expressed in disintegrations per minute per gram of carbon (dpm/gC). The measurements can be compared with normal values for wholesome products by the use of tables.

5. Glycerol

a. AOAC Methods (38).   Here the method 930.35R (surplus method), already mentioned for other cases, is used.

b. Enzymatic Methods.   As in the case of ethanol, Plessi et al. (66) determined the glycerol content in the traditional Modena balsamic vinegar and in apple vinegar by means of successive reactions which lead to the formation of pyruvate which then reacts with the NADH in excess. The NADH consumed is stoichiometric with the amount of glycerol in the sample.

Troncoso Gonzales and Guzman Chozas (67) suggested an enzymatic method which they compared with a chromatographic-spectrophotometric one. In particular, they separated the glycerol using TLC, which was then extracted and made to react with periodic acid. Then there was the reaction with chromotropic acid and the color which developed was read at 540 nm. The enzymatic method is the one described in Ref. 51. From a statistical analysis of the data obtained with the two methods (*F* and *t* test) the authors concluded that there are no significant differences between the two methods, even though they believe the enzymatic method to be more accurate.

c. Other Methods. In the literature several methods for the determination of glycerol can be found, such as, for example, a method of separation from interfering substances with ionic exchangers, oxidation with periodic acid, and colorimetric dosing of the formic aldehyde produced (65).

A chromatographic-spectrophotometric method was used by Troncoso Gonzales and Guzman Chozas (67) on the basis of a former work of Sauvage and Wagstaffe (68) [see section on enzymatic method (V.B.5.b)].

6. Alcohols in Volatile Fractions

Annovazzi Rizzo (69) researched the composition of the volatile fraction of various wine vinegars by means of a liquid–liquid extraction of the volatile fraction and injection into HRGC (OV-351 column). Esanol, *trans*-3-esenol, *cis*-3-esenol, 2,3-butanediol, benzilic alcohol, and β-phenylethyl alcohol are quantified. The levels of concentration vary from 0.1 mg/L to more than 75 mg/L. The extraction takes place on a filtered sample (0.45 μm) to which the internal standard is added (cyclohexanone). Then a solvent is added (dichloromethane/pentane 2:1 v/v) and the extraction is protracted for 12 hr in a water bath at 50°C (Fig. 5). The extracted volatile fraction is concentrated and then analyzed in HRGC.

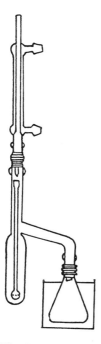

Fig. 5 Liquid–liquid extraction apparatus. (Courtesy of G. Annovazzi Rizzo, Lab. Sanita' Pubblica, USSL no. 51 Novara, Italy.)

For a description of the method suggested by Gerbi et al. (55), see section on alcohols in volatile fractions (V.A.9).

C. Beers and Malt Beverages

1. Alcohol

a. AOAC Methods (38). Determination of alcoholic content (by volume) is described in method 935.21. A distillation is carried out, followed by a measurement of density with a pycnometer.

The determination of alcoholic content (by weight) is described in method 935.22. A distillation is carried out, followed by a measurement of density with a pycnometer. An alternative method is described which requires a measurement with an immersion refractometer (method number 970.15). However, the beer must first be decarbonated.

2. Methanol (+ Ethanol)

a. Biosensors. A type of biosensor for "flow systems" is based on the enzyme alcohol oxidase of the yeast *Hansenula polymorpha* trapped in a gelatin matrix. The detection is by a pO_2 sensor (partial O_2 pressure). This sensor is also suitable to detect ethanol. The ethanol could be measured in concentrations ranging from 0.5 mmol/L up to 15 mmol/L, whereas methanol can be measured from 10 mmol/L to 300 mmol/L (70).

3. Ethanol

a. AOAC Methods (38). Here a GC method is provided for (method number 984.14); it is necessary to decarbonate the sample and to use *n*-propanol as internal standard.

b. HPLC Methods. Cieslak and Herwig (71) perfected a method for the quantification of ethanol using a hydrogen form cation exchange column. They use a 0.01 N H_2SO_4 water solution as mobile phase, at 85°C, RI detector, and obtain the separation of ethanol, glycerol, and glucose in 10 min. Quantification was made with the use of an external standard. The samples to be analyzed are subjected only to ultrasonic degassing. The authors also ascertain an excellent agreement with distillation methods.

c. Enzymatic Methods. In Boehringer's enzymatic method (51), approx. 5–10 ml of the sample first has to be decarbonated via stirring with a glass rod. Then the sample is suitably diluted and enzymatic determination of the ethanol present performed by means of a UV spectrophotometer, using the reagents in the kit.

In the simplified determination, the sample as such is diluted and one proceeds to determination by UV spectrophotometer.

d. Biosensors. A recent review Schwedt and Stein (36) mentions the possibility of determining ethanol in fermentation processes including beer brewing. The enzyme is alcohol oxidase embedded in a silicone rubber membrane. The detection is by an optical sensor (O_2-sensitive fluorescing indicator). The authors state that it is possible to calibrate the biosensor in the range of 50 to 500 mmol/L of ethanol. Furthermore, the detection limit is 10 mmol/L of ethanol and the accuracy is ±4% at 100 mmol/L. Another biosensor for ethanol in alcoholic beverages is based on an alcohol dehydrogenase, NADH oxidase enzyme covalently coimmobilized to Sepharose 4B with an O_2 electrode detector (72).

4. Fusel Alcohols

An example of a determination of fusel alcohols (propanol, isobutanol, and isoamyl alcohol) in the beer fermentation process is provided by Garcia et al. (73) in order to establish a kinetic model at different constant temperature and pH values. The yeast was *Saccharomyces carlsbergensis*. It is an HSGC method with a GC fitted with an FID and a packed column. A sample (100 ml) is degassed first, then an internal standard of *n*-butanol is added. After an incubation at 40°C for 1 hr, 3 ml of the headspace is injected into GC for a temperature-programmed run.

5. Glycerol

   a. AOAC Methods (38). In this method a dichromate oxidation is performed (method number 930.18); this is a surplus method.

   b. HPLC Methods. See Ethanol, Cieslak and Herwig (71).

   c. Enzymatic Methods. In Boehringer's enzymatic method (51), the carbon dioxide must first be removed from the sample of beer (5–10 ml) by stirring it with a glass rod for about 1 min and then diluted according to the estimated glycerol content. A description of the principle of the method can be found in the section on wines [see section on glycerol, enzymatic methods, in Wines and musts (V.A.8.d)].

D. Distilled Liqueurs

1. Alcohol

   a. AOAC Methods (38). The alcoholic content (by volume) is determined by distillation and later measurement with a pycnometer (method number 942.06). Different steps are taken if the sample contains more or less than 60% alcohol by volume.

   An alternative method, applicable to spirits containing less than 600 mg/100 ml of extract, is a hydrometer method (method number 957.03). This method gives the corrections to be applied according to the content of extract.

   Other possibilities are a densitometric method (method number 982.10), a refractometric one (method number 950.04), and one using Williams tubes (method number 926.03).

   The alcoholic content (by weight) is determined by distillation and measurement of the density with a pycnometer (method number 945.07).

2. Methanol

   a. AOAC Methods (38). Here three methods are possible. In the chromotropic acid colorimetric method (method number 958.04), the alcoholic distillate is used and a photometric measurement taken at 575 nm. In the immersion refractometer method (method number 905.01), too, the alcoholic distillate is used. Finally, in a GC method (method number 972.11), a preliminary determination is carried out to ascertain whether *n*-butanol is present. If *n*-butanol is absent, a solution of it may be made and used as internal standard. Otherwise a calibration curve must be prepared with standard methanol solutions.

   b. Other Methods. Similar to the above-mentioned, the dosage of methanol in brandies from pears, blackberries, and kirsch is measured through the colorimetric reaction with chromotrophic acid (reading at 570 nm) (74).

3. Isopropanol and Tertiary Butyl Alcohol (Qualitative)

a. AOAC Methods (38). A qualitative method for ascertaining the presence of these alcohols (besides acetone and other ketones) makes use of a solution of mercuric sulfate, which in the presence of these compounds in the conditions of the method gives rise to a white or yellow precipitate (method number 935.16).

4. Higher Alcohols

a. AOAC Methods (38). These are two GC methods (method numbers 968.09 and 972.10) for the determination of isobutanol, isoamylol, and *n*-propanol, which require the preparation of standard solutions at different concentrations, according to the concentration of alcohols in the sample.

b. Other Methods. A GC method (with capillary column) is described in Ref. 74. This is an improved version of the "Méthodes Officielles d'Analyse des Alcools et Eaux de Vie" for the determination of 2-butanol, 1-propanol, isobutanol, 2-methyl-1-butanol; and 3-methyl-1-butanol. Standard solutions of these alcohols are prepared at five different concentrations for each one. They vary between 10–200, 20–1500, 20–100, 5–60, 15–80, and 50–200 mg/L, respectively. The method uses 4-methyl-2-pentanol (5g/L) as internal standard.

Usseglio-Tomasset and Matta (75) made a collaborative study (seven laboratories) on GC analysis of higher alcohols in "grappa." The method chosen makes use of a packed column with Carbopack 80, 0.2% of Carbowax 1500. 1-Pentanol was used as the internal standard. The authors state that with this method one can expect, in the analysis of grappas, an approximation of <10%, estimated on the results obtained from a standard solution of higher alcohols.

5. Alcohols in Volatile Fractions

Pino et al. (76) compared the efficiency of two methods of extraction of the volatile fraction and, precisely, of the batch and continuous solvent extraction methods. They use pentane, pentane–diethyl ether (2:1), and dichloromethane to make the comparisons. From their research it appears that extraction with pentane gives the best results, especially with a continuous procedure for 6 hr. Of the alcoholic components studied are 1-butanol, 3-methyl-1-butanol, and 2-phenylethanol.

6. Other Determinations

Simal et al. (77) suggested a GC method for the simultaneous determination of ethanol, methanol, propanol, isobutanol, and amylic alcohols (as well as ethyl acetate) in alcoholic drinks. They conduct the analysis with a Porapak Q column and obtain the separation of the various analytes in about 50 min. *n*-Butanol is used as the internal standard. The authors subject to analysis five "aguardientes" and obtain a CV = 3% for ethanol, isobutanol, and amylic alcohols, and CV = 6% for methanol and propanol.

In Ref. 78 a few examples are given of separations of various components present in alcoholic drinks (Scotch whisky, bourbon, cherry brandy) at concentrations between 40 and 60 ppm each, by means of Carbopack B AW/6.6% PEG 20 M column. The alcohols separated in the drinks analyzed are methanol, ethanol, *n*-propanol, isobutanol, and amylic alcohols (Fig. 6a–6c).

Manitto et al. (79) analyze a great number of distilled spirits of different origin with the intention of quantifying the concentration of 2-butanol and the percentage of the predominating enantiomeric form. The anlyses were performed by bidimensional gas chromatography ("heart cut" technique coupled with GC analysis with 2,3,6-trimethyl-β-cyclodextrin 5% OV 1701-OH capillary column). They investigate raw spirits (wine distillate, grape pomace distillate, and lees

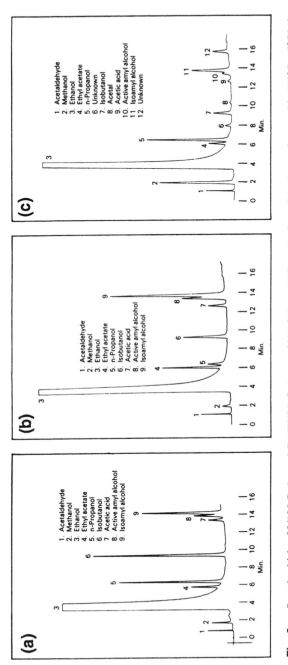

Fig. 6a Scotch whisky. 80/120 Carbopack B AW/5% Carbowax 20M, 2.0 m x 1/4 in. o.d. x 2 mm i.d. glass, col. temp. 70–170°C at 5°C/min, flow rate 20 ml/min, N_2, Det. FID, Sens. 8 x 10^{-11} AFS, sample 0.5 μl Scotch whisky. (Courtesy of Supelco Inc., Bellefonte, PA.)

Fig. 6b Bourbon. 80/120 Carbopack B AW/5% Carbowax 20M, 2.0 m x 1/4 in. o.d. x 2 mm i.d. glass, col. temp. 70–170°C at 5°C/min, flow rate 20 ml/min, He, Det. FID, Sens. 4 x 10^{-10} AFS, sample 0.5 μl bourbon whiskey. (Courtesy of Supelco Inc., Bellefonte, PA.)

Fig. 6c Cherry brandy. 80/120 Carbopack B AW/5% Carbowax 20M, 2.0 m x 1/4 in. o.d. x 2 mm i.d. glass, col. temp. 70–170°C at 5°C/min, flow rate 20 ml/min, He, Det. FID, Sens. 4 x 10^{-10} AFS, sample 0.5 μl cherry brandy. (Courtesy of Supelco Inc., Bellefonte, PA.)

distillate) and commercial beverages (brandy, grappa, Calvados, Armagnac, cognac, whiskey, rum, cherry and bilberry brandy).

In all samples they found a marked predominance of the R form of 2-butanol (50–80%). This conclusion constitutes a strong support to the current assumption that lactic acid bacteria are responsible for the formation of most of the secondary alcohol.

E. Cordials and Liqueurs

1. Alcohol

a. AOAC Methods (38).   By weight or volume, a distillation and a measurement of density with a pycnometer are carried out (method number 935.19). In another method (by volume), a distillation by means of a Vigreaux column and a later densitometric measurement are carried out (method number 983.12).

2. Methanol

This method requires the liquid–liquid extraction of an aqueous phase containing methanol and successive distillation (method number 940.06). The distillate may be used either with the chromotropic and colorimetric method (method number 958.04) or with the immersion refractometer method (method number 905.01).

3. Ethanol

a. Enzymatic Methods.   The only special instruction given by the Boehringer enzymatic method (51) is to bring the concentration of ethanol to the values of application of the kit. In the case of particular liqueurs (e.g., egg-based ones), the method provides for a dilution (1:100 w/v) and then a refrigeration, followed by filtration to separate the fatty matters. The next steps are the same as for other products. The procedure is the same in the case of brandies.

4. Glycerol

a. AOAC Methods (38).   This is the surplus method, number 940.10.

F. Oils and Fats

For the definitions of oils, fats, alcohols, etc., in lipids, please consult Gunstone and Herslof (80).

1. Fatty Alcohols

Fatty alcohols may be present either free or combined, mainly with fatty acids in wax esters.

a. EEC Methods (39).   The total content of alkanols (free and combined) may be determined by using the method described in Refs. 81 and 82. In this method an internal standard is added (1-eicosanol); then the fatty matter is saponified with methanolic KOH and finally the unsaponifiable is extracted with diethyl ether. The alcoholic fraction is separated by means of TLC on basic plates (KOH) of silica gel and derivatized into trimethylsilyl ethers and analyzed by HRGC (Fig. 7). The results are expressed in mg/kg (ppm).

b. On-line Coupled LC-GC.   The possibility of determining easily the content of free alkanols is offered by an *on-line coupled LC-GC* system described by Grob (83). A very interesting application for olive oils is illustrated by Grob et al (84), who make use of a fully automated LC-GC "Dualchrom" system (Carlo Erba, Milan, Italy).

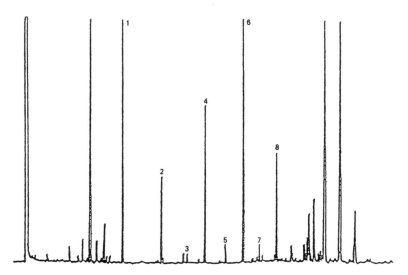

Fig. 7 Chromatogram of the alcoholic fraction of a virgin olive oil. 1, eicosanol; 2, docosanol; 3, tricosanol; 4, tetracosanol; 5, pentacosanol; 6, hexacosanol; 7, heptacosanol; 8, octacosanol. (Courtesy of the Office for Official Publications of the European Communities. *Official Journal of the European Communities*, no. 176/26 of July 20, 1993.)

The aliphatic alcohols are first heat-derivatized with pivalic anhydride added directly to the oil. The solution is properly diluted with *n*-hexane and injected for the LC-GC run (Fig. 8). Naturally with this method it is possible to determine the total content from the unsaponifiable or, in theory, from the content of wax esters or by suitable transesterification.

2. Glycerol

a. Gas Chromatographic Methods. Christie (85) reports a GC method for the determination of glycerol in simple lipids. With this method, methyl pentadecanoate is added as the internal standard and a reaction with lithium aluminum hydride is made. There is then a reaction with acetic anhydride, then the removal of the reagents by evaporation and the dissolution in ethyl ether for the GC analyses.

3. Alcohols in Volatile Fractions

The determination of volatile components in fatty matter is not controlled by any regulations. In literature it is possible to trace many works which deal with this subject, also in relation to modifications induced by heating to which these substances are often subjected.

As an example we mention the work of Snyder and Mounts (86) on five oxidized vegetable oils (low–erucic acid rapeseed, corn, soybean, sunflower, and high oleic sunflower). They use the technique of multiple headspace extraction (MHE) and condition 0.5 g of the sample at 90°C for 30 min in a sealed vial. After pressurizing it for 30 sec, they inject it into GC for 30 sec, in splitless.

Of the alcoholic components identified, they find up to over 9 ppb of pentanol, up to almost 12 ppb of 2-pentenol and about 3 ppb of octen-3-ol.

In a recent work, Snyder and King (87) compare the volatile fractions of a soybean oil extracted

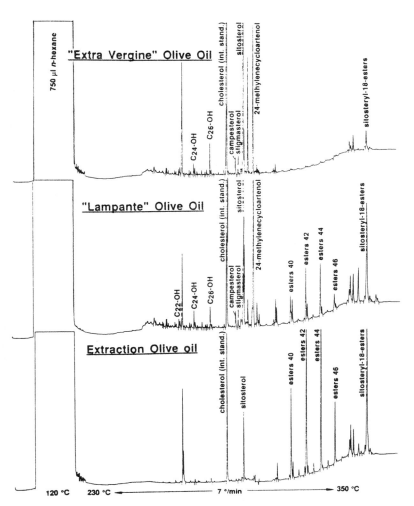

Fig. 8 LC-GC-FID chromatograms for typical olive oils. Nearly complete absence of wax esters (esters 40–46) and very low concentrations of steryl esters indicate a high-quality extra virgin oil. The concentration of free stigmasterol is low. C_{24-26}-OH, fatty alcohols. In lampante oils, more wax esters and steryl esters are found. The concentration of stigmasterol increases more than campesterol if the oil was prepared from olives of low quality. Run at the same sensitivity, chromatograms of solvent-extracted oils are completely overloaded. The refined extraction oil was diluted 1:5 before running the chromatogram shown. Wax ester and steryl ester concentrations are very high. [From *JAOCS, 67*: 626 (1990).]

with hexane with one extracted with an SC-CO₂ system and use a dynamic HSGC method and a thermal desorption of a polymer adsorbant (Tenax TA). In their survey they recognize ethanol, 1-propanol, butenol, 1-pentanol, butendiol, 3-pentenol, 1-hexanol, 1-octen-3-ol.

Again Snyder and King (88), with the same technique adopted in Ref. 87, investigate the alcoholic volatile fraction of canola oil, corn, soybean, and sunflower, where they recognize the presence of octadecanol and hexadecanol.

Selke and Frankel (89) quantify the major volatile products of a soybean oil oxidized in the dark, at room temperature.

With the technique of dynamic headspace they concentrate the volatile compounds on porous polymer (Tenax), which are then thermally desorbed and introduced in splitless into the injection port of a capillary gas chromatograph.

Identification and quantification of the compounds is performed by GC-MS. From the alcohols they identify 1-penten-3-ol.

4. Triterpenic Alcohols

In all of the presented methods, saponification of the fatty matter is used. The most common triterpenols are cycloartenol, 24-methylencycloartanol, α-amyrin, β-amyrin, cyclobranol, and butyrospermol.

In the method described in Ref. 81, use is made of a TLC separation which permits the aliphatic alcohols to be isolated from the rest of the unsaponifiable. The band of alkanols is normally scraped together with the triterpenic alcohols which come immediately after and from which they cannot be clearly separated as the rf values are not very different. It is therefore apparent that in GC separation, after the alkanols, the triterpenic alcohols elute.

Strocchi et al. (90, 91) conducted a study on the modifications that triterpenic alcohols undergo during the refinement process of olive oil and identify a new one (24-methyl-5α-lanosta-9 (11), 24-dien-3β-ol). They saponify in a nitrogen atmosphere and preserve the unsaponifiable in benzene at -30°C.

A further TLC separation is then performed on a basic plate of silica gel with which they isolate the triterpenols. The separation of each triterpenic alcohol is carried out by HRGC (Fig. 9).

Soulier et al. (92) isolated eight triterpenic alcohols from sal and illipe butters besides other compounds. They separate these compounds after saponification of the fatty matter and fractionation of the unsaponifiable on an aluminum oxide (hydrated at 5%). Two successive HPLC

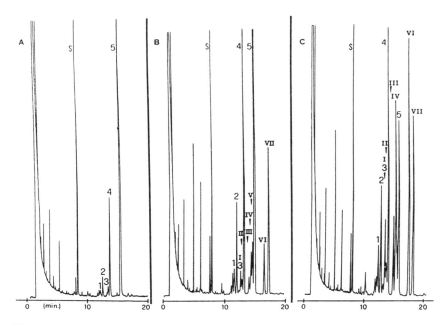

Fig. 9 Gas chromatograms of the TMS derivatives (stationary phase SE-52) of triterpenic alcohols from (A) neutral olive oil; (B) bleached olive oil (1% activated earth at 90°C for 30 min); (C) bleached olive oil (1.5% activated earth at 120°C for 30 min). 1, β-amyrin; 2, butyrospermol; 3, α-amyrin; 4, cycloartenol; 5, 24-methylenecycloartanol; I–VII degradation products; s, internal standard. (Courtesy of Riv. It. Sostanze Grasse, Milano, Italy.)

separations and a TLC-AgNO₃ permit the isolation of highly purified fractions. Through the use, among others, of the ¹H NMR and MS techniques, they identify nine triterpene alcohols (β-amyrin, butyrospermol, cycloartenol, α-amyrin + lupeol, 24-methyleneparkeol, 24-methylenecycloartan-ol, ψ-taraxasterol, and cycloartanol), among which 24-methyleneparkeol and ψ-taraxasterol are indicated for the first time in sal and illipe butters (Table 4). Moreover, the authors suggest an analytical method for recognizing the presence of these fats in foodstuffs.

5. Erythrodiol and Uvaol

The methods presented make use of saponification of the fatty matter. Erythrodiol is the $3\beta,17\beta$-dihydroxy-Δ^{12}-oleane, whereas uvaol is the $3\beta,11\beta$-dihydroxy-Δ^{12}-ursene.

a. EEC Methods (39). The determination of the two triterpenic dialcohols contained in certain types of fatty matters (considerably high concentration in solvent-extracted olive oil, grape pip oil, etc.) is normally performed according to the EEC method described in Ref. 81. As was already described for alkanols, the unsaponifiable is subjected to a TLC separation. The relative band is usually scraped together with the band of sterols and, in the following HRGC separation, after the sterols, there is elution of erythrodiol and uvaol.

6. Other Determinations

Bianchi and Vlahov (93) determined the distribution of "minor components" of olive oil in the various parts of the olive fruit and quantify the content of both alkanols and terpenols.

Kawanishi et al. (94) perfected the separation and quantification of free primary alcohols in various anatomical parts from the germ, kernel, seed coat, shell, and skin (peel) of various nuts, seeds, fruits, and cereals. In particular, they quantify the content of hexacosanol, octacosanol, and triacontanol in the oil from hazel, peanut, almond, and pine seeds (Tables 5–7).

The authors adopt the TLC technique on reverse phase, developing the plates with chloroform/methanol 1:2 to separate the alkanols from the hydrocarbons, esters, and glycerides in the oils. Before GC analysis, the alkanols are derivatized to trimethylsilyl ethers.

Recently, Amelio et al. (95) described a method which may find routine applications and which makes use of SPE for the separation and cleanup of the unsaponifiable from olive oil, from which the aliphatic alcohols are separated by means of HPLC (besides sterols and the two triterpenic dialcohols erythrodiol and uvaol) (Fig. 10; Table 8). The alkanols are then derivatized and analyzed by means of HRGC as in Ref. 81. The use of an autosampler and a fraction collector for use with HPLC permits considerable automatization of the analysis.

G. Dairy Products

In this section some applications to common cases will be considered. In dairy products the presence of alcohols is mainly linked to the volatile fractions.

1. Alcohol

a. AOAC Methods (38). In method 983.12, which may be applied to liqueurs, dairy, and alcoholic dairy products, a distillation with a Vigreaux column is performed. The percentage of alcohol is determined by a densitometric method.

2. Ethanol

a. Enzymatic Methods. In the Boehringer enzymatic method (51), ethanol is determined in products derived from milk (ricotta cheese, kefir). After homogenization, dilution, and clarifica-

Table 4   Sal and Illipe Butters Triterpene Alcohols Data

1 $R_1 = R_2 = R_3 = CH_3$, 9 Cyclo
2 $R_1 = R_2 = R_3 = CH_3$, Δ 9(11)
8 $R_1 = H$, $R_2 = R_3 = CH_3$, $\angle 8$
9 $R_1 = H$, $R_2 = R_3 = CH_3$, 9Cyclo
10 $R_1 = R_2 = R_3 = H$, $\Delta 5$
11 $R_1 = R_3 = H$, $R_2 = CH_3$, $\Delta 7$

Name	GLC RRT	Mass spectra of TMS derivatives and some diagnostic ions
Cycloartenol 1f (TMS)	1.63	498 (4), 483 (5), 408 (83), 393 (46), 365 (58), 339 (26), 297 (5), 286 (14), 255 (4), 241 (4)
24-Methylenecycloartanol 1e (TMS)	1.83	512 (2), 497 (3), 422 (38), 407 (19), 379 (31), 353 (8), 339 (2), 323 (2), 300 (12), 297 (8), 255 (3), 241 (3)
24-Methyleneparkeol 2e (free)	1.79	440 (22), 425 (100), 407 (38), 397 (21), 356 (8), 341 (11), 339 (4), 323 (8), 313 (88), 300 (6), 295 (10), 288 (5), 285 (6), 273 (22), 271 (8), 259 (18), 255 (11), 243 (13)
Butyrospermol 3 (free)	1.53	426 (20), 411 (100), 408 (5), 393 (31), 365 (2), 344 (2), 325 (3), 313 (8), 297 (4), 295 (3), 286 (4), 273 (7), 271 (9), 255 (9), 255 (7), 241 (8)
Lupeol 4 (TMS)	1.67	498 (17), 483 (5), 408 (7), 393 (7), 369 (14), 365 (2), 325 (3), 306 (3), 299 (4), 279 (10), 257 (6), 231 (16), 218 (37), 216 (9), 203 (36), 189 (56), 95 (100)
α-amyrin 5 (free)	1.67	426 (5), 411 (1), 218 (100), 207 (23), 203 (22), 189 (18)
β-amyrin 6 (TMS)	1.53	498 (6), 483 (1), 393 (1), 279 (5), 257 (2), 218 (100), 203 (38), 189 (20)
ψ-taraxasterol 7 (TMS)	1.97	498 (9), 483 (1), 408 (5), 393 (3), 369 (7), 365 (1), 326 (1), 279 (5), 257 (3), 231 (7), 218 (6), 203 (11), 189 (100)

Source: JAOCS, 67: 388 (1990).

Table 5   Composition of Alcohols Extracted from Nuts, Seeds, and Cereals

Material	Part[a]	Relative percent			Source
		Hexacosanol	Octacosanol	Triacontanol	
Hazel	G	39.2	60.8		Fiskobirlil, Turkey
	K	17.1	63.6	19.3	
	SC	29.7	70.3		
	S	28.6	51.0	20.4	
Cashew	G	25.0	35.0	40.0	Vitaya Lakehmi, India
Peanut	G	18.5	50.6	30.9	Liangyougongsi, China
	K	29.5	70.5		
	SC	25.3	52.5	22.2	
	S	10.7	23.9	65.4	
Almond	G	47.3	52.7		Sun Giant Inc., U.S.A.
	K	21.5	65.1	13.4	
	SC	28.6	54.4	16.8	
Pine seed	K	25.6	74.4		Liangyougongsi, China
	S	10.7	76.1	13.2	
Sunflower seed	G	24.5	57.9	17.6	Siegeo Sun Products Ltd., U.S.A.
	K	25.9	74.1		
	SC	29.3	60.2	15.5	
Maize	G	17.8	82.2		Industrial Sechi SA, Peru
	K	33.2	55.4	11.4	
Wheat	G	17.6	20.5	61.9	local market, Japan
Rice	G	13.5	35.4	51.1	local market, Japan
Prune	flesh	25.3	65.9	8.8	Sunsweet Growers Inc., U.S.A.
	seed	42.3	57.9		
Raisin	flesh	16.2	41.9	41.9	Sun Maid Growers Inc., U.S.A.
Hazel	oil	46.4	53.6		
Peanut	oil	35.3	50.3	14.4	
Almond	oil	56.3	43.7		
Pine seed	oil		29.4	70.1	

[a]G, germ; K, degermed kernel; SC, seed coat; S, shell; oil, obtained by press of whole nut.

Source: *JAOCS, 68*: 869 (1991).

tion of the sample with Carrez solutions, the solution is brought up to volume and filtered. Enzymatic determination of ethanol by means of a UV spectrophotometer is performed as has already been seen for other products.

3.   Alcohols in Volatile Fractions

Imhof and Bosset (96) quantified the volatile flavor compounds in pasteurized milk and fermented milk, applying the standard addition method. Instrumental analysis is performed by dynamic headspace and GC-MS.

In particular, the samples were purged with ultrapure nitrogen gas. The analytes were trapped

Table 6   Octacosanol Contents of Fruits

Material	Part	Extract[a] (mg/g)	Octacosanol content (µg/g extract)	Octacosanol content (ppm)	Source
Plum *Prunus domestica*	flesh	3	1415	7.1	Sunsweet Growers Inc., U.S.A.
	seed	77	233	20.1	
Raisin *Vitis labrusca*	flesh	1	708	0.9	Sun Maid Growers Inc., U.S.A.
Apple *Malus pumila var. dulcissima*	peel	58	663	221.8	local market, Japan

[a]Petroleum ether-extractable material.

Source: *JAOCS, 68*: 869 (1991).

Table 7 Octacosanol Contents of Nuts, Seeds, and Cereals

Material	Part[a]	Extract (mg/g)	Octacosanol content (μg/g extract)	Octacosanol content (ppm)
Hazel	G	659	5.6	3.7
Corylus avellana	K	630	10.7	6.7
	SC	153	11.6	1.8
	S	3	30.6	0.09
	W			P 6.0
				E 5.6
Cashew	G	291	4.8	1.1
Anacardium occidentale				
Peanut	G	421	9.2	3.9
Arachis hypogaea	K	471	14.5	6.8
	SC	91	23.4	2.1
	S	1	13.1	0.01
	W			P 5.6
				E 5.2
Almond	G	490	7.8	3.8
Prunus dulcis	K	526	12.4	6.5
	SC	118	15.6	1.8
	W			P 6.2
				E 5.6
Pine seed	K	717	5.3	3.8
Pinus koriensis	W			P 4.5
				E 4.7
Sunflower seed	G	516	8.0	4.1
Helianthus annuus	K	539	22.7	12.2
	SC	123	30.0	3.7
Maize	G	460	17.8	8.2
Zea mays var. indentata	K	5	11.5	0.06
Wheat	G	153	14.1	2.2
Triticum aestivum				
Rice	G	218	20.9	4.6
Oryza sativa				
Coconut	K			P 12.6
Cocos nucifera				E 7.9

[a]G, germ; K, degermed kernel; SC, seed coat; S, shell; W, whole; P, press; E, extraction.
Source: *JAOCS, 68*: 869 (1991).

in a stainless steel tube packed with Carbosieve SIII (0.05 g) and Carbopack B,69–80 mesh (0.20 g) (Tekmar). Detection was performed by a mass-selective detector (MSD) by electron impact (EI). The authors identified, among the other compounds, 2-propanol (78 μg/kg, with a linear range (L.R.) (μg/kg) = 0–700), 1-propanol (18 μg/kg, L.R. = 0–1000), 1-butanol (not detected, L.R. = 0–1000), 3-methylbutanol-1 (2μg/kg, L.R. = 0–300), 3-pentanol (n.d., L.R. = 0–600), pentan-2-one-4-ol (7 μg/kg, L.R. = 0–600), 1-pentanol (15 μg/kg, L.R. = 0–600) (Fig. 11).

Moio et al. (97) recognized 5% of aliphatic alcohols (10 different molecules) in the neutral volatile component of fresh water buffalo milk. The technique is the same for mozzarella cheese. In this study, the CV is calculated to be <7%.

A total concentration of alcohols equal to 0.054 ppm is determined. Of the primary alcohols pentanol (0.01 ppm), hexanol (0.003 ppm), octanol (0.006 ppm), decanol (0.003 ppm), dodecanol and tetradecanol (0.003 ppm) are identified. The alcohol present in greatest quantity is 2-phenylethanol (0.015 ppm), while of the secondary alcohols 1-octen-3-ol was present with 0.012 ppm. Of the triterpenic alcohols 0.003 ppm of geraniol was identified (Fig. 12).

In another study, Moio et al. (98) investigated the neutral volatile compounds in bovine, ovine, caprine, and water buffalo raw milks. The extraction technique is the same as in Ref. 97. For the majority of components the authors found a CV <5% for the extraction procedure yield. Furthermore, they identified 2-ethyl-1-hexanol, 2-phenylethanol, geraniol, dodecanol, and tetradecanol for the first time in these milks.

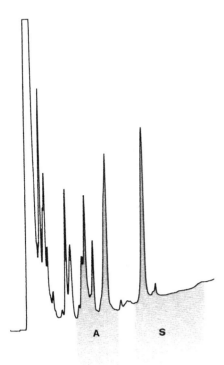

Fig. 10 HPLC of extra virgin olive oil unsaponifiable and collected fractions. A = fraction containing alkanols (6 and 9 min); S = fraction containing sterols (10 and 18 min). [From *J. Chromatogr.*, *606*: 179 (1992).]

Moio et al. (99, 100) obtained the alcoholic fraction of the mozzarella cheese flavor along with the other aroma components, by distillation under vacuum [7 hr under a pressure of 10^{-1} torr, in a modified Forss and Holloway device (101, 102)] and collected them in traps cooled by liquid nitrogen. They extracted the distillate by dichloromethane and the extracted was concentrated first using a Kuderna–Danish and a Dufton column and then analyzed by HRGC and HRGC-MS [electron impact (EI) and chemical ionization (CI) with NH_3 as the reagent gas].

Table 8 Statistical Analysis of Alkanol Data (Student's t-Test)

Method	Parameter	Alkanol[a]							
		a	b	c	d	e	f	g	h
Conventional (NGD method)	Mean (mg/kg)	83.695	7.190	142.187	9.362	130.476	6.992	50.584	429.456
	S.D. (mg/kg)	187.982	12.896	269.367	15.099	214.025	10.581	75.458	780.711
This work	Mean (mg/kg)	82.691	7.183	137.675	9.183	128.109	6.251	51.296	422.313
	S.D. (mg/kg)	175.717	11.945	249.974	13.236	198.039	8.943	65.204	718.363
	t	0.038	0.003	0.121	0.088	0.080	−0.184	−0.070	0.066

[a] a = 1-Docosanol (C_{22}); b = 1-tricosanol (C_{23}); c = 1-tetracosanol (C_{24}); d = 1-pentacosanol (C_{25}); e = 1-hexacosanol (C_{26}); f = 1-heptacosanol (C_{27}); g = 1-octacosanol (C_{28}); h = total alkanols.

Source: J. Chromatogr., *606*: 179 (1992).

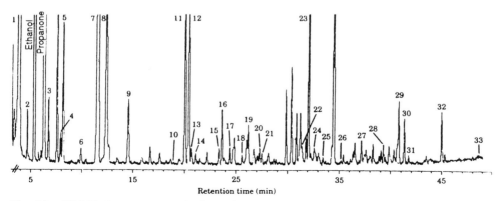

Fig. 11 GC-MS chromatogram of a fermented milk product (yogurt). 3, 2-propanol; 6, 1-propanol; 10, 1-butanol; 14, 3-pentanol; 15, pentan-2-one-4-ol; 16, 3-methylbutanol-1; 18, 1-pentanol. [The alcoholic fraction only has been referenced in the present work. From *Food Sci. Technol., 27*: 265 (1994).]

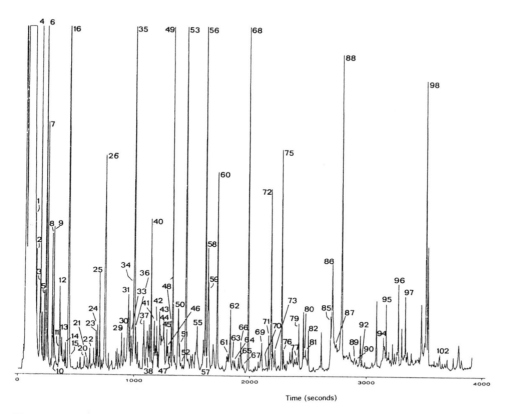

Fig. 12 High-resolution gas chromatography on a DB5 column (for operating conditions refer to the methods section) of the neutral volatile components extracted from raw water buffalo milk after 7 hr at low vacuum distillation (10^{-1} Torr) using a Forss and Holloway device. 12, pentanol; 20, hexanol; 30, 1-octen-3-ol; 38, 2-ethyl-1-hexanol; 44, octanol; 50, 2-phenylethanol; 61, geraniol; 63, decanol; 77, dodecanol; 89, tetradecanol. [The alcoholic fraction only has been referenced in the present work. From *Ital. J. Food Sci., 1*: 43 (1993).]

The authors quantify 19 alcoholic components (in total 0.9 ppm, 20% of the total aroma components), among which are two terpenols: linalool (0.02 ppm) and geraniol (0.04 ppm).

In bovine mozzarella cheese, the typical fermentation alcohols are 3-methyl-1-butanol (0.5 ppm) and 2-phenylethanol (0.2 ppm) (Figs. 13 and 14).

Bosset and Liardon (103) and Liardon et al. (104) investigated the neutral volatile components of the aroma of Swiss gruyere cheese.

They extract the volatile fractions using a vacuum gas-stripping apparatus for 13 hr at room temperature under reduced pressure (0.1–4 kPa) by means of a flow of high-purity nitrogen and condensation in a series of cold traps. The fractions are then extracted with diethyl ether and concentrated in a rotary evaporator. There follows an HRGC analysis with FID and NPD detectors.

The neutral volatile fractions were identified by means of GC-MS. In particular, the authors quantified 16 primary alcohols (C2–C9), 6 secondary alcohols, and 5 ketoalcohols. This study is also conducted with the aim of making known the transformations of the aroma during ripening (Figs. 15 and 16; Tables 9 and 10).

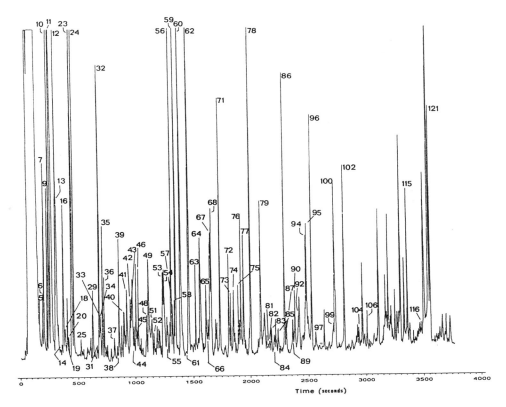

Fig. 13 High-resolution gas chromatography on a DB5-fused silica capillary column of the neutral volatile components of water buffalo mozzarella cheese. For the conditions of gas chromatographic analysis, refer to the methods section. 12, 3-methyl-1-butanol; 13, 2-methyl-1-butanol; 16, 1-pentanol; 18, 3-methyl-2-buten-1-ol; 29, hexanol; 34, 2-heptanol; 36, 2-butoxyethanol; 38, 2-methoxyethoxy-2-ethanol; 41, heptanol; 42, 1-octen-3-ol; 49, 2-ethyl-1-hexanol; 51, benzyl alcohol; 54, octanol; 57, linalool; 58, 2-nonanol; 60, 2-phenylethanol; 64, 1-nonanol; 72, geraniol; 74, 1-decanol. [The alcoholic fraction only has been referenced in the present work. From *J. Food Sci., 1*: 57 (1993).]

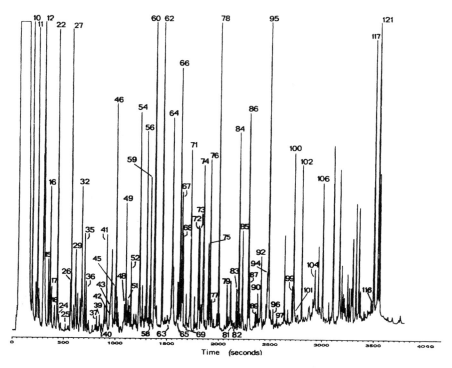

Fig. 14 High-resolution gas chromatography on a DB5-fused silica capillary column of the neutral volatile compounds of bovine mozzarella cheese. For the conditions of the HRGC analysis, refer to the methods section. 12, 3-methyl-1-butanol; 13, 2-methyl-1-butanol; 16, 1-pentanol; 18, 3-methyl-2-buten-1-ol; 29, 1-hexanol; 34, 2-heptanol; 36, 2-butoxyethanol; 38, 2-methyoxyethoxy-2-ethanol; 41, 1-heptanol; 42, 1-octen-3-ol; 49, 2-ethyl-1-hexanol; 51, benzyl alcohol; 54, 1-octanol; 57, linalool; 58, 2-nonanol; 60, 2-phenylethanol; 64, 1-nonanol; 72, geraniol; 74, 1-decanol. [The alcoholic fraction only has been referenced in the present work. From *Ital. J. Food Sci.*, *3*: 215 (1993).]

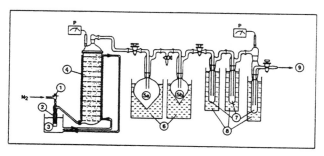

Fig. 15 Vacuum gas-stripping apparatus for cheese aroma isolation. (1) Needle valve for fine regulation of carrier gas flow (N_2); (2) gas washer with sintered glass (H_2O; $V \approx 0.2$ L; $T = 30°C$); (3) water bath with outer circulation ($T = 30°C$); (4) extraction tower filled with grated cheese ($V \approx 8.3$ L; $T \approx 20–25°C$; $P \approx 0.1–4$ kPa; cf. Fig. 16); (5a) cooling trap ($V \approx 3$ L); (5b) cooling trap ($V \approx 2$ L); (6) cooling mixture: Alcohol + solid CO_2 ($T \approx -77°C$); (7) 3 cooling traps ($V \approx 0.5$ L each). The third one prevents oil diffusion from the rotary pump; (8) N_2 liquid ($T \approx -196°C$); (9) high vacuum pump (high-speed rotary pump, model ED 100[*], $P =$ two Pirani vacuum gauge heads (PRCT 10-C) with one Pirani vacuum gauge (model M-12).[*] [*]Edwards high vacuum, Manor Royal, Crawley, Sussex, England. [From *Food Sci. Technol.*, *15*: 143 (1982).]

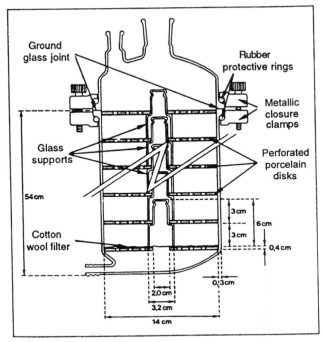

Ground
glass joint

Rubber
protective rings

Metallic
closure
clamps

Glass
supports

Perforated
porcelain
disks

54 cm

3 cm
6 cm
3 cm
0.4 cm

Cotton
wool filter

0.3 cm

2.0 cm

3.2 cm

14 cm

Fig. 16 Detail of construction of the extraction tower. [From *Food Sci. Technol., 15*: 143 (1982).]

H. Fruits, Vegetables, and Juices

1. Alcohol

a. AOAC Methods (38). Determination is conducted on 50 g of original material (method number 920.150). The alcoholic content is determined by volume from specific gravity.

2. Ethanol

a. Enzymatic Methods. The Boehringer enzymatic method (51) requires the sample of juice to be diluted in order to obtain a suitable concentration of ethanol for analysis. Decolorization of brightly colored juices is also required, as described for the analysis of glycerol. In the case of cloudy juices, the use of Carrez solutions is required.

3. Fatty Alcohols

Kawanishi et al. (94) also investigated the alcoholic content of various anatomical parts of various nuts, seeds, and cereals (germ, degermed kernel, seed coat, shell): hazel, cashew, peanut, almond, prune, raisin, apple, coconut, wheat, and rice. The method of analysis has already been briefly described in the section on oils and fats (V.F.6).

Donchev and Velcheva (105, 106) carried out a survey on hydrocarbon and primary alcohols of epicuticular waxes of watermelon and squash because it was found that the cuticular lipid composition can deter herbivorous insects. They extracted the cuticles by light petroleum for 30 hr and then by diethyl ether for 15 hr. Solvents were evaporated under vacuum and the liquid fraction was separated from the solid fraction. The latter was dissolved in acetone. The hydrocarbons and primary alcohols were isolated by PTLC on silica gel using light petroleum–ethyl

Table 9 Identification and Relative Quantification
of Alcohols in a Pool of Ripe Swiss Gruyère Cheese

Component Name	Relative Abundance[1]		
	Outer[3]	Middle[3]	Central[3]
Ethanol	0	168	169
1-Propanol	0	40	44
1-Isobutanol	0	18	18
1-Butanol	0	65	44
3-Methylbutan-1-ol	11	179	76
Pent-3-en-1-ol (T)[2]	0	20	10
1-Pentanol	13	20	10
1-Hexanol	0	15	6
2-Phenylethan-1-ol	12	3	0
1-Hexadecanol (T)	8	5	7
2-Propanol	10	0	0
2-Butanol	0	74	55
3-Methylbutan-2-ol	16	0	0
2-Pentanol	21	25	13
2-Heptanol	52	26	6
2-Nonanol	19	0	0
1-Phenylethan-1-ol	7	8	0

Alcohols identified in other Gruyère samples[4]
1-Heptanol
1-Octanol
1-Nonanol
Furfuryl alcohol
Benzyl alcohol

[1]The concentration defined as the peak area of each component of the flavor mixture has been normalized relative to the peak area of the BHT, a stabilizer of ether, chosen as equal to 10 and used as an internal standard and corrected for the different weights of the 3 zones.
[2]T: tentative identification from the mass spectrum (MS).
[3]See zone fractioning of samples in Part I.
[4]g mode and zone fractioning in Part III.
Source: *Food Sci. Technol., 15*: 143 (1982).

ether–acetic acid (90:10:0.5). Following this, identification of the compounds was performed by GC-MS (Table 11).

4. Glycerol

   a. Enzymatic Methods.   The Boehringer enzymatic method (51) requires that the juice be diluted to bring the concentration of glycerol to values suited to the analysis and cloudy juices to

Table 10 Identification and Relative Quantification[1]
Ketoalcohols in a Pool of Ripe Swiss Gruyère Cheese

Component Name	Relative Abundance[1]		
	Outer[3]	Middle[3]	Central[3]
Butan-2-ol-3-one	60	n.d.	0
3-Methylbutan-3-ol-2-one	56	0	0
Pentan-3-ol-2-one	59	10	0
Pentan-2-ol-3-one	34	7	0
Pentan-2-ol-4-one	11	0	0

[1]Subscripts are defined in Tab. 9.
Source: *Food Sci. Technol., 15*: 143 (1982).

Table 11 Content (%) of Primary Alcohols, Isolated from Liquid and Solid Waxes and Identified by GC-MS

Peak nº	Mol. mass	Number of carbons	Mass identification	Liquid wax			Solid wax		
				Watermelon	Squash		Watermelon	Squash	
				Varieties					
				Cr. sweet	C. Pepo	C. max	Cr. sweet	C. pepo	C. max
1	214	14	n-Tetradecanol	1.0	—	—	—	—	—
2	228	15	n-Pentadecanol	0.5	—	—	—	—	—
3	242	16	n-Hexadecanol	1.7	—	—	—	—	—
4	256	17	n-Heptadecanol	0.7	—	—	—	—	—
5	270	18	n-Octadecanol	4.6	—	—	—	—	—
6	284	19	n-Nonadecanol	4.3	—	—	—	—	—
7	298	20	n-Eicosanol	2.1	—	—	—	—	—
8	312	21	n-Heneicosanol	0.7	—	6.8	—	—	6.8
9	326	22	n-Docosanol	6.2	30.5	12.3	—	4.2	12.3
10	340	23	n-Tricosanol	1.7	—	8.2	—	—	8.2
11	354	24	n-Tetracosanol	16.8	30.0	11.0	9.5	7.7	11.0
12	368	25	n-Pentacosanol	5.3	—	2.7	6.7	—	2.7
13	382	26	n-Hexacosanol	39.7	23.2	34.3	48.6	22.6	34.3
14	396	27	n-Heptacosanol	6.0	—	1.4	10.7	—	1.4
15	410	28	n-Octacosanol	8.7	16.3	23.3	22.4	29.8	23.3
16	424	29	n-Nonacosanol	—	—	—	2.1	8.9	—
17	438	30	n-Triacontanol	—	—	—	—	12.5	—
18	452	31	n-Hentriacontanol	—	—	—	—	3.6	—
19	468	32	n-Dotriacontanol	—	—	—	—	10.7	—

Source: Riv. Ital. Sostanze Grasse, Milano, Italy.

be filtered. In the case of brightly colored juices, the method of decolorization with polyamide or polyvinylpolypyrrolidone (PVPP) is described.

5. Alcohols in Volatile Fractions

Eckert and Ratnayake (107) studied the effect of volatile compounds in inducing germination of conidia of *Penicillium digitatum* on oranges having a torn exocarp. In their other studies they analyzed the surrounding atmosphere in a glass jar (19.4 L) containing ~4 kg of wounded oranges. First the volatiles were cold-trapped and dissolved in methanol–NaCl–saturated water. After extraction by isooctane, the solution was injected in a GC fitted with an FID. The authors quantify the concentration of ethanol at 0.1 µg/ml (Fig. 17). The authors also identify ethanol in ripening apricots.

A survey on volatile constituents in fresh unpasteurized orange juices is conducted by Moshonas and Shaw (108) by means of dynamic HSGC (purge and trap injector) with cryofocusing on capillary tubing kept at -130°C with liquid nitrogen. The identification of volatiles was performed by GC-MS. As regards the alcoholic fraction they quantified component concentrations as little as <1 ppb.

Studies have been made on disrupted potatoes by Krulick and Osman (109) and on raw and cooked potatoes by Maga (110) in order to find out the effect of such manipulations on their composition and their volatile fraction. In the first case, the analytical techniques make use of extractions with solvent, LC, TLC, and GC. Identification of the peaks was confirmed by means of GC-MS. The variations of composition are mainly due to microorganisms and the increase of 1-hexacosanol and 1-octacosanol after only 4 days in incubation at 25°C in the dark is particularly significant. In Maga's paper no chemical analytical methods are described but a wide review about desirable and nondesirable volatile and nonvolatile flavor compounds in raw and processed potatoes is presented. Several tables list many identified compounds, among which alcohols are present. For each of them a bibliographical reference is given (207 references are quoted).

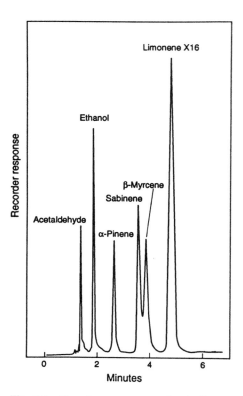

Fig. 17 Gas chromatogram of volatile organic compounds in the atmosphere surrounding wounded oranges. Limonene peak shown was attenuated 16 times. [From *Phytopathology, 84*: 746 (1994).]

6. Triterpene Alcohols

As an example is mentioned a study made by Akihisa et al. (111) on nine leguminosae seeds (*Cajanus cajan, Cicer arietinum, Dolichos biflorus, Lathyrus sativus, Lens culinaris, Pisum sativum, Vigna mungo, Glycine max, Vigna cylindrica*). They recognize 25 triterpenic alcohols including one new compound (24Z-ethylidene-24-dihydroparkeol) by using several TLC and HRGC separations. Identification was supported by MS and 400-MHz ^{1}H NMR.

I. Honey

1. Primary Normal Alcohols

    a. Enzymatic Methods.   Huidobro et al. (112) investigated the content of normal primary alcohols in unspoiled honey, expressing the content as ethanol, as may be seen from the application of the Boehringer enzymatic method (51) which was modified by them. Methanol is excluded from the results of this determination. The samples for analysis are used without clarification or neutralization. The interferences are taken into account by running alcohol dehydrogenase–free control solution against sample and blank. The spectrophotometric reading is taken at 340 nm using a solution of aldehyde dehydrogenase in the reference cuvette. As it can be seen from Tables 12 and 13, the precision and recovery are satisfactory for practical applications.

    The Boehringer enzymatic method (51) requires clarification with Carrez solutions of the sample for analysis.

Table 12 Study of the Precision of the Enzymatic Method
to Determine Apparent Ethanol (mg/kg) in Honey

	sample 8	sample 7	sample 1	sample 25
mean	13.5	35.3	50.1	141.8
SD[a]	0.2348	0.1729	0.1703	0.3127
CV%[b]	1.74	0.48	0.34	0.22

[a] SD, standard deviation. [b] CV%, coefficient of variation percent.
Source: Ref. 112.

2. Glycerol

Also in this case, Huidobro et al. (113) made useful modifications to the Boehringer method (51), adapting it to the specific needs of their analysis. The precision of the method is illustrated in Table 14, where the data relative to the repetition of three samples is collected.

J. Aromas

For a general examination of the problem, readers can refer to Chap. 24. For an overview of the biochemical mechanisms of formation of alcohols in aromas, readers can consult Eskin's volume (114). Refer to Maarse and Belz (115) for further information about analysis of alcohols in aromas.

1. Alcohol

a. AOAC Methods (38). The general method (method number 973.23) is as follows: After a suitable dilution of the sample for analysis and addition of *n*-propanol as the internal standard, a GC analysis is performed.

Table 13 Study of Recovery of the Enzymatic
Method to Determine Apparent Ethanol (mg/kg)
in Honey

present	added	found	recovery,%
	10	16.7	99.0
	10	16.8	100.0
	10	16.7	99.0
	30	36.8	100.0
	30	36.7	99.7
	30	36.8	100.0
6.8			
	50	56.9	100.2
	50	56.9	100.2
	50	56.7	99.8
	150	157.4	100.4
	150	157.9	100.7
	150	159.7	101.9
n			12
mean			100.1
SD[a]			0.763
CV%[b]			0.76

[a] SD, standard deviation. [b] CV%, coefficient of variation percent.

Source: Ref. 112.

Table 14   Study of the Precision of the Enzymatic
Method to Determine Glycerol in Honey

	50.0	154.6	366.2
	49.7	154.2	366.2
	50.0	154.0	366.5
	50.6	154.0	366.9
	49.4	154.5	366.0
	49.2	154.9	365.9
	49.5	155.6	366.8
	50.8	154.8	366.4
	50.3	155.6	365.5
	50.5	154.8	365.4
mean	50.0	154.7	366.2
SD[a]	0.5457	0.5724	0.4984
CV%[b]	1.09	0.37	0.14

[a] SD, standard deviation. [b] CV%, coefficient of variation percent

Source: Ref. 113.

For anise and nutmeg extracts (method number 973.23), the same applies as for the latter quoted method.

Almond extracts (method number 950.44 A, B) require a distillation and determination of density with pycnometer or, alternatively, a GC method (method number 950.44 C).

In cassia, cinnamon, and clove extracts (method number 950.45 A,B), a distillation and determination of density with pycnometer are required or, alternatively, a GC method (method number 950.45 C) is applied.

In lemon, orange, and lime extracts (method number 920.135), the method is applicable to extracts consisting only of oil, alcohol, and water. It is a specific gravity method. Alternatively, a GC method may be used (method number 975.24).

In peppermint, spearmint, and wintergreen extracts (method number 920.147 A, B), the method requires a distillation and measurement of density with a pycnometer. Alternatively, a GC method may be used (method number 920.147 C).

In vanilla extract (method number 920.128), determination of the density of the alcoholic distillate by pycnometer is carried out.

2. Methanol

In lemon, orange, and lime extracts (method number 920.136): this is a surplus method.

In vanilla extract (method number 920.132), a colorimetric method which makes use of chromotropic acid (reading at 575 nm) is applied. Alternatively, the Zeiss immersion refractometer may be used on the alcoholic distillate obtained according to 920.128.

3. Isopropanol

In cassia, cinnamon, and clove extracts (method number 950.45 D), distillation and titration are required.

In lemon extract (in the absence of acetone, method number 942.07), a titrimetric method is carried out.

In lemon and orange flavors (also in the presence of acetone, method number 952.07), a determination is performed by a titrimetric method.

In peppermint, spearmint, and wintergreen extracts (method number 920.147 D), a titrimetric method similar to that already mentioned for lemon extract is carried out.

4. Glycerol

For vanilla extract (method number 920.129), analysis is performed on samples containing 0.1–0.4 g of glycerol, choosing between two different methods (11.010 or 11.011, AOAC Official Methods, 12th ed.), according to the amount of sugar present. These are surplus methods.

In lemon, lime, and orange extracts (method number 909.01), the same procedures as for the preceding case apply.

5. Other Determinations

Simply as examples, a few other determinations on essences in common use in foodstuffs are mentioned.

The Italian Official Pharmacopaea 1991 (116) describes in detail the methods for the identification of essences: in the case, for example, of rosemary (*Rosmarini aetheroleum*), the tests require that the quantitative determination of the monoterpenic alcohol borneol (free + esterified) be carried out. The free borneol is quantified by difference between the ester index after acetylation with acetic anhydride (total borneol) and the ester index of the neutralized ethanolic solution of the essence (combined borneol). As for the essence of thyme (*Thymi aetheroleum*), linalool is also quantified by chromatography.

K. Miscellaneous

1. Alcohol in Fermented Sausage

Johansson et al. (117) investigated some of the transformations which take place during maturation of a fermented sausage with *Pediococcus pentosaceus* and *Staphylococcus xylosus* as starter cultures. As far as the volatile fraction is concerned, the authors attribute the probability of the origin from smoke and seasoning (onion/garlic and pepper), while others come from the activity of muscle enzymes and bacteria. Frozen sausage was ground and immediately extracted by a helium current for 1 hr. Volatile compounds were collected onto a Tenax trap of an automated headspace apparatus.

Thereafter, the injection of the volatile compounds was obtained by means of desorbing at 250°C. The gas chromatograph was coupled with a mass spectrometer and equipped with a nonpolar capillary column. In about 100 volatile compounds, 6 were identified as alcohols (ethanol, 2-methylpropanol, 1,3-butanediol, 1,3-dimethoxypropan-2-ol, furfurylalcohol, 2-butoxyethanol).

2. Ethanol in Canned Salmon

a. AOAC Methods (38). This is a HSGC method (method number 986.12). Determination is performed on the aqueous phase after preparation of a calibration curve.

3. Ethanol in Marmalade

a. Enzymatic Methods. The Boehringer enzymatic method (51) requires use of 10–20 g of product to which is added a little water and which is neutralized with KOH. This is taken to volume (100 ml) and, if necessary, decolored with 2% of polyamide or PVPP, then filtered. The procedure may be found in the general instructions.

4.   Ethanol in Chocolate Sweets Containing Alcohol

   a.   Enzymatic Methods.   The Boehringer enzymatic method (51) requires that a certain amount of the filling be drawn off (0.5 ml if liquid or a quantity which may give a concentration in alcohol less than 0.12 g/L) and diluted with water. The solution thus prepared can be analyzed by the usual method.

5.   Glycerol in Marzipan

   a.   Enzymatic Methods.   The Boehringer enzymatic method (51) requires that 1 g of marzipan be taken for analysis and ground with approx. 2 g of sand. Water is then added and the mixture is heated at 60°C to extract the glycerol. The surnatant is transferred and filtered, and analysis is continued on the clear solution as the method prescribes.

6. Glycerol in Eggs

   a.   AOAC Methods (38).   In this case the method (no. 948.13) is also titrimetric. It requires that the glycerol be oxidized with periodate and the excess of periodate be titrated.

SYMBOLS

b.p. = boiling point
CV% = Coefficient of variation percent = standard deviation/arithmetic means x 100
DEG = diethylene glycol
ECD = electron capture detector
EI = electron impact
FID = flame ionization detector
FPD = flame photometric detector
GC-MS = gas chromatography–mass spectrometry
NMR = nuclear magnetic resonance
HPLC = high-performance (or pressure) liquid chromatography
HRGC = high-resolution gas chromatography
HSGC = headspace gas chromatography
i.d. = internal diameter
LC = liquid chromatography
L.R. = linear range
LSD = light scattering detector
MHE = multiple headspace extraction
MS = mass spectrometry
MSD = mass selective detector
NAD = Nicotinamide adenine dinucleotide
NADH = reduced nicotinamide adenine dinucleotide
NPD = nitrogen phosphorus detector
PTLC = pressure thin layer chromatography
PVPP = polyvinyl polypyrrolidone
rf = response factor
RI = refractive index
SC CO_2 = supercritical carbon dioxide
SFC = supercritical fluid chromatography
SFE = supercritical fluid extraction
TLC = thin-layer chromatography
TLC-$AgNO_3$ = argentation thin-layer chromatography
TMS = trimethylsilyl ether

Carrez 1 (solution) = 3.60 g K_4 [Fe(CN)$_6$] · 3H$_2$O/100 ml
Carrez 2 (solution) = 7.20 g ZnSO$_4$ · 7H$_2$O/100 ml
Surplus method = a satisfactory AOAC method but not in current use for various reasons

ACKNOWLEDGMENTS

Many thanks to my colleagues Dr. R. Rizzo and Mr. F. Varazini; Dr. M. Sacco and Dr. D. Conti, Ponti S. p. A., Ghemme, (NO), Italy; Dr. M. Castino, Stazione Sperim. per l'Enologia, Asti, Italy; Dr. A. Zagato, Ist . Scientifico Scotti–Bassani, Milano, Italy; Mrs. A. Tassara, Ist . Chimica Generale, Università, Genova, Italy; Dr. L. Maxwell.

BIBLIOGRAPHY

1. S. A. Kupina, *Am. J. Enol. Vitic., 35*: 59 (1984).
2. E. A. Crowell and J. F. Guymon, *Am. J. Enol. Vitic., 14*: 214 (1963).
3. M. Groat and C. S. Ough, *Am. J. Enol. Vitic., 29*: 112 (1978).
4. L. M. Klingshrin, J. R. Liu, and J. F. Gallander, *Am. J. Enol. Vitic., 38*: 297 (1987).
5. S. Shimura, T. Hasegawa, S. Takano, and T. Suzuki, *Nutr. Rep. Int., 36*: 1029 (1987).
6. T. Nagata, Y. Katsuta, A. Sugimoto, and T. Sato, *Breed. Sci., 44*: 427 (1994).
7. S. Budavari, *Merck Index*, 11th ed. Merck & Co., Rahway, NJ, 1989.
8. A. Razungles, E. Tarhi, R. Baumes, Z. Gunata, C. Tapiero, and C. Bayonove, *Sciences des Aliments, 14*: 725 (1994).
9. R. E. Major, *LC-GC Int., 4*: 10 (1991).
10. D. Knapp, *Handbook of Analytical Derivatization Reactions*, John Wiley and Sons, New York, 1979.
11. A. E. Pierce, *Silylation of Organic Compounds*, Pierce Chemical, Rockford, IL, 1982.
12. E. Martin, *4. e Congrès de Chimie industrielle*, Bordeaux, 1924.
13. R. J. Bouthilet and W. Lowrey, *JAOAC, 42*: 634 (1959).
14. A. Antonelli and A. Amati, *Vignevini, 7/8*: 35 (1989).
15. K. J. Hyver and P. Sandra, *High Resolution Gas Chromatography*, 3rd ed., Hewlett-Packard, Avondale, PA, 1989.
16. K. Grob, *Split and Splitless Injection in Capillary GC*, 3rd ed., Huethig, Heidelberg, 1993.
17. K. Grob, *On-column Injection in Capillary Gas Chromatography*, 2nd ed., Huethig, Heidelberg, 1991.
18. M. S. Klee, *GC Inlets: An Introduction*, 3rd ed., Hewlett-Packard, Avondale, PA, 1991.
19. R. Buffington and M. K. Wilson, *Detectors for Gas Chromatography: A Practical Primer*, Hewlett-Packard, Avondale, PA, 1991.
20. L. R. Snyder and J.J. Kirkland, *Introduction to Modern Liquid Chromatography*, 2nd ed., John Wiley and Sons, New York, 1979.
21. L. R. Snyder, J. L. Glajch, and J. J. Kirkland, *Practical HPLC Method Development*, John Wiley and Sons, New York, 1988.
22. R. Pecina, G. Bonn, E. Burtscher, and O. Bobleter, *J. Chromatogr., 287*: 245 (1984).
23. R. F. Frayne, *Am. J. Enol. Vitic., 37*: 281 (1986).
24. E. Le Fur, P. X. Etievant, and J.-M. Meunier, *J. Agric. Food Chem., 42*: 320 (1994).
25. J. W. Dolan and L. R. Snyder, *Troubleshooting LC Systems*, Humana Press, Clifton, NJ, 1989.
26. D. Nizery, D. Thiebaut, M. Caude, R. Rosset, M. La Fosse, and M. Dreux, *J. Chromatogr., 467*: 49 (1989).
27. J. W. King, *J. Chromatog. Sci., 28*: 9 (1990).
28. C. M. White, *Modern Supercritical Fluid Chromatography*, Huethig, Heidelberg, 1988.
29. E. Stahl, *Thin Layer Chromatography*, 2nd ed., Springer-Verlag, Berlin, 1988.
30. F. Geiss, *Fundamentals of Thin-Layer Chromatography*, Huethig, Heidelberg, 1987.
31. A. Messina and D. Corradini, *J. Chromatogr., 207*: 152 (1981).
32. R. Wintersteiger and G. Wenninger-Weinzierl, *Fresenius Z. Anal. Chem., 309*: 201 (1981).
33. M. L. Martin, J.J. Delpuech, and G.J. Martin, *Practical NMR Spectroscopy*, Heyden, London, 1980.
34. G. J. Martin, M.L. Martin, F. Mabon, and M.J. Michon, *J. Agric. Food Chem., 31*: 311 (1983).

35. G. J. Martin, M. Benbernou, and F. Lantier, *J. Inst. Brew., 91*: 242 (1985).
36. G. Schwedt and K. Stein, *Lebens. Unters. Forsch., 199*: 171 (1994).
37. O. S. Wolfbeis and H.E. Posch, *Fresenius Z. Anal. Chem., 332*: 255 (1988).
38. *Official Methods of Analysis*, Assoc. Off. Anal. Chemists, 15th ed. (1990), 3rd Supplement (1992), and 4th Supplement (1993).
39. *Official Journal of the European Communities*, L130, 15 May 1992.
40. *Official Journal of the European Communities*, L262, 27 September 1976.
41. P. Gerhards, *ZFL, 43*: 178 (1992).
42. *Metodi ufficiali di analisi per i mosti, i vini, gli agri di vino (aceti) e i sottoprodotti della vinificazione*, Min. Agric. e Foreste della Repubblica Italiana, D.M. 12 Marzo 1986, G.U. n° 61 del 14 Luglio 1986 (1986).
43. M. A. Lage Yusty, C.M. Garcia Jares, J. Simal Lozano, and M.E. Alvarez Piñeiro, *Anal. Bromatol., 41*: 375 (1989).
44. P. Pfeiffer and F. Radler, *Z. Lebensm. Unters Forsch, 181*: 24 (1985).
45. H. Belghith, J.-L. Romette, and D. Thomas, *Biotechnol. Bioeng., 30*: 1001 (1987).
46. R. Pecina, G. Bonn, E. Burtscher, and O. Bobleter, *J. Chromatogr., 287*: 245 (1984).
47. Bio-Rad Laboratories, *Guide to Aminex HPLC Columns for Food and Beverage Analysis*, Hercules, CA, 1994.
48. L. S. Conte, N. Frega, and G. Lercker, *Vignevini, 1–2*: 23 (1986).
49. I. McMurrough and J. R. Byrne, *HPLC analysis of raw materials and alcoholic beverages in brewing and related industries*, in *Food Analysis by HPLC* (L. M. L. Nollet, ed.), Marcel Dekker, New York, 1992.
50. A. Caputi jr., E. Christensen, N. Biedenweg, and S. Miller, *JAOAC Int., 75*: 379 (1992).
51. Boehringer Mannheim Biochemica GmbH., *Biochemical Analysis*, Mannheim, 1989.
52. E. Gómez, J. Laencina, and A. Martinez, *J. Food Sci., 59*: 406 (1994).
53. E. Gómez, A. Martinez, J. Nuñez, and J. Laencina, *Anal. Bromatol., 43*: 59 (1981).
54. M. A. Sefton, I. Leigh Francis, and P. J. Williams, *J. Food Sci., 59*: 142 (1994).
55. V. Gerbi, G. Zeppa, and A. Carnacini, *Ital. J. Food Sci., 4*: 259 (1992).
56. L. F. Di Cesare and R. Nani, *Ital. J. Food Sci., 4*: 279 (1992).
57. M. P. Romero, A. Casp, and J. M. Carrasco, Efficiency of extraction of some wine volatile compounds with dichloromethane-pentane, Proc. Int. Symp. on the Aromatic Substances in Grapes and Wines, (A. Scienza and G. Versini, eds.), S. Michele all' Adige (Italy), 1989, p. 371.
58. H. Stashenko, C. Macku, and T. Shibamato, *J. Agric. Food Chem., 40*: 2257 (1992).
59. K. Umano and T. Shibamoto, *J. Agric. Food Chem., 35*: 14 (1987).
60. S. V. A. Uzochukwu, E. Balogh, O. G. Tucknot, M. J. Lewis, and P. O. Ngoddy, *J. Sci. Food Agric., 64*: 405 (1994).
61. V. Mazzoleni and O. Colagrande, *Ital. J. Food Sci., 2*: 129 (1993).
62. S. Mannino and G. Amelotti, *Sci. Tecnologia degli Alimenti, 1–2*: 172 (1971).
63. I. Moret, G. Scarponi, and P. Cescon, *J. Sci. Food Agric., 35*: 1004 (1984).
64. *Recueil des méthodes internationales d' analyse des vins et des moûts*, Office International de la Vigne et du Vin, Paris, 1990.
65. P. Giannessi and M. Matta, *Trattato di scienza e tecnica enologica*, Vol. 1, Edizioni AEB, Brescia, Italy, 1987.
66. M. Plessi, A. Monzani, and D. Coppini, *Agric. Biol. Chem., 52*: 25 (1988).
67. A. M. Troncoso Gonzales and M. Guzman Chozas, *Belg. J. Food Chem. Biotechnol., 43*: 112 (1988).
68. R. I. Sauvage and P. J. Wagstaffe, *Ann. Fals. Exp. Chim. Toxicol., 66*: 246 (1973).
69. G. Annovazzi Rizzo, *Caratterizzazione di diversi tipi di' Aceto di Vino' in base alle loro componenti volatili*, Thesis, Univ. Cattolica del Sacro Cuore, Milano, 1990.
70. H. Belghith, J. L. Romette, and D. Thomas, *Biotechnol. Bioeng., 30*: 1001 (1987).
71. M. E. Cieslak and W. C. Herwig, *Am. Soc. Brew. Chem., 40*: 43 (1982).
72. J. Ukeda, Y. Fujita, M. Sawamura, H. Kusunose, and Y. Tazuke, *Anal. Sci., 9*: 779 (1993).
73. A. I. Garcia, L. A. Garcia, and M. Diaz, *Process Biochem., 29*: 303 (1994).
74. *Evolution des alcools supérieurs et du méthanol dans les eaux de vie de fruits*, Rev. Fr. Oenol, n° 126. (1990).
75. L. Usseglio-Tomasset and M. Matta, *Boll. Chim. Lab. Prov., 34*: 185 (1983).
76. J. Pino, A. Villarreal, and E. Roncal, *Die Nahrung, 38*: 307 (1994).

77. J. Simal, G. Salgado, and J. M. Rey, *Semaña Vitinicola, 40 (20–48)*: 4355 (1985).
78. *Improved Resolution of Alcoholic Beverage Components*, GC Bulletin 790B, Supelco, Bellefonte, PA, 1982.
79. P. Manitto, F. Chialva, G. Speranza, and C. Rinaldo, *J. Agric. Food Chem., 42*: 886 (1994).
80. F. D. Gunstone and B. G. Herslof, *A Lipid Glossary*, The Oily Press, Ayr, Scotland, 1992.
81. *Official Journal of the European Communities*, L248, Vol. 34, 5 September 1991.
82. *Official Journal of the European Communities*, L22, Vol. 36, 30 January 1993.
83. K. Grob, *On-Line Coupled LC-GC*, Heuthig, Heidelberg, 1990.
84. K. Grob, M. Lanfranchi, and C. Mariani, *J. Am. Oil Chem. Soc., 67*: 626 (1990).
85. W. W. Christie, *Gas Chromatography and Lipids*, The Oily Press, Ayr, Scotland, 1989.
86. J. M. Snyder and T. L. Mounts, *J. Am. Oil Chem. Soc., 67*: 800 (1990).
87. J. M. Snyder and J. W. King, *J. Sci. Food Agric., 64*: 257 (1994).
88. J. M. Snyder and J. M. King, *J. Am. Oil Chem. Soc., 71*: 3 (1994).
89. E. Selke and E. N. Frankel, *J. Am. Oil Chem. Soc., 64*: 5 (1987).
90. A. Strocchi and A. Savino, *Riv. Ital. Sostanze Grasse, 66*: 11 (1989).
91. A. Strocchi and G. Marascio, *Riv. Ital. Sostanze Grasse, 68*: 10 (1991).
92. P. Soulier, M. Farines, and J. Soulier, *J. Am. Oil Chem. Soc., 67*: 6 (1990).
93. G. Bianchi and G. Vlahov, *Fat Sci. Technol., 96*: 2 (1994).
94. K. Kawanishi, K. Aoki, Y. Hashimoto, and A. Matsunobu, *J. Am. Oil Chem. Soc., 68*: 11 (1991).
95. M. Amelio, R. Rizzo, and F. Varazini, *J. Chromatogr., 606*: 179 (1992).
96. R. Imhof and J.O. Bosset, *Leben.-Wiss. u.-Technol., 27*: 65 (1994).
97. L. Moio, J. Dekimpe, P. X. Etievant, and F. Addeo, *Ital. J. Food Sci., 1*: 43 (1993).
98. L. Moio, J. Dekimpe, P. Etievant, and F. Addeo, *J. Dairy Research, 60*: 199 (1993).
99. L. Moio, J. Dekimpe, P. X. Etievant, and F. Addeo, *Ital. J. Food Sci., 1*: 57 (1993).
100. L. Moio, J. Dekimpe, P. X. Etievant, and F. Addeo, *Ital. J. Food Sci., 3*: 215 (1993).
101. D. A. Forss and G. L. Holloway, *J. Am. Oil Chem. Soc., 44*: 572 (1967).
102. P. X. Etievant and C. L. Bayonove, *J. Sci. Food Agric., 34*: 393 (1983).
103. J. O. Bosset and R. Liardon, *Leben.-Wiss. u.-Technol., 17*: 359 (1984).
104. R. Lairdon, J. O. Bosset, and B. Blanc, *Leben-Wiss. u.-Technol., 15*: 147 (1982).
105. Ch. Donchev and M. Velcheva, *Riv. Ital. Sostanze Grasse, 70*: 29 (1993).
106. M. Velcheva and Ch. Donchev, *Riv. Ital. Sostanze Grasse, 69*: 95 (1992).
107. J. W. Eckert and M. Ratnayake, *Phytopathology, 84*: 746 (1994).
108. M. G. Moshonas and P. E. Shaw, *J. Agric. Food Chem., 42*: 1525 (1994).
109. S. Krulick and S. F. Osman, *J. Agric. Food Chem., 27*: 212 (1979).
110. J. A. Maga, *Food Rev. Int., 10 (1)*: 1 (1994).
111. T. Akihisa, Y. Kimura, K. Roy, P. Ghosh, S. Thakur, and T. Tamura, *Phytochemistry, 35*: 1309 (1994).
112. J. F. Huidobro, M. Estrella Rea, P. C. Branquinho de Andreade, M.P. Sanchez, M.T. Sancho, S. Muniategui, and J. Simal-Lozano, *J. Agric. Food Chem., 42*: 1975 (1994).
113. J. F. Huidobro, M. Estrella Rea, P.C. Branquinho de Andreade, M.P. Sanchez, M.T. Sancho, S. Muniategui, and J. Simal-Lozano, *J. Agric. Food Chem., 41*: 557 (1993).
114. N. A. M. Eskin, *Biochemistry of Foods*, 2nd ed., Academic Press, San Diego, 1990, p. 133.
115. H. Maarse and R. Belz, Isolation, separation and identification of volatile compounds in aroma research, D. Reidel, Dordrecht.
116. *Farmacopea Ufficiale Italiana della Repubblica Italiana Droghe. Vegetali e preparazioni*, 1st. Poligrafico dello Stato, Repubblica Italiana, 1991.
117. G. Johansson, J.-L. Berdagué, M. Larsson, N. Tran, and E. Borch, *Meat Sci., 38*: 203 (1994).

17

Determination of the Fat-Soluble Vitamins in Foods by High-Performance Liquid Chromatography

George F. M. Ball
Independent Consultant, Windsor, Berkshire, England

I. INTRODUCTION

The fat-soluble vitamins comprise vitamins A, D, E, and K. Also included are those carotenoids which act as precursors for vitamin A (provitamin A carotenoids). Recommended Dietary Allowances (RDA) have been published in the United States for vitamins A, D, E, and K (1). Other countries and international bodies have compiled similar recommendations. In the United States and Canada fluid milk is fortified by law with vitamin D_3 to a level of 400 IU/quart (10 μg/0.95 L) to meet the RDA of 10 μg. Other commodities such as margarine, milk products, breakfast cereals, dietetic foods, and infant formulas are commonly fortified with vitamins A, D, and E; fortification with vitamin K is confined mainly to infant formulas. The addition of vitamins to a particular processed food is intended to provide a specific proportion of the RDA. General principles for the addition of nutrients to foods have been established by the Codex Alimentarius Commission (2) and the U.S. Food and Drug Administration (3).

Vitamin assays in foods are carried out for a variety of purposes: to implement regulatory enforcement; to check compliance with contract specifications and nutrient labeling regulations; to provide quality assurance for supplemented products; to provide data for food composition tables; to study changes in vitamin content attributable to food processing, packaging, and storage; and to assess the effects of geographic, environmental, and seasonal conditions. The monitoring of foods fortified with vitamins A and D to ensure that neither inadequate nor excessive quantities are added is particularly important because, in addition to the serious pathological consequences of deficiency, these two vitamins are toxic at excessive intakes (4,5). The fortification of animal feeds is another potential route for entry of abnormal quantities of nutrients into human foods, and hence animal feeds should be monitored too—at least for the highly potent vitamin D.

To assess the nutritional value of a food or diet with respect to a particular vitamin, it is necessary to determine the biological activity of the vitamin within the food or diet. Biological activity refers to the vitamin's inherent potency in meeting a specific nutritional requirement and can only be performed directly by means of an animal bioassay using rats or chicks, which serve as models for human bioassays. The result obtained from a bioassay reflects the combined *in vivo* response of the various structurally related forms (vitamers) of a particular vitamin, its precursors, and biologically active metabolites, taking into account the complexities of intestinal digestion, absorption, and transport of the vitamin to the target cells. The biological activity of a vitamin is not synonymous with its bioavailability. Bioavailability refers to the extent of intestinal absorption and metabolic utilization of all dietary forms of the vitamin (6). In the experimental context, bioavailability is defined as the concentration of biologically active forms of the vitamin (as determined by an animal bioassay) divided by the total concentration of the vitamin (as determined by a physicochemical assay after an extraction procedure designed to liberate all of its bound forms) (7). Nonradioactive stable isotopes have been used in isotopic labeling methods to study vitamin bioavailability in human subjects (8).

Bioassays are too time-consuming and expensive to be employed routinely, and the fat-soluble vitamins are determined by physicochemical methods. The current method of choice is high-performance liquid chromatography (HPLC). In nutritional studies, the results of an HPLC assay must be interpreted in light of the vitamin's bioavailability (as far as it is known), the extraction technique employed in the assay, the compounds actually measured, and the biological potencies of these compounds.

II. CHEMICAL AND BIOLOGICAL NATURE OF THE FAT-SOLUBLE VITAMINS

A. Vitamin A and the Provitamin A Carotenoids

Naturally occurring sources of vitamin A are represented by preformed vitamin A (retinol, retinyl esters, and retinaldehyde) and provitamin A carotenoids. An adequate intake of dietary vitamin A is essential for normal vision, growth, cellular differentiation, reproduction, and the integrity of the immune system. Ziegler (9,10) reviewed epidemiological studies which have correlated the intake of carotenoid-rich fruits and vegetables with protection from some forms of cancer. These epidemiological associations, however, do not show cause and effect, and long-term intervention trials with β-carotene supplements are in progress (11).

1. Structure and Biological Activity

a. Preformed Vitamin A. The parent vitamin A compound, retinol, has the empirical formula $C_{20}H_{30}O$ and a molecular weight (MW) of 286.44. Its structure comprises a cyclohexenyl ring attached at the carbon-6 position to a polyene side chain whose four double bonds give rise to *cis–trans* isomerism. The predominant isomer, all-*trans*-retinol, possesses maximal (100%) vitamin A activity and is frequently accompanied in foodstuffs by smaller amounts of 13-*cis*-retinol, which possesses 75% relative activity in the rat (12). Other *cis* isomers of retinol also occur, but they are of low potency and their contribution to the total vitamin A activity of a foodstuff is generally negligible. In most of the foods which contain vitamin A the retinol forms esters with long-chain fatty acids, particularly palmitic acid. Retinaldehyde, which occurs in fish roe and hens' eggs as well as in the eye, exhibits about 90% of the biological activity of all-*trans*-retinol (13).

b. Provitamin A Carotenoids. Carotenoids are yellow, orange, red, or violet pigments that are responsible for the color of many vegetables and fruits. They are divided chemically into two groups: the hydrocarbon carotenoids are known as carotenes and the oxygenated derivatives are termed xanthophylls. The nutritionally most important carotenoid, β-carotene ($C_{40}H_{56}$, MW = 536.88), is composed of two molecules of retinol joined tail to tail; thus the compound possesses maximal (100%) provitamin A activity. The structures of all other provitamin A carotenoids incorporate only one molecule of retinol, hence theoretically contribute 50% of the biological value of β-carotene. In addition to the C_{40} carotenoids, shortened molecules known as apocarotenoids exist, some of which exhibit provitamin A activity (14).

Of the estimated 500 or so carotenoids that occur throughout nature, about 60 possess provitamin A activity in varying degrees. In most vegetables and in many fruits, β-carotene constitutes more than 85% of the total provitamin A activity. Notable exceptions are carrots and oranges, which contain both β-carotene and α-carotene in the ratio of 2:1 (15). β-Cryptoxanthin is the predominant provitamin A carotenoid in orange juice (16) and in some varieties of sweet corn (17). In many fruits and vegetables the concentrations of provitamin A carotenoids are low relative to the concentrations of inactive carotenoids. For example, lutein is the most abundant carotenoid in green leafy vegetables (18), lycopene predominates in tomatoes (19), and capsanthin is the major pigment in red peppers (20). Other inactive carotenoids found in fruits and vegetables include violaxanthin, neoxanthin, zeaxanthin, zeta-carotene, phytoene, and phytofluene (21). Milk products, egg yolk, shellfish, and crustacea also contain carotenoids, which are derived from the animal's diet.

In plant and animal tissues the carotenoids are usually found associated with lipid fractions in noncovalent association with membranes and lipoproteins, and they accumulate, together with chlorophylls, in the chloroplasts of green leaves (22). They also occur as very fine dispersions in aqueous systems, such as orange juice. Carotenoids exist primarily in the all-*trans* configuration, but small amounts of *cis*-isomeric forms have been found in fresh fruits and vegetables. Traditional food processing and preservation methods, especially canning, induce *cis-trans* isomerization. The main *cis* isomers of β-carotene that have been found in fresh and processed fruits and vegetables are 13-*cis* and 9-*cis* (23), which have provitamin A activities relative to all-*trans*-β-carotene of 53% and 38%, respectively (24). A third isomer, 15-*cis*-β-carotene, has also been reported to occur in several fruits and vegetables (23). The main portion of the xanthophylls exists as mono or bis esters of saturated long-chain fatty acids, such as myristic, lauric, and palmitic acids (25,26).

2. Dietary Sources

All natural sources of vitamin A are derived ultimately from provitamin A carotenoids, which are synthesized exclusively by higher plants and photosynthetic microorganisms. Animals are able to convert ingested provitamins to retinol by enzymatic action in the intestinal wall. Thus provitamin A carotenoids are obtained mostly from plant sources and preformed vitamin A from animal sources. Typical vitamin A values in animal products (expressed in μg retinol equivalents/100 g) are calf liver 8300–31,700; whole milk 38–55; butter 674–1062; cheddar cheese 260–440; and eggs 159–210 (27). Much of the dietary vitamin A intake in certain populations is provided by vegetables, among which carrots and green leafy vegetables are particularly rich sources. Typical provitamin A contents in vegetables (μg retinol equivalents/100 g) are carrots, 1355; spinach, 589; and tomatoes, 105 (27).

Foods are supplemented with vitamin A in the form of synthetic retinyl acetate or retinyl palmitate. Synthetic β-carotene, β-apo-8′-carotenaldehyde (apocarotenal), the ethyl ester of β-apo-8′-carotenoic acid (apocarotenoic ester), and the nonprovitamin carotenoid canthaxanthin have been found acceptable as food color additives by the joint FAO/WHO Expert Committee on Food Additives (28). Commercial preparations of these compounds are gelatin-coated and/or antioxidant-stabilized to protect them during processing and to extend the product shelf life (29).

3. Expression of Dietary Values

The vitamin A value of foods has traditionally been expressed in international units (IU). One IU is defined as the amount of vitamin A activity contained in 0.344 μg of all-*trans*-retinyl acetate, which is equivalent to 0.300 µg of all-*trans*-retinol. In 1965, an expert committee decided to abandon the IU, proposing instead that the vitamin A value be designated in terms of retinol equivalents (RE) expressed in micrograms of retinol (30). The RE is defined as the amount of retinol plus the equivalent amount of retinol that can be obtained from the provitamin A carotenoids. It is purely a dietary concept and is not an equivalency in the usual chemical sense.

RE = µg retinol + µg β-carotene/6 + µg other provitamin A carotenoids/12.

For comparison with values in the older literature, the IU values can be converted to retinol equivalents.

To convert IU into RE on the basis of retinol, 1 RE = 1 µg retinol and 1 IU = 0.3 µg retinol. Therefore, 1 RE = 1/0.3 = 3.33 IU vitamin A activity from retinol.

Since the IU was based on studies that did not take into account the poor absorption and biological availability of carotenoids in foods, the equivalency of retinol and β-carotene in the IU system differs from that in the RE system. Thus in the RE system, 1 µg retinol ≡ 6 µg β-carotene, whereas in the IU system, 1 µg retinol ≡ 2 µg β-carotene.

To convert IU to RE on the basis of β-carotene, one must first multiply the IU by a factor of 3 (6/2) to make the equivalency the same as that of the RE system, and then multiply by 3.33. Therefore, 1 RE = 3 × 3.33 = 10.0 IU vitamin A activity from β-carotene. RE = number of IU from retinol/3.33 + number of IU from β-carotene/10. The units for expressing vitamin A activity are presented in Table 1.

For food labeling purposes, which require the actual amounts of vitamin A in the food rather than the nutritional value, data obtained by physicochemical assay are expressed on a weight basis. In plant-derived foods that contain predominantly carotenoids, the units are β-carotene equivalents expressed in micrograms of β-carotene. By definition, 1 β-carotene equivalent is equal to 1 µg of all-*trans*-β-carotene or 2 µg of other, largely all-*trans*, provitamin A carotenoids in the foods. In animal-derived foods that contain predominantly preformed vitamin A, the units are either micrograms of retinol or retinol equivalents (32).

B. Vitamin D

1. Structure and Biological Activity

Vitamin D is represented by cholecalciferol (vitamin D_3) and ergocalciferol (vitamin D_2), which are structurally similar secosteroids derived from the ultraviolet (UV) irradiation of sterols. In vertebrates, vitamin D_3 ($C_{27}H_{44}O$, MW = 384.62) is produced *in vivo* by the action of sunlight on 7-dehydrocholesterol in the skin. Vitamin D_2 ($C_{28}H_{44}O$, MW = 396.63) is produced in plants, fungi, and yeasts by the irradiation of ergosterol. 25-Hydroxyvitamin D_3 is

Table 1 Units for Expressing Vitamin A Values

Compounds[a]	µg/IU	IU/µg	µg/RE	RE/µg
Retinol	0.300	3.33	1.000	1.000
Retinyl acetate	0.344	2.91	—	—
Retinyl palmitate	0.55	1.82	—	—
β-Carotene	0.6 (1.8)[b]	0.56	6	0.167
Mixed provitamin carotenoids	1.2 (3.6)	0.28	12	0.083

[a] All-trans isomers.
[b] The value of 1.8 µg/IU agrees with the convention of considering 6 µg β-carotene as 1 µg RE (31).

the major circulating metabolite and 1,25-dihydroxyvitamin D_3 is the most potent metabolite for the stimulation of intestinal calcium transport and the mobilization of calcium from the bone. An adequate supply of vitamin D is essential in the prevention of rickets in children and of osteomalacia in adults. Many human populations depend on dietary sources of vitamin D because of insufficient biosynthesis of the vitamin due to inadequate skin exposure to sunlight. The biological potencies of vitamins D_2 and D_3 in humans are essentially equal.

2. Dietary Sources

Plants are extremely poor sources of vitamin D and the consumer must rely on animal products for a dietary supply of the vitamin in the natural form. Even so, only trace amounts are found. The richest natural food sources (expressed in µg/100 g), apart from fish liver oils, are fatty fish, such as herring, 22; sardines and pilchards, 8; and tuna, 6 (33). Other significant food sources are butter, 0.8; and cheddar cheese, 0.3 (27). Eggs normally contain about 1.8 (33), but eggs from hens receiving a vitamin D supplement will have a considerably higher vitamin D content. Except for eggs and fatty fish, a serving of food containing only natural sources of vitamin D would probably supply less than 1 µg of vitamin D (34).

The vitamin D activity in animal products is contributed by both vitamin D_3 and its metabolite, 25-hydroxyvitamin D_3. Typical values of the hydroxylated metabolite (µg/100 g) are bovine muscle, 0.2–0.3; bovine liver, 0.3–0.5; bovine kidney, 0.5–1.0 (35); and chicken egg yolk, 1.0 (36). 25-Hydroxyvitamin D_3 is of nutritional importance as it is about five times more active than vitamin D_3 and occurs in significant amounts. In milk, for example, it accounts for 72% of the total vitamin D activity as estimated by the calcium transport assay (37).

3. Expression of Dietary Values

One IU of vitamin D is the activity of 0.025 µg of crystalline cholecalciferol. An expert committee in 1970 recommended that the intake of vitamin D be expressed as µg of cholecalciferol rather than as IU (30).

C. Vitamin E

1. Structure and Biological Activity

Eight vitamers of vitamin E occur in nature: four tocopherols and four tocotrienols. Tocopherols are methyl-substituted derivatives of tocol and comprise a chroman-6-ol ring attached at C-2 to a C-16–saturated phytyl side chain. Tocotrienols are analogous structures whose side chains contain three double bonds. The tocopherols and tocotrienols are designated as α, β, γ, and δ, according to the number and position of the methyl groups in the chromanol ring. The α form is 5,7,8-trimethyl, β is 5,8-dimethyl, γ is 7,8-dimethyl, and δ is 8-methyl. Many of the diverse deficiency syndromes observed in experimental animals deprived of vitamin E can be explained by the vitamin's role as a lipid antioxidant in stabilizing subcellular membranes.

The vitamin E molecule contains three chiral centers, giving rise to eight possible diastereomeric forms. In nature, the polyene side chains of the tocotrienols have the all-*trans* configuration. Naturally occurring α-tocopherol (formerly known as *d*-α-tocopherol) should be designated *RRR*-α-tocopherol to describe its stereochemistry. Totally synthetic α-tocopherol (formerly known as *dl*-α-tocopherol) should be designated all-*rac*-α-tocopherol. The most biologically active vitamer is *RRR*-α-tocopherol ($C_{29}H_{50}O_2$, MW=430.7). The *RRR* forms of the various tocopherols (T) and tocotrienols (T3) possess the following biological potencies (percent activity) as measured by the fetal resorption assay in the rat: α-T (100), β-T (25–40), γ-T (1–11), δ-T (1), α-T3 (29), β-T3 (5) (Ref. 38).

2. Dietary Sources

Vitamin E is distributed widely in nuts, seeds, fruits, vegetables, and grasses, in which all eight vitamers are represented. Major sources of vitamin E in Western diets are the cereal seed oils and the margarines and other products made from them. The indigenous concentrations of vitamin E in fatty foods are much higher than those of vitamins A and D. Typical U.S. values (expressed in mg of total vitamin E/100 g) are wheat germ oil, 166–300; soybean and cotton-seed oil stick margarine, 45; and mayonnaise, 58. The chief animal sources, in which α-tocoph-erol predominates, are high-fat products such as eggs, 1.1; milk, 0.1; butter, 1.6; and liver, 0.7 (39). A U.S. survey of total vitamin E activity in frequently consumed foods analyzed by HPLC was published in 1989 (40).

The acetate ester of α-tocopherol, rather than the free alcohol, is used as a food supplement on account of its greater stability. Both RRR-α-tocopheryl acetate and totally synthetic all-rac-α-tocopheryl acetate are commercially available, the former having a biological activity of 1.36 IU/mg and the latter 1.00 IU/mg (38). The RRR-α-tocopheryl acetate is obtained by extraction from vegetable oils. Since it is not isolated without some chemical processing, it cannot legally be called natural but can be described as derived from natural sources.

3. Expression of Dietary Values

One IU of vitamin E is the activity of 1 mg of all-rac-α-tocopheryl acetate, but the activity is now expressed as mg of α-tocopherol equivalents (30). When calculating the total vitamin E activity of mixed diets in the United States, the mg of β-tocopherol, γ-tocopherol, and α-tocotrienol can be multiplied by factors of 0.5, 0.1, and 0.3, respectively, and added to the mg of α-tocopherol to give the total mg of α-tocopherol equivalents. If only α-tocopherol in a mixed diet is reported, its amount in milligrams can be multiplied by 1.2 to account for the other vitamers present (1).

D. Vitamin K

1. Structure and Biological Activity

Vitamin K is the generic name for 2-methyl-1,4-naphthoquinone and all 3-substituted derivatives of this compound that exhibit antihemorrhagic activity in animals fed a vitamin K–deficient diet. The vitamin is necessary for the synthesis of prothrombin and a number of other proteins involved in blood clotting. Phylloquinone ($C_{31}H_{46}O_{21}$, MW = 450.68), also known as vitamin K_1, has a phytyl side chain with a *trans* double bond at the 2' position and is synthesized by green plants in the chloroplasts. Synthetic preparations of phylloquinone invariably contain a proportion of the 2'-*cis* isomer, which has little or no antihemorrhagic activity. Vitamin K_2 refers to a series of homologs with unsaturated side chains called menaquinones, which are synthesized by bacteria. The commonly occurring menaquinones have side chains composed of 6–10 isoprene units and are designated menaquinone-6 (MK-6) to menaquinone-10 (MK-10) accordingly (41). The underivatized 2-methyl-1,4-naphthoquinone structure, known as menadione (vitamin K_3), is not a natural product, but menadione sodium bisulfite and other water-soluble derivatives are synthe-sized commercially for use as animal feed supplements. Menadione is toxic to infants at excessive dose levels (5) and hence is not used in human medicine or as a food supplement.

2. Dietary Sources

Green leafy vegetables are the outstanding sources of vitamin K, with other vegetables provid-ing useful amounts. Typical published values (μg phylloquinone/100 g raw edible portion) are cabbage, 145; spinach, 400; kale, 817; and broccoli, 205 (42). Soybean oil, which is added to infant formulas, is a valuable natural source of vitamin K, containing 450–630 μg phylloquinone/100 g (43). Examples of animal sources (same units) are ox liver, 94 (44) and milk, 0.3; butter, 7; and eggs, 2 (42).

3. Expression of Dietary Values

Vitamin K is widely distributed in natural foods and the human requirement for it is small. Moreover, intestinal bacteria normally produce menaquinones in amounts sufficient to maintain normal blood clotting function, even in the absence of a dietary intake of vitamin K. Apart from clinical and neonatal cases, a deficiency of vitamin K in humans is therefore rare. For these reasons, an official standard for vitamin K has not been established.

   a.   Scope of Analytical Techniques.   The vitamins were originally isolated and characterized using bioassays to monitor their activity. The bioassays were too time consuming and expensive to apply to routine food analysis, and colorimetric methods were later developed for vitamins A, D, E, and K. These methods require open column chromatography or thin-layer chromatography to isolate the vitamins from interfering substances. The Carr–Price colorimetric method introduced in 1926 still forms the basis of the AOAC method for determining vitamin A in mixed feeds, premixes, and foods (45). During the 1960s, gas chromatography (GC), using so-called packed columns, was widely applied to the determination of vitamins D and E in foods, but cleanup by open column or thin-layer chromatographic techniques was still necessary, as was derivatization to increase the vitamins' thermal stability and volatility. The development of fused silica open tubular capillary columns has revived the interest in GC, leading to a number of recent applications for the determination of vitamin E (46–49). However, since the mid-1970s to the present day, the method of choice for determining the fat-soluble vitamins in foods has undoubtedly been HPLC, owing to the technique's ability to chromatograph the vitamins without the need for derivatization, the nondestructive operation, and the greater separation and detection selectivity.

   At present, there is no universally recognized standard method for determining any of the fat-soluble vitamins that can be applied to all types of food. In this chapter, selected representative published HPLC methods are tabulated and key practical features discussed. The selected method must always be modified to suit the composition of the sample to be analyzed.

E.   Vitamin A and the Provitamin A Carotenoids

Of the preformed vitamin A commonly found in foods, only all-*trans*-retinol and smaller amounts of 13-*cis*-retinol, both in esterified form, are usually present in significant quantities. For the analysis of vitamin A–fortified foods, HPLC can be applied to determine either the total retinol content or the added retinyl ester (acetate or palmitate), depending on the extraction technique employed. The vitamin A activity of plant foods is usually based on the HPLC determination of the three most ubiquitous provitamins, namely, α- and β-carotene and β-cryptoxanthin. The most satisfactory methods are those that separate all-*trans*-β-carotene from its *cis* isomers.

F.   Vitamin D

Most of the published HPLC methods for determining vitamin D in foods are concerned with estimating the vitamin D content in fortified products such as milk in various forms, infant formulas, and margarine. In fortified foods, the amount of naturally occurring vitamin D (if any) is usually relatively very small, and it is only necessary to determine the vitamin D that is added. Even so, fortification levels are very low (e.g., 7.5–12.5 μg/100 g in milk powder) (50), and the determination of vitamin D is by no means a simple task.

A bioassay will account for the activity of previtamin D as well as vitamin D and its various active metabolites. A valid estimate of the vitamin D value of a food should therefore represent "potential vitamin D," i.e., the sum of the vitamin D and previtamin D contents.

When determining naturally occurring vitamin D in animal products for nutritional evaluation purposes, 25-hydroxyvitamin D_3 should be included, as this metabolite contributes significantly to the total biological activity, particularly in milk. 25-Hydroxyvitamin D_3 is present in dairy products, eggs, and meat tissues in sufficient concentration to permit its determination by HPLC using an absorbance detector. In bovine milk the concentration of this metabolite is <1 ng/ml (51) and hence it is usually determined by a competitive protein binding assay after fractionation of the extracted sample by HPLC (52).

G. Vitamin E

The nutritional evaluation of the vitamin E–rich vegetable oils and the products made from them using a nonbiological assay necessitates the determination of the individual tocopherols and tocotrienols, as these vitamers vary widely in biological activity. HPLC is ideally suited for this purpose, and the overall vitamin E value of such foods can be estimated by applying appropriate factors based on relative biological activities. For the analysis of those animal products which are known to contain predominantly α-tocopherol, only this vitamer need be determined. In vitamin E–fortified foods it is usually sufficient to determine either the added α-tocopheryl acetate or the total α-tocopherol.

H. Vitamin K

Among the various vitamin K compounds, only phylloquinone is accounted for in routine food analysis. Milk-based and soy protein–based infant formulas for the full-term infant are supplemented with synthetic preparations of phylloquinone, which invariably contain about 10% of the biologically inactive *cis* isomer (53). The analytical method must therefore exclude *cis*-phylloquinone in the measurement.

III. SAMPLE PREPARATION

The lipid fraction of foods containing the fat-soluble vitamins is composed mainly of triglycerides, with much smaller amounts of sterols, carotenoids, phospholipids, and minor constituents. All of these substances exhibit similar solubility properties to the fat-soluble vitamins and therefore constitute a potential source of interference. A proportion of the indigenous fat-soluble vitamin content of a food is bound up with a lipoprotein complex, and hence the fat–protein bonds must be broken in order to release the vitamin. The vitamin preparations added to foods are often contained within a protective gelatin coating, which will need to be dissolved at the start of the analysis.

It is essential for a successful assay that the vitamins be quantitatively extracted from the food matrix in a form that can be measured by the particular analytical technique to be used. An effective extraction procedure serves to homogenize and concentrate the sample, isolate the vitamin from its association with protein, eliminate as far as is possible known interfering substances, and destroy any indigenous enzyme activity. The vitamin-rich fraction thus obtained may require some form of cleanup before the fat-soluble vitamins can be measured.

The fat-soluble vitamins are photosensitive. Therefore all operations with vitamin solutions and vitamin-containing materials should be carried out in subdued light or in low actinic amber glassware. Ideally, the laboratory or a dedicated room within the laboratory should have the windows (if any) covered with effective blinds, and artificial lighting should be provided by

F40GO "gold" (or equivalent) fluorescent lamps, which exclude radiation wavelengths of less than 500 nm.

A. Extraction Procedures

Factors to consider in selecting a suitable extraction procedure are (a) the analytical information required, (b) the nature of the food matrix, (c) the form in which the vitamin occurs naturally or is added, (d) the nature and relative amounts of potentially interfering substances, (e) the stability of the vitamin towards heat and extremes of pH, and (f) the selectivity and specificity of the analytical technique to be used.

1. Alkaline Hydrolysis

Alkaline hydrolysis (saponification) provides an effective means of removing the preponderance of neutral lipid (mainly triglycerides) from the food sample and is an economical way of digesting a relatively large amount of material for analysis. This extraction procedure can be utilized in assays for vitamins A, D, and E, but it is not expedient for vitamin K compounds, which are rapidly decomposed in alkaline media. Conventional "hot" saponification involves refluxing the suitably prepared sample with ethanolic KOH solution and added antioxidant for 30 min. The hydrolysis reaction attacks ester linkages and releases the fatty acids from the glycerol moiety of glycerides and phospholipids, and from esterified sterols and carotenoids. The reaction also liberates indigenous vitamins from any combined form in which they may exist (e.g., lipoprotein complex) and breaks down chlorophylls into small water-soluble fractions. In addition, it dissolves any gelatin that might have been present in the vitamin premix added to supplemented foods. The sterols, carotenoids, fat-soluble vitamins, and so forth, which constitute the unsaponifiable fraction, are extractable from the saponification digest by liquid–liquid extraction using a water-immiscible organic solvent, after adding water to the digest to facilitate the separation of the aqueous and organic phases. The fatty acids, which are precipitated as their potassium salts (soaps), and the glycerol are not extractable under alkaline conditions. The combined solvent extracts are evaporated to dryness and the residue is redissolved in a small volume of a suitable solvent for chromatographic analysis or further purification. The evaporation step concentrates the sample extract, and this may be obligatory in the analysis of unfortified foods. However, it is time-consuming and provides a further possible opportunity for loss of the vitamin analyte. A practical advantage of saponification is that the esters of vitamins A and E present in the sample are hydrolyzed to their respective alcohols, which allows the estimation of total retinol or total α-tocopherol. The liberation of the unstable retinol and tocopherols from their relatively stable esters during saponification demands that protective measures against light and oxygen be implemented throughout the analysis.

Starchy foods, such as breakfast cereals, can be digested with the enzyme takadiastase before saponification so as to avoid the formation of lumps (54).

2. Enzymatic Hydrolysis

Enzymatic hydrolysis is an effective alternative to saponification for removing triglycerides in vitamin K determinations. Barnett et al. (55) published a method for the simultaneous determination of vitamins A, D, E, and K in milk-based and soy protein–based infant formulas and dairy products fortified with these vitamins. In this procedure, an amount of sample containing approximately 3.5–4.0 g of fat is digested for 1 hr with lipase at 37°C and pH 7.7. This treatment effectively hydrolyzes the glycerides but only partially converts retinyl palmitate and α-tocopheryl acetate to their alcohol forms; vitamin D and phylloquinone are unaffected. The hydrolyzate is made alkaline in order to precipitate the fatty acids as soaps, then diluted with ethanol and extracted with pentane. A final water wash yields an organic phase containing primarily the fat-soluble vitamins and cholesterol.

3. Direct Solvent Extraction

A nonhydrolytic extraction procedure using a suitable solvent system can be applied to the determination of naturally occurring vitamin A esters and supplemental α-tocopheryl acetate. Among the many published procedures, the Rose–Gottlieb method is particularly suitable for extracting the total fat from milk products and infant formulas. It entails treatment of the reconstituted milk samples with ammonia solution and alcohol in the cold and extraction with a diethyl ether/petroleum ether mixture. The alcohol precipitates the protein, which dissolves in the ammonia, allowing the fat to be extracted with the mixed ethers. The method is suitable for the extraction of vitamins A and D but not for vitamins E and K, which are labile under alkaline conditions.

4. Supercritical Fluid Extraction

Supercritical fluids are compounds that are held above their critical temperature and pressure. Under these conditions they have properties intermediate between those of liquids and gases (56). For CO_2, the most widely used extractant for SFE, the critical temperature is 31°C and the critical pressure is 73 atm (57). The low viscosity and high diffusion rate of supercritical CO_2 provide a high efficiency of extraction, during which the test material is maintained in an inert atmosphere and protected from light. By adjusting the conditions of temperature and pressure, the selectivity of the extraction can be optimized, thereby reducing the need for a complex cleanup stage before analysis (58).

B. Cleanup Procedures

Cleanup or fractionation procedures that have been used in fat-soluble vitamin assays include sterol precipitation, open column chromatography, solid phase extraction, and high-pressure gel permeation chromatography. HPLC has been used on a semipreparative scale in vitamin D and vitamin K assays to obtain purified fractions of sample extracts. This technique is discussed in Sec. IV.B.3.

1. Precipitation of Sterols

The bulk of the sterols can be removed from the unsaponifiable fraction of the sample by precipitation from a freezing methanolic solution followed by filtration.

2. Open Column Chromatography

The more recent applications of open column chromatography in vitamin assays utilize liquid–solid (adsorption) chromatography using gravity flow glass columns dry-packed with magnesia, alumina, or silica gel. Such columns enable separations directly comparable with those obtained by thin-layer chromatography to be carried out rapidly on a preparative scale.

3. Solid Phase Extraction

Solid phase extraction refers to the use of disposable prepacked cartridges to clean up the sample matrix (59). Concentration of the sample extract is achieved in the cleanup mode if the elution volume is less than the volume of sample extract applied to the cartridge. The full range of silica-based polar and nonpolar stationary phases encountered in HPLC column packings is commercially available, but only silica and reverse phase C_{18}-bonded silica have so far found application in fat-soluble vitamin assays.

4. Gel Permeation Chromatography

High-pressure gel permeation chromatography, employing up to four 300×7.8 mm internal diameter (i.d.) μStyragel 100-Å columns connected in series, has been used in vitamin A, D, and E assays to remove triglycerides in the analysis of oils and margarine (60), breakfast cereals (61), and infant formulas (62,63). μStyragel 100-Å is a semirigid gel composed of 10-μm particles of polystyrene crosslinked with divinylbenzene and has an average pore size of approximately 40 Å ($1 \text{ Å} = 10^{-10}$ m) (64).

C. Vitamin A

The universally accepted method of extracting vitamin A from almost any type of food commodity or diet sample is saponification. Exceptions are fish roe and hens' eggs, which contain 90% and 10%, respectively, of their vitamin A activity in the form of retinaldehyde (13). Retinaldehyde is destroyed by alkali treatment and hence its determination requires special treatment (65).

A 50:50 mixture of diethyl ether and light petroleum ether (boiling range 40–60°C) has been recommended by the European project COST 91 (European Cooperation in Scientific and Technological Research) (66) for the extraction of vitamin A from the diluted saponification digest. The soaps are soluble in diethyl ether, so the ether extract has to be washed with successive portions of distilled water until the washes are free from alkali (colorless on addition of phenolphthalein). The washing steps must be performed by inverting the separating funnel gently several times to avoid the formation of stable emulsions, which are produced when soaps, water, and hydrophobic solvent are shaken in the absence of ethanol. The use of hexane is advantageous in that soaps are not extracted, but the ethanol–water–soap mixture behaves more like a hydrocarbon solvent, and the addition of water has less effect on the phase separation. Therefore, the minimum number of extractions needed to achieve a quantitative recovery of vitamin A depends on the concentration of fatty acids in the digest (67).

Zaher and Smith (68) developed a rapid saponification method for the extraction of vitamin A from milk and other fluid dairy products which avoids the need for several extractions and washings using separating funnels. To a 50-ml stoppered centrifuge tube is added sequentially 2 ml of sample, 5 ml of absolute ethanol containing 0.1% (w/v) ascorbic acid or 1% (w/v) pyrogallol, and 2 ml 50% (w/v) KOH. The tubes are stoppered, agitated carefully, and placed in a water bath at 80°C for 20 min. During this period, the tubes are agitated periodically to ensure complete digestion of the fat. After saponification, the tubes are cooled with running water and placed in an ice water bath prior to the addition of 20 ml of diethyl ether/petroleum ether (50:50) containing 0.01% (w/v) butylated hydroxytoluene (BHT). The tubes are again stoppered and vortex-mixed vigorously for 1 min, then allowed to stand for 2 min, and again vortexed for 1 min. To each tube is added 15 ml of ice-cold water, and the tubes are inverted at least 10 times. After centrifugation, 10 ml of the upper organic layer is accurately removed by pipet into a tube, and the solvent is evaporated to dryness in a stream of nitrogen or under vacuum at 40°C using a rotary evaporator. The residue is redissolved in 1 ml of methanol (for milk samples) for HPLC injection.

In fortified fluid milks, in which the vitamin A ester (palmitate or acetate) in the form of an oily premix is thoroughly dispersed in the bulk product, the total vitamin A can be extracted directly with hexane. The hexane solution, after removal of the polar material, is then injected into the liquid chromatograph. Thompson et al. (69) developed such a method in which a 2.0-ml milk sample is mixed with 5.0 ml of absolute ethanol in a 15-ml stoppered centrifuge tube and allowed to stand in the dark for 5 min. The milk constituents in this mixture are suspended in 71% aqueous ethanol, which denatures the proteins and fractures the fat globules. Hexane (5.0 ml) is added and the tube contents are vortexed for 30 sec, then allowed to stand for 2 min. The mixing and standing procedure is repeated twice. Distilled water (3 ml) is added and the tube is inverted several times, then centrifuged. The hexane (upper) layer is removed using a Pasteur pipet and analyzed by adsorption HPLC.

D. Provitamin A Carotenoids

Plant material should be extracted as rapidly as possible after it is obtained so as to avoid oxidative degradation of the carotenoids. If the material cannot be extracted immediately, it should be stored in a freezer or freeze-dried. The plant material should be undamaged, as rapid enzymatic degradation of carotenoids occurs immediately when the leaves are cut. Fresh and frozen fruits and vegetables should be blanched to inactivate enzymes before blending to a purée consistency, adding a known volume of water if necessary. Sun-dried fruits such as apricots, peaches, and prunes should be ground in a food chopper and a representative sample allowed to rehydrate by steeping in a known volume of water for several hours before blending. Fruits canned in thick syrup should be washed two or three times with water before blending.

Food samples, such as papaya, that contain β-cryptoxanthin ester among other xanthophyll esters require saponification to hydrolyze these esters and simplify the analysis. Saponification is also advisable for the analysis of green leafy vegetables, as the alkali treatment breaks down chlorophylls which could otherwise sensitize photoisomerization of the carotenoids (70). Kimura et al. (71) tested six widely used saponification procedures and recommended a procedure in which the carotenoids are dissolved in petroleum ether, an equal volume of 10% methanolic KOH is added, and the mixture is left standing overnight (about 16 h) in the dark at room temperature. This treatment was shown to retain β-carotene and β-apo-8'-carotenaldehyde, while completely hydrolyzing β-cryptoxanthin ester.

For the determination of carotenoids in fruits and vegetables, which contain a large percentage of water, direct solvent extraction using a water-miscible organic solvent such as acetone or tetrahydrofuran (THF) is appropriate. THF not only readily solubilizes the carotenoids and the chlorophylls, but it also prevents the formation of emulsions by denaturing the associated proteins (18). This solvent is known to promote peroxide formation; thus it must be stabilized with an antioxidant such as BHT. The extraction may be carried out in the presence of anhydrous sodium sulfate as a drying agent. The addition of magnesium carbonate to the extraction system serves to neutralize traces of organic acids that can cause destruction and structural transformation of carotenoids.

In an extraction procedure described by Khachik and Beecher (72), homogenized vegetables are blended with anhydrous sodium sulfate (200% of the weight of the test portion of vegetable), magnesium carbonate (10% of the weight of the test portion), and THF. The extract is filtered under vacuum, and the solid materials are reextracted with THF until the resulting filtrate is colorless. Most of the solvent is removed on a rotary evaporator at 30°C, and the concentrated filtrate is partitioned between petroleum ether and water to remove the majority of contaminating nonterpenoid lipids. The water layer is washed with petroleum ether several times, and the resulting organic layers are combined, dried over anhydrous sodium sulfate, and evaporated to dryness. The residue is taken up in a small volume of the HPLC solvent for analysis.

Marsili and Callahan (73) compared an ethanol/pentane solvent extraction procedure with a supercritical CO_2 extraction procedure for the HPLC determination of α- and β-carotene in vegetables. A combination of static and dynamic modes of extraction with ethanol modifier at 338 atm and 40°C was necessary in order to achieve optimum recovery with the SFE procedure. The extracted material was recovered by depressurization of the CO_2 across a solid phase trap and rinsed from the trap into a 2-ml vial with HPLC grade hexane for injection. β-Carotene results obtained using the SFE procedure averaged 23% higher than results using the solvent extraction process.

In methods for determining α- and β-carotene and β-cryptoxanthin in orange juice, xanthophylls that were more polar than cryptoxanthin were removed from the unsaponifiable fraction by either open column chromatography on magnesia (74) or solid phase extraction using C18-bonded silica (75).

E. Vitamin D

Saponification is obligatory for the determination of vitamin D in fatty foods because of the vast excess of triglycerides present. Saponification at elevated temperatures results in the thermal isomerization of vitamin D to previtamin D (76). It has been reported (77) that saponification of milk by refluxing for 30 min at 83°C in the presence of pyrogallol results in a 10–20% loss of added vitamin D due to thermal isomerization. This creates a problem in HPLC methods because the vitamin D peak is well separated from the previtamin D peak, and in food analysis it is virtually impossible to measure the latter because of interference from coeluting substances. Several workers have avoided this problem by employing "cold" saponification, i.e., alkaline digestion at ambient temperature overnight with slow constant stirring. Whatever the saponification temperature, it is necessary to perform the reaction in an inert atmosphere. Indyk and Woollard (78) avoided vitamin D losses of 10–20% by flushing the saponification vessel with oxygen-free nitrogen and then sealing the vessel before cold saponification.

For the determination of vitamin D in fortified milks, margarine, and infant formulas, Thompson et al. (79) extracted the unsaponifiable matter three times with hexane in the presence of a 6:4 ratio of water to ethanol. The combined hexane layers were then washed successively with water, 5% aqueous KOH, water, and 55% aqueous ethanol to remove polar material, including 25-hydroxyvitamin D.

The sterols, carotenoids, and other constituents of the unsaponifiable fraction that could interfere in the chromatographic determination of vitamin D must be removed from the solvent extract. A variety of cleanup techniques has been employed, including sterol precipitation, thin layer chromatography, open column chromatography, and solid phase extraction.

Open column chromatography, using gravity flow glass columns dry-packed with neutral 5% or 8% water-deactivated alumina, has been used to purify the unsaponifiable matter obtained from various foods (79–81). The chief aim is to remove sterols and carotenes, but vitamins A and E will also be removed, if present. Thompson and Plouffe (82) collected a vitamin D–containing fraction using an eluant of 1% isopropanol in hexane, followed by a 25-hydroxyvitamin D–containing fraction using 5% isopropanol in hexane. The alumina obtained from commercial sources needs to be activated by heating to drive out the water and then partially deactivated so that it contains 5% or 8% by weight of added water. It is advisable to test the activity by ensuring that vitamin D is recovered completely in the first eluate.

Solid phase extraction using silica cartridges has been used to clean up the unsaponifiable residue obtained from various food samples. Typically, the residue dissolved in hexane is loaded onto a preconditioned silica cartridge, which retains vitamin D onto the highly polar adsorbent. Nonpolar material has a greater affinity for the solvent and hence is unretained. The silica bed is then washed with a solvent that is sufficiently polar to remove further interfering material without displacing the vitamin D. The vitamin D is eluted with a slightly more polar solvent, thus achieving isolation of the vitamin from its less polar coextractants. This technique was used by Bui (54) in the analysis of high-fat vitamin D–fortified milk powders and diet foods. The bulk of the sterols was removed by washing the cartridge with 3 ml of hexane/ethyl acetate (85 + 15), after which the vitamin D was eluted with 5 ml of hexane/ethyl acetate (80 + 20). Sliva et al. (83) dissolved the unsaponifiable residue in dichloromethane/isopropanol (99.8 + 0.2) rather than in hexane and applied this solution to a silica solid phase extraction cartridge. In this case, the cartridge was washed with 2.0 ml of dichloromethane/isopropanol (99.9 + 0.2), and vitamins D_2 and D_3 were eluted with 7.0 ml of dichloromethane/isopropanol (99.8 + 0.2).

Silica solid phase extraction cartridges facilitate the separation of vitamin D from its more polar 25-hydroxy metabolite. In an HPLC method devised by Mattila et al. (36) for determining vitamin D_3 and 25-hydroxyvitamin D_3 in egg yolk, the unsaponifiable residue is dissolved in 10 ml of hexane, passed through a 0.45-μm membrane filter, and applied to a preconditioned silica cartridge. The silica bed is washed with 20 ml of hexane followed by 50 ml of 0.5% isopropanol in hexane, and the washings are discarded. Vitamins D_2 and D_3 are eluted with 35 ml of 0.5% isopropanol in hexane and the silica bed is washed with a further 50 ml of the same solvent. Finally, the corresponding 25-hydroxy metabolites are eluted with 40 ml of 6% isopropanol in hexane.

An alternative strategy is to dissolve the unsaponifiable residue in a polar solvent such as ethanol and then to pass this solution through a reverse phase/solid phase extraction cartridge. Polar compounds will pass through the sorbent, whereas less polar constituents will be retained. A solvent of intermediate polarity is used to wash off the material that is more polar than vitamin D. The vitamin D is eluted with a less polar solvent, leaving the nonpolar material retained on the sorbent. Thus reverse phase/solid phase extraction has the potential for removing material that is more polar than vitamin D as well as material that is less polar. This technique is employed in the method of Reynolds and Judd (84) for the analysis of vitamin D–fortified skimmed milk powder. The unsaponifiable residue is dissolved in 2 ml of ethanol, then diluted with 1 ml of water and applied to a C18-bonded phase cartridge. The cartridge is washed with 15 ml of methanol/THF/water (1 + 1 + 2) followed by elution of the vitamin D with 5 ml of methanol.

A simplified approach, known as matrix removal, is to pass the test extract once through the sorbent under conditions that selectively retain the matrix interferences but not the vitamin D. Cohen and Wakeford (85) applied this technique to the analysis of vitamin D–fortified instant nonfat dried milk. Test portions of milk powder were extracted with stabilized dichloromethane and, after evaporation to near dryness, the residue was dissolved in isooctane/dioxane (95:5) and passed through a Sep-Pak silica cartridge.

F. Vitamin E

Either direct solvent extraction or saponification can be employed in vitamin E assays, depending on the nature of the food matrix and the analytical information required. Saponification at normal refluxing temperatures causes significant losses of tocotrienols, despite the presence of an antioxidant (86), and therefore direct solvent extraction is the preferred technique if it is required to determine the tocotrienols. Vegetable oils may be simply dissolved in hexane and analyzed directly by normal phase HPLC (87–90). Solid food samples require a more rigorous initial extraction before partitioning of the total lipid fraction into an organic solvent. Balz et al. (91) modified the Röse–Gottlieb method to extract vitamin E from infant formulas. Dipotassium oxalate solution (35% w/v) was substituted for ammonia to avoid alkalizing the medium, and *tert*-butylmethyl ether was substituted for diethyl ether because of its stability against the formation of peroxides.

In the analysis of fortified foods that contain significant amounts of naturally occurring vitamin E, saponification provides a convenient means of hydrolyzing the α-tocopheryl acetate to its alcohol form, allowing both the supplemental and natural α-tocopherol to be measured together as a single peak by HPLC. It should be noted that if totally synthetic all-*rac*-α-tocopheryl acetate is the supplement used, its hydrolysis product, all-*rac*-α-tocopherol, is less biologically active than naturally occurring *RRR*-α-tocopherol, making it impossible to calculate a potency value for the total vitamin E. This problem does not arise if the supplement used is *RRR*-α-tocopheryl acetate. Vitamin E is prone to oxidation in an alkaline medium, and the saponification must be performed in the presence of an antioxidant and with nitrogen flushing. About 25 mg of pyrogallol or 250 mg of sodium ascorbate may be sufficient as antioxidant per 1 g of sample to be saponified (92). In an IUPAC method for determining vitamin E in margarine, a mild saponification procedure (10 min at 26°C) is recommended to ensure a quantitative recovery of α-tocopherol from α-tocopheryl acetate (93). For the efficient extraction of tocopherols from the saponification digest using hexane, the ethanol strength must be below 40% (94).

G. Vitamin K

Shearer (95) subjected milk-based and soy protein–based infant formulas and dairy products to lipase digestion, after which the hydrolyzate was made alkaline, then diluted with ethanol and extracted with hexane. Vegetables, fruits, cereals, meats, and fish were ground in a mortar with fine quartz granules before extraction with acetone. After the addition of water and hexane to the acetone extract, the nonpolar phylloquinone partitioned entirely in the upper hexane phase, leaving polar impurities in the acetone/water phase. The hexane extracts obtained after either lipase digestion or acetone extraction were loaded onto preconditioned silica solid phase extraction cartridges. Nonpolar lipids were removed by elution with hexane and discarded, after which the phylloquinone-containing fraction was eluted with hexane/diethyl ether (97:3). Under these elution conditions, polar lipids such as sterols, fatty acids, and phospholipids remain on the adsorbant, but many lipids of similar polarity to phylloquinone (e.g., cholesteryl esters) coelute with phylloquinone.

Booth et al. (96) subjected various food samples (vegetable juice, whole milk, raw spinach leaves, plain bagel, and raw ground beef) to an initial solvent extraction in isopropanol/hexane (3 + 2) followed by sonication. The extracts were evaporated to dryness, redissolved in hexane, and applied to preconditioned silica solid phase extraction cartridges. Following a wash with 100% hexane, the phylloquinone-containing fraction was eluted with hexane/diethyl ether (93 + 3). The milk samples were further purified by a liquid phase reductive extraction step, in which the residue obtained from the evaporated eluate was redissolved in hexane, followed by the addition of 1% (w/v) zinc chloride solution, 3% acetic acid, and 97% acetonitrile. Zinc metal was then added and the mixture vortexed to reduce phylloquinone and dihydrophylloquinone (internal standard) to their acetonitrile-soluble hydroquinones. After centrifugation, the upper hexane layer, which contained the contaminating lipids, was removed and discarded. Water and hexane were added to the remaining solution to reoxidize the hydroquinones to their hexane-soluble quinones. The beef samples were further purified by reverse phase C18 solid-phase extraction to separate phylloquinone from the large quantity of saturated fats present. After evaporation, the residue from the eluate of the silica cartridge was dissolved in isopropanol while heating at 45°C for 10 min. The C18 cartridges were preconditioned by successive washes with methanol/dichloromethane (80:20), followed by 100% methanol and 100% water. After application of the test solution to the preconditioned packing, the cartridge was washed with methanol/water (95:5) followed by 100% acetonitrile, and the phylloquinone-containing fraction was eluted with methanol/dichloromethane (80:20).

Schneiderman et al. (53) extracted phylloquinone from powdered infant formulas using supercritical CO_2 at 8000 psi and 60°C for 15 min. The extracted material was readily recovered by depressurization of the CO_2 across an adsorbant trap, then washed from the trap with a small volume of dichloromethane/acetone (50:50). The resultant solution was evaporated to dryness and the residue was dissolved in mobile phase for direct HPLC analysis. Trial experiments gave recoveries of 92% of phylloquinone (1 ml of a 1 µg/ml standard solution) from a Chromosorb W matrix.

IV. ANALYSIS BY HPLC

A. The Column

The majority of published HPLC techniques used in fat-soluble vitamin assays have utilized 5- or 10-µm particles of silica or derivatized silica packed into stainless steel tubes of typical length 25 cm and standard i.d. 4.6 mm. Radially compressed cartridge-type columns (Waters Chromatography Division) manufactured from heavy wall polyethylene of dimensions 10 cm × 8 mm i.d. have also found application. The insertion of a short guard column between the injector and analytical column protects the latter against loss of efficiency caused by strongly retained sample components and from pump or valve wear particles. The column-packing material is held in the column by fine-porosity frits of stainless steel or some other material. Rabel (97) discussed the care and maintenance of HPLC columns. Membrane filtration and/or centrifugation of all test extracts is important for the removal of particulate or high molecular weight species which might otherwise enter the guard or analytical column.

Improved detection sensitivity can be achieved by reducing the column i.d., and small-bore columns of between 1.0 and 2.5 mm i.d. are available for use in specifically designed liquid chromatographs having an extremely low extracolumn dispersion. Small-bore columns do not give fast analyses, or more efficiency and resolution than conventional columns of 4.6 mm i.d., but their low flow rates lead to a significant reduction in solvent costs and the cost of waste solvent disposal. The enhanced sensitivity is a consequence of the decreased column volume, which reduces band broadening in the column. The minimum detectable mass is directly proportional to the square of the column radius (98); therefore, in theory, a 2.1-mm-i.d. column will provide a mass sensitivity about five times greater than that of a 4.6-mm-i.d. column of the same length. In other words, a comparable peak size will be obtained from a 1-µl injection of a given solution on a 25 cm × 2.1 mm i.d. small-bore column as from a 5-µl injection volume on a 25 cm × 4.6 mm i.d. conventional column. Therefore, the advantage of high mass sensitivity from small injection volumes only becomes apparent when there is a limited amount of sample as in, for example, clinical or forensic analysis.

Mulholland (99) optimized an HPLC system to match the low dispersion requirements of a 2.1-mm-i.d. small-bore column. A 1-µl injection rotor, a specially designed 1-µl flow cell, and minimal external tubing were selected. The system was linked with diode array detection for the determination of vitamins A, D, and E using a 3 cm × 2.1 mm guard column and a 10 cm × 2.1 mm i.d. analytical column of the cartridge type. The high mass sensitivity was demonstrated with an on-column loading of 10 ng of α-tocopheryl acetate, which could be clearly detected and identified.

Food analysts have been reluctant to convert to small-bore HPLC. However, several vitamin methods utilize 3.2-mm-i.d. columns, which in a properly designed system give a twofold increase in sensitivity over 4.6-mm-i.d. columns.

B. Chromatographic Modes

Either normal phase or reverse phase HPLC can be employed for fat-soluble vitamin assays, with the choice depending on the extraction/cleanup procedures employed, what vitamers are required to be measured, and whether the system is to be used for semipreparative or analytical separations.

1. Normal Phase HPLC

This chromatographic mode refers to the use of adsorption chromatography, in which the polar surface of microparticulate silica (or other adsorbant) constitutes the stationary phase, and to normal bonded phase chromatography typified by nitrile- or amino-bonded stationary phases. The nonpolar mobile phase is typically hexane containing a small percentage (usually <5% by volume) of a more polar solvent (e.g., isopropanol) to act as a modifier. The separation of solutes in normal phase chromatography is based on differences in their polar functional groups, which enable compounds to be separated into classes or groups of similar type. In addition, adsorption chromatography provides a powerful means of separating geometrical (cis–trans) isomers, with the separation mechanism being attributed to a steric fitting of solute molecules with the discrete adsorption sites.

An operational disadvantage of adsorption chromatography is the problem of slow equilibration toward water, which is a very strong moderator in deactivating the silica surface. All organic solvents (unless specifically dried) contain an inherent amount of water in the ppm range that is sufficient to affect solute retention. Isohydric mobile phases avoid the long equilibration times usually required with silica columns when changing the eluant, since the eluant is in equilibrium with the adsorbant with respect to water (100).

The columns used in normal phase HPLC can tolerate relatively heavy loads of triglyceride and other nonpolar material, as such material is not strongly retained and can easily be washed from the column after a series of injections. Procedures have been devised for the determination of vitamins A and E in which the total lipid fraction of the food sample is extracted with a nonaqueous solvent, and any polar material that might be present is removed. An aliquot of the nonpolar lipid extract containing the fat-soluble vitamins is then injected into the liquid chromatograph without further purification. Direct injection of the lipid material is possible because the material is dissolved in a nonpolar solvent that is compatible with the nonpolar mobile phase.

2. Reverse Phase HPLC

Reverse phase systems usually employ a bonded octadecylsilane (ODS) stationary phase in which the functional group is an octadecyl (C18) hydrocarbon. Differences in selectivity and chromatographic performance between ODS columns obtained from different manufacturers are attributable to the characteristics of the silica particle (size, shape, and porosity), the coverage of the bonded phase and the surface configuration, and the percentage of accessible active acidic silanol groups after completion of bonding procedures. Fully endcapped stationary phases are almost completely hydrophobic and exhibit true reverse phase properties. Nonendcapped phases contain a percentage of accessible silanol groups that impart some secondary normal phase adsorption characteristics. The surface configuration depends on the type of silanizing agent used in the bonding reaction during manufacture. A monofunctional agent gives rise to a monomeric phase (i.e., a monomolecular layer), while a di- or trifunctional agent promotes crosslinking and results in a polymeric phase.

Separation in reverse phase chromatography depends on the solute's hydrophobicity, which increases with the number of carbon atoms in the molecule. In general, reverse phase columns are usually operated with semiaqueous mobile phases consisting typically of mixtures of methanol (or acetonitrile) and water. However, nonaqueous reverse phase (NARP) chromatography is preferred for the lipophilic carotenes and vitamin K compounds. A typical NARP eluant consists of a polar basis (usually acetonitrile), a solvent of lower polarity (e.g., dichloromethane) to act as a solubilizer and to control retention by adjusting the solvent strength, and, occasionally, a small amount of a third solvent with hydrogen bonding capacity (e.g., methanol) to optimize selectivity. The advantage of nonaqueous mobile phases over semiaqueous eluants is the increased solubility of low polarity compounds, which are therefore eluted early and with less risk of precipitation. To compensate for the increased affinity of the low-polarity compounds for the mobile phase, a highly retentive stationary phase, such as Zorbax ODS (20% carbon loading w/w), is required.

The removal of triglycerides from the sample before injection is essential when using reverse phase chromatography. Triglycerides are insoluble in water and only sparingly soluble in methanol and acetonitrile. If injected, they may not be completely eluted from the column and the retained material would impair chromatographic efficiency, peak shape, and reproducibility. In the absence of nonpolar material in the test solution, reverse phase chromatography exhibits improved reproducibility of solute retention times compared with normal phase chromatography. This is largely because retention in reverse phase chromatography is little affected by small variations in the mobile phase composition and, unlike adsorption chromatography, no significant effect is seen from slight changes in water content. Reequilibration of reverse phase columns is easier and quicker compared with adsorption (silica) columns, owing to the weaker forces involved in the interactions between the solute and the nonpolar stationary phase. Elution with several column volumes of methanol is usually sufficient to restore the column to its original condition.

3. Two-Dimensional HPLC

For determining the trace amounts of naturally occurring vitamins D and K in foods, an adequate separation for quantitation cannot be achieved on a single column. This is because prior chromatographic cleanup techniques fail to remove lipoidal substances that are of similar polarity to the vitamin in question, and these substances interfere in the HPLC analysis. It has therefore been found necessary in many cases to perform the HPLC in two systems. A true two-dimensional combination, involving normal phase and reverse phase columns, can be expected to provide better selectivity than the use of two similar columns. The first system (semipreparative HPLC) is designed to isolate a fraction of the sample extract that contains the vitamin analyte and the internal standard. Ideally, the analyte and internal standard should be unresolved from one another so that they can be collected by reference to a single peak in the chromatogram. If the vitamin/internal standard peak is marked by coeluting peaks, the obvious method of collecting the fraction is to refer to the retention time of a vitamin standard. However, this method is not reliable because in normal phase HPLC (the separation mode invariably employed) it is difficult to maintain constant retention times during a run. Shearer (101) overcame this problem in a vitamin K assay procedure by injecting a phylloquinone standard together with a small amount of the sample extract. By comparing the chromatogram thus obtained with a chromatogram produced from the sample extract alone, the phylloquinone peak could be seen superimposed on the background peaks. Thus the elution of vitamin K was related to the UV-absorbing "fingerprint" given by contaminants in the sample extract. The fraction containing the vitamin of interest and the internal standard is collected in a small tapered tube. A fraction collection with a drop counter system, rather than a volume collection system, is recommended for this purpose (102). The solvent is carefully evaporated off under a stream of nitrogen, and the residue is then dissolved in a small volume of a suitable solvent for the second (analytical) stage of HPLC.

C. Detection Systems

Three types of in-line HPLC detector have been utilized for monitoring the concentration of the vitamins in the column effluent: absorbance, fluorescence, and electrochemical detectors. Photodiode array detection, which permits simultaneous absorbance detection at several wavelengths and continuous memorizing of spectra during the evolution of a peak, has proved invaluable for the assessment of peak purity and identity. All of the fat-soluble vitamins, including carotenoids, exhibit some form of electrochemical activity. Both amperometry and coulometry have been applied to electrochemical detection. In amperometric detectors only a small proportion (usually <20%) of the electroactive solute is reduced or oxidized at the surface of a glassy carbon or similar nonporous electrode, while in coulometric detectors the solute is completely reduced or oxidized within the pores of a graphite electrode. A coulometric detector can function as an on-line reactor and this unique property has been exploited in methods for determining vitamin K. A stringent requirement for electrochemical detection is that the solvent delivery system should be virtually pulse-free.

D. Preparation of Sample Extracts for HPLC

The residues obtained after evaporation of the prepared test extracts should be dissolved in an appropriate known volume of a suitable solvent—ideally the mobile phase to be used for the separation. This is important, as different solubilities of a given solute in the injection and elution solvents, together with the interaction between two different solvents, can result in the production of peak artifacts that could be erroneously interpreted. Chromatographic artifacts and peak distortion due to sample–solvent interactions have been reported in the separation of carotenoids by reverse phase HPLC (103). Before injection, the test solution should be passed through a 0.45-μm membrane filter to remove any microparticulate matter.

E. Quantification

The fat-soluble vitamins and carotenoids are highly susceptible to oxidation and degradation, so commercially available preparations to be used for standards are often less pure than expected. Consequently, the concentrations of standard solutions must be calculated after spectrophotometry, using published absorptivities ($A_1^{1\%}{}_{cm}$) or molar absorptivities (ε) at the wavelength of maximum absorption (λ_{max}).

Several different quantification procedures for vitamin A have been described in the literature, some using retinol directly as a standard and some using retinyl acetate, which is converted to retinol by saponification. The latter approach is generally preferred because crystalline all-*trans*-retinyl acetate is commercially available in high purity and is free from *cis* isomers. Commercial sources of retinol are oily preparations and are at best only about 70% pure. There are two ways of preparing a retinol standard from retinyl acetate.

1. A relatively large amount (typically 25 mg) of retinyl acetate is saponified and extracted, and the residue is dissolved in isopropanol to give a stock solution of retinol, which can be stored for 2–3 months in a refrigerator. This stock solution is diluted with isopropanol to give a suitable working standard solution, whose concentration is determined spectrophotometrically ($A_1^{1\%}{}_{cm} = 1830$ at λ_{max} 325 nm) immediately before use as an external standard in the HPLC procedure.
2. An accurately prepared standard solution of retinyl acetate (i.e., a solution of known concentration) is taken through the saponification and extraction procedure along with each batch of samples, and the resultant retinol solution is used as an external standard without spectrophotometric standardization. This technique, which is recommended by COST 91 (66), compensates for losses of vitamin A incurred during the saponification and subsequent manipulations (i.e., the calculated vitamin A value is recovery-corrected).

For carotenoid analysis, β-apo-8′-carotenaldehyde is frequently used as an internal standard. However, Marsili and Callahan (73) found that internal standardization was unnecessary and preferred the external standard calibration method. External standard calibration is also generally used for vitamin E analysis. The more recent HPLC methods for determining vitamin D use vitamin D_2 as an internal standard for quantifying vitamin D_3, and vice versa. In addition to compensating for vitamin D losses during the extraction and cleanup, internal standardization overcomes the problem of thermal isomerization which takes place during hot saponification. Vitamin K analysis, which also involves a rigorous sample workup, uses internal standard calibration.

F.   Vitamin A

Methods for determining vitamin A, and sometimes β-carotene as well, are summarized in Table 2.

Although retinol and its esters exhibit native fluorescence, the quantum efficiency is low (109) and UV detection is used in most HPLC systems for determining vitamin A. Retinol and retinyl esters have equal molar absorptivities ($ε = 52,275$ at $λ_{max}$ 325 nm) (110). It is advisable to scan the absorbance of standard solutions over the wavelength range 250–400 nm rather than to determine absorbance only at 325 nm, so as to disclose possible decomposition.

Thompson et al. (69) described a method using adsorption HPLC in which supplemental retinyl palmitate can be determined in hexane extracts of milk (whole, semiskimmed, and skimmed) and margarine without the need for saponification. The added β-carotene in margarine is eluted near the solvent front but can be quantified because of the known absence of other carotenoids. The nonpolar lipid material that accumulates on the column is eluted to waste with 25% diethyl ether in hexane after a series of injections. A major advantage of this method is that the vitamin A is maintained in its stable ester form throughout the assay and is protected by the lipids to the point of chromatographic separation. Furthermore, there is no need to evaporate extracts to dryness. The relative standard deviation (RSD) values obtained for retinol and β-carotene in replicate ($n = 10$) samples of milk and margarine were reported to be less than 5%, which indicated that the method has good precision.

Using silica adsorption HPLC, Woollard and Indyk (111) analyzed the *cis-trans* isomers of retinol in dairy products and discussed their significance in biopotency estimates.

The removal of triglycerides from the food sample by saponification presents the opportunity to utilize reverse phase HPLC with its attendant ease of operation. The unsaponifiable matter is conventionally extracted into a solvent, e.g., diethyl ether/petroleum ether (50:50) or hexane, that is incompatible with the semiaqueous mobile phase. It then becomes necessary to evaporate the unsaponifiable extract to dryness and to redissolve the residue in a small volume of methanol (if methanol is the organic component of the mobile phase). Egberg et al. (108) avoided the evaporation process in the analysis of breakfast cereals, margarine, and butter by acidifying the unsaponifiable matter with acetic acid in acetonitrile to precipitate the soaps. An aliquot of the filtered extract could then be injected, after dilution with water, onto a reverse phase column eluted with a compatible mobile phase (65% acetonitrile in water).

Reverse phase chromatography, using semiaqueous mobile phases, is capable of separating all-*trans*-retinol from 13-*cis*-retinol, albeit somewhat poorly. Further separation of retinol *cis-trans* isomers is not achieved, but this is of little concern as other *cis* isomers are not usually present in detectable amounts. In practice, it is convenient to adjust the elution conditions to obtain a single retinol peak that encompasses both isomers and to ignore the lower biological potency of the minor 13-*cis* isomer. Alternatively, the peak areas of the all-*trans* and 13-*cis* isomers can be measured separately, then multiplied by their respective relative potencies of 100% and 75%, and added. A typical chromatogram of an extracted sample of milk is shown in Fig. 1.

G.   Provitamin A Carotenoids

Methods for determining the principal provitamin A carotenoids are presented in Table 3.

Carotenoids, by virtue of their intense color, exhibit very strong absorption in the visible region

Table 2 HPLC Methods for the Determination of Vitamin A Compounds and Carotene in Food

Analyte	Food sample	Sample preparation[a]	Quantitative HPLC			Ref.
			Column	Mobile phase[a]	Detection	
		Normal Phase Chromatography				
Retinyl palmitate	Fortified fluid milk (whole, semiskimmed, skimmed)	Deproteinize 2-ml sample with EtOH in a centrifuge tube, extract once with hexane, add water, centrifuge	LiChrosorb Si-60 5 μm, 0.32 × 25 cm	Hexane/diethyl ether, 98:2	UV 325 nm (vit A palmitate)	69
Retinyl palmitate, carotene	Margarine	Dissolve sample in hexane, shake with 60% aqueous EtOH, centrifuge	LiChrosorb Si-60 5 μm, 0.32 × 25 cm	Hexane/diethyl ether, 98:2	Vis. 453 nm (carotene)	69
Total vitamin A (as 13-*cis*- and all-*trans*-retinol)	Milk, infant formulas	Saponify (ambient), extract with hexane/diethyl ether, 85:15	Apex silica 3 μm, 0.45 × 15 cm	1–5% 2-PrOH in heptane	UV 340 nm	104
Total vitamin A (as 13-*cis*-, 9,13-di-*cis*-, 9-*cis*- and all-*trans*-retinol, carotene)	Cheese	Saponify (ambient), extract with hexane	LiChrosorb Si-60 5 μm, 0.4 × 25 cm	Hexane/methyl ethyl ketone, 90:10	UV 340 nm (retinol) Vis. 450 nm (carotene)	105
		Reverse Phase Chromatography				
Total vitamin A (as retinol)	All food types	Saponify (hot), extract with diethyl ether/petroleum ether, 50:50	μBondapak C18 10 μm, 0.39 × 30 cm	MeOH/H2O, 90:10	UV 325 nm	66

Table 2 (Continued)

Analyte	Food sample	Sample preparation[a]	Quantitative HPLC			Ref.
			Column	Mobile phase[a]	Detection	
Total vitamin A (as retinol)	Unfortified fluid dairy products	Saponify (hot) 2-ml sample in a centrifuge tube, extract once with diethyl ether/petroleum ether (50:50), centrifuge	Nova-Pak C$_{18}$ 0.39 × 15 cm	MeOH/H$_2$O, 95:5	UV 325 nm	68
Total vitamin A (as retinol), β-carotene	Fortified fluid milk (whole, skimmed), infant formulas, margarine	Saponify (hot) in centrifuge tube, extract with hexane (×5)	LiChrosorb RP-18 10 μm, 0.32 × 25 cm	MeOH/H$_2$O, 90:10 (retinol) MeOH/H$_2$O, 99:1 (β-carotene)	UV 325 nm (retinol) Vis. 453 nm (β-carotene)	106
Total vitamin A (as retinol), α- and β-carotenes	Selected foods of animal origin and processed foods	Saponify (hot), extract with hexane	μBondapak C$_{18}$ 10 μm, 0.39 × 30 cm	MeCN/MeOH/EtOAc, 88:10:2	UV 313 nm (retinol) Vis. 436 nm (carotenoids)	107
Total vitamin A (as 13-*cis*- and all-*trans*-retinol)	Breakfast cereals, margarine, butter	Saponify (hot), precipitate soaps with HOAc in MeCN, dilute with H$_2$O	Vydac 201 TP C$_{18}$ 10 μm, 0.32 × 25 cm	MeCN/H$_2$O, 65:35	UV 328 nm	108

[a]Solvent proportions by volume.
Notes and abbreviations: see Footnote to Table 9.

Fig. 1 HPLC of retinol in the unsaponifiable fraction of unfortified raw milk. Column packing material, Nova-Pak C18; mobile phase, methanol/water (95:5); UV detection, 325 nm. Peaks: (1) BHT (added to the extracting solvent as an antioxidant); (2) retinol. (From Ref. 68.)

of the spectrum. Absorption maxima and $A^{1\%}_{1\ cm}$ values for a large number of carotenoids have been compiled by De Ritter and Purcell (117). The $A^{1\%}_{1\ cm}$ value for β-carotene dissolved in hexane or petroleum ether is 2592 at λ_{max} 453 nm. A wavelength of 488 nm was used by Landen et al. (118) to detect β-carotene in unsaponified vegetable extracts because at this higher wavelength there is minimum absorbance for chlorophyll degradation products.

In normal phase HPLC, using silica or polar bonded-phase stationary phases, the carotenes are eluted as an unresolved group near the solvent front, while the polar xanthophylls are highly retained. Where silica columns have been utilized to determine vitamin A in margarine (69) and cheese (105), the sole carotenoid, β-carotene, can be determined at the same time. Calcium hydroxide (lime) columns facilitate the separation of all-*trans*-β-carotene and its 9-, 13-, and 15-*cis* isomers (119). Lime columns are not commercially available; therefore slight differences in packing techniques result in column-to-column variability.

Reverse phase HPLC is the preferred mode for the determination of provitamin A carotenoids in foods because the carotenes are strongly retained and the separation of α- and β-carotene (unresolved by normal phase systems) is easily achieved. In addition, there is little risk of on-column carotenoid degeneration, as experienced with polar stationary phases, owing to the weak hydrophobic forces on which the separation mechanism is based. The xanthophylls are eluted well before the carotenes, the latter requiring strong mobile phases containing little or no water for their elution. Carotenes are only sparingly soluble in typical reverse phase solvents (methanol and acetonitrile), and the inclusion of water further reduces this solubility. Nelis and de Leenheer (120) used isocratic NARP-HPLC with a Zorbax ODS column and an eluant of acetonitrile/ dichloromethane/methanol (70:20:10) to separate nine carotenoids spanning a wide polarity range. This classic separation was achieved by virtue of the fact that the Zorbax ODS material supplied at that time was nonendcapped. The carotenes were retained by hydrophobic interaction with the ODS ligands, while retention of the xanthophylls was controlled by secondary interactions involving residual silanol groups. It should be noted that Zorbax ODS is now manufactured as a fully endcapped material, the nonendcapped form being no longer available (121).

Table 3 HPLC Methods for the Determination of Provitamin A Carotenoids in Food

Analyte	Food Sample	Sample preparation[a]	Quantitative HPLC			Ref.
			Column	Mobile phase[a]	Detection	
			Reversed Phase Chromatography			
α- and β-Carotenes, β-cryptoxanthin	Fruits, vegetables	Extract sample with THF + Na$_2$SO$_4$ + MgCO$_3$	Partisil ODS 5 μm, 0.46 × 25 cm	MeCN/THF/H$_2$O, 85:12.5:2.5	Vis. 470 nm	15
α-Carotene, all-*trans*-β-carotene (separated from *cis*-β-carotenes)	Fruits, vegetables	Extract sample with THF	Vydac 218 TP54 C$_{18}$ 5 μm, 0.46 × 25 cm	MeOH/MeCN/THF, 56:40:4	Vis. 470 nm	112
α-Carotene, all-*trans*-β-carotene (separated from *cis*-β-carotenes)	Raw and cooked vegetables	Extract sample with THF + Na$_2$SO$_4$ + MgCO$_3$	Spheri-5 ODS 5 μm, 0.46 × 22 cm	MeCN/CH$_2$Cl$_2$/MeOH, 70:20:10	Vis. 450 nm	113
α- and β-Carotenes	Vegetables	Extract sample with THF + Na$_2$SO$_4$ + MgCO$_3$ + nonapreno-β-carotene (internal standard), partition into petroleum ether and water	Spheri-5 RP-18 5 μm, 0.46 × 22 cm	MeCN/MeOH/CH$_2$Cl$_2$, 55:22:23	Vis. 470 nm	72
β-Carotene	Vegetables	Saponify (hot), extract with diisopropyl ether	Hypersil-ODS 3 μm, 0.46 × 25 cm	MeCN/MeOH/CHCl$_3$/H$_2$O, 250 + 200 + 90 + 11	Vis. 445 nm	114
β-Cryptoxanthin, α-carotene, all-*trans*-β-carotene (separated from *cis*-β-carotenes)	Orange juice	Saponify (hot), extract with hexane. Magnesia column chromatography	Vydac 201 TP54 C$_{18}$ 5 μm, 0.46 × 25 cm	MeOH/CHCl$_3$, 94:6	Vis. 475 nm	74
α- and β-Carotenes	Olive oil	Saponify (ambient), extract with diethyl ether	Supelcosil LC-18 5 μm, 0.46 × 15 cm	MeCN/2-PrOH/1,2-dichloroethane, 92.5:5.0:2.5	Vis. 458 nm	115
β-Cryptoxanthin, γ-, α-, and β-carotenes	Malaysian vegetables and fruits	Saponify (hot), extract with hexane	μBondapak C$_{18}$ 10 μm, 0.39 × 30 cm	MeCN/MeOH/EtOAc, 88:10:2	Vis. 436 nm	116
α- and β-Carotenes	Vegetables	Supercritical fluid extraction	Vydac 201 TP54 C$_{18}$ 5 μm, 0.46 × 25 cm	MeOH/MeCN/CH$_2$Cl$_2$/hexane, 65:27:4:4	Vis. 450 nm	73

[a]Solvent proportions by volume.
Notes and abbreviations: see footnote to Table 9.

The separation of all-*trans*-β-carotene from its principal 9-*cis* and 13-*cis* geometrical isomers has been achieved using silica-based polymeric C18 bonded phase column packings such as Vydac TP (74,112,122–125) and Spheri-5 ODS (113,126,127). Under the isocratic NARP conditions employed, 15-*cis*-β-carotene coelutes with the 13-*cis* isomer (119,127). Vydac TP and Spheri-5 ODS column packings have very different particle characteristics and surface coverages. Vydac 218 TP (a fully endcapped material) has a medium carbon loading of 9% w/w but an unusually large mean pore diameter of 300Å corresponding to a low specific surface area of 80 m²/g. Vydac 201 TP is a nonendcapped version of 218 TP with otherwise similar characteristics. Spheri-5 ODS (a fully endcapped material) has a mean pore diameter of 80 Å, a specific surface area of 200 m²/g, and a carbon loading of 14% w/w. It appears that the polymeric surface configuration, which is common to both Vydac TP and Spheri-5 ODS, is responsible for the separation of the *cis-trans* isomers of β-carotene, endcapping being unimportant. Indeed, Lesellier et al. (126) demonstrated that a polymeric (as opposed to a monomeric) ODS phase is necessary for the isocratic separation of all-*trans*-α- and β-carotenes from their respective *cis* isomers using NARP chromatography under usual conditions. The separation of the major carotenoids in an extracted sample of frozen orange juice concentrate is shown in Fig. 2.

Owing to reported losses of carotenoids by rapid decomposition through oxidation in the presence of a stainless steel column frit (128), it is recommended to purchase HPLC columns with inert metal-free frits for carotenoid analysis (129,130).

H. Vitamin D

All vitamin D–active compounds possess a broad UV absorption spectrum with a λ_{max} at 264–265 nm and a molar absorptivity (ε) of 18,300 in ethanol or hexane (131). Reported

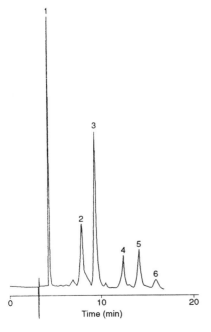

Fig. 2   HPLC of carotenoids in the unsaponifiable fraction of frozen orange juice concentrate after magnesia column cleanup. Column packing material, Vydac 201 TP54 C18; mobile phase, methanol/chloroform (94:6); Vis. detection, 475 nm. Peaks: (1) Sudan I (internal standard); (2) zeinoxanthin; (3) β-cryptoxanthin; (4) all-*trans*-α-carotene; (5) all-*trans*-β-carotene; (6) *cis*-β-carotenes. (From Ref. 74.)

on-column detection limits using absorbance detection range from 1 to 10 ng (54,132,133). Vitamin D does not naturally fluoresce. Hasegawa (134) reported an HPLC method using electrochemical detection in the redox mode, which was applied to the determination of vitamin D in medical nutritional products. The on-column detection limit was estimated to be about 200 pg.

Normal phase HPLC, using either silica or polar-bonded stationary phases, separates, isocratically, vitamin D_3 or D_2 from their respective previtamin and inactive isomers (135). Vitamin D (D_2 + D_3), 25-hydroxyvitamin D_2, and 25-hydroxyvitamin D_3 can be separated from one another and from other hydroxylated metabolites (136), but vitamins D_2 and D_3 cannot be resolved from one another. For nutritional evaluation purposes, it is not essential to distinguish between naturally occurring vitamins D_2 and D_3 because the two vitamers exhibit similar biological potency in man, and their thermal isomerization rates are virtually equal.

In methods that employ saponification at elevated temperatures and that utilize normal phase HPLC in the final separation, compensation must be made for the isomerization of vitamin D to previtamin D. The potential vitamin D can be calculated from measurement of the vitamin D peak by applying a correction factor on the basis that the heat treatment results in a constant ratio of previtamin D to vitamin D (137,138). Another technique is to saponify a standard solution of authentic vitamin D in parallel to the sample, then use this solution as an external standard in the quantification (102). Agarwal (139) circumvented the isomerization problem by converting previtamin D and vitamin D to a common derivative, isotachysterol, by treatment of the unsaponifiable residue with acidified butanol. The isotachysterol was detected at its λ_{max} of 301 nm, which provided greater sensitivity and selectivity compared with the detection wavelength of 265 nm used for vitamin D.

The vitamin D methods presented in Table 4 have been selected because, apart from the method of Agarwal (139), they all use reverse phase chromatography for the quantitative HPLC. Reverse phase HPLC, as in the normal phase mode, separates vitamins D_2 or D_3 from its respective previtamin and inactive isomers (135). Unlike normal phase HPLC, the reverse phase mode, using nonendcapped stationary phases (83), can separate vitamin D_2 from D_3, permitting the measurement of the two adjacent peaks. The 25-hydroxylated metabolites of vitamins D_2 and D_3 can be separated from one another using a Vydac 201 TP column (36). The separation of vitamin D_2 from vitamin D_3 and 25-hydroxyvitamin D_2 from 25-hydroxyvitamin D_3 allows the D_2 form of the vitamin or its 25-hydroxylated metabolite to be used as an internal standard for quantifying the corresponding D_3 form. Internal standardization is desirable, in view of the multistep procedure, to compensate for any losses of vitamin D that may be incurred. When using vitamin D_2 as an internal standard, the previtamin D formation can be neglected, as the isomerization rates of vitamin D_3 and D_2 are practically equal. Therefore the previtamin D/vitamin D ratio between the two forms at any given temperature should be constant. The use of a nonaqueous mobile phase overcomes the problem of prolonged retention of strongly hydrophobic compounds experienced with semiaqueous mobile phases and allows a lower flow rate to be used.

Normal phase/reverse phase HPLC is the ideal combination for semipreparative and analytical separations in two-dimensional HPLC as vitamins D_2 and D_3 coelute during the semipreparative stage, allowing a narrow retention window to be collected for analysis using internal standardization (144). By this means, Johnsson et al. (145) obtained a vitamin D_3 detection limit of 0.1 µg/kg for milk and milk products. The RSD for replicates of fortified milk samples was 0.53% ($n = 7$). A typical semipreparative chromatogram and an analytical chromatogram of a purified extract of margarine spiked with vitamin D_2 are shown in Figs. 3 and 4, respectively.

Indyk and Woollard (140) reported that removal of the cholesterol from the unsaponifiable fraction of fortified whole-milk powder by methanolic precipitation and filtration was an adequate cleanup procedure, making semipreparative HPLC unnecessary. This simplified procedure was

Table 4 HPLC Methods for the Determination of Vitamin D in Food

Analyte	Food sample	Sample preparation[a]	Semipreparative HPLC[a]	Quantitative HPLC[a]	Ref.
Vitamin D_2 or D_3	Infant formulas	Saponify (ambient), extract with hexane. Convert vitamin D to isotachysterol with acidified butanol	Supelcosil LC-18-DB 5 μm, 0.46 × 25 cm MeCN/MeOH, 90:10 UV 301 nm	*Normal Phase* Spherisorb silica 5 μm, 0.2 × 25 cm Hexane/EtOAc/MeOH, 97 + 2.5 + 0.05 UV 301 nm	139
Vitamin D_2 or D_3	Fortified fluid milk (skimmed), whole milk powder, milk powder with soybean, chocolate milk powder, diet food	Digest starchy samples with takadiastase before saponification. Saponify (hot), extract with petroleum ether. Silica solid-phase extraction (high-fat samples only)		*Reverse Phase* Hypersil ODS 5 μm, 0.4 × 12 cm (two columns connected in series) 0.5% H_2O in MeOH UV 265 nm	54
Vitamin D_2 or D_3	Fortified whole milk powder	Add vitamin D_2 or D_3 to sample as internal standard. Saponify (ambient), extract with petroleum ether/diethyl ether, 90:10. Precipitate sterols from a methanolic solution		Radial-Pak cartridge containing either Resolve C_{18} or Nova-Pak C_{18} 5 μm (two cartridges connected in series) MeOH/THF/H_2O, 93:2:5 UV 254 and 280 nm (dual)	140
Vitamin D_2 or D_3	Infant formulas	Add vitamin D_2 or D_3 to sample as internal standard. Saponify (ambient), extract with petroleum ether/diethyl ether, 90:10. Silica solid-phase extraction		Two Radial-Pak cartridges as above MeOH/THF/H_2O, 92:2:6 UV 254 and 280 nm (dual)	141
Vitamin D_3	Milk (unfortified)	Add vitamin D_2 to sample as internal standard. Saponify (ambient), extract with hexane/diethyl ether, 90:10. Silica solid-phase extraction	Radial-Pak cartridge containing Resolve silica 5 μm Hexane/2-PrOH, 99:1 UV 265 nm	Radial-Pak cartridge containing Resolve C_{18} 5 μm, column temperature 30°C MeOH/THF/H_2O, 93:2:5 UV 265 nm	142
Vitamin D_3	Milk	Saponify (ambient), extract with petroleum ether/diethyl ether, 90:10. Silica solid-phase extraction		Vydac 201 TP54 C_{18} 5 μm, 0.46 × 25 cm MeCN/MeOH, 90:10 UV 254 nm	143

Table 4 (*Continued*)

Analyte	Food sample	Sample preparation[a]	Semipreparative HPLC[a]	Quantitative HPLC[a]	Ref.
Vitamin D_3	Infant formulas	Add vitamin D_2 to sample as internal standard. Saponify (hot), extract with hexane. Silica solid-phase extraction		Vydac 201 TP54 C_{18} 5 μm, 0.46 × 25 cm, column temperature 27°C MeCN/MeOH, 91:9 UV 265 nm	83
Vitamin D_2 or D_3	Margarine, fats, and oils	Add vitamin D_2 to sample as internal standard. Saponify (hot), extract with diethyl ether	LiChrosorb Si-60 7 μm, 0.46 × 25 cm Hexane/2-PrOH/THF, 98:1:1 UV 264 nm	ChromSphere C_{18} 8 μm, 0.3 × 10 cm MeCN/CHCl₃/MeOH, 91:6:3 UV 264 μm	144
Vitamin D_2 or D_3	Fortified fluid milk (whole)	Saponify (ambient), extract with hexane	Supelcosil LC-Si 5 μm, 0.46 × 15 cm Cyclohexane/hexane, 50:50 containing 0.5% 2-PrOH UV 254 nm	Radial-Pak cartridge containing Resolve C_{18} 5 μm or Spherisorb ODS 10 μm MeCN/MeOH, 90:10 UV 254 nm	79
Vitamin D_2 or D_3	Margarine, infant formulas	As above, but with alumina column chromatography before semipreparative HPLC	As above, but concentration of 2-PrOH changed to 0.25%	As above	79
Vitamin D_3	Margarine, oil, fortified milk	Add vitamin D_2 to sample as internal standard. Saponify (hot), extract with hexane	Polygosil 60 5 μm, 0.8 × 30 cm, column temperature 30°C Isooctane/CHCl₃/THF/isobutanol, 94:3:2:1 UV 254 nm	Vydac 201 TP54 C_{18} 5 μm 0.46 × 25 cm, column temperature 30°C MeCN/MeOH/CHCl₃, 82:12:6 UV 265 nm	145
Vitamin D_3	Infant formulas	Add vitamin D_2 to sample as internal standard. Saponify (hot), extract with petroleum ether/diethyl ether, 90:10	Polygosil 60 5 μm, 0.8 × 25 cm Isooctane/isobutanol, 99:1 UV 265 nm	Hypersil ODS 5 μm, 0.46 × 25 cm (two columns connected in series) 100% MeOH UV 265 nm	146

Compound	Sample	Sample preparation			Ref.
Vitamin D3	Nutritionally complete liquid diet	Saponify (hot), extract with diethyl ether. Add vitamin D2 as internal standard to the extracted ether solution	Nucleosil 50-5 5 μm, 0.46 × 25 cm Hexane/2-PrOH, 99.5:0.5 UV 265 nm	Hitachi gel 3056 reverse phase column 5 μm, 0.4 × 25 cm MeCN/MeOH/50% HClO4 (970 + 30 + 1.2) containing 0.057 M NaClO4 Dual-cell electrochemical detector (redox mode) + 0.65 V (oxidation) −0.20 V (reduction)	134
Vitamin D3 and 25-OH-D3	Egg yolk	Add vitamin D2 and 25-OH-D2 to sample as internal standards. Saponify (ambient), extract with diethyl ether/petroleum ether, 50:50. Silica solid-phase extraction to obtain a vitamin D fraction and a 25-OH-D fraction	*Vitamin D fraction:* μPorasil silica 10 μm 0.39 × 30 cm Hexane/THF/2-PrOH, 91:1:1 UV 264 nm *25-OH-D Fraction:* μPorasil silica 10 μm Hexane/2-PrOH, 97:3 UV 264 nm	*Vitamin D3:* Vydac 201 TP54 C18 5 μm. 0.46 × 25 cm MeOH/H2O, 94:6 UV 264 nm *25-OH-D3:* Vydac 201 TP54 C18 5 μm MeOH/H2O, 83:17 UV 264 nm	36
Vitamin D3 and 25-OH-D3	Meat and fat from livestock fed normal and excessive quantities of vitamin D	Saponification (hot), extract with hexane/CH2Cl2, 85:15. Alumina column chromatography to obtain a vitamin D fraction and a 25-OH-D fraction	*Vitamin D Fraction:* Apex silica 3 μm, 0.45 × 15 cm Cyclohexane/hexane, 50:50 containing 0.25% 2-PrOH UV 254 nm *25-OH-D Fraction:* Radial-Pak cartridge containing Resolve C18 5 μm Dry 100% MeOH UV 254 nm	*Vitamin D3:* Radial-Pak cartridge containing Resolve C18 5 μm Dry 100% MeOH UV 254 nm *25-OH-D3 by Normal Phase HPLC:* Apex silica 3 μm Heptane/2-PrOH. 96:4 UV 254 nm	82

aSolvent proportions by volume.
Notes and abbreviations: see footnote to Table 9.

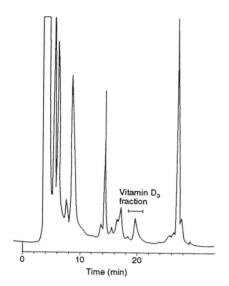

Fig. 3 Semipreparative HPLC of the unsaponifiable material obtained from a typical food sample showing the collected vitamin D_3 fraction. Column packing material, Polygosil 60 silica; mobile phase, isooctane/chloroform/THF/isobutanol (94:3:2:1); UV detection, 254 nm. (From Ref. 145.)

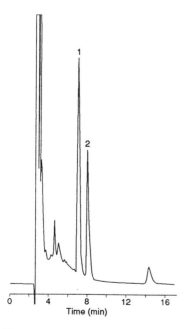

Fig. 4 Analytical HPLC of the vitamin D_3 fraction from a saponified margarine sample isolated by semipreparative HPLC. Column packing material, Vydac 201 TP54 C18; mobile phase, acetonitrile/methanol/chloroform (82:12:6); UV detection, 265 nm. Peaks: (1) vitamin D_2 (internal standard); (2) vitamin D_3. (From Ref. 145.)

made possible by the use of two analytical columns connected in series, which adequately separated vitamins D_2 and D_3 from one another and from vitamins A and E. The analysis of infant formulas (141) required cleanup by silica solid phase extraction to remove the minor tocopherols and tocotrienols, which constituted potential sources of interference. Tandem column chromatography was also used by Bui (54) in a simplified procedure for determining vitamin D in various fortified milks and milk products.

Mattila et al. (36,147) described a two-dimensional HPLC procedure for determining vitamin D_3 and 25-hydroxyvitamin D_3 in egg yolk. Samples were saponified in the presence of vitamin D_2 and 25-hydroxyvitamin D_2 as internal standards, and the extracted unsaponifiable matter was cleaned up by silica solid phase extraction. Eluates containing vitamin D and 25-hydroxyvitamin D were collected separately and, after evaporation, each fraction was subjected to semipreparative adsorption HPLC followed by reverse phase analytical HPLC. The reliability of the method was checked with recovery tests and repeatability tests (see Table 5) and by monitoring the purity of the analyte peaks with a diode array detector.

I. Vitamin E

As shown in Table 6, HPLC methods for determining the individual vitamin E analogs invariably employ fluorescence detection. Besides being more selective, the sensitivity of a fluorescence detector toward nonesterified vitamin E analogs under normal phase chromatographic conditions is at least 10 times greater than that of a variable-wavelength absorbance detector (148). The relative fluorescence responses of the tocopherols at 290 nm (excitation) and 330 nm (emission), as measured by HPLC peak area, were α-T, 100; β-T, 129; γ-T, 110; and δ-T, 122 (148). Fortunately, the fluorescence responses of the corresponding tocotrienols are very similar and therefore tocotrienol standards are not needed for calibration purposes. The fluorescent intensities of the vitamin E analogs are highly dependent on the solvent. Polar solvents such as diethyl ether and alcohols provide greater intensities compared with hexane. The fluorescence is negligible when the compounds are dissolved in chlorinated hydrocarbons (160).

The fluorescence activity of α-tocopheryl acetate is very weak, and the excitation and emission maxima of 285 and 310 nm are even closer together than those for α-tocopherol (295 and 330 nm). Only those spectrofluorometers equipped with a high-energy (150-W) xenon lamp and narrow-band monochromators are capable of stimulating and measuring the weak fluorescence

Table 5 Method Validation Data of HPLC Method for Vitamin D and 25-Hydroxyvitamin D (36, 147)

Data	Vitamin D_2	Vitamin D_3	25-OH-D_2	25-OH-D_3
Content of pooled egg yolk sample (μg/100 g)[a]	—	4.0, 5.6	—	0.98 ± 0.069
Overall mean recovery (%)	70 ($n = 20$)	70 ($n = 20$)	71 ± 5.4 ($n = 8$)	71 ± 3.2 ($n = 8$)
Recovery, calculated on the basis of the internal standard (%)	—	94 ($n = 4$)	—	98 ± 4.6 ($n = 8$)
RSD[b] (%)	—	3.6 ($n = 4$)	—	6.0 ($n = 4$)
Detection limit[c] (ng/injection)	2	2	2	2

[a]To convert these results to a whole-egg basis, divide by 2.9.
[b]Day-to-day repeatability.
[c]Detection limits defined as a signal three times the height of the noise level.

Table 6 HPLC Methods for the Determination of Vitamin E Compounds in Food

Analyte	Food sample	Sample preparation[a]	Quantitative HPLC Column	Mobile phase[a]	Detection	Ref.
Normal Phase Chromatography						
α-, β-, γ-, δ-T, α-, β-, γ-T3	Cereals, flour foods (unfortified)	Extract sample with boiling 2-PrOH, reextract with acetone, partition into hexane	LiChrosorb Si-60 5 μm, 0.32 × 25 cm	0.2% 2-PrOH or 5% diethyl ether in 50% H2O-saturated hexane	Fluorescence: ex. 290 nm em. 330 nm	148
Total α-T, β-, γ-, δ-T, α-, β-, γ-T3	Infant formulas	Saponify (hot), extract with diethyl ether	As above	As above	As above	148
α-, β-, γ-, δ-T	Infant formulas	Saponify (hot), extract with hexane	LiChrosphere Si-60 5 μm, 0.46 × 12 cm	1% 2-PrOH and 0.5% EtOH in hexane	Fluorescence: ex. 292 nm em. 320 nm	149
α-, β-, γ-, δ-T	Various foodstuffs, dairy products, infant formulas	Saponify (hot), extract with petroleum ether/diisopropyl ether (3 + 1)	Radial-Pak cartridge containing Resolve silica 5 μm	Hexane/2-PrOH, 99:1	Fluorescence: ex. 295 nm em. 330 nm	150
α-Tocopheryl acetate	Infant formulas	Disperse sample in nonaqueous solvent mixture (DMSO/DMF/CHCl3, 2 + 2 + 1) containing 0.1% (w/v) ascorbic acid. Partition total lipid fraction into hexane, centrifuge	Radial-Pak cartridge containing Resolve silica 5 μm	0.08% 2-PrOH in hexane	UV 280 nm	151
Total α-T	Animal feeds, human foods (unfortified)	Saponify (hot), extract with diethyl ether	LiChrosorb Si-60 5 μm, 0.46 × 12.5 cm	0.1% 2-PrOH in hexane	Fluorescence: ex. 293 nm em. 326 nm	152
α-, β-, γ-, δ-T, α-, β-, γ-, δ-T3	Vegetable oils and fats (unfortified)	Dissolve sample in hexane	LiChrosorb Si-60 5 μm, 0.4 × 25 cm	0.5% 2-PrOH in hexane	Fluorescence: ex. 290 nm em. 330 nm	93
Total α-T, β-, γ-, δ-T, α-, β-, γ-, δ-T3	Margarines and fats (fortified)	Saponify at 26°C, extract with diethyl ether	As above	As above	As above	93
α-, β-, γ-, δ-T	Fats and oils	Saponify (ambient), extract with diethyl ether	LiChrosorb Si-60 5 μm, 0.4 × 25 cm, column temperature 44°C	0.4% 2-PrOH in hexane	UV 295 nm	153

Compounds	Food type	Extraction	Column	Mobile phase	Detection	Ref.
α-, β-, γ-T, α-, β-T3	Meat and meat products	Saponify (ambient), extract with hexane	LiChrosorb Si-60 5 μm, 0.4 × 25 cm	Hexane/diisopropyl ether, 93:7	Fluorescence: ex. 292 nm em. 324 nm	154
α-, β-, γ-T	Fish and fish products	As above	As above	As above	As above	155
Total α-T, β-, γ-T, α-, β-T3	Infant formulas	As above	As above	As above	As above	156
α-, β-, γ-, δ-T	Cereal products	As above	As above	As above	As above	157
α-, β-, γ-, δ-T3 α-, β-, γ-, δ-T	Rice bran	Saponify (hot), extract with hexane	Supelcosil LC-Si 5 μm, 0.46 × 25 cm	Isooctane/EtOAc/HOAc/DMP 98.15:0.9:0.85:0.1	Fluorescence: ex. 290 nm em. 330 nm	158
				Reverse Phase Chromatography		
α-T	Liver	Saponify (hot), extract with hexane, wash hexane extract with H_2SO_4 (600 ml/L)	Partisil ODS	Hexane/2-PrOH, 99:1	Fluorescence: ex. 210 nm em. 325 nm	159
α-, (β + γ)-, δ-T	All food types	Saponify (ambient), extract with hexane, filter through anhydrous sodium sulfate	Zorbax ODS 5 μm. 0.46 × 25 cm	MeCN/CH_2Cl_2 containing 0.001% triethylamine/MeOH, 700 + 300 + 50	Fluorescence: ex. 290 nm em. 330 nm	40
α-Tocopheryl acetate α-, β-, γ-, δ-T α-, β-, γ-, δ-T3, plastochromanol-8	Breakfast cereals, infant formulas	*Cereals:* add water and MeOH to ground sample, shake and sonicate. Extract with tBME/petroleum ether (10 + 14) *Formulas:* reconstitute sample with water, add dipotassium oxalate solution (35% w/v) and EtOH, extract with tBME/petroleum ether (25 + 35).	LiChrospher 100 diol 5 μm 0.4 × 25 cm	Two-step gradient composed of hexane and an increasing concentration of tBME. 0–4 min hexane, 4–5 min up to hexane/tBME (97:3), 5–41 min isocratic, 41–42 min up to hexane/tBME (95:5). 42–60 min isocratic	Fluorescence: ex. 295 nm em. 330 nm	91

[a] Solvent proportions by volume.
Notes and abbreviations: see footnote to Table 9.

(91). Using such a spectrofluorometer, the fluorescence activity of α-tocopheryl acetate was shown to be less than 0.5% of that of α-tocopherol under defined normal phase conditions (161). Nevertheless, the detection limit of 0.002 μg/g for the ester (8 ng on-column) represents a fourfold gain in sensitivity compared with absorbance detection and is ample to permit its measurement in fortified foods, which contain 5–500 μg/g (162).

In some fluorescence detectors a pronounced short-wavelength excitation maximum at 205 nm can be obtained, which provides a 20-fold increase in fluorescence intensity over the 295-nm maximum for α-tocopherol (163). A disadvantage of short-wavelength excitation is a marked loss of selectivity and an aggravation of quenching effects (162), and hence the longer wavelength (295 nm) excitation is usually employed in food analysis.

Normal phase HPLC is capable of separating isocratically all of the eight nonesterified tocopherols and tocotrienols that may be found in seed oils. The vitamers are eluted in the order of α-T, α-T3, β-T, γ-T, β-T3, γ-T3, δ-T, and δ-T3. Oils and fats can be simply dissolved in hexane and analyzed directly by normal phase HPLC using either silica or polar-bonded stationary phases. α-Tocopheryl acetate is eluted before α-tocopherol under normal phase conditions, so that the acetate and palmitate esters of vitamin A can potentially interfere when present in complex foods such as infant formulas. In the analysis of infant formulas, Woollard and Blott (151) achieved the complete separation of α-tocopheryl acetate from vitamin A esters by means of a Radial-Pak cartridge containing Resolve spherical silica and a mobile phase composition of hexane containing 0.08% isopropanol. The radially compressed cartridge allowed the flow rate to be stepped up from 2 ml/min to 10 ml/min, without incurring a high back pressure, after elution of the α-tocopheryl acetate (7–8 min) as a means of rapidly clearing the lipid material from the cartridge. Chromatograms of the lipid fraction of a fully filled infant formula obtained using dual absorbance and fluorescence detection are depicted in Fig. 5. The fluorescence trace, obtained using a filter-type fluorometer, allows the indigenous α-tocopherol to be conveniently estimated, whereas the UV trace is used to quantify the α-tocopheryl acetate. Recovery studies indicated that the solvent extraction method removed an average 93% (91–96%) of added α-tocopheryl acetate and 92% (91–94%) of α-tocopherol. Supplemental retinyl acetate could be assayed simultaneously with either added or indigenous vitamin E using the appropriate detection mode. With the aid of a dual-monochromator spectrofluorometer, α-tocopheryl acetate and indigenous α-tocopherol could be determined simultaneously with wavelengths of 280 nm (excitation) and 335 nm (emission) (161), but the increased selectivity eliminated detection of the vitamin A esters.

An improvement in the column stability and reproducibility for analysis of vitamin E vitamers on a silica column was reported using a mobile phase composition of isooctane/ethyl acetate/acetic acid/2,2-dimethoxypropane (98.15:0.90:0.85:0.10) (158). The acetic acid component reduced retention times of the late-eluting E vitamers presumably by competing with water and polar material for binding to silanol groups in the stationary phase. 2,2-Dimethoxypropane reacts with water to form acetone and methanol, and its inclusion stabilized retention times and reduced the need for column regeneration.

Baltz et al. (91) utilized a diol stationary phase and a two-step gradient composed of hexane and an increasing concentration of *tert*-butylmethyl ether to determine α-tocopheryl acetate, the eight unesterified tocopherols and tocotrienols, and also plastochromanol-8 in infant formulas and breakfast cereals.

Fluorescence detection is usually obligatory when the total lipid fraction is analyzed, as absorbance detection reveals peaks of lipid origin that interfere with the peaks of the vitamin E analogs. However, if the sample is saponified, absorbance detection can sometimes be used since the unsaponifiable fraction is free from interfering lipid material. The individual E vitamers exhibit relatively low intensities of UV absorption. Published $A_{1cm}^{1\%}$ values for the tocopherols in ethanol, with their absorption maxima, are α-T, 75.8 at 292 nm; β-T, 89.4 at 296 nm; γ-T, 91.4 at 298 nm;

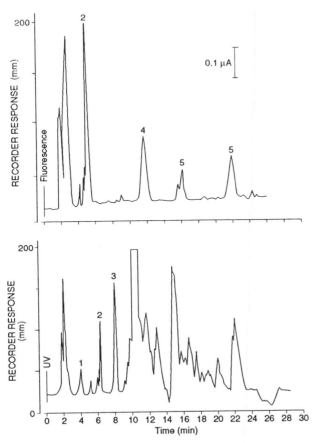

Fig. 5 HPLC of the lipid fraction from a vitamin E–fortified fully filled infant formula containing corn, soya, and coconut oils. Column packing material, Resolve silica; mobile phase, 0.08% isopropanol in hexane. Dual-mode detection: fluorescence, 295 nm excitation with a 335-nm cutoff emission filter (upper trace); UV, 280 nm (lower trace). Peaks: (1) retinyl palmitate; (2) retinyl acetate; (3) α-tocopheryl acetate; (4) α-tocopherol; (5) unidentified vitamin E analogs. (From Ref. 151.)

and δ-T, 87.3 at 298 nm (164). The absorptivity of α-tocopheryl acetate is lower still, with an $A_{1cm}^{1\%}$ value of only 40–44 at the λ_{max} of 285.5 (38). The differences in the spectral characteristics implies that individual standards must be run if each vitamer is to be accurately quantified. Because of these limitations, absorbance measurement is usually confined to the determination of total α-tocopherol in vitamin E–supplemented foods and animal feeds.

A preferred means of analyzing vitamin E–supplemented foods is to saponify the sample and determine the total α-tocopherol by means of reverse phase HPLC and fluorescence detection. The elution order of the vitamin E analogs is δ-T3, β+ plus γ-T3, α-T3, δ-T, β+ plus γ-T, and α-T. The positional β and γ isomers do not separate on reverse phase columns of standard dimensions. Several researchers (149,165) advocate extraction of the unsaponifiable matter with hexane. If the reverse phase HPLC system is operated with a semiaqueous mobile phase, the hexane solution containing the unsaponifiable material must be evaporated to dryness and redissolved in a polar solvent that is compatible with the mobile phase. The use of NARP-HPLC with a predominantly hexane mobile phase eliminates this second evaporation step (159).

Reverse phase HPLC is compatible with electrochemical detection, as the supporting electrolyte is readily soluble in the semiaqueous or methanolic mobile phase. The electrochemical detection of vitamin E in saponified animal feeds was reported to be 20 times more sensitive than fluorescence detection (166). Electrochemical detection cannot measure α-tocopheryl acetate owing to the absence of the oxidizable hydroxyl group, but this is of no concern in the determination of total vitamin E after saponification. It is possible, by adding the electrolyte postcolumn, to use electrochemical detection with normal phase HPLC, as demonstrated by Hiroshima et al. (167) in the analysis of wheat germ oil.

J. Vitamin K

Phylloquinone, with a molar absorptivity (ε) of around 19,000 in the 240- to 270-nm region, can be detected photometrically at levels down to 500 pg on-column. Photometric detection is satisfactory for the analysis of green leafy vegetables and vitamin K–fortified infant formulas, after purification of the sample to remove interfering lipids, but it lacks the required sensitivity for the analysis of other foods. The reversible oxidation–reduction between the quinone and hydroquinone (quinol) forms of phylloquinone allows this vitamin to be detected electrochemically with a 10-fold increase in sensitivity over photometric detection and an improved selectivity. The first combined HPLC-electrochemical measurements of vitamin K used the reductive mode, but this technique suffered from interference from the reduction of oxygen. A redox method was later developed that eliminated interference from the reduction of oxygen and provided an increase in the selectivity and sensitivity of detection. The coulometric detector employed in this method is equipped with a dual-electrode cell in which first vitamin K is reduced upstream at the generator electrode and then the hydroquinone is reoxidized downstream at the detector electrode.

The hydroquinone reduction products of vitamin K, unlike the quinone forms, are highly fluorescent, and electrochemical (168,169) or chemical (96) reduction of vitamin K prior to fluorometric detection has been utilized. Indyk (170) reported that a commercial fluorescence detector facilitates the photochemical reduction and simultaneous fluorescence detection of phylloquinone during normal passage of the column effluent through the flow cell.

HPLC methods for determining phylloquinone in foods are summarized in Table 7.

The separation of *cis*- and *trans*-phylloquinone cannot be achieved using reverse phase liquid chromatography, but *cis*-phylloquinone elutes before the *trans* isomer under normal phase conditions. Hwang (171) employed adsorption column chromatography as a cleanup step following extraction of reconstituted infant formulas with dichloromethane/isooctane (2 + 1). Quantitative HPLC was also performed in the adsorption mode using photometric detection at 254 nm. Both the *cis* and *trans* isomers of phylloquinone could be measured in standards and in liquid formulas, but matrix interferences prevented measurement of the *cis* isomer in powdered formulas. The overall precision of the method for determining *trans*-phylloquinone was 3.3% RSD at concentrations ranging from 30 to 225 µg/L, as fed, and the recovery was 98 ± 4%. The recovery was calculated from the average of 33 determinations at spiking levels ranging from 44 to 92 µg/L, as fed, covering all of the products tested. The detection limit of the assay at a signal/noise ratio of 3 was 0.3 ng for both isomers, equivalent to 2 µg/L of the analyte in the as-fed product.

In a method developed by Shearer (95) for determining phylloquinone, food samples are extracted and subjected to solid phase extraction (as described previously) followed by adsorption HPLC as a semipreparative purification step. The retention window for collecting the fraction containing the phylloquinone and internal standard excludes *cis*-phylloquinone. Reverse phase chromatography is used for the final analytical stage because this mode is effective in separating phylloquinone from closely related structures, including the internal standard. With UV absorbance detection, a nonaqueous mobile phase of dichloromethane in methanol is preferred to a

Table 7 HPLC Methods for the Determination of Vitamin K Compounds in Food

Analyte	Food sample	Sample preparation[a]	Semipreparative HPLC[a]	Quantitative HPLC[a]	Ref.
				Normal Phase	
*trans-*Phylloquinone	Infant formulas	Mix sample with concentrated NH_4OH/MeOH, extract with CH_2Cl_2/isooctane (2 + 1). Silica column chromatography		Apex Silica 5 µm, 0.45 × 25 cm Isooctane/CH_2Cl_2/2-PrOH (70 + 30 + 0.02) UV 254 nm	171
				Reverse Phase	
Phylloquinone	Infant formulas	Digest sample with lipase, extract with pentane		µBondapak C_{18} 10 µm. 0.4 × 25 cm MeOH/MeCN/THF/H_2O, 39:39:16:6 UV 254 nm	172
Phylloquinone	Soy bean oil	Digest sample with lipase, extract with pentane. Alumina column chromatography		Supelcosil LC-18 5 µm, 0.46 × 15 cm MeOH/MeCN/H_2O, 88:10:2 UV 270 nm	173
Phylloquinone	Milk (human and bovine)	Deproteinize sample with EtOH at 4°C with added synthetic phylloquinone homolog (internal standard). Extract with hexane aided by sonication and vortexing. Centrifuge, evaporate hexane layer to dryness	µPorasil silica 10 µm, 0.8 × 30 cm Heptane/EtOAc, 99:1 UV 254 nm	Resolve C_{18} 5 µm, 0.39 × 15 cm 100% MeCN UV 270 nm	174
Phylloquinone	Infant formulas	Supercritical fluid extraction		µBondapak C_{18} 10 µm, 0.39 × 15 cm MeCN/CH_2Cl_2/aqueous 0.025 M $NaClO_4$, 90:5:5 Amperometric detection (reductive mode), silver electrode, -1.1 V. vs saturated calomel reference electrode	53

Table 7 (*Continued*)

Analyte	Food sample	Sample preparation[a]	Semipreparative HPLC[a]	Quantitative HPLC[a]	Ref.
Phylloquinone	Various foods (vegetable juice, whole milk, raw spinach leaves, plain bagel, raw ground beef)	*Vegetable juice:* add dihydrophylloquinone (internal standard) to sample. Extract with 2-PrOH/hexane (3 + 2) and water aided by sonication and vortexing. Silica solid phase extraction *Milk:* as described for vegetable juice followed by liquid phase reductive extraction for removal of lipids *Spinach:* homogenize and grind sample with sodium sulfate. Extraction and cleanup as for vegetable juice *Bread:* add K$_{1(25)}$ (internal standard) to ground sample. Extraction and cleanup as for vegetable juice *Beef:* as described for bread followed by reverse phase C$_{18}$ solid phase extraction		Hypersil ODS 3 μm, 0.46 × 15 cm MeOH/CH$_2$Cl$_2$, 90:10 containing 10 mM ZnCl$_2$, 5 mM HOAc, and 5 mM NaOAc Fluorescence: ex. 244 nm em. 418 nm Postcolumn chemical reactor column packed with zinc metal	96
Phylloquinone	All food types	*Vegetables, fruits, cereals, meat, and fish:* extract sample with acetone, filter, partition the phylloquinone into hexane *Fats, oils, and dairy products:* digest sample with lipase, extract with hexane. Purify hexane extracts obtained from either extraction technique by silica solid phase extraction. The internal standard is phylloquinone 2,3-epoxide (unlabeled for UV detection and tritium-labeled for coulometric detection)	A. Partisil-5 silica, 0.46 × 25 cm 50% water-saturated CH$_2$Cl$_2$/hexane, 15:85 UV 254 nm B. Spherisorb-5 nitrile, 0.46 × 25 cm 50% water-saturated CH$_2$Cl$_2$/hexane, 3:97 UV 254 nm	A. Hypersil ODS, 0.46 × 25 cm CH$_2$Cl$_2$/MeOH, 15:85 UV 270 nm B. Spherisorb-5 C$_8$, 0.46 × 25 cm MeOH/50 mM acetate buffer pH 3.0 (97:3) containing 0.1 mM EDTA Dual-electrode coulometric detection (redox mode), porous graphite electrodes, −1.5 V (generator electrode) + 0.05 V (detector electrode)	95

[a]Solvent proportions by volume.

Notes and abbreviations: see footnote to Table 9.

semiaqueous mobile phase because it provides an increased solubility of lipophilic compounds. To compensate for the increased affinity of vitamin K for the mobile phase, a highly retentive ODS stationary phase is required. NARP-HPLC cannot be used with electrochemical detection because the mobile phase is not polar enough to dissolve the electrolyte needed to conduct a current. With electrochemical detection, therefore, a semiaqueous mobile phase is used in conjunction with a less retentive octyl-bonded (C8) stationary phase. The addition of 0.1 mM EDTA to the mobile phase prevents the reduction of metal ions at the generator electrode. With UV detection, the phylloquinone is quantified by the usual method of peak height ratios using phylloquinone 2,3-epoxide as the internal standard. The epoxide is electrochemically inactive and hence does not appear as a peak when electrochemical detection is used. In this case, quantification is accomplished by the technique of radioisotopic dilution using tritiated phylloquinone 2,3-epoxide as the internal standard. A chromatogram showing the analytical separation of the vitamin K_1 fraction from a sample of brown rice isolated by semipreparative HPLC is shown in Fig. 6.

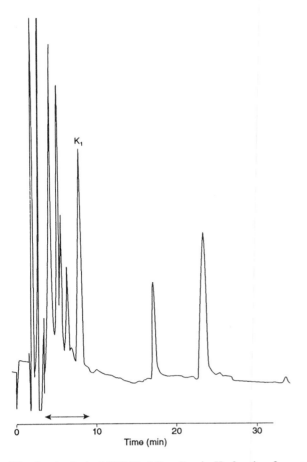

Fig. 6 Analytical HPLC of the vitamin K_1 fraction from an extracted sample of brown rice isolated by semipreparative HPLC. Column packing material. Spherisorb-5 C_8; mobile phase, methanol/50 mM acetate buffer pH 3.0 (97:3) containing 0.1 mM EDTA; dual-electrode coulometric detection (redox mode), -1.5 V (generator electrode), +0.05 V (detector electrode). The arrows signify the fraction containing tritiated phylloquinone 2,3-epoxide (internal standard) and phylloquinone (analyte) that is collected for quantification by radioisotopic dilution. (Courtesy of M. J. Shearer.)

In a method proposed by Booth et al. (96) for the determination of phylloquinone in various food types, extracted samples are subjected to solid phase extraction on silica followed, in the case of meat or milk samples, by further purification by reverse phase/solid phase extraction or liquid phase reduction extraction, respectively (as described previously). The residues obtained after evaporation of the eluates or reductive extracts are dissolved in 100% dichloromethane, followed immediately by the addition of 10 mM zinc chloride solution, 5 mM acetic acid, and 5 mM sodium acetate. This final test solution is analyzed by reverse phase HPLC using a mobile phase composed of methanol/dichloromethane (90:10) containing 10 mM zinc chloride, 5 mM acetic acid, and 5 mM sodium acetate. The fluorescent hydroquinone reduction products of the injected phylloquinone and internal standard are produced on-line using a postcolumn chemical reactor packed with zinc metal. $2',3'$-Dihydrophylloquinone, a synthetic analog of phylloquinone, is a suitable internal standard for the analysis of vegetable juice, whole milk, and spinach. Another synthetic analog, $K_{1(25)}$, is used for the analysis of bread and beef because of the presence of a chromatographic peak corresponding to the retention time of $2',3'$-dihydrophylloquinone. The precisions of the method as calculated from 10 successive determinations on a sample on the same day (intraday) and the same sample on three different days (interday) are presented in Table 8.

Schneiderman et al. (53) employed reverse phase HPLC with reductive mode electrochemical detection for the determination of phylloquinone extracted from powdered infant formulas using SFE. The recovery of phylloquinone from a milk-based formula was 95.6% (RSD of 7.4%) and from a soy protein-based product, 94.4% (RSD of 6.5%).

K. Multiple Vitamin Analyses

HPLC methods for the successive or simultaneous determination of two or more vitamins are summarized in Table 9.

In a method developed by Reynolds and Judd (84) for the determination of vitamins A and D in skimmed milk powder, an aliquot of the unsaponifiable fraction is first analyzed for vitamin A by reverse phase HPLC. The larger remaining portion of the unsaponifiable fraction is then purified by reverse phase solid phase extraction and analyzed for vitamin D using the same HPLC column and mobile phase as used for the vitamin A. The RSDs ($n = 6$) for vitamins A and D were 2.0% and 3.5%, respectively.

Delgado Zamarreño et al. (176) proposed a rapid method for the simultaneous determination of vitamins A, D_3, and E in saponified milk and milk powders using reverse phase HPLC and amperometric detection. The high sensitivity of the electrochemical detection allows the determi-

Table 8 Intraday ($n = 10$) and Interday Precision ($n = 3$) of HPLC Method for Phylloquinone

Food sample	Intraday precision		Interday precision	
	K_1 (µg/100 g)	RSD (%)	K_1 (µg/100 g)	RSD (%)
Vegetable juice	5.29	6.6	5.74	12.3
Milk[a]	0.21	12.1	0.32	13.8
Spinach[a]	482.9	6.3	202.0	7.4
Bread	0.36	10.9	0.39	10.8
Ground beef	2.98	9.5	2.57	12.6

[a]Intraday precisions for milk and spinach samples were determined from samples different from those used to determine interday precisions.
Source: Ref. 96.

Table 9 HPLC Methods for Multiple Vitamin Determinations in Food

Analyte	Food sample	Sample preparation	Quantitative HPLC			
			Column	Mobile phase[a]	Detection	Ref
Total retinol, vitamin D_2 or D_3 (measured concurrently in separate chromatograms)	Fortified skimmed milk powder	Saponify (hot) in presence of vitamin D_2 or D_3 as internal standard, extract with petroleum ether/diethyl ether (50:50), dissolve residue in 25 ml MeOH. For retinol: remove 5-ml aliquot for HPLC. For vitamin D: evaporate remaining 20-ml MeOH solution. C_{18} solid phase extraction	Spherisorb ODS 5 μm, 0.46 × 25 cm	MeOH/H_2O, 97.5:2.5	For retinol: UV 325 nm. For vitamin D: UV 265 nm	84
Total retinol, vitamin D_2 or D_3 (measured concurrently in separate chromatograms)	Fortified fluid milk (whole, semiskimmed, skimmed)	Saponify (ambient), extract with hexane, dissolve residue in 6 ml hexane. For retinol: remove 1-ml aliquot for HPLC. For vitamin D: evaporate remaining 5-ml hexane solution. Alumina column chromatography	Vydac 201 TP C_{18} 10 μm, 0.32 × 25 cm	For retinol: MeOH/H_2O, 90:10. For vitamin D: MeCN/MeOH, 90:10	For retinol UV 325 nm. For vitamin D: UV 265 nm	81
Retinyl palmitate, α-tocopheryl acetate (measured simultaneously in same chromatogram)	Breakfast cereals	Extract sample with a solvent mixture of CHCl3, EtOH and H_2O at 50°C	μPorasil 10 μm, 0.4 × 30 cm	Hexane/CHCl3 containing 1% EtOH, 85:15	UV 280 nm	175
Total retinol, vitamin D_3, total α-tocopherol (measured simultaneously in same chromatogram)	Milk, milk powder	Saponify (hot), extract with hexane	Spheri-5 RP-18 5 μm, 0.46 × 22 cm	MeOH/H_2O, 99:1 containing 0.1 M LiClO4	Amperometric detection (oxidative mode), glassy carbon electrode, + 1.05 V, vs silver–silver chloride reference electrode	176

[a]Solvent proportions by volume.

Chromsphere (Chrompack); Apex (Jones Chromatography); Nucleosil (Macherey-Nagel); LiChrosorb (Merck); Spheri (Perkin-Elmer); Spherisorb (Phase Separations); Zorba: (Rockland Technologies), Vydac (Separations Group); Hypersil (Shandon); Supelcosil (Supelco); Porasil, Resolve, μBondapak, Nova-Pak (Waters Associates); Partisi (Whatman).

MeOH (methanol); EtOH (ethanol); 2-PrOH (isopropanol); HOAc (acetic acid); NaOAc (sodium acetate);EtOAc (ethyl acetate); HClO4 (perchloric acid); LiClO4 (lithium perchlorate); NaClO4 (sodium perchlorate); MeCN (acetonitrile); DMP (2,2-dimethoxypropane); CH_2Cl_2 (dichloromethane); CHCl3 (chloroform); THF (tetrahydrofuran); DMI (dimethylformamide); DMSO (dimethylsulfoxide); tBME (tert-butylmethylether); EDTA (ethylenediaminetetraacetic acid).

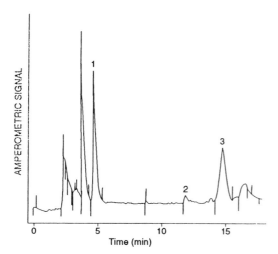

Fig. 7 HPLC of vitamins A, D$_3$, and E in the unsaponifiable fraction of milk. Column packing material, Spheri-5 RP-18; mobile phase, methanol/water (99:1) containing 0.1 M lithium perchlorate; amperometric detection (oxidative mode), +1.05 V. Peaks: (1) retinol; (2) vitamin D$_3$; (3) α-tocopherol. (From Ref. 176.)

nation of vitamin D$_3$ in unenriched fluid milk, which is not possible using UV detection without a preconcentration step. A typical chromatogram showing the indigenous vitamins in fluid bovine milk is depicted in Fig. 7. The on-column detection limits were of the order of 0.07, 4, and 0.2 ng of vitamins A, D$_3$, and E, respectively.

V. FUTURE DEVELOPMENTS

HPLC instrumentation and column technology have undergone major advances since the early 1970s when HPLC made its debut in the field of vitamin analysis. Yet sample preparation in food analysis continues to rely largely on manual wet chemical techniques, which are time consuming, labor intensive, require considerable analytical skill, and constitute the major source of error in the assay procedure. There is also the serious problem of environmental pollution and the exposure of laboratory personnel to toxic chemicals.

The ideal sample preparation technique would be nontoxic and capable of direct coupling to the measurement apparatus, allowing complete automation of the analysis. A potential technique that could meet this dual requirement is SFE, which has been used to extract carotenes from sweet potatoes (177) and vegetables (73), tocopherols from wheat germ (178), and phylloquinone from infant formulas (53). The extraction of vegetable and animal fats with supercritical CO$_2$ has been reviewed (179). For analytical purposes SFE can be coupled to GC or HPLC, but it is most conveniently coupled to supercritical fluid chromatography (SFC). Now that instrumentation and column technology have been improved, SFE-SFC should be investigated as a possible alternative to HPLC for determining the fat-soluble vitamins in foods.

REVIEWS

G. F. M. Ball, *Fat-Soluble Vitamin Assays in Food Analysis*, Elsevier Applied Science, London (1988).
    (Review of physicochemical methods for determining the fat-soluble vitamins in foods and animal feeds.)

G. F. M. Ball, in *Food Analysis by HPLC* (L. M. L. Nollet, ed.), Marcel Dekker, New York, 1992, p. 275. (Review of fat-soluble vitamin assays in foods by high-performance liquid chromatography.)

D. B. Rodriguez-Amaya, *J. Micronutr. Anal.*, 5: 191 (1989). (Review of provitamin A determination in plant foods.)

E. Lesellier, A. Tchapla, C. Marty, and A. Lebert, *J. Chromatogr.*, 633: 9 (1993). (Review of the analysis of carotenoids by high-performance liquid chromatography and supercritical fluid chromatography.)

REFERENCES

1. Food and Nutrition Board, *Recommended Dietary Allowances*, 10th ed., National Academy of Sciences, Washington, DC, 1989.
2. Codex Alimentarius Commission, *General Principles for the Addition of Essential Nutrients to Foods*, *Alinorm 87/26*, Appendix 5, Food and Agriculture Organization, Rome, 1987.
3. United States Food and Drug Administration, *Code of Federal Regulations*, Title 21, Part 104.5, Food and Drug Administration, Washington, DC, 1987.
4. D. R. Miller and K. C. Hayes, in *Nutritional Toxicology*, Vol. 1 (J. N. Hathcock, ed.), Academic Press, New York, 1982, p. 81.
5. S. T. Omaye, in *Nutritional and Toxicological Aspects of Food Safety* (M. Friedman, ed.), Plenum Press, New York, 1984, p. 169.
6. J. F. Gregory III, *Ann. N.Y. Acad. Sci.*, 585:86 (1990).
7. J. F. Gregory III, in *Nutritional Evaluation of Food Processing*, 3rd ed. (E. Karmas and R. S. Harris, eds.), Van Nostrand Reinhold, New York, 1988, p. 719.
8. J. F. Gregory III, *Food Technol.*, 42(10):230 (1988).
9. R. G. Ziegler, *J. Nutr.*, 119:116 (1989).
10. R. G. Ziegler, *Am. J. Clin. Nutr.*, 53:251S (1991).
11. A. Bendich and J. A. Olson, *FASEB, J.*, 3:1927 (1989).
12. S. R. Ames, *J. Assoc. Offic. Anal. Chem.*, 49:1071 (1966).
13. L. M. Sivell, N. L. Bull, D. H. Buss, R. A. Wiggins, D. Scuffam, and P. A. Jackson, *J. Sci. Food Agric.*, 35:931 (1984).
14. J. C. Bauernfeind, *J. Agric. Food Chem.*, 20:456 (1972).
15. J. L. Bureau and R. J. Bushway, *J. Food Sci.*, 51:128 (1986).
16. I. Stewart, *J. Agric. Food Chem.*, 25:1132 (1977).
17. C. Y. Lee, P. E. McCoon, and J. M. LeBowitz, *J. Agric. Food Chem.*, 29:1294 (1981).
18. F. Khachik, G. R. Beecher, and N. F. Whittaker, *J. Agric. Food Chem.*, 34:603 (1986).
19. B. Tan, *J. Food Sci.*, 53:954 (1988).
20. M. I. Mínguez-Mosquera and D. Hornero-Méndez, *J. Agric. Food Chem.*, 41:1616 (1993).
21. F. Khachik, G. R. Beecher, M. B. Goli, and W. R. Lusby, *Pure Appl. Chem.*, 63:71 (1991).
22. T. W. Goodwin and G. Britton, in *Plant Pigments* (T. W. Goodwin, ed.), Academic Press, London, 1988, p. 61.
23. C. A. O'Neil and S. J. Schwartz, *J. Chromatogr.*, 624:235 (1992).
24. D. B. Rodriguez-Amaya, *J. Micronutr. Anal.*, 5:191 (1989).
25. F. Khachik and G. R. Beecher, *J. Agric. Food Chem.*, 36:929 (1988).
26. T. Philip and T.-S. Chen, *J. Food Sci.*, 53:1703 (1988).
27. H. Crawley, in *The Technology of Vitamins in Food* (P. Berry Ottaway, ed.), Blackie, Glasgow, 1993, p. 19.
28. H. T. Gordon and J. C. Bauernfeind, *CRC Crit. Rev. Food Sci. Nutr.*, 18:59 (1982).
29. B. Borenstein, in *Nutritional and Safety Aspects of Food Processing* (S. R. Tannenbaum, ed.), Marcel Dekker, New York, 1979, p. 217.
30. J. G. Bieri and M. C. McKenna, *Am. J. Clin. Nutr.*, 34:289 (1981).
31. J. A. Olson, in *Handbook of Vitamins: Nutritional, Biochemical and Clinical Aspects* (L. J. Machlin, ed.), Marcel Dekker, New York, 1984, p. 1.
32. Food Labelling Regulations, *Statutory Instrument 1984*, no. 1305, as amended, H. M. Stationery Office, London, 1984.

33. R. Passmore and M. A. Eastwood, in *Davidson and Passmore Human Nutrition and Dietetics*, 8th ed. Churchill Livingstone, New York, 1986.
34. D. B. Parrish, *CRC Crit. Rev. Food Sci. Nutr.*, *12*:29 (1979).
35. K. T. Koshy and A. L. VanDerSlik, *J. Agric. Food Chem.*, *25*:1246 (1977).
36. P. Mattila, V. Piironen, E. Uusi-Rauva, and P. Koivistoinen, *J. Food Comp. Anal.*, *6*:250 (1993).
37. L. E. Reeve, N. A. Jorgensen, and H. F. DeLuca, *J. Nutr.*, *112*:667 (1982).
38. L. J. Machlin, in *Handbook of Vitamins: Nutritional, Biochemical, and Clinical Aspects* (L. J. Machlin, ed.), Marcel Dekker, New York, 1984, p. 99.
39. P. J. McLaughlin and J. C. Weihrauch, *J. Am. Diet. Assoc.*, *75*:647 (1979).
40. C. J. Hogarty, C. Ang, and R. R. Eitenmiller, *J. Food Comp. Anal.*, *2*:200 (1989).
41. J. W. Suttie, in *Fat-Soluble Vitamins: Their Biochemistry and Applications* (A. T. Diplock, ed.), Heinemann, London, 1985, p. 225.
42. S. L. Booth, J. A. Sadowski, J. L. Weihrauch, and G. Ferland, *J. Food Comp. Anal.*, *6*:109 (1993).
43. Y. Haroon, M. J. Shearer, S. Rahim, W. G. Gunn, G. McEnery, and P. Barkhan, *J. Nutr.*, *112*:1105 (1982).
44. R. E. Olson, in *Modern Nutrition in Health and Disease* (M. E. Shils and V. R. Young, eds.), Lea and Febiger, Philadelphia, 1988, p. 328.
45. AOAC, *Official Methods of Analysis*, 15th ed., Association of Official Analytical Chemists, Inc., Arlington, VA, 1990, p. 1045.
46. C. Marks, *J. Am. Oil Chem. Soc.*, *65*:1936 (1988).
47. F. Ulberth, *J. High-Resol. Chromatogr.*, *14*:343 (1991).
48. S. Kmostak and D. A. Kurtz, *J. Assoc. Offic. Anal. Chem. Int.*, *76*:735 (1993).
49. J. M. Snyder, S. L. Taylor, and J. W. King, *J. Am. Oil Chem. Soc.*, *70*:349 (1993).
50. D. C. Woollard, *Food Technol. Aust.*, *39*:250 (1987).
51. K. T. Koshy and A. L. VanDerSlik, *J. Agric. Food Chem.*, *27*:650 (1979).
52. B. W. Hollis and N. E. Frank, *Meth. Enzymol.*, *123*:167 (1986).
53. M. A. Schneiderman, A. K. Sharma, K. R. R. Mahanama, and D. C. Locke, *J. Assoc. Offic. Anal. Chem.*, *71*:815 (1988).
54. M. H. Bui, *J. Assoc. Offic. Anal. Chem.*, *70*:802 (1987).
55. S. A. Barnett, L. W. Frick, and H. M. Baine, *Anal. Chem.*, *52*:610 (1980).
56. S. B. Hawthorne, *Anal. Chem.*, *62*:633A (1990).
57. R. E. Majors, *LC-GC Int.*, *4*(3):10 (1991).
58. R. M. Smith, in *Supercritical Fluid Chromatography* (R. M. Smith, ed.), Royal Society of Chemistry, London, 1988.
59. R. E. Majors, *LC-GC*, *4*(10):972 (1986).
60. W. O. Landen, Jr. and R. R. Eitenmiller, *J. Assoc. Offic. Anal. Chem.*, *62*:283 (1979).
61. W. O. Landen, Jr., *J. Assoc. Offic. Anal. Chem.*, *63*:131 (1980).
62. W. O. Landen, Jr., *J. Assoc. Offic. Anal. Chem.*, *65*:810 (1982).
63. W. O. Landen, Jr., *J. Assoc. Offic. Anal. Chem.*, *68*:183 (1985).
64. R. V. Vivilecchia, B. G. Lightbody, N. Z. Thimot, and H. M. Quinn, *J. Chromatogr. Sci.*, *15*:424 (1977).
65. D. B. Parrish, *CRC Crit. Rev. Food Sci. Nutr.*, *9*:375 (1977).
66. G. Brubacher, W. Müller-Mulot, and D. A. T. Southgate (eds.), *Methods for the Determination of Vitamins in Food. Recommended by COST 91*, Elsevier, London, 1985.
67. J. N. Thompson, *J. Assoc. Offic. Anal. Chem.*, *69*:727 (1986).
68. M. Zahar and D. E. Smith, *J. Dairy Sci.*, *73*:3402 (1990).
69. J. N. Thompson, G. Hatina, and W. B. Maxwell, *J. Assoc. Offic. Anal. Chem.*, *63*:894 (1980).
70. T. W. Goodwin and G. Britton, in *Plant Pigments* (T. W. Goodwin, ed.), Academic Press, London, 1988, p. 61.
71. M. Kimura, D. B. Rodriguez-Amaya, and H. T. Godoy, *Food Chem.*, *35*:187 (1990).
72. F. Khachik and G. R. Beecher, *J. Agric. Food Chem.*, *35*:732 (1987).
73. R. Marsili and D. Callahan, *J. Chromatogr. Sci.*, *31*:422 (1993).
74. F. W. Quackenbush and R. L. Smallidge, *J. Assoc. Offic. Anal. Chem.*, *69*:767 (1986).
75. J. F. Fisher and R. L. Rouseff, *J. Agric. Food Chem.*, *34*:985 (1986).
76. K. H. Hanewald, F. J. Mulder, and K. J. Keuning, *J. Pharm. Sci.*, *57*:1308 (1968).
77. J. N. Thompson, W. B. Maxwell, and M. L'Abbé, *J. Assoc. Offic. Anal. Chem.*, *60*:998 (1977).

78. H. Indyk and D. C. Woollard, *N.Z. J. Dairy Sci. Technol.*, *19*:19 (1984).
79. J. N. Thompson, G. Hatina, W. B. Maxwell, and S. Duval, *J. Assoc. Offic. Anal. Chem.*, *65*:624 (1982).
80. J. F. Muniz, C. T. Wehr, and H. M. Wehr, *J. Assoc. Offic, Anal. Chem.*, *65*:791 (1982).
81. A. F. Wickroski and L. A. McLean, *J. Assoc. Offic. Anal. Chem.*, *67*:62 (1984).
82. J. N. Thompson and L. Plouffe, *Food Chem.*, *46*:313 (1993).
83. M. G. Sliva, A. E. Green, J. K. Sanders, J. R. Euber, and J. R. Saucerman, *J. Assoc. Offic. Anal. Chem. Int.*, *75*:566 (1992).
84. S. L. Reynolds and H. J. Judd, *Analyst*, *109*:489 (1984).
85. H. Cohen and B. Wakeford, *J. Assoc. Offic. Anal. Chem.*, *63*: 1163 (1980).
86. C. K. Chow, H. H. Draper, and A. S. Csallany, *Anal. Biochem.*, *32*:81 (1969).
87. P. Taylor and P. Barnes, *Chem. Ind.*, *20*:722 (1981).
88. A. J. Speek, J. Schrijver, and W. H. P. Schreurs, *J. Food Sci.*, *50*:121 (1985).
89. C. G. Rammell and J. J. L. Hoogenboom, *J. Liq. Chromatogr.*, *8*:707 (1985).
90. E.-L. Syväoja, V. Piironen, P. Varo, P. Koivistoinen, and K. Salminen, *J. Am. Oil Chem. Soc.*, *63*:328 (1986).
91. M. K. Balz, E. Schulte, and H.-P. Thier, *Fat Sci. Technol.*, *95*:215 (1993).
92. I. D. Desai, in *Vitamin E: A Comprehensive Treatise* (L. J. Machlin, ed.), Marcel Dekker, New York, 1980, p. 67.
93. W. D. Pocklington and A. Dieffenbacher, *Pure Appl. Chem.*, *60*:877 (1988).
94. T. Ueda and O. Igarashi, *J. Micronutr. Anal.*, *3*:15 (1987).
95. M. J. Shearer, in *Encyclopaedia of Food Science, Food Technology and Nutrition*, Vol. 7 (R. Macrae, R. K. Robinson, and M. Sadler, eds.), Academic Press London, 1993, p. 4804.
96. S. L. Booth, K. W. Davidson, and J. A. Sadowski, *J. Agric. Food Chem.*, *42*:295 (1994).
97. F. M. Rabel, *J. Chromatogr. Sci.*, *18*:394 (1980).
98. R. P. W. Scott, *J. Chromatogr. Sci.*, *23*: 233 (1985).
99. M. Mulholland, *Analyst*, *111*:601 (1986).
100. L. R. Snyder and J. J. Kirkland, *Introduction to Modern Liquid Chromatography*, 2nd ed., John Wiley and Sons, New York, 1979.
101. M. J. Shearer, *Adv. Chromatogr.*, *21*:243 (1983).
102. T. Kobayashi, T. Okano, and A. Takeuchi, *J. Micronutr. Anal.*, *2*:1 (1986).
103. F. Khachik, G. R. Beecher, J. T. Vanderslice, and G. Furrow, *Anal. Chem.*, *60*:807 (1988).
104. J. N. Thompson and S. Duval, *J. Micronutr. Anal.*, *6*:147 (1989).
105. B. Stancher and F. Zonta, *J. Chromatogr.*, *238*:217 (1982).
106. J. N. Thompson and W. B. Maxwell, *J. Assoc. Offic. Anal. Chem.*, *60*:766 (1977).
107. E.-S. Tee and C.-L. Lim, *Food Chem.*, *45*:289 (1992).
108. D. C. Egberg, J. C. Heroff, and R. H. Potter, *J. Agric. Food Chem.*, *25*:1127 (1977).
109. R. E. Lawn, J. R. Harris, and S. F. Johnson, *J. Sci. Food Agric.*, *34*:1039 (1983).
110. A. C. Ross, *Anal. Biochem.*, *115*:324 (1981).
111. D. C. Woollard and H. Indyk, *J. Micronutr. Anal.*, *2*:125 (1986).
112. R. J. Bushway, *J. Agric. Food Chem.*, *34*:409 (1986).
113. F. Granado, B. Olmedilla, I. Blanco, and E. Rojas-Hidalgo, *J. Agric. Food Chem.*, *40*:2135 (1992).
114. A. J. Speek, C. R. Temalilwa, and J. Schrijver, *Food Chem.*, *19*:65 (1986).
115. B. Stancher, F. Zonta, and P. Bogoni, *J. Micronutr. Anal.*, *3*:97 (1987).
116. E.-S. Tee and C.-L. Lim, *Food Chem.*, *41*:309 (1991).
117. E. De Ritter and A. E. Purcell, in *Carotenoids as Colorants and Vitamin A Precursors* (J. C. Bauernfeind, ed.), Academic Press, New York, 1981, p. 815.
118. W. O. Landen, Jr., R. R. Eitenmiller, and A. M. Soliman, *Proc. World Congr. Food Sci. Technol.*, Sept. 28 to Oct. 2, 1987, *Singapore*, Institute of Food Science and Technology, 1989, p. 43.
119. C. A. O'Neil, S. J. Schwartz, and G. L. Catignani, *J. Assoc. Offic. Anal. Chem.*, *74*:36 (1991).
120. H. J. C. F. Nelis and A. P. de Leenheer, *Anal. Chem.*, *55*:270 (1983).
121. N. E. Craft, *Meth. Enzymol.*, *213*:185 (1992).
122. R. J. Bushway, *J. Liq. Chromatogr.*, *8*:1527 (1985).
123. F. W. Quackenbush, *J. Liq. Chromatogr.*, *10*:643 (1987).
124. F. Khachik, G. R. Beecher, and W. R. Lusby, *J. Agric. Food Chem.*, *37*:1465 (1989).
125. M. H. Saleh and B. Tan, *J. Agric. Food Chem.*, *39*:1438 (1991).

126. E. Lesellier, C. Marty, C. Berset, and A. Tchapla, *J. High Resol. Chromatogr.*, *12*:447 (1989).
127. F. Granado, B. Olmedilla, I. Blanco, and E. Rojas-Hidalgo, *J. Liq. Chromatogr.*, *14*:2457 (1991).
128. D. W. Nierenberg and D. C. Lester, *J. Nutr. Growth Cancer*, *3*:215 (1986).
129. W. A. MacCrehan, *Meth. Enzymol.*, *189*:172 (1990).
130. K. J. Scott, *Food Chem.*, *45*:357 (1992).
131. B. E. Miller and A. W. Norman, in *Handbook of Vitamins: Nutritional, Biochemical and Clinical Aspects* (L. J. Machlin, ed.), Marcel Dekker, New York, 1984, p. 45.
132. H. Cohen and B. Wakeford, *J. Assoc. Offic. Anal. Chem.*, *63*:1163 (1980).
133. P. J. van Niekerk and S. C. C. Smit, *J. Am. Oil Chem. Soc.*, *57*:417 (1980).
134. H. Hasegawa, *J. Chromatogr.*, *605*:215 (1992).
135. E. J. deVries, J. Zeeman, R. J. E. Esser, B. Borsje, and F. J. Mulder, *J. Assoc. Offic. Anal. Chem.*, *62*:129 (1979).
136. G. Jones and H. F. DeLuca, *J. Lipid Res.*, *16*:448 (1975).
137. D. C. Sertl and B. E. Molitor, *J. Assoc. Offic. Anal. Chem.*, *68*:177 (1985).
138. E. J. de Vries and B. Borsje, *J. Assoc. Offic. Anal. Chem.*, *65*:1228 (1982).
139. V. K. Agarwal, *J. Assoc. Offic. Anal. Chem.*, *72*:1007 (1989).
140. H. Indyk and D. C. Woollard, *N.Z. J. Dairy Sci. Technol.*, *20*:19 (1985).
141. H. Indyk and D. C. Woollard, *J. Micronutr. Anal.*, *1*:121 (1985).
142. A. Kurmann and H. Indyk, *Food Chem.*, *50*:75 (1994).
143. S. A. Renken and J. J. Warthesen, *J. Food Sci.*, *58*:552 (1993).
144. M. Rychener and P. Walter, *Mitt. Gebiete Lebensm. Hyg.*, *76*:112 (1985) (in German).
145. H. Johnsson, B. Halén, H. Hessel, A. Nyman, and K. Thorzell, *Int. J. Vitam. Nutr. Res.*, *59*:262 (1989).
146. E. J. M. Konings, *Neth. Milk Dairy J.*, *48*:31 (1994).
147. P. Mattila, V. Piironen, C. Bäckman, A. Asunmaa, E. Uusi-Rauva, and P. Koivistoinen, *J. Food Comp. Anal.*, *5*:281 (1992).
148. J. N. Thompson and G. Hatina, *J. Liq. Chromatogr.*, *2*:327 (1979).
149. S. Tuan, T. F. Lee, C. C. Chou, and Q. K. Wei, *J. Micronutr. Anal.*, *6*:35 (1989).
150. H. E. Indyk, *Analyst*, *113*:1217 (1988).
151. D. C. Woollard and A. D. Blott, *J. Micronutr. Anal.*, *2*:97 (1986).
152. U. Manz and K. Philipp, *Int. J. Vitam. Nutr. Res.*, *51*:342 (1981).
153. F. Zonta and B. Stancher, *Riv. Ital. Sostanze Grasse*, *60*:195 (1983).
154. V. Piironen, E.-L. Syväoja, P. Varo, K. Salminen, and P. Koivistoinen, *J. Agric. Food Chem.*, *33*:1215 (1985).
155. E.-L. Syväoja, K. Salminen, V. Piironen, P. Varo, O. Kerojoki, and P. Koivistoinen, *J. Am. Oil Chem. Soc.*, *62*:1245 (1985).
156. E.-L. Syväoja, V. Piironen, P. Varo, P. Koivistoinen, and K. Salminen, *Int. J. Vitam. Nutr. Res.*, *55*:159 (1985).
157. V. Piironen, E.-L. Syväoja, P. Varo, K. Salminen, and P. Koivistoinen, *Cereal Chem.*, *63*:78 (1986).
158. T.-S. Shin and J. S. Godber, *J. Am. Oil Chem. Soc.*, *70*:1289 (1993).
159. C. G. Rammell, B. Cunliffe, and A. J. Kieboom, *J. Liq. Chromatogr.*, *6*:1123 (1983).
160. J. N. Thompson, P. Erdody, and W. B. Maxwell, *Anal. Biochem.*, *50*:267 (1972).
161. D. C. Woollard, A. D. Blott, and H. Indyk, *J. Micronutr. Anal.*, *3*:1 (1987).
162. J. N. Thompson, in *Trace Analysis*, Vol. 2 (J. F. Lawrence, ed.), Academic Press, 1982, p. 1.
163. L. J. Hatam and H. J. Kayden, *J. Lipid Res.*, *20*:639 (1979).
164. P. Schudel, H. Mayer, and O. Isler, in *The Vitamins: Chemistry, Physiology, Pathology, Methods*, 2nd ed., Vol. 5 (W. H. Sebrell, Jr., and R. S. Harris, eds.), Academic Press, New York, 1972, p. 168.
165. V. Piironen, P. Varo, E.-L. Syväoja, K. Salminen, and P. Koivistoinen, *Int. J. Vitam. Nutr. Res.*, *54*:35 (1984).
166. T. Ueda and O. Igarashi, *J. Micronutr. Anal.*, *1*:31 (1985).
167. O. Hiroshima, S. Ikenoya, M. Ohmae, and K. Kawabe, *Chem. Pharm. Bull.*, *29*:451 (1981).
168. B. K. Ayi and A. M. Burgher, in *Production, Regulation, and Analysis of Infant Formula: A Topical Conference*, May 14–16, 1985, Virginia Beach, VA, Association of Official Analytical Chemists, VA, 1985, p. 83.
169. J. P. Langenberg, U. R. Tjaden, E. M. de Vogel, and D. Is. Langerak, *Acta Aliment.*, *15*:187 (1986).
170. H. Indyk, *J. Micronutr. Anal.*, *4*:61 (1988).

171. S.-M. Hwang, *J. Assoc. Offic. Anal. Chem.*, *68*:684 (1985).
172. M. P. Bueno and M. C. Villalobos, *J. Assoc. Offic. Anal. Chem.*, *66*:1063 (1983).
173. F. Zonta and B. Stancher, *J. Chromatogr.*, *329*:257 (1985).
174. B. Fournier, L. Sann, M. Guillaumont, and M. Leclercq, *Am. J. Clin. Nutr.*, *45*:551 (1987).
175. W. A. Widicus and J. R. Kirk, *J. Assoc. Offic. Anal. Chem.*, *62*:637 (1979).
176. M. M. Delgado Zamarreño, A. Sánchez Pérez, C. Gómez Pérez, and J. Hernández Méndez, *J. Chromatogr.*, *623*:69 (1992).
177. G. A. Spanos, H. Chen, and S. J. Schwartz, *J. Food Sci.*, *58*:817 (1993).
178. M. Saito, Y. Yamauchi, K. Inomata, and W. Kottkamp, *J. Chromatogr. Sci.*, *27*:79 (1989).
179. M. T. G. Hierro and G. Santa-María, *Food Chem.*, *45*:189 (1992).

18

Water-Soluble Vitamins

L. Faye Russell

Centre for Food and Animal Research, Ottawa, Ontario, Canada

I. INTRODUCTION

A. Criteria for Method Selection

This chapter is intended as a guide for the analyst in search of a method to quantitate one or more of the water-soluble vitamins in foods. It is not intended as a comprehensive review of all recently published methods in the field but rather is a summary of selected methods. The criteria applied to the selection of the methods reviewed in this chapter are as follows:

1. Selected methods must present results obtained from "real" food samples.
2. The selected methods must lend themselves to at least some degree of automation and unattended operation.
3. Selected methods must include some type of separation step in order to avoid potential interference by other food components.
4. Selected methods must present specific extraction as well as quantitation procedures.
5. Selected methods must provide some indication of reliability in order to provide the reader with a means of evaluating and comparing different analytical procedures.
6. Selected methods must not duplicate techniques summarized in recent review articles.
7. Microbiological methods for analysis of the water-soluble vitamins will not be discussed in the context of this chapter.

B. Chapter Organization

Each vitamin is treated individually and all vitamins are organized in a similar manner. Because nutritional regulations are specific to each jurisdiction, no attempt will be made to catalogue them beyond recognition that an increasingly nutrition-literate consuming public is demanding more and better nutritional information and labeling of their food. The need for reliable vitamin methodology is being fueled by the role of the vitamins in health promotion (e.g., the protective

CFAR Contribution No. 2301

effect of the "antioxidant" vitamins against chronic disease such as cancer) rather than just prevention of nutritional deficiencies.

The chemistry of each vitamin will be discussed only insofar as it impacts on its analysis. The biochemistry, metabolism, and biological functions of the vitamins will be summarized briefly with references to in-depth reviews where available.

The AOAC International "standard" or reference (chemical) methods will be discussed briefly because they are the most common yardsticks against which other methods are judged. References to more detailed discussions of these methods will be cited where applicable.

C. Extraction Techniques

As a general rule, it is very difficult to reliably extract and quantitate multiple vitamins or vitamers from complex food systems due to their diverse physical and chemical properties. The greatest challenge in vitamin analysis is usually their extraction from the food matrix. This is particularly true of naturally occurring vitamins, which are often bound to other food constituents such as proteins or carbohydrates. Extraction conditions need to be chosen with attention to the labile nature of the vitamins in order to release them from the food matrix without degradation or loss. The analyst is also cautioned against indiscriminate "mixing and matching" of extraction and quantitation methods, since the extraction conditions are known to affect subsequent separation and quantitation procedures.

D. HPLC Methods

For a variety of reasons, high-performance liquid chromatography (HPLC) has become the most popular technique for determining the water-soluble vitamins. Reverse phase HPLC is particularly well suited to the separation and quantitation of nonvolatile, water-soluble analytes such as the vitamins. HPLC methods also provide a means of separating the vitamins from the interfering artifacts that occur naturally in complex biological systems such as food. HPLC is readily automatable through the use of autosamplers and even robotics. Additional advantages of HPLC with respect to biological, microbiological, and manual chemical methods include increased speed of analysis; increased precision; reasonable accuracy; and increased specificity including simultaneous separation of multiple vitamins, vitamers, and metabolites (1).

The tables of HPLC methods have been organized by type of column and detector, and by analyte, since these are often the practical criteria which govern the analyst's choice of method. Similarly, the tables of flow injection analysis (FIA) methods are subdivided according to the type of chemistry involved.

E. Method Validation

One the biggest challenges in selecting a method for vitamin analysis is often determining the relative merits of seemingly diverse methodologies. At present there is no specific standard against which all methods are compared. Indicators of assay reliability for which the analyst should look include measures of precision, accuracy, reproducibility, recoveries from spiked food samples, linearity of calibration, limits of detection, comparisons with existing and recognized methods, results of collaborative or interlaboratory trials, and/or measures of peak purity (2,3).

The analyst is cautioned against application of a method to a food substrate for which it has not been validated. The food substrate can have a very significant effect on the type of extraction and handling required to achieve complete extraction and prevent vitamin degradation and loss. If it becomes necessary to apply an existing method to a new food substrate, a full validation of the method for the new substrate should be conducted.

II. VITAMIN C

A. Chemistry and Biochemistry

Vitamin C is widely known for its role in the prevention of scurvy (4–6). However, the precise extent and mechanisms of its biological activity are still being investigated. Based on the ease with which the C vitamers can transfer electrons and act as a redox couple, vitamin C is now thought to participate in biological processes as diverse as digestion and absorption, endocrinology, immunology, oncology, collagen formation, cataract prevention, and detoxification. More detailed information on the biochemistry of vitamin C is available in two recently published reviews (4,5).

Fruit and vegetables are the most common food sources of vitamin C. Vitamin C (Fig. 1) exists naturally as two equally biologically active vitamers, L-ascorbic acid (AA) and dehydroascorbic acid (DHAA) (4,6–9). These vitamers are often bound to protein in natural products. Although isoascorbic acid (IAA) is often added to food as an antioxidant, this form has only 5% of the vitamin activity of AA. Its presence in food may therefore result in overestimation of the total vitamin C content. In addition, trace amounts of its oxidation product, dehydroisoascorbic acid (DHIAA), may be detected in food.

Vitamin C's most important chemical properties from an analytical perspective relate directly to its marked lack of stability. As one of the most labile of the vitamins, it is often used as an indicator of overall vitamin stability in foods. AA and DHAA in solution are readily oxidized, especially upon exposure to elevated temperatures, metal ions, dissolved oxygen, alkaline pH, light or naturally occurring degradative enzymes (4,6–9). While the oxidation of AA to DHAA can be reversed, that of DHAA is irreversible and produces biologically inactive products. As a result, care must be exercised throughout extraction and analysis to prevent degradation and loss of vitamin C.

For detection purposes, AA and DHAA are both UV absorbers (4,8,10). The absorption maximum of DHAA, 210–230 nm, makes it particularly susceptible to interferences from a number of endogenous food components and limits the choice of analytical reagents and solvents. The UV absorption maximum of AA varies from 245 nm to 265 nm depending on the pH of the sample extract. Neither AA nor DHAA is naturally fluorescent. AA can be detected electrochemically based on its reducing capacity. In contrast, DHAA is electrochemically inactive. Conse-

L-Ascorbic Acid   L-Dehydroascorbic Acid

Fig. 1 Vitamin C.

quently, chemical derivatization is often necessary to achieve the sensitivity needed to detect the endogenous vitamin C in food.

B. Methods of Analysis

1. Extraction Techniques

Although the specifics of extraction procedures may vary, they generally adhere to the same basic principles needed to prevent vitamin degradation and loss during extraction and analysis. The most popular extraction reagents are acids, which protect the C vitamers from oxidation and hydrolysis, as well as precipitating protein (4,8,10). Most extractants include metaphosphoric, trichloroacetic, or oxalic acid alone or in combination with other acids or short-chain alcohols, such as methanol or ethanol. Antioxidants and metal chelators such as ethylenediaminetetraacetic acid (EDTA) or diethylenetriaminepentaacetic acid may be added to extractants to protect against oxidation promoted by metal cations (e.g., iron or copper). Additional precautions include purging sample solutions with inert gases to reduce exposure to dissolved oxygen, conducting the extraction at reduced temperatures, and limiting the exposure of samples to white light.

2. Methodology Reviews

Golubkina and Prudnik (8) reviewed the extraction and quantitation of total vitamin C and its individual vitamers in foods. This included a discussion of colorimetric, fluorometric, enzymatic, and chromatographic methods, as well as the use of indicator paper. Polesello and Rizzolo (11,12) prepared comprehensive reviews of HPLC methods published between 1981 and 1989 for quantitating the various C vitamers in a variety of foods. In addition to the chromatographic conditions, they also summarized the extraction conditions and the types of food tested. A similar review by Lambert and De Leenheer (13) covered selected references published in the decade 1980–1990. Washko et al. (10) prepared a review of vitamin C analyses in foods and biological samples, including a valuable discussion of the principles, advantages, and pitfalls of various extraction protocols. Colorimetric/spectrophotometric, fluorometric, gas chromatographic (GC), HPLC, and enzymatic analyses of vitamin C were also covered. This review included a synopsis of factors to consider when selecting an assay for vitamin C. Parviainen and Nyyssönen (4) prepared a review of selected methods for quantitating the C vitamers in foods, biological tissues, and pharmaceuticals, including paper chromatography, thin-layer chromatography (TLC), HPLC, and GC applications. However, little attention was devoted to extraction methods. Velíšek and Davídek (14) reviewed the GC methods for quantitating the C vitamers in foods.

3. Reference Methods

For general food use, the AOAC International currently recommends a semiautomated (segmented flow injection) total vitamin C determination based on oxidation of the AA to DHAA, derivatization of the DHAA with o-phenylenediamine, and fluorescence detection of the resulting quinoxaline derivative (15). The AA content can be determined by difference if a second analysis is conducted, omitting oxidation of the AA in the sample. A similar manual method is recognized for use with vitamin preparations. For determination of AA in juices or infant formula, the AOAC International lists two titrimetric methods based on the reduction of 2,6-dichloroindophenol indicator dye by AA.

   These methods generally require less specialized equipment or training, but they are labor-intensive and time consuming to conduct. All of these methods are susceptible to interferences from other compounds in food samples due to the lack of a separation step.

4.   Recent Developments in Methodology

a.   *HPLC Methods.*   HPLC is the most popular alternative to the standard methods. New HPLC methods continue to be developed for the analysis of total vitamin C and its individual vitamers. Tables 1–6 summarize selected HPLC methods for determining vitamin C that were published between 1990 and mid-1994. For methods published prior to this, refer to Sec. II.B.2, "*Methodology Reviews.*"

b.   *Flow Injection Analysis.*   Fruit juice appears to be a simple enough matrix that its vitamin C content can be reliably determined without first separating the vitamin from interfering artifacts. This may be due in part to processing or fortification, which results in the vitamin C content of juice occurring in the free (unbound) form, in relatively large concentrations, and potentially as a single vitamer. However, the decision to assay a single vitamer by any technique should be weighed carefully in light of the findings of Hoare et al. (16) that the vitamin C content of orange juice was consistently underestimated when the DHAA content was ignored. Tables 7 and 8 summarize FIA methods developed since 1990 for use with fruit juice.

III.   THIAMINE (VITAMIN B₁)

A.   Chemistry and Biochemistry

Historically, thiamine has been known as the vitamin whose deficiency produces beriberi (6,38). Although its effects on nerve function and its antipolyneuritic properties have been recognized for years, the exact means by which thiamine exerts its neurological function is still under investigation. Kawasaki (38) recently reviewed the absorption, metabolism, and physiological function of thiamine.

Food sources of thiamine include meat, fish, whole cereal grains, fortified cereal and bakery products, nuts, legumes, eggs, yeast, fruits, and vegetables (6). In nature vitamin B_1 exits as the nonphosphorylated vitamer and its phosphorylated esters (13,38). Thiamine, the nonphosphorylated vitamer, is the predominant form in plant products. The phosphorylated forms include thiamine monophosphate (TMP), thiamine pyrophosphate (TPP), and thiamine triphosphate (TTP), and are bound to protein (Fig. 2). The diphosphate ester TPP is generally regarded as the physiologically active form and functions as a coenzyme. It is the predominant vitamer in animal products. Total thiamine is the sum of thiamine plus its phosphorylated esters.

The thiamine esters are all relatively stable in the dried state if stored at low temperature in the dark (38). In solution thiamine is stable at pH 2–4, and TPP is stable at pH 2–6 at low temperature. All of the thiamine vitamers are unstable at elevated temperatures and under alkaline conditions.

Detection of thiamine can be problematic due to its relatively low concentrations. To improve the sensitivity of detection, the thiamine vitamers are generally converted to their thiochrome derivatives, which are fluorescent at pH >8 (13,38). The thiochrome derivatives of the thiamine vitamers are relatively stable at room temperature and pH >9, and all exhibit similar fluorescence spectra.

B.   Methods of Analysis

1.   Extraction Techniques

The thiamine vitamers are generally determined as their thiochrome esters (13,38). Before the vitamers can be derivatized they must be released from the food matrix, usually by acid hydrolysis, e.g., using 0.1 N hydrochloric or sulfuric acid at 100°C for 10–30 min. If total thiamine is to be determined, the phosphate esters are subjected to enzymatic hydrolysis using takadiastase and/or

Table 1 HPLC Methods for Quantitating C Vitamers in Foods (C18 columns, UV Absorbance Detection)

Analyte(s)	Type of food analyzed	Extraction conditions	HPLC columns[a]	HPLC mobile phase and flow rate	Detection conditions	Method verification[b]	Ref.
AA, DHAA, & IAA simultaneously.	Oranges; orange juice; beer; kiwi; tomato; camu-camu.	Extraction with methanol + water (5 + 95 v/v); addition of IAA as internal standard when absent from sample; solid phase extraction/ cleanup with C18 Sep-Pak (Waters); derivatization of DHAA with 1,2-phenylenediamine hydrochloride.	*Precolumn:* C18 (Waters). *Analytical:* μBondapak C18 (300 × 3.9 mm, 10 μm; Waters).	*Isocratic:* 5 mM cetrimide in methanol + 50 mM phosphate buffer, pH 4.59 (5 + 95 v/v). 1.8 ml/min.	*AA & IAA:* UV absorbance, 261 nm *DHAA (derivatized):* UV absorbance, 348 nm.	Internal standardization when IAA absent from sample, otherwise external standard-ization. Linear range = 0–20 mg/100 ml. Reproducibility—CV ± 0.80–1.85% for DHAA standards, ± 0.59–1.67% for AA standards, ± 0.49–1.74% for IAA standards (n = 6).	17
AA & DHAA simultaneously.	Citrus fruit juice; fresh fruit and vegetables (caramondin, mango, papaya, sweet pepper, tomato).	Extraction with 0.05 N phosphate buffer; solid phase extraction/cleanup using C18 Sep-Pak (Waters).	*Analytical:* Spheri-5 RP-18 (110 × 4.6 mm, 5 μm; Brownlee Labs) plus two Polypore H (110 × 4.6 mm + 220 × 4.6 mm; Brown-lee Labs) in series.	*Isocratic:* 2% potassium phosphate buffer, pH 2.3. 0.4 ml/min.	*AA:* UV absorbance at 260 nm *DHAA:* differential UV absorbance spectrophotometry at 215 and 260 nm.	External standardization. Reproducibility—CV ± 1.5% for AA, ± 8.3% for DHAA (n = 6) using grapefruit. Recoveries = 91% for AA, 97% for DHAA from juice.	18

Application	Sample	Extraction	Column[a]	Mobile phase	Detection	Comments[b]	Ref.
AA & total vitamin C as AA separately. DHAA by difference.	Citrus juice.	AA: Extraction with ethanol + metaphosphoric acid (50 + 50 v/v); Total AA: Extraction with ethanol + metaphosphoric acid (50 + 50 v/v); reduction of DHAA to AA with 1.2% sodium hydrosulfide.	Analytical: Cosmosil 5C18 (150 × 4.6 mm; Nacalai Tesque) at 40°C.	Isocratic: 0.2% (w/v) metaphosphoric acid. 1.0 ml/min.	UV absorbance, 243 nm.	External standardization. Linear range = 0–100 mg/L, r = 0.9999. Reproducibility—CV ± 3.0–5.9 g/L for AA & DHAA in mandarin and lemon juice (n = 7). Recoveries = 99.3–100.2% for AA, 99.6–101.0% for DHAA from 4 types of citrus juice.	19
AA	Babaco (Carica pentagona Heil).	Aqueous extraction.	Analytical: Lichrosorb-NH2 (250 × 4.6 mm, 5 µm).	Isocratic: acetonitrile + 5 mM phosphate buffer, pH 3.5 (40 + 60 v/v). 1 ml/min.	UV absorbance, 268 nm.	External standardization. Linear range = 20–80 mg/L. Reproducibility—CV ± 5.14% for AA in babaco. LoD = 4.4 mg/L. Recoveries = 65.9% (mean) from babaco.	20
AA	Babaco; berries; currants; feijoa; passion fruit; tamarillos; medlar; persimmon.	Extraction with metaphosphoric acid + acetic acid.	Precolumn: Newguard RP-18 (Brownlee Labs). Analytical: Spherisorb ODS (250 × 4.6 mm, 5 µm).	Isocratic: water acidified with sulfuric acid to pH 2.2. 0.4 ml/min.	UV absorbance, 254 nm.	External standardization. Linear calibration, r = 0.9995. Reproducibility—CV ± 1.7% for AA in babaco (n = 10). LoD = 0.8 mg/L. Recoveries = 94.8% (mean) from babaco.	20 21
AA	Margarine; butter.	Two procedures reported to be equally effective, based on extraction into warm water or hexane.	Analytical: Spherisorb-ODS (150 × 4.6 mm, 3 µm; Tracer Analytical).	Isocratic: water acidified with sulfuric acid to pH 1.95. 0.7 ml/min.	UV absorbance, 254 nm.	External standardization. Linear range = 5–100 µg/ml, r > 0.999. Recoveries = 96–101% (n = 6).	22

[a]Column specifications expressed as (length × id, particle size; manufacturer).
[b]r = correlation coefficient; n = number of determinations. LoD = limit of detection; SNR = signal-to-noise ratio; CV = coefficient of variation.

Table 2 HPLC Methods for Quantitating Total Vitamin C in Foods (C18 Columns, Fluorescence Detection)

Analyte(s)	Type of food analyzed	Extraction conditions	HPLC columns[a]	HPLC mobile phase and flow rate	Detection conditions	Method verification[b]	Ref.
Total vitamin C as DHAA	Canned vegetables; potato chips; fruit juice; infant formula.	Extraction with meta-phosphoric acid + acetic acid; oxidation of AA to DHAA using Norit; derivatization of total vitamin C as DHAA with *o*-phenyl-enediamine.	*Analytical* Bondapak C18 (300 × 3.9 mm; Waters).	*Isocratic:* methanol +water (55 + 45, v/v). 1.0 ml/min.	Fluorescence, 350/430 nm (ex/em).	External standardization. Linear range = 0.138–3.000 μg/ml. Reproducibility—*CV* ± 1.67% in infant formula, ± 1.08% in canned corn (*n* = 5). LoD = 10 ng/100 μl injection. Comparable results from HPLC and AOAC International microfluorometric method for all foods tested. Recoveries = 91–108%.	23

[a]Column specifications expressed as (length × id, particle size; manufacturer).
[b]*r* = correlation coefficient; *n* = number of determinations. LoD = limit of detection; SNR = signal-to-noise ratio; *CV* = coefficient of variation.

Table 3 HPLC Methods for Quantitating C Vitamers in Foods (C18 Columns, Electrochemical Detection)

Analyte(s)	Type of food analyzed	Extraction conditions	HPLC columns[a]	HPLC mobile phase and flow rate	Detection conditions	Method verification[b]	Ref.
AA & IAA simultaneously	Yeast.	Extraction with metaphosphoric acid + cold perchloric acid.	*Precolumn:* ODS-10 (40 × 2.6 mm; BioRad). *Analytical:* Spherisorb ODS-2 (250 × 4.6 mm, 5 µm; Rainin) at 35°C.	*Isocratic:* methanol + 0.08 M acetate buffer, pH 4.2 containing 0.1 mM EDTA and 1.0 mM tetrabutylammonium phosphate (5 + 95 v/v).	Electrochemistry, +0.72 V vs. Ag/AgCl reference electrode, glassy carbon working electrode.	External standardization. Linear range = 0–60 µg/g yeast (dry wt). Reproducibility—CV ± 2.3% for IAA, ± 1.2% for AA. Recoveries quantitative for both vitamers.	24
AA & IAA simultaneously, total AA & total IAA simultaneously. DHAA and DHIAA determined by difference.	Processed meats; ground beef; applesauce; baby food (peaches); baby food (rice cereal).	*AA + IAA:* extraction with cold metaphosphoric acid. *Total AA + total IAA:* extraction with cold metaphosphoric acid; reduction of DHAA & DHIAA to AA and IAA with homocysteine.	*Precolumn:* Spheri-5-RP-18 (30 × 4.6 mm; Brownlee). *Analytical:* Three Supelcosil LC-18-DB (250 × 4.6 mm, 5 µm; Supelco) in series.	*Isocratic:* 0.08 M acetate buffer, pH 5.4 containing 5 mM tetrabutylammonium hydrogen sulfate and 0.15% metaphosphoric acid 0.4 ml/min.	Amperometry, + 0.6 V (oxidative) vs. Ag/AgCl reference electrode, glassy carbon working electrode.	External standardization. Nonlinear calibration curve, range = 0.5–2.5 ng. LoD = 0.5 ng at SNR > 2. Reproducibility—CV ± 1.9% for AA, ± 2.0% for IAA. Recoveries = 94–107% for AA, 94–110% for IAA.	25

[a]Column specifications expressed as (length × id, particle size; manufacturer).
[b]r = correlation coefficient; n = number of determinations. LoD = limit of detection; SNR = signal-to-noise ratio; CV = coefficient of variation.

Table 4 HPLC Methods for Quantitating C Vitamers in Foods (Polymer Columns, Fluorescence Detection)

Analyte(s)	Type of food analyzed	Extraction conditions	HPLC columns[a]	HPLC mobile phase and flow rate	Detection conditions	Method verification[b]	Ref.
AA, DHAA, IAA, & DHIAA simultaneously; total vitamin C by addition (AA + DHAA).	Fresh and processed vegetables and fruit; orange juice; ready-to-eat breakfast cereals; pizza; processed meat.	Extraction with metaphosphoric acid + acetic acid containing EDTA; addition of IAA (or AA) as internal standard when absent from sample; clean-up as required with hexane to remove lipid material or *n*-butanol to remove starch.	*Precolumn:* PLRP-S guard cartridge (Polymer Laboratories). *Analytical:* two PLRP-S in series (150 × 4.6 mm + 250 × 4.6 mm, 5 µm, 100 Å pore; Polymer Laboratories) at 4°C.	*Isocratic:* 0.2 M phosphate buffer, pH 2.14. 0.8 ml/min	On-line postcolumn reactions: oxidation of AA and IAA to DHAA and DHIAA with CuCl₂; derivatization of DHAA and DHIAA with *o*-phenylenediamine. Fluorescence detection at 350/430 nm (ex/em).	Internal standardization with IAA or AA if absent from sample. External standardization if both AA and IAA are present in food. Linear range =3–2000 ng. Recoveries = 80–113% for total vitamin C (AA + DHAA) in a variety of food samples. Good agreement with results of AOAC International reference method.	26 27 28
AA & DHAA simultaneously.	Dairy products.	Extraction with oxalic acid + perchloric acid; derivatization of DHAA with 4-ethoxy-1,2-phenylenediamine; solid phase extraction/clean-up with C18 and cation exchange columns in series; oxidation of AA to DHAA with bromine; derivatization of DHAA with 4-methoxy-1,2-phenylenediamine; solid phase extraction/cleanup with C18 and cation exchange columns in series.	*Analytical:* PRP-1 (150 × 4.1 mm, 5 µm; Hamilton).	*Isocratic:* acetonitrile + 5 mM 1-propanesulfonate in 200 mM phosphate buffer, pH 9 (16+ 84 v/v). 1 ml/min.	Fluorescence, 375/475 nm (ex/em).	LoD = 50 fmol AA, 70 fmol DHAA/5 µl injection at SNR = 3. Reproducibility—CV ± 11% for AA ± 9% for DHAA in cheeses (*n* = 6). Recoveries = 103–119% for AA, 89–91% for DHAA in milk (*n* = 5).	29

[a]Column specifications expressed as (length × id, particle size; manufacturer).
[b]r = correlation coefficient; n = number of determinations. LoD = limit of detection; SNR = signal-to-noise ratio; CV = coefficient of variation.

Table 5 HPLC Methods for Quantitating Total Vitamin C in Foods (Polymer Columns, Fluorescence Detection)

Analyte(s)	Type of food analyzed	Extraction conditions	HPLC columns[a]	HPLC mobile phase and flow rate	Detection conditions	Method verification[b]	Ref.
Total AA & total IAA simultaneously.	Fresh foods (cereals, pulses, nuts, fruit, vegetables, fish, meat, eggs, milk); processed foods (beverages, cereals, fruit, vegetables, fish, meat, milk).	Oxidation of AA and IAA to DHAA and DHIAA by indo-phenol; DHAA and DHIAA derivatized with *o*-phenylene-diamine; solid phase extraction/cleanup with C18 Sep-Pak (Waters).	*Precolumn:* PLRP-S (5 × 3 mm; Polymer Laboratories). *Analytical:* PLRP-S (150 × 4.6 mm, 5 μm, 100 Å pore; Polymer Laboratories).	*Isocratic:* methanol + 80 mM phosphate buffer, pH 7.8 (7 + 3 v/v). 0.8 ml/min.	Fluorescence, 355/425 nm (ex/em).	External standardization. LoD = 0.125 μg/g at SNR = 3. Reproducibility— CV ± 5.2% for total vitamin C in cucumber (*n* = 7). Recoveries = 70–110% for AA, 72–101% for IAA from a variety of fresh and processed foods (*n* = 3).	30

[a]Column specifications expressed as (length × id, particle size; manufacturer).
[b]*r* = correlation coefficient; *n* = number of determinations. LoD = limit of detection; SNR = signal-to-noise ratio; *CV* = coefficient of variation.

Table 6 HPLC Methods for Quantitating Individual C Vitamers in Foods (Ion Exchange, Ion Exclusion, and Amino Columns; UV Absorbance Detection)

Analyte(s)	Type of food analyzed	Extraction conditions	HPLC columns[a]	HPLC mobile phase and flow rate	Detection conditions	Method verification[b]	Ref.
AA & total AA separately. DHAA determined by difference.	Potatoes; strawberries.	AA: Extraction with metaphosphoric acid. DHAA: Extraction with metaphosphoric acid; reduction of DHAA to AA using 30 mM homocysteine.	Precolumn: MicroGuard cation H+. Analytical: Aminex HPX-87H (300 × 7.8 mm, 9 μm; Bio-Rad).	Isocratic: 4.5 mM sulfuric acid. 0.5 ml/min.	UV absorbance, 245 nm.	External standardization. Linear range = 1–20 ng/μl, r = 1.00. LoD = 1 ng/μl using a 10-μl injection. Reproducibility—CV ±8.40% for AA, ± 7.66% for total AA in strawberries (n = 4). Recoveries—96–99% for AA in potatoes and strawberries.	31
AA & total AA separately. DHAA determined by difference.	Orange juice.	AA: Extraction with methanol + citric acid + EDTA. Total AA: Reduction of DHAA to AA with 0.8% homocysteine; extraction with methanol + citric acid + EDTA.	Analytical: Zorbax NH2 (250 × 4.6 mm, 5 μm; Dupont).	Isocratic: methanol + 0.25% phosphate buffer, pH 3.5 (40 + 60 v/v). 1.0 ml/min.	UV absorbance, 244 nm.	External standardization. Linear range = 1–100 mg/L. Recoveries = 99% for AA from orange juice.	16

[a]Column specifications expressed as (length × id, particle size; manufacturer).
[b]r = correlation coefficient; n = number of determinations. LoD = limit of detection; SNR = signal-to-noise ratio; CV = coefficient of variation.

Table 7 Flow Injection Analysis Methods for Quantitating AA in Fruit Juice (Chemiluminescence-Based Chemistry)

Analyte(s)	Type of food analyzed	Extraction conditions	Flow injection chemistry	Detector	Method verification[a]	Ref.
AA	Orange juice; apple juice.	Aqueous extraction.	Ascorbate anion suppresses formation of the chemiluminescent product when luminol reacts with alkaline hydrogen peroxide in the presence of peroxidase.	Fluorescence.	External standardization. Linear range = 1–16 µM. Reproducibility—CV <2% for AA standard solution ($n = 4$). Good agreement with results of standard spectrophotometric method.	32
AA	Tomato juice; orange juice; lemon juice; banana juice.	Extraction with metaphosphoric acid.	AA reduces iron (III) ion to iron (II); iron (II) reacts with luminol in hydrogen peroxide to produce a chemiluminescent product.	Chemiluminescence, photomultiplier tube (PMT) with 400V power source. No wavelength selection applied.	External standardization. Linear range = 0.01–100 nmol/L. Reproducibility—CV ± 1.4% for AA standard solution ($n = 10$). LoD = 1 pmol/L or 30 amol per 30-µl injection at SNR = 2. Recoveries = 95–106% for AA from tomato, orange, lemon and banana juice. Potential interference from oxalate, sulfide, and sulfate anions (causing overestimation).	33

[a] r = correlation coefficient; n = number of determinations. LoD = limit of detection; SNR = signal-to-noise ratio; CV = coefficient of variation.

Table 8 Flow Injection Analysis Methods for Quantitating AA in Fruit Juice (Immobilized Ascorbate Oxidase–Based Chemistry)

Analyte(s)	Type of food analyzed	Extraction conditions	Flow injection chemistry	Detector	Method verification[a]	Ref.
AA	Lemon juice; orange juice; black currant drink; cucumber juice.	Perchloric acid extraction.	Requires two measurements per extract: *a.* Amperometric signal of AA + electrochemically active artifacts in sample. *b.* Oxidation of AA to electrochemically inactive DHAA by passage through Sephadex 4B column containing immobilized ascorbate oxidase. Decrease in amperometric signal (a–b) related to initial concentration of AA.	Amperometric, + 0.6 V, glassy carbon working electrode.	External standardization. Linear range = 0–400 ng/ml, r = 0.9994. LoD = 4.0 ng/ml (2σ) in phosphate buffer. Reproducibility—CV ± 1.0% for AA in standard solution (n = 10). Recoveries = 91–100% for AA from black currant and lemon juice.	34
Total vitamin C as AA	Orange juice; grapefruit juice; mashed potato powder; freeze-dried Brussels sprouts.	Extraction with 15 mM phosphate buffer, pH 5.0, containing 1 mM dithiothreitol.	Dithiothreitol in extraction solution reduces DHAA to AA. Requires two measurements per extract to determine total AA: *a.* Amperometric signal of AA + electrochemically active artifacts in sample; *b.* Oxidation of AA to electrochemically inactive DHAA by passage through packed bed reactor containing ascorbate oxidase immobilized on aminopropyl controlled-pore glass beads. Decrease in amperometric signal (a–b) related to initial concentration of AA.	Electrochemistry, + 850 mV vs. Ag/AgCl reference electrode, glassy carbon wall-jet working electrode.	External standardization, Linear range = 1–200 μg/ml, r = 0.995. Good agreement with results by HPLC-UV and AOAC International methods. Note: IAA will be measured as total vitamin C causing overestimation.	35
AA	Five varieties of tropical fruit juice.	Extraction with 0.1 M citrate-phosphate buffer, pH 7.0.	AA oxidized to DHAA by passage through packed bed reactor containing ascorbate oxidase immobilized on alkylamine glass beads; decrease in oxygen concentration resulting from AA oxidation related to AA content of extract.	Electrochemistry, potentiometric oxygen electrode (36).	External standardization. Linear range = 0.05–3.0 mM. Reproducibility—CV ± 1.0% for AA in standard solution (n = 10).	37

[a] r = correlation coefficient; n = number of determinations. LoD = limit of detection; SNR = signal-to-noise ratio; CV = coefficient of variation.

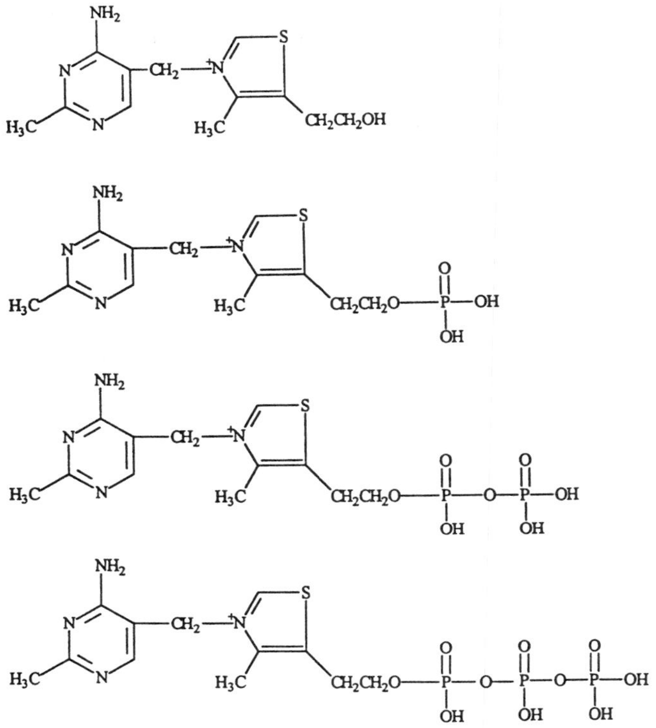

Fig. 2 Thiamine (vitamin B$_1$).

β-amylase. Typical hydrolysis conditions involve incubation at 37°C overnight or for 3 hr at 50°C. Excess starch or protein in the food matrix may also be removed by enzymatic hydrolysis using β-amylase or papain, respectively. Highly fluorescent thiochrome derivatives of the thiamine vitamers are formed by alkaline oxidation with cyanogen bromide or potassium ferricyanide. Potassium ferricyanide is usually the reagent of choice due to the toxicity of cyanogen bromide. If necessary, excess derivatizing reagents can be removed by extraction into isobutanol, or by solid phase extraction or column chromatography using C18 or strong cation exchange materials.

2. Methodology Reviews

Three recent reviews (11–13) provide comprehensive coverage of the HPLC methods published in the 1980s. These papers summarize the quantitation of the thiamine vitamers, and total thiamine alone and in conjunction with a number of other B vitamins, as well as extraction conditions, chromatographic conditions, detection conditions, and the types of food tested. Kawasaki (38) included paper chromatography, TLC, column chromatography, HPLC, and GC techniques in his review of selected methods for quantitating the thiamine vitamers in foods, biological tissues, and pharmaceuticals. However, extraction methods received little emphasis in this review. Velíšek and Davídek (14) reviewed the GC methods for thiamine analysis.

3. Reference Methods

All of the methods recommended by AOAC International are based on the formation of fluorescent thiochrome derivatives of thiamine (39). In samples likely to contain several thiamine vitamers, total thiamine is quantitated by dephosphorylating the thiamine esters with enzymatic hydrolysis prior to derivatization. Column chromatographic cleanup on cation exchange resin is used to purify the extracts and remove excess derivatizing reagents prior to fluorescence quantitation of thiochrome. A recent modification of the AOAC International method reported that the use of C18 instead of cation exchange material for the column chromatographic cleanup produced similar results in less analysis time (40).

4. Recent Developments in Methodology

a. HPLC Methods. The recent work on thiamine methodology is split between its determination alone and its concomitant determination with other B vitamins. Methods which simultaneously quantitate thiamine along with other B vitamins are reviewed in Sec. XI "Simultaneous Determinations of Multiple Vitamins." The HPLC methods developed for total thiamine and the thiamine vitamers exclusive of other vitamins are reviewed in Tables 9–12.

b. Flow Injection Analysis. An FIA method was recently developed for total thiamine in chicken (41) and is summarized in Table 13. The FIA results compared favorably with a similar HPLC method (41) in which the individual thiamine vitamers (thiamine, TMP, and TPP) were simultaneously separated using either a Shodex DM614 column or a Hypersil APS column (Chrompak B.V., Netherlands).

IV. RIBOFLAVIN (VITAMIN B₂)

A. Chemistry and Biochemistry

Although riboflavin deficiency is not associated with a specific deficiency syndrome per se, it produces pellagra-like symptoms including skin lesions around the mouth, nose, and ears (6,48, 49).

Endogenous vitamin B_2 occurs in foods in three principal forms: riboflavin, and its physiologically active coenzyme forms, flavin mononucleotide (FMN) and flavin adenine dinucleotide (FAD) (Fig. 3) (48,49). The vitamin B_2 coenzymes act as cofactors (covalently bound prosthetic groups) or (noncovalently bound) coenzymes for an array of respiratory enzymes. Examples of the physiological functions of the flavin coenzymes include dehydrogenation, hydroxylation, monooxygenation, and electron transfer. The biochemistry and physiology of the B_2 vitamers are discussed in greater detail in a recent review (48).

Food sources of vitamin B_2 include yeast, dairy products, meat, and eggs (2,48,49). The coenzymes are usually the predominant forms of vitamin B_2 in foods and are protein-bound. The notable exception is milk, in which the predominant vitamer is free riboflavin.

The single most important chemical characteristic affecting vitamine B_2 analysis is its photosensitivity. Degradation of vitamin B_2 by ultraviolet (UV) and visible light produces irreversible photoreduction to lumiflavin or lumichrome as well as loss of vitamin acitivity (49). The coenzyme vitamers are also subject to hydrolysis by endogenous phosphatases in a number of foods. These enzymes are generally inactivated by thermal processing and are only of concern in the analysis of fresh products. All of the B_2 vitamers are unstable at alkaline pH and the coenzymes are degraded below pH 5.0. Extraction and analysis of the coenzymes therefore needs to be done at pH 5.0–7.0, while total riboflavin analysis is best carried out at acidic pH.

For detection purposes, the B_2 vitamers have similar UV absorbance spectra and all fluoresce naturally (49). It is therefore feasible to detect the B_2 vitamers spectrophotometrically.

Table 9 HPLC Methods for Quantitating Total Thiamine in Foods (C18 Columns, Fluorescence Detection)

Analyte(s)	Type of food analyzed	Extraction conditions	HPLC columns[a]	HPLC mobile phase and flow rate	Detection conditions	Method verification[b]	Ref.
Total thiamine (thiamine + TMP).	Soy-based products (whole bean, okara, whey tofu).	Extraction by autoclaving at 20 psi for 15 min in hydrochloric acid at pH = 2.0; protein removal by precipitation at pH 4.5 and centrifugation; oxidation to thiochrome using 1% potassium ferricyanide in 15% sodium hydroxide; neutralization with conc. phosphoric acid.	*Precolumn:* RP-18 (3 cm; Brownlee). *Analytical:* Ultrasphere C18 (150 × 4.6 mm, 5 μm; Beckman).	A: acetonitrile B: 0.01 M acetate buffer, pH 5.5 C: water *Gradient:* - $A + B$ (13 + 87, v/v) at $t = 0$ min. 1.2 ml/min. - isocratic at $A + B$ (13 + 87, v/v) until $t = 6$ min. 1.2 ml/min. - step gradient to $A + C$ (90 + 10, v/v) at $t = 6$ min. 2.3 ml/min. - isocratic at $A + C$ (90 + 10, v/v) until $t = 11$ min. - step gradient to $A + B$ (13 + 87, v/v) at $t = 11$ min. 1.2 ml/min. - isocratic at $A + B$ (13 + 87, v/v) until $t = 18$ min. 1.2 ml/min.	Fluorescence, 364/436 nm (ex/em).	External standardization. Linear calibration, $r = 0.997$. LoD = 1.25 ng thiamine. Recoveries = 43–98% from soy products. Good agreement with results obtained by AOAC International method.	42

Table 9 (*Continued*)

Analyte(s)	Type of food analyzed	Extraction conditions	HPLC columns[a]	HPLC mobile phase and flow rate	Detection conditions	Method verification[b]	Ref.
Total thiamine.	Cheese; fortified ready-to-eat breakfast cereal; pork; potato; wheat flour.	Acid hydrolysis with 0.1 M hydrochloric acid; enzymatic hydrolysis with β-amylase and takadiastase; protein precipitation with trichloroacetic acid (cheese); oxidation with alkaline 0.25% potassium ferricyanide; sample cleanup with C18 solid phase extraction.	*Precolumn:* Not specified. *Analytical:* Novapak C18 (150 × 3.9 mm, 4 μm; Waters).	*Isocratic:* methanol + 50 mM phosphate buffer, pH 7.0 (30 + 70, v/v). 1.0 ml/min.	Fluorescence, 445/522 nm (ex/em).	External standardization. Linear range = 1.5–25 ng/10-μl injection. LoD = 10 pg or 30 fmol thiamine as thiochrome at SNR ≥ 2. Reproducibility—CV ± 3.4% for thiochrome. Recoveries = 90–105% TPP as total thiamine from cheese, cereal, pork, and flour.	43

[a] Column specifications expressed as (length × id, particle size; manufacturer).
[b] *r* = correlation coefficient; *n* = number of determinations. LoD = limit of detection; SNR = signal-to-noise ratio; *CV* = coefficient of variation.

Table 10 HPLC Methods for Quantitating Total Thiamine in Foods (Phenyl Columns, Fluorescence Detection)

Analyte(s)	Type of food analyzed	Extraction conditions	HPLC column[a]	HPLC mobile phase and flow rate	Detection conditions	Method verification[b]	Ref.
Total thiamine.	Milk.	Enzymatic hydrolysis of protein with trypsin and thiamine phosphate esters to thiamine with claradiastase; derivatization of thiamine to thiochrome using ferricyanide (derivatization stopped with sodium sulfite); thiochrome extracted with 1-butanol.	*Analytical:* Nucleosil Phenyl (150 mm, 5 μl).	*Isocratic:* methanol + aceto-nitrile + isobutanol + water (80 + 10 + 10 + 5 v/v/v/v). 0.7 ml/min.	Fluorescence, 375/430 nm (ex/em).	External standardization. Recoveries = 95% thia-mine as thiochrome from milk.	44

[a]Column specifications expressed as (length × id, particle size; manufacturer).
[b]r = correlation coefficient; n = number of determinations. LoD = limit of detection; SNR = signal-to-noise ratio; CV = coefficient of variation.

Table 11 HPLC Methods for Quantitating Total Thiamine in Foods (TMS Columns, Fluorescence Detection)

Analyte(s)	Type of food analyzed	Extraction conditions	HPLC columns[a]	HPLC mobile phase and flow rate	Detection conditions	Method verification[b]	Ref.
Total thiamine.	Rice.	Extraction by refluxing at 60°C for 30 min in 0.1 M hydrochloric acid + methanol (60 + 40 v/v); homogenization; centrifugation.	*Analytical:* Zorbax TMS (250 × 4.6 mm; DuPont) at 55°C.	*Isocratic:* phosphate-perchlorate buffer, pH 2.5. 0.4 ml/min.	On-line postcolumn derivatization with alkaline potassium ferricyanide to convert thiamine to thiochrome. Fluorescence, (375/435 nm ex/em).	External standardization. Linear range = 0.35 ng thiamine; $r = 0.999$. Reproducibility—$CV \pm$ 2.8% for total thiamine as thiochrome in 6 varieties of rice ($n = 6$). Recoveries = 94–101% thiamine as thiochrome from brown rice. Good agreement between results by HPLC and AOAC International methods ($r = 0.958$).	45

[a]Column specifications expressed as (length × id, particle size; manufacturer).
[b]r = correlation coefficient; n = number of determinations. LoD = limit of detection; SNR = signal-to-noise ratio; CV = coefficient of variation.

Table 12 HPLC Methods for Quantitating Total Thiamine in Foods (Amino Columns, Fluorescence Detection)

Analyte(s)	Tye of food analyzed	Extraction conditions	HPLC columns[a]	HPLC mobile phase and flow rate	Detection conditions	Method verification[b]	Ref.
Total thiamine	Fortified ready-to-eat break-fast cereal; cereal bars; dessert; yeast.	Extraction with hydrochloric-perchloric acid; hydrolysis of thiamine phosphate esters to thiamine with takadiastase; derivatization of thiamine to thiochrome using alkaline 1% HgCl2; extraction of thiochrome with isobutanol.	*Analytical:* Lichrosorb NH2 (250 × 4.6 mm, 5 μm; SFCC, France).	*Isocratic:* methanol + dichloromethane (10 + 90 v/v). 1.0 ml/min.	Fluorescence, 365/440 nm (ex/em).	External standardization. Linear range = 12–200 ng TPP/ml. Good agreement with results from micro-biological analyses.	46

[a]Column specifications expressed as (length × id, particle size; manufacturer).
[b]r = correlation coefficient; n = number of determinations. LoD = limit of detection; SNR = signal-to-noise ratio; CV = coefficient of variation.

Table 13 Flow Injection Analysis (FIA) Methods for Quantitating Total Thiamine in Foods (Fluorescence Detection)

Analyte(s)	Type of food analyzed	Extraction conditions	Flow injection chemistry	Detector	Method verification[a]	Ref.
Total thiamine	Chicken breast.	Extraction with 2% trichloroacetic acid at 100°C for 30 min.; centrifugation.	Derivatization of thiamine to thiochrome using alkaline 0.04% potassium ferricyanide.	Fluorescence, 365/440 nm (ex/em).	External standardization. Linear range = 0.1–10,000 ng/ml. Reproducibility—$CV \pm 1.52\%$ for total thiamine as thiochrome in chicken breast ($n = 5$). Recoveries = 99.7% thiamine as thiochrome from chicken breast.	41
Total thiamine	Pork; bread; rice; meat pie; milk; fortified breakfast cereal.	Extraction with 25% potassium chloride and 10% takadiastase in 0.02 M acetic acid at 70°C for 1 hr; centrifugation.	Oxidation of thiamine to thiochrome with alkaline ferricyanide; extraction from aqueous extract into isobutanol; thiochrome detected fluorometrically.	Fluorescence, 390/475 nm (ex/em).	External standardization. Recoveries = 98.3% from foods tested. Results agreed well with those obtained from AOAC International method ($r = 0.99$).	47

[a] r = correlation coefficient; n = number of determinations. LoD = limit of detection; SNR = signal-to-noise ratio; CV = coefficient of variation.

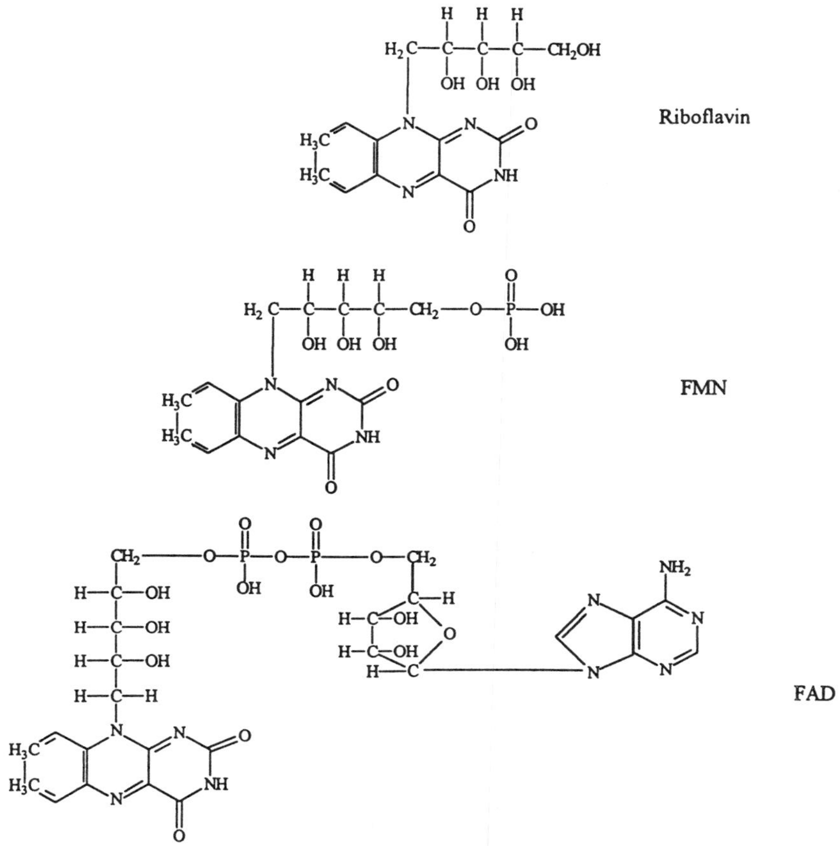

Fig. 3 Riboflavin (vitamin B$_2$).

B. Methods of Analysis

1. Extraction Techniques

Due to their photosensitivity, it is imperative that vitamin B$_2$ analysis be conducted under subdued light using low actinic glassware (13,48,49). For total riboflavin analysis it is necessary to release the coenzymes from the protein to which they are bound and to hydrolyze the coenzymes to riboflavin. This is usually accomplished using dilute mineral acids at 100–121°C for 30 min with or without additional enzyme hydrolysis of the coenzyme vitamers. Milder conditions are required for analyses in which information on the coenzyme vitamers is required. A shorter hydrolysis with cold trichloroacetic acid and no enzymatic hydrolysis has been used for these purposes.

For methods that lack a separation step, endogenous fluorescent substances in the food samples are oxidized to nonfluorescing compounds using potassium permanganate/hydrogen peroxide (13,48,49). Sodium metabisulfite converts riboflavin to its nonfluorescent leuco form and permits measurement of the background fluorescence. Solid phase extraction may also be used for sample purification prior to vitamin B$_2$ quantitation.

2. Methodology Reviews

The HPLC methods published between 1980 and 1990 for determining total riboflavin and the individual B_2 vitamers in foods are reviewed in four recent publications (11,12,13,49). These reviews summarize the extraction and chromatographic conditions and the types of foods tested. Single and multivitamin determinations are presented. Nielsen (48) covered paper chromatography, TLC, open column chromatography, and HPLC of total riboflavin and the individual vitamers in foods, pharmaceuticals, and biological samples. However, extraction conditions received little attention.

3. Reference Methods

The AOAC International currently recommends two microbiological and two fluorometric methods for assaying total riboflavin in foods (50). The fluorometric techniques, which include a manual and a semiautomated procedure, are both based on the same principles (49). The B_2 vitamers are extracted and converted to riboflavin by heating in the presence of mineral acids (e.g., autoclaving in 0.1 N hydrochloric acid at 121°C for 30 min). Protein is precipitated and the extracts are oxidized briefly with potassium permanganate to destroy interfering fluorescent food components. Excess permanganate is decolorized with hydrogen peroxide and the fluorescence of riboflavin measured. Sodium hydrosulfite (dithionite) is then used to reduce riboflavin to its nonfluorescent leuco form and the blank or background fluorescence measured.

Due to the lack of a separation step in the fluorometric methods, overestimation due to the presence of interfering artifacts frequently occurs (51). Recent work (51) indicated that the standard methods are not suitable for samples with high fat content or active degradative enzyme systems (e.g., phosphatases).

4. Recent Developments in Methodology

a. HPLC Methods. Recent HPLC methods for riboflavin are divided between determinations of total riboflavin and simultaneous determinations of the individual B2 vitamers. Tables 14–17 summarize the HPLC methods for the riboflavin in foods that have been developed since 1990. For methods published prior to this, see Sec. IV.B.2, "*Methodology Reviews.*" The simultaneous quantitation of riboflavin with one or more other vitamins is summarized in Sec. XI, "Simultaneous Determinations of Multiple Vitamins."

One paper (52) reported significant quantities of isomeric artifacts in the commercial coenzyme standards and presented a technique for estimating the extent of these impurities. Since the coenzymes are the predominant B_2 vitamers in most foods, failure to account for these impurities during calibration and quantitation of the individual vitamers can lead to significant analytical errors.

b. Flow Injection Analysis. Table 18 summarizes two recent FIA analyses for total riboflavin in food. One method (51) utilizes the AOAC International extraction procedure with an FIA-style modification for detection. The other is based on the natural fluorescence of riboflavin without correction for fluorescent artifacts.

V. NIACIN (VITAMIN B3)

A. Chemistry and Biochemistry

Niacin is known as the pellagra-preventing vitamin (6,13,58). Physiologically it acts as a hydrogen carrier and as a cofactor for a number of dehydrogenases. Niacin participates in oxidation–reduction systems, playing an essential role in the enzyme systems responsible for carbohydrate, fatty

Table 14 HPLC Methods for Quantitating Total Riboflavin in Foods (C18 Columns, UV Absorbance Detection)

Analyte(s)	Type of food analyzed	Extraction conditions	HPLC columns[a]	HPLC mobile phase and flow rate	Detection conditions	Method verification[b]	Ref.
Total vitamin B_2 as riboflavin.	Ready-to-eat baby foods (vegetables, milk, meat, liver and/or fish).	Extraction by autoclaving with 0.1 M hydrochloric acid at 103.5 kPa for 20 min; enzymatic hydrolysis with claradiastase (for B_2 coenzymes) and papain (for protein) at pH 4.5 for 20 hr at 37°C; protein precipitation with trichloroacetic acid at 100°C for 10 min; sample purification/clean-up with C18 solid phase extraction.	*Precolumn:* Nucleosil C18 (50 × 4.6 mm, 10 μm) *Analytical:* Nucleosil C18 (250 × 4.6 mm, 5 μm) at 41°C.	*Isocratic:* acetonitrile + 0.01 M potassium phosphate buffer, pH 7.0 (10.5 + 89.5 v/v). 1.0 ml/min.	UV absorbance, 268 nm.	External standardization. Linear range = 0.05–0.5 μg/ml (r = 0.998). LoD = 30 ng/ml at SNR = 3. Reproducibility—CV ± 10.5% in foods tested (n = 4). Recoveries = 90 ± 7% from foods tested (n = 4). Results generally agreed well with those obtained by microbiological method (r = 0.987).	53
Total vitamin B_2 as riboflavin.	Milk; nondairy imitation milk.	Protein precipitation and extraction with 10% lead acetate at pH 3.2.	*Precolumn:* C18 Bondapak guard (5 μm). *Analytical:* Spherisorb ODS 2 (150 × 3.9 mm, 5 μm).	*Isocratic:* methanol + dilute acetic acid (30 + 70 v/v). 1.5 ml/min.	UV absorbance, 270 nm.	External standardization. Linear range = 0.0–50.0 μg (r = 0.99732). Reproducibility—CV ± 1.6–5.5% in foods tested (n = 6). Recoveries = 92.20% from milk (n = 2).	54

[a]Column specifications expressed as (length × id, particle size; manufacturer).
[b]r = correlation coefficient; n = number of determinations. LoD = limit of detection; SNR = signal-to-noise ratio; CV = coefficient of variation.

Table 15 HPLC Methods for Quantitating Total Riboflavin in Foods (C18 Columns, Fluorescence Detection)

Analyte(s)	Type of food analyzed	Extraction conditions	HPLC columns[a]	HPLC mobile phase and flow rate	Detection conditions	Method verification[b]	Ref.
Total vitamin B$_2$ as riboflavin.	Soy-based products (whole bean, okara, whey, tofu).	Extraction in hydrochloric acid (pH = 2.0) by autoclaving at 20 psi for 15 min; protein removal by precipitation at pH 4.5 and centrifugation.	*Precolumn:* RP-18 (3 cm; Brownlee). *Analytical:* Ultrasphere C18 (150 × 4.6 mm, 5 μm; Beckman).	A: acetonitrile B: 0.01 M acetate buffer, pH 5.5 C: water *Gradient:* - A + B (13 + 87 v/v) at t = 0 min. 1.2 ml/min. - isocratic at A + B (13 + 87 v/v) until t = 10 min. 1.2 ml/min. - step gradient to A + C (90 + 10 v/v) at t = 10 min. 2.3 ml/min. - isocratic at A + C (90 + 10 v/v) until t = 15 min. 2.3 ml/min. - step gradient to A + B (13 + 87 v/v) at t = 15 min. 1.2 ml/min. - isocratic at A + B (13 + 87 v/v) until t = 22 min. 1.2 ml/min.	Fluorescence, 436/535 nm (ex/em).	External standardization. Linear calibration, r = 0.990. LoD = 0.50 ng thiamine. Recoveries = 48–95% from soy products. Results significantly lower than those obtained by AOAC International method. AOAC International method suspected of overestimating total riboflavin content due to fluorescent impurities.	42

| Total vitamin B2 as riboflavin. | Dairy products; liver; beef; pork; fish;cauliflower; orange; potato; tomato; bakery products; baby food; casserole; coffee; beer. | Extraction by autoclaving in 0.1 M sulfuric acid at 121°C for 20 min; adjustment to pH 4.5 with 2.5 M acetate buffer; enzymatic hydrolysis of coenzymes to riboflavin using claradiastase at 45°C overnight; sample cleanup/concentration with C18 solid phase extraction (Sep-Pak cartridges; Waters). | *Precolumn:* Spherisorb S5 ODS2 guard (5 µm; Phase Sep.) *Analytical:* Spherisorb S5 ODS2 (250 × 4.6 mm, 5 µm; Phase Sep.) | *Isocratic:* methanol + water (35 + 65 v/v). 1.0 ml/min. | Fluorescence, 445/525 nm (ex/em). | External standardization. Linear range = 0.19–190 ng/50-µl injection. Reproducibility—CV ± 3.4% in Baltic herring, ± 6.2% in cheese. Recoveries = 99% for riboflavin, 95% for FAD, and 87% for FMN from foods tested. Results agreed well with those obtained by AOAC International method, except for crispbread. AOAC International method presumed to overestimate total riboflavin in crispbread due to presence of fluorescent artifacts. | 55 |

[a]Column specifications expressed as (length × id, particle size; manufacturer).
[b]*r* = correlation coefficient; *n* = number of determinations. LoD = limit of detection; SNR = signal-to-noise ratio; *CV* = coefficient of variation.

Table 16 HPLC Methods for Quantitating B2 Vitamers in Foods (C18 Columns, Fluorescence Detection)

Analyte(s)	Type of food analyzed	Extraction conditions	HPLC columns[a]	HPLC mobile phase and flow rate	Detection conditions	Method verification[b]	Ref.
Riboflavin, FMN, FAD, and sorboflavin simultaneously.	Dairy products.	Extraction/protein solubilization with 6% formic acid + 2 M urea; centrifugation; addition of sorboflavin as internal standard; sample cleanup with silica-based C18 solid phase extraction.	*Analytical:* LC-18 (75 × 4.6 mm; 3 μm; Supelco).	*Isocratic:* 100 mM phosphate buffer, pH 2.9. 1 ml/min.	Fluorescence, 450/530 nm (ex/em).	Internal standardization using sorboflavin. LoD = 3 nmol/L for FAD, 2.5 nmol/L for FMN and riboflavin at SNR = 3. Reproducibility—CV ± 6% for FAD and riboflavin, ± 12% for FMN from milk. Recoveries = 105.8% for FAD, 99.7% for FMN, and 100.2% for riboflavin from cheese.	56
Riboflavin, FMN, and FAD simultaneously.	Milk.	Inactivation of endogenous pyrophosphatase in raw milk by boiling for 3 min; extraction and deproteinization by enzymatic hydrolysis with pronase at 45°C for 1 hr; adjustment to pH 5.5 with phosphate buffer; centrifugation.	*Analytical:* Capcell Pak C18 (250 × 4.6 mm, 5 μm; Shiseido) at 40°C.	*A:* methanol + water (90 +10 v/v). *B:* 10 mM phosphate buffer, pH 5.5 *Gradient:* - A + B (35 + 65 v/v) at *t* = 0 min. - linear gradient to A + B (95 + 5 v/v) at *t* = 8 min. - isocratic at A + B (95 + 5 v/v) for 5 min. 0.8 ml/min.	Fluorescence, 462/520 nm (ex/em).	External standardization. Linear calibration for all vitamers (*r* = 0.999). LoD = 5 pmol for FAD, 2 pmol for FMN and riboflavin. Recoveries = 95–102% for riboflavin, FMN, and FAD from milk.	57

[a]Column specifications expressed as (length × id, particle size; manufacturer).
[b]*r* = correlation coefficient; *n* = number of determinations. LoD = limit of detection; SNR = signal-to-noise ratio; *CV* = coefficient of variation.

Table 17 HPLC Methods for Quantitating B2 Vitamers in Foods (Polymer-Based Columns, Fluorescence Detection)

Analyte(s)	Type of food analzyed	Extraction conditions	HPLC columns[a]	HPLC mobile phase and flow rate	Detection conditions	Method verification[b]	Ref.
Riboflavin, FMN, FAD, and 7-ethyl-8-methyl riboflavin simultaneously.	Liver; beef; chicken breast; milk; eggs; ready-to-eat breakfast cereal; fast-food hamburger.	Addition of 7-ethyl-8-methyl riboflavin as internal standard; extraction with methanol, methylene chloride, and 100 mM phosphate buffer, pH 5.5; centrifugation.	*Precolumn:* PLRP-S (5 × 3 mm; Polymer Labs). *Analytical:* two PLRP-S 100 Å columns in series (150 × 4.6 mm + 250 × 4.6 mm, 5 μm; Polymer Labs) at 40°C.	A: acetonitrile. B: 0.1 sodium azide in 10 mM citrate-phosphate buffer, pH 5.5. *Gradient:* - $A + B$ (3 + 97 v/v) at $t = 0$ min. 1.2 ml/min. - linear gradient to $A + B$ (6 + 94 v/v) at $t = 43$ min. 1.2 ml/min. - linear gradient to $A + B$ (14 + 86 v/v) at $t = 51$ min. 1.0 ml/min. - isocratic at $A + B$ (14 + 86 v/v) until $t = 70$ min. 1.0 ml/min. - linear gradient to $A + B$ (3 + 97 v/v) at $t = 80$ min. 1.0 ml/min. - isocratic at $A + B$ (3 + 9 v/v) until $t = 90$ min. Convex flow rate gradient to 1.2 ml/min at $t = 90$ min.	Fluorescence, 450/522 nm (ex/em).	Internal standardization using 7-ethyl-8-methyl riboflavin as internal standard. LoD = 0.21 ng for riboflavin; 0.89 ng for FMN, 11.15 ng for FAD at SNR = 3. Reproducibility—CV ± 0.5–3.0%. Recoveries = 92–115% from all foods tested. Results agreed well with AOAC International fluorometric method, except for eggs, breakfast cereal, and fast-food hamburger.	52

[a]Column specifications expressed as (length × id, particle size; manufacturer).
[b]r = correlation coefficient; n = number of determinations. LoD = limit of detection; SNR = signal-to-noise ratio; CV = coefficient of variation.

Table 18 Flow Injection Analysis Methods for Quantitating Total Riboflavin in Foods (Fluorescence-Based Chemistry)

Analyte(s)	Type of food analyzed	Extraction conditions	Flow injection chemistry	Detector	Method verification[a]	Ref.
Total riboflavin.	Pork; bread: rice; meat pie; milk; fortified ready-to-eat breakfast cereal.	Extraction with 25% potassium chloride and 10% takadiastase in 0.02 M acetic acid at 70°C for 1 hr; centrifugation.	Extraction from aqueous extract into isobutanol; riboflavin detected fluorometrically.	Fluorescence, 390/475 nm (ex/em).	External standardization. Recoveries = 98.3% from foods tested. Results agreed well with those obtained from AOAC International method (r = 0.99).	47
Total riboflavin.	Liver; beef; chicken breast; milk; egg; fortified ready-to-eat breakfast cereal; bread.	Extraction by autoclaving in 0.1 N hydrochloric acid at 121–123°C for 30 min; protein precipitation by adjustment to pH 6.0–6.5 with sodium hydroxide followed by acidification to pH 4.5 with dilute hydrochloric acid; filtration; adjustment to pH 6.8 with sodium hydroxide; oxidation of fluorescent impurities with potassium permanganate; decolorization of excess permanganate with hydrogen peroxide (based on AOAC International extraction procedure [50]).	Fluorescence measured on 100 μl aliquot of extract in distilled water mobile phase (carrier) at a flow rate of 1.0 ml/min; reduction of riboflavin to non-fluorescent leuco form with sodium hydrosulfite in sodium acetate.	Fluorescence, 440/565 (ex/em).	External standardization. Recoveries = 78–104% from raw liver and model systems. Results agreed well those obtained by HPLC [52].	51

[a] r = correlation coefficient; n = number of determinations. LoD = limit of detection; SNR = signal-to-noise ratio; CV = coefficient of variation.

acid, and amino acid metabolism. The biochemical and physiological functions of niacin are reviewed in more detail by Shibata and Shimono (58).

The niacin vitamers found in foods include nicotinic acid, nicotinamide, nicotinoyl esters (niacytin), and the coenzyme forms nicotinamide adenine dinucleotide (NAD^+), nicotinamide adenine dinucleotide phosphate ($NADP^+$), and their reduced forms, NADH and NADPH (Fig. 4) (58,59). The physiologically available forms in humans are nicotinic acid, nicotinamide, NAD^+, and $NADP^+$. Tryptophan can also be converted to niacin in the body (2,6). The niacin activity of food may therefore be expressed in "niacin equivalents," which is the sum of the niacin content plus 1/60 of the tryptophan content (in mg). The present chapter will be restricted to the niacin vitamers and will not cover tryptophan methodology. Food sources of niacin include cereals and seeds, which contain primarily nicotinic acid, and fish, pork, chicken, and beef, which contain primarily nicotinamide (13). The niacin content of roasted coffee beans is primarily nicotinic acid formed during the roasting process due to removal of the methyl group from trigonelline (1-methylnicotinic acid).

In general, niacin is one of the more stable vitamins (13,58). Aqueous solutions of nicotinic acid and nicotinamide can both be autoclaved for short periods without decomposition (e.g., at 100°C for 10 min). Nicotinic acid's stability is pH-independent. In contrast, nicotinamide is stable in solution at neutral pH but is converted to nicotinic acid by heating in 1 N acid or alkali. NADH and NADPH are stable for short periods in alkali but readily decompose under acidic conditions, while the reverse is true for NAD^+ and $NADP^+$. More detailed information on the chemistry of the niacin vitamers is available in a recent review by Shibata and Shimono (58).

For detection purposes, the niacin vitamers are all UV absorbers (13,58).

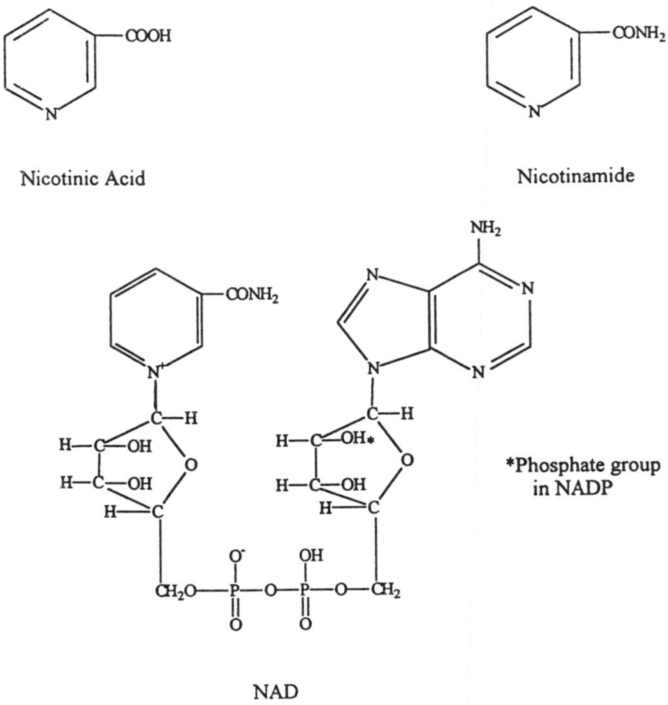

Nicotinic Acid   Nicotinamide

*Phosphate group in NADP

NAD

Fig. 4   Niacin (vitamin B_3).

B. Methods of Analysis

1. Extraction Techniques

Due to the relative stability of niacin, either acid or alkaline hydrolysis can be used to convert nicotinamide to nicotinic acid prior to determination of total niacin as nicotinic acid (2,13). Acid hydrolysis is usually used to estimate the biologically active niacin, while alkaline hydrolysis releases nonbioavailable vitamers and permits determination of the total niacin content. Since alkaline hydrolysis is usually much faster than acid hydrolysis, the latter is commonly supplemented with enzymatic hydrolysis. Takadiastase, papain, and clarase have all been used for this purpose. Organic solvents, such as methanol, can be used for the extraction of free nicotinic acid.

Nicotinamide determinations require more exacting extraction conditions in order to prevent hydrolysis of the amide functional group (13). Typical extraction procedures for nicotinamide involve aqueous extractants and the use of dilute acids, e.g., sulfuric or hydrochloric acid, to release the vitamers from protein.

2. Methodology Reviews

The HPLC methods published between 1980 and 1990 for niacin in foods are reviewed in three recent publications (11–13). These reviews summarized the extraction and chromatographic conditions used as well as the types of foods tested. Single and multivitamin determinations were presented. Shibata and Shimono (58) covered paper chromatography, TLC, open column chromatography, HPLC, and GC techniques, but not extraction conditions, for the niacin vitamers in foods, pharmaceuticals, and biological samples. Velíšek and Davídek (14) reviewed the GC methods of niacin analysis.

3. Reference Methods

The AOAC International currently lists one manual and two automated methods for determination of total niacin (nicotinic acid + nicotinamide) in foods (60). The niacin vitamers are extracted with acid or alkaline hydrolysis, which also facilitates the conversion of nicotinamide to nicotinic acid. Total niacin is determined as nicotinic acid based on cleavage of the pyridine groups by cyanogen bromide and formation of a colored product by reaction with sulfanilic acid. The colored endproduct is determined spectrophotometrically at 470 nm. These methods suffer from several disadvantages including low sensitivity and the use of harmful and unstable reagents (particularly cyanogen bromide and sulfanilic acid) (58).

4. Recent Developments in Methodology

a. HPLC Methods.   The microbiological methods for determining total niacin are still widely used (60). However, these methods tend to be time- and labor-intensive. Reproducibility can also be problematic (58). HPLC is a viable alternative and a number of methods have therefore been developed.

Most of the HPLC methods for niacin in foods concentrate on total niacin rather than the individual vitamers. Tables 19 and 20 summarize the HPLC methods for the niacin vitamers in foods that have been developed since 1990. For methods published prior to this, see Sec V.B.2, *"Methodology Reviews."* The simultaneous quantitation of niacin with one or more other vitamins is summarized in Sec. XI, "Simultaneous Determinations of Multiple Vitamins."

Table 19 HPLC Methods for Quantitating Total Niacin in Foods (C18 Columns, UV Absorbance Detection)

Analyte(s)	Type of food analyzed	Extraction conditions	HPLC columns[a]	HPLC mobile phase and flow rate	Detection conditions	Method verification[b]	Ref.
Total niacin as nicotinic acid.	Beef; semolina; cottage cheese.	Extraction by autoclaving for 15 min at 120°C in calcium hydroxide; adjustment to pH 6.5–7.0; precipitation of excess calcium with oxalic acid; solid phase extraction/cleanup with C18 Sep-Pak.	*Precolumn:* C18 Guard-Pak (Waters) *Analytical:* LC-18-DB (150 × 5.6 mm, 5 μm; Supelco).	*Isocratic:* acetonitrile + water containing 1.0 ml phosphoric acid and 1.0 g sodium dodecyl sulfate (23 + 77 v/v). 1.5 ml/min.	UV absorbance, 254 nm.	External standardization. Linear range = 0.004–0.040 mg nicotinic acid/100 ml. Good agreement with results obtained by microbiological method.	61
Total available niacin as nicotinic acid.	Beans; chickpeas; lentils; green beans; pork muscle.	Extraction by autoclaving for 15 min at 121°C in hydrochloric acid; enzymatic hydrolysis at 48°C for 3 hr using takadiastase; solid phase extraction/cleanup on ion exchange Dowex 1-X8 (BDH) column; vacuum evaporation; dissolution in methanol + 0.1 M acetate buffer, pH 4.7–4.9 (1 + 9 v/v).	*Precolumn:* C18 Porasil (40 × 3.2 mm); Waters). *Analytical:* μBondapak C18 (300 × 3.9 mm, 10 μm; Waters) or Spherisorb ODS2 (300 × 3.9 mm, 10 μm; Spheris). At 13.5–24°C depending on sample type.	*Isocratic:* methanol + 0.01 M acetate buffer, pH 4.66 containing 5 mM tetrabutylammonium bromide (1 + 9 v/v). 1.4–1.5 ml/min.	UV absorbance, 254 nm.	External standardization. Linear range = 40–120 μg nicotinic acid/ml (r = 0.999). LoD = 2.66 ng nicotinic acid at SNR = 3. Reproducibility—CV ± 1.4–9.2% for foods analyzed (n = 5–9).	59

[a] Column specifications expressed as (length × id, particle size; manufacturer).
[b] r = correlation coefficient; n = number of determinations. LoD = limit of detection; SNR = signal-to-noise ratio; CV = coefficient of variation.

Table 20 HPLC Methods for Quantitating Total Niacin in Foods (Ion Exchange and NH$_2$ Columns, UV Absorbance Detection)

Analyte(s)	Type of food analyzed	Extraction conditions	HPLC columns[a]	HPLC mobile phase and flow rate	Detection conditions	Method verifications[b]	Ref.
Total niacin as nicotinic acid.	Vinegar; jam.	Extraction with 20% sodium, hydroxide at 100°C for 30 min; neutralization with 25% hydrochloric acid + methanol (1 + 1 v/v); solid phase extraction/cleanup using AG1-X8 anion exchange resin (Bio-Rad) and Toyopack IC-SP M cation exchange cartridge (Toso).	*Precolumn:* NH2P (50 × 4.6 mm; Asahi Kasei). *Analytical:* Asahipak NH2P-50 (250 × 4.6 mm, 5 μm; Asahi Kasei).	*Isocratic:* acetonitrile + 0.075 M sodium acetate (60 + 40 v/v). 0.5 ml/min.	UV absorbance, 261 nm.	External standardization. Linear range = 10–100 ng nicotinic acid. LoD = 1 ng niacin. Reproducibility—CV ± 1.9–4.7% for fortified vinegar. Recoveries = 91–95% from fortified vinegar (n = 5). Results in reasonable agreement with those from microbiological method (vinegar and jam samples).	62
Total niacin as nicotinic acid.	Fortified foods (pasta, bread, ready-to-eat breakfast cereals, infant formula); unfortified foods (beef soup, tuna).	Extraction by autoclaving at 121–123°C for 45 min in sulfuric acid; adjustment to pH 6.0–6.5 using 7.5 N sodium hydroxide; adjustment to pH 4.5 with sulfuric acid; filtration; adjustment to pH 0.5–1.0 with sulfuric acid; solid phase extraction/cleanup with Florisil column.	*Analytical:* PRP-X100 (250 × 4.1 mm; Hamilton).	*Isocratic:* dilute acetic acid. 1.5 ml/min.	UV absorbance, 254 nm.	External standardization. Linear range = 0.24–0.80 μg nicotinic acid/ml. LoD = 0.11 μg niacin/ml. Reproducibility—CV ± 2.4% for nicotinic acid in whey-based infant formula; ± 2.9% for nicotinic acid in macaroni (n = 10). Recoveries = 99.8 ± 7.7% from foods tested (n = 15). Results from microbiological method averaged 11% higher than those by HPLC.	63

[a]Column specifications expressed as (length × id, particle size; manufacturer).
[b]r = correlation coefficient; n = number of determinations. LoD = limit of detection; SNR = signal-to-noise ratio; CV = coefficient of variation.

VI. VITAMIN B₆ (PYRIDOXINE)

A. Chemistry and Biochemistry

Although no specific vitamin B_6 deficiency syndrome has been demonstrated in humans, this vitamin is known to participate in a wide variety of physiological functions (6,64). Vitamin B_6 is heavily involved in α-amino acid biosynthesis and catabolism as a coenzyme for the aminotransferases. As a coenzyme for a number of decarboxylases it plays an important role in the synthesis of neurotransmitters and brain function. Vitamin B_6 has also been implicated in the treatment of disease states ranging asthma to autism, cancer, and gestational diabetes. The biochemical, physiological, and pharmacological functions of vitamin B_6 are reviewed in greater detail by Ubbink (64).

There are six nutritionally active B_6 vitamers that occur naturally in foods: pyridoxine (PN), pyridoxal (PL), pyridoxamine (PM), and the corresponding 5′-phosphate esters pyridoxine phosphate (PNP), pyridoxal phosphate (PLP), and pyridoxamine phosphate (PMP) (Fig. 5) (6,13,64). Vitamin B_6 also occurs in plant-based foods as a PN glycoside, which has approximately 60% of the bioavailability of PN in humans (65). The predominant B_6 vitamers in meat and fish are PMP and PLP, while that in foods of plant origin, such as cereals, fruit, and vegetables, is PN. PLP is readily converted to PMP during cooking and food processing.

Vitamin B_6 analysis is complicated by several factors, including its occurrence as six vitamers of differing structures (13,64). For extraction purposes, the B_6 vitamers are strictly water-soluble,

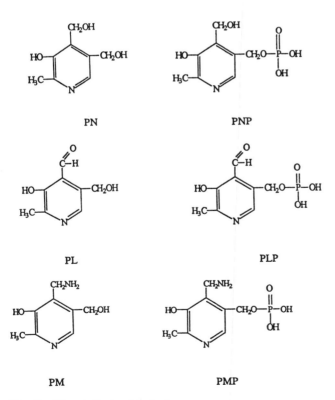

Fig. 5 Vitamin B_6 (pyridoxine).

which severely restricts the use of organic solvents for extraction, purification, or enrichment prior to quantitation. All of the B_6 vitamers are highly light-sensitive, necessitating the use of subdued lighting and low actinic glassware during their extraction and analysis.

Detection of the B_6 vitamers can be problematic because of the low concentrations at which they occur in food (13,64). The sensitivity and specificity of the detection method is therefore critical. The six principal vitamers are all UV absorbers with similar spectra in 0.1 M hydrochloric acid. This is not the case at higher pH. The B_6 vitamers are also naturally fluorescent but this is complicated by the fact that different vitamers show different fluorescence spectra of differing intensities. Derivatization has been used to enhance and standardize the fluorescence properties of these vitamers. Greater detail on the spectral properties of the B_6 vitamers is available in the review by Ubbink (64).

B. Methods of Analysis

1. Extraction Techniques

Total vitamin B_6 has traditionally been determined by extraction at elevated temperatures (autoclaving) in the presence of strong mineral acids. These conditions result in the hydrolysis of the B_6 phosphate esters and the glycosylated forms (13,64,66). Nondegradative extraction of all six vitamers usually involves the use of perchloric or sulfosalicylic acid. However, variations in extraction efficiency of these acids have been reported. Sulfosalicylic acid also interferes with the fluorescence of the B_6 vitamers and must be removed prior to extraction. The glycoside forms in plant materials have traditionally been determined after β-glycosidase hydrolysis to release the bound PN.

2. Methodology Reviews

Polesello and Rizzolo (11,12) and Lambert and De Leenheer (13) reviewed the HPLC methods for vitamin B_6 that were published in the decade 1980–1990. In addition to the chromatographic conditions, they also summarized the extraction conditions and the types of food tested. Ubbink (64) prepared a review of selected methods for quantitating the B_6 vitamers in foods, biological tissues, and pharmaceuticals, as well as for TLC, HPLC, and GC applications. However, little attention was devoted to extraction methods. Velíšek and Davídek (14) reviewed the GC methods for vitamin B_6 analysis.

3. Reference Methods

The methods currently recommended by the AOAC International are both microbiological (67). When using these methods, care should be taken to ensure that the microorganisms used respond to the various vitamers in the same way as humans (1).

4. Recent Developments in Methodology

  a. HPLC Methods.   Much of the recent HPLC work in vitamin B_6 has been directed to the simultaneous determination of the six principal vitamers and the PN glycoside content of plant foods. Tables 21–23 summarize recent (since 1990) HPLC methods for the B_6 vitamers in the absence of other B vitamins. For methods published prior to 1990, see Sec.VI.B.2, "*Methodology Reviews.*" The simultaneous quantitation of vitamin B_6 with one or more other vitamins is summarized in Sec. XI, "Simultaneous Determinations of Multiple Vitamins."

Table 21 HPLC Methods for Quantitating Total Vitamin B6 in Foods (C8 Columns, Fluorescence Detection)

Analyte(s)	Type of food analyzed	Extraction conditions	HPLC columns[a]	HPLC mobile phase and flow rate	Detection conditions	Method verification[b]	Ref.
Total vitamin B6 as pyridoxol.	Unfortified foods (yeast, wheat germ, ready-to-eat breakfast cereal).	Extraction with 0.05 M acetate buffer, pH 4.5; hydrolysis of phosphate esters with acid phosphatase at 37°C overnight; desamination of PM to PN using 1 M glyoxylic acid + ferrous sulfate catalyst at 37°C overnight; reduction of PN to PL with alkaline sodium borohydride.	*Precolumn:* RP-18 guard column (10 μm; Merck). *Analytical:* Lichrospher 60 RP Select B (C8) (250 × 5 mm, 5 μm; Merck).	*Isocratic:* acetonitrile + 0.05 M phosphate buffer containing 5 mM sodium heptane sulfonate (4 + 96 v/v), pH 2.50. 1.0 ml/min.	Fluorescence, 290/395 nm (ex/em).	External standardization. Linear range = 0–405 μg PL/g. LoD = 0.02 μg PL/g. Reproducibility—CV ± 2.2–7.9% for foods tested (n = 6). Recoveries = 90–95% for PL, PM, and PN from foods tested.	68

[a]Column specifications expressed as (length × id, particle size; manufacturer).
[b]r = correlation coefficient; n = number of determinations. LoD = limit of detection; SNR = signal-to-noise ratio; CV = coefficient of variation.

Table 22 HPLC Methods for Quantitating B6 Vitamers in Food (C8/C18 Columns, Fluorescence Detection)

Analyte(s)	Type of food analyzed	Extraction conditions	HPLC columns[a]	HPLC mobile phase and flow rate	Detection conditions	Method verification[b]	Ref.
PL, PM, PN, PLP, PMP, PNP, & PNG[c] simultaneously. Total vitamin B6 by addition.	Milk; orange juice; cabbage; broccoli.	Perchloric acid extraction; addition of internal standard, 4'-deoxypyridoxine.	*Analytical:* TSK ODS-80 Tm (250 × 4.6 mm; Tosoh).	*Isocratic:* methanol + 50 mM phosphate buffer, pH 3.5 containing 120 mM sodium perchlorate (1 + 99 v/v). 1.0 ml/min.	*Postcolumn reaction:* 1.0 M phosphate buffer, pH 7.5 containing 1 mg/ml sodium bisulfite. *Fluorescence:* 305/390 nm (ex/em).	Internal standardization using PNG as internal standard. LoD = 0.2 ng for PM and PMP; 0.4 ng for PNP, PL, & PN; 0.6 ng for PLP & PNG. Reproducibility—CV ± 0.9–3.6% for all analytes tested (n = 6). Recoveries = 85–105% for all analytes from foods tested.	69
PL, PM, PN, PLP, PMP, PNP, and 4'-deoxypyridoxine. Total vitamin B6 by addition.	Fresh carrots; rice bran; oat bran; wheat bran; whole soybean flour.	*Nonglycosylated Vitamers:* addition of 4'-deoxypyridoxine as internal standard; extraction with 5% sulfosalicylic acid and methylene chloride; solid phase extraction/cleanup and removal of sulfosalicylic acid using AG2-X8 (BioRad) anion exchange column. *Glycosylated Vitamers:* extraction as above; optional saponification with 0.7 m KOH; hydrolysis of PNG with β-glucosidase + acid phosphatase at 37°C for 2.5 hr; centrifugal ultrafiltration to remove protein.	*Analytical:* Ultrasphere IP (250 × 4.6 mm, 5 μm; Beckman).	A 33 mM phosphoric acid containing 8 mM octane sulfonic acid, pH 2.2. B 2-propanol + 33 mM phosphoric acid (17 + 83 v/v). *Gradient:* - A+ B (83 + 17 v/v) at t = 0 min. slightly concave gradient to A + B (69 + 31 v/v) at t = 11.0 min. - linear gradient to A + B (0 + 100 v/v) at t = 15 min. - isocratic at 100% B until t = 30 min. - requilibration with A + B (83 + 17 v/v) for 30 min. 1.0 ml/min.	Fluorescence, 295/405 nm (ex/em).	Internal standardization using 4'-deoxypyridoxine as internal standard. Good agreement with results of another study using AOAC International acid catalyzed hydrolysis and microbiological assay.	66

[a] Column specifications expressed as (length × id, particle size; manufacturer).

[b] r = correlation coefficient; n = number of determinations. LoD = limit of detection; SNR = signal-to-noise ratio; CV = coefficient of variation.

[c] PNG = 5'-O-(β-D-glycopyranosyl)pyridoxine.

Table 23 HPLC Methods for Quantitating B6 Vitamers in Food (Anion Exchange Columns, Fluorescence Detection)

Analyte(s)	Type of food analyzed	Extraction conditions	HPLC columns[a]	HPLC mobile phase and flow rate	Detection conditions	Method verification[b]	Ref.
PN, PM, PL, PMP, PLP, & 3-hydroxypyridine simultaneously. Total vitamin B6 by addition.	Raw and fried chicken.	Addition of 3-hydroxy-pyridine as internal standard; extraction with 5% sulfosalicylic acid and methylene chloride (lipid extraction); removal of sulfosalicylic acid on anion exchange resin (Dowex AG-2-X8; BioRad).	*Precolumn:* Bio-Gel HPHT (50 × 4 mm; BioRad). *Analytical:* two ion exchange (diethylaminoethyl functionality on G500 PW) columns in series (750 × 7.5 mm; BioRad).	*Isocratic:* 0.02 M glycine buffer, pH 9.8 containing 0.12 M sodium chloride. 0.8 ml/min. *Column wash:* 0.02 M glycine containing 0.4 M sodium chloride (used for 10 min after each chicken analysis).	*Post-column reaction:* 0.75 M phosphate buffer, pH 4.90 containing bisulfite to enhance fluorescence sensitivity of PM, PMP and PLP. *Fluorescence time program:* 330/400 nm (ex/em) for PM and PMP; 310/400 nm for PN, PL, and internal standard; 330/400 nm (ex/em) for PLP.	Internal standardization using 3-hydroxy-pyridine as internal standard. Linear calibration, $r = 0.98$–0.99. LoD = 0.79 ng for PM, 0.78 ng for PMP, 2.68 ng for PLP, 1.26 ng for PM and 1.24 ng for PL. Recoveries = 83.3–104.6% from raw and fried chicken breast and thigh ($n = 5$).	70

[a]Column specifications expressed as (length × id, particle size; manufacturer).
[b]r = correlation coefficient; n = number of determinations. LoD = limit of detection; SNR = signal-to-noise ratio; CV = coefficient of variation.

VII. FOLACIN

A. Chemistry and Biochemistry

Folacin deficiency produces a megaloblastic anemia very similar to that produced by vitamin B_{12} deficiency. This results from the interdependence of folacin and vitamin B_{12} in at least two biochemical pathways (71). There is also evidence that folacin plays a role in neurological function. In vivo folacin functions as a carrier of one-carbon units and acts as a coenzyme in a number of biochemical reactions including the synthesis of nucleic acids, amino acids, and proteins (71,72). Greater detail on the biochemistry and physiology of folacin is available from recent reviews (72,73).

Folacin is available from a diverse array of plant- and animal-based foods including green leafy vegetables, liver, brewer's yeast, beans, and fermented dairy products (6,72). The folacin vitamers constitute a large group of related compounds having the same basic structure as folic acid (pteroylglutamic acid) but differing in their state of reduction and the number of glutamate residues attached (Fig. 6) (71–73). In theory there are approximately 150 different folacin vitamers (excluding polyglutamyl forms) of differing biological activity. These vitamers can be subdivided into three basic groups: the oxidized monoglutamate vitamers such as folic acid, and the two principal reduced forms, 5,6-dihydrofolate and 5,6,7,8-tetrahydrofolate. The N-5 or N-10 position of the tetrahydrofolates can contain one-carbon substituents such as 5-methyl-, 5-formyl-, 5,10-methylene-, 5,10-methenyl, and 5-formiminotetrahydrofolate. Folic acid does not occur in nature but is the vitamer of choice for fortification. In nature the dihydro- and tetrahydrofolates usually occur as polyglutamates with five or six glutamate residues conjugated via γ-peptide links.

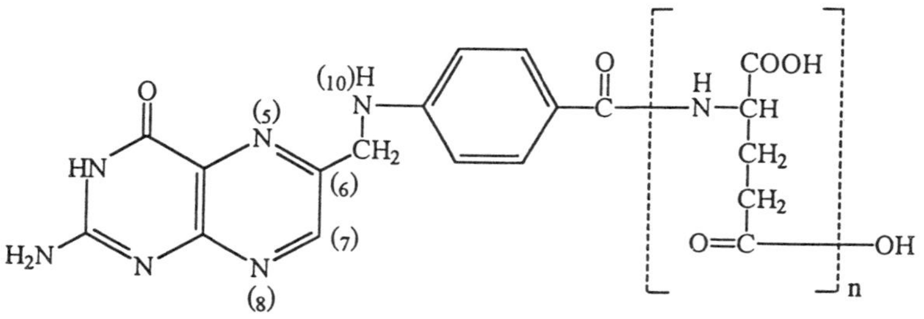

Folic Acid (PteGlu)

State of Reduction
      7,8-dihydro, e.g. H_2PteGlu
      5,6,7,8-tetrahydro, e.g. H_4PteGlu

One Carbon Substituent
      5-methyl, e.g. 5-CH_3-H_4PteGlu
      5-formyl, e.g. 5-HCO-H_4PteGlu
      10-formyl, e.g. 10-HCO-H_4PteGlu
      5,10-methylenyl, e.g. 5,10-CH_2- H_4PieGlu
      5,10-methenyl , e.g. 5,10-CH=H_4PteGlu

Conjugation
      n= number of glutamate residues
      n=1, e.g. 5-CH_3- H_4PteGlu
      n=5, e.g. 5-CH_3- H_4PteGlu$_5$

Fig. 6 Folacin.

The folacin vitamers in foods are relatively labile, being extremely sensitive to heat, light, oxidation, and extremes of pH (13,72,74). The extraction of these vitamins is further complicated by the low concentrations at which they occur naturally and the widespread presence in food of endogenous enzymes that interconvert the vitamers and/or degrade the polyglutamyl chain. In addition, the folacin vitamers can occur bound to protein. Special care must therefore be taken to ensure quantitative extraction. The chemical properties of the folacin vitamers are discussed in greater depth by Hawkes and Villota (72).

For detection purposes, all biologically active folates exhibit characteristic UV absorbance spectra that are virtually identical (73,75–77). Under sufficiently acidic conditions the tetrahydrofolates fluoresce naturally. Oxidative cleavage of the monoglutamates can be used to produce highly fluorescent pterins. Amperometric detection of the folacin vitamers tends to suffer from high background interference by other constituents of biological systems.

B. Methods of Analysis

1. General Comments

Current folacin methodology can be divided into three broad categories: microbiological, (radio-) protein binding, and chromatographic. In a recent interlaboratory comparison of microbiological, radioprotein binding and enzyme protein binding, and HPLC assays for food folacin analysis (78), the microbiological methods were found to give the most consistent results among the 15 participating laboratories. The results from radioprotein binding assay of freeze-dried Brussels sprouts were generally 50–60% higher than either microbiological or HPLC results on the same food substrate and also tended to be more variable. Although the mean folate content obtained by enzyme protein binding assay agreed well with HPLC and microbiological results, there was considerable between-laboratory variation. This study confirmed that the major problem with protein binding assays in food was differing responses of the various vitamers to the binding protein. The HPLC results for overall folate content agreed reasonably well with those found microbiologically, although there were significant discrepancies in the proportions of the individual folacin vitamers reported. It was concluded that future work should focus on improving assay methodology.

2. Extraction Techniques

Sample preparation for folacin analysis generally involves extraction of the folates from the sample matrix followed by enzymatic hydrolysis of the polyglutamate forms to monoglutamates using conjugase (pteroylpolyglutamyl hydrolase) (73,75,79). Most food extraction procedures begin with homogenization and a protein precipitation step, such as mild acidification, heating, addition of organic compounds, such as trichloroacetic acid, or enzyme (e.g., papain) hydrolysis. The specific technique used for protein precipitation will depend on the expected folacin content and the food matrix. The extractants often contain antioxidants such as ascorbic acid, 2-mercaptoethanol, or dithiothreitol in order to preserve the vitamers in their natural oxidation state. Extractants may also be degassed or flushed with an inert gas to exclude oxygen. A recent study (79) indicated that an extractant pH of 7.85 provided the most efficient recovery of naturally occurring folacin in tissues and that a second extraction of the residue produced by the initial extraction significantly increased the completeness of the extraction procedure.

For hydrolysis to monoglutamate forms, the source of the conjugase enzyme is critical as are the conditions used for deconjugation (73,75,79). Chicken pancreas conjugase yields a diglutamate product that produces ambiguous chromatographic results. Hog kidney conjugase appears to be more suitable, although there have been reports of its inhibition by naturally occurring compounds in biological samples. In addition, the importance of optimizing deconjugation time and hog kidney conjugase concentration for each type of sample has been demonstrated (80).

Sample purification prior to chromatographic quantitation is often accomplished with solid phase extraction on cation or anion exchange resins (73,75).

3. Methodology Reviews

The most popular methods for folacin analysis are microbiological, radiometric, and chromatographic techniques. Hawkes and Villota (72) recently prepared a comprehensive review of folacin in food including biological, microbiological, radioassay, column chromatographic, HPLC, and electrochemical methods for determining total folacin and the folacin vitamers. This review concentrated on assay techniques with little emphasis on folacin extraction. Polesello and Rizzolo (11,12) prepared comprehensive reviews of HPLC methods published between 1981 and 1989 for quantitating the various folacin vitamers in a variety of foods. In addition to the chromatographic conditions, they also summarized the extraction conditions and the types of food tested. A similar review by Lambert and De Leenheer (13) covered selected references published in the decade 1980–1990. Mullin and Duch (73) prepared a review of selected methods for quantitating the folacin vitamers in foods, biological tissues, and pharmaceuticals, including paper chromatography, TLC, HPLC, and microbiological applications. However, little attention was devoted to extraction methods.

4. Reference Methods

The AOAC International reference methods for free or total folacin in foods are all microbiological procedures (81). Analysts are cautioned about degradation of folacin occurring while frozen samples are thawing prior to extraction. In addition, different test microorganisms, including *L. casei*, used in the AOAC International methods, have been shown to respond differently to the various folacin vitamers (1,82,83). Folacin methodology in general continues to be problematic.

5. Recent Developments in Methodology

a. *HPLC Methods.* In general, the microbiological assays for folacin are time- and labor-intensive and do not permit differentiation between "free" (monoglutamate forms) and total folacin (84). Radiometric assays are problematic when analyzing complex substrates due to the varying affinities of binding proteins to the folacin vitamers. Many analysts have therefore turned to HPLC as a means of determining the individual folacin vitamers. To date, the number and diversity of structure and properties among the folacin vitamers have prevented development of a single comprehensive method for the simultaneous determination of the folacin vitamers. Increasing use is being made of detection methods having higher specificity such as fluorescence and electrochemistry as a means of compensating for the low concentrations of endogenous folacin in foods. Tables 24 and 25 summarize developments in folacin methodology since 1990. For methods published prior to 1990, see Sec.VII.B.3, "*Methodology Reviews.*" The simultaneous quantitation of total folacin with one or more other vitamins is summarized in Sec. XI, "Simultaneous Determinations of Multiple Vitamins."

VIII.   VITAMIN B₁₂ (CYANOCOBALAMIN)

A.   Chemistry and Biochemistry

A deficiency of vitamin B_{12} produces pernicious anemia, similar to that from folacin deficiency (13,90). This is due to the interdependence of vitamin B_{12} and folacin in at least two biochemical

Table 24 HPLC Methods for Quantitating Folacin in Foods (C18 Columns, Electrochemical Detection)

Analyte(s)	Type of food analyzed	Extraction conditions	HPLC columns[a]	HPLC mobile phase and flow rate	Detection conditions	Method verification[b]	Ref.
5-Methyltetra-hydrofolate	Citrus juice.	Aqueous extraction; hydrolysis of polyglutamates to monoglutamates using hog kidney conjugase at 37°C for 90 min.	*Precolumn:* NovaPak C18 (75 × 3.9 mm, 4 μm; Waters). *Analytical:* Zorbax ODS (250 × 4.6 mm; 5 μm; DuPont).	*A:* methanol + phosphate-acetate buffer, pH 5.0 containing 5 mM tetra-butyl ammonium dihy-drogen phosphate (TBAP) (10 + 90 v/v). *B:* methanol + phosphate-acetate buffer, pH 5.0 containing 5 mM (TBAP) (30 + 70 v/v). *Elution program:* - concentration and clean-up (elution of ascorbic acid) using precolumn with A for 1.25 min. - 10-port valve switched to allow backflush of fo-lates from precolumn to analytical column with B. 1.0 ml/min.	*Electrochemistry:* +200 mV vs. Ag/AgCl, glassy carbon working electrode for folacin. *UV absorbance:* 254 nm for de-tection of ascorbic acid eluting from precolumn.	External standardiza-tion. Linear range = 0.25–1.25 μg 5-methyl tetrahydrofolate/ml, $r =$ 0.9992. Reproduci-bility—$CV \pm 2.8\%$ in citrus juice ($n = 14$). Results generally higher than those from microbiological method without deconjugation and standardized with folic acid.	85

[a]Column specifications expressed as (length × id, particle size; manufacturer).
[b]r = correlation coefficient; n = number of determinations. LoD = limit of detection; SNR = signal-to-noise ratio; CV = coefficient of variation.

Table 25 HPLC Methods for Quantitating Folacin in Foods (C8/C18 Columns, UV Absorbance Detection)

Analyte(s)	Type of food analyzed	Extraction conditions	HPLC columns[a]	HPLC mobile phase and flow rate	Detection conditions	Method verification[b]	Ref.
(Fortified) folic acid.	Infant formula; liquid diet.	Aqueous extraction and hydrolysis with protease-papain at 40°C for 1hr.	*Precolumn:* Bio Series SAX guard (DuPont). *Analytical:* Bio Series SAX (strong anion exchange) (80 × 6.2 mm; Du Pont) and Zorbax RX (C8) (250 × 4.6 mm; Du Pont) at 35°C.	A: acetate-sulfate buffer pH 5.30. *B:* acetonitrile + A (360 + 640 v/v). *Linear gradient with column switching:* - at t = 0 min—100% A at 1.5 ml/min through SAX column. - at t = 8 min—add C8 column to flow path, *i.e.,* 100% A at 0.8 ml/min (through both columns). - at t = 20 min—remove SAX column from flow path, *i.e.,* 100% A at 0.8 ml/min (through C8 column). - linear gradient to A + B (86.1 + 13.9 v/v) at t = 20.1 min. 0.8 ml/min through C8 column. - isocratic at A + B (86.1 + 13.9 v/v) until t = 33 min. 0.8 ml/min through C8 column. - Linear gradient to 100% B at t = 40 min. 0.8 ml/min through C8 column. - Linear gradient to 100% A at t = 44 min. 0.8 ml/min through C8 column.	UV absorbance, 345 nm.	External standardization. LoD = 6 g folic acid /L. Reproducibility—CV ± 3.6%. Recoveries = 95.5–100.2% folic acid from milk- and soy-based infant formula.	86

Analyte	Matrix	Sample preparation	Column	Mobile phase	Detection	Comments	Ref.
Total folacin as: folic acid, dihydrofolic acid, tetrahydrofolic acid, 10-formyltetrahydrofolic acid, 5-methyltetrahydrofolic acid, and 5-formyltetrahydrofolic acid simultaneously.	Egg yolk; lima beans; baker's yeast; bovine liver; soybean; wheat germ; cabbage; lettuce; orange juice; banana.	Extraction with bis-tris buffer, pH 7.8 containing 2% ascorbate and 10 mM mercapto-ethanol by auto-claving at 120°C for 30 min; sample cleanup/folacin purification by affinity chromatography.	*Analytical:* Econosphere C18 (100 × 4.6 mm, 5 µm; Alltech).	A: aqueous solution containing 5 mM tetrabutyl ammonium phosphate, 0.5 mM dithioerythritol, and 25 mM sodium chloride. B: acetonitrile + water (65 + 35, v/v) containing 5 mM tetrabutyl ammonium phosphate, 0.5 mM dithioerythritol and 25 mM sodium chloride. *Gradient:* - A + B (90 + 10 v/v) at $t = 0$ min. - isocratic at A + B (90 + 10 v/v) until $t = 5$ min. - linear gradient to A + B (64 + 36, v/v) at $t = 15$ min. - linear gradient to A + B (50 + 50, v/v) at $t = 35$ min. - linear gradient to A + B (40 + 60, v/v) at $t = 52$ min. 1 ml/min.	UV absorbance (diode array): 258 and 350 nm.	External standardization. Reproducibility—$CV \pm 5$–19% for total folate in egg yolk, baker's yeast, lima beans and bovine liver. Good agreement with results obtained by microbiological assay ($r = 0.966$).	87–89

[a] Column specifications expressed as (length × id, particle size; manufacturer).

[b] r = correlation coefficient; n = number of determinations. LoD = limit of detection; SNR = signal-to-noise ratio; CV = coefficient of variation.

pathways (71). There are two known biochemical reactions for which humans require vitamin B_{12}: the methylation of homocysteine to form methionine (involving the formation of tetrahydrofolate from 5-methyltetrahydrofolate) and the formation of succinyl coenzyme A from methylmalonyl coenzyme A (90). As a result, vitamin B_{12} is implicated in protein and DNA synthesis and cellular membrane integrity (91). More detailed information on the biochemistry and physiology of vitamin B_{12} is available from a recent review (90).

The best food sources of vitamin B_{12} are animal products, in particular organ meats such as liver, kidney, and heart (6,13). Plant products contain very little vitamin B_{12}, which is an important consideration for strict vegetarians.

The B_{12} vitamers are a group of related organometallic compounds having a common corrinoid structure and varying in the substituent bound to the central cobalt atom (Fig. 7) (13,90). The

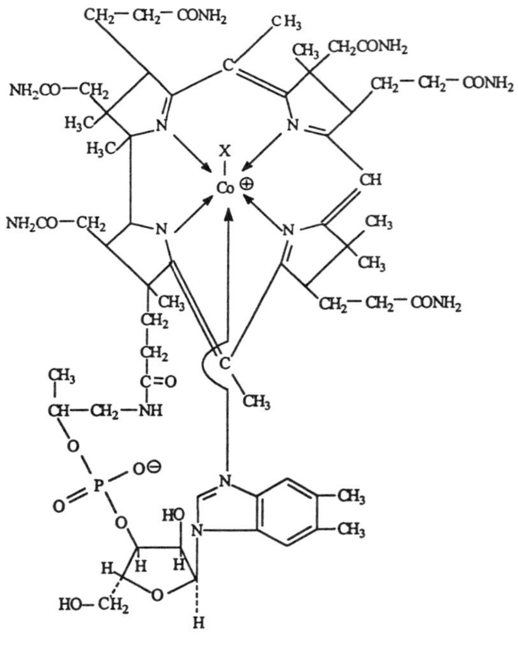

Cyanocobalamin   -X = -CN

Hydroxocobalamin   -X = -OH

Methylcobalamin   -X = $-CH_3$

Adenosylcobalamin -X =

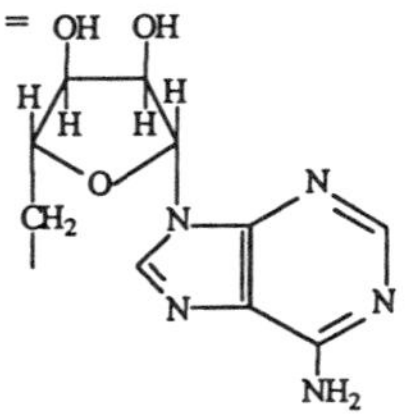

Fig. 7   Vitamin B_{12} (cyanocobalamin).

principal B_{12} vitamers in food include hydroxocobalamin (HOCbl), methylcobalamin (MeCbl), and adenosylcobalamin (AdoCbl). Cyanocobalamin (CNCbl) does not occur naturally but is used in food fortification due to its relative stability. The coenzyme forms, MeCbl and AdoCbl, are the biologically active vitamers.

Cyanocobalamin is the most stable of the B_{12} vitamers (90). It is stable in aqueous solution between pH 4 and 7 and can be heated at 120°C. However, it is susceptible to degradation and loss of vitamin activity under alkaline conditions. Exposure to light leads to HOCbl formation.

In general, the corrinoids are susceptible to loss of amide groups from their side chains under mildly acidic conditions (90). AdoCbl and MeCbl exhibit extreme photosensitivity, which necessitates the use of subdued lighting and low actinic glassware during vitamin B_{12} analyses. However, these vitamers are relatively stable in neutral aqueous solution in the dark and can be heated for 20 min at 100°C. AdoCbl is unstable under acidic and alkaline conditions while MeCbl is stable in the presence of dilute acid or alkali. Lindemans (90) discusses the chemistry of the B_{12} vitamers in greater detail.

For detection purposes, the B_{12} vitamers are all UV absorbers (90). Unfortunately, their absorption spectra vary from vitamer to vitamer and detection is complicated by the low concentrations at which they normally occur in foods.

B. Methods of Analysis

1. Extraction Techniques

Since animal products are the food sources of vitamin B_{12}, these vitamers must be extracted from protein-rich matrices under conditions which will not lead to their destruction (13). As a result, extraction procedures often involve digestion with protease followed by extraction into hot (80°C) ethanol. Aqueous dimethylsulfoxide containing ammonium pyrrolidine dithiocarbamate and citric acid to stabilize the B_{12} vitamers has also been used for extraction. For quantitation as total vitamin B_{12}, heating in a cyanide- or sulfate-containing buffer has been used to extract the endogenous B_{12} vitamers and convert them to the more stable cyano- or sulfitocobalamin form (92). Further purification may be achieved via adsorption onto charcoal or chromatography on anion exchange resin and/or solid phase extraction on C18 cartridges.

Specific extraction conditions have been shown to significantly affect the efficiency of the overall extraction process (92–94). It has therefore been suggested that the extraction conditions be optimized for each type of food sample.

2. Methodology Reviews

The HPLC methods published between 1980 and 1990 for cyanocobalamin alone or in combination with other water-soluble vitamins in foods are reviewed in three recent publications (11–13). Extraction and chromatographic conditions as well as the types of foods tested are discussed. Lindemans (90) recently summarized the paper chromatography, TLC, open column chromatography, HPLC, and GC–mass spectrometry techniques for determining vitamin B_{12} in tissue, plasma, and pharmaceutical samples.

3. Reference Methods

The methods currently validated by the AOAC International for vitamin B_{12} are all microbiological in nature (95).

4. Recent Developments in Methodology

a. Ligand Binding Assays. Although the B$_{12}$ vitamers are UV absorbers, this type of detection, even in combination with HPLC separations, does not provide sufficient sensitivity and selectivity for use at the low concentrations of the endogenous B$_{12}$ vitamers in foods. As a result, recent developments have concentrated on ligand binding assays for vitamin B$_{12}$ in foods. Table 26 summarizes a recently published radioisotope dilution assay for total vitamin B$_{12}$ in milk and Table 27 summarizes an enzyme-linked protein binding assay for cyanocobalamin in fortified breakfast cereal.

IX.   BIOTIN

A.   Chemistry and Biochemistry

Biotin deficiency is relatively rare and is characterized by muscle aches, skin rashes, mild depression, slight anemia, and increased serum cholesterol (6,97). Physiologically, biotin participates in gluconeogenesis, fatty acid synthesis, propionate metabolism, and leucine catabolism by acting as the prosthetic group for a number of carboxylase enzymes (13,97). More detailed information on the biochemistry and physiology of biotin are available from a recent review (97).

Food sources of biotin include liver, rice, egg yolks, chocolate, tomatoes, and yeast extract (6,13,97). While biotin is found in the seeds of many cereals and oilseeds, it is largely unavailable to animals in this form. In addition, the bioavailability of biotin can be reduced by significant intake of avidin, a biotin-binding protein found in egg white. Biotin can occur in food as free biotin but is usually bound to protein (e.g., carboxylase enzymes).

In contrast to the other water-soluble vitamins, biotin (Fig. 8) is relatively stable (6,13,97). It is stable to extended periods of heating in strong acid, e.g., autoclaving at 120°C for 2 hr in 4 M sulfuric acid. Biotin is subject to oxidation to biotin sulfoxides in dilute solutions, a circumstance in chromatographic separations that can be prevented by carefully degassing solutions.

Direct detection in biotin is problematic (13,97). Biotin is not a UV absorber; it does not fluoresce naturally, nor is it electrochemically active. Refractometry, which is a notably nonspecific means of detection, is the only means for direct detection of biotin. For practical purposes, derivatization to spectrophotometrically active products or radioassay techniques are usually required to detect biotin in foods.

B.   Methods of Analysis

1. Extraction Techniques

Since endogenous biotin is predominantly protein-bound and is relatively stable, it is extracted from food under fairly harsh conditions. Normal extraction conditions for total biotin would involve autoclaving at 120°C for 2 hr in the presence of 4 N sulfuric acid (13). Enzymatic hydrolysis (e.g., with papain) can also be used to release biotin from food proteins. Sample cleanup can involve adsorption of biotin on charcoal or ion exchange chromatography.

2. Methodology Reviews

Polesello and Rizzolo (11) and Lambert and De Leenheer (13) briefly summarized HPLC methods for determining the biotin content of foods. Gaudry and Ploux (97) covered nonchromatographic techniques as well as column chromatography, paper chromatography, TLC, GC, HPLC, and LC–mass spectrometry for biotin in biological and pharmaceutical samples.

Table 26 Radioisotope Dilution Assay for Quantitating Total Vitamin B$_{12}$ in Food

Analyte(s)	Type of food analyzed	Extraction conditions	Assay conditions	Detector	Method verification[a]	Ref.
Total vitamin B$_{12}$ as cyanocobalamin.	Milk.	Extraction and conversion to cyanocobalamin by autoclaving at 121°C for 25 min in 0.4 M acetate buffer, pH 4.5 containing 2000 ppm sodium cyanide.	As specified for Vitamin B$_{12}$/Folate Dual Radioassay Kit (Amersham): *i.e.*, incubation for 15 min at room temperature with ^{57}Co-labeled cyanocobalamin tracer solution and protein denaturation reagent; incubation with protein binding solution for 45 min; incubation with charcoal for 15 min; centrifugation.	Gamma counter, for ^{57}Co.	External standardization. Reproducibility—$CV \pm 1.2\%$ in milk. Recoveries = $108 \pm 3\%$ from milk (n = 6). Results were significantly higher than those obtained by microbiological assay using standard extraction method. No significant difference when compared with microbiological assay using current extraction method.	92

[a] r = correlation coefficient; n = number of determinations. LoD = limit of detection; SNR = signal-to-noise ratio; CV = coefficient of variation.

Table 27 Enzyme-Linked Protein Binding Assay for Quantitating Vitamin B_{12} in Food

Analyte(s)	Type of food analyzed	Extraction conditions	Assay conditions	Detector	Method verification[a]	Ref.
Fortified cyanocobalamin.	Fortified ready-to-eat breakfast cereal.	Extraction into methanol + citrate-phosphate buffer, pH 7.0 (1 + 1, v/v); centrifugation; pellet reextracted twice as per above procedure; adjustment to pH 7.0.	Incubation overnight at 4°C in the presence of R-protein-horseradish peroxidase conjugate in microtitration plate wells with cyanocobalamin-keyhole limpet hemocyanin conjugate immobilized on well walls; plate washed with phosphate-buffered saline; incubation for 30 min with peroxidase substrate; reaction stopped with 2 M sulfuric acid.	Optical density microplate reader, 450 nm.	External standardization. LoD = 9 pg cyanocobalamin per well. Reproducibility—CV ± 9.4–7% in fortified ready-to-eat breakfast cereal. Recoveries = 71–93% from unfortified breakfast cereal (n = 5).	96

[a] r = correlation coefficient; n = number of determinations. LoD = limit of detection; SNR = signal-to-noise ratio; CV = coefficient of variation.

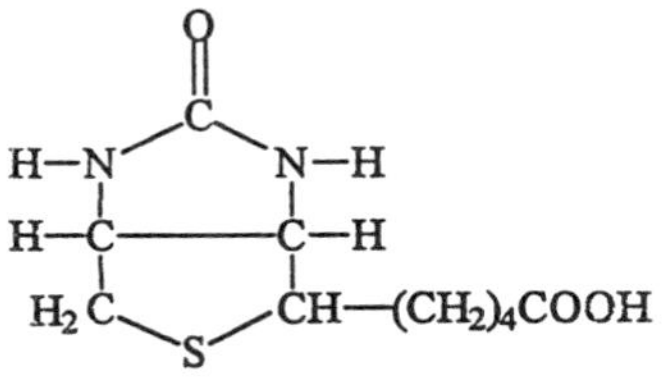

Fig. 8 Biotin.

3. Reference Methods

There are no "official" AOAC International methods for determination of biotin (6). A microbiological method that has yet to be collaboratively studied is recommended for use with food.

4. Recent Developments in Methodology

a. HPLC Methods. A limited amount of work has been done on biotin determination in foods. This is especially true for HPLC methods. Table 28 summarizes a recent HPLC determination of biotin in royal jelly. A simultaneous determination of niacin and biotin in almonds is summarized in Sec. XI, "Simultaneous Determinations of Multiple Vitamins."

X. PANTOTHENIC ACID

A. Chemistry and Biochemistry

Pantothenic acid deficiency appears only in cases of gross malnutrition, and insufficiency is characterized by a burning sensation in the soles of the feet (6).

Free pantothenic acid occurs rarely in nature. The biologically active coenzyme forms, coenzyme A (CoA or CoASH) and acyl carrier protein (ACP), are the principal vitamers in food (13,99). Only the (D+) or (R) enantiomer occurs naturally and possesses biological activity. CoA is integral in the metabolism of carbohydrates, lipids, and nitrogen-containing compounds, participating in the citric acid cycle, and the synthesis of fatty acids, phospholipids, and steroids (cholesterol). Acetyl-CoA acts by mediating the interchange of two-carbon units between donor and acceptor molecules. ACP participates in the biosynthesis of long-chain fatty acids.

Pantothenic acid is present in small quantities in most foods (6,13,99). Rich food sources of pantothenic acid include meats, especially organ meats, fish, yeast, eggs, cheese, whole-grain products, and legumes. Lesser amounts have been found in milk, fruits, and vegetables.

Pantothenic acid (Fig. 9) is relatively labile, being hydrolyzed in strong acids or alkalis, or by thermal processing (13,99). Pantothenic acid is most stable at pH 4–5. Direct detection is feasible by spectrophotometry or refractometry. More detailed information on the chemistry and biochemistry of pantothenic acid is available from a recent review (99).

B. Methods of Analysis

1. Extraction Techniques

Extraction of pantothenic acid from its bound forms has been accomplished with enzymatic hydrolysis using papain, clarase, takadiastase, or intestinal phosphatase (13). The pantothenic acid in meat samples can also be extracted using cold perchloric acid.

Table 28 HPLC Methods for Quantitating Biotin in Foods (C18 Columns, UV Absorbance Detection)

Analyte(s)	Type of sample analyzed	Extraction conditions	HPLC columns	HPLC mobile phase and flow rate	Detection conditions	Method verification[b]	Ref.
Biotin.	Royal jelly.	Extraction of freeze-dried sample by sonication in 10 mM phosphate buffer, pH 3.5; centrifugation.	*Analytical:* Bondclone 10 C18 (300 × 4.6 mm; 10 μm).	*Isocratic:* acetonitrile + 10 mM phosphate buffer, pH 3.5 (10 + 90 v/v).	UV absorbance, 210 nm.	External standardization. Linear range = 50–300 μg/ml. LoD = 0.6 μg/ml. Recoveries = 99–100% from royal jelly samples.	98

[a]Column specifications expressed as (length × id, particle size; manufacturer).
[b]*r* = correlation coefficient; *n* = number of determinations. LoD = limit of detection; SNR = signal-to-noise ratio; *CV* = coefficient of variation.

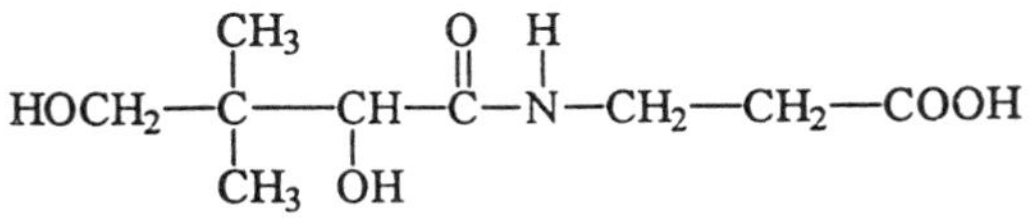

$$\underset{\displaystyle \overset{\displaystyle CH_3}{|}}{\overset{\displaystyle CH_3}{|}}$$

HOCH_2—C——CH—C–N–CH_2—CH_2—COOH

(structure: HOCH_2—C(CH_3)_2—CH(OH)—C(=O)—N(H)—CH_2—CH_2—COOH)

Fig. 9 Pantothenic acid.

2. *Methodology Reviews*

There is a dearth of methods for the determination of pantothenic acid specifically in food. Polesello and Rizzolo (12) and Lambert and De Leenheer (13) briefly discussed HPLC analyses for pantothenic acid in food. Velíšek (99) reviewed nonchromatographic methods and open column chromatography, paper chromatography, TLC, HPLC, and GC primarily for the pantothenic acid in biological and pharmaceutical samples. Velíšek and Divídek (14) reviewed the GC methods for pantothenic acid analysis of foods.

3. *Reference Methods*

The methods sanctioned by the AOAC International for pantothenic acid analysis are all microbiological in nature (100). Analysts are cautioned that food spoilage prior to analysis may lead to inflated pantothenic acid levels (6).

4. *Recent Developments in Methodology*

a. Gas Chromatography-Mass Fragmentography. Very little method development has been done specifically for pantothenic acid in food. Table 29 summarizes a recent GC method that was applied to several food products.

XI. SIMULTANEOUS DETERMINATION OF MULTIPLE VITAMINS

A. Methods of Analysis

1. *Extraction Techniques*

Multivitamin analyses of foods are feasible in specialized cases, such as the determination of fortified (free) rather than naturally occurring vitamins (often bound to other food components), or the determination of vitamins in a relatively simple food system such as juice. Simultaneous separation of multiple water-soluble vitamins is far from insurmountable. The major stumbling block in multivitamin analyses in foods continues to be the nondegradative extraction of the vitamins due to the considerable diversity in their chemical properties, their physical properties, and the conditions that stabilize or destabilize them.

2. *Methodology Reviews*

Several recent reviews provide a comprehensive overview of the HPLC methods that simultaneously determine multiple vitamins in foods (2,11–13). Finglas and Faulks (2) discussed methods for the simultaneous analysis of thiamine, riboflavin, and niacin in foods. The others (11–13) covered methods for various combinations of thiamine, riboflavin, niacin, vitamin B_6, vitamin B_{12}, pantothenate, and folacin.

Table 29 Gas Chromatography–Mass Fragmentography Method for Quantitating Pantothenic Acid in Foods

Analyte(s)	Type of food analyzed	Extraction conditions	GC columns[a]	Carrier gas and flow rate	Detection conditions	Method verification[b]	Ref.
Pantothenic acid, hopantenic acid & 5-[(2,4-dihydroxy-3,3-dimethyl-1-oxobutyl)amino] pentanoic acid simultaneously.	Rice; green tea; dried yeast.	Aqueous extraction at 70°C for 30 min; centrifugation; sample cleanup on ion exchange resin (H type, MCI GEL CK08P); extraction of aqueous column eluate with chloroform, ethyl acetate (×2); addition of internal standard to ethyl acetate phase; evaporation to dryness at 40°C under nitrogen; trimethylsilylation of residue.	Wide-bore fused silica coated with DB-17 (15 m × 0.53 mm; J&W Scientific) at 200°C.	Helium. 15 ml/min. Injection port and separator at 250°C.	Mass fragmentography with multiple ion detector.	Internal standardization using 5-[(2,4-dihydroxy-3,3-dimethyl-1-oxobutyl) amino] pentanoic acid as internal standard. Linear range = 5–100 ng/ml. LoD = 1 ng/ml.	101

[a]Column specifications expressed as (length × id, particle size; manufacturer).
[b]r = correlation coefficient; n = number of determinations. LoD = limit of detection; SNR = signal-to-noise ratio; CV = coefficient of variation.

3. Recent Developments in Methodology

a. HPLC Methods. HPLC is the method of choice for multivitamin determinations of foods. Recent HPLC methods for the simultaneous determination of more than one vitamin are summarized in Tables 30–34 and are organized by vitamin.

XII. FUTURE DIRECTIONS

A. Reference Materials

One of the major problems in vitamin methodology is the lack of standardization during method verification. This makes it difficult if not impossible to assess the validity of individual methods and to compare the different procedures with one another. Although standard reference materials have been developed and are in use for the minerals, the lability of the vitamins has retarded the development of similar materials for the vitamins.

Reference or control materials for method development and quality control purposes should be drawn from a matrix that will mimic real samples as closely as possible, i.e., from food materials rather than standard solutions or pharmaceuticals (107). It is preferable that the analyte of interest be endogenous in the selected reference material in order to mimic the form and compartmentalization of the vitamins in real food samples. Ideal reference materials should be homogeneous in composition, should cover the range of concentrations of the analytes in food samples, and should be accompanied by complete instructions for proper storage, handling, and use. The analyte(s) should be stable in the reference material and their concentration in the reference material should be certified using definitive method(s). A summary of reference materials, their development and uses has been prepared by Wolf (107).

Reference materials are now starting to be developed for vitamins in foods. The results of a recent study indicated that thiamine and vitamin C were relatively stable in whole-milk powder, pork muscle, and freeze-dried haricot beans that were packed under dry nitrogen gas in plastic/aluminum laminated heat-sealed bags and stored at $-18°C$ and $+4°C$ for 24 months (108). However, the authors noted difficulty in distinguishing between variability due to instability in storage and variability in the analytical methodology.

Total diet composites representing the "average" food intake for 25- to 30-year-old males from four geographic regions within the United States were recently tested for thiamine, riboflavin, vitamin B_6, vitamin B_{12}, niacin, pantothenic acid, folic acid, and biotin (109). The food composites were obtained from retail outlets, prepared as ready-to-eat foods and stored in domestic quality self-sealing plastic bags at $-20°C$ for 3–6 weeks before analysis. Based on the consistency of the repeated analyses of vitamin content in food composites collected in different geographic locations and at different times of year, the suggestion has been made that these materials be considered as candidate reference materials for the vitamins listed above. In contrast, the vitamin C content of these composites was found to be quite variable possibly due to its instability during handling and storage.

B. Microcolumn HPLC

Despite the popularity of conventional HPLC methods for vitamin analysis, there are a number of disadvantages inherent in its use (1). These include high setup and operating costs; a requirement for sophisticated technical expertise to operate the instrumentation; extensive sample cleanup and/or analyte derivatization required prior to HPLC separation; lower limits of detection (depending on the method of detection) with respect to microbiological methods in particular; and no information on bioavailability provided.

Table 30 HPLC Methods for Quantitating Thiamine and Riboflavin in Foods

Analyte(s)	Type of food analyzed	Extraction conditions	HPLC columns[a]	HPLC mobile phase and flow rate	Detection conditions	Method verification[b]	Ref.
Thiamine & riboflavin simultaneously.	Broccoli; cereal; flour.	Extraction by autoclaving at 121°C for 30 min in 0.1 N hydrochloric acid; adjustment to pH 4.5 with 2 N sodium acetate; filtration; oxidation of thiamine to thiochrome with 1.0% potassium ferricyanide in 15% sodium hydroxide; sample cleanup using solid phase extraction on C18 Sep-Pak cartridges (Waters).	*Analytical:* μBondapak C18 (300 × 3.9 mm; Waters).	*Isocratic:* methanol + 5 mM ammonium acetate buffer, pH 5.0 (28 + 72 v/v). 1.5 ml/min.	Fluorescence, time-programmed wavelength switching using: - 370/435 nm (ex/em) for first 10 min (for thiochrome) - switch to 370/520 nm (ex/em) for rest of analysis (for riboflavin).	External standardization. Linear calibration up to 100 ng on-column of each vitamin. LoD = 0.05 ng on-column for both thiamine and riboflavin at SNR = 12:1. Good agreement with results obtained by AOAC International method and for AACC check samples.	102

[a]Column specifications expressed as (length × id, particle size; manufacturer).
[b]r = correlation coefficient; n = number of determinations. LoD = limit of detection; SNR = signal-to-noise ratio; CV = coefficient of variation.

Table 31 HPLC Methods for Quantitating Thiamine, Riboflavin, and Pyridoxine in Food

Analyte(s)	Type of food analyzed	Extraction conditions	HPLC columns[a]	HPLC mobile phase and flow rate	Detection conditions	Method verification[b]	Ref.
Thiamine, riboflavin, supplemented pyridoxine, & m-hydroxy-benzoic acid simultaneously.	Milk- & soy-based infant formula.	Extraction with perchloric acid; filtration; addition of internal standard.	*Analytical:* NovaPak C18 (150 × 3.9 mm; Waters).	*Isocratic:* acetonitrile + phosphate-ammonium hydroxide buffer, pH 3.60 containing 0.90 g/L 1-hexane-sulfonic acid (9.0 + 91.0 v/v) 1.0 ml/min.	*Thiamine & riboflavin*—UV absorbance, 254 nm. *Pyridoxine*—fluorescence, 295/395 nm (ex/em).	Internal standardization using *m*-hydroxybenzoic acid. Linear range : 0.2–1.4 μg/ml for thiamine and riboflavin; 0.2–1.2 μg/ml for pyridoxine. LoD = 0.15 μg/ml for thiamine; 0.09 μg/ml for riboflavin; 0.01 μg/ml for pyridoxine. Reproducibility—CV ± 0.9% for thiamine; ± 1.4% for riboflavin; ± 1.8% for pyridoxine in infant formula ($n = 10$). Recoveries = 102% for thiamine and riboflavin from infant formula; 101% for pyridoxine from infant formula. Good agreement with results from AOAC International manual method for thiamine, and AOAC International automated and microbiological methods for riboflavin. Results were lower than AOAC International microbiological method for total vitamin B6.	103

[a] Column specifications expressed as (length × id, particle size; manufacturer).
[b] r = correlation coefficient; n = number of determinations. LoD = limit of detection; SNR = signal-to-noise ratio; CV = coefficient of variation.

Table 32 HPLC Methods for Quantitating Riboflavin, Niacin, Pyridoxine, and Folacin in Food

Analyte(s)	Type of food analyzed	Extraction conditions	HPLC columns[a]	HPLC mobile phase and flow rate	Detection conditions	Method verification[b]	Ref.
Riboflavin, niacin, vitamin B6 & folic acid simultaneously.	Eggs.	Extraction with 0.1 N sulfuric acid; enzymatic hydrolysis with takadiastase and papain.	*Analytical:* μBondapak C18 (300 × 3.9 mm; Waters).	*Isocratic:* methanol + aqueous 2.4 mM PIC-B (hexane sulfonic acid), pH 4.2 (22 + 78 v/v). 0.3 ml/min.	UV absorbance, 254 nm.	External standardization. Reproducibility—*CV* ± 7.9% and 8.1% for riboflavin; ± 10.9% and 14.5% for pyridoxine; ± 29.4% and 27.6% for folic acid; ± 25.5% and 24.1% for niacin in egg yolk and albumen, respectively. Recoveries = 102.1% and 102.5% for riboflavin; 96.6% and 98.4% for pyridoxine; 103.3% and 94.5% for folic acid; 92.7% and 92.4% for niacin from egg yolk and albumin, respectively.	104

[a]Column specifications expressed as (length × id, particle size; manufacturer).
[b]r = correlation coefficient; n = number of determinations. LoD = limit of detection; SNR = signal-to-noise ratio; CV = coefficient of variation.

Table 33 HPLC Methods for Quantitating Niacin and Pyridoxine in Foods

Analyte(s)	Type of food analyzed	Extraction conditions	HPLC columns[a]	HPLC mobile phase and flow rate	Detection conditions	Method verification[b]	Ref.
Fortified nicotinamide and pyridoxine simultaneously.	High protein meal powder; milk fortified with malt and cocoa.	Extraction with 1 M sulfuric acid at 100°C for 30 min.	*Precolumn:* Pellicular ODS (40 × 4.6 mm; Whatman). *Analytical:* Partisil ODS (250 × 4.6 mm, 10 μm; Whatman) or Ultrasphere ODS (250 × 4.6 mm, 5 μm; Beckman) if artifacts interfered with pyridoxine on Partisil column.	*Isocratic:* acetate buffer containing 1-heptane sulfonic acid. 2.0 ml/min.	UV absorbance at 260 nm for nicotinamide in series with fluorescence at 296/396 nm (ex/em) for pyridoxine.	External standardization. LoD = 0.1 mg/100 g for nicotinamide; 0.01 mg/100 g for pyridoxine. Reproducibility—CV ± 1.4% for nicotinamide; ± 1.6% for pyridoxine in liquid meal supplement (n = 6). Recoveries = 97% for nicotinamide and 99% for pyridoxine from food samples tested. Results agreed well with those from microbiological methods.	105

[a] Column specifications expressed as (length × id, particle size; manufacturer).
[b] r = correlation coefficient; n = number of determinations. LoD = limit of detection; SNR = signal-to-noise ratio; CV = coefficient of variation.

Table 34 HPLC Methods for Quantitating Niacin and Biotin in Foods

Analyte(s)	Type of food analyzed	Extraction conditions	HPLC columns[a]	HPLC mobile phase and flow rate	Detection conditions	Method verification[b]	Ref.
Niacin and biotin simultaneously.	Almonds.	Fat extraction with *n*-hexane (Soxhlet); vitamin extraction by sonicating in 0.1 M hydrochloric acid at 75°C for 15 min; adjustment to pH 4.6 using 0.1 M sodium hydroxide; solid phase extraction/cleanup on strong cation exchange cartridge (Baker-10-SPE).	*Precolumn:* Supelcosil LC-8-DB (20 × 4.6 mm, 3 μm; Supelco). *Analytical:* Supelcosil LC-8-DB (33 × 4.6 mm, 3 μm; Supelco) at 35°C.	*Isocratic:* methanol + 5 mM hexanesulfonate containing 0.1% triethylamine, pH 2.8 (15 + 85 v/v). 1.0 ml/min.	UV absorbance, 200 nm	External standardization. Linear range = 2–10 μg of niacin and biotin. Results showed large variability. Recoveries = 97.6% for niacin; 79.8% for biotin from standard solutions.	106

[a]Column specifications expressed as (length × id, particle size; manufacturer).
[b]r = correlation coefficient; n = number of determinations. LoD = limit of detection; SNR = signal-to-noise ratio; CV = coefficient of variation.

Table 35 Capillary Electrophoresis Methods for Quantitating Vitamin C in Fruit Juice

Analyte(s)	Type of food analyzed	Extraction conditions	Type of analysis and column[a]	Analytical conditions	Detection conditions	Method verification[b]	Ref.
AA & IAA simultaneously.	Lemon juice; orange juice.	Juice expressed and filtered.	Capillary Zone Electrophoresis. Coated column (20 cm × 25 μm; BioRad).	Separation buffer = 0.1 M phosphate, pH 5.0. Separation voltage = 8 kV. Injection—on cathodic side by electromigration at 8 kV for 8 sec.	UV absorbance, 265 nm.	External standardization. Linear range = 0–1 mg/ml for peak height vs. concentration; $r = 0.99$ for both AA & IAA ($n = 10$). LoD = 0.25 μg/ml for AA & IAA at SNR > 2. Reproducibility—$CV \pm 5\%$ for AA in fruit juice ($n = 5$). Recoveries—97–101% AA from lemon and orange juice ($n = 5$).	112
AA & IAA simultaneously.	Orange juice; apple juice; grapefruit juice; vegetable juice; cranberry cocktail; white wine.	IAA added as internal standard; metaphosphoric acid (100 g/L) extraction.	Capillary zone electrophoresis. Polyimide-coated column (37 cm × 75 μm; Polymicro Technologies). Length to detector = 30 cm.	Separation buffer = 100 mM tricine, pH 8.8. Separation voltage = 11 kV (297 V/cm). Pressure injection.	UV absorbance, 254 nm.	Internal standardization with IAA. Linear range = 1.6–480.0 μg/ml; $r = 0.9984$.	113

[a]Column specifications expressed as (length × manufacturer).
[b]r = correlation coefficient; n = number of determinations. LoD = limit of detection; SNR = signal-to-noise ratio; CV = coefficient of variation.

The concept of microcolumn HPLC is not new, but dedicated instrumentation is only now being developed. Widespread acceptance of this technique will definitely have to wait for development of appropriate hardware. Conversion of conventional equipment to microcolumn use has proven less than successful, necessitating considerable capital expenditure to set up microcolumn applications. However, there are distinct advantages associated with microcolumn HPLC. These include reduced solvent volume, rapid analysis, high resolution, and smaller sample volume. As concern increases over environmental issues, the significant reduction in solvent used and the concomitant reduction in waste solvent disposal costs that can be realized with microcolumn applications may offset the setup costs to some extent. To date, most of the microcolumn work has been applied to pharmaceuticals, which are less complex than food matrices. However, microcolumn HPLC methods for vitamins in foods may well become an area for future development.

C. Capillary Electrophoresis

Capillary electrophoresis (CE) is fast becoming a viable alternative to conventional and microcolumn HPLC. It provides many of the same advantages as microcolumn HPLC: low solvent volume, rapid analysis, high resolution, and small sample volume. CE techniques are just now leaving the developmental stages and being applied to "real world samples." While the literature contains many CE applications in the pharmaceutical field, similar progress has not been realized in food analysis, largely due to the complex nature of food systems and the corresponding complexity that nondegradative vitamin extraction entails. Not surprisingly, vitamin C analysis of fruit juice is one of the first food samples to which CE has been applied. Table 35 summarizes recent advances in this area. CE is expected to be an area for future development with respect to the micronutrients in foods as CE-compatible extraction conditions are developed.

D. Analytical Automation

Most of the analytical methods now being developed are or can be semiautomated, usually by using an autosampler. This concept has been carried one step further through the use of HPLC-robotics systems to fully automate the extraction portion of the analytical procedure. A robotics-HPLC analysis of the individual C vitamers was recently applied to the analysis of a variety of fresh fruits and vegetables (110,111). The total vitamin C content of a variety of fresh fruits and vegetables and juices was automated by combining a robotic extraction with flow injection analysis. The robotics-FIA results were reported to be reproducible within 1.5% and agreed reasonably well with the AOAC International (reference) method. Recoveries ranged from 70–112% in a variety of samples. With increasing demands for nutritional information, the trend toward analytical automation in the analysis of water-soluble vitamins is expected to continue.

REVIEWS

D. M. Sullivan and D. E. Carpenter, eds., *Methods of Analysis for Nutritional Labelling.* AOAC International, Arlington, VA, 1993.

A. P. De Leenheer, W. E. Lambert, and H. J. Nelis, eds., *Modern Chromatographic Analysis of Vitamins*, 2nd ed., Marcel Dekker, New York, 1992.

L. M. L. Nollet, ed., *Food Analysis by HPLC*, Marcel Dekker, New York, 1992.

A. Polesello and A. Rizzolo, *J. Micronutr. Anal.*, 2: 153 (1986).

A. Polesello and A. Rizzolo, *J. Micronutr. Anal.*, 8: 105 (1990).

REFERENCES

1. R. R. Eitenmiller, *J. Food Qual.*, *13*: 7 (1990).
2. P. M. Finglas and R. M. Faulks, *J. Micronutr. Anal.*, *3*: 251 (1987).
3. G. R. Beecher, and J. T. Vanderslice, Determination of nutrients in foods: factors that must be considered, in *Modern Methods of Food Analysis* (K. K. Stewart and J. R. Whitaker, eds.), AVI, Westport, 1984, pp. 29–55.
4. M. T. Parviainen and K. Nyyssönen, Ascorbic acid, in *Modern Chromatographic Analysis of Vitamins*, 2nd ed. (A. P. De Leenheer, W. E. Lambert, and H. J. Nelis, eds.), Marcel Dekker, New York, 1992, pp. 235–260.
5. M. B. Davies, J. Austin, and D. A. Partridge, Biochemistry of vitamin C, in *Vitamin C: Its Chemistry and Biochemistry*, Royal Society of Chemistry, Letchworth, UK, 1991, pp. 74–96.
6. J. W. DeVries, Water-soluble vitamins, in *Methods of Analysis for Nutritional Labelling* (D. M. Sullivan and D. E. Carpenter, eds.), AOAC International, Arlington, VA, 1993, pp. 131–135.
7. M.-L Liao and P. A. Seib, *Food Chem.*, *30*: 289 (1988).
8. N. A. Golubkina and O. V. Prudnik, *J. Anal. Chem. USSR*, *44*: 1091 (1989).
9. M. B. Davies, J. Austin, and D. A. Partridge, Inorganic and analytical aspects of vitamin C chemistry, in *Vitamin C: Its Chemistry and Biochemistry*, Royal Society of Chemistry, Letchworth, UK, 1991, pp. 115–146.
10. P. W. Washko, R. W. Welch, K. R. Dhariwal, Y. Wang, and M. Levine, *Anal. Biochem.*, *204*: 1 (1992).
11. A. Polesello and A. Rizzolo, *J. Micronutr. Anal.*, *2*: 153 (1986).
12. A. Polesello and A. Rizzolo, *J. Micronutr. Anal.*, *8*: 105 (1990).
13. W. E. Lambert and A. De Leenheer, Quantitative determination of water-soluble vitamins using HPLC, in *Food Analysis by HPLC* (L. M. L. Nollet, ed.), Marcel Dekker, New York, 1992, pp. 341–369.
14. J. Velíšek and J. Davídek, *J. Micronutr. Anal.*, *2*: 25 (1986).
15. D. M. Sullivan and D. E. Carpenter, eds., Vitamin C, in *Methods of Analysis for Nutritional Labelling*, AOAC International, Arlington, VA, 1993, pp. 561–568.
16. M. Hoare, S. Jones, and J. Lindsay, *Food Aust.*, *45*: 341 (1993).
17. S. Zapata and J.-P. Dufour, *J. Food Sci.*, *57*: 506 (1992).
18. M. O. Nisperos-Carriedo, B. S. Buslig, and P. E. Shaw, *J. Agric. Food Chem.*, *40*: 1127 (1992).
19. M. Sawamura, S. Ooishi, and Z.-F. Li, *J. Sci. Food Agric.*, *53*: 279 (1990).
20. A. Romero-Rodriguez, L. Vazquez Oderiz, J. Lopez Hernandez, and J. Simal Gandara, *Sci. Aliments*, *12*: 593 (1992).
21. M. A. Romero Rodriguez, M. L. Vazquez Oderiz, J. Lopez Hernandez, and J. Simal Lozano, *J. Chromatogr. Sci.*, *30*: 433 (1992).
22. J. M. Irache, I. Ezpeleta, and F. A. Vega, *Chromatographia*, *35*: 232 (1993).
23. K. Y. Dodson, E. R. Young, and A.-G. M. Soliman, *J. AOAC Int.*, *75*: 887 (1992).
24. H. S. Kim, P. A. Seib, and O. K. Chung, *J. Food Sci.*, *58*: 845 (1993).
25. W. A. Behrens and R. Madere, *J. Liq. Chromatogr.*, *15*: 753 (1992).
26. J. T. Vanderslice and D. J. Higgs, *J. Nutr. Biochem.*, *4*: 184 (1993).
27. J. T. Vanderslice, D. J. Higgs, J. M. Hayes, and G. Block, *J. Food Comp. Anal.*, *3*: 105 (1990).
28. J. T. Vanderslice, and D. J. Higgs, *J. Chromatogr. Sci.*, *22*: 485 (1984).
29. N. Bilic, *J. Chromatogr.*, *543*: 367 (1991).
30. H. Hayashi and A. Miyagawa, *J. Food Hyg. Soc. Japan*, *31*: 44 (1990).
31. W. D. Graham and D. Annette, *J. Chromatogr.*, *594*: 187 (1992).
32. J. M. Kim, Y. Huang, and R. D. Schmid, *Anal. Lett.*, *23*: 2273 (1990).
33. A. A. Alwarthan, *Analyst*, *118*: 639 (1993).
34. G. M. Greenway and P. Ongomo, *Analyst*, *115*: 1297 (1990).
35. S. Daily, S. J. Armfield, B. G. D. Haggett, and M. E. A. Downs, *Analyst*, *116*: 569 (1991).
36. E. T. A. Marques, F. A. C. Spencer Neto, and J. L. Lima Filho, *Biochem. Ed.*, *19*: 77 (1991).
37. I. D. H. C. Marques, E. T. A. Marques, A. C. Silva, W. M. Ledingham, E. H. M. Melo, V. L. da Silva, and J. L. Lima Filho, *Appl. Biochem. Biotechnol.*, *44*: 81 (1994).
38. T. Kawasaki, Vitamin B1: thiamine, in *Modern Chromatographic Analysis of Vitamins*, 2nd ed. (A. P. De Leenheer, W. E. Lambert, and H. J. Nelis, eds.), Marcel Dekker, New York, 1992, pp. 319–354.

39. C. M. Sullivan and D. E. Carpenter, eds., Thiamine, in *Methods of Analysis for Nutritional Labelling*, AOAC International, Arlington, VA, 1993, pp. 535–543.

40. A. M. Alyabis and K. L. Simpson, *J. Food Comp. Anal.*, *6*: 166 (1993).

41. J. B. Fox, S. A. Ackerman, and D. W. Thayer, *J. AOAC Int.*, *75*: 346 (1992).

42. S. M. Fernando and P. A. Murphy, *J. Agr. Food Chem.*, *38*: 163 (1990).

43. V. Ollilainen, L. Vahteristo, A. Uusi-Rauva, P. Varo, P. Koivistoinen, and J. Huttunen, *J. Food Comp. Anal.*, *6*: 152 (1993).

44. R. Gauch, U. Leuenberger, and U. Müller, *Z. Lebensm. Unters. Forschung*, *195*: 312 (1992).

45. H. Ohta, M. Maeda, Y. Nogata, K.-I. Yoza, Y. Takeda, and Y. Osajima, *J. Liq. Chromatogr.*, *16*: 2617 (1993).

46. A. Laschi-Loquerie, S. Vallas, J. Viollet, M. Leclercq, and V. Fayol, *Int. J. Vit. Nutr. Res.*, *62*: 248 (1992).

47. G. R. Skurray, *ASEAN Food J.*, *5*: 123 (1990).

48. P. Nielsen, Flavins, in *Modern Chromatographic Analysis of Vitamins*, 2nd ed. (A. P. De Leenheer, W. E. Lambert, and H. J. Nelis, eds.), Marcel Dekker, New York, 1992, pp. 355–398.

49. L. F. Russell and J. T. Vanderslice, *J. Micronutr. Anal.*, *8*: 257 (1990).

50. C. M. Sullivan and D. E. Carpenter, eds., Riboflavin, in *Methods of Analysis for Nutritional Labelling*, AOAC International, Arlington, VA, 1993, pp. 439–447.

51. L. F. Russell and J. T. Vanderslice, *Food Chem.*, *43*: 79 (1992).

52. L. F. Russell and J. T. Vanderslice, *Food Chem.*, *43*: 151 (1992).

53. É Barna, *Acta Aliment.*, *21*: 3 (1992).

54. A Muñoz, R. Ortiz, and M. A. Murcia, *Food Chem.*, *49*: 203 (1994).

55. V. Ollilainen, P. Mattila, P. Varo, P. Koivistoinen, and J. Huttunen, *J. Micronutr. Anal.*, *8*: 199 (1990).

56. N. Bilic, and R. Sieber, *J. Chromatogr.*, *511*: 359 (1990).

57. C. Kanno, K. Shirahuji, and T. Hoshi, *J. Food Sci.*, *56*: 678 (1991).

58. K. Shibata and T. Shimono, Nicotinic acid and nicotinamide, in *Modern Chromatographic Analysis of Vitamins*, 2nd ed. (A. P. De Leenheer, W. E. Lambert, and H. J. Nelis, eds.), Marcel Dekker, New York, 1992, pp. 235–260.

59. A. Vidal-Valverde and A. Reche, *J. Agric. Food Chem.*, *39*: 116 (1991).

60. C. M. Sullivan, and D. E. Carpenter, eds., Niacin, in *Methods of Analysis for Nutritional Labelling*, AOAC International, Arlington, VA, pp. 315–325 (1993).

61. T. A. Tyler and J. A. Genzale, *JAOAC*, *73*: 467 (1990).

62. S. Hirayama and M. Maruyama, *J. Chromatogr.*, *588*: 171 (1991).

63. G. W. Chase, W. O. Landen, A.-G. M. Soliman, and R. R. Eitenmiller, *J. AOAC Int.*, *76*: 390 (1993).

64. J. B. Ubbink, Vitamin B₆, in *Modern Chromatographic Analysis of Vitamins*, 2nd ed. (A. P. De Leenheer, W. E. Lambert, and H. J. Nelis, eds.), Marcel Dekker, New York, 1992, pp. 399–439.

65. J. F. Gregory, P. R. Trumbo, L. B. Bailey, J. P. Toth, T. G. Baumgartner, and J. J. Cerda, *J. Nutr.*, *121*: 177 (1991).

66. J. F. Gregory and D. B. Sartain, *J. Agric. Food Chem.*, *39*: 899 (1991).

67. C. M. Sullivan, and D. E. Carpenter, eds., Pyridoxine, in *Methods of Analysis for Nutritional Labelling*, AOAC International, Arlington, VA, 1993, pp. 315–325.

68. M. Reitzer-Bergaentzle, E. Marchioni, and C. Hasselmann, *Food Chem.*, *48*: 321 (1993).

69. K. Tadera and Y. Naka, *Agric. Biol. Chem.*, *55*: 563 (1991).

70. S. J. Olds, J. T. Vanderslice, and D. Brochetti, *J. Food Sci.*, *58*: 505 (1993).

71. R. E. Davis and D. J. Nicol, *Int. J. Biochem.*, *20*: 133 (1988).

72. J. G. Hawkes and R. Villota, *CRC Crit. Rev. Food Sci. Nutr.*, *28*: 439 (1989).

73. R. J. Mullin, and D. S. Duch, Folic acid, in *Modern Chromatographic Analysis of Vitamins*, 2nd ed. (A. P. De Leenheer, W. E. Lambert, and H. J. Nelis, eds.), Marcel Dekker, New York, 1992, pp. 261–283.

74. C. L. Krumdieck, T. Tamura, and I. Eto, *Vitamins Hormones*, *40*: 45 (1983).

75. J. F. Gregory, *JAOAC*, *67*: 1015 (1984).

76. P. M. Keagy, I. Folacin. Microbiological and Animal Assays, in *Methods of Vitamin Assay* (J. Augustin, B. P. Klein, D. Becker, and P. B. Venugopal, eds.), John Wiley and Sons, New York, 1985, pp. 445–471.

77. J. F. Gregory, II. Folacin. Chromatographic and Radiometric Assays, in *Methods of Vitamin Assay* (J. Augustin, B. P. Klein, D. Becker, and P. B. Venugopal, eds.), John Wiley and Sons, New York, 1985, pp. 473–496.

78. P. M. Finglas, U. Faure, and D. A. T. Southgate, *Food Chem.*, *46*: 199 (1993).
79. J. F. Gregory, R. Engelhardt, S. D. Bhandari, D. B. Sartain, and S. K. Gustafson, *J. Food Comp. Anal.*, *3*: 134 (1990).
80. R. Engelhardt and J. F. Gregory, *J. Agric. Food Chem.*, *38*: 154 (1990).
81. C. M. Sullivan and D. E. Carpenter, eds., Folacin, in *Methods of Analysis for Nutritional Labelling*, AOAC International, Arlington, VA, 1993, pp. 265–272.
82. D. M. Goli and J. T. Vanderslice, *J. Micronutr. Anal.*, *6*: 19 (1989).
83. D. M. Goli and J. T. Vanderslice, *Food Chem.*, *43*: 57 (1992).
84. A. Hahn, J. Stein, U. Rump, and G. Rehner, *J. Chromatogr.*, *540*: 207 (1991).
85. D. R. White, H. S. Lee, and R. E. Krüger, *J. Agric. Food Chem.*, *39*: 714 (1991).
86. B. T. Jacoby and F. T. Henry, *J. AOAC Int.*, *75*: 891 (1992).
87. E. Seyoum and J. Selhub, *J. Nutr. Biochem.*, *4*: 488 (1993).
88. G. Varela-Moreiras, E. Seyoum, and J. Selhub, *J. Nutr. Biochem.*, *2*: 44 (1991).
89. J. Selhub, *Anal. Biochem.*, *182*: 84 (1989).
90. J. Lindemans, Cobalamins, in *Modern Chromatographic Analysis of Vitamins*, 2nd ed. (A. P. De Leenheer, W. E. Lambert, and H. J. Nelis, eds.), Marcel Dekker, New York, 1992, pp. 469–513.
91. B. T. Goulding, *Chem. Britain*, *26*: 950 (1990).
92. I. Andersson, R. Lundqvist, and R. Öste, *Milchwissenschaft*, *45*: 507 (1990).
93. B. G. Österdahl, K. Janné, E. Johansson, and H. Johnsson, *Int. J. Vit. Nutr. Res.*, *56*: 95 (1986).
94. M. S. Marcus, M. D. Prabhudesai, and M. S. Wassef, *Am. J. Clin. Nutr.*, *33*: 137 (1980).
95. C. M. Sullivan, and D. E. Carpenter, eds., Cyanocobalamine, in *Methods of Analysis for Nutritional Labelling*, AOAC International, Arlington, VA, 1993, pp. 195–200.
96. S. C. Alcock, P. M. Finglas, and M. R. A. Morgan, *Food Chem.*, *45*: 199–203 (1992).
97. M. Gaudry and O. Ploux, Biotin, in *Modern Chromatographic Analysis of Vitamins*, 2nd ed. (A. P. De Leenheer, W. E. Lambert, and H. J. Nelis, eds.), Marcel Dekker, New York, 1992, pp. 441–467.
98. N. Pranee, H. Aung Kyaw, C. Jittra, and P. Chaveewon, *ASEAN Food J.*, *6*: 163 (1991).
99. J. Velíšek, J. Davídek, and T. Davídek, Pantothenic acid, *Modern Chromatographic Analysis of Vitamins*, 2nd ed. (A. P. De Leenheer, W. E. Lambert, and H. J. Nelis, eds.), Marcel Dekker, New York, 1992, pp. 515–560.
100. C. M. Sullivan and D. E. Carpenter, eds., Pantothenic acid, in *Methods of Analysis for Nutritional Labelling*, AOAC International, Arlington, VA, 1993, pp. 327–330.
101. K. Banno, M. Matsuoka, S. Horimoto, and J. Kato, *J. Chromatogr.*, *525*: 255 (1990).
102. A. Sims and D. Shoemaker, *J. AOAC Int.*, *76*: 1156 (1993).
103. G. W. Chase, W. O. Landen, R. R. Eitenmiller, and A.-G. M. Soliman, *J. AOAC Int.*, *75*: 561 (1992).
104. S. Ötles and Y. Hisil, *Ital. J. Food Sci.*, *5*: 69 (1993).
105. D. I. Rees, *J. Micronutr. Anal.*, *5*: 53 (1989).
106. A. Rizzolo, C. Baldo, and A. Polesello, *J. Chromatogr.*, *553*: 187 (1991).
107. W. R. Wolf, Reference materials, in *Methods of Analysis for Nutritional Labelling* (D. M. Sullivan and D. E. Carpenter, eds.), AOAC International, Arlington, VA, 1993, pp. 111–122.
108. P. C. H. Hollmann, J. H. Slangen, P. M. Finglas, P. J. Wagstaffe, and U. Faure, *Fresenius' J. Anal. Chem.*, *345*: 236 (1993).
109. J. T. Tanner, G. V. Iyengar, and W. R. Wolf, *Fresenius' J. Anal. Chem.*, *338*: 438 (1990).
110. J. Quán de Serrano, L. de González, and N. W. Solomons, *Food Chem..*, *47*: 87 (1993).
111. J. T. Vanderslice and D. J. Higgs, *J. Micronutr. Anal.*, *1*: 143 (1985).
112. B. L. Ling, W. R. G. Baeyens, P. Van Acker, and C. Dewaele, *J. Pharm. Biomed. Anal.*, *10*: 717 (1992).
113. E. V. Koh, M. G. Bissell, and R. K. Ito, *J. Chromatogr.*, *633*: 245 (1993).

Literature review completed in August, 1994.

19

Analysis for Organic Acids

D. Blanco Gomis
University of Oviedo, Oviedo, Spain

J. J. Mangas Alonso
Centro de Investigation Aplicada y Technologia Agroalimentaria (CIATRA), Villaviciosa (Asturias), Spain

I. INTRODUCTION

Because they play a pivotal role in maintaining the quality and nutritional value of a variety of foods, organic acids are among the most frequently assayed substances in this type of substrate. The organic acids present in foods originate from biochemical processes; from their addition as acidulants, stabilizers, or preservatives; or from the activity of some microorganisms (particularly yeasts and bacteria). They contribute to the sensory properties of foods. In fact, the role of organic acids in the taste and aroma of alcoholic beverages is well documented (1); the sour, sharp, and irritating flavors of cider are ascribed to lactic acid, total acidity, and acetic acid, respectively. Malic and acetic acid bear a negative though significant correlation to sweet taste and scented flavors; also, they taste equiacidic with solutions containing sucrose (e.g., 0.7% D-tartaric acid with a 0.8% concentration of this acid plus 2% sucrose) (2). Acetic acid and the global hedonic score for fermented beverages also bear a significant negative mutual correlation.

Organic acids may exhibit other sensory properties. For example, citric acid possesses sour-and-sweet sensory notes, and succinic acid has a salty-bitter taste. On the other hand, the typical taste and flavor of Emmental cheese can be ascribed to propionic acid and a few other acids; also, raw meat smells much like lactic acid, which arises from postmortem anaerobic glucolysis and determines the pH of meat, its final properties and microbial stability. Table 1 lists the reported threshold concentrations for various organic acids in different media (3–5).

Sour-tasting compounds are all acidic. The acidity of mineral acids is determined by their hydrogen ion concentration; that of organic acids, however, cannot be defined in such simple terms. Thus, a solution of acetic acid tastes more sour than one of a mineral acid at the same pH (4); however, the mineral acid tastes more sour than the organic acid in equimolar solutions. The threshold concentrations of organic acids are pH-dependent, yet the relationship is usually quite complicated. The threshold concentrations of formic, malic, and succinic acid increase with increasing pH, whereas those of acetic, butyric, and lactic acid exhibit the opposite trend (6).

Organic acids are added to some foods in order to prevent sedimentation or darkening; thus

Table 1 Typical Threshold Concentrations of
Organic Acids in Various Media

Acid	Conc. (meq/L)
Acetic	1.800
Tartaric	1.200
Citric	2.300
Malic	0.500
Lactic[a]	0.400
Pyruvic[a]	2.800
Phenylacetic[a]	0.007
Propionic[b]	0.270
Valeric[b]	0.005
Isobutyric[b]	0.090
Butyric[b]	0.040
Acetic[b]	0.430
Isovaleric[b]	0.007

[a]Flavor threshold in beer.
[b]Flavor threshold in whiskey.

citric, malic, phosphoric, and ascorbic acid are used to reduce or delay enzymatic browning. Melanin formation following production of a quinone is a pH-dependent process (the rate of the former process increases with increasing pH) (7). Polyhydroxy acids such as citric and malic acid inhibit the polyphenol oxidase system by chelating the copper prosthetic group. On the other hand, nonenzymatic browning is inhibited by more acidic conditions, where the carbonylamino reaction is hindered. However, incorporation of sucrose into an acid medium fosters nonenzymatic browning.

Malolactic fermentation decreases the acidity of some alcoholic beverages including wine and cider. Malic acid is converted to lactic acid, which possesses a weaker taste. This improves the flavor of these beverages and increases their biological stability. The biological advantage of malolactic conversion lies in the fact that it provides the energy required to raise the protonmotive force (Δp) and boost ATPase activity via an increase in the intracellular pH. Also, malolactic fermentation is typically associated with buttery, lactic, nutty, yeasty, oaky, and sweaty flavors, as well as with fruitier notes in wines (8).

Several organic acids including malic, pyruvic, and oxalacetic acid take part in sugar metabolism in fruits. The vacuole malate pool and the malate produced by mitochondrial enzymes are used to synthesize sugars via gluconeogenesis (9). On the other hand, some organic acids in apples, i.e., propionic and butyric acid, are converted to ethylene, which acts as a trigger for ripening (10).

Some organic acids undergo major changes during ripening. Thus, the malic and citric acid contents, and those of major organic acids in most fruits, decrease concomitantly with a decrease in the starch content and an increase in sugar concentrations. The sugar-to-acid and sugar-to-dry matter ratios are frequently used as indicators of grape and orange ripeness.

According to the food laws and regulations of the Food and Drug Administration (FDA), organic acids can be used as acidulants (e.g., citric, fumaric, malic, and sorbic acid), antimicrobial additives (e.g., propionic acid), and sequestrants (e.g., tartaric acid) (11). Most fatty acids of 1–14 carbon atoms possess bacteriostatic or fungistatic properties. According to European Union directives, organic acids used as food additives are legally classified as acidulants and/or acidity correctors (12). They are added to foods at widely variable concentrations (from 200 to 30,000 ppm). Lactic, citric, phosphoric, adipic, carbonic, fumaric, malic, and succinic acid, as well as glucono-δ-lactone, are typically used as acidulant additives.

The high relevance of organic acids to food technology has fostered the development of a host of methods for their determination; most such methods are volumetric, electrochemical, enzymatic, or chromatographic (paper, thin-layer, gas, or high-performance liquid chromatography). However, many of them are inappropriate for the fast, sensitive, reliable determination of the organic acids usually encountered in foods and beverages. Thus, volumetric and spectrophotometric methods are far from specific, which entails implementing time-consuming preliminary separations (precipitation, extraction, or ion exchange). On the other hand, like most biological compounds, organic acids can be assayed enzymatically provided specific enzymes for their catalytic degradation are available. The specificity of these methods enables application to such complex media as foods and beverages with little sample pretreatment. However, enzyme assays call for separate kits for each individual acid, which detracts from throughput and raises costs per test in multideterminations. While still to a limited extent and at an early stage of development in this field, electrochemical methods are increasingly being used in connection with biosensors, particularly those involving immobilized oxidases and dehydrogenases, which are widely used for sensing some organic acids (especially lactic and malic acid) in foods. As far as chromatographic techniques are concerned, paper and thin-layer chromatography are scarcely sensitive and provide information of a qualitative rather than quantitative nature. Gas chromatography is occasionally used for determining organic acids. However, because most of them are not volatile enough, they require derivatization (conversion of the carboxyl group to a methyl or trimethylsilyl ester function), which complicates sample pretreatment. Therefore, gas chromatographic methods are not as straightforward, expeditious, or reliable as required for this type of analysis.

High-performance liquid chromatography (HPLC) has simplified the analysis for various food constituents including organic acids. In fact, it allows the fast, sensitive, and nearly specific determination of organic acids in foods and beverages, and involves uncomplicated sample treatment. A number of ion exchange, ion exclusion, ion pair, and reverse phase chromatographic methods have to date been developed for the separation and determination of organic acids in a variety of samples of scientific and technological interest. The method of choice in each case is dictated by the types of acid to be determined and their proportions as well as by the nature of the matrix that contains them.

II. NATURE OF ORGANIC ACIDS

Organic acids are characterized by their carboxyl group (-COOH), which dissociates into a proton and the conjugate base, and endows them with their acid properties. The acid properties of other compounds present in foods arise from other functions; thus, the acidity of ascorbic acid is due to its two enol groups. Like other families of organic compounds, organic acids can be classified according to the type of carbon chain (aliphatic, alicyclic, aromatic, heterocyclic) they bear, their extent of unsaturation (saturated and unsaturated) and substitution (substituted and unsubstituted), and the number of functional groups they contain (monocarboxylic, dicarboxylic, etc.). The lowest monocarboxylic aliphatic acids (C1–C4) are pungent, highly volatile liquids, whereas those with five or more carbon atoms are oily, slightly water-soluble liquids. Dicarboxylic acids are colorless crystalline solids with melting points in the region of 100°C. Alicyclic acids, which include at least one nonbenzene cyclic hydrocarbon skeleton, are less water-soluble than the previous ones. As such, all of these acids form more or less soluble metal salts and esters; the latter are sufficiently volatile for gas chromatography and have spectral absorbing or fluorescent properties that make them amenable to HPLC analysis. All organic acids occur naturally in a variety of plant and animal substrates. Table 2 lists some of the organic acids most frequently assayed in foods, as well as their dissociation constants (pK), on which their isolation relies heavily.

III. SAMPLE PRETREATMENT

There is currently no available procedure for determining organic acids directly in their matrices, whichever their physical state, without altering or destroying the matrix. In fact, all available methods for the determination of organic acids call for dissolved analytes. Consequently, solid samples require a prior extraction. In addition, the typically high qualitative and quantitative complexity of the substrates to be analyzed (beverages, fruits, vegetables, sweeteners, dried foods, meat products, dairy products, processed foods, *etc.*, which contain a number of organic acids in variable proportions) demands complex sample pretreatment in order to remove potential interferents and also occasionally increase the solute concentration and hence precision, accuracy, and sensitivity. How complex the pretreatment need be will be determined by the nature of the sample to be processed and the determination method to be subsequently applied.

Enzymatic and chromatographic methods (particularly HPLC methods) are those calling for the least pretreatment on account of their high sensitivity and selectivity. As a rule, beverages entail less complex pretreatment than do foods; such pretreatment usually involves the mere dilution in a solution of appropriate pH for enzyme activity to develop, or in the mobile liquid chromatographic phase, to an extent proportional to the solid content, viscosity, and acid concentration of the substrate, as well as to the sensitivity of the sensing system used. The dilution is usually followed by centrifugation or passage through a filter of 0.45 or 0.22 μm pore size. Turbid samples should first be cleared with Carrez reagent (180 mg of $K_4Fe(CN)_6 \cdot 3H_2O$, 360 mg of $ZnSO_4$, and 40 mg of NaOH per gram of sample). A high protein and fatty acid content in the sample can be removed by precipitation with perchloric, sulfuric, phosphotungstic, or trichloroacetic acid. Carbonated drinks should be degassed for 5 min in an ultrasonic bath. Such a straightforward procedure allows the organic acids present in fruit juice (13), wine (14), beer (15), sausage (15), and milk (16) to be determined enzymatically; those in apple juice (17), wine (18,19),

Table 2   Dissociation Constants of Organic Acids in
Aqueous Solutions at 25°C

Monocarboxylic		Di- and tricarboxylic			
Acid	pK	Acid	pK_1	pK_2	pK_3
Unsubstituted					
Formic	3.75	Oxalic	1.20	3.67	
Acetic	4.53	Succinic	4.22	5.70	
Propionic	4.87	Fumaric	3.09	4.60	
Butyric	4.83	Maleic	1.91	6.33	
Isobutyric	4.84	Malic	3.46	5.21	
Substituted		Mucic	3.08	3.63	
Glycolic	3.70	Saccharic	3.01	3.94	
Lactic	3.83	Citric	2.79	4.30	5.65
Pyruvic	2.39	Isocitric	3.28	4.71	6.39
Glyoxylic	2.98	Oxalacetic	2.22	3.89	13.00
Alicyclic					
Quinic	3.58				
Shikimic	4.76				

fermented vegetables (20), and vinegar (21) to be quantified by reverse phase chromatography with UV detection; and those in soft drinks to be measured by ion exclusion chromatography (22). The potential interference of colored compounds with the enzymatic methods for organic acids can be overcome by treating the sample with 1% (w/v) polyvinylpyrrolidone, active carbon, or polyamide. Likewise, the adverse effects of some substances (e.g., phenol compounds) on the column lifetime can be lessened by replacing the guard column after every 100–200 injections. Alternatively, some workers purify samples prior to their chromatographic injection, thereby also improving resolution. Solid–liquid chromatographic procedures involving nonpolar phases or ion exchange resins are the most frequently used for this purpose. The former use Sep-Pak C_{18}, Bond-Elut C_{18}, or similar types of cartridges (18,23), through which a sample volume of 5–10 ml is passed in order to retain neutral and highly lipophilic compounds in the solid phase. The latter procedures involve retaining organic acids on a strong ion exchange resin from which they are subsequently eluted with 1 M HCl (24,25). If more thorough sample cleanup is required (26) or if the carbohydrate, polyphenol, and acid fractions are to be isolated, it is more practical to combine two cartridges (one of each type) and pass the sample through both. The eluate will contain the carbohydrates, which are not retained by either cartridge; the polyphenols will be retained on the nonpolar support and can be reclaimed by elution with methanol; and the organic acids will be retained on the ion exchanger and can be reclaimed as described above.

On account of their high water solubility, organic acids in semisolid and solid samples can be most readily determined merely by cutting up an adequate portion (5–100 g) and grinding the resulting pieces for 3–5 min at a high rate in a blender containing enough water (occasionally hot water, at 60°C), whether acidified or not, and 70–80% ethanol or acetonitrile, depending on the particular type of sample and on whether other constituents (e.g., carbohydrates, phenol compounds) are to be extracted simultaneously. The acids should be completely extracted, so the slurry typically requires stirring for about 2 hr. After centrifugation and filtration through Whatman paper (no. 3 or 4), the filter cake is washed once or twice with fresh portions of extractant and the extracts are concentrated in a rotary vacuum evaporator; the sample is thus made ready for analysis after filtration, deproteination, or solid–liquid extraction as previously described for beverages. This general treatment procedure, with slight modifications as required, has been used for the determination of organic acids in fruits and vegetables (27), fruits (28), cheese (27), and bread and meat (27) by enzymatic methods; as well as apple (29), mango fruit (30), cotton leaves (31), cheese (32), sweet potatoes (33), and meat and dairy products and pickles (34), among others, by HPLC. For gas chromatography, precipitation of nonvolatile acids as their lead salts used to be a common method of isolation (35–39). Currently, however, anion exchange resins are preferentially used before esterification. In fact, Kuksis and Prioreschi (40) compared the recoveries obtained with various isolation procedures and concluded that the ion exchange method developed by Busch et al. (41), which uses a Dowex-1 formate resin and elution with aqueous formic acid, was the most suitable for the joint separation of organic acids.

Judging from the available literature, determinations of organic acids in foods by use of electrochemical methods involve a variable extent of cleanup. Thus, some authors determine lactic and malic acid in white wines and fruit juices with no sample pretreatment (42), whereas others use gelatin in order to remove polyphenols (43). Also, treatments with Celite and subsequent filtration are performed in quantifying fumaric acid in fruit juices and solid foods (44).

Obviously, classical methods of analysis for organic acids are the most demanding as regards sample pretreatment owing to their typically poor sensitivity and selectivity. The determination of total and volatile acidity in coarse liquid samples entails the prior removal of carbon and sulfur dioxide with barium oxide or mercuric oxide. Solid foods, however, are better applied one of the above-described procedures for enzymatic methods (27).

Colorimetric methods entail purifying the sample by passage through an anion exchange column (usually packed with acetate or carbonate resin) prior to determining organic acids. In some instances, a cation exchange column is also used prior to fixation of the acids. In this way, interferences from amino acids such as glutamic acid and some metals such as lead, which is intended to remove polyphenols and oxypoly acids (two known types of interferents with the determination of lactic acid), are circumvented. Organic acids retained on ion exchangers can be eluted with acetate, carbonate, or formate solutions, among others, following washing of the resin with water in order to remove sugars. Using a concentration gradient of the eluant phase affords selective separation of the different organic acids present in the sample—elution of strongly retained acids (e.g., citric and tartaric) requires using a sodium hydroxide solution. On the other hand, interferences with the analytical determination of succinic acid (mainly in the form of other acids) can readily be avoided by selective precipitation or oxidation (with permanganate ion). For acids with a high complex formation ability (e.g., citric acid), sample treatment can be as simple as precipitation of the acid with alcoholic-ammoniacal barium, followed by dissolution with sodium sulfate. These sample cleanup procedures have been used for the separation of various organic acids in fruit juices (45), wines (46), maple syrup (47), and vinegar (48), among others.

IV. DERIVATIZATION

Chemical derivatization is a powerful tool for processing samples when one or more of its components do not meet the requirements of the determination technique of choice. Many organic acids in foods are inadequately volatile for direct gas chromatographic separation, so they must be converted into a volatile, thermostable derivative lending itself readily to sweeping by the carrier gas at the operating temperature used. While this problem does not affect the other determination techniques, which call for dissolved samples, it may be advantageous to include a derivatization reaction in any case so as to improve identifiability and detectability. This is particularly true of the absorptiometric determination of carboxylic acids on account of their structural similarities and the typically low molar extinction coefficients of their chromophores, on which their sensing usually relies owing to the lack of more sensitive properties (e.g., fluorescence, electrochemical activity). Moreover, chemical derivatization may improve the separation of components that cannot be readily resolved in the underivatized sample.

Of the different available alternatives (on-and off-line, pre- and postcolumn), off-line pre-column derivatization is the most commonly used type with organic acids.

Derivatization reactions for the GC determination of organic acids typically involve alkylation or silylation (most often to a methyl, propyl, butyl, or benzyl ester). The esterifying agents most frequently used for derivatization to a methyl ester are diazomethane in ether and boron trifluoride in methanol. While diazomethane is an effective esterifying agent, its high reactivity and toxicity demand that strict safety precautions be exercised (49). There has been a recent shift to using lower alcohols (particularly propanol and butanol) as esterifying agents in combination with a strong acid (H_2SO_4 or HCl) that serves as a catalyst. The growing use of butyl and propyl esters can be ascribed to their lower water solubility and volatility in addition to the easier handling of the resulting esters (50,51). Most reported procedures for the derivatization of organic acids involve reactions that require strictly anhydrous conditions when in fact many samples contain some water. Therefore, in order to avoid time-consuming separations, direct esterification with aqueous butanol and concentrated sulfuric acid can be assayed (52,53). During derivatization, the addition of anhydrous sodium sulfate allows for completion of the reaction. These esterification procedures have been used in the determination of various organic acids in peaches (54), eggs (50,55), cocoa beans (56), and grape juice and wine (57), as well as in the simultaneous determination of D- and L-malic acid in apple juice (58).

The silylation of organic acids, by which carboxyl groups are converted to trimethylsilyl (TMS) ester functions and hydroxyl groups are transformed into TMS ether groups, was formerly carried out with trimethylchlorosilane or hexamethyldisilazane in pyridine as the esterifying agent. However, adverse reports of unstable components resulting in multiple chromatographic peaks led to a search for alternative reagents. Some of the problems were solved by Horning et al. (59), who treated organic acids with methoxylamine hydrochloride in pyridine, followed by silylation with bis(trimethylsilyl)trifluoroacetamide. Subsequently, other authors used a mixture of bis(trimethylsilyl)trifluoroacetamide with hexamethyldisilazane (60,61) or trimethylchlorosilane (62). Fabig et al. (49), in a study of the determination of the organic acids present in mycorrhizal roots involving various methylation and silylation agents, concluded that trimethylsilylation with N-methyl-N-trimethylsilyltrifluoroacetamide was the most suitable procedure. Based on the evidence on silylation kinetics reported by Englmair (63), acetone is advantageous over other solvents such as pyridine. There are conflicting reports concerning the determination of acids as TMS derivatives prepared from their lead salts. Some authors (56,64) found the method unsuitable for quantitative determination of acids in cocoa beans and grape juice, respectively, whereas others (65–68) obtained quantitative results for acids in fruits and wines. These esterification procedures have been used in the determination of various organic acids in wine (66,69), vanilla extract (70), fruits and fruit products (65,71), cheese (72), potatoes (61,73), and potato wastewater (74), among others.

Off-line precolumn derivatization in HPLC involves separating the esters obtained from the organic acids by reverse phase chromatography, which amply surpasses solvophobic chromatography (i.e., the use of undissociated acids as such) and is compatible with elution techniques thanks to the broader lyophilicity range encompassed by the derivatized compounds.

Of the different available derivatization methods for the absorptiometric determination of organic acids, those using organic compounds containing phenacyl (75,76,160), naphthacyl (77), and p-nitrobenzyl groups (78,79,162) are worth special note, as are those using 4-methyl-7-methoxycoumarin (80), 4-methyl-6,7-dimethoxycoumarin (81), and 9-methylanthracene (82) with fluorometric detection. Some derivatives containing one of these functions in addition to one or two electroactive groups (e.g., $-NO_2$) have also been used with electrochemical detection (83,84).

As a rule, the derivatization of organic acids with phenacyl bromide, o-(4-nitrobenzyl)-N,N'-diisopropylurea and some other reagents is carried out in a solvent such as acetonitrile, dioxane, or acetone, in the presence of a catalyst or phase transfer reagent (most often 18-crown-6 ether or one of its derivatives, although tetrabutylammonium hydroxide is preferred for dicarboxylic acids) if an adequate yield is desired (80). The reaction must be carried out at a high temperature (80–100°C) and allowed to develop for 60–75 min. It is advisable to remove any excess reagent prior to injecting the sample in order to improve the appearance of the resulting chromatogram by passing it through a Sep-Pak silica or similar cartridge, or, as recommended by Badoud and Pratz (162), by adding a small amount of a strong ion exchange resin to the sample in a 50:1 resin-to-reagent ratio and allowing it to stand for 15 min prior to injecting a portion of the supernatant into the chromatograph. Caccamo et al. (161) recommend keeping the reaction medium at pH 7 with 0.08 M phosphate buffer in order to ensure a high esterification yield (particularly in citric acid) when phenacyl bromide is used. This reagent was also selected by Marcé et al. (85) for the determination of carboxylic acids in wine following an optimization study (simplex and simultaneous modeling) involving various reagents and experimental conditions.

The esters produced by the above derivatizing reagents usually enable the sensitive determination of their parent acids in foods, with detection limits occasionally comparable to those provided by intrinsically more sensitive techniques such as electrochemistry. Thus, the UV (254 nm) and electrochemical detection (1.1 V) limits of the p-nitrobenzyl derivatives of lactic, formic, and acetic acid are the same and equal to 1.8, 0.9, and 1.1 ng, respectively (83).

V. TITRIMETRIC METHODS

The total acidity due to organic acids is usually quantified by titrating a sample aliquot with an alkali solution of accurately known concentration. The titration endpoint is detected by means of a pH indicator (indicator titration method) or, alternatively, a potentiometric device (potentiometric titration method).

The *indicator titration method* uses 0.5% phenolphthalein as the endpoint indicator and, most often, NaOH as the titrant (KOH is also occasionally employed for this purpose).

The *potentiometric titration method* monitors pH by means of a glass-calomel electrode system. The titration is usually finished at pH 8.2 (Association of Official Analytical Chemists method), yet the endpoint can be assigned to a pH 7.0 (e.g., in the European Community official method).

Total acidity is usually expressed as the number of milliters of a 1 N solution of the alkaline titrant needed to neutralize 100 g of the food concerned or the percentage of lactic acid in the sample. Alternatively, it can also be given as the percentage of tartaric, citric, malic, or acetic acid (the most abundant of all the organic acids in the food concerned), among others.

Total acidity can be determined in an automatic fashion by using the unsegmented flow injection analysis (FIA) technique (86). The sample, occassionally passed through a dialyzer in order to retain potential interferents—mainly suspended and colored matter—, is injected into a carrier stream consisting of a dilute alkaline solution containing a pH indicator (phenolphthalein, bromocresol green, + methyl red). The resulting recording reflects a decrease in the absorbance over the wavelength range 500–550 nm as a result of the neutralization of a given amount of alkali by the acids in the sample, the peak width being a logarithmic function of their combined concentrations.

Volatile acidity can be defined as the amount of acetic fatty acid present in the sample, whether protonated or ionized, with exclusion of lactic and sorbic acid—occasionally added to foods as an antifungal agent—, carbon dioxide, and sulfur dioxide. The acids are separated by distillation on a steam distillation apparatus or a cash electric still. The distillate is titrated with an alkaline solution of accurately known concentration using phenolphthalein as the acidimetric indicator; the titration is finished 15 sec after the pink color of the indicator starts to appear. Volatile acidity is usually expressed as the w/v percentage of acetic acid, which is the major acid in distillates from fermented foods.

Volatile acidity determinations can also be automated with the aid of segmented flow injection techniques (87). The sample is diluted in an acid solution and then mixed with an H_2O_2 solution in order to oxidize sulfur dioxide. The mixture is transferred to a distillation unit and the resulting gas phase rectified. Then the distillate is reacted with bromophenol blue at pH 4.66; the pH decrease produced by the acids in the distillate gives rise to an increase in the absorbance at 450 nm.

VI. COLORIMETRIC METHODS

Spectrophotometric, photometric, and, in general, colorimetric techniques are scarcely used for determining organic acids at present owing to their lack of sensitivity and selectivity. These techniques rely on the Beer–Lambert law, according to which the measured absorbance of a solution depends on the concentration of the absorbing species it contains at its wavelength of maximum absorption. As a result of the natural width of molecular absorption spectra and absorbance additivity, the analytical signal is likely to be perturbed by absorption from other substances present in complex samples and selectivity to be diminished as a result. Also chromophoric groups in the molecular structures of the organic acids of interest have low molar extinction coefficients and hence poor sensitivity. For all of these reasons, application of molecular spectrophotometric techniques to organic acids entails sample cleanup and purification, as well as the use of a color-enhancing reaction that depends on the particular application (see Table 3). As a rule, detection is carried out in the visible spectral region (88–91), yet some acids are detected outside it, e.g. sorbic acid, which is measured in the range 252–255 nm (92). Regarding the sensitivity and reproducibility of spectrophotometric methods as applied to organic acids, it is worth noting the sensitivity of p-hydroxydiphenyl to lactic acid, which ranges from 1 to 10 ppm, with a relative standard deviation (RSD) less than 14.5% and recoveries between 101% and 106%; the determination is markedly influenced by the presence of acetaldehyde. Also, vapor extraction followed by purification via liquid–liquid partitioning with diethyl ether and UV detection allows sorbic acid to be recovered by 95% with an RSD less than 4% (92). Likewise, the FIA determination of tartaric acid by the absorption at 490 nm of the complex it forms with vanadate ion features a linear range from 20 to 400 ppm, with an RSD below 0.8%, an average recovery of 102.5%, and a throughput of 50 samples/hr (90).

VII. ENZYMATIC METHODS

Enzymatic methods use specific enzymes that catalyze given biochemical reactions (e.g., phosphorylations, oxidations, reductions, decarboxylations) involving the analyte (an organic acid in this case); one of the species that takes part quantitatively in the enzymatic process is detected from its absorbing (UV-Vis) properties.

The typically high selectivity of enzymatic methods has turned some of them into references by which other analytical methods (e.g., chromatographic and electrochemical methods) are judged.

As a rule, enzymatic methods for organic acids use nicotinamide adenine dinucleotide (NAD$^+$/NADH) or nicotinamide adenine dinucleotide phosphate (NADP$^+$/NAPDH) as the measured species; the absorption maximum at 340 nm is employed for this purpose in the latter case. Usually, the rate of the enzymatic process is boosted by means of a *trapping reaction*. For example, hydrazine is used as a nucleophilic reagent in order to form a hydrazone by interaction with the carbonyl group of a keto acid; such is the case with oxaloacetate hydrazone, originating in the oxidation of L(-)-malate and pyruvate hydrazone, resulting from the oxidation of lactate (93). The kinetics of the process is controlled by the hydrazone and ketone concentrations. In some cases, another biochemical system is coupled to the primary enzyme system in order to reduce the concentrations of the products resulting from the main enzymatic process. Such is the case with the formation of acetyl coenzyme A by means of phosphotransacetylase in the determination of acetic acid (94) or the use to the glutamate/oxaloacetate/transaminase (GOT) system to decrease the concentration of the oxaloacetate produced in the oxidation of L(-)-malate by malate dehydrogenase (14).

However, the use of a trapping reaction with L(-)-malate can be avoided by replacing malate dehydrogenase with NAD(P)-linked malic enzyme (13), which catalyzes the oxidative decarboxylation of L(-)-malate—the process is thus shifted to the formation of pyruvic acid.

In some cases, detection of NADP(H) at 340 nm is rendered difficult by the strong absorption of some matrix component in this spectral region. Such is the case with the analytical determination of L(-)-malate and acetate in red wines, which may contain a high concentration of anthocyanins and phenolic substances relative to the organic acids present. The use of an electron carrier (e.g., 8-dimethylamino-2,3-benzophenoxazine) to transfer charge from NADH to, for example, 3-(4,5-dimethylthiazolyl-2)-2,5-diphenyltetrazolium bromide (MTT), allows the formation of a complex (MTT-formazan) that absorbs at 578 nm, where no matrix component is bound to interfere at alkaline pH values, i.e., those where the enzyme activity is maximal (14,95). One other enzymatic reaction with detection in the visible spectral region is that used in the determination of D-gluco-

Table 3 Spectrophotometric Methods for Determination of Organic Acids

Analyte	Sample type	Sample preparation	Detection	Ref.
Lactic acid	Wines	Oxidative decarboxylation with $CuSO_4$, $Ca(OH)_2$, and concentrated sulfuric acid	Vis—560 nm (reaction with *p*-hydroxydiphenyl)	46
	Fruit juices and wines	Isolation over an anion exchange resin with acetate as counterion. Elution with 0.5 M sodium sulfite. Oxidation of lactic acid to acetaldehyde with 0.1 M cerium sulfate.	Vis—570 nm (reaction with piperidine)	45
	Dairy products	Deproteination with 20% phosphotungstic acid, extraction with ether for 3 hr	Reaction with ferric chloride	91
Citric acid	Cheeses	Extraction in hot water (60–70°C) and deproteination with 40% (v/v) trichloroacetic acid	Vis—428 nm (reaction with pyridine)	88
	Wines	Precipitation with 20% barium chloride and concentrated ammonium hydroxide. Dissolution with hot 7.1% sodium sulfate and decolorization with 0.4% activated carbon	Vis—420 nm (reaction with Grier's reagent: sulfanilic acid plus sodium nitrite)	89
Malic acid	Wines	Percolation through a cation exchange resin with hydrogen as counterion and decolorization with 0.3% activated carbon. Isolation over an anion exchange resin with formate as counterion and elution with 1 N formic acid	UV—370 nm (reaction with 2, 7-naphthalenedisulfonic acid)	89
	Maple syrup	Percolation through a cation-exchange resin (Dowex 50, 60–80 mesh) with hydrogen as counter-ion. Isolation over an anion-exchange resin with carbonate as counter-ion, and selective elution with 0.25 N and 1 N $(NH_4)_2CO_3$	UV—390 nm (reaction with 2, 7-naphthalenedisulfonic acid)	47
Tartaric acid	Wines	Dilution and addition of glacial acetic (4% v/v)	Vis—490 nm (reaction with vanadate)	90
	Vinegars	Isolation over an anion-exchange resin with acetate as counter-ion. Elution with 7.1% sodium sulfate.	As above	48

nate. The analytical procedure employed involves the formation of Prussian blue by reduction of the ferricyanide resulting from the oxidation of D-gluconate by gluconate dehydrogenase (GADH) (15).

Table 4 lists the enzymatic reactions most frequently used for the determination of organic acids.

The high sensitivity and selectivity of enzymatic methods has lately fostered their use for quantifying organic acids in foods (Table 5). These analytical assets do not always warrant application to this type of substance, however, since the reagents needed are expensive and unstable, and the procedures involved are somewhat time consuming. By way of example, the enzymatically achieved RSD is less than 2.5% for malic acid (13), 7% for acetic acid (94), 4% for lactic acid (16), and 3.5% for citric acid (96). As regards sensitivity, acetic acid can be determined at concentrations as low as 0.4 ppm; malic, lactic, and succinic acid at ~1 ppm; and gluconic and citric acid at 3.30 and 2.64 ppm, respectively (27).

VIII. ELECTROANALYTICAL METHODS

Foremost among the classical electrochemical methods applied to organic acids are redox titrations. However, these methods are usually subject to severe interferences. For example, the determinations of L-ascorbic and D-isoascorbic or erythorbic acid are interfered by reductants potentially present in the sample (e.g., Fe^{2+}, Sn^{2+}, Cu^{1+}, reducing sugars, reductones, sulfur-containing compounds such as sulfides, phenol compounds, water-soluble vitamins) (97). Some of the most commonly used reagents for redox titrations include dichloro-2,6-phenolindophenol, ferrocenium trichloroacetate, *N*-bromosuccinimide, potassium ferricyanide, potassium bromate, iodine, ammonium cerium nitrate, copper nitrate, and potassium dihaloiodates. Removing interferences with electrochemical titrations is a slow, rather complex process, which detracts from usability.

Table 4 Enzyme Complexes[a] Used for Determination of Organic Acid in Foods

Analyte	Enzyme system	Trapping reaction
L(−)Malate	L-MDH or NAD(P)-linked ME	GOT or hydrazine/hydrazone
Acetate	AK → PK → L LDH(NADH); ACS → L-M-DH(NADH)	P-transacetylase/CS
D-Gluconate or D- gluconic acid δ-lactone	GK → 6-PGDH [NADPH] or 2KGR [NADPH] or 5KGR [NADPH] or GADH	
Citrate	CL → L-MDH → L-LDH; A → IDH[NAD(P)H]	
Formate	FDH[NADH]	
D- and L-lactate	DH and L- LDH	GPT
D-Isocitrate	IDH	
Fumarate	F → L-MDH	GOT
Succinate	SCS → PK → L-LDH	
Galacturonate	I → D[NADH]	

[a]MDH, malate dehydrogenase; LDH, lactate dehydrogenase; 2KGR, 2-ketogluconate reductase; 5KGR, 5-ketogluconate reductase; AK, acetate kinase; PK, pyruvate kinase; CS, citrate synthetase; GADH, gluconate dehydrogenase; GK, gluconate kinase; 6-PGDH, 6-phosphogluconate dehydrogenase; IDH, isocitrate dehydrogenase; GOT, glutamate oxaloacetate transaminase; GPT, glutamate pyruvate transaminase; FDH, formate dehydrogenase; F, fumarase; A, aconitase; ACS, acetyl coenzyme A synthetase; SCS, succinyl coenzyme A synthetase; I, isomerase; D, dehydrogenase; ME, malic enzyme.

Table 5 Enzymatic Analysis Methods for Organic Acids in Foods

Analyte	Sample type	Sample preparation	Detection	Ref.
L(-)-Malate	Fruit juices	Dilution and pH adjustment in the neutral region	UV—340 nm (NADPH)	13
	Wines	As in 13	Vis—570/578 nm (MTT-formazan)	14
	Fruits and vegetables	Extraction with hot water and dilution (if required)	UV—340 nm (NADH)	27
Acetate	Fruit juices	Use of gelatin and poly-vinylpyrrolidone (in the event of strong phenolic interference) and pH 8.0	As above	94
	Wines	As in 94	UV—340 nm (NADH) or Vis—570/578 nm (MTT-formazan) with Meldola blue as mediator of electron transfer	94, 95
	Fruits, vegetables, and cheeses	Extraction with hot water. Deproteination with perchloric acid and defatting at a low temperature in an aqueous medium	As above	27
D-Gluconate or D-gluconic acid δ-lactone	Wines, beers, and fruit juices	Use of H_2O_2 to oxidize reductants and catalase to quench H_2O_2. The lactone requires prior alkaline hydrolysis	Vis—660 nm (Prussian blue) or UV—340 nm (NADPH)	15
	Sausages	Extraction with boiling water and defatting; then as above	As above	15
Citrate	Wine and fruit juices	Dilution and decolorization of colored samples with 1% polyamide or polyvinyl-pyrrolidone	UV—340 nm (NADH)	27, 96
	Bread, meat, cheese, vegetables and fruits	Extraction with hot water (60°C); deproteination and defatting with perchloric acid and treatment at a low temperature in an aqueous medium, respectively.	UV—340 nm (NADH)	27
Lactate	Wines and fruit juices	Dilution and decolorization with 1% polyamide or polyvinylpyrrolidone if necessary. Carrez solution can be used for clearing	UV—340 nm (NADH)	27, 93
	Milk	Clearance with 6 N HCl and 6 N NaOH	As above	16
	Cheese and meat	Extraction with hot water and deproteination and defatting as above	As above	27

Succinate	Wines and fruit juices	Dilution and adjustment of pH 8.0 and decolorization with 1% polyvinylpyrrolidone if necessary	As above	27, 28
	Meat, vegetables, and cheeses	Extraction with hot water. Defatting at a low temperature in an aqueous media and deproteination with perchloric acid	As above	27, 28
Fumarate	Wines, fruit juices, and solid foods	As for malic acid	As above	27
Formate	Wines and fruit juices	Dilution and decolorization with 2% activated carbon if necessary	As above	27
	Vegetables and meats	Extraction with hot water and defatting by storage at a low temperature. Deproteination with trichloroacetic acid	As above	27

MTT = [3-(4,5 dimethyl thiazolyl-2)-2,5 diphenyl tetrazolium bromide]; Meldola blue = [8-dimethyl-amino-2,3-benzophenoxazine].

Some organic acids with redox properties (e.g., L-ascorbic and fumaric) have also been determined by polarography. This electrochemical technique requires strict sample cleanup. On the other hand, ac polarography allows the two isomers of ascorbic acid to be determined separately. Also, fumaric acid can be determined by conventional polarography (its reduction potential is −1.15 V) (44).

Combinations of enzymatic and electrochemical methods are potentially of a high analytical interest since many electrochemical devices can be used as flow-through detectors for the automated determination of organic acids. As a rule, electrochemical determinations of organic acids and other biochemicals such as glucose use oxidases that are incorporated into an amperometric transducer sensitive to the oxygen concentration (the Clark electrode). By way of example, lactate oxidase (LOD) is used as "receiver" in the determination of lactic acid; in fact, the enzyme catalyzes the oxidation of lactic acid to acetic acid in the presence of oxygen, with formation of hydrogen peroxide. Lactate monooxygenase (LMO) (98), another catalyst for the oxidation of lactic acid to acetic acid in the presence of oxygen (with formation of water in this case) has also been used for this purpose. Changes in the oxygen content during oxidation are monitored with the above-mentioned amperometric transducer. If LOD is used as the receiver, then lactic acid can be determined by amperometrically detecting the hydrogen peroxide produced in the biochemical reaction (the anodic current is recorded at a potential of 600 mV vs. the reference electrode) (99).

One of the most serious problems of oxygen-sensitive amperometric sensors based on immobilized oxidases are posed by low oxygen tensions. Two effective solutions in this respect are provided by a two-dimensional electrode (oxygen can diffuse into the membrane from two directions, whereas the analyte can diffuse from one only) or an electron transfer mediator (100). The latter functions to ensure effective charge transfer between the transducer and receiver. Ferrocene [bis(η^5-cyclopentadienyl)iron, Fecp$_2$] and its derivatives (Fecp$_2R$) are typical charge transfer mediators. These electroactive substances act as electron acceptors for flavoproteins. For example, succinate dehydrogenase (SDH) catalyzes the oxidation of succinic acid to fumaric acid; the charge transfer reactions involved are illustrated in Scheme 1.

$$
\begin{array}{cccc}
 & Fecp_2R & SDH_{ox} & succinate \\
Electrode\ surface \leftrightarrows e^- \leftrightarrows & \uparrow\downarrow & \uparrow\downarrow & \downarrow \\
 & Fecp_2R^{+\cdot} & SDH_{red} & fumarate
\end{array}
$$

Scheme 1

Some analytical determinations of organic acids with biosensors use dehydrogenases such as L-lactate dehydrogenase (L-LDH), D-lactate dehydrogenase (D-LDH), and L-malate dehydrogenase (L-MDH) (42,101), which are employed for determining L-lactic, D-lactic, and L-malic acid, respectively.

Dehydrogenase-catalysed oxidations involve the formation of NADH as shown in Scheme 2.

$$NAD^+ \rightarrow NADH$$

$$R-COOH_{red} \rightarrow RCOOH_{ox}$$

Scheme 2

The amount of NADH formed in the biochemical reaction is directly proportional to the concentration of organic acid in the sample. The NADH can be detected electrochemically in various ways. One involves using an electroactive mediator and amperometric detection of the oxygen by use of a Clark electrode. Such is the case with the determination of malic acid by oxidation with malate dehydrogenase using vitamin K_3 and diaphorase as mediators; the reduced form of the vitamin is oxidized by dissolved oxygen in the medium, the decay in the O_2 concentration being monitored by means of a Clark oxygen electrode (43). L-Malate can also be measured by means of an oxygen electrode containing immobilized malate dehydrogenase and NADH oxidase. The former enzyme catalyses the oxidation of L-malic acid to oxalacetic acid, with formation of NADH; the latter reoxidizes NADH to hydrogen peroxide from oxygen, the process being monitored amperometrically (42).

Direct detection of NADH from the anodic current produced by application of an oxidation potential to an electrode is rendered difficult by the usually high potentials to be used (~800 mV vs. the Ag/AgCl electrode), which are also bound to oxidize other electroactive substances potentially present in the medium (e.g., ascorbate, which can produce an anodic current 20 times as high as that of NADH).

This shortcoming can be circumvented by using a chemically modified electrode. Thus, by using a graphite surface modified by absorbing N,N-dimethyl-7-amino-1,2-benzophenoxazine, the oxidation potential for NADH can be lowered (to 0 mV vs. Ag/AgCl or 50 mV vs. SCE) and the interference of potentially oxidizable electroactive substances be avoided (101).

Using *conducting salts* comprising two planar organic molecules with an extended π-electron system as electrode material for oxidizing NADH has the advantages over alternative materials such as platinum that no intermediate radical is formed and that a lower overvoltage can be used.

By way of example, the determinations of L-malate and L-lactate with an electrochemical biosensor feature an RSD of 0.92–1.3% for concentrations between 0.01 and 5 mM (43,98,42). Table 6 gathers some applications of this technique to the determination of organic acids.

IX. AUTOMATION OF ANALYSES

The role of robots in analytical laboratories is currently limited to repetitive operations such as weighing, injections into chromatographic columns, and sample preparation, mainly. Owing to the need for careful programming and feedback control, the use of robots is only justified when tedious, repetitive operations must be performed over long periods. On the other hand, flow

Table 6 Electrochemical Methods for Determination of Organic Acids

Analyte	Sample type	Sample preparation	Immobilized enzyme	Detection	Ref.
Fumaric acid	Fruit juices	Shaking with Celite and subsequent filtering		Polarographic. Peak potential –1.15 V	44
	Solid foods	Extraction with methanol and filtering		As above	44
L-Malate	White wines and fruit juices	No sample pretreatment required	Malate dehydrogenase and NADH oxidase	Amperometric, by means of a selective Clark oxygen electrode	42
	Wines	Separation of polyphenols with gelatin (2.5%, 4:1), and filtering and diluting with 0.05 M pyrophosphate buffer and vitamin K$_3$	Malate dehydrogenase and diaphorase (use of vitamin K$_3$ as mediator)	As above	43
L-Lactate	Wines and fruit juices	As above for red wine. For white wines and fruit juices no pretreatment needed	Lactate oxidase	As above	42, 43
	Cells		Lactate oxidase	Amperometric. Potential used for detecting H$_2$O$_2$ formed: +0.6 V vs. SCE	99
	Blood plasma	Unnecessary	Lactate monooxygenase	Amperometric, by means of a selective Clark oxygen electrode	98
	Liquid foods	As above	Lactate dehydrogenase and glutamic-pyruvic transaminase	Amperometric. Detection of NADH over an electrode chemically modified with Meldola blue, at 0 V vs. Ag/AgCl	101

operations are easier to automate, miniaturize, and control in space and time, so that they are inherently more flexible. In fact, typical organic acid determinations by enzymatic, colorimetric, and electrochemical methods are usually implemented via segmented flow analysis (SFA) or flow injection analysis (FIA). No doubt the use of automated analytical techniques of both types results in dramatically increased throughput and reduced analytical costs through decreased reagent consumption.

While segmented flow systems effectively limit sample dispersion as a result of the surface tension forces at the liquid–gas interface, they are more laborious to set up and give rise to higher analytical costs than do flow injection systems. In addition, they feature a more modest throughput (the analytical signal is of the steady-state type rather than of the transient type as in FIA), give rise to pulses arising from air compressibility, require degassing before the analyte reaches the flow cell, and may produce distorted signals from potentiometric detectors due to static charge building up on air bubbles.

FIA systems can be used in two ways as regards enzyme introduction into the manifold in a straightforward process involving a single enzyme complex. Thus, the enzyme solution can be mixed with the carrier by means of a membrane dialyzer, which facilitates sample dilution and filters off interferents that may raise background absorption, after the pH is adjusted with a reaction coil. Alternatively, the enzyme can be used in a chemically immobilized form (e.g., on controlled pore glass, CPG).

Typical applications of these automated techniques to food analysis were described in the previous sections.

X. CHROMATOGRAPHIC TECHNIQUES

A scan of recent literature on the topic clearly reveals that organic acids in fruits, juices, fermented beverages, vegetables, dairy products, roots, and foodstuffs in general are being more and more frequently determined by use of gas–liquid chromatography, and high-performance liquid chromatography preferentially over other methods of analysis (enzymatic, titrimetric, colorimetric, electrochemical), which are usually laborious, time consuming, and error-prone. In addition, they do not lend themselves readily to the simultaneous determination of all acids present in a sample.

A. Gas Chromatography

With the advent of gas-liquid chromatography (GLC), analytical chemists were equipped with a powerful tool for examining organic acids which they exploited extensively during the 1960s and 1970s. Subsequently, they shifted to HPLC, which entails no derivatization of the starting compounds. In the beginning, workers naturally employed chromatographic columns packed with a solid support such as Gas Chrom Z (100–200 mesh) or Diatoport S (60–80 mesh), which was coated with a liquid phase, typically diethyleneglycol adipate (DEGA) or succinate (DEGS), SE-30, OV-1, etc. Currently, open tubular and capillary columns coated with a stationary phase such as DB-1 or Carbowax 20M are preferred for this purpose. Detection is usually carried out with a flame ionization detector (FID) or a high-performance mass spectrometric detector in order to facilitate identification of several organic acids in the sample. Table 7 shows some representative applications of the GLC technique to the determination of organic acids. Additional applications not listed in the table include the determinations of fixed wine acids (102); aconitic acid in sugarcane and sorghum syrups (103,104); lactic, acetic, propionic, and butyric acid in beet juice and molasses (105–107); the effect of cultivars and fertilizers on nonvolatile acids (critic, malic, 2-pyrrolidone-5-carboxylic, and phosphoric acid) in potato tubers (108); lactic, oxalic, sorbic, benzoic, succinic, fumaric, glutaric, malic, tartaric, pimelic, shikimic, citric, quinic, palmitic, caffeic, and ferulic acid in apples—the citric-quinic and caffeic-ferulic mixtures cannot be resolved—(109); lactic and 3-hydroxybutyric acid in egg noodles and cookies in a collaborative study involving 10 laboratories (100); lactic acid and cyclohexanecarboxylic acid as an indicator of olive spoilage (111); over 50 volatile and nonvolatile acids with an MS detector (112); and malic, isocitric, and citric acid in lemon juice (113).

As a rule, GLC provides good sensitivity and selectivity. Recoveries from real samples are close to 100% and coefficients of variation are usually below 5% but reach up to 15% in some instances. Its most serious pitfall is the need for clean-up and derivatization, which should be carefully performed in order to ensure reliable results.

B. Liquid Chromatography

As noted in the Introduction, the choice of a given technique (e.g., ion exchange, ion exclusion, ion pair, or reverse phase chromatography of undissociated or derivatized substrates) for the separation of organic acids in foods is essentially dictated by the nature of the acids and that of the matrix. The most usual method for the determination of organic acids in foods and beverages is ion exclusion chromatography, followed by reverse phase chromatography. Notwithstanding their extensive application in other fields (and also, formerly, in food analysis), ion exchange and

Table 7 Gas–Liquid Chromatographic Methods for Determination of Organic Acids

Acids	Sample	Derivatizing agent	Column	Carrier gas and temperature program	Ref.
Succinic, malic, citric	Peaches	BF$_3$/methanol	6 ft × 5 mm stainless steel 20% diethylene glycol adipate polyester on Diatoport-S	He (90 ml/min) 155–215°C (7.5°C/min)	114
Lactic, succinic	Eggs	BF$_3$/propanol	10 ft × 4 mm glass 10% diethylene glycol succinate on Gas Chrom 7	He (80 ml) 130°C	50
Lactic, oxalic, succinic, malic, citric	Cocoa beans	Hexamethyldisilazane + trimethylchlorosilane	6 ft × 1/8 in. stainless steel 5% SE-30 on Chromosorb G or diethylene glycol adipate (10%) on Gas Chrom A	N$_2$ (30 ml) 90–260°C (6°C/min) or 90–190°C (6°C/min)	56
Tartaric, citric, adipic, gluconic lactone	Baking powder	Methyltrimethylhepta-fluor butyramide + trimethylchlorosilane	2.5 m × 1/8 in. OV-17 on Vara-port-30	N$_2$ (25 ml/min) 130–300°C (6°C/min)	115
Succinic, fumaric, malic, tartaric, citric, *trans*-aconitric	Fruit juices	Bis(trimethylsilyl) acetamide or trifluoroacetamide	160 × 4 mm glass 4% OV-1 on Gas Chrom Q	N$_2$ (25 ml/min) 170°C	71
Succinic, lactic	Cheese	Methanol/(H$_2$SO$_4$)	1.83 × 0.002 m stainless steel 10% diethylene glycol adipate/20% H$_3$ PO$_4$ on Gas Chrom A	He (35 ml/min) 80–190°C (6°C/min)	72
Succinic, lactic, tartaric, citric, citramalic, phosphoric	Wine	Hexamethyldisilazane/diazomethane	4 m × 1/8 in. stainless steel 10% OV-17 or 7% QF on Chrom Q	H$_2$ 80–180°C	116

Table 7 (*Continued*)

Acids	Sample	Derivatizing agent	Column	Carrier gas and temperature program	Ref.
Succinic, fumaric, malic, tartaric, citric, quinic, syringic, glycolic	Fruits and juices	Hexamethyldisilazane + trimethylchlorosilane	6 ft × 1/4 in. stainless steel 3.8% SE-30 on Diatoport S	He (50 ml/min) 90–240°C (6°C/min)	65
Piruvic, β-hydroxy-butyric, fumaric, succinic, malic, pyrrolidone-carboxilic	Eggs and egg products	Methanol/HCl	50 m × 0.3 mm Fluorad FC-431	H₂ (3 ml/min) 5–200°C (6°C/min)	55
Succinic, lactic, citramalic, malic, tartaric, citric	Wine	Methanol/HCl	1 m × 3 mm glass FFAP on Gas Chrom Q	N₂ 40–180°C (4°C/min)	57
Succinic, malic, citric, quinic	Fruits and sweet potatoes	Bistrimethylsilyltri-fluoracetamide trimethylchlorosilane	15 × 0.25 mm fused silica DB-1	H₂ (42 cm/sec) 150–192°C (4°C/min)	61
D- and L-malic	Apple juice	Butanol/HCl	30 × 0.25 mm fused silica Carbowax 20M	He (1 ml/min) 50–130°C (8°C/min) 200°C (3°C/min)	58

ion pair chromatographic methods are gradually being superseded as the previous two alternatives provide better, faster, and more reliable results.

The detectors most frequently used in HPLC are of the refractive index (RI), conductivity, and UV-Vis detectors type. While RI detectors are probably the most flexible, they are somewhat sensitive to changes in pressure, temperature, and the composition of the mobile phase, which demands strict control and the use of isocratic elution. However, they can be of use for determining other components of interest including carbohydrates and alcohols simultaneously in a single chromatographic analysis. Conductivity detectors were originally employed in ion chromatography for the determination or inorganic ions and, later, organic acids. Their inherent pitfalls have deterred potentials users from applying them to food analyses as solute conductivity measurements require the prior elimination of the eluant background conductivity using a conventional suppressing column or a more modern alternative such as a cation exchange membrane loaded with tetrabutylammonium ion, a hollow-fiber suppressor with a neutral or alkaline enhancer, or a redox suppressor. UV-Vis detectors are no doubt the most frequently used at present for determining organic acids in foods. For underivatized organic acids, detection at 206–220 nm usually poses no serious problem in the determination of major organic acids. By using a photodiode array detector, optimizing the detection and quantitation conditions, and checking chromatographic peak purity, determining organic acids is fairly easy as a single sample injection provides all the spectral information required for a given wavelength range during elution, which is collected as a data matrix for subsequent processing.

HPLC is no doubt the most suitable choice for determining organic acids on account of its expeditiousness (a chromatogram typically takes about 15 min to record), sensitivity (detection limits, calculated as three times the standard deviation of background noise, range from 100 ng for species low molar extinction coefficients such as succinic acid and 1 ng for shikimic acid, and can be lowered by a factor of up to 4 using columns of 2 mm i.d.), selectivity (recoveries are usually close to 100%), and reliability (coefficients of variation are typically less than 5%). In addition, the analytical costs of HPLC are fairly low; in fact, the cost of determining six organic acids in wine or cider amounts to only about $3 (reagent and solvent consumption, filters, and column wear included).

C. Ion Exchange and Ion Exclusion Chromatography

The ready ionization of organic acids has long been exploited for their isolation by ion exchange chromatography, which involves the use of an ion exchange resin as stationary phase. Palmer and List (117) set the pace for their isolation from foods. This separation technique is rarely used nowadays, however, as it is clearly outperformed by ion exclusion chromatography, which usually relies on the use of sulfonated styrene-divinylbenzene cation resin. The term ion exclusion chromatography was coined by Wheaton and Bauman (118) to name the exclusion phenomenon undergone by coions on exchangers as a result of electrostatic repulsion forces, which allowed electrolytes to be readily separated from nonelectrolytes. This operational principle, based on the finding of Harlow and Morman (119) that the elution time of an acid from a Dower 50W × 12 column in its hydrogen form depends on its dissociation constant(s) (pK), molecular weight, and water solubility (Tanaka et al. (120) using a strong cation exchange column found the capacity factors of various acids to depend linearly on their first dissociation constant; the retention volume of the weak acids increased with their first pK), was exploited for the separation of organic acids in a variety of matrices including foods. The retention mechanism involved can be explained in the form of Donnan partition equilibria between the protons in the solute and the fixed hydrogen ions in the resin. The relative retention of a highly crosslinked ion exchange resin is also conditioned by hydrophobic interactions and steric effects (121). Ion exclusion chromatography is generally quite efficient; however, it has one major shortcoming: some neutral compounds such as sugars and polyphenols, two usual components of foods that possess similar retention times to those of organic acids, are also retained by cation exchange resins. As a result, samples must be pretreated prior to their chromatographic separation. The column most frequently used for this purpose is the Aminex HPX-87H 300 mm × 7.8 mm i.d. model from Biorad Laboratories. The mobile phase is usually 0.0008–0.185 N sulfuric acid. The column temperature is maintained at 50–75°C and a UV (206–230 nm) or refractive index (RI) detector is normally employed. This technique fails to resolve shikimic and succinic acid (18,19); galacturonic and tartaric acid (24); pyruvic, citramalic, and malic acid (24); and quinic and malonic acid (122). Also, the peaks for succinic, glycolic, and lactic acid, and those of propionic and glutamic acid, are extensively overlapped (122). Table 8 summarizes the most interesting applications of this technique to the determination of organic acids in foods.

D. Reversed Phase Chromatography

Most stationary phases used in bonded phase chromatography in its reverse phase mode are based on octadecylsilane functionality (C_{18} columns). The mobile phases typically used in this context are water, aqueous buffers of a given pH and ionic strength, and mixtures of water and a miscible organic modifier such as methanol or acetonitrile.

Horvath et al. (123,124) showed nonpolar stationary phases such as octadecyl silica to allow the separation of relatively polar substances including organic acids by using an aqueous phase at an appropriate pH to prevent the ionization of the species to be resolved. Under these conditions, retention is governed by hydrophobic interactions between the hydrocarbon rest of the solutes and the hydrocarbon chains of the stationary phase. The foundation of this hydrophobic chromatography lies in the solvophobic theory (125), according to which capacity factors, and hence the retention of the solutes to be resolved, depends essentially on the pH of the mobile phase, which influences their ionization. In addition, the capacity factor of ionic substances increases with the salt concentration or ionic strength of the mobile phase—the effect on acids was studied by Jandera et al. (126). On the other hand, the capacity factor decreases with the concentration of organic modifier (methanol or acetonitrile) and increases with the size of the aliphatic chain or the aromatic skeleton of the acid. Both are expanded on esterification, which results in highly hydrophobic compounds that can readily be resolved in this way—so much so that they require the use of an organic modifier in the mobile phase for elution—and allows gradient elution to be implemented and chromatographic resolution to be substantially improved as a result. The column type most frequently used for this purpose is a C_{18} 5-μm, 250 × 4 mm i.d. one, through which a 10^{-3}–10^{-1} M KH_2PO_4/H_3PO_4 mobile phase at pH 2.2–4.2 is passed. Underivatized acids require less than 5% organic modifier, whereas derivatized acids call for 35–85% of acetonitrile or methanol and isocratic or gradient elution. Marcé et al. (127) optimized the separation of phenacyl derivatives of organic acids in wine using the surface response modeling method (127). Table 9 shows selected applications of this technique to the determination of organic acids in foods.

XI. CONCLUDING REMARKS

From the above exposition it follows that none of the analytical techniques routinely used for determining organic acids in foods possess all of the ideal analytical features. Thus, classical methods are scarcely sensitive and selective; as noted in the section concerned with sample treatment, their implementation entails time-consuming clean-up and, occasionally, preconcentration, which obviously detract from precision and throughput. All of these raise analytical costs,

Table 8 Ion Exchange and Ion Exclusion Chromatographic Methods for Determination of Organic Acids

Acids	Sample	Column	Mobile phase	Detection	Ref.
Acetic, lactic, succinic, gluconic, and pyroglutamic	Vinegars	TSK gel IC-Anion-PW (50 × 4.6 mm) 40°C	0.45 mM OABS, pH 3.5	UV, 302 nm	128
Oxalic, citric, isocitric, gluconic, lactic, malic, quinic, acetic, fumaric, succinic, formic, and glycolic	Beer, wine, and fruit juices (concentrates)	Aminex HPX 87H (300 × 7.8 mm) 65°C	0.005 M H_2SO_4, pH 3.02	UV, 210 nm	129
Seventeen acids involved in Krebs cycle	Dairy products	ION 300 (300 × 7.8 mm) 25–55°C	0.0065–0.008 N H_2SO_4	RI	130
Acetic, lactic, succinic, malic, tartaric, and citric	Wine and grape juice	ION 300 (300 × 7.8 mm) 74°C	0.013 M H_2SO_4	RI	131
Acetic, lactic, succinic, malic, tartaric, and citric	Wine	ION 300 (300 × 7.8 mm) 30–80°C	0.01–0.05 N H_2SO_4	RI	132
Citric, lactic, and acetic	Cocoa	Aminex HPX 87H (300 × 7.8 mm)	0.013 N H_2SO_4	UV 215 nm + RI	133
Acetic, lactic, succinic, malic, formic, and citric	Vinegars	HPICE / AS 1	0.01 N HCl	Conductimetric	134
Tartaric, malic, citric, pyruvic, succinic, lactic, acetic, shikimic, galacturonic, and citramalic	Wine	Aminex HPX 87H (300 × 7.8 mm) 45°C	0.0075 M H_2SO_4	UV, 210 nm	24
Citric, succinic, acetic and some fatty acids	Wine	Dionex HPICE-AS 1 (250 × 9 mm)	1–2 mM H_2SO_4	Conductimetric	135
Citric, malic, lactic, and acetic	Cucumber juice	Aminex HPX 87H (300 × 7.8 mm) 60°C	0.013 N H_2SO_4	UV, 210 nm + RI	136
Succinic	Orange juice	Aminex HPX 87H (300 × 7.8 mm)	0.036 M H_2SO_4	UV, 210 nm	137
Citric, malic, lactic, tartaric, and acetic	Sparkling wine fermenting	Aminex HPX 87H (300 × 7.8 mm) 65°C	H_2SO_4, pH 2.15–2.25	UV, 210 nm	23

Table 8 (Continued)

Acids	Sample	Column	Mobile phase	Detection	Ref.
Pyruvic, succinic, lactic, formic, acetic, and pyroglutamic	Cheese	Aminex HPX 87H (300 × 7.8 mm)	0.1 N H_2SO_4	UV, 220 nm	32
Citric, malic, and succinic	Sweet potatoes	Aminex HPX 87H (300 × 7.8 mm) 75°C	0.0008 N H_2SO_4	UV, 214 nm	33
Citric, malic, tartaric, lactic, formic, acetic, and succinic	Wine and grape juice	Aminex HPX 87H (300 × 7.8 mm)	0.01 N H_2SO_4	RI	26
Citric, malic, tartaric, lactic, formic, acetic, fumaric, quinic, glycolic, and succinic	Coffee	Aminex HPX 87H (300 × 7.8 mm) 50°C	0.185 N H_2SO_4	UV, 210 nm	122
Citric and malic	Cotton leaves	Aminex HPX 87H (300 × 7.8 mm) 30°C	0.014 N H_2SO_4	UV, 210 nm + RI	31
Citric, malic, tartaric, lactic, formic, acetic, oxalic, succinic, and some fatty acids	Carrots fermenting	Aminex HPX 87H (300 × 7.8 mm) 40°C	0.013 N H_2SO_4	RI	20
Citric	Carbonated beverages, fruit juices, and cheeses	Aminex HPX 87H (300 × 7.8 mm)	0.009 N H_2SO_4	UV, 210 nm	22
Acetic	Vinegars, olives, mayonnaise, and other sauces	Aminex HPX 87H (300 × 7.8 mm)	0.009 N H_2SO_4	UV, 210 nm	21
Oxalic, citric, malic, succinic, fumaric, and pyroglutamic	Potatoes	Aminex HPX 87H (300 × 7.8 mm) 41°C	0.018 N H_2SO_4	UV, 220 or 230 nm	138
Lactic, acetic, formic, and some fatty acids	Silages	Aminex HPX 87H (300 × 7.8 mm)	0.0025 N H_2SO_4	UV, 210 nm	139
Lactic, acetic, formic, pyruvic, orotic, hippuric, and some fatty acids	Dairy produce	Aminex HPX 87H (300 × 7.8 mm) 65°C	0.009 N H_2SO_4	UV, 220 and 275 nm	140

OABS, O-amino benzene sulfonic acid; RI, refractive index.

Table 9 Reverse Phase Chromatographic Methods for Determination of Organic Acids

Acids	Sample	Column	Mobile phase	Detection	Ref.
Citric and malic	Calamondin orange juice (Philippines citric)	Spheri-5RP-18 (110 × 4.6 mm) + 2 columns Polypore H (110 × 4.6 and 220 × 4.6 mm)	2% KH_2PO_4, pH 2.3	UV, 215 and 260 nm	141
Tartaric, citric, malic, sorbic, and benzoic	Juices, nectars, and soft drinks	Spheri-5RP-18 (220 × 4.6 mm) 5 μm	Water (HPLC) pH 2.2, H_2SO_4	UV, 214 and 230 nm	142
Quinic, malic, and citric	Kiwi and babaco	Spherisorb ODS-2 (250 × 4.6 mm) 5 μm	Water (HPLC) pH 2.2, H_2SO_4	UV, 214 nm	143
Citric, malic, and oxalic	Fresh fruits and commercial orange juice	Spheri-5RP-18 (110 × 4.6 mm) + 2 columns Polypore H (110 × 4.6 and 220 × 4.6 mm)	2% KH_2PO_4, pH 2.3	UV, 215 nm	144
Pyruvic, lactic, acetic, citric, formic, orotic, and uric	Dairy produce	Beckman C_8 (250 × 4.6 mm) 5 μm	0.5% $(NH_4)_2HPO_4$, pH 2.24:CH_3CN (99.55:0.45)	UV, 214 and 285 nm	145
Formic, acetic, pyruvic, uric, orotic, citric, and lactic	Cheese	Beckman C_8 (250 × 4.6 mm) 5 μm	0.5% $(NH_4)_2HPO_4$, pH 2.24:CH_3CN (99.98:0.2)	UV, 214 nm	146
Quinic, malic, and citric	Tropical fruits	Spherisorb ODS 2 (250 × 4.6 mm) 5 μm	Water (HPLC) pH 2.2, H_2SO_4	UV, 214 nm	147
Oxalic, glycolic, malic, shikimic, lactic, citric, and fumaric	Cane juice	Sup Rs C_{18} 5 μm 32°C	2% $NH_4H_2PO_4$, pH 2.18	UV, 214 nm	148
Malic, shikimic, lactic, citric, succinic, tartaric, acetic, fumaric, citramalic, and glucuronic	Grape juice	Beckman ODS Ultrasphere C_{18} (250 × 4.6 mm) 5μm	0.02% HCOOH (v/v), pH 3.07	UV, 190 nm	149
Malic, shikimic, lactic, citric, succinic, tartaric, acetic, and fumaric	Wines and fruit juices	Spherisorb ODS 2 (250 × 4.6 mm) 5 μm 35°C	Na_2SO_4 (0.1 M)/H_2SO_4, pH 2.45–2.50	UV, 210 nm	150
Oxalic	Carambola (*Averrhoa Carambola L*)	Zorbax NH_2 (250 × 4.6 mm)	NaH_2PO_4(0.15 M), pH 2.4	UV, 206 nm	151
Malic, citric, and succinic	Orange and grape juices	(a) Zorbax NH_2 (250 × 4.6 mm); (b) Hamilton PRP-1 (150 × 4.1 mm) 10 μm	a) NaH_2PO_4 (0.075 M), pH 4.4 b) $HClO_4$ (0.03 N), pH 1.7	UV, 206 nm	152

Table 9 (*Continued*)

Acids	Sample	Column	Mobile phase	Detection	Ref.
Citric	Orange and grape juices	Waters RC-100 (100 × 8 mm) 5 μm	2% $NH_4H_2PO_4$, pH 2.7	UV, 206 nm + RI	153
Oxalic	Rhubarb	LiChrosorb RP-8 C_8 10 μm	0.5% KH_2PO_4/0.005 M TBA/H_3PO_4, pH 2	UV, 220 nm	154
Malic, citric, oxalic, tartaric, succinic, and fumaric	Tropical fruits	Waters RCM-100 C_{18} (10 cm)	2% $NH_4H_2PO_4$/H_3PO_4, pH 2.4	UV, 254 nm + RI	155
Tartaric, malic, and citric	Orange, lemon, apple, and grape juices	Bondapak C_{18} (300 × 4 mm)	H_3PO_4(aq), pH 2.2	RI	156
Quinic, malic, and citric	Blueberry juice	Bondapak C_{18} (300 × 4 mm)	2% KH_2PO_4/H_3PO_4, pH 2.4	RI	157
Quinic, malic, and citric	Apple juice	Supecosil LC-18 (250 × 4.6 mm) 5 μm	KH_2PO_4(0.2M)/H_3PO_4, pH 2.4	UV, 214 nm	158
Tartaric, malic, shikimic, lactic, acetic, succinic, citric, and citramalic	Wine	3 column RP C_{18} (250 × 4.6 mm) 5 μm	0.15% H_3PO_4/25% CH_3CN	UV, 210 nm	24
Acetic, succinic, malic, and tartaric	Vinegars	LiChrosorb RP-18 10 μm	Aqueous octylammonium salicylate	Conductimetric	159
Citric, malic, tartaric, oxalic and ketoglutaric	Mango	Waters Radial-Pak C_{18} 10 μm	0.2M KH_2PO_4/H_3PO_4, pH 2.4	UV, 214 nm	30
Malic, citric, quinic, succinic, lactic and shikimic	Apple juice and cider	Spherisorb ODS-2 C_{18} (250 × 4 mm) 5 μm	0.01 M KH_2PO_4/H_3PO_4 pH 2.25	UV, 206 nm	29
Lactic, acetic, tartaric, malic, citric and succinic	Wines	Merck Lubar RT 240-4 RP-18 5 μm	H_2O/CH_3CN (65:35)[a]	UV, 254 nm	160
Lactic, acetic, tartaric, malic, citric and succinic	Wines	C_{18} (250 × 4.6 mm) 5 μm	H_2O/CH_3CN (Linear gradient elution)[a]	UV, 254 nm	161
Lactic, acetic, quinic, glycolic, citric, and others	Wines and fruit juices	a) Nucleosil-5RP-18 (250 × 4 mm) b) Hypersil ODS (100 × 2.1 mm)	H_2O/CH_3CN (Linear gradient elution)[b]	UV, 265 nm	162

[a]Prior derivatization with phenacylbromide.
[b]Prior derivatization with o-(4-nitrobenzyl)-N,N'-diisopropylurea (PNBDI).
TBA, tetrabutylammonium hydrogen sulfate.

which should in principle be quite low on account of the inexpensive instrumentation and reagents needed.

Colorimetric methods also lack selectivity; however, they can be made sensitive via derivatization reactions. Overall, analyses are sluggish and costly but can be expedited by automation. Like classical methods, colorimetric methods are therefore inadvisable for determining individual organic acids.

Enzymatic methods are no doubt the most selective of all. They feature acceptable sensitivity and precision, but are slow and expensive—frequently unaffordable for routine analyses.

As shown by the literature published on the topic in the last 10 years, chromatographic techniques in general and HPLC in particular—the latter requires no derivatization and involves minimal sample treatment—are no doubt the most frequently chosen methods for determining organic acids in foods. The high sensitivity and selectivity of HPLC allows the effective determination of major organic acids. On the other hand, minor and especially trace organic acids pose a twofold problem. Thus, currently available detectors are not sufficiently sensitive to them, so that a derivatization reaction or preconcentration is usually needed. Mass spectrometers are still very expensive—unaffordable by most laboratories—and difficult to interface to liquid chromatographs. Also, the increased number of potentially detectable species in some foods makes available chromatographic columns inadequate. The problem can be addressed by using commercially available packed capillary columns (40–500 μm i.d.). Adapting ordinary equipment for this purpose is fairly easy and can provide increased efficiency and mass sensitivity, and allow ready coupling to other separation and detection techniques (163–167). In addition, mobile phases are circulated at very low flow rates (2–6 μl/min), which results in dramatically decreased operational costs. The sample volume used in each analysis ranges from 60–100 nL to ordinary volumes (e.g., 5 μl) provided the eluting power of the solvent containing the sample is much lower than that of the mobile phase. The technique is currently at a developing stage and no application to the determination of organic acids has so far been reported. Our group has used it to quantify quinic, malic, shikimic, and citric acid in apple juice with the aid of a C_{18} 30×320 μm i.d. column, 10^{-2} M K_2HPO_4/H_3PO_4 at pH 2.7 as the mobile phase ($q = 2$ μl/min), and UV detection at 206 nm. The detection limits thus achieved for an injected volume of 60 nl ranged from 2.9 ng for citric acid to 0.04 ng for shikimic acid, the coefficient of variation being less than 5% in all instances.

One other promising alternative is capillary electrophoresis (CE), which was born to the marriage of the high separation power of electrophoresis and the instrumentation and automation concepts of chromatography. Sample components are separated by charge and size. The CE technique offers a number of advantages including high simplicity, efficiency (the number of theoretical plates is one order of magnitude greater than that of chromatographic columns), and throughput (168,169). In addition, analytical costs are markedly lower that those of chromatographies. As with capillary liquid chromatography (CLC), there is virtually no reference to the CE determination of organic acids in foods, but only to citric, tartaric, malic, succinic, acetic, and lactic acids in wines using indirect UV detection (170). In any case, this technique has some sensitivity, reproducibility, and quantitation weaknesses. CLC and CE are complementary and demand major improvements as regards the reproducibility of the sampling system and the sensitivity of detection. To the authors' minds, these miniaturized systems will be routinely used in food analyses in the very near future.

REVIEWS

R. S. Kirk and R. Sawyer, *Pearson's Composition and Analysis of Foods*, Longman, New York, 1991, p. 82.

D. B. Gomis, *Food Analysis by HPLC* (L. M. Nollet, ed.), Marcel Dekker, New York, 1992, p. 371.

H. H. Hill and D. G. MacMinn (eds.), *Detectors for Capillary Chromatography*, John Wiley and Sons, New York, 1992.
Official Methods of Analysis of Association of Official Analytical Chemists, Arlington, VI, 1984, 1990.

REFERENCES

1. J. J. Leguerinel, C. M. Cleret, P. Bourgeois, and P. Mafart, *Sciences des Aliments*, *7*: 223 (1987).
2. S. Noordeloos and C. W. Nagel, *Am. J. Enol. Vitic.*, *23*: 139 (1972).
3. K. H. Plattig, in *Sensory Analysis of Foods* (J. R. Piggott, ed.), Elsevier, London, 1988, p. 10.
4. G. A. Harrison, *J. Inst. Brew.*, *76*: 486 (1970).
5. P. Salo, L. Nykanen, and H. Suomalainen, *J. Food Sci.*, *37*: 394 (1972).
6. M. A. Amerine, E. B. Roessler, and C. S. Ough, *Am. J. Enol. Vitic.*, *16*: 29 (1965).
7. N. A. Eskin, in *Biochemistry of Foods*, Academic Press, San Diego, 1990, p. 417.
8. T. Henick-Kling, in *Wine, Microbiology and Biotechnology*, (G. Fleet, ed.), Hardwood Academic, Switzerland, 1993, p. 314.
9. G. Kelly and E. Latzko, in *The Biochemistry of Plants: A Comprehensive Treatise* (P. K. Stumpf and E. E. Conn, eds.), N. E. Tolbert, 1980, p. 199.
10. N. A. Eskin, in *Biochemistry of Foods*, Academic Press, Inc., New York, p. 87 (1990).
11. L. Aurand, A. Woods, and M. Wells, in *Food Composition and Analysis*, van Nostrand Reinhold, New York, 1987, p. 14.
12. European Community, Directive of the Council Meeting of 21 December. *Offic. J. Eur. Community*, *40*: 27 (1988).
13. S. Suye, N. Yoshihara, and S. Inuta, *Biosci. Biotech. Biochem.*, *56*: 1488 (1992).
14. A. Lonvaud-Funel, B. Doneche, and D. Bleuze, *Connaissance Vigne Vin*, *14*: 207 (1980).
15. M. Ameyama, K. Tayama, E. Shinagawa, K. Matsushita, and O. Adachi, *Agric. Biol. Chem.*, *42*: 2347 (1978).
16. C. Mosso, E. Verzotti, D. Marangon, and A. Viretto, *Industrie Alimentari*, *31*: 450 (1992).
17. D. Blanco Gomis, M. J. Morán Gutiérrez, M. D. Gutiérrez Alvarez, and A. Sanz Medel, *Chromatographia*, *24*: 347 (1987).
18. A. Schneider, V. Gerbi, and M. Redoglia, *Am. J. Enol. Vitic.*, *38*: 151 (1987).
19. R. F. Frayne, *Am. J. Enol. Vitic.*, *37*: 181 (1986).
20. R. Andersson and B. Hadlund, *Z. Lebensm. Unters. Forsch.*, *176*: 440 (1983).
21. S. H. Ashoor and J. Welty, *J. Assoc. Off. Anal. Chem.*, *67*: 885 (1984).
22. S. H. Ashoor and M. J. Knox, *J. Chromatogr.*, *299*: 288 (1984).
23. M. T. Guillén, L. Lugar, M. Martí, J. M. Gómez, and M. C. Torre, *Alimentaria*, 51 (1987).
24. J. P. Goiffon, A. Blanchere, and C. Reminiac, *Analusis*, *13*: 218 (1985).
25. M. C. Polo, F. Barahona, and I. Cáceres, *Connaissance Vigne Vin.*, *20*: 175 (1986).
26. J. D. McCord, E. Trousdale, and D. Y. Ryu, *Am. J. Enol. Vitic.*, *35*: 28 (1984).
27. G. H. Boeringer and F. R. Mannhein, *Boehringer Mannhein S. A.* (1975).
28. J. Battle and J. Bouvier, *Revue Française D'Œnologie. Cahier Scientifique*, *101*: 38 (1986).
29. D. Blanco Gomis, M. J. Morán Gutiérrez, M. D. Gutiérrez Alvarez, and J. J. Mangas Alonso, *Chromatographia*, *25*: 1053 (1988).
30. A. P. Medlicott and A. K. Thompson, *J. Sci. Food Agric.*, *36*: 561 (1985).
31. J. D. Timpa and J. J. Burke, *J. Agric. Food Chem.*, *34*: 910 (1986).
32. G. Panari, *Milchwissenshaft*, *41*: 214 (1986).
33. D. H. Picha, *J. Agric. Food Chem.*, *33*: 743 (1985).
34. S. H. Ashoor and J. Welty, *J. Chromatogr.*, *287*: 452 (1984).
35. R. L. Brunelle, R. L. Schoeneman, and G. E: Martin, *J. Assoc. Off. Anal. Chem.*, *50*: 329 (1967).
36. G. E. Martin, J. G. Sullo, and R. L. Schoeneman, *J. Agric. Food Chem.*, *19*: 995 (1971).
37. J. Fitelson and G. L. Bowden, *J. Assoc. Off. Anal. Chem.*, *51*: 1224 (1968).
38. E. Fernández-Flores, D. A. Kline, and A. R. Johnson, *J. Assoc. Off. Anal. Chem.*, *53*: 17 (1970).
39. T. S. Rumsey, C. H. Noller, J. C. Burns, D. Kalb, D. L. Hill, and C. L. Rhykerd, *J. Dairy Sci.*, *47*: 1418 (1964).
40. A. Kuksis and P. Prioreschi, *Anal. Biochem.*, *19*: 468 (1967).

41. H. Busch, R. B. Hurlbert, and V. R. Potter, *J. Biol. Chem.*, *196*: 717 (1952).
42. F. Mizutani, S. Yabuki, and M. Asai, *Anal. Chim. Acta*, 245: 145 (1991).
43. S. Yoshioka, H. Ukeda, and K. Matsumoto, *Electroanalysis*, 4: 545 (1992).
44. AOAC, *Food Additives: Direct. Acidulants*, 348 (1975).
45. M. Rebelein, *Deutsche Lebens-mittel Rudshan*, *59*: 131 (1963).
46. G. Pilone and R. Kunkee, *Am. J. Enol. Vitic.*, *21*: 12 (1970).
47. AOAC, *Official Methods of Analysis, Sugars and Sugars Products*, 598 (1984).
48. MAPA, *Métodos Oficiales de Análisis de Vinagres, 1*: 213 (1986).
49. B. Fabig, K. Vielhauer, A. M. Moawad, and W. Achtnich, *Z. Pflanzenerähr. Bodenk.*, *152*: 261 (1989).
50. H. Salwin and J. F. Bond, *J. Assoc. Off. Anal. Chem*, *52*: 41 (1969).
51. I. Molnár-Perl and M. Morvai, *Chromatographia*, *23*: 760 (1987).
52. I. Molnár-Perl and M. Pintér-Szakács, *J. Chromatogr.*, *365*: 171 (1986).
53. I. Molnár-Perl, M. Morvai, and M. Pintér-Szakács, *Mugy. Kem. Foly.*, *94*: 520 (1988).
54. K. C. Li and J. G. Woodroof, *J. Agric. Fd. Chem.*, *16*: 534 (1968).
55. S. Littmann, E. Schulte, and L. Acker, *Z. Lebensm Unters Forsch*, *175*: 101 (1982).
56. W. Weissberger, T. E. Kavanagh, and P. G. Keeney, *J. Food Sci.*, *36*: 877 (1971).
57. A. Bertrand and R. Triquet-Pissard, *Ann. Fals. Exp. Chim.*, *73*: 623 (1980).
58. V. K. Agarwal, *J. Assoc. Off. Anal. Chem.*, *71*: 466 (1988).
59. M. G. Horning, E. A. Boucher, A. M. Muss, and E. C. Horning, *Anal. Lett.*, *1*: 713 (1968).
60. H. G. Förster, C. Thomas, J. Rathmann, and M. Müller, *Z. Chem.*, *28*: 413 (1986).
61. M. E. Parish, R. J. Braddock, and T. R. Graumlich, *J. Food Sci.*, *51*: 431 (1986).
62. G. W. Chapman and R. J. Horvat, *J. Agric. Fd. Chem.*, *37*: 947 (1989).
63. P. Englmair, *J. Chromatogr.*, *194*: 33 (1980).
64. W. W. Wagener, C. S. Ough, and M. A. Amerine, *Am. J. Enol. Vitic.*, *22*: 167 (1971).
65. E. Fernández-Flores, D. A. Kline, and A. R. Johnson, *J. Assoc. Off. Anal. Chem.*, *53*: 17 (1970).
66. G. E. Martin, J. G. Sullo, and R. L. Schoeneman, *J. Agric. Food Chem.*, *19*: 995 (1971).
67. J. J. Ryan and J. A. Dupont, *J. Agric. Food Chem.*, *21*: 45 (1973).
68. D. A. Heatherbell, *J. Sci. Food Agric.*, *25*: 1095 (1974).
69. R. L. Brunelle, R. L. Schoeneman, and G. E. Martin, *J. Assoc. Off. Anal. Chem.*, *50*: 329 (1967).
70. J. Fitelson and G. L. Bowden, *J. Assoc. Off. Anal. Chem.*, *51*: 1224 (1968).
71. D. W. Baker, *J. Assoc. Off. Anal. Chem.*, *56*: 1257 (1973).
72. C. D. Harvey, R. Jenness, and H. A. Morris, *J. Dairy Sci.*, *64*: 1648 (1981).
73. D. R. Lynch and M. S. Kaldy, *Can. J. Plant Sci.*, *65*: 793 (1985).
74. P. J. W. Rovers and T. Kasprzycka-Guttman, *Food Chem.*, *45*: 283 (1992).
75. S. T. Ingalis, P. E. Minkler, C. L. Hoppel, and J. E. Nordiander, *J. Chromatogr.*, *299*: 365 (1984).
76. E. Vioque, M. P. Maza, and F. Millán, *J. Chromatogr.*, *331*: 187 (1985).
77. M. J. Cooper and M. W. Anders, *Anal. Chem.*, *46*: 1849 (1974).
78. W. Steiner, E. Muller, D. Frohlich, and Battaglia, *Mitt. Geb. Lebensm. Hyg.*, *75*: 37 (1984).
79. Z. L. Bandi and E. S. Reynolds, *J. Chromatogr.*, *329*: 57 (1985).
80. W. Elbert, S. Breitenbach, A. Heftel, and J. Hahan, *J. Chromatogr.*, *328*: 111 (1985).
81. R. Farinotti, P. Siard, J. Bourson, J. Kirkiacharian, B. Valeur, and G. Mahuzier, *J. Chromatogr.*, *269*: 81 (1983).
82. H. Lingeman, A. Hulshoff, W. J. M. Underberg, and F.B.J. M. Offermann, *J. Chromatogr.*, *290*: 215 (1984).
83. Y. Roorda, C. Geonnet, and J. L. Rocca, *Analusis*, *10*: 409 (1982).
84. N. Kubab, R. Farinotti, and G. Mahuzier, *Analusis*, *14*: 125 (1986).
85. R. M. Marcé, M. Calull, J. C. Olucha, F. Borrull, and F. X. Rius, *Anal. Chim. Acta*, *242*: 25 (1991).
86. M. Tortajada, A. Maquieira, and R. Puchades, *Am. J. Enol. Vitic.*, *44*: 118 (1993).
87. M. Dubernet, *Connaissance Vigne Vin*, *3*: 297 (1976).
88. MAPA, *Métodos Oficiales de Análisis. Leche y Productos Lácetos*, *1*: 475 (1986).
89. M. A. Amerine and C. S. Ough, in *Análisis de Vinos y Mostos*, Acribia, Zaragoza, Spain, 1976, pp. 43–44.
90. F. Lázaro, M. D. Luque de Castro, and M. Valcárcel, *Analyst*, *111*: 729 (1986).
91. AOAC, *Official Methods of Analysis, Dairy products*, 279 (1984).

92. F. di Sipio, *Industrie Alimentari*, *31*: 103 (1992).
93. J. Lima and A. Rangel, *Am. J. Enol. Vitic.*, *43*: 58 (1992).
94. L. P. McCloskey, *Am. J. Enol. Vitic.*, *31*: 170 (1980).
95. B. Doneche and P. J. Sánchez, *Connaissance Vigne Vin*, *19*: 161 (1985).
96. G. Henniger and L. Mascaro, *J. Assoc. Off. Anal. Chem.*, *68*: 1024 (1985).
97. C. Hasselmann and P. Diop, *Sciences des Aliments*, *3*: 161 (1983).
98. D. Weigelt, *Analyst*, *112*: 1155 (1987).
99. R. Renneberg, G. Trott-Kriegeskorte, M. GLietz, V. Jäger, M. Pawlowa, G. Kaiser, U. Wollenberger, F. Schubert, R. Wagner, R. D. Schmid, and F. W. Scheller, *J. Biotechnol.*, *21*: 173 (1991).
100. J. Frew, H. Allen, and O. Hill, *Anal. Chem.*, *59*: 933 (1987).
101. L. Gorton and A. Hedlund, *Anal. Chim. Acta*, *213*: 91 (1988).
102. R. L. Brunelle, R. L. Schoeneman, and G. E. Martin, *J. Assoc. Off. Anal. Chem.*, *50*: 329 (1967).
103. C. L. Mehltretter and J. G. Otten, *Int. Sug. J.*, *73*: 235 (1971).
104. A. R. Johnson and E. Fernández-Flores, *J. Assoc. Off. Anal. Chem.*, *52*: 559 (1969).
105. J. F. T. Oldfield, R. Parslow, and M. Shore, *Int. Sug. J.*, *72*: 35 (1970).
106. M. Kiely and P. O'Drisceoil, *Int. Sug. J.*, *73*: 135 (1971).
107. M. Kiely and P. O'Drisceoil, *Int. Sug. J.*, *73*: 196 (1971).
108. S. J. Jadhav and T. W. Andrew, *Can. Inst. Food Sci. Technol. J.*, *10*: 13 (1977).
109. M. Morvai and I. Molnár-Perl, *J. Chromatogr.*, *520*: 201 (1990).
110. S. Littmann-Wienstedt, *Dtsch. Lebensm.-Rundsch*, *85*: 183 (1989).
111. A. Montano, A. De Castro, L. Rejano, and A. H. Sánchez, *J. Chromatogr.*, *594*: 259 (1992).
112. C. J. Belke and A. J. Irwin, *J. Am. Soc. Brew. Chem.*, *50*: 26 (1992).
113. I. Molnár-Perl, M. Morvai, M. Pinter-Szakacs, and M. Petro-Turza, *Agric., Food Chem. Consum., Proc. 5th Eur. Conf. Food Chem.*, *2*: 644 (1989).
114. J. G. Woodroof, *J. Agric. Food Chem.*, *16*: 534 (1968).
115. K. Zürcher and H. Hadorn, *Mitt. Gebiete Lebensm. Hyg.*, *66*: 379 (1975).
116. A. Bertrand, *Ann. Fals. Exp. Chim.*, *67*: 253 (1974).
117. J. Palmer and D. M. List, *J. Agric. Food Chem.*, *21*: 903 (1973).
118. R. M. Wheaton and W. C. Bauman, *Ind. Eng. Chem.*, *45*: 228 (1953).
119. G. A. Harlow and D. H. Morman, *Anal. Chem.*, *36*: 2438 (1964).
120. K. Tanaka, T. Ishizuka, and H. Sunahara, *J. Chromatogr.*, *174*: 153 (1979).
121. K. Kinara, S. Rokushika, and H. Hatano, *J. Chromatogr. 410*: 103 (1987).
122. G. H. D. Van der Stegen, and J. Van Duijn, *Colloq. Sci. Int. Cafe*, *12*: 238 (1987).
123. C. Horvath, W. Melander, and I. Molnar, *J. Chromatogr.*, *125*: 129 (1976).
124. I. Molnar and C. Horvath, *Clin. Chem.*, *22*: 1497 (1976).
125. C. Horvath, W. Melander, and I. Molnar, *Anal. Chem.*, *49*: 142 (1980).
126. P. Jandera, J. Churacek, and J. Bartosova, *Chromatographia*, *13*: 485 (1980).
127. R. M. Marcé, M. Calull, J. C. Olucha, F. Borrull, and F. X. Rius, *Anal. Chim. Acta*, *259*: 237 (1992).
128. A. Yamamoto, A. Matsunaga, and E. Mizukami, *Eisei Kagaku*, *36*: 332 (1990).
129. H. Klein and R. Leubolt, *J. Chromatogr.*, *640*: 259 (1993).
130. G. Doyon, G. Gaudreau, D. St.-Gelais, Y. Beaulieu, and C. J. Randall, *Can. Inst. Food Sci. Technol. J.*, 24 (1991).
131. M. Calull, R. M. Marcé, J. C. Olucha, and F. Borull, *J. Chromatogr.*, *590*: 215 (1992).
132. M. Calull, R. M. Marcé, and F. Borull, *J. Chromatogr.*, *589*: 151 (1982).
133. K. I. Tomlins, D. M. Baker, and I. J. McDowell, *Chromatographia*, *29*: 557 (1990).
134. D. Chabert, B. Giorgio, C. Tisse, M. Guérrère, and J. Estienne, *Ann. Fals. Exp. Chim.*, *80*: 259 (1987).
135. J. Haginaka, J. Wakai, H. Yasuda, and T. Nomura, *J. Chromatogr.*, *447*: 373 (1988).
136. M. J. Lázaro, E. Carbonell, M. C. Aristoy, J. Safón, and M. Rodrigo, *J. Assoc. Off. Anal. Chem.*, *72*: 52 (1989).
137. P. E. Shaw and C. W. Wilson, *J. Sci. Food Agric.*, *14*: 153 (1987).
138. R. J. Bushway, J. L. Bureau, and D. F. Megann, *J. Food Sci.*, *49*: 75 (1984).
139. A. Canale, M. E. Valente, and A. Ciotti, *J. Sci. Food Agric.*, *35*: 1178 (1984).
140. R. T. Marsili, H. Ostapenko, R. E. Simmons, and D. E. Green, *J. Food Sci.*, *46*: 52 (1981).
141. M. O. Nisperos-Carriedo, E. A. Baldwui, M. G. Moshonas, and P. E. Shaw, *J. Agric. Food Chem.*, *40*: 2464 (1992).

142. M. J. López Hernández, J. Simal Lozano, and M. A. Romero Rodríguez, *Anal. Bromatol.*, *XLI-1*: 65 (1989).

143. M. A. Romero Rodríguez, V. M. González Rodríguez, and M. A. Lage-Yusti, *Anal. Bromatol.*, *XLII-2*: 299 (1990).

144. M. O. Nisperos-Carriedo, B. S. Ruslig, and P. E. Shaw, *J. Agric. Food Chem.*, *40*: 1127 (1992).

145. A. E. Bevilacqua and A. N. Califano, *J. Food Sci.*, *54*: 1076 (1989).

146. A. E. Bevilacqua and A. N. Califano, *Food Chem.*, *43*: 345 (1992).

147. M. A. Romero Rodríguez, M. L. Vázquez Orderiz, M. J. López Hernández, and J. Simal Lozano, *J. Chromatogr. Sci.*, *30*: 433 (1992).

148. A. Danielle, C. Myrtil, and A. Parfait, *Int. Sugar J.*, *90*: 28 (1988).

149. M. Llorente, B. Villarroya, and C. Gómez-Cordovés, *Chromatographia*, *32*: 555 (1991).

150. R. M. Marcé, R. Callul, R. M. Manchobas, F. Borrull, and F. X. Rius, *Chromatographia*, *29*: 54 (1990).

151. C. W. Wilson, P. E. Shaw, and R. J. Knight, *J. Agric. Food Chem.*, *30*: 1106 (1982).

152. P. E. Shaw and C. W. Wilson, *J. Sci. Food Agric.*, *34*: 1285 (1983).

153. P. E. Shaw, B. S. Buslig, and C. W. Wilson, *J. Agric Food Chem.*, *31*: 182 (1983).

154. B. Libert, *J. Chromatogr.*, *210*: 540 (1981).

155. P. E. Shaw and C. W. Wilson, *J. Sci. Food Agric.*, *32*: 1242 (1981).

156. D. Bigliardi, S. Gherardi, and M. Poli, *Industria Conserve*, *54*: 209 (1979).

157. E. D. Coppola, E. Conrad, and R. Cotter, *J. Assoc. Anal. Chem.*, *61*: 1490 (1978).

158. E. D. Coppola and M. S. Starr, *J. Assoc. Anal, Chem.*, *69*: 594 (1986).

159. M. C. Gennaro, *J. Chromatogr.*, *449*: 103 (1988).

160. E. Mentasti, M. C. Gennaro, C. Sarzanini, C. Baiocchi, and M. Savigliano, *J. Chromatogr.*, *332*: 177 (1985).

161. F. Caccamo, G. Carfagnini, A. Di Corcia, and R. Samperi, *J. Chromatogr.*, *362*: 47 (1986).

162. R. Badoud and G. Pratz, *J. Chromatogr.*, *360*: 119 (1986).

163. M. N. Novotny and D. Ishii (eds.), *Microcolumn Separations*, Elsevier, Amsterdam (1985).

164. J. P. Chervet and C. J. Meijvogel, *LC-GC*, *4*: 33 (1991).

165. A. Berloni, A. Cappiello, G. Famiglini, and P. Palma, *Chromatographia*, *39*: 279 (1994).

166. B. Lin Ling, W. R. G. Baeyens, and C. Dewaele, *J. High Resolut. Chromatogr.*, *14*: 169 (1991).

167. Milos Krejcí, *Trace Analysis with Microcolumn Liquid Chromatography*, Chromatographic Science Series, Vol. 59, Marcel Dekker, New York, 1992.

168. M. V. Pickering, *LC-GC*, *7*: 752 (1990).

169. W. Steuer, Y. Grant, and F. Erni, *J. Chromatogr.*, *507*: 125 (1990).

170. Waters Technical Bulletin, Organic acids analysis in foods and beverages by capillary electrophoresis, 1990.

20

Organic Bases

Maria Carla Gennaro, Donatella Giacosa, Claudia Abrigo, Paola Biglino, and Ennio Campi
University of Torino, Torino, Italy

I. INTRODUCTION

This chapter is devoted to the methods of identification and determination of organic bases in food. In agreement with Debruyckere and Van Peteghem (1), a presentation of the literature data will follow, rather than a criterion based on the chemical structure of the amines, a classification generally accepted in food science that is based on the significance in food of the presence of different classes or organic bases.

A general review concerning the determination of amines in food and containing 52 references has been published by Buchberger and Weiss (2). Other reviews concern amine analysis in dairy products (3, 67 references) or particularly consider isotachophoresis methods (4). Most of the published manuscripts and reviews concern specific classes of organic bases. According to the above-mentioned criterion, the following two major classes are distinguished:

1. Amines
2. Alkaloids

II. AMINES

In accordance with the concept that in food science the classification of the species of interest is based on their importance and presence in food more than on their chemical structure (1), the following subclasses of amines are here identified and treated:

1. Biogenic amines
2. Unsulfonated aromatic amines
3. Heterocyclic amines
4. Nitrosamines and volatile alkylamines

For each of these classes a brief overview is given and the data published are organized in tabular form, in which for the considered analytes all of the data available concerning analytical

745

method, experimental condition, detection limit, kind of food investigated, and amount found for each species, as well as the sample pretreatment principal steps, are reported.

A. Biogenic Amines

Among biogenic amines some aromatics such as histamine, tyramine, phenylethylamine, tryptamine, and some aliphatics such as spermine, spermidine, putrescine, and cadaverine are referred.

Biogenic amines are naturally present in the central nervous system where they play an important role as neurotransmitters. They are receiving increasing interest because they are often present in some foods, such as cheese, meat, fish, and poultry, and because they are reported to be psychoactive and vasoactive, responsible for diseases with symptoms similar to those of intoxication, such as headache, nausea, depression, hypo- or hypertension, schizophrenia, cardiac palpitation, up to clinical shock.

The major toxic biogenic amine is histamine, which has been associated with scombroid poisoning and for which the U.S. Department of Agriculture has proclaimed an upper safe limit of 50 mg/100 g of food. In turn, p-tyramine and 2-phenylethylamine are the principalagents responsible for migraine.

Biogenic amines in food are formed during the ripening process through the enzyme degradation of proteins to amino acids and the subsquent decarboxylation processes, operated by bacterial enzymes. In particular, microbial spoilage in food can lead to production of decarboxylases, which u nder appropriate conditions can convert amino acids to their corresponding amines. So, for example, cadaverine and putrescine are formed from lysine and ornitine and histamine derives from histidine. Therefore the determination of the precursor amino acids is very important. Biogenic amine content, and in particular putrescine and cadaverine content, is a useful indicator of spoilage of food and can be advantageously used for the assessment of meat quality.

The extraction of amines from the matrix represents a critical step of the analysis. Many procedures have been proposed making use of different extraction steps and different solvents, such as perchloric, trichloroacetic, and hydrochloric acids, methanol, or other organic solvents.

For biogenic amine determination chromatographic methods are generally employed, in particular high-performance liquid chromatography (HPLC), ion pair, ion exchange, gas-liquid chromatography, and thin-layer chromatography. The detectioin is spectrophotometric or fluorometric of the fluorescent derivatives. Pre- or postcolumn derivatization reaction are in fact often employed to improve sensitivity. Dansyl chloride (5-dimethylaminoaphthalene-1-sulfonyl chloride), dabsyl chloride (4-dimethylaminoazobenzene-4′-sulfonyl chloride), o-phthaldialdehyde (OPA), fluorescamine, and ninhydrin are used as derivatizing agents.

In the following tables, the manuscripts published from 1980 are reviewed. For the biogenic amines considered in each manuscript are listed, when available, the method used, the experimental conditions, the detection limit, the kind of food considered and the amount there found, the essential steps concerning the extraction, and the sample treatment before the analysis.

Reviews have also been published concerning the extraction and analysis in food of biogenic amines, their significance and determination (5 with 46 references, 6, 7). A review by Pfannhauer and Pechanek (8) with 123 references deals with the mechanisms of formation and degradation of biogenic amines in fish, cheese, wine, meat, sausages, as well as the toxicological aspects and the principal methods of analysis. Methods for separation and determination of agmatine, cadaverine, histamine, putrescine, spermidine, and tyramine in beer are summarized in Table 1 by the American Society of Brewing Chemists Inc. (9) and the possibilities of isotachophoresis in the analysis of biogenic amines in milk and milk products (10) and in food and drugs (11) are reported.

Table 1 Biogenic Amines

Analyte	Analytical method	Experimental conditions	Detection limit	Applications and amount	Sample treatment	Ref.
Biogenic amines (histamine, tryptamine, 2-phenylethylamine, putrescine, cadaverine) with 1,6-diamino-hexane and tryptophan.	HPLC with the use of coulometric array detection (16 porous graphite working electrodes at increasing potentials) of o-phthalaldehyde derivatives.	ODS HR80 PTFE as the stationary phase. Mobile phase: sodium acetate, acetonitrile, tetrahydrofuran (pH = 6.5) at different ratios and in gradient elution. Potential between 0 and 1200 mV (increments of 80 mV at each electrode) vs. solid-state Pd reference electrode.	Histamine: 16 ppb, tryptophan 21 ppb, tryptamine 18 ppb, tyramine 16 ppb, phenlylethylamine 22 ppb, putrescine 25 ppb, cadaverine 21 ppb, 1,6-diaminohexane 12 ppb	Gewurztraminer, Barolo and Port wines	0.22 μm filtration and derivatization reaction with o-phthaladehyde.	12
Biogenic amines (tyramine, putrescine, cadaverine, tryptamine, 2-phenylethylamine).	HPLC with precolumn derivatization with 9-fluorenylmethyl-chloroformate and fluorometric detection at 315 nm (excitation 265 nm).	Reverse phase stationary phase. Mobile phases: sodium acetate (pH = 4.4) and acetonitrile in gradient elution.	around 1 ppm	Wine, cheese, fish.	Acidification, borate buffer addition, 0.9–fluorenyl-methyl chloroformate derivatization, addition of a mixture of heptylamine, acetonitrile, and HCl to remove the excess of the reagent.	13
Biogenic amines: (cadaverine, histamine, 2-phenylethylamine, putrescine, spermidine, spermine, tyramine, tryptamine, 5-hydroxytryptamine).	Ion pair HPLC with postcolumn derivatization with o-phthalaldehyde in methanol, boric acid, KOH, Brij-35 solution, mercaptoethanol. Fluorescence detection (445 nm, excitation 340 nm).	Novapak C-18 as the stationary phase. Mobile phase: sodium acetate, sodium octane sulfonate (pH = 4.5) with methanol or methanol-acetonitrile in gradient elution.	0.30–0.65 mg/L	Beer, hops, malt, wort.	$HClO_4$ extraction.	14

Table 1 (*Continued*)

Analyte	Analytical method	Experimental conditions	Detection limit	Applications and amount	Sample treatment	Ref.
Biogenic amines (cadaverine, histamine, putrescine, spermidine).	Ion exchange and pulsed amperometric detection. Au working electrode, stainless-steel counter electrode and Ag/AgCl as the reference electrode.	IonPac CS10 cation exchange and IonPac CG10 guard column. Mobile phase: acetonitrile, $HClO_4$, NaClO$_4$ in gradient elution.	putrescine: 5 ng, cadaverine and histamine 12 ng, spermidine 25 ng	Fish.	Homogenization, trichloracetic acid–HCl–heptane extraction, centrifugation (5000 g, 15 min). Reextraction with trichloroacetic acid, cleaning up on LC-18 SPE cartridge and elution with HCl.	15
Biogenic amines.	RP-HPLC and fluorometric detection at 455 nm (excitation 340 nm) of *o*-phthaldialdehyde derivatives.	Nucleosil 100 7C18 as the stationary phase. Mobile phase: hexanesulfonic acid, Kdihydrogenophosphate, acetonitrile in gradient elution.	0.5 mg/kg	Fermented sausages.	Homogenization with $HClO_4$, centrifugation (9000 g, 10 min, 0°C), filtration, dilution 1/10 v/v.	16
Biogenic amines (together with polyols and amino acids).	Capillary GC with FID detection.	Fused silica column of CP-Sil 5CB, He as carrier gas and T programming (10°C/min) from 160 to 300°C.			Derivatization with ethanol, acetonitrile, pyridine, ethyl chloroformate. CHCl$_3$ and NaHCO$_3$ addition. Organic phase extraction.	17
Biogenic amines.	HPLC and spectrophotometric detection at 254 nm of dansyl derivatives.	Spherisorb ODS-2 column as the stationary phase. Mobile phase: ammonium acetate, acetonitrile in gradient elution.	1–5 mg/kg	Minced meat, cheese, tuna fish, dry sausages.	Homogenization with $HClO_4$, centrifugation, filtration, reextraction. $HClO_4$ addition, incubation at 40°C with NaOH, NaHCO$_3$, and dansyl chloride in acetone for 45 min. NH$_4$OH addition, 30 min digestion, acetonitrile, centrifugation, filtration.	18

Analyte	Method	Column/mobile phase	Detection limit	Sample	Sample preparation	Ref.
Biogenic amines and metabolites.	HPLC and electrochemical detection, with a vitreous carbon electrode at +0.75 V vs. Ag/AgCl reference electrode.	ODS silica column and mobile phase of NaH_2PO_4, EDTA, octyl sulfate, methanol, pH = 3.5.		Brain of fish and amphibians.	Homogenization in ice cool $HClO_4$ and dihydroxy-benzylammonium bromide as the internal standard.	19
Biogenic amines (agmatine, putrescine, spermine, spermidine).	HPLC with post-column derivatization with o-phthalaldehyde. Fluorometric detection.	Polyaminepak (strong cation exchange resin) column. Mobile phase: sodium citrate, acetonitrile.		Vegetables.	Homogenization with $HClO_4$, cooling (T = 0°C), centrifugation.	20
Biogenic amines.	Miniaturized disposable amperometric thick-film biosensors.	Enzyme immobilization by crosslinking using glutaraldehyde at a Pt electrode.	0.06 μM of putrescine	Food quality control, fish, fruit juice, wine.		21
Biogenic amines (histamine, tyramine, putrescine, phenylamine, cadaverine, isoamylamine).	HPLC with precolumn derivatization (o-phthalaldehyde) and fluorescence detection at 440 nm (excitation at 230 nm).	Nucleosil 100-5C18 column. Mobile phase: acetonitrile-acetic acid in gradient elution.	1.0 mg/L histamine and tyramine, 0.5 mg/L putrescine, phenethylamine, cadaverine, and isoamylamine.	Wine.	Derivatization (heptylamine as the internal standard) with o-phthalaldehyde in borate buffer, pH = 10.4.	22
Histamine and tyramine.	Fluorometric determination.	Santos method (71).		Chihuahua cheese.		23
Biogenic amines: (cadaverine, histamine, 2-phenylethylamine, putrescine, spermidine, spermine, tyramine, tryptamine).	RP-HPLC with spectrophotometric detection (254 nm) of dansyl-derivatives, 1-7-diaminoheptane as internal standard.	Spherisorb 3S TG RP as the stationary phase. Mobile phase: acetonitrile, water, phosphate buffer solution pH = 7, gradient elution.	Ranging between 0.02 ng (putrescine and tyramine) and 0.08 ng (tryptamine).	Ground (parmigiano-reggiano) cheese. Amount determined: cadaverine and spermidine 0.01 mg/100g, 2-phenylethylamine, putrescine, and spermine below 0.10 mg/100g histamine 0.45 mg/100 g and tyramine 4.20 mg/100g.	Homogenization (HCl, diamino-heptane), centrifugation (14,000g, 20 min, 4°C), reextraction of aqueous residuum, Na_2CO_3 addition (pH > 12), extraction (n-butanol, chloroform) HCl addition, evaporation to dryness, HCl dissolution, derivatization (dansyl chloride, $NaHCO_3$), evaporation, acetonitrile dissolution.	24

Table 1 *(Continued)*

Analyte	Analytical method	Experimental conditions	Detection limit	Applications and amount	Sample treatment	Ref.
Biogenic amines: (agmatine, cadaverine, histamine, 2-phenylethylamine, putrescine, spermidine, spermine, tyramine, tryptamine).	HPLC with spectrophotometric detection (254 nm).	Lichrospher 100 RP-18 column. Water-methanol mixture as the mobile phase in isocratic or gradient elution.		Canned fish: tuna, bonito, mackerel, squid, anchovy.	Homogenization (trichloroacetic acid), filtration, addition of NaOH and benzoyl chloride, addition of saturated NaCl, extraction (ethyl ether), evaporation, methanol addition.	25
Biogenic amines: (cadaverine, ethylamine, histamine, isoamylamine, methylamine, 2-phenylethylamine, putrescine, tyramine, tryptamine).	HPLC with fluorometric detection of phthalaldehyde derivatives (445 nm, excitation 356 nm).	RP-18 as stationary phase and mobile phase of acetonitrile, octan-2-ol, Na_2HPO_4 in gradient elution.		Wines.	SO_2 addition, precolumn derivatization with phthalaldehyde-2-mercaptoethanol.	26
Biogenic amines (histamine, 2-phenetylamine, spermine, spermidine, tryptamine) with alkylamines and benzylamine.	Chemometric (PRISMA method) optimization of IMGE (isoselective multisolvent gradient elution) RP-HPLC method.	Novapak C_{18} as the stationary phase. Multisolvent (tetrahydrofuran, acetonitrile, methanol, water) and ammonium acetate as the mobile phase.		Animal feed, spoiled foodstuff.	$HClO_4$ dissolution, centrifugation, NaOH and saturated $NaHCO_3$ addition, dansylation reaction (acetone as the solvent).	27
Biogenic amines (histamine, 2-phenylethylamine, tyramine, tryptamine).	Ion interaction HPLC and spectrophotometric detection at different wavelengths.	Spherisorb ODS-2 as the stationary phase. Mobile phase: aqueous solution of octylammonium salicylate and octylammonium o-phosphate.	Histamine 900 ppb, 2-phenylethylamine 500 ppb, tyramine 400 ppb, tryptamine 20 ppb	Red wines. Found: 2-phenyethylamine 72 ppm, tryptamine 4 ppm.	Dilution (1/10 v/v), 0.20 μm filtration.	28

Biogenic amines.	HPLC analysis with three-component mobile phase optimization through PRISMA model.	Novapak C_{18} as the stationary phase $T = 37°C$. Mobile phase: mixture of tetrahydrofuran, acetonitrile, methanol, and ammonium acetate.		Fish feed.	$HClO_4$, dissolution, centrifugation, addition of NaOH and $NaHCO_3$, dansylation reaction (in acetone), digestion, NH_4OH addition.	29
Biogenic amines (cadaverine, 2-histamine, 2-phenylethylamine, putrescine, tyramine).	HPLC with spectrophotometric (440 and 570 nm) and fluorimetric detection (485 nm, excitation 254 nm). Ion chromatography.	HPLC method: Hypersil ODS as the stationary phase. $T = 35°C$. Mobile phase of ethanol, acetonitrile, buffer pH = 8. Ion chromatography: column DC-6A resin (Sarasin), $T = 65°C$. Gradient elution with NaCl in citrate buffer (pH = 5.65)–ethanol–Brij.	Tyramine: 5 ppm (UV detection) and 1 ppm (fluorescence). Histamine: 5 ppm (UV detection) and 20 ppm (fluorescence).	Cheese.	Extraction, dansyl derivatization for HPLC analysis. Homogenization with trichloroacetic acid for ion chromatography.	30
Biogenic amines (putrescine, cadaverine, spermidine, spermine, histamine).	HPLC and detection at 254 nm.	SGX-C18 column, H_2O–methanol as the mobile phase for putrescine and cadaverine; methanol–1,6-dioxane-H_2O for spermidine, spermine, histamine.	1 mg/kg	Silage.	For putrescine and cadaverine: $HClO_4$ extraction, addition of hexane-1,6-diamine, $HClO_4$, agitation, addition of benzoyl chloride, digestion. Addition of NaCl, extraction with ethyl ether. For histamine, spermidine, and spermine: acid extraction.	31
Biogenic amines and metabolites.	HPLC with electrochemical detection, vitreous carbon electrode, potential range between 0.72V and –0.85 V.			Wine, beer.	Deproteinization by ultrafiltration, cleaning with Amberlite CG-50, $HClO_4$ as eluant.	32

Table 1 (*Continued*)

Analyte	Analytical method	Experimental conditions	Detection limit	Applications and amount	Sample treatment	Ref.
Biogenic amines (tyramine and 2-phenylethylamine).	Ion-interaction RP-HPLC with spectrophotometric detection at 230 nm.	Spherisorb ODS2 as the stationary phase, octylammonium salicylate or octylammonium o-phosphate as the mobile phase, pH = 6.4.		Chocolate.	Homogenization with ice-cold $HClO_4$ and EDTA, extraction, centrifugation.	33
Biogenic amines (cadaverine, histamine, isoamylamine, phenylethylamine, putrescine, spermidine, spermine, tyramine, tryptamine).	HPLC with UV (254 nm) and fluorescence detection (490 nm and excitation 360 nm) of dansyl derivatives.	LiChrospher RP-18 as the stationary phase. Mobile phase of ethanol, acetonitrile, Tris, acetic acid buffer (pH = 8).	0.5 ppm in liquids, 5.0 ppm in solids	Wine, cheese, meat.	Addition of 1,7-diaminoheptane (internal standard), extraction with acetonitrile and $HClO_4$. Dansylation (in the presence of ethylamine). Extraction in ethylacetate. Evaporation to dryness. Dissolution in acetonitrile.	34
Tyramine.	HPLC with amperometric (glassy carbon electrode, at 0.7 V vs. Ag/AgCl reference electrode) and fluorometric detection at 305 nm (excitation 225 nm).	LiChrosorb RP-Select B stationary phase. Mobile phase: phosphate buffer pH = 7.5, EDTA and acetonitrile.		Camembert cheese.	Perchloric acid and dichloromethane addition, homogenization. NaOH addition, dilution, 0.45 μm filtration.	35
Biogenic amines.	HPLC with fluorimetric detection at 455 nm (Excitation 345 nm) of the *o*-phthaladehyde derivatives.	Shim-Pak CLC-ODS stationary phase, $T = 50°C$. Mobile phase: Na-hexane sulfonate, $NaClO_4$, methanol pH = 3.0.		Herring.	Dried herring homogenized with trichloro acetic acid, centrifuged, diluted.	36

Analyte	Detection	Column / phase	Detection limit	Sample	Procedure	Ref.
Biogenic amines (cadaverine, histamine, putrescine, spermine, spermidine, tyramine).	Normal phase HPLC. Fluorometric detection at 470 nm (excitation 335 nm) of dansyl derivatives.	Ultrasphere Si-column as the stationary phase. Ethylacetate-hexaneethanolamine as the mobile phase.	0.1 mg/kg	Fish, fish products.	Homogenization with trichloroacetic acid, heating at 60°C (15 min), filtration, addition of NaOH (pH = 9), extraction in butanol and HCl. Dansylation (dansyl chloride, 40°C, NaHCO$_3$, acetone). Extraction of dansyl derivatives into benzene, Na$_2$SO$_4$, filtration, evaporation to dryness, dissolution in benzene.	37
Biogenic amines (cadaverine, histamine, putrescine, tyramine, spermidine, spermine).	HPLC analysis. T = 40°C. Diode array detector: 254 nm.	Hypersil ODS as the stationary phase. Mobile phase: acetic acid, acetonitrile, methanol, in gradient elution.	mg/kg: cadaverine 0.2 histamine 0.3 putrescine 0.2 tyramine 0.3 spermidine 0.3 spermine 0.5	Poultry broiler chicken, monitoring of microbial spoilage.	Homogenization with HClO$_4$, chilling, filtration, extraction with HClO$_4$, KOH addition (pH = 5.5), dilution, T = 4°C, overnight digestion, filtration. Cleaning-up with Amberlite CG-50 for spermidine and spermine analysis.	38
Biogenic amines (cadaverine, histamine, putrescine, tyramine).	Fast-atom bombardment mass spectrometry (fast Xe bombardment, 5–8 KeV).	Positive-ion spectra recorded.		Beef joints, fermented sausages.	Addition of heptane or hexanediamine as internal standard. Dansylation reaction. Dissolution in acetonitrile, concentration, addition of glycerol.	39
Biogenic amines (cadaverine, histamine, putrescine and precursor arginine, histidine, lysine).	Ion pair HPLC with UV detection (254 nm).	Octylsilane column. Mobile phase of tetrabutylammonium phosphate in methanol-acetonitrile in linear gradient elution.	4 ppm for each amino acid and 2 ppm for each amine	Fish, meat.	Extraction with methanol, three elutions through pH = 7-buffered Amberlite IRC-50 Plus cation-exchange column, two elutions through pH = 4.25-buffered column of the same material, elution with methanolic HCl, evaporation at 65°C, addition of saturated NaHCO$_3$-methanol, derivatization with dansyl chloride.	40

Table 1 (Continued)

Analyte	Analytical method	Experimental conditions	Detection limit	Applications and amount	Sample treatment	Ref.
Biogenic amines.	Liquid chromatography with post-column reaction with ninhydrin (125°C).	LC-5000 amino acid analyzer with column of BTC2710 resin, T = 65°C. Elution with citrate buffer solution (pH 5.45–5.70) over 65 min.	1 ppm and 0.1 ppm with preconcentration	Wines.	Direct analysis.	41
Biogenic amines.	B2E linked-scan mass spectrometry.	MAT-CH5-DF mass spectrometry modified for collisionally induced dissociation, He as collision gas. The fragment ion at m/e 169 used for precursor ion search		Wine.	Addition of Na_2CO_3 and dansyl chloride in acetone, digestion, evaporation, dissolution in ethyl acetate.	42
Biogenic amines and amino acids.	GLC (after derivatization with heptafluorobutyric anhydride-propan-2-ol) and FID detection.	Flexible column of cross-linked SE-54, temperature programming 75–225°C		Cheddar cheese (high-quality and late-gassing).	Homogenization, extraction, purification on columns of AG 50 W-X8 cation exchange resin, HCl elution, evaporation to dryness. Dissolution in pyridine, benzene, trifluoroacetic anhydride, heating 60–65°C. Cooling. Benzene addition.	43
Biogenic amines (2-phenylethylamine, methylhistamine, histamine, tryptamine, m- and p-tyramine, 3-methoxytyramine, 5-hydroxytryptamine, cadaverine, putrescine, spermidine, spermine).	GLC with ECD detector.	Fused silica column coated with SE-54, programmed (150–240°C), He as the carrier gas.	5–20 pg injected	Cheese and chocolate: the amounts found for all of the amines in four different kinds of cheese and in four kinds of chocolate are given. They vary in wide ranges.	Homogenization in ice cold $HClO_4$, 2-(4-chlorophenyl)ethylamine (internal standard) and EDTA mixture. Centrifugation. Addition of solid $KHCO_3$. Extraction with bis(2-ethylhexyl)-hydrogenophosphate in $CHCl_3$. Purification by back extraction in HCl, benzene, acetonitrile, pentafluorobenzoyl chloride.	44

Analytes	Detection	Conditions	Detection limit	Sample	Extraction	Ref.
					Evaporation. Toluene dissolution.	
Biogenic amines (histamine, 2-phenylethylamine, tyramine, tryptamine and precursors histidine, 2-phenylalanine, tyrosine, tryptophan).	Ion pair HPLC and spectrophotometric detection at 215 and 260 nm.	CPSpher C18 Chrompak as the stationary phase, $T = 40°C$. Mobile phase: Na heptane-sulfonate, KH_2PO_4, H_3PO_4, methanol (or acetonitrile)	10 mg/kg	Gouda cheese.	Extraction in trichloroacetic acid, cooling at 3°C, centrifugation, removal of fatty layer, dilution.	45
Biogenic amines (cadaverine, histamine, putrescine, tyramine, tryptamine).	HPLC, with post-column derivatization (ninhydrin, hydrindantin) and thermosensitized reaction ($T = 145°C$). Spectrophotometric detection at 546 nm.	Nucleosil C18 as the stationary phase. Mobile phase: ninhydrin, hydrindantin, dimethylsulfoxide, sodium acetate buffer (pH = 5), and Na dodecyl sulfate in different ratios.	2 mg/kg in cheese for each of the amines, 0.8 mg/kg in sauerkraut, 0.3 mg/kg in wine	Cheese, wine, sauerkraut, chocolate, fish (tuna).	Homogenization with trisodium citrate, digestion, addition of trichloroacetic acid, centrifugation (10,000g, 4°C), filtration (0.45 μm), dilution.	46
Biogenic amines (histamine, histidine, tyramine, tryptophan) and precursors.	HPLC with spectrophotometric detection at 220 nm.	μ-Bondapack C_{18} as the stationary phase. Mobile phase: butane, pentane, hexane, octane; or camphor-sulfonate (Na salts) in methanol-H_2O-acetonitrile or acetonitrile-H_2O, pH = 3.0.		Cheese.	Homogenization with HCl, cleanup, multistep extraction.	47
Adrenaline noradrenaline, spermidine, spermine, tyramine.	Automated ion exchange method.			Mild and old cheese.	Trichloroacetic acid, $HClO_4$ and methanol extractions. Comparison of recovery yields.	48
Biogenic amines (cadaverine, histamine, phenylethylamine, putrescine, tyramine).	HPLC and fluorometric detection, identification with field desorption mass spectrometry.	Spherisorb ODS (5 μm) stationary phase.		Cheese, fish, yogurt, chocolate, beverages.	Water extraction, centrifugation, HCl extraction, centrifugation, Zerolit 236 column (Na^+ form) passage, HCl elution, evaporation to dryness, treatment with fluorescamine.	49

Table 1 (*Continued*)

Analyte	Analytical method	Experimental conditions	Detection limit	Applications and amount	Sample treatment	Ref.
Biogenic amines.	HPLC analysis with fluorometric detection at 450 nm (excitation 340 nm) of phthalaldehyde derivatives.	Ultrasphere ODS stationary phase. Mobile phase: Na acetate (pH = 6.6), tetrahydrofuran, methanol in gradient elution.		Wine.	Shaking with poly(vinyl-pyrrolidone), filtration up to 0.45 μm.	50
Cadaverine, histamine, putrescine.	HPLC and fluorescence detector.			Putrefaction of food fish products, raw fish.		51
Biogenic amines (histamine, tyramine, tryptamine).	Thin-layer chromatography, fluorescence detection (536 nm, excitation 365 nm).	Chromatogram development with $CHCl_3$, methanol, aqueous NH_3 plate air-dried and methanolic 7-chloro-4-nitrobenzofuran sprayed.	10 mg/kg for each amine	Uncooked german sausages.	Homogenization in HCl, centrifugation $T = 4°C$, filtration, pH adjustment to pH = 10.0 (Na_2CO_3). NaCl saturation. Butanol multiextraction. Centrifugation.	52
Biogenic amines.	Three-step gradient elution from a cation exchange column. Detection with ninhydrin and spectrophotometry at 570 nm.		<1 ppm	Fresh, matured, and spoiled pork and beef. Amount found: <10 ppm in fresh meat.	Acid buffer solution extraction.	53
Biogenic amines.	Cation exchange chromatography.			Ground beef, minced beef, as indication of freshness and quality.	HCl extraction.	54
Biogenic amines.	Automated ion exchange chromatography.			Fresh and processed meat, meat products.	Comparison between thrichloroacetic acid, $HClO_4$ and methanol extractions.	55

Compounds	Method	Column/Conditions	Detection limit	Sample	Sample preparation	Ref.
Biogenic amines (cadaverine, histamine, putrescine).	Capillary gas chromatography with FID detection and mass fragmentation quantification.	WG-11 capillary column, injection T = 210°C, oven T = 80°C (1.5 min) 220°C, carrier gas: helium. Inlet pressure 3.2 atm.	8 ng (GC), 1 ng (GC-MS)	Fresh pork meat. Found: cadaverine 0.18 mg/100 g, histamine 0.2–0.27 ng/100 g, putrescine 0.06 mg/100 g.	$HClO_4$ double homogenization, ion exchange chromatography (weakly acidic cation exchanger, H^+ form), elution with HCl, evaporation or precipitation with sodium tetraphenylborate (pH = 3), 1,3-diaminopropane and 1,8-diaminooctane. Redissolution with HCl and diethyl ether. Extraction with diethyl ether. Tri-fluoroacetylation of amines. Histamine converted into N-trifluoro-acetyl-N-ethoxycarbonylhistamine.	56
Biogenic amines (agmatine, cadaverine, ethanolamine, histamine, 2-phenylethylamine, putrescine, spermidine, spermine, tyramine, tryptamine).	RP-HPLC and fluorometric detection of orthophthaldialdehyde derivatives (excitation 340 nm, emission 460nm) of the dansyl derivatives (excitation 360 nm, emission 460 nm) and of fluorescamine derivatives (excitation 395 nm, emission 460 nm).	μ-Bondapak C_{18} column. Mobile phase: methanol, acetonitrile, acetic acid, and buffer in different concentration ratios in isocratic and gradient elution.	pmol injected in 10 ml: agmatine 0.5, cadaverine 6.0, ethanolamine 13.0, histamine 5.0, 2-phenylethylamine 7.5, putrescine 5.5, tyramine 5.0, tryptamine 5.0	Red must and Villard Noir wine.	Derivatization with dansyl chloride, fluorescamine; and o-phthalaldehyde in presence of mercaptoethanol. Extraction in ethyl acetate.	57
Biogenic amines (histamine, 2-phenylethylamine, tyramine, tryptamine).	Thin-layer chromatography of dansyl derivatives with detection at 350 nm and fluorescence spectrophotometry.	Silica gel G and benzenetriethylamine as the mobile phase.		Soy sauce.	HCl acidification, dilution dansylation reaction (in acetone, Na_2CO_3), digestion, glutamic acid addition, evaporation to dryness, dissolution in benzene.	58

Table 1 (*Continued*)

Analyte	Analytical method	Experimental conditions	Detection limit	Applications and amount	Sample treatment	Ref.
Biogenic amines (ethanolamine, histamine, tyramine, phenylethylamine, putrescine, isoamylamine, cadaverine).	HPLC and fluorescence detection.	RP-18 Ultrasphere as the stationary phase. Mobile phase: methanol, Na acetate, tetrahydrofuran in nonlinear gradient elution.		Wine.	No pretreatment.	59
Biogenic amines (histamine, 2-phenylethylamine, putrescine).	HPLC of dansylated amines and detection at 254 nm.	Ultrasphere ODS as the stationary phase. Methanol, acetonitrile, *o*-phosphoric acid in gradient elution as the mobile phase.	2–5 mg/100 g of sample	Tuna, swiss cheese.	Extraction in methanol, multiple sequential extractions in butanol and HCl, evaporation to dryness, dansylation reaction.	60
Biogenic amines (cadaverine, putrescine).	Ion exchange chromatography.			Cheese, fish, meat.		61
Biogenic amines.	Gas–liquid chromatography.			Mushroom, beans.		62
Biogenic amines (histamine, putrescine, tyramine).	HPLC.			Vinegar, fruit juice.		63
Cadaverine, histamine, putrescine.	Automated ionexchange chromatography.			Cheese, fish.		64
Biogenic amines (histamine, putrescine, spermidine, spermine).	HPLC.			Fish, sardine, tuna.		65
Biogenic amines (agmatine, histamine, putrescine).	Ion exchange thin-layer chromatography and densitometry.			Fish, meat.		66

Compound	Method	Conditions	Detection limit	Sample / Found	Sample preparation	Ref.
Histamine Tyramine.	HPLC Ion pair HPLC of o-phthalaldehyde derivatives and fluorometric detection (emission 425 nm, excitation 338 nm).	Nova-Pak C18 stationary phase. Mobile phase: methanol, disodium hydrogen phosphate, heptanesulfonic acid, pH = 3.0	1–5 ppm in cheese	Dairy products, infant formulas, yogurt, cheese. Found in cheese 25–223 ppm.	Homogenization with methanol at T = 60°C, cooling, methanol addition, filtration.	67 68
Tyramine.	GLC analysis with flame-ionization detector.	OV-17 and OV-210, DMC5 Chromosorb W as the stationary phase. T = 175°C. N_2 as the carrier gas.	0.2 mg/g for packed soups and 40 µg/g for liquid soups	Canned and packed soups of chicken, vegetables, wheat flour, leek, glutamate, skimmed milk, hydrolized proteins, pepper, yeast. Found in tyramine <2–4 mg/pint for chicken and pea soups, <4 mg/pint for can liquid soups.	Homogenization with HCl, centrifugation, amphetamine sulfate (internal standard) addition, evaporation to dryness. Dichloromethane addition, evaporation. Addition of trichloroacetic anhydride, dichloromethane, evaporation to dryness.	69
Tyramine and N-methyltyramine.	Gas–liquid chromatography with FID detection and GC–mass spectrometry (ionizing 70 eV, accelerating 3.5 kV).	GL: linear T programming. Glass column silanized with dimethyldichlorosilane in toluene and n-butanol, $CHCl_3$ as coating solvent, T = 275°C. Nitrogen as carrier gas. GC-mass: glass column 1.5% Ov-17, 0.2% SP-1000, helium. Oven T: 150–250°C.	2 µg	Fermented milk and cheese, soybean paste, soya sauce, Japanese sake, beer. Found in soya sauce 143–882 µg/g, in sake 0.21–0.51 ppm, in beer 1.06–1.30 ppm, in fermented milk 0.41–2.33 ppm, in soybean paste 0.21–169.5 mg/kg, in cheese 45.0–138.4 mg/kg.	Homogenization with $HClO_4$. Centrifugation. Purification by elution through Amberlite CG-120 column, conversion to N,O-bis(ethyloxycarbonyl) derivatives in ethyl chloroformate.	70

Table 1 (*Continued*)

Analyte	Analytical method	Experimental conditions	Detection limit	Applications and amount	Sample treatment	Ref.
Aliphatic amines (dimethylamine and trimethylamine) and biogenic amines.	HPLC with UV detection at 207 nm.	IMP-HPX-72-0 BioRad column in OH form and NaOH as the mobile phase.		Seafood (to assess spoilage), cod, mackerel.	$HClO_4$ homogenization, KOH addition to pH = 7, filtration 0.22 μm.	72
Alkylamine and biogenic amines[+].	HPLC of phthalaldeyde derivatives and fluorometric detection at 425 (excitation at 335 nm).	Fine-SIL C18 column. Mobile phase: methanol-NaH_2PO_4 in different ratios, in isocratic and gradient elutions.	1 μM/ml of extract	Fresh tea shoots, made tea.	Fresh tea shoots: steamed, frozen, dried, pulverized, extracted in water (T = 50°C), filtered, derivatized with o-phthalaldehyde-2-mercaptoethanol in borate buffer solution (pH = 10).	73
Aliphatic amines (together with tyramine and histamine).	Enzyme (momoamine oxidase) electrode.			Meat (freshness testing).		74
Dimethylamine and trimethylamine (together with biogenic amines, urea, and ammonia).	HPLC, UV detection at 208 and 214 nm.	BioRad HPX-72-0 column and mobile phase of NaOH.		Seafood.		75

Source: Ref. 8.

B. Unsulfonated Aromatic Amines

A study in the 1970s estimated that the annual individual consumption of food in the USA WAS 645 KG, 5.5 g of which represented synthetic color. A similar evaluation in the UK states that the amount of food which contains added color represents nearly half of the total diet and the average amount of synthetic colors consumed is around 10 g per head per year (766–78). The kind and the amounts of food and beverages with added color in amounts up to 120 mg/kg of product are surprisingly high.

Besides the possible toxicity of the dye itself, synthetic food colors often contain toxic impurities. These can be contained in the reagents or can be formed as side products during chemical synthesis. In particular, in azo dyes, such as FD&C Yellow No. 6 (or sunset yellow, 1-(4′-sulfo-1′-phenylazo-)2-naphthol-6-sulfonic acid disodium salt), FD&C Red No. 2 (or amaranth, 3-hydroxy-4-[(4-sulfo-1-naphthalenyl)-azo-]-2,7-naphthalenedisulfonic acid trisodium salt), or FD&C Yellow No. 5 (ortartrazine, 4,5-dihydro-5-oxo-1-(4-sulfophenyl)-4-[(4-sulfophenl)azo]-[1]H-pyrazole-3-carboxylic acid trisodium salt), carcinogenic aromatic amines can be present, in particular naphthylamine, benzidine, and aniline, together with other toxic species like 4-aminodiphenyl and aromatic polycyclic hydrocarbons. Owing to impurities contained in the reagents or due to formation in side processes during the chemical synthesis of the dyes, aromatic amines and in particular naphthylamine and benzidine, together with 4-aminodiphenyl and aromatic polycyclic hydrocarbons, can be present in many dyes and mainly in azo dyes. Carcinogenicity and toxicity must be checked therefore not only for dyes but for impurities contained therein. Also toxicity of colors used in food packaging must often be considered along with the possibility of release of toxic species to the contained food.

Another possibility coming from the addition of color to food or beverages must be considered. Admitted colors can in fact undergo degradation reactions which lead to the formation of aromatic amines, in particular aniline, together with naphthionic acid and sulfanilic acids. These processes were shown to be particularly favored by light and by the presence of ascorbic acid. Therefore they can easily develop in bottled soft drinks which are displayed in bright sunlight, as, for example, in shop windows. The problem is of particular imoortance in countries which experience strong sunlight (79).

Methods for the analysis of aromatic amines in synthetic azo dyes and in food and beverages colored with these dyes have been developed. HPLC or gas chromatographic methods are generally employed, often with the help of derivatization reaction with R salt (disodium-3-hydroxynaphthalene-2,7-disulfonate) or pyrazolone T (4,5-dihydro-5-oxo-1-(4-sulfophenyl-)-[1]H-pyrazole-3-carboxylic acid) and fluorometric detection. The methods are presented in Table 2, which gives, when available, the method used, the experimental conditions, the detection limit, the kind of food considered and the amount found in the different foods investigated for aromatic amines, as well as the essential steps concerning the extraction and the sample treatment before the analysis.

Some results indicated the presence of aromatic amines in commercial food dyes as well as in soft drinks and hard candies. Aniline, for example, was found in orange beverages up to 12.6 ng/ml and in hard candies up to 9.2 ng/g and naphthylamine at concentrations up to 8,25 ng/ml in orange beverages and up to 10.6 ng/g in hard candies (80).

Table 2 Unsulfonated Aromatic Amines

Analyte	Analytical method	Experimental conditions	Detection limit	Applications and amount	Sample treatment	Ref.
Unsulfonated aromatic amines (aniline, 1- and 2-naphthylamine, 2- and 4-aminobiphenyl).	Ion pair HPLC with spectrophotometric detection at 426, 512, and 625 nm.	Supelco LC-18 column. mobile phase: methanol, tetrabutylammonium iodide in gradient elution.	<0.3 ng/ml	Soft drinks, hard candies. Aniline was found in orange beverage up to 12.6 ng/ml. 1-Naphthylamine up to 8.25 ng/ml and 2-naphthylamine up to 1.12 ng/ml in grape product. In hard candies aniline was found up to 9.2 ng/g and 1-naphthylamine up to 10.6 ng/g.	Preconcentration on C18 cartridge column, elution with methanol, water. Dilution. pH adjustment to 2.1–2.6 with H_2SO_4. Na_2SO_4 addition. pH adjustment to 8.5 with NaOH. NaCl addition, $CHCl_3$ extraction, H_2SO_4 addition, diazotization, and coupling reaction with disodium-2-naphthol,3,6-disulfonate. Dissolution in methanol.	80
Unsulfonated aromatic amines: 1- and 2-naphthylamine.	Ion pair HPLC and spectrophotometric detection (522 nm).	Supelco LC-18 as the stationary phase. Mobile phase of methanol and tetrabutylammonium phosphate.	8 ng/g	Commercial samples of food color amaranth. Found in up to 435 μg/g of 1-naphthylamine and 214 μg/g of 2-naphthylamine (5% in the free state).	Na_2SO_4 addition, NaOH, NaCl, and $CHCl_3$ addition, diazotization, coupling reaction with 3-hydroxynaphthalene-2,7-disulfonic acid and ditionite.	81
Aniline, p-cresidine.	HPLC with 240–700 nm diode array. Detection at 254 and 546 nm.	Bio-Sil C18 column equilibrated with ammonium acetate and acetonitrile. Mobile phase: water, ammonium acetate, acetonitrile in linear gradient elution.	5 ppb	FD&C Red. No. 40 p-cresidine was found in 26 samples: amount 0–2115 ppb (average 40 ppb); aniline 5–169 ppb.	NaOH addition to water solution, multiple-step extraction on Extrelut QE columns with $CHCl_3$, H_2SO_4 addition, evaporation. H_2SO_4 addition, diazotization, and coupling reaction with R-salt or pyrazolone for aniline.	82

Compound	Method	Sample/matrix	Results	Extraction/sample preparation	Ref.
Unsulfonated aromatic amines (aniline, benzidine) with 4-aminobiphenyl and 4-aminoazobenzene.	HPLC with UV detection at 254 and 510 nm. Microsorb C18 column. Mobile phase of water-acetonitrile-ammonium acetate in gradient elution.	In food color FD&C Yellow No. 6. Found in 34 certified samples: aniline 5.3–422 ppb (average 98), 4-aminobiphenyl 0.2–23 ppb (average 3.5 ppb), 4-aminoazobenzene 0.5–1098 ppb (average 4.0 ppb). Benzidine absent.	Benzidine 2.3 ppb, aniline 34.2 ppb, 4-aminobiphenyl 4.6 ppb, 4-aminoazobenzene 4.2 ppb.	Two extraction methods employed: (A) NaCl and NaOH addition, CHCl$_3$ extraction, H$_2$SO$_4$ addition, CHCl$_3$ removed. (B) Warm water and NaOH addition. Extraction in dry Extrelut QE column. CHCl$_3$ extracts treated as in method A. Diazotization and coupling reaction with R salt or pyrazolone-T.	83
Aromatic amines (benzidine, aniline) together with 4-aminobiphenyl and 4-aminoazobenzene.	HPLC and spectrophotometric detection at 254 and 510 nm. Novapak C18 stationary phase.	In 25 commercial samples of food color FD&C Yellow No. 5. Benzidine was found in 5 samples under 1 ng/g.		Aqueous-alkaline solution extracted into CHCl$_3$, solvent evaporation, extract dissolution in diluted acid. Diazotization and coupling reaction with R salt.	84
Aromatic amines (cyclohexylamine, dicyclohexylamine, aniline) and cyclamate.	Gas chromatography. 10% Carbowax column 2.5% NaOH or Chromosorb PNAW.	Beverages, confectionery.	All the samples contained measurable levels of aniline (mean 67 ng/g, maximum amount 487 ng/g).	Dissolution, sulfosalicylic acid addition, centrifugation, heating with HCl and H$_2$O$_2$, CH$_2$Cl$_2$ extraction.	85
Aromatic amines (dichloroaniline, 2,4-dimethylaniline).	HPLC analysis with electrochemical detection. Vitreous working electrode and Ag/AgCl reference electrode. Applied voltage 0.9–1.0V. μ-Bondapak C$_{18}$ stationary phase. Phosphate buffer (pH = 6.0) and acetonitrile as the mobile phase.	Fatty foods and food packaging, vegetable oils.		Dissolution in hexane, extraction with HCl, hexane, acetonitrile.	86
Aniline, aromatic amines.	Gas–liquid chromatography.	Fats and fatty oils.			87

Table 2 (*Continued*)

Analyte	Analytical method	Experimental conditions	Detection limit	Applications and amount	Sample treatment	Ref.
Aniline, biphenyl-amine, benzidine, naphthylamine, *o*-toluidine.	Microchemical detection.			Cereal, brans, wheat, oat.		88
Aromatic amines.	HPLC			Food packaging material.		89
Aromatic amines.	Gaschromatography.			Food.		90
Unsulfonated aromatic amines (aniline, 1-naphthylamine, biphenyl-2-amine, biphenyl-4-amine, *p*-cresidine).	HPLC and spectrophotometric detection at 512 nm.	Gradient mobile phase aqueous-methanol.		Commercial samples of tartrazine, Sunset Yellow FCF, and Allura Red. Maximum content: aniline 519 ppm, 1-naphthylamine 5.0 ppm, biphenyl-2-amine 2.0 ppm, biphenyl 4-amine 210 ppm.	Na_2SO_4 addition, extraction with $CHCl_3$, diazotization, and coupling with R salt.	91
Aniline and *N*-1-naphthylethylenediamine.	Differential pulse polarography and spectrophotometry.	Dropping Hg electrode, Pt counterelectrode, and saturated calomel electrode. Drop time: 1 sec, scan rate 5m V/sec, pulse height 50 mV.	Degradation occurs after 1 hr of light treatment.	Degradation of food color in food by light with formation of aromatic amines.	Acetate buffer pH = 5.5	79

C. Mutagenic Heterocyclic Amines

Frying or broiling of meat and other protein-rich foods may generate heterocyclic aromatic amines, mainly differently substituted quinolines and quinoxalines, together with amino acid pyrolyzates or aminoimidazoazaarenes. All of these species may be potential human carcinogen and mutagen agents. The quantification of heterocyclic amines in cooked food is therefore essential for human risk assessment.

Heterocyclic amines together with pyridine, indole, and imidazole derivatives are also present in cooking fumes. For their analysis HPLC, gas chromatography, and gas-mass methods have been employed and extraction procedures from food as well as from cooking fumes have been described.

Solid phase extraction procedures have generally been developed and methods based on the combined extraction and adsorption on coupled columns of diatomaceous earth and cation exchange resins have been used. A review was published citing 28 references reporting chromatographic methods for their determination (92).

In Table 3 the published methods are presented and, when available, reported for every manuscript considered: the method used for the heterocyclic amines determination, the experimentalk conditions, the detection limit, the kind of food considered and the amount of amines found therein, as well as the essential steps concerning the extraction and sample treatment prior to analysis.

Table 3 Mutagenic Heterocyclic Amines

Analyte	Analytical method	Experimental conditions	Detection limit	Applications and amount	Sample treatment	Ref.
Mutagenic heterocyclic amines: quinoxaline derivatives.	Gaschromatography-mass spectrometry with EIMS detector at 70 eV.	Fused silica column coated with HP-5, temperature programming from 20°C to 280°C.	For MeIQ$_x$ 0.1 ng/g	Cooking fumes of minced beef and pork fried at 210°C.	Beef and pork were fried at 210°C. By mounting a glass funnel above the pan, the fumes drawn through glass fiber filters and XAD-2 sorbent tube. Desorption with methanol-NH$_3$ solution. Eluates evaporated to dryness and residue dissolved in HCl and extracted with CH$_2$Cl$_2$. Aqueous layer mixed with Na$_2$CO$_3$ and amines extracted in CH$_2$Cl$_2$. Derivatization with 3,5-bistrifluoromethylbenzyl bromide.	93
Heterocyclic amines.	HPLC with UV detection at 263 nm, electrochemical detection at 950 mV, and fluorometric detection at 450 nm (excitation at 360 nm).	LiChrosorb RP-Select B column and mobile phase of ammonium acetate (pH = 4.5), methanol, acetonitrile in isocratic and gradient elution.		Food flavors, meat extract.	Cleanup according to the Gross method (94). The spray-dried product was Soxhlet-extracted, placed on Kieselgur, and extracted with ethyl ether. Cleanup with affinity chromatography on Cu-phthalocyanine complex.	95

Compound	Method	Conditions	Detection limit	Sample	Extraction	Ref.
Heterocyclic amines (quinoline and quinoxaline derivatives).	HPLC with diode array spectrophotometric and fluorometric detection.	TSK gel ODS-80 and Supelguard LC-8-DB precolumn. Mobile phase: triethylamine, acetonitrile, phosphoric acid (pH = 3.2 and 3.6) in gradient elution.	1 ng/g	Pan-broiled meat, fish, salmon. Quinoxaline up to 4.7 ng/g of fried fish pan-broiled 200°C; 4.6 ng/g oven-cooked at 200°C and <1 ng/g barbecued at 270°C (indole derivatives: maximum amount in barbecued at 270°C).	Solid phase extraction in Extrelut column–CH_2Cl_2 filled and Bond-Elut PRS cartridge. Sequential and selective extractions with HCl, methanol, water, ammonia, ammonium acetate.	96
Heterocyclic amines.	RP-HPLC analysis and electrochemical detection E = 0.9 V with vitreous-carbon electrode vs. Ag/AgCl.	Synchropak SCD-100 column, short alkyl chain ligand, end-capped as the stationary phase. Mobile phase of acetonitrile-ammonium acetate buffer (pH = 5.5).	80–475 pg	Food products.		97
Mutagenic heterocyclic amines.	HPLC and mass spectrometry.	Supelco LC-CN or Supelco LC-18-DB stationary phase. Mobile phase of ammonium acetate (pH = 6.8), acetonitrile, methanol in gradient elution.	Low ppb level	Cooked beef products, beef extract, fried beef.	Methanol extraction, cleaning on XAD-2, ethyl acetate elution.	98
Heterocyclic amines (quinolines, quinoxalines, glutamic acid pyrolysates).	Gas chromatography.			Wines, grapes.		99

D. Nitrosamines and Volatile Aliphatic Amines

Developments in environmental carcinogenesis agree in indicating that N-nitroso compounds are a major class of chemical carcinogens which can lead to a wide variety of tumors.

In particular nitrosamines can be formed in vitro by reaction of nitrite with secondary or tertiary amines, as well as also in vivo in the stomach or small intestine, following the simultaneous ingestion of nitrite and precursor amines (100). Carcinogenic nitrosamines have been found in normal feces, confirming their endogenous formation in the gastointestinal tract. It follows that all of the naturally occurring alipatic amines in daily foods might play a significant role in the incidence of some forms of human cancer. These are generally aliphatic amines, in particular methyl-, dimethyl-, butyl, and penthylamine, whichare widely distributed in various food. Many of them are present in seafood or are formed by the breakdown of nitrogen-containing compounds. They are associated with undesirable organoleptic changes to the food. Anyway, even if offensive odors represent an important problem in food as well as in environmental chemistry, aliphatic amine major toxicity is not direct but derives from their capability to form nitrosamines. For this reason the presence in food of volatile aliphatic amines is very important and these two classes of amines are often treated together.

The presence of volatile N-nitrosamine in foods was first evidenced in cured meats and was recently reported in beer and other beverages. N-Nitrosodimethylamine and N-nitrosopyrrolidine are generally diffused in beer. It was shown that they derive from the malt used in the production of beer and were found in other products derived from malt, like vinegar and whiskey. (On the contrary, grape-based beverages like wine or brandy seem to be nitrosamine-free.

Headspace gas chromatographic as well as HPLC methods were employed for volatile and nonvolatile aliphatic and nitrosamines. Derivatization methods with dansyl or dabsyl chloride are generally employed with fluorometric detection in order to improve method sensitivity.

Reviews have been published reporting determination methods for volatile amines (101, with 30 references) and for nitrosamines in food and wine (102, with 259 references). Published manuscripts concerning aliphatic amines and nitrosamines from 1980 are reviewed in Table 4. Available data concerning method used, experimental conditions, detection limit, kind of food considered, and amount there found as well as the essential steps concerning the extraction and the sample treatment before the analysis are reported.

Table 4 Nitrosamines and Volatile Aliphatic Amines

Analyte	Analytical method	Experimental conditions	Detection limit	Applications and amount	Sample treatment	Ref.
Primary amines.	Fluorometry.	Chemical derivatization with naphthalene-2,3-dicarboxaldehyde and o-phthalaldehyde in the presence of borate buffer and a surfactant solution of 2% Triton X-100.		Tea, beer, drugs.		103
Methylamine, dimethylamine, diethylamine.	Ion chromatography.	Silica gel column, modified with sulfo groups and unmodified Silasorb 600. Elution with HNO_3 + isopropanol.	10 ppm			104
Primary and secondary amines.	Gas chromatography with chemiluminescence detector (modified thermal energy analyzer).	Column 10% OV-101 on Volasphere A2 (120–140 mesh). He as carrier gas. Temperature programming 210–224°C.	10 ppb	Food.	Mineral oil vacuum distillation, derivatization with benzenesulfonyl chloride, fractionation.	105
Methylamine, dimethylamine, trimethylamine.	Headspace gas chromatography with FID or N-selective detection.	Column packed with 4% Carbowax 20M and 0.8% KOH on CarbopackB (60–80 mesh). $T = 120$°C.		Fish, shrimps.	$HClO_4$ extraction, KOH addition, $T = 60$°C (20 min) in sealed vials.	106
Primary amines and cysteine.	HPLC with electrochemical, fluorometric, and spectrophotometric detections.	Optimization of the mobile phase by PRISMA model.		Garlic.	Extraction in methanol, cleaning-up on a Bond-Elut C18 or SCX column. Derivatization pre- and postcolumn with phthaldialdehyde 1,1-dimethylethanethiol.	107
Total aliphatic amines.	Flow injection analysis and detection at 460 nm of derivatives.	Carrier solution:NaOH. Reagent solution: Na 1,2-naphthoquinone 4-sulfonate, $CHCl_3$ extraction.	Ethylamine: 0.5 ppm	Alcoholic drinks (beer, wine).	On-line liquid extraction.	108

Table 4　(*Continued*)

Analyte	Analytical method	Experimental conditions	Detection limit	Applications and amount	Sample treatment	Ref.
Trimethylamine.	Titrimetric determination.	Steam distillation, Kjeldhal method, and titrimetry.		Fish (haddock, cod, scorpion fish).	Homogenization with trichloroacetic acid. Centrifugation. Filtration. Addition of NaOH and formaldehyde. Kjeldhal distillation, H_3BO_3 addition.	109
Lower chain alkylamines.	Gas chromatography and thermoionic detection.	Chromosorb 103, temperature programmed between 100 and 160°C (20°C/min). Ar as the carrier gas.	50 μg for dimethylamine	Beer, malt.	For malt: heating 70–80°C with NaOH, condensation in HCl, evaporation, NaOH dissolution. For beer: acid extraction.	110
Volatile amines and nitrosamines.	TLC with spectrophotometric and fluorescence detection of dansyl derivatives.	Silufol TC plate and hexane-$CHCl_3$, ether or $CHCl_3$, benzene, ethylacetate, acetic acid as the mobile phase.			Distillation (HCl or NaOH). Dansylation reaction. Extraction in benzene.	111
Nitrosamines.	Spectrofluorometry at 480 nm (excitation 390 nm).			Cured meat products.	Homogenization with ethyl ether and concentration in Kuderna-Danish apparatus, reduction to secondary amines with Ni-Al alloy. Reaction with fluorescamine.	112
Aliphatic amines (dimethylamine, trimethylamine) and biogenic amines.	HPLC with UV detection at 207 nm.	IMP-HPX-72-0 BioRad column in OH form and NaOH as the mobile phase.		Sea-food (to assess spoilage), cod, mackerel.	$HClO_4$ homogenization, KOH addition to pH = 7, filtration 0.22 μm.	72

Analyte	Method	Concentration	Conditions	Sample	Procedure	Ref.
Alkylamines and biogenic amines.	HPLC of phthalaldehyde derivatives and fluorometric detection at 425 nm (excitation at 335 nm).	1 μM/ml of extract	Fine-SIL C18 column. Mobile phase: methanol-NaH$_2$PO$_4$ in different ratios, in isocratic and gradient elutions.	Fresh tea shoots, made tea.	Fresh tea shoots are steamed, frozen, dried, pulverized, extracted in water ($T = 50°C$), filtered, derivatized with o-phthaldialdehyde-2-mercaptoethanol in borate buffer solution (pH = 10).	73
Dimethylamine and trimethylamine.	Gas chromatography with N and P thermoionic detection.	1 mg/ml of each amine	Glass column packed with Chromosorb 103 (80–100 mesh). Programmed T (140–190°C). He as the carrier gas.	Seafood.	HClO$_4$ extraction, $T = 60°C$, benzene + KOH, centrifugation.	113
Volatile alkyl-amines.	Gas chromatography.			Grapes, wine.		114
Alkylamines.	HPLC analysis.			Beer.		115
Ammonia and amines.	Differential pulse polarography.			Food colors.		116
Secondary amines.	HPLC.			Meat, fish.		117
Primary and secondary amines.	Gas chromatography.			Grain, beer, malt, barley.	Pentafluorobenzamide derivatization.	118
Volatile bases.	Steam distillation.			Fishery products.		119
Long-chain quaternary amines.	Mass spectrometry.			Beef, tallow.		120
Methylamine, ethylamine, isopropylamine, isoamylamine.	Combined TLC and GC.			Pure-milk bacterial cultures, milk products, yogurt, milk powder.		121
Dimethylamine (together with morpholine and piperidine).	Capillary gas chromatography with flame photometry detection after derivatization.			Ham, cod, spinach, fish.		122
Aliphatic amines (together with tyramine and histamine).	Enzyme (momo-amine oxidase) electrode.			Meat (freshness testing).		74

Table 4 (*Continued*)

Analyte	Analytical method	Experimental conditions	Detection limit	Applications and amount	Sample treatment	Ref.
Primary and secondary amines (methylamine, ethylamine, ammonia, diethylamine, dimethylamine).	HPLC.			Marine organisms.	Derivatization with dabsyl chloride.	123
Aliphatic amines.	HPLC.			Fish (freshness indication).	Dabsyl chloride derivatization.	124
Volatile amines, nitrosamines.	Thin-layer chromatography.			Food of animal origin.		125
Volatile amines.	Gas chromatography and mass spectrometry.			Tomato, tomato products.		126
Total volatile amine content.	Enzyme determination of NH_3 and total volatile amine determination with NH_3-selective electrode.	Ammonia selective Orion-95-10 electrode.		Meat tissue, fresh beef, chicken, processed ham, frozen shrimps.	Homogenization, methanol stirring, filtration. Preconcentration with Dowex 50 W-X8 resin.	127
Volatile amines and trimethylamine.	Solid state gas sensor, sintered bulk semiconductor SnO_2, electrically heated ($T > 200°C$) and indirect conductometric detection.	Measurement of transient decrease in resistance of the sensor.		Fish.	Multiple extraction in $HClO_4$, trichloroacetic acid, NaOH in reaction chamber with air bubbled.	128

Analyte	Method	Column/conditions		Sample	Extraction	Ref
Volatile amines (methylamine, dimethylamine, trimethylamine, ammonia).	Headspace gas chromatography.	Chromosorb 103 column.		Raw fish, fish products.	KOH addition.	129
Volatile amines and ammonia.	HPLC and fluorescence detection 500 nm (370 nm excitation).	μ-Bondapak C_{18} as the stationary phase and acetonitrile, buffer phosphate (pH = 7.2) as the mobile phase, in gradient elution.	0.01 mM in NH_3	Meat, tissues, beef.	Acetonitrile extraction, dansyl derivatization (in acetone and $KHCO_3$ pH = 8).	130
Total volatile amines.	Steam distillation and titration, headspace GC for trimethylamine with N-selective FID detector.	The volatile amines are collected in 3% H_3BO_3 and titrated with HCl. GC WCOT fused silica column coated with CP-WAX 51, N_2 as the carrier gas.		Fish, fish products.	Extraction, deproteinization, homogenization ($HClO_4$), filtration, NaOH addition, steam distillation, collection in H_3BO_3.	131
Volatile amines (dimethylamine and trimethylamine).	GC with FID and N/P-FID detectors.	Precolumn Chromosorb 103, Carbopack column coated with Carbowax 20M and KOH.		Seafood.		132
Dimethylamine and trimethylamine (together with biogenic amines, urea and ammonia).	HPLC, UV detection at 208 and 214 nm.	BioRad HPX-72-0 column and mobile phase of NaOH.		Seafood.		75

III. ALKALOIDS

The true alkaloids are basic compounds of vegetable origin characterized by the presence of nitrogen in a heterocyclic structure and by a marked physiological activity (133). Most of these compounds are water-insoluble or slightly soluble, poisonous, and characterized by optical activity.

Alkaloid may be classified following different criteria based on botancial, pharmacological, or chemical considerations, presenting each classification's advantages and disadvantages. When planning food analysis, the botanical point of view is very useful to preview the hypothetical contamination which can be found in the different commercial foods as a function of the starting ingredients.

On the other hand, chemical considerations permit the choice of the best analytical approach. A chemical classification subdivides the alkaloids in derivatives of imidazole, indole, isoquinoline, lupinane, phenylalkylamine, purine, pyridine, quinoline, steroid, tropane, and other heterocyclic nitrogen-containing nucleus (134).

We have chosen to treat the alkaloids here considered according to their presence in different kinds of foods. The following classes are proposed:

1. Methyl- and hypoxanthynes, purines, pyrimidines
2. Cinchona alkaloids
3. Spicy food alkaloids
4. Solanum alkaloids
5. Ergot alkaloids
6. Various alkaloids

Books on this topic (133, whiich deals with alkaloids in general, and 135, with the analysis of the alkaloids) have been published, as have reviews (136, with 182 references, dealing with HPLC analysi of alkaloids; 102, with 259 references, which presents methods for the determination of contaminants in foods; and 137, on application of HPLC techniques in food analysis).

A. Methyl- and Hypoxanthines, Purines, Pyrimidines

1. Methylxanthines

Of particular importance in food chemistry are the methylxanthines, mainly 3,7-dimethylxanthine (theobromine) and 1,3,7-trimethylxanthine (caffeine), which are naturally present in cacao, coffee, and tea. Caffeine is also present in many beverages, such as cola beverages.

Caffeine is also used in medicine because it acts as a diuretic and as a stimulant to the central nervous and cardiovascular systems (138, 139). Taking into account that the therapeutically recommended dose in humans is around 200 mg/day and that a drinking cup of coffee contains about 83 mg of caffeine (140), it is easy to see how a beverage can become a drug. It has been shown that among regular consumers of caffeine even a brief abstinencemay induce anxiety. Among sportsmen example of abuses of caffeine as a doping agent have been reported (138). In addition, some studies have demonstrated that caffeine can be teratogenic in rats. In 1980 the U.S. Food and Drug Administration suggested that pregnant women reduce their caffeine intake.

Similar properties are also referred for theobromine, which is particularly present i n cocoa and its derivatives.

All of these considerations have led to the development of many analytical methods for the identification and determination of caffeine and theobromine. Reviews have been published (141–143) and many manuscripts generated. For the less recent ones only the references are reported here. They concern HPLC methods for determination in coffee and cola (144,145), in decaffeinated coffee (146–148), of theobromine, caffeine, and theophylline in instant coffee and beverages (149), in cocoa, cocoa products, and chocolate (150–154), and dietary and other beverages (155–160). In oparticular, HPLC methods with electrochemical detection have been employed in the analysis of caffeine in cocoa (151, 152, 161, 162) and in milk powders. Other instrumental techniques for caffeine analysis were used, such as conductometry (in tea leaves, 163), mass spectrometry (in cocoa, 164), gas chromatography (in beverage, 165, and in chocolate, 166), spectrophotometry (in coffee, 167, and nonalcoholic beverages, 168), thin-layer-chromatography (in cocoa products, 169–171) or electrochemistry (172).

2. Hypoxanthines, Purines and Pyrimidines

Other amines which play a relevant role in food chemistry are purines and pyrimidine bases and their derivatives. Both purine and pyrimidine nucleotides can be acquired through the diet and particularly through meat. Most dietary purines are oxidized by enzymes in the intesinal mucosa to uric acid, which is the excretory product of purines in humans. The disease known as gout is associated with high levels of uric acid in serum and the result is the deposition of urate salts in various tissues. The analysis of purine and pyrimidine bases and their degradation products can give useful information to detect decomposition of food or to evaluate degradations undertaken in some treatments, such as storing or heating (1).

An important nucleoside is adenosine triphosphate, which is converted by degradation in hypoxanthine and other breakdown products, whose determination in meat gives an indication of the degree of spoilage.

HPLC methods have been employed and electrochemical biosensors developed for hypoxanthine determination.

The more recent manuscripts are presented in Table 5. When available, for every manuscript considered the method used for determination, experimental conditions, detection limit, kind of food considered and the amount of amines therein found are reported, as are the essential steps concerning the e xtraction and sample treatment before the analysis.

Table 5 Methyl- and Hypoxanthines, Purines, Pyrimidines

Analyte	Analytical method	Experimental conditions	Detection limit	Application and amount	Sample treatment	Ref.
Caffeine, together with food additives.	Capillary zone electrophoresis (CZE) and HPLC with spectrophotometric detection (214 nm).	CZE: fused silica capillary (60 cm × 75 mm) column. Mobile phase: sodium phosphate buffer pH 11. Running voltage 15 kV. HPLC: Novapak C18 column and mobile phase of phosphate buffer and acetonitrile.	0.1–2.0 mg/L	Foodstuffs and beverages.	Beverages: sonication and water dilution. Powders: water solution and 0.2 µm filtration.	173
Caffeine, theobromine, and theophylline.	Semimicro-HPLC with UV detection at 207 nm.	Column: Inertsil ODS-2. $T = 40°C$. Mobile phase: methanol, water, phosphate buffer at pH = 3 (v/v ratios = 12/33/5).	0.1–2.5 µg/ml	Foods.	Extraction of pulverized samples with water and ethanol by sonication. Evaporation to dryness. Dissolution in phosphate buffer. Cleanup on Sep-Pak C18 cartridge.	174
Caffeine and additives.	HPLC method with UV detection at 230 nm.	Column: µ-Bondapak C_{18}. Mobile phase: 20 mM ammonium acetate and methanol.	10 ng	Beverages.	NH_3 addition and filtration.	175
Caffeine.	GC-MS with solventless microextraction.	Column: HP-5 (0.17 µm, 25m × 0.32 mm). Temperature program from 70°C to 280°C in different steps. EIMS: 70 eV.	2 µg/ml	Beverages.	Methanol addition. Caffeine is adsorbed onto 1-cm length of bare fused silica fiber protruding from a syringe needle for 5 min and then desorbed in the injection port (split or splitness) of GC-MS system.	176

Analyte	Method	Conditions	Concentration	Sample/Purpose	Notes	Ref.
Caffeine, theobromine, theophylline.	HPLC with UV detection at 274 nm.	Column: LiChrospher 100 RP-18. Mobile phase: 5.0 mM octylamine phosphate pH = 6.4.	Caffeine: 0.4 ppm. Theobromine: 0.15 ppm. Theophylline 0.3 ppm.	Tea, coffee, cola beverages. Found: caffeine 6 ppm in decaffeinated tea, 137 in decaffeinated coffee, 158 in diet cola, 433 in tea, 1760 in espresso coffee. Theobromine: 1.7 ppm in diet cola and 17–26 in the other beverages.	Water dilution and 0.2 μm filtration.	177
Caffeine.	Second-derivative spectrophotometry at 298.6 nm.	Methods for correction of background.		Coffee, tea.	To minimize tannin interference, treatment with 0.1 M NaOH (coffee) or Cu (II) acetate (tea).	178
Caffeine.	D, C, and N isotopic analysis. Principal components and discriminant analysis.	$^{13}C/^{12}C$ and $^{15}N/^{14}N$ ratios of caffeine crystals evaluated by MS after combustion. $^2H/^1H$ ratio evaluated by NMR after combustion and Zn reduction.		To discriminate between natural and synthetic caffeine. To evaluate authenticity and to differentiate American and African coffees.	Caffeine extraction with boiling water, $PbCO_3$ and MgO addition, filtration, $CHCl_3$ extraction, solvent evaporation, caffeine precipitation with light petroleum, and crystallization from methanol.	179
Caffeine.	HPLC method with spectrophotometric detection at 254 nm.	Column: Ultrasphere ODS. Mobile phase: water, methanol mixture.	0.02 mg/ml	Beverages.	Addition of benzyl alcohol in aqueous 50% methanol.	180
Caffeine.	High-performance gel filtration chromatography with spectrophotometric detection at 280 nm	Column: TSK-G 3000 SW with a LKB guard column. Mobile phase: 0.05% NaN_3 in water.	30 μg/ml	Green and roasted coffee. Instant coffee.	Coffee beans extracted with water, $T = 80°C$. The extracts as well as the instant coffee samples filtered, cleared (Carrez solution), centrifuged. The supernatant dissolved in water.	181

Table 5 (*Continued*)

Analyte	Analytical method	Experimental conditions	Detection limit	Application and amount	Sample treatment	Ref.
Caffeine with aminoacids and polyphenols.	Near-infrared reflectance spectroscopy.	Multiple linear regression analysis.		Green tea.		182
Caffeine, theobromine, theophylline.	HPLC with spectrophotometric detection at 272 nm and spectrophotometry: first derivative at 248 nm for theobromine, second derivative at 278 nm for theophylline, and zero-order at 275 nm for caffeine.	Column: C-8 Lichrocart. Mobile phase: 0.01-M sodium acetate, acetonitrile, acetic acid pH = 4.0.	5 μg/ml	Tea, coffee, cream chocolate.	Methanol extraction.	183
Caffeine.	Spectrophotometry at 272.5 nm.	Second-derivative method.		Raw and roasted coffee.	Boiling with MgO, cooling, filtration, acidification (HCl) to pH = 4, cleanup on a column of Sep-Pak C18, methanol elution.	184
Caffeine, together with flavonols, theogallin, gallic acid.	HPLC.	Column: Hypersil ODS with a Nucleosil C18 guard column. Mobile phase: water, acetic acid, and acetonitrile in gradient elution.		Tea.	Boiling in water, filtration, addition of acetonitrile or methanol at $T = 40$°C. Cleanup on RP-18 cartridges and elution with acetonitrile.	185
Caffeine.	IR and NMR spectroscopies.			Coffee beans.	Preparative SFC separation (CO_2, at 5 ml/min), $T = 40$°C, on silica gel column. Addition of methanol. Separation of the supernatant, ethanol addition.	186

Analyte	Method	Conditions	Detection	Sample	Sample preparation	Ref.
Caffeine together with food additives.	Capillary electrophoresis.	$pH = 9.1$		Nonalcoholic beverages.		187
Caffeine together with preservatives, sweeteners, colors.	HPLC method.	Column: Altex Ultrasphere-TM-ODS. Mobile phase: ammonium acetate, methanol.		Beverages and foods.		188
Caffeine and theophylline, together with sweeteners and additives.	HPLC method and UV detection.	Column: Superspher RP-Select B. Mobile phase: KH_2PO_4, acetonitrile ($pH = 4.2$–4.4 for sorbic and benzoic acids).	4 µg/L	Lemonades, fruit juices, dairy products. Yogurt, salad dressing.	Dilution, addition of $K_4Fe(CN)_6$ and $ZnSO_4$, filtration for beverages and milk shakes. For dessert, powders, yogurts, salad-dressing: digestion with water, extraction on C18 cartridge and elution with KH_2PO_4 and acetonitrile mixture.	189
Caffeine.	Spectrophotometry at 258, 272.5, and 285 nm.	Method of Morton and Stubb (190). Second-derivative method (data at 272.5 and 292.5 nm).		Coffee.	Addition of MgO, boiling in water (45 min), filtration, pH adjustment to pH = 4.0 with HCl, extraction on a minicolumn of Sep-Pak C18, elution with ethanol, dilution.	191
Caffeine.	RP anion exchange HPLC with UV, electrochemical, and fluorescence detection.			Plants and plant extracts.		192
Caffeine.	Hydrophobic HPLC and spectrophotometric detection at 214 nm.	Column: Fractogel TSK HW-40, $T = 35$ and 40°C. Mobile phase: KH_2PO_4, Na_2HPO_4 in different ratios and pH (pH = 4.3 and 6.9).	Caffeine: 3 ng/ml	Cola beverages.		193

Table 5 (*Continued*)

Analyte	Analytical method	Experimental conditions	Detection limit	Application and amount	Sample treatment	Ref.
Caffeine.	Thin-layer chromatography with dual-wavelength detection. Scanning of the chromatogram in reflectance mode operating at 275 and 350 nm.	Plate: Silica gel GF254. Mobile phase: $CHCl_3$-acetone.		Coffee, beverages.	Sonication in $CHCl_3$, filtration, evaporation, dilution with $CHCl_3$.	194
Caffeine.	GLC with FID detection and HPLC with spectrophotometric detection at 274 nm.	GC: 3% OV-17 Chromosorb WHP column, T program from 120 to 220°C, N_2 as carrier gas. HPLC: μBondapak C18 column. Mobile phase: water, acetonitrile, and acetic acid in gradient elution.	0.2 μg (GLC) 0.3 μg (HPLC)	Coffee tablets and beverages, chocolate, coffee.	Water and NH_3 addition, Soxhlet extraction with $CHCl_3$ and isopropane, filtration, evaporation, dilution with $CHCl_3$. For HPLC analysis distillation and dissolution in the mobile phase.	195
Caffeine theobromine.	Spectrophotometry with detection at 275.9 nm for caffeine and 272.2 nm for theobromine.		5μg/ml	Cocoa beans.	Extraction in boiling water, addition of Pb acetate; filtration, $NaHCO_3$ addition, pH adjustment to pH = 12.6 with NaOH, $CHCl_3$ extraction for caffeine, and HCl addition for theobromine.	196
Caffeine.	Gas chromatography with N-P detection.	Glass column of 6% OV-101 Chromosorb W (1.2 m × 2 mm). $T = 205$°C. N_2 as the carrier gas.	0.04%	Green tea.	Extraction in boiling water, filtration, caffeine extraction with toluene and H_2SO_4.	197

Analyte	Method	Conditions	Detection limit	Matrix	Sample preparation	Ref.
Caffeine.	Colorimetric method at 537 nm.	Absorbance measurement of tetramethylpurpuric acid (orange-red).	7.7 μM	Coffee.	Extraction in boiling water, filtration, extraction with $CHCl_3$. Addition of 3 M HCl and 15% H_2O_2 to obtain the tetramethylpurpuric acid.	198
Theobromine, theophylline, caffeine, preservatives, and antioxidants.	HPLC with UV detection.	Column: LiChrospher 60 RP-Select B.		Coffee, tea, cocoa products.	Alkaline treatment with MgO. Extraction in acetate buffer solution and oxalic acid, ethanol, propan-2-ol, and acetonitrile.	199
Caffeine.	SFC method with detection at 272 nm.	Column: LiChrosorb Si 60, with T program 40–28°C. Mobile phase: methanol in CO_2 at different ratios.		Coffee, tea infusion.	Extraction from coffee beans with CO_2.	200
Caffeine together with water-soluble vitamins and sodium benzoate.	HPLC with UV detection 210 nm.	Column: Nucleosil C18. T = 40°C. Mobile phase: water, acetonitrile, triethylamine, heptanesulfonate (pH = 2.8).	60 μg/ml	Beverages.	Addition of acetanilide (internal standard) and dilution with the mobile phase.	201
Caffeine.	Spectrophotometry at 276.5 nm.	Second-derivative spectrometry.	0.4 μg/ml	Beverages, cola drinks. 23 samples were found to contain from 0.47 to 108.8 mg/ml.	Extraction under alkaline conditions into $CHCl_3$.	202
Caffeine.	HPLC analysis and spectrophotometric detection at 275 nm.	LiChrosorb Si-60 column. Mobile phase: water, CH_2Cl_2, ethanol.	1 μg/ml	Coffee, black tea, green tea.	Water extraction and cleanup on Extrelut column.	203
Caffeine.	Spectrophotometry at 274 nm and first derivative at 287 nm and gas chromatography with FID detection.	GC column: 1.5 m × 4 mm, 3% of E-30 methylsilicone on Diatomite C-AW DMCS, T = 200°C, N as the carrier gas.	14 μg/ml	Coffee, tea, soft drinks.	Infusion of tea, coffee. Ethanol extraction for GC.	204

Table 5 *(Continued)*

Analyte	Analytical method	Experimental conditions	Detection limit	Application and amount	Sample treatment	Ref.
Caffeine.	Ion pair HPLC analysis with spectrophotometric detection at 280 nm.	Conventional and microbore C18 column. Mobile phase of methanol and tetramethylammonium bromide in different ratios.		Coffee-flavored foods, coffee-cream filled pralines, wafers and low-calorie products.		205
Caffeine.	Gas chromatography with N, P detectors.	GC column: 2 m × 3mm of 4% SE-30 plus 6% of OV-210 on GasChrom Q. N_2 as the carrier gas.	10 pg	Black, green and oolong tea, coffee, cola drinks.	NH_3 and water addition, $CHCl_3$ extraction.	206
Caffeine, together with different food additives.	RP-HPLC and spectrophotometric detection at 254 nm.	Column μBondapak C_{18}. Mobile phase: methanol, acetic acid, water and acetate buffer in different ratios.	10 ng	Ready-to-serve beverages, ice candy, ice-cream, fruit squash, tomato sauce.	Direct analysis after water extraction.	207
Caffeine and other additives.	Chemometric methods.	Principal component analysis and hierarchical cluster analysis.		Black, green, and oolong tea with two varieties in each category. Definition of a quality index.		208
Caffeine.	HPLC with spectrophotometric detection at 270 nm.	μBondapak C_{18} column (with a precolumn of polyvinylpyrrolidinone).		Tea extracts.		209
Caffeine.	Spectrophotometry at 273 nm.			Tea.	Water extraction, sodium phosphate buffer solution (pH = 8.6) addition, $CHCl_3$ extraction, filtration through anhydrous Na_2SO_4, dilution with $CHCl_3$.	210
Caffeine, theobromine and theophylline.	HPLC and spectrophotometric detection at 273 nm.	Spherisorb 800 or LiChrosorb RP-18 as the stationary phase. Mobile phase: methanol, CH_2Cl_2, 0.01M Na acetate, acetonitrile.		Coffee, tea, mate, cola and nonalcoholic beverages.	Aqueous NH_3 addition and $CHCl_3$ extraction.	211

Compound	Method	Conditions	Results	Sample	Procedure	Ref.
Caffeine.	Spectrophotometric determination at 520 nm.	The absorbance of the red-violet murexide solution is measured at 520 nm ($\varepsilon = 2275$ mol^{-1} cm^{-1} L).		Coffee, tea.	Heating on water bath, gradual addition of saturated aqueous bromine solution until persistence of Br color. Evaporation to 1–2 drops, anhydrous Na$_2$SO$_4$ addition. Washing with 1% 2-picoline solution in trichloroethylene and formation of a red-violet murexide compound.	212
Caffeine.	HPLC with spectrophotometric detection at 214 nm and 254 nm.	Column packed with Cyclobond I, β-cyclodextrin-bonded to irregular silica gel. Mobile phase: methanol–1% triethylammonium acetate pH = 4.5.		Nonalcoholic beverages, sweetener Equal, Diet Pepsi, Diet 7-Up, Diet soft drinks, Diet Coke.		213
Caffeine and separation from trigonelline.	Liquid (detection at 264 nm) and TLC (detection at 254 nm) chromatographies.	Baker octadecyl column and elution with CHCl$_3$, TLC on silica gel GF 254, with aqueous 92% methanol as the mobile phase.		Coffee.	Addition of MgO and water, HCl extraction, filtration, dilution.	214
Caffeine, theobromine, theophylline.	HPLC with spectrophotometric detection at 280 nm.	Column: Spherisorb ODS and RSil C18 HL precolumn. Mobile phase: 0.02 M NaH$_2$PO$_4$ (adjusted to pH = 7 with NaOH) in methanol.	3.6 < 5.8%	Guarana (*Paullinia cupana* var. *sorbilis*) seeds. Found: caffeine: 3.6–5.8% theobromine 0.03–0.17%, and theophylline 0.02–0.06%.	Seeds are air-dried, roasted, and ground. Boiling water extraction under reflux. Filtration. Dilution.	215

Table 5 (*Continued*)

Analyte	Analytical method	Experimental conditions	Detection limit	Application and amount	Sample treatment	Ref.
Caffeine.	HPLC with spectrophotometric detection and GC with FID and MS detection.	HPLC: column Nucleosil C-18. GC: column 1% of OV-17 on Supelcoport.		Orange, citrus, lemon. No caffeine content found in orange juices.	Extraction of frozen samples in 80% ethanol, filtration, pH adjustment to pH = 2.0, centrifugation. Adjustment of the supernatant solution to pH = 9.0, butanol extraction, azeotropic distillation of the combined butanolic phases.	216
Caffeine and theobromine.	Spectrophotometry at 276 nm (caffeine) and 272.5 nm (theobromine).	Preparative column (Celite in NaOH, and Celite in H_2SO_4) extraction with $CHCl_3$.		Tea leaves (thea Sinensis, Assamica).	Removal of polyhydric phenols, lowering of sugar and glycoside content with Pb acetate solution. NH_3 addition, cleanup in columns of Celite-NaOH and Celite-H_2SO_4, $CHCl_3$ elution for caffeine.	217
Caffeine.	Supercritical-fluid chromatography directly coupled with supercritical fluid extraction and HPLC with spectrophotometric detection at 272 nm.	SILC18 column. Mobile phase water-methanol.		Roasted coffee beans.	Grounding, water addition, extraction on cartridge. SFC, T = 48°C extract with CO_2 extract trapped on activated charcoal cartridge. Elution of caffeine with methanol.	218
Caffeine.	Spectrophotometry at 250 and 290 nm.			Coffee beans.	Boiling in water in the presence of MgO, filtration, $K_4Fe(CN)_6$ and Zn acetate addition, filtration.	219

Analyte	Method	Column / conditions	Detection	Sample	Sample preparation	Ref.
Caffeine, theobromine, theophylline.	Microbore HPLC with spectrophotometric detection at 280 nm.	Column: Whatman Micro-B of ODS-3. Mobile phase: water, methanol, acetic acid.	150 pg/ml of caffeine, 100 pg/ml for theobromine and theophylline	Cocoa.	Aqueous extraction.	221
Caffeine.	HPLC with spectrophotometric detection at 254 nm.	Column: RP MCH-10. Mobile phase: acetic acid 1% and acetonitrile (4:1).	50 ppm	Tea.	Addition of MgO and water, boiling 1 hour. Cooling, filtration, addition of 8-chlorotheophylline (internal standard), dilution.	222
Caffeine.	HPLC with spectrometric detection at 214 nm.	Column: Partisil-10SCX and mobile phase containing 0.1 M $NH_4H_2PO_4$.	50 ng	Soft drinks.	Decarbonation, filtration.	223
Caffeine with other additives.	HPLC with spectrometric detection at 214 nm.	Column: μ-Bondapak C_{18}. Mobile phase: H_3PO_4, triethylamine, water, pH = 4.3.	0.05 mg/ml	Cola beverages.	Water dilution and degassing.	224
Caffeine.	Silica gel column chromatography and gas chromatography with FID and MS detection.	GC: Carbowax 20M, T program from 50°C to 240°C, with N_2 as the carrier gas.		Indian broken tea, Darjeeling orange pekoe tea.	Infusion in boiling water, cooling in ice bath; filtration and reverse osmosis on cellulose acetate membrane.	225
Caffeine, theobromine, theophylline.	HPLC method with spectrophotometric detection at 275 nm.	Column: LiChrosorb RP-8. Mobile phase: water, methanol, phosphate buffer 0.2 M, pH = 5.0 (36:9:5), at $T = 45$°C.	10 μg/g of caffeine, 5 μg/g of theobromine and theophylline	Cocoa, chocolate, coffee, tea, cola beverage.	Cleanup on a Sep-Pak C-18 cartridge.	226
Caffeine.	HPLC method with spectrophotometric detection at 254 nm.	Column: μBondapak C_{18} equipped with a Bondapak C_{18}/Corasil guard column. Mobile phase: acetic acid, water, propan-2-ol (10:87:3).		Beverages.	Water dissolution. 0.45-μm filtration.	227
Caffeine and theobromine.	Paper chromatography and UV spectrophotometric determination at 274–275 nm.	The paper strips are developed in light petroleum for 1.5 hr to remove fats and 3–4 hr in butanol-saturated NH_3 solution.		Cocoa	Application of the sample to paper strips moisted with 12.5% NH_3.	228

Table 5 (*Continued*)

Analyte	Analytical method	Experimental conditions	Detection limit	Application and amount	Sample treatment	Ref.
Caffeine.	Spectrophotometry at 276 nm.			Green and roasted coffee. Dried and liquid coffee extracts.	Clarification of the aqueous ammoniacal extract by passage through two columns of diatomaceous earth (Celite 545) and respectively NaOH and H_2SO_4. The acid column is washed with ethyl ether and caffeine eluted with $CHCl_3$.	229
Theobromine.	Gas chromatography with FID detection.	Column: Gas-Chrom Q (1 m × 3mm) of 1% of cyclo-hexanemethanol succinate.	50 ppm	Chocolate.	Homogenization, boiling in water with $NaHCO_3$ solution containing isoamylic alcohol, centrifugation, CH_2Cl_2 extraction.	230
Caffeine, theophylline, theobromine.	Spectrophotometry at 500 nm.		20 ppm	Coffee, tea, cola-type beverages.		231
Hypoxanthine.	Amperometric flow injection method and peroxidase electrode.	Combined use of immobilized enzyme reactor and a peroxidase electrode.	0.5 µM	Pork meat.	Injection of the sample solution into a carrier solution (0.1 M pyrophosphate buffer, pH = 8.7) then splitting of the flow in two reactors containing putrescine oxidase and xanthine oxidase immobilized on controlled-pore glass.	232
Hypoxanthine.	Electrochemical method.	Use of nucleoside phosphorylase and xanthine oxidase immobilized on chitosan beads.	µM levels	Fish, for evaluation of fish freshness.	The sample of fish muscle exudate was heat-treated and filtered.	233

Analyte	Method	Conditions	Detection	Matrix	Sample preparation	Ref.
Hypoxanthine.	Electrochemical biosensor.	Enzyme electrodes. Enzymes are immobilized on alkylamine glass beads.		Fish.		234
Hypoxanthine.	Amperometric enzymatic method [at 0.7 V (Pt vs. Ag)] and HPLC methods.	HPLC: LC-8 column and mobile phase of 0.01 M KH_2PO_4, pH = 4.5.		Cooked and raw fish.		235
Hypoxanthine.	Radioimmunoassay (RIA) with radioactivity measurement by liquid scintillation and spectrophotometric method at 290 nm.	RIA method: addition of barbitone buffer, bovine serum albumine affinity-purified antibodies, vortexing, ammonium sulfate addition, vortexing.	15–300 μM	Fish, white bait, turtle.	Homogenization with $HClO_4$, centrifugation at 0°C, pH adjustment of the supernatant to pH 6.5–7.0 with KOH, centrifugation.	236
Hypoxanthine.	Colorimetric method at 593 nm.	Addition of nucleoside oxidase.		Fish, assessment of freshness.		237
Hypoxanthine.	Capillary electrophoresis and detection at 250 nm.	The capillary was 72 cm, effective capillary length 22 cm; the applied voltage = 30 kV.		Fish.	The sample was homogenized with trichloroacetic acid and centrifuged; the supernatant was neutralized with 2M NaOH and diluted with 10 mM phosphate buffer pH 7.8.	238
Hypoxanthine.	Hypoxanthine sensor based or oxygen microelectrode.	Buffer 0.1 M K phosphate pH 7.3.	6.7 μM	Fish.	$HClO_4$ extraction and dilution with buffer solution.	239
Hypoxanthine.	Electroanalytical method with flow injection and immobilized enzyme reactors.	The system contains immobilized enzymes on LiChrosorb-NH_2. A series of catalyzed reactions forms H_2O_2 which is detected with a peroxidase-bovine serum albumin membrane electrode.		Fish.		240
Hypoxanthine.	Semiconductor biosensor.			Fish.		241

Table 5 (*Continued*)

Analyte	Analytical method	Experimental conditions	Detection limit	Application and amount	Sample treatment	Ref.
Hypoxanthine.	Flow injection method.	The system incorporates immobilized urate oxidase, alkaline phosphatase, and purine nucleoside phosphorylase-xanthine oxidase reactors and a vitreous carbon electrode.	0.5 μM	Fish freshness.		242
Hypoxanthine and xanthine.	Enzyme reaction and fluorometric detection.	Coupled reaction of xanthine oxidase-peroxidase-homovanillic acid.		Fish.		243
Hypoxanthine	HPLC method with spectrophotometric detection at 254 nm.	Column: μBondapak C_{18}. $T = 30°C$. Mobile phase: KH_2PO_4 0.04 M, K_2HPO_4 0.06 M buffer.		Fish.	$HClO_4$ homogenization, centrifugation, pH adjustment to pH 6.5–6.8 filtration.	246
Hypoxanthine.	Enzyme sensor.	Membrane (cellulose triacetate, 4-aminomethyloctane, 1.8-diamine, glutaraldehyde) with immobilized xanthine oxidase and nucleoside phosphorylase.	0.5 mM	Fish.		247
Hypoxanthine and breakdown products.	Thin-layer chromatography.	Plates: silica gel 60 F254.		Meat.		248
Hypoxanthine.	Enzyme sensor.	Immobilized xanthine oxidase membrane and oxygen probe.	0.06 mM	Meat, fish.		249
Hypoxanthine and inosine.	HPLC method.			Fish.		250
Hypoxanthine.	HPLC method.			Refrigerated fish.		251
Xanthine.	Amperometric enzyme electrode.	The system is formed by electrodes containing different enzymes which through multistep reactions lead to enzyme deactivation and current fall.		Food.		252

Analytes	Method	Conditions	Matrix	Notes	Ref.
Guanosine, deoxyguanosine, adenine, adenosine, deoxyadenosine, purine nucleosides.	HPLC chromatography with spectrophotometric detection at 254 nm and TLC.	Separation on dextran-based gels Sephadex G10, G25, and LH20, Sephasorb HP Ultrafine columns. Mobile phase: Na acetate pH = 4.0. HPLC: Zorbax C8 and Versapack C18 columns TLC: cellulose F254.	Beer and wort.		253
Purine derivatives, inosine, nucleic acids.	HPLC with detection at 254 nm.	μBondapak C_{18} and LiChrosorb RP-8 columns. Mobile phase KH_2PO_4 (pH = 6.0).	Food.	Acid hydrolysis for total purine enzymatic determination. Enzymatic hydrolysis for purine bases and inosine HPLC determination.	254
Purine bases.	Cation exchange chromatography and spectrophotometric detection at 260 nm.	T program 45–70°C. Elution with Na citrate buffers pH = 3.0, = 4.25, = 5.45.	Carbohydrate-rich foods.	Hydrolysis with trifluoroacetic and formic acids (15 min, $T = 240$°C), CH_2Cl_2 extraction.	255
Purine bases.	Ion chromatography.	Computer programs for calculating the relative compositions of nucleotides, nucleic acids, purines.	Food.	Trifluoroacetic and formic acids hydrolysis.	256.
Orotic acid, allantoin, uric acid, uridine, hypoxanthine, guanine, cytidine, xanthine, pseudopurine, pyrimidines.	HPLC with spectrophotometric detection at 254 nm.	μBondapak C_{18} column. Mobile phase: $NH_4H_2PO_4$ pH = 6.	Milk.		257

Table 5 (*Continued*)

Analyte	Analytical method	Experimental conditions	Detection limit	Application and amount	Sample treatment	Ref.
Purine bases.	HPLC with spectrophotometric detection. The ratio of the absorbances at 245 and 254 nm is used to calculate adenosine content.	Column Bondapak C_{18} or Zorbax ODS. Mobile phase: phosphate buffer (pH = 3.5) and tetrahydrofuran in different ratios.		Chocolate.	Defatting extraction.	258
Purine bases.	Automated cation exchange amino acid analyzer. Detection at 260 nm.	Beckman M82 and Hamilton HP-B-80 columns. Elution with citrate buffer at different gradient elutions, pH, and temperatures.	200 µg	Protein-rich foods, liver.	Hydrolyzation with trifluoroacetic and formic acids for 90 min at $T = 240°C$, under pressure in a PTFE-coated vessel. Cooling, addition of 6-chloropurine as the internal standard, 0.2 µm filtration.	259
Purine alkaloids, trigonelline, theophylline theobromine, caffeine.	HPLC.			Instant coffee, beverages, tea.		260
Purine derivatives.	Enzymatic method.			Coffee, tea, cola.		261
Purine alkaloids.	HPLC.	Prepacked column.		Food, tea, cola, nonalcoholic beverages, bearberry leaves.		262
Purine derivatives, nucleotides.	HPLC.			Malt wort, beer.		263

B. Cinchona Alkaloids

The cinchona alkaloids are used in the pharmaceutical and the food and drink industries.

Quinidine is widely used as an antiarrhythmic drug. Quinine is a very important source of bitter flavoring in the food industry but is also widely used as an antimalarial drug at doses of 1000 mg/day, as well as in analgesic preparations, at doses of 10–400 mg (264). Side effects produced by assumptions of this compound, such as arrhythmia, hypotension, vcomiting, and neurological complications, suggest avoidance of use by children and women during pregnancy and lactation, even though only a few countries oblige by labeling the contents of quinone in soft drinks.

Quinine after oral intake is completely resorbed in the liver. The half-life in plasma is about 4–6 hr andafter 24 hr the plasma concentration is negligible. This compound is mainly metabolized in the liver, where about 80% of the quinine is transformed into a series of phgenolic and nonphenolic metabolites, then excreted in urine (265).

These alkaloids have been determined by a variety of analytical techniques, such as HPLC, capillary gas chromatographic, and isotachophoretic methods.

Table 6 reports, for the analytes considered, the analytical techniques used in the literature and, where reported, the analytical conditions, detection limit, source analyzed, and sample pretreatment.

Table 6 Alkaloids

Analyte	Analytical method	Experimental conditions	Detection limit	Applications and amount	Sample treatment	Ref.
Cinchona alkaloids Quinine, quinidine, cinchonine, cinchonidine.	TSP-LC-MS GC with FID detection.	Column: RLS-903, 0.15 mm 30 m × 0.3 mm. Mobile phase: helium, flow rate 5.0 ml/min. Isothermal analysis at 280°C.		Plant cell cultures. Soft drinks.	Samples are made alkaline with 10% sodium hydroxide, then the alkaloids are extracted with chloroform.	266 267
Quinine.	Isotachophoresis with conductometric and spectrophotometric detection at 254 and 280 nm.	Capillar 200 × 0.2 mm in PTFE, morpholine-thansulfonic acid (MES) pH = 6.0 with 0.05% Mowiol as leading electrolyte, 0.005 M creatinine-MES as terminating electrolyte, 25 mA. End voltage 10 kV.	5 ppm	Bitters and tonics. Found: bitters 30–85 ppm, tonics 40–85 ppm.	Sample degassed by sonication.	265
Cinchona alkaloids Quinine (Qn), quinidine (Qd), cinchonine (Cn), cinchonidine (Cd), dihydroquinidine, dihydroquinine, dihydrocinchonine, dihydrocinchonidine, quinamine, cinchonamine, corynantheal.	HPLC with spectrophotometric detection at 220 nm.	Column: LiChrosorb RP-8 Select B. Mobile phase: 15% CH_3CN in phosphate buffer pH = 3.0.		Bark of *Cinchona* species. Found: Cn 0.25%, Cd 1.23%, Qd 0.34%, Qn 1.20% of original dry sample.	Bark samples were dried at 80°C to constant mass. Bark grounded is mixed with calcium hydroxide and 5% sodium hydroxide for half 1 hr, then extracted in Soxhlet apparatus for 7 hr with toluene as extraction solvent. Extract was evaporated to dryness under a stream of helium at room temperature and diluted with mobile phase.	264

C. Spicy Food Alkaloids

Piperine occurs in concentrations of about 3–5% in ground pepper, which is used for culinary purposes. Its stereoisomers chavicine, isochavicine, and isopiperine also occur in pepper, but in very small amounts, together with other minor constituents of similar chemistry. Pipetine is the alkaloid responsible for the pungency of pepper, so that its determination is an important parameter in food quality control (268).

Capsaicinoids (vanillyl amides of fatty acids) are also responsible for the *hotness* of spicy foods. These compounds produce an irration to the nerve endings responsible for heat and pain sensation in the mouth (269). Only limited information is available about toxicology of capsaicinoids. While capsaicin was shown to produce in rabbits histological degeneration in heart, liver, and kidney and in guinea pigs effects sim,ilar to those of histamine, for humans it seems to be nontoxic, being eliminated in an unmetabolized form in urine (270).

HPLC, gas chromatography, and thin-layer chromatography methods are generally employed with different detectors for the analysis of these alkaloids.

Table 7 lists the literature data for the analytes considered, the analytical techniques used, snd, when reported, the experimental conditions, application, and sample pretreatment.

Table 7 Alkaloids of Spicy Foods

Analyte	Analytical method	Experimental conditions	Detection limit	Applications and amount	Sample treatment	Ref.
Capsaicinoids (nor-dihydrocapsaicin, capsaicin, dihydrocapsaicin, minor capsaicinoids).	HPLC with spectrophotometric (280 nm) and fluorimetric detection (λ_{exc} 220 nm λ_{em} 320 nm).	Stationary phase: Supelco C18 Stainless steel column, 250 × 4.6 mm, 5 μm. Mobile phase: methanol/water (3:2).		Oleoresin, hot sauces.	For oleoresin: methanol addition, shaking, centrifugation, and injection of the filtered supernatant. Tincture: methanol addition, filtration. Another procedure consists in solid phase extraction: silica gel activation with hexane, sample is weighed on the cartridge, rinsed with hexane and eluted with methanol.	269
Capsaicinoids (capsaicin).	TLC with spectrophotometric detection at 730 nm and vanillin as the internal standard.	Stationary phase: activated silica gel plates. Mobile phase: cyclohexane 70%, chloroform 20%, acetic acid 10%.		*Capsicum annuum, Capsicum frutescens.*	Maceration of fruits in ethanol for a week, concentration of supernatant under reduced pressure. Derivatization with Folin–Denis reagent.	271
Capsaicinoids (nordihydrocapsaicin, capsaicin, dihydrocapsaicin).	GC with ECD detection of pentafluorobenzyl derivatives, vanillylamide of octanoic acid as internal standard.	30 m × 0.326 mm i.d. fused silica capillary column with DB-5 chemically bonded stationary phase. Injector at 250°C, split-less mode. Detector at 350°C. Temperature program from 200 to 265° with nitrogen as carrier gas at a flow rate of 1.5 ml/min.	10 pg (capsaicin).	*Capsicum* species (40–60 mg/100 g original weight) and oleoresin extracts (2.5–3.5% w/w in original product) as capsaicinoid content.	Acetone extraction with Soxhlet apparatus, purification with petroleum ether, reextraction with diethyl ether, evaporation, dissolution in acetone. Derivatization: acetone solutions are added with anhydrous potassium carbonate and pentafluorobenzylbromide in excess.	272

Capsaicinoids [nordihydrocapsaicin (ND), capsaicin (C), dihydrocapsaicin (DC)].	GC, FID detection of o-methyl derivatives. Vanillylamide of octanoic acid as internal standard.	2.1 m × 2 mm i.d. packed glass column with 5% SP-2100 on Chromosorb W AW DMCS 100–120 mesh as the stationary phase. Temperature program from 200 to 250°C with nitrogen as carrier gas at a flow rate of 30 ml/min.	Fresh pepper (mg/g in dry wt): 0.25–0.65 ND, 1.2–2.2 C, 0.6–1.4 DC (mg/100 g in original wt) 3.8–12.2 ND, 20.4–41.8 C, 16.5–26.4 DC. Canned peppers (dry wt in mg/g): 0.1–0.3 ND, 0.3–4.2 C, 0.6–1.4 DC (mg/100g in original wt) 0.6–4.9 ND, 3.1–65.4 C, 16.5–26.4 DC. Spices (mg/g in dry wt): 0.05–0.20 ND, 0.2–1.6 C, 0.1–0.8 DC (mg/100 g in original wt) 4.4–20 ND, 18.0–105 C, 10.8–76.0 DC. Sauces (mg/g in dry wt): less than 0.35 ND, less than 2.6 C, less than 1.5 DC (mg/100 g in original wt), less than 1.4 ND, less than 10.8 C, less than 6.15 DC. Resins (mg/100g fresh wt): 90–800 ND, 180–1200 C, 200–1200 DC.	Samples grounded extraction in a Soxhlet apparatus with acetone, evaporation, addition with 0.01 M NaOH methanol-water (70:30) solution, extraction with light petroleum, reextraction with water-methanol (50:50), addition of HCl, extraction with diethyl ether, evaporation and dissolution in methanol. Derivatization with trimethylanilinium hydroxide. The molecular ratio required for quantitative derivatization is between 6:1 and 12:1 of derivatizing agent.	273
Capsaicinoids.	GC, FID detection of methyl derivatives.	Column: 2.1 m × 2 mm i.d. of 5% SP2100 on Chromosorb W AWW-DMCS (100–120 mesh). Temperature programming from 200 to 245°C at 4°C/min.	*Capsicum.*	Extraction of freeze-dried grounded sample in Soxhlet apparatus with acetone; evaporation to dryness, dissolution in methanol and light petroleum, evaporation of methanol layer, dissolution in ethyl ether and water, evaporation of ether layer, dissolution in tetrahydrofuran, derivatization with trimethylanilinium hydroxide.	274

Table 7 (*Continued*)

Analyte	Analytical method	Experimental conditions	Detection limit	Applications and amount	Sample treatment	Ref.
Capsaicinoids, gingerols.	RP-HPLC with spectrophotometric detection at 280 nm N-(4-hydroxy-3-methoxybenzyl) octanamide as the external standard.	Stationary phase: Spherisorb ODS-2 5 μm, 250 × 4.6 mm. Mobile phase: acetonitrile/water 1% acetic acid (1:1 for determination of capsaicinoids, 13:1 for determination of gingerols).		Chilies, ginger.	Ground dried chilies or dried ginger extraction in Soxhlet apparatus with methanol, concentration, filtration. Oleoresin capsicum dissolution in methanol-tetrahydrofuran. Oleoresin ginger dissolution in methanol.	275
Vanillyloctanamide, nordihydrocapsaicin, capsaicin, dihydrocapsaicin, vanillyldecanamide, homocapsaicin, homodihydrocapsaicin.	HPLC with spectrophotometric detection.	Stationary phase: C18 Rad-Pak column 5 μm mobile phase: methanol 63% water 37%, flow rate 3.5 ml/min.		*Capsicum* fruit, (*Capsicum* oleoresins).	Dehydrated ground capsicum or fresh capsicum dissolution in acetonitrile, dilution with double distilled water, solid phase extraction on C18 Sep-Pak cartridge, conditioned with acetonitrile and double-distilled water. Elution with acetonitrile containing 1% acetic acid.	276
Capsaicin, nordihydrocapsaicin, piperine, dihydrocapsaicin.	HPLC with spectrophotometric detection.	Stationary phase: C18 250 × 4.6 mm. Mobile phase: 0.95 g sodium pentanesulfonate in 160 ml water, 335 ml methanol, 5.0 ml acetic acid.	25 ppm.	Black pepper. *Capsicum* fruit, oleoresins.	Spices are ground till to pass in US 20 mesh (oleoresins warmed). Refluxation for 1 hr with acetone, then dilution with the mobile phase.	277
Capsaicinoids.	Spectrophotometry at 275 nm.			Jalapeno pepper.	Pepper pods are injected with methanol, then after 30 min flushed and absorbance measured.	278
Capsaicin (C), nordihydrocapsaicin (ND), dihydrocapsaicin (DC).	HPLC with spectrophotometric detection.	Stationary phase: C18, 10 μm 250 × 4.6 mm. Mobile phase: 1% acetic acid in 40% acetonitrile-60% water (v/v).	0.5 mg of C and DC. 0.6 mg of ND.	Red pepper. Found nordihydrocapsaicin 0.004–0.030%, capsaicin 0.047–0.277%, dihydrocapsaicin 0.017–0.152%.	25 g of grounded red pepper extraction by heating at 65–75°C for 5 hr in 200 ml of methanol 95%. Surnatant filtered is injected on HPLC.	279

Compound	Technique	Conditions	Detection	Sample	Sample preparation	Ref.
Capsaicinoids (capsaicin, dihydrocapsaicin, nordihydrocapsaicin, drocapsaicin, homodihydrocapsaicin, homocapsaicin).	HPLC with spectrophotometric (at 280 nm) and fluorometric detection with λ_{ex} at 280 nm, λ_{em} at 316 nm.	Stationary phase: 5 μm. Ultrasphere ODS 250 × 4.6 mm. Mobile phase: 0.05 M AgNO$_3$ in 60% methanol, 40% water, pH 2.5 with acetic acid.	500 ppb.	Animal feed. Oleoresins. Found in 2.27% as total capsaicinoids.	Oleoresins: sample dissolved in hexane, liquid–liquid extraction with acetonitrile, evaporation under vacuum at 30°C, dissolution in methanol and analysis. Animal feed: liquid–liquid multistep extraction with methanol/0.1 N HCl (8:1), dichloromethane, benzene, acetone, methanol, water.	270
Capsaicinoids.	Paper chromatography.					280
Capsaicin, dihydrocapsaicin, piperine.	Micellar electrokinetic capillary chromatography with spectrophotometric and electrochemical detection.	Capillary column 65 cm × 25 (electrochemical detection) or 100 (spectrophotometric detection) μm. Mobile phase: 0.01 M borate buffer (pH 9.2), 5 mM phosphate buffer (pH 7.2), or 0.02 M 2-morpholineethane sulfonic acid buffer (pH 6.2) with various organic modifier and 0.05 M SDS as surfactant. Detectors: microelectrodes at various applied potentials for electrochemical detection.		*Capsicum* and (*capsicum* oleoresins (determination of total capsaicinoids). Spicy food.		281
Piperine, chavicine, isochavicine, isopiperine.	Microscale NP-LC. UV detection at 252 nm. Anthraquinone as the internal standard.	Column: polyphenol RSil 10 μm, 250 × 0.32 mm. Mobile phase: hexane 40%, tetrahydrofuran 60%, at $T = 50°C$.		Ground pepper. Found in 3–5% w/w of piperine in black and white pepper. 0 in cayenne pepper. 25–40% w/w in pepper extracts.	Samples, added of mobile phase, are stirred for 30 min in a vessel protected from light and the supernatant is injected.	268

Table 7 (*Continued*)

Analyte	Analytical method	Experimental conditions	Detection limit	Applications and amount	Sample treatment	Ref.
Piperine.	HPLC with UV detection at 343 nm.	Stationary phase: C-18 column. Mobile phase: aqueous 77% methanol.		*Piper nigrum.*	Ethanol extraction.	282
Piperine.	RP-HPLC with spectrophotometric dual-wavelength detection at 280 and 343 nm or 254 and 364 nm. *N*-(4-Hydroxy-3-methoxybenzyl)octanamide as the internal standard.	Column: Spherisorb ODS-2 5 μm 250 × 4.6 mm. Mobile phase: aqueous 1% acetic acid/acetonitrile (1:1) at 1.5 ml/min detection.		Pepper, pepper oleoresins.	Ethanol extraction.	283
Piperine, piperettine, piperlonguminine, piperyline, piperanine, piperettyline.	HPLC with spectrophotometric detection at 340 nm.	Column: C18 ion pair material 5 μm (250 × 4.6 mm). Mobile phase: 0.19% Na pentane sulfonate solution in methanol-water-acetic acid (67:32:1).	170 ppm	*Piper nigrum.* Found in (% dry basis) 3.05–5.00 of piperine, 0.21–0.64 of piperettine.	Ground pepper heating and acetone refluxing, dilution with aqueous methanol.	284
Piperine, isopiperine, chavicine, isochavicine, piperettine, piperyline.	Spectrophotometry.	Scanning from 342 to 345 nm.		Pepper oleoresins, soluble pepper seasoning.	1,2-Dichloroethane extraction.	285

Compound	Method	Conditions	Detection limit	Sample / Found	Sample preparation	Ref.
Capsaicin, piperine.	HPLC. UV and electrochemical detection.	Column: Altex Ultrasphere ODS (5 μm, 150 × 4.6 mm) or Varian MCH-5. Mobile phase: 45% acetonitrile in 0.3% phosphate buffer at pH 4.0.	0.06 ppm of capsaicin	Spices, spicy food.	Methanol extraction, filtration.	286
Piperine.	RP-HPLC. UV detection at 345 nm. Phenazine as internal standard.	Column: C8 5 μm, 250 × 7.0 mm. Mobile phase: 63% water, 30% acetonitrile, 7% tetrahydrofuran; flow 2.0 ml/min.		White pepper: found in 2.20–4.35%. Black pepper: found in 2.10–5.03%.	Ethanol addition, boiling for 45 min.	287
Piperine.	TLC with densitometric detection at 345 nm.	Stationary phase: Kieselgel F254. Mobile phase: ethyl acetate, dichloromethane.	5 ng	Black pepper: found in 5.0%. White pepper: found in 3.9%.	Dichloromethane extraction and dilution.	288
Piperine.	Spectrophotometry at 345 nm. HPLC with UV detection at 280 or 345 nm.	Column: μBondapak CN 300 × 3.9 mm. Mobile phase: (1:1) water/methanol.		Black pepper: found in 3.24–5.05%. Green pepper: found in 48.3–55.2%. White pepper: found in 41.4–47.2%.	Ground pepper is extracted in Soxhlet apparatus with diethylether for 20 hr, dried under vacuum, and redissolved in methanol.	289
Piperine.	Spectrophotometry.			Foods.		290
Piperine	Capillary Gas Chromatography with FID detection.	Column: 25 m × 0.5 mm OVl. Inj. volume 0.5 mL at 100°C. 250°C isothermal analysis flow 4.6 ml/min.	Less than 0.027 mg	Commercial pepper; found in 3.7–4.4% w/w. Pepper extracts; found in 29.4–31.7% w/w.	Ground pepper or pepper extract addition to dichloromethane, shaking, settling, injection of supernatant.	267
Piperine.	HPLC.			Pepper, pepper extracts.		291

D. Solanum Alkaloids

Glycoalkaloids are natural toxins, present in all parts of plants of the *Solanum* species. They are considered to be a natural reaction of plants against parasites and plant diseases (292).

Tubers of *Solanum tuberosum* with a high total glycoalkaloid content produce a bitter taste or burning sensation. Usually medium contents in tubers vary up to 15 mg/100 g of fresh product, but wild species of *Solanum* contain higher amounts of glycoalkaloid.

Occasionally deaths have been observed for ingestion of foods with a high total glycoalkaloid content, and this anomalous amount could be due likely to unappropriate storage procedures (293).

Glycoalkaloids consist of a C27 steroidal skeleton to which are attached sugar units. Chromatographic analyses of this compounds can be performed by gas chromatography, as aglycones or derivatized glycoalkaloids, as well as by HPLC methods (292).

Table 8 reports the analytes and the analytical methods with, where reported, the pretreatment of the sample and the detection limits.

Table 8 Solanum Alkaloids

Analyte	Analytical method	Experimental conditions	Detection limit	Applications and amount	Sample treatment	Ref.
α-Solanine. α-chaconine.	HPLC with spectrophotometric detection at 202 nm.	RoSil C8 or Nucleosil-5 C18 columns. Mobile phase: acetonitrile/diammonium phosphate buffer or acetonitrile/water.		Potato tubers. Found in: 72 ppm of α-solanine and 46 ppm of α-chaconine.	Sodium 1-heptanesulfonate and acetic acid extraction, filtration. Elution through a C18 Sep-Pak cartridge, conditioned with methanol, sodium 1-heptanesulfonate. Elution with water-acetonitrile mixture to remove interferents and glycoalkaloids elution with acetonitrile and diammonium phosphate buffer.	292
Solanum glycoalkaloids.	Positive ion fast atom bombardment mass spectrometry with copper probe (PI FAB-MS).	Probe tip bombardment with a beam of Xe atoms produced by an Ion-Tech 11-NF atom gun at 9 kV.		Potatoes.		294
Glycoalkaloids: α-solanine, α-chaconine, β_1-solanine, β_2-solanine, β_1-chaconine, β_2-chaconine.	RP-HPLC with spectrophotometric detection at 200 nm.	RP C-18 column. Mobile phase: acetonitrile and ammonium phosphate pH = 3.5 in different ratios.		Potatoes. Found in: α-solanine 6 μg/mg, α-chaconine 6.5 μg/mg, β_2-solanine 0.2 μg/mg, β_1-chaconine 0.05 μg/mg, and β_2-chaconine 0.1 μg/mg.	Tetrahydrofuran, water, acetonitrile, acetic acid addition. Centrifugation, filtration, evaporation to 4 ml, centrifugation, heptanesulfonic acid washing, filtration. Cleanup with solid phase extraction.	295
C-27 steroidal alkaloids, solanum species: solanidine, demissidine, solasodiene, tomatidine, solasodine, solanthrene, 5-α-cholestane.	High-resolution gas chromatography, FID detection.	CP-Sil-5 capillary column. Carrier gas: helium. Temperature 28°C; isothermal for 28 min, increased at 8°C/min to 320°C.		Potatoes. Amount determined as solanidine glycosides in 10 potato genotype: 1-47 mg/100 g of fresh tuber.	Samples were homogenized and glycoalkaloids extracted and hydrolyzed with HCl 2M and CCl$_4$. The CCl$_4$ phase washed with NH$_4$OH, evaporated to near dryness is redissolved in methanol-toluene.	296

Table 8 (*Continued*)

Analyte	Analytical method	Experimental conditions	Detection limit	Applications and amount	Sample treatment	Ref.
Total glyco-alkaloids.	Thin-layer chromatography.	Chromatographic plates coated with Kieselgel G. Development with water, ethanol, acetic acid mixture.	10 μg as α-solanine pure.	Potato tubers. Found in: 7–120 mg/100 g of fresh weight.	Liquid nitrogen–frozen samples (−20°C) treated in special presses, centrifugated. The clear supernatants spotted on the chromatographic plates.	293
Glycoalkaloids (α-chaconine and α-solanine).	HPLC. Spectrophotometric detection at 215 nm.	Carbohydrate column. Solvent: tetrahydrofuran, water, and acetonitrile.	0.4 μg/ml.	Potato tubers. Amount found in: 4–23 mg/100g of α-chaconine and 2–13 mg/100g of α-solanine.	Addition of methanol, chloroform. Filtration and concentration. HCl addition, sonification, centrifugation, addition to the supernatant of concentrated ammonium hydroxyde. 70°C water bath (30 min) and refrigeration overnight. Dissolution in tetrahydrofuran, water, acetonitrile.	297
Glycoalkaloids.	Colorimetry at 620 nm.	Complex with bromothymol blue.		Potato tubers: 5.9–44 mg/100 g of fresh weight.	Sample macerated with disintegrator is added with H_2SO_4 and filtered under vacuum. The extract is heated at 90°C with air condensed; the hot solution is made alkaline and extracted with chloroform.	293
Solanum alkaloids: solanidine, leptinidine, acetyl leptinidine.	Capillary gas chromatography with internal standard and FID.	Nonpolar megabore fused silica column.	Underivatized alkaloid: around 3 ng.	*Solanum tuberosum.* Found in tubers: solanidine 590 μg/g of dry weight, in leaves: solanidine 6288 μg/g. *Solanum chacoense*: in tuber: solanidine 2877 μg/g, leptinidine 5990 μg/g, and acetyl leptinidine 8909 μg/g.	Freeze-dried sample is added with tomatine. HCl and methanol extraction. Neutralization and centrifugation. Partition of supernatant with benzene, dryness evaporation, chloroform redissolution. Silica gel Sep-Pak elution. Pyridine and ethyl acetate, hexane elution. Evaporation to dryness. Chloroform redissolution.	298

E. Ergot Alkaloids

Ergot alkaloids are peptide alkaloids containing lysergic and isolysergic acids, produced by fungi of *Claviceps* species and other related genera. These compounds are responsible for some poisonings, exerting many pharmacological effects such as vasoconstriction, reproductive disorders, and agalactia (299). Nowadays strict grading standards do no permit that grain which contains appreciable quantities of ergot to reach commercial food channels (300).

Cleaning and milling of grain removes most of the ergot that might otherwise pass in the flour.

The toxic effects exerted by ergotamine and other related alkaloids are exploited in pharmacological research. Vasoconstructive effects could be used to control migraine headaches (3012) as well as to control postpartum bleeding.

Ergot alkaloids can be casually found in foods, due to negligence in storage. Many analytical techniques are used to analyze these compounds, most of them based on chromatographic methods as thin-layer chromatography, gas chromatography, and HPLC with various detectors and tandem MS/MS as well as immunoenzymatic techniques (299).

Table 9 reports literature data for the compounds analyzed and where reported, the analyical conditions, the amount found andthe pre-treatment steps.

Table 9 Ergot Alkaloids

Analyte	Analytical method	Experimental conditions	Detection limit	Applications and amount	Sample treatment	Ref.
Ergot alkaloids.	HPLC and spectro-photometric (at 330, 280, 254 nm), fluorometric (408 nm, excitation 254 nm), amperometric (vitreous carbon electrode operating at 1.2 V vs. Ag/AgCl reference electrode) detection.	C18 column. Mobile phase: phosphate buffer and 1-octane-sulfonic acid.	With spectro-photometric detection 1–20 μg/ml, fluorescence 5–500 ng/ml, amperometric 5–50 ng/ml.	Orange juice, vegetable juice, milk, cola beverages.	For orange juices and vegetables: juice extraction using Analytichem C18 solid phase (SPE). For milk samples: protein precipitation with HCl vortex mixing, centrifugation, 1-octanesulfonic acid extraction. For cola: methanol addition, vortex mixing, filtration.	302
Ergot alkaloids: α-ergocryptinine, ergocorninine, α-ergocryptine, ergocornine, ergocristine.	HPLC with UV detection at 280 nm.	Hitachi gel n. 3111:0 column. Eluent: n-hexane, ethanol, triethylamine.	Around 1 μg.	Ergot.	Ether defatting. Multiple extractions with 1,2-dichloroethane, aqueous solution pH = 10, tartaric acid, H_2SO_4, NaOH, chloroform. Evaporation at 40°C.	303
Ergot alkaloids produced by *Claviceps* species.	Spectrophotometry (detection at 490 nm).	Polystyrene ELISA plate wells coated with lysergic acid-poly (L-lysine) conjugate.	10 ng/g.	Wheat, grass seed.	Incubation with phosphate buffer (pH = 7.4) and Tween 20 (PBST) for 30 min. Addition to ELISA wells with antiserum. Addition of goat anti-rabbit IgG-peroxidase conjugate. Incubation and addition of o-phenylenediamine dihydrochloride in citrate buffer and H_2O_2. Color halted with 3M H_2SO_4.	304

Analyte	Method	Detection	Sample / Found	Extraction	Ref.
Ergot alkaloids: ergosine, ergotamine, α-ergokryptine, ergocristine, ergometrine.	Liquid chromatography with fluorometric detection (excitation at 235 nm with a kV 370 emission at 254 nm).	0.5 ng for ergosine, ergotamine, ergocornine, α-ergocryptine, and ergocristine; 0.1 ng for ergometrine.	Flour, Found in: ergosine 10.8–14.2 µg/kg; ergotamine 1.4–36.9 µg/kg; ergocornine 9.1 µg/kg, -ergokryptine 10.3 mm/kg, ergocristine 2.7–62.2 µg/kg, and ergometrine 0.27–10.4 µg/kg.	Extraction with a mixture of CH_2Cl_2, ethylacetate, methanol, and ammonium hydroxide. Decantation. Filtration. Evaporation to dryness. Ether and methanol dissolution, HCl extraction. Acid layers washed with hexane, made alkaline, extracted in methylene chloride. Organic extract evaporated, methanol dissolved and filtered.	300
Ergot peptide alkaloids.	Quadrupole MS-MS pulsed positive and negative ion modes.	Isobutane as reagent gas. Ar as target in collisionally activated dissociation experiments. Direct introduction of sample.			305
Ergot alkaloids (ergometrin, ergometrinin, ergosin, ergotamin, ergocornin, ergokryptin, ergocristin, ergotaminin, ergocorninin, ergokryptinin, ergocristinin).	HPLC. Fluorometric detection.	Column: Hypersil ODS 3µm, 250 × 4.6 mm. Mobile phase acetonitrile water (47:53) containing 0.1 g of ammonium carbonate detection: λ_{ex} = 327nm, λ_{em} = 398 nm. nm.	Less than 20 ppb of each alkaloid.	Cereals, cereal products. Found a total content below the law limits (0.05% w/w) in both Middle European as well as Canadian products.	Dichloromethane-ethyl acetate–methanol–aqueous ammonia (50:25:5:1) extraction, evaporation, redissolution in toluene-methanol (49:1). Extrelut column pretreated with tartaric acid, cleanup by hexane: isopropyl ether (1:1), elution with dichloromethane after basification with gaseous ammonia. Eluate evaporation and redissolution in mobile phase, filtration.	306

Table 9 (*Continued*)

Analyte	Analytical method	Experimental conditions	Detection limit	Applications and amount	Sample treatment	Ref.
Ergot alkaloids (ergometrin, ergometrinin, ergotamin, ergosin, ergocornin, α-ergokryptin, β-ergokryptin, ergocristin, ergosinin, ergotaminin, ergocorninin, α-ergokriptinin, β-ergokryptinin, ergocristinin).	HPLC with fluorometric detection (λ_{ex} = 327 nm, λ_{em} = 398 nm), TLC with colorimetric detection with *p*-dimethylaminobenzaldehyde and HCl and GC-MS with mass spectrometric detection in chemical ionization with methane as the reactant gas, ionization energy 70 eV, emission current 300 mA, analyzing mass range between 100 and 600 amu.	HPLC: column Shandon Hypersil ODS 3 μm 250 × 4.6 mm; mobile phase: acetonitrile, water (47:53), containing 0.1 g/L of ammonium carbonate. TLC: plate Kieselgel 60, 15 cm, thickness 0.5 mm. Mobile phase: methanol dichlormethan (1:9). GC-MS: column DB-1 15 m × 0.32 mm, film thickness 0.25 μm; temperature programming from 180 to 300°C; helium as carrier gas at a flow of 2.5 ml/min.	Less than 20 ppb for each alkaloid in standard solution; ranging from 25 to 100 ppb in samples.	Bread, brown bread, confectionery products, bran products, cereals, corn flakes, linseeds, sesame seeds, infant formula.	Products are grounded and, when necessary, defatted, then extracted with dichloromethane–ethyl acetate–methanol–conc. aq. ammonia (50:25:5:1). Extracts filtered are evaporated, dissolved in hexane, and cleaned with an Extrelut-modified method.	307

299

Air-dried samples are extracted by shaking briefly with methanol-phosphate buffer containing 0.05% of Tween 20 and settled at –2°C. Supernatant is decanted into polypropylene centrifuge tubes and centrifuged at 14,000g, then supernatant is analyzed.

Cereal grains.

10 ng/g

Microtier plates (Immu-nol 2) coated with ergotamine-ovalbu-min conjugate in carbonate buffer (pH 9.6) are incubated at 37°C for 45 min, then added of sample in phos-phate buffer, of di-luted rabbit antiergot- amine antiserum, incubated for further 15 min. Detection conditions: bound antibody is determined using Bio-Rad goat anti-rabbit IgG-peroxidase conjugate, hy-drogenum peroxide and o-phenylenedi-amine.

ELISA, spectrophoto-metric detection at 490 nm.

Ergotamine, ergosine, ergocristine, ergotaminine, ergosinine.

F. Various Alkaloids

Other alkaloids not considered in the previous sections can be present in foods. So, for example, pyrrolizidine alkaloids oproduced by plants of the genera *Senecio, Crotalaria, Heliotropium, Erechtites, Trichodesma, Amsinckia* are carcinogenic, mutagenic, and teratogenic. They are also present in a wide variety of plants and forage which could be ingested by animals and can therefore be found with their metabolites in milk of some herbivora (308) or be present as impurities in herbal teas and honeys (309). The toxic effect of these alkaloids acts mainly on the liver.

Other analytes, such as those produced by *Lupinus* specie, represent the only antinutritional compoonent of seeds which, on the other hand, are rich in proteins and soluble dietary fibers (310).

Carbendazin and benamyl, systematic bvenzimidazole fungicides, are used for treatment on fruit and vegetables during refrigeration. Since benamyl metabolizes quite rapidly to carbendazin, tolerance ranges from 0.2 to 35 ppm (311).

Table 10 reports these analytes, the literature data concerning the experimental conditions adopted, and, when available, the detection limits, applications, and sample pretreatments.

Table 10 Other Alkaloids

Analyte	Analytical method	Experimental conditions	Detection limit	Applications and amount	Sample treatment	Ref.
Pyrrolizidine alkaloids.	Gas chromatography with ^{63}Ni electron capture detection.	Pyrex column packed with 7% OV-101 on HP Chromosorb W or fused silica SP-2100 capillary column. Carrier gas: nitrogen. Temperature: injector 275°C, detector 300°C.	0.1 ppm	Goat's milk. Found: 0.33–0.81 ppm.	Conversion with alkali to retronecine. Filtration. Cleanup by ion exchange, reverse phase, and porous polymer chromatography.	308
Pyrrolizidine alkaloids (senecionine, anacrotine, rosmarinine jacoline, jaconine, crocandine, fulvine, ehretinine, lycopasanine, heliotrine).	Negative-ion-chemical-ionization-mass spectrometry (NI-CI-MS).	Hydroxide is the reactant ion produced by introduction of nitrous oxide into the ion source via the capillary gas chromatographic inlet line. Methane added through standard CI reagent gas inlet. Ionization energy 140 eV, emission current 0.5 mA.		Herbal teas, milk, honey.		309
Lupanine, 13-hydroxylupanine hemisuccinate ester, *cis*-hexahydrophthalate ester, α-isolupanine, angustifoline, 17-hydroxylupanine, sparteine, cystisine, multiflorine, epilupenine.	ELISA, spectrophotometric detection at 405 nm.	Plate wells are coated with (+)-13-hydroxylupanine hemihexahydrophthalate bovine serum albumin conjugate in carbonate buffer (pH = 9.6). Incubation, addition with polyclonal antialkaloid antibodies from sheep antisera, washing, addition of donkey anti-sheep IgG-alkaline phosphatase, incubation, washing, addition of disodium 4-nitrophenyl phosphate hexahydrate in glycine buffer, pH = 10.	Less than 60 μg/g	Lupin seed.	Milled lupin seed flour is extracted with 5% trichloroacetic acid, neutralized with 10% $NaHCO_3$ and diluted with Tris-buffered saline (pH 7.4).	310
Carbendazin.	RP-HPLC analysis with fluorometric detection at 310 nm (excitation 286 nm).	Ultracarb 30 ODS column. Mobile phase: acetonitrile, methanol, water, monoethanolamine.	15 ng/g	Blueberries. Found: from nondetectable to 155 ng/g.	Methanol addition, filtration, NaCl addition, pH adjustment to pH = 9.5, dichloromethane extraction, evaporation to dryness. Methanol, acetonitrile, and water addition. Sequential column elutions through acid and basic alumina.	311

Table 10 (*Continued*)

Analyte	Analytical method	Experimental conditions	Detection limit	Applications and amount	Sample treatment	Ref.
Quinolizidine alkaloids, dihydrofuro-[2,3-b]-quinolinium alkaloids.	Overpressured layer chromatography (OPLC) with fluorometric detection (λ_{ex} = 310 nm, λ_{em} = 420 nm) or with Dragendorff reagent.	Stationary phases: alumina 60 F254 (type E) or Silica gel G60. Mobile phases: ethyl acetate or ethyl acetate–formic acid–water.		Linseed, plants.	Powdered lupin seeds are made alkaline with aqueous ammonia, extracted with ethyl ether, evaporated, and dissolved in methanol. Freeze-dried plant material is extracted with methanol and the extract purified by shaking with alumina.	312
Alkaloids.	TLC with spectrophotometric detection or with a modified Dragendorff reagent or ethanolic 0.01% 9-hydrazinoacridine.	Stationary phase: Silica gel R. Mobile phase: CCl$_4$-acetone-methanol-ethyleneamine (75:15:2).		Waters.		313
Angyrine, lupanine, aphylline, 5,6-dehydrolupanine, aphillidine, 4-hydroxylupanine.	GC with FID and MS detection.	Column: DB-1 (15 m × 0.25 mm). T program from 150 to 220°C at 6°C/min; helium as carrier gas, injector and detector temperature: 300°C.		*Lupinus latifolius.* Found in: 1.33% of total alkaloids in plant dry weight.	Ground samples, mixed with aqueous ammonia in methanol, are extracted overnight in Soxhlet apparatus with dichloromethane, concentrated by evaporation, partitioned in sulfuric acid, basified, reextracted in dichloromethane, brought to dryness under a rotary vacuum evaporator, dissolved in ethanol.	314

Analyte	Method	Conditions	Amount/detection limit	Sample	Extraction/cleanup	Ref.
Alkaloids.	Tandem MS.			Food and other biological materials.		315
Tropane alkaloids (hyoscine, hyosciamine).	HPLC and spectrophotometric detection at 210 nm.	Lichrosorb RP-18 column. Mobile phase: acetonitrile, methanol, ammonium acetate. Internal standard bamifylline.	12.05 ng hyoscine, 13.25 ng hyosciamine	Feeds, eggs.	CH_2Cl_2 extraction for solid samples and cleanup with C18 Bond-Elut cartridge for liquid samples. Evaporation to dryness, methanol addition.	316
Nitrosodimethylamine precursors: N-methyltyramine, hordenine, gramine.	HPLC with spectrophotometric detection at 280 nm.	For N-methyltyramine and hordenine: μBondapack phenyl-column and mobile phase of sodium monobasic phosphate and methanol at pH = 3. For gramine: Spherisorb ODS column and mobile phase of trichloroacetic acid and methanol.	N-Methyltyramine 0.2 μg/g, hordenine 0.2 μg/g, and gramine 0.1 μg/g	Barley, clean malt, malt byproducts. Found in malt: N-methyltyramine 28.8 μg/g, hordenine 13.1 μg/g. In barley: N-methyltyramine 18.5–40.4 μg/g and hordenine 8.9–29.9 μg/g.	Grinding. Ether, acid, basic extractions. Cleanup.	317
Alkaloids.	TLC separation and near-infrared excited Raman spectroscopy and HPLC.	Stationary phase: ODS-2 Spherisorb.		Chilly.	Grinding extraction with anhydrous Na_2SO_4, ethyl acetate, $NaHCO_3$.	318

REVIEWS

J. Buchberger and G. Weiss, *Lebensmittelchem., Lebensmittelqual, 4* (*Milchproteine*): 195–204 (1991).

D. Gonzalez-Llano, C. Polo, and M. Ramos, *Lait, 70*: 255–277 (1990).

H. Klein and R. Teichmann, *Ernaehrung* (*Vienna*) *10*: 608–613 (1986).

W. J. Hurst, *J. Liq. Chromatogr., 13*: 1–23 (1990).

A. Askar and H. Treptow, "*Biogenic Amines in Foods Occurrence, Significance and Determination.*", Ulmer Publ.; Stuttgart, W. Germany, 197 (1986).

W. Pfannhauer and U. Pechanek, *Z. Gesamte Hyg Ihre Grenzgeb., 30*: 66–76 (1984).

American Society of Brewing Chemists-Inc., *J. Am. Soc. Brew. Chem., 41*: 110–111 (1983).

H. Klein and W. Stoya, *Lebensmittelchem., Lebensmittelqual., 4* (*Milchproteine*): 125–129 (1991).

H. Klein and W. Stoua, *Mitt. Geb. Lebensmittelunters Hyg., 79*: 413–432 (1988).

M. G. Knize, J. S. Felton, and G. A. Gross, *J. Chromatogr., 624*: 253–265 (1992).

R. V. Golovnya, *Naturwiss. Tech. 2*: 173–200 (1988).

Ya. L. Kostyukovskii and D. B. Melamed, *Zh. Anal. Khim., 38*: 1865–1912 (1983).

S. W. Pelletier, *Alkaloids*; *Chemical and Biological Perspectives*, Vol. 1, John Wiley and Sons, New York, 1983.

M. Popl, J. Faehnrich, and V. Tatar, *Chromatographic Analysis of Alkaloids*, Marcel Dekker, New York, 1990, p. 675.

B. D. Kabulov, A. L. D'-yakonov, and S. V. Zalyalieva, *Chem. Nat. Compd.* (*Engl. Transl*), *27*: 521–540 (1992).

J. M. Hall, *Cereal Foods World, 28*: 254–255 (1983).

C. Israelian, *Bios* (*Nancy*), *14*: 41–43 (1983).

W. J. Hurst, W. R. Kreiser, and R. A. Martin jun., *Manuf. Confect., 60*: 39–41 (1980).

M. A. Busturia, *Fluess. Obst, 59*: 123–125 (1992)

REFERENCES

1. G. Debruyckere and C. Van Peteghem, *Food Analysis by HPLC*, Marcel Dekker, New York, 1992, pp. 643–671.
2. J. Buchberger and G. Weiss, *Lebensmittelchem. Lebensmittelqual, 4* (*Milchproteine*): 195–204 (1991).
3. D. Gonzalez-Llano, C. Polo, and M. Ramos, *Lait, 70*: 255–277 (1990).
4. H. Klein and R. Teichmann, *Ernaehrung* (*Vienna*) *10*: 608–613 (1986).
5. W. J. Hurst, *J. Liq. Chromatogr., 13*: 1–23 (1990).
6. A. Askar and H. Treptow, *Biogenic Amines in Foods*: *Occurrence, Significance and Determination*, Ulmer, Stuttgart, 1986, p. 197.
7. H. Treptow and A. Askar, *Ernaehrung* (*Vienna*), *14*: 9–11, 14–17 (1990).
8. W. Pfannhauer and U. Pechanek, *Z. Gesamte Hyg Ihre Grenzgeb, 30*: 66–76 (1984).
9. American Society of Brewing Chemists Inc., *J. Am. Soc. Brew. Chem., 41*: 110–111 (1983).
10. H. Klein and W. Stoya, *Lebensmittelchem Lebensmittelqual.., 4* (*Milchproteine*): 125–129 (1991).
11. H. Klein and W. Stoya, *Mitt. Geb. Lebensmittelunters. Hyg., 79*: 413–432 (1988).
12. G. Achilli, G. P. Cellerino, and G. Melzi d'Eril, *J. Chromatogr. A661*: 201–205 (1994).
13. J. Kirscbaum, B. Luckas, and W.-D. Beinert, *J. Chromatogr. A661*: 193–199 (1994).
14. M. L. Izquierdo-Pulido, M. C. Vidal-Carou, and A. Marine-Font, *JAOAC, Int., 76*: 1027–1032 (1993).
15. R. Draisci, S. Cavalli, L. Lucentini, and A. Stacchini, *Chromatographia, 35*: 584–590 (1993).
16. B. Straub, M. Schollenberger, M. Kicherer, B. Luckas, and W. P. Hammes, *Z. Lebensm. Unters. Forsch., 197*: 230–232 (1993).
17. P. Husek, *J. Microcolumn Sep., 5*: 101–103 (1993).
18. S. Eerola, R. Hinkkanen, E. Lindfors, and T. Hirvi, *JAOAC, Int., 76*: 575–577 (1993).
19. R. Ellis and G. D. Lawrence, *Microchem. J., 47*: 55–59 (1993).
20. H. Ohta, Y. Takeda, K> I> Yoza, and Y. Nogata, *J. Chromatogr.., 628*: 199–204 (1993).
21. U. Bilitewski, G. C. Chemnitius, P. Rueger, and R. D. Schmid, *Sens. Actuators, B, B7*(*1–3*):351–355 (1992).

22. P. Lehtonen, M. Saarinen, M. Vesanto, and M. . Riekkola, Z. *Lebensm. Unters. Forsch.*, *194*:434–437 (1992).
23. M. E. Diaz-Cinco, O. Fraijo, P. Grajeda, J. Lozano-Taylor, and E. Gonzalez-de-Mejia, *J. Food Sci*, *57*: 355–356, 365 (1992).
24. S. Moret, R. Bortolomeazzi, and G. Lercker, *J. Chromatogr.*, *591*: 175–180 (1992).
25. G. C. Yen, and C. L. Hsieh, *J. Food Sci.*, *56*: 158–160 (1991).
26. C. Tricard, J. M. Cazabeil, and M. H. Salagoity, *Analusis*, *19*: M53–M55 (1991).
27. H. Vuorela, P. Lehtonen, and R. Hiltunen, *J. Liq. Chromatogr.*, *14*: 3181–3195 (1991).
28. M. C. Gennaro and C. Abrigo, *Chromatographia*, *31*: 381–386 (1991).
29. K. Outinen, H. Vuorela, R. Hinkkanen, and R. Hiltunen, *Acta Pharm. Fenn.*, *99*: 99–108 (1990).
30. U. Buetikofer, D. Fuchs, D. Hurni, and J. O. Bosset, *Mitt. Geb. Lebensmittelunters. Hyg.*, *81*: 120–133 (1990).
31. M. Krizek, *Arch. Anim. Nutr.*, *41*: 97–104 (1991).
32. N. Nic-a'-Bhaird, J. M. McCrodden, A. M. Wheatley, M. C. Harrington, J. P. Sullivan, and K. F. Tipton, *Biomed. Chromatogr.*, *4*: 229–233 (1990).
33. M. C. Gennaro and C. Abrigo, *Analusis*, *19*: 151–154 (1991).
34. R. Etter, S. Dietrich, and R. Battaglia, *Mitt. Geb. Lebensmittelunters. Hyg.*, *81*: 106–119 (1990).
35. K. Takeba, T. Maruyama, M. Matsumoto, and H. Nakazaka, *J. Chromatogr.*, *504*: 441–444 (1990).
36. S. Suzuki, K. Kobayashi, J. Noda, T. Suzuki, and K. Takama, *J. Chromatogr. 508*: 225–228 (1990).
37. A. Lebedzinska, K. I. Eller, and V. A. Tutel'-yan, *Zh. Anal. Khim.*, *44*: 928–931 (1989).
38. R. E. Schmitt, J. Haas, and R. Amado, *Z. Lebensm. Unters. Forsch.*, *187*: 121–124 (1988).
39. J. Eagles and R. A. Edwards, *Biomed. Environ. Mass Spectrom.*, *17*, 241–244 (1988).
40. F. V. Carlucci and E. Karmas, *JAOAC*, *71*: 564–568 (1988).
41. E. Mueller-Seitz and E. H. Reimerdes, *Lebensmittelchem. Gerichtl. Chem.*, *42*: 10–11 (1988).
42. H. Walther, U. P. Schlunegger, and F. Friedli, *Biomed. Environ. Mass Spectrom.*, *14*, 229–233 (1987).
43. L. C. Laleye, R. E. Simard, C. Gosselin, B. H. Lee, and R. N. Giroux, *J. Food Sci.*, *52*, 303–307 (1987).
44. G. B.Baker, J. T. F. Wong, R. T. Coutts, and F. M. Pasutto, *J. Chromatogr.*, *392*, 317–331 (1987).
45. M. A. J. S. Van-Boekel, and A. P. Arentsen-Stasse, *J. Chromatogr.*, *389*, 267–272 (1987).
46. H. M. L. J. Joosten, and C. Olieman, *J. Chromatogr.*, *356*: 311–319 (1986).
47. S. F. Chang, J. W. Ayres, and W. E. Sandine, *J. Dairy Sci.*, *68*, 2840–2846 (1985).
48. J.. A. Zee, R. E. Simard, and L. L'Heureux, *Lebensm. Wiss. Technol.*, *18*, 245–248 (1985).
49. D. L. Ingles, J. F. Back, D. Gallimore, R. Tindale, and K. J. Shaw, *J. Sci. Food Agri.*, *36*, 402–406 (1985).
50. K. Mayer and G. Pause, *Lebensm. Wiss. Technol.*, *17*, 177–179 (1984).
51. Y. Tonogai, Y. Ito, and M. Harada, *Shokuhin Eiseigaku Zasshi*, *25*, 41–46 (1984).
52. S. Ramantanis, C. P. Fassbender, and S. Wenzel, *Arch. Lebensmittelhyg.*, *35*, 80–82 (1984).
53. B. Rogowski, and I. Doehla, *Lebensmittelchem. Gerichtl. Chem.*, *38*: 20–21 (1984).
54. N. Sayem-El-Daher, R. E. Simard, J. Filion, A. G. Roberge, *Lebensm. Wiss. Technol.*, *17*: 20–23 (1984).
55. J. A. Zee, R. E. Simard, and L. L'Heureux, *J. Food Prot.*, *46*: 1044–1049, 1054 (1983).
56. J. Slemr and K. Beyermann, *J. Chromatogr. 283*: 241–250 (1984).
57. C. Buteau, C. L. Duitschaever, and G. C. Ashton, *J. Chromatogr.*, *284*: 201–210 (1984).
58. K. D. H. Chin, and P. E. Koehler, *J. Food Sci.*, *48*: 1826–1828 (1983).
59. C. Droz and H. Tanner, *Schweiz. Z. Obst. Weinbau*, *119*: 75–77 (1983).
60. J. Y. Hui and S. L. Taylor, *JAOAC*, *66*: 853–857 (1983).
61. U. Pechanek, W. Pfannhauser, and H. Woidich, *Z. Lebensm. Unters. Forsch.*, *176*: 335–340 (1983).
62. S. Yamamoto, H. Itano, H. Kataoka, and M. Makita, *J. Agric. Food Chem.*, *30*: 435–439 (1982).
63. U. Pechanek, G. Blaicher, W. Pfannhauser, and H, Woidich, *Chromatographia*, *13*: 421–427 (1980).
64. U. Pechanek, G. Blaicher, W. Pfannhauser, and H. Woidich, *Z. Lebensm. Unters. Forsch. 171*: 420–424 (1980).
65. H. Schmidtlein, *Lebensmittelchem. Gerichtl. Chem. 33*: 81–83 (1979).
66. S. Pongor, J. Kramer, and E. Ungar, *J. High Resolut Chromatogr. Chromatogr. Commun.*, *3*: 93–94 (1980).
67. D. Froehlich and R. Battaglia, *Mitt. Geb Lebensmittelunters. Hyg.*, *71*:, 38–44 (1980).
68. Th. B. A. Reuvers, M. Martin de Pozuelo, M. Ramos, adn R. Jimenez, *J. Food Sci.*., *51*: 84–86 (1986).
69. R. W. Daisley, and H. V. Gudka, *Communi. J. Pharm. Pharmacol. 32*: 77 (1980).

70. S. Yamamoto, S. Wakabayashi, and M. Makita, *J. Agric. Food Chem.*, *28*: 790–793 (1980).
71. F. J. Santos, *Rev. Agroquim. Tecnol. Aliment. 25*, 362 (1985).
72. T. A. Gill, and J. W. Thompson, *J. Food Sci.*, *49*: 603–606 (1984).
73. T. Tsushida and T. Takeo, *J. Sci. Food Agric.*, *35*: 77–83 (1984).
74. I. Karube, I. Satoh, Y. Araki, S. Suzuki, and H. Yamada, *Enzyme Microb. Technol.*, *2*: 117–120 (1980).
75. R. B. H. Wills, J. Silalahi, and M. Wootton, *J. Liq. Chromatogr.*, *10*, 3183–3191 (1987).
76. J. Walford (Editor), *Development in Food Colours,* Applied Science, London (1980).
77. R. J. Taylor, *Food Additives*, Wiley, Chichester (1980).
78. *Food Colors*, National Academy of Science, Washington, DC (1971).
79. A. G. Fogg and A. M. Summan, *Analyst*, *108*: 691–700 (1983).
80. F. E. Lancaster and J. F. Lawrence, *Food Add. Contam. 9* : 171–182 (1992).
81. F. E. Lancaster and J. F. Lawrence, *Food Add. Contam.*, *6*: 415–423 (1989).
82. N. Richfield-Fratz and J. E. Bailey jum., *J. Chromatogr.*, *405*: 283–294 (1987).
83. N. Richfield-Fratz, J. E. Bailey jun.; and C. J. Bailey, *J. Chromatogr.*, *331*: 109–123 (1985).
84. J. E. Bailey jun. and C. J. Bailey, *Talanta*, *32*: 875–882 (1985).
85. H. Limacher and H. Tanner, *Alimenta*, *24*: 33–35 (1985).
86. C. Wuethrich, U. Baumann, R. Gysin, and B. Marek, *Mitt. Geb. Lebensmittelunters. Hyg.*, 75: 494–505 (1984).
87. J. J. Sanchez-Saez, P. Calvo-Anton, and M. C. Miramar-Blazquez, *Bol. Cent. Nac. Aliment. Nutr.*, 6–8: 14–15 (1981).
88. R. G. Fulcher, T. P. O'Brien, and S. Wong, *Cereal Chem.*, *58*: 130–135 (1981).
89. U. Baumann and B. Marke, *Mitt. Geb. Lebensmittelunters Hyg.*, *71*: 468–483 (1980).
90. I. L.Zhuravleva, R. V. Golovnya, L. F. Palamarchuk, and Z. N. Kishkovskii, *Zh. Anal. Khim.*, *35*: 193–195 (1980).
91. F. E. Lancaster and J. F. Lawrence, *Food Add. Contam.*, 8: 249–264 (1991).
92. M. G. Knize, J. S. Felton, and G. A. Gross, *J. Chromatogr.*, *624*: 253–265 (1992).
93. S. Vainiotalo, K. Matveinen, and A. Reunanen, *Fresenius' J. Anal Chem. 345*: 462–466 (1993).
94. G. A. Gross, *Carcinogenesis*, *11*: 1597 (1990).
95. R. Schwarzenbach and D. Gubler, *J. Chromatogr.*, *624*: 491–495 (1992).
96. G. A. Gross and A. Grueter, *J. Chromatogr.*, *592*: 271–278 (1992).
97. S. M. Billedeau, M. S. Bryant, and C. L. Holder, *LC-GC Int.*, *4*: 38, 40–41 (1991).
98. R. J. Turesky; H. Bur, T Huynh-Ba, H. U. Aeschbacher, and M. Milon, *Food Chem. Toxico.l*, 26: 501–509 (1988).
99. C. E. Daudt and C. S. Ough, *Am. J. Enol. Vitic.*, *31*, 356–359 (1980).
100. J.-K. Lin, and C.-C, Lai, *Anal. Chem. 52*: 630–635 (1980).
101. R. V. Golovnya, *Naturwiss. Tech. 2*: 173–200 (1988).
102. Ya.. L. Kostyukovskii and D. B. Melamed, *Zh. Anal. Khim.*, *38*: 1865–1912 (1983).
103. Z. Li, H. Lin, G. Zeng, J. Gou, and R. Yu, *Fenxi Ceshi Xuebao*, *12*: 51–54 (1993).
104. L. P. Tikhomirova, M. A. Shnaider, L. D. Londar, and N. N. Spiridonova, *Izv. Vyssh. Uchebn. Zaved., Khim. Khim. Tekhnol.*, *33*: 22–23 (1990).
105. B.. Pfundstein, A. R. Tricker, and R. Preussmann, *J. Chromatogr.*, *539*, 141–148 (1991).
106. R. Kruse and J. Stockemer, *Arch. Lebensmittelhyg.*, *40*, 87–89 (1989).
107. S. J. Ziegler, B. Meier, and O. Sticher, *Dtsch. Apoth. Ztg.*, *129*: 318–322 (1989).
108. H. Koizumi and Y. Suzuki, *Anal. Sci.*, *4*: 537–538 (1988).
109. P. Malle and S. H. Tao, *J. Food Prot.*, *50*: 756–760 (1987).
110. J. Culik, V. Kellner, B. Spinar, and F. Frantik, *Kvasny-Prum.*, *33*: 283–286 (1987).
111. Ya. L. Kostyukovskii, and D. B. Melamed, *Zh. Anal. Khim.*, 42: 924–928 (1987).
112. M. D. Ruiz-Lopez, M. C. Lopez-Martinez, and R. Garcia-Villanova, *Analusis*, *15*, 43–46 (1987).
113. R. C. Lundstrom and L. D. Racicot, *JAOAC*, *66*: 1158–1163 (1983).
114. C. E. Daudt and C. S. Ough, *Am. J. Enol. Vitic.*, *31*: 356–359 (1980).
115. G. M. Murray and M. J. Sepaniak, *J. Liq. Chromatogr.*, *6*: 931–938 (1983).
116. A. G. Fogg and A. M. Summan, *Analyst* (London), *108*: 691–700 (1983).
117. M. Bellatti and G. Parolari, *Meat Sci.*, *7*: 59–65 (1982).
118. B. D. Ripley, B. J. French, and -L. V. Edgington, *JAOAC*, *65*: 1066–1072 (1982).

119. J. M. Gallardo, M. Lopez-Benito, L. Pastoriza, and P. Gonzalez, *Inf. Tec. Inst. Invest. Pesq. (Barcelona)*, *65*: 15 (1979).
120. R. J. Cutter, G. Hansen, and T. R. Jones, *Anal. Chim. Acta* , *136*: 135–142 (1982).
121. J. Hladik, L. Curda, and Z. Pokluda, *Prum. Potrovin*, *33*: 101–104 (1982).
122. T. Hamano, Y. Mitsuhashi, and Y. Matuski, *Agric. Biol. Chem.*, *45*, 2237–2243 (1981).
123. J. K. Lin, and C. C. Lai, *Anal. Chem.*, *52*: 630–635 (1980).
124. C. C. Lai, and J. K. Lin, *K'-o-Hsueh-Fa-Chan-Yueh-K'an*, *8*: 941–955 (1980).
125. W. Fritz, and W. J. Uhde, *Nahrung*, *24*:, 471–477 (1980).
126. M. Petro-Turza, I. Szarfoldi-Szalma, and G. Vigh, *Acta Aliment.*, *10*: 407–415 (1981).
127. N. Parris, and T. A. Foglia, *J. Agric. Food Chem.*, *31*: 887–889 (1983).
128. R. M. Storey, H. K. Davis, D. Owen, and L. Moore, *J. Food Technol.*, *19*: 1–10 (1984).
129. Y. Tonogai, Y. Ito, and M. Harada, *Shokuhin Eiseigaku Zasshi*, *25*: 149–157 (1984).
130. N. Parris, *J. Agric. Food Chem.*, *32*: 829–831 (1984).
131. J. Stockemer and R. Kruse, *Arch. Lebensmittelhyg.*, *36*: 116–117 (1985).
132. R. I. Perez-Martin, J. M. Franco, P. Molist, and J. M. Gallardo, *Int. J. Food Sci. Technol.*, *22*: 509–514 (1987).
133. S. W. Pelletier, *Alkaloids: Chemical and Biological Perspectives*, Vol. 1, John Wiley and Sons, New York, 1983.
134. G. Eigenmann and I. Ubaldini, *Nuovo dizionario di Merceologia e Chimica Applicata*, Vols. 1–7, U. Hoepli, Milano, 1977.
135. M. Popl, J. Faehnrich, and V. Tatar, *Chromatographic Analysis of Alkaloids*, Marcel Dekker, new York, 1990, p. 675.
136. B. D. Kabulov, A. L. D'-yakonov, and S. V. Zalyalieva, *Chem. Nat. Compd. (Engl. Transl.)*, *27*: 521–540 (1992).
137. J. M. Hall, *Cereal Foods World*, *28*: 254–255 (1983).
138. D. B. Haughey, R. Greenberg, S. F. Schaal, and J. J. Lima, *J. Chromatogr.* *229*: 395–397 (1982).
139. C. Van der Meer and R. E. Haas, *J. Chromatogr.* *182*: 121–124 (1980).
140. F. T. Delbecke and M. Debackere, *J. Chromatogr.* *278*: 418–423 (1983).
141. C. Israeian, *Bios* (Nancy), *14*: 41–43 (1983).
142. W. J. Hurst, W. R. Kreiser, and R. A. Martin jun., *MC, Manuf. Confect.*, 60:39–41 (1980).
143. M. A. Busturia, *Fluess. Obst*, *59*: 123–125 (1992).
144. J. L. Blauch and S. M. Tarka jun., *J. Food Sci.*, *48*: 745–747, 750 (1983).
145. R. M. Smith, *Food Chem.*, *6:* 41–45 (1981).
146. S. H. Ashoor, G. J. Seperich, W. C. Monte, and J,. Welty, *JAOAC*, *66*: 606–609 (1983).
147. A.. Ferrara, P. Reale, M. G. Calaminici, and T. Iaccarino, *Boll. Chim. Unione Ital. Lab. Prov.*, *33*: 55–60 (1982).
148. A. Duplatre, C. Tisse, and J. Estienne, *Analusis*, *9*: 404–405 (1981).
149. L. C. Trugo, R. Macrae, and J. Dick, *J. Sci. Food Agric.*, *34*: 300–306 (1983).
150. L. Bianco and M. Marucchi, *Boll. Chim. Unione Ital. Lab. Prov.* *33*: 165–168 (1982).
151. J. W. De-Vries, K. D. Johnson, and J. C. Heroff, *J. Food Sci.*, *46*: 1968–1969 (1981).
152. H. Hadorn, *Rev. Choc. Confect. Bakery*, *5*: 26–28 (1980).
153. U. Juergens, K. Von-Grunherr, and G. Blosczyk, *Lebensmittelchem. Gerichtl. Chem.*, *34*: 109–114 (1980).
154. A. Herrmann, E. Loetscher, and M. Wagmann, *Mitt. Geb. Lebensmittelunters. Hyg.*, *73*: 121–128 (1982).
155. J. W. Weyland, H. Rolink, and P. A. Doornbos, *J. Chromatogr.*, *247*: 221–229 (1982).
156. S. J. Reid and T. J. Good, *J. Agric. Food Chem.*, *30*: 775–778 (1982).
157. E. C. P. Gillyon, *Chromatogr. Newsl.*, *8*: 50–51 (1980).
158. J. Van-Duijn and G. HD. Van-der-Stegen, *J. Chromatogr.*, *179*: 199–204 (1979).
159. B. B. Woodward, G. P. Heffelfinger, and D. I. Ruggles, *JAOAC*, *62*: 1011–1019 (1979).
160. U. Juergens and R. Riessner, *Dtsch. Lebensm. Rundsch.*, *76*: 39–42 (1980).
161. G. Sontag and K. Kral, *Mikrochim. Acta*, II: 39–52 (1980).
162. W. R. Kreiser and R. A. Martin, jun., *JAOAC*, *63*: 591–594 (1980).
163. D. C. Woollard, *N. Z. J. Dairy Sci. Technol.*, *17*: 63–68 (1982).
164. K. Matsumoto, R. Tateishi, and Y. Osajima, *Nippon Shokuhin Kogyo Gakkaishi.*, *28*: 269–274 (1981).

165. K. Pfeilsticker and A. Leyendecker, *Z. Lebensm. Unters. Forsch.*, *171*: 174–177 (1980).
166. A. Klatsm and P. Zala, *Elelmez-Ip.*, *34*: 336–338 (1980).
167. H. Ishida, H. Sekine, S. Kimura, and S. Sekiya, *Shokuhin Eiseigaku Zasshi*, *27*: 75–80 (1986).
168. C. Thomcyzk, *Lebensmittellindustrie*, *26*: 561–563 (1979).
169. G. Lehmann, I. Haug, and R. Schloesser, *Z. Lebensm. Unters. Forsch.*, *172*: 87–89 (1981).
170. K. Mizuki and M. Deki, *Kanzei Chuo Bunsekihoho*, *18*: 53–57 (1978).
171. J. Stangelberger, I. Steiner, G. Kroyer, and J. Washuettl, *Lebensmittelchem. Gerichtl. Chem.*, *37*, 70–71 (1983).
172. E. Laub, M. Zimmer, *Lebensmittelchem. Gerichtl. Chem.*, *33*: 117–118 (1979).
173. M. Jimidar, T. P. Hamoir, and D. L. Massart, *J. Chromatogr.*, *636*: 179–186 (1993).
174. H. Terada, A. Suzuki, H. Tanaka, and K. Yamamoto, *Shokuhin Eiseigaku Zasshi*, *33*: 347–354 (1992).
175. F. Wu and P. Zhang, *Sepu 10*: 311–312 (1992).
176. S. B. Hawthorne, D. J. Miller, J. Pawliszyn, and C. L. Arthur, *J. Chromatogr.*, *603*: 185–191 (1992).
177. M. C. Gennaro and C. Abrigo, *Fresenius' J. Anal. Chem.*, *343*: 523–525 (1992).
178. O. Lau, S. Luk, O. Cheng, and T. P. Y. Chiu, *Analyst 117*: 777–783 (1992).
179. D. Danho, N. Naulet, and G. J. Martin, *Analusis*, *20*: 179–184 (1992).
180. Z. Tan, C. He, and S. Fan, *Fenxi-Huaxue*, *19*: 1402–1404 (1991).
181.. L. C. Trugo, C. A. B. De-Maria, and C. C. Werneck, *Food Chem.*, *42*: 81–87 (1991).
182. X. Xia and N. Ding, *Fenxi-Huaxue*, *19*: 945–948 (1991).
183. M. A. Abuirjeie, M. S. El-Din, and I. I. Mahmoud, *J. Liq. Chromatogr.*, *15*: 101–125 (1992).
184. A. Cepeda, P. Paseiro, J. Simal, and J. L. Rodriguez, *An. Bromatol.*, *42*: 241–249 (1991).
185. S. Kuhr and U. H. Engelhardt, *Z. Lebensm. Unters. Forsch.*, *192*: 526–529 (1991).
186. P. Elisabeth, M. Yoshioka, Y. Yamauchi, and M. Saito, *Anal. Sci.*, *7*: 427–431 (1991).
187. T. Schlabach and J. Powers, *Int. Lab.*, 26–28 (1991).
188. Y. Ren, Z. Gao and B. Huang, *Shipin-Yu-Fajiao-Gongye 1*: 72–785 (1990).
189. U. Hagenauer-Hener, C. Frank, U. Hener, abnd A. Mosandl, *Dtsch. Lebensm. Rundsch.*, *86*: 348–351 (1990).
190. A. B. Morton and R. R. Stubb, *Analyst (London)*, *71*: 348 (1964).
191. A. Cepeda, P. Paseiro, J. Simal, and J. L. Rodriguez, *Alimentaria (211)*: 23–27 (1990).
192. A. Crozier and A. M. Monteiro, *Chromatogr. Anal.*, *11*: 5–7 (1990).
193. E. Mueller and H. Jork, *Dtsch. Lebensm. Rundsch.*, *86*: 243–247 (1990).
194. T. Li, *Yaowu-Fenxi-Zazhi*, *10*: 366–367 (1990).
195. F. J. Muhtadi, S. S. El-Hawary, and M. S. Hifnawy, *J. Liq. Chromatogr.*, *13*: 1013–1028 (1990),
196. S. Li, J. Berger, and S. Hartland, *Anal. Chim. Acta*, *232*: 409–412 (1990).
197. B. Guo and H. Wan, *J. Chromatogr.*, *505*: 435–437 (1990).
198. D. I. Kuznetsov and L. I. Semenova, *Pishch. Promst.*, *9*: 65–66 (1989).
199. U. Hagenauer-Hener, U. Hener, F. Dettmar, and A. Mosandl, *Kontakte (Darmstadt)*: *1*, 24–29 (1989).
200. D. YUpnmoor and G. Brunner, *GIT Fachz. Lab.*, *33*: 311–314, 316–317 (1989).
201. Y. Kitada, M. Sasaki, Y. Yamazoe, Y. Maeda, M. Yamamotom and H. Nakazawa, *Bunseki Kagaku*, *37*: 561–565 (1988).
202. P. Deng, H. Li, A. Lu, and Y. Dai, *Shipin Kexue (Beijing)*, *98*: 51–55 (1988).
203. A. Kunugi, T. Aoki, and S. Kunugi, *Shokuhin Eiseigaku Zasshi*, *29*: 136–140 (1988).
204. E. M. Abdel-Moety, *Z. Lebensm, Unters. Forsch.*, *186*: 412–416 (1988).
205. A. Martelli and F. Belliardo, *Riv. Soc. Ital. Sci. Aliment.*, *16*: 121–129 (1987).
206. J. Chen, *Sepu*, *5*: 367–369 (1987).
207. M. Veerabhadrarao, M. S. Narayan, O. Kapur, and C. S. Sastry, *JAOAC*, *70*: 578–582 (1987),
208. X. Liu, P. Van Espen, F. Adams, S. H. Yan, and M. Vanbelle, *Anal. Chim. Acta*, *200*: 421–430 (1987).
209. R. S. Ramakrishm, M. Dias, S. Palamakumbura, and M. Jeganathan, *Can. J. Chem.*, *65*: 947–950 (1987).
210. N. R. Skhiladze, *Chem. Nat. Compd. (Engl.-Transl.)*, *22*:: 491 (1987).
211. M. F. Vergnes and J. Alary, *Talanta*, *33*: 997–1000 (1986).
212. A. K. Agrawal and S. K. Sharma, *J. Inst. Chem. 58*: 146 (1986).
213. H. J. Issaq, D. Weiss, C. Ridlon, S. D. Fox, and G. M. Muschik, *J. Liq. Chromatogr.*, *9*: 1791–1801 (1986).
214. G. Lehmann, J. Ganz, and K. Schmidt, *Dtsch. Lebensm. Rundsch.*, *82*: 43–46 (1986).

215. F. Marx, K. Pfeilsticker, and J. G. S. Maia, *Dtsch. Lebensm. Rundsch.*, *81*: 390–392 (1985).
216. I. Stewart, *J. Agric. Food Chem.*, *33*: 1163–1165 (1985).
217. R. A. K. Srivastava, *Indian Drugs*, 22: 438–440 (1985).
218. K. Sugiyama, M. Saito, T. Hondo, and M. Senda, *J. Chromatogr.*, *332*: 107–116 (1985).
219. P. Paseiro and J. Simal, *An. Bromatol.*, *36*: 97–108 (1985).
220. K. Fukuhara, Y. Matsuki, and T. Nanbara, *Shokuhin Eiseigaku Zasshi*, *26*: 208–212 (1985).
221. W. J. Hurst, K. P. Snyder, and R. A. jun. Martin, *J. Chromatogr.*, *318*: 408–411 91985).
222. M. Dulitzky, E. De-la-Teja, and H. F. Lewis, *J. Chromatogr.*, *317*: 403–405 (1984).
223. C. J. Argoudelis, *J.-Chromatogr.*,*303*: 256–262 (1984).
224. T. A. Tyler, *JAOAC*, *67*: 745–747 (1984).
225. P. Schreier and W. Mick, *Z. Lebensm. Unters. Forsch.*, *179*: 113–118 (1984).
226. H. Terada and Y. Sakabe, *J. Chromatogr.*, *291*: 453–459 (1984).
227. N. G. Webb and D. D. Beckman, *JAOAC 67*: 510–513 (1984).
228. A. M. Sjoberg and J. Rajama, *J. Chromatogr.*, *295*: 291–294 (1984).
229. British Standards Institution, *British Standard*, *BS 5752*: *Part 3* (ISO 4052-1983) (1983).
230. H. Ishida, H. Sekine, S. Kimura, and S. Sekiya, *Shokuhin Eiseigaku Zasshi*, 27: 75–80 (1986).
231. M. S. Karawya, A. M. Diab, and N. Z. Swelem, *Anal. Lett.*, *17*: 77–88 (1984).
232. T. Yao, M. Satomura, and T. Wasa, *Anal. Chim. Acta*, *261*: 161–165 (1992).
233. H. Okuma, H. Takahashi, S. Yazawa, S. Sekimukai, and E. Watanabe, *Anal. Chim. Acta* , *260*: 93–98 (1992).
234. H. Okuma, H. Takahashi, S. Sekimukai, and E. Watanabe, *Nippon Shokuhin Kogyo Gakkaishi*, *38* 1019–1024 (1991).
235. J. H. T. Luong, K. B. Male, and M. D. Huynh, *J. Food Sci.*, *56*: 335–337, 340 (1991).
236. B. Roberts, B. A. Morris, and M. N. Clifford, *Food Chem.*, *42*: 1–17 (1991).
237. Y. Isono, *Agric. Biol. Chem.*, *54*: 2827–2832 (1990).
238. A. L. Nguyen, J. H. T. Luong, and C. Masson, *Anal. Chem.*, *62*: 2490–2493 (1990).
239. M. Suzuki, H.Suzuki, I. Karube, and R. D. Schmid, *Anal. Lett.*, *22*: 2915–2927 (1989).
240. T. Yao and Y. Matsumoto, *Electroanalysis (NY)*, *1*: 173–176 (1989).
241. Y. Shimizu, Y. Takao, and M. Egashira, *J. Electrochem. Soc.*, *135*: 2539–2540 (1988).
242. T. Yao, Y. Matsumoto and T. Wasa, *Bunseki Kagaku*, *37*: 236–241 (1988).
243. Z. Qian and S. Ye, *Huazue-Shiji*, *8*: 361–362, 378 1986).
244. W. Tiemeyer, *Ernaehrung (Vienna)*, *10*: 388–391 (1986).
245. B. G. Burns, and P. J. Ke, *JAOAC*, *68*: 444–448 (1985).
246. J. M. Ryder, *J. Agric. Food Chem.*, *33*: 678–680 (1985).
247. E. Watanabe, K. Toyama, I. Karube, and H. Matsuoka, *S. Suzuki, Appl. Microbiol. Biotechnol.*, *19*: 18–22 (1984).
248. M. Keuler, G. Brauner-Glaesner, and R. Ristow, *Fleischwirtschaft*, *63*: 1877–1880 (1983).
249. E. Watanabe, K. Ando, I. Karube, H. Matsuoka, and S. Suzuki, *J. Food Sci.*, *48*: 496–500 (1983).
250. E. H. Lee, T. Ohshima, and C. Koizumi, *Nippon-Suisan-Gakkaishi*, *48*: 255 (1982).
251. J. J. Warthesen, P. T. Waletzko, and F. F. Busta, *J. Agric. Food Chem.*, *28*: 1308–1309 (1980).
252. W. J. Albery, P. N. Bartlett, and A. E. G. Cass, *Philos. Trans Roy. Soc. London, Ser. B. 316*: 107–119 (1987).
253. C. J. Dale and T. W. Young, *J. Inst Brew.*, *94*: 33–37 (1988).
254. M. Colling and G. Wolfram, *Z. Lebensm Unters. Forsch.*, *185*: 288–291 (1987).
255. E. Lassek and A. Montag, *Z. Lebensm Unters Forsch.*, *184*: 361–365 (1987).
256. W. Herbel and A. Montag, *Z. Lebensm. Unters. Forsch.*, *184*: 8–10 (1987).
257. W. Tiemeyer, *Ernaehrung (Vienna)*, *10*: 388–391 (1986).
258. P. H. Ritter, J. E. Burnett, R. M. Sheeley, W. J. Hurst, and R. A. Martin, Jr., *J. Liq. Chromatogr.*, *9*: 887–891 (1986).
259. W. Herbel and A. Montag, *Z. Lebensm. Unters. Forsch.*, *178*: 81–85 (1984).
260. L. C. Trugo, R. Macrae, and J. Dick, *J. Sci. Food Agric.*, *34*: 300–306 (1983).
261. F. Vojir and F. Petuely, *Lebensmittelchem. Gerichtl. Chem.*, *36*: 73–79 (1982).
262. L. Kraus, N. Linnenbrink, and R. Richter, *Kontakte (Darmstadt)*, *(1)*:, 20–21 (1982).
263. A. A. Qureshi, W. C. Burger, and N. Prentice, *J. Am. Soc. Brew. Chem.*, *37*: 153–160 (1979).
264. D. V.McCalley, *Analyst*, *115*: 1355–1358 (1990).

265. J. C. Reijenga, G. V. A. Aben, A. A. G. Lemmens, T. P. E. Verheggen, C. H. M. M. De-Bruijn and F. M. Everaerts, *J. Chromatogr., 320*: 245–252 (1985).
266. F. A. Mellon, J. R. Chapman, and J. A. E. Pratt, *J. Chromatogr., 394*: 209–222 (1987).
267. M. Verzele, G. Redant, S. Qureshi, and P. Sandra, *J. Chromatogr., 199*: 105–112 (1980).
268. M. Verzele, F. Van-Damme, G. Schuddinck, and P. Vyncke, *J. Chromatogr., 471*: 335–346 (1989).
269. T. H. Cooper, J. A. Guzinski, and C. Fisher, *J. Agric. Food Chem., 39*: 2253–2256 (1991).
270. E. K. Johnson, H. C. jun. Thompson, and M. C. Bowman, *J. Agric. Food Chem., 30*: 324–329 (1982).
271. N. C. S. De-Siqueira, L. A. Mentz, L. Ene, C. Chaves, C. B. Alice, and G. A. A. B. Silva, *Rev. Cienc. Farm. (Araraquara, Braz.), 10*: 107–111 (1988).
272. A. M. Krajewska and J. J. Powers, *J. Chromatogr., 457*: 279–286 (1988).
273. A. M. Krajewska and J. J. Powers, *J. Chromatogr., 409*: 223–233 (1987).
274. A. M. Krajewska and J. J. Powers, *JAOAC, 70*: 926–928 (1987).
275. A. B. Wood, *Flavour Fragrance J., 2*: 1–12 (1987).
276. V. K. Attuquayefio and K. A. Buckle, *J. Agric. Food Chem., 35*: 777–779 (1987).
277. K. M. Weaver, R. G. Luker, and M. E. Neale, *J. Chromatogr., 301*: 288–291 (1984).
278. K. S. Rymal, R. D. Cosper, and D. A. Smith, *JAOAC, 67*: 658–659 (1984).
279. P. G. Hoffman, M. C. Lego, and W. G. Galetto, *J. Agric. Food Chem., 31*: 1326–1330 (1983).
280. N. C. Rajpoot and V. S. Govindarajan, *JAOAC, 64*: 311–318 (1981).
281. M. Y. Khaled, M. R. Anderson, and H. M. McNair, *J. Chromatogr. Sci., 31*: 259–264 (1993).
282. H. Li, G. Lu, and J. Zhou, *Zhongcaoyao, 20*: 300–301 (1989).
283. A.B. Wood, M. L. Barrow, and D. J. James, *Flavour Fragrance J., 3*: 55–64 (1988).
284. K. M. Weaver, M. E. Neale, and A. Laneville, *JAOAC, 71*: 53–55 (1988).
285. T. Lupina and H. Cripps, *JAOAC, 70*: 112–113 (1987).
286. G. H. Chiang, *J. Food Sci., 51*: 499–503 (1986).
287. A. W. Archer, *J. Chromatogr., 351*: 595–598 (1986).
288. H. Glasl and M. Ihrig, *Dtsch. Lebensm. Rundsch., 80*: 111–113 (1984).
289. M. Rathnawathie and K. A. Buckle, *J. Chromatogr., 264*: 316–320 (1983).
290. L. R. Josimovic, *Glas. Hem. Drus. Beograd., 45*: 301–308 (1980).
291. M. Verzele and S. Qureshi, *Chromatographia, 13*: 241–243 (1980).
292. R. J. Houben and K. Brunt, *J. Chromatogr., 661*: 169–174 (1991).
293. D. T. Coxon and P. G. Jones, *J. Sci. Food Agric., 32*: 366–370 (1981).
294. K. R. Price, F. A. Mellon, R. Self, G. R. Fenwick, and S. F. Osman, *Biomed. Mass Spectrom., 12*: 79–85 (1985).
295. M. Friedman and C. E. Levin, *J. Agric. Food Chem., 40*: 2157–2163 (1992).
296. W. M. J. Van Gelder: *J. Chromatogr. 331*: (1985) 285–293.
297. R. J. Bushway, E. S. Barden, A. M. Wilson, and A. A. Bushway, *J. Food Sci., 45*: 1088–1089 (1980).
298. D. R. Lawson, W. A. Erb, and A. R. Miller, *J. Agric. Food Chem., 40*: 2186–2191 (1992).
299. R. A. Shelby and V. C. Kelley, *J. Agric. Food Chem., 38*: 1130–1134 (1990).
300. P. M. Scott and G. A. Lawrence, *J. Agric. Food Chem., 28*: 1258–1261 (1980).
301. E. R. Barnhart and B. Huff, eds., Physicians Desk Reference, 39th ed., Medical Economics, Oradell, NJ, (1985), p. 1415.
302. L. A. Lin, *J. Chromatogr., 632*: 69–78 (1993).
303. A. Yoshida, S. Yamazaki, and T. Sakai: *J. Chromatogr. 170*: 399–404 (1979).
304. R. A. Shelby and V. C. Kelley, *J. Agric. Food Chem., 40*: 1090–1092 (1992).
305. R. D. Plattner, S. G. Yates, and J. K. Porter, *J. Agric. Food Chem., 31*: 785–789 (1983).
306. C. Klug, W. Baltes, W. Kroenert, and R. Weber, *Z. Lebensm. Unters. Forsch., 179*: 245–246 (1984).
307. C. Klug, W. Baltes, W. Kroenert, and R. Weber, *Z. Lebensm, Unters. Forsch., 186*: 108–113 (1988).
308. M. L. Deinzer, B. L. Arbogast, D. R. Buhler, and P. R. Cheeke, *Anal. Chem., 54*: 1811–1814 (1982).
309. P. A. Dreifus, W. C. Brumley, J. A. Sphon, and E. A. Caress, *Anal. Chem., 55*: 1036–1040 (1983).
310. B. N. Greirson, D. G. Allen, N. F. Gare, and I. M. Watson, *J. Agric. Food Chem., 39*: 2327–2331 (1991).
311. R. J. Bushway, H. L. Hurst, J. Kugabalasooiar, and B. Perkins, *J. Chromatogr. 587*: 321–324 (1991).
312. J. Pothier, G. Petit-Paly, M. Montagu, N. Galand, J. C. Chenieux, M. Rideau, and C. Viel, *J. Planar Chromatogr. Mod. TLC, 3*: 356–358 (1990).
313. C. Marutoiu, L. Roman, V. Coman, P. Bodoga, and M. Vlassa, *Rev. Chim., 41*: 281–282 (1990).

314. J. E. Meeker and W. W. Kilgore, *J. Agric. Food Chem. 35*: 431–433 (1987).

315. F. W. McLafferty, *Biomed. Mass Spectrom.*, *8*: 446–448 (1981).

316. I. N. Papadoyannis, V. F. Samanidou, G. A. Theodoridis, G. S. Vasilikiotis, G. J. M. Van-Kempen, and G. M. Beelen, *J. Liq. Chromatogr.*, *16*: 975–998 (1993).

317. B. Poocharoen, J. F. Barbour, L. M. Libbey, and R. A. Scanlan, *J. Agric. Food Chem.*, *40*: 2216–2221 (1992).

318. S. Keller, T. Lochte, B. Dippel, and B. Schrader, *Fresenius' J. Anal. Chem.*, *346*: 863–867 (1993).

21

Phenolic Compounds

Hyoung S. Lee and Bill W. Widmer
Citrus Research and Education Center, Florida Department of Citrus, Lake Alfred, Florida

I. INTRODUCTION

Several reviews have been published which give excellent discussions of chromatography for phenolic analysis. Ribereau-Gayon (1) presented general principles for extraction and separation of phenolic compounds by paper, thin-layer, and column chromatography. Procedures for hydrolysis of bound forms and identification procedures based on R_f values and ultraviolet-visible (UV-Vis) absorption spectra were included. Van Sumere and co-workers (2) reviewed methods for separation of phenolics during the period 1965–1977. During this period analytical methods that were developed made use of the newer and more efficient chromatographic instruments utilizing the principles of gas chromatography (GC) and high-performance liquid chromatography (HPLC). GC applications for phenolics and related compounds such as simple phenols, phenolic aldehydes, phenolic acids, phenolic glycosides, phenolic amides, stilbenes, stilbene glycosides, lignins, coumarins, furanocoumarins, ellagic acid, and flavonoids in plants are well tabulated (2). Since the use of GC is limited to volatile samples, reagents used and the volatile derivatives formed for phenolic analysis are included. For HPLC analysis, the column adsorbants, mobile phase solvents, and phenolic compounds separated are presented in their review.

Hardin and Stutte (3) reviewed chromatographic analysis techniques of phenolic acids and flavonoid compounds. Their review includes early work utilizing paper chromatography from 1959 through use of the more advanced techniques of GC and HPLC up until 1980. Their review also includes procedures for separation of benzoic and cinnamic acid derivatives, isoflavones, flavanones, and flavonols by HPLC. Derivatization methods and stationary phases for GC analysis of benzoic, cinnamic, flavone, and isoflavones were also discussed. Herrmann (4) reviewed analysis of phenolic compounds, with emphasis on hydroxycinnamic acid derivatives, which had been carried out by HPLC. Extraction procedures and modern methods of extract purification through polyamide and anion exchange prior are described along with HPLC procedures for analysis of extracts. Their review also included a tabulation of phenolic acids and their amounts found in pome and stone fruit from 1980 to 1986. Harborne (5) discussed phenolic structure and listed general procedures from extraction to individual identification of plant phenolics. In the same book, Van Sumere (6) summarized the procedures for extraction, separation, and quantitative determination of simple phenols, phenolic acids, and related compounds.

Macheix et al. (7) published spectral properties of various phenolic compounds as well as procedures for extraction, purification, separation, and identification of phenolic compounds mainly occurring in fruits. This study was limited to analysis with HPLC since GC is relatively little used today for the analysis of phenolic compounds. In the recent symposium of phenolic compounds in foods, Khurana (8) reviewed HPLC resolution of the natural phenolic components such as anthocyanins, flavones, carotenoids, beet pigments, curcumins, mangiferin, and gingerolsis in foods. Several mechanisms such as reverse phase, normal phase, and anion exchange to resolve phenol and 1-naphthol on C-18, C-8, weak cation exchange, and amino phases were briefly described. Also, the analysis of phenolic components produced from degradation of natural products during food processing was discussed. For occurrences of hydroxycinnamic acid, hydroxybenzoic acids and derivatives in various foods were well documented by Reschke and Herrmann (9), Macheix and Fleuriet (10), and Herrmann (4).

For coumarins, Ibrahim and Barron (11) discussed methods used for isolation and characterization of coumarins and related compounds by chromatographic methods. Distribution of coumarins in citrus oils, antimicrobial activity against yeast and molds (12), and biogenesis and structural diversity of coumarins in the Rutaceae family have been reviewed (13). Also, discussion of potential value of coumarins as taxonomy markers and their possible functions is included.

Because flavonoids are commonly occurring components in foods and nonfood plants, they have been studied extensively. The structure and properties of many compounds have been worked out and are the subject of many books. The first book strictly devoted to flavonoid chemistry was published in 1962, edited by Geissman (14). Since then among the more significant contributions have been works by Harborne in 1967 with *Comparative Biochemistry of the Flavonoids* (15); Maybry et al. in 1970 with *The Systematic Identification of the Flavonoids* (16); and Harborne et al. in 1975 with a two-volume set entitled *The Flavonoids* (17). This latter volume has twice been updated, in 1982 and 1988, on the advances in flavonoid research since the previous publication (18,19). Two other publications worthy of mention are *Techniques of Flavonoid Identification* by Markham (20) and *Methods in Plant Biochemistry: Plant Phenolics* edited by Dey and Harborne (5). While these publications are not confined exclusively to flavonoids in food, the methods of analysis apply to plant materials in general and as such are applicable to plant food products.

Most of the previous work has documented the natural occurrences of phenolic compounds. Due to their chemical complexity in foods this chapter is limited mainly to the analysis of simple phenolics (such as hydroxycinnamic and hydroxybenzoic acids), coumarins, and polyphenols (flavonoids) in foods. However, discussions about anthocyanins are not inlcuded with flavonoids because this subject is covered in Chap. 23. Also, old references for applications in foods, which were already tabulated in previous works (2,4,6,8) in the context of detailed discussion of the topics, were omitted.

II. PROPERTIES AND OCCURRENCE

Phenolic compounds are a diverse class of chemicals containing a hydroxyl group on a benzene ring. In addition, true phenolics are defined as being derived from the shikimic acid or polyketide pathways. Some phenolics are complex molecules derived from the condensation of two or more components from both pathways. Phenolic compounds can be further divided into separate component classes listed in Table 1. The occurrence of phenolic compounds is widespread throughout nature but with considerable variation occurring between plant species. As a result their occurrence has been used to help determine taxonomic relationships.

Table 1   Phenolic Classes in Plants

Phenolic class	Carbon skeleton
Simple phenols, Benzoquinones	C_6
Phenolic acids	C_6-C_1
Acetophenones, phenylacetic acids	C_6-C_2
Hydroxycinnamic, phenylpropenes, coumarins, isocoumarins, chromones	C_6-C_3
Naphthoquinones	C_6-C_4
Xanthones	$C_6-C_1-C_6$
Stilbenes, anthraquinones	$C_6-C_2-C_6$
Flavonoids, isoflavonoids	$C_6-C_3-C_6$
Lignans, neolignans	$(C_6-C_3)_2$
Biflavonoids	$(C_6-C_3-C_6)_2$
Lignins	$(C_6-C_3)_n$
Condensed tannins	$(C_6-C_3-C_6)_n$

Source: Adapted from Ref. 5.

When analyzing phenolic compounds it is helpful to keep in mind that certain substitution patterns occur more frequently in nature than others. For example, 2-hydroxybenzoic (salicylic acid), protocatechuic (3,4-dihyroxybenzoic acid), vanillic (4-hydroxy-3-methoxybenzoic acid), and syringic acid (3,5-dimethoxy-4-hydroxybenzoic acid) are almost universal in their occurrence. Gallic (3,4,5-trihydroxybenzoic acid), salicylic, gentisic (2,5-dihydroxybenzoic acid), and 2,3-dihydroxybenzoic acids are of common occurrence; while 2-hydroxybenzoic acids with a methoxy group in the 4, 5, or 6 position occur rarely (21).

A.   Phenolic Acids

The term *phenolic acids* encompasses the seven carbon benzoic acids (C6–C1) and nine carbon cinnamic acids (C6–C3). *p*-Hydroxybenzoic acid, protocatechuic acid, vanillic acid, gallic acid, and syringic acid are the principal benzoic acids. Salicyclic acid and gentisic acid, which possess an OH group ortho to the carboxylic acid function, are also often found in plants. Gallic acid occurs as quinic acid esters and is often found in plants as its dimer, ellagic acid.

p-Coumaric, caffeic, ferulic, and sinapic acids are the most common cinnamic acids. Cinnamic acids, since they possess a double bond, are capable of existing in the two isomeric forms *cis*- and *trans*-cinnamic acid. The naturally occurring cinnamic acids have the more stable trans configuration. However, cis and trans isomers can undergo transformation into the other form under certain conditions.

Phenolic acids occur naturally in a wide range of bound forms. They usually occur in the form of esters rather than glycosides. They may be conjugated with organic acids, sugars, amino compounds, lipids, terpenoids, or other phenolics (22). The structures of the principal benzoic and cinnamic acids are presented in Figs. 1 and 2.

B.   Coumarins

The coumarins possess an oxygen heterocyclic of six atoms fused with a benzene ring. They are considered along with the cinnamic acids, since they also possess the C6–C3 configuration. Coumarins are lactones of *O*-hydroxycinnamic acid. The basic structure is as shown in Fig. 3.

Simple coumarins occur naturally as their β-*O*-D-glucosides and umbelliferone, esculetin,

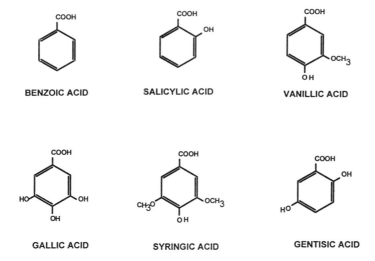

BENZOIC ACID   SALICYLIC ACID   VANILLIC ACID

GALLIC ACID   SYRINGIC ACID   GENTISIC ACID

Fig. 1 Structure of common benzoic acids.

scopoletin, daphnetin, and isofraxidin are the most common coumarins found in plants (Fig. 4). Besides simple coumarins, several natural products are known to have a furano- and pyranocoumarin, which is a furan or pyran ring fused with the benzene ring of a coumarin as shown in Fig. 3. Psoralen in cold-pressed citrus oils is one good example of a furanocoumarin. Coumarins have been reported in a large number of plants and are reported to be rich in citrus peel oil. Coumarins are essentially nonvolatile, found in cold-pressed citrus peel oils but not in distilled oils.

C. Flavonoids

The flavonoids are an important group of phenolic components occurring ubiquitously in plants. Flavonoids are C15 compounds arranged C6–C3–C6 with the central group usually linked with

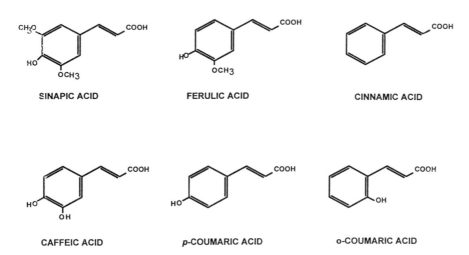

SINAPIC ACID   FERULIC ACID   CINNAMIC ACID

CAFFEIC ACID   p-COUMARIC ACID   o-COUMARIC ACID

Fig. 2 Structure of common cinnamic acids.

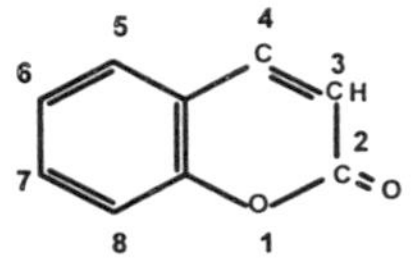

COUMARIN

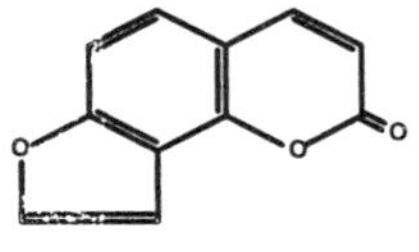

FURANOCOUMARIN

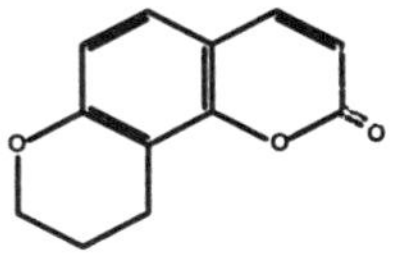

PYRANOCOUMARIN

Fig. 3 Basic structure of coumarins.

oxygen. The flavonoids can be further broken into classes shown in Fig. 5. These include the flavones, flavonols, flavanones, dihydroflavonols (flavanols), anthocyanins, chalcones, and iso-flavonoids. In addition, there are biflavonoids such as the theaflavins and thearubigins, which are a polymer of two or more flavanol molecules joined together. Flavonoids may be present as the aglycone or in a glycoside form bound to various sugars at a carbon (C-glycosides) or hydroxyl (O-glycosides). There are also flavonoids bound to other phenolics such as cinnamic acids.

III. IMPORTANCE IN FOODS

Phenolics are important because of their contribution to the sensory quality of fresh fruits and processed products including color, astringency, bitterness, and flavor. In particular, the phenolic

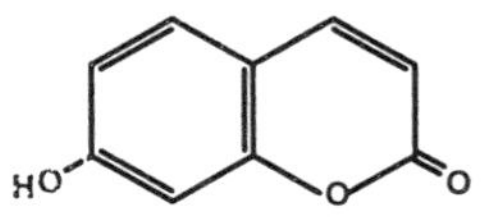

UMBELLIFERONE

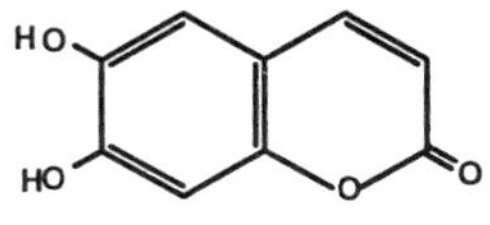

ESCULETIN

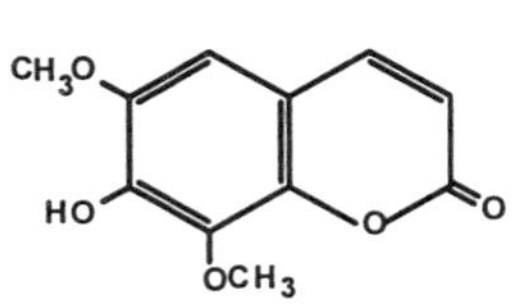

ISOFRAXIDIN

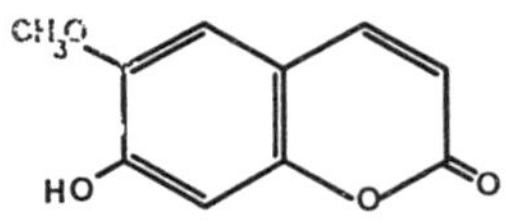

SCOPOLETIN

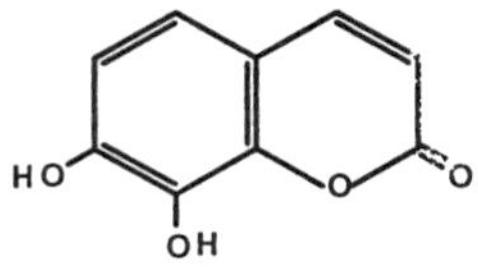

DAPHNETIN

Fig. 4 Structure of common coumarins.

Table 2   Structure of Flavonoids

Compound	Structure
Flavone	
Apigenin	5, 7, 4'-OH
Chrysin	5, 7-OH
Tectochrysin	5-OH; 7-OMe
Geraldone	7, 4'-OH; 3'-OMe
Acacetin	5, 7-OH; 4'-OMe
Genkwanin	5, 4'-OH; 7-OMe
Luteolin	5, 7, 3', 4'-OH
Scutellarein	5, 6, 7, 4'-OH
Isoscutellarein	5, 7, 8, 4'-OH
Diosmetin	5, 7, 3'-OH; 4'-OMe
Ladanein	5, 6-OH; 7, 4'-OMe
Salvigenin	5-OH; 6, 7, 4'-OMe
Tricin	5, 7, 4'-OH; 3',5'-OMe
Thymusin	5, 6, 4'-OH; 7, 8-OMe
Xanthomicrol	5, 4'-OH; 6, 7, 8-OMe
Pebrillin	5, 6-OH; 7, 8, 4'-OMe
Sorbifolin	5, 6-OH; 7, 3', 4'-OMe
Gardenin B	5-OH; 6, 7, 8, 4'-OMe
Gardenin D	5, 3'-OH; 6, 7, 8, 4'-OMe
Thymonin	5, 6, 4'-OH; 7, 8, 3'-OMe
Sideritoflavone	5, 3', 4'-OH; 6, 7, 8-OMe
Isosinensetin	5, 7, 8, 3', 4'-OMe
Sinensetin	5, 6, 7, 3', 4'-OMe
Nobiletin	5, 6, 7, 8, 3', 4'-OMe
Tangeretin	5, 6, 7, 8, 4'-OMe
Rhiofolin	Apigenin-7-*O*-neohesperidoside
Isorhiofolin	Apigenin-7-*O*-rutinoside
Diosmin	Diosmetin-7-*O*-rutinoside
Neodiosmin	Diosmetin-7-*O*-neohesperidoside
Vitexin	8-*C*-glucosylapigenin
Isovitexin	6-*C*-glucosylapigenin
Flavonol	
Galangin	3, 5, 7-OH
Kaempferol	3, 5, 7, 4'-OH
Quercetin	3, 5, 7, 3', 4'-OH
Isorhamnetin	Quercetin-3-OMe
Rhamnetin	Quercetin-7-OMe
Rutin	Quercetin-3-O-rutinoside
Myricetin	3, 5, 7, 3', 4', 5'-OH
Quercetagetin	3, 5, 6, 7, 3', 4'-OH
Gossypetin	3, 5, 7, 8, 3', 4'-OH
Heptamethoxyflavone	3, 5, 6, 7, 8, 3', 4'-OMe
Isoflavone	
Daidzein	7, 4'-OH
Genistein	5, 7, 4'-OH
Glisoflavone	5'-*C*-Isopentenyl; 7', 3', 4'-OH; 5-OMe

Chalcone

Butein	3, 4, 2', 4'-OH
Licochalcone B	3, 4, 4'-OH; 2-OMe
Phloridzin	a,b-Dihydro, 4, 4', 6'-OH; 2'-O-glucoside

Flavanones

Pinocembrin	5, 7-OH
Naringinin	5, 7, 4'-OH
Butin	7, 3', 4'-OH
Isosakuranetin	5, 7-OH; 4'-OMe
Eriodictyol	5, 7, 3', 4'-OH
Hesperitin	5, 7, 3'-OH; 4'-OMe
Eriocitrin	Eriodictyol-7-O-rutinoside
Neoeriocitrin	Eriodictyol-7-O-neohesperidoside
Narirutin	Naringinin-7-O-rutinoside
Naringin	Naringinin-7-O-neohesperidoside
Prunin	Naringinin-7-O-glucoside
Hesperidin	Hesperitin-7-O-rutinoside
Neohesperidin	Hesperitin-7-O-neohesperidoside
Didymin (neoponcirin)	Isosakuranetin-7-O-rutinoside
Poncirin	Isosakuranetin-7-O-neohesperidoside

Flavanol

Pinobanksin	3, 5, 7-OH
Dihydroquercetin	3, 5, 7, 3', 4'-OH
(+) catechin 2 (R)	3 (S), 5, 7, 3', 4'-OH
(-) epicatechin 2 (R)	3 (R), 5, 7, 3', 4'-OH
Gallocatechin	3 (R), 5, 7, 3', 4', 5'-OH
Epigallocatechin	3 (S), 5, 7, 3', 4', 5'-OH
Catechingallate	3-Galloylcatechin
Epicatechin gallate	3-Galloylepicatechin

Isoflavanone

Glycyrrhizoflavanone	5, 7, 3', 4'-OH

compounds such as hydroxycinnamic acids, catechin, procyanidins, flavonol glycosides, and anthocyanins in grapes contribute to the characteristic taste (astringency, bitterness), flavor, oxidative browning, and color changes during the aging of wine (23–25). Even free phenolic acids appear to be the precursor for vinyl phenols and off-flavors formed in citrus products during storage (22).

Since the qualitative and quantitative phenolic compositions are important factors in the browning potentials of apples and certain grape cultivars, it is highly desirable to know the changes that they undergo during maturation as well as the relative amounts of individual phenolic compounds in apples, grapes, and the phenolic distribution in flesh and skin (24,26). Furthermore, some of these compounds possess beneficial pharmacological properties and antibacteriological effects (27) and are used for therapeutic purposes (28). In citrus fruits, phenolic compounds continue to claim attention, not only for their remarkable taste properties but for their biological activities such as anticarcinogenic effects, antiinflammatory properties, and inhibitory activities against histamine release (29).

Coumarins are important because they are used to impart sweetness to perfumes, industrial deodorants (30), wine and brandies (31). Coumarin compounds are generally distributed throughout the citrus species and contribute to the quality of cold-pressed citrus oils. Citrus oils pressed directly from citrus peels are used in perfumery and for the flavoring of candies, soft drinks, and baked goods. Natural coumarins of citrus fruits also appear to play important roles as natural protective agents in citrus fruits. Coumarins were especially useful as taxonomic markers in *Citrus* species (32) to distinguish *Citrus* hybrids. They also have antifungal activity against serious postharvest decay organisms of citrus fruits (33). Numerous physiological activities of coumarins on both plants and animals have received increased attention. In particular, the possible toxic effects of coumarins on animals such as antibacterial activity, vasodilatory and diuretic effects, anticoagulant properties, hepatotoxicity, and respiratory stimulation have been attributed to coumarins (13). The relationship between structure and hypertensive activity of coumarins is also well defined (34).

Flavonoid content in foods is also of interest because on their potential impact of flavor, appearance, and keeping quality of a food. Structural information on some flavonoids encountered in foods is given in Table 2 and Fig. 5. While some flavonoids have no flavor impact, some are bitter or astringent. Naringin, a flavanone glycoside (Table 2), is one of two bitter components primarily responsible for the bitterness in grapefruit. Accurate methods to measure bitter substances in food and juice products are important so processing parameters can be manipulated to control and limit bitterness in these products. Naringin levels are carefully monitored during processing and blending of grapefruit juice to avoid production of a product with excessive bitterness. In the official rules of the State of Florida Department of Citrus (Section 20-64.003), grapefruit juice contains less than 600 ppm naringin. Compliance with the law relating to naringin content in grapefruit juices shall be determined by the Davis method, which is based on the colorimetric method described in the U.S. Department of Agriculture Handbook 6.15.1–6.15.2. Also, food control authorities in Germany suggest guide values of flavonoid glycosides, calculated as naringin, in grapefruit not to exceed 1500 mg/L. Narirutin, another 7-*O*-glycoside of naringinin that is not bitter, differs from naringin only in the disaccharide attached to the parent flavanone. Hesperidin, a tasteless flavanone glycoside, is a significant component of the cloud in lemon and orange juice. Naringin has the disaccharide neohesperidose (2-*O*-L-rhamnosyl-D-glucose) attached at the 7-*O*-hydroxyl position while narirutin contains rutinose (6-*O*-L-rhamnosyl-D-glucose) attached at the 7-*O*-hydroxyl position (35). Flavanone-7-*O*-neohesperidosides are bitter (Table 2: neoeriocitrin, neohesperidin, and poncirin), while the 7-*O*-rutinosides are not (eriocitrin, hesperidin, didymin). Citrus methoxylated flavones are another group of bitter components; however, the levels found in citrus juices are below their flavor thresholds. Flavanols in tea are important components contributing to the bitterness and astringency of tea (36).

Flavonoids in foods can also influence the appearance of food products. Chalcones are yellow pigments while anthocyanins can be orange, red, blue, or violet pigments and are responsible for the characteristic color of many fruits and their food products. Anthocyanins are responsible for the color in cranberry, cherry, raspberry, blackberry, grape, strawberry, radish, and cabbage, to name a few (37). Measurements of anthocyanins in some fruits and their products are used to help determine maturity and quality (38,39). Flavonoids can also have a positive or negative impact on appearance through oxidation and polymerization reactions which form pigments.

IV. SAMPLE PREPARATION

Analysis of phenolics in a raw or processed food matrix begins with extraction, unless one is only interested in a crude estimate of total phenolic content. The extraction procedure depends on the type of food to be analyzed, the phenolic compounds one is interested in measuring, and the analytical procedure to be used. A wide range of sample preparation techniques have been used to extract and purify the phenolic compounds for analysis in foods. However, the first step is almost

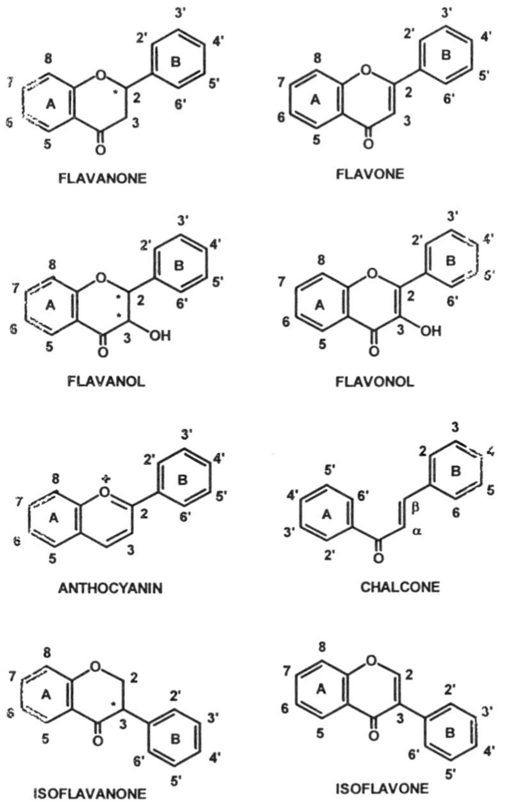

FLAVANONE

FLAVONE

FLAVANOL

FLAVONOL

ANTHOCYANIN

CHALCONE

ISOFLAVANONE

ISOFLAVONE

* chiral center

Fig. 5 Structure and substituent numbering of the different flavonoid classes.

always to first crush, mill, macerate, or grind the food to increase the sample surface area, allowing better contact of the extracting solvent with the sample. If only a portion of the sample is to be extracted, this also helps in mixing the sample to ensure that the extracted portion is representative of the entire sample. Since many phenolic compounds occur in a bound form as glycosides or esters, the sample preparation may include alkaline or acid hydrolysis to free the bound phenolics either before or after the solvent extraction step. The hydrolysis step is omitted if one wishes to measure phenolics in the bound forms.

Crude extracts can sometimes be analyzed as they are or, depending on the analytical technique, the extract often will require concentration to facilitate detection of minor components. After exhaustive extraction for a minor component, components may be present at a very dilute level. Concentration is almost always done at low temperatures and under reduced pressures to minimize the component degradation. Temperatures should be kept below 40°C since at higher temperatures various reactions such as hydrolysis, isomerization, internal redox, and polymerization are more likely to occur (41). Interferences can also arise from coelution of one or more components with those of interest. Therefore the extract may also require a purification step to eliminate components that may interfere with the analysis.

For GC analyses, nonvolatile components in an extract can quickly contaminate the injection port liner. These can then cause adsorption of components in subsequent injections. Adsorbed nonvolatile components may also be pyrolized to volatile components that elute and interfere with detection. Very few flavonoids have sufficient volatility without being labile; therefore samples must be derivatized with various silanizing agents to increase their volatility for analysis. Because of this limitation, HPLC is used much more frequently than GC for flavonoid analysis of food extracts.

A. Extraction

Hydrolysis followed by solvent extraction or direct extraction without hydrolysis are the main techniques used to prepare samples for analysis of phenolic acids. Hydrolysis of bound acids into free acids allows the determination of total phenolic acid content. The most common solvents used are ethyl acetate (28,40–42), diethyl ether (28,41), methanol or aqueous methanol (21,43–45). For liquid–liquid extractions, neither diethyl ether or ethyl actate is able to extract all the phenolic compounds completely, due to the differences in polarity between components (41). The free phenolic acids as well as their glycosides are normally known to be extracted with alcohol–water mixtures (46), but it is necessary to extract the sample several times or reflux for 1 hr to obtain satisfactory results. Ethyl acetate has a slightly better extraction rate for acids and aldehydes of low relative molecular mass. However, extraction reproducibility is greater with diethyl ether. Therefore, diethyl ether is considered more appropriate for extraction of phenolic acids from foods. In situations where one also wants to extract compounds of higher relative molecular mass, such as catechin, proanthocyanidins, and hydroxycinnamic acid esters, ethyl acetate is a more appropriate solvent.

Some phenolic acids are very sensitive to light and oxidation. They are subject to isomerization and degradation during processing and sample preparation. As in any extraction procedure of sensitive compounds, considerable precautions must be taken to prevent isomerization and oxidation of phenolic acids. This is particularly true of caffeoyltartaric acid (caftaric acid) (47,48). In the absence of antioxidants, there can be considerable degradation and formation of an enzymic oxidation product, s-gluthathionylcaftaric acid (49). Addition of 1% ascorbic acid (25,47,50,51), sodium metabisulfite (40,52–54), or cysteine (55) before blending, or immediate immersion in methanol (24) will help prevent oxidative changes during sample preparation of fruit and vegetable products.

The presence of a vinyl group in the side chain of cinnamic acids makes this class of phenolic acids susceptible to cis–trans isomerism changes by UV light (56). On exposure of the trans isomer (naturally occurring form) to UV radiation or daylight, an equilibrium mixture of the cis and trans configuration is formed in MeOH solutions. Conversion of *trans*-ferulic acid in wheat extracts stored for 48 hr, even when refrigerated and in the dark, have been observed (57). An acidic sample matrix can help to stabilize and prevent isomerization. The nondissociated acids have a higher stability than the acids in ionized form. Thus, it is recommended that samples be analyzed within 24 hr of extraction (57).

Polar flavonoids occurring in the form of glycosides or bound to some other polar acyl group (e.g., organic acid) can sometimes be extracted with hot water (46,58,59). Usually, however, organic solvents are necessary for complete extraction. The most common solvents used are methanol (MeOH), ethanol (EtOH), isopropanol (IPA), and acetone (Ace). For less polar aglycones and highly methoxylated aglycones, which frequently are confined to the exterior surfaces of fruits and vegetables, less polar solvents such as the chlorinated hydrocarbons (CH_2Cl_2 and $CHCl_3$), benzene (Bz), diethyl ether (Et_2O), and ethyl acetate (EtOAc) are used. Sometimes the apolar components can be extracted efficiently by just dipping the fruit or plant part in solvent for several minutes.

For fruit juices containing anthocyanin pigments, sodium bisulfite ($Na_2S_2O_2$) can be added before extraction to discolor anthocyanin monomers occurring in the juice and keep them in the aqueous phase (40). In wine and pigmented fruit juices, anthocyanins and tannins both absorb at 280 nm and can interfere with other phenolics during chromatographic analysis. They can also irreversibly adsorb to the stationary phase under some conditions through polymerization and quickly destroy an expensive column. For these reasons they should be eliminated prior to analysis of the other phenolics. To accomplish this in extracts of bilberry juice, Azar and co-workers (40) extracted 50 ml of juice six times with 50 ml of EtOAc following addition of sodium bisulfite. Extracts were then dried over anhydrous Na_2SO_4, concentrated in a rotary evaporator under vacuum at 30°C, redissolved in MeOH, and stored at 0°C under nitrogen until analyzed.

Salagoity-Auguste and Bertrand (42) fractionated the phenolic compounds in wine into neutral (flavonoid) and acidic fractions before HPLC analysis. Fractionation was based on the fact that phenolic acids are completely ionized at pH 7.0 and nondissociated at pH 2.0; thus neutral phenolics can be extracted at pH 7.0 while nondissociated phenolic acids can be extracted at pH 2.0 after removal of the neutral fraction. This method is known to have very low efficiency for procyanidins (47), however, and has been adopted in many studies in order to avoid interferences between phenolic compounds. For example, a simple extraction procedure to obtain both phenolic acids and neutral phenolics such as catechins, procyanidins, and flavonols from apple juices (60) is described. Apple juice (10 ml) is adjusted to pH 7.0 using 0.1 M NaOH, extracted three times with 10 ml EtOAc, the extracts pooled and centrifuged at $100g$ for 10 min. The top EtOAc layer contains the neutral phenolics. Phenolic acids and their derivatives remain in the aqueous phase. The aqueous portions are combined and adjusted to pH 2 with HCl (0.1 M). The nondissociated phenolic acids are then extracted with EtOAC the same as the neutral fraction.

When phenolics are isolated from beverages containing alcohol, such as beer or wines, it is first necessary to remove the ethanol by degassing (61) or evaporation (23,27). The presence of alcohol when extracting with another organic solvent such as EtOAc, ether, or chlorinated solvents will lower the extraction efficiency by changing the partitioning characteristics of the solvent, resulting in an incomplete extraction and/or interfering compounds being extracted along with the phenolics of interest. However, care must be taken when the alcohol is removed to minimize incorporation of heat and air, which could degrade certain phenolic compounds such as flavanols, anthocyanins, and some phenolic acids. Degassing can be obtained by simply transferring the beer 25 times (61).

In some circumstances it is requisite to remove nonpolar substances such as fatty substances and pigments (chlorophyll, carotenoids). These components can frequently adsorb irreversibly to the column packing materials under the conditions of phenolic analysis and quickly ruin the column. Removal of interfering waxes and lipids can usually be accomplished by first dipping the fruit or plant in hexane, since most flavonoids have a very limited solubility in this solvent. Often a liquid sample or crude extract is partitioned with hexane, isooctane, or petroleum ether to remove lipids, carotenoids, and other nonpolar substances (21,43,52,58,61–73). For the aglycones and glycosides of flavanones, flavonols, and proanthocyanins, the extract may alternatively be washed with diethyl ether (74,75) or CH_2Cl_2 (76–78) to remove any lipids and carotenoids.

B. Hydrolysis

Since phenolic acids usually occur in conjugated forms and are seldom found in the free state, the determination of the free phenolic acids of foods necessitates hydrolysis of the phenolics to release any bound phenolics, whereas the hydrolysis step is omitted when the separation and determination of the different phenolic esters in foods are required. Both acid or alkaline hydrolysis are commonly used for the determination of phenolics. Enzymatic hydrolysis with β-glucosidase was also applied (79). β-Glucosidase will hydrolyze most classes of phenolic glucosides, except for the anthocyanins. For identification of hydrocinnamoyl derivatives, hydrocinnamoylquinatesterase (HCQE) was applied to hydrolyze the phenolic esters (67).

Acid hydrolysis, which has the purpose of rupturing glycosidic linkages, can be performed by heating in a water bath at 100°C with 2 N HCl for 2 hr or more (43). Hydrolysis of benzoic and cinnamic acid esters with alkali can be performed with 2 N NaOH at room temperature for 4–24 hr (21,22,80–83) or for 90 min at 60°C (43) under N$_2$. Optimum hydrolysis can be achieved by holding for 6 hr at room temp in 2 N NaOH (82). An acidification step is necessary following alkaline hydrolysis as many phenolic compounds are readily oxidized in alkaline solutions. It is also important that alkaline hydrolysis reactions be carried out in an inert atmosphere (N$_2$) to avoid oxidation.

A typical alkaline hydrolysis procedure for fruits and fruit juices can be followed (22). For fruit samples, 1 kg of fruit was extracted with H$_2$O/EtOH/MeOH (2:1:1) in a blender for 5 min, left to stand for 60 min, filtered, centrifuged, hydrolyzed (2 M NaOH, 24 hr at room temperature), acidified, extracted three times with 300 ml EtOAc, and concentrated. For fruit juice, 250 ml of juice was hydrolyzed (2 M NaOH, under N$_2$ gas, 4 hr, room temperature), and then acidified, centrifuged, and extracted two times with 400 ml EtOAc.

To obtain the aglycones from the cooled solution, diethyl ether and ethyl acetate are both known to be suitable solvents (6). Those concentrated EtOAc or ether extracts should be redissolved in a small volume of methanol (40,80,83), in aqueous methanol (28), or in water (84) for further analysis by reverse phase HPLC (RP-HPLC).

Some losses of caffeic acid can be expected due to the reactive nature of the O-dihydroxyphenols. The recoveries of phenolic acids extracted from juice were as follows: caffeic 61%, coumaric 54%, ferulic 55%, and sinapic 57% (22). Hydrolysis with hydrochloric acid (1–2 hr in 1 N HCl) yields poor recovery of phenolic acids, and ferulic acid was decomposed into a substance with a shorter retention time on HPLC, whereas treatment with NaOH led to a good recovery of ferulic acid (82).

C. Sample Cleanup

Sample cleanup procedures are generally now being used to cleanup extracts prior to further analysis by HPLC or GC. The cleanup stage is a critical part of a method, removing potential interfering components, and necessarily varies according to the type of food matrix to be analyzed. These include liquid–liquid partitioning with a nonmiscible solvent and open column chromatography on Sephadex LH-20, polyamide, Amberlite XAD-2, prep-HPLC, and solid phase extraction using commercially available disposable cartridges.

The most important factors to consider in selecting the cleanup adsorbant are efficiency, recovery, and contamination. More than 90% of flavonoids tested were recovered from Amberlite XAD-2 by eluting with MeOH (45). Cleanups on charcoal and polyamide have been found to lead to irreversible adsorption and poor recoveries (85). The cleanup and elution solvents necessary for analysis of specific foods will also vary with the food and analyte to be analyzed. This is due to the fact that each food contains a unique matrix of components along with those of interest.

For example, fruit jams (45) containing large amounts of sugars and pectins, are a difficult matrix from which to extract phenolic compounds. Sample cleanup methods based on adsorption by nonionic polymeric adsorbants such as Amberlite XAD-2 resin have been found to work best in isolation of phenolic components from the sugars and other polar compounds present (86,87). One successful method using methanol extracts of fruit jams (blend jams with 80% MeOH for 18 hr) first calls for redissolving the extract in water acidified to pH 2.0. The extract is then loaded onto a column (40 × 2 cm) packed with Amberlite XAD-2 resin. Flavonoid compounds remain in the XAD-2 resin column (40 × 2 cm) while sugars, pectins, and other polar compounds elute with the aqueous solvent. The flavonoid fraction can be then eluted with 300 ml methanol after washing the column with 100 ml water (pH 2.0) and 300 ml water (pH 7.0) (45).

For honey (88), a two-step cleanup is necessary, involving open column chromatography on Amberlite XAD-2 to retain the various phenolic compounds while the sugars and other polar compounds elute with the aqueous solvent, followed by further purification of the flavonoids by preparative column chromatography on Sephadex LH-20 column. For fruit products of orange, grapefruit, grape, and lemon, ionic adsorbants such as polyamides (89), polyvinylpyrrolidone (PVP) (85), and Sephadex LH-20 (26,50) were used to isolate and purify the various phenolic compounds. Besides open column preparative chromatography, for biliberry juice, semipreparative HPLC on C18 bonded phase (LiChrosorb RP-18, 7 μm) prior to the analytical HPLC can be applied (40).

Another approach for efficient sample cleanup is solid phase extraction using small, disposable cartridges. Since the introduction of disposable solid phase extraction (SPE) cartridges (small packed chromatography columns) with HPLC packings, SPE has become the preferred method for cleanup of crude extracts prior to analysis. The full range of silica-based polar and nonpolar stationary phases in small cartridges is commercially available and solid phase extraction on C18-bonded phase is mostly used for isolation of phenolics (23–25,51,74,83), replacing the use of PVP or Sephadex for purification steps. Solid phase extraction on disposable cartridge is proving to be an invaluable technique for cleaning samples prior to the phenolic analysis by HPLC in beer (90), wine (23,27,91), fruits (24,25), and vegetables (71). For example, fractionation of phenolic acids from neutral phenolic compounds (such as isoflavonoids) in soybean products prior to HPLC analysis can be achieved by the use of small, disposable, prepacked cartridges. This method (82) includes ethanol extraction of phenolics, mild hydrolysis of phenolic acid esters, and separation of phenolic acids from isoflavonoids by C18 Sep-Pak. The C18 Sep-Pak cartridges were preconditioned with 5 ml MeOH/water for neutral, and further preconditioning with additional 5 ml 0.01 N HCl for acid.

The fractionation of grape phenolics into acidic and neutral groups can be accomplished by passing deproteinated grape juice through a preconditioned C18 Sep-Pak cartridge (24,47,50). A typical preconditioning of C18 Sep-Pak cartridges can be done by sequentially passing 2 ml, dropwise, of methanol and distilled water for neutral Sep-Pak and passing 2 ml of 0.01N instead of distilled water for acidic Sep-Pak. For fractionation, 5 ml of deproteinated juice was adjusted to pH 7.0 with NaOH and passed through the preconditioned neutral Sep-Pak to absorb the neutral phenolic compounds. The effluent portion was then adjusted to pH 2.5 with HCl and passed through a second acidic Sep-Pak to absorb acidic phenolics. The absorbed fractions were eluted with methanol from their respective cartridges.

This technique can also provide selective extraction of phenolic compounds other than anthocyanins in red (Concord and de Chaunac) grapes (25) by preconditioning the C18 cartridge sequentially with 5 ml EtOAc, 5 ml MeOH, and 5 ml 0.01 N HCl, and then phenolic compounds were selectively eluted with 20 ml EtOAc. The recoveries were 98–99% for anthocyanins; nearly 100% (97.1–103.3%) for caffeoyl tartrate; 86–91% for procyanidin B3; and 77% for rutin, which showed lower recovery than the other two phenolics, possibly due to low solubility in EtOAc and low concentration in the grapes.

Separation of flavonols from other phenolic compounds with a high recovery can be achieved by further modification (23). Four fractions including phenolic acid (fraction I), procyanidins, catechins, and anthocyanin monomers (fraction II), flavonols (fraction III), and anthocyanin monomers (fraction IV) are obtained by passing wine through a preconditioned C18 Sep-Pak cartridge and eluting sequentially with methanol (fraction I), 16% acetonitrile at pH 2.0 (fraction II), ethyl acetate (fraction III), and methanol (fraction IV).

Sendra and co-workers (92) made a 40-fold concentrated extract of methoxyflavones from orange and mandarin juices utilizing commercially available C18 Sep-Paks (Waters, Milford,

MA). The Sep-Pak was conditioned with acetonitrile followed by water prior to addition of a 20-ml sample. Polar components were washed off with water followed by 30% aqueous acetonitrile (ACN). Elution of the methoxyflavone extract was then accomplished with 5 ml 50% aqueous ACN. Prior to analysis, the extract was concentrated and redissolved in 95% EtOH, filtered, and analyzed by gradient HPLC (92). The separation of seven citrus methoxyflavones is shown in Fig. 6. C18 SPE procedures to isolate flavanoids from grape (24,25,50), apple (67,93,94), grapefruit (95), and vegetables (68,71) have also been published.

V. ANALYSIS OF PHENOLIC COMPOUNDS

Due to the large number and the chemical complexity of phenolic compounds in plants, the analytical procedures for individual phenolic compounds have been relatively difficult and complicated. Analysis for phenolic compounds in food products can vary from simple colorimetric tests for detection to use of sophisticated instrumentation for separation, quantitation, and characterization of individual components.

A. Colorimetric Method

Colored phenolics such as anthocyanins are the red, blue, and purple pigmented constituents and therefore detection is usually easy. Treatment with ammonia will change anthocyanins from red to yellow and chalcone and aurone pigments will change from orange to red; no color change means that only carotenoid pigments are present. Ammonia or addition of strong alkali will cause color development or a color change in most phenolic compounds. Colorless phenolic compounds can usually be detected in extracts by their UV absorbance as well. Other tests causing a color response with all or most phenolics include the Folin reagents (96,97), Gibbs reagent (98), diazonium salts (99,100), vanillin (101), potassium ferricyanide–ferric chloride (102), titanium chloride (103), and Prussian blue (104). The advantages and disadvantages of measuring total phenolics by various methods were given a critical examination by Swain and Goldstein (105). Their recommendation of Folin reagent for the determination of the total concentration of phenolic

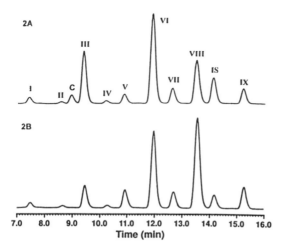

Fig. 6 HPLC separation of orange (a) and mandarin (b) methoxylated flavones by gradient elution. Conditions are given in the text. Peaks: I = isosinesetin, II = hexamethoxygossypetin, III = sinensetin, (IV = tetramethoxyisoscutellarein, V = hexamethoxyquercetagetin, VI = nobiletin, VII = tetramethoxyscutellarein, VIII = heptamethoxyflavone, IX = tangeretin. (From Ref. 92.)

compounds is still in use today, but the Folin–Denis reagent was replaced by the Folin–Ciocalteau reagent. They also recommended vanillin where catechins and proanthocyanidins predominate.

As an example of total phenolic measurement of grape juice, 100 µl of grape juice was filtered (0.45 µm), mixed with 900 µl of distilled water, 5 ml of 0.2 N of Folin-Ciocalteau reagent from Sigma, and 4 ml of saturated sodium carbonated solution (75 g/L), and then the absorbance at 765 nm was measured after 2 hr of reaction (106). Quantitation was based on the standard curve of gallic acid and total phenolic content expressed as mg of gallic acid per liter. For total phenolics and flavanols in *citrus flavedo* (107), *citrus flavedo* tissue (1 g) was homogenized in 20 ml warm MeOH for 5 min, stirred for 60 min, and filtered. The filtrates were used for the determination of total phenols by the Folin–Denis method and flavanols by the vanillin reagent (101).

The method using titanium chloride was developed after this and reportedly gives results similar to those of Folin reagent. The Prussian blue method developed by Budino et al. (104) was developed still later and reportedly is 3 times more sensitive than the titanium chloride method and 20 times more sensitive than vanillin reagent. A color test specific for lignins involves addition of phloroglucinol and hydrochloric acid. Development of a red color indicates that lignin is present in an extract or tissue.

For total coumarins, coumarins were extracted by homogenizing 1 g of citrus flavedo tissue in 20 ml of MeOH/CHCl$_2$/H$_2$O (2:1:0.8) solution for 5 min with a ultraturrax, filtered, and diluted with CHCl$_3$ and H$_2$O. The CHCl$_3$ phase was removed and benzene (1:10 vol) was added. The extract was evaporated to dryness and the residue dissolved in CHCl$_3$. Then coumarin derivatives were quantified by measuring the absorbance at 340 nm (107).

B. Enzymatic Method

Various enzymatic assays using phenol hydroxylase, tyrosinase, or polyphenol oxidase as well as biosensors (oxygen electrode, carbon electrode, or chemically modified electrode) have been reported for phenolic compounds, but no practical application of these methods has gained acceptance to date. Recently, enzymatic assay (108) using fungal laccase for total phenolic content of natural beverages showed linear correlation with the Folin-Ciocalteau method. An enzymic assay for total phenolic is based on measurement of the laccase oxygen demand, i.e., either monitoring the fall in the oxygen tension after addition of enzyme to a diluted juice sample in an endpoint batch assay or monitoring the current due to reduction of products after injection of sample into a flow injection analysis (FIA). Enzymatic methods can provide accurate endpoint measurement, but absolute values may vary with fruits and possibly with cultivar and may suffer from interferences.

C. Paper Chromatography

The classical chromatography techniques used to isolate and separate phenolic constituents in food are preparative paper chromatography (PC), thin-layer chromatography (TLC), and column chromatography (CC) on cellulose, silica, or polyamide phase. In the earliest method, paper chromatography, a sample is placed as a spot on cellulose paper and a mobile phase passed over it. To improve resolution between components the paper was often developed in two dimensions. Mixtures of butanol/acetic acid/water (BuOH/HOAc/H$_2$O) or tertiary butanol (TBuOH)/ HOAc/H$_2$O were commonly used in one dimension followed by development with acetic acid at a perpendicular angle. While testing for phenolics in plant materials and crude extracts with color reactions is useful as an initial screen to obtain an estimation of amounts present, they are susceptible to interferences from other components present in the matrix. It is always preferable to carry out some form of separation prior to detection. One simple and inexpensive test used by

Roux et al. (109) to determine the component classes present is two-dimensional paper chromatography (2D-PC).

The spotted paper is developed in a dimension with n-butanol/acetic acid/water (4:1:5) followed by development in 15% aqueous acetic acid perpendicular to the first development. Spots may be detected with short- and long-wave UV light with further information gained by fuming with ammonia. While this technique has the advantage of being very inexpensive, development times are long and the technique has largely been replaced by more efficient techniques as they were developed, such as TLC, CC, HPLC, and capillary electrophoresis (CE). However, due to its simplicity, very low cost, and effectiveness, PC is still in occasional use today. A few examples of recent PC applications of food phenolics are summarized in Table 3.

Paper electrophoresis has served as a very useful complementary technique to paper chromatography. Differences in the degree of ionization of the phenolic hydroxyl groups or other functional groups in the molecule can cause variations in the electrophoretic mobility (5). Ionization requires use of alkaline buffers, and paper electrophoresis in acetate buffer at pH 5.2 was useful for separating phenolic acids from simple phenols, which are immobile under this condition (110).

D. Thin-Layer Chromatography

Thin-layer chromatography (TLC) separations of flavonoids on cellulose can be performed in the same manner as paper chromatography. The principles of separation in TLC are similar to paper chromatography. The main difference is that a sheet of paper is used for the inert phase for paper chromatography. Although the costs of TLC are slightly higher than PC, commercially available plates on plastic and glass backings with much more uniform layer than paper are also available coated with other adsorbants, making TLC more versatile than PC. Separations performed by TLC have the convenience and simplicity of PC but have faster development times and better resolution. Available adsorbants are the same as those available for column and high-performance liquid chromatography. They include silica and polyamide for normal phase separations and cyanopropyl, octyl, and octadecyl reverse phase materials. Plates are also available in several thicknesses for preparative or analytical work. For this reason TLC is still in common use for preparative separations where only small amounts are needed and as a rapid and low-cost screening method to determine the flavonoid or phenolic classes present (Table 4).

There are also advantages to screening an extract by TLC prior to HPLC. One is detection of contaminants that may irreversibly adsorb and ruin the stationary phase of an expensive column. Solvent conditions necessary for a successful separation may be determined rapidly through simultaneous development of several plates in different solvents. The component classes present may also be rapidly determined by running several plates simultaneously and through the use of selective spray reagents that are available.

There are few examples of TLC applications for quantitative analysis of phenolic acids in the recent literature (25,40,50,62,71). Generally, TLC of phenolics is carried out by normal phase chromatography on cellulose or silica layers eluted with a mixture of hydrocarbon carrier (toluene, dioxane, or benzene) and polar organic modifier (acetone, butanol, ethanol, or acetic acid). Two-dimensional TLC can separate free phenolic acids according to the nature of substituents on the phenolic ring as well as separating glucose esters from glycosides and quinic esters of phenolic acid. For example, normal phase TLC with BuOH/pyridine/H_2O (14:3:3) on cellulose plate (0.1 mm) allowed separation of glucose esters ($0.55 < R_f < 0.75$) from glycosides and quinic esters ($0.1 < R_f < 0.3$) of phenolic acids extracted from bilberry juice (40). For coumarins, the solvent systems of chloroform/methanol (97:3) and butanol/acetic acid/water (40:12.5:29.5) were the most effective eluting solvent systems for separation of most of 30 known coumarins on silica gel (30).

Table 3 Paper Chromatography of Phenolics in Foods

Component	Source	Extraction	Analysis method	Detector	Ref.
Flavonol glycosides of quercetin, isorhamnetin, kaempferol, myricetin, apigenin	Bee pollen	MeOH, Hex, SPE on Sephadex LH-20	Whatman No. 1 (2D) BuOH/HOAc/H2O (4:1:5) and HOAc	UV	250
Flavonol glycosides	Tea, black	EtOH or MeOH	Whatman No. 1 (2D) BuOH/HOAc/H2O and 2% HOAc	UV 275 & 374 nm	205
Flavonol: myricetin, kaempferol, quercetin, rutin	Vegetable: fresh and pickled *Allium cepa* L. (red and white onion), *Cucumis sativus* L (cucumber), *Capsicum annuum* L. (yellow pepper), *Brassica oleracea* L (cauliflower), *Dqucus carota* L. (carrot), *Curcurbita pepo* L. (squash) Tea:consisting of *Thea sinensis*, *Matricaria chamonilla* (camomile), *Hibiscus sabdariffa* L., *Mentha piperita* L. (peppermint), *Malva sylvestris* L. (mallow), *Salvia triloba*, *S. officinalis* L. *S. sclarea* L. (sage)	MeOH, Et2O	Whatman No. 1 and Whatman 3MM, (1 & 2D) TBOH/HOAc/H2O (3:1:1) and Bz/HOAc/H2O (6:7:3), addition of NaOMe, AlCl3 +HCl, NaOAc, or NaOAc+H3BO3	UV 366 nm with NH4	59
Flavone: luteolin	Vegetable: peanut hull	Hex, CHCl3, Ace, alcohol	Silica gel Bz/HCO2Et/HCO2H (70:25:5)	UV 254	70

BuOH, butanol; Bz, benzene; Et2O, diethyl ether; EtOH, ethanol; Hex, hexane; HOAc, acetic acid; MeOH, methanol; NaOMe, sodium methoxide; NaOAc, sodium acetate; SPE, solid phase extraction; TBOH, tertiary butyl alcohol; UV, ultraviolet.

Table 4 Thin-Layer Chromatography of Phenolics in Foods

Component	Source	Extraction	Analysis method	Detector	Ref.
p-Hydroxybenzoic, o-pyrocatechuic, α-, β-, & γ-resorcylic, protocatechuic, gallic, isovanillic, vanillic, syringic, o-, m-, & p-coumaric, caffeic, ferulic, sinapic	Avocado (Persea americana Mill.)	Freeze-dried, grind, MeOH extraction, N_2, alkali hydrolysis (2 N NaOH, 24 hr, N_2 at room temp.), acidify to pH 2.0	Silica gel 1) Bz/MeOH/HOAc (90:16:8) 2 Bz/MeOH/HOAc (35:4:2)	Fluorescence	21
Ferulic, p-coumaric, sinapic	Bean, navy (Phaseolus vulgaris, var. seafarer)	Hexane extraction, alkaline hydrolysis (0.5 N NaOH, 20°C × 16 hr, N_2), acidify to pH 2.0, Et_2O/EtOAc (1:1) extraction	Silica gel & cellulose 1 $CHCl_3$/HOAc/H_2O (4:1:1) 2) toluene/HOAc/ (9:1) 3) 5% HOAc	UV light 254 & 366 nm NH_3 & I_2 vapor	62
Caffeic, ferulic, chlorogenic, gallic, syringic, protocatechuic, p-, m-, & o-coumaric, vanillic, p- & m-hydroxybenzoic acids	Bilberry juice (Vaccinium myrtillus)	EtOAc + $NaHSO_3$ extraction, conc. at 30°C	Cellulose 1) 2% HOAC in H_2O 2) HOAc/H_2O/Bz (22:1.2:60) 3) BuOH/pyridine/H_2O (14:3:3)	UV light spray with $C_6H_6N_2O_2$, 15% $NaHCO_3$	40
Scoparone, umbelliferone, scopoletin	Citrus fruits	40% EtOH or EtOAc extraction, filter, conc., dissolve with EtOH/MeOH (1:1)	Silica gel EtOAc/$CHCl_3$ (60:40)	FL 340/425 nm	33
7-Hydroxycoumarin derivatives	Citrus juice oil (Citrus hassaku)	Steam distillation, crystallization, CC on silica gel & Sephadex LH-20, TLC	Silica gel 30% Et_2O in CH_2Cl_2	NMR, UV, IR	181
Flavanol: catechin, epicatechin, epicatechin polymers	Chocolate, cocoa (Theobroma cacao)	CH_2Cl_2, EtOAc, SPE Sephadex LH-20	Silica gel 1) toluene/Et_2O/HCO_2H (3:6:1) 2) $CHCl_3$/Et_2O/HCO_2H (1:7:1)	vanillin-HCl MS, ^{1}H NMR C^{13} NMR	78

Compounds	Source	Extraction	Chromatography	Detection	Ref
Scopoletin, isoarnottinin, 8-hydroxy-5-methoxypsoralen, apiumetin, dihydroxanthyletin, lomatin, ammirol, 8-(2"-acetoxy-3"-hydroxy-3"-methylbutoxy) psoralen	Fruit: (*Ammi majus*)	Hot EtOH, 10% KOH × 1 hr, acidification, CHCl$_3$ extraction, CC on silica gel & Sephadex LH-20	Silica gel (Si-60F-254)	NMR, UV, IR, MS	180
Flavanols: catechin, epicatechin Flavonol: myricetin, quercetin, kaempferol, flavonol glycosides (14)	Fruit: apple, apricot, grape, orange, peach, pear, pineapple (juice)	ether, EtOAc	Cellulose MN300. 2% HCO$_2$H & IPA/NaOH/H$_2$O (8:1:1)	UV	75
Flavanols: catechin, epicatechin, catechingallate	Fruit: grape, white and peach	EtOH, C18 SPE, CC on Sephadex LH–20	Baker Si 250FC (HPTLC) 1) toluene/Ace/HCO$_2$H (30:30:10) 2) toluene/EtOAc/HCO$_2$H (50:40:10)	UV, Folin–Ciocalteau reagent	25 24 50 170
Hydrocinnamic tartrate	Grape, white (5 cultivars)	Blend with 1% AsA, add EtOH to remove protein, SPE on C18, CC on Sephadex LH–20	Silica gel (TLC) or Baker Si-250F (HPTLC) 1) toluene/Ace/HCO$_2$H (30:30:10) 2) toluene/EtOAc/HCO$_2$H (50:40:10)	UV light, fluorescence, & Folin–Ciocalteau	24
Tartaric esters of hydrocinnamic acids	Grape, white (21 cultivars)	Blend with 1% AsA, add EtOH to remove protein, EtOH extraction, SPE on C18	Silica gel (TLC) or Baker Si-250F (HPTLC) 1) toluene/Ace/HCO$_2$H (30:30:10) 2) toluene/EtOAc/HCO$_2$H (50:40:10)	UV light, fluorescence, & Folin–Ciocalteau	50
Flavanone: 8-Isopentenylnaringenin, sophoraflavanone	Herb: Hop (*Humulus lupulus* L.)	Et$_2$O, 10% Na$_2$CO$_3$	Dowex 1-X4, 80% MeOH	^{1}H NMR, MS	236

Table 4 (*Continued*)

Component	Source	Extraction	Analysis method	Detector	Ref.
Flavone, chalcone	Herb: licorice (*Glycyrrhiza eurycarpa*)	Hex, Bz, Et$_2$O	Silica gel 1) Bz:Et$_2$O (15:1), (3:1), & (1L1) 2) CHCl$_3$/EtOAc (7:1) 3) Bz/H$_2$O/Et$_2$O (97:3)	UV, MS, ^{1}H NMR, 13c NMR	65
Isoflavan: kanzonols	Herb: licorice (*Glycyrrhiza uralensis* & *G. glabra*)	Bz & Hex	Silica gel 1) Bz/Et$_2$O (5:1), & (20:1) 2) Hex/Ace (4:1) 3) CHCl$_3$/Hex (1:1) 4) CHCL$_3$/EtOAc (20:1) 5) toluene/Ace/HCO$_2$H (36:6:1)	UV, MS, ^{1}H NMR 13c NMR	64
Flavone: glisoflavone, glycyrrhisoflavone, Flavonol: kaempferol 3-*O*-methyl ether	Herb: licorice (*Glycyrrhiza*)	Hex and EtOAc	Kieselgel PF254, 1) CHCl$_3$/Ace/HCO$_2$H 2) CHCl$_3$/MeOH 3) Hex/EtOAc	UV, FT-IR, MS, ^{1}H NMR, and 13c NMR	262 66
Flavonol and isoflavone	Herb: licorice (*Glycyrrhiza*)	EtOAc	Silica, solvent not given	MS, UV, ^{1}H NMR	243
Flavone: (14) Flavone: pinobanksin, pinocembrin	Honey	Acidified pH 2.0, SPE on XAD-2	Silica, toluene/HOAc (4:1)	UV 360 nm	88

Compounds	Source	Sample preparation	Chromatography	Detection	Ref
Flavone: galangin, chrysin, tectochrysin; Flavonol: quercetin, kaempferol; Flavonone: pinocembrin, pinobanksin	Honey	Hex, EtOAc, acid hydrolysis	RP-18 F$_{254s}$ 0.25 mm (Merck), 1 M HOAc in 60% MeOH, 1% diphenylboric acid–ethanolamine spray	UV fluorescence	69
Gallic, protocatechuic, PHBA, syringic, vanillic, caffeic, p-coumaric, & ferulic	Palm products (*Phoenix cactilifera*) dates & legmi	Crush, acidify to pH 2.0, Et$_2$O extraction	Silica gel (Si-60F-254) 1) Bz/dioxane/HOAc (90:25:4) 2) Bz/MeOH/HOAc (45:8:3)	UV, 254/366 nm, FeCl$_3$	111
Ellagic, tannic, gallic, syringic, caffeic, ferulic, cinnamic, gentisic, salicyclic, resorcyclic, sinnapic, PHBA, vanillic, benzoic derivatives	Standards	—	Silica gel (Si-60F-254) Pentane-ACN/Et$_2$O (4:1) + 1.25% HCO$_2$H	UV	251
Gallic, chlorogenic; Flavanol: catechin, epicatechin, epigallocatechin	Vegetable: edible corns (*Colocasis esculenta* & *Xanthosoma sagitifolium*)	Blend with cold Ace/H$_2$O (85:15) for 5 min, wash with Hex, SPE on C18	Silica gel BuOH/HOAc/H$_2$O (4:1:5)	UV & 6 reagents	71
Caffeic, caftaric	Wine	Hydrolysis (H$_2$SO$_4$, N$_2$, pH 1.0), filter	Whatman No. 1 BuOH/HOAc/H$_2$O (2:1:1)	UV & ninhydrin spray	145

Ace, acetone; ACN, acetonitrile; AsA, ascorbic acid; BuOH, butanol; Bz, benzene, CC, column chromatography; EtOAc, ethyl acetate; Et$_2$O, diethyl ether; EtOH, ethanol; FL, fluorescence; FT-IR, Fourier transform–infrared; HCO$_2$-Et, ethyl formate; Hex, hexane; HOAc, acetic acid; HPTLC, high-performance thin-layer chromatography; IPA, isopropanol; IR, infrared; MS, mass spectrometry; MeOH, methanol; NMR, nuclear magnetic resonance; PHBA, *p*-hydroxybenzoic acid; SPE, solid phase extraction; TLC, thin-layer chromatography; UV, ultraviolet.

Tentative identifications can be made by comparing R_f values of unknowns with those of standards. Further tests can confirm by examining the color under UV light and after fuming with ammonia (40,62), and characteristic color response toward various spray reagents. For example, leave the sprayed plates in an oven at 50°C for 30 min, observe, and then fume with ammonia. The four cinnamic acids and two benzoic acids (salicylic and gentisic acids) substituted in the ortho position have a blue fluorescence turned green when fumed with ammonia vapor, when illuminated with Wood's light (366nm). Either coumarin or cinnamic acid derivatives (71) fluoresce before and after ammonia vapor treatment and the lack of color in visible light. Spray reagents of Folin–Ciocalteau (25,71) were used as a general reagent for phenolic compounds, which can appear as blue bands with Folin–Ciocalteau (71) and ferric chloride (71,111) reagents.

Under UV light, coumarins fluoresced violet to blue whereas psoralens were reddish brown to yellow (112). Under alkaline conditions the lactone ring of coumarin opens, producing a compound that fluoresces bright yellow with long-wavelength UV radiation (113). Spraying with 2 N NaOH (113) or ferric reagents [$FeCl_3$-$K_3Fe(CN)_6$] (112) and heating can enhance the visual detection of coumarins from plates.

Table 4 contains a list of some recent TLC separations of phenolic compounds from extracts of fruit, fruit juices, honey, chocolate, hops, pepper, and licorice. A few applications of high-performance TLC (HPTLC) are also included in Table 4. HPTLC is a development of TLC carried out using small particles, usually about 5 µm in diameter (18). One of the major problems involved in separation of these phenolic compounds is their similarity in chemical characteristics, which requires more resolving power of chromatography. Furthermore, many phenolics show similar UV absorption spectra with maxima in a narrow range of 280–320 nm (21). Thus, many traditional separation techniques such as PC, TLC, and CC are being replaced by HPLC. For more detailed information on normal phase TLC, the reader can refer to Stahl (114) and Markham (115). For information on reverse phase applications, the reader can refer to Heimler (116).

E. UV-Vis Absorption Spectroscopy

The utility of UV absorption data was not widely recognized until the 1950s due to reliability problems with early instruments. Then the information gained by UV absorption spectra was found to be increased by recording spectra over several experiments. Solutions containing the unknown and one of several shift reagents were performed in succession. Shift reagents are used that cause pronounced bathochromic (longer wavelength) shifts depending on the type and pattern of substitution. Comprehensive reviews on structure analysis of flavonoids by UV-visible (UV-Vis) spectroscopy were subsequently published with comprehensive listings of absorption data for many flavonoids along with shift data for various shift reagents (117–120). The review by Jurd (117) summarized work done up to 1960 while the works published by Harborne (15), Mabry et al. (16), Markham (118), and Averett (119) provide updates of Jurd's review. A more recent review by Vorin (120) appeared in 1983.

The solvents commonly used to record flavonoid spectra have been methanol or ethanol. With the more recent coupling of UV to HPLC instrumentation for detection, acetonitrile, acetonitrile–water, and alcohol–water mixtures with or without small amounts of acid as mobile phase have become common solvents in which flavonoid spectra are recorded. The information gained by use of shift reagents may also be utilized in HPLC analysis through postcolumn addition prior to UV-Vis detection. Shift reagents in common use include sodium methoxide (NaOMe), aluminum chloride ($AlCl_3$), aluminum chloride in dilute hydrochloric acid ($AlCl_3$/HCl), sodium acetate (NaOAc), and sodium acetate/boric acid (NaOAc/H_3BO_3). Absorption maxima are influenced by pH as well as the solvent used to record the spectra, so that one should record both the pH and the solvent when recording absorption spectra.

Flavonoid spectra typically have two major absorption bands as can be seen for the absorbance spectrum of tangeretin, a polymethoxylated flavone (Fig. 7). Band I typically occurs in the 300- to 400-nm range and in general terms can be said to arise from the absorption of the B-ring portion of the flavonoid molecule. Band II typically occurs in the range of 240–290 nm and comes from absorption of the A ring. Addition of substitutents to an aromatic ring capable of electron donation such as hydroxyl or methoxyl groups usually induce strong (10–15 nm) bathochromic shifts for the absorption band associated with that ring. When conjugation occurs between the two rings as in the flavones, flavonols, isoflavones, chalcones, and aurones, electron-donating substituents on one ring may affect absorption maxima, shape, and intensity of both bands. In the following discussion, the described shifts in absorption maxima are bathochromic (to longer wavelength) unless specifically stated as being hypsochromic (shorter wavelength).

Flavone and flavonol absorbance spectra in methanol typically have two major absorption bands. In flavones, band I is most intense at 320–350 nm. In flavonols, band I occurs between 350–380 nm due to a 20- to 30-nm shift from the 3-hydroxyl substituent. This shift is absent in flavonols when the 3-hydroxyl position is blocked as in 3-glycosides. Both have a second strong band in the 250- to 290-nm region. Addition of NaOMe causes ionization of free hydroxyl groups (3,7,4′). Substituted hydroxyls that are acylated or glycosylated are not affected. A shift of 40–65 nm in band I with decreased sensitivity indicates a 3-hydroxyl (flavonol) while no increase or decrease in intensity is indicative of a 4′-hydroxyl. A shift occurring with spectrum deterioration over time is indicative of 3,4′-dihydroxyl or 3,3′4′-trihydroxyl substitution. Spectrum deterioration can also mean a trihydroxyl system on either ring. Formation of a small peak at 320–330 nm indicates 7-hydroxyl substitution (121,122).

The weak base NaOAc ionizes only the most acidic hydroxyl groups. A shift of 5–20 nm in band II indicates 7-hydroxyl substitution. Methyl or O-methyl groups adjacent to this position can block the reaction. Alkali-sensitive systems such as 5,6,7-, 5,7,8-, or 3,3′,4′-trihydroxyl substitution patterns are unstable in base and absorption spectra will degrade over time (121). *Ortho-*

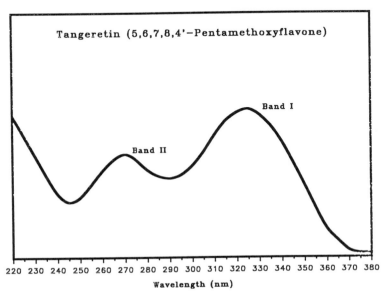

Fig. 7   The absorption spectrum of tangeretin typical of flavonoids with conjugation between the A and B rings.

dihydroxyl substitution in the B ring will show shifts of 12–36 nm in band I on addition of NaOAc/H_3BO_3 relative to spectra in methanol. Problems encountered when using NaOAc and NaOAc/H_3BO_3 have been discussed by Rosler et al. (123).

Aluminum chloride forms complexes with *ortho*-dihydroxy substitution systems as well as 4-keto-5-hydroxy and 4-keto-3-hydroxy substitution systems, which induce shifts in band I. The ketohydroxy complexes are stable in acid while *ortho*-dihydroxy substitution patterns are not. The latter are indicated in A and B rings by 20- to 25- and 30- to 40-nm shifts, respectively, when spectra in AlCl$_3$/HCl and AlCl$_3$ are compared. The *ortho*-dihydroxyl-AlCl$_3$ complex is sensitive to water in ethanol but not in methanol; therefore use of ethanol with this reagent is to be avoided (124). A shift of 35–55 nm in AlCH$_3$/HCl indicates 5-hydroxyl substitution while a 50- to 60-nm shift is evidence of either 3-hydroxyl or 3,5-dihydroxyl substitution (122).

Flavanones, dihydroflavonols, and dihydrochalcones have one strong absorption maxima between 270 and 295 nm with a small peak or shoulder occurring between 300 and 360 nm. The UV absorbance spectrum of the flavanone glycoside hesperidin in Fig. 8 is typical. The use of shift reagents can help determine substitution patterns in the A ring, but due to lack of conjugation between the A and B rings, substitution changes in the B ring show little or no differences in absorption. More sophisticated methods are necessary to determine substitution patterns in this ring. Isoflavones and isoflavanones have spectra similar to the flavanones, except that the strong absorption band usually occurs at shorter wavelengths between 240 and 270 nm with a shoulder or small peak in the range of 300–340 nm (125).

In NaOMe and NaOAc, isoflavones, flavanones, dihydroflavonols with 5,6,7- or 5,7,8-trihydroxyl substitution systems will give absorption spectra that deteriorate over time. An isoflavone with 3′,4′-dihydroxy substitution will also show deterioration over time. A band II 35- to 40-nm shift in NaOMe and NaOAc is indicative of 5,7-dihydroxyl substitution (2′,4′,6′-hydroxyls in dihydrochalcones) in flavanones and dihydroflavanols while a shift of 60 nm indicates a 7-hydroxyl substitution only. In isoflavones the presence of a 7-hydroxy (2′,4′-hydroxyls in dihydro-

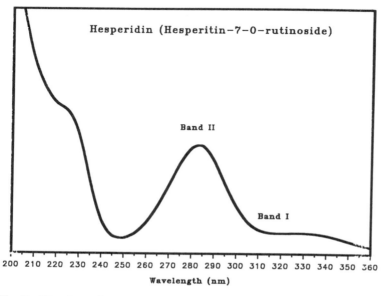

Fig. 8 The absorption spectrum of hesperidin typical of flavanones and other flavonoid classes without conjugation between the A and B rings.

chalcones) system is associated with a shift of 6–20 nm in band II. Some flavanones may give a chalcone absorption spectrum with a maximum around 400 nm due to conversion in base (121).

The same chelation complexes of AlCl$_3$ with *ortho*-dihydroxy, 4-keto-5-hydroxy, and 4-keto-3-hydroxy substitution systems are formed with isoflavones (no 4-keto-3-hydroxy), flavanones, and dihydroflavonols causing bathochromic shifts in band II. However, with lack of conjugation between the ring systems, *ortho*-dihydroxyl groups can only be detected in the A ring and show up as an 11- to 30-nm shift in AlCl$_3$ compared to absorbance maxima in AlCl$_3$/HCl. A free 5-hydroxyl group is indicated by a 10- to 14-nm shift in AlCl$_3$/HCl for isoflavones and a 20- to 25-nm shift for flavanones and dihydroflavonols. Absence of a 5-hydroxyl group is indicated by a shift of 30–38 nm in AlCl$_3$ (121).

Aurones have a strong band I absorption in the range of 370–430 nm according to Markham and Maybry (121). Two other bands also are present at 250–350 nm and 225–270 nm. Chalcones have an absorbance pattern with one band of moderate intensity typically occurring between 220 and 270 nm and a stronger band in the range of 340–390 nm. Sodium acetate (NaOAC) induces either a large bathochromic shift in band I or the appearance of a broad shoulder at high wavelength when there is a free hydroxyl group in the 4 and/or 4' position for chalcones. In aurones with an equivalent hydroxyl substitution pattern (4' and 6 position), the same shift behavior is observed in band I. AlCl$_3$ will detect the presence of a free 2'-hydroxyl substitution in chalcones and 4-hydroxyl substitution in aurones (same as the 5 position in flavone) with observance of a large bathochromic shift in band I (40–60 for chalcones and 60–70 for aurones). Reactions with *ortho*-dihydroxyl groups are distinguished through comparisons of spectra in AlCl$_3$ and AlCl$_3$/HCl and with the use of acetate and H$_3$BO$_3$. Substitution of an *ortho*-dihydroxyl system on the B ring gives a larger bathochromic shift than when present on the A ring.

With NaOMe, a bathochromic shift of 60–100 nm and an intensity increase is indicative of 4'-OH in aurone (80–95 nm) and 4-OH in chalcone (60–100 nm); aurones with 4'-OH give an 80- to 95-nm bathochromic shift with increase in intensity; aurones with 6-OH give a smaller (60–70 nm) shift; aurones with 6,4'-diOH give a smaller shift still; chalcones with 4-OH give a 60- to 100-nm bathochromic shift with increase in intensity; chalcones with 2- or 4'-OH give a 60- to 100-nm shift with no intensity increase; chalcones with 4- and 2- or 4'-OH give a smaller shift of 40–50 nm (121).

Anthocyanins characteristically have strong absorption in the region of 500–550 nm. Of the available shift reagents, only AlCl$_3$ has been used to detect free *ortho*-dihydroxyl systems. When present, a bathochromic shift from 15 to 50 nm in band I is observed while those without adjacent hydroxyl are unchanged.

F. Countercurrent Chromatography

Countercurrent chromatography (CCC) is a liquid–liquid partition technique using two immiscible solvents. Most often one phase is stationary as in other chromatographic techniques where one phase is immobilized on an adsorbent. The second phase is delivered so that it passes through the first phase and separation occurs based on partition coefficients of solutes between the two phases. The two phases must be immiscible. Putman and Butler (126) applied CCC to separate condensed tannin extracted from sorghum grain. Separation of tannins by CCC is partially due to differences in the length of the polymeric tannin molecules. CCC may be divided into two types of practical use: droplet CCC (DCCC) and centrifugal partition chromatography (CPC). While DCCC has a higher capacity (several grams vs. 1 gram) than CPC, it relies on gravity to retain the stationary phase and separations can take up to 2 days (127). CPC relies on centrifugal force to retain the stationary phase and thus solvent flow of the mobile phase can be much faster. Yoshida et al. (128) showed application of CPC to the purification of hydrolyzable tannins extracted from Sinkiang licorice, which are difficult to purify by gel column chromatography because they are readily hydrolyzed during the development.

The main advantages of CCC over other chromatography procedures is that no irreversible adsorption ever occurs because separation relies on liquid–liquid partioning with no solid adsorbant being used. Therefore all of the introduced sample can be recovered. Furthermore, the advent of high-speed CPC led to considerable improvements in terms of separation efficiency, analysis time, and handling of the apparatus compared with DCCC. The use of CCC for flavonoid separation has been limited (Table 5), most likely due to the higher capital cost over column chromatography equipment. It is well suited, however, for preparative separations of crude mixtures or final purification of semipurified product. While the resolution is not very high, 1 to several grams of sample may be loaded onto commercial instruments for recovery. For more detailed information, the reader should refer to Marston and Hostettmann (127), Conway (129), and Mandava and Ito (130).

G. Gas Chromatography

Gas chromatography (GC) has only limited application in the analysis of phenolics due to their limited volatility, the main disadvantage being the additional amount of sample preparation required in order to ensure volatility of the sample. Failure to adequately remove nonvolatile components from the extract will result in rapid contamination of the injection port as nonvolatile components collect and are pyrolyzed in the injector. This can result in adsorption of sample on subsequent injections, possible contamination of the analytical column, and interferences caused by volatile pyrolysis products formed from components trapped in the injector.

Even though GC offers high sensitivity for a wide range of phenolic compounds, fewer analyses of the phenolic acids have been reported (Table 6). It may be due to the fact that phenolic acids are not easily volatilized; analysis of nonvolatile phenolic acids by GC requires the derivatization. The most common derivative of phenolic compounds is trimethylsilyl (TMS) derivative. It is an easy technique and TMS esters of phenolic acid methyl esters are most suitable for quantification with flame ionization detection, (FID). However, TMS derivatives are extremely sensitive to moisture; thus, exclusion of water is a prerequisite for this derivative.

More recently (131), phenolics have been converted to derivatives amenable to GC by treatment with methyl or ethyl chloroformate (MCF or ECF). The reaction, catalyzed by pyridine, proceeds even in the presence of water, but most problematic with these derivatives is multiple peak formations of benzoic and 4-methoxybenzoic acids. Each can give up to four peaks on the chromatogram. In recent work with capillary columns (132), low polarity columns (HP-1, DB-5) are shown to be more effective than higher polarity columns (DB-17, DB-1701, DB-210) for resolution of mixed phenolic compounds.

It has been demonstrated that the semivolatile methoxylated flavones and flavone aglycones can be separated by GC without derivatization (133). Gaydou et al. (134) demonstrated separation of 27 polymethoxylated flavones (PMFs) using a capillary column (50 m × 0.32 mm i.d.) coated with OV-1 (0.15-μm phase thickness). This procedure applied for major PMFs in orange peel oils and the results were comparable to those determined for similar samples using HPLC. However, flavonoid glycosides first require hydrolysis into the corresponding sugar and aglycone residue followed by derivatization, usually into the TMS derivative to increase volatility of flavanone aglycones (74). It was found that heating of the reaction mixture to 60°C was necessary to yield quantitative conversion of flavanone to the TMS chalcone derivative. Otherwise, a mixture of flavanone and chalcone derivatives were obtained (74).

Advantages of GC analysis include an improved separation of closely related isomers and simple coupling to mass spectroscopy (MS) detectors for identification through the fragmentation

Table 5 Column Chromatography of Phenolics in Foods

Component	Source	Extraction	Analysis method	Detector	Ref.
Flavonol: catechin, epicatechin, epicatechin polymers	Chocolate, cocoa (*Theobroma cacao*)	CH_2Cl_2, EtOAc, SPE Sephadex LH-20	Sephadex LH-20, Fractogel TSK HW-40 (F), MCI-gel CHP-20P	Vanillin-HCl, MS, ^{1}H-NMR, ^{13}C NMR	78
Flavone: glucosylsudachitin	Fruit: *Citrus sudachi* peel	Hot H_2O	Silica	UV, ^{1}H NMR	46
Flavone: glucosylapigenin, limocitrin (3-β-D-glucose, 3-β-D-rhamnose); Flanonol: rutin, narcissin; Flavanone: narirutin	Fruit: *Citrus unshiu* peel	Hot H_2O, Hex, and BuOH	$CHCl_3$/MeOH/HOAc/H_2O (70:15:12:3)	^{1}H NMR, UV	58
Flavonol: catechin, epicatechin, epicatechin gallate	Fruit: grape seed (*Vitis vinifera*)	70% Ace, $CHCl_3$, EtOAc, hydrolysis	Sephadex LH-20, EtOH & H_2O/HOAc	UV 280	76
Methoxyflavone:	Fruit: tangerine (*Citrus unshiyu*)	Hex and Bz	Silica L (40/100), 1) Hex/EtOAc 2) Bz/Ace 3) Hex/BuOH 4) $CHCl_3$/MeOH	UV	63
Flavone: apigenin 6,8-di-*C*-glucoside; Flavanone: liquiritigenin glycosides	Herb: licorice (*Glycyrrhiza uralensis*)	EtOAc, *n*-BuOH	Sephadex LH-20, MCI gel CHP20P, Avicel cellulose 1) EtOH/H_2O 2) MeOH/H_2O	UV, MS, ^{1}H NMR, ^{13}C NMR	271
Flavone: glisoflavone, glycyrrhizoflavone, kaempferol 3-*O*-methyl ether; Flavonol and isoflavone	Herb: licorice (*Glycyrrhiza*)	Hex and EtOAc	$CHCl_3$/MeOH/H_2O MCI gel CHP-20P, 60% MeOH	UV 280, IR, MS, ^{1}H NMR, ^{13}C NMR	66
	Herb: licorice (*Glycyrrhiza*)	EtOAc	$CHCl_3$/MeOH/H_2O (7:13:8) and EtOAc	MS, UV, ^{1}H NMR	243
Flavone, chalcone	Herb: licorice (*Glycyrrhiza eurycarpa*)	Hex, Bz, Et_2O	Silica Bz/Et_2O	UV, MS, ^{1}H NMR, ^{13}C NMR	65
Isoflavan: kanzonols	Herb: licorice (*Glycyrrhiza uralensis & G. glabra*)	Bz & Hex	Silica 1) Bz/Et_2O 2) Hex/Ace	UV, MS, ^{1}H NMR, ^{13}C NMR	64
Flavone: 7,8-dihydroxyflavone, chrysin; Flavonol: galangin; Isoflavone: daidzein	Standards	—	PTFE (0.85 mm i.d. 50 ml–1200 rpm & 1.6 mm i.d. 340 ml–100 rpm, $CHCl_3$/MeOH/H_2O (4:3:2) vs. Hex/EtOAc/MeOH/H_2O (1:1:1) vs. *n*-BuOH	UV 254 nm	272
Flavonol: acylated kaempferol and quercetin glycosides	Vegetable: cabbage (*Brassica oleracea*)	70% MeOH, $CHCl_3$, EtOAc	Polygosil C18, Polyclar AT, Sephadex LH-20, H_2O/MeOH	UV, MS, ^{1}H NMR, ^{13}C NMR	136

Ace, acetone; BuOH, butanol; Bz, benzene; EtOAc, ethyl acetate; Et_2O, diethyl ether; EtOH, ethanol; Hex, hexane; HOAc, acetic acid; IR, infrared; MS, mass spectrometry; MeOH, methanol; NMR, nuclear magnetic resonance; SPE, solid phase extraction; UV, ultraviolet.

Table 6 Gas Chromatography of Phenolics in Foods

Component	Source	Extraction	Analysis method	Detector	Ref.
Psoralen, xanthotoxin, bergapten, isopimpinellin, angelicin	Celery	For root & crown: homogenize with 80% meOH, adjust to PH 8.0, Et2O extract. For petiole tissue: homogenize with H2O, Et2O extract, SPE C18 & silica	DB-225 (0.25 μm), 30 m × 0.25 μm i.d., He carrier. Oven: temp. program, 120–180°C	MS (75 eV)	172
Chlorogenic, quinic	Coffee	DMSO, 90°C, 60 min, TMS derivatives	25 m × 0.25 mm i.d. CP-SIL5-CB, He carrier. Oven: temp. program, injector & detector: 310°C	FID, GC-MS, GC-IR	252
Flavanone: naringinin, hesperetin, eriodictoyl	Fruit juice: orange, lemon, grapefruit, and grape	Et2O followed by EtOAc, dried, acid hydrolysis in HCl, TMS derivative	50 m × 0.25 mm i.d. RSL 200 biphenyl (0.2 μm film), He carrier	MS ion scan, m/z 545–633 at 1 s/sec	74
Flavone: 27 methoxylated	Fruit: orange peel oil	Dilution in EtOAc	50 m × 0.32 mm i.d., OV-1, 280°C, isothermal	FID	134
Flavone, flavonol	Standards	—	25 m × 0.25 mm, 0.25 μm film OV-1 DF (Macherey-Nagel, helium carrier, Injector and detector 300°C, column 270°C, isothermal, EI 70eV	MS scan m/z 50–350 EI 70 eV	135
Flavone: 39 methoxylated	Standards	—	25 m × 0.32 mm i.d. 280°C isothermal	MS	270
Flavonol: quercetin, kaempferol, rhamnodiglucocides	Tea, black (*Camellia sinensis* L. O. Kuntze)	MeOH, SPE on polyamide column, benzoylation, enzyme hydrolysis and silization of sugars	23 m × 0.25 mm i.d., OV-1. Oven: temp. program	MS	59
Flavonol: acylated kaempferol and quercetin glycosides	Vegetable: cabbage (*Brassica oleracea*)	70% MeOH, CHCl3, EtOAc, acid hydrolysis and silization of sugar residues	25 m × 0.32 mm HP ultra-1-methyl silicone, He carrier. Oven: temp. program	MS	136

DMSO, dimethyl sulfoxide; EI, electron impact; EtOAc, ethyl acetate; Et2O, diethyl ether; FID, flame ionization detection; GC-MS, gas chromatography-mass spectrometry; GC-IR, gas chromatography–infrared; MS, mass spectrometry; MeOH, methanol; SPE, solid phase extraction; TMS, trimethylsilyl.

pattern (133,135). The coupling of MS to HPLC is much more involved as removal of the liquid mobile phase requires a sophisticated interface. Analysis of derivatized sugar residues from hydrolysis of a glycoside (59,136) also allows identification of the glycoside substituent, although information on point of attachment is lost during hydrolysis. Several examples of recent GC separations of the phenolics can be found in Table 6.

H. High-Performance Liquid Chromatography

Since the application of reverse phase high-performance liquid chromatography (RP-HPLC) in 1976 by Wulf and Nagel (137) with chemically bonded silica gel (μBondapak C_{18}) column with isocratic elution with H_2O/MeOH/HOAc (13:6:1) for separating phenolic compounds, HPLC is the most commonly used chromatographic method for the variety of phenolic compounds in foods. HPLC offers higher resolution and efficiencies over PC, TLC, and CC through the use of tightly packed columns with small particles. There are no restrictions on sample volatility and derivatization requirements as there are in GC.

HPLC also provides a separation efficiency similar to that of GC for phenolic acids without derivatization and with relatively short analysis time, and the components of interest can easily be quantified. A wide range in selectivity is available through the many solvent combinations and column packings available. Accurate high-pressure pumps allow fast analysis times with a high degree of reproducibility, particularly when the column is placed in a temperature-controlled oven. Scale-up of the separation for preparative isolations with large columns is facilitated with vendors offering packed columns in many sizes. Connection to an automatic fraction collector allows precision peak collection capability over multiple injections. HPLC is also easily coupled to a variety of detectors allowing selective or universal detection of components, some of which provide characterization information about eluting components. The combination of HPLC with photodiode array detection (PDA) can provide additional spectral information on each of the resolved components.

The main disadvantage to HPLC is the high capital cost of equipment and columns. Others include the higher level of training necessary to operate equipment; samples must be filtered prior to injection in order to avoid column plugging, and a column can easily be destroyed through irreversible adsorption of components that may be present in a crude extract. It is for these last two reasons that short, easily replaceable precolumns have become popular. They are placed before the more expensive analytical column to trap particulates and components that may adsorb tightly to packing material. This inexpensive precolumn then can be replaced as necessary and extend the life of the expensive analytical or preparative column.

With development and an increase in the use of HPLC methods, an accurate measurement of total phenolics can be obtained, especially when only a few different flavonoid types are present. Individual components are isolated and then quantitated according to individually determined response factors. Total phenolic content of an extract can then be accurately determined through summation provided all of the phenolic components are measured by the HPLC method.

1. Stationary Phase

In most cases, the reported systems for the separation of phenolics and their glycosides in foods were carried out by reverse phase chromatography on silica-based C18-bonded phase columns. Occasionally, silica columns bonded with C8 or C6 were applied in analysis of phenolic acid standards and coumarins (135), and for ferulic acid in wheat straw (139), respectively.

The average particle diameter of HPLC packings is typically 3–10 μm. Columns of smaller particle size (e.g., 3 μm) usually provide a larger number of plates per unit time than do columns with larger particles. However, 3-μm columns are somewhat more difficult to work with and become plugged more easily.

Bailey et al. (140) evaluated three different types of reverse phase columns such as Hypersil octyl (wide pore, lower surface area, less retentive than narrow-pore phases); Hypersil ODS (narrow pore, high surface area packing); and PRP-1 (styrene-divinylbenzene copolymer with no surface hydroxyl groups or surface metals) for tea phenolics. Hypersil ODS column with citrate buffer showed very good resolution of a wide range of tea phenolic compounds. Chromatography with polymer phase column (PRP-1), which has none of the secondary retention encountered with silica-based phases, provided lower resolving power and different selectivity than silica-based ones.

2. Mobile Phase

Most of the solvent systems used in analytical HPLC include binary gradient elutions and, occasionally, isocratic elution for beer (90), cereal (141), grape (51), orange juice (80), wheat straw (139), and wheat (57). Gradient or isocratic elution using solvents of aqueous acetic, formic, or phosphoric acids with MeOH or ACN as organic modifier is common. The range of solvent strengths used in gradient elutions and time required for analytical separation is dependent on the number and type of phenolics in the mixture. Oftentimes multiple-step gradients are employed with complex mixtures. Generally, more sophisticated gradients are required for crude extracts of foods containing many types of phenolics while isocratic methods can be employed for partially purified extracts or crude extracts containing only a few components of similar polarity.

The nature of the solvent such as solvent strength and viscosity of organic modifier has an important influence on chromatographic separation. Acetonitrile (ACN) gave better resolution in shorter times than methanol (MeOH), and tetrahydrofuran (THF) produced even better resolution (81) for cinnamic acids in orange juice (80,81). ACN was found to give better and sharper peak shapes resulting in higher plate number than did MeOH (142). An H_2O/ACN gradient is often employed to reduce the elution time and to sharpen the peak shape for the late eluting phenolics and to improve quantitation at low concentrations. However, MeOH seemed preferable to ACN because of its nontoxic nature and because a higher percentage of organic solvent could be used in the mobile phase to prevent deterioration of the reversed phase column (57). THF has higher solvent elution strength value (eluotropic strength) than ACN, and ACN has a higher value than MeOH. THF often was chosen as the organic modifier instead of methanol or acetonitrile for phenolics (44,80,81,138). In some cases, substitution of MeOH with THF improves resolution. Mobile phase containing THF could substantially improve the chromatographic resolution between four cinnamic acids compared to solvents employing MeOH or ACN (80).

The pH and ionic strength of the mobile phase are known to influence the retention of phenolics on the column depending on the occurrence of protonation, dissociation, or a partial dissociation (142). A change in pH that increases the ionization of a sample could reduce the retention in a reversed phase separation. Thus, small amounts of acetic acid (2–5%), phosphoric or trifluoro-acetic acid (TFA) (0.1%) are included in the solvent system to suppress ionization of phenolic and or carboxylic groups and hence improve resolution and reproducibility of runs.

Since good pH control for buffers is limited to a pH range of 3.7–5.6 for acetate buffers, 2.0–8.0 for citrate buffers, and 1.8–3.5 and 5.8–8.0 for phosphate buffers, the pH of the aqueous mobile phase should be adjusted to optimize the analysis time and separation depending on the buffer used. For example, the pH of aqueous mobile phase was adjusted to 2.5 or 2.6 for phosphate buffer (84,106,143), pH 4.0 for ammonium acetate buffer (90), and pH 5.4 for citrate buffer (57). Ionic strength of these buffers varied from 10 to 100 mM (84,90,91,106,143–146). Buffer strength should be high enough to enhance the separation; however, varying buffer concentrations did not significantly affect the resolution or the retention times (57).

Ion pair HPLC is often used with samples that contain ionizable compounds to provide more retention and higher selectivity than are afforded by the column and organic solvents alone. For fruit juices with little cleanup procedure, an ion-pairing agent such as 10 mg/L of sodium dodecyl sulfate (SDS) was added to control the retention of the amines, phenolic acids, and flavonoids (144).

3. Separation

Elution of phenolics for RP-HPLC is in order of decreasing polarity. Polarity is increased most by hydroxyls at the 4 position, followed by those at the 2 and 3 positions. Methoxy and acrylic substitution reduce polarity and increase retention times (21). Loss of polar hydroxy groups and or addition of methoxy groups can decrease the polarity within each class of benzoic and cinnamic acid. Also, the presence of the ethylenic side chain in the cinnamic acids can reduce their polarity compared with similarly substituted benzoic acids (60). The elution order for benzoic acids is as follows: gallic > α-resorcylic > protocatechuic > γ-resorcylic > gentisic > p-hydroxybenzoic > syringic > vanillic > m-hydroxybenzoic > β-resorcylic > benzoic > salicyclic acid. Salicylic acid was retained longer than its para isomers, which may be due to intramolecular hydrogen bonding (137). Under the same conditions, the elution order for cinnamic acids is caffeic > p-coumaric > sinapic > ferulic > m-coumaric > o-coumaric acid (111). Ferulic acid eluted after p-coumaric, which indicated that the methoxy ($-OCH_3$) substituent is nonpolar as it increases retention.

With RP-HPLC, the most polar coumarin glycosides elute first and the nonpolar furanocoumarins last. Table 7 summarizes the chromatographic retention for common phenolic acids in foods.

A change in temperature normally has only a minor effect on band spacing in RP-HPLC and essentially no effect in normal phase separations. Temperatures of 25°C are more common but elevated temperatures between 30°C and 40°C are often applied for phenolics and derivatives in apples (52), carrots (53), apple juice (60,144), bilberry juice (40), and for cis-trans isomers of caffeic and p-coumaric acids in wines (23).

Chromatographic methods enabling the separation of geometric isomers of phenolic acids in foods involved reverse phase separation based on C18 columns. The cis and trans isomers of caftaric, coutaric, and p-coumaric acid present in grape and wine can be separated by binary gradient elution (23,106), or simply by isocratic elution with 16% aqueous MeOH at pH 2.6 (51). The cis isomers of cinnamic are present in grapes and they are known to elute before the corresponding trans isomers in RP-HPLC (47).

RP-HPLC based on C8 column has also shown potential for the separation of geometric isomers of caffeic, ferulic, and isoferulic acids (138) using aqueous THF as a mobile phase, but this analysis is demonstrated only with standards which are a simple matrix compared to the food materials of a complicated matrix.

Optimization of mobile phase composition offers a shortcut to much higher productivity for analysts. Many procedures for computer-assisted HPLC method development have been described. Several manufacturers (LDC Analytical, LC Resources, etc.) offer software to control the pumps and to save data in order to make optimization of separation easy and for fast-method development, but many users often prefer to apply a simple optimization procedure (57,80). An optimization study of four cinnamics using reverse phase column with isocratic elution with various isocratic mobile phases was discussed (80). The best separation for sinapic, coumaric, ferulic, and caffeic acids was achieved by using THF as modifier in the water or by using THF, methanol, and water for ternary combination (80). The mobile phase consisting of 21% THF/79% water (with 2% HOAc) produced the best separation and allowed the most accurate quantitation of the four hydrocinnamic acids in orange juice.

Table 7 Chromatographic Retention* Along with UV and Fluorescence Data of Phenolic Acids

Phenolic acids	r	R_f^a	R_f^b	Absorption maxima**		Fluorescence maxima***	
				EtOH	MeOH	Excitation	Emission
Benzoic acids							
Benzoic	1.723	0.80	0.72		228, 272	301	412
Salicylic	2.102	0.63	0.62	235, 305	232, 300	301	412
m-Hydroxybenzoic	0.901	0.56	0.55				
p-Hydroxybenzoic	0.760	0.55	0.51	265	252	291	340
Protocatechuic	0.578	0.28	0.26	260, 295	258, 292	304	365
α-Resorcylic	0.536	0.49	0.52		249, 307		
β-Resorcylic	1.010	0.50	0.46		256, 294		
γ-Resorcylic	0.619				246, 306		
Gentisic	0.739	0.32	0.28	237, 335	332	357	503
Gallic	0.432	0.11	0.13	272	270	412	496
Vanillic	0.828	0.54	0.53	260, 290	257, 287	308	360
Syringic	0.796	0.48	0.55	271	272	320	372
Cinnamic acids							
o-Coumaric	2.364				275, 325		
m-Coumaric	1.796				235, 277		
p-Coumaric	1.437	0.52	0.51	222, 290–310	235, 310	328	372
Caffeic	1.000	0.26	0.33	235, 290–320	246, 322	344	433
Ferulic	1.567	0.62	0.60	232, 290–317	240, 328	338	413
Sinapic	1.455	0.57	0.62		246, 336	340	430

r = relatve retention time vs. caffeic by HPLC.
R_f^a = Bz/dioxane/HOAc (90:25:4); R_f^b = Bz/MeOH/HOAc (45:8:3).
*Adapted from Ref. 111.
**Adapted from Ref. 1, 6, and 21.
***Adapted from Ref. 6 and 81.

The variety of detectors available for HPLC analysis which provide additional information about the eluant as it exists the column greatly facilitates unknown characterization. Detailed information is described in a latter part of this chapter. A discussion of the different gradient and isocratic methods for HPLC analysis of phenolics is beyond the scope of this chapter, but many of the commonly used HPLC methods are listed in Tables 8 and 9 of which several will be discussed.

I. Capillary Electrophoresis

While the separation technique of capillary electrophoresis (CE) was first described in 1967 by Hjerten (147), commercial instrumentation only became available within the last few years. The high efficiency of CE for separation of molecules based on mass and charge was first demonstrated in an article published in 1981 by Jorgenson and Lukacs (148) and was limited to the separation of polar components.

The principle of CE is based on a buffer-filled capillary tube whose ends are in buffer-filled reservoirs containing electrodes. Sample is applied by various means to one end of the tube and a charge applied between the electrodes. Either positively or negatively charged species can be selected by changing electrode polarity. Neutral components travel together through the capillary with the electroosmotic flow of buffer induced by the potential difference between electrodes. Charged sample molecules are separated as they migrate through the capillary (against the flow)

Table 8 High-Performance Liquid Chromatography of Phenolic Acids and Coumarins in Foods

Component	Source	Extraction	Analysis method	Detector	Ref.
Esculin, umbelliferone, scopoletin, 4-methyl-umbelliferone	Alcoholic beverages	—	Micropak MCH-5 300 × 4.0 mm, (5 μm), 40°C; A: 5% HOAc in H_2O; B: ACN/H_2O/HOAc (80:15:5)	FL 340/454 nm	31
Chlorogenic, cryptochlorogenic, p-coumaroylquinic isomers	Apple (*Malus* sp.) 11 cultivars (cv)	Homogenized in cold 65% EtOH + 0.5% $Na_2S_2O_5$, petr. ether to remove pigments, EtOAc extraction	150 × 4.6 mm (3 μm) Rosil C18, 35°C; A: H_2O + H_3PO_4 (pH 2.6); B: ACN: MeOH (10:15); 10% B–50% B in 50 min	UV & PDA	52
Chlorogenic	Apple (cv scugog)	Blend with EtOH:HOAc:H_2O (10:1:9) for 5 min at full speed, CC on Sephadex LH-20, PC on Whatman #44, MeOH:HOAc:H_2O (10:1:9)	Spheri 10 RP-18 (10 μm); A: 5% HCO_2H in H_2O; B: MeOH; 17% B–55% B in 39 min	PDA	26
Chlorogenic	Apple (*Malus* Sp.) (8 cv)	Homgenize with MeOH, evaporate to dryness in vacuo at 30°C	Nova pak C18; A: THF; B: 0.1% TFA in H_2O; 20% A–100% A in 20 min	UV 270 nm	44
Chlorogenic, PHBA, p-coumaric, ferulic, sinapic, protocatechuic	Apple (cv golden delicious)	Blend with 80% MeOH for 2 min, SPE on C18	250 × 4.6 mm Econosil C18 (5 μm); A: 5% HOAc; B: ACN; 12% B–20% B in 10 min	UV 254, 265, 272 & 280 nm	94
Chlorogenic, chlorogenic isomers, caffeic, p-coumaric, p-coumarylquinic	Apple juice products	Acidification (pH 2.0), EtOAc extraction, evaporate to dryness	Supelco LC-18 (5 μm); A: 70 mM KH_2PO_4, pH 2.5; B: MeOH; 2% B–40% B in 45 min	PDA	84
Chlorogenic, p-coumaric & derivatives, protocatechuic, phenolic acid derivatives	Apple (petit jaune & judor, Bedan & Kermerrien) & juice	Adjust pH 7.0 with 0.1 M NaOH, EtOAc extraction, enzymatic, centrifuge, acidify aq. phase to pH 2.0	250 × 4.6 mm (5 μm) Spherisorb ODS2, 40°C; A: 0.2% 1 M HCl in H_2O; B: 0.2% 1 M HCl in MeOH; 0% B–50% B in 85 min; Enzymatic: laccase (pH 5.0, 25°C, 5–15 min)	UV 260 nm & 310 nm; oxygen electrode	60
Chlorogenic, coumaric, caffeic	Apple juice	Adjust to pH 6.0 with 4 M NaOH, SPE on C18	LiChrosphere 100-RP-18 (5 μm); 10% MeOH in 0.5% aq. HOAc, 10%–100% in 50 min	PDA	108

Table 8 (*Continued*)

Component	Source	Extraction	Analysis method	Detector	Ref.
Chlorogenic, caffeic, p-coumaric, ferulic	Apple juice (6 varieties)	Filter (0.45 μm)	MCH-10 Micropak C18 (10 μm) A: 0.1 M KH_2PO_4 B: MeOH 10% B–40% B in 20 min	UV 324nm	146
Phenolic acids (16), PHBA, o-pyrocatechuic, α,β,γ-resorcylic, protocatechuic, gallic, isovanillic, vanillic, syringic, o,m,p-coumaric, caffeic, ferulic, SA	Avocado (*Persea americana* Mill.)	Freeze-dried, grind, MeOH extraction, N_2, alkali hydrolysis (2 N NaOH, 24hr, N_2 at room temp), acidify to pH 2.0	250 × 4.6 mm, Adsorbosphere ODS (8 μm) A: 5% HOAc in H_2O B: ACN 10% B–100% B in 60 min	UV 254, 275, & 300 nm	21
Caffeic, p-coumaric, vanillic, ferulic, sinapic	Beer	Acidify with 50 μl HOAc/ml beer, SPE on C18	100 × 4.6 mm Brownlee RP-18-ODS (5 μm) 15% MeOH + 0.1 M $NH_4C_2H_3O_2$ (pH 4.0)	ED	90
Gallic, protocatechuic, PHBA, caffeic, syringic, p-coumaric, ferulic, sinapic, vanillic	Beer	Degas by 25 times transferring, EtOAc extraction	250 × 4.6 mm, Nucleosil C18 (10 μm) A: 3.5% HOAc in H_2O B: MeOH 0% B–50% B in 30 min	ED	61
Gallic, protocatechuic, PHBA, caffeic, syringic, p-coumaric, ferulic, sinapic, vanillic	Beer	EtOAc extraction	250 × 4.6 mm, Nucleosil C18 (10 μm) A: 3.5% HOAc in H_2O B: MeOH 0% B–50% B in 30 min	ED & UV 254 nm	56
Caffeic, ferulic, chlorogenic, gallic, syringic, protocatechuic, gallic, p-, m-, & o-coumaric, vanillic, p- & m-hydroxybenzoic acids	Bilberry juice (*Vaccinium myrtillus*)	EtOAc + $NaHSO_3$ extraction, conc. at 30°C	250 × 4 mm (5 μm) LiChrosorb RP-18, 30°C A: 2% HOAc in H_2O B: 2% HOAc/H_2O/ACN (2:68:309) 7% B–100% B in 50 min	UV 280 nm	40
PHBA, caffeoylquinic, dicaffeoylquinic, & hydrocinnamic esters	Carrot (cv Karotan premia)	Homogenize with 80% EtOH + 0.5% $NaHSO_3$, 4°C, 15 min, EtOAc + 20% NH_4SO_4 + 2% $(HPO_3)_n$ extraction	150 × 4.6 mm Rosil C18 (3 μm) 35°C A: H_2O, pH 2.6 B: MeOH/ACN (1.5:1) 12% B–70% B in 70 min	UV 265 & 325 nm, PDA	53

Compound(s)	Sample	Preparation	Column/Conditions	Detection	Ref.
Psoralen, xanthotoxin, bergapten, isopimpinellin, angelicin	Celery	For root & crown: homogenize with 80% MeOH, adjust to pH 8.0, Et$_2$O extract. For petiole tissue: homogenize with H$_2$O, Et$_2$O extract, SPE C18, & silica	Supelco silica 250 × 4.6 mm i.d. (5 µm), 0.045% MeOH in CHCl$_3$ & Spherisorb 5-phenyl 250 × 4.6 mm i.d. (5 µm) MeOH/H$_2$O (23:27)	UV 250 nm, UV 325 nm	172
Ferulic, p-coumaric, hydrocinnamic acids	Cereal (maize, wheat, barley, & rapeseed)	Microwave digestion (750 W × 90 s) with 4 M NaOH, cool, pH 1.0 with 6 M HCl, EtOAc extraction	10 × 4.6 mm Hichrom ODS (5 µm), H$_2$O/HOAc/BuOH (83:12:5)	UV 275 nm	141
5- & 8-Methoxypsoralen	Citrus oils	Dilute with EtOAc	YMC PVA-LIL 250 × 4.6 mm, isooctane/heptane (93:7)	UV 260 nm, PDA 210–340 nm	225
Total chlorogenic	Coffee (green arabica)	Extract with hot H$_2$O for 15 min, clarification with Carrez solution, centrifuge, filter (0.45 µm)	GFC: TSK-G 3000-SW 300 × 8 mm i.d., 0.05% sodium azide in H$_2$O	UV 280 nm	162
Chlorogenic, isochlorogenic, protocatechuic, ferulic	Coffee pulp (coffee arabica)	80% MeOH + 5 mM cysteine & MeOH + EtOAc extraction	HC ODS/SIL-X, 25°C A: MeOH B: 2% HOAc in H$_2$O 12% A–50% A over 27 min	UV 280 nm	55
Coumarins	Fruits, citrus	Centrifuge with DMSO-MeOH, filter (0.2 µm)	250 × 4 mm i.d. C18 A: 0.01 M H$_3$PO$_4$ B: MeOH 20% B–100% B in 55 min	UV 285 nm, NMR	79
Scoparone, umbelliferone, scopoletin	Fruits, citrus	40% EtOH or EtOAc extraction, filter, conc. Dissolve with EtOH/MeOH (1:1)	Zorbax Rx-C8 250 × 4.6 mm (5 µm) A: 0.1% HOAc in H$_2$O B: 0.1% HOAc in ACN 10% B–50% B in 65 min	FL 340/425 nm	33
Caffeic, coumaric, ferulic, & sinapic	Fruits (orange & grapefruit)	Blend with H$_2$O/EtOH/MeOH (2:1:1), alkaline hydrolysis, (2 N NaOH, 24°C, 24 hr), acidify, EtOAc extraction	LiChrospher 100-RP-18 15 ml/L aq. HOAc in MeOH (77:23)	UV 300 nm	22
Hydrocinnamic acid esters	Fruits (orange, grapefurit, & lemon)	Blend with MeOH, conc., CC on polyamide column	Shandon ODS-Hypersil A: MeOH B: 2% HOAc in H$_2$O 10% A–30% A in 45 min	PDA, NMR, MS	89
Caffeic, chlorogenic acids	Fruit jams (apricot, peach, plum, strawberry)	Blend with 80% MeOH × 18 hr, BuOH ext, CC on Amberlite XAD-2	LiChrochart 100-RP-18 (5 µm) A: H$_2$O/HCO$_2$H (95:5) B: MeOH 5% B–80% B in 35 min	UV 280 nm &350 nm	45

Table 8 (*Continued*)

Component	Source	Extraction	Analysis method	Detector	Ref.
Coumarins	Fruit juice	Centrifuge (10,000g × 4°C × 5 min)	150 × 4.6 mm i.d. MS gel C18 (5 μm), 35°C A: 0.1 M HaH$_2$PO$_4$ + 10 mM SDS B: ACN/0.1 M NaH$_2$PO$_4$ + 0.005% SDS/MeOH (60:30:10) 6% B–100 % B in 45 min	16 serial coulometric	253
PHBA, protocatechuic, gallic, p-coumaric, ferulic, caffeic, sinapic, chlorogenic, phenolic aldehydes, & coumarins	Fruit juice (apple, orange, grape, pear, peach, apricot, & pineapple)	Et$_2$O & EtOAc extraction	Novapak C18 A: 2% HOAc in H$_2$O B: MeOH/HOAc/H$_2$O (30:2:68) 0% B–85% B in 45 min	UV 280 nm & 240 nm	28
Phenolic acids & coumarins	Fruit juice (apple, orange & grapefruit)	Centrifuge (10,000g × 4°C × 5 min), filter (0.22 μm)	150 × 4.6 mm i.d. MS gel C18 (5 μm), 35°C A: 0.1 M NaH$_2$PO$_4$ + 10 mM SDS B: ACN/0.1 M NaH$_2$PO$_4$ + 0.005% SDS/MeOH (60:30:10) 6% B–100% B in 45 min	16 Serial coulometric	144
Cis/trans-Caftaric & *cis/trans*-coutaric	Grape (*Vitis* & non-*vitis* sp.)	Mix with AsA & K$_2$S$_2$O$_5$, press, filter (0.45 mm)	C18, 16% aq. MeOH, pH 2.6	UV 320 nm	51
cis/trans-Caftaric, *cis/trans*-coutaric, gallic, ferulic, caffeic, p-coumaric, cinnamic esters	Grape (Thompson seedless)	Filter (0.45 μm)	Supelco LC-18 (5 μm) A: 0.07M KH$_2$PO$_4$, pH 2.5 B: MeOH 2% B–40% B in 45 min	PDA 280 & 320 nm	106
Hydroxycinnamic acid tartrates	Grape, red (Concord & deChauna)	Homogenize with 80% MeOH & 1% AsA for 2 min followed by SPE on C18	Radial-Pak C18 acidic: 5% HOAc in H$_2$O neutral: 40–100% ACN	PDA	25
Hydroxycinnamic tartrate	Grape, white (5 cv)	Blend with 1% AsA; add EtOH to remove protein, SPE on C18, CC on Sephadex LH-20	Radial-Pak C18 acidic: 5% HOAc in H$_2$O neutral: 40% ACN–100% ACN	PDA	24
Tartaric esters of hydroxycinnamic acids	Grape, white (21 cultivars)	Blend with 1% AsA; add EtOH to remove protein, EtOH extraction, SPE on C18	Radial-Pak C18 acidic: 5% HOAc in H$_2$O neutral: 40%–100% ACN	UV 320 nm & 280 nm	50
Tartaric esters of hydroxycinnamic acids	Grape, white & red	Homogenize with 1% AsA, fractionation on C18 as acidic & neutral	100 × 8 mm, C18 Radial-Pak acidic: 5% HOAc in H$_2$O neutral: 40%–100% ACN	PDA	47

Compounds	Source	Sample Preparation	HPLC Conditions	Detection	Ref.
Vanillic	Olive (Hojiblanca & Cacerena cv) & storage brine	Blend with 80% EtOH + 1% $Na_2S_2O_5$ & EtOAc extraction for brine	Spherisorb ODS-2 (5 μm) H_2O/H_3PO_4/ACN 0%–50% ACN in 40 min	UV 280 nm & PDA scan 200–380 nm	54
Caffeic, ferulic, coumaric, & sinapic acids	Orange juice	Alkali hydrolysis (2 N NaOH, room temp., 4 hr), acidify to pH 4.5, EtOAc extraction	250 × 4.6 mm (5 μm), Adsorbosphere HS C18 THF/2% HOAc-H_2O (21:79)	PDA	80
Ferulic, caffeic, coumaric, sinapic	Orange juice	Centrifuge, SPE on C18	250 × 4.6 mm (5 μm) Bondasil C18 THF/ACN/2% HOAc-H_2O (12:5:83) to THF/2%HOAc in H_2O (35:65) in 28 min	UV 280 nm, 324 nm, & FL	81
Gallic, protacatechuic, PHBA, syringic, vanillic, caffeic, p-coumaric, & ferulic	Palm products (*Phoenix dactilifera*), dates & legmi	Crush, acidify to pH 2.0, Et_2O extraction	μBondapak C18, dioxane/2% HOAc (15:85)	UV 280 nm	111
Chlorogenic, neochlorogenic, caffeic	Peach (15 cv)	MeOH extraction & fractionation on C18	Radial-Pak C18 acidic: 5% HOAc in H_2O neutral: 40%–100% ACN	PDA & MS	170
Chlorogenic, caffeic, p-coumaric, ferulic	Potato tubers (*Solanum tuberosum*)	Slice, mixing with 85% MeOH, conc., CC on Corasil	Zorbax ODS 250 × 4.6 mm (5 μm) A: HOAc/H_2O (2:98) B: HOAc/ACN/H_2O (2:30:68) 10% B–100% B in 30 min	UV 280 nm	163
PHBA, gentistic, vanillic, caffeic, syringic, p-coumaric, ferulic, sinapic, salicylic, isoferulic, o-coumaric	Soybean products	Reflux with 80% EtOH for 1 hr, hydrolysis, (2N NaOH, room temp., 6 hr), SPE on C18	Ultrasphere-ODS C18 A: 5% HOAc in H_2O B: 5% HOAc in MeOH 0% B–100% B in 55 min	UV 280, 254 & 200 nm	82
cis/trans-Caffeic, cis/trans-ferulic, cis/trans-isoferulic, & coumarins	Standards	—	250 × 4.0 mm LiChrosorb RP-8 (5 μm) THF/5% HOAc (13:87)	UV 290 nm	138
Benzoic acids, cinnamic acids, phenols, & phenolic aldehydes	Standards	—	300 × 3.9 mm Novapak C18 A: H_2O/HOAc (98:2) B: H_2O/MeOH/HOAc (68:30:2) 0% B–80% B in 59 min	PDA	166
Chlorogenic acid	Sunflower seed meal	Defat, Soxhlet extraction with 80% EtOH, redissolve in 1 M NaOAc buffer	50 cm × 2 mm i.d. Pellidon polyamide, 0.1 M citric acid	ED	254

Table 8 (*Continued*)

Component	Source	Extraction	Analysis method	Detector	Ref.
Gallic	Tea, black (clone SFS 204)	Liquor	Hypersil ODS (5 μm) 250 × 4.9 mm i.d., A: 2% HOAc B: ACN 8% B–31% B over 50 min	PDA 280–460 nm	140
3,4-Dihydrocoumarin, 6-methylcoumarin, coumarin	Vanilla flavorings	Dilute with 50% MeOH, filter (0.45 μm)	μBondapak C18, 300 × 3.9 mm, (10 μm) MeOH/H₂O (40:60) + 1% HOAc	UV 275 nm	220
Gallic, chlorogenic	Vegetable (cormels) (*Colocasis* & *Xanthosoma* sp.)	Blend with cold Ace/H₂O (85:15) for 5 min, wash with Hex, SPE on C18	μBondapak C18 (5 μm) A: 0.02% HOAc in H₂O B: ACN 9% B–31% B in 50 min	PDA scan 220–600 nm	71
Ferulic acid	Wheat, ground	Acid hydrolysis (0.1 M H₂SO₄ × 30m), α-amylase (55°C × 60 m) treatment	100 × 4.6 mm Hypersil ODS (5 μm), 12% MeOH in 0.01 M citrate buffer (pH 5.4)	PDA 320 nm & FL 312 nm/418 nm	57
Ferulic, 4-coumaric, gallic, 4-hydroxy-benzoic, vanillic, syringic, & phenols	Wheat straw	Hydrolysis with MeOH/H₂O/H₂SO₄ (16.6:83:0.4) for 30 min at 70°C	120 × 4.6 mm Viosfer MeOH: 0.1% THF (15:85)	HPLC/MS	176
p-Coumaric, vanillic, & phenols	Wheat straw, lignin	Alkaline hydrolysis (2 M NaOH, nitrobenzene 160°C × 2 hr), CH₂Cl₂ extraction	120 × 4.6 mm Viosfer C6 (5 μm) MeOH/0.1% HClO₄ (15:85)	ED & UV 280 nm	139

Compound	Source	Sample preparation	Column & conditions	Detection	Ref.
Caffeic, caftaric	Wine	Hydrolysis (H₂SO₄, N₂, pH 1.0), filter	250 × 4.6 mm Brownlee Lab C18 A: 80% ACN/0.01 M NH₄H₂PO₄ (14:86) B: 80% MeOH/0.01 M NH₄H₂PO₄ (17:83) 17% B–50% B in 30 min	UV 320 nm, NMR	145
Esculetin, umbelliferone, scopoletin, methyl-umbelliferone	Wine & brandies	—	Ultrasphere ODS (5 µm) A: 3% HOAc in H₂O B: 3% HOAc in ACN	UV 313 nm, FL 340/425 nm	224
Gallic	Wine & grape seeds	Remove EtOH under vacuum, SPE on C18	250 × 4.6 mm, Altex C18 (5 µm) A: 0.05 M (NH₄)₂PO₄ B: ACN 10% B–20% B over 30 min	ED & UV 280 nm	91
Caffeic, gallic, syringic, ferulic vanillic, & m-, p- & o-coumaric acids	Wine & sherry	SPE on C18	165 × 4.6 mm & 250 × 2.1 mm (5 µm) Spherisorb ODS-2, gradient MeOH/KH₂PO₄	UV 230 & 280 nm	27
cis/trans-Caffeic, cis/trans-p-coumaric, ferulic, hydroxycinnamic tartrates	Wine, red	Remove EtOH by rotary evaporation, SPE on C18	Brownlee C18, 32°C A: ACN B: H₂O, pH 2.6 3% A–100% A in 45 min	UV 280, 313 & 365 nm	23
Gallic, cinnamic acid derivaties	Wine, red	Direct injection	Novapak C18 150 × 3.9 mm i.d. (4 µm) A: 50 mM NH₄H₂PO₄, pH 2.6 B: 20% A + 80% ACN C: 0.2 M H₃PO₄, pH 1.5 100–0% A; 0–80% B; 0–20% C in 60 min	PDA	143

Ace, acetone; ACN, acetonitrile; AsA, ascorbic acid; CC, column chromatography; DMSO, dimethylsulfoxide; ED, electrochemical detector; EtOAc, ethyl acetate; Et₂O, diethyl ether; EtOH, ethanol; FL, fluorescence; GFC, gel filtration chromatography; Hex, hexane; HOAc, acetic acid; HPLC, high-performance liquid chromatography; MS, mass spectrometry; MeOH, methanol; NMR, nuclear magnetic resonance; PC, paper chromatography; PDA, photodiode array; PHBA, p-hydroxybenzoic acid; SPE, solid phase extraction; TFA, trifluoroacetic acid; THF, tetrahydrofuran; UV, ultraviolet.

Table 9 High-Performance Liquid Chromatography of Flavonoids in Foods

Component	Source	Extraction	Analysis method	Detector	Ref.
Flavone: apigenin glycoside Flavonol: glycosides of quercetin, isorhamnetin, kaempferol, myricetin	Bee pollen	MeOH, Hex, SPE on Sephadex LH-20	LiChrospher 100 C18, 5 μ, H_2O/MeOH (65:35) in 20 min	UV 350 nm	72,73
Flavanol: catechin, epicatechin Flavonol: rutin	Coffee pulp (coffee arabica)	MeOH, EtOAc	250 × 2.6 mm, HC ODS/SIL-X A: 2% aq HOAc, B: MeOH 15–50% B in 27 min	UV 280 nm	55
Flavonol	Fruit: apple (*Malus domestica*)	15% HOAc in MeOH	220 × 4.6 mm, Aquapore C18 A: 10% aq HOAc B: ACN 5–20% B in 15 min	UV 350 nm	255
Flavanol: catechin Flavonol: rutin, quercetin, quercetin glycosides	Fruit: apple (*Malus* sp.)	MeOH	Novapak C18 A: aq. TFA B: ACN 20–100%% B in 20 min	UV 270 nm	44
Flavanol: catechin, epicatechin, gallocatechin, dihydroquercetin Flavonol: quercetin glycosides (7)	Fruit: apple (*Malus* sp.)	HOAc in MeOH	220 × 4.6 mm, Aquapore C18 A: aq. HOAC B: ACN 0–20% B in 20 min, 30°C	UV 350 nm	39
Flavanol: catechin, epicatechin	Fruit: apple (*Malus* sp)	EtOH, Hex	150 × 4.6 mm, Rosil C18 A; aq. H_3PO_4 B: ACN/MeOH (2:3) 0–50% B in 50 min, 35°C	PDA	52
Flavanol: (+) catechin, (−) epicatechin	Fruit: apple (*Malus sylvestris*)	80% MeOH in H_2O, SPE on C18	250 × 4.6 mm, Econosil C18 A: aq. 5% HOAc B: ACN 12–20% B in 10 min	UV 254 & 280 nm	94
Flavanol: epicatechin Flavonol: quercetin glycosides (glucoside, galactoside, xyloside, arabinoside, rhamnoside)	Fruit: apple (*Malus* sp.)	MeOH, Hex, SPE C18	100 × 8 mm, C18 Radial-Pak A: aq. 5% HOAc B: ACN 0–40% B in 42 min	PDA scan MS	93,256
Flavanol: catechin, epicatechin Flavonol: myricetin, quercetin, kaempferol Flavonol glycosides (14)	Fruit: apple, apricot, grape, orange, peach, pear, pineapple (juice)	Et_2O, EtOAc	300 × 3.9 mm, Novapak C18 H_2O/MeOH/HOAc (57.5:37.5:5) & H_2O/THF/HOAc (80:17.5:2.5) A: aq. 2% HOAc B: H_2O/MeOH/HOAc (68:30:2) 0–85% B in 65 min	PDA 254, 280, 340, 365 NM	75

Compound	Sample	Extraction	Column, conditions	Detection	Ref.
Flavanol: catechin, epicatechin; Flavonol: rutin, quercetin glycoside	Fruit: apple, grape, pear (juices)	Filtration	250 × 4.6 mm, 5 m, Supelcosil C18; A: 70 mM KH_2PO_4 (pH 2.5); B: MeOH; 2–40% B in 60 min	PDA 280 & 320 nm	186, 106, 184
Flavanone: neoeriocitrin, naringin, neohesperidin; Flavonol: quercetin and kaempferol glycosides and glucuronides, rutin	Fruit: apricot, peach, plum, strawberry, sour orange, apple, pear (jams)	BuOH, SPE Amerlite XAD-2; Hydrolysis	125 × 4.0 mm, LiChroCART C18, 5 μ; A: aq HCO_2H; B: MeOH; 20–80% B in 35 min	PDA 280 & 350 nm	227
Flavonol: quercetin, kaempferol, myricetin; Flavone: apigenin, luteolin	Fruit: blueberry, cranberry, bilberry	3% TFA, EtOAc, hydrolysis	250 × 4.6 mm, Whatman ODS-3, 5μ, aq. HCO_2H/ACN/MeOH (65:10:25); 250 × 4.6 mm, YMC ODS-AM, 5μ; A: aq. 1% HOAc; B: ACN; C: MeOH (75:10:15)–(50:20:30) in 60 min	UV 340 & 360, MS, ^{1}H NMR	169
Flavone: sinensetin, isosinensetin, nobiletin, tetra-*O*-methylisoscutellarein, tangeretin	Fruit: *Citrus* (38 var.)	MeOH	250 × 4.0 mm, 5 μ, Si-60, Merck; A: Hep; B: EtOH; 15–30% B in 12 min	UV 340 nm	257
Flavanone: glycosides	Fruit: *Citrus* (jam)	80% MeOH	125 × 4.0 mm, LiChroCART C18, 5 μ, Merck; A: aq. HCO_2H; B: MeOH; 20–80% B in 25 min	UV 280 nm	228
Flavanone: eriocitrin, naringin, hesperidin; Flavone: tangeretin	Fruit: *Citrus unshiu* and *C. grandis*	MeOH/DMSO (1:1), SPE on C18	250 × 4.0 mm, LiChrospher 100 C18; A: aq. 10 mM H_3PO_4; B: MeOH; 30–100% B in 100 min	PDA, UV 200–360 nm	29
Polyphenols, tannins	Fruit: grape (*Vitis vinifera*)	MeOH and 1% HCl	Sephadex G & LH120, Shodex S-802/S, and Ultrahydrogel 120, H_2O, H_2O/HOAc, or H_2O/MeOH/HOAc	UV 280 nm	258

Table 9 (*Continued*)

Component	Source	Extraction	Analysis method	Detector	Ref.
Flavanol: catechin, epicatechin, epicatechin gallate	Fruit: grape seed (*Vitis vinifera*)	70% Ace, CHCl$_3$, EtOAc, hydrolysis	250 × 4.6 mm, 5 μ Brownlee Spheri-5 C18; A: H$_2$O/H$_3$PO$_4$ (pH 2.6); B: ACN; 11–100% B in 30 min	UV 280 nm	76
Flavanol: catechin, epicatechin, catechingallate	Fruit: grape, white and peach	EtOH, C18 SPE, CC on Sephadex LH-20	100 × 8.0 mm, C18 Radial-Pak; A: 5% aq. HOAc; B: ACN; 0–40% B in 40 min	PDA, MS	25,24, 50,170
Flavanone: naringin, prunin	Fruit: grapefruit (*Citrus paradisi*)	MeOH, SPE C18	250 × 4.0 mm, C18; A: 10 mM aq. H$_3$PO$_4$; B: MeOH; 25–100% B in 55 min	PDA	95
Flavone: luteolin-7-glucoside, apigenin-7-glucoside; Flavonol: quercetin-3-rutinoside and 3-rhamnoside	Fruit: olive (*Olea europaea*)	MeOH	100 × 4.6 mm, Microsphere C18; A: aq. HOAc; B:MeOH; 0–100% B in 60 min	UV 350 nm	209
Flavone: luteolin-7-glucoside	Fruit: olive storage brine	80% EtOH; EtOAc	250 × 4.0 mm, Spherisorb ODS-2; A: H$_2$O/H$_3$PO$_4$; B: ACN; 5–50% B in 40 min	PDA scan 200–380 nm	54
Flavanone: neoeriocitrin, narirutin, naringin, hesperidin, neohesperidin	Fruit: orange, grapefruit, sour orange (juice)	Filtration, SPE on C18	150 × 4.6 mm or 100 × 4.6 mm, 3 μ, C18; aq. HOAc/ACN (82:18); aq. NaH$_2$PO$_4$ (pH 6.2)/ACN (80:20) in 12–18 min	UV 280 nm	193,198, 195
Flavone: isosinensetin, sinensetin, hexamethyl-*O*-gossypetin, tetramethyl-*O*-isoscutellarein, tetramethyl-*O*-scutellarein, nobiletin, heptamethoxyflavone, tangeretin	Fruit: orange (*Citrus sinensis*) & Mandarin	SPE with C18	200 × 2.1 mm, Hypersil C18; A: H$_2$O; B: ACN; C: THF; A/B/C (75.8:21.7:2.5)–(56.8:40.7:2.5) in 16 min, 45°C	PDA 230–400 nm	92

Compounds	Sample	Extraction	HPLC conditions	Detection	Ref.
Flavanone: narirutin, naringin, hesperidin, neohesperidin, didymin; Flavone: Sinensetin, nobiletin, heptamethoxyflavone, tetra-O-methyl-scutellarein, tangeretin	Fruit: orange (juice)	Hex, MeOH, CH$_2$Cl$_2$, SPE on C18	250 × 4.6 mm, C18, 5 μ, A: aq. 1% HOAc, B: ACN/HOAc (99:1), 20–50% B in 10 min	UV 280 nm	196,197
Flavanone: narirutin, naringin, hesperidin	Fruit: orange (juice)	Filter	150 × 4.6 mm, 5 μ, C18, A: aq. 100 mM NaH$_2$PO$_4$ (pH 3.4), B: ACN/100 mM NaH$_2$PO$_4$/MeOH (60:30:10)	Coulometric array, ED	144
Flavanone: neoeriocitrin, narirutin, prunin, naringin, hesperidin, neohesperidin	Fruit: orange & grapefruit (*Citrus aurantium, C. sinensis, C. paradisi*)	DMSO	250 × 4.0 mm, μBondapak C18, 5 μ, H$_2$O/MeOH/ACN/HOAc (15:2:2:1), aq. H$_3$PO$_4$/MeOH (1:3), 6–100% B in 45 min	UV 280 nm, ^{1}H NMR, ^{13}C NMR	178
Flavone: sinensetin, hexa-O-methylquercetagetin, nobiletin, heptamethoxyflavone, tetra-O-methylscutellarein, tangeretin	Fruit: orange oil, Mandarin	CH$_2$Cl$_2$	250 × 4.6 mm, LiChrosorb Si-60, HEP/IPA (60:40) + 0.1% H$_2$O in 25 min	UV 280 nm	226
Flavanone: eriocitrin, neoeriocitrin, narirutin, narirutin, neohesperidin	Fruit: orange, sour orange, grapefruit (juice)	DMF	250 × 4.6 mm, 5 μ, C-18 UHS-Alltech, aq. HOAc/ACN/THF (81:16:3) in 32 min	UV 280 nm	133
Flavanol: catachin, epicatechin	Fruit: peaches	MeOH	80 × 4.6 mm, Pecosphere C18, Perkin-Elmer, A: 0.1% aq. H$_3$PO$_4$, B: MeOH, 5–95% B in 30 min	PDA	225
Flavanol catechin; Flavonol: rutin	Fruit: plum	Hex, EtOH/EtOAc	250 × 4.6 mm, 5 μ, Spherisorb ODS-2, A: 5% aq. HOAc, B: ACN/H$_2$O/HOAc (80:15:5), 0–22% B in 50 min	UV 280 & 313 nm	210
Flavonol: quercetin, kaempferol, glycosides, glucuronides, acylated glycosides	Fruit: raspberry, red (juice)	SPE, polyamide-6	250 × 4.6 mm, 5 μ, Spherisorb ODS-1, A: 1% aq. HOAc, B: ACN, 16–100% B in 77 min	PDA 360 nm	183

Table 9 (*Continued*)

Component	Source	Extraction	Analysis method	Detector	Ref.
Flavone: rhiofolin, neodiosmin Flavanone: neoeriocitrin, naringin, narirutin, hesperidin, neohesperidin, poncirin, prunin, naringinin, hesperetin	Fruit: sour orange (*Citrus aurantium*)	MeOH, dioxane/MeOH (1:1), NaOH or KOH in MeOH, pyridene, DMF, or DMSO	250 × 4.0 mm, μBondapak C$_{18}$ H$_2$O/MeOH vs. H$_2$O/H$_3$PO$_4$/MeOH vs. H$_2$O/HOAc/MeOH vs. H$_2$O/HOAc/ACN vs. gradient systems	PDA 280 nm	192
Flavanones: chiral separation of hydroxy- and methoxyflavanones, prunin, naringin	Fruit: tomato skin and standards	MeOH/H$_2$O (70:30)	250 × 4.6 mm, cellulose triacetate, Chiralcel OA A: Hex/IPA (9:1) B: MeOH/IPA (2:1) 0–60% B in 20 min	UV 254 or 280 nm	260
Flavonol: rutin, quercetin	Grain: flour Other: chocolate, soft drinks, ham	MeOH, SPE on C18	80 × 2.1 mm, 3 μ, Capcell Pak C18-SG 120 H$_2$O/MeOH/0.5 M oxalic acid (36:13:1) in 18 min	UV 350 nm	261
Flavone glycoside: isovitexin	Grain: rice (*Oryza sativa*)	MeOH, SPE on XAD-2	250 × 2.0 cm, Develosil ODS-10 A: H$_2$O B: MeOH 0–80% B in 90 min	UV 280 nm, UV scan, MS, ^{1}H NMR, ^{13}C NMR	177
Flavones, flavanones, dihydroflavanols	Grain: sorghum	H$_2$O/EtOH (1:1) Ace/H$_2$O (7:1)	250 × 4.6 mm, μBondapak C18 A: aq. HOAc, B: MeOH 0–100% B in 45 min	PDA scan 230–360 nm	239
Flavonol: di- and triglycosides	*Herb: Calendula officinalis, Tilia cordata, Ginko biloba*	MeOH	220 × 4.6 mm, Aquapore RP-300 IPA/THF/H$_2$O (10:5:85) or NH$_4$OAc (pH 4.5)/IPA/THF (10:5:85) in 30 min	MS scan *m/z* 260–800, negative and positive ion mode	175
Flavone: glisoflavone, glycyrrhisoflavone, kaempferol 3-*O*-methyl ether	Herb: licorice (*Glycyrrhiza*)	Hex and EtOAc	250 × 4.0 mm, LiChrospher C18 ACN/H$_2$O/HOAc (8:11:1), 40°C	UV 280 nm, IR, MS, ^{1}H NMR and ^{13}C NMR	66, 262
Flavone: pebrillin, gardenin, 5,6-dihydroxy, 7,8,3',4'-tetramethoxyflavone	Herb: *Mentha × piperita*	Et$_2$O, CHCl$_3$	250 × 4.6 mm, C18 Nucleosil A: H$_2$O/MeOH/HOAc (11:9:1) B: MeOH/HOAc (20:1) 2–65% B in 50 min	UV 340 nm, PDA	187, 263, 264

Compound	Sample	Sample preparation	Column and conditions	Detection	Ref.
Flavone (14) Flavonone: Pinobanksin, pinocembrin	Honey	Acidified pH 2.0, SPE on XAD-2	100 × 4.0 mm, 5 μ, LiChroCART C18 MeOH/H$_2$O/HCO$_2$H (50:47:3) in 30 min	UV 340 nm UV-Vis spectra	88
Flavone: chrysin, tectochrysin Flavonol: galangin, quercetin, kaempferol Flavanone: pinocembrin, pinobanksin	Honey	Hex, EtOAc, acid hydrolysis	200 × 3.0 mm, LiChrosorb RP-18 A: aq. H$_3$PO$_4$ (pH 2.6) B: ACN 0–70% B in 50 min	PDA, postcolumn derivatization, KOH	52
Flavone: hydroxy- and hydroxymethoxyflavones	Honey	Acidified to pH 2.0, SPE with Amberlite XAD-2	125 × 4.0 mm, 5 μ, RP-18 LiChroCART A: 5% aq. HCO$_2$H, B: MeOH, 30–80% B in 52 min	PDA 340 and 290 nm	155
Flavanol: catechin, epicatechin Flavonol: myricetin, quercetin, kaempferol	Juice: wine, red	EtOAc	150 × 4.0 mm, Micropak MCH5 C18 A: aq. HClO$_4$ (pH = 2.5) B: MeOH 0–98% B	UV 280 nm	202
Flavanone: hesperidin, naringin, neohesperidin, hesperetin-7-O-glucoside, and dihydrochalcones	Standards	DMSO	250 × 4.0 mm, 5 μ, μBondapak C18 H$_2$O/HOAC/ACN from (80:0.5:20) to (75:0.5:25) in 7 min	UV 280 nm	259
Flavonones: chiral separation of methoxy- and hydroxyflavanones	Standards	—	250 × 4.6 mm, 1) Chiralcel OD 2) Chiralpak OP+ 3) ChiraSpher 4) Cyclobond I 5) Cyclobond I acetyl 6) Cyclobond II a) MeOH B) Hex/IPA C) Hex/MeOH/IPA d) Hex/dioxane	UV 254 nm, Polarimetric	265
Flavanone Flavanol Flavone	Standards	—	125 × 4.0 mm, LiChroCART C18 A: aq. 5% HCO$_2$H B: MeOH, 40–80% B in 45 min	UV 200–360 nm	150
Flavone, flavonol, flavanone (141 standards)	Standards	—	250 × 4.6 mm, LiChrosorb C18 A: aq. 5% HCO$_2$H B: MeOH 0–80% B in 34 min, 35°C	UV 280 nm	158

Table 9 (*Continued*)

Component	Source	Extraction	Analysis method	Detector	Ref.
Theaflavins Flavonol: glycosides of quercetin & kaempferol	Tea, black (77 commercial samples)	Hot H_2O	Condition not listed	UV 380 nm	219
Flavonol: quercetin & kaempferol rhamnodiglucosides	Tea, black (Camellia sinensis L. O. Kuntze)	MeOH, SPE on polyamide column, benzoylation, enzyme hydrolysis and silization of sugars	50 × 4.6 mm, 5 μ, Hypersil C18, aq. 0.2% HOAC/ACN (85:15) in 40 min	PDA MS	59
Theaflavins, thearubigins Flavonol: glycosides	Tea, black (Lattakari assam)	CHCl$_3$, EtOAc	100 × 4.6 mm, Hypersil C18, Shandon A: aq. 0.5% HOAc B: H$_2$O/ACN/HOAc (69.5:30:0.5) 0–100% B in 35 min	Vis scan 370–550 nm	77
Flavone: 5,7-dimethoxyapigenin	Vegetable: asparagus (*Asparagus officinalis*)	MeOH/CHCl$_3$/H$_2$O (12:5:3), SPE on cellulose and C18	220 × 6.0 mm, C18 A: H$_2$O B: MeOH 55–65% B in 5 min	UV, ^{1}H NMR	266
Flavonol: catechin, epicatechin, Flavonol: rutin, quercetin	Vegetable: beans, broccoli, carrots, peas, spinach	MeOH, Hex, SPE C18	100 × 8.0 mm, C18 Radial-Pak A: aq. 5% HOAc, B: ACN 0–40% ACN in 40 min	PDA	68
Flavonol: 7,3'4'-trihydroxy, 7,4'-dihydroxy, geraldone Flavonol: kaempferol, 9 glycosides of kaempferol, quercetin	Vegetable: broad bean (*Vicia faba*)	MeOH, Et$_2$O, BuOH	LiChrospher 100–C18, 5 μ, aq. 5% HCO$_2$H/MeOH (1:1) and (7:3) in 18 min	UV 340 nm	203
Flavonol: acylated kaempferol and quercetin glycosides	Vegetable: cabbage (*Brassica oleracea*)	70% MeOH, CHCl$_3$, EtOAc, acid hydrolysis, and sugar silization	100 × 4.0 mm, Superpac ODS2, 5 μ, H$_2$O/MeOH/TFA (79.9:20:0.1)	UV, MS, ^{1}H NMR, ^{13}C NMR	136
Isoflavone: glycosides	Vegetable: chickpea (*Cicer arietinum* L.)	MeOH/Ace	250 × 4.0 mm, LiChrosorb RP-8 or RP-18 A: aq. HOAc B: ACN 20–80% B in 40 min	UV 287 nm	267

Compound	Sample	Extraction	HPLC conditions	Detection	Ref.
Flavanol: catechin, epicatechin, epigallocatechin	Vegetable: edible corms (*Colocasis esculenta* & *Xanthosoma sagitifolium*)	Ace/H₂O, Hex, SPE on C18	300 × 3.9 mm, μBondapak C18, 5 μ; A: aq. HOAc; B: ACN; 9–31% B in 50 min	PDA scan 220–600 nm	71
Flavonol: kaempferol glycosides	Vegetable: endive (*Cichorium endiva* L.)	EtOH, Pet, EtOAC	200 × 4.0 mm, 5 μ, Chrompak C18; A: H₂O; B: ACN; 0–40% BB in 35 min	UV 280, 325, 260 nm	67
Flavonol: quercetin, kaempferol, myricetin; Flavone: apigenin, luteolin	Vegetable: fresh and processed (28) fruit: (9)	Hydrolysis & extraction HCl + MeOH/H₂O (1:1)	50 × 3.9 mm, 4 μ, Novapak C18, aq. 25 mM KH₂PO₄ (pH 2.4)/ACN (75:25) aq. 25 mM KH₂PO₄ (pH 2.4)/MeOH (55:45) in 25 min	UV 370 nm, PDA 220–450 nm	244, 216
Flavanol: catechin, epicatechin	Vegetable: lettuce (*Lactuca sativa*)	MeOH	250 × 4.0 mm, BioSil C18; A: aq. 100 mM (NH₄)₃PO₄ (pH 2.8); B: MeOH; 16–64% B in 30 min		242, 268
Flavone: luteolin	Vegetable: peanut hull	Hex, CHCl₃, Ace, alcohol	125 × 4.0 mm, LiChrospher C18; H₂O/MeOH (75:25)	UV 254 nm	70
Isoflavone: daidzein, genistein	Vegetable: soy (*Glycine max* L.)	80% MeOH	300 × 4.6 mm, Aquapore C8; A: aq. 100 mM TFA; B: ACN; 0–46.4% B in 21 min	UV 262 nm, MS ^{1}H NMR	179
Isoflavone: daidzein, genistein	Vegetable: soybean (*Glycine max* L.)	MeOH	250 × 4.6 mm, TOSOH C18; A: H₂O; B: MeOH; 50–100% B in 20 min	UV 254 nm	208
Polyphenols, tannins	Wine	Filtration	100 × 4.6 mm, 3 μ, C18; A: aq. 50 mM NH₄HPO₄ (pH 2.6); B: ACN; 0–80% B in 55 min	PDA UV-Vis	269

Table 9 (*Continued*)

Component	Source	Extraction	Analysis method	Detector	Ref.
Flavanol: catechin, epicatechin Flavonol: rutin	Wine	Filtration	150×3.9 mm, 4 μ, C18 Novapak, A: 50 mM $NH_4H_2PO_4$ (pH 2.6) B: ACN 0–64% B in 60 min	PDA, UV-Vis	143
trans-Resveratrol	Wine	EtOAc, UV radiation	ODS-II Whatman, H_2O/ACN/HOAc (70:29:0.1)	UV 280 nm	37
Flavanol: catechin, epicatechin, catechin-gallate Flavonol: myricetin, quercetin, myricetin glycosides	Wine, red	C18 SPE at pH 7.0	220×4.6 mm, C18 A: H_2O (pH 2.6) B: ACN 3–100% B in 45 min, 32°C	UV 280 nm	23

Ace, acetone; ACN, acetonitrile; BuOH, butanol; DMF, dimethylformamide; DMSO, dimethylsulfoxide; ED, electrochemical detector; EtOAc, ethyl acetate; Et_2O, diethyl ether; EtOH, ethanol; Hep, heptane; Hex, hexane; HOAc, acetic acid; IPA, isopropanol; MS, mass spectrometry; MeOH, methanol; NH_4OAc, ammonium acetate; NMR, nuclear magnetic resonance; PDA, photodiode array; Pet, petroleum ether; SPE, solid phase extraction; TFA, trifluoroacetic acid; THF, tetrahydrofuran; UV, ultraviolet; UV-Vis, ultraviolet-visible.

based on mass-to-charge ratio. Flavonoids are weak acids and thus become charged at high pH. For any given buffer strength, a higher pH buffer will increase the separation and migration time of a flavonoid mixture. The most common method of detection is UV adsorption, i.e., a small transparent portion at the end of the capillary functions as the detector cell. Separation of neutral and weakly polar components by CE was first described by Terabe et al. (149) who added a surfactant to the buffer. The surfactant forms micelles around neutral molecules and the charged micelles migrate based on micellar electrokinetic capillary chromatographic (MECC) migration.

The number of flavonoid separations published is limited due to the recent introduction of commercial instrumentation (Table 10). The first publications utilizing commercial instrumentation flavonoid separations utilize MECC to demonstrate the separation and efficiency capabilities of the technique with standard compounds (150–154). Ng and co-workers (151) optimized analytical parameters for separation in less than 20 min of a mixture of seven flavonoids (flavone, two flavonols, flavonol glycoside, two flavanones, flavonone glycoside) of widely differing polarity (151). This demonstrated the capability of MECC to quickly separate many different classes of flavonoids. Morin et al. (152,153) demonstrated the ability of MECC to separate flavonol 3-O-glycosides with differences only in the flavonoid aglycone (152) and only in the sugar moiety (153). Delgado and co-workers (150) compared MECC to HPLC for the separation of a standard mixture of flavonoid found in honey. The time required for the MECC separation was less than half of that required for HPLC.

While separation of a standard mixture allows one to optimize the conditions for the components of interest, plant and food extracts are complex mixtures, making development of a successful analytical method more difficult. Ferreres and co-workers successfully used MECC to differentiate honey from different pollen sources based on flavonoid composition (86). Separation of a standard flavonoid mixture and honey extract are shown in Fig. 9a and b. The MECC separation of honey extracts is significantly shorter (30 min vs. 60 min) than the corresponding HPLC separation utilizing gradient elution (155). Bjergegaard and co-workers (156) measured the separation efficiencies for flavonol glycosides separated from extracts of rape seed (*Brassica napus*), yellow mustard (*Sinapis alba*), and broccoli (*Brassica oleracea*). Calculated efficiencies ranged from 77,000 to 227,000 theoretical plates per meter of capillary.

McGhie (157) quantitatively analyzed flavone and flavone glycoside components in sugarcane extracts by CE with a borate buffer without surfactant. Their results demonstrated that CE is a viable alternative to analysis by either PC or HPLC. The analysis time required for CE was much shorter than that required for HPLC (158). Cancalon and Bryan (159) also used a borate buffer to measure flavanone glycosides and other constituents in citrus juices to distinguish citrus products. The time required for analysis of the flavanones by CE was also shorter than that required for analysis by HPLC.

The advantages to analysis by CE and MECC include shorter analysis times than HPLC, higher resolution, and the requirement of very small sample amounts. The disadvantages are that collected UV spectra often are noisy because of the small amount injected and detection limits for components are higher. The analysis of nonpolar components is still better accomplished by GC or HPLC; however, its use is increasing rapidly for the analysis of flavonoids and other constituents in food that are easily separated based on charge.

J. Size Exclusion Chromatography

Size exclusion chromatography (SEC), which refers to gel chromatography, gel filtration (GFC), and gel permeation chromatography (GPC), is a method for obtaining information about the molecular weight and molecular weight distributions of synthetic polymers and naturally occurring

Table 10 Capillary Electrophoresis of Phenolics in Foods

Component	Source	Extraction	Analysis method	Detector	Ref.
Flavanone: didymin, hesperidin, narirutin, naringin, neohesperidin	Fruit juice: (*Citrus* sp.)	Dilution in H$_2$O	35 mM borax (pH 9.3) buffer optimal	UV scan 200–360 nm	159
Flavone: apigenin, luteolin, chrysin, galangin; Flavonol: 8-methoxykaempferol, myricetin, quercetin, kaempferol; Flavanone: pinobanksin, naringinin, hesperitin, eriodictyol, pinocembrin	Honey	Acidified to pH 2, SPE with XAD-2 and Sephadex LH-20	200 mM sodium borate (pH 8), 50 mM SDS, 5, 10, or 15% MeOH, 20 kV	UV 280 nm	86
Flavonol: aglycone and glycosides	Standards	MeOH/H$_2$O (70:30)	Borate–phosphate buffers with acetyltrimethylammonium bromide or cholate and taurine	UV 350 nm	156
Flavonol: glycosides	Standards	—	Boric acid-NaOH or phosphate-borate buffer, 40°C or ambient	UV scan 200–360 nm	153
Flavone: diosmin, isorhoifolin, diosmin-7-O-glucose, linarin, diosmetin; Flavanone: hesperidin, hesperitin	Standards	—	Phosphate or borate (pH 10.5) or Tris-HCl (pH 7.1) buffers, 50 mM SDS, 60°C, 25 kV	UV scan 200–360 nm	152
Flavone: chrysin; Flavonol: quercetin, morin, rutin; Flavanone: hesperitin, naringinin, naringin	Standards	—	50 mM phosphate/50 mM borate pH 7.5 and 42 mM SDS	UV 210 nm	151
Flavanone:; Flavonol:; Flavone:	Standards	—	200 mM boric acid/50 M SDS pH 8.5, 20.9 kV, 27.3 µA, 25°C	UV 200–360 nm	150
Flavone: apigenin, luteolin, schaftoside, isoschaftoside, swertiajaponin, orientin, isoorientin tricin	Sugarcane	80% ACN	25 mM borate buffers pH 9.5 + 20% MeOH, 35°C, optimal	UV 395 nm	157

ACN, acetonitrile; MeOH, methanol; SDS, sodium dodecyl sulfate; SPE, solid phase extraction; UV, ultraviolet.

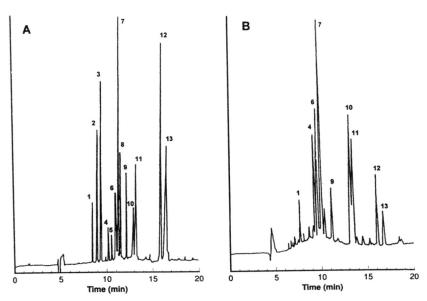

Fig. 9 Separation of standard flavonoid mixture (a) and heather honey extract (b) by MECC using 200 mM borate buffer (pH 8.0), 50 mM SDS, and 10% MeOH. Detection at 280 nm. Peaks: 1 = pinobanksin, 2 = naringinin, 3 = hesperitin, 4 = 8-methoxykaempferol, 5 = myricetin, 6 = quercetin, 7 = leutolin, 8 = eriodictyol, 9 = pinocembrin, 10 = kaempferol, 11 = apigenin, 12 = chrysin, 13 = galangin. (From Ref. 86.)

macromolecules. Separation by molecular size occurs by repeated exchange of the solute molecules between the bulk solvent of the mobile phase and the stagnant liquid within the pores of the packing.

Since Somers's (160) initial work on separation of condensed flavonoid pigments from wine, this technique has been applied to many phenolics such as condensed and hydrolyzable tannins in various food products. For low molecular weight phenolics such as benzoic and cinnamic acids, Lattanzio and Marchesini (161) described that molecular weight has less influence on elution than the presence of phenolic –OH groups, unsaturated carbonyl or carbonyl groups in the solute molecule. However, the effect of the molecular weight on retention is seen from a separation of glycosides of flavonoids. Recently, analytical high-performance liquid chromatography with size exclusion column (HPSEC) of coffee chlorogenics was described by Trugo et al. (162). A wide range of chlorogenic acid isomers are known to be present in coffee, at least five major groups of compounds (163). Extraction and purification of chlorogenic isomers has been accomplished by use of Carrez solutions and separated on a gel filtration stationary phase, TSK-Gel 3000-SW porous spherical and rigid gel, which has a separation range from 1000 to 300,000 of molecular weight. The measurement of total chlorogenic acid (the sum of individual isomers of chlorogenic acid), instead of measuring individual chlorogenic isomers, can be achieved in 20 min with a flow rate of 0.5 ml/min. Results were comparable with (correlation = 0.998) the official Association of Official Analytical Chemists (AOAC) procedure but with lower values. Since SEC technique cannot separate the compounds with similar molecular masses, it is interesting to see the separation between chlorogenic acid and caffeine, which have molecular weights of only 354.31 and 194.2, respectively, using this kind of column. Separation for chlorogenic and caffeine on this column might be based on gel filtration and probably hydrophobic interactions (162).

VI.   DETECTION AND IDENTIFICATION

A.   UV-Vis Detection

Phenolics absorb well in the UV region and the most commonly used detector for HPLC is a variable-wavelength UV or UV-Vis detector. These detectors offer a high degree of sensitivity and are capable of collecting data from one or more wavelengths simultaneously. No single wavelength is ideal for monitoring all classes of phenolics since they display absorbance maxima at different wavelengths (60). UV detection has the disadvantage of not being as sensitive or selective as fluorescence detection and interfering peaks are more common. By selecting the proper wavelength, it is sometimes possible to eliminate interference from a coeluting component if the interfering component does not adsorb at a wavelength adsorbed by those of interest. For maximum sensitivity, usually a wavelength near the maxima is desired; however, in practice, since absorption maxima can differ greatly between compounds (Table 7), wavelengths are set for the best overall detection of all components. Occasionally, in a preparative separation, the wavelength may be set well away from maxima to decrease sensitivity and avoid detector saturation. The values for the absorption maxima of major phenolic acids reported previously are summarized in Table 7.

Most of the benzoic acid derivatives displayed their maxima at 246–262 nm with a shoulder at 290–315 nm, except gallic and syringic acid which have absorption maxima at 271 and 275 nm, respectively (21). The cinnamic acids absorb in two regions in the UV, with one maximum occurring in the range 225–235 nm and the other in the range 290–330 nm (1). The four commonly occurring cinnamic acid derivatives—ferulic, sinapic, caffeic, and *p*-coumaric acids—have maxima at 300 nm (21). Chlorogenic and *p*-coumaric acid exhibited absorption maxima at 310 nm with a shoulder between 270 and 290 nm, and flavonoids for 275–285 nm (60). Figures 10 and 11 present the spectra of common benzoic and cinnamic acids measured in MeOH solution.

The common phenolic acids have been detected at a single wavelength of 254 nm (21,56,94) or 280 nm (40,55,91,111,139,162), by dual monitoring at 254 and 280 nm (94) or at 280 and 320 nm (50,81,106), and by multiple-wavelength monitoring at 254, 275, and 300 nm (21,82). With detection at 320 nm, cinnamic acid derivatives can be detected without any interference from benzoic acid derivatives, which have a higher response at 254 nm. However, detection at 280 nm is the best among the three for the determination of both classes of phenolic compounds (57).

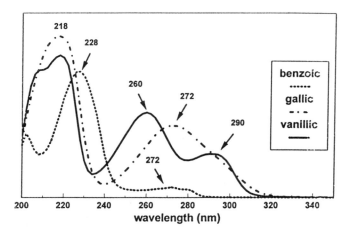

Fig. 10   The absorption spectrum of benzoic acids.

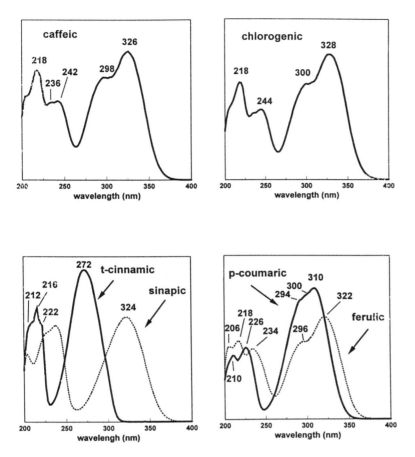

Fig. 11   The absorption spectrum of cinnamic acids.

Because there is a lack of commercial standards for geometric isomers, care needs to be taken when identifying phenolic isomers. Mixture of cis and trans isomers can be prepared by exposure of the standard solutions of *trans*-cinnamic acids to diffused daylight for 2 hr (138). Transformation of trans to cis isomer can occur under UV light. Occasionally, additional peaks were noticed in the latter sections of the HPLC with UV and EC chromatograms. These additional peaks are due to the presence of cis isomers of the cinnamic acid derivatives (56). The possibility of cis and trans isomer forms of cinnamic acid derivatives arises because of a vinyl side chain (56).

B.   Fluorescence Detection

Fluorescence detectors are also used for phenolics (57,81) but have not been applied widely to the detection of phenolics. In beer, there was no advantage found for the detection of phenolic acids by fluorescence over UV detection (164). But in the analysis with orange juice, fluorescence detection offers some major advantage over UV detection, in terms of enhanced selectivity and greater sensitivity (Fig. 12). Dual fluorescence and absorbance detection has been used to simultaneously detect hydroxycinnamic acids and flavanone glycosides in orange juice (81) and for ferulic acid in ground wheat and flour (21). In orange juice, the hesperidin and narirutin, major flavanone glycosides, do not fluoresce. Therefore, they will not interfere with strongly fluorescing

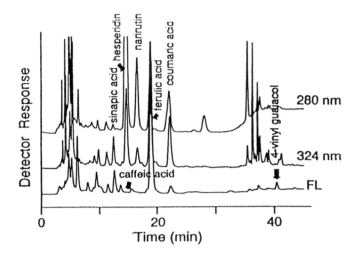

Fig. 12 HPLC for phenolics from Hamlin orange juice using UV and fluorescence detections. (From Ref. 81.)

hydroxycinnamic acids. For ferulic acid in wheat, detection at 280 nm provided the maximum sensitivity for ferulic acid, but detection at 320 nm was preferred because of greater selectivity and ease of quantitation (57). Also, fluorescence with excitation at 312 nm and emission at 418 nm provided good sensitivity and selectivity for ferulic acid. Because of the presence of the ethylene side chain in cinnamic acids, fluorescence can be used for selective detection of cinnamic acid derivatives (57).

In most of the work with quantitation of coumarins, fluorescence detection was employed because of sensitivity and specificity. However, monitoring of the fluorescence at one wavelength was not sufficient to observe all of the coumarin derivatives. Latz and Ernes (165) studied a wide range of fluorescence maxima (314–346 nm for excitation and 420–517 nm for emission) for coumarins as to the possible identity of the individual components and to selectively enhance the fluorescence of coumarins in citrus oils.

C. Photodiode Array Detection

Photodiode array detection, which can yield data in both the time and spectral domains, has led to considerable improvement in HPLC food analysis for identification purpose and demonstrated the usefulness of qualitative information in phenolic analysis based on the absorption spectrum (26–28,47,83).

PDA has three major advantages for HPLC analysis: multiple wavelength detection, peak identification, and peak purity determination. Since PDA can record the characteristic UV spectra of the different phenolics as they elute from the column, characterization and providing of information on the purity of a peak can be facilitated through comparison of the spectra at the front, apex, and tail of each peak. Furthermore, the rapid calculation of absorbance ratios between different wavelengths is possible. Bartolome et al. (166) demonstrated that the ratios A260/A320 nm and A270/A300 nm can partially differentiate the structural features, such as functional group conjugated with aromatic ring, degree of substitution, and position of substituents in some phenolic compounds. For phenolic compounds in apple juices, the absorbance ratios were measured to verify and identify the major phenolic acids and flavonoids in apple juice (60).

A component and standard with identical spectra and retention characteristics provide an additional degree of confirmation that component and standard are identical. Addition of a second pump to deliver postcolumn shift reagents can also be used to observe and record spectral shifts for each eluting component providing additional information. Previously, the sole domain of GC-MS or LC-MS, peak identification and purity checking, can now be done as part of the HPLC analysis at a lower cost. Commonly, identification of phenolic compounds in HPLC analysis was often performed by comparing the retention times and spectral characteristics of their peaks with those of standards. For the peak purity check, spectra from upslope, downslope, and peak apex are used to compare. However, for isomers, UV-Vis spectral information alone cannot be used for positive identification because coeluting isomers can lead to spectra representing a mixture of isomers. Additional means of identification should be used in interpreting HPLC separation, such as mass spectrometry and nuclear magnetic resonance.

D.   Electrochemical Detection

Electrochemical detection (ED) is very sensitive for compounds that can be oxidized or reduced at low-voltage potentials. ED is becoming increasingly important for the determination of very small amounts of phenolics, as it shows enhanced sensitivity and selectivity. ED has been applied in the detection of phenolic compounds in beer (56,61,90), wine (91), beverages (167), and olive oils (168). This procedure involves separation of sample constituents by liquid chromatography prior to their oxidation at a glassy carbon electrode in a thin-layer electrochemical cell. In order to find the optimum amperometric detection potential for the phenolic compounds, preliminary study using cyclic voltammetry is necessary. Voltammetric study can be performed at a glass carbon working electrode with a scan rate of 50 mV/sec (56). Hayes et al. (56) found that a detection potential of +0.9 V appears suitable for the detection of all phenolic compounds in beer except p-hydroxybenzoic acid. For p-hydroxybenzoic acid, a detector potential of approximately +1.20 V was shown to be necessary for detection (56).

Electrochemical detection has the advantage of higher sensitivity than UV detection and also offers greater selectivity (6,168), which is very useful when analyzing real samples as it reduces matrix effects and consequently improves the quantitation and identification of analyte peaks (Fig. 13). The detection limit for ED is about two times higher than UV in sinapic, ferulic, p-coumaric, and caffeic acids (56). Furthermore, ED is almost insensitive to the changes in mobile phase conditions associated with gradient elution. Thus, steady baselines can be achieved at high detector sensitivity settings. Besides sensitivity, ED detection in conjuction with PDA can be used in series to classify on the basis of conjugation pattern and hydroxyl substitution of wine and grape flavonoids (167).

However, fouling of the working electrode by adsorbed phenols or their oxidation products was encountered (168), necessitating the removal of the electrode for cleaning purposes. Precautions should be taken not to cause damage of the electrode surface while cleaning and to prevent trapping of air bubbles in the cell while replacing the electrode in the detector cell.

E.   Coulometric Array Detection

Coulometric array detection allows increased resolution for HPLC analysis of phenolic based on differences in their voltametric properties. The basis for differences in ease of oxidation within and among compound classes corresponds to the patterns of aromatic substituents. Gamache et al. (144) used 16 serial coulometric detectors to generate voltammetric data for phenolic compounds in fruit juice beverages. Differentiation within this group is based on variations in the substitution pattern of alkoxyl alkyl, hydroxyl, and glycosidic groups on the aromatic rings. Gamache et al. (144) described that the rank of ease of oxidation for flavonoids is a flavonoid having the following aromatic substitution: catechol > O-hydroxyl, methoxyl > monohydroxyl > methoxyl. Compounds

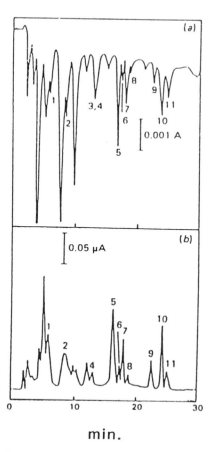

Fig. 13 HPLC for phenolics in beers using (a) UV and (b) electrochemical detection. Peaks: 1 = gallic; 2 = protocatechuic; 3 = p-hydroxybenzoic; 4 = (+)catechin; 5 = vanillic; 6 = caffeic; 7 = syringic; 8 = (−)-epicatechin; 9 = p-coumaric; 10 = ferulic; 11 = sinapic acid. (From Ref. 61.)

having catechol groups such as caffeic acid, catechin, and chlorogenic acid all responded at lower channels of 120 mV dominant oxidation potential. This technique may be utilized to detect adulteration and source classification in many fruit juice products.

F. Mass Spectrometry, Nuclear Magnetic Resonance, and Fourier Transform Infrared

Numerous papers have relied on only UV-Vis spectra for their identification of phenolics, but for positive identification purposes, mass spectrometry (MS), nuclear magnetic resonance (NMR), and fourier transform infrared (FT-IR) could be advantageous detection techniques, especially to validate HPLC or GC determination of phenolics in foods. However, the cost and complexity of their instrumentation limits the widespread utilization of this detection mode.

With phenolics, IR spectra have been less frequently used for identification. However, after GC, the combination of the two techniques (IR and MS) can be a powerful tool for identification of individual phenolics, especially for positional isomers. Williams et al. (132) demonstrated the GC-IR-MS techniques for identification and quantitation of 50 phenolic compounds. For coeluting compounds, spectral substraction (IR) and selected ion techniques (MS) were effective. Mass spectra of positional isomers are usually virtually identical but IR spectra of such isomers can be very different. In view of high discriminating properties of IR spectrometry, the technique might become a valuable tool in addition to GC-MS.

Mass spectrometry is another detection mode that can provide detection of all phenolic compounds in foods. After GC or preparative HPLC, the MS technique has more frequently been employed for structural identification of phenolics in many foods and plants (66,67,136,169–172) because of its sensitivity, selectivity, and ability to provide structural information. With GC-MS for polymethoxylated flavones, Berahia et al. (173) identified and proposed new fragmentation pathways in addition to the retro-Diels-Alder reaction, which gives a characteristic fragment (B^{+0}) from the phenyl group of the flavone skeleton.

LC-MS can represent a fast and reliable method for structural analyses for nonvolatile compounds such as phenolic compounds (174,175), especially for low molecular weight plant phenolics (176), but the limited resolving power of LC hinders its widespread application for phenolics as compared to GC-MS. GC-MS also has the advantage over HPLC-MS of a much better limit of detection. A variety of different ionization modes can be used to provide reproducible ion formation. After preparative HPLC, LC-MS techniques with plasma spray interface have been employed for identification of some flavones in orange and mandarin essential oils (174). Pietta et al. (175) showed that LC thermospray (TSP) MS is an excellent technique for analysis of flavonol glycosides from medicinal plants. Since flavonol glycosides are thermolabile compounds, fragmentation patterns were used to identify the structure. TSP ionization gave parent species with few and diagnostic fragment ions, thus allowing structure elucidation as well as discrimination between different glycosylation sites. Since TSP provides information on molecular ions, sugar sequence, and glycosylation site, it can be of high value for the identification of phenolic compounds occurring in foods. For wheat, Galletti et al. (176) described that HPLC-MS has potential as a technique for determining the presence of useful marker compounds which are indicative of the progress of the lignification during plant growth. Moreover, simultaneous detection by flow splitting or connecting serially with PDA can also provide confirmatory data based on UV spectra. The LC-MS method is reliable and rapid although not particularly sensitive (detection limits were of the order of 1 µg for phenolic acids) and has an important future for analysis of food phenolics.

NMR (^{1}H and ^{13}C) spectroscopy is one of the most valuable methods for identification and characterization of structural analysis. Many applications with proton-NMR (^{1}H NMR) and carbon-13 NMR (^{13}C NMR) demonstrated effectiveness of NMR spectroscopy in the study of acylated flavonol glycosides from cabbage leaves (136), *C*-glycosyl flavonoid from rice hull (177), flavanone glucosides in *Citrus* species (178), isoflavones in soybean foods (179), major flavonol glycosides from blueberry, bilberry, and cranberry (169), enzymatic oxidation products of phenolics in wine (145), coumarins from the fruits of *Ammi majus* (180), coumarins from the juice oil of whole fruits of *Citrus hassaku* (181), phenolic analysis in Rutaceae species (79), and for a new group of bound phenolic acids from oat (182). Their structures were elucidated mainly on the basis of one- and two-dimensional NMR spectroscopy, and in addition to EI or FAB mass spectra, NMR spectrum was recorded to support further confirmation of the positions of the substituents in the phenolics.

VII. APPLICATIONS IN FOOD ANALYSIS

Chromatographic analysis for phenolics may be a useful tool in such areas as characterization, chemotaxonomic study, verification of authenticity, detection of adulterations, varietal and source classification, study of the process and subsequent storage-related variability, and even study of food phenolics in terms of health benefits.

A. Fruits and Fruit Products

Analysis of phenolics in fruits, fruit juices, and wine is often performed by gradient procedures due to the presence of several phenolic classes. The nonanthocyanin flavonoids in most fruits consist mainly of flavonols and flavanols with trace amounts of flavones (183,184). Glycosides are the predominant forms present. These most often are separated by RP-HPLC on C18 columns as stated above with gradients consisting of acidified H_2O and ACN, MeOH, or EtOH. As can be seen from Tables 8 and 9, many different sample preparation procedures have been employed ranging from simple filtration of juice products to solvent extraction and extraction by SPE.

SPE fractionation and RP-HPLC techniques were applied to the study of the seasonal and cultivar variations in the amount of phenolic compounds during ripening of grape berries (24). Neutral and acidic grape phenolics can be fractionated using a solid phase extraction technique and analyzed with different mobile phases (24,25,47,50). The fractionation technique was simple and rapid for the analysis of the major phenolic compounds in grapes, especially for phenolics other than anthocyanins in red grape. Acidic phenolics were determined by isocratic elution with 5% HOAc in water and neutral phenolics were determined by binary gradient elution using acetonitrile (24,25,47,50). Hydroxycinnamic acid tartrates, flavan-3-ols, and oligomeric procyanidins were the major important phenolics in grape. The quantitative and qualitative differences of phenolics in grape berry flesh and skin could help to explain the differences in browning potential of grapes in relation to cultivars and seasonal variations, as well as juices processed with a different degree of extraction (24).

For analysis of flavonols in red raspberries, Rommel and Wrolsted (183) found conventional materials to be ineffective in removing unwanted components from extracts. They resorted to packing small polyamide columns for prefractionation prior to analysis. They obtained a fraction with MeOH containing flavonol glycosides and aglycones and a second fraction of NH_3/MeOH containing flavonol glucuronides, flavonol-*C*-glycosides, and aglycones. Using a gradient of aqueous acid and ACN over 60 min, they found that deactivated C18 columns with endcapping did not perform as well as the Spherisorb ODS-1 column having a low carbon load and no endcapping. Evidently, the hydroxy and silanol interactions play an important role in the separation of these components.

Spanos and co-workers analyzed pear (185), apple (186), and grape (106) juices for phenolic acids and flavonol glycosides by simple filtration of the juice while anthocyanins were isolated by SPE on Sephadex LH-20. They used a gradient of 2–40% MeOH in 70 mM phosphate buffer (pH 2.5) on a Supelcosil C18 column. Phenolic acids eluted in the first 50 min, followed by rutin, quercetin galactoside, quercetin glucoside, phlorizidin, quercetin xyloside, quercetin arabinoside, quercetin rhamnoside, and three glycosides of isorhamnetin. Using the same mobile phase, the anthocyanins in the Sephadex LH-20 fraction were separated along with (+)-catachin and (–)-epicatachin. Figure 14 represents HPLC chromatograms of Thompson seedless grape juice phenolic acids and flavonoids.

Since some apple cultivars are difficult to identify from samples of fruit, quantitative HPLC analysis of polyphenolic compounds in the fruit with corresponding analysis of the polyphenol profile may be useful in taxonomic studies of apple cultivars for classification and identification. HPLC was applied to determine the hydroxycinnamic acid esters with glucose or aldaric acids in the peel and pulp of orange, grapefruit, and lemon (88). Citrus contains specific flavanone glycosides that are not common in the plant kingdom. Analysis for these glycosidic compounds is suggested as they are useful taxonomic markers. Thus, precise quantitative data on the

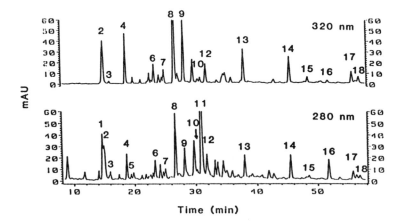

Fig. 14 HPLC chromatogram of grape juice phenolic acids and flavonols. Peaks: 1 = tyrosin; 2 = oxidized cinnamic; 3 = gallic; 4 = oxidized cinnamic; 5 = HMF; 6 = oxidized cinnamic; 7 = *cis*-caftaric; 8 = caftaric; 9 = glutathionylcaftaric; 10 = *cis*-coutaric; 11 = trytophan; 12 = *p*-coutaric; 13 = caffeic; 14 = *p*-coumaric; 15 = ferulic; 16 = hydroxybenzoic ester; 17 = rutin; 18 = quercetin glycoside. (From Ref. 106.).

occurrence of flavonoids in citrus are needed for chemotaxonomic study. Furthermore, in early chemotaxonomic studies, several flavanone glycosides unique to citrus and even to specific cultivars were examined in relation to taxonomic classification (29). HPLC provides polyphenol profiles in the peel and cortex of the apple to be used to characterize apple cultivars by multivariate statistical technique (44).

While the natural occurrence of some phenolic compounds is common in many different plants, some are specific to one class of plant or even a single species. Even with flavonoids that are of common occurrence, amounts present can differ greatly from one plant species to another allowing differentiation through analysis. Based on this knowledge gathered through analysis of authentic species, a database of chemical composition ranges has been acquired for many processed foods and raw materials. With a database, flavonoid patterns can be analyzed through pattern recognition techniques and chemotaxomonic correlations drawn between related plants (44,79,187–191). Applications of analytical HPLC in classification of some fruit cultivars are included in Tables 8 and 9.

Flavonoids in citrus are unique in that they consist primarily of flavanone glycosides. They are 7-*O*-rutinosides and 7-*O*-neohesperidosides of eriodictyol, naringinin, hesperitin, and isosakuranetin. Effective separation of all glycosides present from other components contained in juice and fruit extracts in a reasonable amount of time requires gradient elution with C18 columns due to the complex nature of the juice matrix and differences in polarity between components (192). A great deal of effort has been spent on analytical methodology for separation of these components in citrus products as they are useful in determining the species of fruit from which a product was made (133,144,193–196). Since it is only necessary to quantitate a few of the flavanone glycosides to establish fruit authenticity, isocratic methods are common. Isocratic separations require no reequilibration time between analyses and therefore are sometimes faster when only a few components are to be analyzed. Extracts may be prepared from MeOH or dimethyl sulfoxide (DMSO), or juice products may be analyzed directly after dilution and/or filtration of the sample. Utilizing C18 columns, isocratic solvents of 80–82% aqueous HOAc with ACN (193,198) were used. One of the authors was able to remove minor components interfering with the analysis of naringin and neohesperidin by utilizing lithium acetate– or phophate-buffered mobile phases at pH 5.0–6.3 and deactivated C18 columns. Use of normal C18 columns containing residual silanol groups caused excessive peak tailing with mobile phases not acidified to suppress silanol acitivity. Mouly et al. (133) partially replaced the ACN with THF to remove interfering components from glycosides of eriodictol. For narirutin, hesperidin, and didymin in blood orange juice, the juice sample was simply diluted 10-fold with MeOH and centrifuged, and the clean supernatant was injected into HPLC (197).

The flavones present in citrus are concentrated on the fruit surfaces and are present in processed juices and edible fruit portions at very low levels. Analysis of the flavones in citrus juice requires extraction and concentration to facilitate detection and quantitation. A number of methods have been published for flavones in citrus products (92,196,197,199–201). Sendra and co-workers (92) have perhaps the best method, utilizing a C18 SPE cartridge to isolate flavones from 30 ml of a citrus juice. The cartridge was washed successively with H_2O and 25% ACN. Flavones were then eluted with 55% ACN in H_2O, concentrated, and then redissolved in 0.6 ml ACN. Extracts were analyzed by a reverse phase on a C18 column with H_2O/THF/ACN mobile phase. The THF concentration was kept constant at 2.5% with gradient elution performed in a two-step linear gradient changing the ACN concentration from 21.7% to 40.7% (Fig. 6).

However, when analyzing a raw food or processed products, it is necessary to know not just flavonoid pattern differences between species and varieties (38,72,75,185,194,202) but also how these patterns are influenced and changed during maturation (39,178,185,203–209), food processing (93,106,183–186,210), and storage (93,185,186,192,207) of the food. For example, HPLC techniques for phenolics were used to study the effect of processing, concentration, and storage on the phenolic composition of Granny Smith apple juice (84). The phenolic compound was also analyzed as a potential precursor for an off-flavor compound in orange juices (81,83). Ferulic acid, a specific hydroxycinnamic acid, has been identified as the odorless precursor to the highly malodorous phenols (81). The importance of such monocyclic acids arise from their ability to undergo decarboxylation either by thermal fragmentation or through the activities of microorganisms. Consequently, highly volatile flavor-active phenols are produced (81). Since 4-vinylguaiacol is such a potent deleterious compound in citrus juices and ferulic acid is its putative precursor, a procedure to quantify ferulic acid would be a useful quality control tool (83).

Phenolic analysis has been used for the detection of economic adulteration and especially to verify the authenticity of fruit juices (146). Fernandez de Simon and co-workers (28) analyzed phenolic compounds (flavonols, chalcones, benzoic acids and aldehydes, cinnamic acids, and their

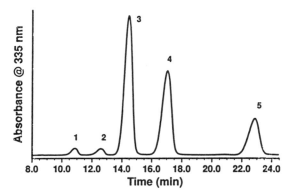

Fig. 15 HPLC separation of methoxyflavones in Valencia orange oil by isocratic elution using H_2O/ACN (60:40) at 1 ml/min on a 150 × 4.6 mm i.d. (3μm) C-18 column. Peaks: 1 = sinensetin, 2 = unknown, 3 = nobiletin, 4 = heptamethoxyflavone, 5 = tangeretin.

derivatives in the form of esters and glycosides) in several commercial juices and nectars to investigate the presence of marker compounds that could be useful in their characterization and differentiation. Analysis for phloretin and isorhamnetin derivatives was used in distinguishing apple and pear juices (211). Flavonol glycosides were suggested as a suitable indicator of adulteration of black currant products with red currants (212) and with blackberries (213). The black currant methanolic extract is purified using polyamide columns and analyzed for flavonoids by HPLC. Anthocyanins are removed by chromatography on Dowex 50W-X4, eluting with 70% MeOH (212). Quercetin-3-O-glucuronide was a suitable indicator to detect adulteration of black currant products with blackberries (213).

Orange juice (*Citrus sinensis*) does not contain either naringin or neohesperidin but grapefruit juice contains both. Thus, the naringin analysis by HPLC can be applied to detect orange juice adulteration with grapefruit juice (214,215). Besides, naringin/neohesperidin ratios can be used to differentiate orange juice which may contain added grapefruit juices from orange juice which may include juices from other naringin-containing cultivars (198). Concentrations of both naringin and neohesperidin can be determined in orange juice by using a single liquid chromatographic isocratic reverse phase system with a C18 column. The detection limit for both compounds is 1.0 ppm (198).

Flavonoids are also used to determine authenticity of raw materials and processed products in terms of species (72,75,216), variety (38,39,44,202), and geographic origin (202,217–219). Through compositional analysis it can be determined if product is authentic or has been adulterated. The flavonoid pattern of orange juice is routinely analyzed as one of several tests to determine if the juice has been blended or adulterated (75,195,198,216). Anthocyanins in wine, cranberry, and grape juice are also part of routine quality tests. Furthermore, this phenolic analysis could be used for the detection of mixtures of fruits in jams. This is especially important when cheaper fruits can be added to more expensive ones in a fraudulent manner. The phenolics present in different commercial jams of apricot, peach, plum, strawberry, sour orange, apple, and pear have been studied and the characteristic compounds for each different jam identified (179).

B. Peel Oils and Flavorings

The high demand for authentic vanilla extract as a flavoring agent has resulted in frequent attempts in adulteration. Thompson and Hoffmann (220) proposed an HPLC method for the quantitation of coumarins as an adulterant in a variety of vanilla flavorings, using a 10-μm μBondapak C_{18} column with MeOH/H_2O (40:60) as the mobile phase. The recovery of coumarin from vanilla extracts spiked at 0.25 mg/ml was 99.9% with the detection limit being about 0.6 ng at a signal-to-noise ratio of 3.

Both normal and reverse phase HPLC have been utilized for coumarin analysis. Normal phase chromatography on a silica gel column in hexane-EtOAC solvent system was known to be superior to reverse phase chromatography with aqueous ACN or MeOH in the resolution of neutral coumarins (221). Phenolic coumarins were resolved by stepwise gradient elution using C18 reverse phase column (5μm) with aqueous MeOH containing HOAC (221). However, normal and reverse phase with a appropriate solvent system complemented each other for the resolution of complex mixtures of coumarins.

For coumarins and psoralens in citrus peel oil, most of the early works were done with normal phase chromatography on silica columns (165,222,223) with isocratic elution with isooctane, heptane, and ethyl acetate. Citrus peel oils can be easily extracted by a hypodermic needle (222). The use of reverse phase methods (31,33,220,224) in citrus oils, fruits, vanilla flavorins, and alcoholic beverages is now more common than that of normal phase methods. Recently, Albanese and Mussinan (225) reported both normal and reverse phase separation for 5-methoxypsoralen

(5-MOP) and 8-methoxypsoralen (8-MOP) in citrus oils. For normal phase HPLC with the polyvinyl alcohol–coated silica column (PVA-SIL, 250 × 4.6 mm i.d.), the effects of alcohol modifier on the separation was significant with this column-packing material. At higher levels of modifier (such as hexanol or 2-propanol), the mobile phase becomes too polar and deactivates the column bed resulting in a loss of separation. Using RP-HPLC on C18 or C8 column with aqueous MeOH or ACN elution, these coumarins suffered unusual peak shape and broad peak shape, which may be due to interaction of the oxygenated psoralen with active sites of the column bed as well as the hydrophobic nature of the analytes (225).

Figure 16 illustrates the separation of nine standard coumarins by an RP-HPLC with an isocratic elution. The HPLC used a Zorbax Rx C8 (250 × 4.6 mm i.d., 5 μm) column maintained at 25°C and eluted with acetonitrile–0.1% HOAc in water (35:65) at 1.0 ml/min. The eluate from the column was passed to a UV detector (UV 330 nm) and then a fluorescence detector (excitation at 340 nm, emission at 425 nm). As for the specificity, some of the coumarins do not have native fluorescence as seen in Fig. 16. Nine coumarins are separated under UV 330 nm (Fig. 16) and three coumarins could not be detected with fluorescence detection. Figure 17a and b are the absorption spectra of coumarins obtained by on-line PDA with HPLC. Since coumarins exhibit strong absorption in the UV region, absorption at approximately 313 nm has been used to estimate the dilution of cold-pressed lemon oil with distilled oil (12). Detailed conditions for coumarin analysis in foods are also included in Table 8.

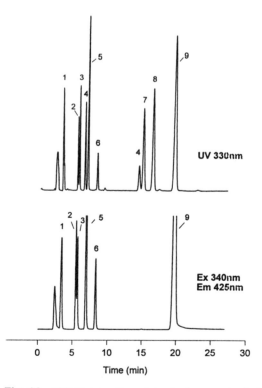

Fig. 16 HPLC separation of standard coumarins by UV and fluorescence detections. Isocratic, ACN: 0.1% HOAc/H$_2$O (35:65), Zorbax RX-C8 (250 × 4.6 mm, i.d. 5 μm). Peaks: 1 = esculin; 2 = umbelliferone; 3 = scopoletin; 4 = 4-hydroxycoumarin; 5 = 4-methylumbelliferone; 6 = scoparone; 7 = 6-methylcoumarin; 8 = 8-methoxypsoralen; 9 = limettin.

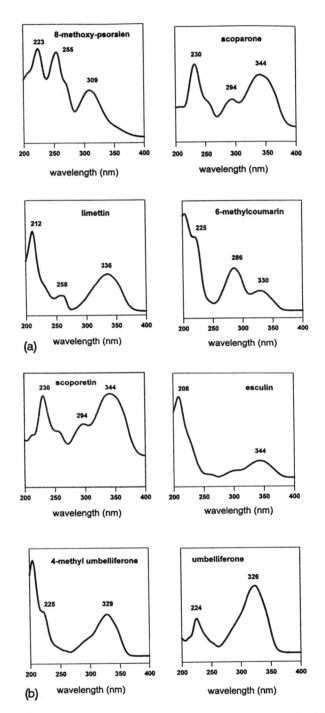

Fig. 17 (a,b) The absorption spectra of coumarins obtained by on-line HPLC-PDA.

Polymethoxylated flavones (PMFs) are characteristic of orange peel oils, and PMF determination has been used to ascertain the geographic origin of industrial peel oils (134). For flavones in citrus peel oils, the authors have analyzed extracts of peel and cold-pressed peel oils diluted in ethanol by reverse phase on various C18 columns with good results. Separations were accomplished with isocratic mobile phases of 38% and 40% ACN in H_2O (Fig. 15). For the dilute citrus oils, gradient elution was preferred to prevent the accumulation of terpenes on the column. With normal phase chromatography, elution order is reversed; terpenes elute with the solvent front and are not a problem. Gaydou and co-workers (226) developed an HPLC method for the determination of flavones in orange and mandarin oils. Using LiChrosorb Si-60 column with heptane/IPA (60:40) as the mobile phase, six flavones (tangeretin, tetra-O-methylscutellarein, heptamethoxyflavone, nobiletin, hexa-O-methyl quercetagetin, and sinensetin) were separated in 25 min. Hadj-Mahammed and Meklati (199) used an HPLC-TSP-MS method for the confirmation of PMFs in Valencia orange peel oil and juice. A C_{18} column (μBondapak, 300×6 mm i.d.) was used with a mobile phase of H_2O/ACN (60:40 v/v) at a flow rate of 1.0 ml/min. Then 20 μl of extract was injected into the HPLC-TSP-MS system and positive ion spectra from m/z 100 to 700 were recorded at 1360 msec. Mass spectrometric identification was done using positive chemical ionization (CI). This technique allowed confirmation of the presence of eight flavones in the peel oils and seven flavones in the juice.

C. Honey

Flavonoid analysis is very useful as an adjunct in studies of the geographic origin of bee pollen and honey (88). Using aqueous HCO_2H/MeOH (50:50), Ferreres and co-workers (88) analyzed 10 flavonols, 6 flavones, and 2 flavanones from a honey extract. The extract was prepared by SPE through XAD-2 to remove the sugars and polar compounds. This was followed by cleanup with Sephadex LH-20 to remove the polymeric brown phenolics and phenolic acid derivatives. Analytical separation time was 30 min on a 100×4 mm, 5-μm, LiChrochart C18 column and many of the components were poorly separated. The latest publication by Ferreres and co-workers (155) on an improved HPLC analysis procedure used a multistep linear gradient with aqueous HCO_2H/MeOH (70:30) as follows: isocratic for 15 min; 60:40 at 20 min; 55:45 at 30 min; 40:60 at 50 min; and 20:80 at 52 min. Separation of the flavonoid extract was improved with poor separation occurring only with the kaempferol 3-OMe and isorhamnetin peaks. Extracts were prepared by passing the sample through XAD-2 resin and were subsequently extracted with Et_2O to remove polymeric components removed previously with Sephadex. Similar analytical procedures have been applied to jams in determining fruit authenticity (227,228).

D. Tea and Coffee

Free gallic acid is the most important phenolic acid in tea, and the amount of gallic acid increases during fermentation. Bailey and co-workers (140) separated the number of tea components including gallic acid in a black tea liquor using a Hypersil ODS column (5 μm) and a PDA detector (280–460 nm). Separation was carried out by a linear gradient of 2% HOAc and acetonitrile, but only a single sample was analyzed without quantitation. Black and green teas also contain many flavonoids including flavanols, flavonols, flavones, theaflavins, and thearubigins (Fig. 18). Aglycones are present in small amounts, but the majority are present in glycoside form. Finger et al. (36) recently published an excellent review on the chromatographic separation of tea constituents.

Flavanols are most abundant in fresh green tea leaves and may compose up to 30% of the dry weight. The major flavanols present are (+)-catechin, (–)-epicatechin, (–)-epicatechin gallate, (–)-epigallocatechin, and (–)-epigallocatechin gallate. Their substitution patterns are given in Table 2. MeOH was the most efficient solvent for extraction of flavanols when 90% aqueous

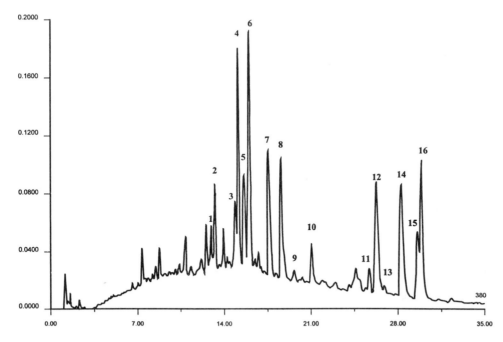

Fig. 18 HPLC separation of flavonol glycosides and theaflavins in decaffeinated black tea liquor using a gradient elution with 100% 0.5% acetic acid to 30% ACN over 35 min. Peaks: 1–8 = flavonol glycosides; 9 = theaflavic acid; 10 = epitheaflavic acid; 11–16 = theaflavins. (From Ref. 77.)

solutions of MeOH, EtOH, and Ace were compared with water-saturated EtOAc (229). Detection of flavanols is most sensitive by UV at approximately 280 nm. A PDA detector may be used to gain spectral information. A reverse phase isocratic method using a C18 column for separation within 35 min of green tea flavanols or flavanols isolated from fermentation studies utilized (1:2:40:157) HOAc/MeOH/DMF/H$_2$O (36). Gradient methods offer advantages with complex samples in being able to faster separate components of widely differing polarity. Kuhr and Engelhardt (230) developed a rapid method for separation of flavanols, caffeine, and phenolic acids in green and black teas from crude extracts that had been further cleaned up by SPE utilizing a C18 cartridge. Using aqueous HOAc and ACN, their separation was accomplished in just 25 min. In black tea, the flavanols have for the most part been oxidatively polymerized to theaflavins and thearubigins. These can be isolated from tea by ethyl acetate extraction or analyzed from whole-tea infusions. Analysis of whole-tea infusions using a small-particle C18 column with 1% aqueous HOAc and ACN, gradient elution, and UV-Vis detection by PDA will separate both the polymerized flavanols and flavonol glycosides with good resolution (77,231).

The flavonol glycosides present in tea are primarily O-linked glycosides while the flavone glycosides occur mostly as C-linked glycosides. The flavonol glycosides are all derived from myricetin, quercetin, and kaempferol. The most important flavones are those derived from apigenin (vitexin and isovitexin). Extraction with methanol will remove both the flavone and flavonol glycosides from tea leaves. Flavonol glycosides have been analyzed by both gradient and isocratic HPLC with good results. The flavonol glycosides can be further isolated from methanol extracts by CC with polyamide. Separation with C18 columns using aqueous HOAc and ACN with isocratic (85:15) (59,232) or gradient elution (77,210) have provided good results (Fig. 18). An advantage to gradient elution is that theaflavins and thearubigins may also be quantified in the same analysis. With the isocratic elution method, 13 mono-, di-, and trisaccharrides were separated in 35 min. The procedure was then scaled up using a large-diameter column and peaks were collected. Identities were confirmed by UV spectra (PDA), ^{1}H and ^{13}C NMR, and GC-MS analysis of the TMS derivatives of sugar residues after hydrolysis (59).

Flavone glycosides occur as C-glycosides and may be isolated from flavonols through hydrolysis of flavonol-O-glycosides. Flavone-C-glycosides are not hydrolyzed and may be isolated from hydrolyzed sugars and flavonol aglycones with SPE using polyamide, then separated by HPLC using conditions similar to analysis for the flavonol glycosides. However, flavones with an unsymmetrical substitution pattern are susceptible to isomerization from Wessely–Moser rearrangement and amounts must be reported as a sum (36). The flavonols and flavonol glycosides in tea are one class of components responsible for the yellow–brown color in green tea. Oxidation of tea flavanols (catechins) during the manufacture of black tea leads to polymerization into orange–red theaflavins and orange–brown thearubigins. Analysis of flavanols and their polymerization products in tea is one basis in the determination of tea quality (77,219,233,234).

The major representatives of phenolic acid on coffee constituents are chlorogenic acid and its isomers (162,163). Chlorogenic acids comprise some groups of compounds mainly formed by quinic acid esterification with either caffeic, ferulic, or p-coumaric acids. Ramirez-Martinez (55) developed an RP-HPLC method for chlorogenic, three isochlorogenics, protocatechuic, and ferulic acid with flavanols from coffee pulp using an ODS/SIL-X column and a fixed wavelength at UV 280 nm. HPLC separation was achieved in 36 min by combining linear gradient elution with isocratic elution. Twelve coffee cultivars were analyzed and chlorogenic acid content averaged about 42.2% of total identified phenolic compounds. In the case of coffee, where phenolic compounds are so varied and some of them exiguous, quantitation of total chlorogenic acid could be facilitated by use of gel filtration chromatography (GFC). Trugo et al. (162) developed high-performance GFC for simultaneous determination of total chlorogenic acid and caffeine in coffee samples by using a TSK-G 3000-SW (300×8 mm i.d.).

E. Cereals

HPLC analysis of ferulic acid can be used as an excellent indicator of endosperm purity in the milling process as a measure of bran contamination in wheat milling fractions. Also, it was shown to be useful in the related areas of wheat breeding and baking (57). Galletti et al. (176) developed an RP-HPLC method for determination of p-coumaric and vanillic acid together with phenols in wheat straw and lignin. Separation was based on a C6 column with an isocratic elution (MeOH–0.1% HClO$_4$, 15:85), and a UV detector set at 280 nm and in series with ED, which led to an increase in sensitivity.

F. Alcoholic Beverages

Spirits are used to store in oak barrels to acquire harmonious organoleptic characteristics. During maturation, certain coumarins are extracted from wood and increase with the length of maturation. Salagoity-Auguste and co-workers (224) used RP-HPLC with water-acetonitrile binary gradient for coumarins in wines and brandies stored in oak barrels. Coumarins were extracted by diethyl ether before injection. Esculetin, umbelliferone, scopoletin, and methylumbelliferone were detected under dual detection by UV 313 nm absorption and fluorescence (ex:340/em:425 nm). These coumarins are known to be very flavor-active and have an impact even though the concentrations found were less than 3 mg/L.

Puech and Moutounet (31) also applied binary gradient elution on a Micropak MCH-5 C18 column with (a) 5% HOAc in water and (b) ACN/water/HOAc (85:15:5) at 40°C to determine esculin, umbelliferone, scopoletin, and 4-methylumbelliferone in different spirits of armagnac, brandy, calvados, cognac, whiskey, scotch whiskey, and rum. Spirits are injected directly with no sample preparation. Scopoletin appeared to accumulate as the spirits aged. The scopoletin content of spirits after 14 years was 6.5 times higher than in new wood, but its content depends on the provenance of the wood, the age of the barrels, and the duration of maturation (31).

RP-HPLC with binary gradient using MeOH and 3.5% HOAc in water (61,56) or isocratic elution with 15% MeOH with 0.1 M $NH_4C_2H_3O_2$ (pH 4.0) with electrochemical detection was applied for major phenolic acids in beer (90). In the MeOH gradient, as the MeOH content increased, resolution between cinnamic acids (such as caffeic) and benzoic acids (vanillic) decreased. Increasing the MeOH content in the mobile phase will favor the solubility of cinnamic derivatives and decrease their retention times more than those of benzoic acid derivatives. Thus, retention times of the cinnamic acid derivatives decreased faster than those for benzoic acid derivatives and consequently the separation between the two groups decreased (56).

Oszmianski et al. (23) demonstrated the separation of phenolic acids from procyanidins and anthocyanins in red wine. Most of the wine phenolics which had been fractionated by SPE could be analyzed by binary gradients (23,91,145,224). For the red wine (143), which was injected directly into the HPLC without sample preparation, a ternary gradient system was employed for gallic and cinnamic acid derivatives.

G. Applications for Health Promotion Benefits

Analysis of food and food plants were also performed to identify components that may offer disease or pest resistance (204,235–240). Some flavonoids have antioxidative activity and analysis helps determine their role in prevention of spoilage during storage (60,70,87,167,177,241,242). Flavonoid analysis of foods is also gaining popularity with the growing evidence of possible health promotion benefits of flavonoids (46,58,66,179,243,244). Methoxylated flavones, flavanols (catechins), and isoflavones are all bioactive components with cancer chemopreventative and antiallergenic acitivity. Polymethoxylated flavones (PMF), sometimes found in certain *Citrus* species, have pharmacodynamic properties, e.g., sinesetin and nobiletin. An analysis of PMF has been used for dietary control in clinical experiments (245). Since 1986, three books have been published (246–248) on research in this area.

For all of these reasons, much time and effort has been spent on development of analytical methodology and in determination of phenolic quantities in foods.

REVIEWS

P. Ribereau-Gayon, *Plant Phenolics*, Hafner, New York, 1972, p. 98.

C. F. Van Sumere, W. Van Brussel, K. Van Casteele, and L. Van Rompaey, *Biochemistry of Plant Phenolics: Recent Advances in Phytochemistry, Vol. 12* (T. Swain, J. B. Harborne, and C. F. Van Sumere, eds.), Plenum Press, New York, 1978, p. 29.

J. M. Hardin and C. A. Stutte, *Food Constituents and Food Residues: Their Chromatographic Determination*, (J. F. Lawrence, ed.), Marcel Dekker, New York, 1984, p. 295.

K. Herrmann, *Crit. Rev. Food Sci. Nutr.*, 28: 315. (1989).

J. B. Harborne, *Methods in Plant Biochem.*, *Vol. 1, Plant Phenolics* (P. M. Dey and J. B. Harborne, eds.), Academic Press, London, 1989, p. 1.

C. F. Van Sumere, *Methods in Plant Biochem.*, *Vol. 1, Plant Phenolics* (P. M. Dey and J. B. Harborne, eds.), Academic Press, London, 1989, p. 29.

J. J. Macheix, A. Fleuriet, and J. Billot, *Fruit Phenolics*, CRC Press, Boca Raton, FL, (1990).

A. L. Khurana, *Phenolic Compounds in Food and Their Effects on Health, Vol. 1, Analysis, Occurrence, and Chemistry* (C. T. Ho, C. Y. Lee, and M. T. Huang, eds.), Am. Chem. Soc., Washington, DC, 1992, p. 77.

A. Reschke and K. Herrmann, *Z. Lebensm. Unters. Forsch.*, *173*: 458 (1981).

J. J. Macheix and A. Fleuriet, *Bull. Liaison-Group Polyphenols*, *13*: 337 (1986).

R. Ibrahim and D. Barron, *Methods in Plant Biochem.*, *Vol 1*, *Plant Phenolics* (P. M. Dey and J. B. Harborne, eds.), Academic Press, London, 1989, p. 75.

W. L. Stanley and L. Jurd, *J. Agric. Food Chem.*, *19*: 1106 (1971).

A. I. Gray and P. G. Waterman, *Phytochemistry*, *17*: 845 (1978).

T. A. Geissman (ed.) *The Chemistry of Flavonoid Compounds*, Pergamon Press, Oxford, 1962.

J. B. Harborne, *Comparative Biochemistry of the Flavonoids*, Academic Press, London, 1967.

T. J. Mabry, K. R. Markham, and M. B. Thomas, *The Systematic Identification of the Flavonoids*, Springer-Verlag, New York, 1970.

J. B. Harborne, T. J. Mabry, and H. Mabry, *The Flavonoids*, Chapman and Hall, London, 1975.

J. B. Harborne and T. J. Mabry, *The Flavonoids: Advances in Research*, Chapman and Hall, London, 1982.

J. B. Harborne, *The Flavonoids: Advances in Research*, Chapman and Hall, London, 1988.

K. R. Markham, *Techniques of Flavonoid Identification*, Academic Press, New York, 1982.

REFERENCES

1. P. Ribereau-Gayon, *Plant Phenolics*, Hafner, New York, 1972, p. 98.
2. C. F. Van Sumere, W. Van Brussel, K. Van Casteele, and L. Van Rompaey, *Biochemistry of Plant Phenolics: Recent Advances in Phytochemistry*, Vol. 12, (T. Swain, J. B. Harborne, and C. F. Van Sumere, eds.), Plenum Press, New York, 1978, p. 29.
3. J. M. Hardin and C. A. Stutte, *Food Constituents and Food Residues: Their Chromatographic Determination*, (J. F. Lawrence, ed.), Marcel Dekker, New York, 1984, p. 295.
4. K. Herrmann, *Crit. Rev. Food Sci. Nutr.*, *28*: 315 (1989).
5. J. B. Harborne, *Methods in Plant Biochem.*, *Vol. 1*, *Plant Phenolics* (P. M. Dey and J. B. Harborne, eds.), Academic Press, London, 1989, p. 1.
6. C. F. Van Sumere, *Methods in Plant Biochem.*, *Vol. 1*, *Plant Phenolics* (P. M. Dey and J. B. Harborne, eds.), Academic Press, London, 1989, p. 29.
7. J. J. Macheix, A. Fleuriet, and J. Billot, *Fruit Phenolics*, CRC Press, Boca Raton, 1990.
8. A. L. Khurana, *Phenolic Compounds in Food and Their Effects on Health*, *Vol. 1*, *Analysis, Occurrence and Chemistry* (C. T. Ho, C. Y. Lee, and M. T. Huang, eds.), Am Chem. Soc., Washington, DC. 1992, p. 77.
9. A. Reschke and K. Herrmann, *Z. Lebensm. Unters. Forsch.*, *173*: 458 (1981).
10. J. J. Macheix and A. Fleuriet, *Bull. Liaison-Group Polyphenols*, *13*: 337 (1986).
11. R. Ibrahim and D. Barron, *Methods in Plant Biochemistry*, *Vol 1*, *Plant Phenolics* (P. M. Dey and J. B. Harborne, eds.), Academic Press, London, 1989, p. 75.
12. W. L. Stanley and L. Jurd, *J. Agric. Food Chem.*, *19*: 1106 (1971).
13. A. I. Gray and P. G. Waterman, *Phytochemistry*, *17*: 845 (1978).
14. T. A. Geissman (ed.), *The Chemistry of Flavonoid Compounds*, Pergamon Press, Oxford, 1962.
15. J. B. Harborne, *Comparative Biochemistry of the Flavonoids*, Academic Press, London, 1967.
16. T. J. Mabry, K. R. Markham, and M. B. Thomas, *The Systematic Identification of the Flavonoids*, Springer-Verlag, New York, 1970.
17. J. B. Harborne, T. J. Mabry, and H. Mabry, *The Flavonoids*, Chapman and Hall, London, 1975.
18. J. B. Harborne and T. J. Mabry, *The Flavonoids: Advances in Research*, Chapman and Hall, London, 1982.
19. J. B. Harborne, *The Flavonoids: Advances in Research*, Chapman and Hall, London, 1988.
20. K. R. Markham, *Techniques of Flavonoid Identification*, Academic Press, New York, 1982.
21. A. M. Torres, T. M. Mau-Lastovicka, and R. Rezaaiyan, *J. Agric. Food Chem.*, *35*: 921 (1987).
22. H. Peleg, M. Naim, R. L. Rouseff, and U. Zehavi, *J. Sci. Food Agric.*, *57*: 417 (1991).
23. J. Oszmianski, T. Ramos, and M. Bourzeix, *Am. J. Enol. Vitic.*, *39*: 259 (1988).
24. C. Y. Lee and A. Jaworski, *Am. J. Enol. Vitic.*, *40*: 43 (1989).
25. J. Oszmianski and C. Y. Lee, *Am. J. Enol. Vitic.*, *41*: 204 (1990).
26. G. Mazza and Y. S. Velioglu, *Food Chem.*, *43*: 113 (1992).
27. G. P. Cartoni, F. Coccioli, L. Pontelli, and E. Quattrucci, *J. Chromatogr.*, *537*: 93 (1991).

28. B. Fernandez de Simon, J. Perez-Ilzarbe, T. Hernandez, C. Gomez-Cordoves, and I. Estrella, *J. Agric. Food Chem.*, *40*: 1531 (1992).
29. Y. Nogata, H. Ohta, K. I. Yoza, M. Berhow, and S. Hasegawa, *J. Chromatogr.*, *667*: 59 (1994).
30. S. Akhtar, M. Ahmad, E. Ali, M. Yamin, and M. K. Bhatty, *Pakistan J. Sci. Ind. Res.*, *31*: 307 (1988).
31. J. L. Puech and M. Moutounet, *J. Assoc. Off. Anal. Chem.*, *71*: 512 (1988).
32. J. H. Tatum and R. E. Berry, *Phytochemistry*, *18*: 500 (1979).
33. G. E. Brown and H. S. Lee, *Phytopathology*, *83*: 1204 (1993).
34. Y. Takehiko, M. Yoshihararu, S. Akiyoshi, I. Yoshitomi, and O. Kozo, *Yukagaku*, *41*: 249 (1992).
35. R. M. Horowitz and B. Gentili, *Arch. Biochem. Biophysic.*, *92*: 191 (1961).
36. A. Finger, S. Kuhr, and U. H. Engelhardt, *J. Chromatogr.*, *624*: 293 (1992).
37. E. H. Siemann and L. L. Creasy, *Am. J. Enol. Vitic.*, *43*: 49 (1992).
38. J. Bakker, P. Bridle, and S. J. Bellworthy, *J. Sci. Food Agric.*, *64*: 31 (1994).
39. C. E. Lister, J. E. Lancaster, K. H. Sutton, and J. R. L. Walker, *J. Sci. Food Agric.*, *64*: 155 (1994).
40. M. Azar, E. Verette, and S. Brun, *J. Food Sci.*, *52*: 1255 (1987).
41. B. Fernandez de Simon, J. Perez-Ilzarbe, T. Hernandez, C. Gomez-Cordoves, and I. Estrella, *Chromatographia*, *30*: 35 (1990).
42. M. H. Salagoity-Auguste and A. Bertrand, *J. Sci. Food Agric.*, *35*: 1241 (1984).
43. T. Kuninori and J. Nishiyama, *J. Chromatogr.*, *362*: 255 (1986).
44. K. B. McRae, P. D. Lidster, A. C. DeMarco, and A. J. Dick, *J. Sci. Food Agric.*, *50*: 329 (1990).
45. F. Thomas-Lorente, C. Garcia-Viguera, F. Ferreres, and F. A. Tomas-Barberan, *J. Agric. Food Chem.*, *40*: 1800 (1992).
46. H. Kumamoto, Y. Matsubara, Y. Iizuka, K. Okamoto, and K. Yokoi, *Agric. Biol. Chem.*, *49*: 2797 (1985).
47. A. Jaworski, and C. Y. Lee, *J. Agric. Food Chem.*, *35*: 257 (1987).
48. J. J. Macheix, J. C. Sapis, and A. Fleuriet, *Crit. Rev. Food Sci.*, *30*: 441. (1991).
49. V. L. Singleton, M. Salgues, J. Zaya, and E. Trousdale, *Am. J. Enol. Vitic.*, *36*: 50 (1985).
50. C. Y. Lee and A. Jaworski, *Am. J. Enol. Vitic.*, *38*: 277 (1987).
51. V. L. Singleton, J. Zaya, and E. K. Trousdale, *Phytochemistry*, *25*: 2127 (1986).
52. M. J. Amiot, M. Tacchini, S. Aubert, and J. Nicolas, *J. Food Sci.*, *57*: 958 (1992).
53. I. Babic, M. J. Amiot, C. Nguyen-The, and S. Aubert, *J. Food Sci.*, *58*: 351 (1993).
54. M. Brenes, P. Garcia, M. C. Duran, and A. Garrido, *J. Food Sci.*, *58*: 347 (1992).
55. J. R. Ramirez-Martinez, *J. Sci. Food Agric.*, *43*: 135 (1988).
56. P. Hayes, M. R. Smyth, and I. McMurrough, *Analyst*, *112*: 1197 (1987).
57. V. Pussayanawin and D. L. Wetzel, *J. Chromatogr.*, *391*: 243 (1987).
58. Y. Matsubara, H. Kumamoto, Y. Iizuka, T. Murakami, K. Okamoto, H. Miyake, and K. Yokoi, *Agric. Biol. Chem.*, *49*: 909 (1985).
59. A. F. Finger, U. H. Engelhardt, and V. Wray, *J. Sci. Food Agric.*, *55*: 313 (1991).
60. E. Delage, G. Bohuon, A. Baron, and J. F. and Drilleau, *J. Chromatogr.*, *555*: 125 (1991).
61. P. J. Hayes, M. R. Smyth, and I. McMurrough, *Analyst*, *112*: 1205 (1987).
62. N. Srisuma, R. Hammerschmidt, M. A. Uebersax, S. Ruengsakulrach, M. R. Bennink, and G. L. Hosfield, *J. Food Sci.*, *54*: 311 (1989).
63. I. D. Chkhikvishvili, N. N. Gogiya, and A. G. Shalashvili, *Appl. Biochem. Microbiol.*, *29*: 451 (1993).
64. T. Fukai, J. Nishizawa, M. Yokoyama, L. Tantai, and T. Nomura, *Heterocycles*, *38*: 1089 (1994).
65. T. Fukai, J. Nishizawa, and T. Nomura, *Phytochemistry*, *35*: 515 (1994).
66. T. Hatano, T. Fukuda, T. Miyase, T. Noro, and T. Okuda, *Chem. Pharm. Bull.*, *39*: 1238 (1991).
67. P. M. Goupy, P. J. A. Varoquaux, J. J. Nicolas, and J. J. Macheix, *J. Agric. Food Chem.*, *38*: 2116 (1990).
68. J. Oszmianski and C. Y. Lee, *J. Agric. Food Chem.*, *38*: 688 (1990).
69. S. Sabatier, M. J. Amiot, and M. Tacchini, *J. Food Sci.*, *57*: 773 (1992).
70. P. D. Duh, D. B. Yeh, and G. C. Yen, *J. Am. Oil Chem. Soc.*, *69*: 814 (1992).
71. T. Agbor-Egbe and J. E. Rickard, *J. Sci. Food Agric.*, *51*: 215 (1990).
72. F. A. Tomas-Barberan, F. Tomas-Lorente, F. Ferreres, and C. Garcia-Viguera, *J. Sci. Food Agric.*, *47*: 337 (1989).
73. F. Tomas-Lorente, M. M. Garcia-Grau, J. L. Nieto, and F. A. Tomas-Barberan, *Phytochemistry*, *31*: 2027 (1992).

74. C. S. Creaser, M. R. Koupaiabyazani, and G. R. Stephenson, *Analyst, 117*: 1105 (1992).
75. B. F. D. Simon, J. Perez-Ilzarbe, T. Hernandez, C. Gomex-Cordoves, and I. Estrella, *J. Agric. Food Chem., 40*: 1531 (1992).
76. J. Oszmianski and J. C. Sapis, *J. Agric. Food Chem., 37*: 1293 (1989).
77. C. Powell, M. N. Clifford, S. C. Opie, M. A. Ford, A. Robertson, and C. L. Gibson, *J. Sci. Food Agric., 63*: 77 (1992).
78. L. J. Porter, Z. Ma, and B. G. Chan, *Phytochemistry, 30*: 1657 (1991).
79. K. Kanes, B. Tiserat, M. Berhow, and C. Vandercook, *Phytochemistry, 32*: 967 (1993).
80. R. L. Rouseff, K. Seetharaman, M. Naim, S. Nagy, and U. Zehavi, *J. Agric. Food Chem., 40*: 1139 (1992).
81. R. L. Rouseff, G. R. Dettweiler, R. M. Swaine, M. Naim, and U. Zehavi, *J. Chromatogr. Sci., 30*: 383 (1992).
82. A. Seo, and C. V. Morr, *J. Agric. Food Chem., 32*: 530 (1984).
83. R. L. Rouseff, T. J. Putnam, S. Nagy, and M. Naim, *Flavor Science and Technology* (Y. Bessiere and A. F. Thomas, eds.), John Wiley and Sons, Chichester, England, 1990, p. 195.
84. G. A. Spanos, R. E. Wrolstad, and D. A. Heatherbell, *J. Agric. Food Chem., 38*: 1572 (1990).
85. K. H. Roseler and R. S. Goodwin, *J. Nat. Prod., 47*: 188 (1984).
86. F. Ferreres, M. A. Blazquez, M. I. Gil, and F. A. Tomas-Barberan, *J. Chromatogr., 669*: 268 (1994).
87. M. Yamada, M. Miyata, Y. Kato, M. Nakamura, M. Nishijima, T. Shibata, and Y. Ito, *J. Food Hyg. Soc. Jap., 34*: 535 (1993).
88. F. Ferreres, F. A. Thomas-Barberan, M. I. Gil, and F. A. Tomas-Lorente, *J. Sci. Agric. Food, 56*: 49 (1991).
89. B. Risch, K. Herrmann, *Z. Lebensm. Unters. Forsch., 187*: 530 (1988).
90. C. E. Lunte, J. F. Wheeler, and W. R. Heineman, *Analyst, 113*: 95 (1988).
91. S. M. Lunte, K. D. Blankenship, and S. A. Read, *Analyst, 113*: 99 (1988).
92. J. M. Sendra, J. L. Navarro, and L. Izquierdo, *J. Chromatogr. Sci., 26*: 443 (1988).
93. S. Burda, W. Oleszek, and C. Y. Lee, *J. Agric. Food Chem., 38*: 945 (1990).
94. A. Bilyk, K. B. Hicks, D. D. Bills, and G. M. Sapers, *J. Liq. Chromatogr., 11*: 2829 (1988).
95. M. A. Berhow and C. E. Vandercook, *Phytochemistry, 28*: 1627 (1989).
96. O. Folin and W. Denis, *J. Biol. Chem., 22*: 305 (1915).
97. O. Folin and V. Ciocalteau, *J. Biol. Chem., 73*: 627 (1927).
98. H. D. Gibbs, *J. Biol. Chem., 72*: 649 (1927).
99. J. B. Pridham, *Anal. Chem., 29*: 1167 (1957).
100. M. G. Bray and W. V. Thorpe, *Methods of Biochemical Analysis*, Vol. 1 (D. Glick, ed.), Interscience, New York, 1954, p. 27.
101. T. Swain and W. E. Hillis, *J. Agric. Food Chem., 10*: 63 (1959).
102. T. White, K. S. Kirby, and E. Knowles, *J. Soc. Leather Trades Chem., 36*: 148 (1952).
103. N. A. M. Eskin, E. Hoehn, and C. Frenkel, *J. Agric. Food Chem., 26*: 973 (1978).
104. R. Budino, D. Tonelli, and S. Girotti, *J. Agric. Food Chem., 28*: 1236 (1980).
105. T. Swain and J. L. Goldstein, *Methods in Polyphenol Chemistry* (J. B. Pridham, ed.), Pergamon Press, Oxford, 1964, p. 131.
106. G. A. Spanos and R. E. Wrolstad, *J. Agric. Food Chem., 38*: 1565 (1990).
107. I. Dubery, *Phytochemistry, 31*: 2659 (1992).
108. S. Cliffe, M. S. Fawer, G. Maier, K. Takata, and G. Ritter, *J. Agric. Food Chem., 42*: 1824 (1994).
109. D. G. Roux, E. A. Maihs, and E. Paulus, *J. Chromatogr., 5*: 9 (1961).
110. J. B. Pridham, *Methods in Polyphenol Chemistry* (J. B. Pridham, ed.), Pergamon Press, Oxford, 1964, p. 111.
111. C. Regnault-Roger, R. Hadidane, J. F. Biard, and K. Boukef, *Food Chem., 25*: 61 (1987).
112. J. F. Fisher and H. E. Nordby, *J. Food Sci., 30*: 869 (1965).
113. F. I. Meredith, C. A. Thomas, and R. J. Horvat, *J. Agric. Food Chem., 34*: 456 (1986).
114. E. Stahl, *Thin Layer Chromatography*, 2nd ed., Springer-Verlag, New York, 1969.
115. K. R. Markham, *The Flavonoids* (J. B. Harborne, T. J. Mabry, and H. Mabry, eds.), Chapman and Hall, London, 1975, p. 1.
116. D. Heimler, *J. Chromatogr., 366*:407. (1986).
117. L. Jurd, *The Chemistry of Flavonoid Compounds* (T. A. Geissman, ed.), Pergamon Press, New York, 1962, p. 107.

118. K. R. Markham, *Techniques of Flavonoid Identification*, Academic Press, New York, 1982, p. 36.
119. J. E. Averett, *Phytochem. Bull.*, *10*: 10 (1977).
120. B. Vorin, *Phytochem.*, *2*: 2107 (1983).
121. K. R. Markham and T. J. Mabry, *The Flavonoids* (J. B. Harborne, T. J. Mabry, and H. Mabry, ed.), Chapman and Hall, London, 1975, p. 45.
122. K. R. Markham, *Methods in Plant Biochemistry, Vol 1, Plant Phenolics* (P. M. Dey and J. B. Harborne, eds.), Academic Press, London, 1989, p. 197.
123. K. H. A. Rosler, D. P. C. Wong, and T. J. Mabry, *J. Nat. Products*, *48*: 837 (1985).
124. L. J. Porter and K. R. Markham, *Phytochemistry*, *9*: 1363 (1970).
125. R. J. Grayer, *Methods in Plant Biochemistry, Vol 1, Plant Phenolics* (P. M. Dey and J. B. Harborne, eds.), Academic Press, London, 1989, p. 283.
126. L. J. Putman and L. G. Butler, *J. Chromatogr.*, *318*: 85 (1985).
127. A. Marston and K. Hostettmann, *J. Chromatogr.*, *658*: 315 (1994).
128. T. Yoshida, T. Hatano, and T. Okuda, *J. Chromatogr.*, *467*: 139 (1989).
129. W. D. Conway, *Countercurrent Chromatography: Apparatus, Theory, and Application*, VCH, New York, 1989.
130. N. B. Mandava and Y. Ito, *Countercurrent Chromatography: Theory and Practice*, Marcel Dekker, New York, 1988.
131. P. Husek, *Chromatographia*, *34*: 621 (1992).
132. D. T. Williams, Q. Tran, P. Fellin, and K. A. Brice, *J. Chromatogr.*, *549*: 297 (1991).
133. P. Mouly, E. M. Gaydou, and J. Estienne, *J. Chromatogr.*, *634*: 129 (1993).
134. E. M. Gaydou, T. Berahia, J. C. Wallet, and J. P. Bianchini, *J. Chromatogr.*, *549*: 440 (1991).
135. T. J. Schmidt, I. Merfort, and G. Willuhn, *J. Chromatogr.*, *669*: 236 (1994).
136. J. K. Nielsen, C. E. Olsen, and M. K. Petersen, *Phytochemistry*, *34*: 539 (1993).
137. L. W. Wulf and C. W. Nagel, *J. Chromatogr.*, *116*: 271 (1976).
138. M. F. M. Borges, F. M. F. Roleira, and M. M. M. Pinto, *J. Liq. Chromatogr.*, *16*: 149 (1993).
139. G. Chiavari, V. Concialini, and G. C. Galletti, *Analyst*, *113*: 91 (1988).
140. R. G. Bailey. I. McDowell, and H. E. Nursten, *J. Sci. Food Agric.*, *52*: 509 (1990).
141. G. J. Provan, L. Scobbie, and A. Chesson, *J. Sci. Food Agric.*, *64*: 63 (1994).
142. G. Marko-Varga and D. Barcelo, *Chromatographia*, *34*: 146 (1992).
143. R. M. Lamuela-Raventos and A. L. Waterhouse, *Am. J. Enol. Vitic.*, *45*: 1 (1994).
144. P. Gamache, E. Ryan, and I. N. Acworth, *J. Chromatogr.*, *635*: 143 (1993).
145. V. F. Cheynier, E. K. Trousdale, V. L. Singleton, M. J. Salgues, and R. Wylde, *J. Agric. Food Chem.*, *34*: 217 (1986).
146. H. S. Lee, and R. E. Wrolstad, *J. Assoc. Off. Anal. Chem.*, *71*: 789 (1988).
147. S. Hjerten, *Chromatogr. Rev*, *9*: 122 (1967).
148. J. W. Jorgenson and K. D. Lukacs, *Anal. Chem.*, *53*: 1298 (1981).
149. S. Terabe, K. Otsuka, K. Ichikawa, A. Tsuchiya, and T. Ando, *Anal. Chem.*, *57*: 834 (1985).
150. C. Delgado, F. A. Tomas-Barberan, T. Talou, and A. Gaset, *Chromatographia*, *38*: 71 (1994).
151. C. L. Ng, C. P. Ong, H. K. Lee, and S. F. Y. Li, *Chromatographia*, *34*: 166 (1992).
152. P. Morin, F. Villard, and M. Dreux, *J. Chromatogr.*, *628*: 153 (1993).
153. P. Morin, F. Villard, M. Dreux, and P. Andre, *J. Chromatogr.*, *628*: 161 (1993).
154. P. G. Pietta, P. L. Mauri, A. Rava, and G. Sabbatini, *J. Chromatogr.*, *560*: 367 (1991).
155. F. Ferreres, F. A. Tomas-Barberan, C. Soler, C. Garcia-Viguera, A. Ortiz, and F. Tomas-Lorente, *Apidologie*, *25*: 21 (1994).
156. C. Bjergegaard, S. Michaelsen, K. Mortensen, and H. Sorensen, *J. Chromatogr.*, *652*: 477 (1993).
157. T. K. McGhie, *J. Chromatogr.*, *634*: 107 (1993).
158. K. V. Casteele, H. G. Geiger, and C. F. V. Sumere, *J. Chromatogr.*, *240*: 81 (1982).
159. P. F. Cancalon and C. R. Bryan, *J. Chromatogr.*, *652*: 555 (1993).
160. T. C. Somers, *Nature*, *209*: 368 (1966).
161. V. Lattanzio and A. Marchesini, *J. Food Sci.*, *46*: 1907 (1981).
162. L. C. Trugo, C. A. B. De Maria, and C. C. Werneck, *Food Chem.*, *42*: 81 (1991).
163. L. C. Trugo and R. Macrae, *Analyst*, *109*: 263 (1984).
164. I. McMurrough and J. R. Byrne, *Food Analysis by HPLC* (L. M. L. Nollet, ed.), Marcel Dekker, New York, 1992, p. 579.

165. H. W. Latz and D. A. Ernes, *J. Chromatogr.*, *166*: 189 (1978).
166. B. Bartolome, M. L. Bengoechea, M. C. Galvez, F. J. Perez-Ilzarbe, T. Hernandez, I. Estrella, and C. J. Gomez-Cordoves, *J. Chromatogr.*, *665*: 119 (1993).
167. S. M. Lunte, *J. Chromatogr.*, *384*: 371 (1987).
168. M. Akasbi, D. W. Shoeman, and A. Saari Csallany, *J. Am. Oil Chem. Soc.*, *70*: 367 (1993).
169. M. P. Tandjung, D. Madhavi, M. A. L. Smith, and M. D. Berber-Jimenez, 1994 IFT Annual meeting abstracts, Institute of Food Technology, Atlanta, Georgia, 1994, p. 37.
170. C. Y. Lee, V. Kagan, A. M. Jaworski, and S. K. Brown, *J. Agric. Food Chem.*, *38*: 99 (1990).
171. R. G. Bailey, H. E. Nursten, and I. McDowell, *J. Sci. Food Agric.*, *64*: 231 (1994).
172. S. Heath-Pagliuso, S. A. Matlin, N. Fang, R. H. Thompson, and L. Rappaport, *Phytochemistry*, *31*: 2683 (1992).
173. T. Berahia, E. M. Gaydou, C. Cerrati, and J. C. Wallet, *J. Agric. Food Chem.*, *42*: 1697 (1994).
174. F. Buiarelli, G. P. Cartoni, F. Coccioli, and E. Ravazzi, *Chromatographia*, *31*: 489 (1991).
175. P. Pietta, R. M. Facino, M. Carini, and P. Mauri, *J. Chromatogr.*, *661*: 121 (1994).
176. G. C. Galletti, J. Eagles, and F. A. Mellon, *J. Sci. Food Agric.*, *59*: 401 (1992).
177. N. Ramarathnam, T. Osawa, and M. Namiki, *J. Agric. Food Chem.*, *37*: 316 (1989).
178. J. Castillo, O. Benavente-Garcia, and J. A. Del Rio, *J. Agric. Food Chem.*, *41*: 1920 (1993).
179. L. Coward, N. C. Barnes, K. D. R. Setchell, and S. Barnes, *J. Agric. Food Chem.*, *41*: 1961 (1993).
180. M. H. A. Elgamal, N. M. M. Shalaby, H. Duddeck, and M. Hiegemann, *Phytochemistry*, *34*: 819 (1993).
181. T. Masuda, Y. Mukari, and N. Nakatani, *Phytochemistry*, *31*: 1363 (1992).
182. F. William, D. C. McLachlan, and B. A. Blackwell, *Cereal Chem.*, *68*: 184 (1991).
183. A. Rommel and R. E. Wrolstad, *J. Agric. Food Chem.*, *41*: 1941 (1993).
184. G. A. Spanos and R. E. Wrolstad, *J. Agric. Food Chem.*, *40*: 1478 (1992).
185. G. A. Spanos and R. E. Wrolstad, *J. Agric. Food Chem.*, *38*: 817 (1990).
186. G. A. Spanos and R. E. Wrolstad, *J. Agric. Food Chem.*, *38*: 688 (1990).
187. B. Voirin, A. Saunois, and C. Bayet, *Biochem. Sys. Ecol.*, *22*: 95 (1994).
188. P. Ranabahu and J. B. Harborne, *Biochem. Sys. Ecol.*, *21*: 715 (1993).
189. R. Hegnauer and R. J. Grayer-Barkmeijer, *Phytochemistry*, *34*: 3 (1993).
190. L. M. Hernandez, F. A. Tomas-Barberan, and F. Tomas-Lorente, *Biochem. Syst. Ecol.*, *15*: 61. (1987).
191. D. J. Guedon and B. Pasquier, *J. Agric. Food Chem.*, *42*: 679 (1994).
192. J. Castillo, O. Benavente-Garcia, and J. A. Del Rio, *J. Liq. Chromatogr.*, *17*: 1497 (1994).
193. R. L. Rouseff, *J. Agric. Food Chem.*, *35*: 1027 (1986).
194. R. L. Rouseff, S. F. Martin, and C. O. Youtsey, *J. Agric. Food Chem.*, *35*: 1027 (1987).
195. W. W. Widmer and S. F. Martin, *Proc. Fla. State Hort. Soc.*, *105*: 149 (1992).
196. G. A. Perfetti, F. L. Joe. Jr., T. Fazio, and W. W. Page, *J. Assoc. Off. Anal. Chem.*, *71*: 469 (1988).
197. H. S. Lee, R. D. Carter, S. M. Barros, D. J. Dezman, and W. S. Castle, *J. Food Comp. Anal.*, *3*: 9 (1990).
198. R. L. Rouseff, *J. Assoc. Off. Anal. Chem.*, *71*: 798 (1988).
199. M. Hadj-Mahammed and B. Y. Meklati, *Lebensm. Wiss. Technol.*, *20*: 11 (1987).
200. B. Heimhubere, R. Galensa, and K. Herrmann, *J. Chromatogr.*, *439*: 481 (1988).
201. R. L. Rouseff and S. V. Ting, *J. Chromatogr.*, *176*: 75 (1979).
202. P. Etievant, P. Schlich, A. Bertrand, P. Symond, and J. C. Bouvier, *J. Sci. Food Agric.*, *42*: 39 (1988).
203. F. A. Tomas-Barberan, M. M. Garcia-Grau, and F. Tomas-Lorente, *J. Agric. Food Chem.*, *39*: 255 (1991).
204. G. Vlahov, *J. Sci. Food Agric.*, *58*: 157 (1992).
205. S. V. Shobha, P. R. Krishnaswamy, and B. Ravindranath, *Phytochemistry*, *31*: 2295. (1992).
206. J. Castillo, O. Benavente-Garcia, and J. A. Del Rio, *Plant Physiol.*, *99*: 67 (1992).
207. M. Barpeled, R. Fluhr, and J. Gressel, *Plant Physiol.*, *103*: 1377 (1993).
208. N. Suganuma and M. Takaki, *Soil Sci. Plant Nutri.*, *39*: 661 (1992).
209. R. Jambunathan, M. S. Kherdekar, and R. Bandyopadhyay, *J. Agric. Food Chem.*, *38*: 545 (1990).
210. J. Raynal, M. Moutounet, and J. M. Souquet, *J. Agric. Food Chem.*, *37*: 1046 (1989).
211. B. Wald and R. Galensa, *Z. Lebensm. Unters. Forsch.*, *188*: 107 (1989).
212. F. Siewek, R. Galensa, and K. Hermann, *Z. Lebensm. Unters. Forsch.*, *179*: 315 (1984).
213. B. Wald, R. Galensa, and K. Hermann, *Flussiges Obst*, *53*: 349 (1986).
214. R. Galensa, F. Siewek, and V. Ara, *Lebensm. Gerichtl. Chem.*, *39*: 5 (1985).
215. R. Galensa, V. Ara, and F. Siewek, *Flussiges Obst.*, *53*: 454 (1986).

216. P. P. Mouly, C. R. Arzouyan, E. M. Gaydon, and J. M. Estienne, *J. Agric. Food Chem.*, *42*: 70 (1994).

217. M. G. L. Hertog, P. C. H. Hollman, and D. P. Venema, *J. Agric. Food Chem.*, *40*: 1591 (1992).

218. K. C. Liu, S. L. Yang, M. F. Roberts, B. C. Elford, and J. D. Phillipson, *Plant Cell Rep.*, *11*: 637 (1992).

219. I. McDowell, J. Feakes, and C. Gay, *J. Sci. Food Agric.*, *55*: 627 (1991).

220. R. D. Thompson and T. J. Hoffmann, *J. Chromatogr.*, *438*: 369 (1988).

221. H. J. Thompson and S. A. Brown, *J. Chromatogr.*, *314*: 323 (1984).

222. J. F. Fisher, and L. A. Trama, *J. Agric. Food Chem.*, *27*: 1334 (1979).

223. C. K. Shu, J. P. Walradt, and W. I. Taylor, *J. Chromatogr.*, *106*: 271 (1975).

224. M. H. Salagoity-Auguste, C. Tricard, and P. Sudraud, *J. Chromatogr.*, *392*: 379 (1987).

225. J. A. Albanese and C. J. Mussinan, HPLC determination of psoralens in citrus oils, 1993 Pittsburg Conference Abstracts, Atlanta, Georgia, 1993, p. 448.

226. E. M. Gaydou, J. P. Bianchini, and R. P. Randriamiharisoa, *J. Agric. Food Chem.*, *35*: 525 (1987).

227. F. Tomas-Lorente, C. Garcia-Viguera, F. Ferreres, and F. A. Tomas-Barberan, *J. Agric. Food Chem.*, *40*: 1800 (1992).

228. C. Garciaviguera, F. A. Tomas-Barberan, F. Ferreres, F. Artes, and F. Tomas-Lorente, *Z. Lebensm. Unters. Forsch.*, *197*: 255 (1993).

229. M. H. Lee, S. C. Chen, and B. W. Min, *J. Chin. Agric. Chem. Soc.* 27: 82 (1985).

230. S. Kuhr and U. H. Engelhardt, *Z. Lebensm. Unters. Forsch.*, *192*: 526 (1991).

231. S. C. Opie, A. Robertson, and M. N. Clifford, *J. Sci. Food Agric.*, *50*: 547 (1990).

232. A. Finger, U. H. Engelhardt, and V. Wray, *Phytochemistry*, *30*: 2057 (1991).

233. P. K. Mahanta and S. Baruah, *J. Sci. Food Agric.*, *59*: 21 (1992).

234. S. Terada, Y. Maeda, T. Masui, T. Suzuki, and K. Ina, *Nippon Shokuhin Kogyo Gakkaishi*, *34*: 20 (1987).

235. D. Wojcik-Wojtkowiak, B. Politycka, and M. Schneider, *Plant Soil*, *124*: 143 (1990).

236. S. Mizobuchi and Y. Sato, *Agric. Biol. Chem.*, *48*: 2771 (1984).

237. K. A. Johnston and R. S. Pearse, *Ann. Applied Biol.*, *124*: 109 (1994).

238. R. Jambunathan, M. S. Kherdekar, and P. Vaidya, *J. Agric. Food Chem.*, *39*: 1163 (1991).

239. I. Mueller-Harvey and J. D. Reed, *J. Sci. Food Agric.*, *60*: 179 (1992).

240. O. Kodama, J. Miyakawa, and T. Akatsuka, *Phytochemistry*, *31*: 3807 (1992).

241. L. R. Howard, K. S. Yoo, L. M. Pike, and G. H. Miller, *J. Food Sci.*, *59*: 110 (1994).

242. K. E. Dangyang and M. E. Saltveit, Jr., *J. Am. Soc. Hort. Sci.*, *114*: 789 (1989).

243. T. Hatano, T. Yasuhara, K. Miyamoto, and T. Okuda, *Chem. Pharm. Bull.*, *36*: 2286 (1988).

244. M. G. L. Hertog, P. C. H. Hollman, and M. B. Katan, *J. Agric. Food Chem.*, *40*: 2379 (1992).

245. R. C. Robbins, *Int. J. Vit. Nutr. Res.*, *46*: 338 (1976).

246. V. Cody, E. Middleton, Jr., and J. B. Harborne, *Plant Flavonoids in Biology and Medicine*, Alan R. Liss, New York, 1986.

247. V. Cody, E. Middleton Jr., J. B. Harborne, and A. Beretz, *Plant Flavonoids in Biology and Medicine II*, Alan R. Liss, New York, 1988.

248. N. P. Das, *Plant Flavonoids in Biology and Medicine III.* Alan R. Liss, New York, 1989.

249. M. S. Ramamurthy, B. Maiti, P. Thomas, and P. M. Nair, *J. Agric. Food Chem.*, *40*: 569 (1992).

250. P. S. Jourdan, C. A. McIntosh, and R. L. Mansell, *Plant Physiology*, *77*: 903 (1985).

251. J. Ferry and R. A. Larson, *J. Chromatogr. Sci.*, *29*: 476 (1991).

252. W. J. Hughes and T. M. Thorpe, *J. Food Sci.*, *52*: 1078 (1987).

253. J. H. Tatum and R. E. Berry, *Phytochemistry*, *16*: 1091 (1977).

254. L. J. Felice, W. P. King, and P. T. Kissinger, *J. Agric. Food Chem.*, *24*: 380 (1976).

255. J. E. Lancaster, J. E. Grant, C. E. Lister, and M. C. Taylor, *J. Am. Soc. Hort. Sci.*, *119*: 63 (1994).

256. W. Oleszek, C. Y. Lee, A. W. Jaworski, and K. R. Price, *J. Agric. Food Chem.*, *36*: 430 (1988).

257. M. Mizuno, M. Iinuma, M. Ohara, T. Tankaka, and M. Iwamasa, *Chem. Pharm. Bull.*, *39*: 945 (1991).

258. J. Cacho and J. E. Castells, *Am. J. Enol. Vitic.*, *42*: 327 (1991).

259. J. A. Del Rio, J. Castillo, and O. Benavente-Garcia, *J. Liq. Chromatogr.* 17: 3461 (1994).

260. M. Krause and R. Galensa, *J. Chromatogr.*, *502*: 287 (1990).

261. H. Terada and M. Miyabe, *J. Food Hygienic Soc. Jap.*, *34*: 385 (1993).

262. T. Hatano, T. Yasuhara, Y. Fukuda, T. Noro, and T. Okuda, *Chem. Pharm. Bull.*, *37*: 3005 (1989).

263. B. Voirin and C. Bayet, *Phytochemistry*, *31*: 2299 (1992).

264. B. Voirin, C. Bayet, and M. Colson, *Phytochemistry*, *34*: 85 (1993).

265. T. C. R. Prickett and J. R. L. Walker, *J. Sci. Food Agric.*, *47*: 53 (1989).
266. U. Mackenbrock, R. Vogelsang, and W. Barz, *Z. Naturforsch. C. J. Biosci.*, *47*: 815 (1992).
267. D. Ke and J. M. E. Saltveit, *Plant Physiol.*, *88*: 1136 (1988).
268. V. L. Singleton and E. K. Trousdale, *Am. J. Enol. Vitic.*, *43*: 63 (1992).
269. T. Berahia, E. M. Gaydou, C. Cerrati, and J. C. Wallet, *J. Agric. Food Chem. 42*: 1697 (1994).
270. S. Yahara and I. Nishioka, *Phytochemistry*, *23*: 2108 (1984).
271. G. C. Chen, *J. Liq. Chromatogr.*, *15*: 2857 (1992).

22

Bittering Substances

Ian McMurrough and David Madigan
Guinness Brewing Worldwide, Dublin, Ireland

I. INTRODUCTION

The measurement of flavor clearly must be of major concern to the food scientist because this feature, arguably above all others, is the criterion by which food quality is judged by the consumer. For comprehensive reviews of flavor measurement the reader is referred to Refs. 1–4 and for specific details on bitterness to Ref. 5. Flavor studies, however, are far from easy, being complicated not only by the myriad of naturally occurring flavor-active substances (2) but by the complex physiological and psychological reactions of humans to these stimuli (3). Indeed, flavor can even be subdivided into taste, smell, texture, temperature, and some less readily defined learned and cultural factors. Taste itself consists of the four key components, sweetness, sourness, saltiness, and bitterness, and these different gustatory sensations are not equally appreciated by the public (3). Beyond doubt, sweetness is the most widely sought-after taste sensation and it is this which is most frequently added to foods to increase palatability (4). Consequently, much has been discovered about the biochemical mechanisms that control the human perception of sweetness, although some mysteries remain. In contrast, much less is known about bitterness and its related sensation of astringency, but the subject was recently reviewed comprehensively (5). Bitterness, in distinction from a lack of sweetness, though widespread in foodstuffs is generally considered a negative factor and one likely to be tolerated less by the young than by the more learned palate. Indeed, rarely are conscious efforts made to increase the bitterness content of foods, and more usually the objectives are to remove bitter compounds. Deliberate bittering is, for the most part, confined to a few beverages (e.g., beer, coffee, tea) wherein a balanced bitterness/astringency adds a clean and refreshing note to a complex profile of other flavor sensations. It is not surprising that it is in this context that most good examples of the application of analytical expertise are found. Since the majority of bitter compounds are either nonvolatile or thermally unstable, high-performance liquid chromatography (HPLC) has become the current method of choice for their separation and quantitation.

II. PROPERTIES OF BITTER SUBSTANCES

The food scientist may encounter a host of heterogeneous bitter substances derived from a wide variety of sources. Some of the bitter principles may occur naturally, in plant or animal tissues, or be derived during processing (e.g., heating) from previously nonbitter components. Many of the most bitter naturally occurring compounds possess therapeutic or other physiological properties and may even be extremely toxic, e.g., strychnine (6). Other constituents may impart bitterness when present at high concentrations but are otherwise unremarkable (e.g., L-phenylalanine). No totally satisfactory explanation based on chemical structure and functional groups has yet emerged for the possession of the bitterness property by certain molecules. It remains to be determined as to what combination of chemical and physical specifications is required to elicit the bitterness response in the appropriate human receptor cell. All that has been established is that bitterness is associated with molecular hydrophobicity, that the configuration and size of the nonpolar region is important, and that molecules require at least one polar group to be bitter (7). Furthermore, it follows that surface activity and solubility properties must also be determinants of specific bittering behavior.

Table 1 lists the detection thresholds of different chemical categories that are considered to be bitter. The list is by no means exhaustive but is intended to include examples of the most frequently encountered and economically important substances (8–20). The diversity of chemical structures that are bitter has been reviewed thoroughly (10). The difference thresholds are approximate measures of the lowest concentrations detectable by taste, when the substance is dissolved either in water or in another weakly flavored aqueous mixture. It is clear that different experimenters have obtained very different results for a number of test substances (e.g., quinine, caffeine). These discrepancies exemplify the difficulties involved in sensory measurement. Dependable sensory analyses are obtained (21) only when sufficient attention is paid to

1. Defining precisely the objective of the test
2. Designing the test to eliminate subjectivity
3. Selecting a suitably sized panel of trained tasters
4. Using the appropriate statistical evaluations to interpret the results

Even though the bitterness thresholds given in Table 1 must be treated with circumspection, it is obvious that some compounds with threshold values less than 0.1 mM must be considered as very bitter. Among the naturally occurring bitter substances the terpenoids and phenolics generally dominate. Even so, some claim that quassin (Fig. 1, Ref. 22) is the most intensely bitter natural substance known (15). Quinine (Fig. 2, Ref. 23) is usually regarded as the standard bitter taste, while phenylthiourea is used to test human sensitivity to bitterness (7). When quinine hydrochloride was used as a standard it was found to be about six times more bitter than hop iso-α acids (Fig. 3, Ref. 24), which in turn were about 10 times more bitter than caffeine (Fig. 4, Ref. 15) when tested at equivalent concentrations (25). Moreover, quinine and especially iso-α acids elicited a lingering bitterness of a different character from the short-lived bitterness of caffeine. Quinine sulfate was used to prove that the bitter character of chlorogenic acid (Fig. 5, Ref. 26) was not significant when the acidic character of the latter was masked (27). For structures that possess both bitter and astringent character, such as proanthocyanidins (Fig. 6, Ref. 28), tasting trials must be specially designed to counteract interferences from palate-blocking effects (29). A properly trained panel can then distinguish between the responses to nonspecific hydrogen bonding, denoted as astringency, and the receptor-specific binding termed as bitterness.

The remainder of this chapter will concentrate on those groups of substances in Table 1 with low bitterness thresholds (<0.2 mM). Those selected for more extensive coverage have significant consequences for the qualities of commercially important extracts of plant materials, e.g., fruit

Table 1 Bitter Taste Detection Thresholds of Various Chemical Categories

Structure	Example	Ref.	Threshold (mmol/L)
L-Form amino acids	L-Tryptophan	8	4
	L-Leucine	9	15
Peptides	L-Leu-L-Leu	9	4
Aliphatic amines	*n*-Butylamine	10	6
	Diethylamine		3
Amides	*n*-Butylamide	10	20
	Benzamide		1
N-containing	Pyridine	10	2
heterocyclics	Purine		5
	Pyrazine		13
	Pyrimidine		18
Urea-type compounds	Phenylthiourea	10	0.002
		11	0.0001
Phenolic acids and	4-hydroxybenzoic acid	10	8
esters	(ethyl ester)		5
Ketones	Methylpropylketone	10	30
	Methylhexylketone		4
Citrus liminoids	Limonin	11	0.08
		12	0.003
Flavanone	Naringin	11	0.2
neohesperidoside		12	0.04
		13	0.2
Quinoline alkaloids	Quinine (sulfate)	11	0.0001
		13	0.03
Methylxanthines	Caffeine	11	0.7
		13	2
		14	0.02
C20-Triterpenoids	Quassin	15	0.0002
Flavanoid polyphenols	(+)-Catechin	16	2
	Procyanidin B3	17	0.07
Hydroxycinnamic acids	Caffeic acid	18	0.06
Hop iso-α acids	Isohumulone	19	0.02
Bitter salts	Magnesium sulfate	20	5

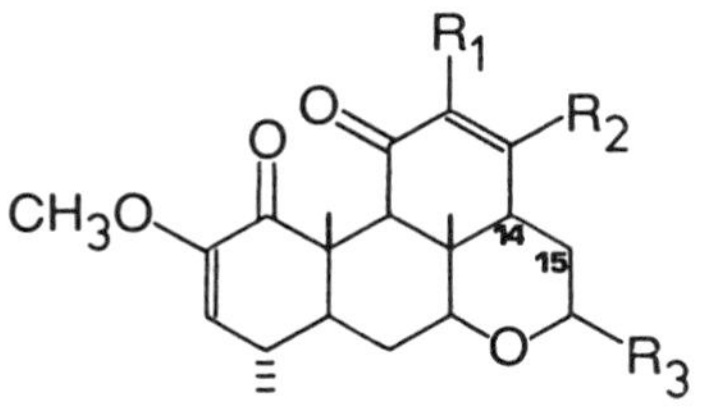

Fig. 1 Structures of quassinoids; quassin (R_1 = OCH$_3$, R_2 = CH$_3$, R_3 = =O), neoquassin (R_1 = OCH$_3$, R_2 = CH$_3$, R_3 = HOH), 14,15-dehydroquassin (R_1 = OCH$_3$, R_2 = CH$_3$, R_3 = =O), 18-hydroxyquassin (R_1 = OCH$_3$, R_2 = CH$_2$OH, R_3 = =O). (Redrawn from Ref. 22.)

Fig. 2 Structure of quinine. (Redrawn from Ref. 23.)

juices, tea, coffee, wine, beer, cocoa, cider, and chicory. Much more comprehensive reviews of bitter substances in fruits, vegetables, meats, and dairy produce than can be undertaken here were recently published (10). Moreover, attention is paid in other chapters in this book to the measurement of certain substances that are important for reasons other than their bitterness (e.g., amino acids, peptides, proteins, organic bases, phenolic compounds, carotenoids, etc.). In the following,

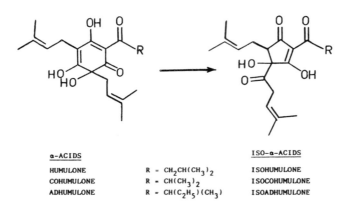

a-ACIDS		ISO-a-ACIDS
HUMULONE	R = CH₂CH(CH₃)₂	ISOHUMULONE
COHUMULONE	R = CH(CH₃)₂	ISOCOHUMULONE
ADHUMULONE	R = CH(C₂H₅)(CH₃)	ISOADHUMULONE

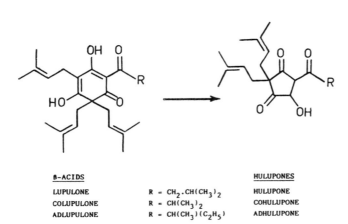

B-ACIDS		HULUPONES
LUPULONE	R = CH₂.CH(CH₃)₂	HULUPONE
COLUPULONE	R = CH(CH₃)₂	COHULUPONE
ADLUPULONE	R = CH(CH₃)(C₂H₅)	ADHULUPONE

Fig. 3 Structures of α acids, β acids, iso-α acids, and hulupones. (From Ref. 24, p. 584, by courtesy of Marcel Dekker Inc.)

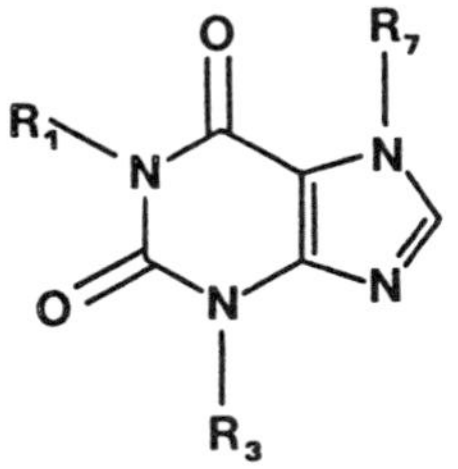

Fig. 4 Structures of some methylxanthines. Caffeine (R_1 = CH_3, R_3 = CH_3, R_7 = CH_3), theobromine (R_1 = H, R_3 = CH_3, R_7 = CH_3). Theophylline (R_1 = CH_3, R_3 = CH_3, R_7 = H). (Redrawn from Ref. 15.)

the focus will be on components whose concentrations in beverages must be controlled, either to maintain a desired level of bitterness or to abate unwanted tastes.

III. CONTROL OF BITTERNESS

The global consumption of alcoholic and nonalcoholic beverages now supports both manufacturing and agricultural activities on a major scale. Whereas the technology for the conversion of the appropriate raw materials to marketable products is not of the highest order of complexity there are nevertheless sufficient technical difficulties to occupy scientists from many disciplines. Bitterness or astringency is often caused in immature fruits by the ubiquitous flavonoid polyphenols (30). Problems of excessive bitterness can frequently be avoided simply by harvesting only sufficiently ripened fruit. However, there are many instances of bitterness in fully ripened produce and of the postharvest development of bitterness.

A. Thermally Induced Bitterness

1. Avocado

The bitterness in avocado is a good example of thermally induced bitterness whereby the degradation of lipids produces olefinic and acetylenic oxygenated aliphatics (31,32).

2. Citrus Fruits

Heating can accelerate lactone ring closure of a nonbitter precursor (limonin monolactone) to form an intensely bitter triterpenoid dilactone, limonin, in citrus fruits (Fig. 7, Ref. 33). The

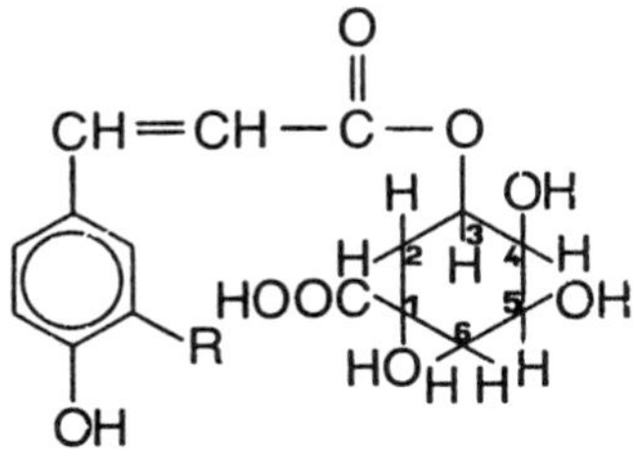

Fig. 5 Structures of some chlorogenic acids: 3-caffeoylquinic acid (R = OH), 3-feruloylquinic acid (R = OCH_3). Other chlorogenic acids that are found in coffee include 4-caffeoylquinic acid, 5-caffeoylquinic acid, 3,4-dicaffeoylquinic acid, 3,5-dicaffeoylquinic acid, 4,5-dicaffeoylquinic acid and 3-caffeoyl, 4-feruloylquinic acid. (See Ref. 26.)

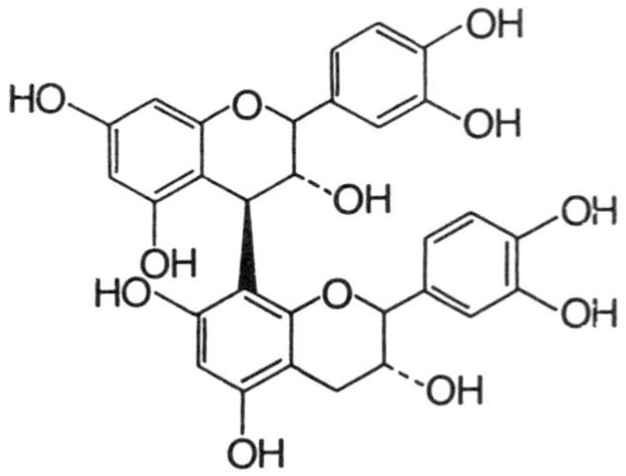

Fig. 6 Structure of procyanidin B2 from apples. Other similar dimeric and trimeric flavanols (proanthocyanidins) are present in hops, barley, and grapes. (Redrawn from Ref. 28.)

chemical transformation that normally takes place slowly in the acidic milieu of the extracted juices can also be accelerated by a lactone hydrolase, even in refrigerated samples, and is referred to as "delayed bitterness" (33). In the juice of the sweet orange the development of limonin to above 6 ppm is an obvious defect, producing an objectionable lingering bitterness, so steps must be taken to minimize such deterioration. Debittering can be achieved biologically, using enzymes or immobilized bacterial cells (34).

3. Maillard Products

There are many instances wherein heating promotes bitterness formation by accelerating either chemical or enzymically mediated change (26). The products of the widely observed Maillard browning reactions include O-containing and N-containing heterocyclics such as furans, imidazoles, pyridines, pyrazines, pyrroles, thiazolidines, thiazolines, and thiophenes. Examples of bitter compounds can be found in each group, but the specific contributions of the many hundreds of identified (and possibly thousands of unidentified) Maillard products to the overall flavor/taste of cooked or roasted foodstuffs remains an awesome challenge.

4. Coffee

The testing of coffee products for bitterness is made difficult because of possible confusion between bitterness and astringency by the public and because of wide variability in individual sensitivities (26). Added to this, much of the perceived bitterness is influenced by the way in which the brew is prepared (e.g., water hardness, temperature) and by personal choice of additives (e.g., sugar, cream). The producers of coffee products can, of course, also be a major influence on bitterness. The roasting of coffee beans may have a pronounced effect on the

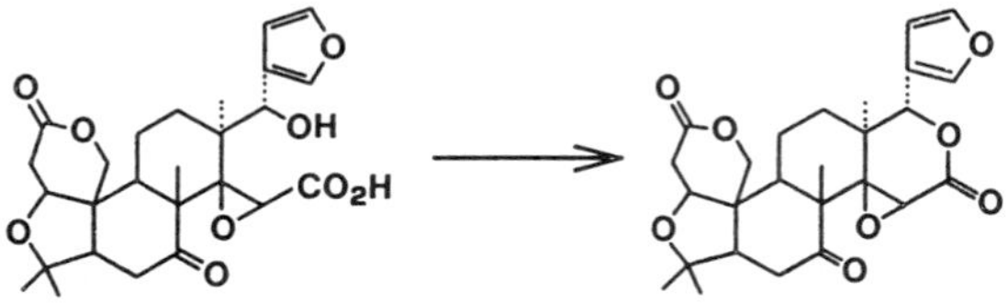

Fig. 7 Formation of limonin from its nonbitter precursor, limonoic A ring lactone. (Redrawn from Ref. 33.)

perceived bitterness of the beverage. Bitterness in coffee increases with the degree of roasting; however, the structures of the bitter compounds remain unknown (10). Research into improved fast-roasting techniques (26) has produced a beverage of reduced bitterness by increasing extractability and aroma volatile levels, thereby decreasing the amount of dry matter initially required to produce the beverage. The roasted beans of lower bitterness were found to contain decreased levels of quinic acid and elevated levels of chlorogenic acids (Fig. 5). This indicated that lower levels of phenolic degradation products of chlorogenic acid were formed, leading to decreased bitterness in the final beverage.

5. Cocoa

The bitter flavor in cocoa is also increased by roasting. The roasting process has been shown to produce a number of bitter diketopiperazines, formed by thermal fragmentation of proteins (35). These compounds were found to interact in a synergistic manner with theobromine (Fig. 4), a bitter purine which occurs naturally in cocoa, thereby further increasing the bitterness of the beverage (35).

6. Roasted Malt

Diketopiperazines have also been found in malt roasted at 210°C for the production of dark beer (36). Whereas the levels measured were below or close to the flavor thresholds of these compounds, the possibility exists (35) that interactions of diketopiperazines can occur which effectively lower bitter taste thresholds. However, in view of the high degree of bitterness imparted by hop-derived compounds to most beers, the significance of diketopiperazines is questionable.

7. Beer

The bitterness of beer derives ultimately from the use of hops. Unlike those instances mentioned earlier wherein application of heat produces undesirable bitterness, heating is an essential feature of the introduction of bitter taste in the traditional brewing process. Bitterness in beer is due to a unique family of homologs, preferably called the iso-α acids (Fig. 3), though the term isohumulones is in common, if incorrect, usage in the brewing world. The iso-α acids (Table 1) are not usually present in the hop plant but their precursors, the α acids (Fig. 3), are always present in fresh hops. The resinous α acids are a family of isoprenyl-substituted phloroglucinol congeners (24) which occur in hops alongside an analogous family of β-acid resins (Fig. 3). Whereas members from both families are weakly water-soluble and weakly bitter, the α acids share the distinction that they alone can isomerize to form intensely bitter water-soluble substances. Iso-α acids are formed from the native α acids during the boiling of worts with hops. This simultaneous extraction and isomerization step is inefficient, however, so that there are longstanding quality control and economic implications. Moreover, the α acids in the hops are not very stable, so the brewer can encounter a gradual deterioration in one of his most important raw materials during storage. To overcome these problems, new hop products have been devised (e.g., pellets, extracts) that have greater stabilities, and some may even contain preisomerized α acids in a concentrated solid or liquid form (24). Even with these recent advances, the brewer must still exert stringent controls to ensure consistency in bittering, and to this end reliable analytical methods are essential.

8. Protein Breakdown

Many foodstuffs require some heat treatment to render them more amenable to storage. The usual goal is the inactivation of native enzymes or the killing of microorganisms that would otherwise cause spoilage. Microbial spoilage is perhaps the major direct cause of bitterness in dairy foods. Some microbial proteases have the ability to hydrolyze milk proteins to bitter amino acids and peptides (37). Protein hydrolyzates, intended as food supplements and flavorings, may also suffer from excessive bitterness when prepared by enzymatic or chemical methods. Such a problem limits the extent to which soy protein can be exploited (38).

9. Fat Breakdown

Another source of microbially produced bitterness in dairy foods can be the fatty acid products of lipolysis (37). Indeed, unacceptably bitter compounds formed by oxidation of unsaturated fatty acids (e.g., linoleic acid) can also arise in soy due to the presence of native lipoxygenase (38). An oxidation can also take place during storage by a nonenzymic autoxidative route, so that stabilization strategies based on defatting with solvents have been devised.

B. Naturally Occurring Bitterness

1. Citrus Fruits

It has been stated (38) that simple phenolic acids and flavonoids might contribute to the overall bitterness of soy, but this remains to be verified. In contrast, there is absolutely no doubt about the contribution of flavonoids to the taste of some citrus varieties. Bitterness in certain totally ripened citrus fruits, notably grapefruit and Seville oranges, is caused by naringin (Table 1, Fig. 8, Ref. 39). In grapefruit juice a naringin content of 300–500 ppm is desirable, to impart the characteristic taste, but more than 700 ppm is far too bitter (33). As with the control of limonin bitterness, late picking of fruit from selected varieties can afford some control over contents of naringin, which can vary fourfold over the season. Both the size of the individual fruits and the climate influence naringin contents and the severity of the process will determine the amounts extracted from the tissues into the juice. Correction of excessive naringin bitterness can be achieved through their removal by adsorbents (33) or through the use of cyclodextrins (39) to form less bitter inclusion complexes.

2. Coffee

As stated earlier, the bitterness observed in coffee results partly from compounds formed during the roasting process. However, coffee does possess significant bitterness as a result of naturally occurring substances in the harvested beans. These include methylxanthines such as caffeine (Fig. 4), and phenolic acids (Fig. 9, Ref. 40), carboxylic acids, and some amino acids and proteins (26). The significance of caffeine in coffee bitterness is interesting in that antagonistic

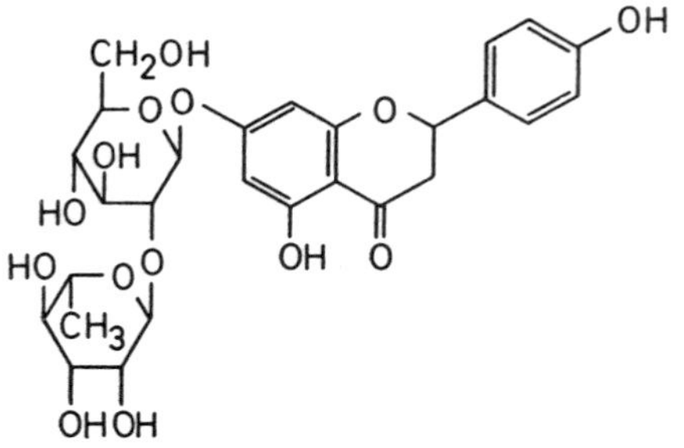

Fig. 8   Structure of naringin (naringenin-7-β-neohesperidoside). (Redrawn from Ref. 39.)

Benzoic acid derivatives:

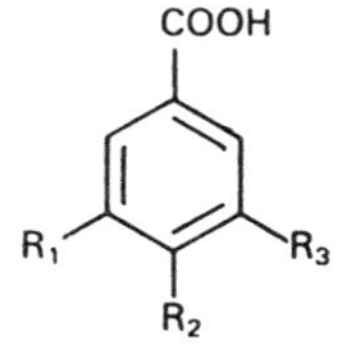

Gallic acid, $R_1 = R_2 = R_3 = OH$
Protocatechuic acid, $R_1 = R_2 = OH$, $R_3 = H$
p-Hydroxybenzoic acid, $R_1 = R_3 = H$, $R_2 = OH$
Vanillic acid, $R_1 = OMe$, $R_2 = OH$, $R_3 = H$
Syringic acid, $R_1 = R_3 = OMe$, $R_2 = OH$

Cinnamic acid derivatives:

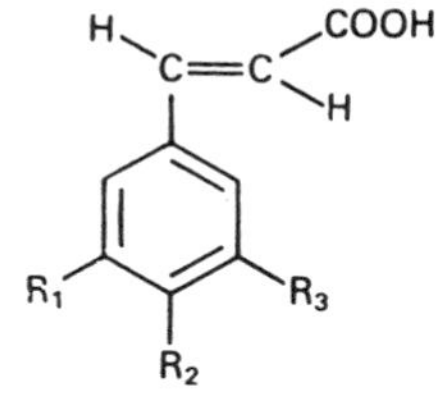

Caffeic acid, $R_1 = R_2 = OH$, $R_3 = H$
p-Coumaric acid, $R_1 = R_3 = H$, $R_2 = OH$
Ferulic acid, $R_1 = OMe$, $R_2 = OH$, $R_3 = H$
Sinapic acid, $R_1 = R_2 = OMe$, $R_3 = OH$

Flavanols:

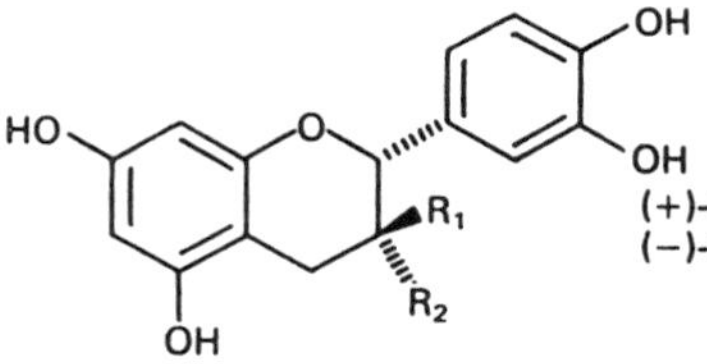

(+)-Catechin, $R_1 = H$, $R_2 = OH$
(−)-Epicatechin, $R_1 = OH$, $R_2 = H$

Fig. 9 Structures of phenolic acids and monomeric flavanols. (Reproduced from Ref. 40.)

interactions between caffeine and polyphenols appear to reduce perceived bitterness. Furthermore the presence of caffeine appears to decrease the astringency associated with polyphenols (26). Of the two species of coffee, arabica and robusta, beans of the former generally contain lower contents of chlorogenic acids and caffeine. Degradation of chlorogenic acids during roasting is known to produce the bitter-tasting quinic acid in amounts many times greater than the taste threshold. Bitterness in coffee is sometimes supplemented by extracts from chicory which contain lactucins (Fig. 10, Ref. 41) as the bitter principles.

3. Tea

The bitterness of tea is associated largely with polyphenolic compounds, which constitute between 10% and 30% of the dry weight of the harvested leaves. These compounds are present

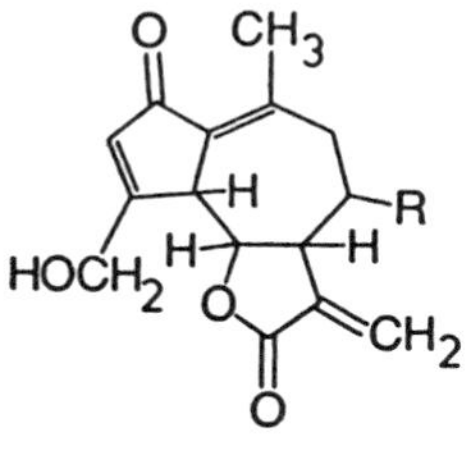

Fig. 10 Structure of lactucin (R = OH) and lactopicrin (R = p-HOC$_6$H$_4$CH$_2$COO), the bitter principles of chicory. (Redrawn from Ref. 41.)

as catechins (Figs. 9 and 11, Ref. 14), some of which are converted to higher molecular weight phenolic compounds, designated theaflavins (Fig. 12, Ref. 14) and thearubigins, during the tea manufacturing process (14). Caffeine plays an important role in determining tea taste, through a combination of its bitter flavor and its interactions with phenolic compounds (42). The bitterness of tea is also affected by some amino acids and saponins (14).

4. Wine

The interplay between bitterness and astringency is nowhere more apparent than in the extraordinarily complex taste of good-quality red wine (43,44). Both taste sensations are caused by phenolic substances and perhaps only the flavanoid polyphenols, referred to as tannins, have any real significance. Wines do contain substantial quantities (120 and 30 ppm, respectively) of both caffeic (Fig. 9) and coumaric acids (45), which arise by hydrolysis of their tartaric esters extracted from the grapes (46), but neither is detectable by wine tasters (45). Similarly, no contribution to taste has been found for chlorogenic acid when present at the normal levels in wine (44). In contrast, the effects of flavanoid tannins are a crucial feature of flavor development in red wines, especially in the more robust classic styles. Typically, the flavanoids in wine are oligomers and polymers based on the monomeric flavan-3-ols (+)-catechin and (−)-epicatechin (Fig. 9). Whereas catechin and its small oligomers are more bitter than astringent (47), the astringent character becomes more prominent than bitterness for molecules containing more than four flavanoid units (48). Young red wines contain a higher proportion of smaller flavanoid oligomers (Fig. 6) to polymers and consequently may be perceived as bitter. As wine ages, the flavanoids polymerize and progressively lose their bitter character while astringency increases. Eventually the polymers so formed are so large that they precipitate and the net effect is a decrease in astringency and the formation of sediment (43). This natural aging process, which can take 10–15 years for wines prepared from high-tannin grape varieties such as Cabernet Sauvignon, requires little more for its successful outcome than a cool cellar and a well-corked bottle. As an alternative to classic production methods some wines are incompletely fermented, so that residual sugars balance any inherent bitterness, and are sold young as ready-to-drink products.

5. Cider

Just as a high tannin content is a feature of some classic wine grape varieties, the most intensively grown apple cultivars for hard cider production in England and France are also distinguished by high concentrations of polyphenols (28). Whereas ciders contain phenolic acids (Fig. 9), phloridzin and flavanoid monomers (Fig. 9) and oligomers (proanthocyanidins, Fig. 6),

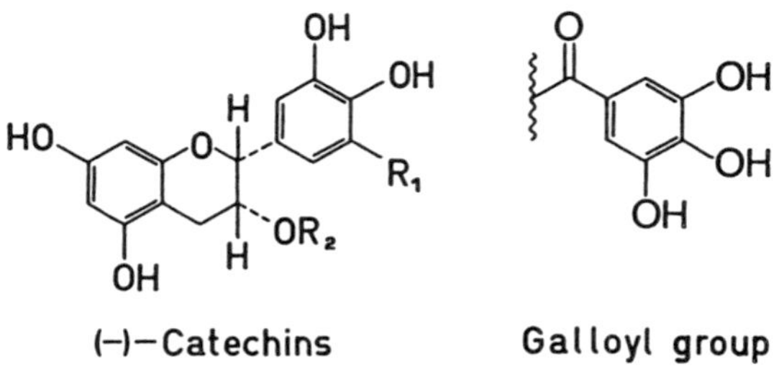

(−)−Catechins   Galloyl group

Fig. 11 Structures of tea catechins; (−)-epicatechin (R_1 = H, R_2 = H), (−)-epigallocatechin (R_1 = OH, R_2 = H), (−)-epicatechin gallate (R_1 = H, R_2 = galloyl group), (−)-epigallocatechin gallate (R_1 = OH, R_2 = galloyl group). (Redrawn from Ref. 14.)

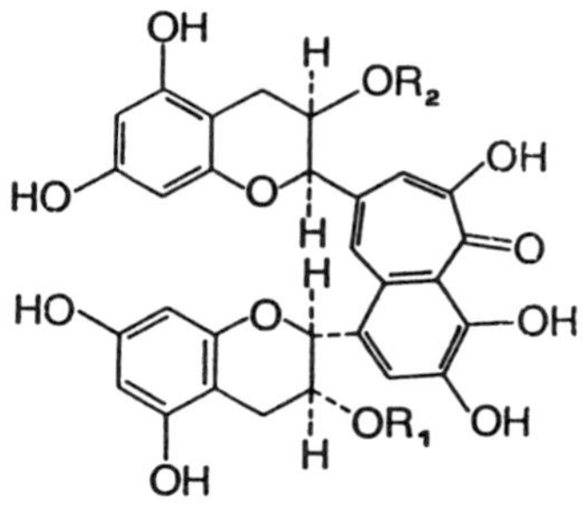

Theaflavins

Fig. 12 Structures of theaflavins; theaflavin ($R_1 = R_2 = H$), theaflavin gallate A ($R_1 = H$, R_2 = galloyl group), theaflavin gallate B (R_1 = galloyl group, $R_2 = H$), theaflavin digallate ($R_1 = R_2$ = galloyl group). (Redrawn from Ref. 14.)

only the latter group is believed to contribute significantly to bitterness and astringency (49). The concentration of these compounds in the final product depends on the efficiency of different extraction processes, the degree of oxidation during processing, and the influence of different clarifying agents (50). Inevitably some oxidation takes place during normal processing, through the action of apple polyphenol oxidase. As a result flavanoid dimers, trimers, and tetramers condense to form colored polymers with an average molecular size equivalent to that of a heptamer. Deliberate aeration of the pulp during apple juice extraction promotes the removal of bitter and astringent flavanoids through their binding to the pomace. Fining with gelatin decreases contents still further by coprecipitation.

IV. SAMPLE PRETREATMENTS

In general, compounds that contribute to the bitterness of foods and beverages are nonvolatile and therefore lend themselves to analysis by colorimetry or HPLC. However, often it is necessary to perform some type of pretreatment to arrive at a sample suitable for analysis. The aim of sample pretreatment is usually to provide a solution of the analyte containing the minimum presence of interferents, which may then be suitable for direct injection onto an HPLC column or analysis by colorimetry.

In the simplest case, pretreatment of the beverage samples may only involve dilution or filtration prior to direct injection onto a chromatographic column, provided that the detection system is sensitive and selective enough not to require either preconcentration of the analyte or removal of interferents (51–53). Should this not be so, pretreatment may involve such steps as precipitation of proteins and other interferents, liquid–liquid extraction, solid phase extraction, pH adjustment, derivatization, or the addition of ion-pairing reagents. In difficult cases, resort to column chromatography or solid phase extraction must be made in order to provide a sufficiently purified solution of the analyte.

The pretreatment of solid samples, such as malt, hops, tea leaves, coffee beans, and fruits, is invariably more complex than that of liquids. Firstly, a representative sample must be taken from the bulk. This stage is critical to the final analysis, a good example being the sampling of hops for analysis of α acids. The α acids are contained in tiny structures near the base of the hop cone known as lupulin glands (54). The lupulin glands are easily dislodged from the cones and tend to fall to the bottom of the sample container, resulting in an uneven α acid distribution. To ensure representative subsampling, the settled lupulin glands should be evenly dispersed throughout the bulk by vigorous shaking. Having taken a representative solid sample, a homogenization and/or particle size reduction procedure is usually necessary to allow efficient extraction of the sample. Extraction is normally carried out with a suitable solvent system such as water, an acid or base or an organic solvent, or a combination of two or more extraction solvents. Increasingly, supercritical fluids are being used for sample extraction. The liquid extract may then require one or more of the treatments described above for liquid samples. A general scheme of analysis by reverse phase HPLC of phenolic compounds, including coumarins, in plant extracts is also given in Ref. 55. Examples of sample pretreatments in the analysis of bitter substances are given in Tables 2–4. A general scheme for pretreatment of samples for the analysis of phenols, phenolic acids, poly-phenols, and some other bittering substances is shown in Fig. 13.

V. HPLC ANALYSIS OF BITTERING SUBSTANCES

A. Beverages

1. Beer

Table 2 lists different methods of beer pretreatment and separation of iso-α acids (Figs. 3 and 14) in beer. The list is by no means exhaustive because very many variations have been tried since HPLC became generally available (24,56–66). Methods given in Table 2 (24,56–61) have been tested by ourselves but no single method has yet been adopted universally by the brewing industry as a standard. Recent reviews (24,62) examine the possible choices in pretreatments, solid phases, mobile phases, calibration protocols, etc. Indeed, the methodology may still be developing and one recent suggestion has been the inclusion of small amounts of EDTA and magnesium acetate in the mobile phase (63). Apart from the lack of consensus on the best separation method to use, the nonavailability of a reliable iso-α acid standard has impeded the use of HPLC in the brewing industry for routine measurement of beer bitterness. Several internal standards have been found to be less than satisfactory and external standardization with mixtures of iso-α acids seems more promising (24). It was recently reported (64) that the dicyclohexylamine salts of the *trans*-iso-α acids are stable for extended periods and this must favorably influence the wider acceptance of HPLC in beer bitterness determination. Moreover, the American Society of Brewing Chemists (ASBC) recently recommended (65) the method (Table 2) devised by Donley (58).

The small amounts of α and β acids (Fig. 3) that are sometimes present in beer can also be measured by HPLC of concentrated isooctane extracts (66).

2. Cider, Apple Juices

An early study of a wide variety of stationary and mobile phases in the analysis of cider procyanidins (67) showed that gel permeation chromatography gave poor resolution of pro-cyanidins, whereas adsorption chromatography led to unacceptably long retention times for the higher procyanidins. Reverse phase separation on octadecyl stationary phases using aqueous methanolic mobile phases at low pH was found to offer the greatest scope for separation (Fig. 15) of procyanidins. Some HPLC methods (68–71) for the analysis of phenolic compounds in cider and apple juices are given in Table 2.

3. Fruit Juices

HPLC has been the method of choice for several years for the measurement of bitterness arising from limonoids and flavanone neohesperidosides. Typical methods (72–76) for determination of these compounds in citrus fruit juices are given in Table 2.

Table 2 HPLC Analysis of Bittering Substances in Beverages

Analyte	Sample type	Sample preparation	Quantitative HPLC			Ref.
			Stat. phase	Mobile phase	Detn.	
Iso-α acids	Beer	1. *Precipitation:* Proteins and dextrins precipitated in 80% CH$_3$OH.	10 μm Nucleosil C$_{18}$ (30 cm × 4 mm)	CH$_3$OH/H$_2$O/85% H$_3$PO$_4$ (70:29:1 v/v/v). Flow = 1 ml/min	UV 270 nm	56
			4 μm Novapak C$_{18}$ (10 cm × 8 mm)	CH$_3$OH/H$_2$O/85% H$_3$PO$_4$ (75:24:1 v/v/v). Flow = 1–2 ml/min	UV 270 nm or 280 nm	24, 57, Fig. 14
		2. *Liquid phase extraction:* Acidified sample extracted with 2 vols 2,2,4-trimethyl-pentane (isooctane)	5 μm RoSil C$_{18}$-D (10 cm × 4 mm)	CH$_3$CN/CH$_3$OH/H$_2$O (25:45:30) containing 0.01M TBAH adjusted to pH 6.8 with H$_3$PO$_4$ (85% w/w). Flow = 1.0 ml/min	UV 270 nm	56
		3. *Solid phase extraction:* Adsorption on precondi-tioned Bakerbond spe Octyl (C8) extraction column. Interferents removed with H$_2$O/H$_3$PO$_4$ (100:0.2) and CH$_3$OH/H$_2$O/H$_3$PO$_4$ (50:50:0.2). Analytes eluted with CH$_3$OH/H$_3$PO$_4$ (100:0.1)	Shim-pack CLC-ODS/H (25 cm × 4.6 mm)	A: 725 ml CH$_3$OH/275 ml H$_2$O containing 2.95 g of TBAP and 17 g of 85% H$_3$PO$_4$. B: CH$_3$OH. Gradient elution from A to B. Column temp. = 50°C. Flow rate = 1.5 ml/min	UV 275 nm	58
		Adsorption on Bond-Elut C$_{18}$ column. Analytes eluted with CH$_3$OH/H$_2$O/H$_3$PO$_4$ (50:50:0.25). Analytes recovered by elution with CH$_3$OH/H$_2$O/H$_3$ PO$_4$ (90:10:0.25)	4 μm Novapak C$_{18}$ 10 cm × 8 mm (in radial compression unit)	CH$_3$OH/H$_2$O/H$_3$PO$_4$ (80:20:1.7 v/v/v) containing 1.3 g/100 ml TBAH Flow = 2 ml/min	UV 270 nm	59
			5 μm Novapak C$_{18}$ 10 cm × 8 mm (in radial compression unit)	1 vial of Waters Assoc. PIC A (TBAP) in CH$_3$OH/H$_2$O (700:300 v/v) Flow = 2 ml/min	UV 280 nm	60
		None:- Direct on-column injection of filtered beer sample	5 μm RoSil C$_{18}$-D 25 cm × 4.6 mm	CH$_3$OH/H$_2$O (80:20) containing 13 g/L of 40% TBAH and 2% of H$_3$PO$_4$ (85% w/v). Flow = 2 ml/min	UV 270 nm	61

Table 2 (*Continued*)

Analyte	Sample type	Sample preparation	Quantitative HPLC — Stat. phase	Quantitative HPLC — Mobile phase	Quantitative HPLC — Detn.	Ref.
Procyanidins and their oxidation products	Ciders and wines	Adsorption on Sephadex LH-20 column. Interferents removed with 20% CH_3OH. Catechin and procyandins then eluted with 60% acetone. Polyphenols are fractionated by counter-current distribution between ethyl acetate and water	Spherisorb Hexyl 5 μm (12 cm × 4.6 mm)	A: H_2O acidified to pH 2.0 or 2.5 by the addition of 0.1% or 0.01% $HClO_4$ respectively. B: CH_3OH step gradient from A to B. Flow = 1.5 ml/min	UV 280 nm	68, 69, Fig. 15
Total phenolics	Apple juice	Apple juice pectin precipitate removed with 1 volume acetone. Analytes adsorbed from concentrated filtrate on Sephadex LH-20 column. Sugars eluted with water. Phenolis eluted with 60% acetone	1. *Normal phase:* Zorbax CN (25 cm × 4.6 mm)	A:THF/hexane/CH_3COOH/ HCOOH/IPA (32:60:4:2:2) B: THF/hexane/CH_3COH/ HCOOH/IPA (64.5:27.5:4:2:2) Gradient from A to B. Flow = 2 ml/min	UV 280 nm	70
			2. *Reverse phase:* μBondapak C18 (30 cm × 3.9 mm)	A: 0.1 M KH_2PO_4/CH_3OH (9:1) B: 0.1 M KH_2PO_4/CH_3OH (6:4) Gradient from A to B. Flow = 2 ml/min	UV 280 nm	70
Procynidins and other phenolics	Apple juice (fish and oxidized)	Fresh apple juice is filtered (0.45 μm) fresh juice). Portion of juice is allowed to brown in presence of solids for 1 hr and is then filtered and centrifuged for analysis (Oxidized juice)	Spherisorb Hexyl 5 μm (10 cm × 5 mm)	A: H_2O acidified to pH 2.5 with 0.1% $HClO_4$. B: CH_3OH C: H_2O neutralized to pH 6.5–7.0 with NH_3. Multiple step ternary gradient	UV 280 nm, Vis. 420 nm	71
Limonin	Grapefruit juice	Juice extracted $CHCl_3$. $CHCl_3$ fractions are evaporated	μPorasil (30 cm × 4 mm)	$CHCl_3$/CH_3CN (95:5). Flow = 2.5 ml/min, Temp = 35°C	R.I.	72, Fig. 16
Limonin	Citrus juice (orange or grapefruit)	Centrifuged juice extracted with $CHCl_3$. $CHCl_3$ extracts evaporated and residue redissolved in CH_3OH	μBondapak CN 10 μm (30 cm × 4 mm)	CH_3OH/H_2O 35:65 or 40:60 v/v. Flow = 1.5 ml/min	UV 210 nm	73

Analyte	Sample	Sample preparation	Column	Mobile phase	Detection	Ref.
Limonin and related limonoids	Citrus juices	Centrifuged grapefruit or orange juice are extracted with $CHCl_3$. $CHCl_3$ extracts evaporated and redissolved in 2 ml mobile phase and filtered (1.2 μm)	Zorbax CN-5 μm (25 cm × 4.6 mm)	Hexane/IPA/CH_3OH (11:12:2 v/v) at 1 ml/min or 6:3:1 v/v/v at 1.5 ml/min for routine analysis)	UV 207 nm	74
Naringin, naringenin	Grapefruit juice	None: Juice is filtered before injection	μBondapak C_{18} 10 μm (30 cm × 4 mm)	H_2O/CH_3CN (80:20 v/v)	UV 280 nm	75, Fig. 17
Naringin, narirutin, hesperidin, neohesperidin	Citrus juices	Analytes adsorbed from centrifuged juice on pre-conditioned Sep-Pak C_{18} cartridge. Flavanone glycosides eluted with CH_3OH	Zorbax C_{18} (25 cm × 4.6 mm)	H_2O/CH_3CN/CH_3COOH (79.5:20:0.5 v/v/v) Flow = 1 ml/min	UV 280 nm	76
Theaflavins	Tea	None: Direct injection of tea infusion following filtration (0.45 μm)	μBondapak C_{18} 10 μm (30 cm × 4 mm)	CH_3COOH/acetone/H_2O (1:60:139 v/v/v) Flow = 2 ml/min	UV/vis 365 nm	77
Theaflavins	Tea	Analytes in tea liquor adsorbed on Sep-Pak C_{18} cartridge, interferents removed with H_2O and 35% v/v CH_3OH. Theaflavins eluted with 80% v/v CH_3OH	Partisil C_{22} 5 μm (10 cm × 4.6 mm)	A: Acetone/H_2O/CH_3OH (69:323:8 v/v/v) pH 2 B:Acetone/H_2O/CH_3OH (109:283:8 v/v/v) pH 2 Gradient from A to B. Flow = 1 ml/min	UV/vis 372 nm	78
Tea phenolics	Tea	Black tea liquor is filtered prior to analysis	Hypersil C_{18} 5 μm (25 cm × 4.6 mm)	A: 2% CH_3COOH B: CH_3CN Gradient: 8% B to 31% B in 50 min. Flow = 1.5 ml/min	Photodiode array	79, Fig. 18
Thearubigins	Tea	Black tea leaves are extracted with boiling distilled water	Hypersil C_{18} 5 μm, pore size 12 nm (25 cm × 4.9 mm)	A: 2% CH_3COOH or 1% citric acid (pH adj. to 2.8 with NaOH) B: CH_3CN Gradient: 8–31% B over 50 min. Flow 1.5 ml/min	UV/Vis 280, 380, 460 nm	80
Flavanols, theogallin, gallic acid	Tea	Boiling water extract of tea purified by solid phase extraction on a pre-conditioned Baker spe C_{18} cartridge. Analytes eluted with H_2O/CH_3CN (85:15 v/v)	Hypersil ODS, 5 μm (25 cm × 4.6 mm).	A: 2% CH_3COOH B: CH_3CN Gradient of B increasing from 12 to 25% in A. Flow = 1 ml/min	UV 278 nm	81

Table 2 (*Continued*)

Analyte	Sample type	Sample preparation	Stat. phase	Quantitative HPLC Mobile phase	Detn.	Ref.
Flavanols	Tea	None: Direct injection of tea infusion following filtration (0.45 μm)	μ-Bondapak C_{18} 10 μm (30 cm × 4 mm)	$CH_3COOH/CH_3OH/DMF/H_2O$ (1:2:40:157 v/v/v) Flow = 2 ml/min	UV 280 nm	77
Gallic acid	Tea	None: Direct injection of tea infusion following filtration (0.45 μm)	μBondapak C_{18} 10 μm (30 cm × 4 mm)	0.02 M, pH 4.5 citrate-phosphate buffer Flow = 2.0 ml/min	UV 254 nm	77
Caffeine, theobromine, theophylline	Tea	None: Direct injection of tea infusion following filtration (0.45 μm)	μBondapak C_{18} 10 μm (30 cm × 4 mm)	$CH_3OH/$ 0.1 M, pH 4.5 citrate-phosphate buffer (20:80 v/v) Flow = 2 ml/min	UV 254 nm	77
Caffeine, theobromine	Tea, coffee, instant cocoa	Beverages prepared to manufacturers instructions. Internal standard (β-hydroxyethyl-theophylline) is added	μBondapak C_{18} (30 cm × 3.9 mm)	CH_3CN/H_2O (8:92 v/v). Flow = 2 ml/min	UV 245 nm	82
Caffeine	Tea, coffee, cola	Aqueous extract adjusted to pH 8–9 with ammonia solution and extracted with $CHCl_3$ containing 5% IPA	μBondapak C_{18} 10 μm (10 cm × 8 mm)	A: 2% CH_3COOH B: CH_3CN Gradient: 10–40% B in 10 min. Flow = 2 ml/min	UV 274 nm	83, Fig. 19
Caffeine, theobromine	Coffee, tea, cola	Beverages diluted with water and filtered (0.22 μm)	LiChrospher 100 RP-18, 5 μm (25 cm × 4.6 mm)	0.005 M octylamine ortho-phosphate, pH 6.4. Flow = 2.0 ml/min	UV 274 nm	84
Caffeine, theobromine, theophylline	Tea, coffee, cola	Tannins removed from tea or coffee infusions by precipitation with lead acetate. Cola drinks are diluted with water	LiChrocart C_8 5 μm (25 cm × 4.6 mm)	0.01 M sodium acetate solution/CH_3CN (85:15), pH 4.0. Flow = 2.0 ml/min	UV 272 nm	85
Phenolic acids and derivatives	Wine	None: Direct injection	Superspher RP-18 (25 cm × 4 mm)	A: CH_3COOH/H_2O (1:99 v/v) B: CH_3COOH/H_2O (5:95 v/v) C: $CH_3COOH/H_2O/CH_3CN$ (5:65:30 v/v/v) Multistep ternary gradient. Flow = 0.5 ml/min	UV 254 nm, 280 nm, 313 nm	90

Analyte(s)	Matrix	Sample preparation	Column	Mobile phase	Detection	Ref.
Phenolic acids	Wine	Analytes adsorbed from concentrated wine sample on preconditioned SepPak C_{18} cartridge. Interferents are eluted with phosphate buffer, and phenolic acids are recovered with tetrahydrofuran	Spherisorb ODS-2.5 µm (16.5 cm × 4.6 mm and 25 cm × 2.1 mm)	A: CH_3OH/phosphate buffer (pH 2.7) (95:5) B: CH_3OH/phosphate buffer (pH 2.7) (5:95) Gradient B to A. Flow = 0.2 ml/min	UV 280 nm	91, Fig. 20
Phenolic acids	Wine	Wine sample is acidified to pH 2 and saturated with NaCl and then extracted with ethyl acetate	Spherisorb ODS-2 (25 cm × 4.6 mm)	$IPA/CH_3COOH/CH_3OH/H_2O$ (2:2:8.7:87.3 v/v/v/v) containing 0.018 M ammonium acetate Flow = 1 ml/min	Amperometric	92
Phenolic acids, flavonols, procyanidins, catechins, tyrosol, tryptophol	Wine	Sample adjusted to pH 7 and then extracted with ethyl acetate to yield neutral phenolic fraction. Remaining aqueous phase is adjusted to pH 2 and reextracted with ethyl acetate to yield acidic phenolic fraction	Micropak MCH 10	A: CH_3OH B: Distilled H_2O adjusted to pH 2.5 with $HClO_4$. Multistep gradient of increasing A in B. Flow = 1.5 ml/min	UV 280 nm	93
Catechins, dimeric and trimeric procyanidins	Wine and grape extracts	Analytes adsorbed on polyamide TLC 6 at pH 7. Phenolic acids removed with water. Catechins eluted with CH_3CN/H_2O (30:70). Procyanidins are eluted with acetone/H_2O (75:25)	Ultrabase 5 µm C_8 or 5 µm Nucleosil C_{18}	A: CH_3COOH/H_2O (10:90) B: bidistilled H_2O Gradient: 10% A to 82% A in 47 min, 82% to 100% A in 8 min. Flow = 0.8 ml/min	UV 280 nm, 313 nm	94, Fig. 21
Flavonol aglycones	Wine	Wine is concentrated and extracted 4 times with diethyl ether	Novapak C_{18} radial compression cartridge	$H_2O/CH_3OH/CH_3COOH$ (65:30:5 v/v/v) Flow = 2 ml/min	UV 365 nm	95

Notes: CH_3OH = methanol, CH_3CN = acetonitrile, H_3PO_4 = orthophosphoric acid, KH_2PO_4 = potassium dihydrogen orthophosphate, THF = tetrahydrofuran, IPA = isopropyl alcohol, CH_3COOH = acetic acid, NaOH = sodium hydroxide, NaCl = sodium chloride, $CHCl_3$ = chloroform, TEAH = tetraethylammonium hydroxide, TBAP = tetrabutylammonium phosphate, TBAH = tetrabutylammonium hydroxide, DMF = dimethylformamide, NH_3 = ammonia, C_2H_5OH = ethanol.

Table 3 Analysis of Bittering Substances in Raw Materials of Beverages

Analyte	Sample	Sample preparation	Quantitative HPLC			
			Stat. phase	Mobile phase	Detection	Ref.
Flavanols, theogallin, gallic acid	Tea leaves	Ground tea extracted with CH$_3$OH and 70% CH$_3$OH. Pooled extracts concentrated and diluted with H$_2$O:CH$_3$CN (90:10 v/v). Analytes in 1 ml of extract adsorbed on preconditioned Baker C$_{18}$ spe. Cartridge containing analytes eluted with H$_2$O:CH$_3$CN (85:15 v/v).	Hypersil ODS, 5 μm (25 cm × 4.6 mm). Guard column: Nucleosil C$_{18}$ (1 cm × 4.6 mm)	A: 2% CH$_3$COOH B: CH$_3$CN Gradient: 88% A for 6 min. Ramp to 75% A in 5 min. Hold 75% A for 15 min. Reset in 8 min. Flow = 1 ml/min	UV 278 nm	81
Caffeine		Ground tea boiled with water and heavy MgO. Mixture is homogenized. Extract filtered and diluted				
Caffeine	Tea leaves, coffee beans	Ground sample treated with NH$_4$OH and then extracted with chloroform containing 5% IPA	μ-Bondapak C$_{18}$ 10 μm (10 cm × 8 mm)	A: 2% CH$_3$COOH B: CH$_3$CN Gradient: 10–40% B in 10 min, 40–10% B in 5 min. Flow: 2 ml/min	UV 274 nm	83
Caffeine, theobromine, theophylline	Cocoa or chocolate cake	Samples are boiled with aqueous Na$_2$CO$_3$ containing amyl alcohol. Aqueous layer is diluted with H$_2$O	LiChrocart C$_8$ 5 μm (25 cm × 4.6 mm)	0.01 M sodium acetate solution/CH$_3$CN (85:15), pH 4.0	UV 272 nm	85
Caffeine	Coffee, coffee products	Coffee or coffee powder is heated with MgO in water. Analytes adsorbed from extract on a preconditioned phenyl solid phase extraction column. Interferents eluted with NH$_3$ solution (0.3 M)/CH$_3$OH (90:10 v/v). Caffeine eluted with CH$_3$OH/H$_2$O/CH$_3$COOH (75:25:1 v/v/v)	C$_{18}$ column with N ≥ 5000 for caffeine.	CH$_3$OH/H$_2$O (30:70 v/v)	UV 272 nm	96

Analyte	Sample	Sample preparation	Column	Mobile phase	Detection	Ref.
α Acids, β acids	Hops, hop pellets and hop extracts	Hops and hop pellets are finely ground and extracted in acidic CH_3OH/diethyl ether. Hop extracts are diluted with CH_3OH	a) Nucleosil C_{18} Hop 5 µm (25 cm × 4 mm) b) Novapak C_{18} 5 µm (RCM 10 cm × 8 mm)	a) $CH_3OH/H_2O/H_3PO_4$ (85:17:0.25 v/v/v) Flow = 0.8 ml/min. b) $CH_3OH/H_2O/H_3PO_4$ (850:170:0.25 v/v/v) Flow = 2 ml/min	a) UV 314 nm b) UV 313 nm	a) 101 b) 102
α Acids, β acids	Hops and hop products	Hops are extracted with CH_3OH and internal standard (Octanal 2,4-dinitrophenylhydrazone) is added. Hop extracts are diluted with CH_3OH	Spherisorb ODS 5 µm (15 cm × 4.6 mm)	180 g citric acid solution (1.25% w/v) made up to 1 L with CH_3OH and adjusted to pH 5.6 with KOH	UV 355 nm	103
α acids, β acids	Hops, hop powders, hop pellets, hop extracts	Hops and hop pellets are extracted with toluene. Internal standard (β-chalkone) is added. Hop extracts are diluted with CH_3OH	Zorbax ODS or Shimpack ODS, 5 µm (25 cm × 4.6 mm). Temp. = 50°C	A: CH_3OH (72.5 ml), H_2O (27.5 ml), 85% H_3PO_4 (1.7 g) and 10% TEAH in H_2O (2.95 g) B: CH_3OH Gradient: 100% A for 5 min, 0-50% B in 15 min, 50% B for 5 min, flow = 1.5 ml/min	UV 270 nm	104
Iso-α acids, α acids, β acids	Worts and beers	Samples are acidified and extracted with isooctane. The isooctane is removed and residue reconstituted in CH_3OH with internal standard (β-phenylchalkone)	Zorbax ODS 5 µm (25 cm × 4.6 mm)	$CH_3OH/H_2O/85\% H_3PO_4$ (72.5:27.5:1 v/v/v) containing 2 mM TEAH	UV 270 nm	105
Iso-α acids, α acids, β acids	Worts and beers	Sample is diluted with CH_3OH, centrifuged, and filtered (0.45 µm)	Nucleosil C_{18} 10 µm (30 cm × 4 mm)	$CH_3OH/H_2O/85\%H_3PO_4$ (75:24.75:0.25 v/v/v) Flow = 2 ml/min	UV 280 nm	106
Iso-α acids	Preisomerized hop extracts	Sample is diluted with internal standard (4-methylbenzophenone) and filtered	Alltech RSL C_{18} 10 µm (25 cm × 4.6 mm)	9.9 ml TBAH in CH_3OH diluted to 320 ml with H_2O adjusted to pH 6.8 with 85% H_3PO_4/CH_3OH (580 ml)/CH_3CN (40 ml)/THF (50 ml). Flow = 1 ml/min	UV 280 nm	107

Table 3 (*Continued*)

Analyte	Sample	Sample preparation	Quantitative HPLC Stat. phase	Mobile phase	Detection	Ref.
Iso-α acids	Preisomerized hop extracts	Sample is diluted with internal standard (β-phenylchalkone)	Shimpack CLC-ODS/HP/N (25 cm $\times$ 4.6 mm)	CH_3OH(725 ml)/H_2O (275 ml)/17 g 85% H_3PO_4/29.5 g 10% aqueous TEAH. pH 3.0–3.5. Flow = 1.5 ml/min	UV 280 nm	108
cis and *trans* isomers of iso-α acids, α acid and β acid contaminants	Preisomerized extracts	Sample is diluted with CH_3OH and filtered	Nucleosil C_{18} 5 μm (25 cm $\times$ 4 mm)	*i)* A: $CH_3CN/H_2O/H_3PO_4$ (60:39:1) B: $CH_3CN/H_2O/H_3PO_4$ (90:9:1) Gradient of A to B. Flow = 1 ml/min. *ii)* A: $CH_3CN/CH_3OH/H_2O/H_3PO_4$ (50:5:49.9:0.1) B: $CH_3CN/CH_3OH/H_2O/H_3PO_4$ (50:45:4.9:0.1) Gradient of A to B. Flow = 1 ml/min	UV 280 nm	24
Iso-α acids	Preisomerized pellets	Pellets are extracted with *n*-butyl acetate containing H_3PO_4	Nucleosil C_{18} 5 μm (25 cm $\times$ 4.6 mm)	$CH_3OH/H_2O/85\%$ H_3PO_4/0.1 M EDTA (81.2:18.4:0.25:0.10 v/v/v/v). Flow = 1 ml/min	UV 270 nm	109

| Phenolic acid derivatives, catechins and procyanidins. | Grape berries | Grape berries with added ascorbic acid are crushed. Juice is deproteinated with C_2H_5OH. Analytes adsorbed from juice at pH 7.0 on pre-conditioned Sep-Pak C_{18} cartridge. Neutral phenolics eluted with CH_3OH. Eluate acidified to pH 2.5 and applied to Sep-Pak. Acidic phenolics eluted with CH_3OH | Waters Radial-Pak C_{18} (10 cm × 8 mm) | a) Acidic phenolics: 5% CH_3COOH in H_2O. b) Neutral phenolics: A: H_2O, B: 40% CH_3CN in H_2O. Gradient: 100% A to 100% B in 50 min. Flow: 1 ml/min | Diode array | 116 |
| Gallic acid and derivatives, catechins, procyanidin dimers and trimers | Grape seeds | Grape seeds are frozen at −40°C, ground, and homogenized in CH_3OH containing ascorbic acid. Water is added and CH_3OH removed by evaporation. Extract is extracted with pet. ether and aqueous phase is fractionated on Sephadex LH-20 and eluted with 96% C_2H_5OH. | Ultracarb C_{18} ODS 20, 4 µm (25 cm × 4.6 mm) | A: 4.5% HCOOH B: CH_3CN:4.5% HCOOH (10:90 v/v) Gradient: 0–50% B in 30 min (convex), 50–100% B in 50 min (linear) Flow = 1.5 ml/min | UV 280 nm | 117 |

Notes: CH_3OH = methanol, CH_3CN = acetonitrile, H_3PO_4 = orthophosphoric acid, KH_2PO_4 = potassium dihydrogen orthophosphate, THF = tetrahydrofuran, IPA = isopropyl alcohol, CH_3COOH = acetic acid, NaOH = sodium hydroxide, NaCl = sodium chloride, $CHCl_3$ = chloroform, TEAH = tetraethylammonium hydroxide, TEAP = tetraethylammonium phosphate, DMF = dimethylformamide, NH_3 = ammonia, C_2H_5OH = ethanol, KOH = potassium hydroxide.

Table 4 HPLC of Bitter Beverage Additives

Analyte	Sample	Sample preparations	Quantitative HPLC			
			Stat. phase	Mobile phase	Detection	Ref.
Lactucin	Chicory roots	Homogenized chicory with added NaCl is extracted continuously with diethyl ether in a liquid–liquid extractor	Novapak C_{18} in Waters RCM 10 cm × 8 mm	H_2O/CH_3OH (50:50). Flow = 2.0 ml/min	UV 258 nm	41
Quinine	Carbonated beverages	Samples are degassed and filtered (0.45 μm)	μBondapak C_{18} 10 μm (30 cm × 3.9 mm)	CH_3OH/CH_3CN/H_2O/CH_3COOH (20:10:70:1). Flow = 1.5 ml/min	UV 254 nm	118 Fig. 24
Quassinoids	*Quassia amara* L.	Quassinoids are isolated from a partially purified commercial mixture by prep. HPLC on a column of Partisil-5	Spherisorb S5-CN, 25 cm × 4.9 mm at 27°C	0.5% CH_3CN in $CHCl_3$. Flow = 0.9 ml/min	UV 256 nm	22 Fig. 25

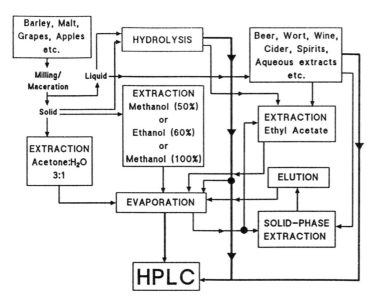

Fig. 13 General scheme for some pretreatments on samples of beverages and their raw materials for HPLC analysis of bittering substances (● either/or). (Redrawn from Ref. 24, p. 626, by courtesy of Marcel Dekker Inc.)

Limonin (Fig. 7) is relatively nonpolar and is therefore readily extracted from aqueous solution into chloroform. Thereafter, for HPLC, relatively polar stationary phases and nonpolar mobile phase mixtures can effect satisfactory mobility (Fig. 16). Naringin (Fig. 8), being a flavanone glycoside, is not readily extractable from fruit juice, even by medium-polarity organic solvents such as ethyl acetate. The mixed functionality of the naringin molecule, however, renders octadecylsilane reverse phases well suited for both its initial extraction (76) and subsequent HPLC (Fig. 17). Indeed, reverse

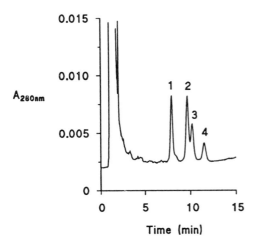

Fig. 14 Separation of iso-α acids in beer sample (pretreated by precipitation with methanol) by elution at 1 ml/min from column (10 cm × 4 mm) of Novapak C18 (4 μm) with methanol/water/orthophosphoric acid (75:24:1). Numbered peaks: 1 = *cis/trans*-isocohumulones, 2 = *cis*-isohumulone, 3 = *trans*-isohumulone, 4 = *cis/trans*-isoadhumulones. (See Refs. 24 and 57.)

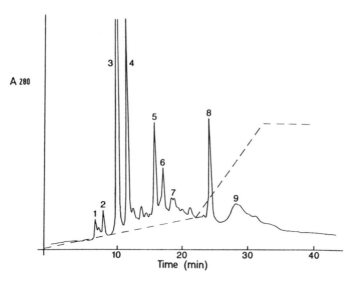

Fig. 15 Separation of a cider tannin extract on Spherisorb hexyl. Solvent gradient (broken line) from 2% B to 25% B in 23 min, 25% B to 98% B in 10 min. Solvent A = H_2O acidified to pH 2.0 with $HClO_4$, Solvent B = CH_3OH. Flow rate, 1.5 ml/min. Peaks: 1 = procyanidin B3, 2 = procyanidin B1, 3 = epicatechin, 4 = procyanidin B2, 5 = procyanidin trimer C1, 6 = procyanidin tetramer(s), 7 = procyanidin pentamer(s), 8 = phloridzin, 9 = oxidized/polymeric procyanidins. (From Ref. 68 with permission.)

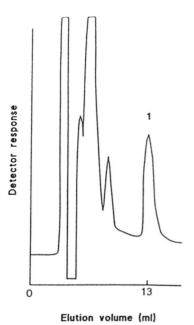

Fig. 16 Separation of limonin (peak 1) in grapefruit juice by elution at 2.5 ml/min from Waters Associates 10 μm Porasil column (30 cm × 4 mm id) with chloroform-acetonitrile (95:5), monitored by differential refractometry. (Redrawn from Ref. 72. Copyright 1975, American Chemical Society.)

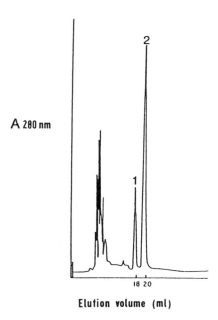

A 280 nm

18 20

Elution volume (ml)

Fig. 17 Separation of naringenin rutinoside (peak 1) and naringin (peak 2) in grapefruit juice by elution at 2.0 ml/min from Waters Associates μBondapak C_{18} column (30 cm × 4 mm id) with water-acetonitrile (80:20 v/v). (Reprinted with permission from Ref. 75, Copyright 1976, American Chemical Society.)

phase HPLC has proven extremely selective for the separation of all flavonoid aglycones, based on differences in hydroxylation patterns, using mixtures of water and organic modifiers (methanol and/or acetonitrile). The presence of glycosidic units increases mobility on reverse phase systems and can add even greater scope for chromatographic selectivity.

4. Tea

Although the terms "astringency" and "bitterness" may not be used universally in the sensory evaluation of tea quality, and despite the fact that these attributes are not routinely measured analytically in tea quality control, many of the compounds present in tea are in fact bitter and/or astringent. Examples include the methylxanthines, theaflavins (Fig. 12), thearubigins, and procyanidins. Reverse phase HPLC on octadecylsilane columns (77–85) is the method of choice for the separation (Fig. 18) of these compounds, most of which are detectable by ultraviolet or visible absorbance. Indeed, ultraviolet/visible detection is made all the more powerful through use of diode array detector systems, which allow much greater characterization of tea chromatograms than could ever be achieved using single-wavelength monitoring. Some other examples of the application of reverse phase HPLC to tea analysis (Fig. 19) are given in Table 2.

5. Wine

Doubtless it is easier by far to appreciate quality in a wine than to define its character entirely by even the most refined of analytical instrumentation (44). Many obstacles still limit the usefulness of HPLC methods in the quantitative expression of some quality features that are readily detectable by eye or palate. Contemporary methodology remains confined to relatively simple molecules with little resolving capacity for compounds such as polymeric flavanoids

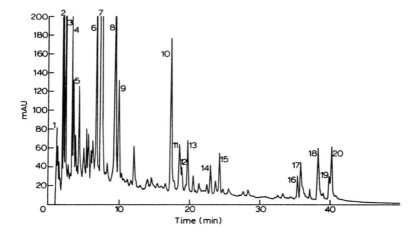

Fig. 18 Chromatogram of black tea monitored at 280 nm. Solvent A, 2% aqueous acetic acid, solvent B, acetonitrile. Linear gradient, 92% A and 8% B to 69% A to 31% B over 50 min. Identified peaks: 2 = theogallin, 3 = gallic acid, 4 = theobromine, 5 = 3-o-caffeoylquinic acid, 6 = 5-o-caffeoylquinic acid, 7 = caffeine, 8 = (−)-epigallocatechin gallate, 9 = p-coumaroylquinic acid, 10 = (−)-epicatechin gallate, 11 = quercetin 3-rhamnosyl glucoside, 12 = quercetin-3-glucoside, 13 = quercetin glycoside, 14 = kaempferol 3-rhamnosyl glucoside, 15 = kaempferol glycoside, 17 = theaflavin, 18 = theaflavin-3-gallate, 19 = theaflavin-3′-gallate, 20 = theaflavin-3,3′-digallate. (Reprinted from Ref. 79, with permission from the *Journal of the Science of Food and Agriculture.*)

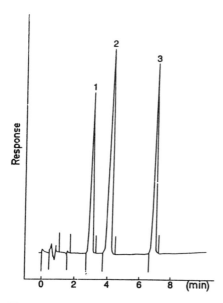

Fig. 19 HPLC chromatogram of standard mixture of caffeine (peak 3), theobromine (peak 1), and theophylline (peak 2) on a column of Waters Associates µBondapak C_{18} (10 cm × 8 mm id) eluted with a gradient of 10–40% acetonitrile in 2% acetic acid in 10 min. Detection is by UV absorbance at 274 nm. (Reprinted from Ref. 83, p. 1021 by courtesy of Marcel Dekker Inc.)

formed during wine maturation (86). Whereas the dependence of bitterness and astringency in red wine on its contents of flavanoids seems certain, a trustworthy quantitative estimate of these sensory properties is beyond the range of existing HPLC methods. The contributions to bitterness of nonflavanoids in wine, such as tyrosol, dipeptides, phenolic acid derivatives, and terpene glyco- sides, appear small compared to the flavanoids (87), and even flavones, flavanones, flavanols, and flavanonols are not of much significance in wines. Consequently, efforts have been concentrated on the HPLC resolution of (+)-catechin, (−)-epicatechin, and their oligomers (Figs. 6 and 9). Recent reviews (86,88) describe HPLC as the method of choice for nonvolatile phenolics, noting typical methodological features such as:

1. Reverse phase octadecyl silica columns.
2. Acidic mobile phases consisting of water-methanol or water-acetonitrile mixtures. The most frequently used acids are phosphoric, perchloric, formic, and acetic acids.
3. Isocratic, gradient, and step-gradient elution.
4. Ultraviolet light and electrochemical detectors.

Direct injection of samples has been possible for the measurement of hydroxycinnamates and anthocyanins in wines, but the HPLC measurement of the more flavor-significant flavanoid oligomers usually requires some sample pretreatment. Different methods of separation and their applications have been extensively reviewed (86,89). Some representative methods (90–95) are given in Table 2 and Figs. 20 and 21.

B. Raw Materials

A distinction between methods suitable for the analysis of components of beverages and their starting materials might seem somewhat superficial, since the analytes are often one and the same. Preparation of the beverages frequently involves little more than the aqueous extraction of a suitably ground or pulverized source material, either with or without heating. Such aqueous extractions are sometimes sufficiently exhaustive when conducted under defined conditions to

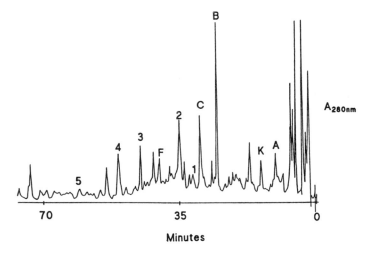

Fig. 20 Chromatogram of wine phenolic acids detected at 280 nm eluted from a column of 5 μm ODS-2 eluted with a gradient of increasing methanol content in pH 2.7 phosphate buffer. Identified peaks: A, gallic acid; B, 4-hydroxybenzoic acid; 1, vanillic acid; 2, syringic acid, caffeic acid; 3, *p*-coumaric acid; 4, *m*-coumaric acid, ferulic acid, C, K, F; 5, unknowns. (Redrawn from Ref. 91 with permission.)

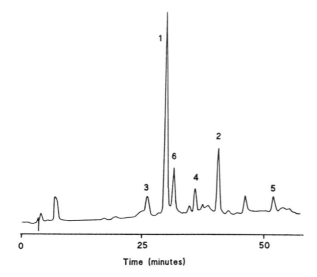

Time (minutes)

Fig. 21 Chromatogram of wine procyanidins eluted from a column (25 cm × 4.6 mm) of Ultrabase C8 with a gradient from 10% A to 82% A over 47 min, and to 100% A over 8 min, at 0.8 ml/min, where A = 10% CH_3COOH and B = H_2O. Identified peaks: 1, procyanidin B1; 2, procyanidin B2; 3, procyanidin B3; 4, procyanidin B4; 5, procyanidin trimer C1; 6, procyanidin trimer 2. (Redrawn from Ref. 94 with permission from the *Journal of the Science of Food and Agriculture*.)

provide quantitative recovery of undegraded analytes from the raw materials and thereby serve as the first step in analysis. For some applications, however, extraction with organic solvents is necessary to recover analytes completely in a sufficiently concentrated solution for HPLC. Even so, apart from minor differences in sample pretreatments the HPLC methods chosen for defined analytes recovered either from the raw materials or from their ultimate beverage products are often interchangeable. Examination of the raw materials can be a good indicator for the quality of the final product. For instance, it has been established that grape skins and seeds are particularly rich in those flavanoids that contribute bitterness and astringency (86). In the case of beer production, however, the bittering substances in hops (α acids) are chemically transformed into the different species that are present in the finished beer (iso-α acids) and so the methods of analysis of these two groups of substances are recognizably different.

1. Tea, Coffee, Cocoa

Table 3 gives a number of methods for the analysis of tea leaves (81,83) and coffee (83,96) or cocoa beans (85). Invariably, the separation and detection methods employed are similar to those which would be applied to the finished beverage. However, the extraction stages in these methods are designed to maximize analyte recovery, in comparison to the usual hot water extractions which give measures of available, but not necessarily total, analyte present. These methods are useful for characterization of raw materials and are perhaps more easily performed reproducibly than hot water extractions.

2. Hops

The measurement by HPLC of the α-acid bittering principles of hops (Fig. 3) and the structurally related though nonbittering β acids has been the subject of intensive studies and consequently has been reviewed comprehensively in recent years (24,62). It is above all the use of hops for bittering that distinguishes beer from other beverages, so that their importance to the brewer cannot be overstated. The main α acids (cohumulone, humulone, and adhumulone; Fig. 3) constitute the main components of the "soft resin fraction" found in the lupulin glands of the dried cones collected from female hop plants. Whereas about 10–25% of the dry weight of hop cones may consist of total resins, only about 2–14% of the weight, depending on the variety, is made up by the α acids. Many other constituents, of varying utility to the brewer, are present in hops (97), some of which complicate the HPLC analysis for α acids. Although the use of hops for brewing has spanned several centuries (98), it has been only in recent times that significant progress has been made in overcoming some major shortcomings. For ease of handling, prod-ucts are now available in the form of packaged pellets or extracts, which are added directly to the wort kettle in accordance with conventional brewing practice (99). Not only do these products occupy less storage space than baled hops but they have the added advantage that their contents of resins are more concentrated and more stable. Other products also designed for kettle addition contain the resins in a preisomerized form. One such product is prepared by addition of magnesium oxide during pelleting so that subsequent heating under vacuum causes efficient transformation of the α acids to iso-α acids (Fig. 3). The isomerization of α acids in wort kettle is normally inefficient with yields (utilization) of soluble iso-α acids typically in the range 25–30%. Greater efficiency in the isomerization is achieved by extracting the α acids from hop tissue in a semipure form using an organic solvent or liquid/supercritical carbon dioxide, and then conducting the isomerization by boiling in potassium carbonate solution (98). Subsequent controlled decreases in pH lead ultimately to a concentrated solution of iso-α acids largely freed from possible contamination with α acids and β acids. Such preisomerized hop extracts are finding increasing usage in the brewing industry for postfermentation bittering.

The brewing analyst and hop product manufacturer therefore have use for HPLC methods for measuring α acids in whole hops, hop pellets, and hop extracts (Fig. 22), and for measuring iso-α acids in preisomerized hop pellets and preisomerized hop extracts (Fig. 23). Clearly, these methods are of importance since commercial pricing of hop materials is based on either α acids or iso-α acids content. Moreover, brewers require reliable measures of the inputs in order to control bitterness of their products. Of lesser routine importance, it is sometimes necessary to measure in

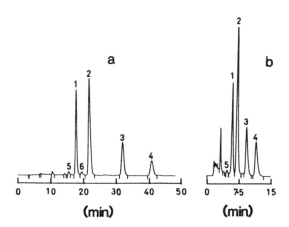

Fig. 22   (a) Separation by recommended EBC (European Brewery Convention) method of α and β acids in commercial hop extract. (b) Separation by recommended IOB (Institute of Brewing) method of α and β acids in extract from hop pellets. Identified peaks: 1, cohumulone; 2, humulone/adhumulone; 3, colupulone; 4, lupulone/adlupulone; 5, deoxycohumulone; 6, deoxyhumulone. (Reprinted from Ref. 24, p. 588, by courtesy of Marcel Dekker Inc.)

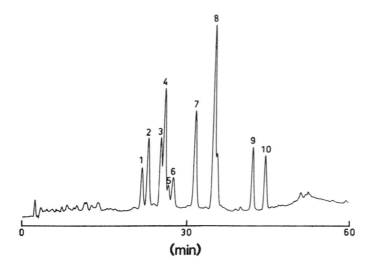

Fig. 23 Separation of iso-α, α, and β acids in a mixture of commercial hop extract and isomerized extract. Elution was at 1 ml/min with a linear gradient of acetonitrile/methanol/water/orthophosphoric acid from 50:5:49.9:0.1 to 50:45:4.9:0.1 in 60 min, detection at 280 nm. Identified peaks: 1, *trans*-isocohumulone; 2, *cis*-isocohumulone; 3, *trans*-isohumulone; 4, *cis*-isohumulone; 5, *trans*-isoadhumulone; 6, *cis*-isoadhumulone; 7, cohumulone; 8, humulone/adhumulone; 9, colupulone; 10, lupulone/adlupulone. (Reprinted from Ref. 24, p. 620, by courtesy of Marcel Dekker Inc.)

detail the utilization of α acids in worts during the brewing process, particularly when the consequences of changes in procedure are examined (24). In contrast, the measurement of β acids (colupulone, lupulone, adlupulone; Fig. 3) is rarely needed by the brewer, but these are normal constituents of hops and their extracts, and are even present in very small amounts in preisomerized extracts (24). For hop growers, however, the relative contents of α acids and β acids can serve as a guide to varietal identification.

The development of methods for the analysis of hop bittering constituents has been traced over three decades (24). Early methods involved countercurrent distribution, thin-layer chromatography on silica gel, and column chromatography on various ion exchangers. Since the early 1970s the possibilities for HPLC captured most of the research interest and despite the advances made by adsorption chromatography on silica gel and anion exchangers the overwhelming potential of reverse phase systems incorporating bonded stationary phases was soon realized. Using columns of octadecylsilica (ODS) all the major α and β homologs and the *cis* and *trans* isomers of the iso-α acids have been resolved (100), and analysis of these compounds, once completely beyond the repertoire of the most accomplished analysts, is now commonplace (101–109). The utility of several commercial ODS columns in this application has been assessed (24) on the basis of hydrophobicity and silanol indices. Those stationary phases judged by these criteria to be most suitable were indeed also specified independently in officially recommended methods (Table 2, Refs. 101, 102, 107–109). Comparisons have also been made of those methods recommended by the European Brewery Convention and the Institute of Brewing (24). For the routine measurement of α-acid and β-acid homologs, mixtures of acidified methanol in water suffice as mobile phases for separation (Fig. 22), but more complex ternary mixtures are needed for the separation (Fig. 23) of iso-α acid isomers (101–109). Other possible variations in experimental protocol, such as detection wavelength, use of ion-pairing reagents, calibration protocols, pH, and so forth, were recently reviewed (24). The only arguments against the universal acceptance of HPLC methods in preference to more traditional methods of analysis are the acceptability of the available calibration standards and a throughput of samples (1–4/hr), which is unsatisfactory to the hop industry at times of peak demand. Possibilities now exist, however, for fast analyses (~5 min) for α acids and β acids on short C_{18} columns (110). For standardization, the use of known mixtures of the α-acid and β-acid homologs is convenient and relatively cheap in the absence of pure standards. Such mixtures are available from the Versuchsstation Schweizerischer Brauereien (Zurich) and the American Society of Brewing Chemists (St. Paul). As another alternative the preparation of 1,2-diaminobenzene complexes of α acids was recently proposed (111). Similarly, known mixtures of iso-α acids such as commercial preisomerized extracts can serve as a standard (56), provided that appropriate storage conditions are used. Other alternatives are the magnesium salts of iso-α acids (112) and the *trans*-isohumulone formed by photoisomerization of pure humulone (113). The most recently proposed alternative (64), which has already generated widespread support, advocates the use of dicyclohexylamine salts of *trans*-isocohumulone, *trans*- isohumulone, and *trans*-isoadhumulone. For more detailed analysis of minor bittering compounds, methods for the preparation of hulupone (Fig. 3) standards have also been described (114). The current state of developments in the HPLC of iso-α acids was concisely reviewed by workers from the University of Ghent, to whom the brewing industry is indebted for much of the progress in this area (115).

3. Grapes

Methods (116,117) for the analysis of phenolics in grape berries and grape seeds are summarized in Table 3. These methods are useful for the selection of high- or low-phenolic-containing varieties, although the results may not be representative of amounts found in the finished product, i.e., wine or grape juice.

C. Bittering Additives

There are few examples in the beverage world in which a distinctly bitter flavor is a goal of the production process, with the obvious exception of the brewing industry. The examples that follow, however, are of compounds that are used specifically with the aim of adding bitterness to the product. Table 4 gives HPLC methods used for the analysis of these substances.

1. Chicory

Chicory is often used as an additive in the coffee industry to reduce production costs while maintaining bitterness. The roots and heads of chicory plants contain the sesquiterpene lactones lactucin and lactucopicrin (Fig. 10), which are responsible for the bitterness of the plant (41).

2. Quinine

Although perhaps best known for its antimalarial activity, quinine (Fig. 2), as its sulfate or hydrochloride salt, is often added as a flavoring agent to carbonated beverages, in particular tonic water. The compound produces a distinctly bitter taste in the beverage. The U.S. Federal Code of Regulations of 1984 imposed an upper limit of 83 mg/L of quinine in soft drinks (118). Therefore it is important for drink producers using this compound to have a sensitive and accurate method for its analysis (Fig. 24).

3. Quassinoids

Quassin (Fig. 1) is said to be the most bitter substance known (15) and thus its use as a bittering agent in beverages or foodstuffs may be ideal (22). Crude quassin, extracted from the wood of the neotropical species *Quassia amara* L. and *Picrasma excelsa*, has been shown (Fig. 25) to contain the quassinoids quassin, neoquassin, and 18-hydroxyquassin (22).

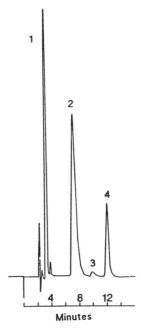

Fig. 24 Chromatogram of low-caloric tonic water on a Waters μBondapak C_{18} 10-μm column (30 cm × 3.9 mm id) eluted with $CH_3OH/CH_3CN/H_2O/CH_3COOH$ (20:10:70:1) at 1.5 ml/min. Detection is at 254 nm. Identified peaks: 1, sodium saccharin; 2, quinine; 3, hydroquinine; 4, sodium benzoate. (Redrawn with permission from Ref. 118. Copyright Springer-Verlag GmbH & Co.)

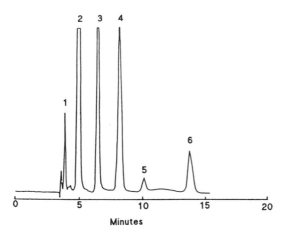

Fig. 25 Separation of a mixture of quassins on a Spherisorb S5-CN column. UV detector at 256 nm. Mobile phase: 0.5% acetonitrile in chloroform. Flow rate: 0.9 ml/min. Temperature: 27°C. Identified peaks: 1, 14,15-dehydroquassin; 2, quassin; 3, neoquassin; 4, neoquassin; 5, 12-hydroxyquassin (tentative); 6, 18-hydroxyquassin. (Redrawn with permission from Ref. 22.)

VI. COLORIMETRIC ANALYSIS OF BITTER SUBSTANCES

From the above examples it is clear that HPLC is widely applied to the analysis of bitter substances across a wide range of industrial applications. There are cases, however, wherein colorimetry is the method of choice, either because HPLC methods for the analytes are not yet sufficiently developed or because colorimetric assays may be simpler and faster to perform. Table 5 summarizes the salient features of such methods for the sample matrices listed below.

1. Beer

The determination of iso-α acids in beer is specifically performed by HPLC, but most breweries choose to use the less specific bitterness units (BU) or European bitterness units (EBU) measurements (119,120) for routine purposes. These analyses are simpler and quicker to perform than HPLC and have the distinct advantage of not requiring a calibration standard. One disadvantage is that beers containing additives (where permitted) which absorb at 280 nm, such as *n*-heptyl-4-hydroxybenzoate, salicylic acid, saccharin, or sorbic acid, will give elevated results using this assay. Colorimetric procedures are in widespread use also for the measurement of total polyphenols (121) and total flavanoids in beer (122).

2. Tea, Nonalcoholic Beverages

Although considerable research interest is being shown in the area of HPLC analysis of tea constituents (see Table 3), there is a widespread use of colorimetry (Table 5) to measure bitter and astringent compounds. The nonspecificity of colorimetry can be an advantage in that groups of bitter substances may be collectively analyzed without the need for complex separations and calculations. However, it has been shown (79) that some colorimetric assays are subject to significant interference. Colorimetry has been successfully used to define and measure nonspecific attributes such as total tannins and relative astringency of teas (123), theaflavins and thearubigins (124,126), and total tannins (127). Colorimetric procedures are also in widespread use for the determination of caffeine in tea and other nonalcoholic beverages (128–130).

3. Citrus Fruits

The analysis of bitter principles in citrus fruits is invariably carried out nowadays using HPLC. However, in the past such analysis was carried out colorimetrically by the Davis method (131). The use of HPLC offers huge advantages in this area in terms of both selectivity and sensitivity, as well as the capability of separating individual compounds.

4. Wine

The analysis of wine phenolics has been greatly enhanced through the use of HPLC (Table 2). As for beer analysis, however, the use of colorimetry can provide valuable information regarding the phenolic content of the beverage, without the need to specify individual compounds. A comparison of spectral and HPLC methods for the analysis of anthocyanins in aging red wine has been carried out (132), and this comparison showed the colorimetric method to give higher values for anthocyanin content. However, discrepancies between chromatographic and colorimetric methods for wine analysis have not precluded the use of the latter for routine analysis. Table 5 gives selected methods (133,134) for the colorimetric analysis of wine phenolics. For a more detailed review of the methods available, the reader is directed to Ref. 133.

Table 5 Colorimetric Analysis of Bittering Substances

Analyte	Sample	Method Summary	Ref.
Bitterness	Beer	Beer is acidified with HCl and extracted with 2 vols. 2,2,4-trimethylpentane (isooctane). Absorbance of isooctane layer at 275 nm is multiplied by 50 to give result in bittering units (BU)	119, 120
Total polyphenols	Beer	Beer is treated with a solution of carboxy-methylcellulose and EDTA. Beer polyphenols are reacted with ferric ions in alkaline solution. Absorbance at 600 nm of the red colored solution is measured against a blank	121
Flavanoids	Beer	Beer is reacted with an acidic solution of DAC. The absorbance of the chromophore produced is measured at 640 nm. The method is standardized with (+)-catechin	122
Total tannins	Tea	Tea or aqueous tannin is complexed with bovine serum albumin (BSA). Mixture is centrifuged and the supernatant assayed for remaining protein with Coomassie blue. Tannic acid equivalence (TAE) is calculated by dividing the concentrations of tannic acid required to completely precipitate the BSA by the concentration of tea required to do the same	123
Theaflavins (TF), thearubigins (TR)	Tea	Aqueous tea extract is extracted with IBMK. Portion of the IBMK layer is diluted with CH_3OH (solution A). Part of the aqueous portion is diluted with H_2O and CH_3OH (solution B). Part of the IBMK layer is shaken with aq. $NaHCO_3$, and part of the washed IBMK layer is diluted with CH_3OH (solution C). $H_2C_2O_4$ and H_2O are added to a portion of the aqueous layer from the first IBMK extraction and the solution diluted with CH_3OH (solution D). The optical densities of solutions A–D are measured at 380 and 460 nm. Appropriate factors are used to calculate thearubigin and theaflavin contents	124
Theaflavins, thearubigins	Tea	Fresh tea liquor is injected onto a preconditioned C_{18} sorbent cartridge. Cartridge is washed with aq. CH_3COOH and a fraction collected. The column is washed with acidic solutions of increasing CH_3OH content and fractions collected. Absorbances of collected solutions are measured at 380, 440, and 460 nm and appropriate factors are used to calculate theaflavin and thearubigin contents. The column is regenerated with aq. CH_3OH followed by CH_3COOH.	125

Theaflavins (TF), thearubigins (TR), highly polymerized substances (HPS)	Tea	Aqueous tea extract is extracted with 25 ml IBMK. A portion of the IBMK layer is diluted with 45% C_2H_5OH (soln. A). IBMK layer is then treated with 2.5% Na_2HPO_4 and part of the treated IBMK is diluted with 45% C_2H_5OH (soln. B.) Part of the aqueous layer from the first extraction is treated with n-butanol, and part of the n-butanol layer is diluted with 45% C_2H_5OH (soln. C) and part of the aqueous layer is treated likewise (soln. D). A_{380} of solns. A–D are measured and multiplied by appropriate factors to calculate TF, TR, HPS, and total liquor color.	126
Tannins	Tea, beer	Diluted beer or tea extract is injected into a stream of Fe(III) via a loop and are subsequently merged with a stream of CH_3COOH/sodium acetate (pH 3.5) and then with another stream containing 1,10-phenanthroline. The mixture passes through a reaction coil, and subsequently through a spectrophotometer where the absorbance is measured at 510 nm relative to that at 680 nm. The method is standardized with gallic acid	127
Caffeine	Nonalcoholic beverages	Sample is extracted continuously with $CHCl_3$ under alkaline conditions with heating. The $CHCl_3$ is reduced to dryness and reconstituted in H_2O. The solution (or a standard solution of caffeine) is reacted with phosphomolybdic acid under acidic conditions with heating. The hot solution is filtered through a Buchner funnel and the precipitate isolated is dissolved in acetone. The absorbance of the solution at 440 nm is compared to that of standard solutions treated as above	128
Caffeine	Nonalcoholic beverages	To sample is added 1.5% $KMnO_4$ solution, reducing solution (Na_2SO_3 and KSCN in H_2O), dilute H_3PO_4, and NaOH. The solution is extracted with $CHCl_3$, the absorbance at 276 nm is measured and compared with that of standard solutions	129
Caffeine	Nonalcoholic beverages	Sample is evaporated to near-dryness on a steam bath and residue is dissolved in NH_4OH, heated on a steam bath, removed, and Celite 545 is added. The sample is purified chromatographically using basic followed by acidic columns of Celite 545. The final eluate is in $CHCl_3$. The absorbance of the eluate is read at 276 nm and compared with standards	130

Table 5 (*Continued*)

Analyte	Sample	Method Summary	Ref.
Flavanones	Citrus fruits	Solution to be tested is added to diethylene glycol in a colorimeter tube. The colorimeter is zeroed using a blue filter. 4 N NaOH is added and the color increase read after 5 min. The method is calibrated with the pure flavanone, or its glycoside, known to be predominantly present in the fruit of interest, e.g., naringin in grapefruit and hesperidin in orange juice	131
Total anthocyanins	Wine	The pH of a wine sample is adjusted to 0.5–0.8 and A_{520} in water or A_{540} in alcohol is measured. The method is calibrated with a suitable pigment, e.g., malvidin-3-glucoside or malvin (malvidin-3,5-diglucoside)	133
Total anthocyanins	Wine	Two tubes are prepared with wine and 0.1% conc. HCl in 95% C_2H_5OH. To one tube is added 2% v/v conc. HCl, pH 6, and to the other pH 3.5 buffer. The absorbance difference between the tubes at 520 nm is compared to that of a solution of a known anthocyanin in 0.1% v/v conc. HCl in C_2H_5OH	133
Total anthocyanins	Wine	Two tubes are prepared containing the following mixture: wine, 0.1% v/v conc. HCl in C_2H_5OH and 2% v/v conc. HCl. To one tube is added H_2O and to the other, 15% $NaHSO_3$. A_{520} is measured and compared with a standard anthocyanin solution	133
Total phenols	Wine	Wine is pretreated with excess acetaldehyde. Appropriate dilutions of standard (gallic acid) or wine are mixed with Folin–Ciocalteau reagent and H_2O. After 0.5–8 min, Na_2CO_3 is added. A_{765} is measured after 2 hr at room temperature. The results are expressed as gallic acid equivalents (GAE)	133
Total phenols	Wine	Sample or standard (catechin) is mixed with H_2O, 0.05 M $FeCl_3$ in 0.1 N HCl and 0.008 M $K_3Fe(CN)_6$. A_{720} nm is read after 20 min	133
Total phenols	Wine	Samples of wine are introduced into a flowing stream of distilled H_2O which is then merged with a stream of Folin–Ciocalteau reagent, followed by a stream of Na_2CO_3 solution. The stream passes through a reaction coil and the absorbance is measured at 760 nm. The method is calibrated with solutions of either gallic or tannic acid.	134

Notes: DAC = *p*-dimethylaminocinnamaldehyde, IBMK = isobutyl methyl ketone.

VII. FUTURE DEVELOPMENTS AND PERSPECTIVES

Many bitter substances exist in nature that have not been included in this chapter. Examples that occur in vegetables, fruits, and herbaceous plants are given in Ref. 5. For instance, several herbs are used to flavor certain specialty wines, such as vermouth (13), and their exclusion from this chapter is warranted only by our self-imposed focus on mainstream beverages. Even then, we have chosen not to deal in depth with the analysis of phenolic substances in beer, while phenolics in tea and wine have occupied much of our attentions. This, we argue, is justified by the overriding importance of iso-α acids to the bitterness of beer compared to the uncertain role of phenolics in this context. For the routine analysis of beer bitterness, HPLC methods still await global acceptance, and their greater precision and accuracy will be weighed against the simplicity and cost effectiveness of more traditional spectrophotometric methods. Indeed, automation of the traditional assay by flow injection instrumentation (135,136) might compete so effectively for the esteem of brewing analysts that HPLC will remain largely a tool for research and special projects. Another example from the recent literature of the application of flow injection analysis is in the analysis of total polyphenols in beer (137). Whereas the technique yields results described as "satisfactory," it has yet to be shown that the values obtained correlate with any aspects of beer quality. In contrast, improved methods of detection for phenolics using electrochemical detectors in amperometric mode (40,51,138) have already shown their worth, albeit in the context of beer colloidal stability (53,139). This method of detection is especially sensitive for easily oxidized substances such as phenolics (51) and hop α and β acids (140). Obviously, the further application of such developments in instrumentation are not confined to the brewing industry, as exemplified by the combined use of coulometric array detection and HPLC for analyzing phenolic and flavonoid compounds in various fruit juices (141).

The next generation of analytical instruments must, like their existing counterparts, be judged by their users on the criteria of both commercial realism and scientific refinement. For the measurement of bitterness in various industries the chosen methods will continue to demonstrate different degrees of instrumental advancement and practical simplicity. Whatever device is dictated by circumstance, it is certain that the final arbiter will remain the human palate.

REVIEWS

H. D. Belitz and H. Weiser, *Food Rev. Int. 1*:271 (1985).

G. E. DuBois, D. E. Walters, and M. S. Kellog, *Flavor Measurement* (Chi-Tang Ho and C. H. Manley, eds.), Marcel Dekker, New York, 1993, p. 239.

H. Heymann, D. L. Holt, and M. A. Cliff, *Flavor Measurement* (Chi-Tang Ho and C. H. Manley, eds.), Marcel Dekker, New York, 1993, p. 113.

R. L. Rouseff, *Bitterness in Foods and Beverages* (R. L. Rouseff, ed.), Elsevier, Amsterdam, 1990, p. 1.

H. G. Schutz, *Flavor Measurement* (Chi-Tang Ho and C. H. Manley, eds.), Marcel Dekker, New York, 1993, p. 95.

S. Van Toller, *Flavor Measurement* (Chi-Tang Ho and C. H. Manley, eds.), Marcel Dekker, New York, 1993, p. 199.

REFERENCES

1. H. G. Schutz, *Flavor Measurement* (Chi-Tang Ho and C. H. Manley, eds.), Marcel Dekker, New York, 1993, p. 95.

2. H. Heymann, D. L. Holt, and M. A. Cliff, *Flavor Measurement* (Chi-Tang Ho and C. H. Manley, eds.), Marcel Dekker, New York, 1993, p. 113.

3. S. Van Toller, *Flavor Measurement* (Chi-Tang Ho and C. H. Manley, eds.), Marcel Dekker, New York, 1993, p. 199.

4. G. E. DuBois, D. E. Walters, and M. S. Kellog, *Flavor Measurement* (Chi-Tang Ho and C. H. Manley, eds.), Marcel Dekker, New York, 1993, p. 239.

5. R. L. Rouseff, *Bitterness in Foods and Beverages* (R. L. Rouseff, ed.), Elsevier, Amsterdam, 1990, p. 1.

6. *The Merck Index* (M. Windholz, ed.), 10th ed., Merck and Co., Inc., Rahway, NJ, 1983, p. 1268.

7. J. A. Maga, *Bitterness in Foods and Beverages* (R. L. Rouseff, ed.), Elsevier, Amsterdam, 1990, p. 35.

8. J. Solms, *J. Agric. Food Chem. 17*: 686 (1969).

9. J. Kirimura, A. Shimizu, A. Kimizuka, T. Ninomiya, and N. Katsuya, *J. Agric. Food Chem. 17*: 689 (1969).

10. H. D. Belitz and H. Weiser, *Food Rev. Int. 1*: 271 (1985).

11. C. H. Brieskorn, *Bitterness in Foods and Beverages* (R. L. Rouseff, ed.), Elsevier, Amsterdam, 1990, p. 15.

12. D. G. Guadagni, V. P. Maier, and J. G. Turnbaugh, *J. Sci. Food Agric. 24*: 1277 (1973).

13. F. Chialva and G. Dada, *Bitterness in Foods and Beverages* (R. L. Rouseff, ed.), Elsevier, Amsterdam, 1990, p. 103.

14. T. Yaminishi, *Bitterness in Foods and Beverages* (R. L. Rouseff, ed.), Elsevier, Amsterdam, 1990, p. 159.

15. R. J. Robbins, M. J. C. Rhodes, A. J. Parr, and N. J. Walton, *Bitterness in Foods and Beverages* (R. L. Rouseff, ed.), Elsevier, Amsterdam, 1990, p. 49.

16. M. Nakagawa, *Bull. Tea Res. Station*, Min. Agr. Forestry, *6*: 65 (1970).

17. I. McMurrough, G. P. Roche, and G. P. Hennigan, *Brewers Digest 59*: 28 (1984).

18. M. Dadic and G. Belleau, Polyphenols and beer flavor, *Proceedings of the American Society of Brewing Chemists* (W. J. Olson, ed.), 1973, pp. 107–114.

19. E. Gienapp and K. L. Schroder, *Die Nahrung 19*: 697 (1975).

20. C. Pfaffmann, *Handbook of Physiology*, Vol. 1, (J. Field, ed.), American Physiological Soc., Washington, D.C., 1959, p. 507.

21. M. C. Meilgaard, G. C. Civille, and B. T. Carr, *Sensory Evaluation Techniques*, 2nd ed., CRC Press, Boca Raton, 1991.

22. R. J. Robins and M. J. C. Rhodes, *J. Chromatogr. 283*: 436 (1984).

23. *The Merck Index* (M. Windholz, ed.), 10th ed., Merck and Co., Inc., Rahway, NJ, 1983, p. 1164.

24. I. McMurrough and J. Byrne, HPLC analysis of bittering substances, phenolic compounds, and various compounds of alcoholic beverages, *Food Analysis by HPLC* (L. M. L. Nollet, ed.), Marcel Dekker, New York, 1992.

25. M. J. Lewis, R. M. Pangborn, and J. Fujii-Yamashita, Bitterness of beer: a comparison of traditional scaling and time-intensity methods, *Proceedings of the 16th Convention of the Institute of Brewing* (Australia and New Zealand Section) (B. J. Clarke, ed.), Sydney, 1980, pp. 165–171.

26. D. A. McCamey, T. M. Thorpe, and J. P. McCarthy, *Bitterness in Foods and Beverages* (R. L. Rouseff, ed.), Elsevier, Amsterdam, 1990, p. 169.

27. C. W. Nagel, I. W. Herrick, and W. R. Graber, *J. Food Sci. 52*: 213 (1987).

28. A. G. H. Lea, *Bitterness in Foods and Beverages* (R. L. Rouseff, ed.), Elsevier, Amsterdam, 1990, p. 123.

29. A. G. H. Lea and G. M. Arnold, *J. Sci. Food Agric. 29*: 478 (1978).

30. E. Haslam, *Plant Polyphenols* (E. Haslam, ed.), Cambridge University Press, Cambridge, 1989, p. 154.

31. B. I. Brown, *J. Chromatogr. 86*: 239 (1973).

32. G. Ben-Et, A. Dolev, and D. Tatarsky, *J. Food Sci. 38*: 546 (1973).

33. A. Puri, *Bitterness in Foods and Beverages* (R. L. Rouseff, ed.), Elsevier, Amsterdam, 1990, p. 325.

34. S. Hasegawa and V. P. Maier, *Bitterness in Foods and Beverages* (R. L. Rouseff, ed.), Elsevier, Amsterdam, 1990, p. 293.

35. W. Pickenhagen, P. Dietrich, B. Keil, J. Polonsky, F. Nouaille, and E. Lederer, *Helv. Chim. Acta 58*: 1078 (1975).

36. S. Sakamura, K. Furukawa, and T. Kasai, *Agric. Biol. Chem. 42*: 607 (1978).

37. R. Schmidt, *Bitterness in Foods and Beverages* (R. L. Rouseff, ed.), Elsevier, Amsterdam, 1990, p. 183.

38. W. E. Marshall, *Bitterness in Foods and Beverages* (R. L. Rouseff, ed.), Elsevier, Amsterdam, 1990, p. 275.

39. P. E. Shaw, *Bitterness in Foods and Beverages* (R. L. Rouseff, ed.), Elsevier, Amsterdam, 1990, p. 309.

40. P. J. Hayes, M. R. Smyth, and I. McMurrough, *Analyst 112*: 1197 (1987).
41. E. Leclercq, *J. Chromatogr. 283*: 441 (1984).
42. D. J. Millin, D. J. Crispin, and D. Swaine, *J. Agric. Food Chem. 17*(4): 717 (1969).
43. A. C. Noble, *Bitterness in Foods and Beverages* (R. L. Rouseff, ed.), Elsevier, Amsterdam, 1990, p. 145.
44. M. A. Amerine and E. R. Roessler, *Wines: Their Sensory Evaluation*, Freeman, New York, 1983.
45. E. Vérette, A. C. Noble, and T. C. Somers, *J. Sci. Food Agric. 45*: 267 (1988).
46. T. C. Somers, E. Vérette, and K. F. Pocock, *J. Sci. Food Agric. 40*: 67 (1987).
47. R. A. Arnold, A. C. Noble, and V. L. Singleton, *J. Agric. Food Chem. 28*: 675 (1980).
48. E. J. L. Robichaud and A. C. Noble, *J. Sci. Food Agric. 53*: 343 (1990).
49. A. G. H. Lea and G. M. Arnold, *J. Sci. Food Agric. 29*: 478 (1978).
50. A. G. H. Lea and C. F. Timberlake, *J. Sci. Food Agric. 29*: 484 (1978).
51. D. Madigan, I. McMurrough, and M. R. Smyth, *Analyst 119*:(5): 863 (1994).
52. I. McMurrough, D. Madigan, and D. Bogan, *Groupe Polyphenols, Proceedings of the 16th International Conference*, Vol. 16, Part 1, Lisbon, INRA, 1992, p. 104.
53. I. McMurrough, R. Kelly, and D. Madigan, *Proceedings of the 24th Congress of the European Brewery Convention*, Vol. 24, Oslo, 1993, p. 663.
54. D. E. Briggs, J. S. Hough, R. Stevens, and T. W. Young, *Malting and Brewing Science, Part 2: Hopped Wort and Beer*, 2nd ed., Chapman and Hall, London, 1981, p. 390.
55. F. Villeneuve, G. Abravanel, M. Moutounet, and G. Alibert, *J. Chromatogr. 234*: 131 (1982).
56. I. McMurrough, M. V. Lynch, F. Murray, M. Kearney, and F. Nitzsche, *J. Am. Soc. Brew. Chem. 45*: 6 (1987).
57. I. McMurrough, M. V. Lynch, F. Murray, and M. Kearney, *Proceedings of the Second Aviemore Conference on Malting, Brewing and Distilling*, Aviemore, Inverness-Shire, 1986, Institute of Brewing, London, 1986, pp. 342–346.
58. J. R. Donley, *J. Am. Soc. Brew. Chem. 50*: 89 (1992).
59. C. Bloomfield, N. J. Arnold, and M. Moir, Estimation of resin acids in hops, extracts and beers using HPLC, *Proceedings of the Second Aviemore Conference on Malting, Brewing and Distilling*, Aviemore, Inverness-Shire, 1986, Institute of Brewing, London, 1986, pp. 336–341.
60. G. K. Buckee, *J. Inst. Brew. 91*: 143 (1985).
61. M. Verzele, C. Dewaele, and M. Van Kerrebroeck, *J. Chromatogr. 244*: 321 (1982).
62. J. M. Cowles, P. L. Ting, A. A. Murakami, and H. Goldstein, *Bitterness in Foods and Beverages* (R. L. Rouseff, ed.), Elsevier, Amsterdam, 1990, p. 251.
63. R. J. H. Wilson, *Ferment 2*: 241 (1989).
64. H. A. Thornton, J. Kulandai, M. Bond, M. P. Jontef, D. B. Hawthorne, and T. E. Kavanagh, *J. Inst. Brew. 99*: 473 (1993).
65. A.S.B.C. Subcommittee report: Iso-α-acids in beer by solid-phase extraction and high-performance liquid chromatography, *J. Am. Soc. Brew. Chem. 51*: 173 (1993).
66. M. Verzele, M. Van Kerrebroeck, C. Dewaele, J. Strating, and L. Verhagen, *J. Chromatogr. 294*: 471 (1984).
67. A. G. H. Lea, *J. Sci. Food Agric. 30*: 833 (1979).
68. A. G. H. Lea, *J. Chromatogr. 194*: 62 (1980).
69. A. G. H. Lea, *J. Sci. Food Agric. 29*: 471 (1978).
70. E. L. Wilson, *J. Sci. Food Agric. 32*: 257 (1981).
71. A. G. H. Lea, *J. Chromatogr. 238*: 253 (1982).
72. J. F. Fisher, *J. Agric. Food Chem. 23*: 1199 (1975).
73. J. F. Fisher, *J. Agric. Food Chem. 26*: 497 (1978).
74. R. L. Rouseff and J. F. Fisher, *Anal. Chem. 52*: 1228 (1980).
75. J. F. Fisher and T. A. Wheaton, *J. Agric. Food Chem. 24*: 898 (1976).
76. R. L. Rouseff, S. F. Martin, and C. O. Youtsey, *J. Agric. Food Chem. 35*: 1027 (1987).
77. A. C. Hoefler and P. Coggon, *J. Chromatogr. 129*: 460 (1976).
78. D. A. Wellum and W. Kirby, *J. Chromatogr. 206*: 400 (1981).
79. R. G. Bailey, I. McDowell, and H. E. Nursten, *J. Sci. Food Agric. 52*: 509 (1990).
80. R. G. Bailey, H. E. Nursten, and I. McDowell, *J. Chromatogr. 542*: 115 (1991).
81. S. Kuhr and U. H. Engelhardt, *Z. Lebensm. Unters. Forsch. 192*: 526 (1991).

82. J. L. Blauch and S. M. Tarka Jr., *J. Food Sci. 48*: 745 (1983).
83. F. J. Muhtadi, S. S. El-Hawary, and M. S. Hifnawy, *J. Liq. Chromatogr. 13*: 1013 (1990).
84. M. C. Gennaro and C. Abrigo, *Fresenius J. Anal. Chem. 343*: 523 (1992).
85. M. A. Abuirjeie, M. Sharaf El-Din, and I. I. Mahmoud, *J. Liq. Chromatogr. 15*: 101 (1992).
86. T. C. Somers and E. Vérette, Phenolic composition of natural wine types, *Modern Methods of Plant Analysis, Vol. 6, Wine Analysis* (H. F. Linskens and J. F. Jackson, eds.), Springer-Verlag, Berlin, 1988, pp. 219–257.
87. J. H. Thorngate, Flavan-3-ols and their polymers, *Beer and Wine Production* (B. H. Gump, ed.), ACS Symposium Ser. 536, 1993, pp. 51–63.
88. V. L. Singleton, Wine Phenols, *Modern Methods of Plant Analysis, Vol. 6, Wine Analysis* (H. F. Linskens and J. F. Jackson, eds.), Springer-Verlag, Berlin, 1988, pp. 173–218.
89. M. E. Evans, *J. Liq. Chromatogr. 6*: 153 (1983).
90. J.-P. Roggero, P. Archier, and S. Coen, *J. Liq. Chromatogr. 14*: 533 (1991).
91. G. P. Cartoni, F. Coccioli, L. Pontelli, and E. Quattrucci, *J. Chromatogr. 537*: 93 (1991).
92. P. J. Woodring, P. A. Edwards, and M. G. Chisholm, *J. Agric. Food Chem. 38*: 729 (1990).
93. M.-H. Salagoïty-Auguste and A. Bertrand, *J. Sci. Food Agric. 35*: 1241 (1984).
94. J. M. R. da Silva, J.-P. Rosec, M. Bourzeix, and N. Heredia, *J. Sci. Food Agric. 53*: 85 (1990).
95. E. Revilla, E. Alonso, and M. I. Estrella, *Chromatographia 22*: 157 (1986).
96. British Standard 5752, Part 12, 1992 Methods of test for coffee and coffee products Part 12. Coffee: determination of caffeine content (routine method by HPLC). British Standards Office, 1992. E. Revilla, E. Alonso, and M. I. Estrella, *Chromatographia 22*: 157 (1986).
97. M. Verzele, *J. Inst. Brew. 92*: 32 (1986).
98. M. Moir, *Ferment 1*: 49 (1988).
99. B. J. Clarke, *J. Inst. Brew. 92*: 123 (1986).
100. M. Verzele and M. De Potter, *J. Chromatogr. 166*: 320 (1978).
101. European Brewery Convention, *Analytica-EBC*, 4th ed., 1977, pp. E123–E124.
102. G. K. Buckee and C. D. Baker, *J. Inst. Brew. 93*: 468 (1987).
103. D. G. Lance, T. E. Kavanagh, and B. J. Clarke, *J. Inst. Brew. 87*: 225 (1981).
104. M. Ono, Y. Kakudo, Y. Yamamoto, K. Nagami, and J. Kumada, *J. Am. Soc. Brew. Chem. 42*: 167 (1984).
105. M. Ono, S. Hashimoto, Y. Kakudo, K. Nagami, and J. Kumada, *J. Am. Soc. Brew Chem. 41*: 19 (1983).
106. I. McMurrough, K. Cleary, and F. Murray, *J. Am. Soc. Brew. Chem. 44*: 101 (1986).
107. American Society of Brewing Chemists, *Methods of Analysis, 8th ed., Hops–9 C. Iso-α-Acids by High Performance Liquid Chromatography*, ASBC, St. Paul, 1992.
108. American Society of Brewing Chemists, *Methods of Analysis, 8th ed., Hops–9 D. Iso-α-Acids by High Performance Liquid Chromatography*, ASBC, St. Paul, 1992.
109. Iso-α-acids in isomerized hop pellets by high-performance liquid chromatography, Report of Subcommittee, *J. Am. Soc. Brew. Chem. 51*: 175 (1993).
110. J. T. Hann, *Lab. Pract. 38*: 59 (1989).
111. C. P. Green and P. Osborne, *J. Inst. Brew. 99*: 347 (1993).
112. M. Moir, *Ferment 3*: 97 (1990).
113. F. R. Sharpe and I. H. L. Ormrod, *J. Inst. Brew. 97*: 33 (1991).
114. P. James, T. Tynan, I. McMurrough, and J. Byrne, *J. Inst. Brew. 96*: 137 (1990).
115. M. Verzele, G. Steenbeke, L. C. Verhagen, and J. Strating, *J. High Resol. Chromatogr. 13*: 826 (1990).
116. A. W. Jaworski and C. Y. Lee, *J. Agric. Food Chem. 35*: 257 (1987).
117. T. Escribano-Bailón, Y. Gutiérrez-Fernández, J. C. Rivas-Gonzalo, and C. Santos-Buelga, *J. Agric. Food Chem. 40*: 1794 (1992).
118. L. P. Valenti, Liquid chromatographic determination of quinine, hydroquinine, saccharin, and sodium benzoate in quinine beverages, *Modern Methods of Plant Analysis, Vol. 8, Analysis of Nonalcoholic Beverages* (H. F. Linskens and J. F. Jackson, eds.), Springer-Verlag, Berlin, 1988.
119. American Society of Brewing Chemists, *Methods of Analysis, 8th ed., Beer–23. Beer Bitterness.* ASBC, St Paul, 1992.
120. European Brewery Convention, *Analytica-EBC, 4th ed. 9.6 Bitterness (International Method) E155*, 1987.
121. European Brewery Convention, *Analytica-EBC, 4th ed. 9.9.1 Total Polyphenols (International Method) E157*, 1987.

122. European Brewery Convention, *Analytica-EBC*, 4th ed. 9.9.2 *Flavanoids (International Method) E159*, 1987.
123. P. J. Rider, A. Der Marderosian, and J. R. Porter, Evaluation of total tannins and relative astringency in teas, *Phenolic Compounds and Their Effects on Health, Vol. 1, Analysis, Occurrence and Chemistry* (C.-T. Ho, C. Y. Lee, and M.-T. Huang, eds.), American Chemical Society, 1992.
124. E. A. H. Roberts and R. F. Smith, *Analyst 86*: 94 (1961).
125. D. L. Whitehead and C. M. Temple, *J. Sci. Food Agric. 58*: 149 (1992).
126. S. N. S. Thanaraj and R. Seshadri, *J. Sci. Food Agric. 51*: 57 (1990).
127. C. Tomàs, M. Celeste, A. Cladera, E. Gómez, J. M. Estela, and V. Cerdà, *Food Chem. 47*(2): 201 (1993).
128. *Official Methods of Analysis of the Association of Official Analytical Chemists*, 15th ed. Method 960.22: Caffeine in nonalcoholic beverages spectrophotometric method, Association of Official Analytical Chemists, Arlington, 1990.
129. *Official Methods of Analysis of the Association of Official Analytical Chemists*, 15th ed. Method 962.13: Caffeine in nonalcoholic beverages spectrophotometric method, Association of Official Analytical Chemists, Arlington, 1990.
130. *Official Methods of Analysis of the Association of Official Analytical Chemists*, 15th ed. Method 967.11: Caffeine in nonalcoholic beverages spectrophotometric method, Association of Official Analytical Chemists, Arlington, 1990.
131. W. B. Davis, *Anal. Chem. 19*: 476 (1947).
132. J. Bakker, N. W. Preston, and C. F. Timberlake, *Am. J. Enol. Vitic. 37*(2): 121 (1986).
133. V. L. Singleton, Wine phenols, *Modern Methods of Plant Analysis, Vol. 6, Wine Analysis* (H. F. Linskens and J. F. Jackson, eds.), Springer-Verlag, Berlin, 1988, pp. 173–218.
134. M. Celeste, C. Tomás, A. Cladera, J.-M. Estela, and V. Cerdà, *Anal. Chim. Acta. 269*: 21 (1992).
135. K. J. Switala and K. G. Schick, *J. Am. Soc. Brew. Chem. 48*: 18 (1990).
136. S. Sakuma, C. Kikuchi, M. Kowaka, and M. Mawatari, *J. Am. Soc. Brew. Chem. 51*: 51 (1993).
137. M. Peris, D. Muller, and A. Maquieira, *Food Chem. 40*: 1 (1991).
138. P. J. Hayes, M. R. Smyth, and I. McMurrough, *Analyst 112*: 1205 (1987).
139. I. McMurrough, R. Kelly, J. Byrne, and M. O'Brien, *J. Am. Soc. Brew. Chem. 50*: 67 (1992).
140. I. McMurrough, J. Byrne, E. Collins, M. R. Smyth, J. Cooney, and P. James, *J. Am. Soc. Brew. Chem. 46*: 51 (1988).
141. P. Gamache, E. Ryan, and I. N. Acworth, *J. Chromatogr. 635*: 143 (1993).

23

Pigments

Reinhard Eder
*Federal Research Institute and College for Viticulture and Pomology,
Klosterneuburg, Austria*

I. CAROTENOIDS

A. Introduction

1. Physical and Chemical Properties

Carotenoids are of widespread occurrence as natural pigments in plants and animals. They contribute the natural yellow, orange, and red colors of many plants and are used extensively as nontoxic natural or nature-identical colorants in foodstuffs (1). Their name is derived from β-carotene, the main representative of their group, which was first isolated from carrots (*Daucus carota*) by Wackenroder in 1831 (2). Carotenoids are isoprenoid polyenes which are formed by head-to-tail linkage of C5 isoprene units, except for one tail-to-tail linkage in the center of the molecule, which makes it symmetrical (3). The C40 carotenoids can be divided into *carotenes*, which are hydrocarbons (e.g., β-carotene, Fig. 1a), and their oxygenated derivates, the *xanthophylls* (e.g., zeaxanthin, Fig. 1b). Fruit xanthophylls are often acylated with fatty acids (4). Besides the C40 carotenoids there are also apocarotenoids, a degraded form with less than 40 carbon atoms in the skeleton, and higher carotenoids where the carbon skeleton contains 45 or 50 carbon atoms, respectively (5).

The characteristic absorption spectrum of each carotenoid is determined by the chromophore, a series of conjugated double bonds. Usually the spectrum shows three absorption bands, which are affected by the length of the chromophore, the nature of the double bond, and the taking out of conjugation of one double bond. Several absorption spectra of some common carotenoids are shown in Fig. 2. A change of solvent may, however, cause a shift of the absorption bands. The extensive double-bond system in the molecule enables the carotenoids to exist in many geometrically isomeric forms (Z-E isomers). Most carotenoids occur naturally in the all-trans form (E isomers), but *cis isomers* (Z isomers) are frequently present in small amounts (6), often being formed as artifacts from all-trans isomers. Cis isomers can be distinguished by a characteristic absorption band ("cis peak") that appears at 300–360 nm (7).

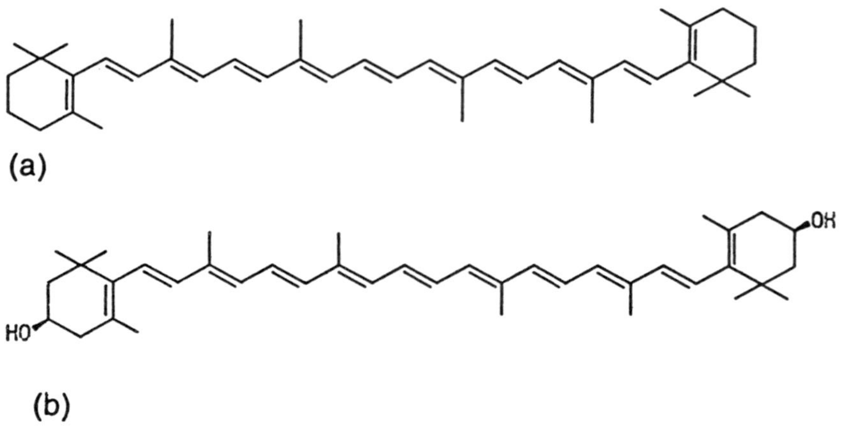

(a)

(b)

Fig. 1   (a) Formula for β-carotene. (b) Formula for zeaxanthin.

Carotenoids are regarded as lipids and therefore as soluble in lipids and fat solvents, such as acetone, alcohol, diethyl ether, and chloroform. In detail carotenes are soluble in apolar solvents like petroleum ether and hexane, while xanthophylls dissolve best in polar solvents such as ethanol and methanol (5). Carotenoids are very unstable pigments, being especially sensitive to light, oxygen, and peroxide. Therefore special precautions have to be taken when handling them.

2. Properties in Food

Carotenoids can be synthesized de novo by higher plants, spore-bearing vascular plants, algae, and photosynthetic bacteria only (8). In addition, carotenoids which have been isolated from

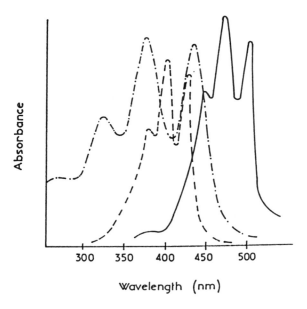

Fig. 2   Absorption spectra of some common carotenoids: (——) lycopene; (- - -) ξ-carotene (in petroleum ether); (-.-.-.-) persicaxanthin (in ethanol). (From Ref. 5.)

animal tissue, e.g., salmon (9), shrimps, egg yolk (10), have been synthesized in plants or bacteria. The distribution of carotenoids in fruits (11–13), vegetables (14,15), leaves (16,17), seeds and animals (18) has been extensively investigated, and good reviews are given by Goodwin (8) and Gross (5), respectively.

In natural tissues carotenoids generally occur as complex mixtures of many substances. In orange juice, for instance, more than 50 carotenoids were identified. Table 1 shows the carotenoid contents of some fruits and vegetables, though it must be considered that they are influenced by climate, site, and maturity (19,20).

3. Arguments for Carotenoid Analysis

Since color is a main parameter used in evaluating the quality of food (21) it is of interest to measure the pigment concentration and distribution in fruits and vegetables. During processing of fruits many of the original carotenoids are degraded. To minimize these undesirable carotenoid losses various processes in food technology have been optimized. For instance, Berset and Marty (22) lowered the β-carotene degradation during extrusion cooking. Another aspect is that forthcoming regulations will restrict the use of foodstuff colorants exclusively to natural compounds, while artificial compounds will disappear progressively from the list of acceptable colorants (23). Therefore the demand for natural food colors is constantly growing. Carotenoids are well-known natural pigments and their application has been comprehensively discussed by Bauernfeind (24); in particular, extracts of anatto, paprika, alfalfa, tagetes, tomato, and carrot have been widely used. The application of carotenoids as natural food colorants is widely permitted within the European community (E160 and E161) and the USA (1). An increasing demand for natural carotenoids is given by the fish industry. Salmonids (e.g., rainbow trout) often do not get a satisfactory red pigmentation when they are bred in farms (25); therefore they have to be fed with synthetic astaxantin, canthaxanthin, or natural compounds such as paprika or dried flowers (26).

There is increasing interest in carotenoids, particularly β-carotene from natural sources because of their role as provitamins A (27) and antioxidants (28). Since the vitamin A potential of cis isomers is less than those of their all-trans counterparts it is important to distinguish and quantify the various forms (29). According to reports and clinical studies, carotenes may be important in the prevention of some forms of cancer (30). Since the National Cancer Institute has recommended an increased intake of food with high carotenoid content (31), more detailed information about the carotenoid composition of foodstuff is necessary.

B. Sample Preparation

Because of the inherent instability of carotenoids some general precautions in handling these substances have to be observed. Any failure to do so may result not only in a low overall quantitative recovery but in possible losses of labile carotenoids, their conversion to other carotenoids, and the appearance of cis-trans isomers as artifacts. Since carotenoids should not be exposed to direct sunlight or to ultraviolet (UV) light, the laboratory should be equipped with suitable blinds to produce diffuse, low-intensity daylight (2) and vessels containing carotenoids should be protected from light by wrapping them with aluminum foil. Heating the carotenoid pigments during the manipulation should be avoided whenever possible. However, it is necessary to remove the solvents used in extraction and purification; consequently, it is important to choose solvents with low boiling points. Moreover, carotenoids, especially when in an adsorbed state, may be oxidized by oxygen or peroxide. Operations like thin-layer chromatography (TLC) or the evaporation of solvents should therefore be carried out in an atmosphere of nitrogen. As acids can initiate oxidative decompositions and isomerizations of carotenoids, they have to be eliminated from all stages of manipulation.

Table 1 Carotenoid Content and Carotenoid Distribution in Fruits and Vegetables (5,19,20,91,92)

	Carotenoid content (mg/kg)	Total carotenoids (%)													
		Phyto-fluene	Phytoen	α-Carotene	β-Carotene	Crypro-xanthin	Lutein	Iso-lutein	Viola-xanthin	Zea-xanthin	Luteo-xanthin	Neo-xanthin	Lyco-pene	Caps-anthin	Capso-rubin
Apple ("Golden Delicious" peel)	13.4	—	—	—	7.9	2.4	9.2	6.8	39.6	3.1	7.6	19.4	—	—	—
Apricot	35.0	6.1	10.0	—	54.0	3.1	2.1	—	—	0.7	0.09	—	1.0	—	—
Banana ("Cavendish," pulp)	1.0	1.2	—	31.5	22.5	0.7	2.5	—	8.6	—	—	—	—	—	—
Pear ("Trévoux", peel)	10.2	—	—	—	6.0	0.4	69.2	1.9	6.6	2.8	1.0	7.0	—	—	—
Grapefruit ("Ruby Red," pulp)	8.2	4.4	16.0	—	27.0	1.5	0.9	—	0.9	—	0.4	—	40.0	—	—
Peach ("Red Haven," mesocarp)	4.7	10.3	—	—	4.5	14.5	2.0	7.6	34.6	6.9	4.7	5.3	—	—	—
Kiwi ("Bruno")	6.5	—	—	—	8.0	3.5	45.9	0.5	1.7	3.0	0.7	13.9	—	—	—
Strawberry ("Tenira")	0.4	—	—	—	3.4	—	20.6	—	8.6	—	1.2	3.1	—	—	—
Orange ("Valencia," pulp)	12.0	4.2	2.0	1.0	2.0	10.6	9.0	2.0	1.8	10.2	11.5	2.5	—	—	—
Papaya	13.8	0.1	0.1	—	29.6	48.2	—	—	3.4	—	—	0.2	—	—	—
Paprika ("Yolo wonder A")	1755.9	—	—	—	15.4	12.3	—	—	7.1	3.1	—	2.0	—	33.3	10.3
Red bell pepper	280.5	—	—	—	11	—	—	—	—	—	—	—	—	60	20
Carrot ("B6274")	94	—	—	34.0	55.3	—	—	—	—	—	—	—	—	—	—
Tomato ("Early Redchief")	950	8.4	19.7	2.3	—	—	—	—	—	—	—	—	66.7	—	—

1. Extraction

Carotenoids should be extracted from tissues as rapidly as possible. If an immediate extraction is not possible, samples should be stored below −18°C until required. In some cases it may also be necessary to cut the material in small pieces or to remove inedible parts. For the extraction the exactly weighted sample and the solvent are transferred into a blender where the sample is simultaneously ground and extracted. Since fresh tissues contain a high percentage of water and carotenoids are liposoluble, the first organic solvent must be miscible with water (e.g., acetone, ethanol, methanol). After one or two extraction steps water-immiscible solvents (e.g., diethyl ether, benzene) can be applied. Dried materials may be also extracted with water-immiscible solvents, but carotenoid recovery is usually better if the tissue is first treated with a little water and then extracted with water-miscible solvents. Prior to the extraction of fruits the addition of antioxidants, e.g., butylated hydroxytoluene (BHT) (5,32), and neutralizing agents, e.g., Tris buffer (5), calcium or magnesium carbonate (33) is recommended. The initial homogenate is filtered under vacuum through a funnel and the residue is recovered for further extractions. The procedure is repeated until no more color can be extracted, but usually two to three such extractions are enough. The carotenoids are then transferred from the water-miscible solvent (e.g., acetone) to an appropriate water-immiscible solvent (e.g., diethyl ether, petroleum ether) by adding enough saturated NaCl solution (5). The two layers are allowed to separate and the top layer, containing the carotenoids, is collected while the bottom layer is reextracted several times until the top layer becomes colorless (34). If chlorophylls are absent the total carotenoid content can be estimated spectrophometrically in the ether phase.

When only small quantities of sample material (0.5 g) are available the tissue is ground in a mortar with anhydrous sodium sulfate and washed with silver sand. This powder can be extracted as described above (2). By using hexane/acetone/ethanol as solvents, Sadler et al. (35) developed a rapid (15 min per extraction), reproducible method with a good recovery rate to extract lycopene and β-carotene from tomato and citrus products. Minguez-Mosquera et al. (36) have used liquid phase distribution with *N,N*-dimethylformamide and hexan or octadecyl (C18) solid phase extraction to obtain fat-free pigments from olive oil samples. Fisher and Rouseff (37) cleaned up saponified orange juice with a C18 solid phase extraction column prior to HPLC analysis. Supercritical CO_2 extraction of β-carotene from sweet potatoes increased the amount of carotenoids extracted but it also resulted in a higher formation of cis isomers (38).

2. Saponification

Saponification is a purification procedure for removing unwanted lipids and chlorophylls. It is omitted when alkali-labile carotenoids (e.g., astaxanthin, fucoxanthin) or carotenoid esters are to be analyzed. To prevent the formation of artifacts produced by aldol condensations between acetone and carotenals, all traces of acetone have to be removed prior to saponification (39).

Saponification is carried out by adding sufficient potassium hydroxide (KOH) solved in methanol or ethanol to give an overall KOH concentration between 5% and 10% in the extract, which is then kept overnight at room temperature under nitrogen (40). Different alternative saponification methods are described: boiling the extract under a stream of nitrogen for 5–10 min (2) up to 30 min (4); keeping it at 40°C for 3 hr (5) or at 60°C for 25 min (41) in every case followed by immediate chilling. Subsequently the cold alkaline solution is mixed with freshly distilled diethyl ether (1:1) in a separating funnel to wash it free of alkali, dried, and evaporated in vacuum.

The two procedures of lipid extraction and saponification can be combined as a prolonged or hot alkaline alcoholic extraction, a method which is satisfactory for certain routine quantitative determinations (42).

3. Removal of sterols

Sterols, which are very abundant in some fruits, can be removed by precipitating them in different solvents. The unsaponifiable matter is dissolved in a minimum volume of methanol, petroleum ether, or acetone. The precipitation is completed overnight at –20°C; following centrifugation the sterol-free supernatant is used for analysis (5).

C. Analysis–Detection–Separation Methods

1. Classical Methods

The total carotenoid content can be analyzed with or without further purification. Formerly the pigment content was determined by colorimetry, which was performed by subjective comparison of colors or by objective measurements of color intensities with a colorimeter or photometer (43). Nowadays spectrophotometry, which allows one to measure the absorption at single wavelengths, is predominant.

   a.   Spectrophotometric Methods.   The extinction (E_{max}) of a known volume of a carotenoid solution is measured at the wavelength of maximal absorbance. The amount of carotenoid (C) dissolved in volume (V) is given by the expression:

$$C = \frac{E_{max} V}{E_{1\%}^{1cm}} \times 100 \tag{1}$$

In the formula $E_{1\%}^{1cm}$, is the specific extinction coefficient of a 1% solution in a 1-cm light pass cell. The absorption maxima of carotenoids in various solvents and their extinction coefficients are given in Table 2 (2). For the determination of a mixture of carotenoids the measurement is carried out at a wavelength of 436 nm, the $E_{1\%}^{1cm}$ of β-carotene = 2500 is used and the results are expressed as β-carotene. Benedek (44) quantified the total pigment content of ground paprika by measuring the optical density at 500 nm and expressing it as capsanthin. By measuring the extinction of 455 and 505 nm the total red and the total yellow pigments of a benzene extract of ground paprika could be distinguished (45). Through reduction of the red polyene ketones with $NaBH_4$ ("reduction method") the red pigments could be differentiated from the total pigment content (46). The detection limit for β-carotene and lycopene is approx. 0.04 µg/g (47). The advantages of the spectrophotometric methods are their simplicity, short analysis time, and

Table 2 Specific Extinction Coefficient (E1cm 1%) of Carotenoids in Stated Solvents

Carotenoid	E1cm 1%	Wavelength (λ)	Solvent
Astaxanthin	1900	473	Acetone
Capsanthin	2072	483	Benzene
Capsorubin	2200	489	Benzene
α-Carotene	2800	444	Light petroleum
β-Carotene	2505	457	Cyclohexan
α-Cryptoxanthin	2636	446	Hexane
Lutein	2236	458	Benzene
Lycopene	3370	487	Benzene
Rubixanthin	2909	462	Acetone
Violaxanthin	2240	454	Benzene
Zeaxanthin	2340	452	Acetone

Source: Ref. 8.

inexpensiveness. However, they do not give any detailed information about the carotenoid composition of the sample.

b. Phase Separation and Countercurrent Distribution. Carotenoids can be purified and characterized according to their polarity by means of solvent partition between immiscible solvents (48). When a petroleum ether pigment extract is mixed in a separating funnel with an equal volume of aqueous methanol (85–95%), the epiphase (petrol ether) will contain the chlorophylls, carotenes, and total esterified xanthophylls. Monoketones and monols will be evenly distributed in both phases, while diols and polyols will be hypophasic (methanol) (5). The liquid–liquid partition was refined by Craig's countercurrent distribution method: Carotenoids were first separated into polarity subgroups by using different solvents, and then each fraction was submitted to column chromatography (49). However, countercurrent distribution has found only a limited application in the field of carotenoid analysis, mostly because of the high costs, the time-consuming procedure, and the possible formation of artifacts.

c. Column Chromatography (CC). Column chromatography is used either for large-scale purifications or to separate an extract into fractions that contain groups of compounds with similar polarity. The carotenoid content of each group can then be quantified photometrically. The principles and applications of this technique for carotenoid analysis have been fully described by Davies (2). Three basic methods of CC are available for the separation of carotenoids:

1. *Zone chromatography where the solvent is run through the column until the zones are separated.* Then the whole column is extruded, sectioned, and the individual colored zones are eluted with a polar solvent.
2. *Stepwise elution chromatography where the sample is first applied to the column in a nonpolar solvent.* After that the column is washed in turn with solvents of increasing polarity (e.g., 0.5%, 1%, 2%, 5%, etc., diethyl ether in light petroleum), causing the pigments to be eluted sequentially.
3. *Gradient elution chromatography where a continuous solvent gradient is used to elute the carotenoids, giving a sharper separation.*

Most procedures are based on adsorption chromatography. The choice of adsorbent and solvent depends, however, on the types of carotenoids to be separated. Some groups of carotenoids and suitable adsorbants are compiled in Table 3; several appropriate solvents for CC are listed in Table 4. The instability of some carotenoids can lead to problems with certain adsorbants. Silica gel or silicic acid, for instance, can cause isomerization of 5,6-epoxycarotenoids. On aluminum some carotenoids (e.g., astaxanthin) can be oxidized and magnesium oxide can catalyze aldol condensation between residual acetone and carotenoid aldehydes. CC is a very time- and material-con-

Table 3 Groups of Carotenoids and Suitable Adsorbants for Column Chromatography

Type of carotenoids	Adsorbants	Ref.
Carotenes	Alumina (activated Brockman grade I or II), silica gel, calcium hydroxide	14, 50
Monohydroxycarotenoids	Aluminum (activated Brockman grade IV or V)	51
Carotenoids of intermediate polarity	Calcium carbonate, magnesium oxid	17, 52
Polar xanthophylls	Cellulose, sucrose	53

Table 4 Suitable Solvents and Carotenoid Groups Separated with Column Chromatography

Solvent system	Type of carotenoids separated
Light petroleum (petrol)	Carotenes
Ether/petrol (5:95)	Carotene epoxides
Ether/petrol (10:90)	Xanthophyll esters
Ether/petrol (20:80)	Monooxycarotenoids
Ether/petrol (1:1)	Dioxcarotenoids, monohydroxy carotenoids
Ether or ethanol/ether (5:95)	Dihydroxy carotenoids, more polar xanthophylls
Ethanol/ether (1:4)	Carotenoid glycosides

Source: Ref. 1.

suming method which favors the isomerization and decomposition of sensitive carotenoids. Furthermore in many cases only a poor resolution of the pigments is achieved; thus this technique is employed less and less.

    d. Adsorbant Chromatography on Paper. A number of adsorbant loaded papers has been applied in carotenoid separation. The most popular are those filled with kieselgur (20%), aluminum oxide (20%), aluminum hydroxide (7.5%), and silica gel (22%). They may be activated by heating to a temperature of 105–150°C before use (54). With circular papers the carotenoid mixture, dissolved in acetone, is applied to the center of the paper and developed radially (2). After the chromatogram is fully developed the R_F values of the colored concentric rings are determined. Nevertheless adsorbant-loaded papers are being substituted by TLC since the latter is more versatile and gives better resolution.

2. Recent Methods

    e. Thin-Layer Chromatography (TLC). The basic technique of TLC has been fully described by numerous authors [55,56,57]. In carotenoid analysis TLC is mainly used for preliminary examinations to give an indication of the number and variety of carotenoids present and to help in the selection of a suitable separation and purification procedure for the given mixture. TLC is also occasionally applied for further purification (on small scale) of carotenoids already separated by CC (2). Gross (58) developed a method for rapid separation of citrus carotenoids by successive TLC on two different adsorbants: silica gel G and magnesium oxide-kieselgur (1:1). Quantification of the individual pigments can be carried out either through direct densitometric determination (59,60) or by scraping the pigment-containing adsorbant from the plate and extracting the carotenoids into an eluant in which the concentration can be measured photometrically [7,21, 61–63]. The carotenoids can be identified by their chromatographic behavior, their absorption spectra, their infrared (IR) spectra, their mass spectra, and by chemical tests on functional groups such as reduction with sodium borohydride for carbonyl groups (2) and acetylation for hydroxyl groups (15). Most TLC methods are based on adsorption, but partition chromatography has also been successfully applied (57). A large number of adsorbants and solvents has been used for the separation of carotenoids; some examples are listed in Table 5. The hR_F values ($hR_F = R_F \times 100$) of carotenes and xanthophylls on four adsorption TLC systems are shown in Table 6. For further details Bolliger (56) and Davies (2), respectively, should be consulted. Since carotenoids are strongly colored the use of special reagents for their detection is usually not necessary. Amounts as small as 0.05 μg β-carotene or 0.01 μg of the deeper red carotenoids may still be detected visually (56). To prevent the rapid fading of carotenoids the developed plates should be sprayed with a solution of liquid paraffin in light petroleum (56). The colorless carotenoid precursor

Table 5 Adsorbants and Solvents Used for TLC

Adsorbant	Solvent	Type of carotenoids	Ref.
Cellulose	Hexane/n-propanol (99:1)	Carotenes, xanthophylles	64
Ca(OH)$_2$	1.5 % p-methylanisol/ petroleum ether	Cis–trans isomers of carotene	7
MgO/Al$_2$O$_3$/cellulose/ CaSO$_4$ (10:6:2:2)	n-Hexane/isopropanol/ methanol (from 100:2:O.2 to 100:0:0)	Cis–trans isomers of carotene	59
Ca(OH)$_2$/silica gel (80:20)	Petroleum ether/benzene (50 + 50)	Carotenals	65
CaCO$_3$/MgO/Ca(OH)$_2$ (30:6:5)	Petrol/benzene/acetone (40:10:1)	Carotenes	66
	Petrol/acetone/ chloroform (5:5:4)	Xanthophylls	66
ZnCO$_3$	Hexane/triethyl- amine/$tert$-amyl alcohol (90:5:5)	Xanthophylls	39
MgO/kieselgur (1:1)	Acetone (4–40%) in petroleum ether	Carotenes, xanthophylls	67
Silica gel G	Isooctane/acetone/diethyl ether (3:1:1)	Carotenes, xanthophylls	60
Silica gel G	Petroleum ether/acetone (60:40)	Xanthophylls	61
Silica gel G	Ethyl acetate/carbon tetrachloride (3:1)	Xanthophylls	40
Silica gel RP-C18	Acetone	Carotenes	68
Al$_2$O$_3$	Benzene	Carotenes	69

Table 6 RF Values of Some Carotenoids on Adsorption TLC

Carotenoids	hR_F value of systems			
	1	2	3	4
α-Carotene	—	84	97	100
β-Carotene	96	69	97	100
7,7′-Dehydro-β-carotene	—	55	97	100
β-Apo-8′-carotenal	—	0	64	100
Isozeaxanthin	63	—	—	—
Bixin	51	0	0	5.5
Canthaxanthin	38	0	0	90
Zeaxanthin	17	0	0	24
Capsanthin	—	0	0	16
Capsorubin	—	0	0	13.5

Source: Ref. 56.
System: 1 = silica gel; n-hexane/ether (30:70)
2 = calcium hydroxide; petroleum/methylene chloride (95:5)
3 = secondary magnesium phosphate; carbon tetrachloride
4 = silica gel G; methylene chloride/ethyl acetate (80:20)

phytofluene can be detected by its characteristic green fluorescence in long-wave UV light (360 nm); the colorless phytoene appears under shorterwave UV light as a weak violet zone which may be easily detected as a dark area against a lighter background after the chromatogram has been sprayed with a solution of Rhodamine 6G (1%) in acetone. The most sensitive method of detecting colorless compounds is the use of iodine vapor, which stains the substances brown. For instance, as little as 0.05 µg of lycopersene can be detected this way (2). Certain functional groups in carotenoids can be stained with special spraying solutions, e.g., carotenoid aldehydes can be detected as violet zones after spraying with ethanolic Rhodamine (1–5%) and 25% ammonium hydroxide. The limit of sensitivity is ~0.03 µg. Carotenoid epoxides can be detected by their blue–green or blue color after exposure to hydrogen chloride (70). In general, TLC methods are reliable, inexpensive, portable, and readily carried out by nontechnical operators, but HPLC methods have the advantage of a shorter analysis time and a higher resolution.

f. High-Performance Liquid Chromatography (HPLC). The technique of HPLC has been widely applied to the study and analysis of carotenoids. Compared with the aforementioned methods, HPLC is characterized by short analysis time, high resolution, good reproducibility, and little structural modification. Generally, carotenoids are identified through their chromatographic behavior or by coelution with authentic standards. In addition, the identity of carotenoids can be confirmed by their UV-Vis absorption spectra, which can be recorded by a stopped flow scanning method or on-line with a photodiode array detector. Functional groups can also be detected when chromatograms are run before and after chemical tests. External or internal standards, e.g., β-apo-8′-carotenal (71), Sudan 1 (72), are used to improve reproducibility. For the stationary phase both normal phase HPLC and reverse phase HPLC are used, the elution being either isocratic or with a gradient. The latter is especially employed for complex extracts containing carotenoids of widely different polarities (1).

NORMAL PHASE HPLC (NP-HPLC). A number of the stationary phases and eluants commonly used for NP-HPLC are summarized in Table 7. The use of pressurized liquid chromatographic separation began with the work of Sweeney and Marsh (73), who separated β-carotene isomers on a mixture of magnesium hydroxide and calcium hydroxide. The efficiency of HPLC technique for separation of complex mixtures of carotenoids was further demonstrated by Stewart and Wheaton (74). They developed a two-step procedure for the separation of the citrus peel carotenoids, fractionating the carotenes on magnesium oxide and the xanthophylls on zinc carbonate. Subsequently, Stewart (75) established an HPLC procedure using a single magnesia column for the determination of α-and β-carotenes and β-cryptoxanthin. To follow the carotenoid changes in citrus juice, however, he used a two-step method whereby the carotenes and cryptoxanthin were analyzed on magnesium oxide and the xanthophylls were chromatographed on silica (76). HPLC on silica was applied to the separation of carotene diols, cis-trans isomers, and diastereoisomers (77) and to the quantitative analysis of carotenoid fatty acid esters from unsaponified fruit samples (72). A typical separation of orange peel carotenoids with NP-HPLC on silica is shown in Fig. 3. The detection limit for β-carotene was found to be ~5 ng [78a]. For the separation of β-carotene isomers an NP-HPLC with calcium hydroxid columns is commonly utilized (78,79), the mobile phase usually containing small amounts of acetone in hexane (80). Vecchi (81,82) managed to separate 11 cis isomers of β,β-carotene on alumina. The separation of all stereoisomers of tunaxanthin was achieved by employing a chiral column of Sumipax OA-2000 (83). Generally, insufficient separation of carotenes, increased formation of isomers (72), as well as poor suitability for gradient elution are the disadvantages of NP-HPLC.

REVERSE PHASE HPLC (RP-HPLC). Recently, reverse phase partition chromatography has become the method of choice for both qualitative and quantitative analysis of carotenoids. The stationary phases commonly used are those with C18-bonded chains (ODS). Their performances are

Table 7 Stationary Phases and Eluants Used for NP-HPLC

Stationary phase	Mobile phase	Type of carotenoids	Ref.
Mg(OH)$_2$ + Ca(OH)$_2$	1.5% p-methylanisole/ petroleum ether	Isomers of β-carotene	73
MgO	A: acetone; B: hexane; g*	Carotene, cryptoxanthin	75
Silica	A: 20% tert-pentyl alcohol/ hexane; B: hexane; g*	Xanthophylls	76
Silica	A: acetone; B: hexane; g*	Carotenes, diols, cis–trans isomers, diastereoisomers	77
Silica	A: hexane; B: ethyl acetate; g*	Carotenoids, carotenoid esters	72
Silica	A: pentane; B; acetone/pentane (80:20); g*	Carotenes, xanthophylls	84
Ca(OH)$_2$	Hexane/0.1–2.0% acetone	Isomers of β-carotene	78
Al$_2$O$_3$	Hexane	Carotene isomers	81
Spherisorb S–5 CN	Hexane/isopropyl acetate/ acetone (76:17:7)	Astaxanthin	26
Spherisorb CN	Hexane/dichloro-methane/methanol/N-ethyldiisopropylamine (60:40:01:0.1)	Tunaxanthins	82
Sumipax OA-2000	Hexane/dichloromethane/ ethanol (54:10:0.1)	Stereoisomers of tunaxanthin	83

g* = gradient elution.

influenced by the extent of endcapping and the carbon loading. Bushway (85) compared various RP and NP-HPLC methods and found that C18 columns provide the quickest and best separation of carotenoids. Polymeric C18 phases exhibited better selectivity for structurally similar carotenoids than monomeric C18 phases, which represent the majority of the commercially available C18 columns (86). These findings were confirmed by Epler et al. (87), who evaluated 65 different HPLC columns under standardized conditions with respect to carotenoid separation and recovery. Numerous mobile phase compositions have been employed for RP-HPLC of carotenoids; some commonly used solvents are listed in Table 8. Acetonitrile-based eluants are applied most frequently, with various organic modifiers (tetrahydrofuran, chloroform, ethyl acetate) added to improve the performance. Methanol-based eluants are also used, both with or without modifiers like tetrahydrofuran, chloroform, water, and hexane. Epler et al. (87) observed that methanol-based solvents typically provided higher recoveries than acetonitrile-based solvents. Separating *Satsuma* carotenoids with RP-HPLC, Noga and Lenz (88) found satisfactory reproducibility of retention time and peak areas and limits of detection from 0.027 ppm for *trans*-violaxanthin to 0.073 ppm for α-carotene. RP-HPLC has been successfully applied to the qualitative and quantitative estimation of carotenes (89), xanthophylls (90), *cis*- and *trans*-carotenoids (80,86), and carotenoid fatty acid esters (51,91). Figure 4 shows an HPLC separation of (A) paprika extract containing carotenoid esters and (B) saponified paprika extract, containing the corresponding carotenoids. RP ion pair chromatography was employed to quantify the chlorophylls and carotenoids in olive oil (36) using the following eluants: (A) methanol/water/ion pair reagent (0.05 M tetrabutylammonium and 1 M ammonium acetate), (B) acetone/methanol (1:1).

3. Further Developments

g. Supercritical Fluid Chromatography (SCF). The use of supercritical fluid chromatography for carotene separation has been examined and optimized, especially in regard to temperature,

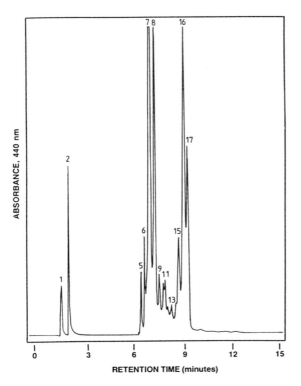

Fig. 3 Normal phase HPLC separation of Valencia orange peel carotenoids. Peaks: 2 = α-cryptoxanthin esters; 5 = lutein diesters; 6 and 7 = violaxanthin diesters; 8 = luteoxanthin diesters; 15 and 16 = violaxanthin monoesters; 17 = luteoxanthin monoesters. The other peaks are not identified. (From Ref. 51.)

pressure, and organic modifiers in the supercritical fluid (98). With an RP column it was possible to resolve an α-carotene cis isomer from an all-trans α-carotene as well as two cis isomers of β-carotene from an all-trans β-carotene. As with HPLC only polymeric C18 columns were able to resolve the cis isomers of α- and β-carotene from the all-trans isomers. SFC offers not only efficient separation but also fast analysis. Indeed the use of SFC with ODS-based columns for the analysis of carotenoid pigments affords a threefold reduction of analysis time compared to HPLC (99). The elution order of carotenoids and their cis isomers was found to be the same as in RP-HPLC. The selectivity of the system could further be increased by adding modifiers (e.g., hexane, methanol, chloroform, methylene chloride) to the CO_2. The correlation found by comparing the selectivity of both SFC and RP-HPLC was very good, with a correlation coefficient of 0.975 (100).

h. High-Performance Liquid Chromatography–Mass Spectrometry (HPLC-MS). HPLC-MS combines traditional HPLC separations with mass spectral identification in a single procedure. The complex carotenoid mixture is separated with RP-HPLC, and the ensuing chromatographic profile is monitored through the total ion current (TIC) response of the mass spectrometer. Usually mass spectra are collected in the electron impact mode (EI) using 40 eV at a source temperature of 240°C (101). Combining these data with the monitored UV-Vis spectra the carotenoids can be identified and quantified. Taylor et al. (101) found a detection limit for carotenoids of ~50 ng. This method is relatively costly in regard to instrumentation and maintenance. These considerations are, however, compensated by its great potential for rapid carotenoid analysis and identifica-

Table 8 Mobile Phases Used for RP-HPLC

Mobile phase	Type of carotenoids	Application	Ref.
A: methanol B: ethyl acetate; g*	Carotenoids, carotenoid esters	Pepper, fruits	92, 51
A: methanol/water (75:25) B: ethyl acetate; g*	Carotenes, xanthophylls	Kiwi	12
A: acetone/water (75:25): B: acetone/methanol (75:25); g*	Carotenes, xanthophylls	Paprika	93
A: acetone B: water; g*	β-Carotene, xanthophylls	Grape, fruits, paprika	61, 94, 95
A: acetonitrile/2-propanol (40:60) B: water; g*	Carotenoids, carotenoid esters, β-carotene	Paprika	91
A: methanol/water (9:1) B: acetone; g*	Carotenes, xanthophylls	Citrus	88
Acetonitrile/ methanol/tetrahydrofuran (40:56:4)	Carotenes, xanthophylls	Vegetable, fruits	85
Acetonitrile/methanol/ methylene chloride/hexane (55:22:11.5:11.5)	Cis–trans isomers of β-carotene	Vegetable	71
A: acetonitrile/water (90:10) B: ethyl acetate; g*	Carotenes, xanthophylls	Tea	96
Acetonitrile/methylene chloride/methanol (82:12:6)	Carotenes	Carrot	97
Methanol/tetrahydrofuran/water (67:27:6)	Lycopene, β-carotene	Tomato; grapefruit	35
Acetonitrile/chloroform (92:8)	α-and β-Carotene	Carrot; tomato	34, 89
Methanol/chloroform (94:6)	β-Carotene	Potato	38

g* = gradient elution.

tion. The separation of asparagus carotenoids with HPLC-MS as well as typical EI mass spectra of some carotenoids were published by Taylor et al. (101).

D. Applications in Food Analysis

The content and distribution of carotenoids in food has been the object of numerous investigations. The carotenoid distribution in plants is complex and shows considerable variations between different species. Furthermore the ripening process in fruits and vegetables leads to significant changes in both the total carotenoid content and the carotenoid distribution. Fruit carotenoids have been reviewed by Goodwin (8), Bauernfeind (24) and, more recently, by Gross (5). Plant and animal carotenoids have been discussed by Krinsky et al. (101). Due to variable chemistry of carotenoids, differences in the matrix composition and new instrumental and methodical developments several different analytical methods have been employed. In Table 9 a compilation of a number of foodstuffs and references to the corresponding analytical method applied are arranged. The general principles of these analytical methods and some consideration concerning their use in carotenoid analysis are discussed above. Subsequently, three analytical procedures are described in detail as representatives for all the other applications in food analysis.

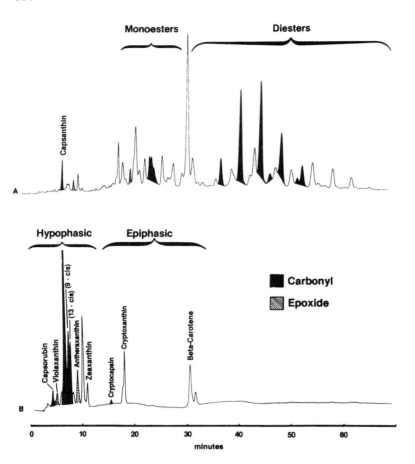

Fig. 4 RP-HPLC separation of (A) paprika extract and (B) saponified paprika extract on a Zorbax C18 column at 460 nm. The solid peaks represent carotenoids containing a ketone group, and the hash-marked peaks represent carotenoids containing an epoxide group. (From Ref. 93.)

Table 9 Some References of Analytical Methods Used for Carotenoid Analysis in Food

Food	CC and spectrophotometry	TLC	HPLC
Apricot	102	103	90
Carrot	104	14	34, 97
Citrus	105	58	76
Grape	106	107	61
Orange	108	40	37
Paprika	44–46	21, 109	41, 97, 95
Pepper	49	110	15, 64
Spinach		111	94
Tomato	112	113	35, 89
Salmonids	25	25	26

1. Spectrophotometric Determination of β-Carotene and Lycopene in Ruby Red Grapefruits (118)

Three to four grapefruits are halved, and the edible meat and juice are collected and blended for 3 min. One hundred milliliters of methanol is added to 100 g of blended sample and this mixture is allowed to stand for ~30 min. Two grams of Hyflo Super-Cel is then added and the samples subsequently filtered through a pad of the same filter aid on a Büchner funnel. The filtrate containing the methanol-water-soluble substances is discarded. The pigments are extracted from the pulp by blending the filter cake with 50 ml of 50% acetone in hexane for 1 min in a microblender cup. After filtration the cake is washed with 20 ml extraction solution. The pulp and pad are blended and twice more washed in the same manner. The combined extracts are placed in a 500-ml separatory funnel and 100 ml of water is added. The aqueous acetone layer is removed and reextracted with hexane until the hexane extract is colorless. Finally, the combined hexane extracts are washed three times with 100 ml water; then the hexane extract is filtered through Na_2SO_4 and made up to a volume of 250 ml. The absorbance of the hexane extract is measured at 451 nm (A_{451} is a maximum for carotene and a near minimum for lycopene) and at 503 nm (A_{503} is a maximum for lycopene and a low absorbance value for carotene). The specific absorbances used for carotene are 250.3 at 451 nm and 51.0 at 503 nm; for lycopene they are 195.5 at 451 nm and 292.7 at 503 nm. Accordingly, the concentrations of the pigments can be calculated as follows:

$$\text{Carotene (mg/100 g)} = 462A_{451} - 309A_{503} \tag{2}$$
$$\text{Lycopene (mg/100 g)} = 395A_{503} - 80.5A_{451} \tag{3}$$

2. Separation of Citrus Carotenoids by TLC (58,115)

The citrus juice is homogenized for 1 min with 1 vol. of petroleum ether and 3 vol. of isopropanol in the presence of a small amount of ascorbic acid. The mixture is diluted with an equal volume of diethyl ether, and sufficiently saturated NaCl solution is added to form two layers. The upper petrol ether/diethyl ether layer, containing all the cartenoids, is washed with distilled water until it is free of alcohol. For saponification the ether layer is combined with an equal volume of 10% KOH in ethanol, covered with nitrogen, and kept overnight at room temperature. The mixture is then washed free of alkali, dried by adding absolute ethanol, and evaporated in vacuum. The nonsaponifiable matter is dissolved in a minimum volume of hot methanol and kept at $-4°C$ overnight. The precipitated sterols are removed by filtration and the methanol is removed by evaporation under reduced pressure.

The saponified pigments are dissolved in a few drops of chloroform and petroleum ether. This solution is then applied as a line on a silica gel G plate (0.4 mm). The chromatogram is first developed with 10% acetone in petroleum ether, with a few crystals of BHT added to the solvent system. When the first band of carotenes has reached the half length of the plate, the plate is transferred to another tank containing 30% acetone in petroleum ether. There the chromatogram is developed until a sufficient separation has been achieved. The phytofluene band should be monitored by UV light at 360 nm and scraped off the plate before chromatography is finished. Figure 5 shows a separation of *Dancy* tangerine carotenoids. The carotenoid content of each band has been determined photometrically. Each band is further rechromatographed on TLC plates of MgO-kieselgur (1:1). Up to 50 μg of pigment may be applied on each plate. The carotenes are developed with 4% acetone in petroleum ether, the monol carotenoids are developed with 10% acetone in petroleum ether, and the diol and polyol carotenoids are separated with 30% acetone in petroleum ether.

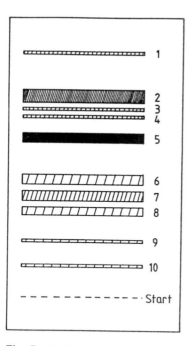

Fig. 5 TLC separation of saponified carotenoid extract of Dancy tangerine juice. Adsorbant: silica gel G, solvent system: acetone/petroleum ether (30:70). Bands: 1 = carotenes, 2 = monol (cryptoxanthin), 3 and 4 = cryptoxanthin epoxides, 5 = citraurin (found only in peels), 6 = diols (lutein, zeaxanthin, and mutatoxanthin), 7 = diol epoxides (2 isomers of autheraxanthin), 8 = diol diepoxides (2 isomers of violaxanthin), 9 and 10 = polyols (neoxanthin isomers). (From Ref. 58.)

3. HPLC Analysis of Carotenoids and Carotenoid Esters in Red Bell Pepper (92)

Red bell peppers without seeds are cut into small pieces and accurately weighed (2–3 g) samples are homogenized with 50 ml methanol and 1 g $MgCO_3$ for 1 min. After filtration through a Bucher funnel under vacuum, the residue is repeatedly extracted with 100 ml of the extracting solvent (acetone/hexane, 80:20) until all of the carotenoids have been extracted. The pooled filtrate is transferred to a separatory funnel and a known amount of the internal standard canthaxanthin added. An equal volume of distilled water is added. The mixture is shaken vigorously and the phases allowed to separate for 5 min. The aqueous phase is filtered through a 0.4-μm nylon filter and used for HPLC analysis. The HPLC separation is carried out on a C18 column (3.9 × 150 mm) with a mobile phase of methanol and ethyl acetate in a linear gradient for 20 min. The detection wavelength is set at 475 nm and the flow rate is 1.8 ml/min. Figure 6 shows the good separations of carotenoids and carotenoid esters which can be obtained with this method.

II. CHLOROPHYLLS

A. Introduction

1. Physical and Chemical Properties

Chlorophylls are the most widely distributed natural pigments, and every organism capable of photosynthesis, including all living plants, algae, and some photosynthetic bacteria, owes its green

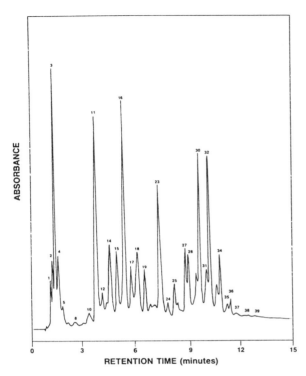

Fig. 6 HPLC determination of carotenoids and carotenoid esters in red bell pepper. Peaks: 2 = capsorubin; 3 = capsanthin; 11 = β-apo-8′-carotenal (Int. Std.); 12, 13, 14, 15, and 18 = capsorubin monoester; 16 = capsanthin $C_{14:0}$; 17, 19 = capsorubin monoester; 23 = β-carotene; 25, 27, 28, 29, 31, 33, 36 = capsorubin diester; 30 = capsanthin $C_{12:0}$ and $C_{14:0}$; 32 = capsanthin $C_{14:0}$ and $C_{14:0}$; 34 = capsanthin $C_{14:0}$ and $C_{16:0}$; 35, 37, 38 = lutein diester. (From Ref. 4.)

color to them. The word *chlorophyll* is derived from the Greek language and means "the green of the leaves." In higher plants chlorophylls are always nonconvalently linked to proteins and accompanied by carotenoid molecules. These chlorophyll-carotenoid proteins are found in organelles known as chloroplasts. The biological function of chloroplasts is to capture light energy, which subsequently is used for the conversion of carbon dioxide to carbohydrates—the process known as *photosynthesis*. As chlorophylls consist of 10 conjugated double bonds they absorb strongly in the blue and red regions of the spectrum, which causes green color (Fig. 8).

In chemical terms chlorophylls are porphyrins like the pigments of blood (hemoglobin) and muscles (myoglobin). The tetrapyrrole ring of chlorophylls is formed by four pyrrole residues joined together by methine groups, the four nitrogen atoms are coordinated with a central magnesium atom (Fig. 7). Unlike hemes, which are iron complexes of porphyrins, chlorophylls contain an additional fifth isocyclic ring as well as a long hydrophobic side chain derived form propionic acid at C7, which is esterified to phytol, a C20 terpenoid alcohol (5). The chlorophyll molecule itself is relatively large, the porphyrin head being about 1.5×1.5 nm in size and the phytol chain being overall about 2 nm in length (116). The additional structural elements of chlorophylls vary widely; more than 50 structures are known already. The most widespread of them are chlorophyll a (chl. a) and chlorophyll b (chl. b), which were isolated for the first time by Sorby in 1873 (117). Chl. b differs from chl. a only in having an aldehyde group in place of the methyl group at the C3 position. However, this small difference in the structure of the two

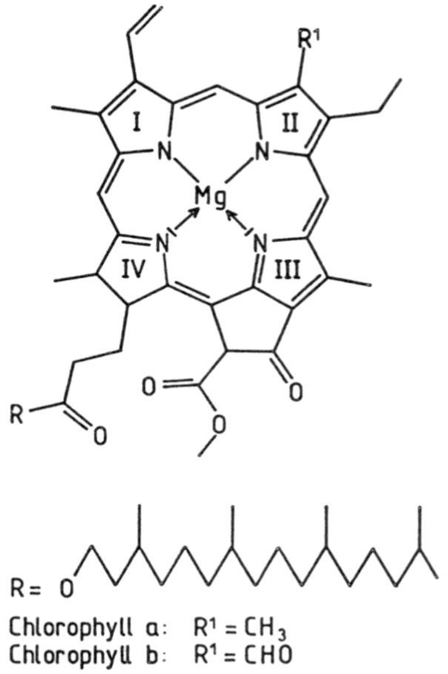

Chlorophyll a: $R^1 = CH_3$
Chlorophyll b: $R^1 = CHO$

Fig. 7 Formula of chlorophylls. R_1: chlorophyll a: $-CH_3$; chlorophyll b: $-CHO$; R: phytol residue ($C_{20}H_{39}O$).

chlorophylls causes significant differences in the absorption spectra and therefore also different green hues. As demonstrated in Fig. 8, the absorption maximum of chl. b is shifted to the green region of the spectrum, so that the color of chl. b is yellow–green while the color of chl. a is blue–green. Chlorophylls c (chls. c) compromise the members of a group of chlorophylls which have a fully unsaturated porphyrin macrocycle. They are most widely distributed in the chromophyte algae (119). Other pigments (e.g., chlorophyll d, chlorophyll a′) have also been isolated and identified; however, they are of less importance. The chemical properties of chlorophylls have been reviewed by Jackson (120). In general, chlorophylls are relatively unstable and are sensitive to light, heat, and oxygen as well as chemical degradation. The phytyl ester may be removed enzymatically by chlorophyllase forming the free acid derivative chlorophyllide. This can also be hydrolyzed by acid or alkali under mild conditions whereby oxidation at C10 may occur (120). The central magnesium ion is immediately removed from the chlorophyll when diluted mineral acids are applied, forming pheophytin, which may further form pheophorbide through hydrolysis (5). Pheophytins and their derivatives lead to an unattractive brown–green color in food. Because of the phytyl residue chlorophylls and pheophytins are lipophilic whereas chlorophyllides and pheophorbides are hydrophilic.

2. Properties in Food

The distribution of chlorophylls among photosynthetic organisms has been listed by Scheer (119) and Jackson (120). Chl. a is present in all organisms capable of oxygenic photosynthesis, in higher plants (green leaves and fruits), ferns, mosses, green algae, and prochlorophytes—the so-called green line (119). Chl. a is usually accompanied by chl. b in a ratio of 3:1. In some algae (brown line, e.g., *Cryptophytes, Diatomes, Pheophytes*) chl. a is complemented by chls. c. The total chlorophyll concentration of fruits and vegetables has been reviewed by Gross (5) and Hermann (118). It can generally be assumed that the deeper the green color of a leaf or a fruit, the higher is

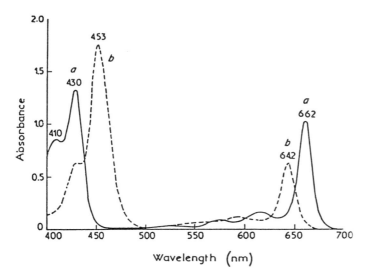

Fig. 8 Absorption spectra of chlorophylls a and b in diethyl ether. Chlorophyll a (—), chlorophyll b (- - -). (From Ref. 5.)

its chlorophyll content. Green leaves of spinach, parsley, or green cabbage can contain up to 2000 mg chlorophyll per kg fresh weight, while, for instance, beans, peas, and cucumbers have contents of about 100 mg/kg (118). In fruits and vegetables the chlorophyll content, and often also the ratio chl. a/b, varies with genus, species, cultivar, and environmental factors (5). Since chlorophylls are generally disintegrating during fruit ripening, the content depends highly on the developmental stage. But there are also some exceptions which retain a high chlorophyll content even in the ripe stage, e.g., certain apple, pear, and gooseberry cultivars. Table 10 shows the average chlorophyll

Table 10 Average Chlorophyll Content in Fruits and Vegetables (mg/kg)

Food	Chlorophyll a	b	Ref.
Bean, green	118	35	123
Cabbage, green	1898	406	123
Cabbage, white	8	2	123
Cucumber	64	24	123
Parsley	890	288	123
Paprika, green	98	33	123
Pea, green	106	22	123
Spinach	946	202	71
Apple ("Golden Delicious," peel)	98	38	124
Grape ("Riesling")	11	4	106
Kiwi ("Bruno")	17	8	125
Pear ("Trévoux", peel, green)	31	13	126
Strawberry (Tenira," green)	5	1	127
Tangerine ("Dancy," green)	249	80	5

content and the ratio chl. a/b of some fruits and vegetables. High amounts of chlorophylls can be found in the deep green peel of avocado [379 mg/kg (121)] and muskmelon [345 mg/kg (122)]. There are, however, pronounced differences between the chlorophyll content of the peel and the pulp of these fruits.

During food processing and storage chlorophylls undergo various alterations which cause grave color changes. Through blanching, preserving, heating, and storage chls. a and b are transformed to pheophytins a and b, which have a brown–green color. And if the preserved food has not been blanched, native enzymes (e.g., chlorophyllase) may also split off the phytol residue and degrade the resulting chlorphyllides. In canned food the central magnesium ion can be replaced with iron or tin ions, which may cause a grey–brown discoloration of the products. By contrast, the insertion of copper and zinc ions produces a green pigment with higher stability and greater tinctorial strength.

3. Arguments for Chlorophyll Analysis: Regulations

Because of their basic role in nature chlorophylls have been the subject of many investigations. For a better understanding of photosynthesis it is necessary that chlorophylls and their derivatives be determined qualitatively and quantitatively. Investigations into their chemical structure, genesis, transformations, and functions in plants supply valuable information for plant physiological studies (128).

As color is an important attribute of food quality it can be used as a criterion for the classification of fruits and vegetables. For instance, with apples the fruit color is an important factor determining the selling price on the U.S. market, the preferred colors being red or yellow, with green fruits selling poorly (129). Numerous analytical methods have been established to quantify the color and to identify the native colorants of food (63,130,131). During the ripening of fruits the green chlorophylls disappear and yellow pigments (e.g., carotenoids, flavonoids) become visible. However, when bananas are ripened at temperatures above 24°C they fail to develop a fully yellow peel because the chlorophyll breakdown is suppressed. Those "green-ripe" bananas are perceived to be of poor quality and consequently fetch a low price (132). The physiology of degreening in maturing seeds is of great importance for the canola industry because the presence of as little as 6% green seeds results in green, poor-selling oil (132). Generally the pigment content and the change in content during ripening can be used as an index for the physiological age of fruits (133). The chlorophyll concentration in apple peels is a potential indicator of maturity at harvest and helps to predict the quality (134). The senescence of leafy vegetables is characterized by a degradation of chlorophylls accompanied by a marked loss of greenness. Both the pathway and the parameters influencing this undesirable alteration have been monitored, but further investigations are required (135).

During food processing (e.g., blanching, heat-drying, storage) the structural characteristics of natural pigments are often modified and the color of the product is usually changed. The determination of the chlorophyll content and its derivatives can therefore indicate the extent of color damage in processed food (136). During the traditional process of canning most chlorophylls are converted to pheophytins and pyropheophytins due to the severe heat treatment. Since the bright green color of chlorophylls is more pleasing to the consumer than the olive brown color of pheophytins and pyropheophytins great efforts are invested into developing better technologies. Recent studies have shown that in comparison to conventional preservation methods aseptic technologies can reduce color losses (137). Furthermore a correlation between the chlorophyll content and the undesirable "grassy" taste of black tea has been found. To minimize this off-flavor in black tea the factors affecting the chlorophyll content (e.g., nitrogen fertilization, plucking criteria) have been investigated (138). Using multiple regression analysis a model for predicting the quality of manufactured black tea from the carotenoid and chlorophyll composition of fresh green tea leaves has been established (139).

Chlorophylls and chlorophyll derivatives are also often used as food colorants to restore the color of freshly harvested crops. Since chlorophylls have been an integral part of vegetable food stuff for animals and humans throughout history, they appear to be harmless. Unfortunately their possible widespread application as a green food colorant is hampered by their inherent instability [116]. Nevertheless the substitution of synthetic pigments used in foods with natural ones would be an important step to healthier food. Mg-chlorophyll (E 140) and Cu-chlorophyll (E 141) are both permitted as food additives, but Cu-chlorophyll has shown a better stability than Mg-chlorophyll in regard to the exposure to light, oxygen, temperature, sodium bisulfite, benzoyl peroxide and acids [140]. The commercial production of coppered chlorophyll has been described in detail by Humphrey (141). Other metal-substituted chlorophylls, e.g., iron(III), zinc(II) chlorophylls, are used for certain kinds of medicaments and have further received attention because of their potential use as electrode materials for photoelectron conversion (142).

B. Sample Preparation

Because chlorophylls are exceptionally labile pigments proper care has to be exercised during extraction and analysis. All manipulations ought to be carried out rapidly in darkness or in dim light to avoid photobleaching; the amount of heating and oxidation must be kept at a minimum; and native enzymes (e.g., chlorophyllase) should be inactivated.

1. Extraction

For the determination of chlorophyll pigments it is preferable to use freshly grown plant material. If storage should become necessary the tissue can be kept in the dark at $0°C$ or below (128). A great variety of extraction methods have been used and comprehensive discussions are given in the literature (5,128,143,144). The criterion for a good extraction procedure is whether it brings all pigments in solution and causes little or no alteration (143). Because of the instability of chlorophylls it is almost impossible to extract food material without formation of alteration products. With the aim of inactivating enzymes, some researchers have immersed plant material in boiling water (blanching or scalding) before extraction. This causes isomerization of chlorophylls, e.g., chls. a and b, forming compounds that are diastereomeric at C13, e.g., chls. a' and b' (128). Boiling also causes the breakdown of chlorophylls to pheophytins if the tissue is somewhat acidic. Under acidic conditions chl. a is generally less stable than chl. b, causing the formation of relatively large quantities of pheophytin a during the extraction of fruit samples (145). To avoid the conversion of chl. a to pheophytin a the original acidity of the tissues must be neutralized with buffers like sodium carbonate, calcium carbonate, and sodium hydrogen phosphate (128). For kiwi fruits Fuke et al. (145) found that the addition of 2.0 g sodium carbonate per 50-g sample is sufficient. However, when the material is extracted under conditions that do not inactivate enzymes, chlorophylls will undergo hydrolysis and oxidation: chlorophyllase will catalyze the hydrolysis of the phytyl group resulting in the formation of chlorophyllides and oxidative enzymes will promote the formation of 13-hydroxychlorophylls (146).

Commonly the food material is disintegrated and chlorophylls are extracted in a blender. Alternatively, small portions may be ground in a mortar with the solvent and sand, then filtered or centrifuged. As up to 90% of the fresh weight of fresh plant tissue are water, water-miscible solvents should be used for extraction. The most frequently used extraction solvents are acetone and methanol. They break the noncovalent linkage between pigments and proteins and bring the chlorophylls in solution (5). Generally cold solvents are used (134) and the acetone or alcohol concentration in the extract should be maintained near 80% or 90% (128). However, acetone and methanol have disadvantages, as acetone is often a poor extractant for fresh plants and alcohols like methanol may cause some allomerization reactions (143). Solvent mixtures (e.g., acetone/methanol) were also used for extracting chlorophylls from diverse sources. The extraction of fresh plant material was improved with a mixture of methanol–petroleum ether (147). For the purification of chlorophylls and carotenoids from olive fruits an extraction procedure with N,N-dimethylformamide and hexane was developed (148).

To obtain clear solutions the extracts are either filtered or centrifuged and the residues reextracted with the same solvents (143). Another possibility is to transfer the chlorophylls from the acetone phase into an ether phase by adding sodium chloride (149) or sodium sulfate (129). More recently, acetone extracts were purified and fractionated with C18 solid phase extraction cartridges, dephytylated pigments were eluted with 70% acetone and phytylated pigments with 90% acetone (150). Chlorophyll extracts should be analyzed as soon as possible, but if they have to be stored they must be kept in the cold and dark.

C. Analysis–Detection–Separation Methods

There are many methods for the estimation of chlorophylls in food extracts. Traditional methods, based on colorimetry, photometry, and fluorometry, are simple and therefore widely applied. Single chlorophylls may be separated and purified with chromatographic methods and determined individually. Nondestructive methods are quite useful for maturity classification of fruits and vegetables.

1. Classical Methods

Colorimetric methods, e.g., subjective comparisons of color intensities or objective instrumental measurements with a colorimeter, are simple to handle and therefore still in frequent use.

a. Spectrophotometric Methods

DETERMINATION OF CHLOROPHYLLS. The absorption spectra of chlorophylls (Fig. 8) show absorption maxima between 600 and 700 nm. Generally two chlorophylls (a and b) are present in food extracts. The photometric measurement is based on the fact that the specific extinctions of chlorophylls are additive. Since chls. a and b have distinct absorption maxima, it is possible to calculate the concentration of each pigment by subtracting the contributed absorption of the second pigment from the maximum extinction of the first pigment. However, incorrect results are obtained if high amounts of pheophytin are present in the extracts (149). For creating equations that solve simultaneously for both chls. a and b, it was necessary to determine the extinction coefficient of each chlorophyll at its own wavelength maximum and at the wavelength maximum of the other chlorophylls present in the mixture. Mackinney (151) was the first to publish specific extinction coefficients for chl. a ($e_{664nm} = 82.04$; $e_{647nm} = 16.75$) and chl. b ($e_{664nm} = 9.27$; $e_{647nm} = 45.60$) in 80% acetone. Based on these results, Arnon (152) developed equations for the simultaneous estimation of chls. a and b in 80% aqueous acetone extracts of higher plant leaves (Table 11).

Those values have been modified by various investigators, but Arnon's equations continue to be frequently used. In 1989 Porra et al. (153) showed that the coefficients published by Mackinney are too low; they determined accurate extinction coefficients in 80% aqueous acetone: chl. a ($e_{664nm} = 85.95$; $e_{647nm} = 20.79$) and chl. b ($e_{664nm} = 10.78$; $e_{647nm} = 51.84$). When the absorbance is measured in other solvents like diethyl ether (149) or 96% ethanol (154) corresponding equations have to be employed (Table 11).

Nomograms were constructed to simplify the measuring of the amounts of chls. a and b from spectrophotometric readings (155). Photometric methods are widely used because of their simplicity and low costs. However, they have some drawbacks:

Table 11 Equations for the Simultaneous Determination of Chlorophylls
with Spectrophotometry

Solvent	Equations	Ref.
80% acetone	Total chlorophyll (mg/L) = $20.2E_{645} + 8.02E_{663}$	153
	Chlorophyll a (mg/L) = $12.7E_{663} - 2.69E_{645}$	
	Chlorophyll b (mg/L) = $22.9E_{645} - 4.68E_{663}$	
Diethyl ether	Chlorophyll a (mg/L) = $10.1E_{662} - 1.01E_{644}$	149
	Chlorophyll b (mg/L) = $16.4E_{644} - 2.57E_{662}$	
	Alternatively:	
	Total chlorophyll (mg/L) = $7.12E_{660} + 16.8E_{642.5}$	148
	Chlorophyll a (mg/L) = $9.93E_{660} - 0.777E_{642.5}$	
	Chlorophyll b (mg/L) = $17.6E_{642.5} - 2.81E_{660}$	
96% ethanol	Total chlorophyll (mg/L) = $20.0E_{649} + 6.1E_{665}$	154
	Chlorophyll a (mg/L) = $13.7E_{665} - 5.76E_{649}$	
	Chlorophyll b (mg/L) = $25.8E_{649} - 7.6E_{665}$	

1. Chlorophyll degradation products have absorption spectra similar to those of the chlorophylls and seriously interfere in the determination.
2. There are considerable differences between the results obtained with the various sets of equations.
3. The sensitivity is relatively poor (156).

Generally these methods are satisfactory for estimations of chlorophyll concentrations in the nmol/ml range, for lower concentrations the more sensitive technique of spectrofluorometry is required (157).

DETERMINATIONS OF CHLOROPHYLLS AFTER THEIR CONVERSION TO PHEOPHYTINS. Because of the instability of chlorophylls under light it would be preferable to convert them to the more stable pheophytins before quantitative measurements (143). Chlorophylls were therefore converted to pheophytin by treatment with HCl (158), ion exchange resins (159), and oxalic acid (160), and subsequently determined colorimetrically or spectrophotometrically.

Vernon (161) devised a method for determining total chlorophyll, total pheophytin, chls. a and b, pheophytins a and b, and the percentage retention of chlorophyll. The principle of the method is to prepare an acetone extract and to take two samples. One is the control sample (no conversion) and in the second the chlorophylls are converted to pheophytins by adding saturated oxalic acid in 80% acetone. Both samples are kept in the dark at room temperature for 3 hr and then their absorbances are measured at 536, 558, 645, 649, 655, 622, 665, 666, and 667 nm. An additional reading at 700 nm is used to correct for any turbidity in the extract. The measurements at 649 and 665 nm are employed for the direct calculation of the chlorophyll concentrations. The concentrations of the formed pheophytins are calculated by the following equations:

$$\text{Total pheophytin (mg/L)} = 6.75A_{666} + 26.03A_{655} \tag{4}$$
$$\text{Pheophytin a (mg/L)} = 20.15A_{666} - 5.87A_{655} \tag{5}$$
$$\text{Pheophytin b (mg/L)} = 31.90A_{655} - 13.40A_{666} \tag{6}$$

For the spectrophotometric determination of chlorophylls after their conversion into pheophytins a detection limit of 5 µg/ml was found (162).

b. Spectrofluorometric Methods. It has been found that chlorophylls and their derivatives fluoresce independently of each other in mixtures and that the concentration of those components could be determined from their fluorescence when irradiated with blue and violet light (163). Fluorometry is a very sensitive method. White et al. (164) stated that chlorophylls, chlorophyllides, pheophytins, and pheophorbides in diethyl ether or 80% acetone mixtures may be detected over a concentration range from 0.001 to 0.3 µmol/L. When a solution contains several different pigments, e.g., chls. a and b, pheophytins a and b, the concentration of each component is calculated by equations. These equations are based on the assumption that the total fluorescence of a pigment mixture is the sum of the fluorescence of each component in the mixture. Various formulas are described in the literature; the equations of White et al. (164) are given as an example:

$$\text{Chlorophyll a} = (0.733F_{405} - 0.0292F_{460}) \times 10^{-3} \ \mu\text{mol/L} \tag{7}$$
$$\text{Chlorophyll b} = (0.292F_{460} - 0.131F_{405}) \times 10^{-3} \ \mu\text{mol/L} \tag{8}$$

In addition, spectrofluorometry is an excellent method for the determination of high chl. a/b ratios (>6). The method is based on the low excitation of chl. a at the wavelength where the excitation of chl. b is maximal (165). Measurements are made under two sets of conditions: For chl. a/b ratios between 6 and 60 the fluorescence of diethyl ether solutions of the pigments are excited at 453 nm at room temperature and the fluorescence emission measured at 646 and 666 nm. For ratios up to 1000 the pigments are dissolved in ethanol and excited at 478 nm and a temperature of 77 K. The fluorescence is recorded at 658 and 678 nm (143).

Spectrofluorometry has been used for the determination of chlorophyll concentrations in leaf extracts and in eluates from paper chromatograms (166). Marine and freshwater biologists have employed fluorometric methods extensively because of the low concentration of chlorophyll and pheophytin present in plankton (167). However, the use of fluorescence methods requires sophisticated spectrofluorometers with automatic correction for photomultiplier and monochromator responses as well as wavelength-dependent variation of the energy output of the light source (157). Those costly specifications of the instrumentation hamper the applicability of fluorometric methods.

   c. Nondestructive Methods.   Sometimes it is convenient to determine the chlorophyll content in the intact sample without having to destroy it. With nondestructive methods serial analysis can be made and the sampling errors are limited (144). The external food color is an important quality parameter and greatly affects the consumer acceptance. Since the color of fruits and vegetables corresponds to the content of pigments, methods estimating the color can be employed for measuring colorants like chlorophyll. During plant development the chlorophyll content of many foods decreases, causing characteristic changes in color. Therefore nondestructive methods measuring the external color present an opportunity to classify the maturity of fruits and to detect the optimum harvest date (167). A great variety of methods have been described, and the recording of absorption, reflectance, and emission spectra and the physical measurement of the color expressed by tristimulus values are widely applied. The relationship between the three attributes hue, lightness (value), and saturation (chroma) is graphically represented in color solids (e.g., Munsell solid, Hunter solid). The retina of the human eye has three types of color receptor molecules, so that every pigment corresponds to each of the primary hues red, blue, and green (5). Tristimulus colorimeters simulate the color receptor molecules of the human eye by using one filter for each primary hue (168). Both Gardner and Hunter laboratory instruments use the Hunter solid with following coordinates: +a (red) to –a (green), +b (yellow) and –b (blue) (169). The hue function (arctan b/a) is highly correlated ($r = +0.973$) with the residual chlorophyll content in dried parsley leaves (136). Correlation studies between the total chlorophyll content and hue function in frozen kiwi fruit slices gave a correlation coefficient of –0.7726 (167). These results demonstrate that tristimulus colorimetry is a suitable method for evaluating the color characteristics and the pigment content of fresh and processed food.

For the detection of the reflectance and/or transmittance characteristics of food, the wavelength is chosen which has the highest correlation between the measurements and the tissue color. The maximum reflectance wavelength of apples (171) was found to be at 678 nm, the reflectance of green leaves was measured at 625 nm with a colorimeter or a spectrophotometer, and it was inversely proportional to the total chlorophyll concentration (172). Changes of the chl. a/b ratio in ripening citrus fruits were monitored by means of reflectance measurement at 648 and 674 nm (173). In the case of ripening tomatoes it has been found that measuring the light transmittance at 500, 540, and 595 nm was more sensitive to small changes in color than light reflectance systems (174). Using a body transmittance spectroscope with a wavelength range between 500 and 900 nm it was possible to distinguish immature and mature green papaya fruits (175).

It has been observed that chlorophyll-containing plant material emits a very low-intensity light for several seconds or even minutes after it has been illuminated by a light source and then placed in darkness (176). The duration and intensity of this delayed light emission (DLE) is positively correlated with the concentration of chlorophyll and related substances (177). Various studies have demonstrated the feasibility of using DLE to estimate the maturity of foodstuffs like tomatoes (178), bananas (179), papaya (180), fresh tea leaves (181), muskmelon (167), and Japanese persimmon (182).

d. Phase Separation and Countercurrent Separation. In earlier times plant pigments were often separated on the basis of their different solubility in various solvents. A valuable application is the separation of phytylated and nonphytylated pigments by shaking an aqueous acetone solution with light petroleum. The phytylated pigments pass into the petroleum layer while the unphytylated pigments remain in the acetone phase (143). The purification of chlorophyll derivates has also been achieved by liquid–liquid partition on Celite columns with diethyl ether, acetone, and HCl (183). Chls. a and b were separated by multiple liquid–liquid partition using a solvents system consisting of petroleum ether/benzene/methanol/formamide (184).

e. Column Chromatography. The preparation and purification of plant pigments by column chromatography has a long tradition. The separation of chls. a and b was achieved for the first time in 1906 by Tswett (185), who passed a leaf extract through a column of precipitated chalk. Column chromatography is a suitable method for the isolation of relatively large quantities; with analytical procedures it is applied to the pretreatment of samples. Many different adsorbants have been used for the separation of chlorophylls, including inorganic compounds such as alumina and magnesia and organic compounds such as powdered sucrose, inulin, starch, and cellulose (5). Inorganic adsorbants are usually less recommended because they often cause decomposition of the labile chlorophylls. Powdered sucrose is the most widely used adsorbant. Both confectioner's sugar containing 3% starch and icing sugar containing 1–5% calcium phosphate have been found satisfactory for separating chlorophylls (143). For the precleaning of samples several separation methods with ion exchangers and molecular sieves are suitable (186). It was found that DEAE-Sepharose CL-6B is effective for separating esterified and nonesterified chlorophylls and also for the separation of chls. a and b (187). The chromatography of chloroplast pigments has been reviewed by Strain and Sherma (188); detailed information is also given by Holden (143).

2. Recent Methods

f. Thin-Layer Chromatography (TLC). TLC has now become widely used for separating and analyzing chlorophylls and has replaced paper chromatography, which is handicapped by its insufficient selectivity and resolution (186). Because of the instability of chlorophylls special precautions have to be taken during sample preparation and chromatography, e.g., low light intensity, development of TLC plates in the dark, pigment concentration and application under a stream of nitrogen, and rapid handling at all stages (143). To protect chlorophylls from oxidation, antioxidants such as ascorbic acid and butylated hydroxytoluene should be added (186). TLC of chlorophylls was described by Shioi (186), Sesták (189), Holden (143), and others. The separation of chlorophylls is mainly based on adsorption, but partition chromatography between liquid–liquid phases is also sometimes used (190). TLC separation can be realized either mono- or two-dimensionally with both NP-TLC or RP-TLC (190). The performance of separation depends mainly on the support and the developing solvents chosen. A variety of supports and developers have been used. The most recommended supports are powdered silica gels, cellulose, sucrose, and, more recently, chemically bonded silica gels; see Table 12 (186). The use of silica gel layers has been described for the determination of chlorophylls in kiwi (145), capsicum (201), and cucumber (202). However, silica gel and kieselgur supports induce the formation of a range of breakdown products (64). Since artifacts are mainly produced under ionizing conditions with polar solvents, the use of these supports should be restricted to nonionizing conditions. The separation of chlorophylls and its derivatives on thin layers of cellulose is easy to handle and no artifacts are formed. Cellulose TLC was used to study the chlorophyll degradation products from pea puree (203), bean leaves (204), and black tea leaves (205). Sucrose, another organic support, was successfully employed for the mono- and two-dimensional separation of chlorophylls (192,197). The use of RP-TLC has increased within the last years; oil-impregnated layers of sugar, kieselgur, and silica gel have been employed with solvent mixtures consisting of methanol and acetone (143). The separation of chlorophylls was improved by using HPTLC plates, which have a particle size

Table 12 Supports and Solvents Used for TLC of Chlorophylls

Supports	Solvents	Ref.
	Monodimensional TLC	
Cellulose	Petrol ether/acetone/1-propanol (90:10:0.45)	191
	Methanol/dichlormethane/water (100:18:20)	193
	Light petroleum/benzene/chloroform/acetone (50:35:10:5)	64
Kieselgur + silica gel	Petrol-ether/2-propanol/water (100:10:0.25)	194
Polyethylene	Acetone/water (90:10)	195
Silica gel	*tert.*-Butanol/n-pentane (1:9)	145
	Benzene/hexane/1-propanol (15:3:2)	140
	Petroleum ether/acetone/dithylamine (10:4:1)	63
Silica gel + 10% ammonium sulfate	Isooctane/acetone/diethyl ether (3:1:1)	196
Sucrose (icing sugar)	1% 2-propanol in light petroleum	197
Chemically bonded silica gel:		
RP-C8	Benzene/*n*-hexane/acetic acid (20:10:0.3)	198
RP-C18	Acetonitrile/methanol/methylene chloride/hexane (15:80:2.5:2.5)	69
	Methanol/acetone/water (30:20:2)	200
	Two-dimensional TLC	
Cellulose	1) Hexane/0.6% 1-propanol	192
	2) Hexane/1.2%–2.4% 1-propanol	
	1) Petrol ether/1-propanol (98.5:1.5)	199
	2) Chloroform/petrol ether/acetone (18:5:81.5:0.5)	199
Sucrose	1) Hexane	192
	2) Hexane/0.4%–1.6% 1-propanol	
	1) Petrol ether/1-propanol (99.2:0.8)	195
	2) Petrol ether/chloroform (80:20)	

of 5–10 μm instead of the 5- to 20-μm size of the plates previously used (186). With chemically bonded silica gel TLC plates (C18) the chlorophyll constituents of several green vegetables were successfully separated (69). The individual pigments were identified according to their R_F values on TLC plates, their absorption spectra, and by cochromatography with authentic standards. For quantitative determination the pigment bands were eluted with suitable solvents and their absorption, reflectance, or fluorescence spectra recorded (143). Since recovery from the plate is insufficient, direct measurements by densitometric, reflectance, or fluorometric methods are recommended. With a dual-wavelength scanner the limit of sensitivity is 1–10 ng per spot and it can be further increased to picograms per spot by fluorescence measurements (186). Compared with HPLC, the sensitivity and reproducibility of TLC is not as good and the duration of one analysis is generally longer. However, TLC is less expensive, is simple to handle, and does not require large quantities of solvents (197).

$\quad$ *g. High-Performance Liquid Chromatography (HPLC).* The first application of HPLC to the analysis of chlorophylls was by Evans et al. in 1975 (206). They separated pheophytins a and b on Corasil II with a mobile phase consisting of ethyl acetate and light petroleum (1:5). In comparison with other analytical techniques HPLC has significant advantages; therefore nowadays it is the method of choice for qualitative and quantitative analysis of chlorophylls in food. The use of HPLC offers several significant advantages over TLC, including better resolution, greater speed, and higher reproducibility and sensitivity (86,207). Furthermore to date no degradation of samples has been reported when separating chlorophylls with RP-HPLC. However, HPLC is handicapped by the high investments necessary and the high costs for each analysis.

$\quad$ Since chlorophylls have distinct spectroscopic properties, absorption or fluorescence spectrophotometry can be employed for their identification and quantification (186). Usually wavelengths between 430 and 440 nm and 645 and 660 nm, respectively (129,150), are used for spectrophotometric detection. With spectrophotometry detection limits of ~80 ng chlorophyll (156) and 1 ng chlorophyll (208), respectively, can be achieved. Fluorescence detection is more sensitive and specific than absorption measurement, therefore allowing the determination of picogram amounts of chlorophylls in the eluant (186). In general, the chlorophylls are excited at 430–440 nm and the emission is measured at 650–670 nm. Both NP and RP methods have been developed for the analysis of chlorophylls and their derivatives. Commonly used eluants for NP- and RP-HPLC as well as adsorbants and references are given in Table 13.

NORMAL PHASE HPLC (NP-HPLC). Various researchers have employed NP-HPLC for the separation of chlorophylls (208,209). Silica gel was the predominantly used absorbant; different solvent mixtures were used for the pigment elution (Table 13). As the separation mechanism of silica gel depends on polarity differences, this support is highly efficient to resolve intermediate to nonpolar pigments with large polarity differences (e.g., chlorophylls). However, it is less effective for polar pigments such as chlorophyllides (186). With NP-HPLC the simultaneous determination of chlorophylls and carotenoids in spinach (208,210) and algeal tissue was possible (156). However, it seems that RP-HPLC, especially when using C18 columns, is superior to NP-HPLC. Pigment degradation and long analysis times are two of the main drawbacks when silica is used as column-packing material (207). Furthermore NP separation requires an additional extraction step to eliminate water from the extract and to transfer the pigments from the acetone solution into hexane, since water in the extract causes problems with the chromatographic resolution and leads to varying retention times (210).

REVERSE PHASE HPLC (RP-HPLC). During the last years RP-HPLC with chemically bonded nonpolar stationary phases (C8 or C18) has been widely applied to the study of chlorophylls and their derivatives (211,212,213). With this system simple solvent mixtures such as methanol/water/

Table 13 HPLC Condition for Analysis of Chlorophylls

Column	Solvent	Detection (nm)	Pigments	Application	Ref.
Normal phase					
Silica gel SS-05	1% isopropanol in hexane	380	chl, pheo, car	Spinach	208
LiChrospher Si 100	0.8–2% 2-propanol in hexane	420, 430	chl	Clover	209
Partisil 10	Petroleum/acetone/dimethylsulfoxide/ diethylamine (25:23.25:1.25:0.25)	440	chl, pheo, chlid, car	Phytoplankton	156
Silica gel Lober 60-B	A: heptane/ether/hexane (60:20:20); B: heptane/ether/hexane (60:20:15)	425	chl, car	Tabacco	217
Silica gel	0.5% 2-methylpropanol in hexane	440	chl	Apple	134
Silica gel	A: 1.7% 2-propanol in hexane; B: 25% 2-propanol in hexane	658	chl, pheo, chlid, pheid	Spinach	210
Reverse phase					
μBondapak C₁₈ Porasil B	A: 80–100% methanol in water; B: 10–75% ether in methanol	440	chl, pheo, car	Spinach	216
μBondapak C₁₈	A: 75% methanol in water; B: ethyl acetate methanol/water (50:37.5:12.5)	654	chl, pheo	Spinach	129
μBondapak C₁₈	A: 1.5% TBAA/water/methanol (10:10:80); B: 20% acetone in methanol	654	chl, pheo	Rice grains	213
Hypersil ODS		440 F430/600	chl, pheid, car	Algae	218
Novapak C₁₈	Acetonitrile/chloroform/water (83:15:2)	440	chl, pheo, car	Bermudagrass	214
Chromsil C₁₈	Acetonitrile/methanol/ethyl acetate (53:40:7)	430	chl, pheo, chlid, car	Peam, celery	213
Radialpak RP	A: 80% methanol in water; B: ethyl acetate ethyl acetate/methanol/water (60:30:10)	660	chl, pheo, chlid, pheid	Canola seeds	150
Zorbax ODS C₁₈	A: 90% acetonitrile in water; B: ethylacetate	436, 658	chl, pheo	Spinach	136
Hypersil ODS		450	chl, pheid, chlid, car	Tea leaves	86
Hypersil ODS	A: 75% methanol in water; B: ethyl acetate	430	chl, pheo, car	Kiwi	12
Spherisorb ODS-2	A: methanol/acetonitrile/dichloromethane/ hexane (15:75:5:5); B: methanol/acetonitrile/dichlormethane/ hexane (15:40:22.5:22,5)	430, 460	chl, car	Beans	215
Spherisorb ODS-2	A: Water/ion pair reagent/methanol (1:1:8) B: acetone/methanol (1:1); ion pair reagent: 0.05 M tetrabutylammonium acetate–1 M ammonium acetate in water	430	chl, chlid, pheo, pheid, car	Olives, olive oil	207

Abbreviations: chl, chlorophylls; chlid, chlorophyllids; pheo, pheophytins; pheid, pheophorbids; car, carotenoids.

ethyl acetate (131,136) can be used (Table 13). The retention mechanism is mainly due to partitioning the components between phases rather than to nonpolar interactions between solute molecules and carbon functional groups in the stationary phase. The elution order of chlorophylls depends on their polarity: The more polar a pigment is, the earlier it is eluted (186). The retention time increases in the following order: chlorophyllides b < chlorophyllides a < pheophorbides < chlorophyll b < chlorophyll a < pheophytin b < pheophytin a (210). The separation of polar pigments such as chlorophyllides and pheophorbides with RP-HPLC is less effective, but the resolution can be improved by the use of ion suppression or ion pairing techniques (34,207). In addition, polyethylene, which has a similar separation mechanism as bonded silica gel, has been employed as an efficient alternative support in the RP mode (186). RP-HPLC has been successfully applied to the determination of chlorophylls and their derivatives in a wide variety of food-stuffs, e.g., spinach (129,136,211), kiwis (12), green olives (207), olive oils (34), green tea leaves (86), canola seeds (150), green peas (213), celery leaves (213), turf bermudagrass (214), and beans (215). In general, the duration of one analysis varies between 10 min (212) and 50 min (135). However, Eskins et al. (216) published a method that required 270 min. The coefficient of variation for the chromatographic determination lies within the range 0.6–1.8% (156). Recovery values of 98.3% and 98.7% can be obtained for chls. a and b, respectively (215). Numerous papers (86,213,216) deal with the simultaneous determination of chlorophylls and carotenoids by means of RP-HPLC (Table 13). This offers the possibility of studying the essential pigments in fruits and vegetables within less than 30 min (12,215).

3. Further Developments

h. Mass Spectrometry (MS). The characterization of chlorophylls by means of MS was hampered for a long time by their thermal instability. With the introduction of desorption-ionization MS it became possible to analyze the chlorophylls without prior removal of the magnesium. Various desorption-ionization methods, e.g., secondary ion mass spectrometry (SIMS), fast atom bombardement (FAB), field desorption (FD), plasma desorption (PD), laser desorption (LD), have been reported and MS is successfully used in chlorophyll research (219). For the characterization of chlorophylls the determination of the molecular ion is often enough. It is, however, commonly complemented by analysis of absorption and nuclear magnetic resonance (NMR) spectra. When chlorophyll mixtures are analyzed without prior separation only quantitative information can be obtained, due to different electron affinities, formation rates, and desorption efficiencies. Mass spectra of single pigments can be monitored on-line through a combination of HPLC and MS. For instance, chls. a and b and the corresponding pheophytins, chlorophyllides, pheophorbides, and pyropheophytins of fresh and heat-treated spinach tissue were separated with RP-HPLC and their molecular weights determined by positive ion fast atom bombardment mass spectroscopy (FAB-MS) (220). Plasma desorption mass spectroscopy (PD-MS) has been successfully used to analyze the formation of alteration products and to study the chlorophyll–protein complex (219).

D. Applications in Food Analysis

Chlorophylls are the most widely distributed natural pigments and occur in all parts of plants, especially leaves. In foodstuffs which originated from higher plants, chl. a is accompanied by chl. b; in algal food it is complemented by chls. c. The chlorophyll content of fruits and vegetables, which varies with genus, species, cultivars, ripening stage, and environmental factors, has been the object of numerous investigations (5,118). As the study of chlorophylls has attracted the attention of many chemists for a long time, various analytical methods have been developed.

References to some of the analytical methods applied to the determination of chlorophylls and their derivatives in some prominent fruits and vegetables are summarized in Table 14.

Frequently used methods are spectrophotometry, TLC, HPLC, and nondestructive methods. As representatives for all of the other applications in food analysis, three analytical procedures are described in detail.

1. Spectrophotometric Determination of Total Chlorophyll and Chl. a and Chl. b in Plants (235)

Weighed plant material (2–10 g) is disintegrated in a blender cup that contains ~0.1 g $CaCO_3$. After the tissue is thoroughly homogenized the extract is filtered through a funnel. The residue is washed with 85% acetone and if necessary a small amount of ether is used to remove the last traces of pigment. The filtrate is transferred into a volume flask and diluted to volume with 85% acetone. Aliquots of 25–50 ml of filtrate are pipetted into a separator containing ~50 ml ether. Subsequently water is added with care until it is apparent that the fat-soluble pigments have completely entered the ether layer. The water layer is drained and discarded and the separator containing the ether solution is placed in the upper rack of a support. Approximately 100 ml water is added to a second separator, which is placed in the same rack below the first separator. A scrubbing tube is set in place and then the ether solution is allowed to run, in small droplets, through the lower separator containing the water. The upper separator and the scrubbing tube are rinsed with some ether. Then the scrubbing tube is placed in the upper separator and the two separators are exchanged, so that the ether-containing separator is now placed in the upper rack. The water in the upper separator is drained and discarded. The ether solution is washed until all acetone is removed (5–10 washings). Then the ether solution is transferred to a 100-ml volume flask, diluted to volume, and mixed. Approximately 5 ml anhydrous (Na_2SO_4 is pipetted into a 60-ml reagent bottle which is filled with the pigment solution. When this solution is optically clear, aliquots are transferred to another dry bottle and the solution is diluted with enough dry ether to give absorption values of 0.2–0.8 at the chosen wavelength of 660.0 nm. The absorption values of the pigment-containing solution are recorded at 660.0 and 642.5 nm with a spectrophotometer. The chlorophyll concentrations are calculated as follows:

$$\text{total chlorophyll} = 7.12A_{660.0} + 16.8A_{642.5} \tag{9}$$
$$\text{chlorophyll a} = 9.93A_{660.0} - 0.777A_{642.5} \tag{10}$$
$$\text{chlorophyll b} = 17.6A_{642.5} - 2.81A_{660.0} \tag{11}$$

Table 14 Some References of Analytical Methods Used for Chlorophyll Analysis in Food

Food	CC and spectrophotometry	Nondestructive methods	TLC	HPLC
Apple	124	221	—	134
Beans	222	223	224, 225	215
Broccoli	226	227	69	69
Cabbage	228	—	69	69
Kiwi	229	170	131, 145	12, 67
Olives	63	—	63, 148	34, 207
Pea	—	—	230, 231	213
Spinach	161, 232, 233	—	196, 233	136, 211
Tea	138	—	234	86, 139

2. Determination of Chlorophylls in Kiwi Fruit by TLC (145)

One to two grams of sodium carbonate is added to 50 g kiwi fruit paste to adjust the pH value to 8–9. Subsequently, chilled acetone is added to the sample to produce an ~80% solution. This solution is mixed in a homogenizor and centrifuged at 40,000-g for 10 min (0–5°C). The residue is mixed with adequate amounts of 80% acetone and centrifuged. This operation is repeated until the green color is completely removed from the residue. The supernatants are collected and transferred to a volumetric flask. Fifty or 100 ml of the solution is transferred to a separatory funnel, and diethyl ether and cold deionized water are added. After vigorously shaking and subsequent standing the aqueous layer is discarded. The washing procedure is repeated 5–10 times to remove all the acetone. The diethyl ether layer is then dehydrated with analytical grade sodium sulfate anhydride. An aliquot is transferred to a beaker and evaporated under a steam of nitrogen. In order to obtain a good separation in the TLC, it is necessary to remove water from the chlorophyll solution completely. Therefore it is necessary to use diethyl ether anhydride and sodium sulfate anhydride to dissolve the chlorophyll residue and to remove the water completely. After the solvent has been evaporated, the residue is dissolved and diluted to 2.0 ml with chromatographic grade acetone. The chlorophyll solution (10 μl) is applied to a silica gel plate (e.g., precoated Kieselgel 60F-254, 10 × 5 cm) and developed immediately with *tert*-butanol/*n*-pentane (1:9). The separated spots of chls. a and b are determined with a dual-wavelength chromatoscanner, using 410 nm for chl. a and 470 nm for chl. b, and 700 nm as the reference wavelength. A TLC separation of chls. a and b and densitometric scans are shown in Fig. 9.

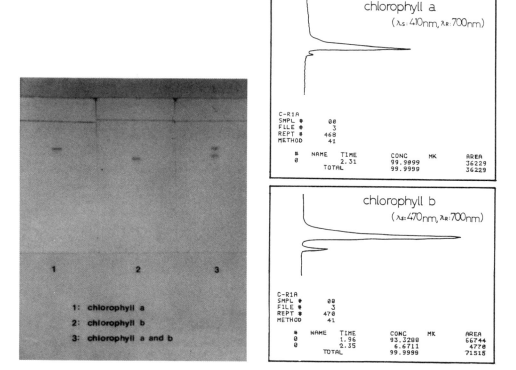

Fig. 9 TLC patterns and densitograms of chlorophyll a and b isolated from kiwi-fruits. (From Ref. 145.)

3. HPLC Analysis of Chlorophylls, Chlorophyll Derivatives, and Carotenoids in Spinach (135)

Pigments are extracted by grinding 2.5-g leaves of spinach in 20 ml cold acetone and 2.5 ml distilled water with a mortar and pestle. Then the homogenate is filtered and rewashed with 80% cold acetone until the residue is colorless. In low light the solution is brought to a final volume of 50 ml. Prior to HPLC analysis the solution is passed through a 0.22-μm filter. Pigments are separated by a C18 column (4.6 × 250 mm) using two solvents: (A) 80% methanol, 20% water, and (B) ethyl acetate in a gradient. A chromatogram of pigments extracted from spinach leaves is shown in Fig. 10.

III. ANTHOCYANINS

A. Introduction

1. Physical and Chemical Properties

Anthocyanins are the largest group of water-soluble pigments in nature. They are present in almost all higher plants and can be found in all parts of the plant, but their most obvious occurrence is in fruits and flowers where they are responsible for the attractive red, violet, and blue colors (236). The term *anthocyanin* is derived from the Greek words for 'flower' and 'blue' and was first introduced in 1835 by Marquart (5). The most important contributions to the isolation, purification, and identification of anthocyanins were made by Willstätter et al. (237). Anthocyanins are regarded as flavonoid compounds since they possess the characteristic $C_6C_3C_6$ carbon skeleton. Natural anthocyanic pigments (anthocyanins) are always glycosides which will separate into the aglycone forms (anthocyanidins) and sugars upon hydrolysis. The anthocyanidins are hydroxylated and methoxylated derivatives of 2-phenylbenzopyrylium or flavylium salts (238). Individual anthocyanins are characterized by the number of hydroxyl groups in the molecule, the degree of methylation of these hydroxyl groups, and the nature, number, and position of sugars attached to the molecule. So far 17 naturally occurring anthocyanidins are known, but only six of them are of widespread appearance and therefore commonly contribute to the pigmentation of plants (239). These common six anthocyanidins—cyanidin (cy), delphinidin (dp), malvidin (mv), pelargonidin

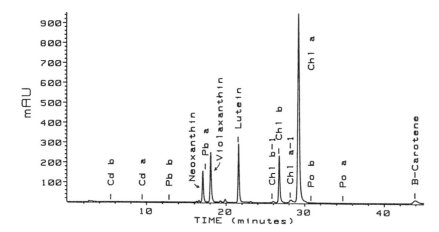

Fig. 10 Chromatogram of chlorophylls, chlorophyll derivatives, and carotenoids extracted from spinach leaves and analyzed by HPLC with a photodiode array detector. Cd, chlorophyllide; Pb, pheophorbide; Chl, chlorophyll; Po, pheophytin. (From Ref. 135.)

(pg), peonidin (pn), and petunidin (pt)—are all C3, C5, C7, and C4′ hydroxylated derivatives (Fig. 11). Since each anthocyanidin may be glycosylated and acylated by various sugars and acids at different positions, the number of anthocyanins is ~15–20 times higher than the number of anthocyanidins. The sugars most commonly bonded to anthocyanidins are glucose, galactose, rhamnose, and arabinose; in some cases also di- and trisaccharides. Glycosylation frequently occurs at C3, C5, and C7, but glycosylation at C3′, C4′, and C5′ has also been observed. The most common anthocyanidin glycosides are 3-monosides, 3-biosides, 3,5-diglycosides, and 3,7-diglycosides. The sugar residues may be further acylated with organic acids like *p*-coumaric acid, caffeic acid, ferulic acid, malic acid, or acetic acid (239).

In aqueous media anthocyanins undergo structural transformations that are pH-dependent. It has been found (236) that four anthocyanin forms exist in equilibrium: the red flavylium cation, the blue quinonoidal base, the colorless carbinol pseudobase, and the colorless chalcone. In strongly acidic media (pH values below 2) anthocyanins exist predominantly in their red-colored form as flavylium cations. At weakly acidic, neutral, and basic pH values the carbinol and the quinonoidal base forms dominate the flavylium cation, so that the color fades and shifts from red to blue (236). A blueing of the anthocyanic color is also achieved by the reaction of anthocyanins containing *ortho*-dihydroxy groups with $AlCl_3$ at pH values 2–4 (bathochromic shift) and through the formation of anthocyanin-flavanol complexes (copigmentation) (5).

As anthocyanins have a long chromophore of eight conjugated double bonds they appear strongly colored in an acid medium. Their absorption spectra are characterized by two separate bands—one in the visible region between 465 and 550 nm and the second, which is less intense, in the UV range between 270 and 280 nm (5). The absorption spectra in the visible region of pelargonidin, cyanidin, and delphinidin in 0.01% HCl-methanol are shown in Fig. 12. It must be noted that the absorption maxima and the corresponding intensities vary with the nature of the solvent; analytical measurements have therefore to be made under standardized conditions. The absorption characteristics of anthocyanin solutions are markedly affected by the structure of the pigment. For instance, an additional hydroxyl group will cause a relatively large bathochromic

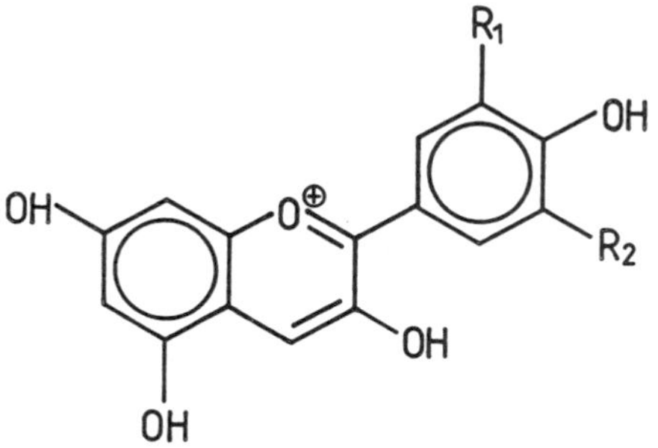

Fig. 11 Formula of anthocyanidins (flavylium cations):

Name	R_1	R_2
Cyanidin (cy)	OH	H
Delphinidin (dp)	OH	OH
Malvidin (mv)	OCH_3	OCH_3
Pelargonidin (pg)	H	H
Peonidin (pn)	OCH_3	H
Petunidin (pt)	OCH_3	OH

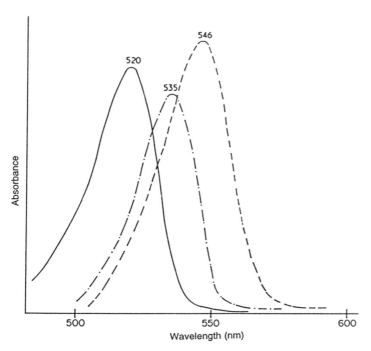

Fig. 12 Absorption spectra in the visible region of (—) pelargonidin, (– – – –) cyanidin, and (– – – –) delphinidin in 0.01% Hcl/methanol. (From Ref. 5.)

shift to longer wavelengths (pelargonidin: λ_{max} = 520 nm, delphinidin: λ_{max} =546 nm) and a change in color from scarlet red (pg) to blue–mauve (dp). In contrast, methylation has little or negligible effect, while glycosylation at C3 produces an absorption maximum shorter by 15 nm. Acylation with cinnamic acids causes a characteristic shoulder in the UV region at 310–335 nm (5).

As all natural anthocyanins suffer from inherent instability, they may be degraded to form colorless or brown-colored, often insoluble products. The major factors affecting their stability are pH, temperature, light, and oxygen, but other factors like enzymes, metal ions, and so forth can also increase the anthocyanin destruction. In general, anthocyanins are most stable in acidic, oxygen-free media under cold and dark conditions.

2. Properties in Food

Many fruits and vegetables owe their attractive blue, violet, or red colors to the presence of anthocyanins. In 70 species of 33 families of angiosperms anthocyanins are found in membrane-bound anthocyanoplasts located within the main cell vacuole (239). Especially the fruits of the Vitaceae (grapes) and the Rosaceae (e.g., apple, cherry, plum, peach, strawberry, blackberry, raspberry) are characterized by their anthocyanin content. Other families containing anthocyanic pigments include the Ericaceae (blueberry, cranberry), Saxifragaceae (red and black currant), Caprifoliaceae (elderberry), Solanaceae (e.g., tamarillo, huckleberry), Gramineae (e.g., maize, barley, rice), Anacardiaceae (e.g., mango), Sapindaceae (e.g., lychee) Rutaceae (e.g., blood orange), Malvaceae (e.g., roselle), Fabaceae (e.g., beancowpea, pea, soybean), Apiaceae (e.g., carrot), Brassicaceae (e.g., cabbage, radish), Solanaceae (tomato, eggplant), Convolvulaceae (e.g., sweet potato), Berberidaceae (e.g., barberry), and others. Of course, many of the mentioned food plants also contain other pigments like carotenoids and chlorophylls. A large number of over 200 different anthocyanins can be found in these plants, but since only six anthocyanidins are common this variation is mainly due to the glycosidic type (5). The color of most fruits is caused by a complex mixture of anthocyanins. For instance, Sapers et al. (240) found as many as 16 anthocyanins in highbush blueberry cultivars; Nagel and Wulf (241) separated 21 anthocyanin components in skins of *V. Vinifera* cv. *Cabernet Sauvignon*. However, simple patterns of only one or two anthocyanins can also be detected, e.g., in passion fruit and peaches. Cyanidin is the most common anthocyanidin in fruits (occurrence in 90% of the fruits examined), followed by delphinidin (35%), peonidin (30%), pelargonidin (20%), and petunidin and malvidin (each 15%) (238). These anthocyanins are also predominant in vegetables and grains (239). Glycosylation at the C3 position is very widespread. Diglycosides are formed either by linkage of sugars to two different hydroxyls (generally C3 and C5 or C7) or by linkage of two sugars to only one position (C3). Triglycosides are not very numerous in fruits. Acylated anthocyanins are fairly often found in fruits [e.g., grapes (242) eggplant and garden huckleberry], and in vegetables [e.g., cabbage, radish, mungbean, celery, chicory, potato]. However, they occur rarely in grains (maize) (239). The total anthocyanin content of different plant tissues may display considerable differences. For example, in fruits it ranged from 0.25 mg/100 g (pear) to 760 mg/100 g (*Vaccinium elliotti*), red currant cultivars varied from 11.9 to 18.6 mg/100 g, cranberry cultivars varied from 46 to 172 mg/100 g (238). The distribution and contents of anthocyanic pigments in fruits, vegetables, and grains have been comprehensively summarized by Mazza and Miniati (239), Macheix et al. (238), Gross (5), and Timberlake and Bridle (243), respectively. Examples showing the total anthocyanin content and their distribution in some fruits and vegetables are given in Table 15.

3. Arguments for Anthocyanin Analysis: Regulations

Color is an important attribute of most fruits and vegetables as well as their products. As the red, violet, and blue colors of many plants are due to the presence of anthocyanins, it is of interest to analyze them. Both quantitative and qualitative aspects are responsible for the color quality of the foodstuff. With the object of improving the chromatic characteristics of fruits the evolution of anthocyanins in fruits and influencing factors (e.g., temperature, water) have been studied extensively (245,246). Unfortunately, anthocyanins degrade easily according to various reaction mechanisms, as has been reviewed by Markakis (247). During processing and storage of anthocyanin-containing food products like juice, wine, jam, syrups, and dried and frozen fruits, the original color deteriorates and changes from natural red or purple to reddish brown (248). Subsequently, problems often occur in fruit wines and juices, caused by the formation of haze and sediments (249). Numerous investigations have already shown that degradation and polymerization are influenced by oxygen, ascorbic acid, light, pH, and temperature. Generally the anthocyanin degradation follows a first-order reaction kinetics. To minimize these undesirable modifications of sensory properties further studies concerning the stability of anthocyanins should be carried out. However, in red and port wines a distinctive conversion of the original fruity red color to a well-developed, aged red wine color is desired (250).

Following the current trend away from synthetic colors, the use of anthocyanins as natural food colorant will increase steadily (251). From the medical point of view anthocyanic pigments are regarded as safe since they have been consumed by humans and animals for countless generations without any apparent adverse effects to health. Furthermore various beneficial therapeutic properties, particularly in ophthalmology and for some blood circulation disorders, are attributed to anthocyanins (252). However, their technical applicability is hampered by their inherent instability and pH dependence. Some of these disadvantages may be overcome by a recently discovered group

Table 15 Total Anthocyanin Content and Major Anthocyanins in Some Fruits (238, 239)

	Anthocyanin content (mg/kg)	% of total anthocyanin distribution																			
		cy3gl	cy3ar	cy3ga	cy3rt	cy3sm	cy3sp	cy3 (2xyru)	cy3 (2glru)	dp3ru	dp3gl	dp3ga	dp3ar	pg3gl	pt3gl	mv3gl	mv3ga	m3ar	pn3ga	pn3ar	pn3gl
Apple (peel)	100–21,600	1–5	4–10	85–94	—	—	—	—	—	—	—	—	—	—	—	—	—	—	—	—	—
Billberry	4600	9	—	19	—	—	—	—	—	—	9	19	—	—	—	tr	—	—	—	—	—
Blackberry	820–1800	69–100	—	—	0–32	—	1	—	—	30	13	—	—	—	—	22	—	—	—	—	—
Black currant	2500	17	—	—	35	—	—	—	—	—	—	—	—	—	—	—	—	—	—	—	—
Blueberry [244]	1600–5030	1–6	—	1–20	—	—	—	—	—	6–19	1–9	6–20	4–15	—	0.5–9	1–21	8–32	5–14	—	—	—
Cherry (sweet)	3500–4500	+	—	—	+	—	—	—	—	—	—	—	—	—	—	—	—	—	—	+	+
Cherry (tart)	288	3–19	—	—	11–27	—	1–16	—	25–77	—	—	—	—	—	—	—	—	—	—	—	—
Cranberry	460–1720	—	13–25	16–25	—	—	—	—	—	—	—	—	—	—	—	—	—	—	23–39	11–21	—
Elderberry	4480	33–66	—	—	—	32–55	—	—	—	—	—	—	—	—	—	—	—	—	—	—	—
Grape (blue)	80–3880	1–6	—	—	—	—	—	—	—	—	5–17	—	—	—	6–12	36–43	—	—	—	—	—
Plum	19–53	37	—	—	45	—	—	—	—	—	—	—	—	—	—	—	—	—	—	—	5–13
Raspberry (red)	230–590	11–45	—	—	5–32	—	20–72	—	0–38	—	—	—	—	—	—	—	—	—	—	—	—
Red currant	119–186	2–10	—	—	8–17	10–31	4–9	28–73	0–28	—	—	—	—	—	—	—	—	—	—	—	—
Strawberry	127	0–50	—	—	—	—	—	—	—	—	—	—	—	50–100	—	—	—	—	—	—	—

Abbreviations: cy, cyanidin; dp, delphinidin; mv, malvidin; pg, pelargonidin; pe, peonidin; pt, petunidin; gl, glucosid; ar, arabinosid; ga, galactosid; xy, xylosid; rt, rutinosid; sm, sambubiosid; sp, sophorosid; xyru, xylosylrutinosid; glru, glucosylrutinosid; tr, traces.
++, major pigment; + minor pigment; tr, traces.

of acylated B-ring-substituted anthocyanins (253). The use of anthocyanins of natural origin for food and beverages is widely permitted within the European Community (E163) and the USA. Numerous plants and plant parts (e.g., grapes, roselle, bilberry) have been suggested as potential commercial sources of anthocyanidins, but only those where the pigment is a byproduct of another valuable product (e.g., juice, wine) are of economical interest (116).

As the individual anthocyanin composition is distinctive for any given plant, the anthocyanin analysis is very useful in distinguishing between species. The chemotaxonomic differentiation is commonly based on qualitative differences (254), but within one cultivar (e.g., grapes) even varieties can be discriminated by quantitative differences (255). The characteristic differences in anthocyanin patterns have also been successfully applied to the detection of adulterations in products of cranberries (256), black currants, (256), blackberries (257), and grapes (258).

B. Sample Preparation

During sample preparation and analysis some precautions have to be taken since the anthocyanins are not stable in neutral or alkaline solutions under high light intensities and at high temperatures.

1. Extraction

As anthocyanins are decomposed in neutral or alkaline solutions, extraction procedures generally involve the use of diluted acids in polar solvents. The most common extraction procedure involves maceration or soaking of plant material in methanol containing a small amount (e.g., 1%) of HCl (259). Acidification with strong acids like HCl serves to maintain a low pH. It may, however, alter the native form of complex pigments by breaking associations with metals, copigments, and proteins (260). To obtain nature-like anthocyanins the use of neutral solvents (e.g., 60% methanol, ethylene glycol, n-butanol, acetone) and mild organic acids (e.g., formic acid, acetic acid) has been recommended (116).

2. Purification

For the purification of crude anthocyanin extracts various methods have been successfully applied. In the past the sample cleanup has been carried out through precipitation with basic lead acetate (261), paper chromatography (262), and solvent–solvent extraction (263). These methods are very time consuming and cause considerable breakdown of anthocyanins. More recently, solid phase extraction with insoluble polyvinylpyrrolidone (PVP) (264), Sephadex G-25 (265), Sephadex LH-20 (266), polyamide (267), ion exchange resins (268), acid alumina (269), and octadecylsilane (270) have been used. The simple and rapid procedures are based on the principle that anthocyanins are bound to the adsorbants while interfering substances (e.g., sugars, acids) are washed from the columns or cartridges. In general, anthocyanins are subsequently eluted from the adsorbant with acidified methanol (271).

C. Analysis–Detection–Separation Methods

For the determination of anthocyanins various methods including spectrophotometry, TLC, and HPLC have been described in the literature (5,262). Depending on the object of the analysis quantitative methods are often divided into three groups: (a) determination of total anthocyanin content in systems containing little or no interfering substances; (b) determination of total anthocyanin content in systems that contain interfering material; and (c) determination of individual anthocyanins (116,262).

1. Classical Methods

   a. Spectrophotometric Methods. As anthocyanins have an absorption maximum between 500 and 535 nm their photometric determination is based on the extinction of light energy, which is defined by the Lambert–Beer law. However, self-association and copigmentation may cause severe deviations in the Lambert–Beer law and may lead to inaccurate estimations. It may be necessary to dilute the samples considerably with acidic buffers (e.g., pH 1.0 buffer solution for samples containing predominantly anthocyanin 3-glucoside). Fresh plant extracts or juices usually contain only few interfering substances and the total anthocyanin content may therefore be determined simply by measuring the sample at a single wavelength. In samples containing a mixture of anthocyanins absolute concentrations may be estimated by the use of weighted average absorptivities and absorbance measurements at weighted average wavelengths (272).

   The single-pH method should not be used for samples that have been heat-treated, stored, or processed because they contain interfering, brown degradation products from sugar or pigment breakdown (116). For such samples improved anthocyanin determination methods have been developed, e.g., differential or subtractive absorption methods, which minimize the interference caused by degradation products. The differential methods take advantage of the fact that the spectral characteristics of anthocyanins change with the pH while those of interfering brown products are stable (273). It can be assumed that the difference in absorbance between two pH values at the same wavelength is a measure of the anthocyanin concentration because the absorbance caused by the interfering material is eliminated through substraction (116). The use of several different pH values has been described in literature (274), but measurement at pH 1.0 and 4.5 is the most common (275,276). After the pH adjustment a period of at least 1 hr is required to allow the different anthocyanin forms to equilibrate. Furthermore with the differential method an anthocyanin degradation index, defined as the ratio of total anthocyanin determined at a single pH value to that determined by the differential method, can be calculated (116). The simple ratio of the absorbance of red pigments (measured in the 500- to 550-nm region) to that of brown pigments (measured in the 400- to 440-nm region) also provides a good index of color deterioration, e.g., in red wine the color hue is defined by the ratio A_{420nm}/A_{520nm} (277). For the subtractive method the total absorbance of the sample at the absorption maximum is first measured. After that the anthocyanins are bleached and the blank reading, which is caused solely by interfering material, is determined. The most common bleaching agents used are sodium sulfite (277,278) and hydrogen peroxide (279). Unfortunately, these bleaching agents also cause a decrease in the absorbance of interfering substances, which may give rise to erroneously high values of total anthocyanin concentration (116).

   With all of these methods the anthocyanin concentration (C) may be calculated by the following equation:

$$C(\text{mg/L}) = \frac{E \times \text{MW} \times 10 \times V}{e}$$

where E is the extinction at an appropriate wavelength or the difference of extinctions at two different wavelengths or after bleaching, MW is the molecular weigth of the pigment, V is the dilution factor, and e is the molar extinction coefficient. Molecular extinction coefficients (the extinction of an anthocyanin solution containing 1 mol/L) of some anthocyanins are listed in Table 16. Methods for determining color density, color hue, polymeric color, total anthocyanin content, degree of ionization, and monomeric anthocyanin color calculated from only a few absorbance readings are given by Wrolstad (280) and Somers and Evans (277), respectively. Due to its simplicity, speed, and low cost, photometry is often applied to anthocyanin analysis. However, it lacks high selectivity and sensitivity.

Table 16 Molecular Extinction Coefficients (e) of Some Anthocyanin Pigments

Pigment	MW[a]	e	Solvent	Wavelength (nm)	Ref.
cy 3-glu	485	25,740	0.1 N HCl	520	281
cy 3-gal	485	46,200	0.1 N HCl in EtOH (15:85)	535	282
cy 3-diglu	647	30,175	0.1 N HCl	520	283
dp 3-glu	501	29,000	1 N HCl in MeOH	543	284
pl 3-glu	469	27,300	1 N HCl in water	496	285
pn 3-glu	499	11,300	0.1 N HCl in MeOH	536	286
pn 3-arab	499	46,100	0.1 N HCl in EtOH (15:85)	532	282
pn 3.5-diglu	661	36,654	0.1 N HCl	520	283
pt 3-glu	515	12,900	0.1% HCl in MeOH	546	286
mv 3-glu	529	28,000	0.1 N HCl	520	281
mv 3.5-diglu	691	37,700	0.1 N HCl	520	281

Source: From Refs. 262 and 272.
[a]Calculated as the chloride salt.

b. Nondestructive Methods. Nondestructive methods like tristimulus colorimetry, light reflectance, or transmittance measurements allow anthocyanin determination in intact samples (e.g., fruits, juices) without any prior chemical manipulation. The principle of tristimulus colorimetry was discussed in the section on chlorophylls and an excellent review was given by Francis and Clydesdale (287). As the three chromatic hues—red, blue, and green—correspond to the three types of receptor molecules in human eyes, tristimulus colorimetry is a good means to describe real, visual colors. Tristimulus measurement has therefore been widely applied to studies concerning stability, deterioration, and acceptability of anthocyanin color in foodstuffs (288), e.g., in grape wine (289,290), black currant syrup (248), strawberry syrup (291), raspberry juice and wine (292), and frozen sour cherries (293). The tristimulus parameter $a*$ (redness) and the total anthocyanin content were correlated, e.g., in young port wine $r = -0.906$ (294) and in sour cherries $r = -0.55$ to -0.77 (293). For apples cv. *Red Delicious* the coefficient of determination (R^2) for the relationship between anthocyanin content [photometric method: $(A_{530} - A_{620}) - 0.1 (A_{650} - A_{620})$; $e = 46,200$] and the chromatic ratio ($a*/b*$) was calculated with 0.59 (295). Thus tristimulus colorimetry can be used with certain reservations to determine the total anthocyanin content of fruits. This nondestructive determination has also been employed to follow changes in individual fruits (e.g., apples) during the growing season or to monitor the effects of various treatments.

Analogous to tristimulus colorimetry a high correlation between reflectance or transmittance measurements and the anthocyanin content of fruits was also found (5). While reflectance measurement has mainly been used to monitor the color development in different fruits such as cherries (296) and grapes (297), transmittance measurement has been successfully adapted for the sorting of fruits into specified maturity categories (298).

c. Electrophoresis. Within special pH ranges anthocyanins exhibit an ionic structure, which gives them mobility in an electric field (271). At first electrophoretic investigations were made on filter paper using different electrolytes like acetate buffers (299), 0.03 M phosphoric acid (300), borate buffers (301), and 0.1 M citric acid at pH 2.0 (302). When Markakis (300) tested different buffer systems at various pH values, he observed that the isoelectric points of anthocyanins were very close to the pH at which these pigments displayed minimum color. Using cellulose acetate film in 0.05 M acetate buffer at pH 4.2, the anthocyanin pigments of various flowers could be separated into several, mainly cationic spots (303). More recently, Tsuda and Fukuba (304), who employed a Triton X-100/AlCl₃–containing electrolyte system, reported successful separations of anthocyanins. However, electrophoresis is seldom used in anthocyanin analysis because it carries various disadvantages such as bad selectivity, low sensitivity, and difficult quantification.

d. Column Chromatography (CC). CC is a well-established method for preparative isolation, sample purification, and, when connected with a photometric unit, for direct estimation of individual anthocyanins. The most common column support materials are cation exchange resins, aluminum oxide, polyamide, and PVP (Polyclar AT). At the beginning cation exchange chromatography was tried with little success because the weak cation exchanger (e.g., Amberlite CG-50) did not fix some anthocyanins while strong cation exchangers (e.g., Amberlite IRA-120) bound some anthocyanins irreversibly, which caused incomplete recovery and variable quantitative results (305). With the goal of improving the recovery very polar solvents such as HCl-water, HCl-methanol, formic acid, and dimethylsulfoxide were used, which in turn caused decomposition of pigments too (5). However, more recently oenocyanin anthocyanins were separated on a sulfoxyethylcellulose cation exchange resin in six fractions (306). Ion exchange resins were also successfully used to remove metal traces from anthocyanins and achieve a preliminary separation (307). The separation of red and black raspberry anthocyanins was obtained with a cellulose powder resin and *n*-butanol/acetic acid/water (5:1:4) as the solvent system (308).

PVP has been widely used for the isolation and purification of individual anthocyanins from plant materials such as strawberries, rhubarb, and raspberries (309). Using insoluble PVP as the stationary phase, Hrazdina (310) successfully separated the 3,5-diglucosides or peonidin, petunidin, cyanidin, delphinidin, and malvidin of grapes.

With polyamide CC it was possible to resolve the complex anthocyanin mixtures of red radish root and red cabbage leaf as well as detect and enrich minor anthocyanins occurring in extremely low amounts (311). Gel filtration on Sephadex with aqueous acetone as eluant was successfully applied to separate condensed and monomeric wine anthocyanins (312). Recently the purification and concentration of anthocyanins was achieved on porous polymers like formophenolic or polyacrylic resins (312). Since the disadvantages of traditional CC (e.g., limited resolution, long analysis time, high chemical consumption) are overcome by HPLC, the quantification of anthocyanin pigments is nowadays seldom done with CC.

e. Gas Chromatography (GC). As anthocyanins exhibit low volatility they have to be derivatized prior to GC analysis. Using trimethylchlorosilane (TMCS) and hexamethyldisilazane (HMDS) for the derivatization, Bombardelli et al. (314) produced volatile nitrogen-containing derivatives which, after injection into the GC system, were transformed into quinoline-like compounds. The GC separation with a column of coiled silanized tubes containing 0.5% OV-101 liquid phase loaded on silanized Chromosorb W HP (100–120 mesh) gave well-separated peaks. The subsequently recorded mass spectra coincided with the corresponding anthocyanin derivatives (315). Despite these excellent results GC analysis is hampered by the required derivatization, so that it is rarely used for anthocyanin analysis.

f. Paper Chromatography (PC). The use of PC was introduced to the study of anthocyanins by Bate-Smith (316) in 1948. He successfully separated anthocyanins of various flowers on filter paper developed with butyl alcohol/acetic acid/water (BAW, 4:1:5) and published R_f values of numerous anthocyanins. Since then PC has been extensively used for the separation, purification, and identification of anthocyanins found in many plant species (317). In particular, Harborne (318) identified more than 100 different anthocyanins and tabulated R_f values in different solvent systems for a great number of pigments. Thus the anthocyanin composition of various fruits and vegetables like grapes (319), strawberries, rhubarb, radish, onions (320), low-bush blueberries and cranberries (321) was elucidated. Usually the separation is carried out on Whatman No. 1 or 3 filter paper. The choice of developing solvent depends on the content of interfering substances (e.g., sugars, degraded pigments) in the sample. If the sample contains relatively little sugar or degradation products, the paper should first be developed in BAW. If the content of interfering substances is high, 1% HCl in water should be used as the first solvent (262). The developing solvents can be divided into two main groups—aqueous and oily—and the compositions of some common solvent systems are listed in Table 17. Generally the aqueous solvents require shorter developing times (4–8 hr) than the oily solvents (14–18 hr), which are therefore more conveniently run overnight (262). Chromatograms are normally developed with the solvent descending the paper, but it has been claimed that more reliable results can be obtained with BAW ascending the paper. After the chromatographic separation, the developed chromatograms should be dried for ~20 min; subsequently, the different anthocyanin spots can be visualized under daylight or UV light (318). The identification of individual anthocyanins is based on their chromatographic and spectrophotometric properties. Further information can be gained through acid or alkaline hydrolysis of the sample prior to PC (5). Ribéreau-Gayon and Ribéreau-Gayon (319) applied PC to discriminate american grape varieties (e.g., V. riparia, V. rupestris), which contain fluorescing

Table 17 Selected Solvents for the Chromatography of Anthocyanins

Abbreviation	Composition	Ref.
BAW	n-Butanol/acetic acid/water (4:1:5), upper phase, aged 4 days for R_f data	316
BAcW	n-Butanol/glacial acetic acid/water (6:1:2), upper phase	322
BEW	n-Butanol/95% ethanol/water (4:2:2.2)	262
BuHCl	n-Butanol/2 N HCl (1:1), upper phase, paper equilibrated 24 hr after spotting and before running, in tank containing aqueous phase of BuHCL mixture	318
BuN	n-Butanol/2% ammonium hydroxides (1:1), paper-dipped in saturated solution of boric acid in water prior to use	262
BFW	n-Butanol/formic acid/water (100:25:60)	320
BfFW	butylformate/formic acid/water (7:2:1)	327
BBPW	n-Butanol/benzene/pyridine/water (5:1:3:3)	262
BBFW	n-Butanol/benzene/formic acid/water (100:19:10:25)	320
Forestal	Glacial acetic acid/conc. HCl/water (30:3:10)	319
Formic	Formic acid/conc. HCl/water (5:2:3)	262
1% HCl	Conc. HCl-water (1:97)	318
HOAc-HCl	Water/glacial acetic acid/12 N HCl (82:15:3)	318
30% HOAc	Glacial acetic acid/water (30:70)	318
50% HOAc	Glacial acetic acid/water (50:50)	262
MAW	Methanol/acetic acid/water (90:5:5)	262
PAW	n-Pentyl alcohol/acetic acid/water (2:1:1)	327
PhOH	Phenol/water (4:1)	318
PrN	n-Propanol/conc. ammonium hydroxide (7:3)	262
PrHCl	Isopropanol/2 N HCl (1:1)	318
TAW	Toluene/glacial acetic acid/water (4:1:5)	262
WHCIP	Water/HCl/propionic acid	323

anthocyanin 3,5-diglucosides, and European *(V. vinifera)* grape varieties, which lack those fluorescing pigments. Fitelson (322) detected adulterations in colored fruit juices and grape juices by PC with *n*-butanol/acetic acid/water (6:1:2) as solvent. Occasionally, two-dimensional PC is employed, but the anthocyanins tend to fade and form large, diffuse spots, so that this method has not been often used (318). As PC has some flaws (e.g., resolution, analysis time, sensitivity) it is increasingly substituted by more sophisticated techniques like TLC and HPLC.

2. Recent Methods

g.   *Thin-Layer Chromatography (TLC).*   TLC has been successfully applied to the separation, isolation, and detection of anthocyanins (5,271). Compared to PC it offers the advantages of shorter analysis time, smaller sample amounts, less diffusion, and better reproducible quantification (317). Inspired by the effectiveness of PC, cellulose and microcellulose thin layers have been extensively used for numerous applications in anthocyanin analysis such as the separation of raspberry anthocyanins (323), the isolation of grape anthocyanins (324), and the detection of adulterations in black raspberry juice (325). Most of the researchers have worked with MN 300 cellulose powder, which was heated to 105°C for 5 min before use. As the anthocyanin composition of grapes is very complex, numerous attempts have been made to improve their chromatographic resolution. Good results have been achieved with mixed thin layers consisting of silica gel and cellulose (1:1) and an additional improvement was reached by two-dimensional or sequential development of the thin-layer plates (326). Figure 13 shows a good separation of anthocyanins from grapes of *V. rotundifolia,* which was achieved by two-dimensional and sequential development, respectively, of mixed silica gel/cellulose (1:1) TLC plates (327). More recently, RP-TLC (C18) was successfully used for the separation of grape anthocyanins and for the detection of various prohibited colorants (328). R_x values of 17 grape anthocyanins and of some natural and artificial colorants were listed. In general, the chromatographic separation is based on the polarity of the anthocyanins, which increases with the number of hydroxyls in the molecule (e.g., delphinidin is more polar than cyanidin). Furthermore the polarity depends on the nature and number of the sugar substituents, with the acylated monoglucosides being less polar than the nonacylated, of which the 3,5-diglucosides are the most polar anthocyanins (5). Table 18 shows R_f (R_x) values of selected anthocyanins separated on three different TLC supports (cellulose; cellulose MN 300 + silica gel G; SIL-RP-C18) and different solvent systems. However, as R_f

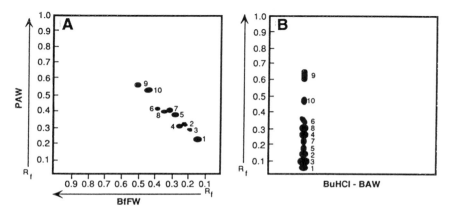

Fig. 13   Separation of grape anthocyanins *(V. rotundifolia)*: (A) Two-dimensional TLC, 1. solvent: PAW, 2. solvent BfFW. (B) Sequential TLC, BuHCl followed by BAW. (From Ref. 327.)

Table 18 R_f (R_x) Values of Selected Anthocyanins with Five Different TLC Systems[a]

Anthocyanin	A	B	C	D	E[b]
cy 3-glu	0.23	0.09	—	—	1.64
cy 3,5-diglu	0.13	0.26	—	—	2.46
cy 3-glu-acetat	—	—	0.44	0.53	0.41
cy 3-rut	0.25	0.26	—	—	—
pl 3-glu	0.29	0.25	—	—	—
pl 3-rut	0.35	0.41	—	—	—
pn 3-glu	—	—	0.42	0.52	1.19
pn 3.5-diglu	—	—	0.17	0.48	2.03
pn 3-glu-acetat	—	—	—	—	0.23
mv 3-glu	—	—	0.28	0.55	1.00
mv 3.5-diglu	—	—	0.13	0.46	1.78
mv 3-glu-acetat	—	—	—	—	0.20
mv 3-glu-cumarat	—	—	0.58	0.73	0.04
pt 3-glu	—	—	—	—	1.31
pt 3.5-diglu	—	—	—	—	2.19
pt 3-glu-acetat	—	—	0.44	0.53	0.26
dp 3-glu	—	—	—	—	1.80
dp 3,5-diglu	—	—	—	—	2.70
dp 3-glu-acetat	—	—	0.44	0.53	0.48
dp 3-glu-cumarat	—	—	0.57	0.57	0.10

[a](A) Cellulose-BAW (323). (B) Cellulose-WHCIP (323). (C) Cellulose MN 300/silica gel G (1:1)-BfFW (329). (D) Cellulose MN 300/silica gel G (1:1)-PAW (329). (E) SIL-RP-C18. Phosphate buffer, methanol, sequential.
[b] = R_x (mv 3-glu) (328).

values in TLC are not as reliable as with PC, the application of reference pigments and the calculation of R_x values should be preferred (271). By measuring the photometric densities of each chromatographic spot the quantities of every anthocyanin can be evaluated (326). A detection limit of 10 mg/L malvidin 3,5-diglucoside was determined with RP-TLC (328).

 h. High-Performance Liquid Chromatography (HPLC). Since the work of Manley and Shubiak (330), who were the first to apply HPLC to anthocyanin analysis, numerous HPLC techniques have been developed for the separation and quantification of anthocyanins and anthocyanidins. Nowadays HPLC has become the method of choice because it offers the advantage of being a rapid, sensitive, and quantitative method. The quantitation of chromatograms is, however, hampered by the lack of availability of pure anthocyanin standards. Although in the last years some anthocyanin standards have been made commercially available, in many cases the preparative isolation of pure anthocyanins is still necessary. On the other hand, it is a common practice that only one anthocyanin standard is used (e.g., malvidin 3-glucoside) and that all other anthocyanin peaks are quantified against this standard. As there are considerable differences in the quantitative absorbance values and the absorption maxima, this practice will give slightly incorrect results (331).

 HPLC separation of anthocyanins has been carried out almost exclusively on C18 columns (271). Occasionally silica columns bonded with octyl (C8) and hexyl (C6) have also been used (332). The average particle diameter of HPLC packings is typically between 3 and 10 μm, with 5 μm being the most frequently used (271). In RP-HPLC the anthocyanins elute according to their polarities (delphinidin < cyanidin < petunidin < pelargonidin < peonidin < malvidin) (271). The elution of pigments is usually achieved by gradient elution, but occasionally isocratic elution is also sufficient, e.g., with fruits which have a relatively simple anthocyanin pattern such as mango (238) or for the separation of anthocyanidin mixtures (333). Solvent systems for the HPLC analysis of anthocyanins always include an acid to reach low pH values (<2.0), which ensure that most of the anthocyanins are present in the red flavylium cation form. Formic acid (up to 10%) is most widely used. It is, however, quite corrosive (334), and perchloric acid (335) and trifluoroacetic acid (336) are equally aggressive. Commonly used alternatives are acetic acid [15%, (337)], phosphoric acid [3–4%, (338)], and phosphate buffer (270). It should, however, be considered that extensive use of solvents more acidic than pH 2 could result in poor reproducibility and short column lifetime due to loss of bonded phases from the surface of the silica stationary phase support. The employment of nonsilica, polymeric columns, which are stable from pH 1 to 13, allows the use of strong acidic solvents, thus causing sharper peaks because the anthocyanins are almost completely in their flavylium cation form (271). Commonly methanol (339) or acetonitrile (340) is used as organic modifier, but some researchers substituted acetone for methanol (341) and obtained similar separations.

The eluting anthocyanins are monitored with UV-Vis or photodiode array detection systems (271). Usually the detection wavelength is set in the range between 510 and 545 nm. Depending on the sensitivity of the detector system and the anthocyanin pigment, a detection limit of 0.5–1.0 mg/L anthocyanin can be assumed generally. The identification of anthocyanins is carried out by cochromatography of known, available standards or based on retention times, which are commonly compared with data from literature. By scanning the absorption spectra on-line, photodiode array detection makes an additional structural characterization of pigments possible (270). Chemically anthocyanins can be identified by acid hydrolysis and subsequent determination of anthocyanidin and sugar moiety or by mild alkaline hydrolysis and subsequent determination of the acyl component (262). Furthermore the elucidation of structure, especially for acylated anthocyanins, can be confirmed by fast atom bombardment mass spectrometry and NMR spectroscopy (342).

When samples contain high quantities of interfering material (e.g., sugars, colloids), a sample cleanup procedure is recommended. A sample preparation with solid phase extraction cartridges (C18) was found to be quite useful and effective (270,340). Prior to HPLC analysis all samples should be filtered through a 0.45-μm or smaller membrane filter.

During the last two decades RP-HPLC analysis has been successfully used to study anthocyanins in numerous foodstuffs. Because of the great number of publications only an incomplete short excerpt from those many applications and corresponding HPLC methods can be given in Table 19. To demonstrate the potential of this technique an RP-HPLC separation of black currant juice is shown in Fig. 14. Despite the high costs of acquisition and maintenance, RP-HPLC will likely remain the best qualified method for anthocyanin analysis in the next years because it offers, compared with other methods, shorter analysis time, better selectivity, higher versatility, and better sensitivity.

3. Further Developments

The aforementioned methods are well qualified for the separation and quantification of monomeric anthocyanins. However, the determination of polymeric anthocyanins is still an open problem. New analytical methods like capillary electrophoresis or supercritical fluid chromatography (SFC) will perhaps make the analysis of these polymeric pigments possible. Due to its advantages like speed and selectivity it can be assumed that HPLC will stay the method of choice for analyses of monomeric anthocyanins in the next years. Better separations of complex anthocyanin mixtures can be expected through the use of micro-HPLC and newly developed stationary phases. The quantification of individual anthocyanins will hopefully be improved by an increased application

Table 19 Some RP-HPLC Methods and Their Applications in Anthocyanin Analysis

Mobile phase	Stationary phase	Sample	Anthocyanins	Ref.
Binary gradient: A: formic acid/water (1:99) B: methanol	Superpac Pep-S (5 μm; 250 × 4 mm)	Grape	Anthocyanidins	343
Binary gradient: A: formic acid/water (10:90) B: formic acid/water/methanol (10:44:45)	Novapak ODS (5 μm; 150 × 4.6 mm)	Grape	Anthocyanidin 3-glucoside, acylated anthocyanins	245
Binary gradient: A: 5% formic acid B: methanol	Micropak MCH-5 C18 (5 μm; 150 × 4 mm)	Red wine	Anthocyanidin 3-glucoside, acylated anthocyanins	344
Binary gradient: A: formic acid/water (10:90) B: formic acid/water/methanol (10:40:50)	Supelcosil LC-18 (3 μm; 100 × 5 mm)	Cranberry (V. oxycoccus)	pn-glu, pn-gal, pn-ara, cy-glu, cy-gal, cy-ara.	339
Binary gradient: A: 4.5% formic acid B: acetonitrile	μBondapak C18 (10 μm; 250 × 4.6 mm)	Grapes	Anthocyanidin 3-glucoside, acylated anthocyanins	345
Binary gradient: A: 10% formic acid B: 10% formic acid/30% acetonitrile/60% water	RP-C18 (5 μm; 250 × 4 mm)	Chokeberry	cy-gal, cy-ara, cy-xyl	346
Binary gradient: A: 10% formic acid B: acetonitrile	HS-5-C18	Blueberry	Anthocyanidin 3-glucoside, 3-galactosides, 3-arabinosides	244
Water/acetonitrile/formic acid (81:9:10)	LiChrospher RP-18	Black currant, strawberry, cherry, raspberry, elderberry	Anthocyanidin glucosides, galactosides, arabinosides, rutinosides	347
Binary gradient: A: acetic acid/water (15:85) B: water/acetic acid/methanol (65.15:10)	LiChrosorb RP-18 5 μm; 250 × 4 mm)	Grape	Anthocyanidin 3-glucoside, acylated anthocyanins	337
Water/acetic acid/methanol (71:10:19)	μBondapak C18 (10 μm; 250 × 4.6 mm)	Apple	cy-gal, cy-ara	348
Binary gradient: A: 15% acetic acid B: acetonitrile	Supelcosil ODS (5 μm; 250 × 5 mm)	Cranberry, roselle, strawberry	Anthocyanidins	340
Binary gradient: A: 15% acetic acid B: acetonitrile	Supelcosil LC-18 5 μm; 250 × 4.6 mm)	Raspberry	cy-soph, cy-glu, cy-glurut, pg-soph, cy-rut	249

Table 19 (*Continued*)

Mobile phase	Stationary phase	Sample	Anthocyanins	Ref.
Phosphoric acid/acetonitrile (80:20)	Dychrom C18	Tart cherry	cy-soph, cy-glurut, cy-glu, cy-rut	338
Binary gradient: A: 4% phosphorc acid B: acetonitrile	PLRP-S (5 μm; 250 × 4.6 mm)	Cranberry, roselle, strawberry, blackberry, cherry, elderberry	anthocyanidin glucosides, galactosides, arabinosides, sambubiosides, sophorosides	340
Binary gradient: A: phosphate buffer, pH 1.8 B: methanol	LiChrosorb RP-18 (5 μm; 250 × 4 mm)	Grape, wine	Anthocyanidin 3-glucoside, acylated anthocyanins	270
Binary gradient: A: 25% acetone/10% formic acid B: 10% formic acid	HS-5-C18	Grape leaves, stems, tendrils	Anthocyanidin glucosides	349
Binary gradient: A: 0.05 M phosphorc acid B: tetrahydrofuan	Nucleosil-5-C18 (5 μm; 150 × 4.6 mm)	Elderberry	cy-samb, cy-glu, cy-sambglu	350
Binary gradient A: 20% methanol/0.0251 M perchloric acid B: 80% methanol/0.0251 M perchloric acid	Sepharon-SGX.018 (150 × 3.2 mm)	Red wine	anthocyanidin 3-glucoside, acylated anthocyanins	335

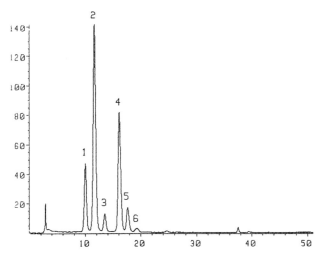

Fig. 14 RP-HPLC separation of black currant juice. Peaks: 1 = delphinidin 3-glucoside; 2 = delphinidin 3-rutinoside; 3 = cyanidin 3-glucoside; 4 = cyanidin 3-rutinoside; 5 = pelargonidin 3-glucoside (= Int. Std.); 6 = pelargonidin 3-rutinoside. (EDER, unpublished).

of pure anthocyanin standards. By direct combination of HPLC with MS a new avenue in on-line characterization and identification of anthocyanins might be opened.

D. Applications in Food Analysis

The content and distribution of anthocyanins in fruits and vegetables has been the object of numerous investigations during the last century (237). Chemical analyses have made it possible to follow the changes in anthocyanin composition that occur during ripening, processing, and storage of fruits (246,335) and to determine factors that affect the stability of these pigments (236). Chemotaxonomic studies have discovered quantitative and qualitative differences in the distribution of individual anthocyanins between cultivars and varieties (340,344). These valuable findings have been applied to the characterization and verification of cultivars (244,338). Usually varieties differ only in the relative amount of anthocyanins present, and since these quantities are determined not only by genetics but by environmental factors, extensive statistical calculations are necessary for the classification of varieties (255). The distinctive anthocyanin profiles have also been a useful tool for the detection of adulterations in fruit juice products which are rich in anthocyanin pigments (271,322). In red wine HPLC analyses of anthocyanin profiles have also been used for geographic classifications (344) and photometric anthocyanin determinations have been applied to measure the chemical age (277). Because of the long history of anthocyanin analysis, various techniques have been employed for food analysis. In Table 20 a number of foodstuffs and references to the corresponding analytical method are compiled.

As examples of all other possible applications, three analytical procedures are subsequently described in detail.

1. Spectrophotometric Determination of the Anthocyanin Content in Blackberry Fruit Juice (276):

pH 1.0 buffer: Combine 125 ml 0.2 N KCl and 385 ml 0.2 N HCl, adjust to pH 1.0 and dilute to 1 L with water.

pH 4.5 buffer: Combine 400 ml 1 M sodium acetate and 240 ml 1 N HCl, adjust the pH to 4.5 with the same solutions and dilute to 1 L with water.

An aliquot of the fruit juice sample is diluted with pH 1.0 buffer so that the absorbance at 510 nm is less than 1.0. Then another aliquot of the fruit juice sample is diluted to the same strength with

Table 20 Some References of Analytical Methods Used for Anthocyanin Analysis in Food

Food	CC and spectroscopy	Nondestructive methods	PC/TLC	HPLC
Blackberry	276	351	352	340
Black currant	291,353,361	248	256,354	291,355
Blueberry	—	356	—	244
Cherry	—	292,351	357	338,340
Cranberry	272,275	358	—	339
Elderberry	359	351	359	340,350
Grape/wine	277,278,309	289,290	319,322,329	270,337
Raspberry	308	—	308,309,324	249
Red radish	310	—	320	—
Rhubarb	—	—	309,320	—
Strawberry	274,291	351	309,320	291,340

pH 4.5 buffer. The absorbance at the wavelength of maximum absorbance (510 nm) and at 700 nm is measured in both samples. Then the absorbance at 700 nm, which is caused by haze, is substracted from that at 510 nm. Subsequently, the difference in absorbance at 510 nm between the pH 1.0 and pH 4.5 samples is determined. The anthocyanin content, estimated as cyanidin 3-glucoside, is calculated by the following equation:

$$c(mg/L) = \frac{A \times MW \times DF \times 1000}{eL}$$

For cyanidin 3-glucoside: $e = 29600$, MW = 445, and L = pathlength = 1.0, DF = dilution factor.

2. Determination of Anthocyanins in Red Wine and Fruit Juice by Means of RP-TLC (328)

Red wine, fruit juice, or methanolic (0.1% HCl) extracts of grapes and fruits are applied as lines on an RP-TLC plate (e.g., SIL-RP C18). The chromatogram is first developed with 20% methanol, 0.25 M HCl until the front has almost reached the top of the plate. The TLC plate is then dried for 30 min in a refrigerator and subsequently developed with the second solvent, 40% methanol, 0.25 M HCl. After the anthocyanins have been sufficiently separated the TLC plate is sprayed with 1% HCl and the pigment spots are evaluated under visible light and UV irradiation. This fast and reliable method can be used for various applications,

3. Separation of Red Wine or Fruit Juice Anthocyanins by RP-HPLC (270):

Prior to HPLC analysis the wine or juice sample is transferred quantitatively onto a C18 solid phase extraction cartridge which has been preconditioned with 2 ml methanol and 2 ml water. The loaded cartridge is washed with 2 ml water and then the cartridge is dried with vacuum. The anthocyanins are slowly eluted with 1 ml methanol containing 0.1% HCl. Subsequently, 10 μl of this methanolic anthocyanin extract is injected into the HPLC system and separated with a LiChrospher 100 RP (5 μm, 250 × 4 mm) analytical column, which is protected by a short (4 × 4 mm) precolumn filled with the same material. The elution of anthocyanins is carried out with a gradient of phosphate buffer, pH 1.8 and methanol, the flow rate is set at 0.8 ml/min and the column temperature at 40°C. The eluting pigments are monitored with a photodiode array detector set at 525 nm. Figure 15 shows a typical chromatogram of the red wine variety "Zweigelt."

IV. BETALAINS

A. Introduction

1. Physical and Chemical Properties

Betalain, a term introduced by Mabry and Dreiding in 1968 (360), includes two classes of water-soluble plant pigments: the red–violet betacyanins and the yellow betaxanthins (361). For a long time the chemistry of these pigments was obscure and they were confusingly named "nitrogenous anthocyanins." But since 1957 the successful elucidation of their structure has begun, Wyler and Dreiding (362) and Schmidt and Schönleben (363) simultaneously isolated and characterized betanin, the red–violet glucoside from *Beta vulgaris* root. Subsequently, the beta-xanthins were discovered in 1964, when indicaxanthin was isolated from prickly pear and its structure fully clarified (364). Further investigations confirmed the assumption that betacyanins and betaxanthins are chemically and genetically closely related substances as both classes have the characteristic betalain chromophore—a protonated 1,7-diazaheptamethin system, which can undergo structural transformations as demonstrated in Fig. 16. The distinction between betaxan-

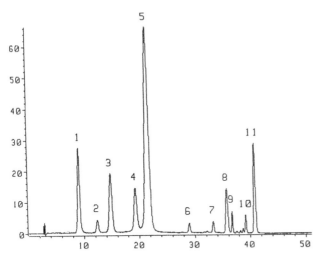

Fig. 15 Separation of red wine anthocyanins cv. *'Zweigelt'* by RP-HPLC. Peaks: 1 = delphinidin 3-glucoside; 2 = cyanidin 3-glucoside; 3 = petunidin 3-glucoside; 4 = peonidin 3-glucoside; 5 = malvidin 3-glucoside; 6 = delphinidin 3-acetylglucoside; 7 = petunidin 3-acetylglucoside; 8 = peonidin 3-acetylglucoside and malvidin 3-acetylglucoside; 9 = delphinidin 3-cumarylglucoside; 10 = petunidin 3-cumarylglucoside; 11 = peonidin 3-cumarylglucoside and malvidin 3-cumarylglucoside. (From Ref. 270.)

thins and betacyanins, however, lies in the different substitution on the dihydropyridine moiety by specific R and R′ groups. While in yellow betaxanthins the conjugation system of 1,7-diazaheptamethin is not extended and the absorption maximum is at ~480 nm, the conjugated chromophore is extended in betacyanin by R and R′ groups comprising a substituted aromatic ring (cyclodopa), which causes an absorption maximum of ~540 nm (3). Until today over 50 betacyanins have been identified, but most of them are based on the two isomeric aglycones betanidin and isobetanidin (the C15 epimer of betanidin), whose chemical structure is shown in Fig. 17. According to the structural formula for betanidin, there are two asymmetrical carbon atoms present in the molecule (C2 and C15). Betanidins are therefore optically active. In acetate buffers betanidin has a specific optical rotation of $[\alpha]_{680}$: +530 ±50°, betanin (pH 5.0) of $[\alpha]_{680}$: +310° and isobetanin of $[\alpha]_{680}$: −840 ±60° (365). In nature all betacyanins are glycosylated, with glucose,

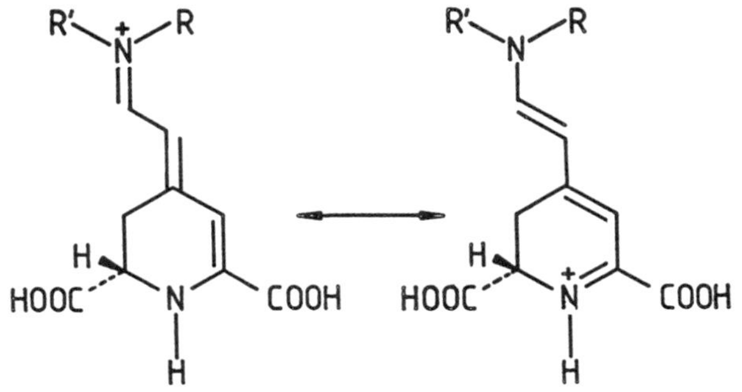

Fig. 16 The betalain chromophore and its structural transformations. (From Ref. 361.)

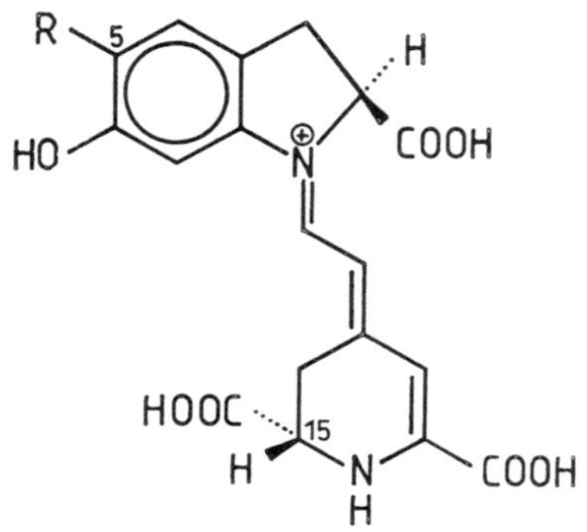

Fig. 17 Formula of betacyanins: R = *O*-glucose, betanin; R = OH, betanidin; R = *O*-glucose-6-sulfate, prebetanin.

sophorose and rhamnose being the most common glycosyl moieties (366). The most prominent betacyanin is betanin, the 5-O-β-glucoside of betanidin, which is the major pigment of red beets, *B. vulgaris* (367). Acylation of betacyanins occurs quite frequently. To date the structures of over 40 acylated derivatives have been elucidated. Numerous acylating moieties have been identified, the most common being sulfuric, malonic, 3-hydroxy-3-methylglutaric, citric, *p*-coumaric, ferulic, caffeic, and sinapic acid (361). In betaxanthins the betalain chromophore is substituted with either an amine or an amino acid (Fig. 18). Some of the occurring substitution patterns for R' are glutamine (vulgaxanthin I; R = H), glutamic acid (vulgaxanthin II, R = H), methionine sulfoxide (miraxanthin I, R = H), aspartic acid (miraxanthin II, R = H), tyramine (miraxanthin III, R = H), L-dopa (dopaxanthin, R = H), and histidine (muscaaurin VII, R = H). The R group is usually a hydrogen, only in indicaxanthin and portulaxanthin the R and R' groups form a prolin or hydroxyproline moiety (361). Betaxanthins usually occur as aglycones; like betacyanins, they are optically active.

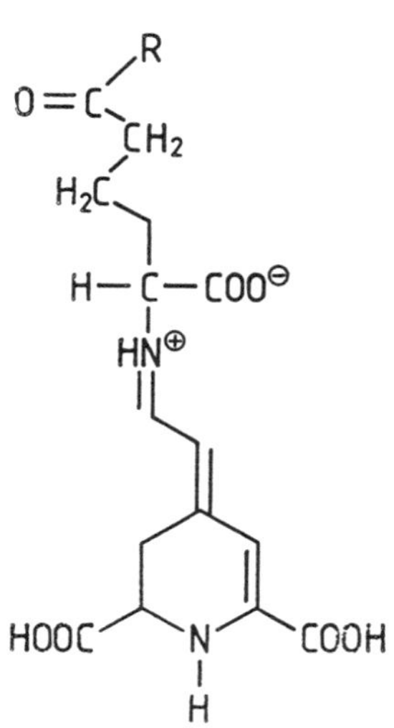

Fig. 18 Formula of betaxanthins: R = NH_2, vulgaxanthin I; R = OH, vulgaxanthin II.

Both betacyanins and betaxanthins are insoluble in organic solvents and in electrophoresis they migrate as anions even at pH values as low as 2.4 (366). In accordance with other natural pigments betalains have a limited stability, they are very sensitive to factors like heat, oxygen, pH, and light (368). The thermolability of betalains has been documented in various publications. Von Elbe et al. (369) showed that the thermally induced degradation of betanin in model solutions followed a first-order reaction kinetics and the half-life of betanin solutions at 50°C and pH 5.0 was 312 ± 30 min. These results were essentially confirmed by Saguy (370), who found that the heat-induced degradation of red beet pigments followed a first-order reaction kinetics with activation energies of 19.2 ± 0.5 and 16.3 ± 0.6 kcal/mol for betanin and vulgaxanthin I, respectively. In addition, the stability, especially the thermal stability, of betalains is strongly influenced by pH. Savolainen and Kuusi (371) established that both betacyanins and betaxanthins were most stable at pH values between 5 and 6, and less stable at lower pH values. Betalains were found to be highly sensitive to oxygen; indeed, even trace levels of residual oxygen increased the pigment degradation (372). In the presence of sufficient oxygen betanin loss followed a first-order reaction kinetics, with an activation energy of 20.4 kcal/mol for heated betanin solutions, but in the absence of oxygen betanin stability was greatly enhanced and degradation occurred according to a 0.5 reaction rate order (372). Finally, betalains are also subject to photochemical degradation. It could be observed that betanin degradation is strongly increased upon exposure to daylight (373) or to fluorescent lighting (368). On the other hand, the stability of betanin was improved by sequestrants like EDTA or citric acid (370) which caused an increase in half-life values by 1.5 (372). Schwartz and von Elbe (374) identified the degradation products of betanin in solution upon heating as betalamic acid and cyclodopa 5-O-glucoside. Of special importance was the finding that this degradation reaction was partially reversible (375). As betalains are very sensitive substances, special precautions, similar to those mentioned for carotenoids, have to be taken in handling them.

2. Properties in Food

The distribution of betalains in plants is restricted to the order Centrospermae (synonym: Caryophyllales), wherein only 10 betalain-producing families have been identified. Betalain pigment accumulations in cell vacuoles of the flowers, fruits, leaves, stalks, and/or roots are responsible for the attractive yellow, red, or violet color of these plant parts (376). Only a few betalain-containing plants are used for human nutrition. In essence these are beet root (*B. vulgaris* L.), prickly pear (*Opuntia ficus indica*), *Basella rubra*, pokeberry (*Phytolacca decandra* L.), and amaranth (*A. tricolor* and *A. caudatus*). In addition, special betalains (e.g., muscapurpurin, muscaaurins I–VII) have been found in mushrooms. They are, however, usually poisonous (361). The main source of betalains in food are beet roots, which have a reported betacyanin level of 35–223 mg/100 g fresh weight and a betaxanthin level of 33–91 mg/100 g fresh weight (377). Among cultivars the colorant properties showed considerable diversity. Beside large differences in the betalain content qualitative discrepancies could be observed (373). While red beet roots (*B. vulgaris* L.) contain both red (betacyanins) and yellow pigments (betaxanthins); the yellow beet root (*B. vulgaris* L. cv. *Burpee's Golden*) has only yellow pigments. The main red pigment in red beet roots is betanin, which accounts for 75–95% of the total betacyanin content. The dominant yellow pigment is vulgaxanthin I, which is also the major pigment of the yellow beet root (378). Other betalains occuring in beets are the betacyanins isobetanin, isobetanidin, prebetanin, and isoprebetanin and the betaxanthin vulgaxanthin II. The color of prickly pear, which is usually consumed either as fresh fruit or as juice, is mainly produced by the betaxanthin indicaxanthin (361,379). Recently, the betalains of the fruit *Basella rubra,* which is used as a vegetable in tropical areas and also as a raw material for cosmetic dyes, could be identified as gomphrenin I (15S-betanidin 6-O-β-glucoside), gomphrenin II (15S-betanidin 6-O-[6'-O-(4-coumaroyl)-β-glucoside), small amounts of the respective R forms (isogomphrenin I and II), and gomphrenin III (15S-betanidin 6-O-[6'-O-feruloyl-β-glucoside]) (380). As considerable amounts of the red–violet betacyanins amaranthin and isoamaranthin are accumulated in mature leaves of some amaranth species (*A. tricolor* and *A. caudatus*), these plants, which are already grown as a grain crop and as a vegetable, are a potential source for the production of natural food colorants (381).

3. Arguments for Betalain Analysis: Regulations

Betalains are the pigments responsible for the color of certain food stuffs (e.g., beet roots, prickly pear). The color of food is a highly important attribute because it is usually the first property consumers observe, and hence this attribute has an excluding character in respect to the other attributes and properties (379). Since color is such an important quality characteristic it is of interest to determine the corresponding pigments and to study their stability (369). To improve the visual color of betalain-containing food (e.g., canned red beets) it is necessary to carry out extensive investigations. For instance, concerning the relationship between variety and content and/or distribution of pigments (373), the relationship between environmental factors and pigment accumulation in plants, and color changes during processing and storage (382).

The use of betalain pigments as food colorants dates back at least one century, when inferior red wines were colored with betalain-containing juices (e.g., red beet root juice). This common practice was, however, soon prohibited and the application of betalain colorants was widely replaced by artificial dyes, which displayed better stability, lower price, and higher purity. But in past years the interest in natural food colorants has been renewed, mainly because of the consumers concern about the safety of some artificial colorants which may be hazardous to human health (383). As a result of this, the number of permitted artificial dyes was markedly reduced and new efforts had to be made to develop natural food colorants (384). However, current legislation restricts the application of betalain colorants to concentrates or powders (E162) obtained from aqueous extracts of beet roots (361). Since only 0.1–0.3% of the fresh weight of beet roots are betalains, effective extraction procedures are required. While pressing yielded a pigment recovery of only 45–50%, the use of a continuous diffusion-extraction procedure increased the yield of betacyanin and betaxanthin to 90% and 80%, respectively (385). Commercial beet colorants contain 0.4–1.0% pigment (expressed as betanin), which are accompanied by various plant metabolites such as sugars (80%), proteins and amine bases, choline, glutamine, and geosmin, which together are responsible for the typical, undesirable taste and smell (361). To improve the color stability and to remove the "off-flavor" commercial beet colorants were purified by ion exchange and gel filtration chromatography (386) or by fermentation (387), but the application of the resulting products is still prohibited (361). More recently, microbiologically purified beet root colorants were also used in various ways for pharmaceutical purposes (388). If betalains are to become a more competitive and viable alternative to synthetic colorants in the near future, further investigation into their biochemistry and genetics is necessary.

B. Sample Preparation

1. Extraction

Before the extraction of betalains can take place the plant material (e.g., beet roots) should be washed, blanched, and cut into small slices. Occasionally they are also lyophilized and powdered (371). Subsequently, the tissue is either pressed to produce a pigment-containing juice or homogenized with extraction solvents. As betalains are highly soluble in water they are usually extracted with water (389) or water-based solvents such as water/0.1% HCl (390) or 0.1 M McIlvaine's citric-phosphate buffer solution (pH 5.0) (378). In some cases Celite was added to the mixture to improve the clarification procedure (390). Other authors employed alcohol-containing solvents for the betalain extraction, e.g., ethanol/water (1:1) (383), or 60% and 80% aqueous methanol (391). The initial homogenate is clarified by either filtration or centrifugation, the filter cake may be reextracted until no more pigments can be extracted. Both the clarified homogenate and the plant juice (e.g., beet juice) can be directly used for electrophoresis, spectrophotometry, or HPLC, although an additional dilution step may be necessary. However, the juice and homogenate contain many interfering substances which may lead to erroneous results, especially with spectrophotometric determination. To avoid such incorrect measurements, additional purification steps (e.g., gel filtration, ion chromatography), which also cause a crude separation of betalain pigments, are commonly carried out (392).

2. Purification

The purification of crude betalain extracts is usually accomplished by chromatographic or electrophoretic methods. These procedures do not only purify but also separate the pigments, thereby allowing the quantitation of total and individual betalains simultaneously (361). Numerous researchers employed a two-step purification procedure. The first step is either ion chromatography with cationic exchange resins [e.g., Dowex 50W-X2, H^+ form (393)] or gel filtration [e.g., Sephadex G-25 (394)]. The second purification step is carried out by chromatography on polyamide (393) and/or PVP [e.g., Polyclar AT (368,395)]. Prior to cation exchange or gel filtration chromatography the pH value of the crude extract has to be adjusted with hydrochloric acid to ~ 3.0 (394). Piatelli et al. (393) developed a purification procedure which was also suitable for the separation of complex betalain mixtures. This method is based on nonionic adsorption of betacyanins and betaxanthins onto cation exchange resins (Dowex 50W-X2, H^+ form) and subsequent elution of the pigments with water. The purified pigment mixture was then separated on polyamide, using a citrate buffer (pH 4.0) as eluant. These chromatographic methods still play an essential role in the identification of unknown betalains because they allow the preparative isolation of pure pigments required for structure elucidation (391).

C. Analysis–Detection–Separation Methods

All of the methods named below have been developed and used for the determination of betalains in beet roots. Some adaptations may be necessary for the analysis of other food materials.

1. Classical Methods

a. Spectrophotometric Determination. The spectrophotometric determination of betalains is based on their absorption characteristics, which follow the Lambert–Beer law. The absorption spectrum of betanin, the main betacyanin of beet roots, shows an absorption maximum at 535–540 nm with an absorptivity value of $E_{1\%}^{1cm} = 1120$ (396). The yellow vulgaxanthin has its maximum absorbance at 476–478 nm with $E_{1\%}^{1cm} = 750$ (393). Nilson (365) demonstrated that the absorption spectra of these pigments are not affected when measured at pH 4.5–7.0, while at pH <4.5 the betalains produced a more violet color and at pH >8.5 the color turned yellow. Since the pH of fresh beet root juice varies between 6.14 and 6.50, the varying acidity in the cell sap will not influence the color characteristics (365). Based on extensive studies with pure betanin and vulgaxanthin I, Nilson (365) developed a spectrophotometric method for the simultaneous determination of red and yellow pigments in beet root juice without an initial separation of these pigments. As the absorption spectrum of a mixture of two pigments is an addition of the absorption spectra of each pigment, it is necessary to substract the absorption of one pigment at the maximum of the other pigment and vice versa. Nilson (365) found that betanin had an absorption at the absorbance maximum of vulgaxanthin I (476–478 nm), while vulgaxanthin I did not absorb at the absorbance maximum of betanin (535–540 nm), which means that the absorption at 535–540 nm is caused only by betanin. To quantifiy the concentration of each pigment he calculated the ratio of $A_{534}/A_{476} = 3.1$ and $A_{538}/A_{600} = 11.5$, respectively. In order to minimize the error it is necessary to dilute the juice until the absorbance at 538 nm lies between 0.2 and 0.5. Furthermore he introduced a correction term for interfering impurities which are present in beet root juice. Considering all of these factors he established the following equations for the calculation of the absorption of the different pigments in beet root juice:

$$x = 1.095(a-c) \tag{14}$$

$$y = b-z-\frac{x}{3.1} \tag{15}$$

$$z = a-x \tag{16}$$

The calculation of the proportion of vulgaxanthin I and betanin from the absorption of beet root juice is shown in Fig. 19. The values a, b, and c are the absorptions of the extract at 538, 476, and 600 nm, respectively; x is the absorption of betanin, y that of vulgaxanthin I, and z the absorption of the impurities. After x and y have been calculated with the help of the equation system, the amounts of betanin and vulgaxanthin I can be estimated using $E_{1\%}^{1cm} = 1120$ for betanin and $E_{1\%}^{1cm} = 750$ for vulgaxanthin I. This method is easy to practice, cheap, and gives highly reproducible results; therefore it is still in common use. The error, which is introduced by calculating all betacyanins as betanin (isobetanin) and all betaxanthins as vulgaxanthin I, is negligible since betanin-isobetanin and vulgaxanthin I together contribute over 95% of the beet root color.

Another quick and accurate method for the determination of all major beet pigments (betanin, vulgaxanthin I, and betalamic acid) and browning substances was introduced by Saguy et al. (395). Their computer-aided determination is based on nonlinear curve fitting of the spectrum with a predicted function of the individual pigments. As this procedure obviates the need for laborious and time-consuming separations, it is useful in the continuous monitoring of time- and temperature-related processes. The results which were obtained with this method showed an excellent agreement with the actual betalain contents, compared with the above-mentioned Nilson method it gave an even better accuracy. Although spectrophotometry has the advantage of simplicity, speed, and inexpensiveness, it is hampered by insufficient selectivity.

 b. Electrophoresis. Both betacyanins and betaxanthins have three carboxyl groups and a quarternary nitrogen atom with a weak positive charge, which together with the carboxyl group

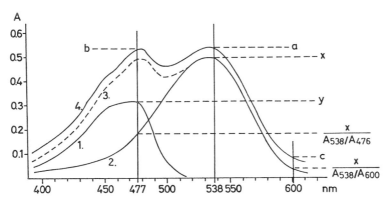

Fig. 19 Calculation of vulgaxanthin I and betanin's proportion of the absorption of beet root juice. Curve 1: Vulgaxanthin's proportion of the absorption. Curve 2: Betanin's proportion of the absorption. Curve 3: Betanin and vulgaxanthin without impurities. Curve 4: Fresh beet juice including impurities. (From Ref. 365.)

in the 2 position gives them amphoteric properties. Therefore betalains occur as negatively charged ions at pH >2.0 and migrate in an electric field. Electrophoretic separation of betalains was commonly achieved by paper electrophoresis using different paper strips (e.g., Whatman No. 1, Beckman No. 319328, S&S 4043B), but with varying success (365). It was found that the separation is highly influenced by the buffer composition and concentration, pH, electroendosmotic flow, temperature, etc. To obtain sharper bands it is recommended that electrophoresis be performed at low temperatures (e.g., 4°C). The voltage gradient is set between 5.6 and 13 V cm^{-1} (371,389) and the current gradient between 0.5 and 2 mA cm^{-1} (365,371). Nilsson (365) obtained his best separations with 0.1 M phosphate buffer, pH 5.5. Some typical examples of electrophoretic separations of beet root juice at different pH values are shown in Fig. 20. Powrie and Fennema (397) found that the separation of prebetanin and vulgaxanthin I was considerably improved when pyridine was included as base in the buffer solutions. In accordance with this finding, von Elbe et al. (389) used a 0.15 M pyridine-citric acid buffer, pH 4.5, for the separation of betacyanins in red table beets. After electrophoresis the papers are immediately dried in a vacuum oven at 30°C or 80–90°C (365,371); occasionally calcium chloride is added (389). The differently colored bands are marked in daylight. Additionally, betaxanthins show a strong yellow–green fluorescence under UV light (365). Pigments can be identified according to their relative migration length (E_b values) and also by coelectrophoresis with standard betalains. For quantitative analyses the band areas of the pigments are scanned with a densitometer and the concentrations are calculated as betanin. When determining the betacyanins in red beets von Elbe et al. (389) found a standard deviation of 0.4 μg within the range 2–9 μg betanin and an average recovery rate of 98%. With the help of

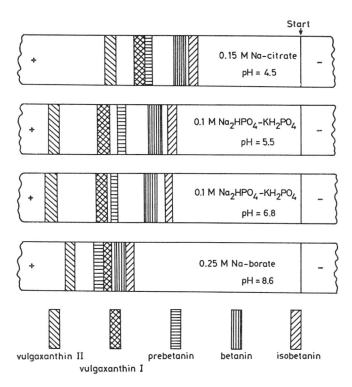

Fig. 20 Electrophoresis at different pH values of juice from fresh beet roots. The band with prebetanin also contains isoprebetanin. (From Ref. 365.)

paper electrophoresis betalains of yellow and golden beet were successfully separated into eight components, the most important of which were vulgaxanthin I and II (371). Because of its simplicity and speed paper electrophoresis has often been used, but it has only limited value where complete betalain separations and identifications are wanted. In earlier times a combination of electrophoresis and subsequent paper chromatography was necessary for the complete separation of betaxanthins. In the case of betacyanins this could be achieved by combining column chromatography on polyamide with subsequent electrophoresis (365). Nowadays analytical HPLC yields excellent separation of natural betalain mixtures.

c. Thin-Layer Chromatography (TLC). While PC was shown to be of limited applicability for betalain analysis, TLC was occasionally used for the separation and identification of these pigments (383). Recently, betacyanins of various plant sources could be separated on microcrystalline cellulose using ethyl acetate/formic acid/water (33:7:10) as the mobile phase (380,391).

d. Column Chomatography (CC). The separation of betalains for both analytical and preparative purposes was achieved with CC using polyamide (365) and/or gel filtration supports (398). Crude plant extracts were frequently precleaned by passing them through an ion exchange resin (e.g., Dowex 50WX2 H^+ form). CC on polyamide was described in detail by Nilsson (365), who separated beet root pigments into three differently colored fractions (vulgaxanthins, betanin, and isobetanin) by using a 0.01 M Na-citrate buffer, pH 5.4 as eluant. The suitability of gel filtration for the separation and quantification of betanin was proven by Adams and von Elbe (398). They found that chromatography on gel filtration supports resulted in a rapid and efficient separation of beet root betalains into numerous pigment bands. Additionally they reported average distribution coefficients (K_{av}) for betanin on Sephadex G-25 or Bio-Gel P-6 from 0.8 to 2.0 at pH 4.0 and 2.0, respectively. The resolution of betanin and betanidin was better on columns packed with Sephadex G-25 than on columns packed with Bio-Gel P-6. In contrast, Bio-Gel P-6 showed a greater loading capacity than Sephadex G-25. When Bio-Gel P-6 was chosen as the support, excellent pigment separations were obtained using a phosphate buffer at pH 3.0. The quantification was carried out by measuring the absorption at the maximum wavelength of each eluting peak. The resulting peak areas were related to standard concentrations of betanin and the ensuing standard betanin equation was linear. Due to long analysis times, low selectivity, and high chemical consumption, traditional CC has mostly been replaced by HPLC.

2. Recent Methods

High-Performance Liquid Chromatography (HPLC). The first successful attempt to use HPLC for the analysis of betalains was carried out by Vincent and Scholz (399). With the help of paired ion chromatography (PIC) on an RP column (μBondapak C_{18}) they separated betacyanins and betaxanthins of red beet. The ion pairing approach was adopted in order to exploit the highly ionic character of these pigments and the durability of the C18 RP column. Using methanol-water at a pH value of 7.5 and tetrabutylammonium as the PIC reagent, nonpolar unprotonated complexes of the solute molecules were generated. Two wavelength settings (538 and 476 nm) were used to selectively monitor the red and yellow pigments, although the red betacyanins do absorb to a certain degree at 476 nm, making 476 nm the wavelength of choice for a comprehensive screening of betalains. According to their chemical structure, yellow betaxanthins were more polar and therefore eluted earlier than betacyanins. For quantitative analyses the HPLC system was calibrated with a purified betanin standard, and a good linearity over a practical work range of 0.005–0.03% (w/v) betacyanin could be obtained. The relative standard deviation among replicate samples was ~2.0%, the detectability limit was found to be ~10–4% (399). Huang and von Elbe (394) separated and purified betanin and amaranthin from amaranth leaves using a combination of gel filtration and PIC. In spite of these good results the majority of HPLC investigations concerning betalains was carried out without the addition of ion pair reagents. RP-C8 (400) or RP-C18 (401) columns were employed almost exclusively as the stationary phase, while the mobile phase commonly consisted of an acid or buffer (e.g., phosphate buffer, acetic acid), which ensures a low pH (~3.0), and an organic modifier (e.g., methanol, acetonitrile) (402). The conditions of some selected HPLC analyses are compiled in Table 21. Generally, the HPLC separation of betalains can be accomplished within ~20 min. Two typical chromatograms of betalains in fermented red beet root extract are shown in Fig. 21. The quantitative estimation of betacyanins is often based on the absorptivity value $E_{1\%}^{1cm} = 1120$ for betanin. However, to achieve more exact results, Schwartz and von Elbe (401) recommended the use of individual absorptivity values (e.g., betanidin hydrochloride $E_{1\%}^{1cm} = 1275$). With the aid of HPLC methods some little

Table 21 Selected Experimental Conditions for HPLC Analysis of Betalains

Column	Solvent	Detection (nm)	Betalain	Application	Ref.
μBondapak C18 (10 μm)	A = methanol/phos phate buffer pH 2.75 (18:82) B = methanol; gradient	535	betcyan	Red beets	383,401
LiChrosorb RP-8	A: water/acetic acid (98:2) B: water/acetic acid/methanol (70:10:20); gradient	436	betxanth	Yellow beets	400
LiChrosorb RP-18 (5 μm. 250 × 4 mm)	A: 1.5% phosphoric acid in water B: phosphoric acid/ acetic acid/ acetonitrile/water (1.5:20:25:53.5); gradient	480	betcyan, betxanth	Aizoaceae, Cactaceae	402
μBondapak C18 (300 × 4 mm)	A: 0.005 M PIC reagent A in water B: 0.005 M PIC reagent B in methanol; gradient	476,538	betcyan, betxanth	Red beets	399
μBondapak C18	A: 8% acetonitrile in 5 mM PIC A solution B: 12% acetonitrile in 5 mM PIC A solution; gradient	546,436, 313	amar, bet	Amarant	394
μBondapak C18	Methanol/phosphate buffer pH 3.0 (1:5)	535	bet	Red beets	368
Nucleosil C18	A: 1.5% phosphoric acid in water B: 1.5% phosphoric acid, 20% ethanol, 25% acetonitrile in water; gradient	540	betcyan	Basella, Gomphrena	380,391

Abbreviations: betcyan, betacyanins; betxanth, betaxanthins; amar, amaranthin; bet, betanin.

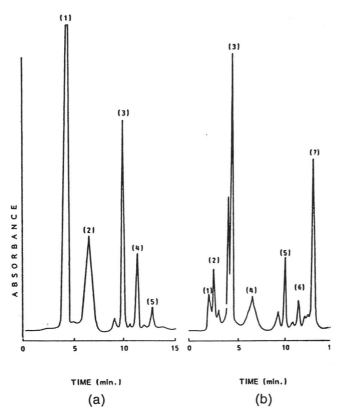

Fig. 21 HPLC separation of betalains in fermented red beet root extract. (a) Detection of betacyanins at 538 nm. Peaks: 1 = betanin, 2 = isobetanin, 3 = betanidin, 4 = isobetanidin, 5 = prebetanin. (b) Detection of betaxanthins at 477 nm. Peaks: 1 = vulgaxanthin I, 2 = vulgaxanthin II, 3 = betanin, 4 = isobetanin, 5 = betanidin, 6 = isobetanidin, 7 = prebetanin. (From Ref. 388.)

investigated betalains could be detected and valuable information concerning the stability and degradation of various betacyanins and betaxanthins could also be gathered (372,401). Due to its speed, high resolving power, good sensitivity, and the fact that preliminary purification is not always necessary, HPLC has become the method of choice for quantitative analyses of individual and total betalains (361).

3. Further Developments

The merits of HPLC, such as high separation capacity, easy quantification, speed, and good reproducibility, make it highly probable that HPLC will remain the predominant method for betalain analyses in the next years. Further improvements of HPLC separations can be expected through the use of newly developed columns (e.g., microcolumn) and stationary phases (e.g., microbore material), which also lower the required sample volume and the consumption of solvents. On-line characterization and identification of eluted pigments will also be improved through the employment of more sophisticated detectors like diodenarray and electrochemical detectors. In addition, the use of spectroscopic methods like MS and NMR for identification of betalains will be extended (391). The applicability of relatively new analytical methods such as supercritical fluid chromatography and capillary electrophoresis for betalain analyses is yet unclear and should be investigated in the future.

D. Applications in Food Analysis

The special pigments of Centrospermae have attracted considerable interest during the past 40 years. In particular, the pigment composition of beet root has been the topic of much investigation. Initially the betalains of beet root were separated and isolated using paper electrophoresis (393) and CC (365). Their chemical structure was elucidated by means of chemical tests and spectroscopic methods (364). The traditional methods for the quantitative determination of betacyanins and betaxanthins in beet root were spectrophotometry, mainly the Nilsson method (365), and to some degree also electrophoresis followed by densitometric evaluation (389). More recently, HPLC has become the method of choice for the separation and quantification of beet root pigments (399). As beet roots are an interesting source for the production of natural food colorants, numerous studies concerning the colorant content and stability of different beet root varieties have been undertaken (373). Among others the influence of pH, temperature, oxygen, and light on betacyanins and betaxanthins has been examined using electrophoresis and spectrophotometry (368, 371). Several researchers calculated the corresponding degradation rates and isolated and identified the main degradation products by means of HPLC (372,374). The extraction of pigments from beet root is often incomplete and therefore new techniques have been tested to increase the pigment recovery (378). Photometric analysis showed, however, that the resulting extract could contain at most only 0.4–1.0% pigment (expressed as betanin). Chromatographic methods were successfully applied to increase the betalain concentration in the colorant and to remove interfering metabolites, which impart an undesirable smell and taste to the pigment extract (387). Using another method the total solids were decreased and the betacyanin concentration increased proportionately by fermentation of the beet root extract (387). The applicability of betalains as natural colorants for such foods as sausage was evaluated using photometric and nondestructive methods (384,403).

As representatives for all the other applications in food analysis, two analytical procedures are described in detail.

1. Spectrophotometric Determination of Betanin (Betacyanins) and Vulgaxanthin I (Betaxanthins) in Beet Juice (365):

Ten grams of extracted beet juice is added to 0.1 M pH 5.0 McIlvaine's citric-phosphate buffer solution and made up to 1000 ml. Then the diluted juice is clarified by filtration through Whatman No. 1 paper. The McIlvaine buffer is used as blank sample and subsequently the absorption at 476, 537, and 600 nm is recorded. The percentage of betanin and vulgaxanthin I is calculated as follows:

$$x = 1.095 \,(A_{537} - A_{600}) \qquad (17)$$
$$y = 0.258 A_{537} + A_{476} - 0.742 A_{600} \qquad (18)$$

where x is the absorption of betanin and y is the absorption of vulgaxanthin I. Thus the concentrations of betanin and vulgaxanthin I can be calculated using an $E_{1\%}^{cm} = 1120$ for betanin and $E_{1\%}^{cm} = 750$ for vulgaxanthin I (380).

2. Determination of Red Beet Betacyanin Pigments by HPLC (401):

One hundred grams of fresh beets from five randomly selected roots is used for analysis. The tissue is blended for 1 min with 100 ml EtOH/H$_2$O (50:50, v/v) under a stream of N$_2$ to lessen oxidative enzymatic reactions. The blender walls are washed with 100 ml of H$_2$O and the mixture is blended for an additional 5 min under N$_2$. Fifty grams of this homogenate is removed, filtered over a 10-g bed of Celite, and washed with 300–400 ml of H_2O until the tissue-Celite mixture is colorless. The filtrate is quantitatively transfered to a 500-ml volumetric flask and brought to volume. A 1:5 dilution of this sample is used for HPLC analysis.

After filtration through a 0.45-µm filter 20 µl of the sample is injected and separated on a reverse phase column (µBondapak C_{18}, 10 µm). A linear gradient is used starting from 100% solvent A = $CH_3OH/0.05$ M KH_2PO_4 (18:82, v/v) adjusted to pH 2.75 with H_3PO_4 and ending with 80% solvent A and 20% solvent B = CH_3OH. The chromatogram is monitored at 535 nm, taking less than 15 min to run. A typical example of an HPLC separation of betacyanins is shown in Fig. 22.

V. MYOGLOBIN

A. Introduction

1. Physical and Chemical Properties

Myoglobin, a monomeric, globular heme protein, is the principal but not the whole source of color in meat (3). Additional contributors to the color of meat are the blood pigments hemoglobin and cytochrome. The proportion of hemoglobin in the meat color depends mostly on the degree of bleeding out of the meat. In well-bled pieces from cattle and pig it will be less than 6% and 8%, respectively (404). The myoglobin molecule consists of a prosthetic group, the heme, and a polypeptide chain, the globin, which is composed of 153 amino acids (405). Due to species-specific changes in the amino acid sequence the molecular weight of myoglobin varies from 16,900 to 17,850 (405,406). The heme prosthetic group is responsible for oxygen binding and confers an intense red or brown color to the protein. The reactive heme group consists of a central iron atom and a large planar ring, the porphyrin (protoporphyrin IX), whose structure is that of four pyrrole rings linked together by methene bridges as shown in Fig. 23 (407). The bonding of the iron to the tetrapyrollic ring structure satisfies four of the six coordination positions of the atom. The fifth is coordinated to the imidazole residue to histidine within the protein structure and the remaining sixth position is capable of binding high-field ligands like O_2, CO, NO, CN, and N_3 (408). The central iron may be found in the oxidation states II or III, whereby at high pH values the iron of the heme is predominately in the ferrous state, while low pH values accelerate the conversion of ferrous iron to the ferric state. Unlike oxygen, which can only bind to the ferrous form of heme

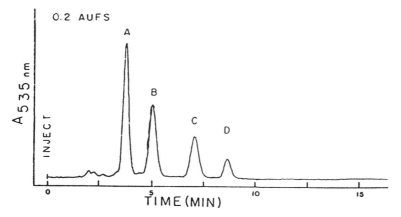

Fig. 22   HPLC separation of betacyanin pigments on a µBondapak C_{18} column. Peaks: A = betanin, B = isobetanin, C = betanidin, D = isobetanidin. (From Ref. 401.)

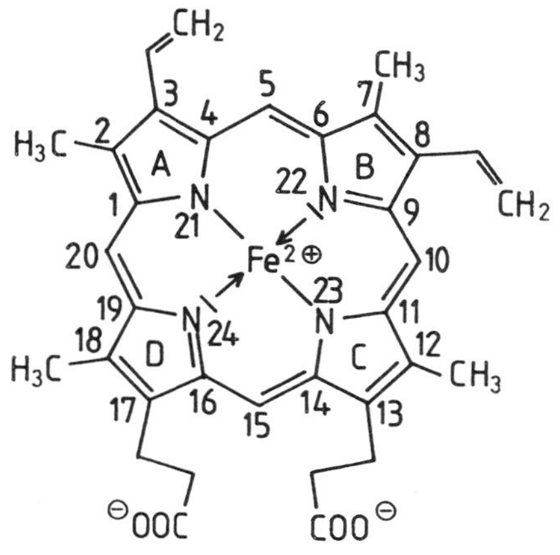

Fig. 23 Structure of heme (Fe^{2+} protoporphyrin).

pigments, CO, NO, CN, and N_3 can bind to either form of the heme pigments, while Cl^- and H_2O can bind only to the ferric form (409). The most functional property of myoglobin is its ability to bind molecular oxygen (O_2) reversibly, in this way creating an intracellular reservoir of oxygen and facilitating the diffusion of oxygen from the plasma membrane to the mitochondria where the oxygen is consumed (410). In addition to enabling the heme group to perform its physiological function as an oxygen carrier, the globin converts the insoluble free heme to a water-soluble complex and protects the heme from oxidation. Therefore the oxidation of the ion can only take place if the protein is denatured. The heme moiety in myoglobin is not a fixed prosthetic group, and therefore reversible dissociation into heme and apoproteins may occur. While the affinity of the heme for the protein is very high at neutral pH (equilibrium constant 10^{12}–10^{15} M), it is considerably lower at more acidic pH values (407). Myoglobin is quite reactive and can undergo several reactions, e.g., autoxidation, reaction with nitrite, and denaturation, all of which are of great importance for the color of fresh and processed meats (411).

2. Properties in Food

Meat color is mainly a result of the concentration of meat pigments, the chemical state of these pigments, and physical characteristics such as fat deposition and diverse surface properties. Dependent on species, breed, sex, age, type of muscle, training, and nature of nutrition, the myoglobin content of meat can vary greatly. A general rule states that the more intensively a muscle is used, the higher is its myoglobin content. For example, the constantly operating muscle of the diaphragm has more myoglobin than the occasionally and less intensively used *M. long. dorsi* (412). The average myoglobin content of this muscle is also a good illustration for the differences between species, e.g., in pig, cattle, and whale it contains 0.10, 0.50, and 0.90% myoglobin, respectively (413). As aforementioned the iron atom of the heme can exist in either the ferrous or the ferric state and can form complexes with certain ligands, all of which greatly affects the color of the meat. The color of fresh meat is defined by the relative amounts of three derivatives of myoglobin: myoglobin (Mb, Fe^{2+}), oxymyoglobin (MbO2, Fe^{2+}), and metmyoglob-in (MetMb, Fe^{3+}). The absorption spectra of these three myoglobin derivatives are shown in Fig. 24. Reduced Mb is the purple pigment of deep muscles and the meat under vacuum (λ_{max} = 555 nm). Upon exposure to air Mb combines with oxygen to form the bright red MbO2 (λ_{max} = 542 nm), which is synonymous with freshness and considered attractive by the consumer. However,

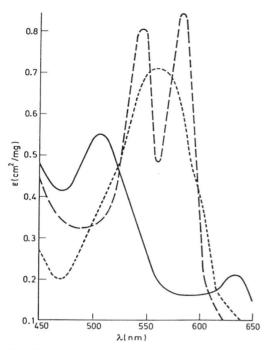

Fig. 24 Absorption spectra of MbO₂ (— — —), Mb (- - - -), and MetMb (—). (From Ref. 3.)

contact of Mb with oxygen also leads to the formation of the oxidized form, MetMb (λ_{max} = 505 and 635 nm), which is brown or gray and rather unattractive (408). These three forms of myoglobin are in a state of equilibrium in meat muscles and are constantly interconverted within a dynamic cycle. The formation of the attractive MbO₂ is favored by a high oxygen tension (pO2 > 4 Torr). The undesired oxidation toward MetMb is predominant when the oxygen tension is ~4 Torr and an oxygen tension below 4 Torr leads to the predominant formation of Mb. In fresh meat reducing substances like NAD^+ or FAD^+ are endogenously produced, and they are responsible for the constant reduction of the brown–gray MetMb to the purple Mb, which means the dynamic cycle can continue as long as sufficient oxygen is present (410). The heating of meat samples, however, causes considerable destruction of the iron-porphin complex and therefore mainly the denatured hemoproteins are present in cooked meats. The dark red color of cured meat is achieved by the reaction of myoglobin with nitrite ions, a process which forms nitrosyl myoglobin (MbNO). After heat denaturation this pigment is converted to the more stable nitrosyl hemochrome, which is pink (411). The chemistry of meat pigments has been the subject of numerous investigation and comprehensive reviews are given by various authors (408,410,411).

3. Arguments for Myoglobin Analysis: Regulations

The red color is regarded as an important parameter of meat and dark fish quality. It is well known that the consumer considers the bright red color of MbO₂ in fresh meat desirable, while the brown color of MetMb is considered undesirable (414). Studies concerning the factors influencing the stability of myoglobin have shown that this pigment is very susceptible to heat, acid denaturation, and autoxidation at freezing temperatures. In addition, it has been detected that the stability of myoglobin from different animals varies considerably (415). Since the formation of MetMb is the major cause of surface discoloration in fresh meat, a better knowledge of the reactions involved in MetMb formation might help to optimize the color of meat products. A traditional method to achieve a more stable and attractive meat color is to cure it. However, due to the toxicity of nitrite and related substances the resulting chemical changes of the myoglobin have been investigated extensively and consequently the permitted amount of nitrite has been considerably reduced (407). The often observed problem that the color of cooked cured meats fades rapidly on exposure to air and light has also been the object of numerous investigations (416). Another argument for myoglobin analysis is the possibility of determining the origin of meat and meat components in meat products (e.g., sausages, burgers). This very important and challenging task in meat hygiene and meat control is based on the fact that myoglobin has a species-specific microheterogeneity. Therefore it has been possible to detect nonpermitted or nondeclared meats by myoglobin analysis employing analytical methods like serology (416), electrophoresis (417), and MS (406). With the intention of determining the degree of bleeding out of the meat (418) or to detect an eventual addition of blood to the meat (419), various analytical methods have been developed to differentiate between myoglobin and hemoglobin. Finally, it has been shown that the analysis of free porphyrins, which result from the decomposition of myoglobin, is a suitable method for the determination of meat spoilage (420).

B. Sample Preparation

1. Extraction

Due to its hydrophilicity myoglobin can be extracted from meat with deionized (421) and/or distilled water (422) or aqueous buffers. Commonly used buffer solutions are 0.075 M or 0.5 M Tris/HCl buffer, pH 7.9 or 8.9 (422,423), and 0.04 M, 0.05 M or 0.067 M phosphate buffer pH 6.8, 7.5, or 6.5 (424–426), respectively. The colored heme moiety, which is set free after dissociation of the heme-protein molecule, is usually extracted with organic solvents like ethanol or acetone (427). A common procedure for the extraction of myoglobin is to trim off the visible fat from the meat, cut the piece into small slices, and homogenize the slices together with one of the solvents mentioned above. To achieve an effective extraction of the meat pigments the homogenate is stirred for 2 or 3 hr at 37°C. Afterward the homogenate is centrifuged and/or filtered and the resulting supernatant beneath the lipid layer is collected (426). To minimize myoglobin decomposition the extraction is carried out at low temperatures (e.g., 4°C). The addition of potassium azide is recommended to prevent bacterial growth (427). The crude meat extract either can be used directly for myoglobin analysis or can be subjected to further purification and concentration procedures.

2. Purification

Three analytical procedures are usually applied to the purification of myoglobin from crude meat extracts:

1. *Fractionating with ammonium sulfate between 55% (w/v) and 90% saturation.* Prior to the fractionation the extract is adjusted to pH 7.0 and then solid ammonium sulfate is added to give a 55% and 90% saturation. The last precipitate is redissolved in distilled water and dialyzed against distilled water (429). In a similar procedure, myoglobin and hemoglobin are separated by precipitating the hemoglobin in an 85% ammonium sulfate solution (419). Hayden (416) successfully removed hemoglobin and sarcoplasmic proteins from myoglobin with 20% lead subacetate, potassium phosphate, and 70% saturated ammonium sulfate.
2. *Dialysis.* Sugden and Saschenbrecker (425) dialyzed the centrifuged meat buffer homogenate against nine volumes of water to precipitate actomyosin and myosin. A two-step procedure

consisting of a first dialysis against distilled water and a second dialysis against a species-specific buffer was developed by Renerre (422).

3. *Column chromatography.* Various column supports have been reported for the purification of myoglobin from crude meat extracts. In particular, gel filtration media like Sephadex G-25, G-50, and G-75 have been widely used (429,430). Satterlee and Zachariah (431) purified metmyoglobin from cytochrome C and nonheme proteins by chromatography on DEAE cellulose. A chromatographic column containing hydroxylapatite was successfully used for the concentration of myoglobin in an extract from heated meat products (432). Myoglobin from turkey was isolated and purified via application of a three-step chromatographic procedure starting with Sephadex G-75, followed by CM cellulose and DEAE cellulose. Due to their low myoglobin content, extracts of some meat samples (e.g., heated meat, sausage) have to be concentrated prior to myoglobin analysis. A simple and common procedure for this is ultrafiltration (417).

C. Analysis–Detection–Separation Methods

1. Classical Methods

a. Spectrophotometric Determination.   The spectrophotometric determination of myoglobin in meat can be achieved by two different methods. The first involves the extraction of pigments from meat and their subsequent analysis by absorption measurement (433,434). In the second method the unextracted pigments are directly assessed by reflectance spectrophotometry (435, 436). As meat samples and meat extracts contain both myoglobin and interfering substances the direct photometric measurement of myoglobin is possible only because of differences in the visible absorption spectra of the various heme derivatives. As an alternative, myoglobin can be purified prior to photometric analysis by chromatography or precipitation. There are some traditional methods for determining the total concentration of myoglobin in meat extracts whereby myoglobin is chemically converted to one specific derivative, e.g., to hematin as in Hornsey's method (437) or to the cyanmet derivative as described by Warriss (438). However, this conversion step leads to a loss of information. Therefore methods which estimate the relative concentration of myoglobin derivatives are to be recommended. Krzywicki (433) determined the total myoglobin concentration, as well as the relative concentrations of MbO_2, MetMb, and Mb, by measuring only the absorbances at 525, 545, 565, and 572 nm. The concentrations were calculated according to the following equations:

$$\text{Mb (\%)} = 0.369R_1 + 1.140R_2 - 0.941R_3 + 0.015 \tag{19}$$
$$\text{MbO}_2 \text{ (\%)} = 0.882R_1 - 1.267R_2 + 0.809R_3 - 0.361 \tag{20}$$
$$\text{MetMb (\%)} = -2.514R_1 + 0.777R_2 + 0.800R_3 + 1.098 \tag{21}$$
$$\text{Total myoglobin} = 0.132A_{525} \text{ mmol/L} \tag{22}$$

where R_1, R_2, and R_3 are the absorbance ratios A_{572}/A_{525}, A_{565}/A_{525}, and A_{545}/A_{525}. Analyses of regression indicated that results obtained with this method have a good conformity with results from the cyanmet method. The standard deviation of the myoglobin determination was 0.01 and that for both oxymyoglobin and metmyoglobin was 0.02. The detection limit was ~10 μmol. Another rapid photometric method for the simultaneous determination of Mb and MetMb was published by Trout (439), who measured the absorbances of beef extracts at 525, 572, and 700 nm and calculated the percentage of MetMb and Mb using the following formulas:

$$\text{MetMb (\%)} = \{1.395 - [(A_{572} - A_{700})\{(A_{525} - A_{700})]\} \times 100 \tag{23}$$
$$\text{Mb (mg/ml)} = (A_{525} - A_{700}) \times 2.303 \times \text{dilution factor} \tag{24}$$

A simple spectrophotometric method useful for the determination of absolute and relative concentrations of MbO_2, MbCO, and MetMb in meat or fish samples was developed by Wolfe et al. (434).

The spectra of samples were recorded twice in a wavelength range of 500–700 nm—the first time after the extraction, when any Mb initially present was in the state of MbO_2, and the second time after saturation of the extract with CO. The calculation was accomplished using five formulas, giving <5% error for the determination of the proportions of MbO_2 and MbCO. The total pigment determination also had a precision of ~5%.

Since extraction procedures are a source of considerable error, reflectance spectrophotometry of meat gives better results. Van den Oord and Wesdorp (435) used differences in the absorbances at 580 and 630 nm to determine the MbO_2 and MetMb content in samples with low Mb content. Franke and Solberg (414) measured the relative concentration of MetMb by adjusting the spectra to a constant value at 750 nm and measuring the absorbance at 632 nm. Eagerman et al. (432) published a rapid procedure for following oxidative or reductive changes in meat myoglobin based on differences in the reflectances at 632 and 614 nm. This method is very sensitive to slight color changes in intact beef slices but does not predict or measure the percentage of MetMb present in the sample. Bevilacqua and Zaritzky (436) measured reflectance characteristics of meat at three wavelengths: 525 nm, which is isobiestic for all three derivatives; 572 nm, which is isobiestic for MbO_2; and 473 nm, which is isobiestic for MbO_2 and MetMb. Working with the Kubelka–Munk theory they developed equations for the evaluation of the relative concentrations of MbO_2, Mb, and MetMb. Subsequently, they applied their new methodology to the determination of the kinetic constants of meat myoglobin oxidation and oxygenation rates in unsterile muscle.

b. Column Chromatography (CC).   CC was mainly devised with the intent of purifying heme pigments (428,431) rather than using it as a quantitative method. The most common techniques for the separation of myoglobin and hemoglobin are gel filtration and ion exchange chromatography (440). In particular, the use of Sephadex has been described by numerous researchers (424,429,440). Bünnig and Hamm (441) determined the hemoglobin and myoglobin content of the skeletal muscle of pigs and cattle by Sephadex G-75 gel chromatography and subsequent photometric measurement of the eluant. Dowiercial et al. (442) separated myoglobin and hemoglobin from sheep using Sephadex G-75 and Sephadex G-25. Then they measured the concentration of each pigment in the collected fractions. A similar method was published by Field et al. (430), who accomplished the fractionation and quantification of meat pigments using a Sephadex G-50 column. Warris (424,440) was the first to combine CC and photometry for the direct determination of myoglobin and hemoglobin. The method consisted of a gel chromatographic separation on Sephadex G-50 followed by an on-line photometric measurement of the eluting pigments. Prior to the chromatographic separation the pigments were converted to the cyanmet forms with sodium cyanide and sodium nitrite. A 0.1 M phosphate buffer, pH 6.8, containing 0.1 M sodium chloride, 0.0005 M sodium cyanide, and 0.002% Hibitane, was used as eluant. The effluent was monitored at 420 nm and the quantity of myoglobin determined by comparing the peak heights with those of standards. A detection limit of 2 μg and a determination limit of 5 μg was observed. The mean recoveries were 100.1 ± 4.0% for ox, 100.5 ± 3.5% for sheep, and 100.1 ± 2.1% for pig myoglobin; the standard deviation of the mean of duplicates over 14 samples was ±0.125 mg/g for an overall mean myoglobin concentration of 4.08 mg/g. Nowadays CC is being increasingly replaced by HPLC, which offers a shorter analysis time and a better selectivity.

2. Recent Methods

c. High-Performance Liquid Chromatography (HPLC).   HPLC was first used for the determination of myoglobin by Powell et al. in 1984 (443). Human muscle, urine, and serum myoglobin was injected directly in an anion exchange column (AX-300) and eluted isocratically with 0.022 M Tris buffer, pH 7.5. Serum and muscle myoglobin gave two peaks in the chromatograms, whereas usually only one peak could be detected in urine. The minimum detection level was 2 mg/L and the mean recovery was 94–95%. Due to the omission of a sample pretreatment this method is very simple and rapid. The HPLC run can be carried out in less than 10 min. Ghrist et al. (444) reported differences in the retention times of HPLC separations of myoglobin and hemoglobin when they used a zirconia-stabilized silica particle packing stabilized with diol bonding. When the aim of the HPLC analysis is the quantification of myoglobin, the heme group of the protein should be converted to the same state of oxidation prior to analysis. Oellingrath and Slinde (445) transformed the myoglobin to its cyanoferric form by adding a small amount of potassium ferricyanide or potassium cyanide to the extract. They tested three HPLC columns for their suitability for the analysis of meat pigments in ground beef extract. The best separation of myoglobin and hemoglobin could be achieved with a hydrophobic interaction column (Bio-Gel TSK Phenyl 5-PW), which was operated with a linear gradient of 1.7 M ammonium sulfate, 0.1 M sodium phosphate buffer, pH 7.0 (eluant A) and 0.1 M sodium phosphate buffer, pH 7.0 (eluant B). Figure 25 shows the chromatograms of a standard solution (myoglobin and hemoglobin) and of a ground beef extract. The time of analysis was shorter with the anion exchange column (DAED Microanalyzer MA-7P) than with the hydrophobic interaction column, the retention times for myoglobin being 0.3 and 7.8 min, respectively (446). The percentage recoveries for both myoglobin and hemoglobin ranged from ~90% to ~110%. Gel filtration chromatography on a Bio-Sil TSK-SW column was not suitable for the separation of heme proteins. However, size exclusion HPLC with a Spherogel TSK-3000 SW was successfully applied to the separation of bovine and porcine muscle proteins including myoglobin (421). One good resolved myoglobin peak could be

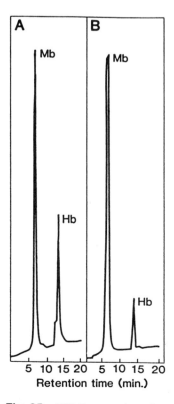

Fig. 25 HPLC separation of myoglobin and hemoglobin on Bio-Gel TSK Phenyl-5-PW. (A) Standard solution. (B) Ground beef extract. (From Ref. 446.)

separated from porcine sarcoplasmic extracts by means of RP-HPLC using trifluoroacetic acid and acetonitrile as eluant (447). In general, HPLC is well reproducible, sensitive, and rapid and has terrific potential as a routine method for the determination of myoglobin in meat samples.

d. Electrophoresis. Electrophoresis is a well-established method for the separation of myoglobin from protein mixtures. Since 1960 protein starch gel zone electrophoresis has been used for identification of fish and animal species (448). It was soon observed that the electrophoretic separation of meat myoglobin gives not one band but a series of bands, which varied from species to species. Consequently, various electrophoretic methods have been developed for the characterization and identification of meat and fish from different species. Höyem and Thorson (449) used disk electrophoresis on polyacrylamide (7.5%) to identify meats from nine animal species by their species-specific myoglobin patterns. Also the adulteration of meat products with whale and horse meat could be detected. Electrophoretic examination of meat juice from various animal species was conducted on a cellulose acetate film by Götze (450). Kim and Shelef (451) employed thin layer agarose gel electrophoresis for the characterization of beef, pork, chicken, and turkey meat. A commonly used method for the identification of animal species is isoelectric focusing (IEF) in polyacrylamide gels ($T = 5\%$, $C = 3\%$) (452). Bauer and Hofman (432) separated myoglobin using IEF at a constant power of 40 V cm^{-1} in pH gradients of 3–10 and 5–8, respectively. Applying this technique they were able to identify 19 meat species in both raw and heated (up to 100°C) materials. In earlier publications myoglobin bands were recognized by their own reddish brown color—the so-called myoglobin method. However, the detection of myoglobin with the myoglobin method is not very sensitive and therefore it was often necessary to concentrate meat extracts from animals which contain low myoglobin concentrations (e.g., rabbit, pig), as well as from heated or pale meats. The sensitivity was markedly increased by adapting a pseudoperoxidase staining procedure for the detection of myoglobin. The pseudoperoxidase staining is based on the ability of hemoproteins to catalyze the oxidation of *o*-dianisidine by hydrogen peroxide, thereby forming an intense brown color marking the myoglobin in the gel (452).

While isoelectric focusing (IEF) and polyacrylamide gel electrophoresis (PAGE) separate myoglobin into a variety of bands, which is the condition for the identification of species, other methods are preferred for the quantification of myoglobin. Hofmann and Blüchel (404) found that SDS electrophoresis in 12% polyacrylamide resulted in only one myoglobin band. The pigment was stained with Coomassie blue R-250 and subsequently the intensity of the spot was measured with a densitometer. Analysis of regression showed a linear correlation between the peak area and the myoglobin content, with a minimum detection level of ~125 ng myoglobin. This method is also suitable for the determination of hemoglobin and other meat proteins like actine and myosin. As the handling of electrophoresis becomes more and more comfortable and automated, this technique will have a great potential for use in routine analysis.

e. Immunochemical Methods. Uhlenhuth (453) was the first to demonstrate that the precipitin reaction could be used to determine the species of origin for blood and muscle tissues. Hashimoto and Yasui (454) prepared rabbit antisera to horse myoglobin to detect horse meat in fresh and canned meat products. This antiserum was superior in antibody titer to an antiserum against cooked horse serum and to an antiserum to a saline extract of horse meat in precipitin analyses. However, there was some cross-reactivity of their antiserum to horse Mb with beef extracts. They concluded that such an antiserum could detect horse meat even in products heated to 70°C and that positive precipitin tests were possible only when some native proteins were present in the product. Hayden (416) used antisera to myoglobin, isolated from ovine, porcine, and equine muscle, in agar-gel diffusion experiments to determine the presence of flesh from these species in beef products. Goat antiporcine Mb serum was used to detect the addition of pork in fresh ground beef at the 3% level and in ground beef heated at 70°C at the 10% level. Rabbit antiequine Mb serum was applied to the detection of horse meat in heated beef sausages at the 3% level. He also detected the presence of lamb in fresh ground beef at the 3% level with monospecific rabbit antiovine Mb serum. Janssen et al. (455) improved the suitability of isoelectric focusing for meat species identification by adapting an immunochemical staining procedure. After isoelectric focusing on an ultrathin polyacrylamide gel the proteins were transferred to nitrocellulose by electroblotting and the blot was incubated with anti–human Mb serum. With this method the detection of less than 10% pork, horse, and sheep meat in a beef-based meat product heated to 120°C for 5 min was possible. Recently, Zijderveld and Koolmees (456) used anti-Mb as a highly specific reagent for the immunohistochemical identification of muscle proteins in comminuted meat products.

3. Further Developments

Some other recently developed analytical methods may offer new possibilities for the analysis of myoglobin. The potential of electrospray mass spectrometry to identify heme pigments from pig, beef, sheep, and horse was demonstrated by Taylor et al. (406). They obtained species-specific spectra corresponding to the molecular weights of the globin portions. Electrospray mass spectrometry is favored for its speed and sensitivity but is hampered by high costs. The on-line combination of HPLC and mass spectrometry is also a very promising method for the selective and rapid determination of myoglobin from protein mixtures. McLuckey et al. (457) separated numerous proteins including myoglobin down to 2.5 pmol per component using a combination of ion spray liquid chromatography and ion trap mass spectrometry. High-performance capillary electrophoresis also has a great potential for the fast separation of charged molecules. Testing the influence of column temperature on the electrophoretic behavior of myoglobin, Rush et al. (458) separated Fe^{3+} from Fe^{2+} myoglobin

D. Application in Food Analysis

Due to its fundamental physiological function as an oxygen carrier and its importance for the meat color, myoglobin has been the object of numerous investigations since the beginning of the eighteenth century (425). Spectrophotometry has been the classical method for determining the relative and the absolute pigment concentrations of myoglobin derivatives in fresh whole and comminuted meats and fish (434). The total myoglobin concentration and relative concentrations of oxidized, oxygenated, and reduced myoglobin derivatives have been determined by simple photometric measurements at four wavelengths (433). Photometric methods were used to study the factors affecting the myoglobin concentration in muscles such as age, activity, species, and nutrition (412). With the aim of improving the color quality of processed meat, the influence of various factors such as pH, sodium chloride content, sodium tripolyphosphate content, cooking temperature, curing salt concentration, and freezing rate were investigated (439). With the application of reflectance spectrophotometry it was possible to determine myoglobin changes in intact meat samples (435). Column chromatography was a valuable method for the isolation and purification of heme proteins from various sources such as beef, pig, sheep, and turkey meats (428). In addition, a gel chromatographic method using Sephadex G-50 was described for the quantitative analysis of myoglobin in muscle of rat, pig, sheep, and ox (424). A SDS electrophoretic method was employed for the separation of myoglobin and hemoglobin. Only one band was found for each component. These bands were stained with Coomassie blue R-250 and their quantity was estimated densitometrically (404). With PAGE and isoelectric focusing myoglobin were separated into a series of myoglobin bands. These electrophoretic patterns were species-specific and were used for the identification of animal species as well as for the quality control of meat products (423). The simple myoglobin method was hampered by its low sensitivity; therefore it is not suitable for samples which contain low myoglobin concentrations like pale or heated meat. The detection level could be lowered by using the pseudoperoxidase (452) and immunochemical (455) stainings. Various HPLC techniques like size exclusion (421), anion exchange (443), hydrophobic interaction (445), and RP-HPLC (447) were applied to separate sarcoplasmic proteins and to determine myoglobin. New analytical methods like HPLC-MS and high-performance capillary electrophoresis are very promising and their suitability for myoglobin analysis will be studied within the next years.

In representation of all other applications in food analysis a simple and rapid HPLC procedure is described in detail.

1. Myoglobin Determination by HPLC (443)

Ten milligrams of skeletal muscle is homogenized in 1.0 ml distilled water and centrifuged at 10,000g for 2 min. Ten microliters of the supernatant is directly injected into an anion exchange column (AX-300, 6.5 μm, 250 × 4 mm). Myoglobin is eluted with 22 mM Tris buffer, pH 7.5, and two peaks are monitored at 405 nm. One HPLC run can usually be accomplished in less than 10 min.

REVIEWS

Belitz, H. D. and W. Grosch, *Lehrbuch der Lebensmittelchemie*, Springer-Verlag, Berlin, 1987.
Goodwin, T. W. (ed.), *Chemistry and Biochemistry of Plant Pigments*, Academic Press, London, 1976.
Gross, J., *Pigments in Fruits*, Academic Press, London, 1987.
Hendry, G. A. F. and J. D. Houghton (eds.), *Natural Food Colorants*, Blackie and Son Ltd., Glasgow, 1992.
Ledward, D.A., in *Developments in Food Proteins, Vol. 3* (B. J. F. Hudson, ed.), Elsevier, London, 1984.
Markakis, P. (ed.), *Anthocyanins as Food Colors*, Academic Press, New York, 1982.
Mazza, G. and E. Miniati, *Anthocyanins in Fruits, Vegetables and Grains*, CRC Press, Boca Raton, 1993.
Nillson, T., *Lantbrukshögskolans Annaler, 36*: 179 (1970).
Scheer, H. (ed.), *Chlorophylls*, CRC Press, Boca Raton, 1991.
Vernon, L. P. and G. R. Seely (eds.), *The Chlorophylls*, Academic Press, New York, 1966.
Walters, C. L., in *Meat* (D. J. A. Cole and R. A. Lawrie, eds.), Butterworths, London, 1975.

REFERENCES

1. G. Britton, in *Natural Food Colorants* (G. A. F. Hendry and J. D. Houghton, eds.), Blackie and Son Ltd., Glasgow, 1992, p. 141.
2. H. Davies, in *Chemistry and Biochemistry of Plant Pigments* (T. W. Goodwin, ed.), Academic Press, London, 1976, p. 38.
3. H. D. Belitz and W. Grosch, *Lehrbruch der Lebensmittelchemie*, Springer-Verlag, Berlin, 1987, p. 196.
4. G. K. Gregory, T.-S. Chen, and T. Philip, *J. Food Sci., 42*: 1071 (1987).
5. J. Gross, *Pigments in Fruits*, Academic Press, London, 1987, p. 87.
6. L. Zechmeister, *Cis–trans Isomeric Carotenoids, Vitamins A and Arylpolyenes*, Springer-Verlag, Wien, 1962.
7. J. Schwartz and M. Patroni-Killam, *J. Agric. Food Chem., 33*: 1160 (1985).
8. T. W. Goodwin, in *Chemistry and Biochemistry of Plant Pigments* (T. W. Goodwin, ed.), Academic Press, London, 1976, p. 38.
9. J. Spinelli and C. Mahnken, *Aquaculture, 13*: 213 (1978).
10. C. A. Bailey and B. H. Chen, *J. Food Sci., 54*: 584 (1989).
11. F. Khachik, G. R. Beecher, and W. R. Lusby, *J. Agric. Food Chem., 37*: 1465 (1989).
12. M. P. Cano, *J. Agric. Food Chem., 39*: 1786 (1991).
13. J. Gross, *Gartenbauwissenschaft, 49*: 18 (1984).

14. J. G. Buishand and W. H. Gabelman, *Euphytica*, *28*: 611 (1979).
15. M. I. Minguez-Mosquera and D. Hornero-Méndez, *J. Agric. Food Chem.*, *41*: 1616 (1993).
16. S. J. Taylor and I. J. McDowell, *J. Sci. Food Agric.*, *57*: 287 (1991).
17. M. C. Santos, P. A Bobbio, and D. B. Rodriguez-Amaya, *Acta Alimentaria*, *17*: 33 (1988).
18. K. L. Simpson, in *Chemistry and Biochemistry of Marine Food Products* (R. E. Martin, G. J. Flick, C. E. Hebard, and D. R. Ward, eds.), AVI, Westport, 1982, p. 115.
19. K. Hermann, *Industr. Obst- u- Gemüseverwert.*, *77*: 250 (1992).
20. K. Hermann, *Industr. Obst- u- Gemüseverwert.*, *78*: 1 (1993).
21. M. I. Minguez-Mosquera, M. Jarén-Galán, and J. Garrido-Fernández, *J. Agric. Food Chem.*, *40*: 2384 (1992).
22. C. Marty and C. Berset, *J. Food Sci.*, *51*: 698 (1986).
23. J. L. Multon (ed.), *Additives et Auxiliares de Fabricacion dans les Industries Agro-Alimentaires*, Collection Sciences et Techniques Agro-Alimentaires, Tec et Doc-Lavoisier Apria, Paris, 1992.
24. J. C. Bauernfeind (ed.), *Carotenoids as Colorants and Vitamin A Precursors*, Academic Press, New York, 1981.
25. T. R. Sommer, F. M. Souza, and N. M. Morrissy, *Aquaculture*, *106*: 63 (1992).
26. H. K. No and T. Storebakken, *Aquaculture*, *97*: 203 (1991).
27. K. L. Simpson, *Proc. Nutr. Soc.*, *42*: 7 (1983).
28. G. W. Burton, *J. Nutr.*, *119*: 109 (1989).
29. C. A. Pesek, J. J. Warthesen, and P. S. Taoukis, *J. Agric. Food Chem.*, *38*: 41 (1990).
30. R. G. Ziegler, *Am. J. Clin. Nutr.*, *53*: 251 (1991).
31. R. R. Butrum, C. K. Clifford, and E. Lanza, *Am. J. Clin. Nutr.*, *48*: 888 (1988).
32. J. Gross, *Gartenbauwissenschaft*, *50*: 88 (1985).
33. K. Helrich, *Official Methods of Analysis*, Association of Official Analytical Chemists, Arlington, 1990, p. 1048.
34. J.-P. C. Hsieh and M. Karel, *J. Chromatogr.*, *259*: 515 (1983).
35. G. Sadler, J. Davis, and D. Dezman, *J. Food Sci.*, *55*: 1460 (1990).
36. M. I. Minguez-Mosquera, B. Gandul-Rojas, and M. L. Gallardo-Guerrero, *J. Agric. Food Chem.*, *40*: 60 (1992).
37. J. F. Fisher and R. L. Rouseff, *J. Agric. Food Chem.*, *34*: 985 (1986).
38. G. A. Spanos, H. Chen, and S. J. Schwartz, *J. Food Sci.*, *58*: 817 (1993).
39. I. Stewart and T. A. Wheaton, *J. Chromatogr.*, *12*: 2947 (1973).
40. J. Gross, M. Gabai, and A. Lifshitz, *J. Food Sci.*, *36*: 466 (1971).
41. Y. Ittah, J. Kanner, and R. Granit, *J. Agric. Food Chem.*, *41*: 899 (1993).
42. G. Britton and T. W. Goodwin, *Meth. Enzymol.*, *18C*: 654 (1971).
43. J. B. Moster and A. N. Prater, *Food Technol.*, *3*: 146 (1957).
44. L. Benedek, *Z. Lebensm.-Unters.-Forsch.*, *107*: 228 (1958).
45. M. Fekete, L. Kozma, and T. Huszka, *Acta Alimentaria*, *5*: 119 (1976).
46. M. Baranyai and J. Szabolcs, *Acta Alimentaria*, *5*: 87 (1976).
47. R. R. Cruse, B. J. Lime, and R. A. Hensz, *J. Agric. Food Chem.*, *27*: 641 (1979).
48. R. Willstätter and H. J. Page, *Justus Liebigs Annalen Chemie*, *404*: 237 (1914).
49. L. A. Curl, *J. Agric. Food Chem.*, *10*: 504 (1962).
50. C. Marty and C. Berset, *J. Food Sci.*, *53*: 1880 (1988).
51. T. Philip and T. S. Chen, *J. Food Sci.*, *53*: 1703 (1988).
52. I. Stewart and T. A. Wheaton, *J. Chromatogr.*, *55*: 325 (1971).
53. S. Liaaen-Jensen and A. Jensen, *Meth. Enzymol.*, *23*: 586 (1971).
54. A. Jensen, in *Carotine und Carotenoide*, Steinkopff, Darmstadt, 1963, p. 119.
55. J. G. Kirchner, *Thin-Layer Chromatography*, Interscience, New York, 1967.
56. H. R. Bolliger, in *Dünnschicht-Chromatographie* (E. Stahl, ed.), Springer-Verlag, Berlin, 1962.
57. K. Randerath, *Dünnschicht-Chromatographie*, Verlag Chemie, Weinheim, 1962.
58. J. Gross, *Chromatographia*, *13*: 572 (1980).
59. T. Johjima, *J. Jpn. Soc. Hort. Sci.*, *62*: 567 (1993).
60. I. Csorba, Z. Buzás, B. Polyák, and L. Boross, *J. Chromatogr.*, *172*: 287 (1979).
61. A. Razungles, C. I. Bayonove, R. E. Cordonnier, and R. L. Baumes, *Vitis*, *26*: 183 (1987).
62. P. Kanasawud and J. C. Crouzet, *J. Agric. Food Chem.*, *38*: 237 (1990).

63. M. I. Minguez-Mosquera, J. Garrido-Fernández, and B. Gandul-Rojas, *J. Agric. Food Chem.*, *38*: 1662 (1990).
64. K. A. Buckle and F. M. M. Rahman, *J. Chromatogr.*, *171*: 385 (1979).
65. A. Winterstein, A. Studer, and R. Rüegg, *Chem. Ber.*, *93*: 2951 (1960).
66. A. Hager and T. Meyer-Bertenrath, *Planta*, *69*: 198 (1966).
67. J. Gross, *Lebensm.-Wiss. u.- Technol.*, *15*: 36–38 (1982).
68. R. Eder, R. Kalchgruber, S. Wendelin, M. Pastler, and J. Barna, *Mitt. Klosterneuburg.*, *41*: 168 (1991).
69. M. Mottier, *Mitt. Gebiete Lebensm. u. Hyg.*, *43*: 118 (1952).
70. B. H. Davies, S. Matthews, and J. T. O. Kirk, *Phytochemistry*, *9*: 797 (1970).
71. F. Khachik, G. R. Beecher, and N. F. Whittaker, *J. Agric. Food Chem.*, *34*: 603 (1986).
72. T. Philip and T.-S. Chen, *J. Chromatogr.*, *435*: 113 (1988).
73. J. P. Sweeney and A. C. Marsh, *JAOAC*, *53*: 937 (1970).
74. I. Stewart and T. A. Wheaton, *J. Chromatogr.*, *55*: 325 (1971).
75. I. Stewart, *JAOAC*, *60*: 132 (1977).
76. I. Stewart, *J. Agric. Food Chem.*, *25*: 1132 (1977).
77. A. Fiksdahl, J. T. Mortensen, and S. Liaaen-Jensen, *J. Chromatogr.*, *157*: 111 (1978).
78. K. Tsukida, K. Saiki, T. Takii, and Y. Koyama, *J. Chromatogr.*, *245*: 359 (1982).
79. L. A. Chandler and S. J. Schwartz, *J. Food Sci.*, *52*: 669 (1987).
80. C. A. O'Neil and S. J. Schwartz, *J. Chromatogr.*, *624*: 235 (1992).
81. M. Vecchi, G. Englert, R. Maurer, and V. Meduna, *Helv. Chim. Acta*, *64*: 2746 (1981).
82. M. Vecchi, G. Englert, and H. Mayer, *Helv. Chim. Acta*, *65*: 1050 (1982).
83. Y. Ikuno, T. Maoka, M. Shimizu, T. Komori, and T. Matsuno, *J. Chromatogr.*, *328*: 387 (1985).
84. N. E. Craft, S. A. Wise, and J. H. Soares, *J. Agric. Food Chem.*, *41*: 208 (1993).
85. R. J. Bushway, *J. Liq. Chromatogr.*, *8*: 1527 (1985).
86. F. W. Quackenbush, *J. Liq. Chromatogr.*, *10*: 643 (1987).
87. K. S. Epler, L. C. Sander, R. G. Ziegler, S. A. Wise, and N. E. Craft, *J. Chromatogr.*, *595*: 89 (1992).
88. G. Noga and F. Lenz, *Chromatographia*, *17*: 139 (1983).
89. M. Zakaria, K. Simpson, P. R. Brown, and A. Krstulovic, *J. Chromatogr.*, *176*: 109 (1979).
90. F. Khachik, G. R. Beecher, and W. R. Lusby, *J. Agric. Food Chem.*, *37*: 1465 (1989).
91. P. A. Biacs, H. G. Daood, T. T. Huszka, and P. K. Biacs, *J. Agric. Food Chem.*, *41*: 1864 (1993).
92. G. K. Gregory, T.-S. Chen, and T. Philip, *J. Food Sci.*, *52*: 1071 (1987).
93. C. Fisher and J. A. Kocis, *J. Agric. Food Chem.*, *3*: 55 (1987).
94. A. Razungles, J. Oszmianski, and J.-C. Sapis, *J. Food Sci.*, *54*: 774 (1989).
95. M. Baranyai, Z. Matus, and J. Szabolcs, *Acta Alimentaria*, *11*: 309 (1982).
96. S. J. Taylor and I. J. McDowell, *J. Sci. Food Agric.*, *57*: 287 (1991).
97. P. W. Simon and X. Y. Wolff, *J. Agric. Food Chem.*, *35*: 1017 (1987).
98. M. C. Aubert, C. R. Lee, A. M. Krstulovic, E. Lesellier, M. R. Péchard, and A. Tchapla, *J. Chromatogr.*, *557*: 47 (1991).
99. E. Lesellier, A. M. Krstulovic, and A. Tchapla, *J. Chromatogr.*, *641*: 137 (1993).
100. E. Lesellier, A. Tchapla, and A. M. Krstulovic, *J. Chromatogr.*, *645*: 29 (1993).
101. R. F. Taylor, P. E. Farrow, L. M. Yelle, J. C. Harris, and I. G. Marenchic, in *Carotenoids Chemistry and Biology* (N. I. Krinsky, M. M. Mathews-Roth, and R. F. Taylor, eds.), Plenum Press, New York, 1990, p. 105.
102. A. L. Curl, *Food Res.*, *25*: 190 (1960).
103. J. Gross and G. Eckhardt, *Phytochemistry*, *20*: 2267 (1981).
104. N. Umiel and W. H. Gabelman, *J. Am. Soc. Hort. Sci.*, *96*: 702 (1971).
105. A. L. Curl, *J. Agric. Food Chem.*, *1*: 456 (1953).
106. A. L. Curl, *J. Food Sci.*, *29*: 241 (1964).
107. J. Gross, *Gartenbauwissenschaft*, *49*: 180 (1984).
108. A. L. Curl and G. F. Bailey, *J. Food Sci.*, *26*: 442 (1961).
109. M. Vinkler and M. Kiszel-Richter, *Acta Alimentaria*, *1*: 41 (1972).
110. B. Camara and R. Moneger, *Phytochemistry*, *17*: 91 (1978).
111. H. G. Daood, B. Czinkotai, A. Hoschke, and P. Biacs, *J. Chromatogr.*, *472*: 296 (1989).
112. A. Ben-Aziz, G. Britton, and T. W. Goodwin, *Phytochemistry*, *12*: 2759 (1973).
113. T. Johjima and H. Ogura, *J. Jpn. Soc. Hort. Sci.*, *52*: 202 (1983).

114. B. J. Lime, F. P. Griffiths, R. T. O'Connor, D. C. Heinzelman, and E. R. McCall, *J. Agric. Food Chem.*, *5*: 941 (1957).
115. J. Gross, M. Gabai, and A. Lifshitz, *J. Food Sci.*, *36*: 466 (1971).
116. G. A. F. Hendry, in *Natural Food Colorants* (G. A. F. Hendry and J. D. Houghton, eds.), Blackie and Son Ltd., Glasgow, 1992, p. 79.
117. H. C. Sorby, *Proc. R. Soc.*, *21*: 442 (1873).
118. K. Hermann, *Industr. Obst- und Gemüseverwert.*, *78*: 42 (1993).
119. H. Scheer, in *Chlorophylls* (H. Scheer, ed.), CRC Press, Boca Raton, 1991, p. 3.
120. A. H. Jackson, in *Chemistry and Biochemistry of Plant Pigments* (T. W. Goodwin, ed.), Academic Press, London, 1976, p. 1.
121. J. Gross and I. Ohad, *Photochem. Photobiol.*, *37*: 195 (1983).
122. M. Flügel and J. Gross, *Angew. Botanik*, *56*: 393 (1982).
123. Y. Izaki, K. Yoshida, K. Hidaka, and K. Toda, *J. Jpn. Soc. Food Sci. Nutrit.*, *39*: 485 (1986).
124. N. Gorin and H. Zonneveld, *Lebensm.-Wiss. u.- Technol.*, *10*: 50 (1977).
125. J. Gross, *Gartenbauwissenschaft*, *47*: 162 (1982).
126. J. Gross, *Gartenbauwissenschaft*, *49*: 128 (1984).
127. J. Gross, *Gartenbauwissenschaft*, *47*: 142 (1982).
128. W. A. Svec, in *Chlorophylls* (H. Scheer, ed.), CRC Press, Boca Raton, 1991, p. 89.
129. P. M. Gorski and L. L. Creasy, *J. Am. Soc. Hort. Sci.*, *102*: 73 (1977).
130. S. J. Schwartz, S. L. Woo, and J. H. von Elbe, *J. Agric. Food Chem.*, *29*: 533 (1981).
131. M. P. Cano and M. A. Marin, *J. Agric. Food Chem.*, *40*: 2141 (1992).
132. H. D. Blackbourn, M. J. Jeger, P. John, and A. K. Thompson, *Ann. Appl. Biol.*, *117*: 147 (1990).
133. A. E. Watada, K. H. Norris, J. T. Worthington, and D. R. Massie, *J. Food Sci.*, *41*: 329 (1976).
134. M. Knee, G. S. Hatfield, and S. M. Smith, *J. Hort. Sci.*, *64*: 403 (1989).
135. N. Yamauchi and A. E. Watada, *J. Am. Soc. Hort. Sci.*, *116*: 58 (1991).
136. C. Berset and P. Caniaux, *J. Food Sci.*, *48*: 1854 (1983).
137. S. J. Schwartz and T. V. Lorenzo, *J. Food Sci.*, *56*: 1059 (1991).
138. L. J. van Lelyveld, C. Fraser, B. L. Smith, and G. Visser, *S.-Afr. Tydskr. Plant Grond*, *7*: 188 (1990).
139. S. Taylor, D. Baker, P. Owuor, J. Orchard, C. Othieno, and C. Gay, *J. Sci. Food Agric.*, *58*: 185 (1992).
140. P. A. Bobbio and M. C. Guedes, *Food Chem.*, *36*: 165 (1990).
141. A. M. Humphrey, *Food Chem.*, *5*: 57 (1980).
142. H. Inoue, M. Imai, T. Naemura, K. Furuya, and Y. Shizur, *J. Chromatogr.*, *645*: 259 (1993).
143. M. Holden, in *Chemistry and Biochemistry of Plant Pigments* (T. W. Goodwin, ed.), Academic Press, London, 1976, p. 1.
144. H. H. Strain and W. A. Svec, in *The Chlorophylls* (L. P. Vernon and G. R. Seely, eds.), Academic Press, New York, 1966, p. 22.
145. Y. Fuke, K. Sasago, and H. Matsuoka, *J. Food Sci.*, *50*: 1220 (1985).
146. F. C. Pennington, H. H. Strain, W. A. Svec, and J. J. Katz, *J. Am. Chem. Soc.*, *89*: 3875 (1967).
147. W. A. Svec, in *The Porphyrins*, Vol 5 (D. Dolphin, ed.), Academic Press, New York, 1978, p. 341.
148. M. Minguez-Mosquera and J. Garrido-Fernandez, *J. Agric. Food Chem.*, *37*: 1 (1989).
149. J. H. C. Smith and A. Benitez, in *Modern Methods of Plant Analysis*, Vol. 4 (K. Paech and M. V. Tracey, eds.), Springer-Verlag, Berlin, 1955, p. 142.
150. A. M. Johnson-Flanagan and M. R. Thiagarajah, *J. Plant Physiol.*, *136*: 180 (1990).
151. G. Mackinney, *J. Biol. Chem.*, *140*: 315 (1941).
152. D. I. Arnon, *Plant Physiol.*, *24*: 1 (1949).
153. R. J. Porra, W. A. Thompson, and P. E. Kreidemann, *Biochem. Biophys. Acta*, *975*: 384 (1989).
154. J. F. G. M. Wintermans and A. de Mots, *Biochem. Biophys. Acta*, *109*: 448 (1965).
155. Z. Sesták, *Biologia Plant*, *8*: 97 (1966).
156. J. K. Abaychi and J. P. Riley, *Anal. Chim. Acta*, *107*: 1 (1979).
157. R. J. Porra, in *Chlorophylls* (H. Scheer, ed.), CRC Press, Boca Raton, 1991, p. 31.
158. J. L. Wickliff and S. Aronoff, *Plant Physiol.*, *37*: 584 (1962).
159. J. R. Wilson and M.-D. Nutting, *Anal. Chem.*, *35*: 144 (1963).
160. G. Mackinney and C. A. Weast, *Ind. Eng. Chem.*, *32*: 392 (1940).
161. L. P. Vernon, *Anal. Chem.*, *32*: 1144 (1960).

162. J. R. Wilson and M. D. Nutting, *Anal. Chem.*, *35*: 144 (1963).
163. R. H. Goodwin, *Anal. Chem.*, *19*: 789 (1947).
164. R. C. White, I. D. Jones, E. Gibbs, and L. S. Butler, *J. Agric. Food Chem.*, *20*: 773 (1972).
165. N. K. Boardman and S. W. Thorne, *Biochim. Biophys. Acta*, *253*: 222 (1971).
166. H. I. Virgin, *Physiologia Plant*, *13*: 155 (1960).
167. W. R. Forbus Jr., G. G. Dull, and D. Smittle, *J. Food Sci.*, *56*: 981 (1991).
168. F. J. Francis, *Hort. Sci.*, *5*: 102 (1970).
169. F. J. Francis, *Hort. Sci.*, *15*: 58 (1980).
170. M. P. Cano, M. A. Marin, and B. De Ancos, *Z. Lebensm.-Unters.-Forsch.*, *197*: 346 (1993).
171. H. Schultz, *Gartenbau*, *23*: 371 (1976).
172. H. M. Benedict and R. Swidler, *Science*, *133*: 2015 (1961).
173. O. L. Jahn and R. Young, *J. Am. Soc. Hort. Sci.*, *101*: 416 (1976).
174. O. L. Jahn, *J. Am. Soc. Hort. Sci.*, *100*: 688 (1975).
175. G. S. Birth, G. G. Dull, J. B. Magee, H. T. Chan, and C. G. Cavaletto, *J. Amer. Soc. Hort. Sci.*, *109*: 62 (1984).
176. B. L. Strehler and W. A. Arnold, *J. Gen. Physiol.*, *34*: 809 (1951).
177. G. Tollin, E. Fumimori, and M. Calvin, *Nature*, *181*: 1266 (1958).
178. W. R. Forbus Jr., S. D. Senter, and R. L. Wilson, *J. Food Sci.*, *50*: 750 (1985).
179. Y. Chuma, K. Nakaji, and M. Ohura, *ASAE Trans.*, *23*: 1043 (1980).
180. W. R. Forbus Jr., S. D. Senter, and T. T. Chan, *J. Food Sci.*, *52*: 356 (1987).
181. K. Nakaji, Y. Sagiura, and Y. Chuma, *J. Soc. Agric. Mach. (Jpn.)*, *39*: 483 (1987).
182. W. R. Forbus Jr., J. A. Payne, and S. D. Senter, *J. Food Sci.*, *56*: 985 (1991).
183. D. W. Hughes and A. S. Holt, *Can. J. Chem.*, *40*: 171 (1962).
184. P. H. Hynninen and N. Ellfolk, *Acta Chem. Scand.*, *27*: 1463 (1973).
185. M. Tswett, *Ber. Deutsch. Bot. Ges.*, *24*: 384 (1906).
186. Y. Shioi, in *Chlorophylls* (H. Scheer, ed.), CRC Press, Boca Raton, 1991, p. 59.
187. T. Omata and N. Murata, *Plant Cell Physiol.*, *24*: 1093 (1983).
188. H. H. Strain and J. Sherma, *J. Chromatogr.*, *73*: 371 (1972).
189. Z. Sesták, *J. Chromatogr. Libr.*, *3*: 1039 (1975).
190. A. Hager and T. Bertenrath, *Planta*, *58*: 564 (1962).
191. M. F. Bacon, *J. Chromatogr.*, *17*: 322 (1965).
192. M. Durand and D. Laval-Martin, *J. Chromatogr.*, *97*: 92 (1974).
193. H. A. W. Schneider, *J. Chromatogr.*, *21*: 448 (1966).
194. A. Hager and T. Meyer-Bertenrath, *Planta*, *69*: 198 (1966).
195. S. W. Jeffrey, *Biochim. Biophys. Acta*, *162*: 271 (1968).
196. I. Csorba, Z. Buzás, B. Polyák, and L. Boross, *J. Chromatogr.*, *172*: 287 (1979).
197. I. Sahlberg and P. H. Hynninen, *J. Chromatogr.*, *291*: 331 (1984).
198. Y. Takeda, Y. Saito, and M. Uchiyama, *J. Chromatogr.*, *280*: 188 (1983).
199. M. Vesk and S. W. Jeffrey, *J. Phycol.*, *26*: 322 (1987).
200. R. Ziegler, A. Blaheta, N. Guha, and B. Schönegge, *J. Plant Physiol.*, *132*: 327 (1988).
201. S. H. Schanderl and D. Y. C. Lynn, *J. Food Sci.*, *31*: 141 (1966).
202. C. A. Rebeiz, M. Yaghi, M. Abou-Haidar, and P. A. Castelfranco, *Pl. Physiol.*, *46*: 57 (1970).
203. K. A. Buckle and R. A. Edwards, *Phytochemistry*, *8*: 1901 (1969).
204. M. F. Bacon, *J. Chromatogr.*, *17*: 322 (1965).
205. P. K. Mahanta and M. Hazarika, *J. Sci. Food Agric.*, *36*: 1133 (1985).
206. N. Evans, D. E. Games, A. H. Jackson, and S. A. Matlin, *J. Chromatogr.*, *115*: 325 (1975).
207. M. I. Minguez-Mosquera, B. Gandul-Rojas, A. Montano-Asqerino, and J. Garrido-Fernández, *J. Chromatogr.*, *585*: 259 (1991).
208. K. Iriyama, M. Yoshiura, and M. Shiraki, *J. Chromatogr.*, *154*: 302 (1978).
209. P. Kuronen, K. Hyvärinen, P. Hynninen, and I. Kilpeläinen, *J. Chromatogr.*, *654*: 93 (1993).
210. F. L. Canjura and S. J. Schwartz, *J. Agric. Food Chem.*, *39*: 1102 (1991).
211. F. L. Canjura, S. J. Schwartz, and R. V. Nunes, *J. Food Sci.*, *56*: 1639 (1991).
212. S. J. Schwartz, J. H. von Elbe, and R. C. Lindsay, *J. Agric. Food Chem.*, *31*: 349 (1983).
213. H. G. Daood, B. Czinkotai, A. Hoschke, and P. Biacs, *J. Chromatogr.*, *472*: 296 (1989).
214. C. A. Bailey and B. H. Chen, *J. Chromatogr.*, *455*: 396 (1988).

215. J. Lopez-Hernandez, L. Vazquez-Oderiz, E. Vazquez-Blanco, A. Romero-Rodriguez, and J. Simal-Lozano, *J. Agric. Food Chem.*, *41*: 1613 (1993).
216. K. Eskins, C. R. Scholfield, and H. J. Dutton, *J. Chromatogr.*, *135*: 217 (1977).
217. D. W. De Jong and W. G. Woodlief, *J. Agric. Food Chem.*, *26*: 1281 (1978).
218. R. F. C. Mantoura and C. A. Llewellyn, *Anal. Chim. Acta*, *151*: 297 (1983).
219. J. E. Hunt and T. J. Michalski, in *Chlorophylls* (H. Scheer, ed.), CRC Press, Boca Raton, 1991, p. 835.
220. R. B. van Breemen, F. L. Canjura, and S. J. Schwartz, *J. Agric. Food Chem.*, *39*: 1452 (1991).
221. M. Knee, *J. Food Technol.*, *15*: 493 (1980).
222. J. Abbas, M.-A. Rouet-Mayer, and J. Philippon, *Lebensm.-Wiss. u.- Technol.*, *22*: 68 (1989).
223. N. Muftugil, *J. Food Proc. Preserv.*, *10*: 69 (1986).
224. H. Zonneveld, W. Klop, and N. Gorin, *Z. Lebensm.-Unters.-Forsch.*, *178*: 20 (1984).
225. M. J. Maunders, S. B. Brown, and H. W. Woolhouse, *Phytochemistry*, *11*: 2443 (1983).
226. K. W. Lebermann, A. I. Nelson, and N. P. Steinberg, *Food Technol.*, *22*: 487 (1968).
227. R. L. Shewfelt, E. K. Heaton, and K. M. Batal, *J. Food Sci.*, *49*: 1612 (1984).
228. P. A. Rouso, H. C. Harrison, and J. P. Palta, *Hort. Sci.*, *21*: 499 (1986).
229. G. L. Robertson, *Food Chem.*, *17*: 25 (1985).
230. K. A. Buckle and R. A. Edwards, *Phytochemistry*, 8: 1901 (1969).
231. A. Attila, *Industr. Obst- u- Gemüseverwert.*, *56*: 335 (1971).
232. E. Glynne-Jones, R. Marshall, R. A. Hahn, and G. Read, *J. Chromatogr.*, *114*: 232 (1975).
233. F. Lajollo, S. R. Tannenbaum, and T. P. Labuza, *J. Food Sci.*, *36*: 850 (1971).
234. P. K. Mahanta and M. Hazarika, *J. Sci. Food Agric.*, *36*: 1133 (1985).
235. K. Helrich (ed.), *Official Methods of Analysis*, Association of Official Analytical Chemists, Arlington, 1990, p. 62.
236. R. Brouillard, *Anthocyanins as Food Colors* (P. Markakis, ed.), Academic Press, New York, 1982, p. 1.
237. R. Willstätter and A. E. Everest, *Justus Liebigs Annalen Chemie*, *401*: 189 (1913).
238. J. J. Macheix, A. Fleuriet, and J. Billot, *Fruit Phenolics*, CRC Press, Boca Raton, 1990.
239. G. Mazza and E. Miniati, *Anthocyanins in Fruits, Vegetables and Grains*, CRC Press, Boca Raton, 1993.
240. G. M. Sapers, A. M. Burgher, J. G. Phillips, S. B. Jones, and E. G. Stone, *J. Am. Soc. Hort. Sci.*, *109*: 105 (1984).
241. L. W. Wulf and C. W. Nagel, *Am. J. Enol. Vitic.*, *29*: 42 (1978).
242. M. Williams, G. Hrazdina, M. M. Wilkinson, J. G. Sweeney, and G. Iacobucci, *J. Chromatogr.*, *155*: 389 (1978).
243. C. F. Timberlake and P. Bridle, *Anthocyanins as Food Colors* (P. Markakis, ed.), Academic Press, New York, 1982, p. 126.
244. J. R. Ballington, W. E. Ballinger, and E. P. Maness, *J. Am. Soc. Hort. Sci.*, *112*: 859 (1987).
245. M. L. González-San José, L. J. R. Barron, and C. Diaz, *J. Sci. Food Agric.*, *51*: 337 (1990).
246. J. P. Roggero, S. Coen, and B. Ragonnet, *Am. J. Enol. Vitic.*, *37*: 77 (1986).
247. P. Markakis, *Anthocyanins as Food Colors* (P. Markakis, ed.), Academic Press, New York, 1982, p. 163.
248. G. Skrede, *J. Food Sci.*, *50*: 514 (1985).
249. A. Rommel, D. A. Heatherbell, and R. E. Wrolstad, *J. Food Sci.*, *55*: 1011 (1990).
250. J. Bakker, N. W. Preston, and C. F. Timberlake, *Am. J. Enol. Vitic.*, *37*: 121 (1986).
251. L. Gao and G. Mazza, *J. Agric. Food Chem.*, *42*: 118 (1994).
252. J. Wilska-Jeszka and K. B. Zajac, *Int. Food Ingredients*, *3*: 11 (1991).
253. F. J. Francis, *Trends in Food Sci. Techn.*, *3*: 27 (1992).
254. V. Hong and R. E. Wrolstad, *J. Agric. Food Chem.*, *38*: 698 (1990).
255. J. P. Roggero, J. L. Larice, C. Rocheville-Divorne, P. Archier, and S. Coen, *R. F. Oe.*, *112*: 41 (1988).
256. H. Woidich and T. Langer, *Fruchtsaft Ind.*, *4*: 234 (1959).
257. R. E. Wrolstad, J. D. Culbertson, C. J. Cornwell, and L. R. Mattick, *JAOAC*, *65*: 1417 (1982).
258. R. Eder, R. Kalchgruber, S. Wendelin, and J. Barna, *Mitt. Klosterneuburg*, *40*: 152 (1990).
259. M. Wilkinson, J. G. Sweeny, and G. A. Iacobucci, *J. Chromatogr.*, *132*: 349 (1977).
260. A. B. Moore, F. J. Francis, and F. M. Clydesdale, *J. Food Protect.*, *45*: 738 (1982).
261. L. R. Mattick, L. D. Weirs, and W. B. Robinson, *Am. J. Enol. Vitic.*, *20*: 206 (1969).

262. F. J. Francis, *Anthocyanins as Food Colors* (P. Markakis, ed.), Academic Press, New York, 1982, p. 182.
263. S. Sakamura and F. J. Francis, *J. Food Sci.*, *26*: 318 (1961).
264. B. H. Barritt and L. C. Torre, *J. Chromatogr.*, *75*: 151 (1973).
265. H. S. Lee and L. Wicker, *J. Food Sci.*, *466 (1991)*.
266. J. H. Kim, G. I. Nonaka, K. Fujieda, and S. Uemoto, *Phytochemistry*, *28*: 1503 (1989).
267. L. Wulf and C. Nagel, *Am. J. Enol. Vitic.*, *29*: 42 (1978).
268. H. Sakellariades and B. S. Luh, *J. Food Sci.*, *39*: 329 (1974).
269. R. Iori, P. G. Pifferi, and A. Vaccari, *Progress in Food Engineering* (C. Cantarelli and C. Peri, eds.), Forster, Küsnacht, Switzerland, 1983, p. 581.
270. R. Eder, S. Wendelin, and J. Barna, *Mitt. Klosterneuburg*, *40*: 68 (1990).
271. H. S. Lee and V. Hong, *J. Chromatogr.*, *624*: 221 (1992).
272. T. Fuleki and F. J. Francis, *J. Food Sci.*, *33*: 72 (1968).
273. T. C. Somers, *Phytochemistry*, *10*: 2175 (1971).
274. E. Sondheimer and Z. I. Kertesz, *Anal. Chem.*, *20*: 245 (1948).
275. T. Fuleki and F. J. Francis, *J. Food Sci.*, *33*: 78 (1968).
276. R. E. Wrolstad, J. D. Culbertson, C. J. Cornwell, and L. R. Mattick, *JAOAC*, *65*: 1417 (1982).
277. T. C. Somers and M. E. Evans, *J. Sci. Food Agric.*, *28*: 279 (1977).
278. Y. Glories, *Conn. Vigne Vin*, *18*: 253 (1984).
279. T. Swain and W. E. Hillis, *J. Sci. Food Agric.*, *10*: 63 (1959).
280. R. E. Wrolstad, *Color and Pigment Analyses in Fruit Products*, Bull. 624, Oregon Agricol. Exp. Station, Corvallis, 1976.
281. J. W. McClure, *Plant Physiol.*, *43*: 193 (1967).
282. C. Zapsalis and F. J. Francis, *J. Food Sci.*, *30*: 396 (1965).
283. G. K. Niketic-Aleksic and G. Hrazdina, *Lebensm.-Wiss. u.- Technol.*, *5*: 163 (1972).
284. S. Asen, N. W. Stuart, and H. W. Siegelman, *Proc. Am. Soc. Hort. Sci.*, *73*: 495 (1959).
285. E. C. Jorgensen and T. A. Geissman, *Arch. Biochem. Biophys.*, *70*: 389 (1955).
286. T. C. Somers, *J. Sci. Food Agric.*, *17*: 215 (1966).
287. P. Francis and F. M. Clydesdale, *Food Colorimetry: Theory and Applications*, AVI, Westport, 1975.
288. A. Kramer, *Food Technol*, *30*: 62 (1976).
289. R. Eder, S. Wendelin, R. Kalchgruber, F. Rosenthal, and J. Barna, *Mitt. Klosterneuburg*, *42*: 148 (1992).
290. M. Drdák, F. Malik, and J. Karovicová, *Weinwissenschaft*, *43*: 129 (1988).
291. G. Skrede, R. E. Wrolstad, P. Lea, and G. Enersen, *J. Food Sci.*, *57*: 172 (1992).
292. A. Rommel, D. A. Heatherbell, and R. E. Wrolstad, *J. Food Sci.*, *55*: 1011 (1990).
293. G. Urbányi and K. Horti, *Acta Alimentaria*, *21*: 307 (1992).
294. J. Bakker, P. Bridle, and C. F. Timberlake, *Vitis*, *25*: 67 (1986).
295. S. Singha, T. A. Baugher, E. C. Townsend, and M. C. D'Souza, *J. Am. Soc. Hort. Sci.*, *116*: 497 (1991).
296. S. R. Drake, E. L. Proebsting, and S. E. Spayd, *J. Am. Soc. Hort. Sci.*, *107*: 180 (1982).
297. A. E. Watada and J. A. Abott, *J. Food Sci.*, *40*: 1278 (1975).
298. W. E. Ballinger, W. F. McClure, W. B. Nesbitt, and E. P. Maners, *J. Am. Soc. Hort. Sci.*, *103*: 629 (1978).
299. E. Bayer, *Angew. Chemie*, *78*: 834 (1966).
300. P. Markakis, *Nature*, *24*: 1092 (1960).
301. N. Ishikura and K. Hayashi, *Botan. Mag. Tokyo*, *76*: 6 (1963).
302. J. Von Elbe, D. Bixby, and J. Moore, *J. Food Sci.*, *34*: 113 (1969).
303. Y. Osawa, M. Koizumi, N. Saito, and T. Kawei, *Phytochemistry*, *10*: 1591 (1971).
304. T. Tsuda and H. Fukuba, *J. Jpn. Soc. Nutr. Food Sci.*, *42*: 79 (1989).
305. R. I. Lin and B. W. Hilton, *J. Food Sci.*, *45*: 297 (1980).
306. G. Spagna and P. G. Pifferi, *Food Chemistry*, *44*: 185 (1992).
307. P. G. Pifferi and A. Vaccari, *Lebensm.-Wiss. u.- Technol.*, *14*: 85 (1980).
308. G. Daravingas and R. F. Cain, *J. Food Sci.*, *30*: 400 (1965).
309. R. E. Wrolstad and B. J. Struthers, *J. Chromatogr.*, *55*: 405 (1971).
310. G. Hrazdina, *J. Agric. Food Chem.*, *18*: 243 (1970).
311. D. Strack and R. L. Mansell, *J. Chromatogr.*, *109*: 325 (1975).
312. T. C. Somers, *Vitis*, *7*: 303 (1968).

313. H. Saquel-Barel, F. Comte, H. Kiepferle, and J. Crouzet, *Lebensm.-Wiss. u.- Technol.*, *15*: 199 (1982).
314. E. Bombardelli, A. Bonati, B. Gabetta, E. M. Martinelli, G. Mustich, and B. Danieli, *J. Chromatogr.*, *120*: 115 (1976).
315. E. Bombardelli, A. Bonati, B. Gabetta, E. M. Martinelli, and G. Mustich, *J. Chromatogr.*, *139*: 111 (1977).
316. E. C. Bate-Smith, *Nature*, *161*: 835 (1948).
317. A. J. Shrikhande and F. J. Francis, *C. R. Food Sci. Nutrit.*, *7*: 193 (1976).
318. J. B. Harborne, *J. Chromatogr.*, *1*: 473 (1958).
319. J. Ribéreau-Gayon and P. Ribéreau-Gayon, *C. R. Acad. Sci.*, *244*: 233 (1957).
320. T. Fuleki, *J. Food Sci.*, *34*: 365 (1969).
321. T. Fuleki and F. J. Francis, *Phytochemistry*, *6*: 1705 (1969).
322. J. Fitelson, *JAOAC*, *50*: 293 (1967).
323. B. H. Barrit and L. C. Torre, *J. Chromatogr.*, *75*: 151 (1973).
324. G. Hrazdina and A. J. Franzese, *Phytochemistry*, *13*: 225 (1974).
325. D. Jorysch and S. Marcus, *JAOAC*, *49*: 623 (1966).
326. R. A. Fong, R. E. Kepner, and D. Webb, *Am. J. Enol. Vitic.*, *22*: 150 (1971).
327. R. G. Goldy, E. P. Maness, H. D. Stiles, J. R. Clark, and M. A. Wilson, *Am. J. Enol. Vitic.*, *40*: 253 (1989).
328. R. Eder, R. Kalchgruber, S. Wendelin, and J. Barna, *Mitt. Klosterneuburg*, *40*: 152 (1990).
329. R. A. Fong, A. D. Webb, and R. E. Kepner, *Phytochemistry*, *13*: 1001 (1974).
330. C. H. Manley and P. Shubiak, *Can. Inst. Food Sci. Technol. J.*, *8*: 35 (1975).
331. J.-P. Roggero, P. Archier, and S. Coen, *Science des Aliments*, *12*: 37 (1992).
332. J. Bakker and C. F. Timberlake *J. Sci. Food Agric.*, *36*: 1325 (1985).
333. M. Wilkinson, J. G. Sweeny, and G. A. Iacobucci, *J. Chromatogr.*, *132*: 349 (1977).
334. G. Darné and Y. Glories, *Vitis*, *27*: 71 (1988).
335. M. Drdák, P. Daucik, and J. Kubasky, *Mitt. Klosterneuburg*, *39*: 228 (1989).
336. J. Taylor, *J. Sci. Food Agric.*, *49*: 487 (1989).
337. O. Lamikanra, *Food Chem.*, *33*: 225 (1989).
338. A. Chandra, M. G. Nair, and A. Iezzoni, *J. Agric. Food Chem.*, *40*: 967 (1992).
339. O. M. Andersen, *J. Food Sci.*, *54*: 383 (1989).
340. V. Hong and R. E. Wrolstad, *J. Agric. Food Chem.*, *38*: 708 (1990).
341. C. W. Nagel and L. W. Wulf, *Am. J. Enol. Vitic.*, *30*: 111 (1979).
342. J. Bakker and C. F. Timberlake, *J. Sci. Food Agric.*, *36*: 1315 (1985).
343. L. Gao and G. Mazza, *J. Agric. Food Chem.*, *42*: 118 (1994).
344. P. Etiévant, P. Schlich, A. Bertrand, P. Symonds, and J.-C. Bouvier, *J. Sci. Food Agric.*, *42*: 39 (1988).
345. E. Hebrero, C. Santos-Buelga, and J. C. Rivas-Gonzalo, *Am. J. Enol. Vitic.*, *39*: 227 (1988).
346. J. Oszmianski and J. C. Sapis, *J. Food Sci.*, *53*: 1241 (1988).
347. J. P. Goiffon, M. Brun, and M. J. Bourrier, *J. Chromatogr.*, *537*: 101 (1991).
348. T. Y. Lin, P. E. Koehler, and R. L. Shewfelt, *J. Food Sci.*, *54*: 405 (1989).
349. R. G. Goldy, W. E. Ballinger, E. P. Maness, and W. H. Swallow, *J. Am. Soc. Hort. Sci.*, *112*: 880 (1987).
350. K. Bronnum-Hansen and S. H. Hansen, *J. Chromatogr.*, *262*: 385 (1983).
351. H. Sämann, *Mitt. Klosterneuburg*, *39*: 202 (1989).
352. J. Fitelson, *JAOAC*, *52*: 649 (1969).
353. J. Koch and E. Haase-Sajak, *Z. Lebensm.-Unters.-Forsch.*, *131*: 347 (1967).
354. A. Mehlitz and B. Matzik, *Flüssiges Obst.*, *24*: 7 (1957).
355. J. Taylor, *J. Sci. Food Agric.*, *49*: 487 (1989).
356. C. S. T. Yang and P. P. A. Yang, *J. Food Sci.*, *52*: 346 (1987).
357. J. Y. Do, S. Potewiratananond, D. K. Salunkhe, and A. R. Rahman, *J. Food Technol.*, *11*: 265 (1976).
358. F. J. Francis, *J. Food Sci.*, *50*: 1640 (1985).
359. A. Porpáczy and M. Lásló, *Acta Alimentaria*, *13*: 109 (1984).
360. T. J. Mabry and A. S. Dreiding, in *Recent Advances in Phytochemistry* (T. J. Mabry, R. E. Alston, and V. C. Runeckles, eds.), Appleton-Century-Crofts, New York, 1968, p. 145.
361. R. L. Jackman and J. L. Smith, in *Natural Food Colorants* (G. A. Hendry and J. D. Houghton, eds.), Blackie and Son Ltd., Glasgow, 1992, p. 183.
362. H. Wyler and A. S. Dreiding, *Helv. Chim. Acta*, *40*: 191 (1957).

363. O. T. Schmidt and W. Schönleben, *Z. Naturf.*, *12b*: 262 (1957).
364. M. Piattelli, L. Minale, and G. Prota, *Tetrahedron*, *20*: 2325 (1964).
365. T. Nilsson, *Lantbrukshögskolans Annaler*, *36*: 179 (1970).
366. M. Piattelli, in *Chemistry and Biochemistry of Plant Pigments* (T. W. Goodwin, ed.), Academic Press, London, 1976.
367. R. A. Harmer, *Food Chem.*, *5*: 81 (1980).
368. E. L. Attoe and J. H. von Elbe, *J. Food Sci.*, *46*: 1934 (1981).
369. J. H. von Elbe, I.-Y. Maing, and C. H. Amundson, *J. Food Sci.*, *39*: 334 (1974).
370. I. Saguy, *J. Food Sci.*, *44*: 1554 (1979).
371. K. Savolainen and T. Kuusi, *Z. Lebensm.-Unters.-Forsch.*, *166*: 19 (1978).
372. E. L. Attoe and J. H. von Elbe, *J. Agric. Food Chem.*, *30*: 708 (1982).
373. G. M. Sapers and J. S. Hornstein, *J. Food Sci.*, *44*: 1245 (1979).
374. S. J. Schwartz and J. H. von Elbe, *Z. Lebensm.-Unters.-Forsch.*, *176*: 448 (1983).
375. A. S. Huang and J. H. von Elbe, *J. Food Sci.*, *50*: 1115 (1985).
376. M. Piattelli, in *The Biochemistry of Plants, Vol. 7, Secondary Plant Products* (P. K. Stumpf and E. E. Conn, eds.), Academic Press, New York, 1981.
377. T. J. Ng and Y.-N. Lee, *Hort. Sci.*, *13*: 581 (1978).
378. R. C. Wiley and Y.-N. Lee, *J. Food Sci.*, *43*: 1056 (1956).
379. C. Sáenz, E. Sepúlveda, E. Araya, and C. Calvo, *Lebensm.-Wiss. u.-Technol.*, *26*: 417 (1993).
380. W. E. Glässgen, J. W. Metzger, S. Heuer, and D. Strack, *Phytochemistry*, *33*: 1525 (1993).
381. Teutonico and D. Knorr, *Food Technology*, *39*: 49 (1985).
382. J. H. von Elbe, S. J. Schwartz, and B. E. Hildenbrand, *J. Food Sci.*, *46*: 1713 (1981).
383. A. Bilyk, *J. Food Sci.*, *44*: 1249 (1979).
384. J. H. von Elbe, *Food Technol.*, *29*: 42 (1975).
385. R. C. Wiley, Y.-N. Lee, J. J. Saladini, R. C. Wyss, and H. H. Topalian, *J. Food Sci.*, *44*: 208 (1979).
386. C. K. Wilkins, *Int. J. Food Sci. Technol.*, *22*: 410 (1987).
387. H. Pourrat, B. Lejeune, F. Regerat, and A. Pourrat, *Biotechnology Lett.*, *5*: 381 (1983).
388. A. Pourrat, B. Lejeune, A. Grand, and H. Pourrat, *J. Food Sci.*, *53*: 294 (1988).
389. J. H. von Elbe, S. H. Sy, I.-Y. Maing, and W. H. Gabelman, *J. Food Sci.*, *37*: 932 (1972).
390. A. Wohlpart and S. M. Black, *Phytochemistry*, *12*: 1325 (1973).
391. S. Heuer, V. Wray, J. W. Metzger, and D. Strack, *Phytochemistry*, *31*: 1801 (1992).
392. J. P. Adams and J. H. von Elbe, *J. Food Sci.*, *41*: 78 (1976).
393. M. Piatelli, L. Minale, and G. Prota, *Phytochemistry*, *4*: 121 (1965).
394. A. S. Huang and J. H. von Elbe, *J. Food Sci.*, *51*: 670 (1986).
395. I. Saguy, I. J. Kopelman, and S. Mizrahi, *J. Food Sci.*, *43*: 124 (1978).
396. M. Piatelli and L. Minale, *Phytochemistry*, *3*: 307 (1964).
397. W. D. Powrie and O. Fennema, *J. Food Sci.*, *28*: 214 (1963).
398. J. P. Adams and J. H. von Elbe, *J. Food Sci.*, *42*: 410 (1977).
399. K. R. Vincent and R. G. Scholz, *J. Agric. Food Chem.*, *26*: 812 (1978).
400. D. Strack and H. Reznik, *Z. Pflanzenphysiol.*, *94*: 163 (1979).
401. S. J. Schwartz and J. H. von Elbe, *J. Agric. Food Chem.*, *28*: 540 (1980).
402. D. Strack, U. Engel, and H. Reznik, *Z. Pflanzenphysiol.*, *101*: 215 (1981).
403. J. H. von Elbe, J. T. Klement, C. H. Amundson, R. G. Cassens, and R. C. Lindsay, *J. Food Sci.*, *39*: 128 (1974).
404. K. Hofmann and E. Blüchel, *Fleischwirtsch.*, *71*: 1290 (1991).
405. J. Belica, *Fleischwirtsch.*, *61*: 1262 (1981).
406. A. J. Taylor, R. Linfort, O. Weir, T. Hutton, and B. Green, *Meat Sci.*, *33*: 75 (1993).
407. D. Ladikos and B. L. Wedzicha, *Food Chem.*, *29*: 143 (1988).
408. M. Renerre, *Int. J. Food Sci. Technol.*, *25*: 613 (1990).
409. D. U. Ahn and A. J. Maurer, *Poultry Sci.*, *69*: 2040 (1990).
410. K. Potthast, *Fleischwirtsch.*, *67*: 50 (1987).
411. J. B. Fox, *J. Agric. Food Chem.*, *14*: 207 (1966).
412. R. A. Lawrie, in *Meat Science* (R. A. Lawrie, ed.), Pergamon Press, Oxford, 1966, p. 271.
413. H. Kolb, G. Heinz, and H.-W. Wiegand, *Fleischwirtsch.*, *70*: 956 (1990).
414. W. C. Franke and M. Solberg, *J. Food Sci.*, *36*: 515 (1971).

415. M.-Ch. Yin and C. Faustman, *J. Agric. Food Chem.*, *41*: 853 (1993).
416. A. R. Hayden, *J. Food Sci.*, *44*: 494 (1979).
417. F. Bauer and K. Hofmann, *Fleischwirtsch.*, *67*: 1 (1987).
418. U. Götze, *Fleischwirtsch.*, *54*: 234 (1974).
419. A. B. Karasz, R. Andersen, and R. Pollman, *JAOAC*, *59*: 1240 (1976).
420. I. Ingr, *Fleischwirtsch.*, *54*: 735 (1974).
421. C. E. Davis and J. B. Anderson, *J. Food Sci.*, *49*: 598 (1984).
422. M. Renerre, *Ann. Technol. Agric.*, *26*: 419 (1977).
423. F. Bauer and K. Hofmann, *Fleischwirtsch.*, *69*: 419 (1989).
424. P. D. Warriss, *Anal. Biochemistry*, *90*: 447 (1978).
425. E. A. Sugden and P. W. Saschenbrecker, *Can. J. Comp. Med.*, *40*: 270 (1975).
426. R. A. Lawrie, *J. Agric. Sci.*, *40*: 356 (1950).
427. J. I. Anderson and D. J. Locke, *Nature*, *175*: 819 (1955).
428. D. G. Cornish and G. W. Froning, *Poultry Sci.*, *53*: 365 (1974).
429. T. Nakanishi and M. Izumimoto, *Agric. Biol. Chem.*, *36*: 1505 (1972).
430. R. A. Field, L. R. Sanchez, J. E. Kunsman, and W. G. Kruggel, *J. Food Sci.*, *45*: 1109 (1980).
431. L. D. Satterlee and N. Y. Zachariah, *J. Food Sci.*, *37*: 909 (1972).
432. F. Bauer and K. Hofmann, *Proceedings 33rd Int. Congress Meat Science and Technology*, p. 364, Helsinki, 1987.
433. K. Krzywicki, *Meat Sci.*, *7*: 29 (1982).
434. S. K. Wolfe, D. A. Watts, and W. D. Brown, *J. Agric. Food Chem.*, *26*: 217 (1978).
435. A. H. A. van den Oord and J. J. Wesdorp, *J. Food Technol.*, *6*: 1 (1971).
436. A. E. Bevilacqua and N. E. Zaritzky, *J. Food Proc. Preserv.*, *10*: 1 (1986).
437. H. C. Hornsey, *J. Sci. Food Agric.*, *7*: 534 (1956).
438. P. D. Wariss, *J. Food Technol.*, *14*: 75 (1979).
439. G. R. Trout, *J. Food Sci.*, *54*: 537 (1989).
440. P. D. Warriss, *Anal. Biochem.*, *72*: 104 (1976).
441. K. Bünnig and R. Hamm, *Z. Lebensm. Unters.-Forsch.*, *155*: 332 (1974).
442. R. Dowiercial, L. Gach, and A. Pisula, *Nahrung*, *22*: 35 (1978).
443. S. C. Powell, E. R. Friedlander, and Z. K. Shihabi, *J. Chromatogr.*, *317*: 87 (1984).
444. B. F. D. Ghrist, M. A. Stadalius, and L. R. Snyder, *J. Chromatogr.*, *387*: 1 (1987).
445. I. M. Oellingrath and E. Slinde, *J. Food Sci.*, *50*: 1551 (1985).
446. I. M. Oellingrath, A. Iversen, and G. Skrede, *Meat Sci.*, *28*: 313 (1990).
447. R. J. McCormick, G. R. Reeck, and D. H. Kropf, *J. Agric. Food Chem.*, *36*: 1193 (1988).
448. R. R. Thompson, *JAOAC*, *43*: 763 (1960).
449. T. Höyem and B. Thorson, *J. Agric. Food Chem.*, *18*: 739 (1970).
450. U. Götze, *Fleischwirtsch.*, *50*: 203 (1970).
451. H. Kim and L. A. Shelef, *J. Food Sci.*, *51*: 731 (1986).
452. F. Bauer and K. Hofmann, *Fleischwirtsch.*, *67*: 861 (1987).
453. P. Uhlenhuth, *Dt. Med. Wschr.*, *27*: 780 (1901).
454. Y. Hashimoto and T. Yasui, *J. Food Sci.*, *42*: 1189 (1957).
455. F. W. Janssen, G. H. Hägele, A. M. B. Voorpostel, and J. A. de Baaij, *J. Food Sci.*, *55*: 1528 (1990).
456. M. H. G. Zijderveld and P. A. Koolmees, *Meat Sci.*, *27*: 55 (1990).
457. S. A. McLuckey, G. J. Van Berkel, G. L. Glish, E. C. Huang, and J. D. Henion, *Anal. Chem.*, *63*: 375 (1991).
458. R. S. Rush, A. S. Cohen, and B. L. Karger, *Anal. Chem.*, *63*: 1346 (1991).

24

Aroma Compounds

Robertino Barcarolo, Cristina Tutta, and Pierino Casson
Instrumental Analyses Laboratory, Dairy and Food Biotechnology Institute of Thiene, Thiene, Italy

I. INTRODUCTION

The study of volatile compounds has proved to be of great importance in determining the characteristics of foodstuffs. Flavor, together with color and texture, represents one of the three most important attributes of a foodstuff (1) and contributes in equal measure with the other two components to its acceptability. The term "flavor" has been universally adopted to indicate the set of olfactory and taste perceptions in organoleptic testing that accompany the tactile sensations involved. It can be considered a combined neurophysiological response to the chemically stimulated senses of taste and smell. By "taste" we mean the sensations registered by the papillae present on the tongue and on the innermost part of the oral cavity. It is connected with the sense of sweet, salty, sour, and bitter. Nevertheless, smell is the sensation that makes the greatest contribution to defining the characteristic flavor of most foods. When a smell has pleasant associations it can be defined as an aroma. There are thousands of compound volatiles present in food that are able, albeit with differing intensities, to trigger neurophysiological responses as a result of their interaction with olfactory epithelial cells present in the nasal cavity.

Flavors can be naturally present in food (2) or they can be generated by precursors as in the case of roasted, toasted, baked, or otherwise cooked products. The reactions to heat that occur during the preparation process in fact bring about numerous chemical and physical changes in the natural components of the foodstuff. Maillard's reactions (3,4) are of fundamental importance in this connection for the development of aromatic compounds (Fig. 1). Flavors can also derive from enzymic processes, from fermentation processes of microbic origin (6) involving a complete synthesis from simple precursors of a specific metabolite, or from bioconversions of specific substrates (Fig. 2).

Given the highly volatile nature and the low stability of many aromatic compounds, the aroma of many foodstuffs can undergo loss of appeal as a consequence of evaporation or through chemical reactions such as autoxidation, polymerization, decomposition, and hydrolysis. Off-flavors (8–10) can be defined as flavors that are not normally associated with the food in question. They may be

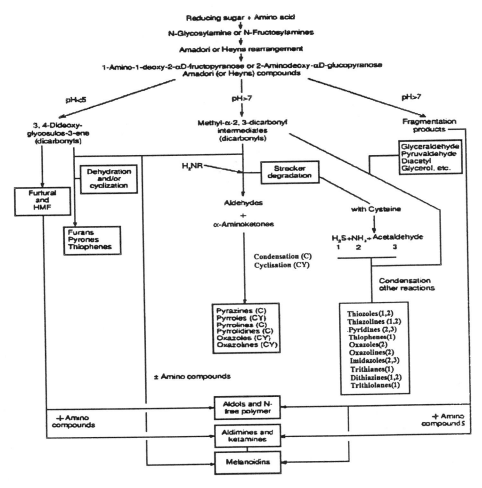

Fig. 1 Maillard reaction scheme. (From Ref. 5.)

of a kind not considered acceptable in any food because they are completely alien to the palate, or else they may simply be flavors that are not associated with the particular product. The compounds that underlie the off-flavors may derive from primary molecules present in the foodstuffs, or they may spread into the food as a result of secondary contamination from pesticides, pollutants, packaging materials, or from other foodstuffs stored in the same place.

Projects specifically directed to a comprehensive categorization of flavors have been in progress for several years and involve various lines of research:

Ascertainment of the chemical makeup of volatile compounds with the aim of identifying a qualitative/quantitative profile of odors for each foodstuff by means of chemicophysical methods

Identification of the formation processes of each of the components of the aroma, i.e., knowledge about the precursors and their modifications during the stages of transformation and processing of the raw material

Biochemical categorization of the specific protein receptors present in the olfactory epithelium and physiological analysis of their interaction with the volatile molecules

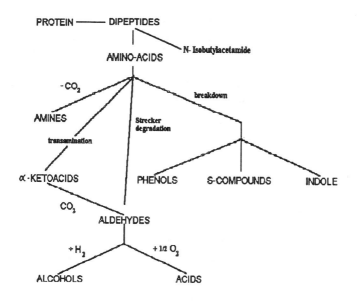

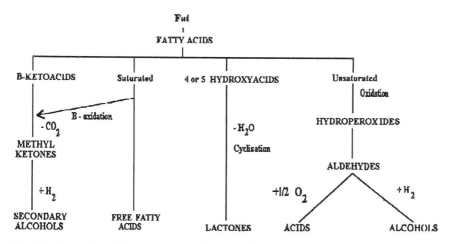

Fig. 2 Formation of flavor compounds from milk fat and protein. (From Ref. 7.)

The knowledge of a chemical-structural nature thereby acquired is an important starting point for further work aimed at various objectives connected with research and production in the area of aromas. In particular it makes it possible to specify more clearly which mechanisms underlie the natural production of aromas and to predict the alterations which are likely to come about in a foodstuff as a consequence of processing and conservation treatment. This knowledge can also be useful in the production of synthetic compounds and the identification of off flavors.

II.   CHARACTERISTICS OF AROMAS

The biochemical approach to distinguishing the molecular composition of aromas does not provide an indication of the likelihood that they will bring about sensory responses. The complexity of a foodstuff's aroma is in point of fact due not only to the fact that it is made up of various compounds but also to the fact that each component possesses a complex smell that derives from several basic olfactory sensations. In the first place, it needs to be borne in mind that in order for a substance to be assigned a smell it must be volatile enough for a sufficient number of its molecules to reach the olfactory epithelium so as to interact with protein receptors. The main physical and molecular parameters by which aromatic substances can be categorized are their boiling point, vapor pressure, dimensions, their shape and the presence in them of functional groups.

Most molecules of aromatic significance have a molecular weight of less than 300; substances with a high molecular weight, given their low volatility and the absence of receptors for them, may be excluded by perception thresholds.

Shape has to be considered in relation to the position of the functional group that can form a bond with the receptor. The structure and steric unamenability of a molecule are factors capable of generating differing olfactory properties. Moreover, the presence of geometrical and optical isomers within a clearly defined class of compounds triggers different smells. It should be remembered, though, that molecules of enantiomeric composition occur naturally in small quantities and their odor threshold values are generally extremely low.

Pelosi (11) proposes a classification of odorous molecules as follows:

Molecules lacking functional groups, such as saturated, unsaturated, and aromatic hydrocarbons
Molecules possessing a single functional group
Molecules with two or more functional groups

In the first category belong molecules capable of creating weak, nonspecific bonds (van der Waals forces) with receptors in the olfactory epithelium. As the molecule is able to establish these bonds regardless of its positioning in relation to the receptor, interactions with more than one receptor are often possible. This accounts for the nonintense and undefined smell that hydrocarbons generally have. Whenever a single functional group is present, this becomes the most important structural element inasmuch as it constitutes the point at which the odorous molecule will form a bond with the receptor. The bonds formed are of the relatively strong and specific hydrogen bond kind, and at least one heteroatom will be involved. Dipole–dipole interactions are nevertheless also possible. In the case of molecules with more than one functional group, each of these may form very specific bonds with different points of the same receptor, which calls for precise structural requirements especially as regards the distance between the functional groups. Another case occurs when only one of the functional groups can form a bond with a specific receptor while the other or others contribute only to the profile of the molecule.

Having established that it is the structural properties of molecules that determine the smell, it is necessary to proceed to an evaluation of the methods that enable us to separate them from the nonvolatile matrix and thus to identify them.

Volatile compounds occur in widely varying concentrations, although most are present in small quantities. The total quantity of aroma compounds naturally present in foodstuffs ranges from a few ppm right up to several hundred ppm (12). The concentration of single types of molecules typically varies from 1 to 100 ppb, although quantities of the order of a few ppt can occur. This wide range of olfactory perception values associated with very simple molecules is one of the surprising aspects of food aromas. It happens, indeed, that molecules present only at trace levels can be more important from a sensory point of view than others that are present in greater quantities. Despite the fact that a large quantity of compounds are usually identified in a foodstuff, generally speaking only a limited number of molecules have a significant effect on the aroma profile. In his analysis, the flavor chemist has to focus his attention on these molecules which, in the terminology used by Jennings and Sevenants (13), have been defined as "character impact" molecules.

III. PREPARATION OF SAMPLES

For the chemical analysis it is necessary, first of all, to obtain a concentration of isolated aromatic substances so as to be able to proceed with the combinations of the purification and separation process and arrive at the identification of the compounds present in the mixture. An increasingly complex and sophisticated analytical methodology is called for to deal with the wide range of thermal stabilities, polarities, volatilities, and solubilities, as well as the chemical reactivity that also characterizes the nonvolatile molecules present in the mixture. In addition to these factors, the presence of interfering compounds (water, ethanol, and so on) means that sophisticated procedures have to be adopted to remove them.

In research, one of the most routine, and therefore often insufficiently considered, aspects is that of the selection and conservation of the sample. It is common knowledge that the sample must be representative but it is also necessary for it to have good conservation qualities, understood not only as a tendency not to become adulterated but above all in terms of its capability of maintaining a constant ratio between the quantities of volatile compounds existing in the vapor phase and in the matrix. It is a recognized fact that the composition of volatile substances in the vapor phase depends on the vapor pressures of the compounds in the pure state at the temperature of the foodstuff. For this reason, in order to conserve their typical characteristics, unpackaged products must be transported to the research laboratories and transferred to suitable containers, the volume of which should not be much in excess of that of the sample so as to avoid allowing the volatile substances to become sparsely spread throughout too large a volume. It is also advisable to remove the air from the container and replace it with an inert gas such as nitrogen or helium. The latter can be considered preferable because it does not have the tendency to become soluble in the liquid phase. Conservation in a freezer may be problematic in the case of certain solid foodstuffs since the formation of ice would cause the internal matrix to break, freeing highly volatile substances that may disperse during the preparation of the sample. It is therefore preferable to proceed straightaway with the analysis, at the same time bearing in mind that this can be conducted by means of various differing techniques (Fig. 3) which can be divided into two different basic research directions: that of the standard approach based on selection extraction and that of headspace analyses (15).

It should be stressed that our aim is not to achieve an exhaustive detailed study of all of the techniques that play a part in the chemical specification of the characteristics of flavors but to identify certain guidelines as a preliminary to subsequent detailed analysis.

IV. SELECTIVE EXTRACTION

The approach based on the selective extraction of the volatile fraction followed by various chromatographic preparation techniques has the advantage of ensuring a sufficiently large sample for subsequent analyses but proves to be somewhat complex in view of the time required for completing the various steps, which can result in deterioration of the sample. Despite the progress that has taken place in the chemistry of foodstuffs, few new techniques have been developed in recent years and none of them allow all of the volatile components of flavor to be completely isolated.

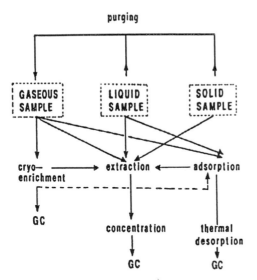

Fig. 3 Some common routes to concentrate volatile compounds from different types of sample. (From Ref. 14.)

A. Isolation and Concentration Techniques

Distillation is, for many foodstuffs, the first form of isolation of the volatile fraction. It can be carried out by means of various techniques (16), the choice of which is governed by the nature of the foodstuff. For products unaffected by heat or for the analysis of specific cooked flavors, steam distillation (SD) at normal pressure, with recovery of the volatiles in refrigerated entrapment systems, offers a very straightforward solution. However, for thermolabile foodstuffs (e.g., fruit) it is suggested that the extraction be conducted under less extreme conditions, operating with reduced temperatures and pressures and incorporating additional refrigerated entrapment devices (Fig. 4) between the sample and the vacuum systems. These refinements lead to a substantial increase in the concentration but increase the time required. Distillation techniques guarantee a high level of extraction, making it possible for further analyses to be carried out. Although operating under conditions of high vacuum accompanied by effective cooling systems, there are still losses of the more volatile compounds. Since multistep procedures are involved, the time required for the analyses can be lengthy, with possible concentration losses over the various stages.

Headspace condensation is a particular steam distillation carried out at ambient temperature. It is a form of extraction that makes use of the condensation brought about by a "cold finger" condenser that comes in contact with the vapors given off by foodstuffs (preferably fresh fruit or vegetables) kept for a long time inside special purpose containers (18).

The volatile fractions separated using distillation techniques are recovered as very dilute aqueous solutions, so they have to be concentrated before being analyzed. Extraction with solvents is the most commonly used method for recovering the aqueous distillate, sometimes preceded by cryoconcentration. The technique of freeze-drying brings about the partial but selective removal of the water by crystallizing it in the form of ice. All the same, this leads to the loss of certain substances by evaporation or by their becoming trapped in the ice crystals. Large quantities of organic solvent are necessary for the extraction process, which can be carried out using either separating funnels or liquid–liquid extractors.

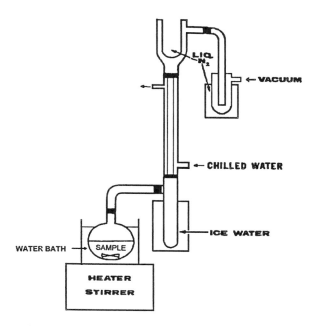

Fig. 4 Distillation apparatus. The flask containing the sample was connected to an ice-cooled trap equipped with a chilled water condenser, two liquid nitrogen traps in series, and a vacuum pump. None of the volatiles reached the second nitrogen trap. (From Ref. 17.)

In the study of volatile compounds in foods and drinks, the most common procedure is a combination of steam distillation at atmospheric pressure and simultaneous extraction with organic solvents (SDE). The Likens and Nickerson continuous extractor (19) is the most popular model in use for this purpose. In this apparatus a solvent and an aqueous homogenate of the sample are placed in two flasks and brought to the boil. The vapors released by the distillation flask are condensed and extracted into the organic solvent in a cooled chamber. The condensate in the aqueous phase and the organic phase form different strata, on account of their insolubility and different densities, in the U-shaped part of the apparatus. The solvent, with the volatile substances extracted, flows into the recovery flask while the water returns to the flask containing the original sample. An improved version of this piece of equipment (20) includes a head condenser to minimize the losses of the more volatile compounds (Fig. 5). There are significant variations in the efficiency of extraction and recovery (21,22), depending on which solvents are involved. The principal criterion guiding the choice of solvent is the degree of selectivity required. Among those most widely in use, mention can be made of dichloromethane, pentane, ethyl ether, which is the most suitable for a successful recovery of the whole volatile fraction because of its low selectivity, and Freon 11. The latter, being nonpolar, extracts little of the lower alcohols (23) and therefore lends itself particularly to the analysis of wines or spirits. An apparatus (24,25) that uses Freon 11 for continuous extraction of terpenes from wines is shown in Fig. 6. Prior to use, the sample has to be frozen (0°C) and the lower part (50 mm) of the extraction unit submerged in ice to avoid the formation of emulsions.

SDE can be considered a valid method for quantitative extractions, with its main limitation being the formation of secondary flavors as a consequence of degradation due to oxidation above all in the lipids. In order to lower the boiling temperature some researchers have recommended working at reduced pressures. However, only a few of the suggested improvements along these

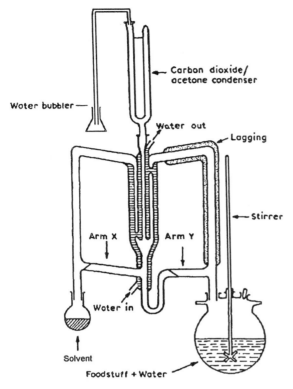

Fig. 5 Improved version of the Likens and Nickerson apparatus including a double-surface water condenser in the condensation area and a head CO_2-acetone condenser. (From Ref. 20.)

lines have proved to be really innovative (26–28). A modification aimed at optimizing the extraction apparatus proposed by Godefroot et al. (29) was created recently (30), and this makes it possible to work at low temperatures (37°C) with high vacuums, thereby avoiding degradation caused by heat (Fig. 7) and ensuring recovery levels comparable to those achieved under normal atmospheric conditions.

Another solution for recovering headspace volatiles formed from heated food samples (31) employs a gas-washing bottle and a liquid–liquid continuous extractor in tandem (Fig. 8). In this apparatus, which involves mild temperature extraction, volatiles from samples are purged into deionized water by a stream of inert gas. The volatiles dissolved by the water are simultaneously and continuously extracted with an appropriate solvent. Liquid–liquid extraction (LLE) takes advantage of the division of the flavors between an aqueous sample and an organic solvent immiscible in water, in accordance with Nernst's law. LLE selectivity depends on the solvent chosen and on other factors such as the pH, ionic forces, water-to-solvent ratio, number of extractions, types of analytes and their concentration (32,33). This technique, which can be used with samples with a low fat content, involves certain disadvantages linked to its low sensitivity, the time required for some extractions, and, above all, the possible formation of emulsions. This last drawback can, however, be removed by adding sodium sulfate or saturating the sample with sodium chloride.

If the SDE or LLE procedures are used, the volatile fraction separated is diluted in an organic solvent that may contain small quantities of water that can be dried using anhydrous sodium sulfate.

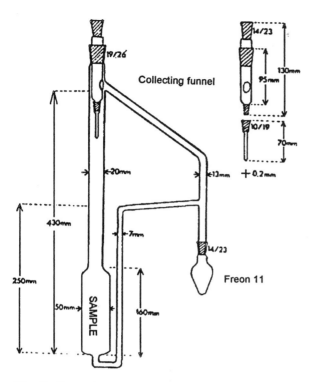

Collecting funnel

Freon 11

SAMPLE

Fig. 6 Extraction apparatus for the continuous Freon 11 extraction of grape juice and wine samples with details of collecting funnel. (From Ref. 25.)

The final organic solution obtained by means of these methods of extraction can be analyzed during GC with no further preparation or it can be concentrated by means of evaporating concentrators (Kuderna-Danish), distillation columns of the Vigreux, Oldershaw, or Dufton types, or microdistillation systems.

An alternative solution for the recovery of aromatic components from aqueous solutions is solid phase extraction (SPE) (34–36). The underlying principle of the technique is the selective entrapment of the organic compounds on adsorbant materials of various kinds (Fig. 9). The desorption of the captured substances under analysis is achieved by selective elution in limited volumes of an appropriate solvent. The adsorption properties of the solid matter depend on its chemical composition (presence of functional groups) and surface structure (porousness and surface area). While the use of adsorbant materials is one of the most effective means of separation, recovery is nevertheless not always complete. Moreover, the nonlinear pattern of trapping that emerges for different quantities of solution fed in and the displacement effects that are encountered can lead to inaccuracy in the quantitative analysis. Figure 10 shows the breakthrough curve for an SPE column.

Over the last few years many studies (37–42) have demonstrated that extraction methods using supercritical fluids (SFE) represent a valid solution in flavor analysis. CO_2 is the most widely used fluid in the extraction of flavors since it is natural, nonflammable, and chemically inert. At temperatures or pressures above its critical point it possesses the mass transfer properties of a gas and the solvation capabilities of a liquid. With its high density and diffusivity, low viscosity, and limited surface tension, it is able to penetrate effectively in complex matrices resistant to access, thus ensuring rapid and efficient transfer of the constituents being analyzed. Compared with the

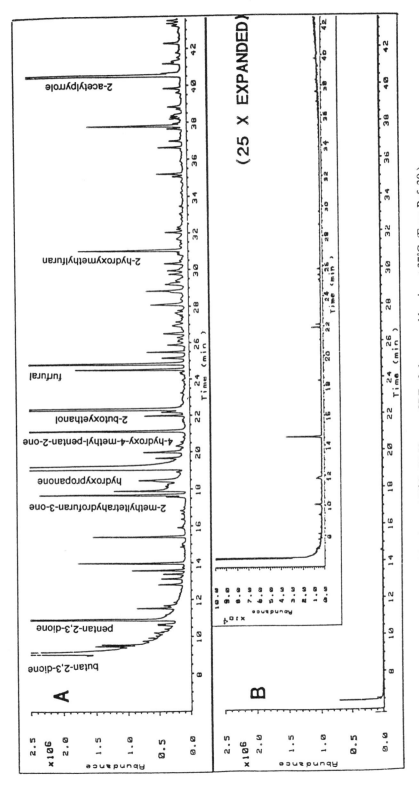

Fig. 7 (A) Classical SDE of glucose and leucine at 105°C. (B) Vacuum SDE of glucose and leucine at 37°C. (From Ref. 30.)

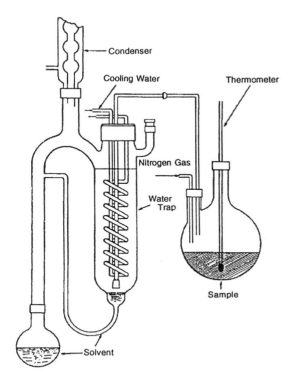

Fig. 8 Apparatus used to collect headspace volatiles from sample containing fatty materials. (From Ref. 31.)

traditional methods, it offers the advantages of good selectivity and a high rate of separation per time unit, does not require high extraction temperatures, and reduces the inconveniences associated with the use of organic solvents. Since the sample does not need to be manipulated nor are concentration procedures called for between the extraction and the chromatographic analysis, potential losses of analytes and heat-induced deterioration are lessened. It needs to be added,

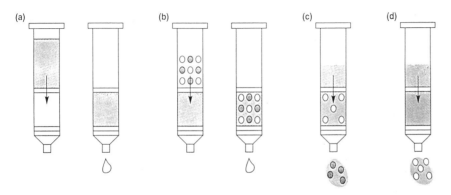

Fig. 9 SPE steps: (a) column conditioning, (b) sample loading, (c) column rinsing, and (d) elution. (From Ref. 37.)

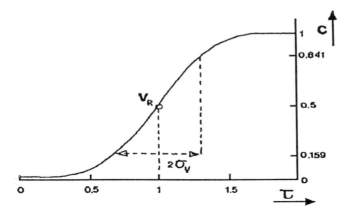

Fig. 10 The breakthrough curve of a SPE column. $c = C/Co$ and $\tau = t/rt$, where c and Co are the solute concentrations in the effluent and in the sample (i.e., the influent), respectively, t and rt are the time and the retention time of the solute, respectively. (V_r = the retention volume of the solute and σ_v = the elution band broadening.) (From Ref. 34.)

however, that while this technique has been found suitable when dealing with natural essences or at any rate with samples with a low fat content, it has not proved to be ideally suited to highly volatile aromatic compounds.

B. Fractionation Techniques

The flavors isolated frequently call for fractionation into differing chemical classes to facilitate their separation and identification. The most commonly used fractionation methods bring about a subdivision into acids, bases, phenols, and neutral classes of compounds. Figure 11 shows an analytical procedure for the separation of the volatile phenol compounds in wine (43). Fractionation is particularly important for foodstuffs that have undergone heat treatments, since they have a complex chemical composition, both because of the number of compounds present and in view of the structurally distinct types represented. It can be carried out directly on a portion of the aqueous distillate, making use of the division of the various classes of compounds on the basis of pH range differences. Differing techniques of preparatory chromatography can be employed for the fractionation as a function of differences in polarity, molecular weight, and the presence of functional groups. Medium-pressure liquid chromatography on silica gel is often used for fractionating extracts obtained by SDE in order to ease subsequent identification analyses since, in addition to removing interfering compounds, it promotes the manifestation of components present in minor quantities.

V. HEADSPACE ANALYSES

The other methodological approach to the analysis of aroma compounds is based on the isolation and examination of the volatile fraction present in the vapor phase above the foodstuff matrix, by using various forms of headspace analysis (HS). As foodstuffs present multicomponent matrixes, the solubility and the bonding capacity of the substances involved varies and the same molecule may occur with different vapor pressures in the various foodstuffs. HS techniques (44,45) are the most suitable kind for the study of very volatile compounds; they often

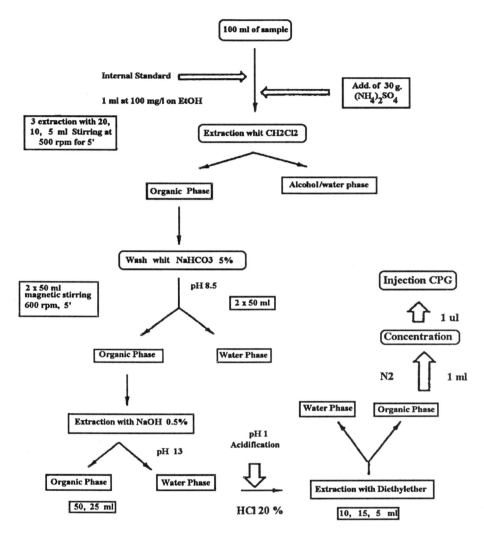

Fig. 11 Procedure for determining volatile phenols concentration in wine. (From Ref. 43.)

call for no more than a simple preparation of the sample and lead to less artifact formation. A comparative chart (46) relating to the extraction and concentration of the volatile compounds in cheese is reproduced in Table 1. The main advantage of the HS approach as compared with that involving total extraction is that the headspace, at room temperatures, presents an aroma profile matching that of natural olfactory perception and thus represents the smell the foodstuff actually has. The problems that are encountered with these procedures have to do with the technical difficulty of measuring small amounts of highly volatile compounds without discriminating against others in the overall series that makes up the fraction of volatiles contained in the headspace. Headspace sampling does not lend itself to the analysis of constituents with a high boiling point in that such components are hardly present at all in the vapor phase at normally operative temperatures. Many different methods are in common use to perform the gas extraction involved and the subsequent transfer into the gas chromatograph can also be achieved by a number of techniques.

Table 1 Summary of the Main Advantages and Disadvantages of Four Techniques for the Extraction and Concentration of Volatile Flavor Compounds of Cheeses as Well as for Their Injection into a Gas Chromatograph

	High-vacuum and low-temperature distillation	Steam distillation (stripping)	Dynamic headspace analysis (Rektorik)	Dynamic headspace analysis (Tekmar)
Amount of sample (g)	2500	250	25	5–25
Duration of analysis	3 days	3 hr	2 hr	1–5 hr
Temperature (°C)	0–10	60	40	40
Pressure (mbar)	0.001	20	Ambient	Ambient
Application for	Research	Research/routine	Research/routine	Research/routine
Advantages	Mild extraction; Concentrated extracts; Multiple injections possible; Distillate can be used for other analyses	Suitable for middle to low-volatile components; Multiple injections possible; Distillate can be used for other analyses; Internal standard possible	Small amount of sample; No organic solvent(s) needed; Extraction and concentration in one step; Internal standard possible	As for Rektorik and additionally: Suitable for high and middle-volatile components; Several adsorbants available, special mixtures possible; Very low detection limit; Good repeatability
Drawbacks	Large amount of sample; Time and lab space consuming; Multiple processing steps; High vacuum; Organic solvent(s); Losses in concentration step; Artifacts and impurities (e.g., joint grease); Suitable only for solid samples	Multiple processing steps; Breakdown reactions of thermolabile components; Organic solvent(s); Losses in concentration step; Artifacts and impurities (e.g., joint grease)	Foam problems with liquid samples; Multiple injection not possible; Only two adsorbant materials available; Desorption temperature not measurable; Production of artifacts in desorption step; Poor repeatability	Foam problems with liquid samples; Multiple injection not possible; Purity of water; Purity of purge gas; Cooling with liquid nitrogen; With split injection not possible

Source: Ref. 46.

A. Equilibrium Headspace Sampling (E-HS)

The simplest form of headspace analysis is that of direct gas chromatographic injection of small volumes of vapor phase compounds given off by foodstuffs (Fig. 12a). In this technique the sample is placed in a closed vial and heated in a bath or oven at a regulated temperature until it reaches an equilibrium with its vapor phase. The distribution of an analyzed component between the two phases is determined by the distribution coefficient (K) in accordance with the following equation: $K = C_s/C_v$, where C_s and C_v are the concentrations of the component in the sample and in its gaseous phase, respectively. To calculate this coefficient, conditions of thermodynamic equilibrium are required that are influenced by the time and by the temperature involved in conditioning the sample. With liquid samples the addition of salts until saturation is reached reduces the distribution coefficient and increases the effectiveness of the sampling. A greater concentration of organic constituents in the vapor phase can also be achieved by increasing either the volume of sample in the vial or the temperature, although the latter can lead to decomposition or deterioration phenomena (47,48). The possible formation of artifacts and the masking of some of the volatile compounds may also derive from the use of stripping salts that can react with the sample. A limiting factor with E-HS is the fact that its sensitivity is low inasmuch as only a part of the vapor phase present in the vial comes to be analyzed. It also needs to be stressed that for quantitative analyses the coefficient K must be included in the calibration procedure for every component and must be constant over a wide range of concentrations. The prerequisite of thermodynamic equilibrium is nevertheless fully satisfied only in the case of liquid samples. The addition of the internal standard to avoid matrix effects therefore lends itself particularly to the quantitative analysis of this type of sample. With solid samples, on the other hand, whenever it is not possible to mix them with an appropriate eluant this calibration procedure cannot be applied.

B. Multiple Headspace Extraction (MHE)

The MHE technique (49–52) was introduced to bypass the problems of diffusion of volatiles from the matrix which are encountered in single static headspace analysis. The procedure is based on

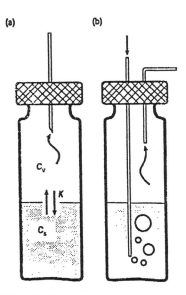

(a)   (b)

Fig. 12 Basic headspace sampling techniques, including (a) equilibrium (static) sampling and (b) dynamic sampling. (From Ref. 47.)

the assumption that by repeating the sampling several times it is possible to extract from the sample the entire amount of each volatile. After each extraction the system is repressurized by adding an inert gas and then reequilibrated before the next step in the process of removal. In this way the pressure in the vial is kept constant so that the removal of the analytes always takes place under isobaric conditions. It was found that the concentration of an analyzed constituent diminishes over successive extractions in an exponential manner (Table 2), which allows quantification on the basis of an extrapolated straight linear regression plotted between the logarithm of the peak areas and the number of analyses conducted (Fig. 13). The quantitative analysis is normally performed through calibration with an external standard and it supplies a valid contribution in predicting the distribution coefficients in a gas–solid system. The MHE procedure can be recommended for very volatile compounds present in homogeneous matrixes or for solutions, properly speaking, of solid samples (53,54). It does not resolve the problems of low sensitivity noted with E-HS and so cannot be used in analyses with extremely low concentrations of the analytes.

C. Dynamic Headspace Sampling (D-HS)

Dynamic extraction of the vapor phase of a foodstuff is carried out using an inert gas for continuous stripping of the volatile components of the matrix (Fig. 12b). In this way thermodynamic equilibrium between the sample and its vapor phase is never reached, so that a high level of extraction efficiency is ensured. The gas is fed in directly either over or through the thermostatically controlled sample for a period of time sufficient to allow extraction of most of the volatiles present. The desorption time can be ascertained by repeating the analysis several times with the same sample. The extraction gas flow has to be sufficiently strong to achieve rapid stripping of the vapor phase (55). Generally the total volume of gas extracted is too great to be able to proceed with direct injection, and concentration methods using adsorbant solids or cryogenic liquids are required before the gas chromatography analysis. The procedure can thus be divided into two phases: (a) extraction and entrapment; (b) desorption and transfer to the GC column (Fig. 14). D-HS offers certain advantages compared with E-HS because the matrix does not influence the analytical results, so that quantification is possible by means of calibration with an external standard with both viscous liquids and solid samples, although in this case the quantitative results

Table 2 A Series of Data Collected for Coffee Components

Peak no.	RT [min]	Area (1)	Area (2)	Area (3)	Area (4)	Slope	Total area
1	3.5	94,012	60,907	40,103	27,043	−0.420	274,000
2	3.7	6,866	4,833	3,408	2,472	−0.344	23,600
3	3.9	24,156	16,360	11,190	7,718	−0.382	76,100
4	4.1	129,939	93,525	67,780	49,243	−0.324	469,000
5	4.5	20,293	14,619	10,665	7,898	−0.320	74,100
6	4.8	78,749	66,403	56,690	48,300	−0.164	521,000
7	5.0	46,933	36,980	29,459	23,425	−0.232	226,000
8	5.2	6,146	4,518	4,071	3,039	−0.232	29,700
9	5.4	1,538	1,127	901	685	−0.272	6,460
10	6.6	562	443	443	382	−0.134	4,490
11	7.3	13,909	12,842	10,642	8,355	−0.153	97,900
12	9.3	2,957	2,462	2,139	1,879	−0.157	20,400
13	9.6	874	643	578	516	−0.194	4,960

Source: Ref. 52.

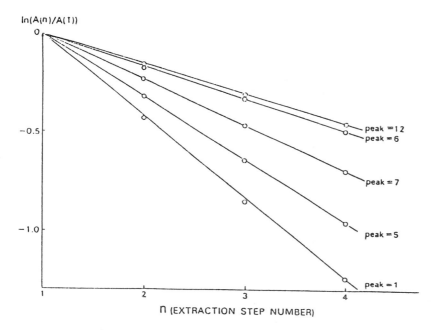

Fig. 13 MHE plot for some of the components of coffee headspace. The slope varies with the component. The total area, ΣA, of each peak summed over an infinite number of extractions is estimated as $\Sigma A = A(1)/[1 - A(n - 1)/A(n)]$ where $A(n)$ is the peak area measured following the nth extraction step. (From Ref. 52.)

have to be correctly interpreted. The quality of the analytical data obtained is strongly influenced by the diffusion processes of the volatiles from the solid matrix. In addition to the nature, form, and size of the sample, other parameters with an important effect on quantification are the temperature and time taken for desorption, and the extraction gas flow rate. A mathematical model was recently proposed for quantifying the volatile compounds in polymers (56).

1. Purge and Trap

A form of D-HS that can be used with liquid samples (or with slurries of solid samples) is known as the purge-and-trap method (57). The sensitivity of this technique is affected by the time and temperature involved in the sampling, the stripping flow, and the solutes' distribution coefficient. The last parameter depends, among other things, on their affinity to water. As a consequence, aromatic compounds that are polar get trapped to a greater extent in the solution, thereby introducing distortions in the context of the volatile fraction as a whole. The presence of solid particles and any tendency that may exist to produce froth can also influence the distribution of solutes. A theoretical model for predicting recovery through dynamic extraction of liquid samples was put forward by Curvers et al. (58). The main drawback with this procedure continues to be the difficulties associated with the large amount of water extracted, which calls for various kinds of additional refinements to avoid problems with both the concentration systems and the analytical columns.

D. Discussion

The preparation of the sample can sometimes call for a series of additional refinements so as to facilitate and/or improve the extraction. In the case of solid samples, crumbling or cutting the

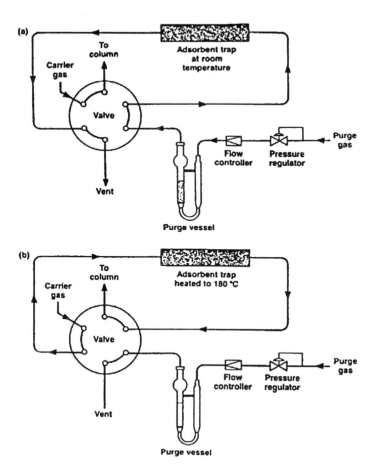

Fig. 14 Flow diagram of a purge-and-trap sampling system in (a) purge mode and (b) desorb mode. (From Ref. 57.)

sample into small pieces to increase the surface area exposed will speed up the transition into the vapor phase. With liquid samples, besides the addition of salts, correction of the pH may be worthwhile in order to encourage precipitation of interfering compounds. Control of the temperature in the conditioning process is the first requirement in research on volatile compounds. In headspace analysis, rises in temperature can bring about sensitivity increases, but they can also provoke artificially created changes and lead to aroma profiles that are not representative of the actual perceived properties of the foodstuffs concerned. It is advisable therefore to equilibrate the sample at low temperatures (30–40°C).

Another parameter to consider is the sampling pressure. It is normal practice to pressurize a vial or other container for the purpose of transferring the volatile fraction to the concentration units. This involves a quantitative skewing against higher boiling substances because of their lower volatilities. Lower boiling substances, on the other hand, have the tendency to remain in the vapor phase unless pressure increases are significant. A workable solution can be found boosting the stripping flow by making use of an auxiliary vacuum system. In this way the flow of carrier gas fed into the vial balances the headspace flow extracted, thereby keeping pressure in the vial equal to that outside.

E. HS Analysis Concentration Techniques

The D-HS techniques are normally considered to be more sensitive than those of E-HS as they make it possible to achieve a thorough extraction of the volatile substances present in the sample. In capillary columns, however, the introduction of large volumes of headspace would result in peak broadening. In order to obtain fuller batches of aromatic compounds, of which only traces are often present, or to extend the detection limits, an effective preconcentration of the stripped volatiles is therefore necessary. The choice of concentration method depends on the chemical makeup and volatility of the analytes and on the structure of the matrix in which they are present. The methods most widely in use are based on the use of cryogenic systems and entrapment on adsorbant solid materials. A few solutions exist which set up a direct on-line connection between the sampling apparatus, the concentration system, and the GC column. Much more frequently there is a break in the process and an off-line concentration of the headspace sample is followed by an on-line desorption.

1. Cryofocusing

Trapping can be carried out by keeping the whole GC column at cryogenic temperatures (59) during the headspace transfer or by inserting a cryogenic trap between the injector and the analytical column. In the first case it is necessary to have a gas chromatograph equipped with suitable accessories making it possible to operate at cryogenic temperatures (e.g., -100°C). These temperatures are attained by the use of liquid nitrogen as coolant, which calls for precise and sensitive heat control of the oven. Complete chilling of the oven calls for a high consumption of liquid N_2 and adds to the GC analysis time, but can prove necessary in the case of incomplete retention of some of the more volatile components in the cryotraps (60).

The use of cryofocusing traps is a very versatile system that lends itself to both laboratory and commercial applications. Liquid nitrogen (-196°C) is the most widely used means of chilling and offers the best retention capabilities for the most volatile compounds, but liquid CO_2 or acetone/ice mixtures can also be used. With cryogenic trapping systems either the first section of the capillary column (e.g., 10–15 cm long) or a short precolumn (which may be open tubular, coated or packed with adsorbant materials) is frozen during the injection. Their efficiency is dependent on many parameters, in particular stripping flow, amount of headspace material transferred, cryofocusing temperature, presence of adsorbant substances, and so on. The principal problem with on-line entrapment derives from the possible obstruction of the trapping system following the formation of ice crystals, which may occur above all in connection with large volumes of headspace materials containing water vapor, not necessarily in high proportions. The presence of adsorbant materials within the traps on the one hand increases their retention capacity, but on the other hand leads to a reduced water tolerance capacity. Thus the problem of water removal becomes of prime importance whenever D-HS analysis techniques are employed. Various solutions have been put forward to avoid this difficulty, such as the use of condensers before the cryofocusing, or the use of selective membranes (Nafion tube), drying tubes, or cryoloops. These solutions nevertheless have certain limitations due to memory effects (condensers) or to the fact that the aromatic fraction contains many water-soluble compounds and the discrimination involved in their removal would cause distorted results. An effective solution in our estimation must depend on optimizing the efficiency parameters discussed above. In particular, sampling at room temperature, when water vapor tension is very low, can help to keep down its presence in the headspace.

2. Adsorption onto Solids

The use of trapping systems containing solid adsorbant materials enables large volumes of headspace to be efficiently concentrated while diminishing the problems due to the presence of water. Another advantage of this method is that it makes it possible to carry out the analysis in two distinct steps, so that the headspace concentration phase is separate from the subsequent gas chromatography analysis. In off-line systems, whenever the analyst does not proceed immediately with the desorption of the trapped substances it is possible to keep the adsorption tubes so as to complete the analysis within the set time. Apart from having a high retention capability, the solid adsorbant materials used must be hydrophobic, thermally stable, not contain impurities, and must release the trapped substances quickly during desorption. In the analysis of volatiles different chromatographic substrates can be used, including porous polymers, silica gels, or adsorbant carbons, although none of these can be said to exhibit all of the ideal characteristics (34,61–63).

The aromatic substances adsorbed can be recovered by heat desorption or by liquid extraction using a solvent. This is necessary for the analysis of extremely thermolabile molecules, although recovery is not always satisfactory and in any case there is considerable dilution of the extract, with consequent loss of sensitivity in the analysis. Heat desorption, unlike solvent extraction, involves one-shot analysis of the whole sampled fraction but guarantees maximum sensitivity without preparation of the sample and interference effects from the solvent. The release of the trapped substances calls for rapid and drastic increases in temperature and leads to the formation of large volumes of vapor. Given the length of time required for the desorption, this vapor cannot be instantaneously transferred to the analytical column and has to be cryogenically refocused prior to the GC analysis. The double thermal cycle introduced represents the main limitation of this analytical procedure in that it can lead to artifact formation. The high temperatures necessary for the desorption can cause breakdown of the adsorbant materials and of the thermally unstable aromatic molecules as well as generating synthetic compounds as a result of reactions between the primary aromatic components. Finally, it should be pointed out that the use of adsorbant trapping systems leads to nonlinear recovery patterns for which reason the quantitative analysis of compounds of which only traces are present can turn out to be inaccurate despite use of the internal standard.

3. Other Preconcentration Techniques

Another possibility for preconcentrating the volatile substances present in the headspaces is provided by the use of open tubular traps. These capillary traps may contain either a thick film of an apolar stationary phase (FTs) or a layer of activated charcoal particles (COTs). The adsorbed organic substances can be thermally desorbed or using a few microliters of a suitable solvent. The use of very short traps (~5 cm × 0.3 mm i.d.) coated with a layer of fine charcoal particles (64) ensures high adsorption capacities and makes thermal desorption (Fig. 15) a possibility in the GC injector, without having to rely on further refocalizations. Nevertheless COTs call for high temperatures for the desorption, manifests a tendency to favor the adsorption of compounds present in higher concentrations, and, when dealing with samples with a high level of humidity, the condensation of water on the charcoal surface facilitates the retention of the partially polar molecules. Alternative solutions call for the use of long traps (~1–2.5 m) with silicon polymer in very thick films (65,66). These materials are inert, permeable, and thermostable and their concentration mechanism is based on partitioning processes. Nonetheless in the analysis of foodstuffs these traps offer limited retention capacities for the most volatile compounds and tend to discriminate against the more polar compounds.

In view of these considerations, we argue that the most efficient solutions in research into volatile compounds consist of using an on-line headspace extraction process, optimizing the sampling parameters (temperatures and pressures), and carrying out the concentration in the single step of cryofocusing.

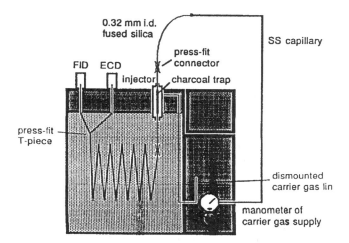

Fig. 15 Gas chromatograph for thermal desorption of charcoal open tubular traps. Carrier gas is fed into the trap by an additional line to which the short fused silica leg of the trap is connected. Thermal desorption occurs in the regular vaporizing injector. (From Ref. 64.)

VI. SEPARATION TECHNIQUES

In research on flavors, the most widespread separation techniques are those involving chromatography. Given the volatility characteristics of aromatic compounds as well as the high resolution and sensitivity involved, gas chromatography is considered an ideal technique for analyzing them. For certain specific purposes liquid chromatography can also make a useful contribution.

A. Gas Chromatography

The complexity of mixtures of aromas calls for highly efficient separation methods. In order to determine qualitatively and quantitatively the specific components, of which there are often only traces present, the best use needs to be made of all the parameters offered by gas chromatography. Gas chromatography analysis can be performed on liquid extracts as well as on headspace. In the first case there is the advantage of being able to use standard configuration GC without modifying instrumentation. However, the presence of a solvent, which constitutes most of the sample injected, does lead to certain substances being masked. Applying the technique to headspace constituents involves desorption directly in the gas chromatograph and frequently requires the application of cryofocusing systems. The intrinsic characteristics of headspace sampling impose a strict evaluation in the choice of analytical columns. Packed columns emerge as most appropriate for the analysis of extremely volatile compounds such as light hydrocarbons and in general for headspaces containing few chemical categories, since their retention capacity is based on processes involving both adsorption and partitioning. For, while chromatographic sensitivity obtained depends primarily on column capacity, it is inversely proportional to the separation efficiency. It should be remembered that the considerable flows of carrier gas required by packed columns hamper the manipulation of the effluents by the mass spectrometer making appropriate interface splitting necessary. Capillary columns are able to offer a high level of efficiency in the separation of most compounds occurring in the sample. A review assessing the importance of the choice of the stationary phase was recently compiled (67). The principal parameters to consider are those of separability, polarity, and thermal stability. Polyethylene glycol (PEG, Carbowax) stationary phases have been much used in the analysis of the constituents of aroma compounds in view of the high degree of separation ensured, despite the low level of reproducibility they offer for analyses with large time lapses between them. With the adoption of analytical techniques like D-HS new difficulties emerge, linked to the water vapor that is frequently present in the sample. This makes it necessary to use thick, polydimethylsiloxane stationary phases, inert to water, so as to avoid shifts in the retention times of some of the compounds. Considering the wide range of polarities among the compounds that occur in aroma extracts, it is possible to choose between completely nonpolar fixed traps (100% polydimethylsiloxane) or intermediate traps (polydimethylsiloxane with variable percentages of phenyl and cyanopropylsilicone). Recently, capillary porous-layer open tubular (PLOT) columns were developed in which the stationary phase is not a liquid but an adsorbant solid deposited on the walls as a thin layer of small particles. These PLOT columns combine the speed and resolution of the capillary columns with the strong separating power of gas–solid chromatography. With the separation of chiral compounds it should be remembered that optical isomers, as opposed to geometric isomers, cannot be separated by normal stationary phases except after the formation of their diastereomeric derivatives (68). Another possibility is provided by the use of optically active stationary phases (69–71).

The analytical efficiency of the columns depends not only on the type and stationary phase used but on the choice of the most appropriate operating conditions. In particular, control of the flow rate, as well as control of injector and oven temperatures, is important for obtaining a good level of separation efficiency.

Introducing the sample in conventional heated vaporization chamber injectors, apart from the biases this may introduce (72), can cause degradation or decomposition of labile compounds (73). These problems can be resolved by making use of cold on-column injectors (74–77) and programmed temperature variations (PTV) (78–81). Both techniques result in improvements in the quantitative analysis. PTV injection in the context of foodstuff analyses is often carried out at subambient temperatures in solvent split mode. By packing the cold glass liner with adsorbant materials (e.g., Tenax) (82–85) it is possible to reduce the loss of the more volatile solutes that evaporate together with the solvent. In any case, in headspace analysis split injection offers low sensitivity because the greater part of the volatiles is vented out through the splitter and does not enter the column. Moreover, difficulties can occur in the quantitative analysis as the mixtures of aroma compounds cover a wide range of volatilities. The commonest technique to ensure a high level of sensitivity is that of splitless injection, which, given the large volumes fed in, requires cryogenic refocusing systems. While D-HS always calls for the use of cryogenic traps, in the case of E-HS, should these not be available, it is possible to make use of alternate reconcentration techniques based on the use of precolumns and blocking of the carrier gas during injection (86).

For increasing separating capacity in the case of mixtures comprising compounds with a wide band of volatilities and polarities, a technique that has attracted a great deal of interest is multidimensional gas chromatography (MDGC) (87–90). This technique involves the use of a number of columns of different types placed in series that can be positioned either in a single oven or in more than one oven with independent temperature regulation. A diagram of the two column-switching systems currently employed is provided in Fig. 16. The first one may be a packed column or a capillary column with a stationary phase of different polarity from that of the second one. In both of these solutions it proves to be indispensable to place a cold trapping system between the two columns as well as a system for equilibrating the flow into the second column to make it compatible with the requirements of any mass spectrometer being used. Finally, let's note that a split can be incorporated between the analytical column and the detector so that part of the effluent can be redirected to an outlet for olfactory testing.

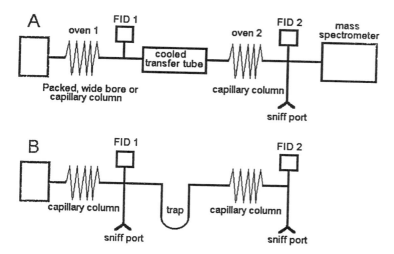

Fig. 16 Two column-switching systems for MDGC. (a) Schematic drawing of the Siemens system (Model Sichromat 2) with intermediate trapping on a cooled transfer tube. The selected fractions trapped (from the first column) are reinjected into the second column by heating the tube. (b) Schematic drawing of the Chrompack MUSIC system. By means of the two splits, 10% of the effluent is led to the detector and 90% to the sniff port. (Modified from Ref. 87.)

B. Liquid Chromatography

Recent research indicates that certain inodorous compounds have an important effect on flavor in acting as precursors or having an interfering effect in the release of volatile molecules. These compounds generally have a high molecular weight and low volatility and can be analyzed in HPLC (91). This technique allows compounds with a wide range of polarities to be analyzed, normally at room temperature, without altering their characteristics. It is used above all for detecting adulterations, but also for isolating and purifying the flavor components (92) which are then identified as IR or MS. Nevertheless HPLC in the analysis of complex mixtures of flavors does not constitute a real alternative to GC in view of its low separating efficiency, lack of sensitive universal detectors, the difficulties involved in the interface with the mass spectrometer, and, above all, the impossibility of proceeding with a parallel sensory analysis.

In spite of all that, direct interfacing with the gas chromatograph (LC-GC) offers new possibilities in the analysis of some classes of foodstuff compounds (93,94), notably essential oils (95).

VII. IDENTIFICATION

Mass and IR absorption spectra, NMR spectra, and Kovats indices on polar and nonpolar columns, together with its odor, are the properties that make a compound identifiable.

A. GC-MS Interface

Techniques combining capillary gas chromatography with mass spectrometry (GC-MS) have come to be extensively employed in research on flavors (96,97). In such a coupled setup the operative characteristics of the chromatograph gas (under pressure) and of the mass spectrograph (vacuum) have to be maintained. In the commonly available GC-MS systems the use of long capillary columns (30–50 m) with narrow internal diameters (0.25–0.32 mm) makes it possible to manage gas flows compatible with the vacuum system without having recourse to the open-split type of interfaces used in the past. The length of the columns must be such as not to obtain a gas flow above 1–1.5 ml/min at the ion source. With headspace samples, given the large amount of water that accompanies the gas flow it is advisable to choose columns with thick films of polydimethylsiloxane (3–5 μm). The use of these columns together with the maintenance of a low oven temperature during sampling and transfer in the presence of the source of vacuum favors rapid progress of water toward the ion source (elution at 50°C), reducing deterioration of the column. The efficiency of a column connected to a mass detector is always inferior to that of an equivalent column connected to a normal-pressure detector, since the vacuum causes an increase in the linear speed of the carrier, making performance fall short of the optimal Van Deemter curve.

When cryofocusing systems are being used, since not all of the volatile substances in the headspace are completely captured in the traps, for the entire duration of the entrapment the carrier is substituted by part of the headspace sample. By adopting a counterflow feed system the analytical column can be kept isolated during sampling and can help, as Fig. 17 shows, to resolve these problems (98).

In order to avoid condensation of substances with a high boiling point and, above all, the decomposition of thermolabile compounds, it is advisable to keep the transfer line at temperatures between 180°C and 200°C.

B. Mass Spectrometry

The main functional parts of the mass spectrometer are the ion source and the spectrum analyzer. Modern GC-MS works by means of electronic impact (EI) sources or chemical ionization (CI) sources; in both cases molecules are hit by a flow of electrons, bringing about their ionization.

The ion sources most commonly in use are of the electronic impact type, with which the molecules get bombarded by a highly charged stream of electrons. The resultant degree of splitting in the molecules is linked to their structure and to the energy of the electron field, which is normally 70 eV. It is in fact essential for the identification of the compounds to obtain spectra under these conditions, since they are the ones adopted by the most common reference libraries. The temperature of the ion source must not be too high, in order to avoid risks of oxidation, nitration, or chemical attack by the effluents of the analytical column. Temperatures of around 150°C can be considered ideal both for reducing background noise and because under high-vacuum conditions all of the aromatic substances are already in either the vapor or gaseous phase. Given their generally low molecular weight, aromatic substances often produce diagnostic fragmentation effects with low m/z values (mass/charge). It is therefore important to carefully eliminate the gases that derive from the air, in particular oxygen and nitrogen, as they can worsen the instrumentation's signal-to-noise (S/N) ratio. In view of the restricted range of molecular weights along aromatic substances, for separation of the ions generated by the ion source it is preferable, given their sensitivity, to use short quadrupole detectors. Recently, excellent results were achieved in food flavor research by ion trap mass detectors (ITD), which make it possible to detect the ions from within the ionization chamber. This means a considerable gain in terms of sensitivity since the ion generation and detection take place in the same physical space. Problems can arise, however, with the resolution of fused low-level peaks since, in order to avoid the formation of M+H fragments, the instrument generally brings about adjustments of some of the operative parameters with consequent lessening of sensitivity. Currently, the spectra obtained with EI by means of ITD are closely comparable to those achieved by quadrupole systems. A further advantage of the latest generation of ITDs is the possibility they offer of trapping an ion fragment and of being able to fragment it further so as to confirm the molecular ion, thereby creating a GC-MS-MS system. The mass spectra obtained through EI are not always able to provide useful information on the molecular weight since the high energy level involved may, in some classes of compounds such as alcohols and esters (99), not lead to the formation of intense molecular ions. In such cases, if the substance being examined is not to be found in the library (not an uncommon circumstance), the solution is to make use of chemical ionization (CI) sources. With this technique the molecules are ionized as a result of a chemical reaction with an ionized low-pressure gas (ammonia, methane, isobutane, etc.) present in excess quantities. The molecule–gas complexes that form have a low energy capacity, which leads to the formation of the molecular ion with splitting of the simplest bonds. The spectra obtained by means of chemical ionization are diagnostic to a high degree but cannot be checked against reference libraries in view of the fact that it is difficult to create the latter since the ions in excess correspond not just to the specific parameters pertinent to the detector but also to the type and partial pressure of the reactant gas.

The intensity of the ionic fragments obtained can be electronically registered as the sum of the intensities of all of the ions (total ion current, TIC) or by means of a programmed selection of certain ions (selected ion monitoring, SIM). The first of these techniques is by far the most widely used in flavor analysis since the composition of the sample is often unknown. Full-scan identification of molecules of which only traces (a few picograms) are present calls for optimization of ionic emission parameters at the source, of lens focalization, and of coordination with the detector. Similarly, the detection side of the apparatus, whether incorporating an electromultiplier or a photomultiplier with conversion dynode, must be able to amplify the greatest number of ions possible. The various relative abundances of the calibration gas (perfluorotributylamine or perfluorokerosene) need to be carefully checked, especially in the 20–250 m/z range. In any case it should be remembered that the mass detector does not provide equal sensitivity across molecules of different types as its responsiveness depends on the stability of the molecules and the intensity of the ions produced. In addition, its sensitivity is linked to the detector's calibration parameters, which can vary even in the course of an analysis (caused, for example, by oxidation on the electrode of oxygenated compounds present in large amounts). Lastly, the scanning speed cannot be brought down to less than one scan ever 0.5 sec (with gaps of 200 m/z), so that few integration points are generated. For quantification of mass it is thus a prerequisite to work with single ions or few ions so as to be able to use a high scanning speed and to carry out frequent recalibrations with standard substances. For accurate analysis, whenever it is possible it is recommended to work with the standard substance deuterated so as to acquire, in the same retention time, both the compound in question and the relative standard.

Ascertainment of the isotopic ratios by means of mass spectrometry (IRMS) is a method that, in addition to its usefulness in testing the authenticity of natural products (100–102) and in investigating certain biosynthetic processes, can be used for quantifying the volatile components present in food (103,104). Table 3 shows the isotopic abundance values for some natural and synthetic aromatic molecules, obtained using three different methods.

C. GC-FTIR

MS analysis cannot always ensure identification of the aromatic compounds, so it should be supplemented by Fourier transform infrared spectroscope (FTIR). In supplying indications of the positioning of the functional groups in the molecules, this is a useful instrument to back up the mass spectrometer (105). However, given budget limitations, few laboratories can afford to do both MS and GC-FTIR, which explains the limited use made of it in flavor analyses (106–108). For this reason, and considering its lower sensitivity and the limited water tolerance of many spectroscopes, as well as the lack of extensive libraries (109), GC-FTIR systems may be considered simply as complementary instruments, but not as alternatives to GC-MS systems.

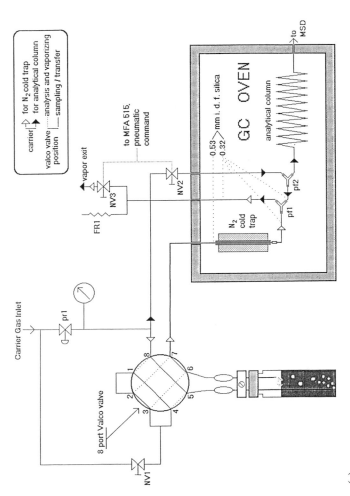

(a)

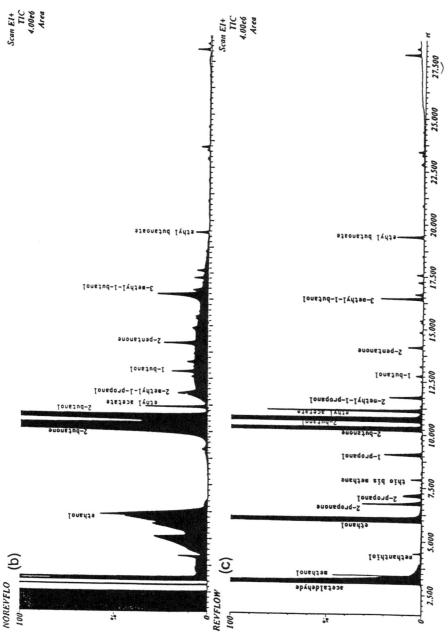

Fig. 17 (a) Schematic diagram of the dynamic headspace cold trap equipment, including the system for reversing the carrier gas flow during sampling. Downstream of the pressure regulator pr1, the carrier gas is split in two: the first stream flows directly through the sampling valve to the cryogenic trap and exits through NV3, while the second stream is used to supply the analytical column and create a reversed flow. (b) Dynamic HS-GC-MS of cheese with reversal of the carrier gas flow. (c) Dynamic HS-GC-MS of cheese without reversal of the carrier gas flow. (From Ref. 98.)

Table 3 Carbon and Hydrogen Isotope Summary for 11
Different Flavor Materials

	Stable Isotope Summary		
	^{14}C, dpm/gC	δ^{13}C, ‰	δD, ‰
Acetaldehyde			
Natural (3)	16.38 ± 0.07	−17.66 ± 4.96[a]	−202 ± 64
Synthetic (5)	0.04 ± 0.15	−28.54 ± 1.65	−76 ± 6
Amyl Acetate			
Natural (8)	15.37 ± 1.00	−17.47 ± 6.05[a]	−292 ± 38
Synthetic (6)	0.04 ± 0.11	−28.51 ± 1.88	−123 ± 19
Mixtures (6)	11.48 ± 0.20	−22.18 ± 4.61[a]	−252 ± 15
Anethole			
Natural (14)	15.76 ± 0.46	−28.48 ± 1.81	−85 ± 12
Synthetic (8)	15.87 ± 1.91	−31.30 ± 0.78	−58 ± 28
Benzaldehyde			
Natural (28)	16.09 ± 0.61	−28.51 ± 1.03	−111 ± 18
Synthetic (12)	0.03 ± 0.12	−25.99 ± 1.29	+600 ± 67
Synthetic (6)	0.01 ± 0.06	−29.07 ± 0.57	−58 ± 15
Cinnamic Aldehyde			
Natural (30)	16.11 ± 0.94	−27.30 ± 0.69	−120 ± 15
Synthetic (8)	0.04 ± 0.08	−25.63 ± 0.76	+515 ± 42
Synthetic (2)	−0.01 ± 0.01	−29.00 ± 0.26	−27 ± 23
Ethyl Acetate			
Natural (13)	15.60 ± 0.70	−19.19 ± 6.05[a]	−242 ± 30
Synthetic (8)	0.09 ± 0.16	−29.16 ± 3.29	−98 ± 55
Ethyl Butyrate			
Natural (14)	15.87 ± 0.61	−15.33 ± 3.86[a]	−238 ± 29
Synthetic (9)	0.06 ± 0.16	−27.04 ± 2.72	−57 ± 44
Mixtures (3)	5.38 ± 0.10	−22.84 ± 1.50[a]	−106 ± 64
Ethyl Caproate			
Natural (11)	15.68 ± 0.75	−29.02 ± 1.63	−246 ± 23
Mixtures (8)	12.33 ± 0.49	−31.78 ± 1.03	−220 ± 16
Linalool			
Natural (13)	16.31 ± 1.08	−27.02 ± 1.27	−297 ± 26
Synthetic (6)	0.02 ± 0.06	−27.69 ± 2.26	−196 ± 59
Methyl Salicylate			
Natural (5)	16.56 ± 0.88	−33.11 ± 0.39	−130 ± 50
Synthetic (10)	0.46 ± 0.48	−29.33 ± 0.77	−71 ± 16
Vanillin			
Beans (15)		−20.05 ± 0.83	−77 ± 15
Lignin (10)	15.25 ± 0.75	−26.89 ± 0.34	−185 ± 12
Guaiacol (2)		−25.50 ± 0.90	−20 ± 4

[a]Extreme σ values due to both C3 and C4 source materials.
Source: Ref. 100.

D. Sensory Evaluation

Although GC-MS analysis is a guaranteed source of qualitative and quantitative information on
the volatile components of foodstuffs, it nevertheless does not provide a response to their
characteristics as aromas. For this reason, in aroma research it is necessary to correlate the
chemical identification data with olfactory response. E-HS techniques are suited to this purpose

because they make it possible to use the same substrate for both analyses, thereby avoiding the problem of variability of samples. The possibility of simultaneous chemical and organoleptic analysis is obtained by splitting GC effluents simultaneously to an FID detector and to an outlet for olfactory testing. The use of nondestructive detectors (photoionization detector, PID) makes it possible for sniffing to take place at the discharge outlet.

The methods proposed (110–112) for assessing aroma value, which depend on the relationship between the concentration of a compound in a product and its odor threshold value, are based on sniffing analysis of diluted extracts. Figure 18 shows a comparison between the GC profile obtained using an FID and that obtained by analyzing increasing dilutions of an organic wine extract by sniffing the gaseous effluent exiting from the column (113). These procedures none-theless will always lead to the formation of subjective judgments, so that more accurate methods need to be found for the quantification of aromas. Recently, by combining a sensory apparatus with techniques of multivariate statistical analysis an odorimetric system has been developed that accords with the perception of the human nose. The technology of the sensors is based on the combined working of metallic semiconductors whose electrical resistance changes when certain odors are present. Besides the standard metal oxide sensors, more selective conducting polymers are available but as yet they are less sensitive and have a shorter working life. Multivariate analysis (using systems of computation that simulate neuron networks) of the responses simultaneously supplied by several sensors makes it possible to arrive at a character-ization of the set of odors for each type of product, carrying out cluster analyses in very short periods of time (114).

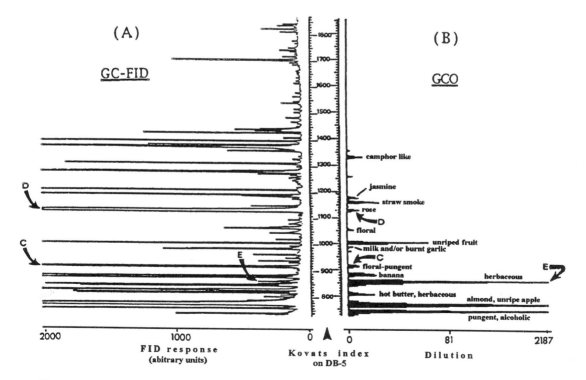

Fig. 18 Relationship between GC-FID analysis and sensory responses during GC sniffing of increasing dilutions of "Biancolella" wine extract. Peak areas in B profile (charm values) are directly proportional to aroma values. (From Ref. 113.)

VIII. APPLICATIONS

In research into aromas it is not possible to identify all of the volatile components present in the food using a single method of analysis. It is therefore necessary to focus the research at each stage on the specific classes of substances that are of interest. Whenever the volatile fraction of greatest interest consists of low-boiling compounds, headspace analysis techniques performed at room temperature are the best ones available. They ensure a high concentration though they may not always allow for accurate quantitative measurements. For the analysis of the heavier fractions it is advisable to turn to other extraction techniques, notably SDE and SPE. As a result, HS analysis and the classic isolation and concentration techniques are to be considered complementary methods in the preparation of samples for GC. Decisions concerning procedures of analysis have to be made on the basis of certain considerations relating to the water content in the matrix to be analyzed.

1. *Low moisture content (<20%).* As a rule, foodstuffs with a low water content are the easiest to analyze even though they generally have very complex matrixes, since they derive from technological transformations involving heat treatments. Such foodstuffs include coffee, cocoa, tea, oven-cooked products, oils, and honey. HS analysis can be carried out either in static or in dynamic mode (Fig. 19), bearing in mind that quantitative specification is difficult in the case of recovery from solid matrixes like these. Given the considerable quantity of high-boiling com-

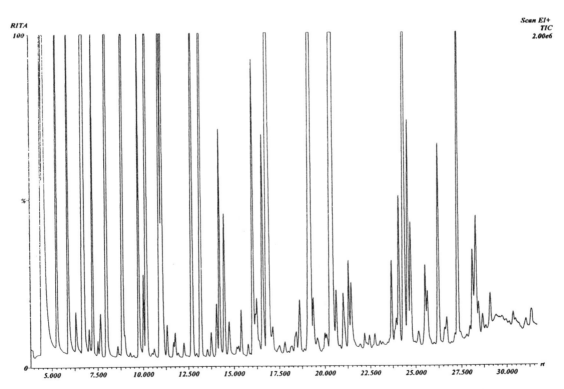

Fig. 19 Dynamic HS-GC-MS analysis of filter-ground coffee. The sample was stripped for 150 sec at room temperature and the HS components condensed in the cold trap at -140°C. GC was performed on a 50 m × 0.32 mm fused silica column coated with 3.5 μm film of JXR. Helium was used as carrier gas at 0.4 bar. MS was scanned from m/z 20 to m/z 450 with a cycle time of 1 sec. The ion source was maintained at 200°C and spectra were obtained by electron impact (70 eV).

Table 4 Concentrations (µg/L) of the Volatile Fraction Components of a Vinegar, an "Asti," and a Red Wine, Determined by LLE (pentanedichloromethane, 7:3) and Hydrophilus Extrelut Resin

	Vinegar		"Asti"		Wine	
	Solvent	Extrelut	Solvent	Extrelut	Solvent	Extrelut
2,3-Butanedione					2540	2646
Isobutyl acetate	1937	2290	63	53	101	443
2-Butanol	593	798	303	799	310	690
Ethyl butyrate			296	265	200	240
1-Propanol	560	729	6672	11424	10235	23091
2-Methyl-1-propanol	11736	10142	11379	16741	27207	64100
Isoamyl acetate	2760	3088	872	1180	1495	1896
1-Butanol			629	766	627	1074
3-Methyl- + 2-Methyl-1-butanol	15914	18149	29816	53796	46123	142313
Ethyl hexanoate			691	564	480	327
Ethyl pyruvate			232	254	38	208
Hexyl acetate + 3-Hydroxy-2-butanone	33537	39407	3753	6294	3291	4196
3-Methyl-1-pentanol			42	59	105	126
Ethyl lactate	1332	1390	6957	8775	7273	10099
1-Hexanol			1292	1263	1389	1268
trans-3-Hexen-1-ol			31	36	27	28
3-Ethoxy-1-propanol	102	119	89	86	199	173
cis-3-Hexen-1-ol			95	87	147	119
cis-Furanlinalooloxide			72	62		
Benzaldehyde	266	97	237	47	227	50
Ethyl-3-hydroxybutyrate			47	44	395	272
Propionic acid			233	177	410	206
(-)-2,3-Butanediol	4893	12666	7116	28858	10839	42747
Linalool			199	168		
Isobutyric acid (+ n.i. in vinegar)	2236	2263	459	294	870	462
(R,S)-2,3-Butanediol	3156	4148	2515	7805	4051	9391
γ-Butyrolactone	3725	4894	1665	2001	2340	3855
Butyric acid			898	492	516	201
Ethyl decanoate			122	319	123	192
Isovaleric acid			216	142	587	149
Diethyl succinate	267	269	346	328	516	227
α-Terpineol			463	392		
Valeric acid (+ n.i.)	14701	13332	385	461	1034	1304
Citronellol			50	42		
Phenethyl acetate + Nerol	93	154	199	147	3914	6490
Hexanoic acid	⁻94	85	5275	4338	1578	984
Geraniol			28	35		
Benzyl alcohol	1146	1198			759	832
2-Phenyl ethanol	20998	22442	12114	14367	26511	45613
Octanoic acid	774	561	7104	8805	2569	2143

n.i.: not identified
Source: Ref. 115.

pounds present, SDE is necessary in order to obtain a complete picture of the aroma profile. With spices and condiments the presence of unvolatile molecules and oleoresins acts in favor of SFE to obtain high quality extracts.

2.   *Intermediate moisture content (20–60%).* Typical foodstuffs belonging to this category are cheeses, meats, sausages and salamis, and fish. If D-HS is used it may be necessary to have recourse to adsorption on solid materials. In the case of direct trapping, in the cryofocusing step alone it is necessary to bring the temperature of the traps down to -80°C so as to avoid complete capture of any alcohols that may be present in excess quantities. Distillation in a vapor current and extraction by solvent are still the techniques most often used for many of these foods and they have to avoid the cosolubilization of nonvolatile components that may lessen the concentration and contaminate the GC-MS apparatus. Interference caused by the fats to be found in the matrix is one of the biggest problems encountered in these cases. With meats and fish, in order to selectively determine compounds containing N, O, and S it is advisable to use an atomic emission detector in the GC analysis.

3.   *High moisture content (>60%).* For the analysis of fresh fruit and vegetables, milk, and liquid foodstuffs in general, the guidelines for analysis set out above remain valid. Headspace analysis can be carried out, if necessary, with the purge-and-trap technique. The parameters conditioning the efficiency of the cryogenic traps have to be optimized to avoid the breakthrough phenomena that are very common in these cases. In addition to HS analysis, in analyzing fruit, vegetables, and their derivatives, use is commonly made of extraction with solvents, SDE (performed in very bland conditions), headspace sampling plus partitioning with solvents, and SPE. This last separation technique can be used to isolate the less volatile components of milk but is ideally suited for applications involving wines, vinegars, and spirits (Table 4). For the analysis of the terpenoid compounds that are determinant in forming the sensory impression of the bouquets of aromatic wines (116), the most widely adopted method (117) involves adsorption of the volatile substances, whether free or glycosidically bonded, onto a nonionic resin (Amberlite XAD-2) and subsequent selective elution with different solvents. While the free forms can be directly determined, the bound ones first have to be enzymatically hydrolyzed to free the aglycons.

REVIEWS

Dumont, J. P. and J. Adda, *Progress in Flavor Research* (J. Adda, ed.), Elsevier, Amsterdam, 1979, p. 245.

Hurrell, R. F., *Food Flavours* (I. D. Morton and A. J. Macleod, eds.), Elsevier, Amsterdam, 1982, pp. 399–437.

IFT, *Food Technol.*, *43*(12): 99 (1989).

Nijssen, B., *Volatile Compounds in Food and Beverages* (H. Maarse, ed.), Marcel Dekker, New York, 1991, pp. 689–735.

Schreier, P., *Developments in Food Flavours* (G. G. Birch and M. G. Lindley, eds.), Elsevier, London, 1986, pp. 89–118.

REFERENCES

1. IFT, *Food Technol.*, *43*(12): 99 (1989).
2. P. Schreier, *Developments in Food Flavours* (G. G. Birch and M. G. Lindley, eds.), Elsevier, London, 1986, pp. 89–118.
3. R. F. Hurrell, *Food Flavours* (I. D. Morton and A. J. Macleod, eds.), Elsevier, Amsterdam, 1982, pp. 399–437.
4. H. E. Nursten, *Developments in Food Flavours* (G. G. Birch and M. G. Lindley, eds.), Elsevier, London, 1986, pp. 173–190.
5. T. Nagodawithana, *Food Technol.*, *46*(11): 140 (1992).

6. IFT, Use of biotechnology to enhance food flavor, Proceedings of Symposia, *Food Technol.*, *46*(11): 121–161 (1992).

7. J. P. Dumont and J. Adda, *Progress in Flavor Research* (J. Adda, ed.), Elsevier, Amsterdam, 1979, p. 245.

8. B. Nijssen, *Volatile Compounds in Food and Beverages* (H. Maarse, ed.), Marcel Dekker, New York, 1991, pp. 689–735.

9. M. J. Saxby, *Food Flavours* (I. D. Morton and A. J. MacLeod, eds.), Elsevier, Amsterdam, 1982, pp. 439–457.

10. F. B. Whitfield, *Developments in Food Flavours* (G. G. Birch and M. G. Lindley, eds.), Elsevier, London, 1986, pp. 249–273.

11. P. Pelosi, *Caratteristiche Olfattive e Gustative degli Alimenti* (G. Montedoro, ed.), Chiriotti, Torino, 1985, pp. 35–81.

12. R. Emberger, *Cereal Food World*, *30*: 691 (1985).

13. W. G. Jennings and M. R. Sevenants, *J. Food Sci.*, *29*: 158 (1969).

14. J. Roerade and S. Blomberg, *HRC*, *12*: 138 (1989).

15. D. A. Cronin, *Food Flavours* (I. D. Morton and A. J. MacLeod, eds.), Elsevier, Amsterdam, 1982, pp. 15–45.

16. C. Weurman, *J. Agric. Food Chem.*, *17*: 370 (1969).

17. M. G. Moshonas and P. E. Shaw, *J. Agric. Food Chem.*, *35*: 161 (1987).

18. L. Schreyen, P. Dirinck, F. van Wassenhove, and N. Schamp, *J. Agric. Food Chem.*, *24*: 1147 (1976).

19. S. T. Likens and G. B. Nickerson, *Am. Soc. Brew. Chem. Proc.*, 5–13 (1964).

20. A. J. MacLeod and S. J. Cave, *J. Sci. Food Agric.*, *26*: 351 (1975).

21. J. Alberola and L. J. Izquierdo, *Flavor of Foods and Beverages; Chemistry and Technology* (G. Charalambous and G. E. Inglett, eds.), Academic Press, New York, 1978, pp. 283–304.

22. C. S. Cobb and M. M. Bursey, *J. Agric. Food Chem.*, *32*: 197 (1978).

23. D. W. Clutton, *J. Chromatogr.*, *167*: 409 (1978).

24. A. Rapp, H. Hastrich, and L. Engel, *Vitis*, *15*: 29 (1976).

25. J. Marais, *S. Afr. J. Enol. Vitic.*, *7*(1): 21 (1986).

26. T. H. Schultz, R. H. Flath, T. R. Mon, S. B. Eggling, and R. Teranishi, *J. Agric. Food Chem.*, *25*: 446 (1977).

27. A. J. Nuñez, J. M. H. Bemelmans, and H. Maarse, *Chromatographia*, *18*: 153 (1984).

28. B. A. Charpentier, M. R. Sevenants, and R. A. Sanders, Proceedings of the 4th International Flavor Conference, Rhodes, 1985 (G. Charalambous, ed.), Elsevier, Amsterdam, 1986, p. 413.

29. M. Godefroot, P. Sandra, and M. Verzele, *J. Chromatogr.*, *203*: 325 (1981).

30. L. Maignial, P. Pibarot, G. Bonetti, A. Chaintreau, and J. P. Marion, *J. Chromatogr.*, *606*(1): 87 (1992).

31. K. Umano and T. Shibamoto, *J. Agric. Food Chem.*, *35*: 14 (1987).

32. F. I. Onuska, *HRC*, *12*: 4 (1989).

33. M. M. Leahy and G. A. Reineccius, *Analysis of Volatiles, Methods, Application* (P. Schreier, ed.), Walter de Gruyter, Berlin, 1984, pp. 19–47.

34. I. Liska, J. Krupcik, and P. A. Leclercq, *HRC*, *12*: 577 (1989).

35. L. Jordan, *LC-GC Int.*, *6*(10): 594 (1993).

36. M. Zief and R. Kiser, *Am. Lab.*, *22*(1): 70 (1990).

37. R. E. Majors, *LC-GC Int.*, *4*(3): 10 (1991).

38. S. B. Hawthorne, M. S. Krieger, and D. J. Miller, *Anal. Chem.*, *60*(5): 472 (1988).

39. S. B. Hawthorne, D. J. Miller, M. S. Krieger, and Z. Freseniuz, *Anal. Chem.*, *330*(3): 211 (1988).

40. D. A. Moyler, *Developments in Food Flavours* (G. G. Birch and M. G. Lindley, eds.), Elsevier, London, 1986, pp. 118–129.

41. M. McHugh and V. Krukonis, *Supercritical Fluid Extraction: Principles and Practice*, Butterworth, Boston, Chap. 10, 1986.

42. J. P. Vidal and H. Richard, *Flavour Science and Technology* (M. Martens, et al., eds.), John Wiley and Sons, New York, 1987, pp. 535–540.

43. P. Chatonnet and J. M. Boidron, *Sci. Alim.*, *8*: 479 (1988).

44. B. V. Ioffe and A. G. Vitenberg, *Headspace Analysis and Related Methods in Gas Chromatography*, John Wiley and Sons, New York, 1982.

45. H. Hachenberg and A. P. Schmidt, *Gas Chromatographic Headspace Analysis*, Heyden, London, 1977.

46. J. O. Bosset and R. Gauch, *Int. Dairy J.*, *3*: 359 (1993).
47. J. V. Hinshaw, *LC-GC Int.*, *3*(6): 20 (1990).
48. Z. Penton, *HRC*, *15*(12): 834 (1992).
49. B. Kolb, P. Pospisil, and M. Auer, *Chromatographia*, *19*: 113 (1984).
50. B. Kolb, *Chromatographia*, *15*: 587 (1982).
51. B. Kolb and P. Pospisil, *Sample Introduction in Capillary Gas Chromatography*, Huethig, Heidelberg, 1985, p. 191.
52. N. Onda, A. Shinohara, H. Ishii, and A. Sato, *HRC*, *14*: 357 (1991).
53. J. Gregoire, *Proceedings of the 6th Int. Symp. Capillary Chromatography*, Riva del Garda 1985 (P. Sandra and W. Bertsch, eds.), Huethig, Heidelberg, 1985, pp. 353–363.
54. A. Venema, *HRC & CC*, *13*: 537 (1990).
55. B. Kolb, *Proceedings of the 13th Int. Symp. Capillary Chromatography*, Riva del Garda 1991 (P. Sandra, ed.), Huethig, Heidelberg, 1991, pp. 369–379.
56. A. Hagman and S. Jacobsson, *Anal. Chem.*, *61*: 1207 (1989).
57. J. V. Hinshaw, *LC-GC Int.*, *3*(2): 22 (1990).
58. J. Curvers, T. Noy, C. Cramers, and J. Rijks, *J. Chromatogr.*, *289*: 171 (1984).
59. J. F. Pankow and M. E. Rosen, *HRC & CC*, *7*: 505 (1984).
60. P. Gagliardi, *Proceedings of the 8th Int. Symp. Capillary Chromatography*, Riva del Garda 1987 (W. Bertsch, ed.), Huethig, Heidelberg, 1987, pp. 390–405.
61. N. H. Mosesman, W. R. Betz, and S. D. Corman, *LC-GC Int.*, *1*(3): 62 (1988).
62. I. Maier and M. Fieber, *HRC*, *11*: 566 (1988).
63. H. J. Schaeffer, *HRC*, *12*: 69 (1989).
64. K. Grob, A. Artho, C. Frauenfelder, and I. Roth, *HRC*, *13*: 257 (1990).
65. S. Blomberg and J. Roerade, *HRC*, *13*: 509 (1990).
66. U. B. Burger and M. Le Roux, *HRC*, *15*: 373 (1992).
67. M. J. Feeney and W. Jennings, *Am. Chem. Soc. Symp. Series*, 176 (1988).
68. A. Mosandl, *Food Rev. Int.*, *4*(1): 1 (1988).
69. W. A. König, *Gas Chromatographic Enantiomer Separation with Modified Cyclodextrins*, Huethig, Heidelberg, 1992.
70. J. V. Hinshaw, *LC-GC Int.*, *6*(10): 606 (1993).
71. H. Frank, *Chromatographic Separation of Optical Isomers*, Huethig, Heidelberg, 1994.
72. K. Grob, *Classical Split and Splitless Injection in Capillary Gas Chromatography*, Huethig, Heidelberg, 1986, pp. 283–306.
73. W. Bretschneider and P. Werkhoff, *HRC*, *11*: 589 (1988).
74. G. Schomburg, H. Behlau, L. Dielmann, F. Weeke, and H. Husmann, *J. Chromatogr.*, *142*: 87 (1977).
75. K. Grob and K. Grob, Jr., *J. Chromatogr.*, *151*: 311 (1978).
76. K. Grob and G. Grob, *HRC & CC*, *1*: 263 (1978).
77. K. Grob, *On-Column Injection in Capillary Gas Chromatography*, 2nd ed., Huethig, Heidelberg, 1991.
78. W. Vogt, K. Jacob, and H. W. Obwexer, *J. Chromatogr.*, *174*: 437 (1979).
79. W. Vogt, K. Jacob, A. B. Ohnesorge, and H. W. Obwexer, *J. Chromatogr.*, *186*: 197 (1979).
80. G. Schomburg, *Proceedings of the 4th Int. Symp. Capillary Chromatography*, Hindelang, 1981 (R. E. Kaiser, ed.), Huethig, Heidelberg, 1981, pp. 371, A921.
81. F. Poy, S. Visani, and F. Terrosi, *J. Chromatogr.*, *217*: 81 (1981).
82. S. Nitz and E. Jülich, *Analysis of Volatiles, Methods, Application* (P. Schreier, ed.), Walter de Gruyter, Berlin, 1984, pp. 151–170.
83. S. Nitz, F. Drawert, and E. Jülich, *Chromatographia*, *18*: 313 (1984).
84. J. Tabera, G. Reglero, M. Herraiz, and P. G. Blanch, *HRC*, *14*: 392 (1991).
85. F. Poy and L. Cobelli, *Sample Introduction in Capillary Gas Chromatography* (P. Sandra, ed.), Huethig, Heidelberg, 1985, p. 93.
86. R. Barcarolo and P. Casson, *HRC 18*: 5 (1995).
87. H. Maarse and F. van den Berg, *Distilled Beverages Flavour: Recent Developments* (J. R. Piggot and A. Peterson, eds.), Ellis Horwood, Chichester, 1989, pp. 1–15.
88. G. Schomburg, *Sample Introduction in Capillary Gas Chromatography* (P. Sandra, ed.), Huethig, Heidelberg, 1985, pp. 235–260.
89. W. Shaowen, W. H. Chatman, and S. O. Farwell, *HRC*, *13*: 229 (1990).

90. S. Nitz, H. Kollmansberger, and F. Drawert, *Bioflavour 1987: Analysis, Biochemistry, Biotechnology* (P. Schreier, ed.), Walter de Gruyter, Berlin, 1988, pp. 123–135.

91. R. Rouseff, Characterization and Measurement of Flavor Compounds (D. D. Bills and C. J. Mussian, eds.), Am. Chem. Soc. Symp. Series, 1985, p. 289.

92. B. F. Kenney, *Food Technol.*, *44*(9): 76 (1990).

93. K. Grob and J. M. Stoll, *HRC & CC*, *9*: 518 (1986).

94. I. L. Davies, K. E. Markides, M. L. Lee, M. W. Raymur, and K. D. Bartle, *HRC*, *12*(4): 193 (1989).

95. F. Munari, G. Dugo, and A. Cotroneo, *HRC*, *13*(1): 56 (1990).

96. P. Schreier, *Characterization, Production and Application of Food Flavours* (M. Rothe, ed.), Akademic Verlag, Berlin, 1988, pp. 23–42.

97. Noever de Brauw and M. C. Ten, *Flavour '81* (P. Schreier, ed.), Walter de Gruyter, Berlin, 1981, pp. 253–286.

98. R. Barcarolo, P. Casson, and C. Tutta, *HRC*, *15*(5): 307 (1992).

99. G. Lange and W. Schultze, *Flavour Fragr. J.*, *2*: 63 (1987).

100. A. R. Culp and E. Noakes, *J. Agric. Food Chem.*, *40*: 1892 (1992).

101. U. Hener, R. Braunsdorf, and A. Mosandl, *Proceedings of the 15th Int. Symp.*, Riva del Garda 1993 (P. Sandra, ed.), Huethig, Heidelberg, 1993, pp. 998–1003.

102. S. Nitz, B. Weinreich, and F. Drawert, *HRC*, *15*: 387 (1992).

103. P. Schieberle and W. Grosch, *J. Agric. Food Chem.*, *35*: 252 (1987).

104. W. Grosch, Sen., H. Guth, and G. Zeiler-Hilgart, *Flavour Sci. Technol., Proceedings of 6th Symp.*, Weurman, 1990, pp. 191–194.

105. W. Herres, *HRGC-FTIR: Capillary Gas Chromatography-Fourier Transform Infrared Spectroscopy*, Huethig, Heidelberg, 1987.

106. C. L. Wilkins, *Anal. Chem.*, *59*: 571 (1987).

107. S. Compton and P. Stout, *LC-GC Int.*, *4*(3): 42 (1991).

108. P. Schreier, *Abh. Akad. Wiss. DDR, Abt. Math., Naturwiss., Tech.* (2, Charact., Prod. Appl. Food Flavours), pp. 23–42 (1988).

109. G. Fischboeck, R. Kellner, and W. Pfannhauser, *Agric. Food Chem. Consum., Proc. 5th Eur. Conf. Food Chem.*, *2*: 631–633 (1989).

110. T. E. Acree, J. Barnard, and D. G. Cunningham, *Food Chem.*, *14*: 273 (1984).

111. D. G. Cunningham, T. E. Acree, J. Barnard, R. M. Butts, and P. A. Braell, *Food Chem.*, *19*: 137 (1986).

112. W. Grosch and P. Schieberle, *Characterization, Production and Application of Food Flavours* (M. Rothe, ed.), Akademie Verlag, Berlin, 1988, pp. 139–151.

113. L. Moio, P. Etievant, and F. Addeo, *Vignevini*, *5*: 71 (1993).

114. T. Aishima, *J. Agric. Food Chem.*, *39*: 752 (1991).

115. V. Gerbi, G. Zeppa, and A. Carnacini, *Ital. J. Food Sci.*, *4*: 259 (1992).

116. A. Rapp, *Wine Analysis* (H. F. Linskens and J. F. Jackson, eds.), Springer-Verlag, Berlin, 1988, pp. 29–66).

117. Y. Z. Gunata, C. L. Bayonove, R. L. Baumes, and R. E. Cordonnier, *J. Chromatogr.*, *331*: 83 (1985).

25

Dietary Fiber

John A. Monro
New Zealand Institute for Crop and Food Research Limited,
Palmerston North, New Zealand

I. INTRODUCTION

Several articles have discussed the current state of dietary fiber analysis in light of the rapid developments that have taken place in recent years. While commenting on methods for analyzing dietary fiber, they usually do not give the detail required to carry out analyses as originally prescribed. However, a number of methods have been extensively summarized in a recent review by Prosky and De Vries (1) in the context of food regulation. Southgate (2) gave a more succinct (4 pages) overview of methods in relation to fiber definition, which has been an important and contentious issue. Asp et al. (3) critically assessed major methods, specifying sources of error associated with various steps and discussing recent studies of precision. A number of analysts prominent in fiber methodology also provided a detailed update of current methods and new developments (4).

Methods for analyzing dietary fiber have proliferated in the last 25 years, making the task of selecting a method, understanding the meaning of the results that it provides, and understanding how they differ from those of another method increasingly difficult. This chapter aims to provide some clarification by describing the range of methods available, their origins and intrelationships, what they measure, and, most importantly, what is involved in putting them into practice.

II. WHAT IS DIETARY FIBER?

Plant cell walls are at the core of the concept of dietary fiber (5,6). But it is a concept with very ragged boundaries around which there has been a great deal of discussion about which components to include, other than cell wall polysaccharides, and about the best methods to account for them. Non–cell wall polysaccharides, amylase-resistant starch, lignin, and any other materials that resist digestion by enzymes of the human digestive tract have been included in the various definitions of dietary fiber (7).

Plant cell walls are, however, directly or indirectly the origin of most of the fiber in the great majority of foods—most fruits, vegetables, nuts, cereals, pulses, and their products—and they are the common denominator in the various ways that it is conceptualized. Substances that have been included in dietary fiber are shown in Table 1.

III. PROPERTIES OF DIETARY FIBER

Knowledge about the structure of plant cell wall polymers is key to understanding the interactions between them that govern various properties of fiber, including its response to analytical conditions and its impact on physiology and health.

A. Chemical Nature of Dietary Fiber

1. Cell Wall Polysaccharides

The young (primary) plant cell wall is a framework of cellulose microfibrils that largely determines the mechanical properties of the wall (8) and of the dietary fiber derived from it (9). The cellulose microfibrils are embedded in a matrix of polysaccharides that show immense variability in composition and structure according to the species, tissue, cell type, and age. As the cell wall differentiates for specialized functions such as support (e.g., vascular fibers) or protection (e.g., seed coats), an organized secondary layer of cellulose microfibrils may be deposited inside the primary wall, often with hydroxycinnamyl alcohols polymerizing throughout the wall to form the vast macromolecular network that is lignin (10,11). It is a theme on which there are countless variations in nature.

2. Polysaccharide Classes

Plant cell wall polysaccharides have traditionally been classified into pectic substances, hemicellulose, and cellulose on the basis of conditions that preferentially extract them (11,12) (Table 1). But the boundaries between the polysaccharide classes are not absolute because the classification is based on extracting with a limited range of reagents and conditions that affect polysaccharides in all three classes to some degree, and often do not totally extract a given polysaccharide. For instance, although pectic substances, by definition, include polysaccharides soluble in cold and hot water and in hot chelating solutions, the latter conditions fail to extract a proportion of pectic-type polysaccharides. Thus some typically pectic polygalacturonic acid polymer resists extraction by hot chelating agents that conversely cause some fragmentation of typically hemicellulosic polysaccharides. Similarly, Klason lignin is so-called because it is not entirely lignin, but is a collection of acid-resistant materials remaining after 72% w/w (or 12 M) H_2SO_4 extraction of cellulose.

Selvendran and O'Neill (12) thoroughly reviewed procedures for preparing cell walls and for extracting and analyzing polysaccharide polymers from them.

3. Structural Features of Fiber Polysaccharides

The monosaccharide composition and structural features of polysaccharides commonly occurring in pectic substances, hemicellulose, cellulose, and some examples of gums and mucilages are shown in Table 2. Polysaccharides in each class are quite heterogeneous, varying enormously through the plant kingdom in their monosaccharide constituents, structure (branchedness, side chains, linearity), and the degrees to which they are associated in the cell wall. The matrix polysaccharides consequently exhibit a spectrum of properties, from the readily soluble members of the "pectic substances" that contribute substantially to "soluble fiber," to more

Table 1 Main Components that May Be Present in Dietary Fiber

Component	Characteristics
Pectic substances	Heteropolysaccharides, predominantly polygalacturonic acid polymers and associated neutral fractions extracted from cell walls with hot water, hot Ca^{2+} chelating solutions (e.g., EDTA, ammonium oxalate), or dilute acid. About 35% of dicot primary wall. Usually a soluble fiber component.
Hemicellulose	Complex polysaccharide mixture extracted from cell walls after pectin, with alkali (e.g., 10% KOH) for polymeric, or acid (e.g., 1.0 M H_2SO_4 100°C) for hydrolyzed. Major component of primary walls. Contributes mainly to insoluble fiber.
Cellulose	Polysaccharide left after extracting pectic substances and hemicellulose. Main structural polymer in plants, high in fibrous tissues with secondary cell walls. Strong alkali-insoluble, 72% (w/w) H_2SO_4-soluble at room temperature.
Lignin	Left after 72% (w/w) H_2SO_4 cell wall extraction (Klason lignin). Extensive macromolecule from phenolic alcohols condensing in plant cell wall. Low in most foods.
Protein	About 10% of primary cell wall. Partly crosslinked with wall polysaccharides. Susceptible to proteases.
Gums and mucilages	Usually complex, heterogeneous polysaccharides of diverse origin–higher plant, algal, bacterial. Mainly nonstructural. Common features: water-soluble, form viscous solutions or gels. Widely used in food industry, but small component of food. Higher plant: guar gum, gum arabic, gum tragacanth, larch gum. Algal: carrageenan, alginate. Bacterial: xanthan gum, whelan, gellan. Usually contribute to soluble fiber.
Resistant starch	Amylase-resistant. Mainly in starchy foods (cereals, pulses). Result of structure of food and/or starch granule impeding enzyme access to starch, or of starch retrogradation due to food processing. High amylose starches partly resistant. May add to insoluble fiber—depending on analysis method and fiber definition used. A small component of most foods.
Cuticular substances	Complex waxes present on protective plant surfaces such as the epidermis. Contribute to insoluble fiber weight unless removed with organic solvent.
Suberin	Long-chain fatty acid derivatives that occur in protective surfaces of the plant such as bark.
Phenolic compounds	Polyhydroxyphenolic compounds such as tannins may condense or react with other food components to form compounds that analyze as lignin.
Maillard products	Compounds that result from the reaction of sugars and proteins in heated foods. May analyze as lignin.
Chitin	Second most abundant natural polysaccharide. From invertebrate exoskeletons. Structure similar to cellulose but based on amino sugars. Not a common dietary fiber.
Man-made	E.g., polydextrose, carboxymethylcellulose, lactulose.
Storage polysaccharides	Storage polysaccharides analyzing as fiber include resistant starch, fructans, galactomannans.

Source: Refs. 3, 8, 10, 11.

Table 2   Dietary Fiber Polysaccharides

Type	Main chain sugars and linkage	Side chain sugars	Structural features
Structural: Cell wall			
(a) Pectic substances			
Rhamnogalacturonan	D-Galacturonic acid α-(1-4) with interspersed L-rhamnose (1-2), (1-4)	Arabinogalactans	DP ≅ 2000. Highly branched at rhamnose units. Carboxyl groups partially methyl-esterified.
Homogalacturonan	D-Galacturonic acid, α-(1-4)	D-Galactose, L-arabinose oligosaccharides	DP > 25. Partially methyl-esterified.
Arabinogalactan	D-Galactose β-(1-3)	D-Galactose, L-arabinose	Probably derived from rhamnogalacturonan.
Arabinan	L-Arabinose, α-(1-3), (1-5)		Highly branched. DP 34, 100.
Galactan	D-Galactose, mainly β-(1-4) also β-(1-6)		DP 30–50.
(b) Hemicellulose			
Xylan	D-Xylose, β-(1-4)	D-Glucuronic acid	Linear. Partly acetylated.
Xyloglucan	D-Glucose, β-(1-4)	D-Xylose, D-glucose, L-arabinose, D-galactose	DP ≅ 200.
Glucuronoarabinoxylan	D-Xylose, β-(1-4)	D-Glucuronic acid, L-arabinose	Arabinose may be covalently linked to lignin via ferulic acid.
Glucomannan	D-Glucose, β-(1-4) D-Mannose, β-(1-4)		Extended chains of similar conformation to cellulose.
Mixed β-glucan	D-Glucose, β-(1-3), (1-4)		Mixed linkage so nonlinear conformation
(c) Cellulose			
β-glucan	D-Glucose, β-(1-4)	Xylose (very small amounts)	DP 6000–7000. Highly crystalline areas.
Mainly nonstructural: Gums and mucilages			
Gum arabic	D-Galactose β-(1-4), (1-6)	D-Galactose, L-arabinose, L-rhamnose	Heteropolysaccharide. Highly branched. DP ≅ 6000.
Guar gum	D-Mannose β-(1-4)	D-Galactose every alternate unit	Numerous side chains.
Gum tragacanth	D-Galacturonic acid	D-Galactose, D-xylose, L-arabinose.	Complex and large. Numerous side chains cation involvement.
Xanthan gum	D-Glucose, β-(1-4)	D-Mannose, D-glucose	Every eighth sugar residue in main chain bears a mannose side chain, and every sixteenth a glucose.

Source: Refs. 8, 10, 11.

resistant and mechanically strong types that typify "hemicelluloses" and are part of "insoluble fiber."

An alternative classification of cell wall polysaccharides in terms of their associations within the cell wall has been suggested; pectic substances covalently associated with galacturonic acid–containing polysaccharides and hemicelluloses noncovalently associated with cellulose (14).

4. Polysaccharide Structure and Physical Properties

Within the variability of cell wall polysaccharides there are general features of structure that limit the extent to which polysaccharide chains become annealed, in turn governing their solubility, the choice of reagent to solubilize them, and their physiological effects as dietary fiber (9). Some structural features that restrict the alignment of polysaccharide chains, limiting formation of noncovalent bonds, and their consequent effect on physical properties are summarized in Table 3.

C. Properties of Dietary Fiber in Food

The physical properties governed by polymer interactions in the cell wall in turn determine the properties of dietary fiber in food and in the gut, and so too its physiological effects. Some of the suggested relationships between fiber properties and health are summarized in Table 4.

The dietary fiber in most noncereal plant foods consists mainly of primary cell walls; it contains a relatively large proportion of matrix polysaccharides (pectic substances and hemicellulose) with little lignin encrustation, so the ratio of soluble to insoluble fiber is high. It does not impart a fibrous texture to food but provides a physical constraint against which the pressure of the turgid cell acts, giving a crisp texture to fresh food (17). It can be softened by cooking, particularly when it contains high levels of methyl-esterified pectic polysaccharides, as they are susceptible to depolymerization during β elimination of the methyl groups (18). Both the soluble and insoluble fiber from primary cell walls is quite susceptible to bacterial fermentation.

Cell walls with secondary thickening are strong, giving a coarse texture to foods in which they occur. Fiber in most foods is derived mainly from primary cell walls, but "roughage" from secondary cell walls is more effective in its action in the large bowel, where lignin confers resistance to bacterial action. For instance, wheat bran is an effective laxative because secondary thickening and lignification ensure that the bran particles remain intact, providing bulk (15) and mechanical stimulation (19) in the large bowel. Cereal refining has been concerned largely with removing secondary cell walls from flour, at a cost to health.

Table 5 gives examples of levels of soluble and insoluble fiber polysaccharides that are likely to be found in some common foods.

IV. WHY MEASURE DIETARY FIBER?

Several factors have given impetus to the development of methods for dietary fiber analysis for use in determining food composition (1):

Recognition of the fundamental difference in digestive capabilities of monogastrics, and of ruminants for which fiber analyses were originally designed

The need to measure the unavailable fraction of foods, to make allowance for its assumed noncontribution to energy intake

Table 3 Effect of Primary Structure on Dietary Fiber Polysaccharide Interactions and Properties

Primary structure variation	Effect on interaction between polysaccharides	Resulting properties and examples
(a) Favoring solubility		
Degree of polymerization: low	Little entanglement. Extended regions of annealment not possible.	Cold-water-soluble pectin. Low viscosity.
Side chains: many Branching: much	Alignment hindered so little noncovalent bonding.	Pectic arabinogalactan.
Conformation: nonlinear and/or irregular	Disordered chains prevent extensive alignment for noncovalent bonding.	Pectic rhamnogalacturonan (kinked at rhamnose units).
Carboxyl groups: free ionized	Electrostatic repulsion prevents chain alignment.	Pectic galacturonan when Ca^{2+} removed by chelator such as EDTA.
(b) Favoring insolubility		
Degree of polymerization: high	Entanglement. Extended alignments possible.	Some depolymerization favors extraction, e.g., rhamnogalacturonan, cellulose
Side chains: few Branching: little	Little hindrance to noncovalent bonding in free linear regions allowing junction zones to form.	Mechanically strong. Water-insoluble. H-bond cleaving agent, e.g., 10% KOH, or acid (e.g., 1 N H_2SO_4) hydrolysis to extract (e.g., hemicelluloses).
Conformation: linear regular	Ordered packing and extensive alignment occurs allowing pseudocrystalline, multichain structures.	Mechanically strong, insoluble, solvent-resistant, e.g., xylans, mannans, cellulose, chitins.
Carboxyl groups: Ca^{2+}-esterified	Ca^{2+}-stabilized junction zones.	To extract need agents that sequester Ca^{2+}, e.g., hot EDTA, hot phosphate, NH_4 oxalate.
Carboxyl groups: methyl-esterified	Interchain hydrophobic bonding possible. Restrict Ca^{2+}-mediated crosslinking.	β elimination of methyl groups depolymerizes (hot neutral to alkaline conditions), e.g., cold water–insoluble—hot water–soluble pectin.

Source: Refs. 9, 10, 11.

Evidence for the benefits to health (Table 4) arising from undigested food residue in the gastrointestinal tract, with dietary fiber no longer being seen as a nutritionally inert food component

A need for regulation, particularly in view of the intense development and marketing of products on the basis of the supposed benefits to health conferred by the dietary fiber in them

Demand for quantification of dietary fiber has come from nutritionists, food regulatory authorities, food producers, and the public.

Table 4 Relationships Between Fiber Properties, Physiological Effects, and Health

Fiber property	Physiological effects	Possible effect on health
Soluble, forming viscous solutions, e.g., β-glucan, pectin, gums, mucilages.	Delayed gastric emptying, maintaining satiety and lowering food intake. Displaces energy-rich foods lowering energy intake. Slows diffusion, increases rate of transit, slowing or reducing rate of absorption of:	↓Obesity (risk factor for diabetes, hypertension, heart disease).
	sugars,	↓Hyperglycemia.
	lipids, cholesterol, bile acids.	↓Plasma lipids, cholesterol, thus heart disease.
	Less enterohepatic recycling increases 2° bile acids (cancer promoting) in colon.	↑Colon cancer.
Soluble, and insoluble with little secondary thickening or lignification. Readily fermented in the colon (e.g., fruit and vegetable cell walls).	Increases colonic fermentation, increasing short chain fatty acids:	
	butyric (stabilises colonic epithelium);	↓Colon cancer.
	propionic (lowers cholesterol synthesis);	↓Heart disease.
	lower pH decreases bile acid solubility in the colon.	↓Colon cancer.
	Increases bacterial conversion of 1° to 2° bile acids (cancer promoting).	↑Colon cancer.
	Increases bacterial mass but fiber destroyed, little net increase in moist fecal bulk. Concentration of carcinogens and promotors may increase.	↑Colon cancer.
Much secondary thickening and lignification. Insoluble. Partly resistant to fermentation in the colon, e.g., wheat bran.	Increases fecal bulk and moisture, reducing:	↓Hemorrhoids, diverticulitis, varicose veins, hiatus hernia.
	local pressure, straining, and	
	bile acid and carcinogen concentrations.	↓Colon cancer.
	Resistant fiber (cell walls) provide: mechanical stimulus for bowel activity eliminating carcinogens and toxins, and large binding surface reducing levels of free carcinogens and toxins.	↓Colon cancer.

Source: Refs. 1, 15, 16, 20.

V. FOOD REGULATIONS

For food regulatory purposes in the United States, Scandinavia, Germany, Switzerland, and other countries, the Prosky (AOAC) method has been accepted (1). Indeed, the method was developed partly in response to the U.S. Food and Drug Administration's need to replace the ruminant-oriented crude fiber and detergent methods with an inexpensive, rapid method for routine analysis of human foods in conventional analytical laboratories.

The European Community Scientific Committee for Food (SCF) has recently decided that for regulatory purposes dietary fiber should be defined as non-starch polysaccharides of cell-wall origin, as are measured by the Englyst procedure (21).

As methods for dietary fiber analysis are continuing to be developed and made more nutritionally valid, it is likely that methods used for regulatory purposes will change, especially as the

Table 5 Dietary Fiber Content of Some Common Types of Food (g/100g edible weight)

Food	Fiber fraction		Total fiber
	Soluble	Insoluble	
Vegetables			
Carrots (raw)	1.5	1.6	3.2
Sweet potato (boiled)	1.53	1.24	2.78
Broccoli (boiled)	1.3	1.9	3.2
Fruits			
Apple (raw flesh)	0.5	0.99	1.53
Banana (raw)	0.86	0.6	1.46
Kiwi fruit (flesh)	1.36	0.26	1.62
Cereals products			
Flour, white	1.02	2.12	3.14
Flour, whole meal	1.56	10.48	12.04
Bread, white	1.65	1.15	2.80
Bread, whole meal	2.13	3.58	5.72

Source: New Zealand Food Composition Database (OCNZ94).

purpose of labeling becomes increasingly to inform the consumer about food components in the context of health.

VI. ANALYSIS OF DIETARY FIBER

By all definitions, dietary fiber is a complex of food polymers. Therefore this chapter does not consider methods for measuring individual components that may be part of dietary fiber, such as β-glucan, except where they are required to obtain a measure of dietary fiber as a whole.

Techniques not yet commonly used for routine analysis of dietary fiber in human foods, such as nuclear magnetic resonance spectroscopy (NMR) and near-infrared reflectance spectroscopy (NIRS), will not be discussed in detail given the limited space available. NIRS is a method with great potential because of its speed and simplicity. However, the equipment is costly and its usefulness depends on the adequacy of the chemical methods (reviewed here) used to calibrate it.

A. Overview

Most of the methods currently used and/or commonly referred to are listed in Table 6. As many of the methods shown in the table are points in a continual evolution, they may also be ordered into a number of lineages, as in Fig. 1, which shows that although methods (e.g., 1CF) originated last century for analysis of fiber in forage, their sustained development for application to human foods has taken place mainly in the last 25 years or so. The methods may be grouped by the means used for digesting the sample to dissociate nonfiber components from fiber (chemical hydrolysis, detergent/chemical hydrolysis, etc.) and by the ways that the fiber is subsequently measured (gravimetric, colorimetric/gravimetric, etc.).

The choice of method for fiber analysis is determined in the first place by the operational definition of dietary fiber being used. The procedure used for extracting and digesting a food sample determines the components that remain to be measured as dietary fiber, and methods for measuring these components determine which of them is actually part of the complex finally called "dietary fiber." Each definition (or synonym) for dietary fiber has its own analytical strategy,

Table 6 Methods for Analysis of Dietary Fiber: Overview

	Method		Procedure to:			
Code[a]	Name	Year	Digest sample	Measure fiber	Fraction measured	Ref.
1CF	Crude fiber	1859	Chem	Grav	CF	22
2NAF	Normal acid fiber	1955	"	"	NAF	23
3ADF	Acid det. fiber	1963	Det/Chem	"	ADF	24
4NDF	Neutral det. fiber	1963	NDet	"	NDF	25
5Sch	Schaller	1977	NDet/enz	"	NDF	26
6R&V	Robertson, Van Soest	1981	"	"	NDF	27
7M&B	Mongeau, Brassard	1986	"	"	NDF, SDF	28
8UED1	Urea enz. dialysis	1989	Urea/enz	"	TDF	29
9UED2	Urea enz. dialysis	1990	"	"	SDF, IDF	30
10W&O	Williams, Olmstedt	1935	Enz	Col/grav/bioch	SDF, Hemi, Cell.	31
11McC	McCance	1936	"	Difference		32
12W&B	Weinstock, Benham	1951	"	Grav	TDF(u)	33
13Hell	Hellendorn	1975	"	Grav	IDF(u)	34
14Elch	Elchazly	1976	"	"	ADF, Cell., KLig	35
15Berl	Berlin	1985	"	"	SDF, IDF	36
16S&W	Schweizer, Wursch	1979	"	"	SDF, IDF(u)	37
17A&J	Asp, Johannson	1981	"	"	"	38
18Lund	Lund	1983	"	"	TDF, SDF, IDF	39
19Fur	Furda	1981	"	"	"	40
20Pro1	Prosky/AOAC	1984	"	"	TDF	41
21Pro2	Prosky/AOAC	1988	"	"	SDF, IDF	42
22Lee	Lee	1990	"	"	TDF	43
23LiA	Li, Andrews	1988	"	"	SDF, IDF	44
24LiC1	Li, Cardozo	1990	"	"	TDF	45
25LiC2	Li, Cardozo	1992	Water	"	TDF	46
26Sou	Southgate	1969	Enz	Col/grav	SNSP, Hemi, Cell.	47
27Sell	Selvendran, Dupont	1980	"	GC/ColU/Grav	TNSP, KLig	48
28T&A	Theander, Aman	1981	"	GC/Pot/Grav	TNDP, SNDP, INDP, KLig	49
29Upp	Uppsala	1986	"	"	"	50
30En1	Englyst	1981	Enz/Chem	GC/ColU	SNCP, INCP, Cell., KLig	51
31En2	Englyst et al.	1982	"	"	TNSP, NCP, INSP	52
32Sel2	Selvendran et al.	1981	DMSO/Enz	GC/ColU/Grav	TNSP, KLig	53
33En3	Englyst, Cummings	1984	"	GC/ColU	TNSP	54
34F&T	Faulks, Timms	1985	"	Col	"	55
35En4	Englyst, Hudson	1987	"	Col	TNSP, INSP	56
36Bril	Brillouet et al.	1988	"	GC/ColU/Grav	SNSP, ICWM	57
37En5	Englyst, Cummings	1988	"	GC/ColU	TNSP, INSP	58
38En6	Englyst et al.	1992	"	"	"	59
39Mon	Monro	1993	"	"	"	60

[a]Methods currently in use or widely referred to are in bold print in this and subsequent tables.

Abbreviations for all tables: ADF = acid detergent fiber; Bioch = biochemical; Cell. = cellulose; CF = crude fiber; Chem = chemical hydrolysis; Col = colorimetric; ColU = colorimetric for uronic acids only; Det = detergent; DMSO = dimethylsulfoxide; Enz = enzymatic; EtOH = ethanol, GC = gas chromatography; grav = gravimetric; Hemi = hemicellulose; ICWM = insoluble cell wall material; KLig = Klason lignin; NAF = normal acid fiber; NCP, SNCP, INCP = total, soluble, and insoluble noncellulosic polysaccharide; NDF = neutral detergent fiber; o/n = overnight; NSP = nonstarch polysaccharide; ONSP = other nonstarch polysaccharides (not cell wall structural or RS, i.e., gums and mucilages, storage polysaccharides, etc.); Pect = pectic substances; Phos = phosphate; Pot = potentiometric; RS = resistant starch; s/n = supernatant; TDF, SDF, IDF = total, insoluble, and soluble dietary fiber; TDF(u), IDF(u) = TDF and IDF uncorrected for ash and/or protein; temp = temperature; TNDF, SNDF, INDF = total, soluble, and insoluble nondigestible polysaccharide; TNSP, SNSP, INSP = total, soluble, and insoluble nonstarch polysaccharide; wt = weight.

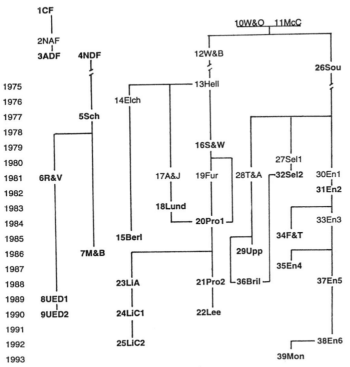

Fig. 1   Evolutionary relationships between the methods for analysis of dietary fiber outlined in Table 6.

depending on the components that it dictates that one should measure (Table 7). Thus, although the methods of Table 6 usually entail a small number of basic analytical stages, which are shown in Fig. 2 in relation to the definitions in Table 7, at each of these stages a selection of preparative and measurement procedures have been used (Table 8), in various combinations, to achieve the objective of the stage.

The different definitions of dietary fiber and associated methods have given rise to a range of dietary fiber preparations (CF, NAF, ADF, etc.; see key to Table 6). These food fractions contain differing proportions of cell wall and other food constituents, as summarized in Table 9.

Understandably, selecting a method from the large number available and knowing that it is measuring a relevant food fraction has been a problem for analysts. There are several reasons why a large number of methods have evolved.

There has not been agreement on a single precise definition of dietary fiber as a basis for a single analytical approach. Components included in various definitions range from cellulose plus lignin, to the full gamut of food components that remain after enzymic digestion (Tables 1 and 9).

Definitions have, in some cases, been pragmatically modified to align with analytical capabilities. For instance, the definition of fiber as "non-starch polysaccharide" (65) arose from "the remnants of plant cells resistant to the digestive enzymes of man" (63), because the components of the former are amenable to precise chemical analysis, whereas the latter can

Table 7 Relationship Between Fiber Definitions and Analytical Methods

	Fiber definition	Analytical objective	Digestion-measurement (methods)[a]
A	The residue of plant food left after extraction with solvent, dilute acid, and dilute alkali (61).	To obtain an empirical measure of the **nonnutritive fraction of animal feeds.**	Chem—grav, e.g. digestion Det/chem—grav (**1CF–3ADF**)
B	Insoluble organic matter indigestible by animal enzymes (62).	"	NDet—grav, NDet/enz—grav (**4NDF–6R&V**)
C	Remnants of plant cells resistant to hydrolysis by the alimentary enzymes of man (63).	To measure **nondigestible food constituents** (mainly polysaccharides and lignin) resistant to activities of the type shown by human digestive enzymes.	Enz/Grav, Urea/enz—grav (**7M&B–9UED2, 12W&B–25LiC2**)
D	Sum of lignin and the plant polysaccharides not digested by the endogenous secretions of the human digestive tract (64).	To measure **nondigestible polysaccharides** (NDP) and **lignin**, resistant to activities of the type shown by human digestive enzymes.	Enz–specific[b] (**10W&O, 11McC, 26Sou–29Upp**)
E	Nonstarch polysaccharides (65).	To measure **nonstarch polysaccharides** (NSP) as an index of plant cell walls in food.	DMSO/enz-specific (**30En1–39Mon**)

[a]See Table 6.
[b]"Specific" includes GC and colorimetric methods.

in theory contain any food component and cannot be measured practically by specific chemical methods. Gravimetric methods have not needed to be as specific in their associated definition.

Methods for each analytical strategy have continually been developed to improve their validity and suitability for routine food analysis and to include new reagents, enzymes, equipment, and procedures for measuring fiber constituents.

Despite the intention of the methods using enzymes to provide a "physiological" measure of fiber as material resistant to the digestive enzymes of the human gut, methods have not been constrained by a defined set of analytical conditions representing the gut lumen, in which the enzymes would normally act (60).

B. Outline of Individual Methods for Fiber Analysis

An outline of each of the methods for fiber analysis in Table 6 is presented in Table 10 (comments on methods), Table 11 (specified pretreatment of samples), Table 12 (specified methods for preparing fiber), table 13 (specified enzyme preparations), and Table 14 (fiber measurement). The summaries in Tables 11–14 are intended to give an idea of steps required to apply the methods and enable changes occurring in their evolution to be seen. The descriptions are based on details available in the original publications, which the reader is urged to consult for information on equipment, manipulations, precautions and various other practical suggestions, before applying any of the methods.

Table 8 Range of Techniques and Reagents Used at Different Stages of Fiber Analysis

Stage of analysis (objective)	Technique and reagents used
Preparation (drying and disintegration) ↓	Oven drying, air drying, freeze drying, no drying. Wet homogenization, grinding, ball milling, sonication (Table 11).
Preextraction (defatting, decolorizing) ↓	None, acetone, chloroform, chloroform-methanol, diethyl ether, ethanol, 90% ethanol, 80% ethanol, ethanol benzene, hexane, petroleum spirit (Table 11).
Digestion (dissolution of starch and other nonfiber food components) ↓	Hot alkali, cold alkali, hot acid, acid detergent, neutral detergent, urea, bile acids, cold water, hot water, DMSO. Buffers: acetate, phos, citrate/phos, MES/Tris, HCl/NaCl, intestinal (Table 12). Amylases, amyloglucosidases, proteases, pullulanase, mixed hydrolases (Table 13).
Separation (recovery of fiber from digestate) ↓	Precipitation, dialysis, filtration, centrifugation, ultrafiltration (Table 12).
Measurement of fiber	Direct weighing, loss in weight on ashing, sample weight minus other components, colorimetry, GLC/colorimetry, HPLC (Table 14).

Source: Refs. 22–60 (see Table 6).

1. Sample Preparation

Preparing samples (Table 11) is an extremely important preliminary to most fiber determinations because thorough extraction of nonfiber components, particularly starch, is necessary to avoid overestimating fiber. Samples must therefore be sufficiently fragmented and defatted for polymers to freely diffuse and/or for enzymes to gain unhindered access to their substrates during digestion.

Pretreatment should, however, be the minimum necessary to enable sample digestion and fiber measurement without interference, as heating, dehydrating, and grinding may lead to reactions that change the solubility of fiber and other food components (7,52), reducing the validity of methods that depend on fiber solubility and/or indigestible residue weight. Hot drying may lead to artifacts such as Maillard products and resistant starch (7,21) and is likely to cause partial degradation of the fiber, thus altering its solubility. Maillard products behave as lignin in fiber analysis, inflating fiber values from gravimetric methods that include lignin in fiber, particularly when there is no correction for residual protein. Very small particle size has been associated with loss of fiber in analysis, especially in methods using chemical hydrolysis and/or detergent (Methods 1CF-7M&B) (66,67). This may be due to increased reactive surface or losses of fine material during filtration.

Dry milling or grinding, before or after defatting, is the most common means of disintegrating samples. A dry, homogeneous powder is convenient for accurately obtaining a representative subsample for analysis, although sample disruption should ideally be achieved with a short period of liquid homogenization or wet ball milling when the purpose of analysis is to measure dietary fiber components in food as it is consumed and digested. Neither milling nor homogenization will have an effect similar to chewing and are therefore not physiological. But thorough tissue disruption is necessary to successfully extract starch and protein, and it allows a measure of the potential effect of the chemical environment of the gut on the solubility of food cell wall polysaccharides free from the variable effect of gross food structure.

Methods associated with definitions of dietary fiber given in Table 7 (A-E)				
A 1CF-3ADF	**B** 4NDF-7M&B	**C** 15Berl- 25LiC2	**D** 28T&A- 29Upp	**E** 32Sel2- 38En6

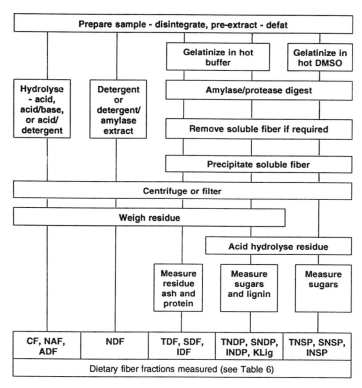

Fig. 2 Relationship between fiber definitions, methods for fiber analysis, steps involved in applying the methods, and dietary fiber preparations obtained. (Methods 8UED1 and 9UED2 are similar to methods under C above, but cold 8M urea used for starch gelatinization. Method 39Mon extracts soluble fiber under physiological conditions before applying methods under C, D, or E above.)

Defatting-depigmenting methods have used a range of solvents (Table 11). Alcohol pretreatments should be avoided when soluble fiber is to be measured because ethanolic dehydration may lead to retrogradation of starch and alter the solubility of some fiber polysaccharides (12).

Samples with a fat content greater than 5–10% are usually defatted. The fat level that can be tolerated without affecting fiber analysis has not been determined but will depend on fat in concert with other properties of a food (3).

2. Sample Digestion

The main objective of sample digestion is to extract and/or degrade food components that are not part of dietary fiber as defined for the method and that would invalidate fiber measurements if not removed. It is most important to extract digestible starch, as it will behave as apparent cell wall polysaccharide, contributing to fiber measured either gravimetrically or as polysaccharide sugars. Resistant starch (Table 1) formed during sample processing can contribute to fiber measured by many of the methods in Table 6, but excluding those using DMSO gelatinization

Table 9 Food Components in Dietary Fiber and Fiber Fractions

Fiber fraction[a]	Food components[a]						Protein residue, artifacts, Maillard products, minerals, etc.[c]
	Lignin	Cell.	Hemi	Pect	ONSP[b]	RS	
	Cell wall polysaccharides						
	Nonstarch polysaccharides						
	Nondigestible polysaccharides						
	Nondigestible food residue						
CF	—	—	—				—
NAF[d]	———	———	———	—			—
ADF[d]	———	———	———				
NDF	———	———	———	——			
TDF(u)	—————————————————————————————————						————
TDF	—————————————————————————————————						————
SDF, SNDP, SNSP				———	———	———	
IDF(u)						———	———
IDF						———	———
TNDP		———	———	———	———	———	
INDP		———	———	———		———	
TNSP		———	———	———	———		
INSP		———	———	———			
ICWM	———	———	———				
KLig	—						—

[a]See key to Table 6.
[b]ONSP assumed to be soluble polysaccharide.
[c]Residual protein, artifacts, Maillard products, etc.
[d]Often includes a significant proportion of hemicellulose and pectin.

(32Sel2-39Mon) before digesting, or gravimetric methods involving acid hydrolysis (1CF, 2NAF, 3ADF) or urea (8UED1, 9UED2).

When dimethylsulfoxide (DMSO) is used to gelatinize starch prior to digestion, in methods measuring fiber as NSP (Table 6, 32Sel2-38En6), it may extract some otherwise insoluble cell wall polysaccharide (3,48), inflating SNSP values at the expense of INSP. The problem is overcome by extracting soluble fiber before DMSO treatment, as in method 39Mon (60).

Protein is also digested in most methods, although a correction may be applied by subtracting residual protein as N × 6.25. But it is an approximate correction that can lead to error when the levels of residual protein are high.

a. Points of Procedure. A number of common sense directions that continually recur throughout instruction to the methods for fiber preparation, outlined in Table 12, include the following:

Always run blanks through the analysis to correct for any contribution of reagents and enzymes to measured fiber.

Table 10 Outline of Methods for Fiber Analysis: Comments on Methods

Method	Comments
1CF	AOAC method 962.09, AACC method 32-10. Empirical. Technically difficult. Most pectin and hemicellulose, some cellulose and lignin destroyed. Nitrogen high in residue. Used to predict forage digestibility. Not suitable for human foods but still used as a statutory method.
2NAF	An attempt to reduce variability in CF due to alkali in method 1CF. Most pectin and hemicellulose lost.
3ADF	Gives low-nitrogen residue mainly lignin + cellulose for predicting forage digestibility. Used as a pretreatment before lignin (ADF lignin; AOAC method 973.18D) and cellulose analysis but some lignin loss, and extraction of hemicellulose and pectin may be incomplete.
4NDF	Gives low-nitrogen residue. Measures "cell walls" in forages but less suitable for foods as most soluble fiber (e.g., pectin) lost due to heat + EDTA. NDF-ADF used to measure hemicellulose. Starch contaminates residue in starchy samples and slows filtration.
5Sch	AACC insoluble dietary fiber method 32-20. NDF method with an amylase treatment removing persistent starch.
6R&V	Like 5Sch, NDF method with amylase treatment.
7M&B	AOAC method 992.6, AACC method 32-06. Like 5Sch and 6R&V, plus analysis of SDF. However, NDF + SDF may not equal TDF by this method.
8UED1	TDF method. Urea hydrates and extracts starch and soluble polysaccharides without heat damage. Lower ash, protein, and resistant starch in the residue than AOAC (21Pro2) method. Inexpensive.
9UED2	Modified 8UED1 to separate SDF and IDF.
10W&O	Archetypal enzymic method for IDF. Three-day digestion, loss of SDF and some hemicellulose in alkaline pH. Measurement of cellulose as glucose and hemicellulose as pentose not valid. Forerunner to 26Sou. Technically difficult.
11McC	TDF indirectly by difference, but specific amylase digestion of starch also forerunner to method 26Sou.
12W&B	Accelerated Remy (1931; pepsin, diastase, pancreatin digestion, 8 days) CF method using fungal enzyme. SDF not recovered.
13Hell	Prototype "physiological" method using mammalian enzymes (pepsin and pancreatin) but conditions used not physiological and SDF not recovered.
14Elch	Single tube method for ADF and its cellulose and lignin content. SDF and most hemicellulose lost.
15Berl	Modified 14 Elch. Unusual as separates SDF and IDF after autoclaving, before starch digestion, and recovers SDF by membrane filtration. Use of citrate/phos buffer allows intermediate separation step to be avoided. Results similar to AOAC method (21Pro2).
16S&W	Modified 13Hell mainly to recover SDF.
17A&J	Modified 13Hell with 80% EtOH precipitation for recovering both SDF and IDF. Replaced by 18Lun.
18Lun	Improved 17A&J for short digestion (Termamyl), and easy fiber collection by filtering on Celite.
19Fur	Another method to recover and measure SDF and IDF, but replaced by (incorporated into) 21Pro1.
20Pro1	Based mainly on 15S&W, 18Lund, and 19Fur; developed as rapid gravimetric method. Requires ash and protein correction (with associated error) on residue. TDF procedure modified slightly in 20Pro2. Filtration may be difficult with viscous digests.

Table 10   (*Continued*)

Method	Comments
21Pro2	Modified 20Pro1 for TDF (AOAC method 985.29 for TDF) by changing NaOH and phos buffer concentration, also by including SDF and IDF separation (AOAC method 991.42 for IDF). Tends to give higher total fiber values than specific methods such as 29Upp.
22Lee	AOAC method 991.43 for TDF, SDF, and IDF, AACC method 32-07. Modified 21Pro1. Changing buffer from phos to MES/Tris eliminates a pH adjustment reducing number of steps, averts phos precipitation into fiber, and decreases volume. Faster than 21Pro2 giving similar results but improved precision.
23LiA	Modified 20Pro1; autoclaving, one enzyme (no protease), one buffer, no pH adjustments. Fast. May heat-degrade fiber. Residue protein in high-protein low-fiber food may add to correction error.
24LiC1	Modified 23LiA measuring SDF and IDF.
25LiC2	20Pro1 reduced to simple water extraction. For fruit and vegetables (low starch and protein).
26Sou	Archetypal method based on fractionating fiber into water-soluble NSP, "hemicellulose," cellulose. Fractionation laborious, still valid, but methods for measuring polysaccharide fractions improved.
27Sel1	For clean cell walls before GC analysis of TNSP. Further developed to prepare clean cell walls for polysaccharide analysis (12).
28T&A	Forerunner to 29Upp; both include resistant starch.
29Upp	Well-researched single tube GC method. Potentiometric analysis of uronic acids slow but colorimetry may replace it. Results similar to but often lower than with 21Pro2 (AOAC) gravimetric method for TDF.
30En1	Like 26Sou but fractions by GC and colorimetry.
31En2	Modified 31En1, with more specific enzymes, GC, and colorimetry only for uronic acids. Excludes resistant starch. Fractionation complex for routine analysis but provides detailed information.
32Sel2	Method for preparing clean, starch-free cell walls, but some loss.
33En3	Simplified 31En2 for total NSP; DMSO plus 16 hr starch digestion, NH_4OH replaces $BaCO_3$ to neutralize hydrolyzates, rapid derivatization for GC. Higher soluble fiber than 21Pro2 probably due to higher pH, and DMSO extraction (72,87).
34F&T	NSP method similar to 33En3 in digestion but faster (uses DMSO/heat-stable amylase). Colorimetric. Results correlate well with GC but slightly higher.
35En4	33En3 made colorimetric, giving similar results.
36Bril	Detergent plus proteases under mild conditions and digestate removal before heating, with subsequent starch digestion similar to 21Pro2, gives clean deproteinized cell wall material (ICWM = INSP + lignin). ICWM gravimetrically, soluble NSP by GC and colorimetry.
37En5	Modified 31En2, similar to 33En3 but adapted to give TNSP and INSP values (SNSP by difference). Hot, pH 7, phos buffer step extracts more pectic substances than pH 5 acetate buffer in 29Upp.
38En6	Modified 37En5 by using Termamyl for rapid starch hydrolysis. Conditions for hydrolyzing polysaccharides, GC and colrimetric analysis of monosaccharides, and derivation of correction factors discussed in detail.
39Mon	Introduces physiological relevance by extracting and separating SDF under physiological conditions before digesting starch. Circumvents effects of heat and DMSO during digestion on polysaccharide distribution between SNSP and INSP. Allows effects of food processing/cooking on fiber solubility to be measured.

Source: Refs. 22–60 (see Table 6).

Table 11 Outline of Methods for Analysis of Dietary Fiber: Specified Pretreatment of Samples

Method	Drying	Disintegrating	Defatting/decolorizing
1CF	Air-dry.	Finely grind.	Diethyl ether or petroleum spirit, if >1% fat.
2NAF	Oven-dry.	Grind 0.8–1 mm.	EtOH: benzene (1:2 v/v), 8 hr, Soxhlet.
3ADF, 4NDF, 5Sch, 6R&V	Air-dry.	Grind 1 mm.	Acetone, 400 ml/100 g sample 1 hr if >10% fat.
7M&B	Freeze-dry.	Grind to pass 20 mesh (0.075 mm^2).	Acetone, room temp 1 hr, if >5% fat.
8UED1, 9UED2	Freeze-dry.	Dry-mill to 0.5–1 mm.	Petroleum ether, 3 × 25 ml/10 g sample if >10% fat.
10W&O	Air-dry.	Ball-mill to pass 20 mesh.	None.
11McC		Grind.	EtOH (90% v/v), 16 hr (o/n).
12W&B	Dry.	Grind.	None.
13Hell	Fresh or dry.	Blend in water or suspend if powdered.	None.
14Elch	Dry.	Grind.	Hot 85% methanol then diethyl ether in Soxhlet.
15Berl	Freeze-dry.	Grind to <150 µM.	Petroleum ether.
16S&W	As is, or dry if defatted.	Homogenize if fresh, finely grind if dry.	Hexane or chloroformmethanol (2:1 v/v) in Soxhlet, if >10% fat.
17A&J	As is or dried.	Blend in water or suspend if dry.	None.
18Lund	Homogenize and freeze-dry.	Mill to 0.3 mm.	Petroleum ether 40 ml/g sample, room temp, 15 min.
19Furda	Dry.	Grind to pass 1 mm.	Diethyl ether 5–6 hr, Soxhlet.
20Pro1	Vacuum oven, 70°C, o/n.	Dry-mill to pass 0.3–0.5 mm.	Petroleum ether 3 × 25 ml/g sample if >5% fat.
21Pro2, 22Lee	Homogenize or freeze-dry.	Reduce to 0.35 mm.	Petroleum ether 3 × 25 ml/g, if >10% fat.
23LiA, 24LiCl1	Freeze-dry.	Mill to pass 40 mesh.	Hexane if >5% fat.
25LiC2	Freeze-dry.	"	None.

Table 11 (*Continued*)

Method	Drying	Disintegrating	Defatting/decolorizing
26Sou	As is.	Finely grind dry after defatting.	Boiling 85% methanol 3 × 25 ml/3–5 g sample, then 3 × acetone or ether.
27Sel1	As is.	Ball-mill in 90% EtOH 100 ml/5 g.	Boiling 90% EtOH, 20 g/100 ml, 90% EtOH (ball-milling), absolute EtOH, diethyl ether.
28T&A	Wet or dry.	Grind or homogenize.	Reflux 80% EtOH, 300 ml/3 g sample, 45 min, wash 2 × 50 ml 80% EtOH. Reflux 150 ml chloroform 30 min, wash 2 × 50 ml chloroform.
29Upp	As is or dry.	Mill if dry, homogenize if wet.	If >6% fat, sonicate 1–3 g in 80% EtOH, 15 min, 3 × 75 ml. Centrifuge. Petroleum ether or hexane 10 min, 2 × 50 ml.
30En1	As is.		None.
31En2	As is if possible, otherwise freeze-dry.	If low water, ball-mill briefly. If high water, homogenize. If freeze-dried, ball mill.	None.
32Sel2	As is.	Blend and ball-mill in 1% Na deoxycholate–25 mM Na metabisulfite or ascorbic acid, 15 hr, 2°C.	Phenol/acetic acid/water (2:1:1, w/v/v), 100 ml/10 g sample, centrifuge, wash with water. Freeze-dry.
33En3	See 31En2.	See 31En2.	Acetone 40 ml/200 mg, 30 min, if >5% fat and/or >10% water.
34F&T	Freeze-dry.	Grind to <1 mm.	Boiling 80% EtOH then boiling acetone, each 3 × 25 ml/5 g.
36Bril	As is or freeze-dry.	Grind to pass 0.5 mm.	None.
35En4, 37En5, 38En6, 39Mon	As is or freeze-dry.	Mill if <10% water, homogenize or freeze-dry and mill if >10% water.	Acetone 40 ml/300 mg dry matter if >5% fat and/or >10–15% water.

Source: Refs. 22–60 (see Table 6).

Table 12 Outline of Methods for Analysis of Dietary Fiber: Specified Methods for Preparing Fiber

Method	Fiber preparation (sample digestion)
1CF	Defat 2 g, add 1 g treated asbestos, boil in 200 ml 1.25% (w/v) H$_2$SO$_4$, 30 min. Filter, wash 4 × 50–75 ml water. Boil in 200 ml 1.25% (w/v) NaOH, 30 min. Filter, wash with 25 ml hot 1.25% H$_2$SO$_4$, 3 × 50 ml water, EtOH. Dry 130°C, 2 hr.
2NAF	Defat 1 g. Reflux in 200 ml 0.5 M H$_2$SO$_4$–0.2 ml Teepol, 1 hr. Filter in sintered glass crucible, wash 3 × 50 ml hot water, EtOH, diethyl ether. Dry 100°C.
3ADF	Reflux 1 g in 100 ml 0.5 M H$_2$SO$_4$-2% cetyltrimethylammonium bromide 60 min. Filter on tared Gooch crucible. Wash hot water, acetone. Dry 8 hr, 100°C.
4NDF	Reflex 0.5–1 g 60 min in neutral detergent (ND: per L; 18.61 g diNa EDTA, 6.81 g Na borate decahydrate, 30 g Na laurel sulfate, 10 ml 2-ethoxyethanol, 4.56 g separately dissolved Na$_2$HPO$_4$) with 2 ml decahydronaphthalene + 0.5 g Na sulfite added just before boiling. Filter in tared Gooch crucible. Wash well with hot water, acetone. Dry 100°C, 8 hr.
5Sch	As for 4NDF, but after filtering off ND, wash with 200 ml hot water, add 100 ml 0.1 M phos. buffer pH 7.0–α-amylase solution, suck some through, plug, digest 37°C, 18 hr. Filter, wash with water and acetone. Dry 100°C.
6R&V	To 0.5–1 g add 50 ml cold ND, boil 30 min, add 50 ml cold ND + 2 ml amylase solution. Boil 30 min, filter. Wash 2 × boiling water, 2 × acetone. Dry 105°C o/n.
7M&B	**NDF:** Reflux 0.5–1 g sample in ND (no Na sulphite) 1 hr, filter in tared crucible, add 10 ml α-amylase solution + 15 ml hot water, stand 5 min, filter, add further 10 ml amylase + 15 ml hot water to plugged crucible, digest 55°C, 60 min. Filter, wash 3 × water, 2 × acetone. Dry 100°C o/n. **SDF:** Heat 0.5 g sample in 20 ml water 100°C 15 min. Cool to 55°C, and 2 ml amyloglucosidase, digest 55°C, 1.5 hr, reheat 100°C 30 min, filter, wash with 10 ml hot water. Add 2 ml amyloglucosidase to filtrate + washing. Digest 55°C 1 hr. Add 4 vol EtOH, stand 1 hr, filter in tared 10–15 μm pore Gooch crucible, wash 2 × 75% EtOH, 2 × acetone. Dry 100°C, o/n.
8UED1	Weigh 0.9-1 g into dialysis tubes pretreated with 10% EtOH, add 30 ml 0.67% heat-stable amylase in 8 M urea, digest room temp 3.5–4.5 hr. Add 0.5 ml protease (Savinase), digest 50°C 2-28 h, with continuous water exchange. Empty into container, add 4 vol EtOH, stand 4 hr. Filter through tared Whatman 54 filter paper. Wash with 80% EtOH. Dry 105°C, 8 hr.
9UED2	As for 8UED1 (above) until 2–28 hr dialysis completed, then filter through tared Whatman 54 filter paper, wash with water. **SDF:** To filtrate plus washing add 4 vol EtOH, stand 4 hr, filter through tared filter paper. Dry ppt on paper 105°C, 8 hr. **IDF:** Wash res from dialysis with 80% EtOH. Dry 105°C, 8 hr.
10W&O	Suspend 0.25 g sample in 25 ml water, autoclave 30 min, add 20 ml bile salt (50 ml 0.2 M KH$_2$PO$_4$, 23.4 ml 0.4 M NaOH, 2.0 g Na taurocholate, 6.6 ml water, pH 8) and 5 ml pancreatin solution, digest 45°C, 3 days. Filter, wash with EtOH, benzene, diethyl ether. Dry 70°C, 2 hr. Hydrolyze in 60% (v/v) H$_2$SO$_4$, 6–10°C 24 hr, dilute to 4% (v/v), reflux 3 hr, filter hydrolysate, neutralize with NaOH. Wash res with water, EtOH, benzene. Dry 110°C.
11McC	Weigh dry EtOH-extracted res. Digest with takadiastase. Measure starch and protein ($N \times 6.25$) in res. Unavailable carbohydrate by difference.
12W&B	Autoclave 1 g sample in 100 ml 0.1 M phos buffer pH 4.9, 10 min (or boil 15 min, cool, reboil 5 min) cool to 50°C, add 0.45 g Rhozyme S (protease-amylase). Digest 50°C, 24 hr (or 1 hr and correct for protein [Kjeldahl $N \times 6.25$] in res). Filter in sintered glass funnel, wash with hot water, ether, dry.

Table 12 (*Continued*)

Method	Fiber preparation (sample digestion)
13Hell	To 50-ml sample slurry add 50 ml 0.2 M HCl + 100 mg pepsin, digest 40°C 18 hr. Neutralize with 4 M NaOH, add 50 ml phos buffer pH 6.8, 300 mg Na dodecyl sulfate. Digest 1 hr, 40°C. Add 4 M HCl to pH 4.5. Centrifuge, filter s/n in tared porosity 3 crucible with 15-mm sand bed, repeat 3 × with water, 3 × acetone, finally adding residue to crucible. Dry 105°C o/n.
14Elch	Digest with amyloglucosidase 55°C, 3 hr, then trypsin 37°C, 18 hr in centrifuge tube. Centrifuge. Dry res. Extract with acid detergent (method 3ADF), wash, dry, and weigh (= ADF). Extract ADF with 72% (w/w) H_2SO_4, 30 h, 4°C, wash, dry res (= "crude lignin"). Oxidise duplicate ADF with potassium permanganate. Dry res (= "crude cellulose").
15Berl	To 2 g defatted sample in 100 ml centrifuge tube add 40 ml water, 2 drops octanol, and autoclave 1 hr, 130°C. Centrifuge. Wash res (R) with 2 × 40 ml water stirring 15 min. Combine s/n plus washings.
	SDF: To s/n + washings add 5 ml buffer (130 ml 0.05 M citric acid + 250 ml 0.1 M Na_2HPO_4 + 50 mg thiomersal) and 5 ml amyloglucosidase solution. Digest 55°C, 3 hr. Add 5 ml 0.48 M Na_2HPO_4 to near pH 7.5, and 10 ml pancreatin/trypsin solution. Digest 37°C o/n (15 hr). Filter on 5000-Dalton cutoff membrane filter (3–4 bar pressure), wash with 25 ml water. Dry the membrane plus SDF 105°C o/n.
	IDF: Suspend res (R) in 40 ml water and digest as for SDF. Centrifuge, wash res with 40 ml water, centrifuge again, transfer res to a tared sintered glass funnel, wash with ethanol, acetone. Dry o/n 105°C.
16S&W	To 50- to 60-ml sample slurry (2.5 g dry) add 5 M HCl to H 1.5, 50 mg pepsin, 200 μl chloroform. Digest 37°C, 20 hr, pH 1.5, + 25 mg more pepsin at 12 hr. At 20 hr add 3 M NaOH to pH 6. Add 25 ml 0.1 M phos buffer pH 6.0–100 mg pancreatin–50 mg amyloglucosidase. Digest 37°C, 18 hr. Centrifuge.
	IDF: Wash res with water, acetone. Dry under reduced pressure 50°C.
	SDF: To s/n and washing add 4 vol EtOH, centrifuge, wash ppt. with 80% EtOH, acetone, ether. Dry under reduced pressure 50°C.
17A&J	Heat 50-ml sample suspension 100°C, 15 min. Cool, add 50 ml 0.2 M HCl–100 mg pepsin. Digest 18 h, 40°C. Neutralize with NaOH. Add 50 ml 0.1M Na phos buffer pH 6.8–100 mg pancreatin. Digest 1 hr, 40°C. Acidify to pH 4.5. Filter and wash in tared sintered glass crucible.
	IDF: Wash res 3 × 50 ml water, 2 × 50 ml acetone, dry 105°C o/n, weigh.
	SDF: Evaporate filtrate + washings to 150 ml, add 4 vol EtOH, centrifuge, wash 3 × 50 ml 80% EtOH, 2 × 50 ml acetone. Dry 50–60°C, 1 hr.
18Lun	Suspend 1 g sample in 25 ml 0.1 M phos buffer pH 6.0. Add 0.1 ml heat-stable amylase, digest 100°C, 15 min. Cool, add HCl to pH 1.5 + 100 mg pepsin, digest 40°C, 60 min. Add 20 ml water, NaOH to pH 6.8, 10 mg pancreatin. Digest 40°C, 60 min. Add HCl to pH 4.5.
	TDF: Add 4 vol 95% EtOH, 60°C, 1 hr. Filter through weighed crucible plus 0.5 g Celite. Dry 105°C o/n.
	SDF, IDF: Filter digest as for total fiber but without prior EtOH, wash res 2 × 10 ml water (res = R, filtrate = F).
	SDF: To F plus washings add 4 vol EtOH, 60°C, 1 hr. Collect ppt on Celite, wash 2 × 10 ml each of 78% EtOH, 95% EtOH, acetone. Dry 105°C o/n.
	IDF: Wash R with 2 × 10 ml each 95% EtOH, acetone. Dry 105°C o/n.

19Fur   Disperse 2 g defatted sample in 200 ml 0.005M HCl pH 2.5, boil 20 min.Cool to
         60°C, add 0.3 g diNaEDTA, 60°C, 40 min. Cool, add 10 mg α-amylase and 10
         mg protease, digest 20–30°C o/n. Filter, wash (few ml water) in tared Gooch
         crucible.
         IDF: Wash res with water, EtOH, acetone. Dry in 70°C, vacuum oven.
         SDF: Filtrate to pH 2–3 with HCl, add 4 vol EtOH, 1 hr. Filter as above, wash
         ppt with 75% EtOH, EtOH, acetone. Dry in 70°C vacuum oven.
20Pro1   Duplicate 1 g samples in 50 ml 0.05 M phos buffer pH 6 + 0.1 ml heat-stable
         α-amylase. Digest 95–100°C, 15 min. Cool, add 0.28 N NaOH tp pH 7.5 + 5
         mg protease. Digest 60°C, 30 min. Cool, add 0.329 M phosphoric acid to pH
         4.5, + 0.3 ml amyloglucosidase. Digest 60°C, 30 min. Add 280 ml 95% EtOH
         at 60°C, stand 1 hr. Filter on tared crucible plus Celite. Wash with 3×20 ml
         74% EtOH, 2×10 ml each of 95% EtOH acetone. Dry in 70°C vacuum or
         105°C air oven. Cool in desiccator and weigh.
21Pro2   Duplicate 1 g samples in 50 ml 0.8 M phos buffer, final pH 6.0, + 0.1 ml heat
         stable α-amylase, 95–100°C, 15 min. Cool to room temp, add 10 ml 0.275 M
         NaOH to pH 7.5. Add 5 mg protease. Digest 60°C, 30 min. Cool, add 10 ml
         0.325 M HCl to pH 4.0–4.6, add 0.3 ml amyloglucosidase. Digest 60°C, 30 min.
         IDF: Filter on tared crucible plus Celite, wash res with 2×10 ml each of water,
         95% EtOH, acetone. Dry as for SDF.
         SDF: Adjust s/n and water washings to 100 g, add 400 ml 95% EtOH at 60°C,
         stand at room temp 1 hr. Collect ppt in tared crucible plus Celite, wash with
         2×20 ml 78% EtOH, 2×10 ml 95% EtOH, 2×10 ml acetone. Dry in 70°C
         vacuum oven or 105°C air oven o/n.
         TDF: Adjust volume of digest to 100 ml, continue as for soluble fiber.
22Lee   Disperse duplicate 1-g samples in 40 ml 0.05 M MES/0.05M Tris buffer pH 8.2.
         Add 50 μl heat stable α-amylase solution, heat 95–100°C, 15 min. Cool to
         60°C, add 100 μl protease solution, digest 60°C, 30 min. Add 5 ml 0.561 N
         HCl, adjust to pH 4.0–4.7 with 1 N NaOH or 1 N HCl at 60°C. Add 300 μl
         amyloglucosidase, digest 60°C, 30 min.
         TDF: Add 225 ml 95% EtOH at 60°C, stand 1 hr room temp. Filter through
         tared crucible plus Celite, wash with 2×15 ml each, 78% EtOH, 95% EtOH,
         acetone. Dry in 70°C vacuum oven or 105°C air oven o/n.
         SDF, IDF: Filter digestate in tared crucible plus celite. Wash with 2×10 ml
         water at 70°C (res = R, filtrate = F).
         IDF: Wash res (R) with 78% and 95% EtOH, acetone and dry, as for TDF.
         SDF: To F + water washings add 4 vol 95% EtOH at 60°C, stand 1 hr room
         temp. Collect ppt on Celite and wash with EtOH-acetone and dry as above.
23LiA   To duplicate 0.5 g samples add 25 ml deionized water, heat 121°C $1\frac{1}{4}$ hr, or
         autoclave 1 hr. Cool to 65–70°C, add 2 ml amyloglucosidase solution. Digest
         55°C, 2 hr. Make to 100 ml with 95% EtOH, stand 1 hr. Filter in tared crucible
         plus Celite. Wash with 2×10 ml each 78% and 95% EtOH, and acetone. Dry
         105°C o/n.
24LiC1   As above (23LiA) but filter after amyloglucosidase digestion (before EtOH).
         IDF: Rinse res with 35 ml 95% EtOH, acetone. Dry 105°C o/n.
         SDF: Combine filtrate and 95% EtOH washing, add 100 ml 95% EtOH, stand 1
         hr, filter, rinse ppt with EtOH and acetone. Dry 105°C o/n.
25LiC2   Extract duplicate 500 mg freeze-dried samples with 25 ml deionized water, stand
         37°C, 90 min. Add 100 ml 95% EtOH, stand at room temp 1 hr. Filter in tared
         crucible plus Celite, wash 2×20 ml 78% EtOH, 2×10 ml 95% EtOH, 10 ml
         acetone. Dry 105°C 2 hr.

Table 12 (*Continued*)

Method	Fiber preparation (sample digestion)

26Sou   Sequentially extract 0.1–0.5 g defatted sample with: (1) Hot water, 10 ml, 80°C, 30 min, centrifuge (res = R1). To s/n add 4 vol EtOH. Centrifuge and add to pellet 0.2 ml 2 M acetate buffer pH 4.5 and 3 ml 5% (w/v) takadiastase. Digest 37°C, 18 hr. Add 4 vol EtOH, centrifuge and hydrolyse pellet with 5% H_2SO_4, 100°C, 2.5 hr (hydrolyzate = H1). (2) Res R1: Add 5 ml hot water, 100°C, 10 min. Cool, add 1.2 ml 2 M acetate buffer pH 4.5, and 5 ml 5% (w/v) takadiastase, digest 37°C, 18 hr. Add 4 vol EtOH, centrifuge (res = R2) and wash with 80% EtOH. (3) Res R2: Add 10 ml 5% (v/v) H_2SO_4, 100°C, $2\frac{1}{2}$ hr. Add equal vol EtOH, centrifuge, wash res 2 × 50% EtOH. (Res = R3, s/n + washes = H2). (4) Res R3: Wash with EtOH, diethyl ether, dry. Weigh tube plus res. Add 5 ml 72% (w/w) H_2SO_4, 0–4°C, 24 hr. Dilute rapidly with water and centrifuge or filter (Res = R4), wash 2 × water, combine s/n and washes (H3). (5) Res R4: Wash res with EtOH once and 3 × diethyl ether. Dry in air, then 95–100°C, 10 min. Weigh. Ignite if ash correction to lignin desired.
H1 = hydrolyzed water soluble NSP
H2 = hydrolyzed insoluble non-cellulose NSP (hemicellulose)
H3 = hydrolyzed cellulose
Res. R4 = lignin + ash

27Sel1   To 50 mg alcohol insoluble residue add 25 ml 20 mM phos-10 mM NaCl, pH 7.0. Heat 85–90°C, 3 hr. Cool to 40°C, add salivary amylase (15U) and pullulanase (3U), digest 37°C, 17 hr. If starch test with I_2/Kl positive, repeat enzyme addition and digest 7 hr more. Dialyze digestate, freeze-dry, and weigh (= cell walls). Hydrolyse subsample of cell walls with 72% (w/w) H_2SO_4 room temp, 3 hr, then dilute to 1 M and hydrolyze 100°C, 3 hr. Hydrolyze another subsample in 72% H_2SO_4, filter, and wash (res = lignin) in tared crucible.

28T&A   TDF: Measured by difference after analysis of total polysaccharide and starch in defatted sample.
IDF, SDF: To defatted res from 3 g sample add 50 ml 0.1 M acetate buffer pH 5–6, 85°C, 30 min. Add 0.1 ml heat stable amylase. Digest 85°C 45 min, boil 15 min, filter, wash 3 × 50 ml boiling water.
IDF: Wash res 3 × 50 ml EtOH, 3 × 50 ml acetone, dry.
SDF: Digest filtrate and water wasings with 0.3 ml amyloglucosidase 60°C, 15 h. Dialyze vs distilled water 48 hr, freeze-dry.
Hydrolyses for GC: To 500 mg defatted sample (for TDF) or 50–100 mg fiber (SDF, IDF) add 5 ml 12M H_2SO_4, room temp, 2 hr. Dilute to 0.358 M, reflux 6 hr. Filter on glass filter, wash thoroughly with water (res = KLig).

29Upp   Suspend 1.5–2.0 g dry preextracted sample (1 g if starch-rich) in 75 ml Na acetate buffer pH 5.0, add 0.1 ml heat-stable amylase. Digest 100°C, 30 min. Cool to 60°C, add 0.5 ml amyloglucosidase. Digest 60°C, o/n.
TDF: Add 4 vol EtOH, 4°C, 0.5 hr. Centrifuge, wash res 2 × 80 ml 80% EtOH, 50 ml acetone, dry in warm air. (Or by method 28T&A for TDF).
SDF, IDF: Centrifuge digestate, wash res (R1) with 2 × 50 ml water.
SDF: Combine s/n with washings, concentrate to 100 ml, dialyze vs tap water (1 × 24 hr), distilled water (1 × 24 hr), freeze dry, and weigh.
IDF: Wash R1 with 2 × 50 ml EtOH, 50 ml acetone, dry in warm air.
Hydrolyses: Grind subsample (150–250 mg) of each fiber preparation, hydrolyse with 3.0 ml 12 M H_2SO_4, 30°C, 1 hr. Add 79 ml water, 5 ml myoinositol (3 mg/ml, internal standard), and autoclave 125°C, 1 hr. Klason lignin = washed filtered res from 12 M H_2SO_4 hydrolysis of IDF.

30En1   Sequentially extract 200 mg sample, separating res and extract by centrifugation:
         (1) 40 ml Na acetate buffer pH 5.0, 90°C, 5 hr. Res = R1, s/n = S1.
         (2) Digest R1 with amyloglucosidase 45°C, 16 hr. Res = R2, s/n = S2.
         (3) Hydrolyse R2 with 0.5 M H_2SO_4. Res = R3, s/n = S3.
         (4) Hydrolyse R3 with 12 M H_2SO_4, 4°C, 24 hr. Filter. Res = R4, filtrate = F1.
         Add 4 vol EtOH to S2, centrifuge and hydrolyse ppt (S2R) with 0.5 M H_2SO_4.
         S1 contains free sugars.
         S2R = water soluble noncellulosic polysaccharides as hydrolyzate.
         S3 = water insoluble noncellulosic polysaccharides as hydrolyzate.
         F1 = cellulose as hydrolyzate.
         R4 = crude lignin.

31En2   Add to three 100- to 300-mg samples 10 ml acetate buffer pH 5.2, 100°C 1 hr.
            Cool to 42° C, add 0.1 ml amylase-pullulanase solution. Digest 40°C 16 hr.
         TNSP: Add 4 vol EtOH, stand 1 hr, centrifuge, wash res 2 × 40 ml 80% EtOH,
            40 ml acetone, dry. Hydrolysis: Add 1 ml 12M H_2SO_4, 35°C, 1 hr. Add 6 ml
            water and 5 ml standard (1 mg/each of erythritol and myoinositol), heat
            100°C, 2 hr.
         INSP: Add to digestate 40 ml 0.2 M phos buffer pH 7.0, 100°C, 1 hr, centrifuge,
            wash res 2 × with 40 ml buffer, 40 ml acetone. Dry. Hydrolyze as for total NSP.
         Noncellulosic NSP: As for Total NSP except hydrolysis in 5 ml 2 M H_2SO_4 + 5
            ml standard, heat 100°C, 2 hr.
         Remove 2 ml from above hydrolyzates for uronic acid determination. Neutralize
            rest with 3 g $BaCO_3$, centrifuge.

32Sel2   Extract res from 10 g sample with 200 ml 90% DMSO, 20°C. Centrifuge and
            add to res 50 ml 20 mM phos buffer–10 mM NaCl pH 7.0 + 0.2 ml
            amyloglucosidase. Digest 37°C, 24 hr. Centrifuge and wash res with water.
            Freeze dry (= cell walls). Hydrolysis: Add 72% (w/w) H_2SO_4, room temp, 3
            hr, then dilute to 1 M and hydrolyze 100°C, 3 hr.

33En3   To 50- to 200-mg sample add 2 ml DMSO, 100°C, 1 hr. Add 8 ml buffer pH 5.2
            and 0.1 ml amylase-pullulanase solution. Digest 45°C, 16–18 hr. Add 40 ml
            EtOH, stand 1 hr room temp. Centrifuge, wash res 2 × 50 ml 85% EtOH, 40
            ml acetone. Dry 65°C. Acid hydrolysis: As for method 35En4 (below).

34F&T   To 200-mg sample add 10 ml 0.1 M Tris-maleate buffer pH 6.7, 100°C, 10 min.
            Add 0.2 ml heat-stable amylase. Digest 100°C, 15 min. Add 40 ml EtOH, 0°C,
            30 min. Centrifuge. To res add 2 ml DMSO, 100°C, 5 min, then 8 ml 0.1 M
            acetate buffer pH 4.6 + 0.1 ml amyloglucosidase. Digest 37°C, 35 min. Add
            40 ml EtOH, 0°C, 30 min, centrifuge. Dry res. Acid hydrolysis: Add 2 ml 12
            M H_2SO_4, 35°C, 1 hr. Add 22 ml water, heat 100°C 2 hr with aliquot removed
            after 1 hr for uronic acid determination. Dilute to volume for colorimetry.
            Hydrolyzate contains total NSP.

35En4   To two 100- to 200-mg dry samples add 2 ml DMSO, heat 100°C, 1 hr. Add 8 ml
            0.1 M acetate buffer pH 5.2, 50°C. Add 0.5 ml α-amylase solution and 0.1 ml
            pullulanase. Digest 42°C, 16 hr.
         TNSP: To first sample add 40 ml EtOH, stand 1 hr, room temp, centrifuge,
            wash res 2 × 85% EtOH, 40 ml acetone, dry res (R1) 65–70°C.
         INSP: To second sample add 40 ml 0.2 M phos buffer pH 7.0, heat 100°C, 1
            hr. Centrifuge, wash res with 50 ml phos buffer, 50 ml 85% EtOH, 40 ml
            aetone. Dry res (R2) at 65–70°C.
         R1 and R2 hydrolysis: Add 2 ml 12 M H_2SO_4, 35°C, 1 hr, then add 22 ml
            water, 100°C, 2 hr.

Table 12   (*Continued*)

Method	Fiber preparation (sample digestion)
36Bril	Homogenize 5 g sample in 180 ml 0.1 M phos buffer (pH 7.5, 1% Na laurel sulfate, 1% 2-mercaptoethanol, 0.02% Na azide). Add 5 ml pronase solution. Digest 40°C 2 hr., centrifuge (res = R1, s/n = S1). Disperse res R1 in 60 ml 90% DMSO, 100°C, 30 min. Add 120 ml 0.1 M acetate buffer pH 5.0, 100°C + 11 μkats (0.5 ml) Termamyl. Digest 100°C, 15 min. Cool to 40°C, add 5 ml amyloglucosidase solution. Digest 40°C, 2 hr. Centrifuge (res= R2, s/n = S2). **INSP**: Wash res R2 with water, suspend in 50 ml EtOH, filter, wash with 50 ml acetone, 50 ml diethyl oxide. Dry in 70°C vacuum oven o/n. **SNSP**: Combine S1 and S2, dialyze against distilled water.
37En5	As for 35En4 but uses colorimetry to measure NSP.
38En6	To two samples of maximum 300 mg dry matter add 2 ml DMSO, 100°C, 30 min. Add 8 ml "enzyme solution 1." Digest 100°C, 10 min, cool to 50°C, add 0.5 ml "enzyme solution 2." Digest 50°C, 30 min then 100°C, 10 min. **TNSP**: To first sample add 40 ml EtOH, 0°C, 30 min. Centrifuge, wash res with 50 ml 85% EtOH, 50 ml EtOH, 20 ml acetone. Dry at 80°C (R1). **INSP**: Add 40 ml phos buffer pH 7.0, heat 100°C, 30 min, room temp 10 min. Centrifuge, wash res with 50 ml water, 50 ml EtOH, 20 ml acetone. Dry completely at 80°C (R2). R1 and R2 hydrolysis: Add 5 ml 12 M H_2SO_4, 35°C, 1 hr, then add 25 ml water, heat 100°C, 1 hr.
39Mon	Wash 100–200 mg (dry) sample with 12 ml gastric buffer (0.1 M HCl + NaOH until pH 2.0), 37°C, 10 min. Centrifuge. To res. add 12 ml gastric buffer + 20 mg pepsin. Digest 37°C, 2 hr. Centrifuge, wash res with 12 ml intestinal buffer (per L: 8 g NaCl, 0.2 g KCl, 0.26 g $MgSO_4 \cdot 7H_2O$, 0.06 g $NaH_2PO4 \cdot 2H_20$, 1 g $NaHCO_3$, 0.2 g CaCl2), centrifuge. To res add 12 ml 0.5% (w/v) pancreatin in intestinal buffer (centrifuged supernatant). Digest 2 hr, 37°C. Centrifuge, wash with 2×12 ml intestinal buffer, freeze-dry. Analyse res and an undigested sample for total fiber. Physiological SDF = TDF in undigested sample – TDF in digested sample.

Source: Refs. 22–60 (see Table 6).

Weigh to 0.1 mg.

Weigh appropriate residues before subsampling to enable fiber in the original food to be calculated.

Treat all samples in exactly the same way.

Mix thoroughly after each addition of reagent.

Constantly agitate during digestion.

Avoid lumps during digestion and washing.

Avoid stranding material on the sides of beakers and tubes.

Thoroughly suspend residues when washing.

Use porosity 2 grade where sintered glass filters are specified.

Cool heat-dried residues in a vacuum desiccator, or use the hot weighing procedure (28,67), to avoid moisture contamination.

Long and variable filtration times can lead to error, so use a sample mass that allows the method to proceed smoothly.

Observe usual precautions when using solvents and corrosive chemicals.

Table 13 Outline of Methods for Analysis of Dietary Fiber: Specified Enzyme Preparations

Method	Enzyme preparation
1CF, 2NAF, 3ADF, 4NDF	Enzymes not used.
5Sch	α-Amylase: Sigma A-6880; use 2.5% w/v in 0.1 M phos buffer pH 7.0.
6R&V	α-Amylase: Detergent stable, Sigma A-6505, type IIIA, from *Bacillus subtilis*; use 2% w/v in water/methoxyethanol (9:1).
7M&B	α-Amylase: Sigma A-6880; use 5% in 0.1 M phos buffer pH 7.0. Amyloglucosidase: Sigma A-9268; use 15% v/v in 2.0 M Na acetate–acetic acid buffer pH 4.5.
8UED1	α-Amylase (heat-stable): Sigma A-0164; use 0.67% in 8 M urea. Or use Novo 240 L, 0.33% v/v in 8 M urea. Protease: Savinase (Novo 8.0 L) use as is.
9UED2	As for 8UED1 but Termamyl 120 L (Novo) used to prepare 0.33% (v/v) heat-stable amylase/8 M urea solution.
10W&O	α-Amylase/protease: Pancreatin,(Merck USP); use 10% w/v in 8.5% NaCl solution.
11McC	α- and β-Amylase: Takadiastase.
12W&B	α-Amylase/protease: Rhozyme S (Rohm and Haas); use as is.
13Hell	α-Amylase/protease: Pancreatin (Merck); use as is. Protease: Pepsin (Merck); use as is.
14Elch	α-Amylase: Amyloglucosidase. Protease: Trypsin.
15Berl	Amyloglucosidase: Serva (13451; 18 U/mg); use 800 mg in 100 ml water. Proteases: Pancreatin (Sigma P1750). Trypsin (Sigma T8128; 1000–5000 BAEE-U/mg). Protease solution: Stir 1 g pancreatin in 90 ml water 30 min, add 200 mg trypsin, stir 3 min, centrifuge, make s/n to 100 ml.
16S&W	(a) Amyloglucosidase: Glucoamylase, Sigma grade II. (b) Amylase/protease: Pancreatin, Sigma grade IV. Use 50 mg (a) + 100 mg (b) in 25 ml 0.1 M phos buffer pH 6.0. Protease: Pepsin (Fluka, 30 U/mg); use as is.
17A&J	α-Amylase: Pancreatin (Merck; 50,000 amylase U/g); use as is. Protease: Pepsin (Merck; 30,000 U/g); use as is.
18Lund	α-Amylase, (heat-stable): Termamyl 60 L (Novo); use undiluted. Amylase/protease: Pancreatin (Sigma, 4 × NF); use as is. Protease: Pepsin NF (Merck), use as is.
19Fur	α-Amylase: ex *Bacillus subtilis* (Sigma Type IIA), 4 × crystallized; as is. Protease: Subtilopeptidase A, (Sigma type VIII) crystallized and lyophilized; use as is.
20Pro1	α-Amylase (heat-stable): Termamyl 120 L (Novo); use undiluted. Amyloglucosidase: Sigma A 9268; use undiluted. Protease: Subtilopeptidase A, (Sigma type VIII), use as is.
21Pro2	α-Amylase (heat stable): Termamyl 120 L (Novo); use undiluted. Amyloglucosidase: Sigma A 9913; use undiluted. Protease: Sigma P 3910; use as is.
22Lee	α-Amylase: Heat stable, (a) Sigma A-3306 or Termamyl 300 L (Novo); use undiluted. Amyloglucosidase: (b) Sigma A 9913; use undiluted Protease: (c) Sigma P3910, 50 mg/ml 0.05 M MES/Tris buffer, pH 8.2. (a) (b) and (c) are sold in a pretested kit (Sigma TDF-C10).
23LiA, 24LiC1	Amyloglucosidase: Sigma A 9913; use 0.15 ml amyloglucosidase + 1 ml 4M Na acetate buffer pH 4.8 + 0.85 ml water/2 ml enzyme solution.
25LiC2	Enzymes not used.
26Sou	Mixed amylase: Takadiastase (Parke Davis); use a 5% (w/v) suspension.

Table 13 (*Continued*)

Method	Enzyme preparation
27Sel1	Amylase (salivary): Sigma; use 15 U/50 mg sample in 25 ml phos buffer. Pullulanase: Boehringer; use 3 U/15 U of salivary amylase as above. Amyloglucosidase: Boehringer; use 20 U/50 ml Na acetate buffer pH 5.0 as an alternative to the salivary α-amylase/pullulanase.
28T&A	α-Amylase: Heat stable, Termamyl 6 KNN/g (Novo); use undiluted. Amyloglucosidase: ex *Aspergillus niger* (*Boehringer 14 U/mg*); use undiluted.
29Upp	α-Amylase: Heat stable, Termamyl 120 L (Novo); use undiluted. Amyloglucosidase: Boehringer (140 EU/ml), ex *Aspergillus niger*, use undiluted.
30En1	Amyloglucosidase: Boehringer 102857, 10 mg/ml; dilute to 1.5 mg/ml acetate buffer pH 5.0 before use.
31En2, 33En3	(a) α-Amylase: Hog pancreatic amylase (Sigma A 4268). (b) Pullulanase: Boehringer 108944. Use (a) and (b) as 5000 units α-amylase plus 5 units pullulanase per ml 0.1M Na acetate buffer pH 5.2.
32Sel2	Amyloglucosidase: Boehringer; use undiluted.
34F&T	α-Amylase: Termamyl 120 L (Novo), use undiluted. Amyloglucosidase: ex *Aspergillus niger* (Boehringer 102857); use undiluted.
35En4, 37En5	α-Amylase: Pancreatin; 2 α-amylase capsules (350 mg pancreatin powder/capsule) + 9 ml water, mix 10 min, centrifuge 10 min, 1500 g, use supernatant. Pullulanase: Boehringer 108944 diluted 1:100 in water just before use.
36Bril	α-Amylase: Termamyl 60L (Novo), 22.8 μKat/ml. Amyloglucosidase: Merck; 20 mg/ml in 0.1M acetate buffer pH 5.0. Protease: Pronase from *Streptomyces griseus* (Boehringer); use 5 mg/ml 0.1M phos buffer pH 7.5.
38En6	α-Amylase: Termamyl (Novo). Use 2.5 ml in 200 ml 0.1 M Na acetate buffer pH 5.2 ("Enzyme solution 1"). α-Amylase/protease: Pancreatin (Paynes and Byrne). Pullulanase: Boehringer 108944; "Enzyme solution 2": 1.2 g pancreatin + 12 ml water, mix 10 min, centrifuge 10 min, 1500 g. Take 10 ml s/n and add 2.5 ml pullulanase.
39Mon	α-Amylase/protease: Pancreatin (Sigma P7000); Dissolve 0.5% w/v in intestinal buffer, centrifuge, use supernatant. Protease: Pepsin (Sigma P7000). Use during physiological extraction of soluble NSP, subsequent steps use TNSP, TNDP, or TDF methods (Table 6) with enzymes specified.

Source: Refs. 22–60 (see Table 6).

b. Fiber Fractions. Many methods are designed to achieve some fractionation of dietary fiber into its components. Separation into soluble (SDF, SNDP, SNSP) and complementary insoluble (IDF, INDP, INSP) fiber fractions is most common but is usually based on solubility under analytical conditions designed to maximize starch digestion rather than under physiological conditions of the gastrointestinal tract.

Digestion conditions vary from method to method, leading to method-dependent differences both in distribution of polysaccharides between fractions (68–71) and in composition within the fractions shown in Table 9. For instance, in raw fruit and vegetables the distribution of polysaccharides between soluble and insoluble fractions may be strongly governed by the effect of extraction conditions on pectic substances. Hot phosphate at pH 7 appears to extract Ca^{2+}, while heat at neutral pH will favor β elimination; the combined sensitivity to buffer species and pH leads

Table 14   Outline of Methods for Analysis of Dietary Fiber: Fiber Measurement

Method	Fiber measurement
1CF	Weigh asbestos pad + res. Ignite at 600°C. Reweigh. Crude fiber = wt loss on ignition - wt loss of asbestos blank.
2NAF	Weigh dried crucible + res. Ash at 550°C, o/n. Measure total N in duplicate res. Normal acid fiber (NAF) = loss in wt on ashing - total N × 6.25.
3ADF	Weigh dried crucible + res. Acid detergent fiber (ADF) = gain in wt of crucible.
4NDF, 5Sch, 6R&V	Weigh dried crucible + res. Ash at 500–550°C, 3 hr. Reweigh. Neutral detergent fiber (NDF) = loss in wt on ashing.
7M&B	IDF (= NDF): Weigh crucible + res. Ash at 525°C, 4 hr. Hold at 100°C o/n, weigh hot. IDF = loss in wt on ashing. SDF: Weigh crucible + res, and blank. Ash and weigh as above. SDF = loss in wt - loss in wt of blank.
8UED1, 9UED2	Weigh filter papers + residues, and filter paper blanks. Analyse a set of duplicates for protein by Kjeldahl method, correcting for N in filter paper. Ash other duplicates and blanks at 525°C for 8 hr. TDF = wt res-(wt protein in res + wt ash in res)-wt blank.
10W&O	Weigh res, ignite, reweigh, loss in wt = lignin. Measure total reducing sugars with copper reagent, cellulose as yeast-fermentable reducing sugar (glucose), and hemicellulose as nonfermentable reducing sugar (pentose) in the hydrolyzate.
11McC	Unavailable carbohydrate = sample wt - (moisture + lipid + starch + protein + ash).
12W&B	Weigh funnel + res. Ash and reweigh. Loss in wt = crude fiber.
13Hell	Weigh dry crucible + res. IDF = increase in wt of crucible.
14Elch	TDF = wt of digested res; ADF = lignocellulose; lignin = wt res after 72% H_2SO_4 hydrolysis of ADF; Cellulose = wt res after potassium permanganate oxidation of ADF duplicate.
15Berl	SDF = wt of dry retentate - protein in retentate (Kjeldahl); IDF = wt of insoluble res - wt res protein - wt res ash.
16S&W	SDF and IDF = wt of dry residues. Composition of fiber fractions: Hydrolyse 100 mg each of the soluble and insoluble fiber preparations, with 10 ml 0.5 M H_2SO_4, 100°C, 3 hr. Determine component sugars by GC, uronic acids by colorimetry. Res after hydrolysis of insoluble res = crude lignocellulose.
17A&J	IDF = wt (crucible + res) - wt crucible. SDF = wt dried EtOH ppt.
18Lun	SDF and IDF: weigh dry residue + crucible, ash at 550°C, 5 hr. Fiber = wt (crucible + res) - wt (crucible + ash) - wt ashfree reagent blank.
19Fur	IDF = wt (res + crucible) - wt crucible. SDF = wt dried EtOH ppt - content of protein and ash.
20Pro1, 21Pro2, 22Lee, 23LiA, 24LiC1, 25LiC2	TDF, IDF, SDF (in duplicate samples and reagent blanks): Weigh crucibles after cooling in dessicator. Determine protein (Kjeldahl N × 6.25) on one set of duplicates. Incinerate second set at 525°C, 5 hr. Apparent fiber = wt res - wt protein - wt ash. Actual TDF, IDF, or SDF = apparent fiber - fiber in blank.
26Sou	NSP measured as sugars in hydrolysates H1, H2, and H3. Measure hexoses, pentoses, and uronic acids in H1 and H2, and hexoses (= cellose in H3) and pentoses in H3. Hexoses by anthrone reaction, pentoses by orcinol/ferric-chloride reaction, uronic acids by carbazole reaction. Weigh res (R4) and ash. Lignin = wt res R4 - wt ash.

Table 14   *(Continued)*

Method	Fiber measurement
27Sel1	Cell walls = dry wt of dialysed digestate. Composition of cell walls by GC of alditol-acetate derivatives of sugars in neutralized cell wall hydrolyzate. Uronic acids by carbazole reaction. Lignin = wt residue after 72% (w/v) H_2SO_4 hydrolysis.
28T&A, 29Upp	Measure neutral sugars in hydrolyzed fractions (SNDP, INDP, TNDP) by GC of their alditol-acetate derivatives. Measure uronic acids in a separate subsample through potentiometric changes due CO_2 released during refluxing in hydroiodic acid under nitrogen, passed through $KI/CdSO_4$ solution, and collected in a cell containing 0.02 M NaOH. Calculate neutral sugar and uronic acids as polysaccharides.
	Klason lignin = dry wt of res from 12 M H_2SO_4 hydrolysis. IDF and SDF = neutral sugar + uronic acid + Klason lignin in fraction. TDF = neutral sugar + uronic acid + Klason lignin - starch.
30En1	NSP in polysaccharide(soluble noncellulosic, water-insoluble noncellulosic, cellulose) fractions calculated from analysis of the hydrolyzates; neutral sugars determined by GC of their alditol-acetate derivatives, uronic acids colorimetrically by the dimethylphenol reaction.
31En2	As for 30En1 (above) but fractions determined are total NSP, noncellulosic NSP, and insoluble NSP. Soluble NSP calculated: soluble NSP = total NSP - insoluble NSP.
32Sel2	Essentially as for 27Sel1.
33En3	As for 31En2 but only total NSP measured (neutral sugars + uronic acids) and improved derivatization for GC.
34F&T	Calculate TDF from sugars measured colorimetrically in hydrolyzate; neutral reducing sugars by *p*-hydroxybenzoic acid hydrazide reaction, uronic acids by dimethylphenol reaction.
35En4	As below (37En5) but sugars in hydrolyzates measured by dintrosalicylate reaction for reducing sugars.
36Bril	Wt insoluble cell walls = wt res - wt ash.
	Soluble NSP; Measure uronic acids in soluble polysaccharide by *m*-phenylphenol colorimetric method. Measure neutral sugars by GC after hydrolysing with 2 M trifluoroacetic acid, 1.25 hr, 120°C, and derivatizing to alditol acetates.
37En5, 38En6, 39Mon	Total NSP and insoluble NSP calculated from sugars measured in hydrolysates (from R1 and R2): neutral sugars by GC (or by HPLC or colorimetry) of their alditol acetate derivatives, and uronic acids colorimetrically by dimethylphenol reaction. Soluble NSP = Total NSP - Insoluble NSP.

Source: Refs. 22–60 (See Table 6).

to disrupted junction zones and depolymerization, increasing the solubility of pectic polyuronide and associated polysaccharides (Tables 1 and 2) compared with methods using hot acetate buffer at pH 5 or physiological conditions (68–70).

The Southgate (26Sou) (47) and early Englyst [30En1 (51), 31En2 (52)] methods base their approach to fiber fractionation on the traditional division of cell wall polysaccharides into the pectic substances, hemicellulose, and cellulose. They therefore provide more detail of fiber composition than other methods, particularly when coupled with GC analysis of the constituent sugars, but are considered exacting and too elaborate for routine use.

3. Enzymes Used

a. Why Use Enzymes? Nondigestibility is a central feature of all of the dietary fiber definitions outlined in Table 7, and even if not stated, it is implied (Table 7 definition E; NSP). Enzymes with similar digestive capacities to those in the mammalian gut are used to provide a measure of fiber consistent with its "physiological" definition as a nondigestible food component. Another reason for using enzymes is that both starch and dietary fiber are predominantly polysaccharide and must be separated before a valid measure of fiber as an individual food component can be obtained. The specificity of enzymes is a valuable tool for the selective digestion of starch from fiber.

The main enzymes used during digestion in dietary fiber analysis are amylases and amyloglucosidases (or pullulanase) to remove starch, and proteases (Table 13). If starch is not thoroughly extracted dietary fiber values will be spuriously high irrespective of method (Table 6) used. Protein extraction is most important in gravimetric methods, although high protein concentrations may also interfere with colorimetry (71).

Amylase and amyloglucosidase are both endo enzymes, but amylase has difficulty hydrolyzing α-(1–4) bonds in the vicinity of α-(1–6) branch points in starch, so that starch fragments (limit dextrins) are produced when amylase alone is used to digest starch. Any limit dextrins precipitated by 80% ethanol will contribute to fiber in methods using an 80% ethanol precipitation step to recover soluble fiber from the digestion/extraction medium. The problem is overcome by including amyloglucosidase or pullulanase to complete the hydrolysis of starch.

b. *Use of Enzymes.* The purity of enzyme preparations should be pretested to ensure that they are not active against fiber under the conditions used, but have the desired activities, by running through the entire procedure with model substrates. For example, enzyme activities and substrates used to detect them, as used by the Prosky (41,42) and Lee (43) (AOAC) methods, have been pectinase (citrus pectin), hemicellulase (strachtan = larch gum), and β-glucanase (β-glucan), with expected substrate recoveries of 95–100%, and amylase (wheat and corn starches) and protease (casein), with 1–2% expected recoveries.

4. Fiber Measurement

Fiber is measured (Table 14) either by nonspecific gravimetry or by specific methods for measuring the fiber polysaccharides as their monosaccharide constituents after hydrolysis. The approach used by different methods was given in Table 6.

a. *Gravimetric Methods.* Gravimetric methods measure fiber as the weight of residue remaining after digestion, usually corrected for protein and ash. Protein is calculated from total nitrogen measured by the Kjeldahl method in one of a duplicate set of samples, using the equation: Protein = $N \times 6.25$. Ash is usually determined by subtracting the weight of the residue holder (most often crucible plus Celite) from the holder plus ash. Protein and ash determinations are also conducted on blanks, so that the corrected fiber content of the blank is obtained. Fiber in the sample can then be calculated as follows: sample fiber = weight of digested residue – residue protein – residue ash – fiber in blank (adjusted for ash and protein in blank).

The gravimetric methods for measuring fiber are relatively easy and precise. They do not require the sophisticated equipment or expertise demanded by GC analysis, nor do they require the skill and control that seems necessary with some of the more sensitive colorimetric methods. Equipment for centrifuging or filtering, ovens for drying and ashing, an accurate balance, water baths, and a pH meter are the main requirements and are available to most analytical laboratories.

Gravimetric methods are nonspecific insofar as they measure any residue components, including resistant starch, that survive the preextraction and digestion stages of analysis, are insoluble in 80% ethanol, and are not covered by the correction for protein and ash. Tannins and tannalized food constituents, Maillard products, and other artifacts from food processing may be present and lead to overestimation of fiber (72) (and of lignin where measured), particularly in samples (such as white bread) containing only small amounts of dietary fiber (73). The contribution of such nonspecific materials has been considered especially important in foods that are intrinsically unusual in containing, for instance, high levels of tannin (carob pods) or having an unusual processing history.

Because of their nonspecificity, it may be difficult to assess the accuracy with which gravimetric methods determine dietary fiber—defined as nondigestible polysaccharide plus lignin—for a particular food. Where direct comparisons have been made between gravimetric and specific methods, the gravimetric methods have often given higher values for dietary fiber than when it is measured as specific food components (74–76) such as NSP.

Correcting for residual protein using $N \times 6.25$ also partially corrects for Maillard products. But the factor 6.25 is derived from a particular set of amino acids and is otherwise arbitrary. Therefore, it is best to aim for a low level of protein in fiber residues to avoid error associated with the correction.

b. Nongravimetric Methods.   Nongravimetric methods for analyzing dietary fiber rely on measuring the sugar constituents of the fiber polysaccharides by GC or HPLC, in combination with colorimetry, or by colorimetry alone (Table 6). Acidic cell wall sugars (galacturonic acid and glucuronic acid) are difficult to measure under the same conditions as neutral sugars during GC and HPLC, and so are measured separately by colorimetry although Theander [28T&A (49) and 29Upp (50)] used potentiometry. The specific colorimetric methods that have been used in dietary fiber analysis are summarized in Table 15 (71,77–80). Southgate provided detailed instructions to carry out a range of colorimetric measurements on food polysaccharides (71).

GC or HPLC provides information on the monosaccharide composition of the fiber but requires chromatographic equipment. Derivatization for GC analysis of sugars is now rapid (57,59), but GC has the drawback of rapid column deterioration, which can be expensive and can lower reliability. An alternative derivatization to aldonitrile acetate rather than alditol acetate derivatives and insertion of a guard column has been proposed to lengthen column life (80). However, now that HPLC has been successfully used for separating the monosaccharides found in fiber hydrolyzates (82–84), it is likely to be the preferred method because, although the columns are more expensive, they have a vastly longer life. For routine use in laboratories without GC or HPLC equipment colorimetry is satisfactory for measuring dietary fiber (55,56).

In GC (50,59) and colorimetric (55,56) analysis of monosaccharides correction factors are necessary to cover losses during hydrolysis, differing derivatization or color yields, and differing GC responses for the individual sugars.

The problem of nonspecificity of reaction in colorimetric methods has required conditions to maximize the desired reaction relative to others, with dichromatic readings for some methods and rigorous exclusion of any extraneous organic matter. Reaction conditions must therefore be strictly controlled. Such methods may therefore be subject to error until practice and skill has been acquired. Therefore, it is advisable to include blanks and standards with each set of unknowns (71).

When measuring fiber polysaccharide colorimetrically from the mixture of monosaccharides in an hydrolyzate, differing color yields of the various monosaccharides may need to be accounted for. When the monosaccharide composition of fiber preparations is not known in advance but the type of food is known, the problem can be largely overcome by using a mixed sugar standard representing the type of sample being analyzed. Englyst and Hudson (56) used a mixture of arabinose/xylose/glucose (3:4:3) for cereals and arabinose/glucose/galacturonic acid (1:2:1) for fruits and vegetables. If the monosaccharide composition of fiber from a particular food is known, a glucose standard may be used with an appropriate correction factor for the food.

All free sugars in the sample, and most of the protein, should have been extracted from the fiber

Table 15 Nongravimetric Methods for Measuring Dietary Fiber
Polysaccharides as Their Sugar Constituents

Method	Ref.
Colorimetric methods:	
Anthrone reaction (hexoses).	37
To 1 ml of test solution (25–200 µg sugar/ml) in a glass-stoppered tube add 10 ml anthrone reagent [10 g thiourea, 0.5 g anthrone/litre 66% (v/v) H_2SO_4]. Equilibrate at room temp, heat 100°C 15 min. Cool and read at 620 nm after 20–30 min.	
Orcinol/ferric chloride reaction (pentoses).	37
To 3 ml test solution (2–10 µg pentose/ml) in a glass-stoppered tube add 3 ml orcinol/$FeCl_3$/HCl reagent (10% orcinol in EtOH: 0.1% (w/v) anhydrous $FeCl_3$ in concentrated HCl; 1:10). Heat 100°C 45 min. Cool and read at 670 nm.	
p-Hydroxybenzoic acid hydrazide (PAHBAH) reaction (reducing sugars). To 1 ml sample containing up to 16 µg/ml sugar add 2 ml 1% (w/v) PAHBAH solution (2% PAHBAH in 0.5 M HClmm $CaCl_2$: 3 M NaOH, 1:1). Heat 100°C 5 min. Cool 15–20 sec. in ice water, read at 410 nm.	77
Dinitrosalicylate (DNS) reaction (reducing sugars).	56
To 1 ml sugar solution (hydrolysate) containing 0.5–2.0 mg/ml sugar add 0.5 ml glucose solution (0.5 mg/ml) and 0.5 ml 3.9 M NaOH. Mix. Add 2 ml DNS solution (10 g 3,5-dinitrosalicylic acid + 16 g NaOH + 300 g Na/K tartarate made to 1 L with water). Heat 100°C, 10 min. Cool. Add 20 ml water, mix. Read absorbance at 530 nm.	
Dimethylphenol (DMP) reaction for (uronic acids).	56, 78
To 0.3 ml of hydrolysate (max. uronic acid 150 µg/ml, dilute with 2M H_2SO_4) add 0.3 ml NaCl-H_3BO_3 (2 g NaCl plus 3 g H_3BO_3 in 100 ml water) and mix. Add 5 ml conc. H_2SO_4 and mix immediately. Heat 70°C, 40 min. Cool. Add 0.2 ml DMP solution (0.1 g 3,5-DMP in 100 ml glacial acetic acid). Read between 10 and 15 min later at 400 and 450 nm. Subtract 400 nm from 450-nm reading. Use 0.91 as the factor for converting monosaccharides to polysaccharides.	
m-Hydroxydiphenyl (MHDP) reaction (uronic acids).	79, 80
To 1 ml solution containing 5–75 µg/ml uronic acids add 6 ml H_2SO_4/tetraborate solution (0.0125 M Na tetraborate in conc. H_2SO_4) with tubes in ice. Immediately mix thoroughly. Heat 100°C 5 min exactly and cool. Add 0.1 ml MHDP solution (0.15% MHDP in 0.5% NaOH). Immediately mix thoroughly. Include sample blanks replacing MHDP solution with 0.5% NaOH. After 15 min read at 520 nm. Subtract sample blank from reading.	
Carbazole reaction (uronic acids)	37
Add 1 ml test solution containing 4–40 µg/ml uronic acid to 5 ml borate/H_2SO_4 (0.025 M Na tetraborate, $10H_2O$ in conc. H_2SO_4) at 4°C. Mix gently, heat at 100°C 10 min. Cool to room temp and add 0.2 ml carbazole solution [0.125% (w/v) carbazole in EtOH], and mix. Read absorbance at 530 nm.	

Table 15   (*Continued*)

Method	Ref.
Gas chromatography:	
Alditol acetate derivatization of sugars for gas chromatography:	59

To 3 ml hydrolyzate add 1 ml internal standard (1 mg allose/ml 50% saturated benzoic acid). With tubes in ice water add 1 ml 12.5 M ammonia solution (may need a little extra) until alkaline. Add 5 μl octan-2-ol and 0.2 ml ammonia-$NaBH_4$ solution (1.2 g $NaBH_4$ in 6 ml 12.5 M ammonia). Mix and heat 40°C, 30 min. Cool and add 0.4 ml glacial acetic acid and mix. Remove 0.5 ml to 30-ml glass tube and add to it 0.5 ml 1-methyl imidazole, 5 ml acetic anhydride, and immediately mix. Leave 10 min, add 0.9 ml absolute EtOH, mix, and stand 5 min. Add 10 ml water, mix, and stand 5 min. Add 0.5 ml bromophenol blue solution (0.4 g/L). With tubes in ice water add 5 ml 7.5 M KOH. After 2 min again add 5 ml 7.5 M KOH. Mix by inversion and after phase separation remove upper phase avoiding lower (blue) phase. Inject 0.5–1.0 μl for conventional GLC of the alditol acetate derivatives.

before it is hydrolyzed for analysis because they can interfere with colorimetric methods. Where sample preextraction (Table 11) and digestion (Table 12) produce clean residues that are predominantly cell wall polysaccharide, and the corrections discussed above are applied, colorimetry can give a quantitative and accurate figure for fiber polysaccharide, judging from its agreement with parallel GC determinations (56).

HYDROLYSIS OF FIBER POLYSACCHARIDES.   Acid hydrolysis of fiber polysaccharides to free monosaccharides is usually achieved in a two-step hydrolysis, unless a 1 N H_2SO_4 treatment is used to extract hemicellulose separately from cellulose, as in methods 26Sou (47), 30En1 (51), and 31En2 (52). In the first step, polysaccharides are hydrolyzed with 72% (w/w) (12 M) H_2SO_4 producing sulfated monosaccharides, and in the second step, the acid is diluted and heated to hydrolyze sulfated to nonsulfated monosaccharides. Some destruction occurs, particularly in the second step (59).

Conditions for optimizing the balance of hydrolysis and destruction over the whole acid treatment have been systematically examined (59,85) with the aim of rapid hydrolysis, minimal destruction, and small correction factors. As a result, Englyst recommends the use of 12 M H_2SO_4, 35°C, 1 hr, followed by dilution to 2 M H_2SO_4, 100°C, 1 hr, and has derived the factors that it is necessary to apply to individual monosaccharides before their quantities are added to give an NSP value (59).

VII.   PROBLEMATIC FIBER COMPONENTS

Soluble fiber, resistant starch, and lignin can be regarded as problematic because there is disagreement about the most appropriate way to measure them and/or whether or not they should be measured as part of dietary fiber at all.

A.   Resistant Starch

Resistant starch (52) (Table 1) is a minor component in most foods (7), but it may contribute significantly to nondigestible polysaccharides in starchy foods that contain little cell wall material, particularly when the physical structure of the food or starch, or retrogradation as a result of food processing, renders it indigestible (86,87).

Gravimetric methods have been criticized for not providing a measure of nonstarch cell wall materials free from resistant starch, as the concept of dietary fiber was originally intended to equate to plant cell walls in food, and inclusion of resistant starch may make fiber values vulnerable to food processing (73). Others (1) argue that when starch behaves as dietary fiber (nondigestible polysaccharide) it should be classified as a fiber component, particularly as it has the physiological properties of fiber.

Formation of resistant starch during analysis should of course be avoided, as it would lead to overestimation of the nondigestible polysaccharide that was present in the food before analysis. Although resistant starch is digested after being gelatinized with hot DMSO in methods measuring fiber as NSP (Table 6), concern has been expressed (3) that the distribution of polysaccharides between soluble and insoluble fiber fractions is affected by the DMSO treatment, although a study of the Englyst procedure did not detect an effect of DMSO on the distribution (87).

Englyst prefers measuring resistant starch as an important food constituent in its own right, with fiber measured separately as nonstarch polysaccharide, so that it can act as a valid index of plant cell walls in a food (79).

Methods are available for measuring resistant starch. They usually involve predigesting samples with α-amylase and then extracting the resistant starch from the residue with 2 M KOH, digesting it with amyloglucosidase, and finally measuring the freed resistant starch as glucose using the glucose oxidase assay (76,88).

B. Soluble Fiber

1. Relevance and Standardization

Interest in measuring soluble and insoluble components of fiber intensified as evidence accumulated that the difference in physical properties of the two fractions might have different consequences for health (Table 4). The problems of lack of standardization and physiological relevance in soluble fiber analysis were recently discussed (60).

Most methods for soluble fiber analysis are simply modifications of total-fiber methodology, with a step inserted after starch digestion to separate soluble fiber from insoluble material by filtering or centrifuging the digest. The soluble fiber is then recovered from the liquid phase, most often by precipitating with ethanol. There are three important consequences of adopting such an approach:

1. Soluble fiber values are of doubtful nutritional relevance when the fiber is solubilized under conditions used for starch gelatinization and digestion (60), unless the conditions are similar to those of the gastrointestinal tract.
2. Soluble fiber values are method-dependent because conditions for extracting starch/soluble fiber are not standardized. Methods differ in both pH and buffer species, both of which can have a powerful effect on solubility of pectic substances (69,70,73), which are a major component of soluble fiber.
3. True differences in soluble fiber values for a food before and after cooking will be largely obliterated in methods that apply a heat treatment when gelatinizing/digesting starch, as the food, in effect, becomes "cooked" during analysis (60).

The method 39Mon (60), in which soluble fiber is extracted under simulated standard GI conditions and recovered before the residue is subjected starch digestion, circumvents the influence of nonphysiological conditions on the distribution of polysaccharides between the soluble and insoluble fractions. The procedure can be put at the front end of any total fiber method to give a "nutritionally valid" measure of soluble fiber and remaining total (i.e., insoluble fiber). Also, the gastrointestinal conditions can be changed to align the method with gut physiology in the group of interest.

If soluble fiber values are to be used to predict fiber soluble in the gut, as has been the case in clinical and nutritional research, it makes sense first to define the relevant internal conditions and to then use these to extract the soluble fiber. The solubility of fiber is likely to be affected by other food components associated with it and by various properties of the food that may be expressed differently under different conditions. Therefore, the only way to be sure that soluble fiber measured during analysis *in vitro* represents that soluble *in vivo* is to extract under conditions dictated by the gut rather than by analysts. Then the idiosyncratic response of a food to analytical conditions will not give rise to problems of nutritional validity.

2. Incomplete Recovery of Soluble Fiber

Some polysaccharides may escape precipitation by 80% ethanol used in recovering soluble fiber. They normally represent only a few percent of NSP (7,86) but can be important in samples containing high levels of storage fructans, such as artichokes and onions. If of interest, they can be precipitated from the 80% ethanol supernatant with 96% acetone.

C. Lignin

Measuring true lignin is difficult and is not attempted in most fiber methods. Instead, those that measure "lignin" use the weight of residue remaining after 72% (w/w) (12 M) H_2SO_4 hydrolysis of the insoluble fiber residue. Such Klason lignin (KLig in Table 6) is an approximation to true lignin, as it may contain a number of the nonspecific residual substances that were discussed above as adding to insoluble fiber when determined gravimetrically. If Klason lignin represents material present in food as consumed and is not created during analysis, it can be regarded as a true fiber component, depending on definition of fiber being used.

Englyst (58,73) argues that as lignin is a minor component of most foods, it is better to measure fiber as nonstarch polysaccharide to avoid including leftovers of doubtful pedigree on the assumption that they are lignin. If lignin is suspected of being a significant component it can, like resistant starch, be measured separately.

As a practical procedure for measuring lignin, the method of Van Soest and Wine (89) has been recommended (71). It involves measuring the loss in weight of acid detergent residue upon potassium permanganate oxidation, giving a measure of lignin by difference.

A colorimetric procedure based on reaction with acetyl bromide is also available (90) but it has not been used in the analysis of dietary fiber per se.

VIII. PRECISION IN FIBER ANALYSIS

Several recent studies have been conducted into the precision of dietary fiber analysis and have been thoroughly reviewed (1,3,76). Overall, within-laboratory coefficients of variation averaged 4.6 (range 2.0–6.6), while between-laboratory coefficients averaged 9.0 (range 4.8–17.2). Variability tends to be greater with low-fiber than high-fiber foods. A high degree of repeatability is achievable if conditions are carefully standardized and sound laboratory practice is applied with reagents, equipment, and procedures.

Because dietary fiber and the food in which it occurs are highly variable complexes, there is no universal analytical standard. Certified reference materials are now available (91) with specified TDF (AOAC) values: Diet material no. 1548 (3.69% TDF) from the U.S. National

Institute of Standards and Technology (Gaithersberg, MD), and CRM 381 (rye flour; 8.22% TDF), CRM 382 (wheat flour; 3.25% TDF), CRM 383 (haricot beans; 11.91% TDF).

IX. LIMITATIONS OF FIBER ANALYSIS

The methods for dietary fiber analysis discussed here give information solely about the quantity of fiber in a food. They do not provide information about polysaccharide properties—degrees of polymerization, distribution of side chains, crosslinks, and so on—that govern physical properties of fiber and, in turn, physiological responses to it. They give no clues to cell wall, cell, or plant tissue morphology, all of which contribute to the effects of fiber in the gut. Emergent properties of the organized systems of cell wall polymers and cells, such as coarseness or resistance to bacteria, cannot be accessed by dietary fiber analyses alone.

It is expecting too much of fiber analysis that it should provide a sure basis for predicting physiological effects, which depend on much more than mere quantity. Fiber analyses are similar to most other analytical procedures in this respect. If such predictions are to be made, a range of physicochemical properties such as viscosity, water holding capacity, particle size, and fermentability must be measured (92,93). Solubility is the only physical variable that is measured in fiber analysis, but even so the conditions of solubilization have seldom been physiological.

X. CHOICE OF METHOD

If fiber analysis is to be carried out for the purposes of food labeling, a method will be prescribed by food regulatory authorities. For the purpose of research, many methods have a place, but they should be selected on their ability to measure variables that are germane to the research. A prerequisite to choosing a method is therefore a clear understanding, firstly, of the research variables, and secondly, of precisely what the method measures. "Dietary fiber" is an imprecise term that has unfortunately become fixed into the vernacular of nutrition. If the relevant composition variable can be identified as a discrete chemical entity (such as nonstarch polysaccharide, NSP), it should be measured as NSP rather than as some less specific complex, such as TDF, that may not give an accurate measure of the chemical component of interest. As far as is possible the aim should be to say clearly what one is talking about and to measure it specifically.

XI. THE FUTURE

The methods outlined above have shown a trend toward improved precision and convenience, with less attention to nutritional validity. There is scope for increased automation to increase the speed of analysis. Rapid, simple methods such as NIRS are likely to become more widely used. However, the usefulness of such methods depends on measurements of fiber that are not only accurate but meaningful. It seems, then, that a task for the immediate future is to answer the questions: How should the term "dietary fiber" be defined? Should it be dispensed with altogether and replaced with terms that refer more specifically to discrete chemical components of food and digesta? If dietary fiber is to be retained as a physiological concept for use in a nutritional context, perhaps the conditions under which it is prepared should be aligned more with biological systems in which it is functional. When such conditions are set by the human body rather than by analysts, both standardization and relevance are likely to be improved.

REVIEWS

Asp, N.-G., T. F. Schweizer, D. A. T. Southgate, and O. Theander, *Dietary Fibre: A Component in Food: Nutritional Function in Health and Disease* (T. F. Schweizer and C. A. Edwards, eds.), Springer-Verlag, London, 1992, p. 57.

Furda, I. and C. J. Brine (eds.), *New Developments in Dietary Fiber*, Plenum Press, New York, 1990.

Prosky, L. and J. W. DeVries, *Controlling Dietary Fiber in Food Products*, Van Nostrand Reinhold, New York, 1992.

Southgate, D. A. T., *Nutr. Food Sci.*, *23*: 15 (1992).

REFERENCES

1.   L. Prosky and J. W. DeVries, *Controlling Dietary Fiber in Food Products*, Van Nostrand Reinhold, New York, 1992.
2.   D. A. T. Southgate, *Nutr. Food Sci.*, *23*: 15 (1992).
3.   N.-G. Asp, T. F. Schweizer, D. A. T. Southgate, and O. Theander, *Dietary Fibre—A Component in Food: Nutritional Function in Health and Disease* (T. F. Schweizer and C. A. Edwards, eds.), Springer-Verlag, London, 1992, p. 57.
4.   *New Developments in Dietary Fiber*, Experimental Medicine and Biology, 270 (I. Furda and C. J. Brine, eds.), Plenum Press, New York, 1990.
5.   K. W. Heaton, *Human Nutrition: Clin. Nutr.*, *37C*: 151 (1983).
6.   R. R. Selvendran, B. J. H. Stevens, and M. S. DuPont, *Adv. Food Res.*, *31*: 117 (1987).
7.   N.-G. Asp, *New Developments in Dietary Fiber* (I. Furda and C. J. Brine, eds.), Plenum Press, New York, 1990, p. 227.
8.   J. R. Colvin, *Plant Carbohydrates II: Extracellular Carbohydrates* (W. Tanner and F. A. Loewus, eds.), Encyclopedia of Plant Physiology, New Series, Vol. 13B, Springer-Verlag, Berlin, 1981, p. 9.
9.   E. R. Morris, *Dietary Fibre: Chemical and Biological Aspects* (D. A. T. Southgate, K. Waldron, I. T. Johnson, and G. R. Fenwick, eds.), Royal Society of Chemistry, Cambridge, U.K., 1990, p. 91.
10. K. Kato, *Plant Carbohydrates II, Extracellular Carbohydrates* (W. Tanner and F. A. Loewus, eds.), Encyclopedia of Plant Physiology, New Series, Vol. 13B, Springer-Verlag, Berlin, 1981, p. 29.
11. P. M. Dey and K. Brinson, *Adv. Carboh. Chem. Biochem.*, *42*: 265 (1984).
12. R. R. Selvendran and M. A. O'Neill, *Methods in Biochemical Analysis*, Vol. 32 (D. Glick, ed.), John Wiley and Sons, New York, 1987, p. 25.
13. R. W. Bailey, A. Chesson, and J. Monro, *Am. J. Clin. Nutr.*, *31*: 577 (1978).
14. M. McNeil, A. G. Darvill, and P. Albersheim, *Fortschr. Chem. Org. Naturst.*, *37*: 191 (1979).
15. D. Kritchevsky, *Ann. Rev. Nutr.*, *8*: 301 (1988).
16. M. A. Eastwood, *Ann. Rev. Nutr.*, *12*: 19 (1992).
17. M. C. Bourne, *J. Texture Stud.*, *10*: 25 (1979).
18. K. W. Waldron and R. R. Selvendran, *Dietary Fibre: Chemical and Biological Aspects* (D. A. T. Southgate, K. Waldron, I. T. Johnson, and G. R. Fenwick, eds.), Royal Soc. Chem., Cambridge, U.K., 1990, p. 44.
19. J. Tomlin and N. W. Read, *BMJ*, *297*: 1175 (1988).
20. M. L. Dreher, *Handbook of Dietary Fiber*, Marcel Dekker, New York, 1987.
21. H. N. Englyst, M. E. Quigley, and G. J. Hudson, *Eur. J. Clin. Nutr. 49 Suppl 3*: 548 (1995).
22. W. Henneberg and F. Stohmann, *J. Landw.*, *3*: 485 (1859).
23. D. M. Walker and W. R. Hepburn, *Agric. Progr.*, *30*: 118 (1955).
24. P. J. van Soest, *JAOAC*, *46*: 829 (1963).
25. P. J. van Soest, *JAOAC*, *46*: 825 (1963).
26. D. R. Schaller, *Food Prod. Dev.*, *11*(9): 70 (1977).
27. J. B. Robertson and P. J. van Soest, *J. Anim. Sci.*, *45*(1): 254 (1977).
28. R. Mongeau and R. Brassard, *J. Food Sci.*, *51*: 1333 (1986).
29. J. L. Jeraci, B. A. Lewis, J. B. Robertson, and P. J. van Soest, *JAOAC*, *72*: 677 (1989).
30. J. L. Jeraci, B. A. Lewis, J. B. Robertson, and P. J. van Soest, *New Developments in Dietary Fiber* (I. Furda and C. J. Brine, eds.), Plenum Press, New York, 1990, p. 227.

31. R. D. Williams and W. Olmstedt, *J. Biol. Chem.*, *108*: 653 (1935).
32. R. A. McCance, E. M. Widdowson, and L. R. B. Shekelton, *Spec. Rep. Series*, Med. Res. Council (London), No. 213: HMSO, London, U.K., 1936.
33. A. Weinstock and G. H. Benham, *Cereal Chem.*, *28*: 490 (1951).
34. E. W. Hellendoorn, M. G. Noordhoff, and J. Slagman, *J. Sci. Food Agric.*, *26*: 1461 (1975).
35. M. Elchazly and B. Thomas, *Z. Lebensm. Unters Forsch.*, *162*: 329 (1976).
36. F. Meuser, P. Suckow, and W. Kulikowski, *Z. Lebensm. Unters Forsch.*, *181*: 101 (1985).
37. T. F. Schweizer and P. Würsch, *J. Sci. Food Agric.*, *30*: 613 (1979).
38. N.-G. Asp and C.-G. Johansson, *The Analysis of Dietary Fiber in Food* (W. P. T. James and O. Theander, eds.), Marcel Dekker, New York, 1981, p. 173.
39. N.-G. Asp, C.-G. Johansson, H. Hallmer, and M. Silijeström, *J. Agric. Food Chem.*, *31*: 476 (1983).
40. I. Furda, *The Analysis of Dietary Fiber in Food* (W. P. T. James and O. Theander, eds.), Marcel Dekker, New York, 1981, p. 163.
41. L. Prosky, N.-G. Asp, I. Furda, J. W. DeVries, T. F. Schweizer, and B. F. Harland, *JAOAC*, *67*: 1044 (1984).
42. L. Prosky, N.-G. Asp, T. F. Schweizer, J. W. DeVries, and I. Furda, *JAOAC*, *71*: 1017 (1988).
43. S. C. Lee and V. A. Hicks, *New Developments in Dietary Fiber* (I. Furda and C. J. Brine, eds.), Plenum Press, New York, 1990, p. 237.
44. B. W. Li and K. W. Andrews, *JAOAC*, *71*: 1063 (1988).
45. B. W. Li and M. S. Cardozo, *New Developments in Dietary Fiber* (I. Furda and C. J. Brine, eds.), Plenum Press, New York, 1990, p. 283.
46. B. W. Li and M. S. Cardozo, *JAOAC*, *75*: 372 (1992).
47. D. A. T. Southgate, *J. Sci. Food Agric.*, *20*: 331 (1969).
48. R. R. Selvendran and M. S. DuPont, *J. Sci. Food Agric.*, *31*: 1173 (1980).
49. O. Theander and P. Äman, *Sw. J. Agric. Res.*, *9*: 99 (1976).
50. O. Theander and E. A. Westerlund, *J. Agric. Food Chem.*, *34*: 330 (1986).
51. H. Englyst, *The Analysis of Dietary Fiber in Food* (W. P. T. James and O. Theander, eds.), Marcel Dekker, New York, 1981, p. 71.
52. H. N. Englyst, H. S. Wiggins, and J. H. Cummings, *Analyst*, *107*: 307 (1982).
53. R. R. Selvendran, S. G. Ring, and M. S. DuPont, *The Analysis of Dietary Fibre in Food* (W. P. T. James and O. Theander, eds.), Marcel Dekker, New York, 1981, p. 95.
54. H. N. Englyst and J. H. Cummings, *Analyst*, *109*: 937 (1984).
55. R. M. Faulks and S. B. Timms, *Food Chem.*, *17*: 273 (1985).
56. H. N. Englyst and G. J. Hudson, *Food Chem.*, *24*: 63 (1987).
57. J. M. Brillouet, X. Rouau, C. Hoebler, J.-L. Barry, B. Carré, and E. Lorta, *J. Agric. Food Chem.*, *36*: 969 (1988).
58. H. N. Englyst and J. H. Cummings, *JAOAC*, *71*: 808 (1988).
59. H. N. Englyst, M. E. Quigley, G. J. Hudson, and J. H. Cummings, *Analyst*, *117*: 1707 (1992).
60. J. A. Monro, *Food Chem.*, *47*: 187 (1993).
61. *AOAC Official Methods of Analysis*, 14th ed., Association of Official Analytical Chemists, Arlington, VA, 1984.
62. P. J. van Soest, *JAOAC*, *49*: 546 (1966).
63. H. Trowell, *Am. J. Clin. Nutr.*, *25*: 926 (1972).
64. H. Trowell, D. A. T. Southgate, T. M. S. Wolever, A. R. Leeds, M. A. Gassull, and D. Jenkins, *Lancet*, *1*: 967 (1976).
65. H. N. Englyst, H. Trowell, D. A. T. Southgate, and J. H. Cummings, *Am. J. Clin. Nutr.*, *46*: 873 (1987).
66. S. N. Heller, J. M. Rivers, and L. R. Hackler, *J. Food Sci.*, *42*: 436 (1977).
67. R. Mongeau and R. Brassard, *Cereal Chem.*, *56*: 437 (1979).
68. H. Graham, M.-B. G. Rydberg, and P. Äman, *J. Agric. Food Chem.*, *46*: 494 (1988).
69. J. A. Monro, *J. Food Comp. Anal.*, *4*: 88 (1991).
70. J. Gooneratne, G. Majsak-Newman, J. A. Robertson, and R. R. Selvendran, *J. Agric. Food Chem.*, *42*: 605 (1994).
71. D. A. T. Southgate, *Determination of Food Carbohydrates, 2nd ed.*, Elsevier, New York, 1991.
72. M. G. E. Wolters, C. Verbeek, J. J. M. van Westerop, R. J. J. Hermus, and A. G. J. Voragen, *JAOAC*, *75*: 626 (1992).

73. H. N. Englyst and J. H. Cummings, *New Developments in Dietary Fiber* (I. Furda and C. J. Brine, eds.), Plenum Press, New York, 1990, p. 205.

74. N. W. Vollendorf and J. A. Marlett, *J. Food Comp. Anal.*, *6*: 203 (1993).

75. R. Mongeau and R. Brassard, *J. Food Comp. Anal.*, *2*: 189 (1989).

76. O. Theander, P. Äman, E. Westerlund, and H. Graham, *New Developments in Dietary Fiber* (I. Furda and C. J. Brine, eds.), Plenum Press, New York, 1990, p. 273.

77. M. J. Koziol, *Anal. Chim. Acta*, *128*: 195 (1981).

78. R. W. Scott, *Anal. Chem.*, *51*: 936 (1979).

79. N. Blumenkrantz and G. Asboe-Hansen, *Anal. Bioch.*, *54*: 484 (1973).

80. P. K. Kintner and J. P. Van Buren, *J. Food Sci.*, *47*: 756 (1982).

81. S. J. Cooper, R. C. Palmer, and D. C. Porter, *Research Report 65. The British Food Manufacturing Industries Association*, Surrey, 1989.

82. K. A. Garleb, L. D. Bourquin, and G. C. Fahey, *J. Agric. Food Chem.*, *37*: 1287 (1989).

83. Y. C. Lee, *Anal. Biochem.*, *189*: 151 (1990).

84. M. E. Quigley and H. N. Englyst, *Analyst*, *117*: 1715 (1992).

85. C. Hoebler, J. L. Barry, A. David, and J. Delort-Laval, *J. Agric. Food Chem.*, *37*: 360 (1989).

86. P. Colonna, V. Leloup, and A. Buléon, *Eur. J. Clin. Nutr.*, *46*(Suppl.2): 517 (1992).

87. J. A. Marlett, J. G. Chesters, M. J. Longacre, and J. J. Bogdanske, *Am. J. Clin. Nutr.*, *50*: 479 (1989).

88. H. N. Englyst, S. M. Kingman, and J. H. Cummings, *Eur. J. Clin. Nutr.*, *46*(Suppl.2): 533 (1992).

89. J. B. Robertson and P. J. Van Soest, *The Analysis of Dietary Fiber in Food*, Marcel Dekker, New York, 1981, p. 123.

90. I. M. Morrison, *J. Sci. Food Agric.*, *23*: 1463 (1972).

91. D. Mugford, *Nutr. Soc. Aust., Occ. Publ.*, *1*: 19 (1993).

92. M. A. Eastwood and E. R. Morris, *Am. J. Clin. Nutr.*, *55*: 436 (1992).

93. J. F. Thibault, M. Lahaye, and F. Guillon, *Dietary Fibre, a Component of Food: Nutritional Function in Health and Disease* (T. F Schweizer and C. A. Edwards, eds.), Springer-Verlag, London, 1992, p. 21.

DATE DUE

ILL# 5532 718

due 3-19-01 to

TN Tech.

GAYLORD

PRINTED IN U.S.A.